International Conference on Energy Engineering and Environmental Protection (EEEP 2016)

IOP Conference Series: Earth and Environmental Science
Volume 52

Sanya, China
21 – 23 November 2016

Editors:

Kuo-Lin Huang
Kyoung-Woong Kim
Jianshe Liu

ISBN: 978-1-5108-3662-4
ISSN: 1755-1307

Printed from e-media with permission by:

Curran Associates, Inc.
57 Morehouse Lane
Red Hook, NY 12571

CURRAN ASSOCIATES INC. proceedings.com

Some format issues inherent in the e-media version may also appear in this print version.

Printed by Curran Associates, Inc. (2017)

For permission requests, please contact the Institute of Physics at the address below.

Institute of Physics
Dirac House, Temple Back
Bristol BS1 6BE UK

Phone: 44 1 17 929 7481
Fax: 44 1 17 920 0979

techtracking@iop.org

Additional copies of this publication are available from:

Curran Associates, Inc.
57 Morehouse Lane
Red Hook, NY 12571 USA
Phone: 845-758-0400
Fax: 845-758-2633
Email: curran@proceedings.com
Web: www.proceedings.com

Table of contents

Volume 52

International Conference on Energy Engineering and Environmental Protection (EEEP2016)

21–23 November 2016, Sanya, China

Accepted papers received: 10 January 2017
Published online: 13 February 2017

Preface

Power and Energy Engineering

Environmental Protection and Pollution Control

Environmental chemistry and Biology

Geographic information and Remote Sensing Science

IOP Conf. Series: Earth and Environmental Science **52** (2017) 011001 doi:10.1088/1755-1315/52/1/011001

Preface

Kuo-Lin Huang[1], Kyoung-Woong Kim[2], Jianshe Liu[3]

Editors, Proceedings of the International Conference on Energy Engineering and Environmental Protection

[1]National Pingtung University of Science and Technology,Taiwan
huangkL@mail.npust.edu.tw
[2]Gwangju Institute of Science and Technology,South Korea,kwkim@gist.ac.kr
[3]Donghua University (DHU),China,liujianshe@dhu.edu.cn

The International Conference on Energy Engineering and Environmental Protection (EEEP2016) was successfully held in Sanya, China during November 21-23, 2016. The conference was an annual forum for researchers and application developers in the area of energy engineering and environmental protection.

This conference proceeding included 112 accepted articles selected from 240 submissions.

We would like to express our gratitude to the reviewers of these manuscripts, who provided constructive criticism and stimulated comments and suggestions to the authors. We are extremely grateful as organizers, technical program committee and editors and extend our most sincere thanks to all the authors for their excellent contribution and work. Our sincere gratitude also goes to the IOP Publishing editors and managers for their helpful cooperation during the preparation of the proceeding.

On behalf of The Organizing Committees of EEEP2016

IOP Conf. Series: Earth and Environmental Science **52** (2017) 011001 doi:10.1088/1755-1315/52/1/011001

EEEP2016-photos from the conference

Reviewer List

A Marchenko
A.Muthamilselvan
Aamir Hussain Memon
Abdolmajid Lababpour
Abdulrahman Alabdulaaly
Adam Cholewa
Adedapo Awolayo
Ah Heng You
Ahmed Awadallah-F
Ahmed Zobaa
Aimen Gheribi
Ajay Bissessur
Alejandro López Belchí
Alex Chang
Alexander Konovalenko
Ali Bahr Ennil
Ali Ihsan Şekertekin
Ali Shekarifard
Ali Sungur
Amen Al-Yaari
Ana Claudia Moreira Teodoro
Ana Lopez
Ana Marqués
Andrzej Szczurek
Angalaeswari S
Angelica Vieira
Anna Bryła
Anne-Claude Romain
Anping Liu
Anshuman Khardenavis
Anton Purnama
Antonio Gadelha
Antonio Lopes
Antonio López-Uceda
Antonio Pulido Alonso
Aouss Gabash
Arash Houshmand
Ariane Liu
Aritra Das
Asu Inan
Athos Agapiou
Balal Yousaf
Bao Nisha
Baris Yilmaz
Bayu Prabowo
Beata Zielińska
Bekas Georgios
Benjamin Montag
Bernard Saw
Bernhard Stoevesandt
Binbin Yue
Bing Wang
Blaž Likozar
Bob Paine
Bofeng Xu
Bojan Đurin
Boleslav Taraba
Bui T.T. Quyen
Byeung Hun Son
Carlos Duque
Cenk Sevim
Chanyut Kalakan
Charlie Lin
Chee Kong Yap
Chen Zhang
Chiara Maccato Unipd
Chris Lacor
Christopher Page
Chuah Lee Siang
Chung Tai-Shung Neal
Chunhui Cao
Chunyong Zhang
Claudio Jose Dos Santos
Corinne Unger
Cs Tan
Csóka Levente
Cui Yang
Czesław Adamiak
Dale R. Lutz
Damian Ginestar
Daniel Dunea
Daniela Mladin
Daniele Contini
Dariusz Czylkowski
Deirdre Dragovich
Dejian Wu
Demin Chen
Dequn Zhou
Deyou Li
Dimitris Kaziolas
Dmitrii A. Chechushkov
Doan Quang Tri
Dong Han
Dong Jiang
Dong Yin
Dongliang Zhong
Dorota Burchart-Korol
Dyche Anderson
Edoardo Moreci
Eduardo Cervi
Eduardo Schettini Costa
Effie Marcoulaki
Elena Pummer
Elisa Ferracin
Elissar Gemayel
Eloi Rufato
Eman S.Mansor
Emil Bayramov
Fabiana Maria Da Silva
Fabio Monforti
Farhad Balali
Farrukh Chishtie
Farshad Torabi
Farzad Mirzaei
Fatemeh Bahadori
Fátima Candeias
Fei Xing
Fei Yang
Feiqiang He
Fengzhong Sun
Florian Betz
Francesco Aristodemo
Francuois Muller
Frede Blaabjerg
Fu Qing
Fumihide Kojima
Gabriele Buttafuoco
Gafsi. Mostefa
Gang Pei
Gao Wei
Gaydaa Al Zohbi
George Alexthopil
Georgios C. Christoforidis
Georgios Sylaios
Gerald Steinmaurer
German Pachurin
Giacomo Valerio Iungo
Gilmar Perbiche Neves
Gm Shafiullah
Golnaz Jozanikohan
Graça Rasteiro
Grazia Leonzio
Grigorios Kyriakopoulos
Grzegorz Chomka
Guandalini Roberto
Guangwei Wang

Guoqian Chen
Győri Zoltán
Gzh Wang
Hadi Ebrahimi Harate
Hai Yun
Haifeng Zhuang
Hailiang Ma
Hamid Sarlak Chivaee
Han Gaoyi
Harold Sun
Hasan Dilbas
Haydar Demirhan
He Shien
Heiko Kollasko
Heli Wei
Himanshu Sekhar Sahu
Hong Minh Hoang
Hongtao Zhu
Hongyan Su
Houde Su
Huajun Song
Huang Muyi
Hui Zhong
Huiying Liu
Huiyong Yin
Ihuoma N. Anyanwu
Inderjeet Singh
Iñigo Ortega
Ioana Maior
Issam Abadlia
Izumi Hirasawa
Jaime-Francisco García-Pérez
Jamuna K
Jan Fulneček
Jan Tkac
Jayalakshmi N S
Jayalakshmi N. Sabhahit
Jenn-Gwo Hwu
Jérôme Mendes
Jianbo Bai
Jiangtao Wang
Jianjun Li
Jie Zhang
Jin Jiang
Jingrui Xie
Jingsong Deng
João Soares
Joaquin Ortega Casanova
Johan René González Moya
Johannes Idsø
John A Rupp
Jordi Palacin
Jose Ismael Acosta Rodriguez
Jose Moreira
José Villar Collado
Josef-Peter Schöggl
Juan Huan
Juan Manuel Trujillo Gonzalez Docente
Julian Antonio Pucheta
Julio Hernandez Blanco
Jumin Zhao
Jun Liao
Jun Xie
Jun Zheng
Junjie Li
Jun-Ming Hong
Junsheng Liu
Junshu Wu
K. S. Siddiqi
Kai Hou
Kairao Wu
Kalyan Sehanobish
Karlis Baltputnis
Karvinen Aku
Ke Dai
Kelvin Swee
Kenneth A. Loparo
Khalid Ahmad
Khandaker Abdullah Al Mamun
Khelifa Abdelkrim
Kingsley Dogah Etornam
Kordun Oleksandr
Krzysztof Pytel
Ks.Suh
Lan Cuo
Laura Porcel Rodríguez
Le Van Si
Lee Seung Jo
Lei Wu
Lei Yan
Lei Ye
Lele Zou
Leslaw Teper
Levente Csoka
Lhoussain El Bahir
Li Wangliang
Li Wei
Li Wenfang
Liancheng Zhu
Lianyou Liu
Lin Liang
Liu Zhongliang
Longhua Guo
Longsheng Lu
Lu Diannan
Luca Podofillni
Luo Yong
M. Rashid
Ma Chunyan
Maciej Dzikuc
Maciej Dzikuć
Maged Marghany
Manoel Regis L.V. Leal
Manuel Cortez
Manuel Oliveira
Marat Khakimyanov
Marcelo G. Cendoya
Marcin Woch
Martin Koller
Maryam Lashgari
Masakazu Niinae
Md. Abdus Salam
Md. Nazmul Islam Sarkar
Meng Wang
Miao Chao
Michelle Kondo
Miguel Angel Espinosa
Miguel Angel Sanz Bobi
Miguel López
Miklas Scholz
Mirna Habuda-Stanić
Mohamed Attia
Mohamed Memmah
Mohammad Abutayeh
Mohammad Nurunnabi
Mohammadreza Mazidi
Mohammed Matouq
Mohan Raj
Mohand Tazerout
Mohd Rizaimy Shaharudin
Mojtaba Hedayati
Mónica Meraz
Monika Maciejewska
Mouloud Denai
Mt Sambandam
Mukhtar Abdulkadir
N. B. Manik
N. N. Rao
Najmeh Neshat

Namor Emil
Nestor Tancredi
Nicola Strisciuglio
Nikola Botoucharov
Nikos Markatos
Nima Soltani
Nor 'Asnilawati Salleh
Nor Erniza M Rozali
Nurul Ihsan Fawzi
O. Le Corre
Olga Eugenia Scarpati
Orestis Schinas
Ouattara Mathilde
Panagiotis Grammelis
Paula Matono
Payam Ghorbannezhad
Peng Bo
Peng Xu
Pengyuan Shen
Ping Ning
Ping Zhou
Pingfang Yan
Pingqi Gao
Pío Aguirre
Piotr Gutwinski
Podofillini Luca
Pouyan Ghabezi
Preye David Orodu
Przemysław Borkowski
Qianfeng Wang
Qianggang Wang
Qibao Qi
Qifeng Zhuang
Qinghong Wang
Qingsong Wang
Qingxiang Zhou
Qingyu Yang
Qiuan Zhu
Qiuxuan Wu
Qixing Wu
Quan Zhang
Rabih Mezher
Rafael Maya-Yescas
Rainer Brüggemann
Ran Zhu
Ratan Joarder
Reik Donner
Renjuan Sun
Ren-Zhou Gui
Ricardo José Lucas Lagoa
Richard Gloaguen
Richard Viskup
Rita De Cássia M. Sales
Robert Devoy
Robert Jandl
Romain Gallen
Rubing Zhang
Ruhollah Taghizadeh
Ruwei Wang
S.-H.Tabatabaei
S.Meignanalakshmi
Saad Saleem
Sabino Veintemillas Verdaguer
Saeed Manshadi
Saheb Koussa Djohra
Samia Khadhar
Samira Keivanpour
Sandro Nizetic
Sasan Ebrahimi
Seed Wang
Semih Ozden
Sen Guo
Shafique Ahmed Junejo
Shan Yao
Shang-Ru Zhai
Shao Hong-Bo
Sheli Chai
Shengchun Yang
Shenshen Su
Shizhong Li
Shouxiang Wang
Shuren Wang
Simona Caprarescu
Siyu Li
Slawomir Wierzbicki
Srdjan Kostic
Stafford W. Sheehan
Stefan Kern
Steve Brown
Sundar Raj Thangavelu
Suo Yang
Suping Cui
Syed Shatir A. Syed-Hassan
Sylvain Laporte
Tahir Yavuz
Taiping Wang
Tamas Janos Katona
Tao Xi
Tatyana Lyubimova
Tau One
Tie-Qiao Tang
Ting Wang
Tomas Orfanus
Tongguang Wang
Toufic Mezher
Tsiberkin Kirill
Umeyr Kureemun
Usep Surahman
V Kumar
Vamshi Krishna
Vladimir A. Plotnikov
Vladimir Obolkin
Vugar Fataliyev
Wai Ming Cheung
Wang Hanxi
Wei Cao
Wei Yang
Wei-Cheng Wang
Weimin Huang
Wenhao Niu
Wenjuan Zhang
Wenwu Yu
Wojciech Skarka
Xiangkai Li
Xianhui Zhao
Xian-Min Ma
Xiao Ziniu
Xiaocheng Li
Xiaohui Zhang
Xiaojun Yao
Xiaoquan Lu
Xiaoyan Liu
Xiaoyu Wang
Xingru Feng
Xue Li
Xuepin Liao
Y S Tsai
Yaling Li
Yan Peng Ban
Yan Zi-Feng
Yann Meyer
Yawei Du
Yi Chang
Yifan Zhang
Yılmaz Yildirim
Yizi Shang
Yonghe Wang
Yonghong Hu
Yongli Huang
Yongqiang Zhao
Yongqing Wang
Yongsheng Zhang

Yuan-Jian Yang
Yusuf Alper Kaplan
Yuting Tan
Zakhar Slukovskii
Zaki Zahran
Zdenko Šimić
Zezhou Wu
Zhang Changxing
Zhang Qi
Zhang Xiaomei
Zhang Yugui
Zhang Zhe
Zhang Zhiyong
Zhe Lei
Zheng Han
Zhi Wu
Zhiming He
Zhixiong Li
Zhoulu Yu
Zong-Xiao Yang

IOP Conf. Series: Earth and Environmental Science **52** (2017) 011002 doi:10.1088/1755-1315/52/1/011002

Peer review statement

All papers published in this volume of *IOP Conference Series: Earth and Environmental Science* have been peer reviewed through processes administered by the proceedings Editors. Reviews were conducted by expert referees to the professional and scientific standards expected of a proceedings journal published by IOP Publishing.

An non-uniformity voltage model for proton exchange membrane fuel cell

Kelei Li[1], Yankun Li[1], Jiawei Liu[2, 3] and Ai Guo[2]

[1] CRRC Qingdao Sifang Co., Ltd, Qingdao, 266111, China.
[2] School of Electrical Engineering, Southwest Jiaotong University, Chengdu, 610031, China.

E-mail: daben2011@foxmail.com

Abstract. The fuel cell used in transportation has environmental protection, high efficiency and no line traction power system which can greatly reduce line construction investment. That makes it a huge potential. The voltage uniformity is one of the most important factors affecting the operation life of proton exchange membrane fuel cell (PEMFC). On the basis of principle and classical model of the PEMFC, single cell voltage is calculated and the location coefficients are introduced so as to establish a non-uniformity voltage model. These coefficients are estimated with the experimental datum at stack current 50 A. The model is validated respectively with datum at 60 A and 100 A. The results show that the model reflects the basic characteristics of voltage non-uniformity and provides the beneficial reference for fuel cell control and single cell voltage detection.

1. Introduction

PEMFC is a new generation device gradually used and recognized in the transport sector with the advantages of no pollution, high efficiency and fast start up [1-3]. In the field of rail transportation, fuel cell has built-in power system without the line traction power system and greatly reduces line construction investment with the tremendous development potential. But some bottleneck problems are also needed to solve such as cost and life in fuel cell commercialization [4-6].

Fuel cell model is classified as mechanism model and simplified model. Their characteristics and performances are studied. With some rational assumptions, the mechanistic model is obtained by mass conservation, energy conservation law, ideal gas equation, heat equation, mass transfer, electrochemical equation and internal cell characteristics. Due to the complex process of the fuel cell, the simplified model ignoring certain factors is put forward to reflect some essential characteristics of the fuel cell.

Mechanism model is on the basis of strict physical relationship. Its form is complex and sometimes some key parameters are difficult to get. Thus this modeling approach has been curbed and reduces its effectiveness and feasibility. Actual phenomenon and the experimental datum are used to describe the system experience model. In order to avoid the complex relationships between physical quantities, simplify the model and improve the feasibility, experimental model has been widely applied. In membrane electrode model, the most classic fuel cell performance model is set up under the steady-state condition of MK-IV PEMFC by Amphlett J C et al. [7] The model can predict the degree of polarization in different working conditions. Mann R F et al. [8] consider the anodic polarization phenomenon and get electricity output characteristic model by introducing membrane resistance coefficient as calculating the ohmic overvoltage. Based on the single cell performance empirical formula, Lee J H et al. [9] divides stack into several units and obtains stack power by adding each unit power. It is a more complete experience model to predict PEMFC voltage characteristics. Li Qi et al. [10] are proposed for engineering analysis about PEMFC experimental model. The key parameters are identified by particle swarm optimization algorithm of the model. Chen Huicui et al. [11] study the dynamic variable loading voltage response process through an equivalent circuit model. Su Zhou et al. [12] add the gas diffusion layers and the membrane electrode assembly dynamic model to the existing model in order to study the membrane water content. In addition, control theories and methods are gradually applied to the fuel cell modeling. These models are more complex and difficult for control strategy.

The important factors affecting the PEMFC operation life include operating conditions, stack temperature, reaction gas distribution, gas excessive coefficient, pressure, frequent start-stop and dynamic loading. The stack consists of many series single fuel cell. Its life depends on the performance of the worst single cell. The individual voltage uniformity is an important factor affecting stack life and used to measure the single voltage difference. The evaluation of fuel cell voltage uniformity has four methods, such as single cell voltage fluctuation rate, single voltage root-mean-square, single voltage fluctuation amplitude and graphical method [13, 14]. The single cell voltage fluctuation rate is the ratio of the standard deviation to the mean cell voltage, called a relative standard deviation. The single cell voltage root-mean-square is the absolute standard deviation. The single cell voltage fluctuation amplitude is difference between the maximum voltage and the minimum voltage. Graphical method is a figure of single cell voltage distribution which can directly reflect cell voltage uniformity.

Individual voltage uniformity is not involved in existing fuel cell models. Based on classic model proposed by Amphlett J C [7] and Lee J H[9], this paper establishes the fuel cell stack model reflecting the individual voltage uniformity. The stack voltage model is verified by 14.4 kW PEMFC test platform. Position coefficients are estimated with the experimental datum at stack current 50 A. The model is validated respectively with datum at 60 A and 100 A. The results show that the model reflects the basic characteristics of voltage non-uniformity and provides the beneficial reference for fuel cell control and single cell voltage detection.

2. PEMFC working principle

The single structure of PEMFC is mainly composed of bipolar plate, diffusion layer, catalyst layer and electrolyte membrane as shown in figure 1. Air (oxygen) and hydrogen respectively enter into cathode and anode of the cell.

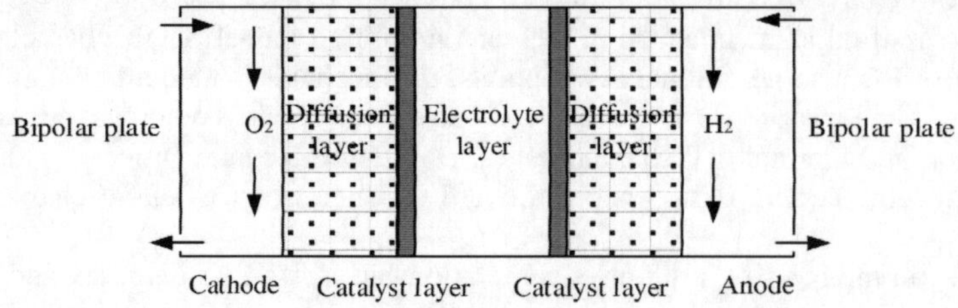

Figure 1. The single structure of PEMFC

The electrolyte membrane separates cathode and anode into two parts. The hydrogen reduces the activation energy and dissociates into hydrogen ions and electrons under the catalyst:

$$H_2 \rightarrow 2H^+ + 2e^- \quad (1)$$

Hydrogen ions move toward cathode through the proton exchange membrane. The electrons move to cathode through the external circuit. In the role of catalyst, oxygen molecules, electrons and hydrogen ions produce water. The reduction reaction is:

$$\frac{1}{2}O_2 + 2H^+ + 2e^- \rightarrow H_2O \tag{2}$$

The overall reaction is an electrochemical reaction of hydrogen and oxygen producing water:

$$H_2 + \frac{1}{2}O_2 \rightarrow H_2O \tag{3}$$

3. Voltage model of PEMFC

Open circuit voltage E is calculated according to the Faraday constant and the energy balance between the reaction gas and products[15]:

$$E = 1.229 - 0.85\times10^{-3}(T_{fc} - T_{atm}) + 4.3085\times10^{-5}T_{fc}[\ln(p_{H_2}) + \frac{1}{2}\ln(p_{O_2})] \tag{4}$$

where T_{fc} is fuel cell temperature, T_{atm} is ambient temperature, p_{H_2} is hydrogen pressure and p_{O_2} is oxygen pressure. When the current passes through the fuel cell, a series of physical and chemical changes occur in the process. There is resistance during each process. In order to make the reaction continuously, it is necessary to consume energy to overcome their resistance, so the voltage will be lost. Depending on the cause of the generated loss, it can be divided into three parts: the activation loss, ohmic loss and concentration loss. Activation loss is the energy loss caused by electron transfer at anode and cathode. Activation overpotential occurs in anode and cathode plates of the fuel cell. However, the hydrogen oxidation reaction is very fast at the anode, but the oxygen reduction reaction at the cathode is very slow. Therefore, voltage drop generated by the activation loss is controlled by the cathodic reaction condition. Activation over potential v_{act} is[16]:

$$v_{act} = v_0 + v_a(1 - e^{-c_1 i}) \tag{5}$$

where i is the current density, v_0 is the voltage drop under zero current density, v_a and c_1 are constants affected by oxygen partial pressure and temperature.

Ohmic loss is caused by the resistances of proton transmission and electric transmission. The current density is proportional to the voltage drop, namely:

$$v_{ohm} = i \cdot R_{ohm} \tag{6}$$

where R_{ohm} is the internal resistance. Resistance mainly depends on the humidity and temperature of the fuel cell membrane. The ohmic resistance is a function of the membrane conductivity σ_{m}:

$$R_{ohm} = \frac{t_m}{\sigma_m} \tag{7}$$

where t_m is the thickness of the proton exchange membrane. Concentration loss is due to the change of the reactant concentration. At high current densities these losses appear as a rapid voltage drop. An approximate equation of voltage drop caused by the concentration loss is [17]:

$$v_{conc} = i\left(c_2 \frac{i}{i_{\max}}\right)^{c_3} \tag{8}$$

where c_2, c_3 and $i_{\max}$ are constants depending on the temperature and partial pressure of the reactants determined by experiment, $i_{\max}$ is the current density caused by a steep voltage drop.

By combining formula (4)-(8), the terminal voltage v_{fc} of fuel cells can be written as:

$$v_{fc} = E - v_{act} - v_{ohm} - v_{conc} \tag{9}$$

The block diagram is classical PEMFC model shown in figure 2(a) from formula (9). Inputs are stack current, stack temperature, anode and cathode pressure. These amounts will calculate open-circuit voltage, activation loss, ohmic loss and concentration loss by the formula (4)~(8). The space position of monomer is different in the stack. The pressure and temperature of the anode and cathode of monomer can't be same, so that the voltage of monomer is not equal. Position coefficient is introduced in order to reflect their non-uniformity, whose block diagram is shown in figure 2(b).

The voltage non-uniformity model is established in MATLAB/Simulink software. The thickness of proton exchange membrane is 125 μm and stack active area is 300 cm^2. By the law of partial pressure, the oxygen partial pressure is 21% of the cathode pressure. Air humidity exceeds 95% before entering the stack. Because stack reaction product is water, the humidity of proton exchange membrane is inferred 100%.

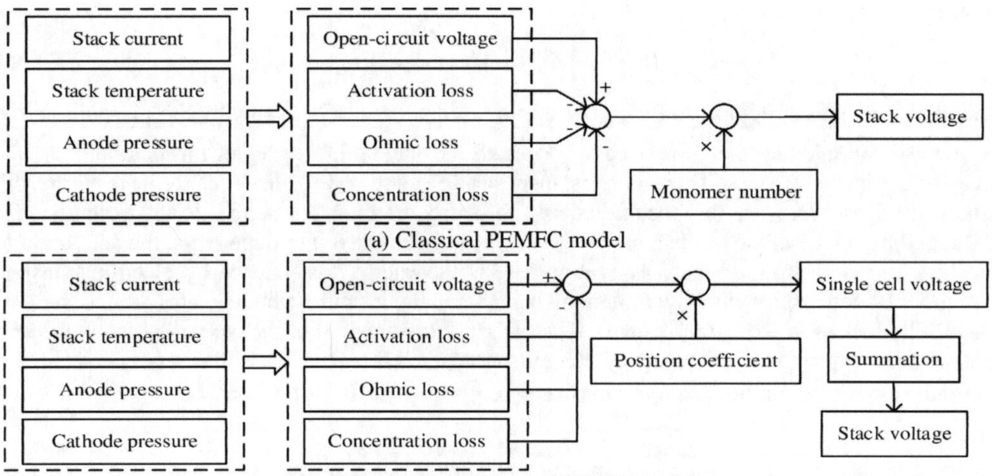

(b) Voltage non-uniformity PEMFC model

Figure 2. PEMFC model

4. Experimental verification

4.1. Experiment platform

The experiment platform is established as shown in figure 3 with Ballard's FCvelocity®-9SSL 14.4 kW liquid-cooled stack. Stack is constituted by 75 cells. The rated current is 300 A and the rated voltage is 48 V. The hydrogen enters into the PEMFC anode from the hydrogen storage tank with 13 Mpa high-pressure through the reducing valve, mass flow controller and humidifier. The open degree of the backpressure valve is adjusted at the anode outlet to meet pressure difference between the cathode and anode at the stack inlet. Compressed air enters into the high pressure gas bomb, then passes through mass flow adjustor and humidifier. Perma Pure FC-125 and FC-300 series humidifiers are used for hydrogen and air respectively. The way is gas-liquid humidifying. The terminal of stack connects electronic load to extract a specified stack current. For data acquisition and control, ZTYC's USB7635BD data acquisition card is applied to connect LabVIEW software. In the system, temperature, pressure, flow rate, current, stack voltage, cell voltage can be real-time monitored.

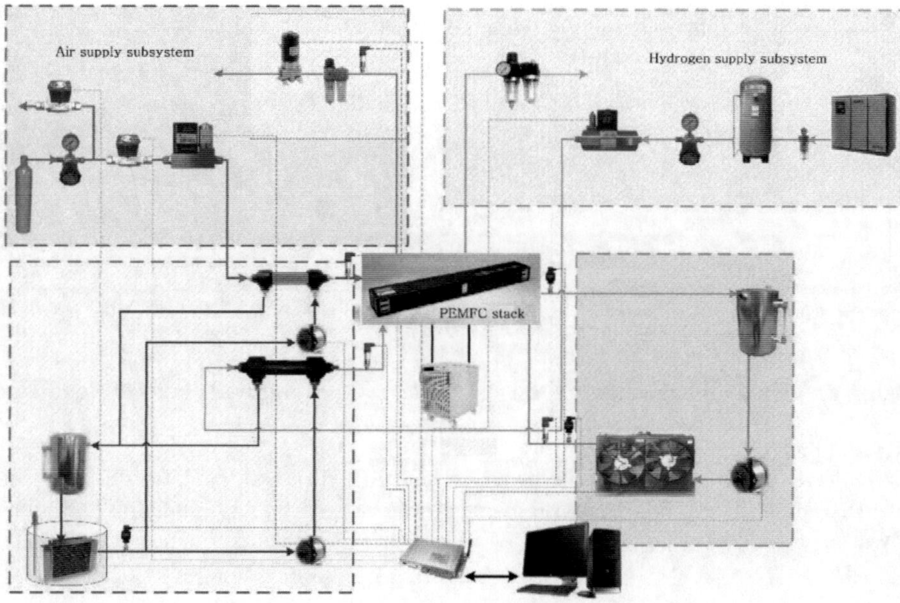

Figure 3. The schematic of PEMFC system

4.2. Experimental verification

The experiment condition is under constant temperature and isopiestic pressure. Firstly, experiment platform runs for half an hour to ensure the stack activity released in full. The temperature at stack outlet of cooling liquid meets expected value. The experimental conditions are set as shown in table 1.

Table 1. The experiment operating conditions

Ambient temperature(℃)	Air inlet pressure(kPa)	Hydrogen inlet pressure(kPa)	Cooling liquid inlet temperature(℃)
27.81	178.42	192.72	58

The relationship between the excess coefficient of fuel and air and the extraction current is shown in Table 2 according to the parameters of Ballard's technical manual[18].

Table 2. Fuel and air excess coefficient

Current(A)	Fuel excessive coefficient	Air excessive coefficient
15	6.3	5.1
30	3.4	2.4
60	2.2	1.8
120	1.9	1.8
180	1.6	1.8

Stack current is gradually loaded from 20A to 160A with step 10A. Every time current changes, stack has been in a steady state. The steady-state voltage values of 75 cells are tested at different current, and average cell voltage is as shown in figure 4. With the stack current increasing, average single voltage decreases gradually. When the stack current is 30A~140A range, cell model voltage mean value matchs with experimental preferably. When the stack current is 10A, 160A, errors were 4.015% and -1.101%. It indicates that the model can reflect the experimental stack cell mean voltage.

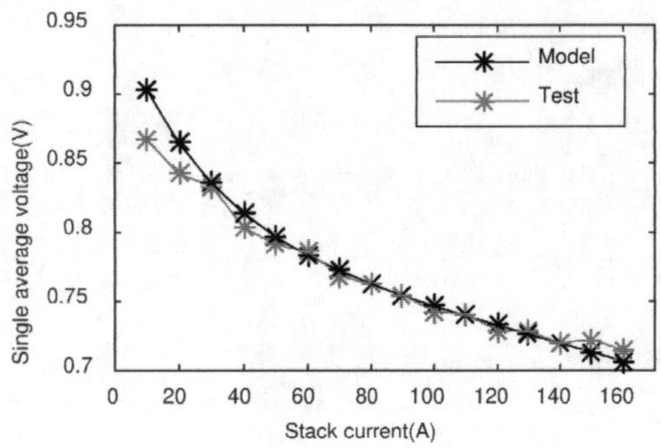

Figure 4. Average voltage between test and model

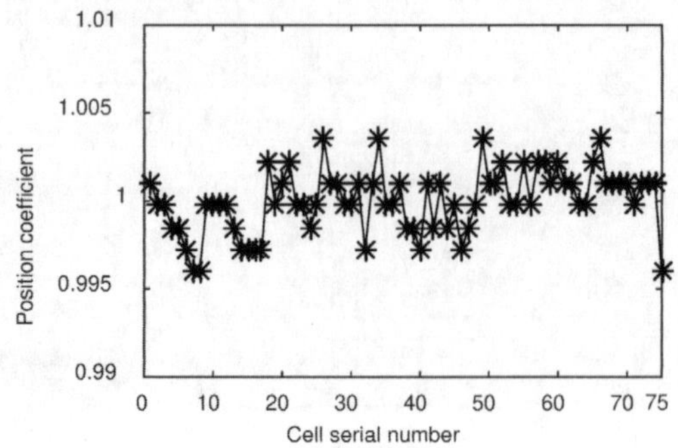

Figure 5. Position coefficient

5. Results and discussion

Individual voltage uniformity is not involved in existing fuel cell models. Position coefficient responses the degree of deviating from the average cell and can be represented as the cell voltage divided by the average cell voltage, namely:

$$\alpha_{p,i} = \frac{V_i}{\overline{V}} \qquad i = 1,\cdots,N \tag{10}$$

where $\alpha_{p,i}$ is the position coefficient, V_i is the single cell voltage, $\overline{V}$ is the stack single average voltage and N is the stack monomer number. The position coefficient is shown in figure 5 with experimental data at 50 A.

Based on the model, the comparison curves of the voltage model and the experimental results are shown in figure 6 and 7. In figure 6, the experimental values is higher than model values when the current is 60 A. However, in figure 7 the model value is higher than the experimental value at 100A. Their trends are basically same. The experimental value is slightly higher than the model value in figure 6 ,but the model value is higher than the experimental value in figure 7. The model is fitted well with the experimental data in cell No. 1-55 and differs slightly in the cell No. 56-75.

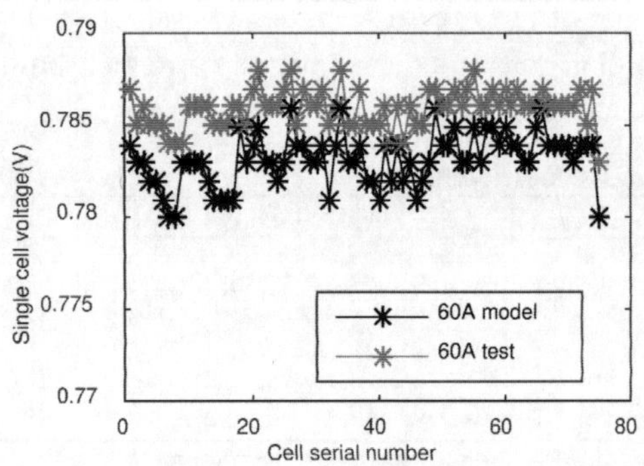

Figure 6. Single cell voltage distribution at 60A

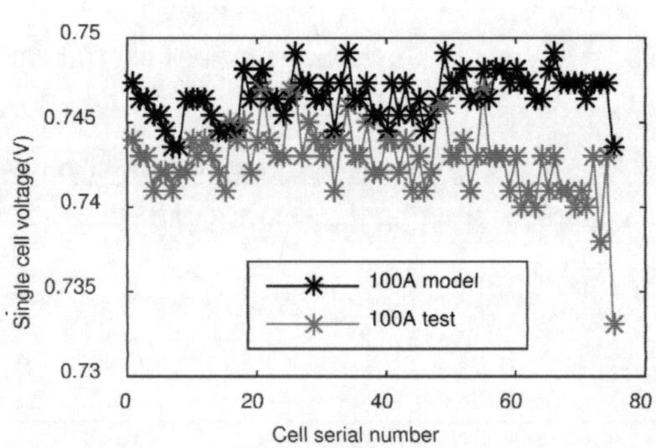

Figure 7. Single cell voltage distribution at 100A

Fuel cell voltage fluctuation rate C_v satisfies:

$$C_v = \sqrt{\frac{\sum_{i=1}^{N}\left(\frac{V_i - \overline{V}}{\overline{V}}\right)^2}{N}} \cdot 100 \tag{10}$$

The fluctuation rates of 60A and 100A are calculated according to equation (11), as shown in Table 3.

Model voltage fluctuation rate is the same, because the location coefficient is only got by 50A numerical calculation. Current will affect the position coefficient. Since the model does not consider its impact, fluctuation rate of model and experiment has a large difference.

Table 3. Voltage fluctuation rate

Stack current(A)	Model	Test
60	0.1823	0.1309
100	0.1823	0.2882

6. Conclusion

Monomer voltage non-uniformity is an important performance index for PEMFC. This paper firstly introduces the working principle of fuel cell and voltage model. The position coefficient is introduced based on the classical model to represent monomer voltage. The position coefficient is calculated with experimental data at 50 A. The model is validated at 60 A and 100 A. The model and experimental values can well reflect the stack cell voltage imbalance at 60 A. At 100A, the model values coincide with the experimental values when the cell numbers are 1-55. The model and experimental values differ at No. 56-75. The model voltage fluctuation rate is same, because the position coefficient is only got by experimental data at 50 A. There is no consideration of the impact on the current. Thus there is a big fluctuation rate difference between model and experiment. In addition, the current impact on location coefficient is the next step to research project.

Appendices

Table A1.

Symbol	Meaning	Unit
T_{fc}	fuel cell temperature	K
T_{atm}	ambient temperature	K
p_{H_2}	hydrogen pressure	Pa
p_{O_2}	oxygen pressure	Pa
i	current density	A/cm^2
v_0	voltage drop under zero current density	V
R_{ohm}	internal resistance	$\Omega \cdot cm^2$
t_m	thickness of the PEM	μm
i_{max}	current density caused by steep voltage drop	A/cm^2
$\alpha_{p,i}$	position coefficient	constant
V_i	single cell voltage	V
$\overline{V}$	stack single average voltage	V
N	stack monomer number	constant

Acknowledgments

We gratefully acknowledge the financial support provided by National Natural Science Foundation of China (51607149).

References

[1] Pachauri R K, Chauhan Y K 2015 A study, analysis and power management schemes for fuel cells *Renewable and Sustainable Energy Reviews* **43** 1301-19.

[2] Kwon O, Shin H, Cheon S, et al. 2015 A study of numerical analysis for PEMFC using

a multiphysics program and statistical method *International Journal of Hydrogen Energy* **40**(35) 11577-86.
[3] Da Costa Lopes F, Watanabe E H, Rolim L G B 2015 A Control-Oriented Model of a PEM Fuel Cell Stack Based on NARX and NOE Neural Networks *IEEE Transactions on Industrial Electronics* **62**(8) 5155-63.
[4] Whiteley M, Fly A, Leigh J, et al. 2015 Advanced reliability analysis of Polymer Electrolyte Membrane Fuel Cells using Petri-Net analysis and fuel cell modelling techniques *International Journal of Hydrogen Energy* **40**(35) 11550-58.
[5] Jian Q, Zhao Y, Wang H 2015 An experimental study of the dynamic behavior of a 2 kW proton exchange membrane fuel cell stack under various loading conditions *Energy* **80** 740-745.
[6] Zhou S, Gao K, Zhi X. 2015 An Improved Dynamic Model for Proton Exchange Membrane Fuel Cell Systems *Journal of Tongji University* **6**(43) 882-87.
[7] J. C. Amphlett, R. M. Baumert, R. F. Mann, et al. 1995 Performance Modeling of the Ballard Mark IV Solid Polymer Electrolyte Fuel Cell I. Mechanistic Model Development *Journal of the Electrochemical Science* **142**(1) 1-8.
[8] Mann R F, Amphlett J C, Hooper M, et al. 2000 Development and application of a generalised steady-state electrochemical model for a PEM fuel cell *Journal of Power Sources* **86**(1-2) 173-180.
[9] Lee J H, Lalk T R 1998 Modeling fuel cell stack systems *Journal of Power Sources* **73**(2) 229-241.
[10] Li Q, Chen W, Wang Y, et al. 2011 Parameter Identification for PEM Fuel-Cell Mechanism Model Based on Effective Informed Adaptive Particle Swarm Optimization *IEEE Transactions on Industrial Electronics* **58**(6) 2410-19.
[11] Chen Huicui, Pei Pucheng 2014 Dynamic model of proton exchange membrane(PEM) fuel cell during load changes *Journal of Tsinghua University* **54**(10).
[12] Zhou Su, Gao Kunpeng, Zhi Xuelei 2015 An improved dynamic model for proton exchange membrane fuel cell systems *Journal of Tongji University* **06** 882-887.
[13] Dai Chaohua, Shi Qing, Chen Weirong 2016 A Review of the single voltage uniformity in Proton Exchange Membrane Fuel Cells *Proceedings of the CSEE* **05** 1289-1302.
[14] Li Y, Zhao X, Liu Z, et al. 2015 Experimental study on the voltage uniformity for dynamic loading of a PEM fuel cell stack *International Journal of Hydrogen Energy* **40**(23) 7361-69.
[15] J. C. Amphlett, R. M. Baumert, R. F. Mann, et al. 1995 Performance Modeling of the Ballard Mark IV Solid Polymer Electrolyte Fuel Cell II. Empirical Model Development *Journal of the Electrochemical Society* **142**(1) 9-15.
[16] Larminie J, Dicks A, McDonald M S 2003 Fuel cell systems explained[M]. Chichester, UK: J. Wiley.
[17] Guzzella L 1999 Control oriented modelling of fuel-cell based vehicles[C]//Presentation in NSF Workshop on the Integration of Modeling and Control for Automotive systems.
[18] Ballard. FCvelocity®-9SSL V4 Product Manual and Integration Guide[M]. 2011.

Benefit and cost competitiveness analysis of wind and solar power inter- continent transmission under global energy interconnection mode

Xiaoxia Wei[1, 4], Jian Ding[2], Jie Liu[3], Tiezhong Wei[3]

1. State Grid Energy Research Institute, Guangyi Building A1206, Guangyi Street 5#, Xicheng district, Beijing, China
2. Beijing Haitong Tiancheng Information Technology Co.Ltd., China
3. Heilongjiang Power Company, China

E-mail: 20583853@qq.com

Abstract. Relying on the global energy Interconnection, considering the energy implementation, carrying out clean energy alternative is mainly to use the clean energy to take place of fossil energy. Under the green development scenario, This research gives the global energy interconnection development model, makes the Artic and the Equation as the connection points, gives the Northern hemisphere interconnection model and equator interconnection model unite the whole world energy. This research also identifies the factors effecting the transmission changes cost, including generation cost, transmission cost and landing cost. And take two continents connection as the prediction example, estimate these two continents cost benefit and variable power-jointed scheme cost competitiveness. It showed that under the global energy interconnection mode, the trans-continent mode had better benefit, and the landing cost is good to be used, can solve the pollution and energy restriction.

1. Introduction

Global energy supply is constrained by resource reserves and rising supply costs. To the amount, fossil energy reserves are limited, non-renewable, large-scale development and utilization will lead to depletion of resources [1, 2]. To the layout, the world's energy resource and consumption are reverse distribution; energy exploitation is centralizing to a few countries. At present, the supply costs of fossil and clean energy is showing a "one rising and one dropping". With the rapid development of clean energy technology, the cost will decrease [3].

The global energy environment is affected by global warming and the ecological environment destruction. The Fossil fuel combustion is the main source of global greenhouse gas emission, which can directly affect land reduction, species extinction, food supply, human health. Sulfur dioxide and other pollutants are leading acid rain and other environmental pollution. In addition, a large number of fossil fuels in mining, transportation, use of the various aspects of the water quality, soil, water and other environmental pollution [4].

The global energy efficiency needs to enhance energy development, configuration and utilization efficiency. At present, the world average oil recovery rate of only 34%; coal recovery rate is from 65% to 70%; the world average level of thermal power coal consumption is about 330 g standard coal / (kilowatt hour), the German average thermal power coal consumption of 290 g standard coal / (kilowatt hour), while some countries coal power is up to 370 g standard coal / (kilowatt hour). Energy efficiency of developed countries is higher than developing countries; per unit GDP of OECD countries are only about 25% of the non-OECD countries [5, 6].

2. Global energy interconnection development model

According to the analysis of global clean energy demand, expected 2050, the electricity delivered amount of the Arctic channel will reach 3 trillion kilowatt-hour /year, and that of the equatorial region will reach 9 trillion kilowatt-hour /year. Electricity accounts for 16% of global electricity demand. In addition, there is a certain proportion of trans-continental power exchange, access to peak load, complementary resources, sharing of spare networking benefits.

2.1. Northern hemisphere interconnection model

Northern Hemisphere Interconnection model is mainly based on the Arctic power transmission channel. On the one hand, it is carrying the Greenland, Norway, Barents Sea, Kara Hai, Bering Strait, while on the other hand, it is a global energy Internet strategy platform connecting Arctic wind power base for electricity delivery needs, on the other hand, it can realize northern hemisphere three continents connection, forming the global energy interconnection strategy platform. This connection can give full play to great grid connection; give many benefits of peak and valley regulation, mutual backup, cross-complementarity. In addition, making use of the time difference between the continents, send the Arctic wind power to other continents, meet the day peak load needs, and improve Arctic wind power utilization efficiency, showed in Table 1.

Table 1. Global Energy Interconnection Development Model

Connection	Sending and receiving	Line setting	2050 transmission capacity
Arctic to Northeast Asia	The Kara Sea wind power base of the Arctic to North China. Bering Strait wind power base to North China, Japan and South Korea.	Adopt overhead lines ± 1100 kV HVDC transmission technology to China's power transmission channels, Adopt UHV DC submarine cable to Japan, South Korea.	Transmission capacity is 1.2 trillion kilowatt-hour / year, channel capacity is 240 million kilowatts.
Arctic to Europe	Greenland wind power to the north of England. Offshore wind power in the Norwegian Sea and the Barents Sea can be sent to Europe via the land channel.	Adopt UHV DC submarine cable connecting Green land South wind power base to Iceland, then leading to the north of England.	Transmission capacity is 0.8 trillion kilowatt-hours / year; channel capacity is 160 million kilowatts.
Arctic to North America	Bering Strait wind power base to the western United States load center. Southern Greenland wind to the east coast of Canada, then the eastern United States to the central load Ottawa.	Adopt ± 1100 kV HVDC transmission technology from Quebec to New York	Transmission capacity is 1trillion kilowatt-hours / year, channel capacity is 200 million kilowatts.

2.2. Equator interconnection model

Equator Interconnection model, is considering the equatorial power transmission channel, carrying North Africa, East Africa, the Middle East, Australia, South America and other equatorial regions of the solar power base of electricity delivery, can achieve the northern hemisphere, the southern hemisphere interconnect the main communication channel. Through the equator Interconnection model, not only can solve the equatorial solar power delivery problems, also can make the southern and northern hemisphere interconnect into realization. Because of the large gap of these continents in the time zone, solar radiation intensity and load have certain simultaneity. The peak time of North Africa sunshine in Europe is the same with Europe load peak. This can make the benefit of solar power generation, showed in Table 2.

Table 2. Figure with Equator Interconnection model

Connection	Sending and receiving	Line setting	2050 transmission capacity
North Africa to Europe	Solar power base of Morocco, Algeria, Tunisia, Libya, Egypt and other countries of North Africa to the South of Europe	Adopt ± 1100 kV HVDC transmission technology	Transmission capacity is 1.5 trillion kilowatt-hours / year; channel capacity is 300 million kilowatts.
The Middle East to South Asia	Solar power base of the Middle East, Saudi Arabia, Oman, the United Arab Emirates in South Asia to the western of India	Adopt UHV DC submarine cable connecting the Strait of Hormuz to reach Iran, then by land UHVDC from Pakistan to the western India Mumbai load center	Transmission capacity is 2.5 trillion kilowatt-hours / year; channel capacity is 500 million kilowatts.
Australia to Southeast Asia	Australian northern solar power base using high-voltage submarine cable cross-sea landing in Indonesia, and then through a short cross-sea distance through Singapore to Thailand	Adopt ± 1100 kV HVDC transmission technology	Transmission capacity is 1 trillion kilowatt-hours / year; channel capacity is 200 million kilowatts.

2.3. Wind, Solar and Other Renewable Energy Transmission Efficiency Model

Global energy interconnection can guarantee the energy supply of economic and social development. Relying on the global energy interconnection, can guarantee the long-term stable energy supply, by widely distributed large potential of clean energy. From now on, if wind and solar power grow at 12.4% per year, by 2050, the non-fossil energy will reach 80% of the world total energy consumption. By then, wind and solar energy will become the leading energy, but its development is less than 5% of the total resources.

Global energy interconnection can reduce energy supply costs. Take Asia and Europe interconnection as example, Asia is the sending area(Xinjiang Province of China, Kazakhstan, the Russian Siberian region and Mongolia)which is compiling natural gas, wind and solar and other clean energy, using ± 1100 kV HVDC transmission technology, sending power to Germany. The channel using hour of DC is 5500 comparing with Germany's receiving offshore wind, the program of the Asia and Europe inter-connection through Russia St.Peterburg relaying transmission is cheaper 30.4% than Germany offshore wind. The maximum power cost is 0.3648 yuan/kW·h. If we adopt the nonstop transmission of Asia and Europe, the maximum power cost is 0.526 yuan/kW·h, which is cheaper 43.8% than Germany offshore wind power. Through the implement of Asia and Europe inter-connection, this connection can reduce Germany power supply cost; the benefit of intercontinental connection is obvious.

Transmission efficiency is mainly to compare the landing cost with generation cost of sending area, analyse its economic efficiency. Considering the development mode of the global energy interconnection, the research on transmission benefit identification is divided into four levels, showed in Figure 1.

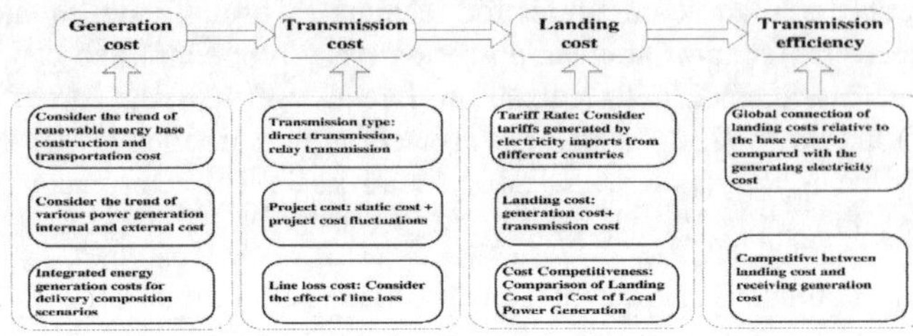

Figure 1. Analysis process of transmission efficiency

3. Identify the factors effecting the transmission changes cost

Table 3. Overview of system-friendly policy tools and impacts on System Value

	Policy tool	Country example	Impact on SV
System service capabilities	Grid codes that require advanced capabilities Advanced design of system services markets	Participation of wind in balancing the grid in Denmark and Spain	By providing system services from renewable energy, more thermal generation can be turned off during times of abundance
Location of deployment	Integrated planning of grid infrastructure and generation	Integrated planning in Brazil	Siting renewable energy generation in locations infrastructure Locational signals in remuneration available boosts
	Locational signals in remuneration schemes	Mexican auction system; differentiation of FIT levels in China	
Technology mix	Technology-specific auctions that reflect the value of each technology as determined in long-term planning	South Africa	Deploying a mix of technologies can lead to a more stable VRE profile and reduce periods of VRE excess, hence boosting SV
	System Value reflected in multi-technology auctions	Mexico	
Economic design criteria	Partial exposure to market prices via premium systems	German and Danish market premium systems, US Tax Credits	Investors are encouraged to choose a technology that generates during times of high electricity prices
Integrated planning, monitoring and revision	An integrated long-term plan for VRE and flexible resources, updated regularly	Integrated system planning in Denmark	Aligned deployment of VRE and flexible resources enhances SV; regular update of the long-term path allows reaping the full benefit of technology innovation

3.1. The generation cost of renewable energy

The sending area's generation cost, is considering renewable energy generation cost and different energy delivery compound mode.(1) Considering the characteristics of resources in each region, identify the trend of the renewable energy base construction and transportation cost.(2) Considering the development trend of power generation technology, identify the trend of various power generation internal and external costs. (3) Identify the costs of different electricity generation, such as hydropower, gas and coal, according to unit cost, coal price, gas price, utilization hours, tax rate, operating period, power consumption rate, annual operation and maintenance cost, and hydropower station submerged compensation. (4) Identify the synthetic generation cost under different power combing scheme, considering the proportion capacity of wind power and solar energy, hydropower regulatory capacity of the flood season, the regulation of gas and electricity capacity.

3.2. The cost trend of wind and solar development

Wind and solar power generation can affect the system cost through solving programs; it is mainly focusing on some certain aspects, showed in Table 3.

In 2015, wind and solar power generation amount accounts 90% to total new power increased. Between 2008 and 2015, the land wind power's average cost decreased about 35%. While the solar power generation cost decreased about 80%. In Figure 2, more evidence showed that, the nest generation wind and solar power generation technique will be more mature, and much more economy, which can push the wind and solar into the widely usage. The Figure gives, costs refer to global average of levelised cost of electricity (LCOE) with country specify assumptions on investment costs and cost of financing. Different costs per country are averaged weighted by annual capacity additions. The sending area's renewable energy bases, established in the area of wind and solar energy, facing the energy bundling sending out, its cost will decrease gradually. The solar energy LCOE will decrease from 0.13dollar/kw·h of 2015, to 0.055dollar/kw·h of 2025, the decreasing ratio is 59%. The onshore wind cost will drop to 0.053dollar/ kw·h of 2025, affected by wind tower, generator, wind farm development and other factors. Consider the renewable energy cost trend, the energy persistence, fast impact the coal, gas and other fossil energy. In the future, the solar and wind power sending costs will have more advantage.

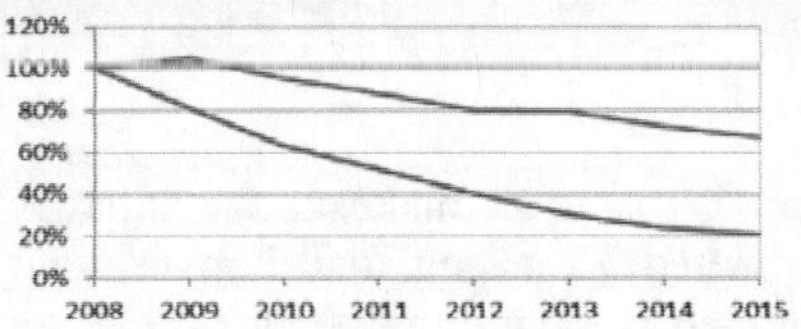

Figure 2. 2008-2015 wind and solar power generation cost curve

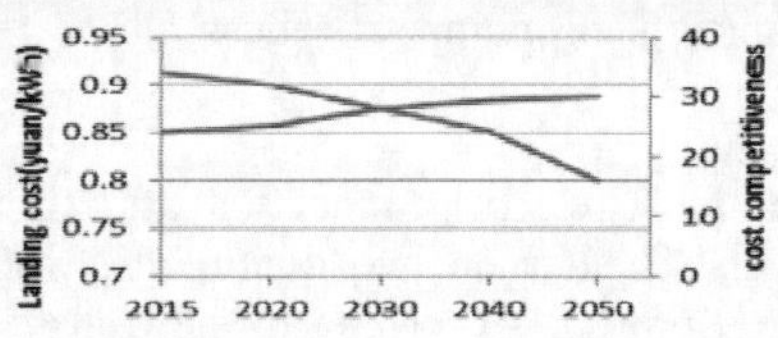

Figure 3. Under scheme two/1 analyse the landing cost and cost competitiveness

3.3. Transmission cost of transmission channel

Transmission costs of transmission channels depend on how to select transnational power transmission mode, the corresponding project cost and the cost of fluctuations in different countries, line loss and the cost of electricity in different countries and settlement changes in the situation and other issues.

(1) The choice of power transmission type, consider power supply and demand of inter-continent and inter-country, analyse the advantage of choosing direct or continuous transmission mode, analysis characteristics of UHV DC transmission project, transmission channel characteristics, utilization hours and its supporting power supply, including the grid to provide the price of transmission services.

(2) Engineering cost, combined with the above project planning to estimate the static cost of construction, project cost fluctuations, the relevant technical costs of the project cost-related.

(3) The line losses cost, should be calculated on the basis of the principle of tariff balance. It is assumed that the cost of landing is the settlement price of the grid companies that provide transmission

services and grid companies in the receiving areas, which is equivalent to the cost of local power purchase.

$$C_{lineloss} = C_{generation} \times K / (1-K) \quad (1)$$

$C_{lineloss}$ is line loss cost, $C_{generation}$ is generation cost, K is line loss.

3.4. The receiving area landing cost

The receiving area landing cost refers to the receiving area getting the sending renewable energy adds the transmission cost.

$$C_{landing} = C_{generation} + C_{line}(C_{lineloss}) + T_{tariff} \quad (2)$$

$C_{landing}$ is landing cost, C_{line} is transmission cost, T_{tariff} is tariff.

(1) Landing Cost should be calculated according to the cumulative principle of cost. Power transmission projects across continents, need to consider cross-border tariffs generated by different countries when electricity imports. Each transit country should be a link, based on DC lines, converter stations and supporting exchange projects in the country's investment were calculated transmission costs and line loss costs, the country's landing costs and based on the country's tariff rate. The resulting tariffs are added together as the cost of the next country's electricity generation.

(2) There are relay points in some countries. In the landing cost calculation, the DC cost of the before relay point and the cost of electricity generated by the surplus electricity are taken as the DC generation cost , which can be used as after the power-weighted average of the respective country.

3.5. The landing cost competitiveness

The landing cost competitiveness is to analysis and comparison, under the basic and global interconnection scenarios, the use of different energy combinations in the form of energy transmission area, compared with the cost of receiving electricity in the renewable energy power generation. The lower the landing cost, the more competitive the corresponding price will be, and it will have the better economical, the transmission efficiency will be obviously.

$$P = \left(1 - C_{landing} / C_{generation}\right) * 100\% \quad (3)$$

P is the landing cost competitiveness.

4. Prediction

Taking the most feasible connection of Asia and Europe as example, make up six intercontinental transmission schemes on two continents, which are estimated according to the investment of DC transmission project. According to the transmission cost, estimate wind and solar energy generation cost, transmission cost, landing cost, showed in Table 4.

4.1. The prediction of power generation cost

The generation cost and power usage hour Consider the proportion of wind and solar capacity, hydropower flood season adjustment, gas power adjustment; it can get different power generation cost, showed in Table 5.

4.1.1. Tariff China's tariff rate ranges from 0% to 35%, Poland ranges from 0% to 15% (average 4.2%), and Germany ranges from 5% to 17%.By the end of 2011, Russia, Belarus, Kazakhstan, the three countries jointly created the Eurasian Economic Commission, on September 1 of 2014, the weighted average tariff rate fell to 7.1%.In this research, it chooses the rate can be seen by Table 6.

4.1.2. The loss rate prediction ± 800, ± 1100 kilovolt converter station power loss is generally between 0.5% to 1%, this research make the value as 0.75% .

Table 4. The investment of sending side power(Million kilowatts, yuan / kWh）

Project design	**Channel introduction**	**Capacity**	**Wind power**	**Solar**	**Hydropower**	**Gas power**	**Generation price**	**Utilization hour**	**Total investment**
pure wind power	power connection at Xinjiang	1100	1700				0.5100	4400	1052.837
Wind+solar		1100	1700	800			0.5559	5600	1091.192
Siberian hydropower and Xinjiang wind power	Power connection at Siberian	1100	800		1100		0.5903	5700	786.935
Siberian gas, Xinjiang wind power and Mongolia solar		1100	800	200	1100		0.5932	6100	825.29
Kazakhstan gas, Xinjiang wind power	Power connection at Ekibastuz	1100	1200			1100	0.4525	6600	2056.671
Kazakhstan gas, Xinjiang wind power, the Mongolian solar		1100	1200	200		1100	0.4562	7000	1775.267

Table 5. The cost analysis of sending side power-jointed

	program design	synthesized generation cost（yuan/kWh）	Synthesized hour（h）
1	Kazakhstan gas / hydro + China Xinjiang wind power	0.4601	6000
2	Kazakhstan gas / hydro + China Xinjiang + Mongolia solar wind power	0.4978	5800
3	China Xinjiang wind power + Russia Siberia hydropower	0.4962	5900

Table 6. Tariffs along the lines

country	Calculation tariff（%）
China	17
Kazakhstan	7.1
Russia	7.1
Belarus	7.1
Poland	4.2
Germany	11

4.2. The transmission cost prediction

According to the transmission scheme and power-jointed program mentioned above, and combine with basic usage hour, can get the relative transmission cost, showed in Table 7.

Table 7. The corresponding transmission cost (yuan/kWh)

	one/1	one/2	two/1	two/2	three/1	three/2
Transmission cost 1	0.126225	0.140363	0.135270	0.149710	0.185018	0.201080
Transmission cost 2	0.078844	0.081095	0.078844	0.081095	/	/

4.3. The landing cost

According to the relative basic usage hour and its calculation method, can get each transmission scheme landing cost and cost competitiveness, showed in Table 8.

Table 8. The relative landing cost and competitiveness analysis (yuan/kWh、%)

	one/1	one/2	two/1	two/2	three/1	three/2
Landing cost	0.9905	1.0566	0.9121	0.9699	1.0768	1.1694
cost competitiveness	17.46	11.95	24.00	19.17	10.27	2.55

In Table 8, the landing cost of two/1 is lowest, and cost competitiveness is highest. While the landing cost of three/2 is highest, and cost competitiveness is the lowest. This showed that the relative landing cost of scheme two is higher than the scheme one. The relay transmission program's landing cost is lower than the direct transmission program. The power scheme two is higher than scheme two and three. So scheme two's line loss cost is lower than other two schemes.

While the scheme one and three covers one more country than scheme two. And because of China is the DC original point, considering the tariff of covering countries, offset the making Kazakhstan as the original point scheme's transmission loss affection. So the landing cost of scheme one and three is higher than two. And the tariff of scheme three is obvious high. Cost competitiveness of direct scheme can not get the same affection with relay scheme. The landing cost of scheme one and two is lower than scheme three.

Figure 3 gives the 2015-2050 landing cost and cost competitiveness. Considering all kinds of factors, as the jointed power basement reset power proportion of the scheme one and two is the same, the after generation power cost of scheme one is higher than scheme two. We can get the result is that; the lower of generation cost, the landing cost affected by generation cost will be great.

5. Conclusion

The global clean energy resources are rich, water resources are more than 10 billion kilowatts, wind energy resources exceeds 1 trillion kilowatt, and solar energy resources exceed 100 trillion kilowatts, surpassing the human society. This research gives the global energy interconnection development model, through the northern hemisphere interconnection model and equator interconnection model can unite the whole world energy. It also identifies the factors effecting the transmission changes cost, including generation cost, transmission cost (line loss) and landing cost. And take Asia and Europe connection as the prediction example, estimate these two continents cost benefit and variable power-jointed scheme cost competitiveness. The research showed that under the global energy interconnection mode, the trans-continent mode had better benefit, and the landing cost is good to be used, can solve the pollution and energy restriction.

Reference

[1] Tian X and Sun Y 2014 *Power System Technology* **39** p 663–668
[2] Wei X, Zhang J and Huang H 2010 *Power System Technology* **39** p 46–50
[3] Yang H, Wang D and Liu D 2011 *Power System Protection and Control* **43** p 91–96
[4] Huang Y, Wang Z, Liu J, Zhang L and Li J 2014 *Electric Power Construction* 32 p 100–103
[5] Xu J 2013 *North china electric power university* p 29
[6] Sun B 2015 *Zhejiang University* p 121

Magnetic seeding sedimentation (MSS) of coal slimes

Xiqing Wu[1], Tao Yue, Liang Dai

School of Minerals Processing and Bioengineering, Central South University, Changsha 410083, China

E-mail: xiqingwu@csu.edu.cn

Abstract. Magnetic seeding sedimentation (MSS), i.e. adding magnetic seeds and pre-magnetization for sedimentation, is a technique especially for sedimentation of fine slimes, improving the sedimentation performance by introducing the magnetic interactions between particles in a suspension and enlarging the apparent size of the fine particles. The fine coal slimes with a size of 66.68%-38μm were investigated by the MSS. Sedimentation tests were conducted, and some measurements, such as laser size analysis, magnetic susceptibility by vibrating sample magnetometer (VSM), were also applied in order to probe the mechanism of the MSS. Based on the tests, measurements and calculations it was demonstrated that the sedimentation of coal slimes increased with the additions of the magnetic seeds, and in the presence of the polyacrylamide, and also there appeared a relatively large apparent size of slimes after additions of magnetic seeds and/or polyacrylamide. So, the reason for the influence of MSS lies in fact that the presence of the polyacrylamide intensified the adsorption of magnetic seeds on the coal particles and the coverage of the magnetic seeds on the coal surface from 0.2% wt. to1.3% wt., resulting in increased magnetic susceptibility of coal particles from $9.13\times10^{-9}m^3/kg$ to $22.17\times10^{-9}m^3/kg$ and thus a low magnetic field strength of pre-magnetization needed for the magnetic agglomeration to happen among the coal particles (the threshold of magnetic field strength for agglomeration) from 602mT to 24mT prior to proper sedimentation.

1. Introduction

Fine minerals with fine-grained particle size are not suitable for effective sedimentation, and further study on the sedimentation of fines has important practical significance. Take coal slimes as an example, slimes produced during the coal washing process, forms a stable suspension easily with difficult sedimentation due to its good hydrophobic surface, fine particle size, less momentum, high clay content, high negative potential, and etc. In order to improve the sedimentation of coal slimes, quite a few of settling techniques have been investigated. Most of them are based on the addition of various organic and inorganic chemical agents, such as the application of new organic flocculants [1-2], the use of mixed organic flocculants [3], the combined application of surfactants and flocculants [4-6], combination of inorganic coagulants and organic flocculants [7], the introduction of magnetic field and electric field [8-10], and etc.

Magnetic seeding flocculation is a technology of separation which is mainly used in magnetic minerals [11-12]. The author [13] patented a new method, so-called 'magnetic seeding flotation' (MSF), can be used for fines flotation of coal slims, enlarging the apparent size of fine particles by magnetic seeding agglomeration (MSA). This method is characterized in that the directional adsorption of magnetic seeds and agents on the target mineral surface, enlarging the apparent size of fine particles.

Based on the method of MSA, this paper is intended to investigate the magnetic seeding sedimentation (MSS) of coal slimes, and to explore the mechanism of magnetic seeding agglomeration (MSA) of mineral slimes as well.

2. Materials and methods

2.1. Materials

(1) Coal slimes: taken from a colliery in Shanxi province, China, containing 44.21% of fixed carbon and 28.41% of ash. Its particle size was 66.80% -38μm, and its specific surface area 1.33m2/g, so it's far below the optimum particle size for flotation. The X-ray diffraction analysis showed that the ash was mainly composed of kaolinite and quartz.

This slimes was mainly used for the investigation of the MSS and size analysis of the magnetic seeding agglomeration (MSA) in this study.

(2) Pure coal: purified from the above coal slimes by the following sub-processes: screening two times respectively with the apertures 0.5mm and 0.074mm to obtain the fraction between the two sizes, then two gravity separations by a shaking table to achieve the purity of containing 2 wt.% ash, and then being ground in a ceramic ball mill for ten minutes after filtering and drying. This sample was mainly used for magnetism analysis.

(3) Magnetic seeds: synthesized in laboratory by the method of air-oxidation of ferrous salts in an open system [14]. The synthesis was conducted in a 200 ml beaker with pure water inside. The water was maintained at 85 ℃ while stirring at a speed of 300 r/min, and then $FeSO_4 \cdot 7H_2O$ and ammonium hydroxide were added in sequence. The synthesis was finished after 30 min stirring, and the formed suspension was kept in store for use, and the average particle size of the magnetic seeds was 2.05μm.

(4) Polyacrylamide（PAM）: anionic type with a molecular weight of 3 million.

2.2. Methods

(1) Sedimentation test

A suspension with the concentration of the slimes 20% wt., with additions of the magnetic seeds, the PAM or the PAM plus magnetic seeds, was conditioned for 3 minutes in a pre-magnetizing blender with the volume 100 ml and the maximum magnetic field intensity 50mT [15], and then was poured into a 100 ml graduated glass cylinder for sedimentation test.

(2) Laser diffraction for agglomerate size analysis

A suspension with the concentration of the slimes 5%wt., without or with additions of the magnetic seeds and/or the PAM, was conditioned for 3 minutes in the above pre-magnetizing blender and then was introduced into a beaker for laser diffraction analysis by the equipment Mastersizer2000.

(3) Vibrating sample magnetometer (VSM) analysis

Hysteresis loops of the pure coal, the pure coal coated by the magnetic seeds, and the pure coal coated by the magnetic seeds at the presence of the PAM, were obtained by vibrating sample magnetometer HH-15.The pure coal coated by magnetic seeds was prepared by mixing the pure coal and the magnetic seeds in the above pre-magnetizing blender for 4 minutes, then the slurry passed through a magnetic separation tube with a magnetic field intensity of 150mT to remove free magnetic seeds, and the passed tailing was then filtered and dried for the VSM [16].

3. Magnetic seeding sedimentation (MSS) tests and size analysis

3.1. Magnetic seeding sedimentation (MSS) tests

The influences of some variables on the magnetic seeding sedimentation were tested, and the results are shown in following figures. It can be seen in Figure 1 that in the presence of the magnetic seeds the polyacrylamide has a stronger sedimentation than the PAM alone, i.e. the MSS is conducive to the sedimentation of the coal slimes.

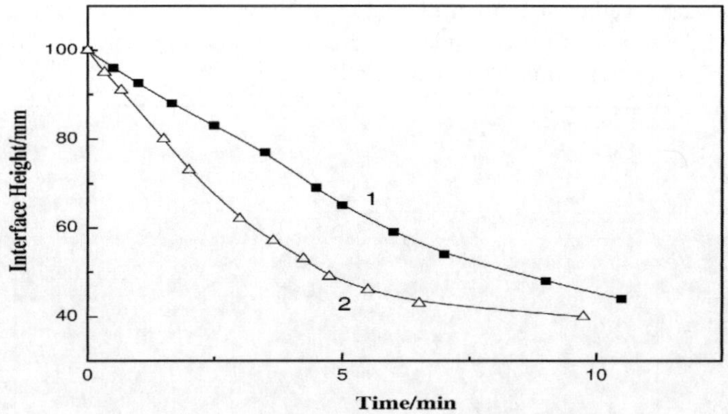

Figure 1. Sedimentation curves of coal slimes
1-60mg/L polyacrylamide, 2-60mg/L polyacrylamide+70mg/L magnetic seeds

Considering that the magnetic field intensity of pre-magnetization is an important parameter in the process of MSS, the influence of the magnetic field intensity on the MSS, under the condition of 60mg/L Polyacrylamide plus 35mg/L magnetic seeds， as shown in Figure 2.

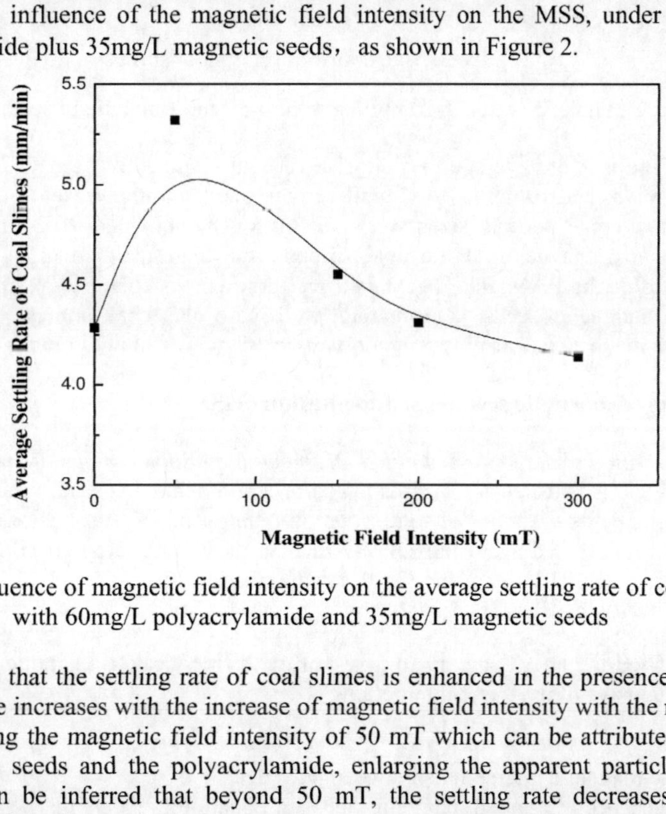

Figure 2. Influence of magnetic field intensity on the average settling rate of coal slimes with 60mg/L polyacrylamide and 35mg/L magnetic seeds

Figure 2 indicates that the settling rate of coal slimes is enhanced in the presence of the magnetic field. The settling rate increases with the increase of magnetic field intensity with the maximum rate of 5.31 mm/min by using the magnetic field intensity of 50 mT which can be attributed to the MSA of the slimes, magnetic seeds and the polyacrylamide, enlarging the apparent particle size for MSS. However, it also can be inferred that beyond 50 mT, the settling rate decreases which may be attributed to the self-agglomeration of magnetic seeds prematurely during the process of pre-magnetization, reducing the effect of MSA, and then reducing the settling rate of coal slimes.

3.2. Size analysis in the MSS

In consideration of the magnetic seeds and the coal particles are both in the range of micro sizes, the laser diffraction method was applied to analyze the sizes of coal slimes in the MSS in order to investigate if the magnetic seeding agglomeration (MSA) happens. For the convenience to observe the size changes, all size distribution curves, under conditions of blank, magnetic seeds, polyacrylamide, and magnetic seeds plus polyacrylamide, are listed in Figure 3.

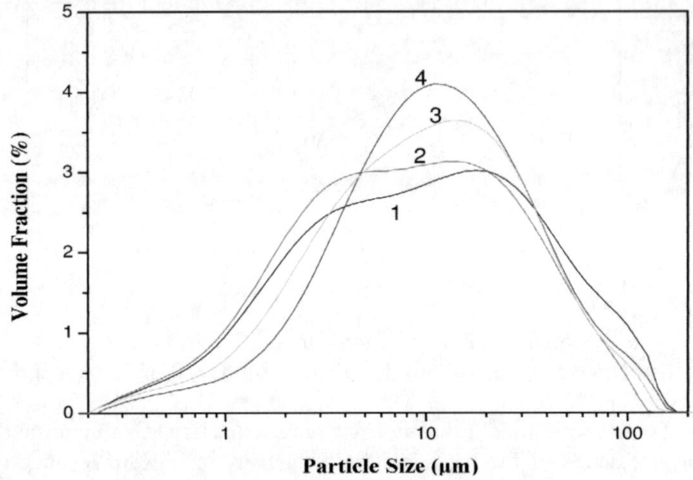

Figure 3. Size distribution of coal slimes

1-blank, 2-50 mg/L magnetic seeds, 3-10 mg/L PAM, 4-50 mg/L magnetic seeds and 10 mg/L PAM

As shown in Figure 3, the average particle size of coal slimes increased to 21.74μm after adding 50mg/L magnetic seeds and 10 mg/L PAM whilst in the blank and the presence of the magnetic seeds or the PAM their average particle sizes were 16.40μm, 18.25μm and 20.24μm, respectively. It is inferred from the size changes that the coal slimes were agglomerated after the additions of the magnetic seeds and/or the PAM. The PAM is a well-known flocculant [17], and by contrast it is a surprise that the magnetic seeds agglomerate well with the coal slimes, and particularly the combination of the magnetic seeds plus polyacrylamide is a powerful agglomerant.

4. The mechanism of magnetic seeding sedimentation (MSS)

4.1. General explanation of the MSS-Expanded DLVO with additional magnetic interaction

Whether the particles in a suspension are dispersed or agglomerated is determined by the interactions between particles, and the variation of total potential energy of the suspension system is a decisive factor on its stability. According to the DLVO theory, the total energy of colloidal particles V_T is summed by electric potential V_R and Van der Waals interaction V_A:

$$V_T = V_R + V_A \qquad (1)$$

Svoboda expanded the DLVO theory to a system with fine weakly magnetic particles, and built a theoretic model inclusive of magnetic interactions [18]:

$$V_T = V_R + V_A + V_M \qquad (2)$$

where V_M is the magnetic interactions between particles.

While in the presence of applied magnetic field, an equation to calculate the magnetic energy was proposed:

$$V_M = -\frac{32\pi^2 a^6 \chi^2 B_0^2}{9\mu_0 r^3} \qquad (3)$$

where a is the radius of particles, χ the volumetric magnetic susceptibility of the particles, B_0 the magnetic induction, μ_o the permeability of vacuum and r the distance between two particles.

Hence, the suspension of the slimes tends to be instable as a result of the pre-magnetization and magnetic seeds covering the surface of particles. In the MSA of the coal slimes, the combined positive effect of the magnetic seeds and polyacrylamide is noteworthy of further study. So, the mechanism of MSA in the MSS was probed through the vibrating sample magnetometer analysis (VSM) and related calculations below.

4.2. Determinations of magnetism by the VSM, coverage of magnetic seeds on coal surface and the critical magnetic field intensity BA for agglomeration

According to equation (3), the magnetic interaction VM is highly related to the volumetric magnetic susceptibilityχ. The hysteresis loops of the pure coal, the pure coal coated by the magnetic seeds, and the pure coal coated by the magnetic seeds at the presence of the PAM, were obtained by the VSM, as shown in Figure 4.

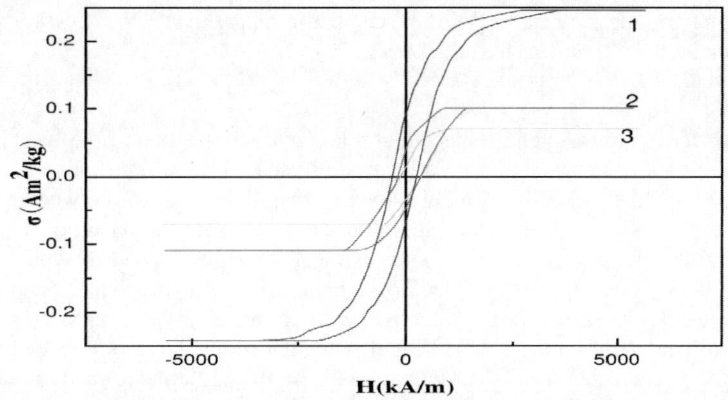

Figure 4. Hysteresis loops of coal

1-Coal-35 mg/L magnetic seeds and 12 mg/L PAM, 2-Coal-35 mg/L magnetic seeds, 3-Coal

As illustrated in Figure 4, surprisingly the pure coal presents a weak magnetism of saturation magnetic induction 0.065 Am^2/kg probably due to contamination of trace of pyrite inside. After the additions of the magnetic seeds and the PAM, its saturation magnetic induction reached 0.244Am^2/kg whilst only addition of the magnetic seeds, its saturation magnetic induction raised to 0.089Am^2/kg. This may be attributed to the combined effect of the magnetic seeds and the PAM. Without doubt, the increase of the magnetism of the coal means a largely increased value of magnetic interactions between particles VM based on the equation (3).

Furthermore, the author [19] translated the difference of their magnetisms into the adsorption or coverage of the magnetic seeds on the surface of coal by an indirect model, and the results can be briefed as follows: in the same dosage of the magnetic seeds 35 mg/L, the coverage of the magnetic seeds on the coal surface is increased from 0.2% wt. to 1.3% wt., namely, the PAM strengthens the adsorption of the magnetic seeds on the surface of coal.

According to equation (3), the magnetic energy is related to the applied magnetic field B_0 (i.e. .the pre-magnetization field intensity here) as well. Watson [20] put forward a formula to calculate the critical magnetic field intensity B_A for flocculation of minerals to happen as follows:

$$B_A = -\left(\frac{2KT\mu_0}{\pi^2 x^2 a^3}\right)^{\frac{1}{2}} \quad (4)$$

where K is Boltzmann constant, its value 1.3806505×10^{-23} J/K; T degree Kelvin, μ_0 the vaccum permeability, its value $4\pi\times10^{-7}N/A^2$; χ is specific magnetic susceptibility of the mineral, and a the radius of mineral particles. The result of calculation is listed in Table 1.

Table 1. Critical magnetic field intensity of coal to flocculation

Sample	χ ($10^{-9}m^3/kg$)	B_A (mT)
Coal	0.89	601.95
Coal with magnetic seeds	9.13	58.67
Coal with magnetic seeds+ PAM	22.17	24.16

As shown in Table 1, the threshold of magnetic field strength for coal to agglomeration decreased greatly by adding the magnetic seeds, especially by simultaneously adding the magnetic seeds and the PAM.

In addition, based on his calculations the author [19] also explained that when the magnetic seeds are placed in a 50mT pre-magnetization field, their specific magnetization is $13.4Am^2/kg$, resulting in a magnetic field intensity on their surface 112.26mT.

5. Conclusions

(1) The MSS (the addition of magnetic seeds under a low applied magnetic field) was beneficial to the sedimentation of coal slimes, and also greatly strengthened with the aid of polyacrylamide.

(2) The mechanism of the MSS lies mainly in the magnetic actions between particles in the suspension. The MSS promoted the agglomeration of slimes, resulting in properly-sized agglomerates right for sedimentation. The average apparent sizes of coal slimes in four agglomeration conditions, such as the blank, magnetic seeds, polyacrylamide and polyacrylamide plus magnetic seeds, were 16.41μm, 18.25μm, 20.24μm and 21.74μm, respectively. The increase of granularity of coal slimes under the MSS facilitated the sedimentation process.

(3) The effect of the polyacrylamide plus magnetic seeds on the agglomeration stood out among them. The PAM strengthened the adsorption of magnetic seeds on the surface of coal particles from mass fraction 0.2% to 1.3%, consequently raising the specific susceptibility from $9.13\times10^{-9}m^3/kg$ to $22.17\times10^{-9}m^3/kg$. So, the MSA under the condition of polyacrylamide plus magnetic seeds reduced the critical magnetic field intensity for the agglomeration of coal slimes from 602 mT to 24 mT.

Acknowledgements

The authors would like to acknowledge the financial support from the National Natural Science Foundation of China (No.51274256).

References

[1] Karmakar N C, Rath S K, Sastry B S, Singh R P 1998 Investigation on flocculation characteristics of polysaccharide - based graft copolymers in coal fines suspension *Journal of applied polymer science* **70**(13) 2619-25.
Pal S, Mal D, Singh R P 2005 Cationic starch: an effective flocculating agent *J. Carbohydrate Polymers* **59**(4) 417-423.

[2] Yu X 1993 Somasundaran P. Enhanced flocculation with double flocculants *J. Colloids and Surfaces A: Physicochemical and Engineering Aspects* **81** 17-23.

[3] Fan A 1998 *The role of polymers and surfactants in flocculation and dispersion of aqueous suspensions* (Columbia University).

[4] Besra, L, Sengupta D K, Roy S K, Ay P 2003 Influence of surfactants on flocculation and dewatering of kaolin suspensions by cationic polyacrylamide (PAM-C) flocculant *J. Separation and Purification Technology* **30**(3) 251-264.

[5] Besra, L, Sengupta D K, Roy S K, Ay P 2002 Studies on flocculation and dewatering of kaolin suspensions by anionic polyacrylamide flocculant in the presence of some surfactants *J.*

International journal of mineral processing **66**(1) 1-28.

[6] Fuerstenau D W, Rosenbaum J M, You Y S 1988 Electrokinetic behavior of coal *J. Energy & fuels* **2**(3) 241-245.

[7] Kochmarsky V 1996 Magnetic treatment of water: possible mechanisms and conditionsfor applications *J. Magnetic and Electrical Separation* **7** 77-107.

[8] Isaka K, Kimura Y, Sumino T 2013 *U.S. Patent No. 8,388,845*. Washington, DC: U.S. Patent and Trademark Office.

[9] Johnson T J, Davis E J 2000 Experimental data and theoretical predictions for the rate of electrophoretic clarification of colloidal suspensions *J. Environmental Science & Technology* **34**(9) 1806-12.

[10] Stratulat R, Calugaru G, Badescu V 2000 Magnetic carriers particles for selective separation in environmental and industrial processes *J. Analele Stiintifice Ale Universitatii, Tomul XLV-XLVJ* **45**.

[11] Terashima Y, Ozaki H, Sekine M 1986 Removal of dissolved heavy metals by chemical coagulation, magnetic seeding and high gradient magnetic filtration *J. Water Research* **20**(5) 537-545.

[12] Wu X Q, Dai C, Dai L 2014 Magnetic seeding flotation: China. Patent No. CN104117432A.

[13] Wu X Q, Yang P W, Cheng Z, Dai L 2013 Surface Magnetization of Hematite by Metal Ions. In *Advanced Materials Research* **785** 1104-10 Trans Tech Publications.

[14] Wu X Q, Yue T, Dai L 2016 Magnetic seeding flotation of fine minerals. In: *28th International Conference of Mineral Processing* Quebec Canada, Sept. 2016.

[15] Wu X Q, Xu P Y, Duan Y F, Hu C, Li G P 2012 Surface magnetization of siderite mineral *J. International Journal of Mining Science and Technology* **22**(6) 825-830.

[16] Lv Y T, Zhao L Y, Shi Q L 2013 Effects of magnetic field on flocculating sedimentation of slime-water. *Journal of Heilongjiang Institute of Science & Technology* **23**(5) 424-426.

[17] Svoboda J 1982 Magnetic flocculation and treatment of fine weakly magnetic minerals *J. IEEE Trans. Magn*. **18**(2) 796-801.

[18] Dai L 2015 *The magnetic seed-polymer agglomeration of coal slimes* (Changsha: School of Minerals Processing and Bioengineering, Central South University).

[19] Watson J 1976 Magnetic separation at high magnetic fields, *In: Sixth International Cryogenic Engineering Conference. IPC Science and Technology Press Ltd.*, Grenoble, France.

IOP Conf. Series: Earth and Environmental Science **52** (2017) 012004 doi:10.1088/1755-1315/52/1/012004

Highly improved hydrogen storage capacity and kinetics of the nanocrystalline and amorphous $PrMg_{12}$-type alloys by mechanical milling

Y H Zhang [1, 2, 3], H W Shang[2], Y Q Li[2], Z M Yuan[2], T Yang[2] and D L Zhao[2]

[1] Key Laboratory of Integrated Exploitation of Baiyun Obo Multi-Metal Resources, Inner Mongolia University of Science and Technology, Baotou 014010, China

[2] Department of Functional Material Research, Central Iron and Steel Research Institute, Beijing 100081, China

E-mail: zhangyh59@sina.com

Abstract. Nanocrystalline and amorphous $PrMg_{11}Ni + x$ wt.% Ni ($x = 100, 200$) alloys were synthesized by mechanical milling. Effects of Ni content and milling duration on the structures, hydrogen storage capacity and kinetics of the as-milled alloys were investigated systematically. The structures were characterized by XRD and HRTEM. The hydrogen desorption activation energy was calculated by using Kissinger method. The results show that increasing Ni content dramatically improves the electrochemical discharge capacity of the as-milled alloys. Furthermore, the variation of milling time has a significant impact on the kinetics of the alloys. As the milling time increased, the high-rate discharge ability (HRD), gaseous hydrogen absorption capacity and hydrogenation rate increased at first but decreased finally, while the dehydrogenation rate always increased.

1. Introduction

The threat of exhaustion of oil has always perplexed human. It is necessary to develop and use clean energy. Hydrogen is an ideal clean energy, however, it is limited in practical use due to lack of safe methods to store hydrogen [1]. $REMg_{12}$-type alloys are regarded to be the most promising alternative to satisfy the requirements for practical application because of their high hydrogen storage capacity [2]. However, some inherent drawbacks, such as sluggish kinetic and poor cyclic stability, have to be improved for the alloys.

Alloying and microstructure modification are effective approaches for improving the hydrogen storage properties of Mg-based alloys [3, 4], such as partially substituting Mg with rare earth elements [5-8] and adding transition metals [9, 10]. In our previous study, the $Mg_{10}NiR$ (R = La, Nd and Sm) alloys possess excellent kinetics by adding rare earth element [11]. Furthermore, the kinetics of the Mg-based alloys is very sensitive to the structures [12]. In the present work, Mg in the $PrMg_{12}$-type alloy was partially substituted by Ni, and mechanical milling was used to fabricate the nanocrystalline and amorphous $PrMg_{11}Ni + x$ wt.% Ni ($x = 100, 200$) alloys. The effects of Ni content and milling time on the hydrogen storage capacity and kinetics have been investigated in detail.

2. Experimental

The experimental alloys were prepared using a vacuum induction furnace. For avoiding the deviation from the stoichiometry, extra 10 percent of Mg was adding into the crucible to make up for Mg losses during the smelting process, which have been proved quite effective and appropriate in our previous study. In addition, helium is used as a protective atmosphere to restrain Mg from volatilization. Besides that, the method of gradually increasing temperature was taken and the process parameters were repeated adjusted to short the melting time. The molten alloy was poured into a copper mould cooled and a cast ingot was obtained. The as-cast alloys were mechanically crushed into powder with a diameter of about 50 μm and mixed with carbonyl nickel powder with a weight ratio of 1:1 and 1:2 respectively. Then the mixed powder was mechanically milled in a planetary-type mill. The milling speed was 135 rpm and the duration time was 5, 10, 20, 40 and 60 h, respectively.

The phase structures were determined by X-ray diffraction (XRD) (D/max/2400). The diffraction, with the experimental parameters of 160 mA, 40 kV and $10°\cdot min^{-1}$ respectively, was performed with $CuK_{\alpha 1}$ radiation filtered by graphite. The powder samples of the as-milled alloys were observed by high resolution transmission electron microscope (HRTEM) (JEM-2100F, operated at 200 kV) and the crystalline states were ascertained by electron diffraction (ED).

The electrochemical properties of the alloys were measured at 303K using a tri-electrode open cell. The alloy electrodes were first charged, then rested for 15 min, and finally discharged to the cut-off voltage of -0.500 V with a constant current density of 60 $mA\cdot g^{-1}$ The electrochemical impedance spectra (EIS) were measured at different temperature using an electrochemical workstation (PARSTAT 2273). The fresh electrodes were fully charged and then rested for 2 h for achieving to the stabilization of the open circuit potential. The EIS of the alloy electrodes were measured at a 50% depth of discharge (DOD), the frequency ranging from 10 kHz to 5 mHz, and the amplitude of signal potentiostatic or galvanostatic measurements being 5 mV. The potentiodynamic polarization curves were measured in the potential range of -1.2 to +1.0 V (vs. Hg/HgO) with a scan rate of 5 $mV\cdot s^{-1}$.

The gaseous hydrogen absorption capacity and kinetics of the alloys were measured by an automatically controlled Sieverts apparatus. The hydrogen absorption was conducted at 3 MPa of hydrogen pressure and 573 K of temperature. The hydrogen desorption was conducted under the pressure of 1×10^{-4} MPa and the same temperature. Hydrogen desorption properties were also measured by using DSC (NETZSCH STA 449F3) at heating rates of 5, 10, 15 and 20 $K\cdot min^{-1}$.

3. Results and discussion

3.1. Microstructure characteristics

The phase components and structure of the as-cast and milled $PrMg_{11}Ni + x$ wt.% Ni (x =0, 100, 200) alloys are identified by XRD analysis, as shown in figure 1. The as-cast $PrMg_{11}Ni$ alloy contains a major phase $PrMg_{12}$ and a secondary phase Mg_2Ni. Mechanical milling makes the diffraction peaks of the $PrMg_{12}$ and Mg_2Ni phases almost disappear, indicating that the crystalline structure has transformed to the nanocrystalline or amorphous structure. The diffraction peaks become broaden is because of a large amount of internal energy were generated in the alloys. In addition, increasing Ni content makes the intensity of diffraction peaks evidently lower and the width clearly increase. It means that increasing Ni content facilitates the glass forming of the alloys. As reported by Abdellaoui et al. [13], increasing Ni content in $REMg_{12}$ alloys can lower the activation energy that crystalline turns to amorphous. Moreover, it can be seen from figure 1 (b) that PrH_3, MgH_2 and Mg_2NiH_4 are formed in the hydrogen absorption reaction, originating from the following reactions [14, 15]:

$$Pr_{12}Mg_{17} + H_2 \rightarrow MgH_2 + PrH_3 \quad (1)$$

$$Mg_2Ni{+}H_2 \rightarrow Mg_2NiH_4 \quad (2)$$

Also, hydrogenation results in the diffraction peaks become broaden, which is ascribed to the lattice internal stress and cell volume expansion generated by hydrogen atoms entering the lattice interstitials. The saturated hydrogen absorption alloys still maintain a nanocrystalline and amorphous structure, meaning that they have good structural stability.

Figure 2 shows HRTEM micrographs and ED patterns of the as-milled $PrMg_{11}Ni + x$ wt.% Ni (x =

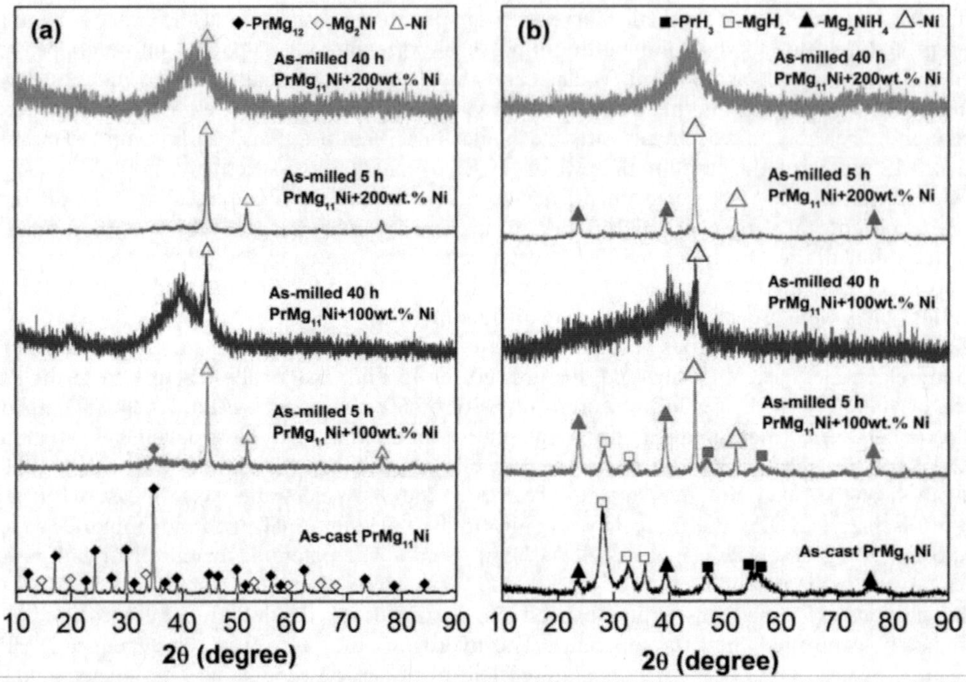

Figure 1. XRD profiles of the as-cast and milled $PrMg_{11}Ni + x\%$ Ni (x = 0, 100, 200) alloys before and after hydriding: (a) before hydriding, (b) after hydriding.

100, 200) alloys. It can be seen that the as-milled alloys exhibit a nanocrystalline and amorphous structure, and the amount of the amorphous phase in the as-milled alloys visibly increases with milling time and Ni content increasing, which is supported by the analysis of Debye-Scherrer rings. Also, it is noted that prolonging milling time from 5 h to 20 h gives rise to the grain sizes of the alloys visibly decrease and the disordered degree of the microstructure increase dramatically. It is ascribed to the generation of a large amount of internal energy during mechanical milling, which causes non-stabilization of the lattice, ultimately generating the fine ultrafine grain or amorphous structure.

3.2. Electrochemical discharge capacity and kinetics

Figure 3 presents the relationship between the discharge capacity of the alloys and the milling time. The as-milled alloys exhibit excellent activation capability, reaching to the maximum discharge capacities at the first cycle. The discharge capacities of the alloys markedly increase with milling time increasing, but it is noteworthy that long time milling (more than 40 h) brings on a slight decrease in the discharge capacity for the x = 200 alloy. As desired, increasing Ni content has a great contribution on the discharge capacity of the alloys. To be specific, increasing Ni content from 100 to 200 wt.% makes the discharge capacity of the as-milled (40 h) alloys grow from 141.8 to 1059.4 $mAh \cdot g^{-1}$ at a discharge current density of 60 $mAh \cdot g^{-1}$. Mechanical milling significantly ameliorates the discharge capacity which is most likely ascribed to the formation of the nanocrystalline structure due to hydrogen concentrations in the grain boundary are much higher than that of the grain interior and amorphous region. In addition, it is believed that hydrogen atoms adhere more strongly to the surface of the nanocrystalline nickel than the polycrystalline nickel, and facilitating the dissociation of hydrogen. As for the positive effect caused by increasing Ni content on the discharge capacity, it is unquestionably ascribed to its high electrocatalytic activity. Mechanical milling can cause the cold welding of nickel particles and alloy particles, forming the highly dispersed metallic Ni nanocrystals through alloy matrix which reduces the stability of the hydride and improves the discharge capacity of

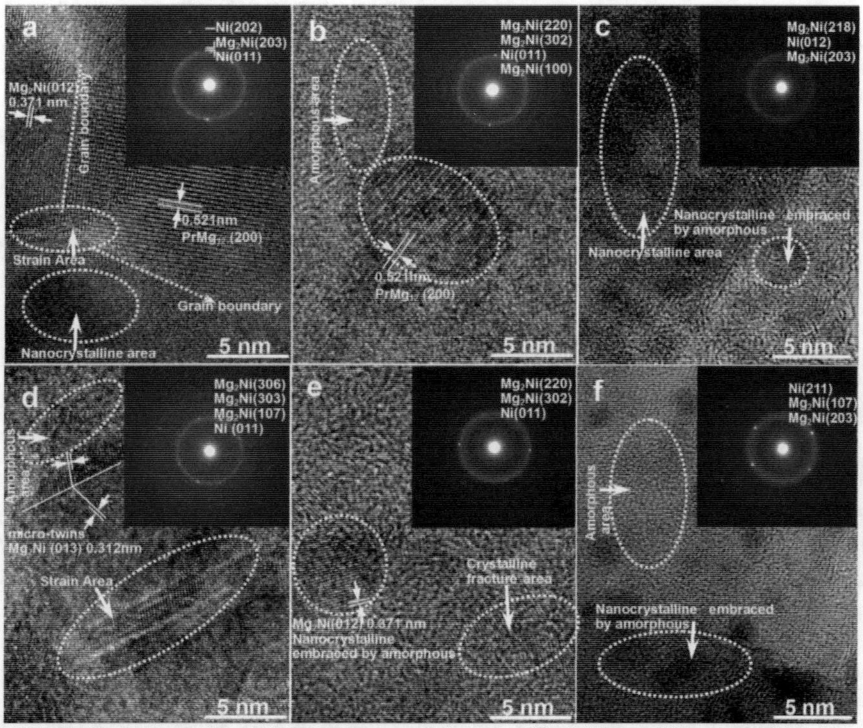

Figure 2. HRTEM micrographs and ED patterns of the as-milled $PrMg_{11}Ni + x$ wt.% Ni (x = 100, 200) alloys: (a), (b) and (c) x = 100 alloy milled for 5, 20 and 40 h; (d), (e) and (f) x = 200 alloy milled for 5, 10 and 20 h.

the alloys. The electrochemical kinetics of an alloy electrode is characterized by high rate discharge ability (HRD), which is defined as: HRD= $C_{600}/C_{60}\times100\%$, where C_{600} and C_{60} are the maximum discharge capacities of the alloy electrode at the current densities of 600 and 60 $mA{\cdot}g^{-1}$, respectively. The milling time dependence of the HRD of as-milled $PrMg_{11}Ni + x$ wt.% Ni (x = 100, 200) alloys is depicted in figure 4. Evidently, the HRDs of the alloys first increase and then decrease with milling time prolonging, and the maximum HRDs of the alloys are 68.53% and 71.98%, respectively. In addition, the x = 200 alloy exhibits a higher HRD than that of the x = 100 alloy for the same milling duration, indicating that increasing Ni content facilitates the improvement of the electrochemical kinetics of alloys. When the ball mill time doesn't exceed 20 h, the HRDs only have a little difference. However, when the ball mill time is more than 20 h, the difference of the HRDs has a sharp increase, suggesting that the catalytic action of Ni on the electrochemical kinetics of alloy electrodes is closely related to the size and the surface state of Ni particles.

The HRDs of an alloy electrode basically depends on the hydrogen diffusion capability in the alloy bulk and the charge-transfer rate on the surface of the alloy electrode. To reveal the mechanism of the electrochemical kinetics of alloys, the effects of Ni content and milling time on the diffusion ability of hydrogen atoms and the charge-transfer rate are studied. Here, we use the limiting current density (I_L) to evaluate the effect of the Ni content and milling duration on the hydrogen diffusion rate of the alloy electrodes. The I_L value can be obtained by measuring the potentiodynamic polarization curve of an alloy electrode, as depicted in figure 5. An inflection can be clearly seen in each anodic polarization curve, which is termed as limiting current density (I_L). It is deemed an oxidation reaction occurring on the surface of the alloy electrode, and the oxidation layer hindering hydrogen atoms from further penetrating. Thereby, the limiting current density (I_L) can be viewed as a critical current density for passivating. Based on the data in figure 5, the relationships between the I_L values of the alloys and milling time can be established, as inserted in figure 5 (a) and (b), respectively. Obviously, the I_L

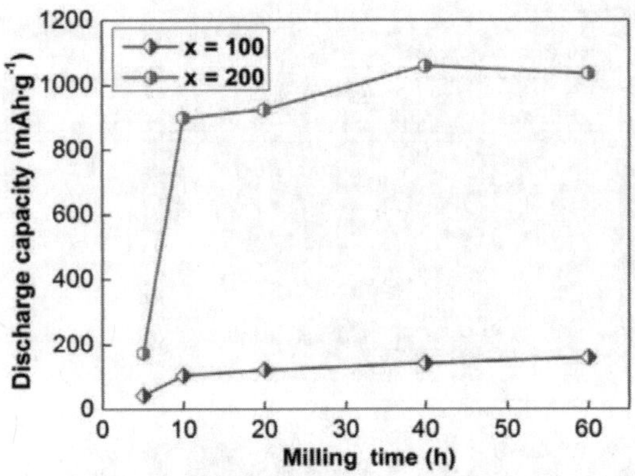

Figure 3. Evolution of the discharge capacity of the as-milled $PrdMg_{11}Ni + x$ wt.% Ni (x = 100, 200) alloys with milling time.

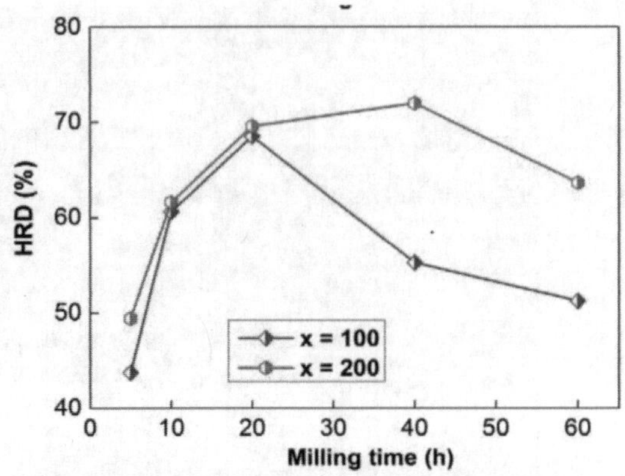

Figure 4. Evolution of the HRD of the as-milled $PrdMg_{11}Ni + x$ wt.% Ni (x = 100, 200) alloys with milling time.

values of the alloys first increase and then decrease with milling time prolonging. A close inspection shows that the milling times corresponding to the maximum I_L values are different with HRDs, which implies that there is other factor affecting electrochemical kinetics of the alloys in addition to the diffusion ability of hydrogen atoms. Apparently, it is charge transfer rate which can be characterized by the electrochemical impedance spectrum (EIS).

The EIS curves of the as-milled (5h) $PrMg_{11}Ni + x$ wt.% Ni (x = 100, 200) alloys were measured at different temperatures (303 K, 313 K and 323 K), as shown in figure 6. There are two distorted capacitive loops at the high and the middle frequency region as well as a line at the low frequency region in each EIS. The smaller semicircle in the high frequency region represents the contact resistance between the alloy powder and the conductive material, and the larger one represents the charge-transfer resistance (R_{ct}) on the alloy surface while the straight line in the low frequency region represents Warburg impendence. Kuriyama considered that the R_{ct} depends on both the reactivity of the alloy surface and reaction area. And the electrochemical reactivity of the alloy surface can be determined with apparent activation enthalpy $\Delta_r H^*$, which can be obtained by the following equation:

$$\ln\left(\frac{T/R_{ct}}{K/\Omega}\right) = C_0 - \frac{\Delta_r H^*}{RT} \quad (3)$$

where R_{ct} is the charge-transfer resistance for the metal hydride electrodes, R is the gas constant, T is

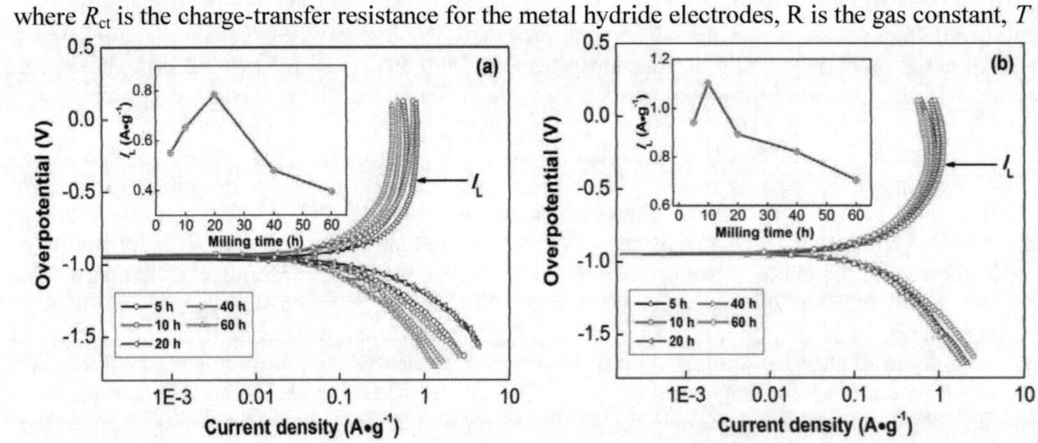

Figure 5. Potentiodynamic polarization curves of the as-milled $PrMg_{11}Ni + x$ wt.% Ni (x = 100, 200) alloys: (a) x=100, (b) x=200.

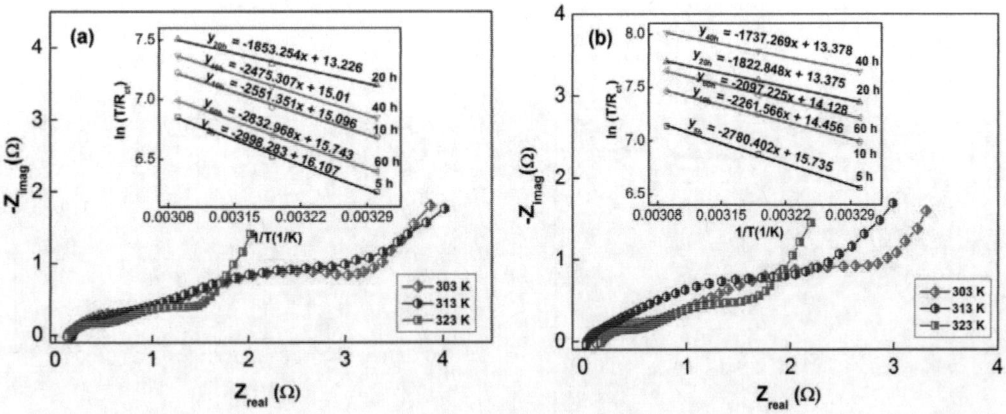

Figure 6. Plots of $\ln(T / R_{ct})$ vs. $1/T$ of the as-milled $PrMg_{11}Ni + x$ wt.% Ni (x = 100, 200) alloys and the EIS of the as-milled (5 h) alloys: (a) x = 100, (b) x = 200.

the temperature and C_0 is a constant in which the surface area is included. The Kuriyama plots of $\ln(T / R_{ct})$ vs. 1/T of the as-milled alloys are nearly linear, as inserted in figure 6. From the slopes of the Kuriyama plots, the $\Delta_r H^*$ can be calculated. The $\Delta_r H^*$ first decreases and then increases with milling time increasing, which is opposite with the change trend of HRDs, implying that the electrochemical kinetics is basically dominated by the charge-transfer ability on the surface of the alloy electrode. Therefore, it can be inferred that the improvement of HRDs of the as-milled alloys are principally ascribed to the decrease of the activation enthalpy by increasing milling time.

3.3. Gaseous hydrogen absorption/desorption capacity and kinetics

Figure 7 shows the variations of the hydrogen absorption capacities of the as-milled $PrMg_{11}Ni + x$ wt.% Ni (x = 100, 200) alloys with the hydrogen adsorption time at 3 MPa H_2 and 573 K. The alloys exhibit a very fast hydrogen absorption rate in the initial stage and the hydrogen content is almost saturated in 100 s. Meanwhile, the variation of milling time has an obvious effect on the hydrogen absorption capacity of the alloys. The relationship between the hydrogen absorption capacity C_{100}^{a} (approximately represents the maximum hydrogen absorption capacity) and the milling time are inserted in figure 7. The decisive factor of the hydrogen desorption reaction can be determined by calculating activation energy. Here, Kissinger method is used to estimate the hydrogen desorption activation energy of the

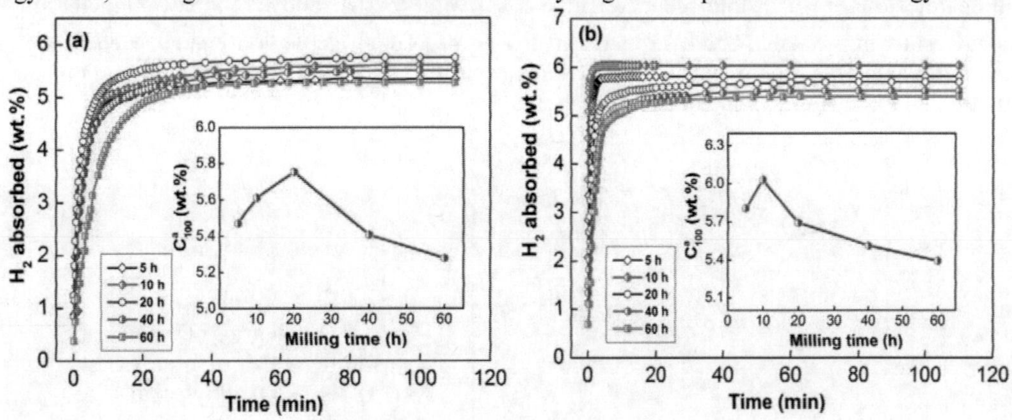

Figure 7. Hydrogen absorption capacity and kinetic curves of the as-milled $PrMg_{11}Ni + x$ wt.% Ni (x = 100, 200) alloys at 573 K: (a) x = 100, (b) x = 200.

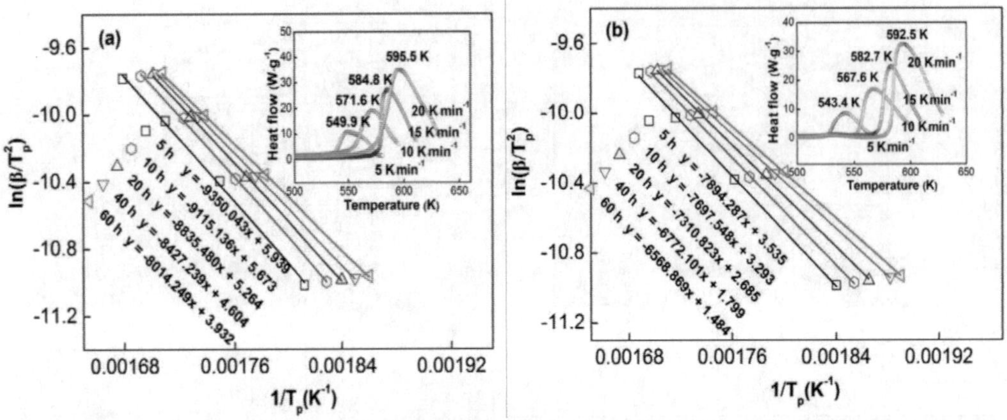

Figure 8. DSC curves of the as-milled (5 h) $PrMg_{11}Ni + x$ wt.% Ni ($x = 100, 200$) alloys at various heating rates and Kissinger plots: (a) $x = 100$, (b) $x = 200$.

alloys to reveal the mechanism of the action of adding Ni and prolonging milling time on the hydrogen desorption kinetics of the alloys. The dehydrogenation activation energy (E_a^{de}) can be calculated using the Kissinger equation [19]:

$$\frac{d[\ln(\beta / T_p^2)]}{d(1/T_P)} = \frac{-E_k^{de}}{R} \tag{4}$$

where E_a^{de} is the activation energy, β is the heating rate, T_p is the absolute temperature corresponding to the maximum desorption rate in the DSC curves, and R is the ideal gas constant. As a representative, the hydrogen desorption DSC curves of the as-milled (5 h) alloys are presented in figure 8. It is noted that there is a clear endothermic peak corresponding to the hydrogen desorption. Simultaneously, the endothermic peak of the $x = 200$ alloy has a drift to low temperature at the same heating rate, suggesting that the hydrogen desorption reaction rate is improved by increasing Ni content. The Kissinger graphs of $\ln(\beta / T_p^2)$ vs. $1/T_p$ of the as-milled alloys can be plotted by using logarithmic transform of the equation (2), as inserted in figure 8, which is nearly linear. Thus, the activation energy of the alloys can be easily calculated from the slops of the plots. The relationship between the activation energy and milling time can be established, as illustrated in figure 9. Apparently, increasing milling time leads to a visible reduction in the E_k^{de} value of the alloys. Moreover, increasing Ni content results in a notable decrease in the hydrogen desorption activation energy. Hence, the real driving force for the improvement of the kinetics is the hydrogen desorption activation energy decreased by increasing Ni content and milling time.

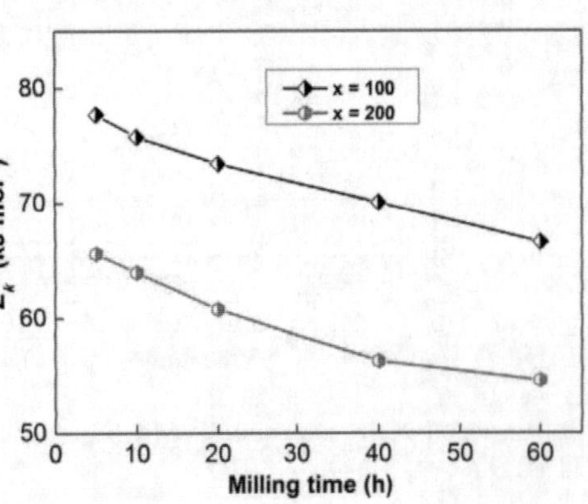

Figure 9. Evolutions of values of the as-milled $PrMg_{11}Ni$ + x wt.% Ni (x = 100, 200) alloys with milling time.

4. Conclusions

The hydrogen storage capacity and kinetics of the as-milled $PrMg_{11}Ni$ + x wt.% Ni (x = 100, 200) alloys has been investigated. The major conclusions are summarized as follows:

(1) Increasing Ni content dramatically enhances electrochemical capacity of the as-milled alloys. Ni content increased from 100 to 200 wt.% gives rise to the discharge capacity of the as-milled (40 h) alloys grows from 141.8 to 1059.4 $mAh \cdot g^{-1}$.

(2) The gaseous hydrogen absorption capacity of the alloys first augments and then declines with milling time prolonging. The maximum value is 5.75 wt.% and 6.03 wt.% for the x=100 and 200 alloys, respectively.

(3) Kissinger method was used to evaluate hydrogen desorption activation energy of the as-milled alloy. $\Delta_r H^*$ obviously decreases with Ni content and milling time increased, which is convinced to be the real driving force of the improvement of the hydrogen desorption kinetics of the as-milled alloys.

(4) The electrochemical kinetics of the alloys first increases and then decreases with milling time increasing, and the maximum HRD values of the as-milled alloys correspond with the minimum $\Delta_r H^*$ values, suggesting the charge transfer ability of the alloy electrode surface is a major governing factor of the electrochemical kinetics of the alloys.

Acknowledgements

This work is financially supported by the National Natural Science Foundations of China (51161015 and 51371094).

References

[1] Vinay B, Aline R, LucAymard, Gholam-Abbas N and Jean-Marie T 2007 *England. Int. J. Hydrogen Energy* **32** 4900.
[2] Schlapbach L and Züttel A 2001 *England. Nature* **414** 353
[3] U Eric A. Lass 2011 *England. Int. J. Hydrogen Energy* **36** 10787
[4] Liu Y F, Pan H G, Gao M X and Wang Q D 2011 *England. J. Mater. Chem.* **21** 4743
[5] Siarhei K, Lars R, Thomas R, Thomas W and Bernd K 2011 *England. Int. J. Hydrogen Energy* **36** 10808.
[6] Song M Y, Kwon S N, Park H R and Hong S H 2011 *England. Int. J. Hydrogen Energy* **36** 13587.
[7] Zou J X, Zeng X Q, Ying Y J, Chen X, Guo H, Zhou S and Ding W J 2013 *England. Int. J. Hydrogen Energy* **38** 2337.
[8] Luo F P, Wang H, Ouyang L Z, Zeng M Q, Liu J W and Zhu M 2013 *England. Int. J. Hydrogen Energy* **38** 10912.
[9] Shang C X, Bououdina M, Song Y and Guo Z X 2004 *England. Int. J. Hydrogen Energy* **29** 73.
[10] Huot J, Pelletier J F, Lurio LB, Sutton M and Schulz R 2003 *Netherlands. J. Alloys Compd.* **348** 319.
[11] Zhang Q A, Jiang C J and Liu D D 2012 *England. Int. J. Hydrogen Energy* **37** 10709
[12] Teresiak R A, Gebert A, Savyak M, Uhlemann M, Mickel C and Mattern N 2005 *Netherlands. J. Alloys Compd.* **398** 156.
[13] Abdellaoui M, Mokbli S, Cuevas F, Latroche M, Percheron-Gue´gan A and Zarrouk H 2003 *Netherlands. J. Alloys Compd.* **356-357** 557.
[14] Roman V D, Andrey A, Poletaev, Jan K S, Boris P T and Volodymyr A Y 2010 *Netherlands. Acta Mater.* **58** 2510.
[15] Ouyang L Z, Huang J M, Fang C J, Zhang Q A, Sun D L and Zhu M 2012 *England. Int. J. Hydrogen Energy* **37** 12358.
[16] Spassov T, Lyubenova L, Köster U and Baró M D 2004 *Netherlands Mater. Sci. Eng., A* **375-377** 794.

An estimation method of the fault wind turbine power generation loss based on correlation analysis

Tao Zhang [1, 2, 3]**, Shourang Zhu** [1, 2]**, Wei Wang** [1, 2]

[1] NARI Group Corporation (State Grid Electric Power Research Institute), Nanjing 211000, China
[2] NARI-TECH Control Systems Ltd., Nanjing 210061, China

E-mail: linyurong@163.com

Abstract. A method for estimating the power generation loss of a fault wind turbine is proposed in this paper. In this method, the wind speed is estimated and the estimated value of the loss of power generation is given by combining the actual output power characteristic curve of the wind turbine. In the wind speed estimation, the correlation analysis is used, and the normal operation of the wind speed of the fault wind turbine is selected, and the regression analysis method is used to obtain the estimated value of the wind speed. Based on the estimation method, this paper presents an implementation of the method in the monitoring system of the wind turbine, and verifies the effectiveness of the proposed method.

1. Introduction

With the development of wind power industry in China, more and more attention has been paid to the study of the economic benefit of wind power. The study of literature [1] shows that the wind turbine failure rate has an important influence on the economic benefit of wind farm. In order to measure the influence, it is necessary to obtain the estimated value of the power generation lost during the wind turbine failure.

2. Problems in the estimation of loss of power generation

According to the theory of wind energy conversion and Bates theory in [2], the active power W of wind turbine is:

$$W = \frac{1}{2} C_p \rho S v^3 \qquad (1)$$

where ρ represents the air density, S stands for the wind turbine swept area, v is wind speed, and C_p indicates the wind energy utilization coefficient. From equation (1), it can be found that the relationship between active power and wind speed is cubic, and it is closely related to the wind sweep area, air density and wind energy utilization coefficient. The blade angle of wind turbine will change with the change of wind speed, which will affect the power coefficient of wind turbine, and the air density will change with time. So it is very difficult to estimate the theoretical output power of the fault wind turbine by Eq.(1).

The initial approach to estimating the power loss of the fault generator is to refer to the normal operation of the wind turbine to estimate the power generation; or to estimate the average wind speed of the wind farm during the outage period and then calculate the power generation capacity. Literature [3] points out that these two methods are easy to implement, but have great error.

With the help of the Weibull distribution of wind speed and wind turbine output power curve in [4], another researchers present an estimation method for wind turbine power generation in [3] and [5], respectively. In [3], the proposed method uses wind tower wind speed as a fault wind turbine speed estimates, and combined with the wind turbine theory output power curve to estimate the power loss. In the literature [5], based on the analysis of the shortcomings of the methods mentioned in the literature [3], the average value of the wind speed of two normal running wind turbines which is nearest to the fault is taken as the wind speed estimation value of the fault wind turbine, and combined with the actual output power curve of the wind turbine to estimate the loss of power generation.

The research of literature [5] has shown that it is reasonable to estimate the loss of power generation by using the actual output power curve of the wind turbine rather than the theoretical output power curve, but in the estimation of wind speed it is simple to use the average value of the wind speed of two normal running wind turbines. Wind speed is affected by terrain, climate, season and other factors. Therefore, the wind speed estimation method is only based on the geographical distance, although there is a certain rationality, but the lack of a strong theoretical basis.

3. Problems in the estimation of loss of power generation

In view of the shortcomings of the proposed method of [5], this paper presents a method to estimate the power generation loss of fault wind turbine. This method is mainly studied in the aspect of wind speed estimation, and an estimation method based on correlation analysis is put forward.

This section will begin with a general description of the estimation method of the loss of wind power generation in this paper. Through the above analysis, it is not difficult to see that the wind speed and the output power curve are the two factors to estimate the loss of power generation. Due to the failure of the wind turbine wind speed is not available, the wind speed of the fault wind turbine must be estimated first, and then combined with the output power curve to obtain the estimated loss of power generation.

In terms of the output power curve will continue to use the research results of the literature [5], namely the use of the actual output power curve. The fitting method and process of the actual output power curve is not the focus of this paper, and the literature [5] has been introduced in detail. The actual output power curve of the fault wind turbine is given directly, which can be described by the following formula according to literature [6]:

$$P_G(v)=\begin{cases} 0 & v<v_S \\ f_1(v), & v_S<v\le v_1 \\ ...., & \\ f_k(v), & v_K<v\le v_E \\ 0, & v>v_E \end{cases} \tag{2}$$

where v_S is cut-in wind speed, v_E is cut-out wind speed, $v_1,v_2,\cdots,v_K$ are actual wind speeds when the power curve fitting is performed, and $f_1,f_2,\cdots,f_K$ are curve functions corresponding to each.

Suppose the fault start time of wind turbine G is t_1, and the fault end time is t_2. The estimated value $\hat{K}_G$ of the electric energy in the wind turbine G fault period can be obtained by the following formula:

$$\hat{K}_G=\int_{t1}^{t2} P_G(\hat{v}_G(t))dt \tag{3}$$

where $\hat{v}_G(t)$ stands for estimated value of wind turbine speed during its fault period.

According to the study of [7], $f_1, f_2,..., f_K$ can be described by polynomial. Therefore formula (3) of the integral calculation is feasible.

The following two sections will be discussed in the acquisition of another element $\hat{v}_G(t)$.

4. Wind velocity correlation

Correlation analysis theory has been applied to many fields of power system according to literature [8] and [9]. In the clean energy field, the correlation analysis is applied to the power prediction by [10].

In this paper, the correlation analysis theory is used to the wind speed estimation of fault wind turbine.

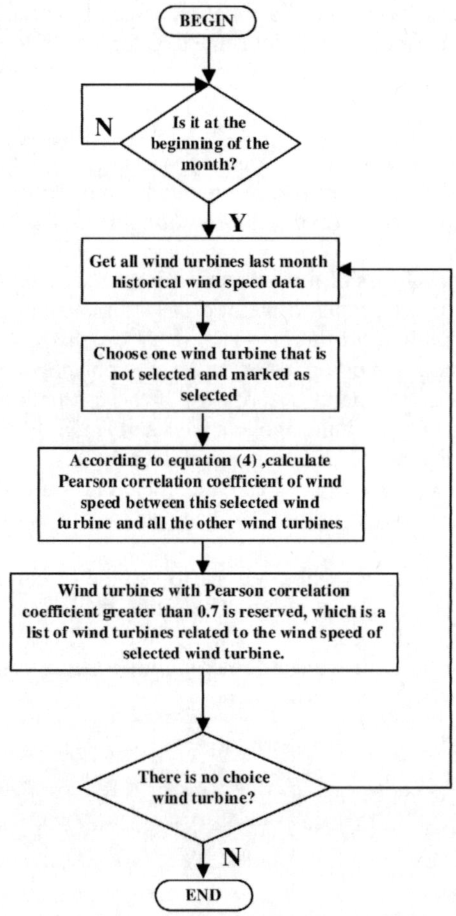

Figure 1. Pearson correlation coefficient calculation of wind speed in wind turbine

By means of correlation analysis, the normal operation wind turbine whose wind speed has a strong linear correlation with the wind speed of fault wind turbine is selected. Then wind speed of fault wind turbine will be estimated by this selected wind turbine.

The Pearson correlation coefficient is proposed as a measure of the correlation between two wind speeds and in different wind turbines, which is based on the research of [11]. Pearson correlation coefficient is a linear correlation coefficient which reflects the degree of correlation between the two variables. The Pearson correlation coefficient r of any two wind turbine A and B can be defined as:

$$r = \frac{q\sum_{i=1}^{q} v_{Ai}v_{Bi} - \sum_{i=1}^{q} v_{Ai}\sum_{i=1}^{q} v_{Bi}}{\left[q\sum_{i=1}^{q} v_{Ai}^2 - (\sum_{i=1}^{q} v_{Ai})^2\right]^{1/2}\left[q\sum_{i=1}^{q} v_{Bi}^2 - (\sum_{i=1}^{q} v_{Bi})^2\right]^{1/2}} \quad (4)$$

Where q represents the number of wind speed values, v_{Ai} and v_{Bi} are two wind turbine wind speed values. Accord to the formula (4), the range of r is between -1 and 1. If the r is positive, it indicates that the wind speed of two wind turbine is positively correlated, that is, the wind speed of a wind turbine becomes larger, and the wind speed of the other wind turbine becomes larger. If the r is negative, it indicates that the wind speed of two wind turbine is negatively correlated, at this time, if the wind speed of a wind turbine becomes larger, the wind speed of the other wind turbine will become smaller.

When the absolute value of r is 1, the wind speed of two wind turbine is completely linear correlation. When the absolute value of r is zero, the wind speed of two wind turbine is completely linear correlation. When the absolute value of r is between 0 and 1, there is a certain linear relationship between the wind speed of two wind turbines. The absolute value of r closer to 1, it shows that the closer the linear relationship. Generally, when the absolute value of r is between 0.7 and 1, we can think of a strongly linear correlation between the wind speed of two wind turbines.

By the above analysis, the acquisition of the correlation coefficient between the wind speeds of two wind turbines depends on the wind speed history data. Wind turbine monitoring system has stored all the running data of the wind turbine, including the wind speed. So it is very convenient to realize the correlation analysis of the wind speed in the wind turbine monitoring system.

Figure 1 shows the implementation of the specific process. Eventually each wind turbine will get a list of wind turbines that are linearly related to its wind speed. The wind turbines in the list can be arranged in the order of the absolute value of the Pearson correlation coefficient.

5. Wind speed estimation

The research of literature [12] describes a regression analysis method based on correlation coefficient. When there is a close linear relationship between the two sequences of variables, a sequence of variables can be calculated by another sequence of variables using the regression equation. According to this method, a wind speed estimation method is designed in this paper.

The analysis in the previous section shows that the linear correlation degree between the wind speeds of two wind turbines can be reflected by the Pearson coefficient. When the wind turbine G fails, the normal operation wind turbine N which has the absolute maximum value of the Pearson coefficient can be selected from the list of wind turbines. In the time of failure of the wind turbine G, the wind speed sequence of the wind turbine N can be measured. The regression analysis method is used to analyze the wind speed sequence of the wind turbine N, and the wind speed estimation values in the wind turbine G fault time period can be obtained.

It is still assumed that the fault start time of the wind turbine G is t_1, and the fault end time is t_2. $\hat{v}_G(t)$ is the estimation value of wind speed of wind turbine G during its fault period. $v_N(t)$ is wind speed of wind turbine N. According to the regression equation, we can get:

$$\hat{v}_G(t) = a + bv_N(t) \quad (5)$$

where a and b stand for correlation coefficient, their acquisition depends on the historical wind speed data as same as Pearson correlation coefficient calculation demand. Specific as follows:

$$b = \frac{\sum_{i=1}^{q}(v_{Ni} - \overline{v}_N)(v_{Gi} - \overline{v}_G)}{\sum_{i=1}^{q}(v_{Ni} - \overline{v}_N)^2}, \quad a = \overline{v}_N - b\overline{v}_G \quad (6)$$

where q is the number of historical wind speed data, $\overline{v}_N$ and $\overline{v}_G$ stand for the average value of wind speed. That there is:

$$\overline{v}_N = \frac{\sum_{i=1}^{q} v_{Ni}}{q}, \quad \overline{v}_G = \frac{\sum_{i=1}^{q} v_{Gi}}{q} \tag{7}$$

By means of equation (5) and equation (7), the estimated value of the wind speed of the fault wind turbine can be obtained. At this point, the wind speed estimation of the two elements of the fault generator loss estimation is solved.

6. Implementation and validation of the estimation method

A remote centralized monitoring system for wind farm has been built by SPIC JIANGSU NEW ENERGY CO. LTD in Yancheng City, Jiangsu province, CHINA. In order to validate the proposed method for estimating the loss of wind power generation, the proposed method is applied to the remote centralized monitoring system, and the Zhengdong wind farm is selected as the experimental object.

First of all, the research of wind speed correlation is verified by experiment. Table 1 presents the Pearson correlation coefficient between five wind turbines. The historical data used in the calculations are from March 2015. This five wind turbines are located on one collector line, and are in a zigzag arrangement.

Table 1 shows that the biggest Pearson correlation coefficient of wind speed with the wind turbine $B2$ is wind turbine $B4$, and the biggest Pearson correlation coefficient of wind speed with the wind turbine $B3$ is wind turbine $B5$. It can be seen that the wind speed correlation of the strongest two wind turbines may not be the most recent. In the following experiments, the wind turbine $B2$ is supposed as the fault wind turbine.

Table 1. Pearson correlation coefficient of wind speed in wind turbine.

	B1	B2	B3	B4	B5
B1	1.000	0.905	0.853	0.872	0.771
B2	0.905	1.000	0.881	0.912	0.798
B3	0.853	0.881	1.000	0.894	0.907
B4	0.872	0.912	0.894	1.000	0.909
B5	0.771	0.708	0.907	0.909	1.000

Figure 2 shows the actual curve of the wind turbine $B2$ and $B4$ wind speed from 0 to 12 April 1, 2015. The dark curve is the $B2$ wind speed, and the light curve is the $B4$ wind speed. From the figure can be seen in about 1 to 3 hours, 4 to 7 hours, 8 to 10, 11 to 12 hours, and so on, the two curves have the same trend.

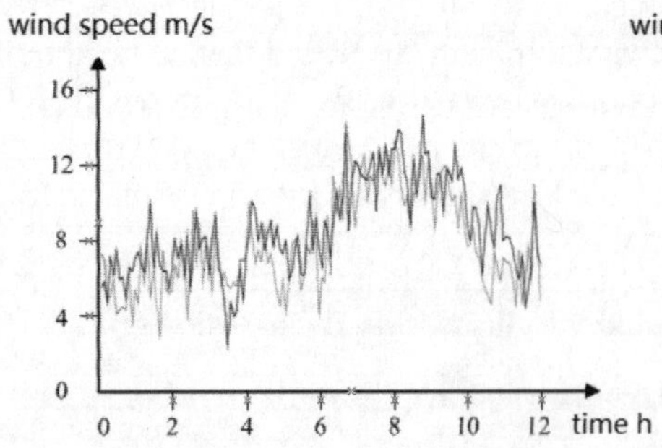

Figure 2. Actual wind speed of wind turbine

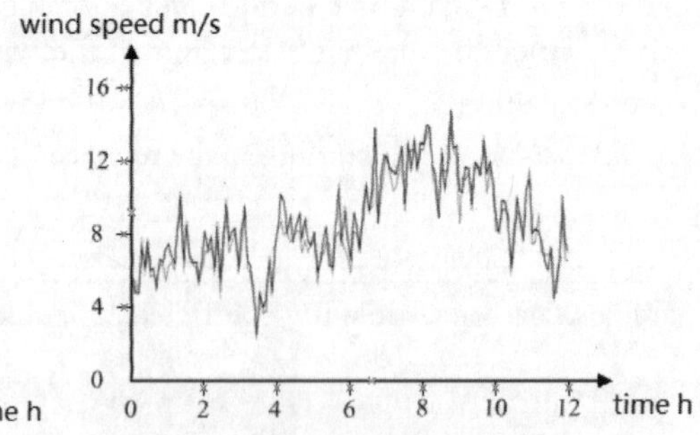

Figure. 3 Actual and estimated wind speed of wind turbine B2

The fault is assumed to occur on the wind turbine $B2$ in this period of time. According to the wind speed of the wind turbine $B4$ and the formula (5) to the formula (7), the estimated wind speed of the wind turbine $B2$ during this period of time is calculated. Figure 3 shows the comparison between the actual wind speed and the estimated wind speed of the wind turbine $B2$.

The dark curve in Figure 3 is still the actual wind speed of the wind turbine $B2$, and the light curve is the estimated wind speed of the wind turbine $B2$ obtained by the method of the wind speed estimation in this paper. From the graph we can see that the curve of wind speed estimation is consistent with the actual curve.

7. Concluding remarks

In this paper, an in-depth study on the estimation of loss of power generation of wind turbine during its fault period is presented, and a method based on correlation analysis is employed. In this method, by analyzing the correlation of wind speed between wind turbines, the normal operating wind turbine which has a strongly linear correlation of wind speed with the fault wind turbine is selected. Then by using the wind speed of the normal operating wind turbine, the fault wind turbine wind speed can be estimate by according to regression analysis method, and combined with the actual output power curve, the loss of power generation is estimated. The experimental results show that the method is accurate and effective.

References

[1] Gao Yang 2014 Analysis of wind power generation capacity Science and *Technique Information* **3** p 220

[2] Yang Gang and Chen Ming 2008 An effective method for calculating power generation quantity for wind turbines *East China Electric Power* **36** p 12

[3] Liu Changhua 2010 Calculating method of power generation loss by wind generating set outage based on excel *INNER MONGOLIA ELECTRIC POWER* **28** p 27

[4] Lv Dong 2012 Discussion and practice of a fast method for calculating the generating capacity of the wind turbine *Science & Technology Information* **25** p 55

[5] Zhang Tao, Zhu Shourang, Wang Wei, et al. The Research on an Estimation Method of Power Generation Loss of the Fault Wind Turbine *Jiangsu Electrical Engineering* **34** p 30

[6] Boettcher F,Peinke J, Kleinhans D, Friedrich R 2007 Handling system driven by different noise sources:implications for power curve estimations *Wind Energy, Springer* p 179

[7] Kang M S. 2007 Generation cost assessment of an isolated power system with a fuzzy wind power generation model *IEEE Trans on Energy Conversion* **22** p 394

[8] Li Wei, Bi Tianshu, Yang Qixun 2011 Fault Phase Selection with Fault Component in same-tower Double-circuit Lines Based on Correlation Analysis *Automation of Electric Power Systems* **35** p 58

[9] Li Yan, Chen De shu, Yuan Rong xiang 2003 Study on simulation and improvement of transient current differential protection based on correlation analysis *RELAY* **31** p 64

[10] Wang Yinsong and Su Ziqing 2015 A Prediction Method for Wind Speed Based on the Adjacent Wind Power Generator Measured Data Correlation Analysis *Journal of North China Electric Power University* **42** p 91

[11] Kusiak A. and Li W Y. 2010 Estimation of wind speed: A data-driven approach *Journal of Wind Engineering and Industrial Aerodynamics* **98** pp 559-567

[12] Jiang Shufa 1999 Application of correlation coefficient in regression analysis *Journal of Shanghai Institute of Electric Power* **15** pp 34-39

Study on the in-situ coupling process of fermentation, extraction and distillation for biobutanol production: process analysis

Fuqiang Jin[1], Xiaodong Zhang[2], Dongliang Hua, Haipeng Xu, Yan Li, Hui Mu

Key Laboratory for Biomass Gasification Technology of Shandong Province, Energy Research Institute of Shandong Academy of Sciences, Jinan 250014, China

E-mail: [1] fuqiangjin@126.com; [2] zhangxd@sderi.cn

Abstract. The transfer process of the in-situ coupling process of fermentation, extraction and distillation for biobutanol production was discussed from a theoretical point of view. The existence of temperature gradient in the extraction section was proved. The force of solute in the extracted liquid was discussed. And the mass transfer mechanism and impetus of the FEDIC process was analyzed. The theoretical analysis could provide a foundation for the following research.

1. Introduction

In recent decades, the decline of petroleum reserves, the surge in population and the degradation of the environment have initiated great interest in the development of renewable biofuels. Among the biofuel alternatives, biobutanol, as the main product of acetone-butanol-ethanol (ABE) fermentation, is one of the most potential replacement fuels [1-3]. However, ABE fermentation suffers a strong end-product inhibition which has adversely affected the economics of commercial production [4-6].

The main limitation of ABE fermentation is the toxic effect of butanol on microorganisms which leads to the low solvent productivity. The cells are completely inhibited at 10-15 g/L of biobutanol [7-9]. It is not economically competitive for butanol recovery by conventional distillation [7, 10]. Therefore, various in-situ butanol removal technologies (such as extraction [10-12], gas stripping [13, 14], adsorption [8] and pervaporation [15,16]) from the fermentation system have been used to mitigate butanol inhibition and improve productivity. Although there has been significant progress in the integration of fermentation and product recovery techniques, there are various challenges before the commercial applications of these technologies. For example, in-situ solvent extraction fermentation has been proposed as one of the most promising approaches for minimizing butanol inhibition and increasing product titer. However, the high price of the extracting agent and the subsequent cost of extracting agent recycling have prevented its application on a large scale [10, 11].

In order to reduce the energy consumption for ABE recovery and extracting agent recycling, we introduced an in-situ coupling process of fermentation, extraction and distillation (abbreviated as FEDIC process) for biobutanol production [17]. In this process, a concentrated ABE solution could be obtained by heating the upper part of the extracted liquid (light phase) at the same time of the extractive fermentation. Since only a small part of the extracted liquid needed to be heated and the corresponding cooling operation was avoided, it is expected that the FEDIC process could reduce the energy consumption. In this paper, the transfer process of the coupling process was discussed.

2. Materials and methods

2.1. Materials

Acetone, butanol and ethanol were analytical grade reagents and used as received. Castor oil (hydroxyl value > 155, acid value: about 1.5 mg KOH/g, density: 0.96 g/ml) was obtained from the local market.

2.2. Experimental Equipment

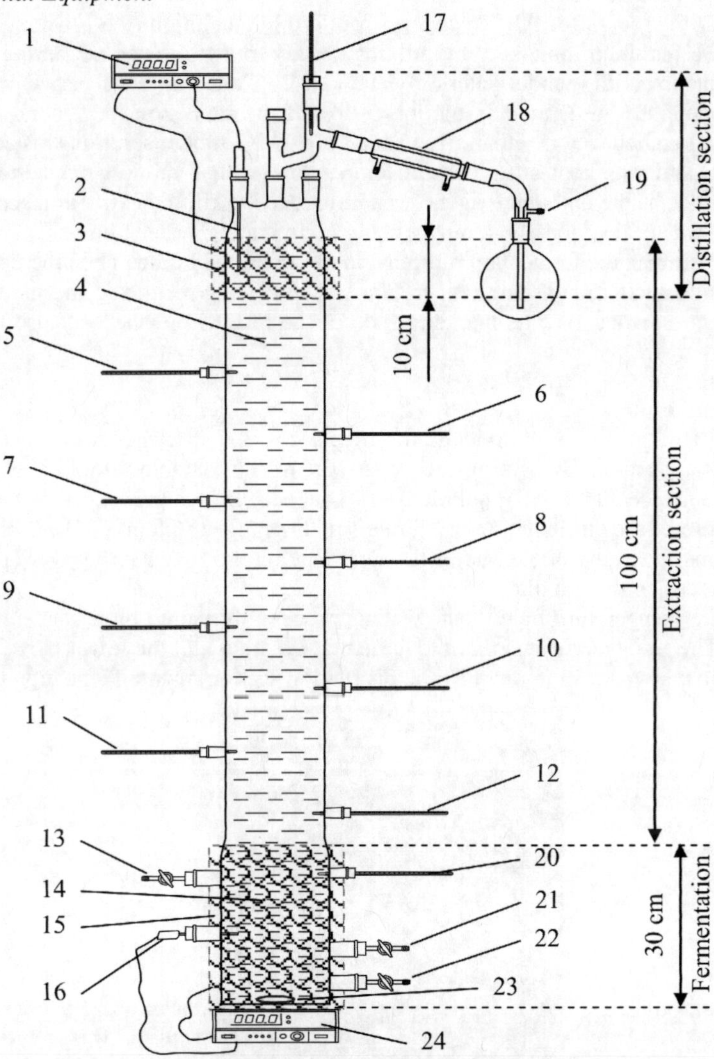

Figure 1. Experimental set-up of the FEDIC process. (1) Temperature controller; (2, 16) Thermocouple; (3, 15) Heating mantle; (4) Extracting agent; (5-12, 17, 20) Thermometer; (13) Feed inlet; (14) Fermentation broth; (18) Condenser; (19) Vacuum system interface; (21) Gas inlet; (22) Outlet port; (23) Magnetic stirring bar; (24) Magnetic stirrer with heating.

As shown in Figure 1, the experimental set-up is a 1.5-meter tall tank with a distillation system at the top and several thermometers (or thermocouples). It is divided into three sections, i.e. fermentation section, extraction section and distillation section. A heating mantle around the fermentation section is to keep the fermentation broth at a constant temperature, and another one around the distillation section is for heating the upper part of the extracted liquid (light phase) to gain the ABE solution. The distillation section was connected with a vacuum system, whose pressure was controlled by a vacuum pump adjusted according to the vacuum degree. The fermentation section had a gas inlet for introducing air to stir the solution in the tank.

2.3. Experimental Procedure

To evaluate the FEDIC process for ABE recovery from a dilute solution, the process was conducted using a model ABE solution obtained by mixing water, butanol acetone and ethanol. Initial concentrations of butanol in the model solution were 12 g/L. Concentrations of acetone and ethanol used for formulation of the ABE model solution followed the ratio typically found in batch ABE fermentations, 6:3:1 (butanol/acetone/ethanol). About 3 L of ABE model solution was added to fill the fermentation section and was kept at a constant temperature with a modest magnetic stirring. And about 10 L of castor oil, as the extracting agent, was introduced to fill the extraction section. Then the experimental set-up was sealed and vacuumized using a circulating water pump. Air was introduced from gas inlet 21 for stirring the fermentation broth and the extracted liquid. Then the upper part of the extracted liquid was heated to a constant temperature by the heating mantle 3, and the temperature of different sections was recorded. The temperature field of the FEDIC process was also measured by a FLIR T400 infrared camera.

3. Results and discussion

The set-up of the FEDIC process was divided into three sections, i.e. fermentation section, extraction section and distillation section. By heating the upper part of the extracted liquid (light phase in the distillation section) a concentrated ABE solution was obtained. In this process, it is a prerequisite to ensure the temperature of the fermentation broth in a suitable range to maintain the biological activity of the microorganisms for cyclic utilization of the fermentation broth, since the overhigh temperature would cause the microorganisms to die.

Figure 2 shows the temperature field of the FEDIC process, measured by an infrared camera after running time 24 h. The results indicated that the fermentation broth and the lower part of the extracted liquid could maintain a low temperature while the distillation section maintained a high temperature.

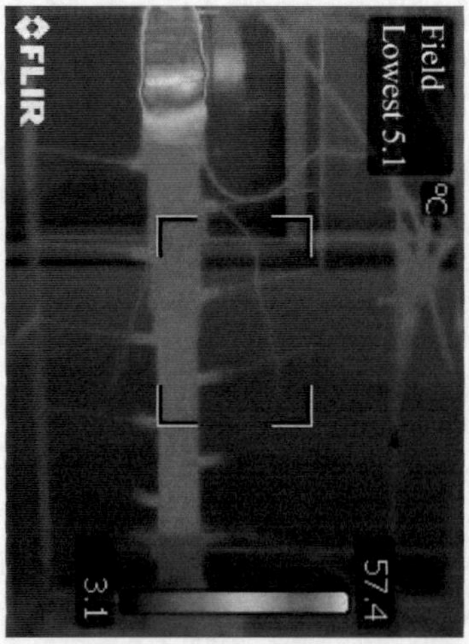

Figure 2. The temperature field of the FEDIC process with magnetic stirring (without gas stirring, DT = 80 °C, BT = 27.5 °C, running time 24 h).

According to the theories of extraction and pervaporation we deduced the mechanism of the FEDIC process, as shown in Figure 3: (1) ABE migrates from the fermentation broth to the liquid-liquid phase interface by molecular diffusion and convection. (2) ABE passes through the phase interface from the fermentation broth to the extracted liquid by molecular diffusion. (3) ABE continues upward diffusing by molecular diffusion and convection. (4) ABE reached high temperature zone and then were vaporized. The bubbles grow, rise to the gas-liquid phase interface and pass through the phase interface. (5) ABE vapor continues upward migrating and is condensed.

Our further experimental results have proved that the fermentation broth and the lower part of the extracted liquid could maintain a low temperature. Thus the FEDIC process was achievable from the point of the temperature field. Here, the existence of temperature gradient in the extraction section was proved using mathematical methods.

The right graph in Figure 3 is the simple graph of the extraction section. T_A and T_B are the temperatures of A zone and B zone, respectively. When the system reaches steady state, there apparently exist $T_A \geqslant T_B$, since the upper part of the extracted liquid was heated. And T_B is higher than room temperature.

Assuming $T_A=T_B$, which means that there does not exist temperature gradient and the heat transfer between A zone and B zone is the equal. And because of $T_B >$ room temperature, B zone must transfer heat to the outside world (heat loss $q>0$) according to the second law of thermodynamics. It will lead to the decreasing of T_B. Thus $T_B \neq T_A$, which contradicts the hypothesis ($T_A=T_B$). It suggests that the assumption is incorrect.

So $T_A>T_B$, the temperature gradient must exist. And it suggests that the heat would transfer downward.

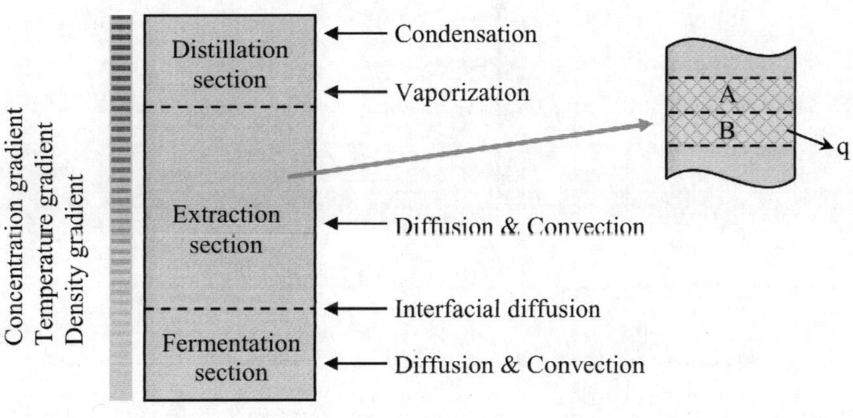

Figure 3. Graph for the process analysis of the FEDIC process.

The mass transfer impetus of ABE include concentration gradient, temperature gradient and density gradient: (1) There must be a concentration gradient since the extracted components were continuously separated out in-situ and keeps in a unsaturated state. (2) The higher the temperature was, the higher the molecular diffusion coefficient were. And the higher the temperature was, the higher the partition coefficient of some extracting agents (such as oleyl alcohol and iso-octanol) were. (3) The density gradient exists since the density of ABE is lower than the used extracting agent. And the density gradient can produce buoyancy to ABE.

Here we mainly discuss the mass transfer in the extracted liquid of the FEDIC process, since the other sections were similar with the conventional extraction and distillation.

The force diagram of ABE in the extracted liquid is shown in Figure 4. ABE is regarded as a small ball with a diameter d. The ball is subject to three forces, namely gravity G, buoyancy F_b and resistance F_r, written respectively as

$$G = mg = \frac{\pi}{6} d^3 \rho g \tag{1}$$

$$F_b = \frac{\pi}{6} d^3 \rho_e g \tag{2}$$

$$F_r = \xi \frac{\pi}{4} d^2 \rho_e \frac{u^2}{2} \tag{3}$$

where m is the mass of the ball (ABE), g is the gravitational acceleration, ρ is the density of the ball, ρ_e is the density of the extracted liquid, ξ is resistance coefficient and u is the instantaneous velocity of the ball.

Thus the force of the ball is described by the equation

$$F = \frac{\pi}{6} d^3 \rho_e g - \frac{\pi}{6} d^3 \rho g - \xi \frac{\pi}{4} d^2 \rho \frac{u^2}{2} \tag{4}$$

Newton's second law gives

$$F = ma \tag{5}$$

where a is the acceleration of the ball.

Thus:

$$F = \frac{\pi}{6} d^3 \rho_e g - \frac{\pi}{6} d^3 \rho g - \xi \frac{\pi}{4} d^2 \rho \frac{u^2}{2} = ma \tag{6}$$

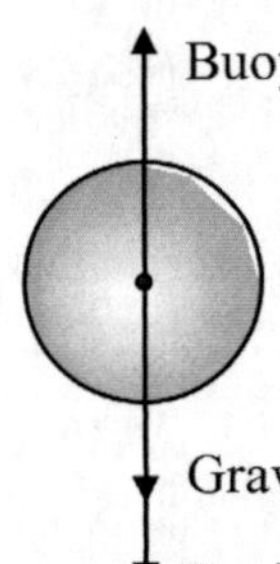

Figure 4. The force diagram of ABE in the extracted liquid.

At the bottom of the extraction section u is very small and F > 0. The ball will accelerate upwards. The resistance F_r increases with the increasing of the instantaneous velocity u. When F = 0, the ball reaches uniform rising state. But for the FEDIC process, it can hardly take place since there are concentration gradient, temperature gradient and density gradient.

In addition, the instantaneous velocity u can be considered to be produced by the force F and concentration gradient, thus

$$u = v_1 + v_2 \tag{7}$$

where v_1 and v_2 are the velocity produced by the force F and concentration gradient, respectively.

$$a = \frac{dv_1}{dt} \tag{8}$$

where t is time.

$$J = c(v_2 - v_l) = -D\nabla c \tag{9}$$

where J is diffusion flux of the component with only consideration of concentration gradient, v_2 is the velocity of the component, v_l is the average velocity of the fluid, and c is the molar concentration of the component. ∇ is the notation for the gradient. D is the diffusion coefficient of the component, also known as the Fick diffusion coefficient.

Thus the mass transfer in the extracted liquid has been preliminarily made clear from what has been discussed above.

In addition, the mass transfer mechanism of the FEDIC process was similar with that of pervaporation. The extracting agent used in the FEDIC process could be regarded as a thick layer of liquid film, while pervaporation usually used a solid-supported membrane. Compared with the pervaporation membrane, the liquid film had several significant advantages such as fast mass transfer, easy automatization, cheapness, convenient operation and low operation cost. So the FEDIC process was a promising separation approach.

4. Conclusion

In this paper, the transfer process analysis of the in-situ coupling process of fermentation, extraction and distillation for biobutanol production was carried out. The existence of temperature gradient in the extraction section was proved. Thus the fermentation broth and the lower part of the extracted liquid could maintain a low temperature while the upper part of the extracted liquid was maintaining a high temperature. And the mass transfer mechanism and impetus of the FEDIC process was analyzed, which proved that the FEDIC process was achievable in theory. However, there are still many unknowns about the FEDIC process. So the comprehensive experimental and theoretical researches should be conducted in the future. In addition, it is also a valuable work to investigate the similar applications in other fermentation process and wastewater treatment.

Acknowledgements

The authors are thankful for the financial support by National Science and Technology Support Program of China (No. 2014BAD02B04), the Natural Science Foundation of Shandong Province, China (No. ZR2015YL016) and Jinan Youth Science and Technology Star Project, China (No. 201406021).

References

[1] Green E M 2011 Fermentative production of butanol — the industrial perspective *Curr. Opin. Biotechnol.* **22** 337-343
[2] Dürre P 2011 Fermentative production of butanol — the academic perspective *Curr. Opin. Biotechnol.* **22** 331-336
[3] Xue C, Zhao X-Q, Liu C-G, Chen L-J and Bai F-W 2013 Prospective and development of butanol as an advanced biofuel *Biotechnol. Adv.* **31**(8) 1575-84
[4] Ezeji T C, Qureshi N and Blaschek H P 2007 Bioproduction of butanol from biomass: from genes to bioreactors *Curr. Opin. Biotechnol.* **18** 220-227
[5] Ezeji T C, Milne C, Price N D and Blaschek H P 2010 Achievements and perspectives to overcome the poor solvent resistance in acetone and butanol-producing microorganisms *Appl. Microbiol. Biotechnol.* **85** 1697-1712
[6] Bankar S B, Survase S A, Singhal R S and Granström T 2012 Continuous two stage acetone-butanol-ethanol fermentation with integrated solvent removal using Clostridium acetobutylicum B 5313 *Bioresour. Technol.* **106** 110-116
[7] Ezeji T C, Karcher P M, Qureshi N and Blaschek H P 2005 Improving performance of a gas stripping-based recovery system to remove butanol from Clostridium beijerinckii fermentation *Biopro. Biosyst. Eng.* **27** 207-214
[8] Qureshi N, Hughes S, Maddox I S and Cotta M A 2005 Energy-efficient recovery of butanol from model solutions and fermentation broth by adsorption *Biopro. Biosyst. Eng.* **27** 215-222

[9] Yeom S H 2011 Simulation study of biobutanol production in a polymer-loaded two-phase partitioning bioreactor (PL-TPPB): Model development *Korean J. Chem. Eng.* **28** 2170-77

[10] Yen H-W and Wang Y-C 2013 The enhancement of butanol production by in situ butanol removal using biodiesel extraction in the fermentation of ABE (acetone-butanol-ethanol) *Bioresour. Technol.* **145** 224-228

[11] Li Q, Cai H, Hao B, Zhang C, Yu Z, Zhou S and Liu C 2010 Enhancing Clostridial Acetone-Butanol-Ethanol (ABE) Production and Improving Fuel Properties of ABE-enriched Biodiesel by Extractive Fermentation with Biodiesel *Appl. Biochem. Biotechnol.* **162**(8) 2381-86

[12] Kurkijärvi A, Lehtonen J and Linnekoski J 2014 Novel dual extraction process for acetone-butanol-ethanol fermentation *Sep. Purif. Technol.* **124** 18-25

[13] Cai D, Wang Y, Chen C, Qin P, Miao Q, Zhang C, Li P and Tan T 2016 Acetone-butanol-ethanol from sweet sorghum juice by an immobilized fermentation-gas stripping integration process *Bioresour. Technol.* **211** 704-710

[14] Xue C, Du G-Q, Sun J-X, Chen L-J, Gao S-S, Yu M-L, Yang S-T and Bai F-W, 2014 Characterization of gas stripping and its integration with acetone-butanol-ethanol fermentation for high-efficient butanol production and recovery *Biochem. Eng. J.* **83** 55-61

[15] Cai D, Zhang T, Zheng J, Chang Z, Wang Z, Qin P-Y and Tan T W 2013 Biobutanol from sweet sorghum bagasse hydrolysate by a hybrid pervaporation process *Bioresour. Technol.* **145** 97-102

[16] Menchavez R N and Ha S H 2013 Ultrasound-enhanced recovery of butanol/ABE by pervaporation *Appl. Biochem. Biotechnol.* **171** 1159-69

[17] Jin F, Hu S, Wang B, Xu H, Zhang J, Wang J, Yang G, Cai F and Zhang X 2011 A coupling process and device of acetone-butanol fermentation and separation *China Patent* 201110039036.7

Oscillation mode analysis considering the interaction between a DFIG-based wind turbine and the grid

Wangping Wu, Da Xie[1], Yupu Lu, Zuyi Zhao, Songtao Yu

The Department of Electrical Engineering, Shanghai Jiao Tong University, Shanghai 200240, China

E-mail: xieda@sjtu.edu.cn

Abstract. Sub-synchronous interactions between wind farms and transmission networks with series compensation have drawn great attention. As most large wind farms in Europe and Asia employ doubly fed induction generator turbines, there has recently been a growing interest in studying this phenomenon. To study the stability of wind turbine with doubly fed induction generator after a small disturbance, a complete small signal system is built in this paper. By using eigenvalue and participation factor analysis, the relation between the modes and state variables can be discovered. Thereafter, the oscillation modes are classified into electrical resonance, sub-synchronous resonance, sub-synchronous oscillation, sub-synchronous control interaction, and low frequency oscillation. To verify the oscillation frequency of each oscillation mode, time-domain simulation based on MATLAB/Simulink is presented. The simulation results justify the effectiveness of the small-signal models.

1. Introduction

With the deterioration of environment and concerns of energy security and fossil energy deficit, wind power becomes the fastest growing power technology [1-3]. Among several wind generation technologies, variable speed wind turbines utilizing doubly fed induction generators (DFIGs) have attracted extraordinary attention because of their advantages over other type of wind turbines [4, 5].An increasing number of wind farms are expected to be connected to lines with series compensation, which however may produce adverse effects, such as sub-synchronous interactions (SSI), with other power system components [6]. As a result, the system generally experience over-voltages and sub-synchronous oscillations, which may cause equipment damage.

Frequency response and eigenvalue techniques are fundamental tools in analyzing small signal stability of multi-machine power systems [7].Small signal stability model is derived and an eigenvalue-based objective function is utilized using particle swarm optimization to optimize the controllers' parameters [8]. [9] develops an approach to analyze the impact of increased penetration of DFIG-based wind turbines on transient and small signal stability of a large power system. In [10], a frequency response-based method is also introduced, as an alternative to the eigen analysis approach, to provide a robust performance evaluation approach. To study the stability of doubly fed induction generators on large power systems , a control strategy to mitigate the impact of reduced inertia is confirmed by time domain and Prony analysis in [11]. However, the references on DFIGs mainly concentrate on certain oscillation modes or optimization of controllers' parameters, they do not analyze all the oscillation modes utilizing detailed mathematic model.

This study is targeted at studying SSI between the grid and DFIG-based wind turbines. Small signal stability model of a wind turbine to a infinite bus system was developed based on previous work [12]. The eigenvalues and their related state variables are then studied by utilizing participation factors. Accordingly, the oscillation modes are classified into five categories: electrical oscillation, sub-synchronous resonance (SSR), sub-synchronous oscillation (SSO), sub-synchronous control interaction (SSCI), and low frequency oscillation.

The major contribution of this paper is: i) it deduces the small-signal model of a whole system from basic flux linkage, voltage and torque equations, including three-mass shaft, induction generator, rotor side converter, grid side converter, DC-link and transmission line; ii) it analyzes all oscillation modes by eigenvalues and participation factors analysis; iii) time-domain simulation is presented to verify the oscillation frequency obtained by the small-signal model.

The rest of the paper is organized as follows: Section 2 introduces the small-signal model of each part and the whole system. Section 3 analyzes the simulation results of small-signal model to identify the oscillation frequency of each eigenvalue and the corresponding oscillation mode. In Section 4, the time-domain simulation is presented to verify the oscillation modes in the small-signal model. Conclusions are drawn in Section 5.

2. Small-signal Modeling of the DFIG

The schematic diagram of the studied system is presented in figure.1, where the wind turbine driven DFIG is connected to the infinite bus system through a transmission line. The one-wind turbine to infinite bus system consists of a three-mass shaft, a induction generator, a rotor side converter(RSC), a grid side converter (GSC), a GSC side reactance, and a transmission line.

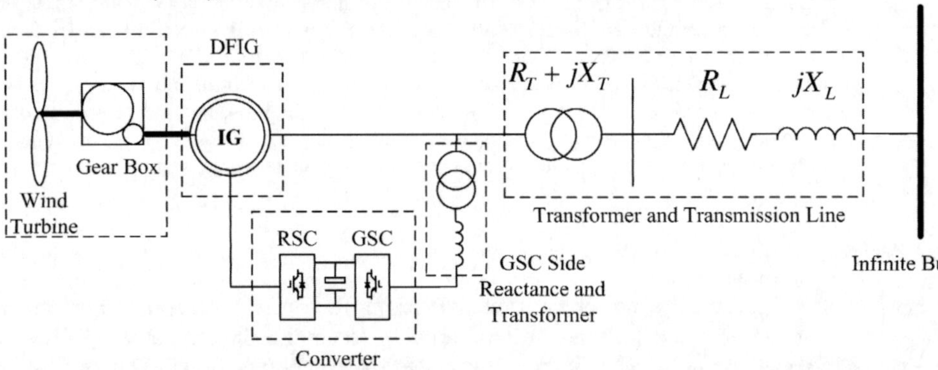

Figure 1. Block diagram of the DFIG-based wind generator model

2.1. Modeling of three-mass shaft

The wind turbine drive train generally comprises a windmill blade, a low-speed shaft, gearbox, a high-speed shaft and a generator rotor, and it is usually represented by two-mass model. However, the two-mass model is not appropriate for transient stability studies, such as torque oscillations, that may interact with the electrical system.A three-mass model is therefore employed in this study.

Within the three-mass model, the inertia of the gearbox is shared between the low-speed shaft (LSS) and high-speed shaft (HSS), and the inertia of the high-speed shaft aggregates the inertia of the generator. Thus, all three masses correspond to the mass moment of the inertia of blades J_1, low-speed shaft J_2 and high-speed shaft J_3, respectively. By taking into account the stiffness and the damping factors for both shafts, the dynamic equations converted into the high-speed shaft can be written as

$$\begin{cases} \dot{\theta}_1 = \omega_{turb} \\ \dot{\theta}_2 = \omega_{gear} \\ \dot{\theta}_3 = \omega_{gen} \\ J_1\dot{\omega}_{rot} = T_{aero} - K_1(\theta_1 - \theta_2) - D_1\omega_{rot} \\ \quad - D_{12}(\omega_{rot} - \omega_2) \\ J_2\dot{\omega}_2 = -K_1(\theta_2 - \theta_1) - K_2(\theta_2 - \theta_3) - D_2\omega_2 \\ \quad - D_{12}(\omega_2 - \omega_{rot}) - D_{23}(\omega_2 - \omega_{gen}) \\ J_3\dot{\omega}_{gen} = -T_{em} - K_2(\theta_3 - \theta_2) - D_3\omega_{gen} \\ \quad - D_{23}(\omega_{gen} - \omega_2) \end{cases} \tag{1}$$

where ω_{turb} is the wind turbine angle speed; ω_{gear} is the low-speed shaft angle speed; ω_{gen} is the high-speed shaft angle speed; θ_1, θ_2, θ_3 are the twist angles of the wind turbine, low-speed shaft, and high-speed shaft, respectively; K_1 and K_2 are the shaft stiffness coefficients; D_1, D_2 and D_3 are the self damping coefficients; D_{12} and D_{23} are the mutual damping coefficients; T_{em} is the electromagnetic torque; T_{aero} is the aerodynamic torque.

2.2. Modeling of induction generator

For modeling the induction generator model, machine equations are represented in synchronously rotating frame where the direction of the d-axis is aligned with the rotor's magnetic flux linkage. The stator, rotor, flux, electromagnetic torque equations are described as follows

$$\begin{cases} u_{ds} = R_1 i_{ds} + \dot{\psi}_{ds} - \omega_1 \psi_{qs} \\ u_{qs} = R_1 i_{qs} + \dot{\psi}_{qs} + \omega_1 \psi_{ds} \\ u_{dr} = R_2 i_{dr} + \dot{\psi}_{dr} - \omega_s \psi_{qr} \\ u_{qr} = R_2 i_{qr} + \dot{\psi}_{qr} + \omega_s \psi_{dr} \end{cases} \tag{2}$$

$$\begin{cases} \psi_{ds} = L_s i_{ds} + L_m i_{dr} \\ \psi_{qs} = L_s i_{qs} + L_m i_{qr} \\ \psi_{dr} = L_m i_{ds} + L_r i_{dr} \\ \psi_{qr} = L_m i_{qs} + L_r i_{qr} \end{cases} \tag{3}$$

$$T_{em} = \frac{3}{2} p L_m (i_{dr} i_{qs} - i_{ds} i_{qr}) \tag{4}$$

where u_{ds} and u_{qs} are d and q axis stator voltages; u_{dr} and u_{qr} are d and q axis rotor voltage; R_1 and R_2 are stator and rotor resistances; ψ_{ds} and ψ_{qs} are the d and q axis stator flux; ψ_{dr} and ψ_{qr} are the d and q axis rotor flux; i_{ds} and i_{qs} are the d and q axis stator currents; i_{dr} and i_{qr} are the d and q axis rotor currents; L_m is the mutual inductance; L_s is the stator self-inductance; L_r is the d and q axis rotor self-inductance; ω_1 is the stator flux angular velocity; ω_s is the slip angular velocity; p is the pole pair.

2.3. Modeling of RSC

The rotor side converter aims to control the output active power to track the input of the wind turbine torque. The decoupled control strategy is used for the DFIG system to regulate its output power, as shown in figure 2.

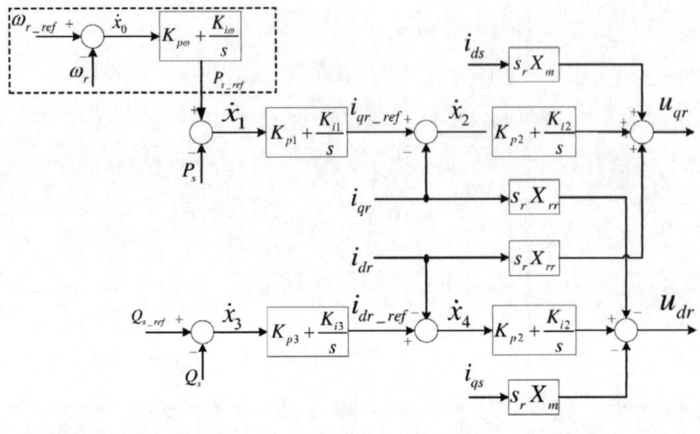

Figure 2. Control block diagram of the rotor side converter

With introducing variables x_0, x_1, x_2, x_3 and x_4, the rotor side converter model can be expressed by the following equations

$$\begin{cases}
\Delta\dot{x}_0 = \Delta\omega_{r_ref} - \Delta\omega_r \\
\Delta\dot{x}_1 = \Delta P_{s_ref} - \Delta P_s = K_{p\omega}(\Delta\omega_{r_ref} - \Delta\omega_r) \\
+ K_{i\omega}\Delta x_0 - (i_{ds0}\Delta u_{ds} + u_{ds0}\Delta i_{ds} + i_{qs0}\Delta u_{qs} + u_{qs0}\Delta i_{qs}) \\
\Delta\dot{x}_2 = K_{p1}(\Delta P_{s_ref} - \Delta P_s) + K_{i1}\Delta x_1 - \Delta i_{qr} \\
= K_{p1}[K_{p\omega}(\Delta\omega_{r_ref} - \Delta\omega_r) + K_{i\omega}\Delta x_0] + K_{i1}\Delta x_1 - \Delta i_{qr} \\
- K_{p1}(i_{ds0}\Delta u_{ds} + u_{ds0}\Delta i_{ds} + i_{qs0}\Delta u_{qs} + u_{qs0}\Delta i_{qs}) \\
\Delta\dot{x}_3 = \Delta Q_{s_ref} - \Delta Q_s \\
= \Delta Q_{s_ref} - (i_{ds0}\Delta u_{qs} + u_{qs0}\Delta i_{ds} - i_{qs0}\Delta u_{ds} - u_{ds0}\Delta i_{qs}) \\
\Delta\dot{x}_4 = K_{p3}(\Delta Q_{s_ref} - \Delta Q_s) + K_{i3}\Delta x_3 - \Delta i_{dr} \\
= K_{p3}\Delta Q_{s_ref} + K_{i3}\Delta x_3 - \Delta i_{dr} \\
- K_{p3}(i_{ds0}\Delta u_{qs} + u_{qs0}\Delta i_{ds} - i_{qs0}\Delta u_{ds} - u_{ds0}\Delta i_{qs})
\end{cases} \tag{5}$$

The output variables of the grid side converter is given by

$$\begin{cases}
\Delta u_{dr} = K_{p2}[K_{p3}(\Delta Q_{s_ref} - \Delta Q_s) + K_{i3}\Delta x_3 - \Delta i_{dr}] \\
\quad + K_{i2}\Delta x_4 - X_m i_{qs0}\Delta s - X_m s_{r0}\Delta i_{qs} - X_{rr} i_{qr0}\Delta s_r - X_{rr} s_{r0}\Delta i_{qr} \\
\quad = K_{p2}[K_{p3}\Delta Q_{s_ref} - K_{p3}(i_{ds0}\Delta u_{qs} + u_{qs0}\Delta i_{ds} - i_{qs0}\Delta u_{ds} - u_{ds0}\Delta i_{qs})] \\
\quad + K_{p2}(K_{i3}\Delta x_3 - \Delta i_{dr}) + K_{i2}\Delta x_4 \\
\quad - X_m i_{qs0}\Delta s_r - X_m s_{r0}\Delta i_{qs} - X_{rr} i_{qr0}\Delta s_r - X_{rr} s_{r0}\Delta i_{qr} \\
\Delta u_{qr} = K_{p2}[K_{p1}(\Delta P_{s_ref} - \Delta P_s) + K_{i1}\Delta x_1 - \Delta i_{qr}] \\
\quad + K_{i2}\Delta x_2 + X_m i_{ds0}\Delta s_r + X_m s_{r0}\Delta i_{ds} + X_{rr} i_{dr0}\Delta s_r + X_{rr} s_{r0}\Delta i_{dr} \\
\quad = K_{p2}\left\{K_{p1}[K_{p\omega}(\Delta\omega_{r_ref} - \Delta\omega_r) + K_{i\omega}\Delta x_0] + K_{i1}\Delta x_1 - \Delta i_{qr}\right\} \\
\quad - K_{p2}K_{p1}(i_{ds0}\Delta u_{ds} + u_{ds0}\Delta i_{ds} + i_{qs0}\Delta u_{qs} + u_{qs0}\Delta i_{qs}) \\
\quad + K_{i2}\Delta x_2 + X_m i_{ds0}\Delta s_r + X_m s_{r0}\Delta i_{ds} + X_{rr} i_{dr0}\Delta s_r + X_{rr} s_{r0}\Delta i_{dr}
\end{cases} \tag{6}$$

2.4. Modeling of GSC

The dc-link voltage is maintained by controlling the grid side converter using the decoupled d-q vector control, with the d-q frame oriented along the stator voltage vector position.

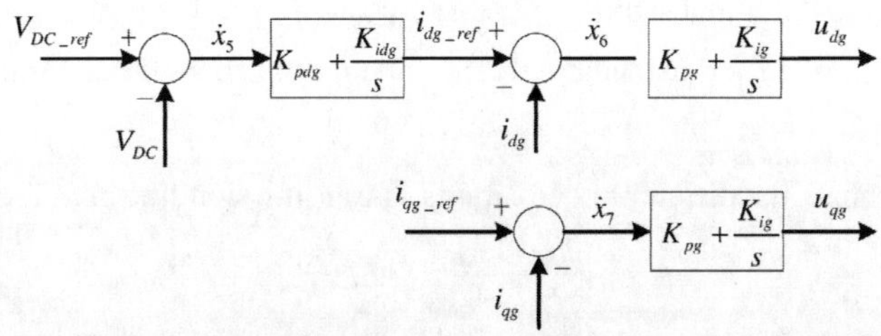

Figure 3. Control diagram of the grid side converter

As shown in figure 3, the dc-link is controlled by i_{dg} while the reactive power is controlled by i_{qg}. With introducing variables x_7, x_8 and x_9, the grid side converter model can be expressed by the following equations

$$\begin{cases}\Delta\dot{x}_5 = \Delta V_{DC\_ref} - \Delta V_{DC} \\ \Delta\dot{x}_6 = \Delta i_{dg\_ref} - \Delta i_{dg} = K_{pdg}(\Delta V_{DC\_ref} - \Delta V_{DC}) + K_{idg}\Delta x_5 - \Delta i_{dg} \\ \Delta\dot{x}_7 = \Delta i_{qg\_ref} - \Delta i_{qg}\end{cases} \tag{7}$$

The output variables of the grid side converter is given by

$$\begin{cases}\Delta u_{dg} = K_{pg}[K_{pdg}(\Delta V_{DC\_ref} - \Delta V_{DC}) + K_{idg}\Delta x_5 - \Delta i_{dg}] + K_{ig}\Delta x_6 \\ \Delta u_{qg} = K_{pg}(\Delta i_{qg\_ref} - \Delta i_{qg}) + K_{ig}\Delta x_7\end{cases} \tag{8}$$

2.5. Modeling of DC-link

The dc-link model can be obtained according to that the active power flow through the converters is balanced, given as follows

$$Cu_{DC}\frac{du_{DC}}{dt} = \frac{3}{2}(u_{dg}i_{dg} + u_{qg}i_{qg} - u_{d2}i_{d2} - u_{q2}i_{q2}) \tag{9}$$

where C is the capacitance of the capacitor; u_{DC} is the dc-link voltage, defining its positive direction as discharge direction.

2.6. Modeling of Grounded Capacitor

The differential equations of grounded capacitor are given by

$$\begin{cases}\frac{du_{gc,x}}{dt} = \omega_1 u_{gc,y} - \omega_1 X_{gc} i_{gc,x} \\ \frac{du_{gc,y}}{dt} = -\omega_1 u_{gc,x} - \omega_1 X_{gc} i_{gc,y}\end{cases} \tag{10}$$

where $u_{gc,x}$ and $u_{gc,y}$ are the x and y axis voltages; $i_{gc,x}$ and $i_{gc,y}$ are the x and y axis currents; X_{gc} is the value of the grounded capacitor.

2.7. Modeling of GSC side Reactance and Transformer

The differential equations of GSC side reactance and transformer are given by

$$\begin{cases}\frac{di_{gx}}{dt} = -\frac{R}{L}i_{gx} + \omega_1 i_{gy} + \frac{1}{L}u_{gx} \\ \frac{di_{gy}}{dt} = -\frac{R}{L}i_{gy} - \omega_1 i_{gx} + \frac{1}{L}u_{gy}\end{cases} \tag{11}$$

where u_{gx} and u_{gy} are the x and y axis voltages; i_{gx} and i_{gy} are the x and y axis currents; R and are L the equivalent resistance and reactance of GSC side reactance and transformer.

2.8. *Modeling of Transmission Line*

Based on the analysis above, the differential equations of transmission line are given as

$$\begin{cases} \frac{di_{Lx}}{dt} = -\frac{\omega_1 r}{X_L} i_{Lx} + \omega_1 i_{Ly} + \frac{\omega_1}{X_L} u_{sc,x} \\ \frac{di_{Ly}}{dt} = -\frac{\omega_1 r}{X_L} i_{Ly} - \omega_1 i_{Lx} + \frac{\omega_1}{X_L} u_{sc,y} \\ \frac{du_{sc,x}}{dt} = \omega_1 u_{sc,y} - \omega_1 X_{sc} i_{Lx} \\ \frac{du_{sc,y}}{dt} = -\omega_1 u_{sc,x} - \omega_1 X_{sc} i_{Ly} \end{cases} \tag{12}$$

where $u_{sc,x}$ and $u_{sc,y}$ are the x and y axis voltages; i_{Lx} and i_{Ly} are the x and y axis currents; X_{sc} is the reactance of the series capacitor, r and X_L are the resistance and reactance of the transmission line.

2.9. *Modeling of the whole system*

To study the system stability after suffering a small disturbance, it is necessary to build the complete small-signal model of the system, as shown in figure. 4.

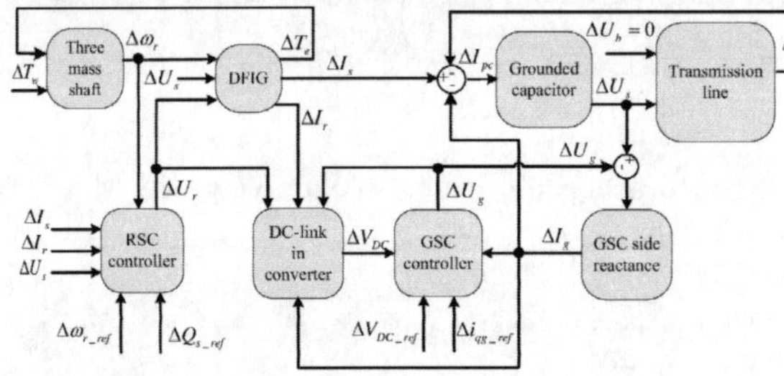

Figure 4. Small signal model of DFIG to infinite bus

By solving the Equations above, the union model can be written in a compact form in

$$\Delta\dot{X} = A\Delta X + B\Delta u \tag{13}$$

where

$$\begin{aligned} \Delta X &= [\Delta X_M \quad \Delta X_G \quad \Delta X_{RSR} \quad \Delta X_{DC} \quad \Delta X_{GSI} \quad \Delta X_{RL} \quad \Delta X_{PC} \quad \Delta X_{TL}]^T \\ &= [\Delta\theta_{turb}, \Delta\theta_{gear}, \Delta\theta_r, \Delta\omega_{turb}, \Delta\omega_{gear}, \Delta\omega_r, \Delta\psi_{qs}, \Delta\psi_{ds}, \Delta\psi_{qr}, \\ &\quad \Delta\psi_{dr}, \Delta x_0, \Delta x_1, \Delta x_2, \Delta x_3, \Delta x_4, \Delta V_{DC}, \Delta x_5, \Delta x_6, \Delta x_7, \Delta i_{gx}, \Delta i_{gy}, \\ &\quad \Delta u_{pc,x}, \Delta u_{pc,y}, \Delta i_{Lx}, \Delta i_{Ly}, \Delta u_{sc,x}, \Delta u_{sc,y}]^T \end{aligned}$$

$$\Delta u = [\Delta T_w \quad \Delta\omega_{r\_ref} \quad \Delta Q_{s\_ref} \quad \Delta V_{DC\_ref} \quad \Delta i_{qg\_ref} \quad \Delta U_b]^T.$$

3. Analysis of the Small-signal Model

The small signal model of the above system is built in MATLAB/Simulink. All parameters of the wind turbine system are provided in Appendix. Based on the small signal model, the eigenvalues of a state matrix can be calculated. The eigenvalues of the system is shown as Table 1.

It can be seen that the system is stable after suffering small disturbance because all of the eigenvalues have negative real parts. There are nine oscillation modes and nine evanescent modes. The participation factors can disclose the relation between the modes and the variables, which are shown in Table 2.

Table 1. Eigenvalues of the small signal system

No.	Eigenvalues	Oscillation frequency, Hz	Damping	No.	Eigenvalues	Oscillation frequency, Hz	Damping
$\lambda_{1,2}$	$-1715\pm2.68\times10^{8}i$	4.26×10^{7}	0	$\lambda_{17,18}$	-11.72±12.08i	1.92	0.6963
$\lambda_{3,4}$	$-3189\pm1.35\times10^{8}i$	2.15×10^{7}	0	$\lambda_{19,20}$	-0.319±3.177i	0.51	0.1
λ_{5}	-942.2	0	1	λ_{21}	-2.74	0	1
$\lambda_{6,7}$	-23.63±498i	79.25	0.0474	λ_{22}	-1.20	0	1
$\lambda_{8,9}$	-51.17±285i	45.37	0.1767	λ_{23}	-1.00	0	1
$\lambda_{10,11}$	-16.75±147i	23.40	0.1132	λ_{24}	-0.0538	0	1
λ_{12}	-139.8	0	1	λ_{25}	-0.008	0	1
$\lambda_{13,14}$	-1.477±77.97i	12.41	0.0189	λ_{26}	-0.0003	0	1
$\lambda_{15,16}$	-8.91±27.45i	4.37	0.3087	λ_{27}	-1.825×10^{-14}	0	1

Table 2. Participation factors of state variables

	$\lambda_{1,2}$	$\lambda_{3,4}$	$\lambda_{6,7}$	$\lambda_{8,9}$	$\lambda_{10,11}$	$\lambda_{13,14}$	$\lambda_{15,16}$	$\lambda_{17,18}$	$\lambda_{19,20}$
Δi_{Lx}	0.0000	0.0134	**0.2106**	0.0008	**0.1902**	0.0030	0.0098	0.0230	0.0012
Δi_{Ly}	0.0034	0.0000	**0.2415**	0.0007	**0.1609**	0.0020	0.0169	0.0151	0.0010
$\Delta u_{sc,x}$	0.0000	0.0000	**0.2521**	0.0050	**0.2302**	0.0014	0.0077	0.0067	0.0004
$\Delta u_{sc,y}$	0.0000	0.0000	**0.2605**	0.0055	**0.2498**	0.0018	0.0046	0.0089	0.0005
$\Delta u_{gc,x}$	0.0000	**0.5000**	0.0000	0.0000	0.0000	0.0000	0.0000	0.0000	0.0000
$\Delta u_{gc,y}$	**0.5000**	0.0000	0.0000	0.0000	0.0000	0.0000	0.0000	0.0000	0.0000
$\Delta\psi_{qs}$	**0.1529**	0.0000	0.0175	**0.1767**	0.0875	0.0022	0.0143	0.0312	0.0019
$\Delta\psi_{ds}$	0.0000	**0.6032**	0.0385	**0.1661**	0.0455	0.0012	0.0023	0.0140	0.0005
$\Delta\psi_{qr}$	**0.3393**	0.0000	0.0032	0.0155	0.0320	0.0013	0.0117	0.0310	0.0015
$\Delta\psi_{dr}$	0.0000	**0.1339**	0.0160	0.0414	0.0412	0.0014	0.0076	0.0244	0.0012
Δx_{3}	0.0000	0.0000	0.0000	0.0000	0.0000	0.0000	0.0000	0.0000	0.0000
$\Delta\theta_{turb}$	0.0000	0.0000	0.0000	0.0000	0.0000	0.0002	0.0008	0.0293	**0.5005**
$\Delta\theta_{gear}$	0.0000	0.0000	0.0000	0.0000	0.0005	**0.4499**	0.0298	**0.1237**	0.0044
$\Delta\theta_{gen}$	0.0000	0.0000	0.0000	0.0001	0.0016	0.0564	0.0196	**0.4601**	**0.2494**
$\Delta\omega_{turb}$	0.0000	0.0000	0.0000	0.0000	0.0000	0.0002	0.0008	0.0293	**0.5005**
$\Delta\omega_{gear}$	0.0000	0.0000	0.0000	0.0000	0.0005	**0.4499**	0.0299	**0.1251**	0.0044
$\Delta\omega_{gen}$	0.0000	0.0000	0.0010	0.0087	0.0372	0.0575	**0.1975**	**1.0560**	0.0201
Δx_{4}	0.0000	0.0000	0.0000	0.0001	0.0002	0.0000	0.0005	0.0111	0.0079
Δx_{5}	0.0000	0.0000	0.0000	0.0084	0.0062	0.0007	**0.7054**	**0.2006**	0.0009
ΔV_{DC}	0.0000	0.0000	0.0004	0.0554	0.0212	0.0015	**0.7167**	**0.1567**	0.0009
Δx_{6}	0.0000	0.0000	0.0000	0.0000	0.0000	0.0000	0.0014	0.0006	0.0000
Δi_{gx}	0.0000	0.0173	0.0102	**0.3833**	0.0432	0.0008	0.0034	0.0124	0.0004
Δi_{gy}	0.0044	0.0000	0.0042	**0.4233**	0.0522	0.0005	0.0808	0.0457	0.0010
Δx_{7}	0.0000	0.0000	0.0000	0.0048	0.0023	0.0001	0.0962	**0.1584**	0.0973
Δx_{0}	0.0000	0.0000	0.0000	0.0001	0.0009	0.0008	0.0265	**0.3211**	**0.2137**
Δx_{1}	0.0000	0.0000	0.0000	0.0000	0.0000	0.0000	0.0001	0.0007	0.0008
Δx_{2}	0.0000	0.0000	0.0000	0.0002	0.0004	0.0000	0.0007	0.0013	0.0012

According to the eigenvalues in Table 1 and the participation factors in Table 2, the oscillation modes can be classified into: electrical resonance, sub-synchronous resonance (SSR), sub-synchronous oscillation (SSO), sub-synchronous control interaction (SSCI), and low frequency oscillation.

3.1. *Electrical resonance*

As shown in Table 2, $\lambda_{1,2}$ are highly sensitive to voltage of the grounded capacitor $\Delta u_{gc,y}$, generator flux $\Delta\psi_{qs}$ and $\Delta\psi_{qr}$. Similarly, $\lambda_{3,4}$ are highly sensitive to voltage of the grounded capacitor $\Delta u_{gc,x}$, generator flux $\Delta\psi_{ds}$ and $\Delta\psi_{dr}$. So $\lambda_{1,2}$ and $\lambda_{3,4}$ are mainly affected by the generator and grounded capacitor. Therefore, they belong to electrical resonance.

However, the corresponding oscillation frequencies of $\lambda_{1,2}$ and $\lambda_{3,4}$ are very high due to the small value of the grounded capacitance. Generally, the grounded capacitor is equal to open circuit and has no effect to the whole system. Therefore, the two oscillation modes are not the emphasis of the system.

$\lambda_{8,9}$ are highly sensitive to generator flux $\Delta\psi_{qs}$ and $\Delta\psi_{ds}$, and current through GSC side reactance Δi_{gx} and Δi_{gy}, so they are mainly affected by the generator and the GSC side reactance. Therefore, they also belong to electrical resonance.

3.2. *Sub-synchronous resonance*

SSR can potentially occur in the fixed-series compensated DFIG-based wind farm even at realistic levels of series compensation [13]. In a series-compensated power system, the natural resonance frequency is given by

$$f_e = f_0\sqrt{\frac{X_{sc}}{X_L}} \tag{14}$$

where f_0 is the synchronous frequency, X_{sc} is the reactance of the series capacitor, and X_L is the total reactance of the transmission line, transformers, and generator.

Therefore, the resonance frequency is influenced by compensation level $k_{com} = \frac{X_{SC}}{X_L}$. In the small signal model simulation, the compensation level is 35%. Therefore, the system resonance frequency is given by

$$f_e = f_0\sqrt{k_{com}} = 50\times\sqrt{0.35} = 29.58\text{Hz} \tag{15}$$

In the system, $\lambda_{6,7}$ and $\lambda_{10,11}$ are highly sensitive to current and voltage of transmission line, i.e., variables Δi_{Lx}, Δi_{Ly}, $\Delta u_{sc,x}$ and $\Delta u_{sc,y}$, so they are mainly affected by the transmission line and the series capacitor. Moreover, the oscillation frequency of $\lambda_{6,7}$ is 79.25Hz, which is approximately equal to $50 + f_e = 79.58\text{Hz}$. The oscillation frequency of $\lambda_{10,11}$ is 23.40Hz, which is approximately equal to $50 - f_e = 20.42\text{Hz}$.

Therefore, the oscillation mode of $\lambda_{6,7}$ belongs to super-synchronous resonance and that of $\lambda_{10,11}$ belongs to sub-synchronous resonance.

3.3. *Sub-synchronous oscillation*

Doubly-fed induction generator wind turbines close to series capacitors are prone to subsynchronous oscillations [14].

$\lambda_{13,14}$ are highly sensitive to twist angle of gearbox $\Delta\theta_{gear}$ and $\Delta\omega_{gear}$, and slightly sensitive to twist angle of generator $\Delta\theta_{gen}$ and $\Delta\omega_{gen}$, so they are mainly affected by the three mass shaft. $\lambda_{17,18}$ are highly sensitive to twist angles of generator and gearbox, i.e., variables $\Delta\theta_{gen}$, $\Delta\omega_{gen}$, $\Delta\theta_{gear}$, and

$\Delta\omega_{gear}$, and also sensitive to DC-link voltage ΔV_{DC}, variables of converter controller such as Δx_0, Δx_5, and Δx_7, so they are mainly affected by the three mass shaft and the converter controller.

Therefore, the oscillation modes of $\lambda_{13,14}$ and $\lambda_{17,18}$ belong to sub-synchronous oscillation.

3.4. Sub-synchronous control interaction

Severe damage to wind turbines due to SSCI has been observed in the case of an interconnection of a wind farm to a series compensated transmission system[15]. The SSCI is not related to mechanical shaft system and may change under different system conditions and converter control algorithms.

In the system, $\lambda_{15,16}$ are highly sensitive to variable of converter controller Δx_5 and DC-link voltage ΔV_{DC}, and also sensitive to generator angle speed $\Delta\omega_r$, so they are mainly affected by the converter controller and rotor speed of generator. Therefore, the oscillation mode of $\lambda_{15,16}$ belongs to sub-synchronous control interaction.

3.5. Low-frequency oscillation

Damping low frequency oscillations of power systems is one of the most important determinants of system stability. Due to the increasing growth of wind farms with high penetration levels in power systems as well as their participation and control, it seems important to increase the damping system oscillations[16]. It occurs in the 0.2 to 2.5 Hz frequency range.

In the system, $\lambda_{19,20}$ are highly sensitive to twist angles of wind turbine and generator $\Delta\theta_{turb}$, $\Delta\theta_r$, $\Delta\omega_{turb}$, and variable of converter controller Δx_0, so they are mainly affected by the three mass shaft and rotor speed control of the generator.

4. Time-domain Simulation Testing

In order to test the oscillation modes of the small signal model, the system shown in figure 1 was implemented in MATLAB/Simulink. The simulation results are shown in figure 5-9.

4.1. Electrical resonance

As analyzed above, the oscillation frequency of $\lambda_{8,9}$ is 45.37 Hz, and the corresponding oscillation mode is mainly affected by the generator and the GSC side reactance. Therefore, we observe the GSC output current and study the harmonic components through spectral analysis using the FFT analysis built-in the Simulink, as shown in figure 5.

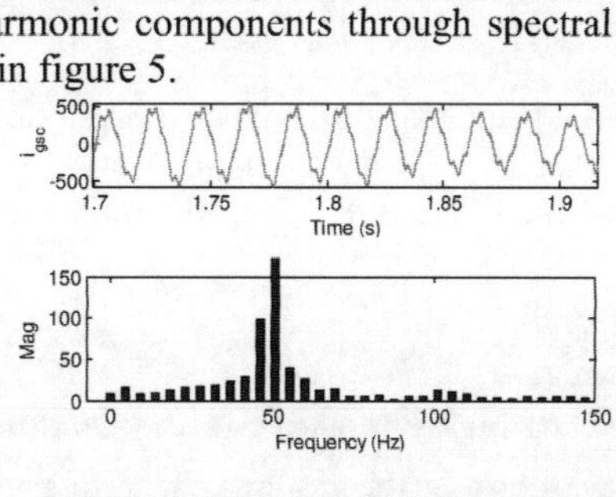

Figure 5. FFT analysis of GSC output current

It is observed from figure 5 that the fundamental frequency of the GSC output current is 50Hz and there exist harmonics in the waveforms. Apparently, the most obvious harmonic is 45Hz, which is consistent with the simulation result of the small signal model.

4.2. Sub-synchronous resonance

In the proposed system, the compensation level is 35%. Therefore, the network resonance mode has a frequency of 29.58Hz ($\approx$ 30Hz), which induces both 20- and 80-Hz oscillations in the grid side. Therefore, we observe the stator output current and study the harmonic components in the waveform, as shown in figure 6.

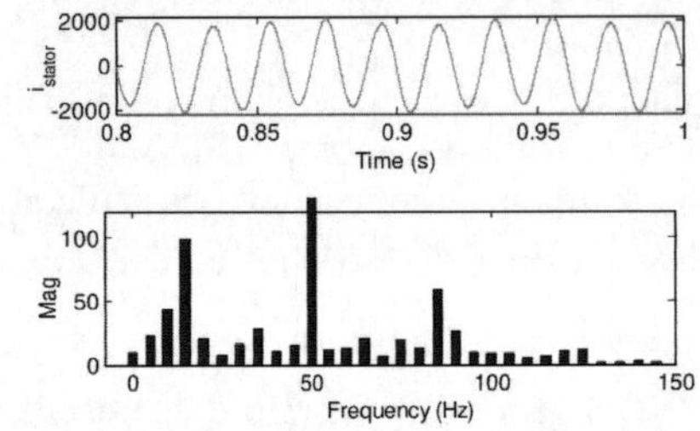

Figure 6. FFT analysis of the stator output current

It is observed from figure 6 that the fundamental frequency of the current is 50 Hz. Harmonics are also observed in the spectrum, most noticeably at 20Hz and 80Hz, which is consistent with the simulation result of the small signal model.

4.3. Sub-synchronous oscillation

The results above indicate that there exist two oscillation frequencies in the system, i.e., 12.41Hz corresponding to $\lambda_{13,14}$, and 1.92Hz corresponding to $\lambda_{17,18}$, both of which are closely related to the three mass shaft. Therefore, we study the waveform of the electromagnetic torque and the corresponding spectral analysis, as shown in figure 7.

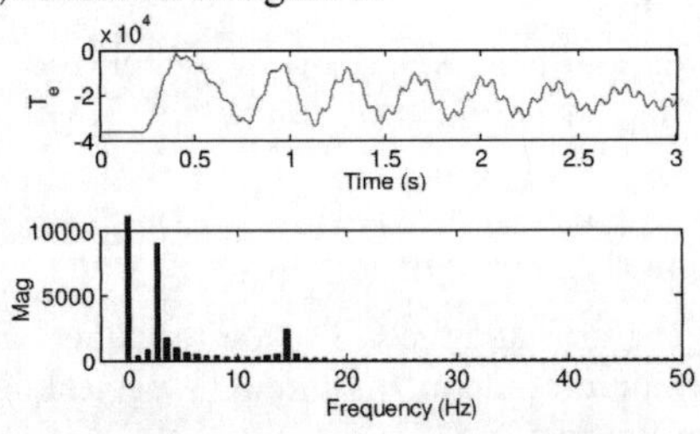

Figure 7. FFT analysis of the electromagnetic torque

It is observed from figure 7 that the electromagnetic torque exhibits a considerable level of harmonic distortion, most noticeably at 2.5Hz and 14Hz, which are consistent with the SSO modes of $\lambda_{17,18}$ and $\lambda_{13,14}$.

4.4. Sub-synchronous control interaction

The SSCI oscillation mode of $\lambda_{15,16}$ is highly sensitive to the DC-link controller and the oscillation frequency is 4.37Hz. Therefore, we observe the dc-link capacitive voltage and study the harmonic components in the waveform, as shown in figure 8.

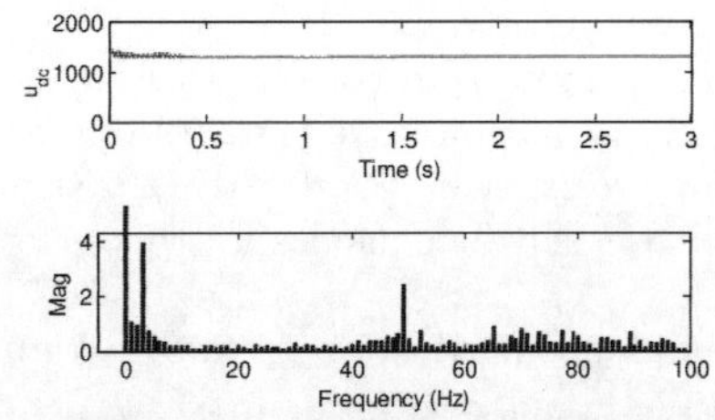

Figure 8. FFT analysis of the dc-link capacitive voltage

It is observed from figure 8 that the dc-link capacitive voltage exhibits a certain level of harmonic distortion, most noticeably at 4.5Hz and 50Hz. While 50Hz is the fundamental frequency of the system, 4.5Hz is consistent with the SSCI oscillation mode of $\lambda_{15,16}$.

4.5. Low-frequency oscillation

The low-frequency oscillation mode of $\lambda_{19,20}$ is mainly affected by the three mass shaft and rotor speed control of the generator. Therefore, we observe the rotor speed of generator and study the harmonic components in the waveform, as shown in figure 9.

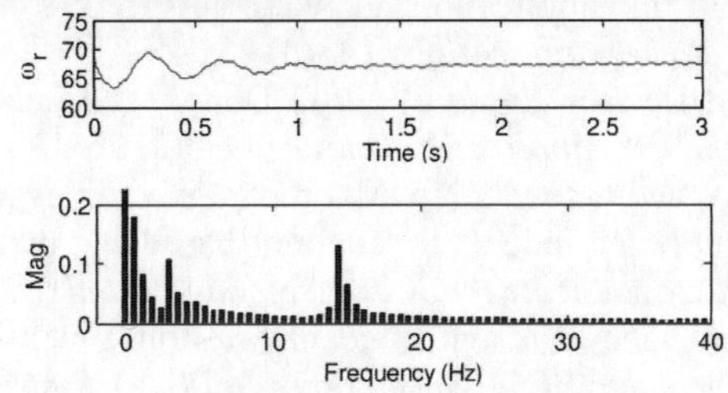

Figure 9. FFT analysis of the rotor speed of generator

It is observed from figure 9 that the generator rotor speed exhibits a considerable level of harmonic distortion, most noticeably at 0.5Hz, 2.5Hz, and 14Hz. While 2.5Hz and 14Hz are corresponding to the SSO modes of $\lambda_{17,18}$ and $\lambda_{13,14}$, 0.5Hz is consistent with the low-frequency oscillation mode of $\lambda_{19,20}$.

5. Conclusions

In this paper, a small signal model of DFIG-based wind turbine connected to power grid is presented. To investigate the vibration modes of the whole system, each sub-model needs to be separately built.Therefore, the dynamic model of three-mass shaft wind turbine, induction generator, and back-to-back converters together with the controllers are derived.Through modal analysis of the union small signal stability model, the model is used to detect and analyze the oscillation modes in the system. Participation factors and eigenvalues analysis reveals that there exist electrical resonance at 45.37Hz, SSR at 79.25Hz and 23.40Hz, SSO at 12.41Hz and 1.92Hz, SSCI at 4.37Hz, and low-frequency oscillation at0.51Hz.The time-domain simulation results in MATLAB/Simulink verified that oscillation modes and corresponding oscillation frequencies detected in the small signal model are correct.

Acknowledgments

This work was supported by the National Natural Science Foundation of China (51277119).

Appendix

Wind turbine: rated wind speed =12 m/s, rated rotor speed=18.1 r/s, blade length= 40.25 m, blades=3. Induction generator: power rating=2 MW, stator voltage=690 V, pole pair=4, frequency=50 Hz, stator resistance R_1=0.0234 Ω, rotor resistance R_2=0.0013Ω, stator dq axis self-inductances $L_{ds}=L_{qs}=L_s$=4.595 mH, rotor dq axis self-inductances $L_{dr}=L_{qr}=L_r$=4.575 mH, mutual inductance L_m=4.5mH.

Three-mass model (transferred to high speed side): J_1=2901.8 $kg\cdot m^2$, J_2=20 $kg\cdot m^2$, J_3=121 $kg\cdot m^2$, $K_1=6.59\times10^4 N\cdot m/rad$, $K_2=9.22\times10^4 N\cdot m/rad$, gearbox ratio N=47:1.

References

[1] Bird L, Bolinger M, Gagliano T, Wiser R, Brown M and Parsons B 2005 Policies and market factors driving wind power development in the United States *Energy Policy* **33(11)** 1397-1407

[2] Söderholm P, Ek K and Pettersson M 2007 Wind power development in Sweden: Global policies and local obstacles *Renewable and Sustainable Energy Reviews* **11(3)** 365-400

[3] Hatziargyriou N and Zervos A 2001 Wind power development in Europe *Proceedings of the IEEE* **89(12)** 1765-82

[4] Pena R, Clare J C and Asher G M 1996 Doubly fed induction generator using back-to-back PWM converters and its application to variable-speed wind-energy generation *IEEE Proceedings Electric Power Applications* **143(3)** 231-241

[5] Muller S, Deicke M and De Doncker R W 2002 Doubly fed induction generator systems for wind turbines *IEEE Industry applications magazine* **8(3)** 26-33

[6] Su J, Shi L, Yao L, Ni Y and Guo Z 2015 A comparative sub-synchronous resonance analysis of grid-connected doubly fed induction generator based and permanent magnet synchronous generator based wind farms *Electric Power Components and Systems* **43(7)** 792-809

[7] Martins N 1986 Efficient eigenvalue and frequency response methods applied to power system small-signal stability studies *IEEE Transactions on Power Systems* **1(1)** 217-224

[8] Wu F, Zhang X P, Godfrey K and Ju P 2007 Small signal stability analysis and optimal control of a wind turbine with doubly fed induction generator *IET Generation Transmission & Distribution* **1(5)** 751-760

[9] Tang J, Achilles R and Agrawal B 1992 Readers Guide To Subsynchronous Resonance-IEEE Committee Report *IEEE Transactions on Power Systems* **7(1)** 150-157

[10] Tabesh A and Iravani R 2008 Small-signal model and dynamic analysis of variable speed induction machine wind farms *IET Renewable Power Generation* **2(4)** 215-227

[11] Gautam D, Goel L, Ayyanar R, Vittal V and Harbour T 2011 Control strategy to mitigate the impact of reduced inertia due to doubly fed induction generators on large power systems *IEEE Transactions on Power Systems* **26(1)** 214-224

[12] Zhang Y, Xie D, Feng J and Wang R 2014 Small-signal modeling and modal analysis of wind turbine based on three-mass shaft model *Electric Power Components and Systems* **42(7)** 693-702

[13] Mohammadpour H A, Shin Y J and Santi E 2014 March SSR analysis of a DFIG-based wind farm interfaced with a gate-controlled series capacitor *In 2014 IEEE Applied Power Electronics Conference and Exposition-APEC 2014* **3**110-17

[14] Leon A E 2016 Integration of DFIG-Based Wind Farms Into Series-Compensated Transmission Systems *IEEE Transactions on Sustainable Energy* **7(2)** 451-460

[15] Nath R and Grande-Moran C 2012 May Study of sub-synchronous control interaction due to the interconnection of wind farms to a series compensated transmission system *In Transmission and Distribution Conference and Exposition (T&D), 2012 IEEE PES* 1-6

[16] Effatnejad R, Zare A, Choopani K and Effatnejad M 2016 March DFIG-based damping controller design to damp low frequency oscillations in power plant industry In 2016 International Conference on Industrial Informatics and Computer Systems (CIICS) 1-5

Analysis on the accommodation of renewable energy in northeast China

Jun Liu[1], Jinfang Zhang, Feng Tian, Zhe Mi

State Grid Energy Research Institute, Future Science & Technology Park, Beijing, P.R.China

E-mail: liujun@sgeri.sgcc.com.cn

Abstract. The accommodation and curtailment of renewable energy in northeast China have attracted much attention with the rapid growth of wind and solar power generation. Large amount of wind power has been curtailed or abandoned in northeast China due to several reasons, such as, the redundancy of power supplies, inadequate power demands, imperfect power structure with less flexibility and limited cross-regional transmission capacity. In this paper, we use multi-area production simulation to analyse the accommodation of renewable energy in northeast China by 2020. Furthermore, we suggest the measures that could be adopted in generation, grid and load side to reduce curtailment of renewables.

1. Introduction

Recently, wind power in northeast China has been largely curtailed with increasing installed capacity. National Energy Administration (NEA) has promulgated series of policies to improve the accommodation of wind power in northeast China, such as "Guidelines to promote the coordinated development of power sectors in northeast China". These policies aimed to control the growth rate and installed capacity of renewable energy in northeast China. However, the curtailment of wind power is even getting worse than ever before.

In the future, it is planned to construct more coal-fired power plants, nuclear plants, which might lead to greater challenge to the accommodation of renewable energies. How to mitigate the curtailment of renewables and enlarge the accommodation range and scale, is important to promote the revitalization of northeast industrial base [1].

2. Development status of renewables in Northeast China

At the end of 2015, the total power installed capacity is about 121GW in northeast China (including Eastern of Inner Mongolia), as shown in Figure 1. Power supply in northeast China mainly consist of coal-fired, wind and hydro, with installed capacity of 84.5GW, 24.7GW and 8GW, respectively, which account for 97% of total installed capacity. The installed capacity of solar power is about 0.8GW in northeast China. Meanwhile, peak load of northeast China power grids is about 54.4GW. There are two cross-regional transmission lines located in northeast China grid and connected with northern China grid. Total transmission capacity is about 5GW.

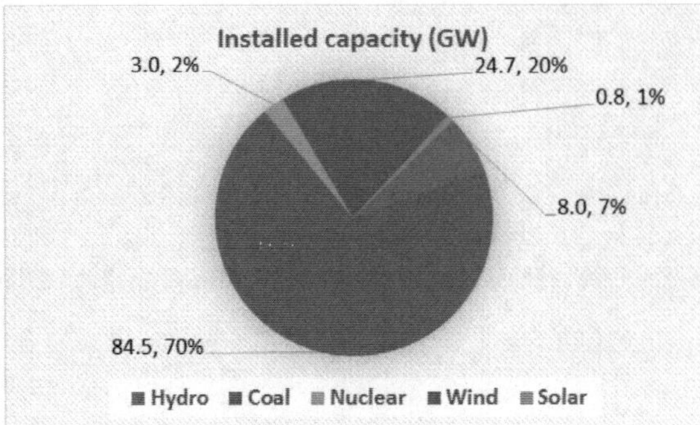

Figure 1. Power supply structure in northeast China

According to statistics, in 2015, total curtailment of wind power in northeast China is up to 8100GWh, accounting for 17.4% of theoretical wind generation. Wind curtailment ratio of Jilin, Heilongjiang, Liaoning and Inner East Mongolia reached 14.6%, 31.1%, 20.5% and 10.3%, respectively. Curtailment of solar power didn't occur in northeast regions.

3. Reasons that lead to curtailment of renewables in northeast China

Main reasons that cause the serious curtailment of renewables in northeast China could be summarized as follows:

First reason is the serious redundancy of power supplies and inadequate power demands in northeast. The growth rate of power demands has been slowed down recently, which exacerbate curtailment of renewable energy. During 12th Five-Year period, total installed capacity of power supply in northeast increased by 37%, 20 percentage points higher than that of load growth. In 2015, the installed capacity is 2.2 times larger than maximum load. It indicates prominent redundancy electricity phenomenon. The utilization hours of coal-fired power plants in northeast China has been declined rapidly since 2006, which decreased from 5900 hours in 2006 to 4079 hours in 2015. At the same time, electricity demand growth slowed down in the northeast. The growth rate of electricity consumption became negative in 2015, which is 2% lower than that of last year. It leads to great difficulties to the accommodation of electricity (including renewables) in local regions.

Second reason is the imperfect power structure. The proportion of flexible power source to total installed capacity is relatively low in northeast China. 70% of the power is supplied by coal-fired power plant, while over 60% of coal-fired plants are combined heat plants (CHPs) in each province of northeast China, especially in Jilin province with a percentage of 75%. In winter, most CHPs didn't participate in deep peaking with mere flexibility, ranges from 15%-20%. However, windy days and hours coincide with heating periods in northeast. As a result, there are great contradictions between heat demands and power balance, the accommodation of renewable energy and minimum operating level of CHPs. In addition, the installed capacity of captive coal-fired power plants (most of them are CHPs, as well) accounted for more than 10%. Load of these captive power plants is relatively fixed and merely participate in peaking load, which gives rise to the decline of power output level of public plant and renewable energy.

Third reason is the cross-regional transmission capacity could not meet the requirement of renewable energy accommodation in large spatial range. At present, the northeast power girds are connected with northern China power grids only by Gaoling back to back DC transmission project and Suizhong power plant grafting project. Suizhong power plant grafting project operated as "point to grid" mode with the capacity of 2GW and mainly deliver coal-fired power generation. It means that actual delivery capacity of northeast China is around 3GW, while the electricity surplus of northeast is more than 20GW.

4. Accommodation of renewables in Northeast China during "13th Five-Year" period

4.1. Development planning of power in northeast during "13th Five-Year" period

Based on power demand and power source structure forecast of "13th Five-Year" period in northeast China, we established a development scenario of renewable energy. Power demand of northeast China will be 450 TWh and the maximum load will reach 70GW at the end of 2020. Zarud-Qingzhou UHVDC (ultra-high voltage DC) transmission lines and Yuanbaoshan power plant grafting project will be put into operation by 2020. Total installed capacity of northeast will be around 150GW at the end of 2020. Each type of power generation is listed as Table 1.

Table 1. Power generation of northeast China by 2020 (Unit: GW)

Generation	2020
Coal-fired	95
Gas-fired	0.3
Hydro	7.9
Pumped Hydro	3.5
Nuclear	6.7
Wind	30
Solar	5
Biomass	2

4.2. Multi-area production simulation model

Aiming to analyse the accommodation of renewable energy in northeast China by 2020, we established a multi-area production simulation model. Production simulation of power system involves the unit maintenance or unit commitment and economic load distribution, power system balance and system peak load balance, etc. Multi-area power production simulation model will take technical and economic characteristics of various types of generation units into consideration, to ensure safe and stable operation of power systems and the continuous power supply.

Objective function:

Main objective function of multi-area production simulation is to optimize the operation cost or pollutant emission of power systems and mitigate the curtailment of renewable energy, listed as following [2-5]:

$$\begin{aligned} LeastCost = {} & \min(E_{wind}) \cap \min(E_{solar}) \cap \min\{E_{hydro} \mid \max(\overline{P}_{hm} + R_{hm})\} \cap \max\{\overline{P}_{pm} + R_{Pm}\} \\ & \cap \max\{P_{Am} + R_{Am}\} \cap \max\{\min(\Delta P_{sm})\} \cap \min\{F(P_T) \mid \max(\min(\Delta P_{sm}) \cap \min(\Delta E_s)\} \end{aligned} \tag{1}$$

Where, E_{wind}, E_{solar} and E_{hydro} represents curtailment of wind, solar and hydro respectively. $\overline{P}_{hm}$, $\overline{P}_{pm}$, $\overline{P}_{Am}$ mean that Hydro, pumped hydro and nuclear power plant should generate as much as possible under the consideration of reserve capacity $\overline{R}_{hm}$, $\overline{R}_{pm}$, $\overline{R}_{Am}$. ΔE_s, ΔPs_m are defined as electricity shortage and power surplus. P_T is the power output of coal-fired power plant. As to coal-fired power plant, the operation cost should be least to minimize the abundancy or insufficient of power supply month by month.

Main constrains [6-7]:

(1) power balance: the output power of plants in local area and transmission power delivered into this area should meet with the load.

(2) electricity balance: the generation of local power plants and electricity delivered into this area should meet with the electricity demands.

(3) reserve: load, rotating and failure reserve are the sum of reserve capacity that each power plant could provide, and larger than the minimum requirements of system or areas.

(4) peaking: peaking load capacity should larger than the sum of failure reserve and load peak-valley difference.

(5) transmission capacity: forward or backward delivering power should not excess the transmission capacity.

(6) maintenance: simultaneous maintenance unit should be under the maintenance field constraint.

(7) emission: air emissions, such as CO_2, SO_2, NO_x and dust, should be within the given limitations.

Optimization flowchart:

First of all, optimal power and electricity exchange between different areas are obtained under the delivery capacity constraints of cross-regional transmission lines, by promoting technical and economic characteristics of each power plants. And then, unit commitment of coal-fired power plants in each area are calculated and optimized according to power and electricity balance. Finally, the location and capacity of each power plant to fulfil load requirements will be determined. The curtailment of renewable energy is obtained, as well [8-9].

4.3. Accommodation of renewables in northeast by 2020

Considering all kinds of unit maintenance, congestion, reserve, the results of multi-area production simulation indicates that electricity surplus amounted to 26GW-28GW of northeast China by 2020 and the curtailment of renewables is about 15.1%. Based on this development scenario, maximal accommodation capacity of northeast China might be around 10GW as the curtailment is limited under 5%. Furthermore, when we mitigate the curtailment under 10%, maximal accommodation scale could be improved to 18GW.

4.4. Measures to mitigate the curtailment of renewables

In the generation side, the minimum technical output rate of the CHPs in heating period could be reduced to release more flexibility and accommodate more renewables. If costly technical retrofitting measures are not carried out in northeast, the minimum technical output rate of CHPs could be adjusted from 80% currently to 70%. As a result, the curtailment of renewable can be reduced to 6% in northeast region by 2020. Meanwhile, operating costs of coal-fired power plant decreased by nearly 1.7%, and carbon emissions decreased by 1.9%, as shown in Figure 2.

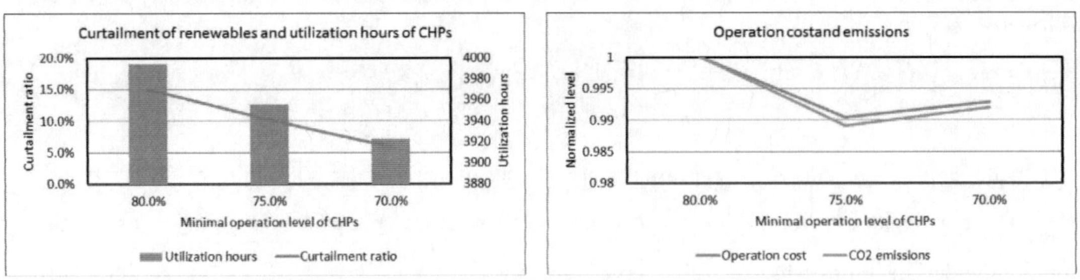

Figure 2. Simulation results of the generation side measures in mitigating renewable curtailment

In the grid side, it is an effective way to mitigate curtailment by increasing cross transmission power. It could release more flexibility of transmission lines during the valley load period, such as nighttime. Simulation results showed that the curtailment of renewables could be mitigated 6 percentage points by increasing delivery power of Gaoling back to back DC transmission project and Zarud-Qingzhou UHVDC transmission lines at nighttime from 50% to 70%, as given in Figure 3. Utilization hours of coal-fired power plants in northeast grids increased from 3994 hours to 4057 hours, while that of Shandong province (receiving side of Zarud-Qingzhou UHVDC transmission lines) might reduce by 98 hours.

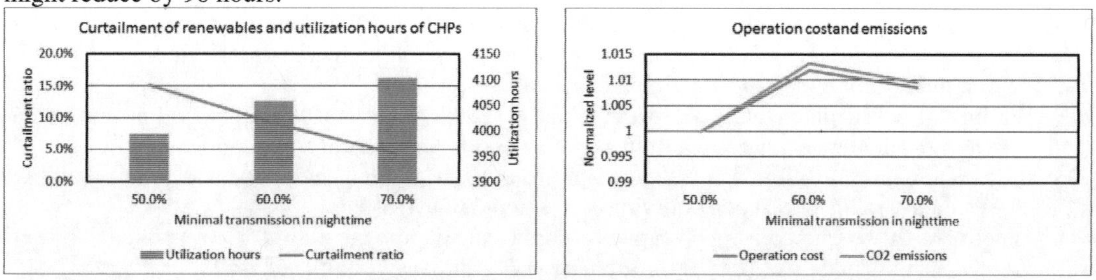

Figure 3. Simulation results of the grid side measures in mitigating renewable curtailment

At the load side, it is important to increase the load level in windy period. There are several ways to increase load level, such as, wind heating, demand side response, TOU (time of use) electricity price mechanisms, etc. It is assumed that 2GW load would be released at load valley period by 2020. As a result, the curtailment of renewables in northeast might be mitigated by 3 percentage points, as shown in Figure 4.

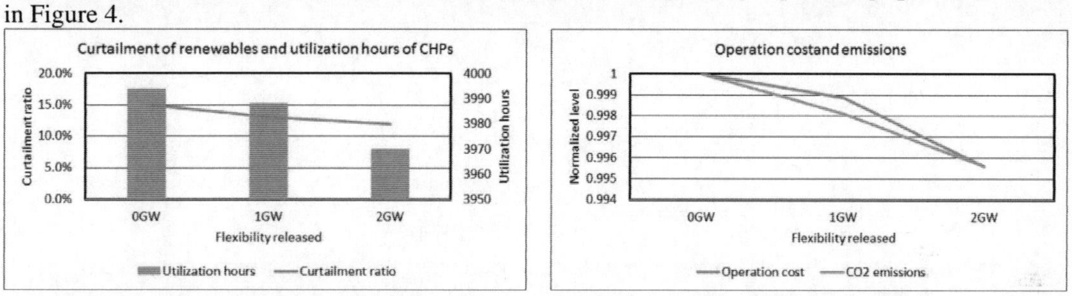

Figure 4. Simulation results of the load side measures in mitigating renewable curtailment

5. Conclusions

According to production simulation on the accommodation of renewable energy in northeast China by 2020, the way to mitigate curtailment could be concluded. In short-term, by promoting the construction of market structure and incentive policies for renewable energy accommodation in northeast regions, under the premise of not significantly increasing power system operation cost to release flexibility of generation, grid and load, the curtailment of renewable could be largely reduced. The measures include moderately reducing minimum output level of CHPs in heating period, adjusting operation curves of cross-regional transmission lines. In the long-term, it is realistic to simultaneously strengthen mitigation measures in generation, grid and load, such as flexibility retrofitting of coal-fired power plants, demand side response, encouraging active loads by incentive policy and beneficial mechanisms. Meanwhile, the inner power grids structure should be planned, optimized and strengthened in northeast China.

References

[1] Holttinen H 2005 Impact of hourly wind power variations on the system operation in the Nordic countries *Wind Energy* **8** 197-218.

[2] Garver L L 1972 Adjusting maintenance schedules to levelize risk *IEEE Trans on Power Apparatus and Systems* **91(5)** 2057-63.

[3] Yare Y, Venayagamoorthy G K 2010 GMS considering uncertainty in wind power in a wind-hydrothermal power system *IEEE Power and Energy Society General Meeting*. Minneapolis, MN USA: IEEE 1-7.

[4] Zhang Jietan, Cheng Haozhong, Hu Zechun, et al. 2009 Power system probabilistic production simulation including wind farms *Proceedings of the CSEE* **29(28)** 34-39(in Chinese).

[5] Zhang Jietan, Wang Maochun, Xu Yourui, et al. 2011 Generation unit maintenance scheduling based on minimum cumulative risk algorithmfor power system containing wind farms *Power SystemTechnology* **35(5)** 97-102 (in Chinese).

[6] Walling R A, Banunarayanan V, Miller N, et al. 2008 Analysisof wind generation impact on ERCOT ancillary services requirements. New York: General Electric Company.

[7] Karki R, Poand H, Billinton R 2006 A simplified wind power generation model for reliability evaluation *IEEE Transactionson Power Systems* **21(2)** 533-540.

[8] Amelin M 2009 Comparison of capacity credit calculation methods for conventional power plants and wind power *IEEE Transactionson Power Systems* **24(2)** 685-691.

[9] Yang Hongming, Wang Shuang, Yi Dexin, et al. 2013 Stochastic optimal dispatch of power system considering multi-wind power correlation *Electric Power Automation Equipment* **33(1)** 114-120.

Gas geochemistry characteristic of shale gas in Longmaxi Formation, SE Sichuan Basin, China

Chunhui Cao[1, 2, 3], Li Du[1], Liwu Li[1], Jian He[1] and Zhongping Li[1]

[1] Key Laboratory of Petroleum Resources, Gansu Province/ Key Laboratory of Petroleum Resources Research, Institute of Geology and Geophysics, Chinese Academy of Sciences, Lanzhou 730000, PR China
[2] Key Lab of Mineral Resources in Western China (Gansu), School of Earth Sciences, Lanzhou University, Lanzhou 730000, China

E-mail: caochunhui@lzb.ac.cn

Abstract. Shale gas samples collected from Lower Silurian Longmaxi Formation of Southern Sichuan Basin in Weiyuan were analysed for stable isotope composition of noble gases and molecular composition, stable carbon isotope composition of hydrocarbons. Results show these shale gases are of organic origin gas, and produced at high-over maturity stage. All the analysed hydrocarbon gases reveal complete inversed isotopic trends from methane to propane, and $\delta^{13}C_1$, $\delta^{13}C_2$ have obviously different between Weiyuan and Changning areas. CO_2 was mainly generated during thermogenic processes of transformation of organic matter, although some gases can contain components from endogenic processes and from thermal destruction of carbonates. He and Ar are mainly product of α-decay of U and Th enriched in crustal materials. A small contribution of mantle origin He was found in the gas reservoirs. Continuous monitoring data indicate $^3He/^4He$ ratio didn't change with the mining time.

1. Introduction

An enormous amount of potential shale gas resources exist in the Chinese Sichuan Basin and the amount of recoverable resources in the area of Weiyuan-Neijiang-Luzhou-Yibing reaches roughly 3 trillion m^3. Of this amount, there are (4.1-10.3) × 10^{12} m^3 of shale gas resource reserves in the Longmaxi Formation stratum, lying at a burial depth of less than 4000 m [1]. Exploratory development of shale gas in the areas of Changning-Weiyuan, Zhaotong, Fushun-Yongchuan and Fuling in the Sichuan Basin has achieved quite a high rate of commercial gas production, where the average gas yield of each trial-produce well now reaches as high as 1.5 × 10^5 m^3/ day.

Methane is the main component of shale gas produced by horizontal well fracturing in Longmaxi Formation of the Sichuan basin, and this type of shale gas has the lowest humidity of all gases on Earth (0.24–0.70%). Results of carbon isotope analysis for shale gas emanating from a well in the Longmaxi Formation in the Changning-Weiyuan area indicate that the carbon isotopic reversal has the following order: $\delta^{13}C_1>\delta^{13}C_2>\delta^{13}C_3$. However, there is an evident difference in the carbon isotope composition between the areas of Changning and Weiyuan; the $\delta^{13}C_1$ and the $\delta^{13}C_2$ values in the Weiyuan area are roughly 8‰ and 6‰ lighter, respectively, than in the Changning area [2, 3]. Differences in the carbon isotope composition of methane between Changning and Weiyuan could possibly be attributed to the following two causes: firstly, it is possible that the shale gas went through different geochemical processes during its generation and evolution in these two areas [4], and secondly, shale gas possibly currently undergoes different physical processes (adsorption/desorption) during the storage and exploitation processes conducted in these two areas.

In recent years, stable isotopes and noble gases have been used to discuss natural gas geochemistry, and have particularly been used to distinguish bacterial and thermogenic gas origins where characterization of parameters relate to the genesis of thermogenic gas (such as primary versus secondary cracking, system openness, relations between gas isotope signatures, and biodegradation) [5-7]. This paper uses noble gas isotopes, gas composition, and stable carbon isotopes to analyze free gas in the Longmaxi shale of SE Sichuan Basin. The original/carbon isotopic difference between the shale gases in SE Sichuan Basin is then discussed. The aim of this paper is to provide theoretical guidance for the exploration and development of shale gas in the Changning and Weiyuan areas, as well as in other areas with marine organic-rich shale.

2. Geological setting

The Sichuan Basin in southeast China is located to the west of the Yangtze Craton and covers an area of more than 18×10^4 km^2. It is a multiple-cycle sedimentary basin, confined in the north by the Micang Mountains (uplift) and the Daba Mountains (fold belt), in the south by the Daxiangling and Lou Mountains (fold belt), in the west by the Longmen Mountains (fold belt), and in the east by the Dalou Mountains (Figure 1). The Caledonian, Hercynian, Indosinian, and Yanshanian orogenies and the Himalayan movement are all recorded in this basin; these tectonic movements resulted in complex deformation and denudation [8, 9]. The Sichuan Basin experienced a transition from marine facies to continental facies sedimentation at the end of the Middle Triassic, and there are six sets of source rocks within the basin: the Lower Cambrian, Lower Silurian, Lower Permian, Upper Permian, Lower Jurassic and Upper Triassic. Of these, the first four sets represent the dominant source rocks [10, 11], and the area is an important oil- and gas-bearing superimposed basin. The Longmaxi Formation of the Lower Silurian is widely distributed throughout the Sichuan Basin and consists of black graptolitic shale in the lower section and nodular limestone in the upper section [12].

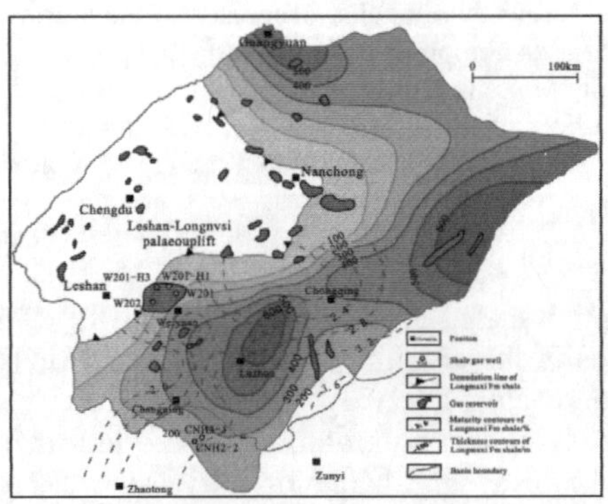

Figure 1. Geological sketch map of Sichuan Basin showing locations of main gas sampling sites, isolines of Ro values, and shale thickness of Longmaxi Formation

3. Samples and experiments

Natural gas samples were collected from the Longmaxi Formation in the Lower Silurian strata, the southwest of the Sichuan Basin. Free gases were collected directly at the well head using a high pressure cylinder (~1000 cm^3, 15 Mpa) with metal valves on both ends, and atmospheric contamination during sampling was minimized by allowing the gas to flush through the system for approximately 5 min [13]. After collecting samples, the pressure of the metal container was retained at higher than 1 atm to prevent possible atmospheric leakage into the container. General information of sampling site locations is given in Figure 1.

Chemical composition, C isotopic compositions and noble gas isotope compositions were analyzed in the Key Laboratory of Petroleum Resources Research, Institute of Geology and Geophysics, Chinese Academy of Sciences at Lanzhou. Chemical composition and isotopic compositions analysis methods were described in Cao et al. (2016). The purification and analysis steps of noble gas and data correction procedures were described in details in previous works [14, 15].

4. Results and discussion

Carbon isotope analysis results of Longmaxi Formation in SE Sichuan basin are shown in Figure 2. $\delta^{13}C_1$=-27.2‰~-36.8‰, $\delta^{13}C_2$= -33.7‰~ -41.9‰; There are obvious differences of carbon isotope composition in Changning and Weiyuan Longmaxi Formation shale gas. Carbon isotopic composition is lighter in Weiyuan than Changning area. $\delta^{13}C_1$= -34.5‰ ~ -36.8‰, $\delta^{13}C_2$= -37.6‰ ~ -41.9‰, in Weiyuan; $\delta^{13}C_1$= -27.2‰ ~ -27.3‰, $\delta^{13}C_2$= -33.7‰ ~ -34.1‰, in Changning. In Weiyuan $\delta^{13}C_1$ is almost 8‰ lighter and $\delta^{13}C_2$ is almost 6‰ lighter than Changning. However, Changning and Weiyuan shale gas carbon isotope distribution patterns are all reversals: $\delta^{13}C_1>\delta^{13}C_2$.

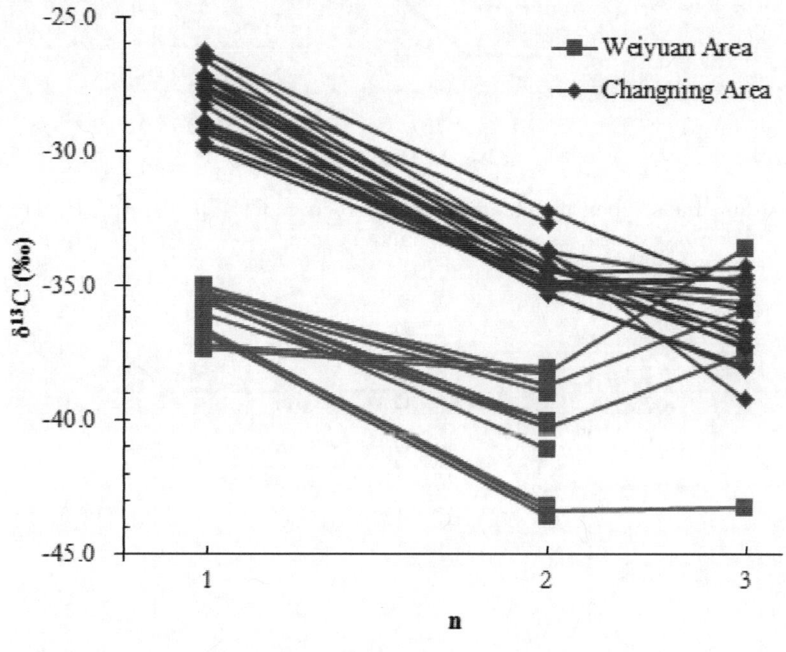

Figure 2. Carbon isotope composition of hydrocarbon from Longmaxi Formation in **SE** Sichuan basin

Molecular and stable isotopic compositions (Figure 3) show that both Weiyuan and Changning shale gases are genetically related to thermogenic activity (thermocatalytic). The genetic correlation determined between the hydrocarbon index and $\delta^{13}C_1$ (Figure 3), and the geochemical analysis of dispersed organic matter revealed the occurrence of mainly mixed type II and III kerogens. A significant proportion of the shale gases in the Weiyuan area are generated from type II kerogen, whereas Changning shale gases are mainly generated from type III kerogen. Weiyuan and Changning shale gases are the driest natural gases in the world [4], and the absence of butanes, pentanes, hexanes, and heptanes in the analyzed gases also indicates that they are genetically connected to the high-over maturity stage.

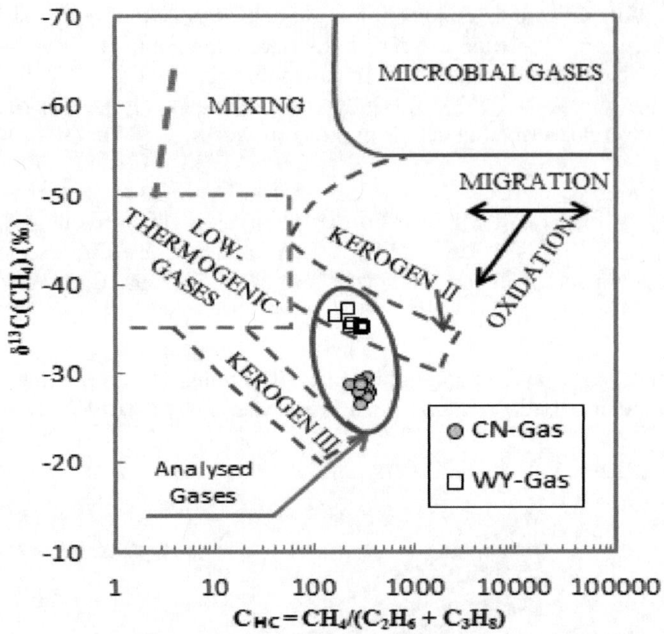

Figure 3. $\delta^{13}C$ of methane versus hydrocarbon index C_{HC} (i.e., $CH_4/ [C_2H_6 + C_3H_8]$) of methane for analyzed natural gases

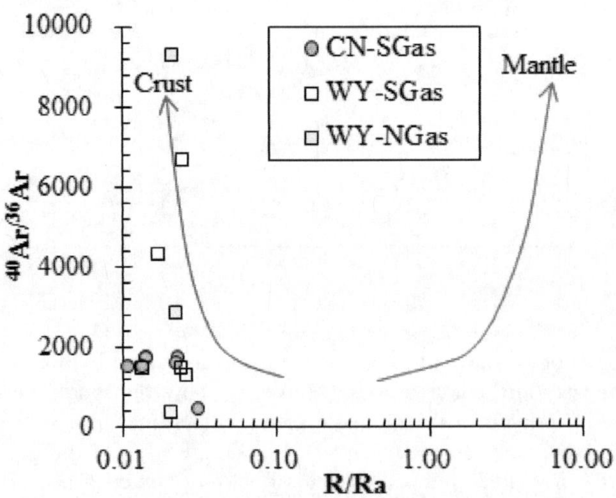

Figure 4. Plot of $^3He/^4He$ vs. $^{40}Ar/^{36}Ar$

Two main trends can be identified in the plot of helium and argon isotopic ratios (Figure 4). One trend has a negative slope and corresponds to the addition of radiogenic ^4He and ^{40}Ar, which is related to the natural radioactivity of sediments and possibly the underlying continental basement. The other trend is a positive slope, which corresponds to mixture with a mantle end-member. However, almost all the samples obtained are under the negative slope line, which suggests that only a small amount of mantle end-member helium and argon are present. Furthermore, there is more ^{40}Ar in Weiyuan shale gas than in Changning shale gas.

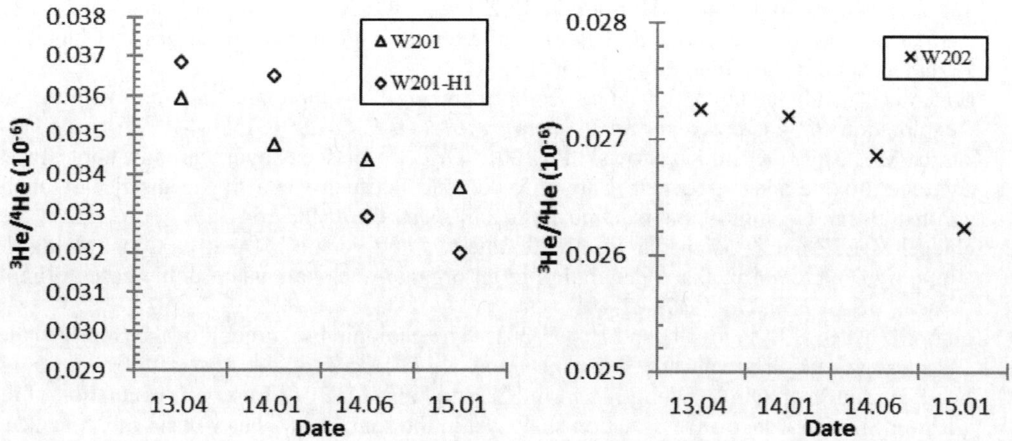

Figure 5. Variations in $^{3}He/^{4}He$ with time

W201 and W201-H1 wells were sampled four times between 2014 and 2015, and well W202 was sampled three times (Figure 5, data of 13.04 is from [4]). There were no evident changes in values of $^{3}He/^{4}He$ throughout the mining time, but the value of $^{3}He/^{4}He$ in W201-H1 tended to be higher than in W201 and W202. This could be because a small amount of He is introduced into the shale when using the horizontal fracturing technique, as fracturing fluids are not added. Despite the yield of shale gas being improved when using the horizontal fracturing technique, to some degree a certain amount of atmospheric constituents are also brought into the shale, which consequently destroys the original characteristics of shale gas.

5. Conclusion

The distribution patterns of carbon isotope compositions are reversals ($\delta^{13}C_1>\delta^{13}C_2$) both in Changning and Weiyuan shale gas. The carbon isotope composition is different in Changning and Weiyuan Longmaxi Formation shale gas. In Weiyuan area, $\delta^{13}C_1$ is almost 8‰ lighter and $\delta^{13}C_2$ is almost 6‰ lighter than Changning area.

Longmaxi shale gases are one of the driest natural gases in the world. They are genetically related to thermogenic gas, and generated from type II and III kerogen. Longmaxi shale gases are genetically high-over maturity gases.

Longmaxi shale gases are mainly Crust origin. Horizontal fracturing technique may change the original chemical characteristic of shale gases.

Acknowledgements

The research has been financially supported by the NSF of China (41502143), the Key Laboratory Project of Gansu Province (1309RTSA041).

ReferencFes

[1] Wang N, Pei L, Lei D F and Zeng B 2015 Analysis of unconventional gas resources distribution and development status in China *Petroleum Geology and Recovery Efficiency* **22**(1) 26-31.

[2] Cao C H, Lv Z G, Li L W and Du L 2016 Geochemical characteristics and implications of shale gas from the Longmaxi Formation, Sichuan Basin, China *Journal of Natural Gas Geoscience* **1** 131-138.

[3] Dai J X, Zou C N, Liao S M, Dong D Z, Ni Y Y, Huang J L, Wu W, Gong D Y, Huang S P and Hu G Y 2014 Geochemistry of the extremely high thermal maturity Longmaxi shale gas, southern Sichuan Basin *Org. Geochem* **74** 3-12.

[4] Tang Q Y, Zhang M J, Cao C H, Song Z, Li Z P and Wu X 2015 The molecular and carbon isotopic constrains on origin and storage of Longmaxi Formation shale gas in Changning area, Sichuan Basin *China Interpret* **3**(2) 35-47.

[5] Prinzhofer A and Battani A 2003 Gas isotopes tracing: An important tool for hydrocarbons exploration *Oil & Gas Science and Technology-Rev. IFP.* **58** 299–311.

[6] Kotarba M J, Nagao K and Karnkowski P H 2014 Origin of gaseous hydrocarbons, noble gases, carbon dioxide and nitrogen in Carboniferous and Permian strata of the distal part of the Polish Basin: Geological and isotopic approach *Chem. Geol.* **383** 164-179.

[7] Schlegel M E, Zhou Z, McIntosh J C, Ballentine C J and Person M A 2011 Constraining the timing of microbial methane generation in an organic-rich shale using noble gases, Illinois Basin, USA *Chem. Geol.* **287** 27-40.

[8] Zhu Y M, Chen S B, Fang J H and Luo Y 2010 The geologic background of the Siluric shale-gas reservoiring in Szechwan, China *J. China Coal Soc.* **35** 1160-64.

[9] Chen S B, Zhu Y M, Qin Y, Wang H Y, Liu H L and Fang J H 2014 Reservoir evaluation of the Lower Silurian Longmaxi Formation shale gas in the southern Sichuan Basin of China *Mar. Petrol. Geol.* **57** 619–630.

[10] Huang J Z, Chen S J, Song J R, Wang L S, Gou X M, Wang T D and Dai H M 1997 Hydrocarbon source systems and formation of gas fields in Sichuan Basin. *Sci. China D.* **40** 32-42.

[11] Zhu Y M, Hao F, Zou H Y, Cai X Y and Luo Y 2007 Jurassic oils in the central Sichuan basin, southwest China: unusual biomarker distribution and possible origin *Org. Geochem.* **38** 1884-96.

[12] Zhao W Z, Yang X P, Steve K and Zhang B M 2006 Reservoir potential of Silurian carbonate mud mounds in the southern Sichuan basin, central China *Acta Geol. Sin.* **80**(5) 684-692.

[13] Gonzalez-Penagos F, Rouchon V, Guichet X and Moretti I 2016 The distribution of thermogenic, bacterial and inorganic fluid sources in the petroleum systems of the Llanos Basin (Colombia) – Insights from the noble gases and carbon stable isotopes *Mar Petrol Geol* **71** 391-403

[14] Zhang M J, Tang Q Y, Hu P Q, Ye X R and Cong Y N 2013 Noble gas isotopic constraints on the origin and evolution of the Jinchuan Ni–Cu–(PGE) sulfide ore-bearing ultramafic intrusion, Western China *Chem. Geol.* **339** 301–312.

[15] Ye X R, Tao M X, Yu C A and Zhang M J 2007 Helium and neon isotopic compositions in the ophiolites from the Yarlung Zangbo River, Southwestern China: The information from deep mantle *Sci. China D.***37** 573-583.

Converter-based accumulation of electric energy generated by microbial biofuel cell

A N Reshetilov[1], A E Kitova[1], A V Dyakov[2, 4], P M Gotovtsev[2], R G Vasilov[2], M A Gutorov[3]

[1] G.K. Skryabin Institute of Biochemistry and Physiology of Microorganisms (IBPM), Russian Academy of Sciences, Pushchino, Moscow Region, Russia
[2] National Research Centre "Kurchatov Institute", Moscow, Russia
[3] Gamma LLC, Zelenograd, Moscow, Russia

E-mail: anatol@ibpm.pushchino.ru

Abstract. Microbial biofuel cell (BFC) was used as a primary energy source for energy storage system. The converter BQ25504 (Texas Instruments) was applied for transformation of electrical energy from microwatt primary sources. The energy storage operation begins if BFC output voltage was higher or equal to 300 mV. In case of stationary operation it was possible to provide energy storage of BFC output voltage equal to 100 mV. The developed system based on converter enables to increase the initial voltage BFC of 0.5 V to 3.1 V; accumulated energy is stored on the various capacitors. Resulting voltage was stable with application of condensers with capacities from 100 μF to 6800 μF. In case of application of 3.1 V and 6800 μF condenser the storage energy was equal to 32.7 mJ. It was enough to provide short time operation of diode L-1154SURDK (2.0 V, 20 mA) and electrical motor M25E-4L (MITSUMI; 3.0 V, 100 mA). Designed system can be applied for energy supply of small electrical devices (for example remote sensors) and autonomous microrobots.

1. Introduction

Recently, increased attention of researchers is paid to development of analytical (biosensor) and energy (biofuel cells) - systems based on biomaterials, which can act as enzymes, cells of warm-blooded organisms or microbial cells. As noted in [1], biosensors and biofuel cells (BFC) devices belong to the same class. In BFC the oxidation of organic compounds that serves as a source of electrical energy; biosensors determine the desired connection and thus can generate current, if applied to their electrodes a potential difference. In both cases the direct conversion chemical energy into electrical energy by biocatalysts.

BFC can be used for electric power low power devices - this point of view is represented in the paper [1]. Low power means enclosed in the range from 10^{-7} to 10^{-3} W, extracted from the one square centimetre of electrode area. In this regard, it may be noted that there is a problem of increasing power BFC for their wider use. In this case BFC can be use not only as power sources for remote sensors and devices, but also they may be used in medicine as an electric power source for microdevices such as pacemakers, micropumps, remote sensors *etc*. It is necessary to mention that operation of all of those devises is operate with its own modes of energy consumption and in some case even periodically, for example based on sensor signal. Additionally, microbial BFC high power can be applied during the municipal wastewater treatment to obtain electricity [2].

The problem of increasing power can be solved by various methods. For example, BFC and biosensors with exoelectrogenic bacteria have the advantage over the cells with intracellular bacteria at equal parameters [3] [4]. Also, series and parallel connection of BFC was used in [5]. Another approach can be based on design of the BFC with application of bacterial cells with exoenzymes on the cell wall [6].

BFC was formed on the basis of living organisms – clams [5]. A set of BFC was connected with a storage capacitor (supercapacitor/ Electric double-layer capacitor) with a capacity of 1 F, who was charged by BFC during about 60 min to 240 mW with stored energy ~ 29 mJ. The stored energy was enough to turn at an angle of 90º to the rotor of the electric motor that has input impedance of 6.5 Ohm. Another example application relates to BFC using them as a source to power biosensor. So, in [7] represent biofuel cell integrated in the contact lens that generates power through the use of glucose, located in the human lacrimal fluid, virtually without changing its contents in the body. BFC generate the open circuit voltage of 0.57 V at develop the power of 1 $\mu w/cm^2$. The authors believe that the use of the diagnostic tool based on the contact lens, containing glucose biosensor would be useful for patients with impaired glucose metabolism. This example can also refer to a regime of periodic use of low-power BFC. In the paper [8] BFC was used for powering up a controller, that sands remote signal trough transmitter. In a review paper [9] attention was paid to the analysis of biofuel cells that have fundamental features - membraneless system that operates without chemical mediator. Many of them have a right to be called micro fuel due to miniature size. A preferred type of enzymes used in these elements - is the cathode compartment - membrane PQQ-dependent dehydrogenases and laccases - in the anode compartment. It seems promising to consider further use BQ25504 type converters to improve and stabilize the voltage in the elements of the type described.

Based on the fact that the BFC is practically useful device whose value would increase significantly in the presence of methods savings generated electricity, in this paper we consider a new approach to the issue of energy storage. Texas Instruments Company manufactures Ultra Low Power Boost Converter with Battery Management for Energy Harvester, designed to work with the micro and milliwatt power sources generated by photovoltaic, thermoelectric, electromagnetic and vibration transducers. Boost Converter based on the chip bq25504; chip transforms received from microwatt power source and transmits it to the storage elements, mainly electrical capacity, including Li-ion batteries, electric double layer capacitors.

The goal of this research was to assess the possibility of applying a voltage converter bq25504 in combination with biofuel cells for energy storage. Despite the development of chips for the unstable power source, accumulation of testing a fuel cell has been described previously. In this work, we examined the accumulation mode, wherein the initial voltage of BFC equal to 350 – 400 mV, the converter rises to the level of 3.1 and stored on the capacitors of various capacities. As a source of electricity has been used microbial BFC with microwatts power output. Developed energy storage system can be used in micropower supply for electronic devices and autonomous micro-robots.

2. Materials and methods

Strain of *Gluconobacter oxydans* VKM B-1280 (All-Russian Collection of Microorganisms) was used in research. Measuring cell were represent by two series-connected BFC with open circuit voltage for each ~ 180 - 250 mV. In series connection the voltage is summed. The volume of the anode chambers was equal and it fits 5 ml. the rods of spectral graphite with diameter of 6 mm were used as the electrodes of the anode and cathode, which are immersed in the solution to a depth of 15 mm. Electrolytic anode and cathode connection ditch single BFC carried out through a hole in the wall of the anode and cathode, a closed membrane MF-4SK ("Plastpolymer", Saint-Petersburg, Russia) with an area of 1.2 cm^2. 25 mm of potassium-sodium phosphate buffer (pH 6.0) with addition of 50 mm of sodium chloride were used as the basic solution. 2,6-dichlorophenolindophenol (DGFIP) for the anode compartment and the hexacyanoferrate (III) potassium (GCF) for the cathode (both the mediator procured from Sigma-Aldrich, Co) were used as mediators. Glucose and ethanol were used as substrates for oxidation by *G. oxydans* cells. On external circuit potentiostats (IPCMicro, "Kronas",

IOP Conf. Series: Earth and Environmental Science **52** (2017) 012010 doi:10.1088/1755-1315/52/1/012010

Russia and Ametek, Inc., USA) were connected. Registration of cyclic voltammograms was performed at scan rate 3 mV/s. Buffer solution was stirred with a magnetic stirrer. Accumulation mode considered for two BFC connected in series.

Immobilization of microorganisms was accomplished by inclusion in the chitosan gel applied to the surface of the electrode. First of all, 20 μl of biomass diluted 2-fold with a buffer solution, was applied to the electrode. The height occupied by the microbial cells on the electrode was 2 cm. The electrode was dried at room temperature. Then 20 μl was applied to a 2% solution of chitosan in 1% acetic acid [10] over the cells, and dried at room temperature.

Converting complex was presented by controller device bq25504, which is connected to the power source in the form of microbial BFC and a measuring device, paired with a personal computer (PC) equipped with software for data logging (Fig. 1). The authors of this study have developed 4-channel measurement device and the PC software. At Figure 1 the designation Vin corresponds to a measurement channel for electric potential supplied from the BFC to the input of the Converter, Vbat measurement channel corresponds converted voltage supplied to the storage element, Vstor corresponds to a channel voltage measurement output bq25504 chip, Vbat_ok channel is a voltage comparison circuit at the output with a predetermined output voltage. The measuring device has high input impedance in order to prevent the shunting effect on the accumulation mode converter. Energy storage mode was studied on the capacitance of the i1000 100 μF capacitor (SamYoung), as well as the capacity of electric double layer capacitors 1 F (Panasonic brand).

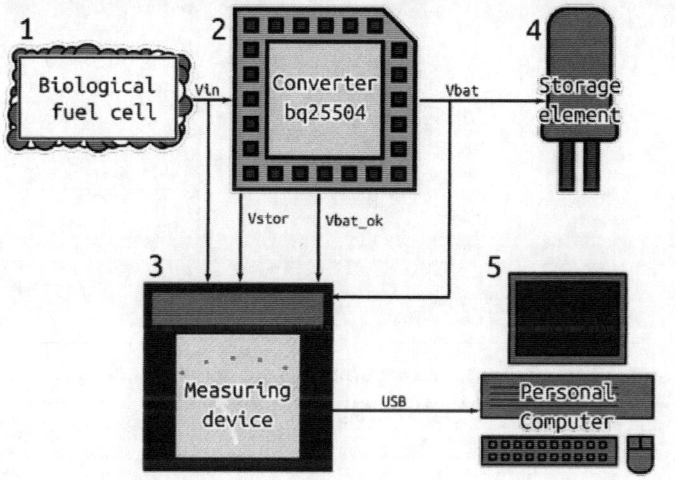

Figure 1. Block diagram of the complex for electrical energy convertion generated by microbial BFC.
1 – BFC, 2 – converter bq25504, 3 - measuring device, 4- storage element, 5 – Personal Computer. Vin –the voltage supplied from the converter to the BFC, Vbat – the converted voltage to the storage element, Vstor – channel voltage measurement at the output of the chip, Vbat_ok – channel is a voltage comparison circuit at the output with a predetermined output voltage.

3. Results and discussion

During first part of research operation of complex with bq25504 converter and chemical energy source was investigated. On the second part bq25504 converter and PV solar cell operation was investigated and in the last part joint operation of bq25504 converter and microbial BFC was provided.

3.1. Chemical energy source "DAEWOO" (AA LR6 1.5 V DAEWOO MEGA ALCALINE).

Processes of energy storage in 1F ionistor are shown at figure 2.

The diagram shows that the input voltage of the converter during the experiment remained constant, thus creating the conditions for rapid conversion. On the drive voltage has reached the value of 3.1 V approximately during 1.5 min and stored on an external capacitor unchanged for 1 min.

Results of experiments with PV solar cell are shown at figure 3. In this case, the solar cell was used as a source with an unstable output voltage varies depending on the lighting. In figure 3 it can be seen exposure uneven the source voltage level due to different levels of light and, as a result, uneven growth of the charge on the capacitor of 1000 μF. After approximately 3 min voltage reaches 3.1 V, and then cut off the device. It should be noted that when a low initial voltage source (350 mV) the charging time of the integrating capacitor has reached three minutes at 1000 μF capacitance.

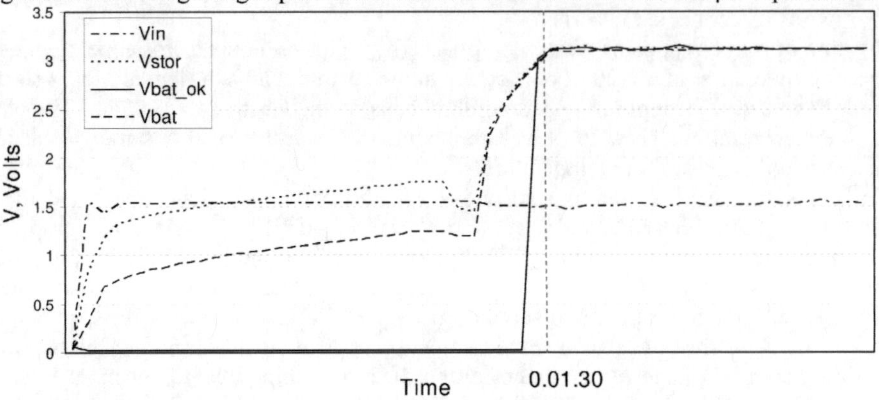

Figure 2. Voltage diagram Vin, Vstor, Vbat_ok and Vbat (explanation in text).

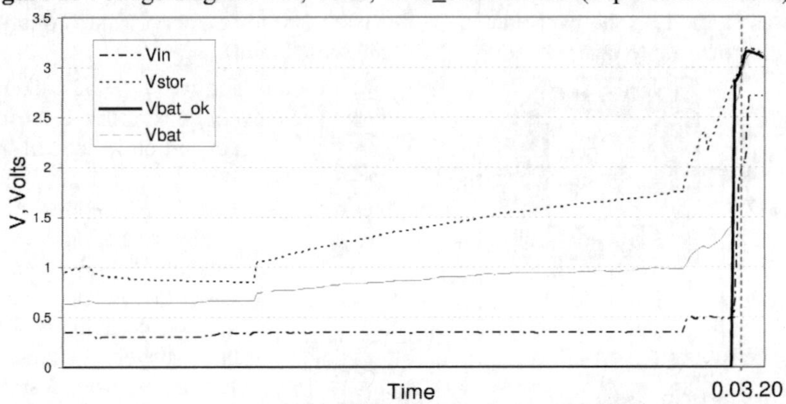

Figure 3. Voltage diagram Vin, Vstor, Vbat_ok and Vbat (experiment with PV solar cell KPS-5130PD7C, Kingbright Electronic).

3.2. Converter energy storage from BFC

On figure 4 cyclic voltammograms for single BFC and two BFC connected in series are demonstrated. It can be seen that the series connection enables to measure the voltage source, which provides start-up condition of the converter voltage higher or equal to 330 mV.

Power output curves for single BFC and two BFC connected in series are shown at figure 5. Series connection of power leads to growth due to the growth of the generated voltage.

3.3. Operation of "BFC – voltage converter – condenser" investigation.

Condenser with 1000 μF capacity was applied as energy storage device. Results of experiments are presented at figure 6.

Figure 6 shows that the voltage of the capacitor after approximately 30 minutes' set at 1.2 or more increases; though final potential value is not reached. Hold Time (indicated by two vertical lines) was around 9 min. The resulting voltage is 3.1 V in this case has been reached, it is possible to explain the parameters of BFC - the area of ion-exchange membrane was initially 0.3 cm^2 and was insufficient to provide the necessary flow of converter input power. This effect is also explained in terms of the value of the internal resistance of the BFC, which in this case was around 3700 Ohms. High internal resistance was due to two factors - the small area of the ion exchange membrane and the initial low ionic strength buffer (25 mM). The following experiment was increased ionic strength of the buffer solution by the addition of sodium chloride (50 mM) and increased the area of the ion exchange membrane (up to 1.2 cm^2), thus reducing the internal resistance of 900 Ohms. Results of this experiment are shown at figure 7. In the experiment, 100 μF capacitor was used as the storage element.

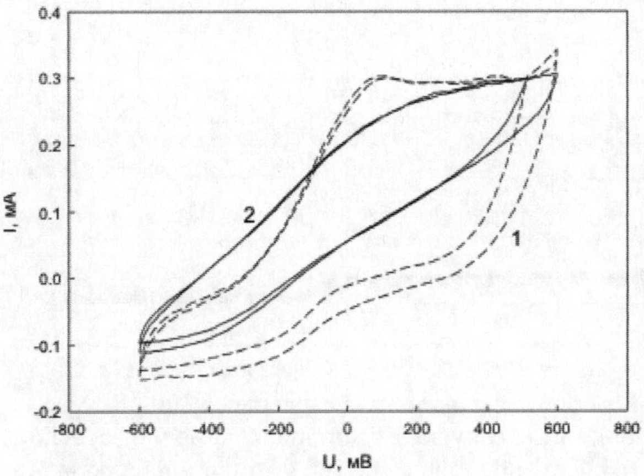

Figure 4. Cyclic voltammograms for single BFC (1) and two BFC connected in series (2).

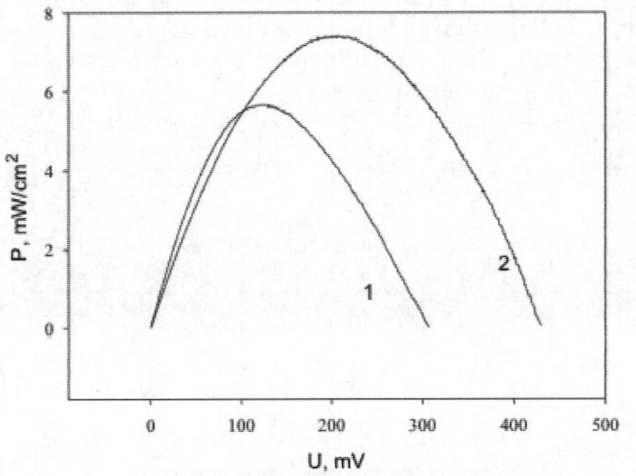

Figure 5. Power output curves for single BFC (1) and two BFC connected in series (2).

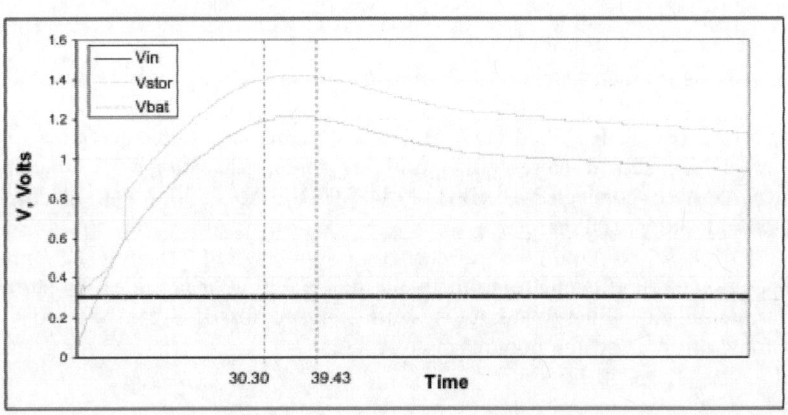

Figure 6. Energy storage experiments results on condenser with 1000 μF capacity.

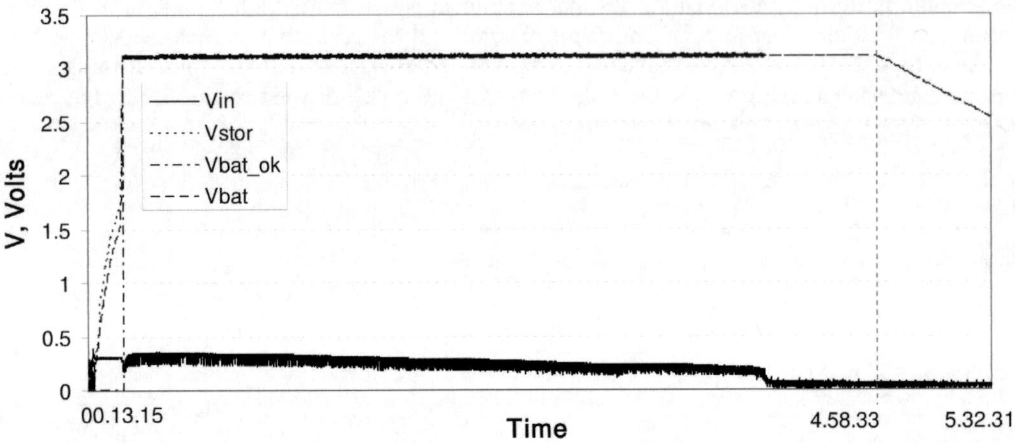

Figure 7. The accumulation of electrical energy to the integrating capacitance 100 μF.

The diagram shows the voltage rise at the output of the converter Vstor and Vbat on the capacitor. 100 μF capacitor for a time of about 13 minutes to charge up to the value 3.1 V. A tenfold reduction in capacitance and the internal resistance of the BFC possible to raise the voltage on the drive to the desired value for a much shorter time is 13 minutes. Voltage storage time was about 5 hours.

The result of the energy storage in the capacitor 1000 μF using BFC that have an ion exchange membrane area was increased and amounted to 1.2 cm^2, shown at Figure 8.

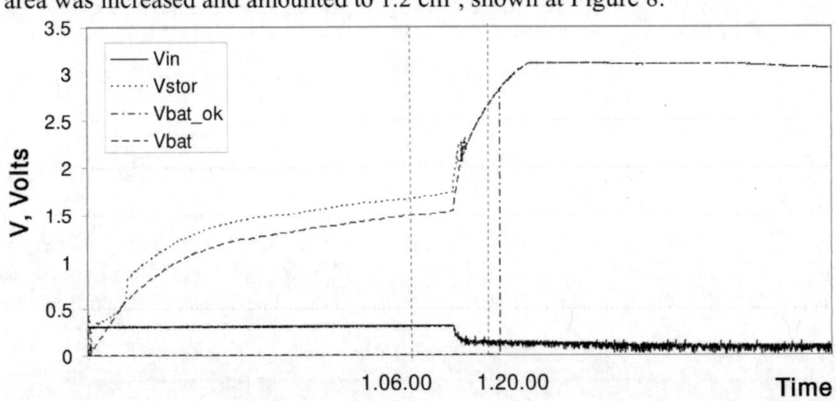

Figure 8. Energy storage in the capacitor 1000 μF using an ion exchange membrane BFC area of 1.2 cm^2.

The possibility of using the described mode of increase of the output voltage values to the BFC was tested, which may be of practical interest. Integrating capacitance to voltage 3.1 V accumulates , and then it turns the two devices connected - LED (L-1154SURDK, 2.0 V, 20 mA) or electric microengine M25E-4L (MITSUMI; 3.0 V, 100 mA).

Charged to 3.1 μF capacitor with 6800 μF capacity provide stored energy to 32.7 mJ and allowed to short-term operation (around 1 minute) to maintain the LED light or provide a few turns of the rotor rotation electric microengine. Energy storage system designed for BFC can be used in micropower supply for electronic devices and autonomous micro-robots.

4. Conclusion

In conclusion, we can note that operation of "BFC-converter-capacitor" system was shown. The process of accumulation depends on the capacity of the final receiver and in low power BFC accumulated voltage may not reach the set value. Conversion of the output voltage occurs more slowly than the above internal resistance BFC. Further research plans include assessing and improving system efficiency, increasing power density and stability of the BFC in time that will result in the design of practically important devices.

Acknowledgements
The work was financially supported by RFBR (grant 15-29-01292 ОФИ_М_2015) and by State tusk of Ministry of Education and Science of Russian Federation № 14.2094.2014/K.

References

[1] Bullen R A, Arnot T C, Lakeman J B, Walsh F C 2006 Biofuel cells and their development *Biosensors and Bioelectronics* **21(15)** 2015-45
[2] Du Z, Li H, Gu T 2007 A state of the art review on microbial fuel cells: A promising technology for wastewater treatment and bioenergy *Biotechnol*. Adv. **25** 464–82. doi:10.1016/j.biotechadv.2007.05.004
[3] Birgit Feketeföldi, Bernd Cermenek, Christina Spirk, Alexander Schenk, Christoph Grimmer, Merit Bodner, Martin Koller, Volker Ribitsch, Viktor Hacker 2016 Chitosan-Based Anion Exchange Membranes for Direct Ethanol Fuel Cells *Journal of Membrane Science & Technology* http://dx.doi.org/10.4172/2155-9589.1000145
[4] Hendrik duToit, Mirella Di Lorenzo 2015 Continuous power generation from glucose with two different miniature flow-through enzymatic biofuel cells *Biosensors and Bioelectronics* **69** 199–205
[5] Szczupak A, Halamek J, Halamkova L, Bocharova V, Alfonta L, Katz E 2012 Living battery – biofuel cells operating in vivo in clams *The Royal Society of Chemistry Energy & Environmental Science* DOI: 10.1039/c2ee23209j
[6] Fan Shuqin, Hou Chuantao, Liang Bo, Feng Ruirui, Liu Aihua 2015 Microbial surface displayed enzymes based biofuel cell utilizing degradation products of lignocellulosic biomass for direct electrical energy *Bioresource Technology* **192** 821–825 http://dx.doi.org/10.1016/j.biortech.2015.05.090
[7] Falk M, Andoralov V, Blum Z, Sotres J, Suyatin D B, Ruzgas T, Arnebrant T, Shleev S 2012 Biofuel cell as a power source for electronic contact lenses *Biosensors and Bioelectronics* **37** 38–45
[8] Kevin MacVittie, Tyler Conlon, Evgeny Katz 2014 A wireless transmission system powered by an enzyme biofuel cell implanted in an orange *Bioelectrochemistry* http://dx.doi.org/10.1016/j.bioelechem.2014.10.005
[9] Magnus Falk, Zoltan Blum, Sergey Shleev 2012 Direct electron transfer based enzymatic fuel cells *Electrochimica Acta* **82** 191–202
[10] Odaci D, Timur S, Telefoncu A 2008 Bacterial sensors based on chitosan matrices *Sensors and Actuators B: Chemical* **134(1)** 89–94

IOP Conf. Series: Earth and Environmental Science 52(2017) 012011 doi:10.1088/1755-1315/52/1/01

Multi-state time-varying reliability evaluation of smart grid with flexible demand resources utilizing *Lz* transform

Heping Jia[1], Wende Jin[2], Yi Ding[1,3], Yonghua Song[1], Dezhao Yu[1]

[1] College of Electrical Engineering, Zhejiang University, Zhejiang, China
[2] State Grid Zhejiang Electric Power Company, Zhejiang, China

E-mail: yiding@zju.edu.cn

Abstract. With the expanding proportion of renewable energy generation and development of smart grid technologies, flexible demand resources (FDRs) have been utilized as an approach to accommodating renewable energies. However, multiple uncertainties of FDRs may influence reliable and secure operation of smart grid. Multi-state reliability models for a single FDR and aggregating FDRs have been proposed in this paper with regard to responsive abilities for FDRs and random failures for both FDR devices and information system. The proposed reliability evaluation technique is based on *Lz* transform method which can formulate time-varying reliability indices. A modified IEEE-RTS has been utilized as an illustration of the proposed technique.

1. Introduction

With the rising challenges of conventional energy crisis and environmental pollutions, the utilization of renewable energies has been rapidly expanding in the last decades [1]. The proportion of installed capacity for renewable energy will continue growing in the years to come [2]. However, renewable energies differ from other traditional power generation methods mainly in its natures of randomicity, intermittence and fluctuation [3,4]. In order to ensure the secure and reliable operation of power systems, some rapid start-up conventional generating units, such as thermal power units, are necessary for accommodating renewable energy [5]. Nevertheless, these techniques may increase operation and maintenance cost as well as greenhouse gases emissions [6].

Nowadays, the development of demand response has not only made better integration of renewable energy but also increased efficiency of power systems [7,8]. In this paper, flexible demand resources (FDRs) are defined as some dispatch-able devices in electric utilization by consumers whose power capacities can be actively participated to achieve demand response [9,10]. Some general components of FDR including electrical vehicles, heat pumps, air conditioners and distributed generations have been utilized to realize bidirectional interaction between electric power providers and electric power consumers in smart grid [11,12]. The potentials of FDRs for different scenarios have been studied in recent years [13,14]. The utilization of FDRs can contribute to more than 20% peak demand reduction [15].

However, the multiple uncertainties from internal and external factors of FRPs make great influences on power system in term of reliability and security. External uncertainties such as weather and operating conditions, internal uncertainties such as random failures and response degree of FDR devices to price signals, incentive mechanisms and dispatch commands are supposed to have significant effects on the operation of FDRs. Furthermore, with the development of cyber physical systems (CPS) [16], the active participations of FDRs are more dependent on information systems. Accordingly, the vulnerability of CPS affects the involvement of FDRs in smart grid [16]. Therefore, in order to achieve reliable operation and effective management of power systems, it is essential to carry out reliability evaluation of FDRs with regard to growing uncertainties.

Extensive researches for reliability assessment of generation, transmission and distribution systems have gained certain progress [17]. However, the reliability model focused on demand side gets relatively less attention [18-20]. The reliability models of consumers in demand side are more related to simulation techniques which only provide inaccurate results [18, 21]. Analytical approaches of reliability such as universal generating function (UGF) have been utilized in power systems with high wind power penetration [22, 23]. The UGF methods can deal with reliability evaluation for the system to overcome the curse of dimensionality compared with state-space based method such as Markov process model and reduce the computational burden. Whereas, UGF methods consider the capacities and probabilities of components as random variables instead of time-varying variables [24]. The general definitions for Lz transform approaches have been proposed in [25] to overcome the shortcomings. Lz transform methods have been formulated in availability evaluation for refrigeration system [26] and water cooling system [27]. In these presented articles, multi-state time-vary reliability models of FDRs are proposed utilizing Lz transform techniques which are extended from UGF methods.

The rest of the paper is organized as follows. Section 2 illustrates multi-state reliability model for a single FDR device in term of potential capacity and random failures. Reliability model of aggregation for multiple FDRs utilizing Lz transform technique is formulated in Section 3. The IEEE RTS [28] is modified to depict the validity and benefits of the proposed approach in Section 4. Section 5 gives the conclusions.

2. Multi-state reliability model for a single FDR device

The power capacity of a FDR at time t is determined by the availability of the FDR at that time. Multi-state model can be utilized to illustrate power capacity of FDR considering potential of FDR and random failures. The corresponding Lz transform to describe the potential of FDR k at bus i can be defined as

$$Lz_{ik}^{f}(z,t) = \sum_{j_{ik}=1}^{K_{ik}} p_{j_{ik}}^{f}(t) \cdot z^{f_{k,j_{ik}}} \tag{1}$$

where $p_{j_{ik}}^{f}(t)$ and $f_{k,j_{ik}}$ are the time-varying probability at time t and the power capacity of FDR k for the state j_{ik} at bus i, respectively. K_{ik} is the state number for FDRs at bus i. $Lz_{ik}^{f}(z,t)$ describes the potential distribution for FDR k. z indicates the Lz transform for a discrete random variable with its state performances and corresponding state probabilities.

The power capacity of a FDR is also determined by its reliability considering random failures of itself and information system. The power capacity is $f_{k,j_{ik}}$ when the FDR in operating state. When the information system or FDR device fails, the power capacity is zero. Therefore, the reliability model of FDR in terms of random failures is utilized by a two-state model. The Lz transform of the capacity for FDR k at bus i can be defined as

$$Lz_{ik}^{re}(z,t) = (1 - up_{j_{ik}}^{re}(t)) \cdot z^{f_{k,j_{ik}}} + up_{j_{ik}}^{re}(t) \cdot z^{0} \tag{2}$$

where $up_{j_{ik}}^{re}(t)$ represents the time-varying failure probabilities of FDR k and information system at bus i.

The multi-state power capacity model of a FDR considering both the potential of FDR and random failures can be obtained by combining (1) and (2). The Lz transform for FDR k at bus i in terms of both potentials and reliability of FDR can be calculated utilizing universal generating operator (UGO) $\otimes_{ser}$ for multi-state component in series.

IOP Conf. Series: Earth and Environmental Science **52** (2017) 012011 doi:10.1088/1755-1315/52/1/012011

$$
\begin{aligned}
Lz_{ik}^{fd}(z,t) &= \otimes_{ser}\left\{Lz_{ik}^{f}(z,t), Lz_{ik}^{re}(z,t)\right\} = \otimes_{ser}\left\{\sum_{j_{ik}=1}^{K_{ik}} p_{j_{ik}}^{f}(t)\cdot z^{f_{k,j_{ik}}}, (1-up_{j_{ik}}^{re}(t))\cdot z^{f_{k,j_{ik}}} + up_{j_{ik}}^{re}(t)\cdot z^{0}\right\} \\
&= \sum_{j_{ik}=1}^{K_{ik}} p_{j_{ik}}^{f}(t)\cdot(1-up_{j_{ik}}^{re}(t))z^{\min(f_{k,j_{ik}}, f_{k,j_{ik}})} + \sum_{j_{ik}=1}^{K_{ik}} p_{j_{ik}}^{f}(t)\cdot up_{j_{ik}}^{re}(t) z^{\min(f_{k,j_{ik}},0)} \\
&= \sum_{j_{ik}=1}^{K_{ik}} p_{j_{ik}}^{f}(t)\cdot(1-up_{j_{ik}}^{re}(t))z^{f_{k,j_{ik}}} + \sum_{j_{ik}=1}^{K_{ik}} p_{j_{ik}}^{f}(t)\cdot up_{j_{ik}}^{re}(t) z^{0} = \sum_{j_{ik}=1}^{K_{ik}} p_{j_{ik}}^{f}(t)\cdot\begin{bmatrix}(1-up_{j_{ik}}^{re}(t))z^{f_{k,j_{ik}}} \\ +up_{j_{ik}}^{re}(t)z^{0}\end{bmatrix}
\end{aligned} \tag{3}
$$

3. Multi-state reliability model of aggregation for multiple FDRs

The power capacity of an individual FDR is limited and the aggregation of multiple FDRs is needed for providing the necessary power system reserves. The reliability model of multiple FDRs after aggregation is illustrated in Figure 1.

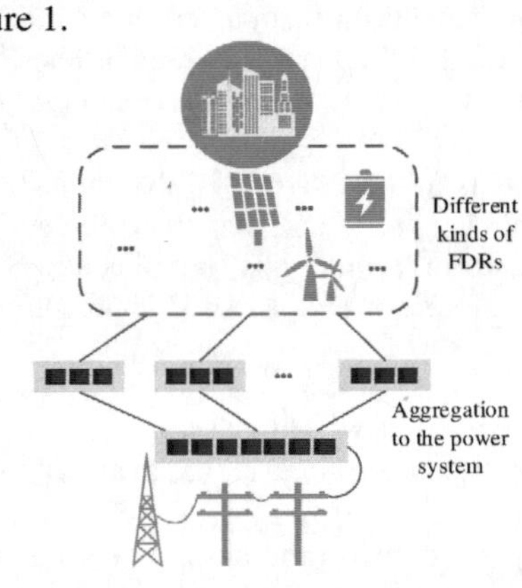

Figure 1. Reliability model of multiple FDRs after aggregation

The multi-state reliability model of aggregating FDRs at bus i can be represented by Lz transform. The power capacity of multiple FDRs can be assumed as statistically independent considering the independent distribution of FDRs. The UGO $\otimes_{par}$ for multi-state component in parallel is utilized to obtain the corresponding Lz transform of multiple FDRs after aggregation. The Lz transform for n_i FDRs at bus i can be defined as the following polynomial:

$$
\begin{aligned}
Lz_{i}^{fdr}(z,t) &= \otimes_{par}\left\{Lz_{i1}^{fd}(z,t),\cdots,Lz_{ik}^{fd}(z,t),\cdots Lz_{in_i}^{fd}(z,t)\right\} \\
&= \otimes_{par}\left\{\sum_{j_{i1}=1}^{K_{i1}} p_{j_{i1}}^{f}(t)\cdot\begin{bmatrix}(1-up_{j_{i1}}^{re}(t))z^{f_{1,j_{i1}}} \\ +up_{j_{i1}}^{re}(t)z^{0}\end{bmatrix},\cdots,\sum_{j_{in_i}=1}^{K_{in_i}} p_{j_{in_i}}^{f}(t)\cdot\begin{bmatrix}(1-up_{j_{in_i}}^{re}(t))z^{f_{n_i,j_{in_i}}} \\ +up_{j_{in_i}}^{re}(t)z^{0}\end{bmatrix}\right\} \\
&= \sum_{j_{i1}=1}^{K_{i1}}\cdots\sum_{j_{in_i}=1}^{K_{in_i}} p_{j_{i1}}^{f}(t)\cdots p_{j_{in_i}}^{f}(t)\cdot\left[\prod_{k=1}^{n_i}(1-up_{j_{ik}}^{re}(t))z^{\sum_{k=1}^{n_i} f_{k,j_{ik}}} + \cdots + \prod_{k=1}^{n_i} up_{j_{ik}}^{re}(t)z^{0}\right] \\
&= \sum_{j_{ia}}^{K_{ia}} p_{j_{ia}}^{a}(t)\cdot z^{fdr_{j_{ia}}}
\end{aligned} \tag{4}
$$

where $p_{j_{ia}}^{a}(t)$ and $fdr_{j_{ia}}$ are the time-varying probabilities and power capacities of aggregating FDRs at bus i for state j_{ia}, respectively. K_{ia} represents the number of states for aggregating FDRs at bus i.

Considering the strong correlation of the FDRs in the same type, the multiple states of aggregation for each type can be obtained by summing up the power capacities for the same type of FDRs:

$$fd_{j_{im}} = \sum_{k=1}^{n_{im}} f_{k,j_{im}} \tag{5}$$

where $fd_{j_{im}}$ represents the power capacity in state j_{im} for aggregation of type m at bus i. $f_{k,j_{im}}$ indicates the power capacity of FDR k in state j_{im}.

The Lz transform for n_{im} FDRs after aggregation can be expressed as:

$$Lz_{im}^{fda}(z,t) = \sum_{j_{im}=1}^{K_{im}} p_{j_{im}}^{f}(t) \cdot z^{\sum_{k=1}^{n_{im}} f_{k,j_{im}}} = \sum_{j_{im}=1}^{K_{im}} p_{j_{im}}^{f}(t) \cdot z^{fd_{j_{im}}} \tag{6}$$

where $p_{j_{im}}^{f}(t)$ represents the time-varying probabilities of the power capacity of aggregating FDR for the state j_{im} at bus i. K_{im} is the number of states for aggregating FDRs.

4. Reliability indices

In order to obtain reliability indices of the systems, the reliability model for conventional generator in [22] is utilized. Reliability indices such as instantaneous availability (IA), expected instantaneous performance (EIP), expected instantaneous performance deficiency (EIPD) and expected energy not supplied (EENS) are formulated in this paper to achieve reliability evaluation for power systems.

$$IA(t) = \sum_{j=1}^{K} p_j(t) \quad (L_j \le G_j) \tag{7}$$

$$EIP(t) = \sum_{j=1}^{K} G_j(t) \cdot p_j(t) \tag{8}$$

$$EIPD(t) = \sum_{j=1}^{K} (L_j(t) - G_j(t)) \cdot p_j(t) \quad (L_j > G_j) \tag{9}$$

$$EENS(t) = \int_0^t \left(\sum_{j=1}^{K} p_j(t) \cdot (L_j(t) - G_j(t)) \right) \cdot dt \quad (L_j > G_j) \tag{10}$$

In (7), (8), (9) and (10), L_j and G_j denote load and generation of the system in state j, respectively.

5. System studies

The modified IEEE-RTS [28] has been utilized to illustrate the reliability model of FDRs. Considering the effects of FDRs in term of accommodating high penetration wind power, a large wind farm with 300 identical 2-MWwind turbine generators is added to the system. The characteristics of the wind turbine generators are represented as multi-state models [29]. The power output and state probability of a wind turbine generator developed in [29] are utilized in this case study. There are four identical 575-MW coal thermal generating units [30] and three 197-MW oil thermal generating units [28] installed in this system. It is assumed that there are 200000 FDRs distributed in the system which can be aggregated to 200 clusters. The state-space diagram for a general FDR is depicted in Figure 2. The state transition rates between two states are shown in Table I. The mean time to failure (MTTF) and mean time to repair (MTTR) of FDR are assumed to be 3650 hours and 50 hours, respectively.

There are two scenarios in the case to make comparisons. Scenario A takes the FDRs into account, while the FDRs are not considered in Scenario B. The operating period of the two scenarios is set to be 100 hours.

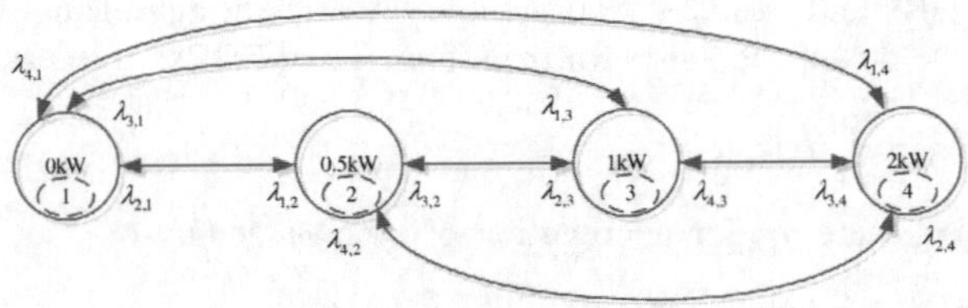

Figure 2. The state-space diagram for a general FDR

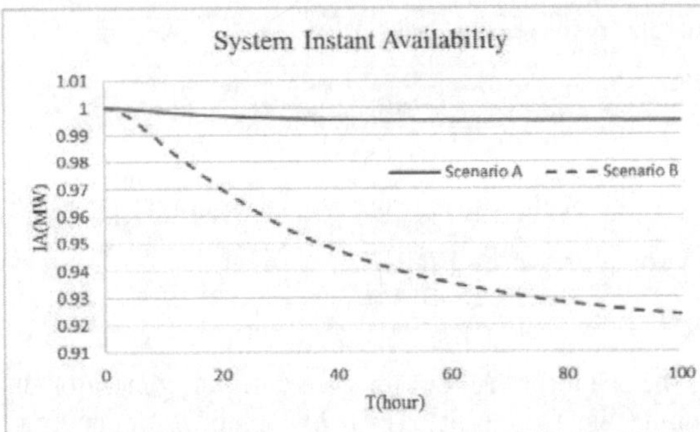

Figure 3. System instant availability for different scenarios

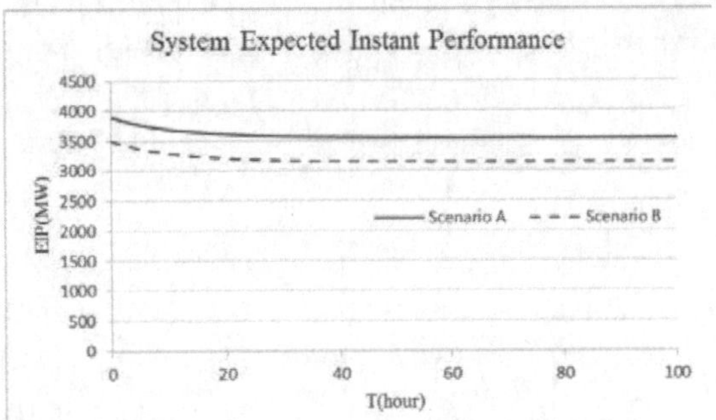

Figure 4. System expected instant performance for different scenarios

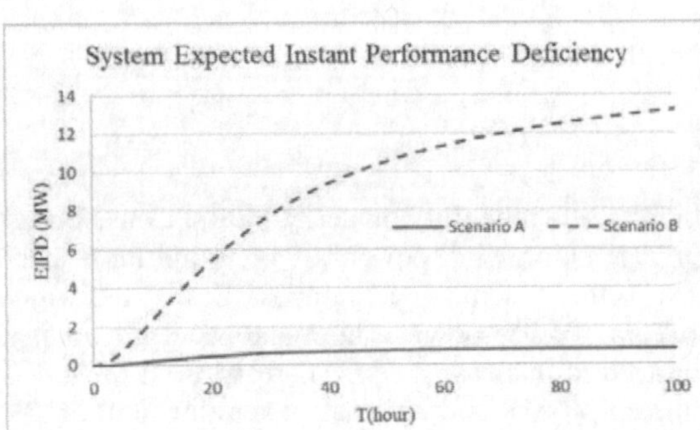

Figure 5. System expected instant performance deficiency for different scenarios

Table 1. The state transition rates for a general FDR

Transition rates(/hour)	0kW	0.5kW	1kW	2kW
0kW	--	0.08	0.0133	0
0.5kW	0.0294	--	0.3235	0.0294
1kW	0.0288	0	--	0.3558
2kW	0.0002	0.0001	0.0007	--

The system time-varying reliability indices including IA, EIP, and EIPD are obtained utilizing the proposed techniques in Figure 3 –Figure 5. In these figures, the solid lines indicate the reliability indices for systems with FDRs, while the dotted line denote reliability indices for systems without the consideration of FDRs. These figures clearly demonstrate the reliability for power systems with FDR is higher than that of systems without FDRs. Moreover, the EENS for 100h of systems with FDRs and without FDRs are 60.29 MWh and 903.05 MWh, respectively. It indicate that in this case the system with FDRs is more reliable than the system without FDRs.

6. Conclusions

With the development of smart grid technologies, flexible demand resources have become active participants in demand response. The uncertainties from FDRs may influence the reliable operation of power system. Moreover, conventional reliability evaluations about generation systems, transmission systems and distribution systems have gained well developments. However, reliability model of FDRs in demand side gets little attention. A multi-state time-varying reliability model of FDRs utilizing *Lz* transform is proposed in this paper. The proposed method can handle reliability evaluation for systems with time-vary probabilities and overcome the curse of dimensionality. The proposed technique is applied to the modified IEEE RTS to validate its feasibility and effectiveness.

Acknowledgments

This work was supported in part by the National Key Research and Development Program of China under Grant 2016YFB0901103, the China NSFC under Grant 51577167 and the State Grid Zhejiang Electric Power Company (52110416001T).

References

[1] Aghaei J and Alizadeh M I 2013 Demand response in smart electricity grids equipped with renewable energy sources: a review *Renewable & Sustainable Energy Reviews* **18**(2) 64-72.

[2] Liu M Quilumba F and Lee W J 2015 A collaborative design of aggregated residential appliances and renewable energy for demand response participation *IEEE Transactions on Industry Applications* **51** 1-1.

[3] Ivana Dusparic, Adam Taylor, Andrei Marinescu, et al. 2015 Maximizing renewable energy use with decentralized residential demand response *2015 IEEE First International Smart Cities Conference* (ISC2).

[4] Wang J, Shahidehpour M and Li Z 2008 Security-constrained unit commitment with volatile wind power generation *IEEE Transactions on Power Systems* **23**(3) 1319-27.

[5] Ortner A, Graf C 2013 Multi-market unit-commitment and capacity reserve prices in systems with a large share of hydro power: a case study *2013 10th International Conference on the European Energy Market* (EEM).

[6] Milligan M, Porter K, Demeo E, Denholm P, Holttinen H and Kirby B, et al. 2009 Wind power myths debunked *IEEE Power & Energy Magazine* **7**(6) 89-99.

[7] Klobasa M 2010 Analysis of demand response and wind integration in Germany's electricity market *Iet Renewable Power Generation* **4**(1) 55-63.

[8] Ma O, Alkadi N, Cappers P, Denholm P, Dudley J and Goli S, et al. 2013 Demand response for ancillary services *IEEE Transactions on Smart Grid* **4**(4) 1988-95.

[9] Briefing Paper: A Review of Current Issues with Long-Term Resource Adequacy 2013 *California Public Utilities Commission Energy Division & Policy and Planning Division.*

[10] Shariatzadeh F, Mandal P and Srivastava A K 2015 Demand response for sustainable energy systems: a review, application and implementation strategy *Renewable & Sustainable Energy Reviews* **45**(4) 343-350.

[11] Yan H, Li B, Chen S, Zhong M, Li D and Jiang L, et al. 2015 Future evolution of automated demand response system in smart grid for low-carbon economy *Journal of Modern Power Systems & Clean Energy* **3**(1) 72-81.

[12] Palensky P and Dietrich D 2011 Demand side management: demand response, intelligent energy systems, and smart loads *Industrial Informatics IEEE Transactions* **7**(3) 381-388.

[13] Federal Energy Regulatory Commission 2008 Assessment of demand response and advanced metering.

[14] Federal Energy Regulatory Commission. A national assessment of demand response potential 2009 The Brattle Group, Freeman Sullivan & Co, and Global Energy Partners.

[15] Cappers P, Goldman C and Kathan D 2010 Demand response in U.S. electricity markets: empirical evidence *Energy* **35**(4) 1526-35.

[16] Sridhar S, Hahn A and Govindarasu M 2012 Cyber–physical system security for the electric power grid *Proceedings of the IEEE* **100**(1) 210-224.

[17] Billinton R, Allan N R 1996 Reliability Evaluation of Power Systems. Second Edition, Plenum

[18] Schneider, K. P., Fuller, J. C., & Chassin, D. P. 2011 Multi-state load models for distribution system analysis *IEEE Transactions on Power Systems* **26**(4) 2425-33.

[19] Gaspar C and Gomes A 2016 A contribution of demand response for the reliability of a power system," *2016 IEEE 16th International Conference on Environment and Electrical Engineering (EEEIC)*, Florence pp. 1-6.

[20] Kopsidas K, Kapetanaki A, Levi V and Milanovic J 2016 Optimal demand response scheduling with real time thermal ratings of overhead lines for improved network reliability *IEEE Transactions on Smart Grid*, in press.(doi: 10.1109/TSG.2016.2542922)

[21] Li G, Bie Z, Hua B, et al. 2012 Reliability evaluation of distribution systems including micro-grids considering demand response and energy storage. *2012 47th International Universities Power Engineering Conference (UPEC).*

[22] Ding Y, Wang P, Goel L, Loh P C and Wu Q 2011 Long-term reserve expansion of power systems with high wind power penetration using universal generating function methods *IEEE Transactions on Power Systems* **26**(2) 766-774.

[23] Ding Y, Singh C, Goel L and Ostergaard J 2014 Short-term and medium-term reliability evaluation for power systems with high penetration of wind power *IEEE Transactions on Sustainable Energy* **5**(3) 896-906.

[24] Lisnianski A and Ding Y 2016 Using inverse Lz -transform for obtaining compact stochastic model of complex power station for short-term risk evaluation *Reliability Engineering & System Safety* **145** 19-27.

[25] Lisnianski A, Frenkel I 2012 Recent Advances in System Reliability *Springer*.

[26] Frenkel I, Lisnianski A and Khvatskin L 2012 Availability assessment for aging refrigeration system by using lz-transform *Journal of Reliability & Statistical Studies* **5**(2) 33-43.

[27] Frenkel I, Khvatskin L, Daichman S, et al. 2013 Assessing Water Cooling System Performance: Lz-Transform Method Availability *2013 Eighth International Conference on Reliability and Security* (ARES).

[28] Tripak O and Kongsiriwong S 1989 A reliability test system for educational purposes-basic data *IEEE Transactions on Power Systems* **4**(3) 319-325.

[29] Dobakhshari A S and Fotuhi-Firuzabad M 2009 A reliability model of large wind farms for power system adequacy studies *IEEE Transactions on Energy Conversion* **24**(3) 792-801.

[30] Lisnianski A, Elmakias D, Laredo D and Haim H B 2012 A multi-state markov model for a short-term reliability analysis of a power generating unit *Reliability Engineering and System Safety* **98**(1) 1-6.

Analysis on capability of load following for nuclear power plants abroad and its enlightenment

Kuan Zheng[1, 4], Fu-qiang Zhang[1], Ting-ting Deng[2], Jin-fang Zhang[1], Weihua Hao[3]

[1] State Grid Energy Research Institute, Beijing, 102209, China
[2] Beijing Jiaotong University, Beijing, 100044, China
[3] State Grid Shandong Electric Power Corporation Maintenance Company, Jinan, 250118, China

E-mail: Kzheng302@126.com

Abstract. With the acceleration adjustment of China's energy structure, the development of nuclear power plants in China has been going back to the fast track. While as the trend of slowing electric power demand is now unmistakable, it enforces the power system to face much greater pressure in some coastal zones where the nuclear power plants are of a comparative big proportion, such as Fujian province and Liaoning province. In this paper, the capability of load following of nuclear power plants of some developed countries with high proportion of nuclear power generation such as France, US and Japan are analysed, also from the aspects including the safety, the economy and their practical operation experience is studied. The feasibility of nuclear power plants to participate in the peak regulation of system is also studied and summarized. The results of this paper could be of good reference value for the China's nuclear power plants to participate in system load following, and also of great significance for the development of the nuclear power plants in China.

1. Introduction

With the acceleration adjustment of China's energy structure, the construction of Nuclear power plants (NPPs) in China has been restarted and went back to the fast track of development. By the end of June 2016, there are 30 nuclear power units being in commercial operation in China with a total installed capacity of 28.6GW, and from January to June in 2016, the cumulative output of the NPPs in commercial operation is 95400GWh, which rose 26.58% compared with those of the same period in 2015 [1]. During the period of the 13th Five Year Plan, we expect that there would be 37.5GW nuclear power to be constructed step by step, and the total installed capacity of nuclear power will reach 58GW by the year of 2020 [2].

There are a few advantages and disadvantages for the rapid development of NPPs. On one hand, it contributes more to a clean energy and reduces the environment stress, and some relevant industries are also driven to develop at the same time. On the other hand, as the proportion of renewable energy power is increasing, the prompt development of NPPs will bring more pressure of the peak regulation for the power system. For this problem, some European and American countries with high proportion of NPPs have owned advanced technology and accumulated rich experience in actual operation, which could provide much reference value for NPPs to participate in the load following and the optimal operation of the power system of China in the future.

2. Regulation capability of NPPs abroad

In 1980s, as a result of improvement of the light water reactor, the output of NPPs can be manoeuvred, and then the system regulators of European and American put forward more and more specific requirements about the adjust and control capability of NPPs. In 1991, the utilities from Great Britain, France, Germany and other countries considered that a more stringent specification would be needed to cover a wide range of designs, and thus the European Utilities Requirements (EUR) were created. The EUR explicitly state that modern reactors should implement significant manoeuvrability capability and, in particular, to be able to operate in the load-following mode [3]:

1) The unit must be capable of continuous operation between 50 and 100% of its rated power (Pr).

The standard plant design shall allow the implementation of scheduled and unscheduled load following operation during 90% of the whole fuel cycle. The unit shall be capable of load-following operation in the range of output from 100% Pr down to the minimum load of the unit. The rate of change of electric output shall be 3% of Pr/min. The unit shall be expected to go through the following number of load scheduled variations, each variation being defined as a transient from full power to minimum load and back to full power:

2 times per day;
5 times per week;
Cumulatively 200 times per year.

2) The unit may be required to participate in emergency load variations:
   Increasing of output: according to requirements for secondary control.
   Decreasing of output:
   Amplitude: down to minimum load of the unit;
   Rate of change: 20% of Pr/min.
3) The unit shall be capable of taking part in the primary control of the grid. This is a prerequisite for connection to the grid. The primary control range shall be ± 2% of the rated power Pr, but higher values may be agreed between system operators and plant operators, though not higher than ± 5% Pr. The unit shall be capable of activating, within 30s, the total primary range of control requested at a quasi-steady frequency deviation of ± 200 mHz, and maintaining supply for at least 15 minutes.
4) The standard plant design shall allow the implementation of a secondary control. The minimum control range for secondary control operation shall be ± 10% of Pr above the minimum load taking into account the control range. The variation rate shall be 1% of Pr/min. Higher values may be agreed between system operator and plant operator, though not higher than 5% of Pr/min.
5) The unit shall be able to contribute to grid restoration. Re-supplying customers will be based on an agreement between the system operator and the plant operator. If the unit participates in re-supplying customers, it should be capable of withstanding sudden load steps up to 10% of Pr.

After decades of development, the modern light-water reactors have already been able to meet all the requirements above. Taking the European EPR and Russian VVER-1000/1200 as examples, the specific parameters of their load following capacity are shown in Table 1.

3. Safety analysis of load following

At present, there is no any disastrous accident of NPP in history, such as the core explosion and the fuel leakage, which are caused by load following. In the development process of more than 60 years of NPPs around the world, there are 3 major nuclear accidents: the accident of Three Mile Island NPP in United States on August 24, 1979, which is ranked level five accident according to the INES scale caused mainly by operation-judgment failure and mechanical failure; the Chernobyl accident in Soviet Union on April 26, 1986, which is ranked level seven accident (the worst accident so far) that happened mainly because of the wrong operation and its design defects; the accident of Fukushima Daiichi NPP in Japan on March 11, 2011, which is ranked level seven accident due to the insufficient design for the tsunami and its aged components and . In conclusion, the key factors leading to those three accidents are design weakness, human errors and scarce capacity for resistance to extreme natural disasters, and it is no relation to the operation mode of NPPs' load following.

Table 1. European nuclear regulatory capacity datasheets

Operating mode	Typical unit parameters			
	EPR		**VVER-1000/1200**	
	Adjustment range	Adjustment rate	Adjustment range	Adjustment rate
Daily load following	60%-100%Pr	5%Pr/min	50%-100%Pr	5%Pr/min
	25%-60%Pr	2.5%Pr/min		
Primary frequency modulation	±2.5%Pr	1%Pr/sec	±2%-±5%Pr	1%Pr/sec
Secondary frequency modulation	When the unit output is between min to 60%Pr: ±4.5%Pr	1%Pr/min	±10%Pr	1%-5%Pr/min
	When the unit output is between 60%-100%Pr: ±10%Pr	2%Pr/min		
Emergency situations	Minimum output-100%Pr	Subject to adjustment	50%-100%Pr	20%Pr/min

According to the research report [3], the risk of NPPs load following include two aspects: extended low power operation (ELPO), the other is frequent load following (FLF). However, the accidents caused by these two aspects could not exceed the nuclear power accident with two-level in general. It could cause temperature rise of moderator if low power operation lasts for a long time, and then the risks of fuel the pellet melting and the radioactive fuel leaking to the primary circuit could increase a lot. Moreover, the frequent load following could lead to greater pressure for storage management of radioactive waste and radioactive waste liquid, as well as the radioactive transportation in the future.

4. Economic analysis of load following

The levelized cost of generating electricity (LCOE) is one of the important indicators to measure the power economy. LCOE is closely related to the cost of power supply, operation and maintenance, fuel, emissions, retire and annual power generation. For nuclear power, as the investment cost is its mainly cost and the fuel cost is usually not more than 20% of the total cost, when NPPs participate in the power system load following, the LCOE changes much more than the other power, which is caused by the drop of load rate (the proportion of annual energy output decrease, etc.). According to the data of international atomic energy agency [4] in 2010, on condition that the discount rate is 5%, the load rate decreases from 90% to 70% and the LCOE of nuclear power increases by 35% approximately, which exceeds the coal power with 10% and gas-fired-electric power with 7%.

Since most of modern NPPs implement strong maneuverability capabilities in their designs, there is no or very small impact of the load following on the acceleration of aging of large equipment [5, 6]. However, there is some influence of the load-following on the aging of some operational components (e.g. valves), and thus one can expect a slight increase of the maintenance costs. Also, for older plants some additional investment could be needed in order to become eligible for operation in the load-following mode.

5. The practical operation experience of load following abroad

France is the largest country on the proportion of nuclear power in the world, 75% of its electricity from nuclear power. Because the share of nuclear power in the national energy mix is high, some

NPPs operate in the load-following mode to respond to the daily variations of the electricity demand (see an example in Figure 1) [7]. However, not all the nuclear power units should participate in the frequent load following during the lifetime. In general, the Électricité De France (EDF) will determine the operation mode and maintenance management of nuclear power units together with the power system dispatching department three years ahead of schedule. Among the 58 nuclear power units within the scope of the jurisdiction of the French EDF, there are 10 nuclear power units which do not participate in the power load following, and the amount of participating in the power load following for each nuclear power unit also ranges a lot. For example, according to the data of 2010, the amount of power output of the nuclear power unit whose load following times is the largest reached about 130 times.

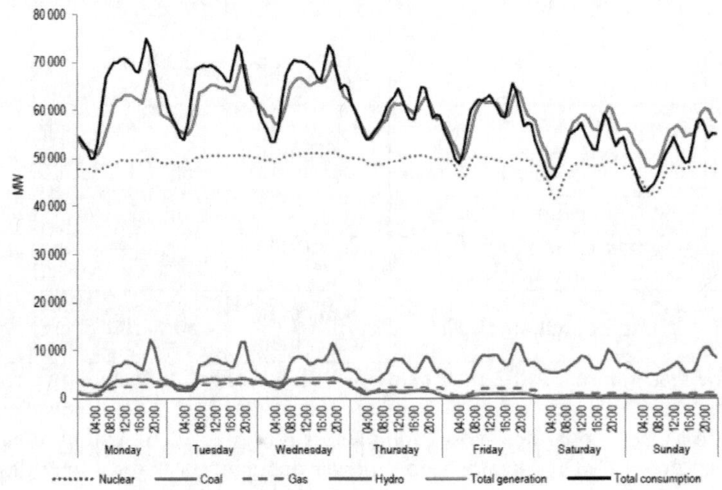

Figure 1. Practical electricity generation in France during 2 weeks in November, 2010.

Although the United States and Japan have the record of nuclear power unit daily load following [8], but since there is enough electricity power with peak regulation capability in these two countries, such as gas-fired power and oil-fired power etc. Therefore, the NPPs of these two countries are operating in the base-load mode in most time. There are some records to show that the PWR nuclear power units of Westinghouse in United States have the practical experience of daily load following operation, and one of the nuclear power units has finished more than 600 times of daily load following operation during the consecutive 4 life times of fuel. Also, the Ikata Nuclear Power Plant of the Shikoku Electric Power Company in Japan has more than 20 years of power load following operation, and the unit No.2 conducted load following in the mode of "12-3-6-3" from October 1987 to February 1988.

Table 2. Practical examples of load following of NPPs in some countries

Country	Reactor type	Control range of load following	Variation rate of output in 1 minute
United States	PWR BWR	50% -100%Pr	About 0.25%-1%
France	PWR	40% -100%Pr	Maximum 5%, Generally 3%
Germany	PWR	50% -100%Pr	5%-10%

6. Conclusion

Compared to the countries with high proportion of NPPs, such as French, the total installed capacity of NPPs in China is small but the development is very fast. During the period of the 13th Five Year Plan, the installed capacity of NPPs in some provinces such as Fujian will increase more than 10% of the total capacity. Analysing the load following capability and the practical operation experience of NPPs abroad indicates that the NPPs can participate in the peak regulation of system by the load following. Although from the economics and safety analysis NPPs is not the best power source to handle the peak regulation of system, the load following operation of NPPs is also of great importance for some provinces in China which are facing big pressure of the peak regulation of system thanks to the fast development of NPPs.

In practice, AP1000, CAP1400 and Hualong No.1 belong to the third-generation reactors of NPPs, which means that the reactors invented by China have the similar load following capability as the French EPR and Russian VVER. However, for China, the performance testing for NPPs, the operation procedures of daily load following and operator training, etc. are lacking. Therefore, for the nuclear power companies and the research institutes should not only learn the technology of load-following of NPPs in France or other countries, but also set the relevant operation specifications and managements for the NPPs operation to change the old opinion that NPPs only operates with base load, what's more, the skills training of the operation personnel for the load following of NPPs is also indispensable to realize the load following of NPPs.

References

[1] The operation statistic of nuclear power from January to June in 2016 in China.China Nuclear Energy Association. http://www.china-nea.cn/html/2016-07/36412.html.

[2] Development plan of nuclear power in China in a long term *National Development and Reform Commission*

[3] Technical and economic aspects of load following with nuclear power plants *Nuclear Energy Agency* http://www.oecd-nea.org/

[4] Operating expearience with nulcear power stations in member states in 2009, IAEA, Vienna, Austria.

[5] Jonas Persson et al. December 2012 Additional costs for load-following nuclear power plants, experiences from Swedish, Finnish, German and French Nuclear power plants ELFORSK rapport **12** 71

[6] Bruynooghe G, et al. 2010 Load-following operating mode at nuclear power plants (NPPs) and incidence on operation and maintenance (O&M) costs. Capability with wind power variability, *JRC Scientific and Technical Reports* SPNR/POS/10 03 004, rev.05

[7] Electrical Energy Statistics for France 2010 RTE. http://clients.rte-france.com/lang/fr/visiteurs/vie/telecharge.jsp.

[8] Yu Wenqi, et al. 2011 A review on load following of pressurized water reactor of nuclear power plants *Energy of China* **33(3)** 38-40.

Multiple external hazards compound level 3 PSA methods research of nuclear power plant

Handing Wang[1], Xiaoyu Liang, Xiaoming Zhang, Jianfeng Yang, Weidong Liu and Dina Lei

Suzhou Nuclear Power Research Institute, Shenzhen, 518000, China

E-mail: wanghanding@cgnpc.com.cn

Abstract. 2011 Fukushima nuclear power plant severe accident was caused by both earthquake and tsunami, which results in large amount of radioactive nuclides release. That accident has caused the radioactive contamination on the surrounding environment. Although this accident probability is extremely small, once such an accident happens that is likely to release a lot of radioactive materials into the environment, and cause radiation contamination. Therefore, studying accidents consequences is important and essential to improve nuclear power plant design and management. Level 3 PSA methods of nuclear power plant can be used to analyze radiological consequences, and quantify risk to the public health effects around nuclear power plants. Based on multiple external hazards compound level 3 PSA methods studies of nuclear power plant, and the description of the multiple external hazards compound level 3 PSA technology roadmap and important technical elements, as well as taking a coastal nuclear power plant as the reference site, we analyzed the impact of off-site consequences of nuclear power plant severe accidents caused by multiple external hazards. At last we discussed the impact of off-site consequences probabilistic risk studies and its applications under multiple external hazards compound conditions, and explained feasibility and reasonableness of emergency plans implementation.

1. Introduction

After the Fukushima nuclear accident, the severe accidents consequences evaluation has also raised more and more attention, especially the nuclear power plants suffered from multiple external hazards compound cases. Some external hazards may occur simultaneously and may affect the plant’ s safety function or result in degradation of the safety function of nuclear power plant. It is an important measure to protect the safety of nuclear facilities. Because of the impact of meteorological, geological hazards under multiple external hazards conditions, the emergency plans implementation feasibility and rationality also need to be explored [1].

Based on the nuclear power plant multiple external hazards compound level 3 PSA methods studies, and the description of the multiple external hazards compound level 3 PSA technology roadmap, important technical elements that different from the internal event level 3 PSA. We took a coastal nuclear power plant as a reference site, and considered the site’s earthquake and tsunami compound hazards and high winds and heavy rain compound hazards. And combined with source term study results of multiple external hazards compound Level 2 PSA, as well as the consideration of extreme meteorological conditions we analyzed multiple external hazards compound caused nuclear power plant severe accident and its impact of plant consequences. Then we calculated some cases, discussed plant consequences probabilistic risk studies and applications under multiple external hazards compound conditions.

2. Technical method

2.1. Evaluation method

Multiple external hazards compound level 3 PSA methods are almost same as commonly level 3 PSA, but some key technical elements need to be considered as multiple external hazards affected factors. First, source term study results come from multiple external hazards compound Level 2 PSA; Second, considering the impact of extreme weather conditions in high winds and heavy rain compound hazards, and it requires the select of a representative weather condition sequence to calculate, and also we need new sampling method to process meteorological conditions in order to verify the different weather conditions effects; Third, considering that the impact of emergency response in earthquake and tsunami compound hazards may results in evacuation failed, and so on[2]. These affected technology elements will be described in section 2 in detail. The definitions for external hazards and extreme conditions are based on the nuclear power plant design basis. For example, seismic external hazard analysis scope is from 0.2g seismic acceleration to above. The high winds external hazard is wind speed more than 20m/h.

The estimation of accident risks requires consideration of both frequencies of accident occurrence and those accidents consequences. NUREG-0396 description for severe accidents consequence risk analysis of probability theory evaluation method [3] is as follows: As the release category i, through calculating dose complementary cumulative frequency distribution (CCDF) curve, it can obtain meteorological conditions probability pi(x) exceeds the specified dose. If the occurrence frequency of release category i is fi, , in the x distance, the conditional probability exceeds a specified dose in considering the entire accident spectrum is as follows:

$$p(x) = \frac{\sum_i p_i(x) f_i}{\sum_i f_i} \qquad (1.1\text{-}1)$$

Multiple external hazards compound level 3 PSA consequence risk analysis results are described in section 3.1. Multiple external hazards extreme weather conditions affected analysis results are described in section 3.2.

2.2. Grid structures

In this research, we taking a coastal nuclear power plant as a reference site, and getting site characteristics of relevant data. Site around of 80km is as an analysis evaluation area, which radially divided into 12 loops.

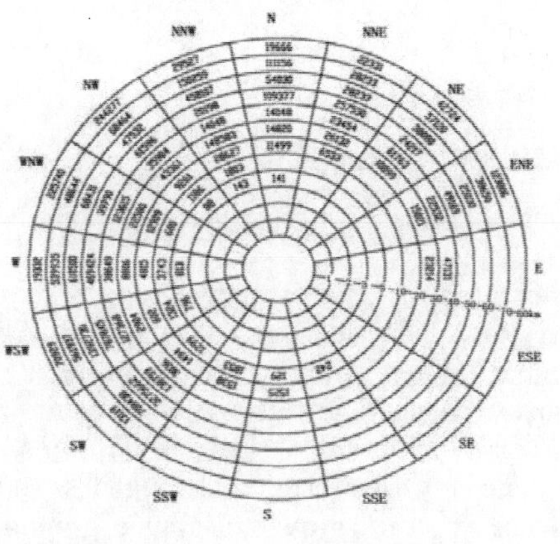

Figure1. Grid Structure Schematic

In the consequence analysis code (MACCS2), the surrounding area of reference site of 80 km was meshed [4]. The surrounding area of reactor is divided into polar coordinate grids, the reactor at the center is located in the coordinate of r=0, θ=0, and every 22.5° azimuth will be a circumferential region which has been divided into 16 sectors. Figure 1 is a grid structure schematic. N represents the north orientation.

3. Multiple external hazards LEVEL 3 PSA Technology Elements impact analysis

3.1. Multiple external hazards compound accident source term

The starting points of multiple external hazards compound consequence analysis is relavent multiple hazards compound level 2 PSA produced radioactive source term and frequency. The radioactive source term is the key parameter to determine the accidents off-site consequences, and the final form of level 2 PSA analysis, as well as the inputs of off-site consequences probability evaluation, which include the radioactive fission product' s releasing volume, releasing time, duration time, releasing height and so on. The radioactive source term includes many different radionuclides, and they are different in physical and chemical natures, as well as in migration behaviors in the environment and risk impacts on society. This severe accident source term adopts multiple external hazards compound Level 2 PSA results.

3.2. Multiple external hazards meteorological processing method and Atmospheric dispersion

Multiple external hazards include extreme weather conditions such as high winds and heavy rain, and we need new sampling method to process meteorological conditions in order to verify the extreme weather conditions effects. Multiple external hazards extreme weather conditions need to make many meteorological groups to examine the impact of high winds and heavy rain. Therefore, we considered to separately extract a number of weather sequences such as heavy rain and high winds as sample box. In addition, constant weather conditions of heavy rain and high winds also were analyzed. External hazards weather conditions processing approach shows in Figure 2.

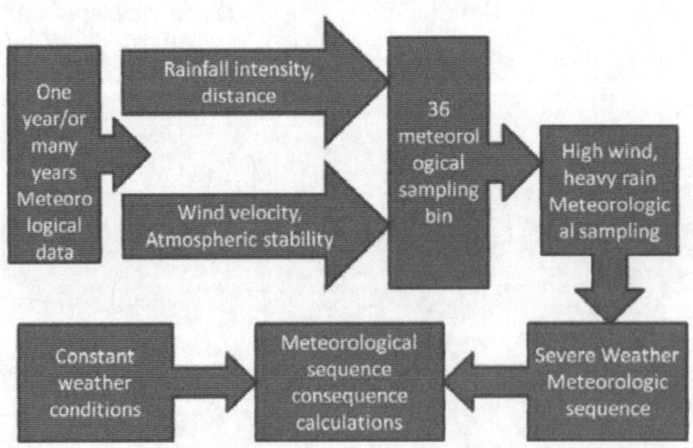

Figure 2. Multiple External Hazards Extreme Weather Conditions Processing method

Took the reference site meteorological data of many years as sampling data, and used category sampling method to sample meteorological sequence. The sampling bin was based on the multiple external hazards extreme weather. When applying Gaussian plume model to calculate multiple external hazards consequence, the atmospheric diffusion parameters also need to be amended. Because the reference site geographical conditions is hilly terrain, and ground roughness is large, so take the use of atmosphere diffusion parameters of larger ground roughness (z0=1m). Diffusion coefficient is related with atmospheric turbulence structure, ground clearance, ground roughness, leak duration time, sampling interval, wind speed and distance leaving the leak source. For other site-related parameters, such as mixing layer height, shielding factor and etc., selecting conservative values through sensitivity analysis [5].

Model assumes pollutant concentration distribution is a Gaussian distribution in space. According to the traditional Gauss formula, when the plume are not limit to the ground and the inversion layer, Gaussian plume model can be expressed as follow:

$$\chi(x,y,z)=\frac{Q}{2\pi u\sigma_y\sigma_z}\exp\left[-\frac{1}{2}\left(\frac{y}{\sigma_y}\right)^2\right]\exp\left[-\frac{1}{2}\left(\frac{z-h}{\sigma_y}\right)^2\right] \tag{2.2-1}$$

Where: χ (x, y, z) is time integral air concentrations at the downwind position (x, y, z), Bq • s/m3; Q is source intensity, Bq; u is the average wind speed, m/s; σ y and σ z is horizontal and vertical standard deviation of plume respectively, m; (x=0, y=0, z=h) is the source term position; h is the release height, m.

3.3. Dose response and health effects

Radioactive substances are released into the atmosphere after a severe accident, and because the atmospheric diffusion, deposition, the body accumulated radiation dose from the plume irradiation and ground deposit irradiation. Dose-response model is converting radioactive substances concentration to the human body cumulative radiation dose. Figure 3 is a dose assessment and health effects evaluation schematic. Considered exposure pathways include: plume immersion radiation, ground deposition irradiation, inhalation irradiation, re-suspended inhalation irradiation, skin deposition irradiation. Early accident radiation exposure considered only the first three. Plume immersion irradiation pathway expected dose could be calculated by the following formula [6].Plume immersion outside irradiation:

$$DC_k=(\sum_i AC_i^C\cdot DFC_{\infty ik})C\cdot F\cdot SFC \tag{2.3-1}$$

Where: k is plume segment through grid caused organs k irradiation dose, Sv; CiC is the plume centerline i nuclide time integrated air concentrations, Bq • s / m3; DFC∞ik is i radionuclide for k organ semi-infinite plume dose conversion factor, Sv • m3 / (Bq • s); C is limited plume correction factor; F is exposure time share when the plume passes, F=te/t0, te is people exposure time (s), t0 is the duration time of the plume through a segment (s); SFC is Plume irradiation shielding factor.

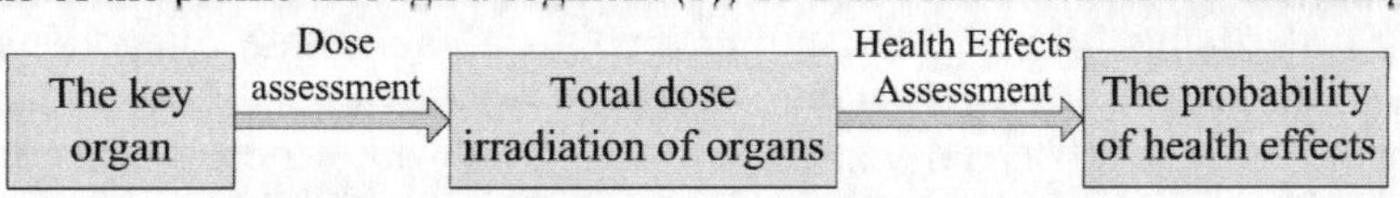

Figure 3. Dose evaluations and health effects schematic

3.4. Multiple external hazards Emergency response

There is difference between multiple external hazards compound level 3 PSA and commonly level 3 PSA about emergency response. In the external hazards compound circumstances, such as earthquake and tsunami compound disaster, its ground transportation will be greatly affected. Early evacuation is more difficult. Most residential are brick houses around the nuclear power plants, shielding performance is better, but these residential brick houses may be affected by earthquake and tsunami, and shielding performance will decrease. Therefore, refer to international general emergency response principles and emergency response measures, it has take into account of the nuclear power plant relatively realistic emergency response measures and evacuation plan in multiple external hazards.

Average speed of the car, under normal weather conditions is 30Km/h, under bad weather conditions is 15Km/h, generally night weather is 25Km/h, bad weather at night is 12.5Km/h. In contrast, NUREG-1150 assumes 4.3Km/h (delayed 1 hour), the technical guide which is a development site guideline adopted 16Km/h in United States, while the average evacuation speed is 7.5Km/h obtained in RSS(Reactor safety research)[7].The average speed of the car is a relevant parameter, because we need to complete evacuation before reactor releasing. The consequence of a severe accident will be more severe if the time of evacuation is longer.Based on the above, and considering multiple external hazards conditions such as earthquake and tsunami compound disaster. Consider that earthquake may cause highway unavailable, and not consider early evacuation but the shielding effects.

4. Calculation results analysis

4.1. The site boundary dose risk

Dose assessment provides the conditional probability distribution, these consequences probability distribution is based on the assumption that the source term caused by accidents has occurred, that the consequences probability distribution is caused by different weather conditions change result in dose change. Table 1 is each release category average dose risk of whole body dose at the site boundary in seven days.

Table 1. Seismic and tsunami compound dose risk in site boundary (7days)

Release categories	Release categories description	Body Average dose（Sv）	Risk (Sv / reactor years)	Percentage in total risk
CFV	Emission was filtered, the core melt is not covered by water	1.80E-01	1.60E-10	0.61%
CFL	Containment late failure, spraying is not operating, the core melt is not covered by water	7.07E+00	1.86E-09	7.13%
CMT	Containment foundation melt wear failure	2.15E+00	2.12E-12	0.01%
CFE	Containment early failure, containment spray is not running	9.34E+01	1.79E-08	68.78%
CIS	Containment isolation failure	1.42E+02	8.51E-11	0.33%
CBP	Containment bypass, no fission products clean, radioactive leak directly into the environment	2.32E+02	6.03E-09	23.14%
Total risk			2.61E-08	100.00%

We can draw the results from Table 1 that when no emergency response, CFE (early failure) and CBP (containment bypass, no cleaning fission products) these two-release categories possess higher dose risk, but lower release frequency. Therefore, when compound external hazard caused radioactive release is inevitable, we should try to lead release categories with serious consequences into less effected categories by necessary mitigating means as much as possible, such as CFV (filter drain) release category which can result in lower dose risk.

4.2. Weather conditions effected analysis

In order to research the impacts of the extreme weather condition influenced consequences, we calculate the results about time integrated air concentration and ground concentration in different wind speeds and rainfalls for new Category meteorological sampling. We can obtain the result that under high winds and heavy rain condition, radioactive concentration is high in within two kilometers of the site. We also researched the radionuclide result that under the constant weather conditions, in use of different wind speeds and rainfalls. The same rainfall is 93mm/h, different wind speeds is10m/s and 30m/s respectively, the air concentration is high in low wind speed. If the wind speed is same, rainfall is different, the air concentration is high in the small rainfall, but the ground concentration is small. These calculations results that using of new Category meteorological sampling method and constant weather conditions are shown in Figure 4 and Figure 5.

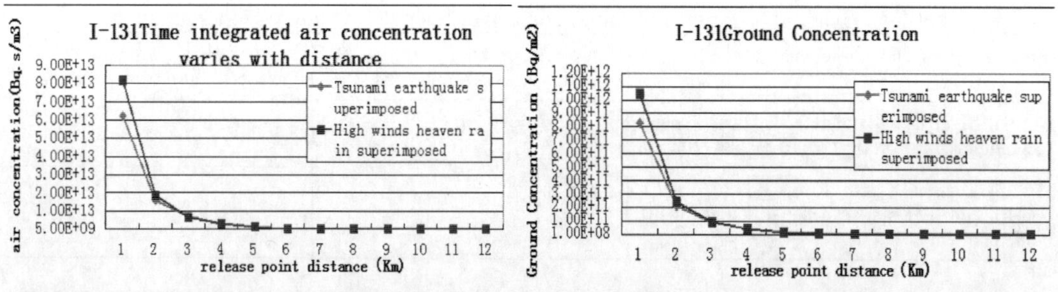

Figure 4. New categories sampling weather conditions air concentration and ground concentration

We can see from the Fig.4. They are time integrated air concentration and ground concentration varies with the distance under the multiple external hazards conditions. The blue line is high winds and heavy rain compound external hazard. The red line is the tsunami and earthquake compound external hazard. The high winds and heavy rain radioactive air concentration and ground concentration are all higher than that of the tsunami and earthquake in the distance within two kilometers. So the extreme weather conditions mainly affect two kilometers distance. Fig.5 shows the results under the constant weather condition.

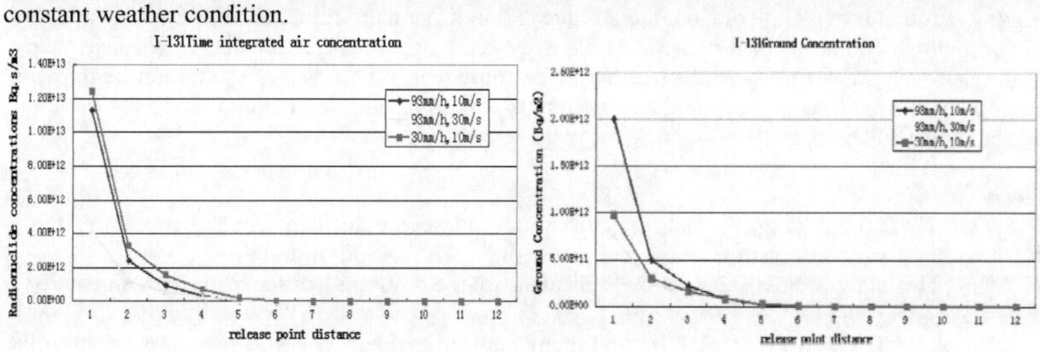

Figure 5. Air concentration and ground concentration in constant weather condition

We can see from the Fig.5 of time integrated air concentration varies with the distance. The same wind speed is 10m/s, different rainfall is 93mm/h and 30mm/h respectively, the smaller rainfall caused more severe consequence and the I-131 time integrated air concentration is higher. The same rainfall is 93mm/h, different wind speed is respectively 10m/s and 30m/s, the lower wind speed I-131 time integrated air concentration is higher.We can also see the results on the ground concentration from the Fig.5. The same wind speed is 10m/s, different rainfall is 93mm/h and 30mm/h respectively, the bigger rainfall caused more severe consequence, and the I-131 ground concentration is higher. The same rainfall is 93mm/h, different wind speed is respectively 10m/s and 30m/s, the lower wind speed caused more severe consequence, and the I-131 ground concentration is higher. That is because the higher wind speed, the lower I-131 time integrated air concentration.

4.3. Emergency response consequences risk analysis

There are differences between multiple external hazards emergency and normal emergency. We can see from Table 2 of multiple external hazards emergency response affection considered. In normal condition considered emergency, the site boundary body dose risk reduces two magnitudes, but in multiple external hazards condition considered emergency, body dose risk only reduces one magnitude. The frequency of multiple external hazards is low, but the consequence risk is higher than the normal conditions (internal event) considered emergency.

Table 2. External hazards compound consequences risk considered emergency

Emergency response Dose type	Site boundary body dose risk (Sv / reactor years)No emergency	Site boundary body dose risk (Sv / reactor years)Emergency
Normal Evacuation (delayed 1 hour, speed 15Km/h, evacuation range 5Km) hidden (10Km)	2.61E-08	5.30E-10
Multiple External hazards Evacuation (delayed 1 hour, speed 1.5Km/h, evacuation range 5Km) hidden (10Km)		4.50E-09

5. Conclusions

We made a reference plant consequence analysis model, studied nuclear power plant severe accident consequences probabilistic evaluation methods under multiple external hazards conditions. And then we get some multiple external hazards consequence risk analysis results and risk insights as follows:

Earthquake and tsunami compound dose risk results show that some release categories have high release frequency, but its plant dose risk is low. Therefore, when multiple external hazards compound radioactive release happens inevitably, we should try to lead the release categories with serious consequences to release categories with less effect by necessary mitigating means as much as possible. This can make dose risk reduction.

High winds and heavy rain compound consequences indicate that radionuclide release has a higher risk in within two kilometers of the site. Multiple external hazards considered emergency consequence risk is higher than the normal considered emergency consequence risk. So we should not be thinking that multiple external hazards occurrence frequency is very low, and then ignore these risks, which is incorrect, but should be overall considering the plant consequences risk.

References

[1] Wang H, Zhu Y, Yang Y, Yang Z 2015 Study on severe accident consequence probability safety assessment method of nuclear power plant *AEST* **49** **(3)** 480-484

[2] Wang H, Huang S. Fukushima D NPS accident source term and consequence assessment 2011 *RPB* **31(3)** 7-11

[3] Yao R Review and progresses in studies of nuclear accident consequence assessment 2009 *RPB***29(1)** 1-10

[4] Collins H E, Grimes B K, Galpin F 1978 NRC: Planning basis for the development of state and local government radiological emergency response plans in support of light water nuclear power plant *NUREG-0396*

[5] Li X, Zhu Z, Zhang J, et al. 1994 The probability of an accident consequence assessment methods, procedures and progress *RPB* **14(4)** 21-22

[6] Chanin D I, Sprung J L, Rollstin J A, Ritchie L T, Jow H-N 1990 Melcor accident consequence code system (MACCS) *NUREG/CR-4691*-Vol.1; SAND--86-1562-Vol.1 ON: TI90009797; TRN: 90-011398

[7] United states nuclear regulatory commission 1990 Severe Accident Risks: An Assessment for Five US Nuclear Power Plants *NUREG-1150*

One-step fabrication of carbon fiber derived from waste paper and its application for catalyzing tri-iodide reduction

Shunjian Xu

Xinyu Institute of New Energy, Xinyu University, Xinyu 338004, China

E-mail: xushunjian@126.com

Abstract. Two carbon fibers were first fabricated by one-step pyrolysis of papers (filter paper and facial tissue), and then employed as catalytic materials for counter electrodes in dye-sensitized solar cells (DSCs) to investigate their potential application. The results show that the microstructure transformation and main weight loss of both the papers are mainly happened in the temperature range of 300–400 °C. After pyrolysis at 800°C, the weight remaining of the filter paper and facial tissue is 1.92% and 4.95%, respectively. The obtained carbon fibers belong to an amorphous carbon consisting of the randomly oriented stacks of graphene sheets. The diameters of both the carbon fibers are about 10 μm, on which there are a certain amount of fine carbon nanofibers. The amorphous microstructure and unique fine nanofibers of the carbon fibers induce more excellent catalytic activity for triiodide ion reduction compared with the biochar (derived from poplar leaf) and the graphite. As a result, the carbon fiber based DSCs display obviously higher efficiency than the biochar or graphite based ones. The conversion efficiency of the DSCs employing the filter paper derived carbon fiber, facial tissue derived carbon fiber, biochar and graphite is 4.72%, 4.70%, 1.33% and 0.77%, respectively.

1. Introduction

Waste papers, common renewable resource, have been traditionally recycled in paper industry as fibrous raw materials [1]. Moreover, waste papers have often been employed to prepare carbon fibers with high-added-value owing to their low ash content and reasonable carbon residue, which have been applied in several fields, such as gas adsorption [2] and reaction formation of SiC ceramics [3]. The invention of dye-sensitized solar cells (DSCs) opens a potential gate for the application of the waste papers derived carbon filers in new energy field.

As a kind of devices to harness the limitless power of the Sun, DSCshave attracted increasing interest for their low cost, easy fabrication, flexibility and moderate efficiency [4]. Typically, a DSC consists of a dye sensitizer, TiO_2photoanode, an electrolyte and a Pt counter electrode. The role of the Pt counter electrode, which consists of Pt film deposited on FTO glass, is to collect electrons arriving from the external circuit and to catalyze the reduction of triiodide ion. However, the scarce resource, high costs and poor stability of Pt could inhibit its large-scale application in the future [5]. To solve this problem, several carbon materials, such as carbon black, graphite, activated carbon, mesoporous carbon, carbon nanotubes and graphene, have been used as catalysts to replace Pt for counter electrode, and their devices show favorable photovoltaic performances [6-8]. Compared with the above carbon materials, the carbon fiber obtained from one-step pyrolysis of waste paper shows simpler process, more abundant raw materials and more environmental friendly. Therefore, to introduce the waste paper based carbon fiber into the DSCs as the counter electrode not only reaches the destination on its application in new energy field, but also displays a potential way with low-cost to replace the Pt counter electrode.

In this research two kinds ofcarbon fibers were firstfabricated by one-step pyrolysis ofwaste papers including filter paper and facial tissue, and then used as catalytic materials for counter electrodes in DSCs to investigate their potential application. Both of the carbon fiber based devices show more excellent photovoltaic performances than the biochar or graphite based ones. The conversion efficiency of the DSCs employing carbon fiber electrodes is around 4.70% while that of the devices consisting of biochar and graphite electrodes is no more than 1.40%.

2. Experimental

2.1. Fabrication and characterizations of carbon fiber

Two papers were individually selected as the precursors to prepare the carbon fibers. The paper waspyrolyzed by slowly heating to 800 ºC for 1h in aflowing N_2 atmosphere, and mechanically milled to obtain carbon fiber powders. The carbon fibers derived from filter paper and facial tissue arelabeled as CF-FP and CF-FT, respectively.

Thermal decomposition behavior of the papers was examined by a simultaneous thermogravimetry–differential thermal analysis (TG–DTA; SDT-Q600, TA, USA) using a heating rate of 10 °C·min^{-1} in N_2.The crystal microstructure of the carbon fiberswas identified by an X–ray diffraction (XRD; D8 ADVANCE, Bruker, Germany) using Cu$K\alpha$ radiation with a wavelength (λ) of 0.154056 nm.The scanning step was 0.02 º, and the scanning area was 15–75º.The morphology of the carbon fibers wascharacterized by a field emission scanning electron microscopy (FE-SEM; JSM-6701F, JEOL, Japan).N_2 adsorption-desorption isotherms of the carbon fiberswere examined on an automatic adsorption instrument (ASAP 2020M, Micromeritics, USA), using nitrogen at 77 K. Surface areas were calculated by Brunner–Emmet–Teller (BET) method from N_2desorption isotherms.

2.2. Configuration and measurements of carbon fiber based cells

The DSC was configured by assembling a N719 sensitized TiO_2photoanode and a carbon fiber counter electrode, between which was separated by using a 50 μm thick spacer. The surface area of both TiO_2 film and carbon fiber film was about 0.7 cm × 0.7 cm. For obtaining the photoanode, a transparent TiO_2film was first coated on the FTO substrate, and then a second layer containing light scattering particles of TiO_2 was coated on the obtained TiO_2 film. After sintered at 450 °C for 30 min, the double-coated TiO_2 film was further treated with $TiCl_4$ aqueous solution followed by heat treatment at 450 °C for 30 min. The photoanode was obtained by further immersing the double-coated TiO_2 film into a N719 ethanol solution at room temperature for 24 h [9].The electrolyte was 0.5M LiI, 0.05M I_2, 0.5M TBP ina solvent of 3-methoxypropionitrile.The carbon fiber electrode was prepared by coating a carbon fiber blended paste on FTO substrate and then heated at 120 °C for 2h. The carbon fiber blended paste was obtained through grinding 200mg carbon fiber powders, 20mg carboxymethyl cellulose (CMC) and 5ml ethylalcohol aqueous solution.

Electrochemical impedance spectroscopy (EIS) measurements were conducted with an electrochemical workstation(CS310, Corrtest, China)at10 mV of amplitude. A symmetric cell consisting of two identical counter electrodes and an electrolyte was used in the EIS measurements.The electrolyte was same as the one used in the fully functional DSCs. Cyclic voltammetry (CV) experiments were performed using a conventional three-electrode system at a scan rate of 50 mV·s^{-1}. The as-prepared electrode, Pt coil, and Pt wire were used as the working electrode, auxiliary electrode, and reference electrode, respectively. The electrolyte was an acetonitrile solution containing 0.1 M $LiClO_4$ as the supporting electrolyte and 10 mMof LiI+ 1 mM of I_2 as the redox couple.Photocurrent-voltage performances of the DSCs were measured under the illumination of a Xenon light source. The light intensity was 100 mW·cm^{-2}.

3. Results and discussion

To explore the thermal decomposition behavior of the filter paper and facial tissue selected for the preparation of the carbon fibers, their TG and DTA curves wereobtained, as shown in Figure 1. It is obvious that both the TG curves are similar. There are obvious weight losses in the temperature ranges of 20–80 °C and 300–400 °C in both the curves. The weight loss in the first temperature range, which is mainly caused by the evaporation of adsorbed water, is 4.57% and 5.36% for the filter paper and facial tissue, respectively. The weight loss in the second temperature range is more serious than that in the first temperature. In this temperature range (300–400 °C), the weight loss of the filter paper is 88.31% while that of the facial tissue is 78.87%. The occurrence of an endothermic peak in the DTA curves of both the papers is accompanied by the serious weight loss. The above results suggest that the microstructure in both the paper is mainly changed in the temperature range of 300–400 °C. Moreover, with increasing temperature to 800°C, the weight loss of both the papers reaches a plateau. The weight remaining of the filter paper and facial tissue at 800°C is 1.92% and 4.95%, respectively.

The XRD patterns of the carbon fibers are demonstrated in Figure 2(a). No obvious difference can be observed between the patterns of CF-FP and CF-FT, meaning the existence of nearly identical microstructure in both the carbon fibers.Two broadening peaks of (002) and (100) associated with graphite characteristic appear in the patterns. According to Bragg formula ($d=\lambda/2\sin\theta$), the layer distanceof the (002) crystal plane ($d_{(002)}$) of the carbon fibers is estimated to beabout 0.3890 nm, which is larger than that of the (002) crystal plane of the graphite (0.3335nm).It is reasonable to conclude that both of the carbon fibers prepared from the pyrolysis of the papers belong to an amorphous carbon consisting of the randomly oriented stacks of graphene sheets. The amorphous microstructure in the carbon fibers is beneficial for catalyzing the reduction of triiodide ion since the active sites in the carbon materials for catalysis are located at the edges of graphene stacks [10].

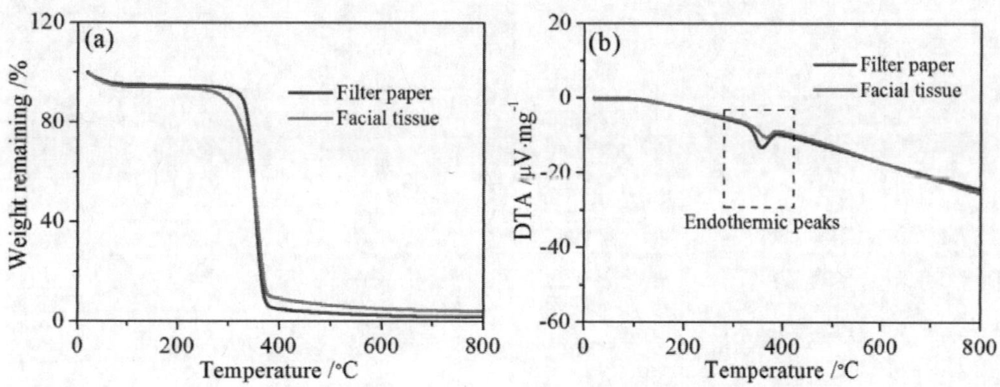

Figure 1. TG (a) and DTA (b) curves of different papers used for the preparation of carbon fibers.

The SEM morphologies of the carbon fibers of CF-FP and CF-FT are illustrated in Figure 3. It is obvious from Figure 3 (a) and (b) that both the carbon fibers show different stack patterns. The carbon fibers in CF-FP derived from filter paper connect each other as a net. In contrast, the carbon fibers in CF-FT derived from facial tissue have oriented distribution with regular wave-structure in the oriented direction. The diameters of the carbon fibers in CF-FP and CF-FT are about 10 μm. The amplified SEM images in Figure 3(c) and (d) further demonstrate that there are a certain amount of fine carbon nanofibers on the surfaces of the 10 μm-sized carbon fibers in CF-FP and CF-FT. The fine nanofibers enlarge thesurface areas of the samples, resulting in the increase of the active sites for catalysis. The N_2 adsorption-desorption isothermsof the carbon fibers are displayed in Figure 2(b). It is obvious that both of the carbon fibers exhibit type II isotherms, which are mainly induced by adsorption-desorption of macroporous or dense materials [11]. The inflection point in the isotherms, as indicated by the arrow, is a cut-off point of the adsorptions of monolayer and multilayer. The above results indicate that the single carbon fiber belongs to dense materials, which limits the surface areas of CF-FP and CF-FT to a certain degree. The value of BET surface area (S_{BET}) of CF-FP and CF-FT is 2.21 and 3.24 $m^2 \cdot g^{-1}$, respectively.

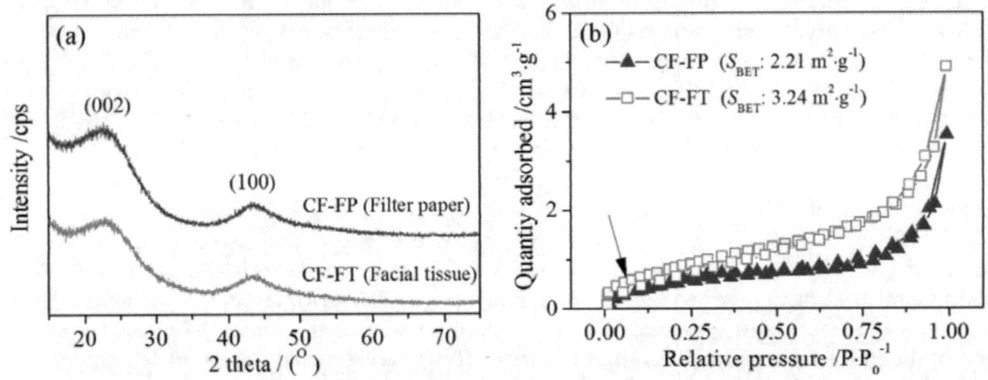

Figure 2. XRD patterns (a) and N_2 adsorption-desorption isotherms (b) of carbon fibers derived from different papers.

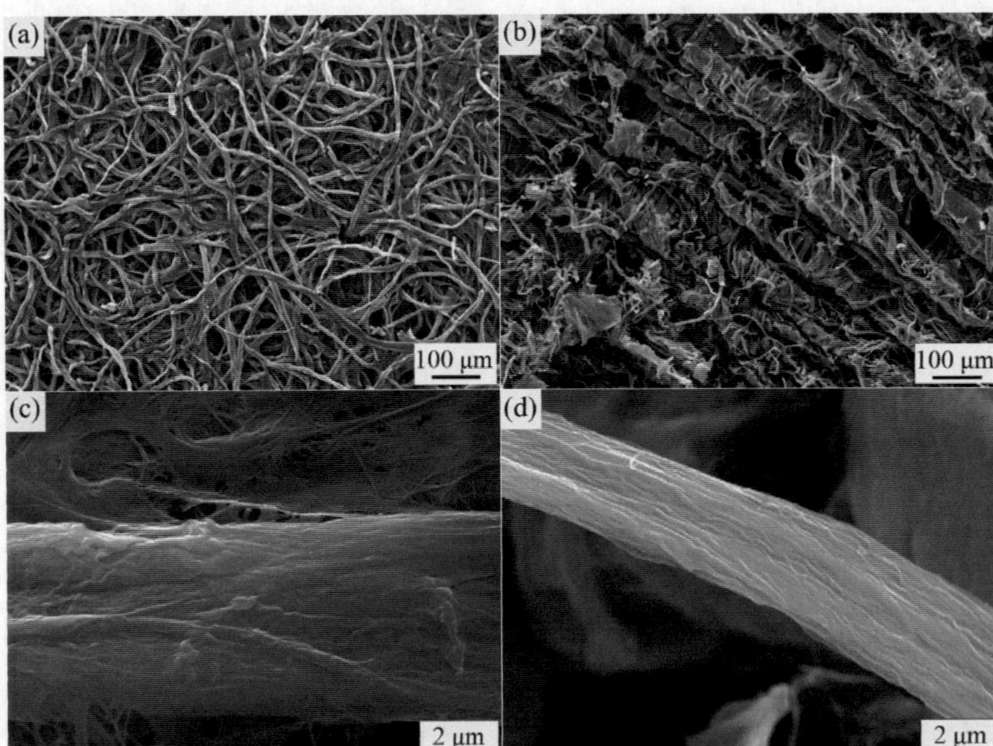

Figure 3. SEM morphologies of carbon fibers derived from different papers. (a) and (c) CF-FP from filter paper; (b) and (d) CF-FT from facial tissue.

To insight into the catalytic activities of the carbon fibers derived from the papers, the EIS and CV were employed. The typical Nyquist plots for the carbon fiber counter electrodes are shown in Figure 4(a). For comparison, Figure4(a) also illustrates the Nyquist plots of biochar and graphite counter electrodes, which were respectively prepared by coating a biochar (derived from poplar leaf) blended paste and a graphite blended paste on the FTO substrates and then annealing at 120 °C for 2h. Two impedance parameters, charge transfer resistance (R_{ct}) and ohmic serial resistance (R_s),can be achieved from plot fitting by using the equivalent circuit, and are summarized in Table 1. The R_{ct} characterizes the catalytic activity for the triiodide reduction, and the $R_s$is mainly related to the sheet resistance of theelectrode [12]. Both of the carbon fiber electrodes present lower R_{ct} than the biochar and graphite electrodes, implying higher catalytic activity toward the triiodide ion reduction. The R_{ct} values of thecarbon fiber electrodes from CF-FP and CF-FT are respectively 4.42 and 4.15 Ω cm^2 while those of the biochar and graphite electrodes are respectively 12.50 and 52.50 Ω cm^2. The more excellent catalytic activity of the carbon fibers could be mainly attributed to their amorphous microstructure and unique fine nanofibers (Figure 2 and 3). Moreover, the almost identical R_{ct} values between the carbon fiber electrodes suggest nearly same catalytic activity toward the triiodide ion reduction. Figure 4(a) also shows that the sheet resistance of each carbon fiber electrode is lower than that of the biochar or graphite electrode. The R_s values of the electrodes prepared from CF-FP, CF-FT, biochar and graphite are 22.61, 23.38, 38.60 and 46.15 Ω cm^2, respectively.

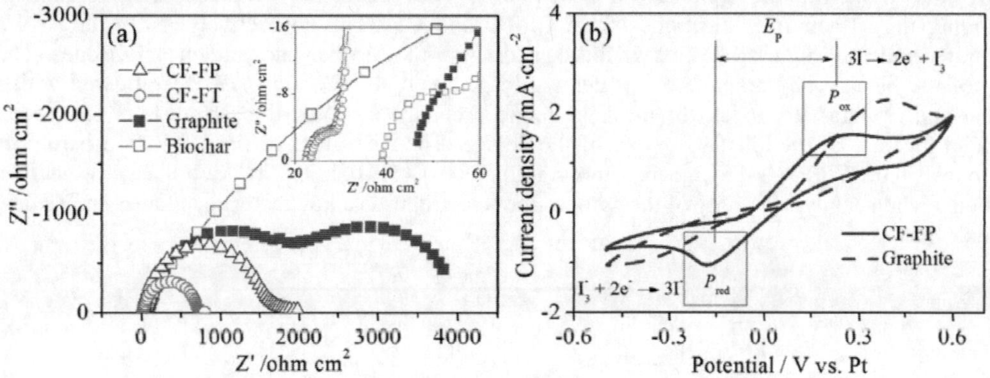

Figure 4. Nyquist plots (a) and CVs (b) of counter electrodes prepared from various carbon materials. The insert in (a) shows the expanded range in high frequency region.

Table 1. Electrochemical performances ofcounter electrodes obtained from various carbon materials and their photovoltaic properties for DSCs*

Carbon	Electrochemical			Photovoltaic			
	R_{ct}/Ωcm^2	R_s/Ωcm^2	E_p/V	J_{sc}/mA·cm^{-2}	V_{oc}/V	*FF*	η/%
CF-FP	4.42	22.61	0.45	15.02	0.697	0.451	4.72
CF-FT	4.15	23.38	—	14.80	0.686	0.463	4.70
Biochar	12.50	38.60	—	10.96	0.488	0.249	1.33
Graphite	52.20	46.15	0.67	9.90	0.447	0.175	0.77

*R_{ct}: Charge transfer resistance; R_s:Ohmic serial resistance; E_p: Peak separation between the P_{ox} and P_{red}; J_{sc}: Short-circuit current density; V_{oc}: Open-circuit voltage; *FF*: Fill factor;η: Conversion efficiency.

The CVs for the carbon fiber (CF-FP) and graphite electrodes are displayed in Figure 4(b). A pair of redox peaks are observed in both the CV curves.The peak in the positive sideis attributed to the oxidation reaction, while the negative oneresults from the reduction reaction[13]. In the reduction peak(P_{red}), the current density of the carbon fiberelectrode is higher than that of the graphiteelectrode. It means that there are a faster reaction rate and a higher catalytic activity for the triiodide ion reduction on the carbon fiber electrode than on the graphiteelectrode [14]. Moreover, the reduction peak for the carbon fiberelectrode shifts more positively compared with the graphiteelectrode, indicating that the reduction reaction takes place at lower overpotential on the former electrode than on the latter electrode [15]. In the oxidation peak (P_{ox}), the carbon fiberelectrode presents lowercurrent density compared with the graphiteelectrode, which is believed tohave an insignificant effect on the catalytic performances [16].The peak separation (E_p) between the P_{ox} and P_{red} is another important parameter to estimate the catalytic properties [17].For clarity, the $E_p$is listed in Table 1. The higher catalytic activity of the carbon fiber electrode is found to be in consistency with its respective lower $E_p$of 0.45V, with reference to that of the graphite electrode(0.67 V).Evidently, the above results are consistent with the EIS analysis (Figure4a).

Figure 5 shows the photocurrent density–voltage (J– V) curvesof the DSCs employingfour kinds of carbon materials as counter electrodes: CF-FP, CF-FT, biochar and graphite. The parameters of the photovoltaic properties are summarized in Table 1. The conversion efficiency (η) of the DSCs employing the carbon fiber electrodes from CF-FP and CF-FT is respectively 4.72% and 4.70%, which are evidently higher than that of the devices with the biochar and graphite electrodes. The η value of the biochar and graphite based cells is respectively 1.33% and 0.77%. Compared with the biochar and graphite electrodes, the most pronounced change by using the carbon fiber electrodes is the rapid increase of the Fill factor (FF) of the devices.The FF values of the DSCs consisting ofthe electrodes from CF-FP, CF-FT, biochar and graphite are 0.451, 0.463, 0.249 and 0.175, respectively. The outstanding catalytic activity of the carbon fiber electrodes is a key factor to enhance the FF and η of the devices.

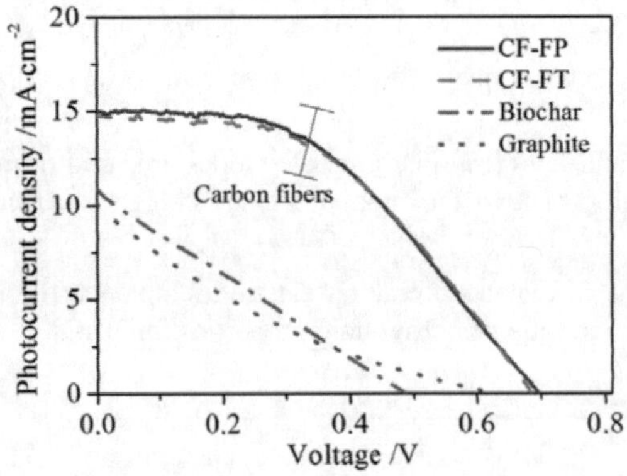

Figure 5. J– V curves of DSCs consisting of counter electrodes based on various carbon materials.

4. Conclusion

During pyrolysis, the microstructure transformation and main weight loss of the selected papers are mainly emerged in the temperature range of 300–400 °C. The weight remaining of the filter paper and facial tissue at 800°C is 1.92% and 4.95%, respectively. The obtained carbon fibersbelong to an amorphous carbon consisting of the randomly oriented stacks of graphene sheets. The diameters of both the carbon fibers are about 10 μm, on which there are a certain amount of fine carbon nanofibers. When both of the carbon fibers are employed as the catalytic materials for the counter electrodes, the corresponding DSCs display obviously higher efficiency than the devices based on the biochar and graphite as counter electrodes. The conversion efficiency of the DSCs employing the filter paper derived carbon fiber, facial tissue derived carbon fiber, biochar and graphite is 4.72%, 4.70%, 1.33% and 0.77%, respectively. The high efficiency is mainly attributed to the unique structural and morphological characteristics of the carbon fibers, such as amorphous microstructure and unique fine nanofibers. Moreover, the low temperature (120 °C) treatment process can be employed to fabricate a low-cost and efficient counter electrode for both glass based DSCs and flexible DSCs

Acknowledgements
This work was financially supported by National Natural Science Foundation of China (51462035), Training Programme Foundation for Young Scientist of Jiangxi, China (20133BCB23035), Natural Science Foundation of Jiangxi (20161BAB206106) and Educational Commission of Jiangxi, China (KJLD13100).

References

[1] Pivnenko K, Eriksson E and Astrup T F 2015 *Waste manage.***45** 134
[2] Xie L P, Lin W G and Yang X M 2006 *New Carbon Mater.* **21** 156
[3] Pan J M, Cheng X N and Yan X H 2013 *Ceram. Int.* **39** 6131
[4] Cebeci C, Kilicarslan F A and Gurbuz O 2016 *Dyes Pigments* **134** 77
[5] Hasin P 2016 *Sol. Energy* **135** 398
[6] Rahman M S, Hammed W A and Yahya R B 2016 *J. Polym. Res.* **23** 192
[7] Yue G, Wu J, Lin J Y, Xiao Y, Tai S Y, Lin J, Huang M and Lan Z 2013 *Carbon* **55** 1
[8] Tetsuya A and Hajime H 2013*Mater. Lett.* **94** 15
[9] Xu S J, Luo Y F, Liu G W, Qiao G J, Zhong W, Xiao Z H, Luo Y P and Ou H 2015 *Electrochim. Acta* **156** 20
[10] Xu S J, Luo Y F and Zhong W 2011*Sol. Energy* **85** 2826
[11] Sing KSW 1985*Pure Appl. Chem.* **57** 603
[12] Roy-mayhew J D, Bozym J, Puncht C and Aksay I A 2010 *ACS Nano* **4** 6203
[13] Wang S F, Rao K K, Yang TCK and Wang H P 2011*J. Alloy. Compd.* **509** 1969
[14] Hu H, Chen B L, Bu C H, Tai Q D, Guo F, Xu S, Xu J H and Zhao X Z 2011*Electrochim. Acta* **56** 8463
[15] Peng S J, Cheng F Y, Shi J F, Liang J, Tao Z L and Chen J 2009*Solid State Sci.* **11** 2051
[16] Veerappan G, Kwon W and Rhee S W 2011*J. Power Sources* **196** 10798
[17] Yue G T, Wu J H, Xiao Y M, Huang M L, Lin J M, FanL Q and Lan Z 2013 *Electrochim. Acta* **92** 64

Study on the statistical characteristics of solar power

Jun Liu

State Grid Energy Research Institute, Future Science & Technology Park, Beijing, P.R.China

E-mail: liujun@sgeri.sgcc.com.cn

Abstract. Solar power in China has grown rapidly recently. It exists variation of solar power due to cloudy and dusty, which is not that much of wind. The way to evaluate the statistical characteristics of solar power is important to the analysis of power system planning and operation. In this study, a multi-scale spatial and temporal framework of evaluating indices was established to describe the variation of its own natural features and the interaction between solar and load, grids. Finally, we have a case study on the variation, comparison, penetration, etc.

1. Introduction

At present, study on consolidation analysis of solar energy resources and operation information of solar power in China is still in initiative stage [1]. The power output of photovoltaic (PV) generation varies with the sunrise and sunset. The change of weather leads variations to PVs, as well. It gives rise to difficulties in power system planning and operation. It has not been formed a recognized and effective evaluation system yet. Evaluation of solar power output should start from the relationship between solar power and the power grids. By combining solar power output with its own statistical characteristics, multi-scale spatial and temporal statistical properties of the solar power generation layer by layer evaluation index system should be established [2-4].

The basic idea of analysis is: the frameworks of the index that reflect the natural characteristics of solar power and the interactive indicators that related with power grids features should be established, firstly. And then, indices classes according to the importance and coverage will be conducted based on core index and supplemented by multi-class derived index. And finally, the statistical characteristics of solar power output in typical regions will be investigated.

2. Statistical characteristics of solar power indicator framework

Statistical characteristics of solar power indicators can be divided into two categories: one is set to reflect solar energy / solar power of own characteristics indicator; the other is an indicator of the interaction between the solar power to the power grids. Index classification summarized as shown in Figure 1.

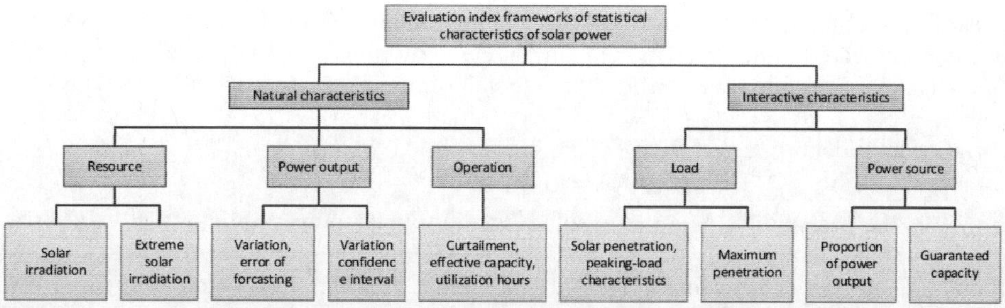

Figure 1. Solar power statistical characteristics evaluation system.

Characteristics of indicators generally include natural resource characteristics, output characteristics and operational characteristics [1]. Resource features indicators are generally based on historical weather data, commonly used in the planning stage; Power output characteristic index based solar power plants based on the real output data, etc. which are usually concerned with stochastic variable characteristics on the power grids; Operating characteristic index consider solar power facilities utilization, a measure of economic solar power operations [5-6].

Indicators reflect the interactive features of the matching relationship between solar power and the load, solar power and traditional power supply. The former mainly noted as the penetration of solar power and solar power generation peaking characteristic indicators; the latter mainly reflect solar power capacity index as a power supply.

3. Indicators of natural characteristics of solar power

3.1. Power output characteristics

Variation and randomness are the main natural characteristics of solar power output / solar energy. As to solar energy resources, it is a manifestation of the natural characteristics of geographical, meteorological and topography factors. Concerned with the solar power output, the feature is based on the characteristics of solar power conversion considering solar energy resources.

Variation refers to changes of the solar power output / solar irradiation intensity, which are determined by specified sequence in multi-spatial and temporal scales, point by point. And the change rates of power output are core indicators of solar to describe variation of solar power.

1) solar power output variation: within the given period, difference of the power output ΔP of two neighboring point.

$$\Delta P = P(t+T) - P(t) \tag{1}$$

2) solar power output change rate: the proportion of solar power output variation to the rated capacity of solar power plants.

$$\rho\% = \frac{P(t+T) - P(t)}{P_{\text{base}}} \times 100\% \tag{2}$$

Where: $P(t+T)$ is defined as solar power output at $t+T$ time; $P(t)$ is solar power output at time t; P_{base} is rated capacity of solar power plant. For the different time scales, T is corresponding to different values.

The variation of solar power generation could be divided by time scale, such as, in seconds, minutes, hours. Meanwhile the variation could be divided by spatial scale, such as, stand-alone plants, clustered power plants and regional generations. With the viewpoint of statistical methods, there are indices of probability distribution of solar power output, forward / backward maximum variation, probability of forward / backward variation of solar power output, etc [7].

In general, variation increases with time scale increasing, while decreases with spatial scales increasing which lead to the overlap and smoothness of power output in different areas. This smoothness could be defined as confidence interval of variation coefficient *R* and a smoothing coefficient *S*.

1) Variation confidence interval *R*:

$$P(X \leq R) = p \tag{3}$$

Where: *P* is the probability; *X* is the absolute value of solar power variation; *p* value is a given probability.

2) Smoothing effect coefficient *S*: standard deviation of solar irradiation sequence difference (levelized with installed capacity).

$$S = \frac{\widehat{\sigma}_{\text{single}} - \widehat{\sigma}_{\text{cluster}}}{\widehat{\sigma}_{\text{single}}} \tag{4}$$

where the subscript "single" and "cluster" corresponds to a single plants and clustered plants. This index is used to quantify the cluster of solar power plants with respect to the degree of smoothing on a single field of solar power variation. Variation in confidence interval *R*, for example, the index obtained by accumulating the probabilities described solar power variation in a specific area within a few minutes, which can provide reference information for the secondary frequency regulations.

3.2. Solar resource features

Resources index mainly depict the richness and distribution of solar energy resources. Solar irradiation is the core index. Commonly derived indicators also include the probability distribution of the solar irradiation and the average solar irradiation. The introduction of resource indicators are as follows:

1) Solar irradiation correlation coefficient: The correlation coefficient refers to a different location weather station metering sequence.

2) Solar energy resources disequilibrium coefficient: I_{ref} is selected from maximum value of average solar irradiation in each solar power plants. Resource coefficient of each solar power plant could be determined as $a_i=I_i/I_{ref}$ by average solar irradiation in each solar power plants. And then, disequilibrium coefficient could be calculated by averaging a_i.

3) Active solar irradiation: solar irradiation within the effective range of the entire statistical period that contribute to power generation.

4) Probability of maximum effective solar irradiation: the occurrence probability of solar irradiation between the start of the photoelectric effect and saturation conditions, which could be noted as that the maximum effective hours and load hours over the years, as well.

5) Probability of extreme variation in solar irradiation: the probability of solar irradiation from less than / greater than the saturated conditions of solar irradiation increase / decrease to above / under the saturated condition.

3.3. Operational characteristics of solar power

Solar power operation indicators are used to evaluate the technology and economy of solar power plant facilities, which are defined as follows:

1) Curtailment / probability of solar power generation: the curtailment / probability when the curtailment of solar power occurs if the power output reaches a given ratio of installed capacity.

2) Monthly / yearly solar power generation.

3) Effective capacity of solar power: under the premise of maintaining reliability level of original power system, traditional power generation could be substituted by solar power generation, which is defined as effective capacity.

4) Annual utilization hours of solar power: annual generation of solar power divided by the rated capacity.

4. Interactive indicators

4.1. Interaction with load

(1) solar penetration

Influence of solar power integration is not only related with installed capacity, but also the scale of power grids.

1) Theoretical penetration $P_{r\_th}(t)$: The ratio of the installed capacity to power load.

$$P_{r\_th}(t)=\frac{P_{base}(t)}{P_{load}(t)} \tag{5}$$

2) Actual penetration $P_{r\_ac}(t)$: Real-time solar output divided by load.

$$P_{r\_ac}(t)=\frac{P(t)}{P_{load}(t)} \tag{6}$$

Where $P(t)$, $P_{base}(t)$, $P_{load}(t)$, is the solar power output, installed capacity and load, respectively Theoretical penetration is commonly used index, which reflects the development scale of integrated solar power of the region. However, the actual power output of solar power generation varies and differs from installed capacity. The integration influence of solar power might be over-estimated when theoretical penetration is used. As a result, actual penetration is suggested to be adopted in the analysis of solar power integration.

(2) Peaking-load characteristics of solar power

Net load is actual load overlapped with the output of solar power, which is treated as negative load. Study on the relationship of the solar power and load mainly based on comparison of original load and net load. Peak-valley difference of net load is noted as key index.

It is assumed P_{imax} and P_{imin} as maximum and minimum values of original load curve, while P_{imax} 'and P_{imin}' are maximum and minimum values of net load. P_{Vi} and P_{Vi} 'denote as peak-valley difference of original and net load. ΔP_{Vi} is the change of peak-valley difference before and after solar integration. Those above-mentioned parameters have the following relationship:

$$\begin{cases} P_{vi}=P_{i\max}-P_{i\min} \\ P'_{Vi}=P'_{i\max}-P'_{i\min} \\ \Delta P_{Vi}=P_{Vi}-P'_{Vi} \end{cases} \tag{7}$$

Where, it indicates anti-peaking when ΔP_{Vi} is negative.In general, seasonal characteristics of the solar power output leads to seasonal peaking requirements. It is necessary to evaluate the influence of solar integration on peaking characteristics of power system season by season.

4.2. Interaction with grids

Integrated solar power will replace conventional power in some extent. However, the special nature features of solar power determines the difference between the level of its actual output and rated capacity. The proportion of solar power output to system load is defined as key indicators to measure the supply capacity of solar power [8-9].

Solar power output ratio (availability): Noted as a percentage of the measured active power solar to installed capacity.

$$\eta\%=P_r(t)/P_{base} \tag{8}$$

Where: $P(t)$ is measured solar power output at time t; while P_{base} is the installed capacity of solar power. There is special significance in two occasions:

① guaranteed capacity: peak load hours, 95% of the solar power generation is more than guaranteed capacity C_g.

② effective output: load valley periods, 95% of solar power output is less than C_e.

Guaranteed capacity reflects the capacity of solar energy that can be substantially ensure the active power during peak load period. It can provide a reference for the calculation of the power balance. Effective output is the general solar power during valley load period.

5. Case study on the evaluation of solar power output characteristic

5.1. Comparison of solar power output with different scale

This section will use power output curves of 3 locations to study the difference of power output of different scales. These regions are named as region A, B and C. Solar output curves of these regions in December 23rd to 29th are selected and normalized as shown in Figure 2.

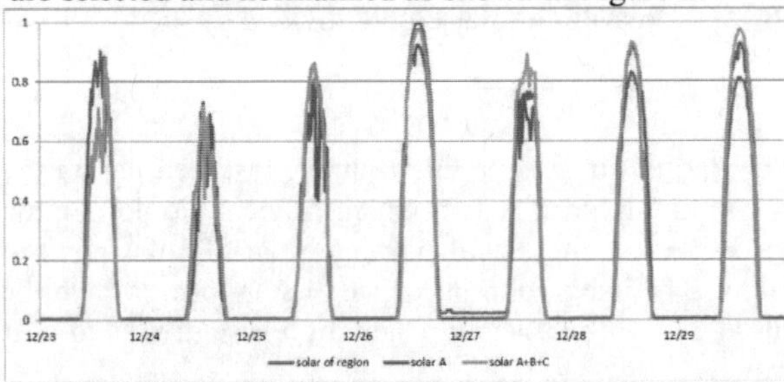

Figure 2. weekly solar power output curves

As shown in Figure 2, the output of the power plant A varies due to weather factors. However, variation of accumulated solar power output curve of these 3 locations has decreased. Furthermore, these variations of solar power in the regional power grids are least because of the complementary effect between the various solar power plants.

By investigating annual solar output data with an interval of 15min, standard deviation of solar in regional power grids is 0.214, while the standard deviation of plant A is 0.304, and plant A+B+C is about 0.260. The smoothing effect coefficient of plant A and plant A+B+C is 0.297 and 0.176, respectively. It means that complementary effect will smooth the solar output curves when the integration capacity increases.

5.2. Variation of solar power under different time scales

The probability density of variation in solar plant A is shown is Figure 3. With increasing time scales, the distribution of probability varies from the "narrow and high" shape to "wide and low", which means the variation enhanced. The probability of forward variation is larger than that of backward variation. As a result, the determination of primary, secondary frequency regulations and reservations should take the statistical results of solar power output in multi-time scales.

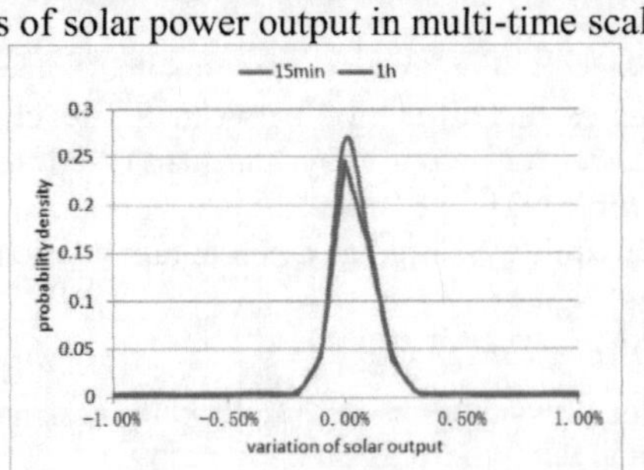

Figure 3. annual solar power output statistical properties of different time scales

Statistical characteristics of the solar power plants in different seasons is shown in Figure 4. The results show solar power output variations of seasonal characteristics. During winter, variation occurs more frequently due to variable solar irradiation. On the contrary, the output in summer varies slight because of high irradiation.

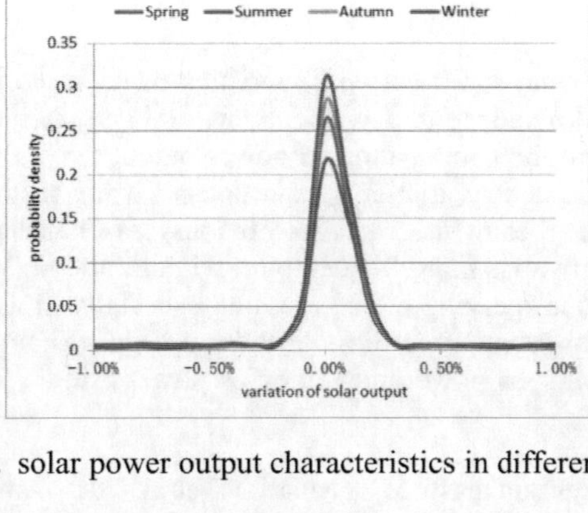

Figure 4. solar power output characteristics in different seasons.

5.3. Comparison on the output of solar and wind

Wind and solar power output is different, as shown in Figure 5. The probability density curve of wind plants is kind of "high and narrow", while the probability curve of solar power plants is relatively "short and wide". This shows that wind power variation is more frequent than that of solar power, and solar power is also relatively large probability that there will be variation circumstances, such as cloudy and dust.

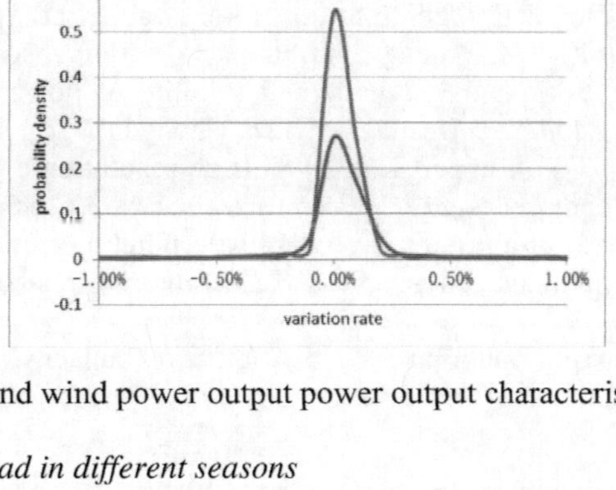

Figure 5. solar power and wind power output power output characteristics of FIG Comparison

5.4. Influence of peaking-load in different seasons

When large-scale solar power generation integrated to the power grids, the peak-valley difference will have a great impact on the power system, as shown in Figure 6. The variation of peak-valley difference is defined as the difference between original load and net load.

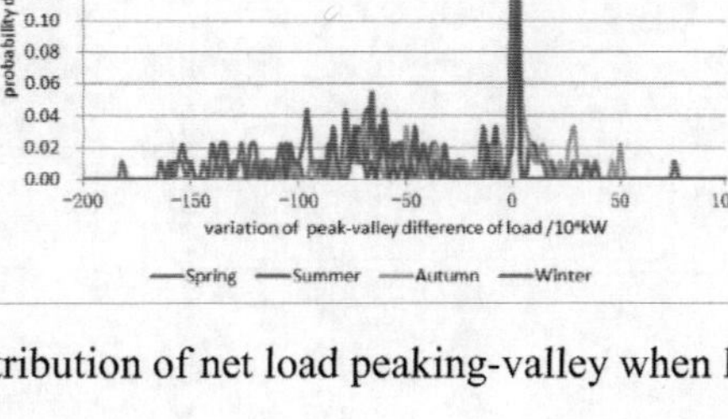

Figure 6. probability distribution of net load peaking-valley when large scale solar integrated

Figure 6 shows that the probability distribution curve shift to the left direction, indicating solar power plant integrated to grid might cause anti-peaking, especially in winter and spring.

6. Conclusions

In this study, we investigated the statistical characteristics of solar power based on an evaluation indices framework. Indices were defined to describe the natural and interactive characteristics of solar power. The results showed that the combination of power output curves of power plants in different locations will get smoothed. As to variation, solar doesn't vary that much of wind, and more frequently in short time scale. Meanwhile, it exists obvious seasonal characteristics of solar power output. Solar power generation varies more frequently in winter due to more cloudy days. Meanwhile, solar power plant might have anti-peaking effect, as wind, especially in winter and spring. It could be attributed to the low demands in some time slices of those seasons. The results of this study could be used to analyze the influence of solar power integration on power grids.

References

[1] Keyhani A, Ghasemi-Varnamkhasti M, Khanali M, et al. 2010 An assessment of wind energy potential as a power generation source *Energy* **35**188-201.

[2] Tan Y T, Kirschen D S, Jenkins N 2004 A model of PVgeneration suitable for stability analysis *IEEE Transactions on Energy Conversion* **19(4)** 748-755.

[3] Li F, Li W, Xue F, et al. 2010 Modeling and simulation of large-scale grid connected photovoltaic system *Proceedings of the Power System Technology of IEEE*, Aviva.

[4] Tan Y T, Kirschen D S 2007 Impact on the power system of a large penetration of photovoltaic generation *IEEE Power Engineering Society General Meeting*, Tampa.

[5] Bebic J. 2008 Power system planning: emerging practices suitable for evaluating the impact of high-penetration photovoltaics. USA: National Renewable Energy Laboratory (NREL).

[6] Yang Weidong, Zhou Xia, Xue Feng 2010 Impacts of large scale and high voltage level photovoltaic penetration on the security and stability of power system *Proceedingsof the Centre of the Picture Industry Meeting of IEEE*, Chengdu.

[7] Villalva M G, Siqueira T D, Ruppert E 2010 Voltage regulation of photovoltaic arrays: small-signal analysis and control design *IET Power Electronics* **3(6)** 869-880.

[8] Park J, Liang W, Choi J, et al. 2009 A probabilistic reliability evaluation of a power system including solar/photovoltaic cell generator *IEEE Power & Energy Society General Meeting*, Calgary, Canada.

[9] Madaeni S H, Sioshansi R, Denholm P 2013 Comparing capacity value estimation techniques for photovoltaic solarpower *IEEE Transactions on Photovoltaics* **3(1)** 407-415.

Characteristics of surface solar radiation in Sino-Singapore Eco-city

M Hou[1], H Huang[2, 4], Y F Shen[1], W Yao[1], G S Wang[3], Q J Bu[1], X L Shan[1], C H Chang[1]

[1]Meteorological Bureau of Binhai New Area, Tianjin 300457, China
[2]Tianjin Climate Center, Tianjin 300047, China
[3]National Marine Data and Information Service, Tianjin 300171, China

E-mail: huanghe04@aliyun.com

Abstract. Using solar observation and meteorological data of the Sino-Singapore Tianjin Eco-city from August 14th 2014 to August 12th 2015, characteristics of solar radiation of the eco-city and characteristics of solar radiation on a tilted surface under different weather conditions were analyzed and assessed. And the accuracy and error sources of isotropic and anisotropic calculation model for solar radiation on a tilted surface were studied. The results show that observed radiation on a horizontal and tilted surface is quite different at monthly, seasonal and annual time scales, so the estimated photovoltaic power generation based on the solar radiation on a horizontal surface is not accurate. Diurnal cycle of solar radiation is affected by different weather conditions and the power stations need to adjust generation strategies according to weather conditions. Accuracy of the two kinds of tilted radiation calculation models is similar and the overall calculation effect is reasonable. The uncertainty of the direct portion segment calculation function is the main cause of calculated errors.

1. Introduction

Sino-Singapore Tianjin Eco-city (eco-city for short, the same below) is located in the northern part of Binhai New Area. It is a major strategic cooperation project between China and Singapore governments to address global climate change, save resources and energy and build a harmonious society. Eco-city gives priority to develop renewable energy, and utilize wind, solar, geothermal and other green energy to reduce the impact of human activities on the environment. Solar energy is an important part of the eco-city construction for making full use of solar energy resources as a supplement to conventional energy sources [1-2]. Eco-city has completed 9.27 MW solar photovoltaic power stations and achieved parallel operation. Traditionally, photovoltaic power stations tilt solar panels to obtain more solar energy. However, the meteorological stations only carry out radiation observations on a horizontal surface, and the power stations are rarely equipped with tilted radiation observations consistent with solar panels inclination. Therefore, the effective radiation adopted by power plant in the actual operation needs to be calculated by the tilted radiation calculation models based on geometrical optics theory [3-14].

To assess solar energy resources in the eco-city, and to provide more accurate professional weather services and countermeasures for the photovoltaic power generation, Meteorological Bureau of Binhai New Area and Meteorological Bureau of Tianjin collaborate and launch experimental observation of eco-city solar energy resources, key prediction technology research of photovoltaic power generation, assessment and services. In this paper, using eco-city solar observation experimental data, solar radiation characteristics and the impact of weather conditions on the solar radiation are analyzed in depth. Results and sources of error of solar radiation on a tilted surface calculated by different models are discussed as well.

2. Observation experiment and data processing

2.1. Site and equipment Introduction
Solar radiation stations are located in the eco-city sewage treatment plant roof surrounded by open space and there are no tall buildings and no white reflector around. Observation station equipped with two solar radiation sensors that one horizontal, one tilted (toward the south, inclination of 21.5 °, the same as solar panels). Models are TBQ-2-B (Beijing Huatron Technology Co., Ltd), spectral range 300 ~ 3000 nm, the radiation range 0 ~ 2000 W / m^2, the degree of stability less than ± 2%, cosine response ≤7% (when solar elevation angle of 10°), and sampling interval 1 min. To ensure the accuracy and consistency of the observed data, two sensors were calibrated and synchronized observation before install. During the observation, remove dust and wipe are periodically done to the sensor glass cover to avoid observational error caused by fouling.

2.2. Data processing and methods

2.2.1. Quality Control. Observation data used in this paper is solar irradiance on a horizontal and tilted surface for the period 14 August 2014 to 12 August 2015 by the hour from 9 to 16. Figure 1a shows scatter plot of irradiance on a horizontal and tilted surface. As a whole there is good agreement between the two, and most of the data points distribute in the vicinity of the fitting trend line. But there are still a fraction of data points deviated larger, so quality control to the observational data are needed.

Quality control methods used in this paper are that linear fit irradiance on a horizontal and tilted surface firstly, and then use fitting formula to calculate the corresponding tilted irradiance while the irradiance on a horizontal surface as independent variables. Furthermore difference between measured and calculated values and the mean range are given. Ultimately the difference 3 times greater than the mean range is eliminated. Figure 1b shows the scatter plot after quality control that it shows a very good agreement.

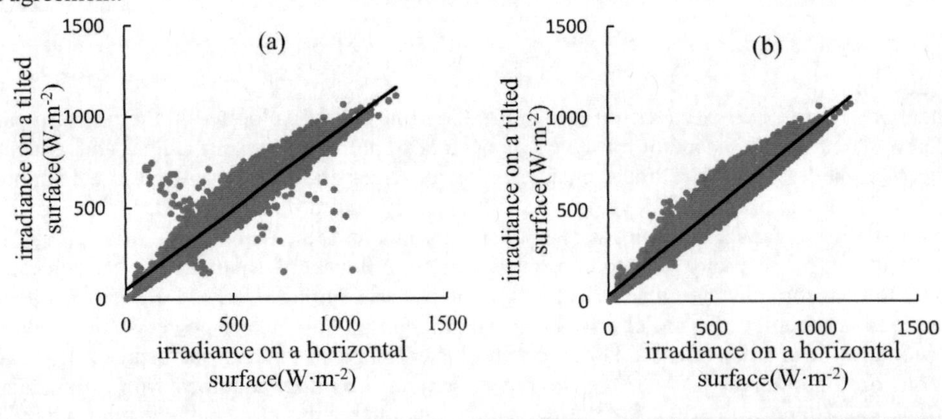

Figure 1. Scatter plot of radiation on a horizontal and tilted surface before (a) and after (b) filtering.

After the quality control, irradiance will be converted to radiation exposure hourly, the total cumulative radiation exposure from 9:00 to 16:00 act as Daily radiation exposure. And then the monthly average radiation, the seasonal radiation exposure and the annual radiation exposure are calculated. In order to reflect the physics regulation and seasonal characteristics of solar radiation better, vernal equinox, summer solstice, autumnal equinox and the winter solstice are used as boundary nodes, namely spring form March 22 to June 22, Summer from June 23 to September 22, fall from September 23 to December 22, winter from December 23 to March 21.

2.2.2. Tilted solar radiation model. Solar radiation on a tilted surface include direct radiation, scattered radiation and reflected radiation which tilted surface received, wherein the calculation method of direct radiation and reflected radiation is relatively simple. But for scattered radiation, calculation model is different because of isotropic and anisotropic difference in the sky. One model used here is isotropic model Klein [9] proposed, another is anisotropic model Hay [10] proposed.

1) Isotropic model, the solar radiation on a tilted surface is calculated as follows

$$H_\theta = H_b R_b + \frac{(H - H_b)(1+\cos\theta)}{2} + \frac{H\rho(1-\cos\theta)}{2} \tag{1}$$

where H_θ is solar radiation on a tilted surface, H_b is direct radiation on a horizontal surface, R_b is ratio of direct radiation on a tilted surface to direct radiation on a horizontal surface, H is solar radiation on a horizontal surface, θ is tilted angle, ρ is ground reflectance. R_b is calculated by

$$\begin{aligned} R_b = \Big[& (\cos\theta\sin W\sin \mathrm{h})\frac{\pi}{180}(k_{ss}-k_{sr}) - (\sin W\cos \mathrm{h}\sin\theta\cos \mathrm{V})\frac{\pi}{180}(k_{ss}-k_{sr}) + \\ & (\cos \mathrm{h}\cos W\cos\theta)(\sin k_{ss}-\sin k_{sr}) + (\cos W\cos \mathrm{V}\sin \mathrm{h}\sin\theta)(\sin k_{ss}-\sin k_{sr}) - \\ & (\cos W\sin\theta\sin \mathrm{V})(\cos k_{ss}-\cos k_{sr})\Big]\left[2(\cos \mathrm{h}\cos W\sin k_s + \frac{\pi}{180}k_s\sin \mathrm{h}\sin W)\right]^{-1} \end{aligned} \tag{2}$$

where W is solar declination, h is geographic latitude, k_{sr} is sunrise hour angle on a tilted surface, k_{ss} is sunset hour angle on a tilted surface, V is azimuth on a tilted surface, k_s is sunset hour angle on a horizontal surface. k_{sr} and k_{ss} are calculated by

$$k_{sr} = -\min\left\{k_{s'}, \left|-\cos^{-1}\left(-\frac{\mathrm{a}}{\mathrm{D}}\right) + \sin^{-1}\left(\frac{\mathrm{c}}{\mathrm{D}}\right)\right|\right\} \tag{3}$$

$$k_{ss} = \min\left\{k_{s'}, \left|\cos^{-1}(-\frac{\mathrm{a}}{\mathrm{D}}) + \sin^{-1}(\frac{\mathrm{c}}{\mathrm{D}})\right|\right\} \tag{4}$$

where $k_s = \cos^{-1}(-\tan \mathrm{h}\tan W)$, $\mathrm{a} = \sin W(\sin \mathrm{h}\cos\theta - \cos \mathrm{h}\sin\theta\cos \mathrm{V})$,
$\mathrm{b} = \cos W(\cos \mathrm{h}\cos\theta + \sin \mathrm{h}\sin\theta\cos \mathrm{V})$, $\mathrm{c} = \cos W\sin\theta\sin \mathrm{V}$, $\mathrm{D} = (\mathrm{b}^2 + \mathrm{c}^2)^{1/2}$,
$W = 23.45\times\sin(360\times\frac{284+d}{365})$, d is the number of calendar days of the date from January 1.

2) Anisotropic model, the tilted solar radiation is calculated as follows

$$H_\theta = H_b R_b + \frac{H\rho(1-\cos\theta)}{2} + (H - H_b)[\frac{H_b}{H_o}R_b + \frac{\left(1-\frac{H_b}{H_o}\right)(1+\cos\theta)}{2}] \tag{5}$$

where H_o is astronomical radiation in the top of the atmosphere.

Using the measured data of tilted surface of eco-city, the accuracy of two models is assessed. But radiation on a horizontal surface needs to be "direct-diffuse separation" due to the lack of observations of direct radiation and diffuse radiation. So the "direct ratio" method Erbs [15] proposed is used to achieve "direct-diffuse separation" of the total radiation.

3) "direct-diffuse separation" method

Direct ratio is defined as the proportion of direct radiation in the total radiation, which is calculated as follows

$$A = \frac{H_b}{H} \tag{6}$$

where A is direct ratio. Direct ratio can represent by atmospheric transparency coefficient. And the atmospheric transparency coefficient refers to the ratio of the observed total radiation and astronomical radiation in the top of the atmosphere. Its relationship with the direct ratio as follows

$$\begin{cases} A = a_0 k_T & k_T < b_0 \\ A = c_0 + c_1 k_T - c_2 k_T^2 + c_3 k_T^3 - c_4 k_T^4 & b_0 \le k_T \le b_1 \\ A = a_1 & k_T > b_1 \end{cases} \tag{7}$$

where k_T is atmospheric transparency coefficient , a_0 、 a_1 、 c_0 - c_4 are fitting coefficients of regression equations, dimensionless , $a_0 = 0.09$, $a_1 = 0.835$, $c_0 = 0.0489$, $c_1 = 0.1604$, $c_2 = 4.388$, $c_3 = 16.638$, $c_4 = 12.366$, b_0 、 b_1 are divided interval coefficients of subsection simulation curve, dimensionless, $b_0 = 0.22$, $b_1 = 0.80$.

3. Results and analysis

3.1. Characteristics of the solar radiation eco-city

Figure 2 presents the seasonal cycle of radiation on a horizontal and tilted surface in eco-city. It indicates that radiation on a horizontal and tilted surface both feature unimodal distribution: the maximum radiation appears in May, and the minimum radiation occurs in January. Both radiations on a horizontal and tilted surface are relative deficit in June, which is possible as the relative humidity is relative high during this season. The radiation on a horizontal and tilted surface is equal in March and October; from April to September, however, the solar radiation is higher than radiation on a tilted surface. The radiation on a tilted surface is higher than that on a horizontal surface for the rest months. Therefore, using observed total radiation on a horizontal surface to evaluate monthly power generation is not accurate. In addition, the annual mean total radiation on a horizontal surface exceeds the total radiation on a tilted surface by 124.2 $MJ\ m^{-2}$ (Table 1). Seasonally, the total radiation reaches peak in spring, and then declines in summer; it continues to decrease in autumn and with slight increase in winter. It suggests that the total radiation on a horizontal surface is higher than radiation on a tilted surface in spring and summer, and vice versa. Taken together, there is large discrepancy between observed radiation on a horizontal and tilted surface on monthly, seasonal and annual time scales. Using observed radiation on a horizontal and tilted surface to evaluate and estimate the monthly, seasonal and annual power generation will result in ineligible bias. Therefore, in order to accurately assess the solar resource and photovoltaic power generation, we suggest to conduct the radiation observation on a tilted surface parallel to the solar panel for at least one year.

Table 1. Annual mean and seasonal mean radiation on a horizontal and tilted surface.

	Radiant exposure on a horizontal surface($MJ{\cdot}m^{-2}$)	Radiant exposure on a tilted surface($MJ{\cdot}m^{-2}$)
Year	5142.0	5017.8
Spring	1744.5	1540.1
Summer	1536.4	1337.8
Autumn	849.0	968.7
Winter	906.0	1064.4

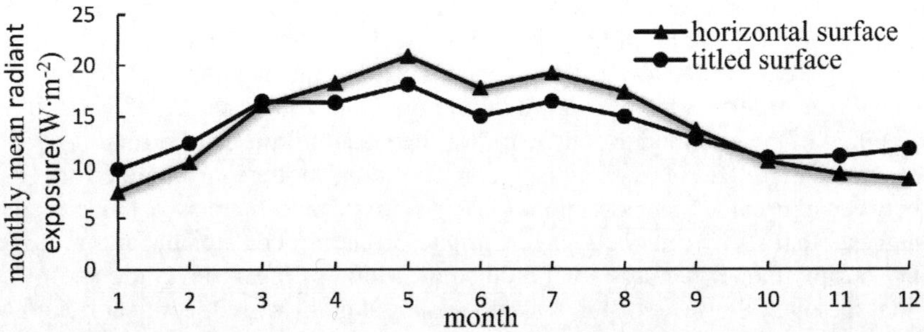

Figure 2. Changes of radiation on a horizontal and tilted surface across the year in eco-city.

Curves in Figure 3 show the diurnal variations of irradiance on a tilted surface on sunny, partly cloudy, cloudy, overcast and rainy days. Sunny day indicates day with no special weather phenomenon, less than thirty percent of daily average total cloud cover, greater than 15 km visibility. Thirty to fifty, sixty to ninety, and greater than ninety percents of daily average total cloud cover signify partly cloudy, cloudy, overcast sky, correspondingly. Rainy day is defined when there is persistent rainfall. It can be found that, on the sunny and partly cloudy days, the irradiance are maintained at a higher level throughout the day with the maximum value at 12 pm and minimum value at 16 pm. The diurnal curve of the irradiance is just like a parabola, which means sufficient solar radiation and large utilization rate because of the small impact to the grid during grid-connected power generation. Under the cloudy and overcast sky, the irradiance is small than it under the sunny and cloudy conditions and its curve is like a zigzag pattern with large fluctuation. In this case, grid-connected power generation has significant impact on the grid. There is smallest irradiance (small than 200 W/m^2) on the rainy day without diurnal cycle. It indicates that utilization rate of solar energy is low. The power station needs to adjust the power supply plan and carry out the maintenance of power grids. In a word, the diurnal variation of the solar radiation has certain randomness and volatility. These features have an effect on the stability of power system. Therefore, the power generation plan needs to be adjusted according to different weather conditions in time. Plans of conventional energy generation and photovoltaic power generation also need to be made. Reduce standby rotation capacity, keep the power system stability, and reduce the waste of resources.

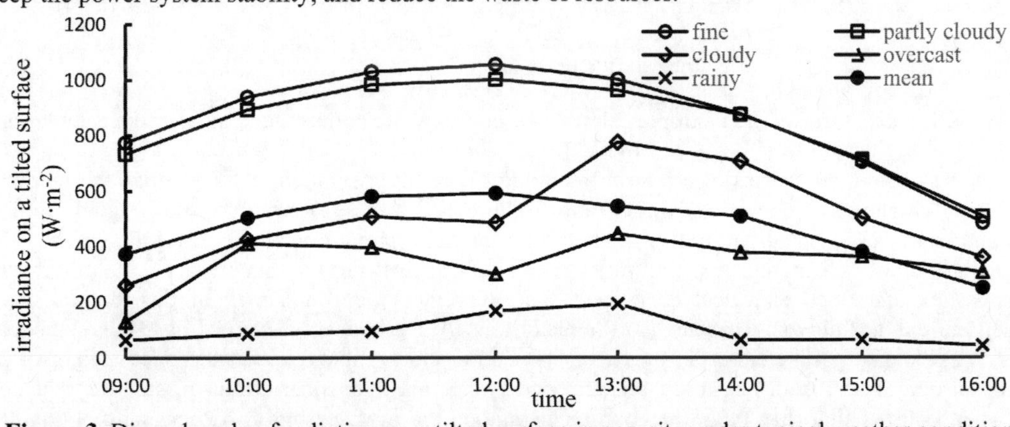

Figure 3. Diurnal cycle of radiation on a tilted surface in eco-city under typical weather conditions.

3.2. Assessment of the tilted solar radiation calculation models

The former analyses suggest that there is discrepancy between radiation on a horizontal and tilted surface on monthly, seasonal and annual time scales, and that it is more reasonable to evaluate power generation using radiation on a tilted surface. Before the power station was founded, however, long-term radiation observation on a tilted surface was lacking in the evaluation of solar radiation and power generation. Previous evaluations were carried out via combining the solar radiation on a horizontal surface from local meteorological institutes with isotropic or anisotropic tilted radiation calculation models. The evaluations using this approach rely on the accuracy of the two models. This study analyzes one-year observations of total radiation on a horizontal and tilted surface in Eco-city, evaluates the accuracy of the two models, and explores the possible causes of errors.

Table 2 presents the correlation coefficients, mean deviation, root mean square error [16-19], and t-statistic [20] between the calculated radiations on a tilted surface from two models and observed radiations. It suggests that the two models have similar correlation coefficients and root mean square error, but the anisotropic model features smaller absolute value of mean deviation and t-statistic. The implication is that the overall effects of the two models are equal, while the anisotropic model shows higher accuracy. Compared with observations, however, both models present large error (Figure 4). For the anisotropic model, although the relative error of 90% results are less than 30%, the partition of results which show less than 10% relative error is only 28%. In addition, 10% results present more than 30% relative error. For the calculated values based on statistics, the absolute difference ranges from -271 W/m^2 to 137 W/m^2.

Table 2. Statistical analysis of tilted radiation calculated values in two models.

	Correlation coefficient	Mean variation(W·m^{-2})	root-mean-square error (W·m^{-2})	t-statistics
Isotropic model	0.96	-1.63	77.45	1.09
Anisotropic model	0.96	-0.05	77.61	0.04

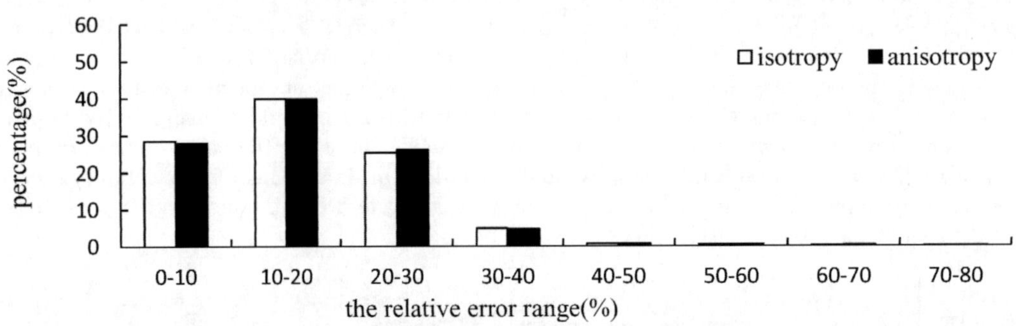

Figure 4. Relative error of the isotropic calculation model (white column) and anisotropic calculation model (black column).

Now we explore the causes of errors in the models, and propose methods to improve the models. This study employs the concept of direct ratio to "direct-diffuse separation" of total radiation on a horizontal surface. As the direct radiation and scatter radiation are not available, we follow Erbs [15] to calculate direct ratio based on the empirical equation of direct ratio and coefficient of atmospheric transmission, and fitted empirical coefficient. The universality and accuracy of this method need further test due to following reasons: (1) the method itself is a conceptual model based on statistical theory; (2) the data sample is small; (3) the time period is short. Table 3 shows the calculated error of the anisotropic model in different seasons and under different atmospheric transmission. It indicates that when k_T is smaller than 0.22, the coefficients among the four seasons is higher than 0.99, while the mean deviation, root mean square error and t-statistic are smallest when k_T is larger than 0.22, suggesting that the accuracy of the model is very high, and the empirical equation of direct ratio is universal. When k_T is between 0.22 and 0.8, the correlation coefficients for all seasons decline, and the root mean square error is 80 W/m^2. Note that the squared error in spring and summer is opposite to that in autumn and winter: spring and summer present positive deviation, indicating that the model overestimates the radiation during these seasons; autumn and winter present negative deviation, indicating that the model underestimate the radiation in these seasons. Therefore, the universality of the model is poor when k_T is between 0.22 and 0.8. To achieve calculated results with higher accuracy, we need to correct the parameters of the direct ratio calculation models for each season. The situation when k_T is higher than 0.8 is similar to that when k_T is between 0.22 and 0.8. Also, when k_T is higher than 0.8, the mean deviation and root mean square error in autumn and winter is higher than other conditions. Taken together, the direct portion segment calculation function and its coefficients in our calculation model are not universal and accurate for different seasons and atmospheric transmission. In the practical application of power stations, it is suggested to assess the universality of the calculation models besides conducting the radiation observation on a horizontal and tilted surface. In addition, adjustments and corrections of calculation coefficients are needed to gain more accurate calculated results of radiation on a tilted surface.

Table 3. Statistical analysis of anisotropic calculation model under different conditions.

	range of k_T	Correlation coefficient	Mean variation($W \cdot m^{-2}$)	root-mean-square error ($W \cdot m^{-2}$)	t-statistics
	<0.22	0.99	19.21	21.85	16.72
Spring	0.22~0.8	0.98	68.95	79.09	35.22
	>0.8	0.93	55.04	67.45	17.18
	<0.22	0.99	20.59	22.47	21.36
Summer	0.22~0.8	0.99	69.39	75.11	50.28
	>0.8	0.92	46.20	62.53	12.11
	<0.22	0.99	15.93	17.41	20.15
Autumn	0.22~0.8	0.96	-54.15	83.26	20.12
	>0.8	0.95	-116.04	125.57	19.04
	<0.22	0.99	14.52	15.96	15.04
Winter	0.22~0.8	0.94	-56.97	86.12	20.42
	>0.8	0.98	-98.65	102.78	30.59

4. Conclusion

Using solar observation and meteorological data over the Sino-Singapore Tianjin Eco-city from 14 August 2014 to 12 August 2015, this study summarizes the characteristics of solar radiation on a horizontal and tilted surface, and explores the tilted radiation calculation models. The conclusions include:

a. Discrepancy between solar radiation on a horizontal and tilted surface in eco-city exists on monthly, seasonal and annual time scales. The radiation observation on a horizontal surface is not accurate enough to evaluate solar resources, and we suggest to conduct radiation observation on a tilted surface parallel to solar panel for at least one year.

b. The characteristics of solar radiation differ under various synoptic conditions. It is suggested to adjust power generation strategies in response to different weathers, especially for cloudy weather which are attributable to the irregularity and instability of power generation.

c. The accuracy of the two radiation calculation models is equivalent, and the calculation effects are acceptable. Error, however, still exists. The universality of the empirical equation of direct radiation ratio and correlation coefficients needs improvement. Besides conducting solar radiation observations on a horizontal and tilted surface, we suggest to assess the universality of tilted calculation models during the practical applications of power stations, and adjust and correct calculation coefficients based on realistic situations.

Acknowledgement
The investigations were supported by Tianjin Meteorological Bureau, project No.: 201610ybxm07

The research was performed with the financial support from Tianjin Science and Technology, project No.: 14ZCZDGX00037

References

[1] Wang X Q, Li B and Wang T L 2010 *Northern Horticulture* **19** 44–46
[2] Wang C Y, Zhang H H and Peng E R 2013 *Modern Agricultural Science and Technology* **11** 231–233
[3] Chong W, Lv W H, Sha Y Z and Bian Z Q 2015 *Acta Energiae Solaris Sinica* **36(12)** 3041–47
[4] Li F, Chen Z H, Cai T, Ma J Y and Xu J 2013 *Acta Energiae Solaris Sinica* **34(6)** 974–983
[5] Li F, Zhao J B, Duan S X, Yan Q Q, Shen Y B and Chen Z H 2015 *Acta Energiae Solaris Sinica* **36**(2) 502–509
[6] Li Q L, Wang P Y and Yang F 2008 *Energy conservation technology* **6** 571–574
[7] Xu L 2001 *Journal of Yunnan University of the Nationalities(Natural Sciences Edition)* **10(4)** 492–493
[8] Yang J H, Mao J J and Chen Z H 2002 *Journal of Shanghai jiaotong university* **36(7)** 1032–36
[9] Klein S A 1977 *Solar Energy* **19(4)** 325–329
[10] Hay J E 1979 *Solar Energy* **23(4)** 301–307
[11] Liu B Y H and Jordan R C 1961 *ASHRAE Journal* **3** 526–541
[12] Reindl D T, Beckman W A and Duffie J A 1990 *Solar Energy* **45(1)** 9–17
[13] Muneer T 1997 *Solar Radiation and Daylight Models for the Energy Efficient Design of Buildings* Oxford (UK: Architectural Press)
[14] Ma C C Y and Iqbal M 1983 *Solar Energy* **31(3)** 313–317
[15] Erbs D G, Klein S A and Duffie J A 1982 *Solar Energy* **28(4)** 293–302
[16] Xing H Y, Chong W, Sha Y Z and Lv W H 2012 *Optical Engineering* **51(4)** 046001-1–6
[17] Ruiz-Arias J A, Alsamamra H, Tovar-Pescador J and Pozo-Vazquez D 2010 *Energy Conversion and Management* **51(5)** 881–893
[18] Skeiker K 2006 *Energy Conversion and Management* **47(4)** 331–345
[19] Chong W, Sha Y Z, Xing H Y and Lv W H 2012 *Acta Optica Sinica* **32(1)** 0112001-1–9
[20] Stone R J 1993 *Solar Energy* **51(4)** 281–291

Experimental study of a novel photovoltaic solar-assisted heat pump /loop heat-pipe (PV-SAHP/LHP) system

Tao Zhang[1, 3], **Gang Pei**[2], **Qunzhi Zhu**[1], **Jie Ji**[2]

[1] School of Energy and Mechanical Engineering, Shanghai University of Electric Power, #2103 Pingliang Road, Shanghai City, China
[2] Department of Thermal Science and Energy Engineering, University of Science and Technology of China, #96 Jinzhai Road, Hefei City, Anhui Province, China

E-mail: taozhang@shiep.edu.cn

Abstract. A prototype of a photovoltaic solar-assisted heat-pump/loop heat-pipe system (PV-SAHP/LHP) was constructed in this paper. The system was a combination of photovoltaic solar-assisted heat pump system (PV-SAHP) and loop heat pipe photovoltaic/thermal (LHP-PV/T) system. The combined system can carry out with two modes but using the same working fluid, and the two modes can switch operation freely. R600a was employed as the working fluid, and system performance under different working mode was presented in this paper. The results show that the day average photothermal efficiency & photovoltaic efficiency can reach to 43.6% & 11.3% under LHP-PV/T working mode compared with that of 57.5% & 12.1% under PV-SAHP working mode; Besides that, a day average COP of 3.66 was obtained under PV-SAHP working mode.

1. Introduction

A photovoltaic/thermal hybrid solar system (or PVT system for simplicity) is a combination of photovoltaic (PV) and solar thermal components/systems, which produce both electricity and heat from one integrated component or system. In other words, PV is used as (part of) the thermal absorber [1]. In order to get more power and heat from PV/T system, it is necessary to cool the PV cell and decrease its temperature. The liquid based photovoltaic thermal collector systems are practically more desirable and effective than air based systems. Temperature fluctuation in liquid based PV/T is much less than the air based PV/T collector which subjected to variation in solar radiation levels [2]. However, there are freezing problem when the normal water-based PV/T systems working in winter. Loop heat pipe (LHP) PV/T systems, which using low-boiling refrigerants as the working fluid, proposed a promising prospect to solve the problem above [3].

Photovoltaic solar-assisted heat pump (PV-SAHP) is a combination of solar energy and heat pump that can improve the quality of the energy available and shows potential for different applications [4]. The evaporator-collector used in such system can absorb both solar and ambient energy [5]. Abou-Ziyan [6] conducted series of experiments with R22 and134a to identify appropriate refrigerant for solar applications. Hawlader et al. [7], Lu et al. [8] and Chyng [9] used solar assisted heat pump (SAHP) to produce hot water. Besides, the PV performance can be improved because of the low evaporating temperature [10].

A novel photovoltaic solar-assisted heat pump/loop heat pipe system (PV-SAHP/LHP), which is a combination of PV-SAHP system and LHP-PV/T system, was conducted in this paper. The combined system can work with two modes, i.e. PV-SAHP independent operation or LHP-PV/T independent operation, but employing the same working fluid. Iso-butane (R600a), one of the natural fluids that with appropriate thermodynamic and physical properties, well energy performance and negligible ozone depletion potential (ODP) and global warming potential (GWP) is suitable as the working fluid [11] and [12]. R600a is widely used in the domestic refrigerator/freezers and residential heat pump (notably in Europe) [13]. The system can switch operation freely according to different climate or temperature need. The photothermal and photovoltaic performance under different operation mode were presented this paper.

2. System design and experiment set up

2.1. System design and setup

System’ s schematic diagram is shown in Figure 1. It comprise two PV/T solar evaporator collectors employing a vacuum magnetron cermet as selective absorbing coating, a water tank with an inner spiral coil pipe, six inspection mirrors, one compressor, one capillary, one gas-liquid separator, one PV control equipment, and the connect pipes. The R600a is employing as the working fluid, and the refrigerant volume-charge is 36.6% of whole system. Detailed information of the system components are listed in table 1.

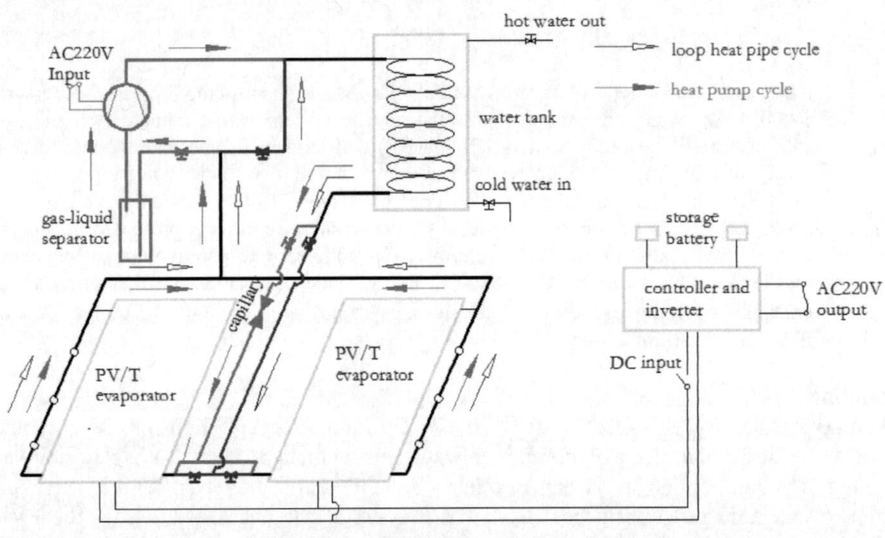

Figure 1. Schematic diagram of PV-SAHP/LHP system

Table 1. Detailed information of the system components

Device & copper tube	Parameter	Remarks
water tank	Φ450×1570mm	150 ±1L
evaporator solar collector	Φ1270×780×1.16mm	PV coverage rate=0.56
vapor rising pipe	Φ28×1000×1mm	/
liquid return pipe	Φ 16×1400×1	/
condensation spiral coil pipe	Φ12×18000×1	/
inspection mirror	/	Emerson
compressor	1/4P	
battery	150AH	
Other pipes	Φ9.52	
Gas-liquid separator		5.9L

2.2. Experimental procedure

The following is the LHP-PV/T's cycle process. First, the incident solar irradiation is partly absorbed by the liquid R600a in the copper tubes behind the PV/T evaporator, to heat the liquid R600a to R600a vapor. R600a vapor is continuously heated when flowing along the copper tubes. Then the R600a vapor flows into spiral coil pipe in the water tank along the vapor rising pipe, where it is cooled by the water and condensed into fluid or gas-liquid and transfers the heat energy to water. Condensed R600a finally flows to the evaporator section through liquid return pipe under gravity for next heating cycle. While in the PV-SAHP cycle process, the heated R600a vapor is flow into the gas-liquid separator instead, and then be compressed by the compressor; the condensed R600a is flow into the capillary instead, and then return to the PV/T evaporator.

The experiments were carried out in Hefei (31.52°N, 117.17°E), a city located in central China. The collector was installed facing south with a tilt angle of 40°, which was a little higher than the city's latitude, because the working tilt angle of the heat pipe was considered. The water tank was a little higher than the collector. The inspection mirrors and the pyranometer were parallel with the collector. Seven thermal T type thermocouples were set in the water tank to monitor temperature variation. Ambient temperature and solar irradiation were also recorded. All the measured parameters were recorded using Agilent 34970A every 30 s. Figure 2 presents the experiment setup.

Figure 2. PV SAHP/LHP system experiment setup

3. Results and discussion

The daily thermal efficiency of solar water heating system can be evaluated using the daily average efficiency, which is defined as[14]:

$$\overline{\eta_t} = {C_p M (T_f - T_i)}\big/{HA} \qquad \text{(For LHP-PV/T; 1a)}$$

$$\overline{\eta_t} = {[C_p M (T_f - T_i) - W_p]}\big/{HA} \qquad \text{(For PV-SAHP; 1b)}$$

Where M is the mass of water in the water tank; Cp is the specific heat of water; H is the total solar irradiation on the collector surface during the experiment; W_p is the power of compressor; T_i and T_f are the initial and final temperatures of water in the water tank, respectively; A is the aperture area of the collector in this paper.

The daily electric efficiency of a hybrid PV/T system can be evaluated using the daily average efficiency, which is defined as[15]:

$$\overline{\eta_e} = {V_{pv} I_{pv}}\big/{HA_{pv}} \qquad (2)$$

Where V_{pv} , I_{pv} and A_{pv} are the output voltage, output current and aperture area of the PV module, respectively.

The coefficient of performance (COP) of a heat pump can be evaluated as [16]:

$$COP = {C_p M (T_f - T_i)}\big/{W_p} \qquad (3)$$

3.1. Performance of LHP-PV/T independent operation mode

In order to show the details of the system daily performance under LHP-PV/T independent operation, a fully representative sample day was chosen. The average solar irradiation, ambient temperature, presented in Figure 3, during the test was 703W/m^2, 30°C, respectively; test time was from 8:09 to 16:09, and the water temperature rising gradually form 19.1°C to 43.4°C during the test. System instantaneous photothermal efficiency and photovoltaic efficiency of every 30 minutes is shown in Figure 4a and Figure 4b, respectively.

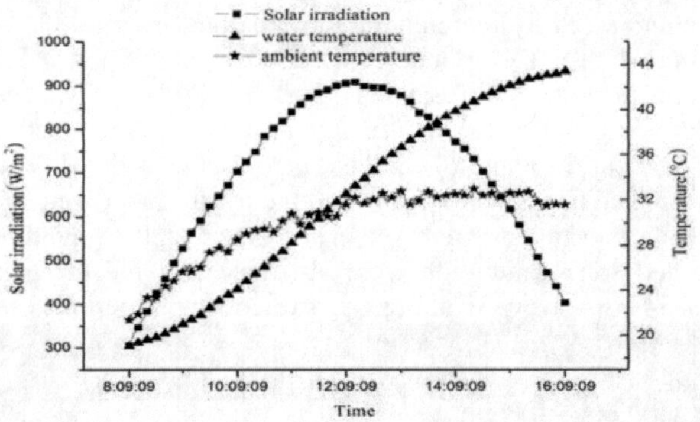

Figure 3. Variation of solar irradiation, ambient temperature, water temperature under LHP-PV/T independent operation mode

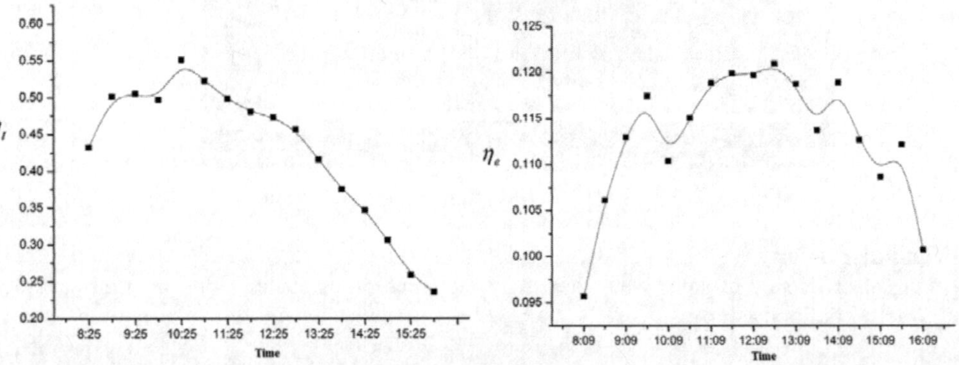

(a) (b)

Figure 4. Variation of photothermal (a) & photovoltaic efficiency (b) under LHP-PV/T independent operation mode

Figure 4a shows that system instantaneous photothermal efficiency presents a trend of increasing first and then decreased. It is because in the beginning of the tests, temperature difference between water and the ambient temperature, positive or negative was small. The incident angle decreased with time, thereby increasing photothermal efficiency. In the middle of the tests, temperature difference between water and the ambient temperature was bigger, making the heat loss bigger. However, there was also a bigger irradiation density and a smaller incident angle, so efficiency decreased slowly. At the end of the tests, the temperature difference between the water and the ambient was big, and also there was big incident angle, so the efficiency decreased sharply. The highest photothermal efficiency during the test was 55.1%, while the average was 43.6%. Figure 4b presents the fluctuation of photovoltaic efficiency during the test; it also has a trend of increasing first and then decreased. The fluctuation reason was similar to the photothermal efficiency, but with different range. The highest photovoltaic efficiency during the test was 12.1%, while the average was 11.3%.

Obtained results indicate that the LHP-PVT system manage to achieve the performance no less than the water based PV/T systems [17-19], and the daily average primary efficiency can reach 71.9%, which offers a promising prospect. Therefore, the results indicate that the LHP-SWH can ensure a satisfactory performance of the LHP-PV/T. Moreover, it must be underlined that the diode characteristic of the LHP can ensure a smaller heat loss at night compared to the conventional water based PV/T systems.

3.2. Performance of PV-SAHP independent operation mode

System performance under PV-SAHP independent operation mode was also studied. Ambient environment, including ambient temperature, solar irradiation, and the water temperature fluctuation of one of the typical day, are shown in Figure 5.The average solar irradiation during the test was 765W/m^2, and the average ambient temperature was 27.1°C, test time was from 8:30 to 14:00. Figure 6 shows the trend of temperature rising and variation of compressor power. The results show that the water temperature is linear rising from 20.5°C to 53.6°C, and the compressor power is rising gradually from 195W to 372W. The condensing pressure rising with the temperature rising, so as the compressor power.

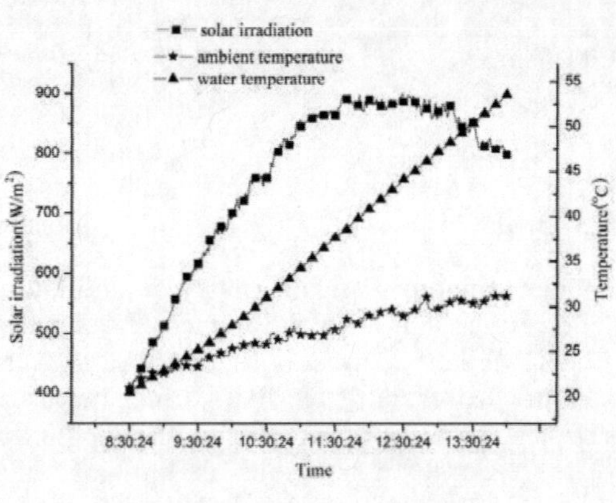

Figure 5. Variation of solar irradiation, ambient temperature, water temperature under PV-SAHP independent operation mode

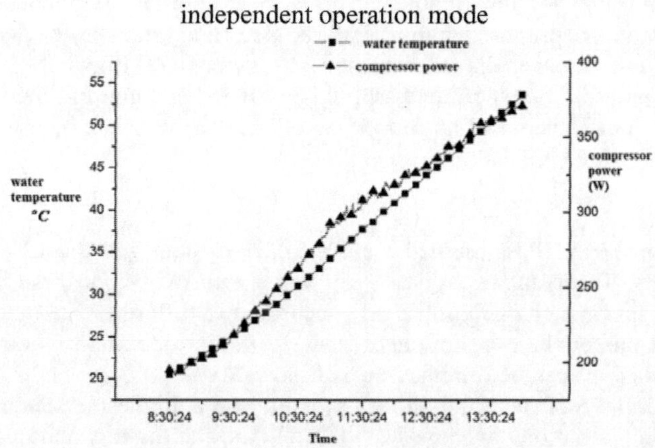

Figure 6. Variation of water temperature and compressor power under PV-SAHP independent operation mode

Figure 7 shows the system instantaneous photothermal efficiency and COP every 30mins under PV-SAHP independent operation mode. It can be seen that the overall trend of COP is increasing first and then decreased, while the photothermal efficiency is declining most of time. The highest and lowest COP during the test was 4.37 and 3.03, respectively. The daily average COP and photothermal efficiency were 3.66 and 57.5%, respectively, which present a satisfactory performance compared with the works in a review [20]. Considering that, the compressor rated power used here is only 184W, the PV/T collector area is small too, and it can be conclude that the hybrid PV-SAHP system can achieve a better performance if the system is bigger.

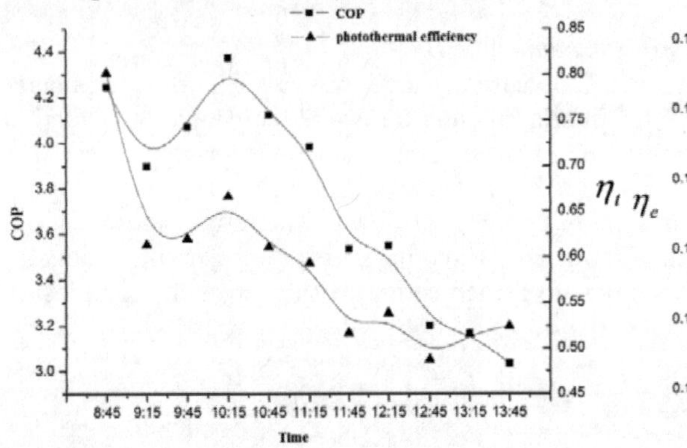

Figure 7. Variation of COP and photothermal efficiency under PV-SAHP independent operation mode

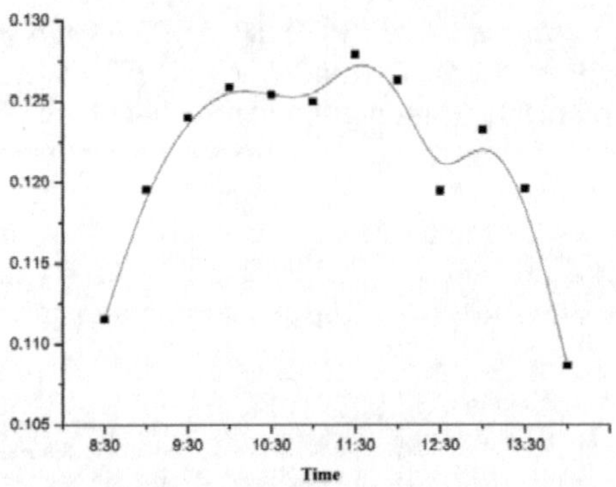

Figure 8. Variation of photovoltaic efficiency under PV-SAHP independent operation mode

Figure 8 presents the instantaneous photovoltaic efficiency every 30mins; it has a similar trend with LHP-PV/T working mode. The results show that the highest instantaneous photovoltaic efficiency during the test can reach to 12.8%, while the lowest is 10.9%. System average photovoltaic efficiency can reach to 12.1%, which is higher than that of LHP-PV/T also. The absorb plate always under a low temperature because of the refrigerants evaporation benefits the PV performance compared with the LHP-PV/T system.

Satisfactory COP, photothermal efficiency and photovoltaic efficiency, one can conclude that the system operates well under the PV-SAHP working mode. And it can also be conclude that the hybrid PV-SAHP/LHP system achieves a remarkable performance under the two independent working modes. However, the essence of the proposed hybrid PV-SAHL/LHP system is the working mode switch according to the ambient environment or domestic requirement. The switching strategy should be deeply studied. Therefore, further experiments should be carried out, and full theoretical model should be established. Based on that, the novel hybrid PV-SAHP/LHP system can be more power saving, and can present a more promising application prospect.

4. Conclusion

A novel system (PV-SAHP/LHP, combined the PV-SAHP system and the LHP-PV/T system) was conducted in this paper. The combined system can work with two modes, but employing the same working fluid. System can switch operation freely according to different climate or temperature need. The photothermal and photovoltaic performance under different operation mode were presented this paper. The major conclusions can be summarized as follows:

1) Under the LHP-PV/T working mode, system had a highest instantaneous photothermal efficiency of 55.1%, while the day average was 43.6%. And the highest photovoltaic efficiency can reach to 12.1% with an average value was 11.3%.

2) Under the PV-SAHP working mode, system had a highest COP of 4.37, while the day average value was 3.66. The average photothrmal and photovoltaic efficiency during the test was 57.5% and 12.1%, respectively, which were both higher than that of LHP-PV/T operation mode.

Acknowledgments
This work was supported by the Shanghai colleges and universities young teachers training funding scheme (ZZsdl15076) and double type teachers' plan of shanghai university of electric power.

References

[1] PVT Roadmap - a European guide for the development and market introduction of PV-thermal technology 2006 PVT forum: <http:// www.pvtforum.org/index.html>.
[2] Daghigh R, Ruslana M H, Sopian K 2011 Advances in liquid based phtotvoltaic/thermal(PV/T) collectors *RENEW SUST ENERG REV* **15** 4157-70.
[3] Pei G, Zhang T, Yu Z, Fu H and Ji J 2010 Comparative study of a novel heat pipe photovoltaic/thermal collector and a water thermosiphon photovoltaic/thermal collector *P I MECH ENG A-J POW* **225(3)** 271-278.
[4] Zakaria Mohd Amin, MNA Hawlader 2010 Development of a multi-function solar assisted heat pump system.In:*ICAE2010—international conference on applied energy Singapore.*
[5] MNA Hawlader, Zakaria Mohd Amin,Ye Shao chun 2009 Solar assisted heat pump: research and development at US CAFEO 27th Singapore.
[6] Abou-ZiyanHZ 1997 Solar-assisted HZ.R22 andR134a heat pump systems for low-temperature Applications *APPL THERM ENG* **17** 455-469.
[7] Hawlader MNA,Chou S K,Ullah MZ 2001 The performance of a solar assisted heat pump water heating system *APPL THERM ENG* **21** 1049-65.
[8] Lu A, ChartersWWS, Chaichana C 2002 Solar heat pump systems for domestic hot water *Sol Energy* **73** 169-175.
[9] Chyng J P 2003 Performance analysis of as solar-assisted heat pump water heater *Sol Energy* **74** 33-44.
[10] Ji Jie, Pei Gang, Chow Tin-tai, Liu Keliang, He Hanfeng, Lu Jianping, Han Chongwei 2008 Experimental study of photovoltaic solar assisted heat pump system *Sol Energy* **82** 43-52.
[11] Kurylo M J 1990 The chemistry of stratospheric ozone: its response to natural and anthropogenic influences *Int. J. Refrig* **13(2)** 62-72.
[12] Molina M J, Rowland F S 1974 Stratospheric sink for chlorofluoromethanes: chlorine atom-catalysed destruction of ozone *Nature* **249(5460)** 810-812
[13] http://www.heatpumpcentre.org/en/aboutheatpumps/heatpumpworkingfluids/Sidor/default.aspx
[14] Zhang Tao, Pei Gang, Zhu Qunzhi, Ji Jie 2016 Investigation on the optimum volume-filling ratio of a loop thermosyphon solar water-heating system *J Sol Energ-T ASME* **384(4)** 041006
[15] Pei Gang, Zhang Tao, Fu Huide, Ji Jie, Su Yuehong 2012 An experimental study on a novel heat-pipe type photovoltaic/thermal system with and without a glass cover *Int J Green Energy* **10(1)** 72-89.
[16] M Yahya, Ahmad Fudholi, Hadyan Hafizh, Kamaruzzaman Sopian 2016 Comparison of solar dryer and solar-assisted heat pump dryer for cassava *Sol Energy* **136** 606-613.
[17] Ji Jie, Lu Jian-Ping, Chow Tin-Tai, He Wei, Pei Gang 2007 A sensitivity study of a hybrid photovoltaic/thermal water-heating system with natural circulation *Appl Energy* **84** 222–237.
[18] He Wei, Chow Tin-Tai, Ji Jie, Lu Jianping, Pei Gang, Chan Lok-shun 2006 Hybrid photovoltaic and thermal solar-collector designed for natural circulation of water *Appl Energy* **83** 199–210.
[19] Chow T T, He W and Ji J 2007 An experimental study of facade-integrated photovoltaic/water-heating system *APPL THERM ENG* **27(1)** 37–45.
[20] Peter Omojaro, Cornelia Breitkopf 2013 Direct expansion solar assisted heat pumps: A review of applications and recent research *RENEW SUST ENERG REV* **22** 33-45.

IOP Conf. Series: Earth and Environmental Science **52** (2017) 012018 doi:10.1088/1755-1315/52/1/012018

Research of nano silver alloy pasteto solar cell

Yizhou Du[1,3],XiaoxaiPeng[3,4], He Wang[2]

[1] Industrial design Department, The IT Electronics Eleventh Design & Research Institute Scientific and Engineering Corporation Limited, Chendu, 610000, China
[2]Physics Department, Xi'an Jiao Tong University, 710000, Xi`an, China
[3]GansuUniversityof Chinese Medicine, China

E-mail:471565156@qq.com

Abstract. From theory and experiment, the method toimprove the performance of silver paste was explored. Through the analysis about the SEM images to the electrode contact section of the crystal silicon solar cell, the different contact modes and resistance calculation methods of the metal paste and silicon were verified. Through the resistance calculation, consideringthe different oxide and silver silicon contact alloy characteristics, two kinds of additive compositions were found that which may have good performance in alloy paste of the solar cell.

1. Introduction

Over the world, more than 85% crystalline silicon solar cells adopt the silver paste as raw material of thick film screen printing production. As precious metal material, silver`s high market price makes it difficult to meet the needs of our use. Thus, systematic studying crystalline silicon solar cell emitter, decreasing the contact resistance of crystalline silicon solar cells are important to improve the efficiency of crystalline silicon solar cells and decrease the cost of PV power generation.

Due to the diversity about material resistance and alloy properties, the crystal silicon solar cell has contact resistance on the contact surface of the emitter and the silver electrode. As one of the most important functional structures of crystalline silicon solar cells, the electrical properties of Ag-Si contact affect the performance of solar cells. There are a lot of researches on the electrical properties of Ag-Si contact, mainly about the contact conduction mode, the carrier transport mechanism, the contact resistance and so on[1-4].

In this letter, the method to improve the performance of silver paste was explored. Through the calculation and analysis of the electrode contact section SEM images of the crystal silicon solar cell, the different contact modes and resistance calculation methods of the metal paste and silicon were verified. Through the resistance calculation, considered the different oxide and silver silicon contact alloy characteristics, two kinds of additive compositions were found that which may have good performance in alloy paste of the solar cell.

This letter will be helpful to the production of crystalline silicon solar cell, reasonably reduce the resistance of silver silicon contact, improve the efficiency of the crystalline silicon solar cell. At the same time, it can also helps to optimize the ratio of material in silver alloy paste, reduce the cost of PV power generation.

2. Metal paste and silicon electric contact model

2.1. Metal paste and silicon contact carrier conduction mechanism

Due to the excellent electrical properties, rich crust reserves and so on, Tin is widely used in electrical contact materials field. At the same time, Sn-Si alloy has a good performance in anti-oxidation, anti corrosion, anti welding, reduce resistance and other aspects. Therefore, the electrical contact properties of tin and silicon will be discussed further in this letter.

In order to facilitate the research, we made the 156*156mm Sn electrode crystal silicon solar cell tablet from "the Yellow River" factory as the sample.

In order to directly study the situation about the Sn-Si contact, we made a series of SEM images to the contact area of the sample, as shown in figure 1.

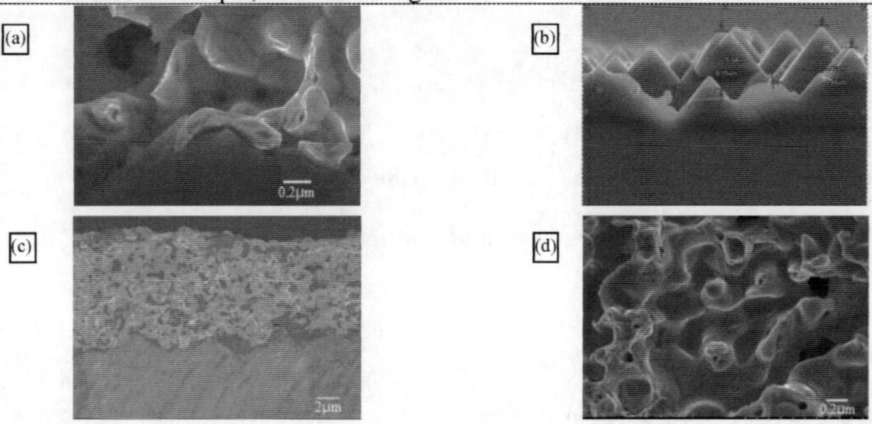

Figure 1. SEM picture to the contact area of the sample (a) 50000 times thepinhole structure (b) 10000 times pyramidal texture of the cell tabletsurface structure(c) 5000 times thick glass section profile (d) 20000 times thin glass cross section profile

According to the SEM images, three main modes of Sn-Si contact can be found. Similar to the three contact modes of screen printed silver paste and silicon contact proposed by Song-Yeu Tsai and Ching-Hsi Lin in 1982 [5].Samples in Sn-Si are mainly performed three kinds of electrical contact mode, the carrier also transfers through the three kinds of modes:

First, the carrier is transported through the direct contact with silicon. This is the direct contact mode of Sn-Si.[6-9]

Second, the carrier is indirectly tunneling through the thin glass layer between the Sn and Si. This is the Sn-thin glass layer-Si indirect contact mode. According to the concept of Tsai Ching-Hsi, Lin Song-Yeu, this model should be the main mode of the screen printing paste and the silicon electric contact carrier[5].

Third, the carrier can be transported directly to the electrode through the thick glass layer which is with the not continuous silicon. This is a Sn-thick glass layer (with a discontinuous metal particle)-Si indirect contact. This contact resistance is very large, so it is not conducive to improve the PV efficiency of solar cells [10], [11].

The three contact modes between metal and silicon are formed during the sintering process, so the sintering process will directly affect the properties of the metal and silicon electrical contact.

Based on the existing research and literature, we can calculate the contact resistance of the three contact modes.

Among them, the contact resistance of Sn-Si direct contact mode can be calculated by the classical contact theory; the contact resistance of Sn-thin glass layer-Si mode can be calculated by the formula of tunneling resistance at low voltage; the contact resistance of Sn-thick glass layer (with a discontinuous metal particle)-Si indirect contact mode can be calculated through the calculation of the electrical conductivity. The electrical resistivity and conductivity under three contact modes can be deduced from Eqs. (1)~(3). [5-9]

$$\rho_c \cong (\frac{1}{E_{00}})\exp(qV_{b0} / E_{00}) \tag{1}$$

$$\rho_{c2} = \frac{h^2 d}{e}(2m\phi)^{1/2}\exp[\frac{4\pi d}{h}(2m\phi)^{1/2}] \tag{2}$$

$$\delta = 2e^2 d^2 v_p N(E_F)\exp(-2\gamma d - w / \kappa T) \tag{3}$$

After that, the total contact resistance can be obtained combined with the contact area ratio of the three modes. (According to the relevant literature research, the contact area ratio of the three contact modes is about 3:3:1. [5], [7])

The calculated value total resistance of Sn-Si sample is $R = 6.86 \times 10^{-5} m\Omega$, which is similar to the actual measured value. It proved that the calculation method of the total resistance applied.

3. Study on a new type of positive silver paste additives

3.1. The selection of additives

Because the effect of additives on the positive silver paste`s property has multiple ways, there will be a variety of factors should be considered in the selection of additives. Due to the electric performance is relatively independent and the interface resistance calculation method are analyzed, the additives is selected mainly based on the electrical properties here.

According to the electrical contact barrier between metal and P type doped silicon, we selects the element as the additive with the low electric contact barrier. Under the condition that the crystal structure and the properties of the alloy are not considered, the ohmic contact resistance formed between the lower electric contact barrier is lower. This judgment can be directly proved by common sense: the electrical contact barrier of the daily known good electrical properties metal will be lower than it of the bad electrical properties metal [7]. In addition, also have to consider the scarcity degree of elements and the market price of oxide. Like gold and other similar elements, although the electrical contact barrier is low, the output around the world is not enough or the price is too high. This makes it not conducive to reduce the cost of solar cell production. Finally, in order to facilitate the test, we also have to consider if there is the relevant oxide particle product with high purity, suitable size and shape. Otherwise it will be difficult to carry out the research. Finally, 9 metal oxide products are selected as the modified additive component. They are SnO_2, PbO, Nb_2O_5, ZrO_2, TiO_2, CeO_2, PdO, Er_2O_3, RuO_2.

3.2. Contact resistance calculation

In order to further explore their electrical properties as well as a reference for the experimental verification, this letter calculates the expected contact resistance of the nine elements.

Although the elements are different, the electrical contact is still mainly by direct contact, Sn-thin glass layer-Si indirect contact, Sn-thick glass(with discontinuous tin particles)-Si indirect contact modes. Therefore, the tin silicon contact model can be directly applied to the calculation of ohmic contact resistance of other elements.

According to the electrical contact barrier between different metal and P type doped silicon, the total contact resistance of different metal electrode crystal silicon solar cell can be calculated by the method mentioned earlier. The results are showed in table 1.

Table 1. Predicted total contact resistance of different metal electrode

	($m\Omega$)		($m\Omega$)		($m\Omega$)		($m\Omega$)
Sn	$6.89*10^{-5}$	Pb	10.78	Nb	12.25	Zr	20.96
Ti	15.70	Ce	153.38	Pd	$1.171*10^{-28}$	Er	152.24
Ru	$1.189*10^{-15}$						

According to the calculation results, the relationship between the total contact resistance of different metal electrode crystal and the electrical contact barrier between different metal and P can be embodied by Matlab, as shown in figure 2.

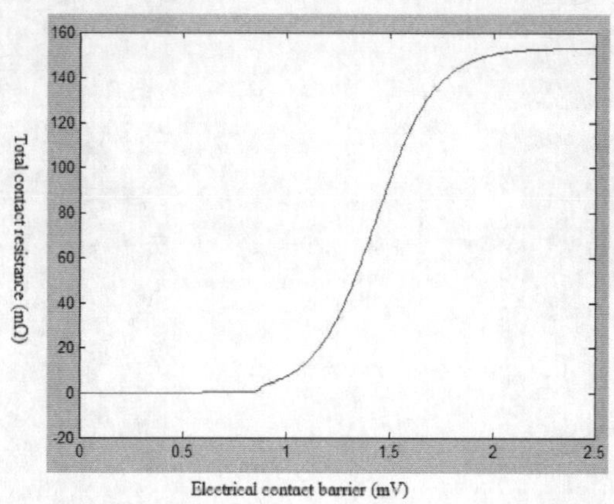

Figure 2. Relationship between contact resistance and electric contact potential barrierbetween the metal and the P type doped silicon

In the figure, the contact resistance is mainly generated when the electric contact barrier is in the range of [5], [10]. It can be learned that, as long as the electrical contact barrier is maintained at less than 1, the total contact resistance and other electrical properties will not generate a large float.

According to the calculated results, the electrical properties of several elements in accordance with the order of Pd, Ru, Sn, Pb, Nb, Ti, Zr, Er, Ce, respectively, the electrical properties of the order decreased. Pd, Ru, Sn, as the best electrical properties of several kinds of metal, can be considered.

3.3. Analysis of crystal properties of alloy

The electrical properties of metals have been discussed above. However, this is a discussion on the premise of not considering the different crystal structure and the special alloy phase. In order to make a more detailed study, further microstructure and properties of the alloys was analyzed.

3.3.1. Ag/Sn alloy At present, there are many researches on the properties of silver tin alloy. Four kinds of micro morphology of SnO2 in Ag-Sn alloy are showed in figure 3.[12]

SnO_2particles have four different morphologies. Each kind of SnO2 particle will form different kind of alloy. With the addition of different metal, sintering process and temperature, the properties of the alloy are also different. Therefore, there are many researches on the Ag-Tin electrical contact materials.

At the present stage, the research is mainly focused on the improvement of metal oxide additives in Ag-Tin alloy. In order to improve the characteristics of alloys in oxidation, corrosion resistance, welding resistance and so on. The appearance of these characteristics is closely related to the crystal structure of the alloy and the contact interface. The appearance of these characteristics is closely related to the crystal structure of the alloy and the contact interface. For example, some of the alloy phase will only form at a fixed temperature. So in the sintering process, it must reach or exceed the temperature and maintain for a certain time.

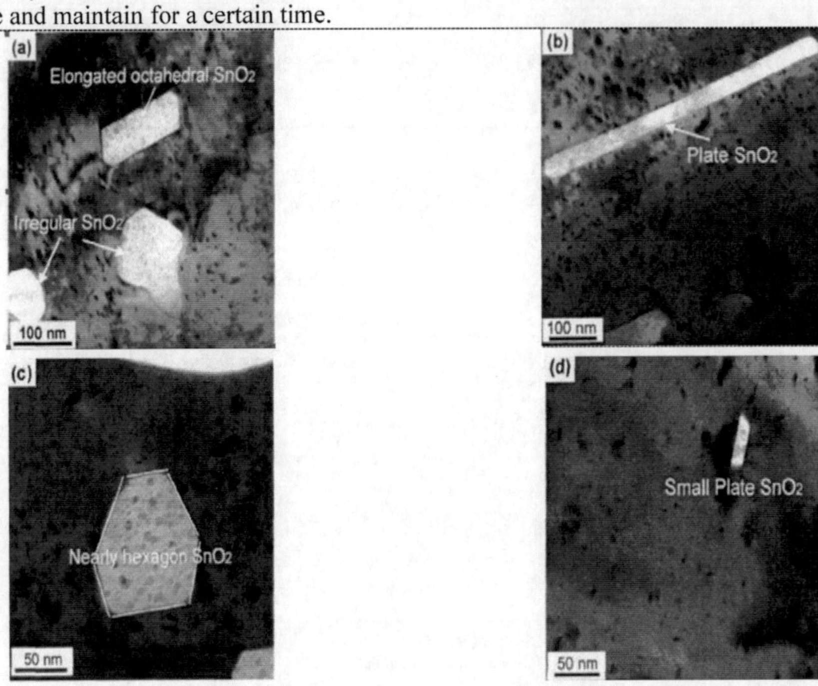

Figure 3. Four kinds of micro morphology of SnO2 in Ag-Sn alloy (a) Long octahedron (b) Long flake form (c) Pseudo - hexagonal (d) Short flake form [12]

3.3.2. Ag/Ru alloy Compared with Sn and other electrical contact material, Ru has no special place in resistance and anti welding etc. However, Ru will form a special alloy with Ag[13]. The micro morphology photo of Ag-Ru alloy are showed in figure 4.

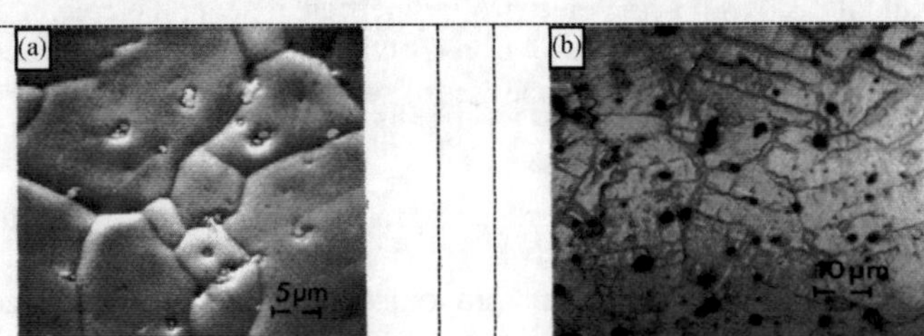

Figure 4. Micro morphology photo of Ag-Ru alloy with different magnification times (a)2000 (b) 1000 [13]

The contact resistance of the Ag-Ru alloy is very small, even in the humid environment the alloy is rusty and oxidized. At the same time, even if the contact force is very small, the contact resistance can still be kept in a small value, as shown in figure 5. [14]

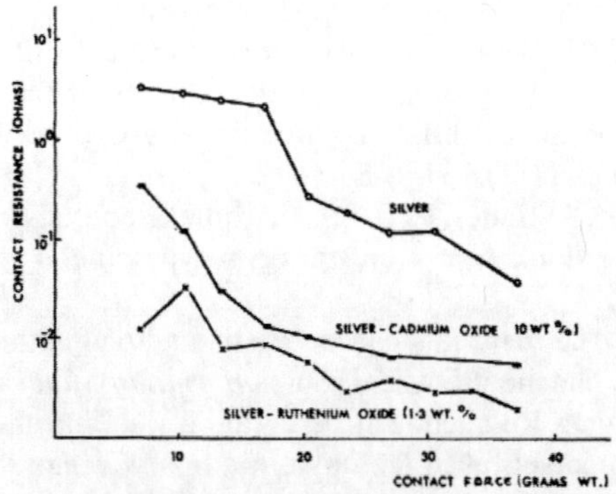

Figure5. The relationship between contact resistance and contact force in the humid environment after 21 hours [14]

This special property of Ag-Ru alloy has special value in actual production. The contact resistance of most electric contact material will increase rapidly because of rust [14].It caused that a lot of electrode contacts will be difficult to use, or cannot be used when they are rusted. Some of them can even caused some danger. As a result, the service life and reliability of these electrical contacts will be reduced, affecting the quality of the products. If the electrical contact material is made of Ag-Ru alloy, the electrical properties of the material will be kept in a damp, rusty and little contact force condition [14]. This will greatly extend the life, the scope of application and the reliability of the electrical contact.

4. Conclusions and outlook

Based on the analysis of the conductive mechanism of crystalline silicon solar cells, the effects of different additions on the properties of nano silver alloy pastes were discussed.

Based on the calculation of resistance and the characteristics analysis about the alloy, it is concluded that Ag-Sn and Ag-Ru alloy may have good performance in alloy paste of the solar cell. However, the actual production and detection of the alloy paste were deficient about the two kinds of additives. In addition, the influence of sintering process is not discussed in this letter either. The next step is to test actual performance about those two kinks of additives in alloy paste of the solar cell, and to discuss the effect of sintering process on the performance.

With the rapid development of PV industry, the demand for silver paste, which has excellent electrical conductivity, printing adaptability and aging resistance, is growing rapidly. Therefore, it is necessary to continue optimizing the alloy paste of solar cells. It will be helpful to reduce the consumption of Ag and further reduce the cost of PV power generation.

References

[1] Zhang YP, Yang YX, Zheng JH 2008 Investigation of Ag-bulk/glass-phase/Si heterostructures of printed Ag contacts on crystalline Si solar cells *Sol Energy Mater Sol Cells* **92** 1011–15

[2] Pysch D, Mette A, Filipovic A, et al. 2009 Comprehensive analysis of advanced solar cell contacts consisting of printed fine - line seed layers thickened by silver plating *Progress in Photovoltaics: Research and Applications* **17(2)** 101-114

[3] Green M A 2011 Ag requirements for silicon wafer - based solar cells *Progress in Photovoltaics: Research and Applications* **19(8)** 911-916

[4] Cabrera E, Olibet S, Glatz-Reichenbach J, et al. 2011 Current transport in thick film Ag metallization: Direct contacts at Silicon pyramid tips *Energy Procedia* **85** 40-545

[5] Yang Hong, Wang He, Lei Xiandao, Chen Chuanke, Wang Minqiang 2014 Interface modelling between the printed thick-film silver paste and emitter for crystalline silicon solar cells *International Journal of Numerical Modelling: Electronic Networks, Devices and Fields*

[6] Schroder DK , Meier DL 1984. Solar cell contact resistance—A review, *IEEE Transactions on Electron Devices IEEE Transactions on Electron Devices* **13(2)** 121-134
[7] Vinod PN 2011 Specific contact resistance measurements of the screen-printed Ag thick film contacts in the silicon solar cells by three-point probe methodology and TLM method *Materials in Electronics* **11(3)** 101-108
[8] Kontermann S , Willeke G , Bauer J 2010 Electronic properties of nanoscale silver crystals at the interface of silver thick film contacts on n-type silicon *Applied Physics Letters* **16(2)** 156-159
[9] Yang H , Wang H 2011 The materials characteristic and degradation of cell efficiency for solar grade silicon made by the metallurgical route *Journal of Materials Science* **14(2)** 132-141
[10] Cabrera E, Olibet S, Glatz-Reichenbach J, et al. 2011 Current transport in thick film Ag metallization: Direct contacts at Silicon pyramid tips? *Energy Procedia* **8**5 40-545
[11] Hörteis M 2009 Fine-line printed contactson crystalline silicon solar cells *University of Konstanz*
[12] Qiao Xiuqing 2013 Morphology control and modification of Sn02 and its application in Ag based electrical contact materials *Zhejiang University*
[13] Hermana D M 2008 Microstructure and properties of a silver–erbium oxide alloy *Journal of Alloys and Compounds* **454(454)** 292–296
[14] Pedder D J 1976 Silver-Ruthenium Dioxide Contacts *Electrocomponent Scienceand Technology* **2** 259-261

Data analysis of the high frequency surface wave radar during typhoon Chan-hom

Cheng Li[1], Hui Wang[1], Jia Gao[1, 4], Huan Li[1], Guosong Wang[1], Song Pan[1], Wenjing Fan[1], Kexiu Liu[1], Chen Zhao[2] and Anxiang Qi[3]

[1] National Marine Data and Information Service, Tianjin, 300171, China
[2] Wuhan University, Wuhan, 430070, China
[3] East China Sea Branch of S.O.A, Shanghai, 200136, China

E-mail: licheng_ouc@163.com

Abstract. Multi-frequency high frequency radar with small circular array was deployed in Zhujiajian and Shengshan to detect the winds, waves, and currents in the overlapping area operationally in 2007. No. 1509 typhoon Chan-hom landed in the coastal areas of Jujiajian, and then moved north by east, passing the radar detection area. This paper compared the radar observed data to anchored-buoy observed data and ASCAT satellite remote sensing large area wind data respectively. The results of comparison indicated that radar basically reflected the real distribution of wind and current of Zhoushan area during typhoon Chan-hom, showing the radar is qualified to detect the winds and currents under complex marine conditions.

1. Introduction

Multi-frequency high frequency radar with small circular array [1] makes use of the specialty that vertical polarized high frequency electromagnetic wave has a lesser attenuation when spread in ocean surface to detect sea surface dynamic parameters, such as winds and currents, and low velocity moving objects at sea beyond visual range [2]. In contrast to traditional marine detection equipment, radar has the advantage of wide coverage, good real-time character, all-weather and cheap to operate. Radar works at different frequency simultaneously, which greatly increase the sea surface dynamic parameters of one echo and has better inversion results, give consideration to long ranging and high detection accuracy. In addition, variant frequency radar has a strong self-correcting capability, anti-electromagnetic interference, and more stable via data fusion [3].

No. 1509 typhoon Chan-hom, which reached super typhoon grade, was formed at 8 PM on 30 June, 2015, and then moved northwestward. It landed in Zhujiajian Town, Zhoushan City, Zhejiang Province at about 16:40 on 11 June, and moved north by east after its landing, passing by Yellow Sea and close to the Korean Peninsula, with its intensity decreasing. In this case, the typhoon center passed by radar detection area, continuous observation data under extreme marine environment was obtained by the radar. Part of the observation data is displayed in this paper, comparison and analysis of radar observation data to buoy observation data and the first Advanced Scatterometer (ASCAT) satellite remote sensing are carried out to analyze the accuracy of detected hydrologic parameters and evaluate the capability of the radar.

2. Data comparison with buoy data

The small array high frequency conversion radar is formed by two intermediate range high frequency radars, located at Sitiao (122.4275° E, 29.8931° N) and Shenshan (122.8358° E, 30.7019° N) respectively, with a distance of about 100 km. The radiation angle of each radar is 90°, as the result of terrain limitation. The radar, with a measuring range of 200km and 0.05° (longitudinal and latitudinal), covers the area south by east to Zhoushan Islands and sweeps once per 10 minutes. A 3-meter integrated anchored-buoy, 30km from Shengshan radar station, was placed in the detection area to detect winds, currents, and waves for three month comparing observation. Figure 1 shows the locations of radar, buoy and track of typhoon Chan-hom.

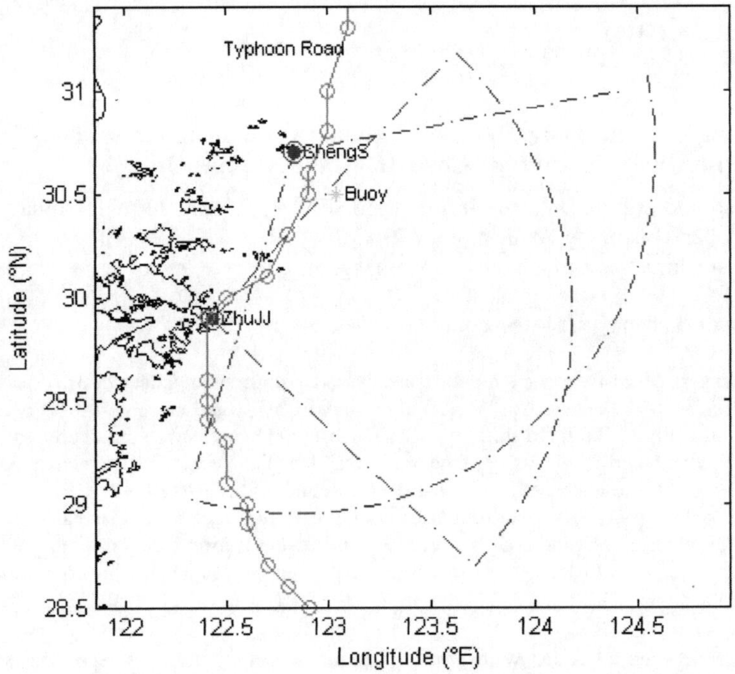

Figure 1. Locations of radar, buoy and track of typhoon Chan-hom

2.1. *Current Comparison*

An acoustic Doppler current profilers is installed beneath the buoy to observe current profiles, and the 3.1 meter depth current data is compared with radar observation data for a tidal period of 25 hours. Figure 2 shows the comparison results of radar radial flow to lytic radial current from ADCP. There is a good consistency between the radar radial currents and ADCP radial currents, showing that the radar observation accuracy is qualified. The comparison result of Shengshan radar station is better than that of Zhujiajian Station, because the buoy is nearer with Shengshan station, and the normal angle is smaller. So you can see that the closer distance from radar, the better the quality of the data, so does the normal angle [4]. The comparison results show that the radar can reflect the real marine dynamics and all of the mean absolute error, root-mean-square error, and coefficient of association are in reasonable range (Table 1). Figure 3 is the comparison result between radar synthetic vector currents and ADCP data. The results show that both the velocity and direction of the currents are of good consistency. The comparison results of flow direction show that the currents are reciprocating, conforms to the characteristic of local flow field.

2.2. Wind Comparison

Wind measuring transducer (YOUNG 05103) is installed on the top of buoy, and starts observe at whole point with a frequency of half-hour. The transducer collects data every 3 seconds as the instantaneous wind velocity and direction for 10 minutes, and then the mean velocity and direction of the 10 minutes are calculated as the mean velocity and direction at the end of the 10 minutes [5]. The comparison time of data is from 1:30, 16 June to 23:30 11 August. The actual sample number of radar data is 8041, 98.4% for the obtain rate of data, although it should be 8170.

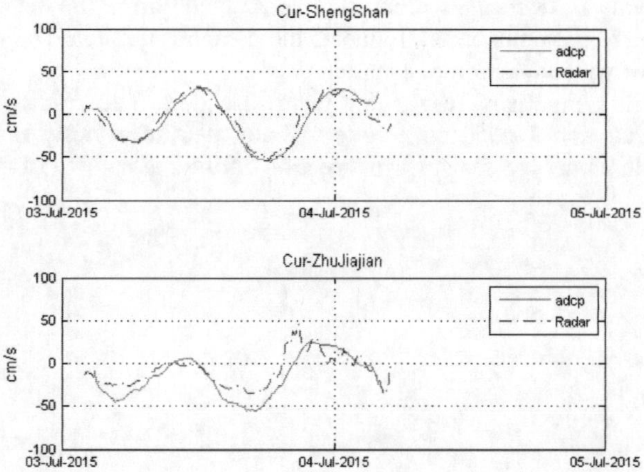

Figure 2. Radial flow comparison

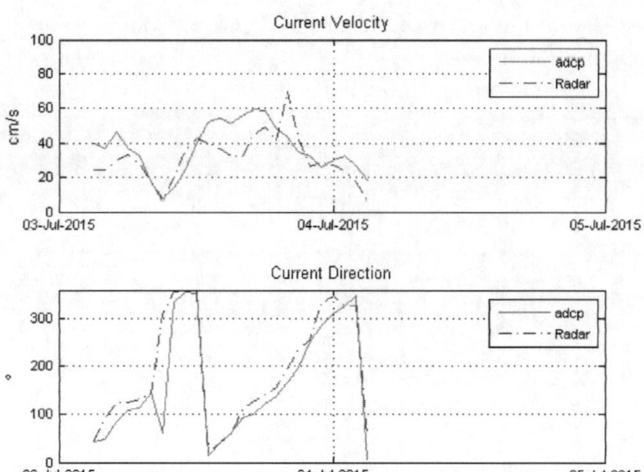

Figure 3. Vector flow comparison

Table 1. Radial flow comparison results

radial flow velocity(cm/s)	mean absolute error	Standard deviation of mean absolute error	root-mean-square error	coefficient of association
Shengshan	2.09	6.55	6.88	0.97
Zhujiajian	8.24	12.19	13.71	0.89

Figure 4 shows the comparison results for radar observation wind data with buoy observation wind data. The results show that the radar obtained the complete records of wind velocity variation trend during the time that typhoon passing the detection area, especially the wind increase and decrease process from 7:00, 9 July to 2:30 12, July 2015. Among the detection time series, the maximum observation wind velocity is 18.5 m/s, which occurs not much during the detection time series, so the root-mean-square error of 3.26 m/s basically meets the demand of radar. The error indicates the radar has a good capability of wind detection to a certain degree.

Figure 5 shows the comparison results for wind direction. The results indicate that the radar observation wind direction has good coherence to buoy data, especially that the radar data better reflects all the changing process of sea surface direction during the landing of typhoon.

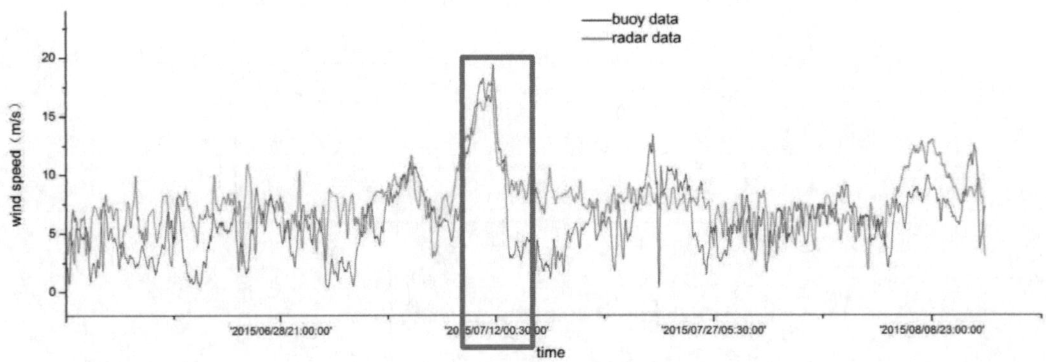

Figure 4. Wind speed comparison

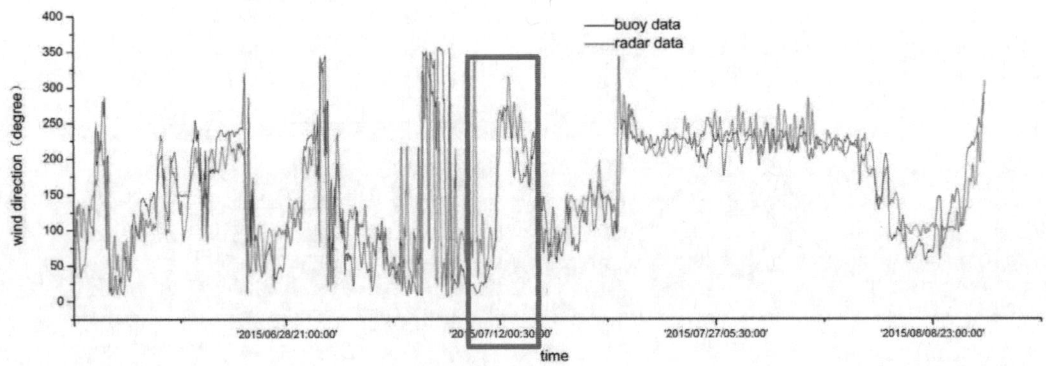

Figure 5. Wind direction comparison

The radar of 7.5~8.5M frequency band is deemed to detect wind field above 5m/s, according to radar detection theory [6]. So the wind data over 5m/s is filtered out to be compared with buoy data. The results (Figure 6) indicate the root-mean-square error is smaller than all the data, showing that the greater the wind speed is, the better radar detection capability. The wind direction is mainly northbound before the arrival of typhoon, while it turned to northwest from west when typhoon passing by, as the buoy was on the eastern way of typhoon. The wind direction observed at the buoy position turned to southeast along with the typhoon heading further north. The observed wind direction is a well matched with the wind shift characteristics of typhoon in northern hemisphere.

In Figure 6, radar observed wind direction changes lags behind the buoy observed wind direction changes, because the buoy directly obtain observed data via transducer, while radar needs to invert ocean scattering echo, which cost some time, to obtain wind direction data.

Overall, radar observed wind speed is greater than buoy observed wind speed, the mean absolute error, root-mean-square error radar are in reasonable range (Table 2). Also the radar can basically reflect the real wind change process and can meet the performance requirements.

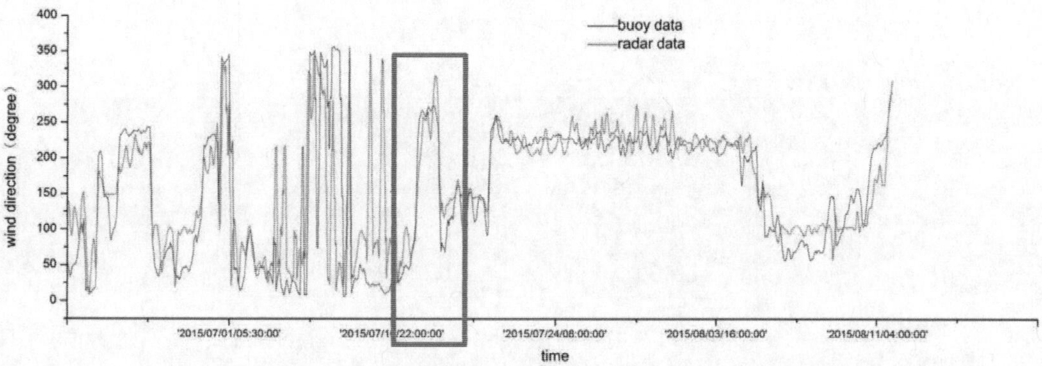

Figure 6. Wind speed comparison (when wind speed greater than 5m/s)

Table2. Wind element errors

Wind element	mean absolute error	Standard deviation of mean absolute error	root-mean-square error	coefficient of association
Wind speed(m/s)	1.60	2.84	3.26	0.53
Wind direction(°) (wind speed 0m/s)	9.36	37.60	38.75	0.66
Wind direction (°) (wind speed >5m/s)	9.43	25.99	27.00	0.66

3. ASCAT satellite wind data comparison

More wind data comparison between radar and satellite is necessary, since the fixed point observation is limited by its small coverage and low spatial resolution.

The first Advanced Scatterometer (ASCAT) was launched on the EUMETSAT MetOp-A satellite in October 2006. It became fully operational in May 2007 and continues to operate today. Scatterometers are essentially radars that transmit microwave pulses down to the Earth's surface and then measure the power that is returned back to the instrument [7]. This "backscattered" power is related to surface roughness. For water surfaces, the surface roughness is highly correlated with the near-surface wind speed and direction. Hence, wind speed and direction are retrieved from measurements of the scatterometer's backscattered power.

Figure 7 and Figure 8 show the large area wind distribution inverted by radar and ASCAT satellite wind data. As can be seen from the charts, radar inverted wind field agrees with ASCAT satellite wind data basically, mirrors the wind field distribution during the typhoon, indicates that the radar is qualified to detect the winds under complex marine conditions and extraordinary weather disasters.

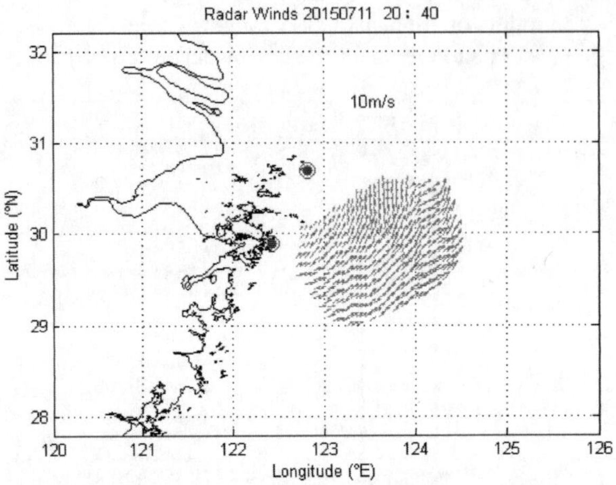

Figure 7. Wind distribution observed by radar

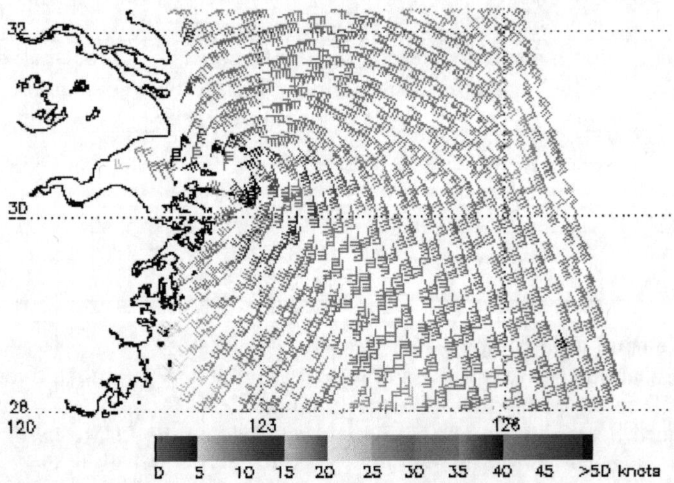

Figure 8. ASCAT wind distribution 20150711-20:40 (NOAA/NMSDIS/Center for Satellite Applications and Research)

4. Conclusions

The winds and currents data obtained by radar during typhoon Chan-hom are analyzed in this paper. In total, radar is qualified to observe winds and currents by the comparison with buoy observed data and ASCAT satellite wind data, proving that radar can detect the winds under complex marine conditions and extraordinary weather disasters.

Three methods of marine surveys are compared in this paper, radar has its advantage over fixed buoy and satellite. Buoy observation is for fixed point, and, cannot meet the demands of mesoscale to small scale marine observation, because it is limited by its small range and weak resistance to destroy. Satellite remote sensing data has the advantage of high sampling rate and wide coverage [8], but there are drawbacks, such as it cannot detect for a continuous long time in a particular area and data loss exists, as a result of the limitation of satellite orbit and period. In addition, satellite observed wind

IOP Conf. Series: Earth and Environmental Science **52** (2017) 012019 doi:10.1088/1755-1315/52/1/012019

speed is valid only if the speed is less than 24m/s, or its accuracy decreases when the typhoon is strong enough. High frequency surface wave radar is able to do large area, real-time, long-term marine dynamic parameters detection, it is the foundational tool of realizing marine real-time monitoring plane, can make up the deficiency of the conventional marine monitoring of China, improve abilities of marine environment real-time monitor in China Sea region.

References

[1] Leigh B 1998 HF Radar-The Key to Efficient Wide Area Maritime Surveillance *EEZ Technology, Edition 3* London: ICG Publishing LTD 115-118.

[2] Wait J R 1966 Theroy of HF ground wave backscatter from sea waves *JGR* **71** 4839-4842.

[3] Moore R K Fung A K 1979 Radar determination of winds at sea *Proc IEEE* **67** 1504-1521.

[4] Lipa B, Barrick D 1986 Extraction of sea state from HF radar sea echo: Mathematical theory and modeling *Radio Sci* **21(1)** 81-100.

[5] Heron M L and Rose R J 1986 On the application of HF ocean radar to the observation of temporal and spatial changes in wind direction *IEEE Journal of Oceanic Engineering* **11(2)** 210-218.

[6] Wu X B, Li L, Li Y, et al. 2012 Experimental Research on significant wave height detecting with RADAR OSMAR071 *Oceanologia Et Limnologia Sinica* **43(2)** 210-216.

[7] Bartalis Z, Wagner W, Naeimi V, et al. 2007 Initial soil moisture retrievals from the METOP-A advanced scatterometer (ASCAT) *Geophysical Research Letters* **34(20)** 122-122.

[8] Figasaldaña J, Wilson J J W, Attema E, et al. 2002 The advanced scatterometer (ASCAT) on the meteorological operational (MetOp) platform: A follow on for European wind scatterometers *Canadian Journal of Remote Sensing* **28(3)** 123-127.

Numerical analysis of the wake of a 10kW HAWT

S G Gong[1], Y B Deng, G L Xie and J P Zhang

School of Mechanical Engineering, Xiangtan University, Xiangtan, Hunan, 411105, China

E-mail: gongsg@xtu.edu.cn

Abstract: With the rising of wind power industry and the ever-growing scale of wind farm, the research for the wake performance of wind turbine has an important guiding significance for the overall arrangement of wind turbines in the large wind farm. The wake simulation model of 10kW horizontal-axis wind turbine is presented on the basis of Averaged Navier-Stokes (RANS) equations and the RNG k-ε turbulence model for applying to the rotational fluid flow. The sliding mesh technique in ANSYS CFX software is used to solve the coupling equation of velocity and pressure. The characters of the average velocity in the wake zone under rated inlet wind speed and different rotor rotational speeds have been investigated. Based on the analysis results, it is proposed that the horizontal spacing between the wind turbines is less than two times radius of rotor, and its longitudinal spacing is less than five times of radius. And other results have also been obtained, which are of great importance for large wind farms.

1. Introduction

Flow analysis of wind turbine wake has been one of the important subjects in the field of aerodynamics. Researchers have studied the influence of wind turbine wake on wind turbine design and performance [1-2], for example, in large offshore wind farm, average power loses due to wind turbine wakes are of the order of 10 to 20% of total power output. The power of wind turbine can be impacted by the velocity distribution of wind turbine wake zone. Specially, the near-wake velocity will affect the aerodynamic loading on the blade, and the far-wake velocity will have direct effect on the output power of downstream wind turbines [3].

The wake distribution of Horns Rev large offshore wind farm was simulated and analyzed by computing through the computational fluid dynamics (CFD), and further compared with measures data of wind farm have been finished. The research results indicate that wind farm models require modification to reduce under-prediction of wake losses while CFD models typically over-predict wake losses [4]. Sarlak, et al [5] studied the wake and power characteristics of a horizontal-axis wind turbine in a wind tunnel, and their results demonstrate how the blockage effect increases with the tip speed ratio. Sarmast, et al [6] has investigated the wake interaction between two model scale wind turbines with span-wise offset by using Large Eddy Simulation. The results show that the rotor characteristics for both rotors are well captured numerically even if the downstream rotor operates into stall regimes. The near-wake velocity field of a model horizontal-axis wind turbine (HAWT) has been calculated with different tip speed ratios, and the distributions of the near-wake velocity behind the HAWA rotor were obtained [7]. Wang, et al [8] has studied the wake of wind turbine under the stochastic wind field, and indicated that the wind turbine wake tip vortex line appeared serious distortion and asymmetry. Zhou, et al. [9] employed Weissinger-L lifting surface model with a time-marching free vortex methodology to compute the effect of steady wind shear and transient dynamic inflow wind on the wake structure and performance characteristics of a HAWT rotor. In experimental aspect, Hu, et al.

[10] utilized wind tunnel experimental model to study the influence of the different incoming wind speed and turbulence intensity on the wind turbine wake. Gao, et al. [11] have carried out the measure of velocity field in near wake of HAWT by using cross correlation Particle Image Velocimetry (PIV) and axis-coder positioning period sampling system at wind tunnel opening. Zhang, et al. [12] have collected the wake stream data as far as 4.5 times of the rotor diameter in wind turbine downstream by using TR-PIV, and found that the dissipation speed of the tip vortex is faster than the center vortex and the leapfrogging phenomenon of the tip vortex. Lignarolo, et al. [13] have investigated the vortical structures of the wake of a HAWT model by stereoscopic PIV, and their research indicated that the wake instability has a strong impact on the momentum deficit recovery of the wake. Many research results about the wind turbine wake and performance have acquired, but for a specific wind turbine, the wind turbine wake and its performance need to be further confirmed.

In this paper, the wake flow for 10 kW HAWT will be simulated by using sliding mesh technology and RNG k-ε turbulence model that apply to rotate fluid analysis. The wake flow field of wind turbine under different wind rotor velocity is studied. The results obtained are considered to be highly helpful in the design and installation of 10kW HAWT.

2. Computing Model

The design parameter of 10kW HAWT is given in the Table 1, and the geometric parameter of blades are shown in the Table 2.

Table 1. design parameter of HAWT

Rated power (kW)	Rated wind speed(m/s)	Cut-in wind speed(m/s)	Cut-out wind speed (m/s)	Number of blades	Rotor diameter(m)	Rated rotor speed (rpm)
10	10	3.2	16	3	8	180

Table 2. geometric parameter of blades

Section	Span(m)	Chord(m)	Torsional angle	Profoil
1	0	0.2	14.6413	Cylinder
2	0.4	0.2	14.6413	Cylinder
3	0.8267	0.5298	14.6413	SG6040
4	1.0533	0.4763	12.0487	SG6040
5	1.2800	0.4169	9.6188	SG6040
6	1.5067	0.3665	7.6366	SG6040
7	1.7333	0.3257	6.0381	SG6040
8	1.9600	0.2923	4.7485	SG6040
9	2.1867	0.2649	3.6922	SG6040
10	2.4133	0.2422	2.8095	SG6040
11	2.6400	0.2229	2.0633	SG6040
12	2.8667	0.2066	1.4184	SG6040
13	3.0933	0.1923	0.8454	SG6040
14	3.3200	0.1796	0.3107	SG6040
15	3.5467	0.1671	-0.2293	SG6040
16	3.7733	0.1516	-0.9051	SG6040
17	4.0000	0.0000	0.6946	SG6040

The three blades model of 10kW HAWT obtained by using data in the Table 2 is shown as Figure 1, and the three-dimension numerical model of wind turbine is also shown as Figure 2. The model in Figure 2 is divided into 3.8 million grids, in which the mesh smoothing is achieved and the local refinement near the surface of blade is adopted. The grid model of computed field and near the surface of blade is shown as Figure 3, Figure4, respectively.

The boundary conditions are given as follow: Inlet velocity is v=10m/s. The rotor rotational speed of blade is 180rpm, and outlet pressure is the average static pressure. The surface of blade is set to no slipping and impermeable boundary condition. The wind flow is defined as the axis free flow under the steady condition. The convergence error is set to 10^{-5}.

For considering the influence of blade rotation effect on airfoil performance, the simulation of 3D aerodynamic performance of 10kW HAWT is achieved by using sliding mesh unsteady computing method and based on the 3D *Navier-stokes equation* and *RNG k-ε turbulence model*. The pressure implicit split operator in SIMPLE algorithm is used to solve the coupling equation of pressure and velocity. The pressure item is discretized by the PRESTO format. The overall computing time is set to 28s, and the time step is 0.0028s.

Figure 1. Geometric model of wind rotor

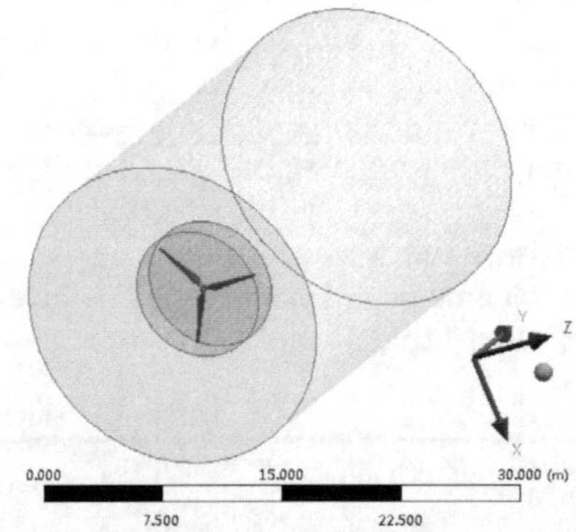

Figure 2. The computing model

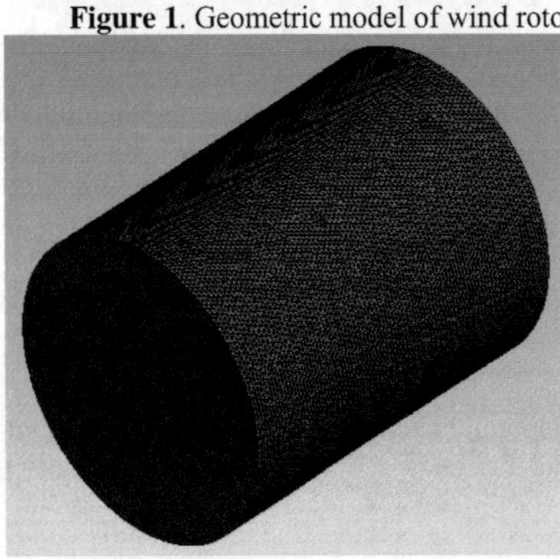

Figure 3. Overall grid of computed field

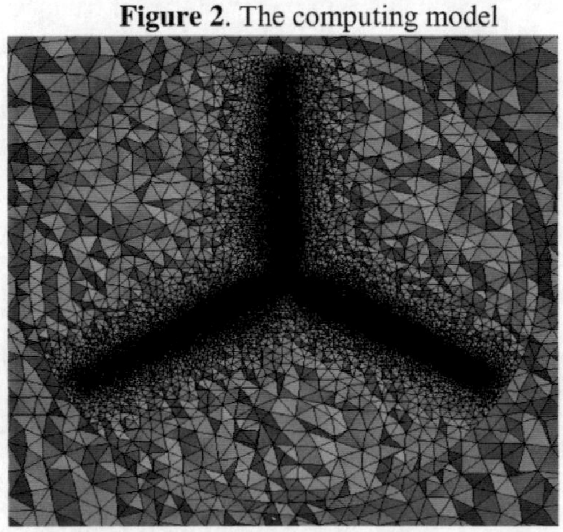

Figure 4. Grid near the surface of blade

3. Simulated results and analysis

3.1. The wake of under rated condition

The wake feature of the wind turbine is obtained by simulated computing under rated condition, that is: rated wind velocity is 10m/s and rated rotor speed is 180rpm. The average wind speed contour slice of the wake field of wind turbine in different along-wind location is shown in Figure 5.

In Figure 5, we can find that the average wind velocity contours begin to change in the 5r (r is radius of wind rotor) location of the wake region, and while the wind come to the seven times rotor radius location, the distribution of velocity contour have a great change. When the distance of wake reaches 9*r*, the contour distribution of wind velocity is close to the one of incoming flow.

The curve of cross-wind average wind speed in the different location of wake field is shown as Figure 6. The cross-wind average wind velocity in the 2*r* and 3*r* location of the wake field has a sharp change. The minimum cross-wind average wind speed in the center of wind rotor is 0.1 times of inflow wind speed, and it is mainly concentrated in range of +/- 1.5*r*. The main reason of change is that the flow is affected by the wind wheel rotation and wheel hub. As the wake area increases to 6*r*, the cross-wind involved area of the wake have been expanded upward to +/- 2.5*r*, but the minimum cross-wind velocity increase gradually 0.8 times of inflow wind. When the wake region continues to increase to 10*r*, the average cross-wind velocity is close to the inflow wind speed, that is, the flow field of whole area has been basically no longer affected by the wind wheel rotation.

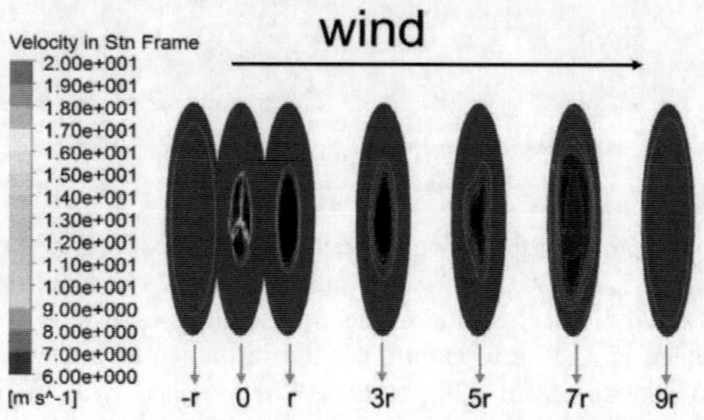

Figure 5. The average wind speed contour slice in the wake field along-wind location

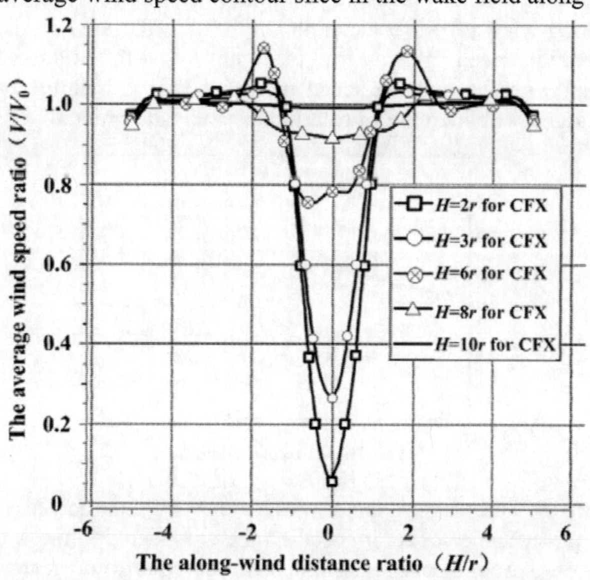

Figure 6. The average wind speed of cross-wind in the wake area

Based on the above analysis, we can obtain that the arrangement between wind turbines in the wind farm considers not only the situation of local wind resource, and also think about the influence of horizontal spacing and longitudinal spacing between the wind turbines, so as to reduce the impacts of the wake on its two sides and downstream wind turbine. Thus, we suggest that the horizontal spacing and longitudinal spacing between the wind turbines is less than 4*r* and 10*r*, respectively, in the installation of 10kW HAWT.

3.2. The wake under different wind rotational speed

In order to study on the influence of wind wheel rotation speed on the wake flow and aerodynamic performance of wind turbine, while the velocity of incoming flow is 10m/s, and the velocity of rotation is 80rpm, 120rpm, respectively, the simulation analysis based on the model shown in Figure 2 is achieved. The change curve of the along-wind average wind speed in the wake area is obtained, and the results are shown in Figure 7.

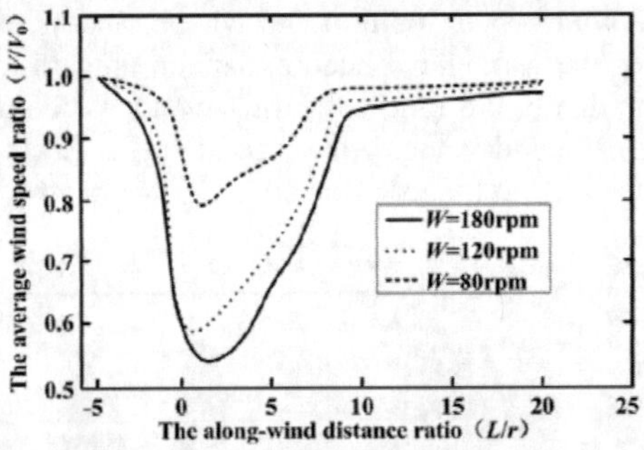

Figure 7. The curve of the along-wind average wind speed in wake area

As we can see in Figure 7, the average wind speed in the same location of the wake area is decreased obviously as the wind wheel rotation velocity increase, especially from 80rpm to 120rpm, the greater the rotation speed is, the more significant the influence of it on the rear wind turbine is. While the rotation speed increase from 120rpm to 180rpm, the growth of average wind velocity becomes slower with rotation speed increase. It indicates that the effect of the rotation speed increase on the wake flow will no longer be visible as the rotation speed of rotor reaches a certain value.

The curve of the along-wind pressure factor in the rotor wake area is shown in Figure 8 under different wind wheel rotation speed. We can find in Figure 8 that the pressure factor after and before wind wheel is increasing with the rotation speed increase. This is mainly reason why the blade tip speed ration will be enlarged with increase of rotational speed, and gradually close to its optimal value.

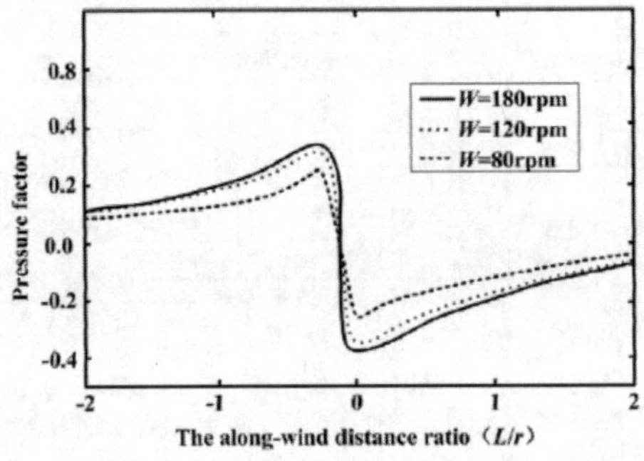

Figure 8. The pressure factor of the along-wind in the wake area

In order to quantify the influence of rotor speed on the cross-wind in the wake area, the cross-wind speed curve in the section that is 3r and 8r distant from the wind rotor in the wake area is obtained, respectively, as shown in Figure 9 and Figure 10.

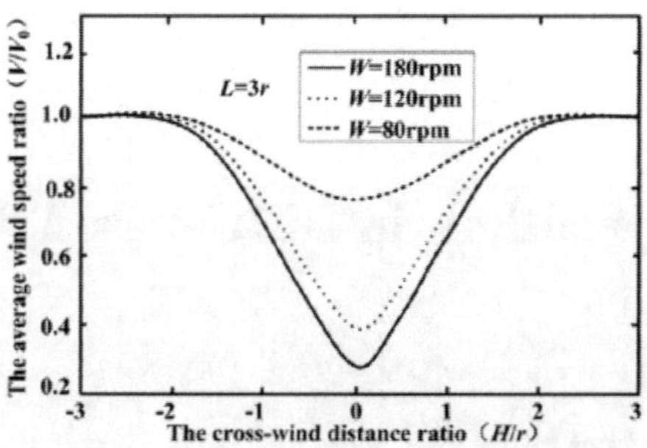

Figure 9. The cross-wind average speed as the distance is 3*r*

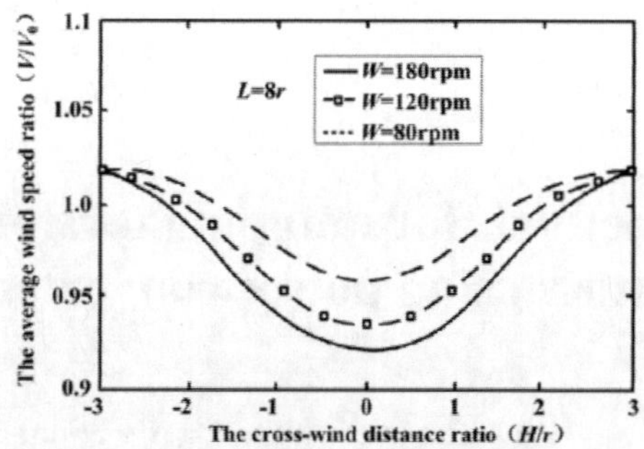

Figure 10. The cross-wind average speed as the distance is 8*r*

As shown in Figure 9 and Figure 10, when the rotor rotational speed increases from 80rpm to 120rpm, that is, the tip speed ratio increases from 3.45 to 5.02, the cross-wind average speed ration in 3*r* section decline significantly, and its minimum value decrease from 0.8 to 0.4. However, when the distance between the cross-section and rotor is 8*r*, the cross-wind average speed ratio decline is not quite obvious, but the affected cross-wind range compared with 3r section is increased. The mean wind speed ration in the given section will go on decreasing as the rotor rotational speed increase to 180rpm, but it decreases at a slowly rate and the affected cross-wind range also remain unchanged.

4. Conclusion

The conclusion can be obtained by simulation analyzing of the wake flow for 10kW HAWT, and it is given as follows.

(1) Based on *Reynolds Averaged Navier-Stokes* equations and RNG k-ε turbulence model for rotational flow analysis, the wake flow analysis under the rated operating condition is achieved. Through the analysis of average wind speed in the wake field, it is proposed that the horizontal spacing between the wind turbines is less than 4*r*, and its longitudinal spacing is less than10*r*.

(2) By simulation analyzing for the wake flow under different rotor rotational speed, the influence of the rotor rotational speed on the wake flow is obtained. It is beneficial to optimize the relation of rotor rotational speed with transverse and longitudinal spacing between the wind turbines.

Acknowledgment

This research is funded by National Natural Science Foundation of China [51375417, 51405415], and the financial support to the first author is gratefully acknowledged.

References

[1] Hu Danmei, Ou Yanghua and Du Zhaohui 2006 *Acta Energi. Sin.* **27** 606-612
[2] Vermeer L J, Screusen J N and Crespo A 2003 *Prog. Aerosp. Sci.* **396** 467-510
[3] Whale J and Anderson C G 2000 *J. Wind Eng. Ind. Aerod.* **84** 1-21
[4] Barthelmle R J, Hansen K, Schepers J G, et al 2009 *Wind Energ.* **12** 431-444
[5] Sarlak H, Nishino T, Martínez tossas L A, et al 2016 *Renew. Energ.* **93** 340-352
[6] Sarmast S, Chivaee H S, Ivanell S, et al. 2014 *J. Phys. Conf.* **524** 012137
[7] Hu Danmei, Zhang Jianping 2010 *Acta Energi. Sin.* **31** 1485-1490
[8] Wang Xiao, Cao Jiufa and Wang Tongguang 2016 *Chin. J. Comput. Mech.* **33** 343-350
[9] Zhou Wenping, Tang Shengli and Lu Hong 2012 *Proc. CSEE* **32** 122-127
[10] Hu Danmei and Du Zhaohui 2009 *J. Hydrodyn.* **21** 285-291
[11] Gao Zhiying, Wang Jianwen, Dong Xueqing, et al. 2011 *Acta Energi. Sin.* **32** 897-900
[12] Zhang Shucheng, Dong Xueqing, Wang Jianwen, et al. 2016 *J.Eng. Thermophy.* **37** 1432-1437
[13] Lignarolo LEM, Ragni D, Krishnaswami C, et al. 2014 *Renew. Energ.* **70** 31-46

Two methods for damping torsional vibrations in DFIG-based wind generators using power converters

Zuyi Zhao, Yupu Lu, Da Xie[1], Songtao Yu, Wangping Wu

[1] Shanghai Jiaotong University, 200240, Shanghai, China

Email: feier199106@sjtu.edu.cn

Abstract. This paper proposes novel damping control algorithms by using static synchronous compensator (STATCOM) and energy storage system (ESS) to damp torsional vibrations in doubly fed induction generator (DFIG) based wind turbine systems. It first analyses the operating characteristics of STATCOM and ESS for regulating power variations to increase grid voltage stability. Then, new control strategies for STATCOM and ESS are introduced to damp the vibrations. It is followed by illustration of their effectiveness to damp the drive train torsional vibrations of wind turbines, which can be caused by grid disturbances, such as voltage sags and frequency fluctuations. Results suggest that STATCOM is a promising technology to mitigate the torsional vibrations caused by grid voltage sags. By contrast, the ESS connected to the point of common coupling (PCC) of wind turbine systems shows even obvious advantages because of its capability of absorbing/releasing both active and reactive power. It can thus be concluded that STATCOM is useful for stabilizing power system voltage fluctuations, and ESS is more effective both in regulating PCC voltage fluctuations and damping torsional vibrations caused by grid voltage frequency fluctuations.

1. Introduction

Due to the increasing environmental concerns regarding to electricity generation, wind energy has gained worldwide attention and taken up a big proportion in energy mix globally [1]. Among various types of wind turbines, variable speed turbines equipped with doubly fed induction generators (DFIGs) have attracted extraordinary attention because of their advantages over other wind turbines. One is that their power converters only need be rated to handle a fraction of the total DFIG power in order to achieve full control, normally below one-third [2][3]. Since a DFIG includes a wound rotor induction generator supplied on the stator side by the grid connection, grid faults such as voltage sags and frequency fluctuations can excite low frequency torsional oscillations of the mechanical components of DFIG-based wind generators. It can fatally result in electric torque ripples as well as shaft torsional oscillations [4]. Moreover, shaft torsional oscillations may cause a considerable reduction in the fatigue life of the shaft system. Therefore, it is extremely important to investigate the torsional vibrations of DFIG in order to improve the fatigue life of systems and ensure reliable wind energy integration.

Static Synchronous Compensator (STATCOM) is a shunt connected flexible AC transmission system (FACTS) device, which is commonly utilized in transmission systems to overcome the limitations of transmission distance [5]. A comprehensive study was conducted in [6] to investigate how STATCOMs could improve the steady-state stability and dynamic properties of wind farms. It is found that they are able to enhance their low-voltage ride-through capability [7] and improve transient voltage stability during grid disturbances [8]. It is also concluded that suitable controllers of STATCOM can provide auxiliary damping during power system oscillations in a series of compensated wind parks [9]. Because of its capability to increase voltage stability margins, STATCOM is a promising solution to mitigate torsional vibrations in wind turbine shafts, particularly during grid voltage disturbances. However, due to low capability of active power output, the performance of

STATCOM can be compromised when applied to wind farms with active power oscillation or bus frequency fluctuations.

Early research has shown that battery storage shunts connected to a wind energy system can avoid the dramatically change of generator voltage and rectifier voltage to a wind gust [10], [11]. Energy Storage System (ESS) is commonly installed within an electricity system to absorb the excess generation or to meet peak loads [12]. Currently, Battery Energy Storage System (BESS) is one of the most used storage technologies on the market [13], which is also employed in this paper. It has been reported that ESS can greatly contribute to improving the power quality and stability margin of a wind farm [14], [15]. Nevertheless, little research has addressed the contributions of ESS to torsional vibrations in the drive-train shaft of a wind turbine system.

In order to fill the research gap, this paper introduces the mathematic model and control strategy of the DFIG system. The principle of STATCOM is analyzed to damp the torsional vibrations. A new damping control algorithm is developed for ESS to deal with torsional vibrations caused by both voltages sags and frequency deviations. The real-time simulation implemented in Matlab/Simulink is conducted to verify the proposed damping strategy. The major contribution of this paper is: i) a model of DFIG-based wind generator suitable for torsional vibrations modeled; ii) a wind turbine equipped with STATCOM to damp torsional vibration caused by voltage fluctuations; iii) a new damping control algorithm for ESS to mitigate torsional vibrations caused by both voltage and frequency fluctuations.

This paper is organized as follows. Section 2 illustrates the principle of STATCOM to damp the torsional vibrations caused by voltage fluctuations. Section 3 introduces the model of ESS and proposes the novel control strategy of vibration damping. The case study in Section 4 shows the damping efficacy of STATCOM for torsional vibrations caused by voltage sags. Finally, conclusions are drawn in Section 5.

2. Torsional damping using STATCOM

The STATCOM is a shunt-connected FACTS device to regulate the voltage at the point of common coupling (PCC) by absorbing or generating reactive power to the grid. It works as a voltage source converter and generates a set of balanced three-phase sinusoidal voltages with rapidly controllable amplitude and phase angle.

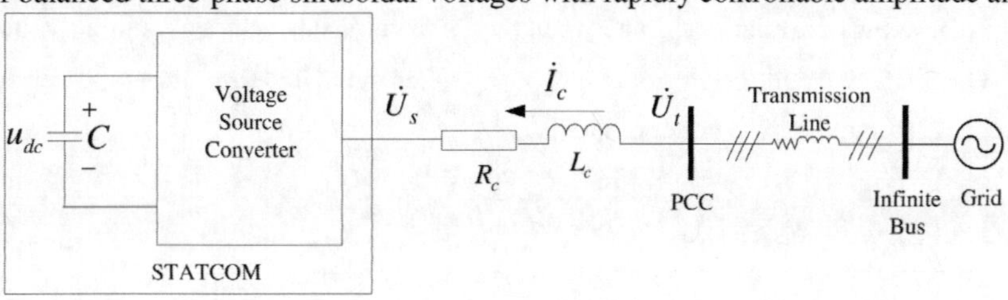

Figure 1. Schematic diagram of STATCOM system connected to the grid

Figure 1 shows the diagram of a STATCOM system connected to the grid, where U_t is the PCC voltage, U_s is the output voltage of the STATCOM, and I_c is the current through the transmission line. In normal conditions, the STATCOM output voltage U_s is synchronous with U_t under voltage control. By neglecting the resistor R_c of the transmission line, the current I_c and the reactive power Q_c exchanged between STATCOM and the grid are

$$\begin{aligned} \dot{I}_c &= \frac{\dot{U}_t - \dot{U}_s}{jX_c} = -j\frac{U_t - U_s}{X_c} \\ Q_c &= \mathrm{Im}\{\dot{U}_t \dot{I}_c\} = \frac{U_t(U_t - U_s)}{X_c} \end{aligned} \tag{1}$$

Therefore, when U_s is lower than the PCC voltage U_t, the STATCOM acts as an inductance, absorbing reactive power from the grid. When U_s is higher than U_t, the STATCOM works as a capacitor generating reactive power to the grid. In steady state, the PCC voltage always leads the output voltage of STATCOM by a small angle to supply the active power consumed due to inverter losses.

In this paper, the control objective of the STATCOM is to support the PCC voltage of wind turbine systems during grid voltage sags, which contributes to torsional vibration damping. The control loops have two objectives: 1) the q-axis loop consisting of an inner current-control loop and an outer voltage-control loop is to achieve flexible output voltage of DFIG; 2) the d-axis loop is to keep the dc-link voltage constant. The control block ensures the stable regulation of the voltages of PCC and dc link.

3. Torsional Damping Using ESS

ESS able to supply active power to the grid and active power modulation is an advantage to introduce additional damping to torsional vibration caused by frequency fluctuation of grid voltage and load variations.

Battery is considered to have high energy density and is a very promising option to realize energy storage system. The diagram of the proposed energy storage system is shown in Figure 2, which consists of two individual power electronic converters. The bidirectional DC/DC converter and PWM converter are required to link the battery and the PCC through coupled inductor X_E to make the system more compact and cost efficient. The two-stage converter device is the key component of ESS to increase grid frequency/voltage stability by absorbing or generating required active/reactive power. The DC/DC converter is responsible to boost the battery voltage and control the delivered or absorbed power, while the PWM converter is used for power factor correction.

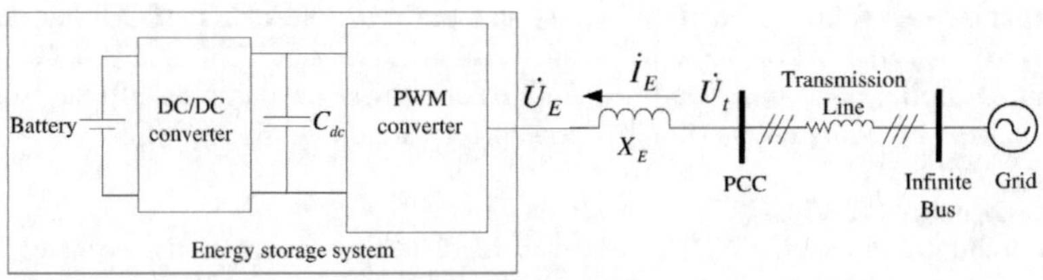

Figure 2. Diagram of Energy Storage System connected to the grid

Figure 3 graphically shows the relationship between the PCC voltage $\dot{U}_t$, the ESS output voltage $\dot{U}_E$, and ESS input current $\dot{I}_E$. The PCC voltage $\dot{U}_t$ leads $\dot{U}_E$ by δ determined by the output power of the ESS. The relationship between $\dot{U}_E$ and the PCC voltage $\dot{U}_t$ is

$$\dot{U}_E + jX_E\dot{I}_E = \dot{U}_t \tag{2}$$

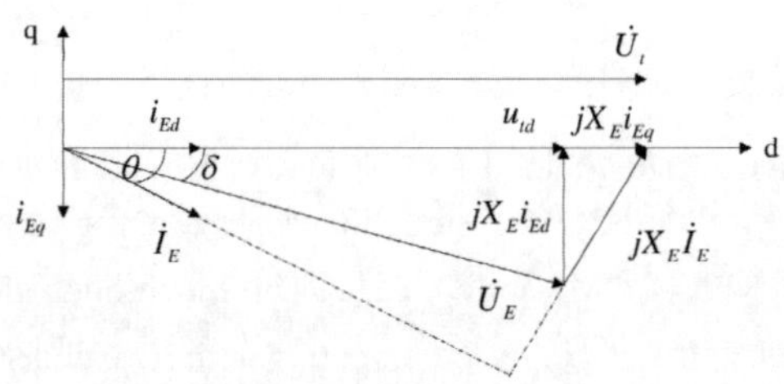

Figure 3. Phasor diagram of the operating principle of ESS

The ESS input current $\dot{I}_E$, which lags the PCC voltage by θ, is decomposed into two components: i_{Ed} in phase with the PCC voltage $\dot{U}_t$ and i_{Eq} in quadrature. The active and reactive power interchanges with the grid are also decoupled and controlled by i_{Ed} and i_{Eq}, which are

$$\begin{aligned} P_E &= u_t i_{Ed} = u_t I_E \cos\theta = \frac{u_t u_E}{X_E}\sin\delta \\ Q_E &= u_t i_{Eq} = u_t I_E \sin\theta = \frac{u_t(u_t - u_E\cos\delta)}{X_E} \end{aligned} \tag{3}$$

When $\dot{U}_E$ lags the PCC voltage $\dot{U}_t$, i.e., $\delta > 0$, the ESS acts as load absorbing active power from the grid. Under this condition, the PWM converter works in rectification mode and charges batteries through the DC/DC converter working in buck mode. When $\dot{U}_E$ leads the PCC voltage $\dot{U}_t$, i.e., $\delta > 0$, the ESS acts as a voltage source generating active power to the grid. Under this condition, the batteries discharge to the grid through the two-stage converter device. The DC/DC converter works in boost mode and the PWM converter works in inverter mode. Similarly, the reactive power exchange between the ESS and the grid can be calculated according to the relationship between u_t and $u_E \cos\delta$ .Therefore, the two-stage converter device can be controlled to realize operation in four-quadrant, and absorbs or generates the desired active/reactive power by adjusting the ESS output voltage $\dot{U}_E$.

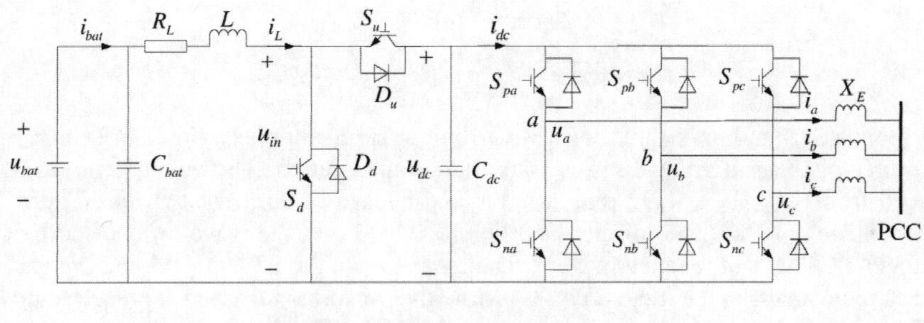

Figure 4. Circuit configuration of ESS

Figure 4 shows the circuit configuration of ESS. The DC/DC converter works in boost/buck mode by controlling the boost switch S_d and the buck switch S_u. In boost mode, the switch S_d is pulse modulated width and the switch S_u is always in off state. In buck mode, the switch S_u is pulse modulated width and the switch S_d is always in off state. The DC/DC converter is to keep the dc-link voltage constant, provided by the capacitor C_{dc}. The PWM converter is to achieve independent control of the active and reactive power exchanged between ESS and the grid. The instantaneous three-phase i_{abc} currents are sampled and transformed to d-q components i_{Ed} and i_{Eq}. The voltage balance across the inductors in voltage vector oriented d-q frame is

$$\begin{aligned} u_{td} - u_{Ed} + \underbrace{\omega L_E i_{Eq}}_{\Delta u_{Ed}} &= L_E \frac{di_{Ed}}{dt} + R_E i_{Ed} \\ u_{tq} - u_{Eq} - \underbrace{\omega L_E i_{Ed}}_{\Delta u_{Eq}} &= L_E \frac{di_{Eq}}{dt} + R_E i_{Eq} \end{aligned} \tag{4}$$

The DC/DC converter control loop consisting of an inner current-control loop and an outer voltage-control loop keeps the dc-link voltage constant. The control loops of PWM converter have two objectives: 1) the q-axis loop with inner current-control loop and outer voltage-control loop achieves flexible regulation of the PCC voltage; and 2) the d-axis loop regulates active power exchange between ESS and the grid according to the frequency fluctuations of infinite bus voltage. Therefore, the control block ensures the regulation of the PCC voltage and the required active power due to frequency fluctuations.

The advantages of ESS operation are that it is: 1) able to support PCC voltage during grid voltage sags, which contributes to the damping of torsional vibrations; and 2) expected to damp the torsional vibrations caused by frequency fluctuations in the power network by absorbing or generating active power during transient state.

4. Torsional Damping Using ESS

4.1. Simulation results of STATCOM control

Grid faults, such as voltage sags and frequency fluctuations, even if far away from the location of the wind turbine system may cause torsional vibrations in the drive train. In order to test the performance of the proposed damping approach, a wind turbine equipped with STATCOM has been modeled and simulated in the Matlab/Simulink, as shown in Figure 5. The objective of the wind turbine system is to enhance the capability of the wind turbine to ride through transient voltage sags in the grid. The STATCOM employs a 7-level series H-bridge topology. The PCC of the wind turbine connects to an infinite bus of 10kV through a 10km-long transmission line. In this case, voltage sags occurring at the infinite bus are treated as small electric disturbances to excite torsional vibrations of the drive train.

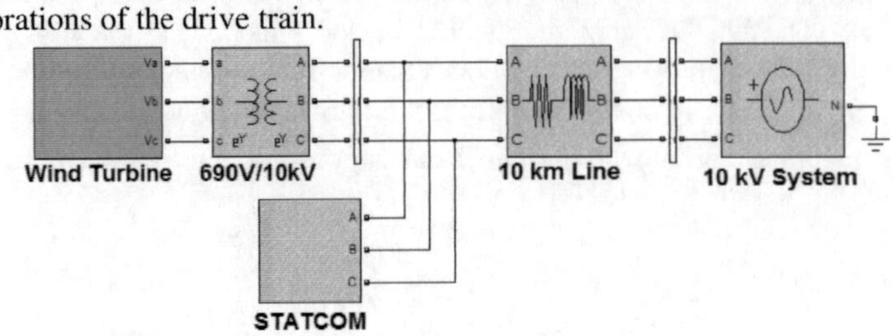

Figure 5. Simulink diagram of wind turbine system equipped with STATCOM

The case of a three-phase grid voltage sag is taken into consideration. The voltage command of the infinite bus is step changed from 1.0 p.u. to 0.92 p.u. at t=1 s and back to 1.0 p.u. at t=1.1 s. Figure 6(a) shows the waveforms of the voltages of PCC and infinite bus. Figure 6(b) shows the waveforms of current outputs of the wind turbine and STATCOM, and reactive power exchange between the STATCOM and the grid. At beginning, the system operates in normal state and the STATCOM neither absorbs nor provides reactive power to the grid. At t=1 s, the infinite bus voltage is decreased by 7% and the STATCOM current increases to compensate the voltage decrease by generating reactive power. The PCC voltage maintains at the normal level, as shown in Figure 6. At t=1.1 s, the infinite bus voltage comes back to the normal state and the corresponding reactive power drops to zero. It indicates the dependence of the PCC voltage on the voltage variation at the infinite bus due to the reactive power generated from the STATCOM.

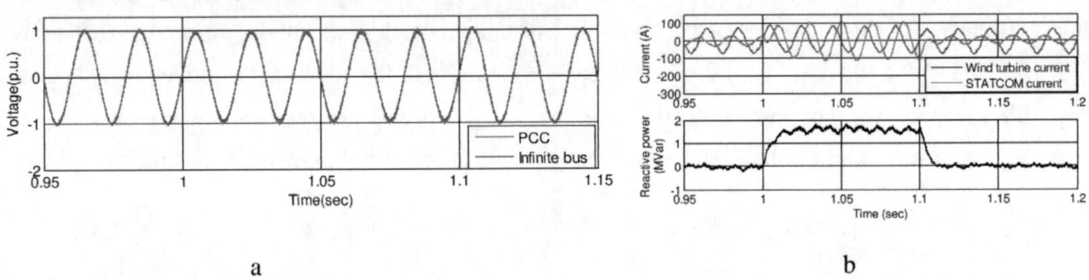

Figure 6. PCC voltage Reactive and power output of STATCOM during during grid voltage sags

Figure 7 shows the low speed shaft torque T_{lss} and the wind turbine angle speed ω_{rot} response with and without the STATCOM. The red curves denote the case without the STATCOM. The vibrations last over 4 s for both low speed shaft torque and wind turbine angle speed. The black curves representing the system equipped with STATCOM present better damping feature, where the torsional vibrations are damped effectively and rapidly in1 s.

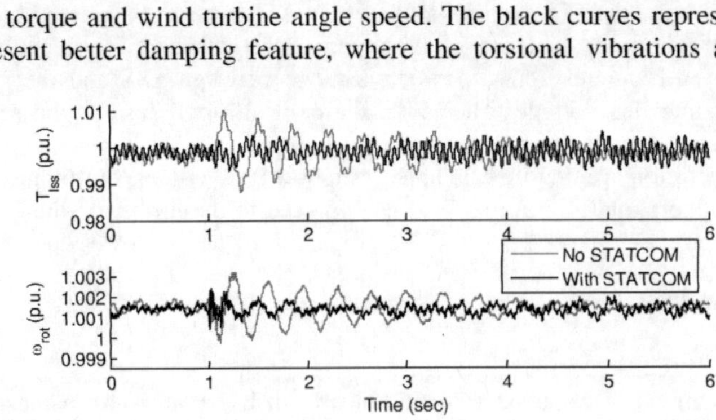

Figure 7. Low speed shaft torque and generator speed during grid voltage sag

4.2. *Simulation results of ESS control*

In order to test the performance of the proposed approach, the wind turbine system with ESS has been modeled and simulated in the Matlab/Simulink, shown in Figure 8.

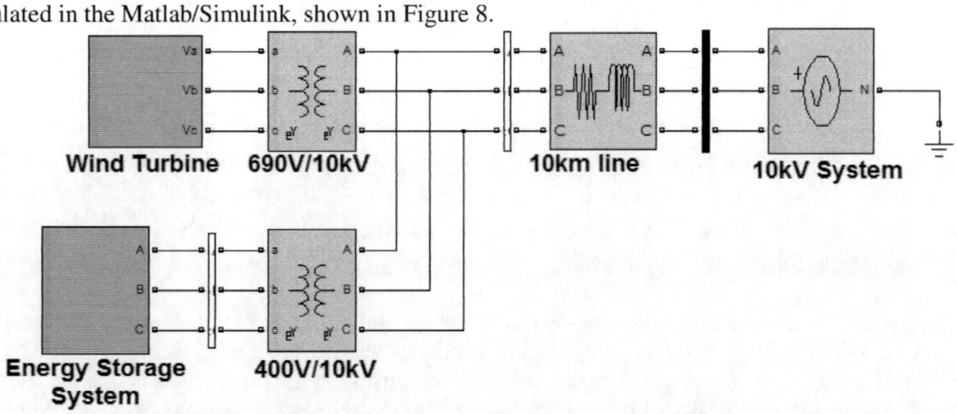

Figure 8. Simulink models of wind turbine system equipped with ESS

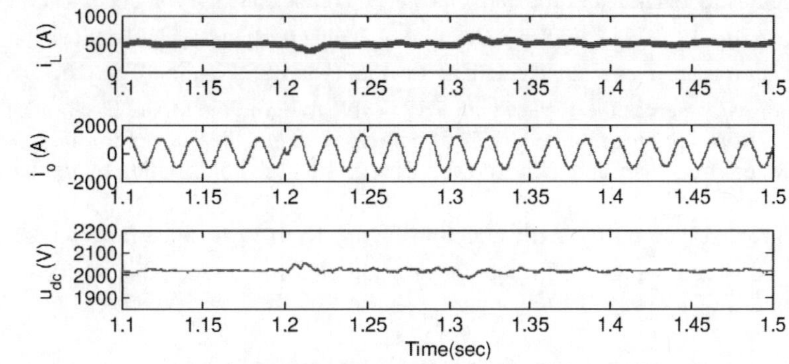

Figure 9. Response of ESS on grid voltage sags

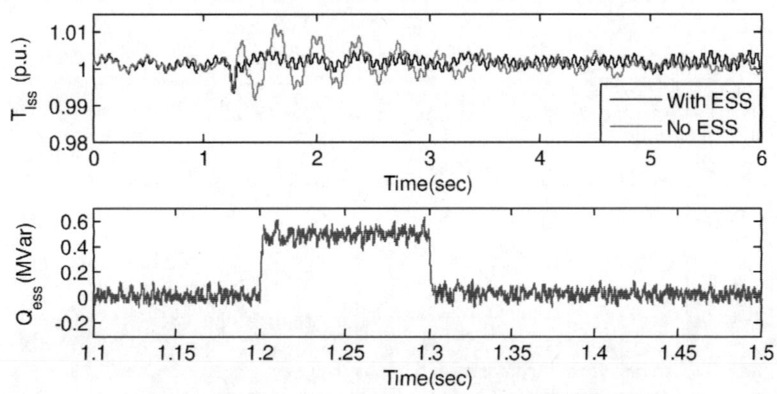

Figure 10. Torsional vibration damping effect due to reactive power output

4.2.1. Simulation analysis of voltage sags In the case of voltage sags, the voltage command of the infinite bus is step changed from 1.0 p.u. to 0.92 p.u. at t=1.2 s and back to 1.0 p.u. at t=1.3 s.

Figure 9 shows the response of ESS on the variations of infinite bus voltage. The battery output current i_L is constant during the voltage vibrations, which suggests that the voltage sags have no effect on active power exchange between the ESS and the grid. The ESS output current i_o increases to generate desired reactive power compensation for the PCC voltage sags, as shown in Figure 10. Initially, the ESS neither absorbs nor provides reactive power to the grid. At t=1.2 s, the infinite bus voltage is decreased by 8% and the ESS compensates the voltage decrease by generating reactive power Q_{ess}. At t=1.3 s, the infinite bus voltage is step changed back to the normal state and the corresponding reactive power drops to zero. Figure 10 shows the low speed shaft torque T_{lss} response with and without the ESS. Without ESS, the vibration lasts over 2 s in low speed shaft torque, whereas the system with ESS is able to damp the torsional vibrations effectively and rapidly within 0.5 s.

4.2.2. Simulation analysis of voltage frequency decrease The capability of the ESS in damping torsional vibrations caused by grid voltage frequency fluctuations is observed under conditions of frequency step changes. In the case of grid voltage frequency decrease, the frequency command of the infinite bus is step changed from 1.0 p.u. to 0.96 p.u. at t=1 s and back to 1.0 p.u. at t=1.2 s. Figure 11 shows response of ESS on variation of the infinite bus voltage. Initially, the ESS neither absorbs nor provides active/reactive power to the grid. The ESS output current i_o increases to generate desired active power to compensate the frequency decrease. Correspondingly, the batteries discharge to keep the dc-link voltage constant during the vibration. Figure 12 shows the low speed shaft torque T_{lss} response with and without the ESS. Without ESS, the vibration lasts over 2 s in low speed shaft torque, whereas the system with ESS is able to damp the torsional vibrations effectively and rapidly within 0.5 s. However, due to the features of the wind turbine, there is no droop characteristic relationship between system frequency and active power. The ESS has no contribution to regulating the frequency of the PCC voltage.

Moreover, the stator active power P_1 is illustrated in Figure 13 to verify the effectiveness of the proposed control approach. For the system without ESS, the power oscillation can be seen due to the torsional vibrations in the drive train, whereas in the case of system equipped with ESS, the power vibrations have been effectively suppressed.

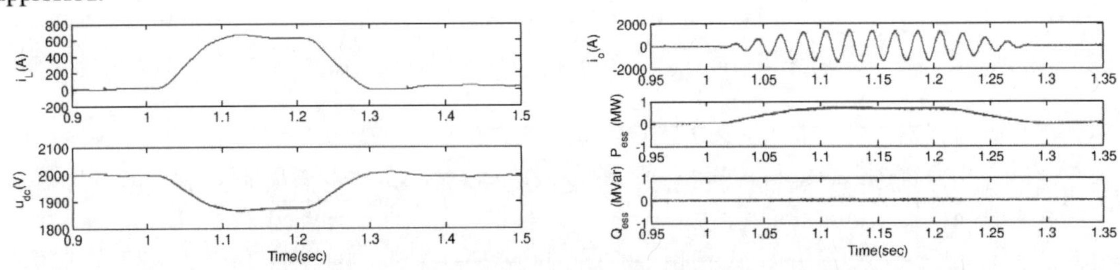

Figure 11. Response of ESS on voltage frequency decrease

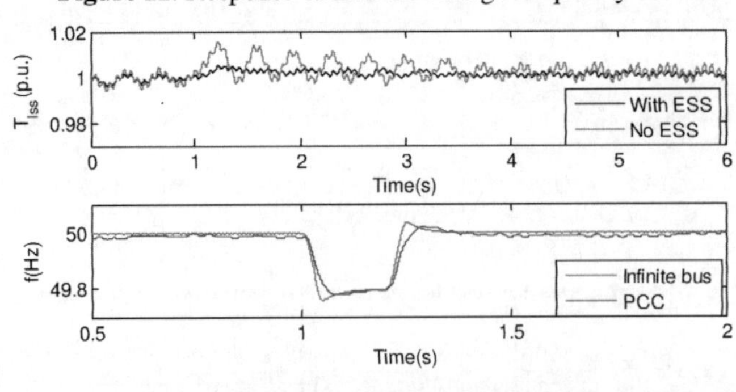

Figure 12. Torsional damping effect for bus frequency decrease

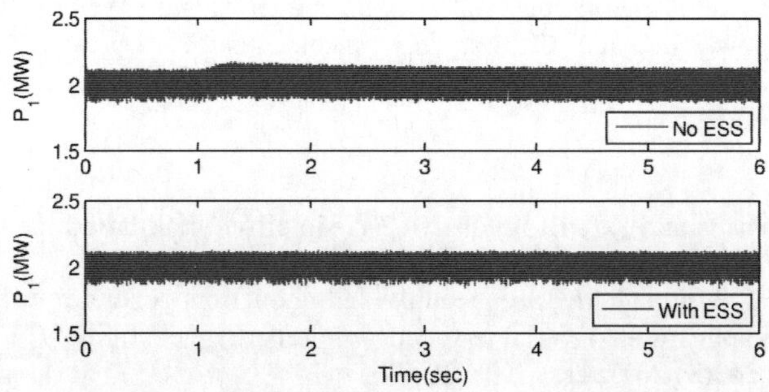

Figure 13. Damping effect of active power fluctuation

4.2.3. Simulation analysis of voltage frequency increase Figure 14 shows the response of ESS on variations of the infinite bus voltage.

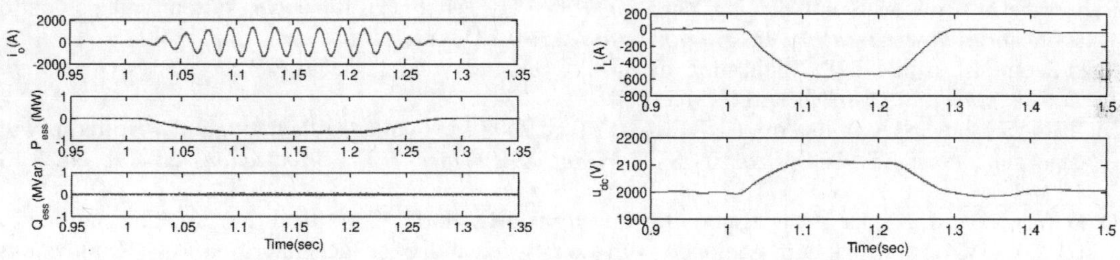

Figure 14. Response of ESS on voltage frequency increase

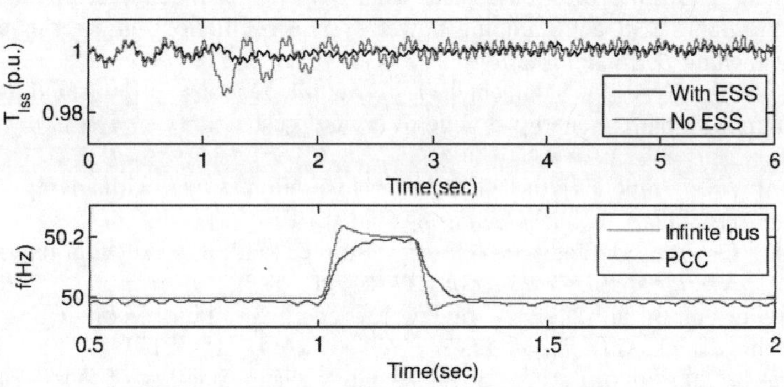

Figure 15. Torsional damping effect for bus frequency increase

Initially, the ESS neither absorbs nor provides active/reactive power to the grid. Its output current i_o increases to absorb active power due to the frequency decrease. Thus, the batteries charge from the capacitor to keep the dc-link voltage constant during the vibration. Figure 15 shows the low speed shaft torque T_{lss} response with and without the ESS. In the case without ESS, the vibration lasts over 2 s in low speed shaft torque, whereas the system with ESS is able to damp the torsional vibrations effectively and rapidly within 0.5 s. Similarly, the ESS has no contribution to regulating the frequency of the PCC voltage.

5. Conclusions

This paper has proposed novel methods to damp the drive train torsional vibrations in a wind turbine system, based on STATCOM and ESS, respectively. The damping control scheme of ESS, including the bidirectional DC/DC converter and PWM inverter, has been proposed as well. Simulation results have shown that the STATCOM is a promising solution to suppress the torsional vibrations by stabilizing the bus voltage of wind systems. Further, ESS making use of the bidirectional DC/DC converter and PWM converter is able to meet the bidirectional active/reactive power demand. Case studies revealed that ESS presents superiority performance over STATCOM on damping torsional vibrations caused by grid voltage sags and is especially irreplaceable for those caused by frequency disturbances.

References

[1] Karki Rajesh, Suman Thapa and Roy Billinton 2012 A simplified risk-based method for short-term wind power commitment *Sustainable Energy, IEEE Transactions* **3(3)** 498-505.

[2] Pena R, Clare J C and Asher G M 1996 Doubly fed induction generator using back-to-back PWM converters and its application to variable-speed wind-energy generation *IEEE Proceedings-Electric Power Applications* **143(3)** 231-241.

[3] Abad Gonzalo, et al. 2010 Direct power control of doubly-fed-induction-generator-based wind turbines under unbalanced grid voltage. *Power Electronics, IEEE Transactions* **25(2)** 442-452.

[4] Badrzadeh Babak and Salman S K 2009 Investigation of the torsional vibration and transient stability margin for doubly-fed induction generators *Sustainable Alternative Energy (SAE), 2009 IEEE PES/IAS Conference on*. IEEE.

[5] Rao Pranesh, Crow M L and Zhiping Yang 2000 STATCOM control for power system voltage control applications *Power Delivery, IEEE Transactions* **15(4)** 1311-17.

[6] Saad-Saoud Z, et al. 1998 Application of STATCOMs to wind farms *IEE Proceedings-Generation, Transmission and Distribution* **145(5)** 511-516.

[7] MOLINAS Marta, SUUL Jon Are, UNDELAND Tore 2008 Low voltage ride through of wind farms with cage generators: STATCOM versus SVC *Power Electronics, IEEE Transactions* **23**. Jg., Nr. 3, S. 1104-17.

[8] Qiao Wei, Ganesh Kumar Venayagamoorthy and Ronald G. Harley 2009 Real-time implementation of a STATCOM on a wind farm equipped with doubly fed induction generators *Industry Applications, IEEE Transactions* **45(1)** 98-107.

[9] El-Moursi, Mohamed S, Birgitte Bak-Jensen and Mansour H. Abdel-Rahman 2010 Novel STATCOM controller for mitigating SSR and damping power system oscillations in a series compensated wind park *Power Electronics, IEEE Transactions* **25(2)** 429-441.

[10] Borowy Bogdan S and Ziyad M Salameh 1997 Dynamic response of a stand-alone wind energy conversion system with battery energy storage to a wind gust *Energy Conversion, IEEE Transactions* **12(1)** 73-78.

[11] Lu Ming-Shun, et al. 2008 Combining the wind power generation system with energy storage equipments *Industry Applications Society Annual Meeting* **IAS'08** IEEE.

[12] Barton John P and David G Infield 2004 Energy storage and its use with intermittent renewable energy *Energy Conversion, IEEE Transactions* **19(2)** 441-448.

[13] Díaz-González Francisco, et al. 2012 A review of energy storage technologies for wind power applications *Renewable and Sustainable Energy Reviews* **16(4)** 2154-2171.

[14] ZHANG Bu-han, et al. 2006 Improvement of Power Quality and Stability of Wind Farms Connected to Power Grid by Battery Energy Storage System *Power system technology* **15** 013.

[15] Li Xiangjun, Dong Hui and Xiaokang Lai 2013 Battery energy storage station (BESS)-based smoothing control of photovoltaic (PV) and wind power generation fluctuations *Sustainable Energy, IEEE Transactions* **4(2)** 464-473.

Torsional vibration characteristic study of the grid-connected DFIG wind turbine

Songtao Yu[1], Da Xie[1, 3], Wangping Wu[1], Chenghong Gu[2], Furong Li[2]

[1] School of Electronic and Electrical Engineering, Shanghai Jiaotong University
[2] Department of Electronic and Electrical Engineering, University of Bath, Bath BA2 7AY, U.K.

Email: 792252002@sjtu.edu.cn; profxzg@hotmail.com

Abstract. This paper studies the torsional vibration characteristics of the grid-connected doubly-fed induction generator (DFIG) wind turbine by small signal analysis method. Firstly a detailed small-signal stability union model of the grid-connected DFIG wind turbine is developed, including the mechanical system and electrical system. To study the dynamic characteristic of the blade, gearbox, low speed and high speed shafts, a three mass shaft model for the mechanical system is adopted. At the same time, small signal models of DFIG, the voltage source converter (VSC) and the transmission line of the electrical system are developed respectively. Then, through calculating the eigenvalues of the state matrix A and the corresponding participation factors, the modal analysis is conducted in the shaft torsional vibration issues. And the impact of the system parameters including the series compensation capacitor, the flat-wave reactor, the PI parameters, especially the speed controller of generator rotor on shaft torsional vibration are discussed. The results show that the speed controller strengthens association between the mechanical system and the electrical system, and also produces a low-frequency oscillation mode.

1. Introduction

As an environmentally friendly resource, wind energy has been developed quickly in recent years [1-2], which has led to the explosive growth of wind turbines connected to the power grid. With obvious disadvantages of variable production, however, the large amount of wind turbines integrated to the grid will bring a great impact of low-frequency oscillation and wind turbine shaft torsional vibration on power system [3]. Worse still, the torsional vibration can cause a reduction in the lifespan of the wind turbine resulted from fatigue, even the shaft fault [4], which contributes critically to the stability and safe operation of power system. Therefore it is of greatly practical significance to study the wind turbine torsional vibration characteristics further.

The wind turbines are mainly categorized into two types: fixed speed and variable speed wind turbines. Doubly fed induction generator (DFIG)-based wind turbine is widely used nowadays due to its variable speed operation and power decoupling controllability [5-7]. DFIG is also one of the most economical types of wind turbines, which permits the regulation of reactive power with variable rotor speed. With the increasing use of DFIG, a multitude of researchers are keen on the study of DFIG. Reference [8] invented an effective rotor position phase lock loop (PLL) to track the maximum power point (MPPT) of DGIG. By developing a new control scheme for the rotor side converter of DFIG, inter-area oscillations caused by the long transmissions are damped in [9]. The impact of increased penetration of DFIGs on the power system are studied in [10] by modelling DFIGs as synchronous machines. The researches mentioned above are mainly concentrated on studying the normal operation of DFIG, the impact of DFIGs on the power system and lack the ability of analysing the dynamic characteristics of the DFIG system.

Thus the modelling methods to study dynamic characteristics of DFIG-based wind turbines during the transient process are gaining increasing interests. References [11-13] developed a lumped-mass, a two-mass and a three-mass drive train model of wind turbine shaft system respectively to model the mechanical system of wind turbines exactly. An available model of DFIG is developed in [14] to study how the model parameters will influence the transient responses of DFIG-based wind plants. But the mechanical system and electrical system are both simplified during the investigation. A two-mass drive train model is adopted in [15] to investigate the machine dynamics of DFIG. Reference [16] use the modal analysis to characterize the small-signal behaviour of DFIG wind turbine, and research the DFIG intrinsic dynamics with the change of the series compensation system parameters, operating points, and some other factors. However, the aforementioned works are mainly studying the impact of the aerodynamic factors or the long transmission lines and neglect the influence from the power grid and the control system. Also, which parts of the wind power system and to what extent will the different system parameters influence the dynamitic character of the mechanical system are unrevealed.

In order to exactly analyze the torsional vibration characteristics of the mechanical system of DFIG, a three-mass shaft model, which consists of blades, a gearbox, a low speed shaft, a high speed shaft and an induction generator rotor in [17] is adopted. The small signal model of the voltage source converter (VSC) is developed, where the three subsystems, including a rotor side converter (RSC), a grid side converter (GSC) and a DC-link are modelled respectively. The transmission line and the flat-wave reactor together with the transformer are modelled as an equivalent RLC (resistor, inductor and capacitor) line and an equivalent RL (resistor and inductor) line respectively. Then, the union small signal model of DFIG consisting of seven modules is obtained. Through conducting modal analysis, the torsional vibration characteristics of the shaft are demonstrated in detail. With the multi-mass shaft model, the dynamic response of the shaft to the electrical and control system parameters is reflected clearly.

The main contributions of this paper are that it: i) develops a small signal model of DFIG-based wind power system for easy shaft torsional vibration analysis; ii) studies the influence of the electrical and control systems on the shaft torsional vibration and finds that only the speed controller will have an impact; iii) provides a new method for parameter design for DFIG.

2. Grid-connected DFIG Wind Turbine Modelling

The schematic diagram of the studied DFIG wind turbine is shown in figure 1. The wind turbine is connected to the induction generator via a gearbox. The stator side of the induction generator is connected to the infinite bus through a transmission line, including the impedance of transformer and cable. The rotor side of the induction generator is fed through a VSC to supply exciting voltage to the generator.

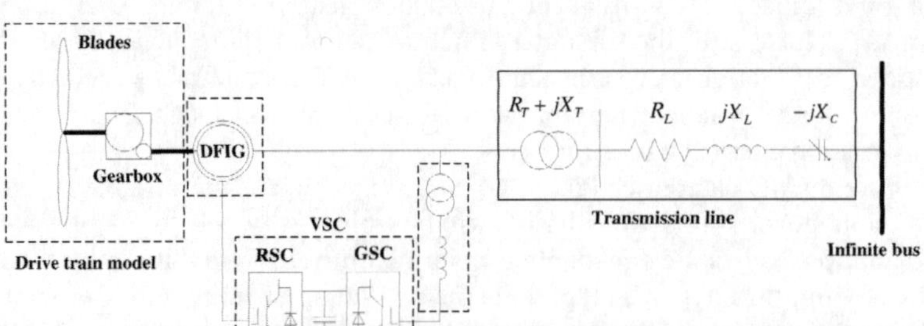

Figure 1. Schematic diagram of the DFIG wind turbine connected to infinite bus system

As shown in figure 1, the model of the grid-connected DFIG wind turbine consists of four parts: the drive train, the induction generator, the transmission line and the converter controller model.

2.1. Drive Train Model

To reflect the dynamic characteristics of the drive train, a three-mass model presented in [17] is used to model the mechanical system. The drive train model can be expressed as

$$\begin{cases} J_1 \frac{d}{dt}\omega_{rot} = T_{aero} - K_1(\theta_1-\theta_2) - D_1\omega_{rot} - D_{12}(\omega_{rot}-\omega_2) \\ J_2 \frac{d}{dt}\omega_2 = -K_1(\theta_2-\theta_1) - K_2(\theta_2-\theta_3) - D_2\omega_2 \\ \qquad - D_{12}(\omega_2-\omega_{rot}) - D_{23}(\omega_2-\omega_{gen}) \\ J_3 \frac{d}{dt}\omega_{gen} = -T_{em} - K_2(\theta_3-\theta_2) - D_3\omega_{gen} \\ \qquad -D_{23}(\omega_{gen}-\omega_2) \\ \frac{d}{dt}\theta_1 = \omega_{rot} \\ \frac{d}{dt}\theta_2 = \omega_2 \\ \frac{d}{dt}\theta_3 = \omega_{gen} \end{cases} \tag{1}$$

where J_1, J_2 and J_3 denote the moment of inertia of the blades, low speed shaft and the high speed shaft respectively; ω_{rot}, ω_2 and ω_{gen} denote the angular velocity of the three parts respectively; θ_1, θ_2 and θ_3 are mechanical rotation angle of the three parts respectively; D_1, D_2 and D_3 are damping coefficient of the three parts respectively and D_{12} the damping coefficient between the blades and the low speed shaft and D_{23} the damping coefficient between the low and high speed shaft; K_1 and K_2 are elastic coefficient between the blades and the low speed shaft, the low speed and high speed shaft respectively; T_{aero} is input wind torque from the blade side; T_{em} is the electromagnetic torque of the generator.

Based on small signal analysis, the standard state space equation of (1) after it is linearized can be expressed as

$$\begin{cases} \dot{X}_{DT} = A_{DT}X_{DT} + B_{DT}u_{DT} \\ Y_{DT} = C_{DT}X_{DT} + D_{DT}u_{DT} \end{cases} \tag{2}$$

where subscript DT denotes the three-mass drive train model.

2.2. DFIG Model

In this paper, DFIG is modelled with the five-order model of an induction generator (IG), which is shown in [18]. Here only the linearized model of DFIG derived using small signal analysis is shown, whose standard state equation can be expressed as follows.

$$\begin{cases} \dot{X}_{DFIG} = A_{DFIG}X_{DFIG} + B_{DFIG}u_{DFIG} \\ Y_{IDFG} = C_{DFIG}X_{DFIG} + D_{DFIG}u_{DFIG} \end{cases} \tag{3}$$

where subscript DFIG denotes the double fed induction generator.

2.3. Transmission Line Model

The transmission is equivalent to a RLC line model, which is modelled the same way shown in Reference [17]. Then the standard state space equation can be written as

$$\begin{cases} \dot{X}_{TL} = A_{TL}X_{TL} + B_{TL}u_{TL} \\ Y_{TL} = C_{TL}X_{TL} + D_{TL}u_{TL} \end{cases} \tag{4}$$

where subscript *TL* denotes the transmission line.

The flat-wave reactor with the small transformer is equivalent to a RL line model, whose state space equation can be obtained through equation (4).

2.4. VSC Model

The VSC of DFIG consists of a RSC and a GSC, which are connected back-to-back via a DC-link. Because of the variable-frequency supply provided by the DFIG converter, the rotor angular frequency and synchronous angular frequency are decoupled, which realizes the operation of wind turbine with variable speed. The equivalent circuit of the VSC is shown in figure 2.

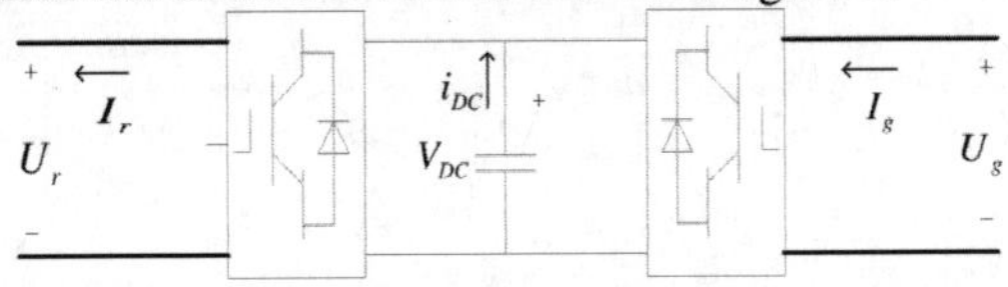

Figure 2. Equivalent circuit of the DFIG converter

2.4.1 DC-Link Model. The power balance equation about the DFIG converter is described by

$$P_{DC} = P_r - P_g \tag{5}$$

where P_r is the rotor-side active power, P_g is the grid-side active power, and P_{DC} is the active power of the capacitor in the dc link.

Then

$$CV_{DC}\dot{V}_{DC} = u_{dg}i_{dg} + u_{qg}i_{qg} - \left(u_{dr}i_{dr} + u_{qr}i_{qr}\right) \tag{6}$$

After conversion by p.u. and linearization of equation (6), the standard state equation of DC-link can be obtained as follows:

$$\dot{X}_{DC} = A_{DC}X_{DC} + B_{DC}u_{DC} \tag{7}$$

where subscript DC denotes the DC-link.

2.4.2 RSC Controller Model. The RSC is responsible for regulating DFIG active power and terminal voltage based on stator flux vector orientation method. The block diagram of RSC controller is shown in figure 3, where u_{qr}^{φ} and u_{dr}^{φ} are used to control the active power and voltage respectively.

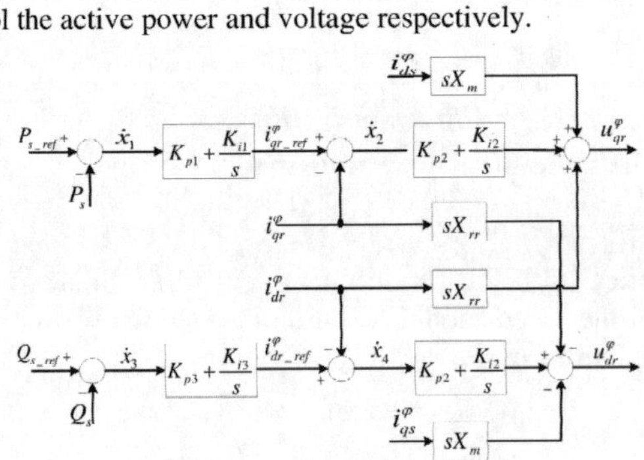

Figure 3. Block diagram of RSC controller

Assume that the converter operates fast enough so that the dynamics can be ignored and the controlled variable closely follows the given value.

During linearization at the steady state point, the following equations are obtained first.

$$\begin{cases} \Delta P_{s\_ref} = 0 \\ \Delta P_s = i_{ds0}^{\varphi}\Delta u_{ds}^{\varphi} + u_{ds0}^{\varphi}\Delta i_{ds}^{\varphi} + i_{qs0}^{\varphi}\Delta u_{qs}^{\varphi} + u_{qs0}^{\varphi}\Delta i_{qs}^{\varphi} \\ \Delta Q_{s\_ref} = 0 \\ \Delta Q_s = i_{ds0}^{\varphi}\Delta u_{qs}^{\varphi} + u_{qs0}^{\varphi}\Delta i_{ds}^{\varphi} - i_{qs0}^{\varphi}\Delta u_{ds}^{\varphi} - u_{ds0}^{\varphi}\Delta i_{qs}^{\varphi} \end{cases} \tag{8}$$

Then the standard state space equation of RSC can be expressed as follows

$$\dot{X}_r = A_r X_r + B_r u_r \tag{9}$$

where subscript r denotes RSC.

2.4.3 GSC Controller Model The GSC is responsible for controlling the DC-link voltage as a constant value and the output reactive power of DFIG. The block diagram of GSC controller is shown in figure 4.

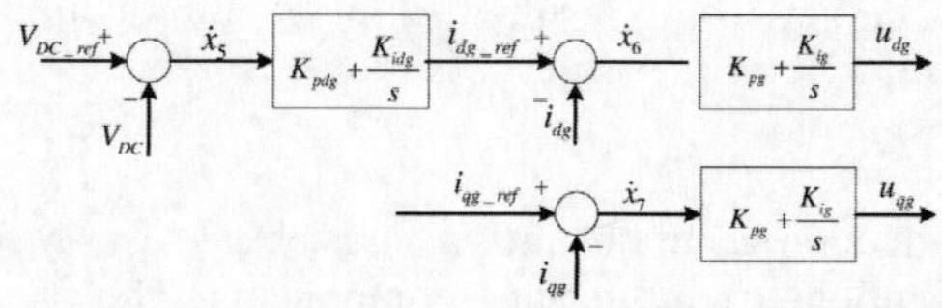

Figure 4. Block diagram of GSC controller

Similar to the RSC, the linearized equations of GSC can be expressed by

$$\begin{cases} \Delta\dot{x}_5 = \Delta V_{DC\_ref} - \Delta V_{DC} \\ \Delta i_{dg\_ref} = K_{pdg}(\Delta V_{DC\_ref} - \Delta V_{DC}) + K_{idg}\Delta x_5 \\ \Delta\dot{x}_6 = \Delta i_{dg\_ref} - \Delta i_{dg} \\ \quad = K_{pdg}(\Delta V_{DC\_ref} - \Delta V_{DC}) + K_{idg}\Delta x_5 - \Delta i_{dg} \\ \Delta\dot{x}_7 = \Delta i_{qg\_ref} - \Delta i_{qg} \\ \Delta u_{dg} = K_{pg}[K_{pdg}(\Delta V_{DC\_ref} - \Delta V_{DC}) + K_{idg}\Delta x_5 - \Delta i_{dg}] + K_{ig}\Delta x_6 \\ \Delta u_{qg} = K_{pg}(\Delta i_{qg\_ref} - \Delta i_{qg}) + K_{ig}\Delta x_7 \end{cases} \tag{10}$$

where $\Delta V_{DC\_ref} = 0$ and $\Delta i_{qg\_ref} = 0$.

The standard state space equation of GSC can be expressed as follows

$$\dot{X}_g = A_g X_g + B_g u_g \tag{11}$$

where subscript g denotes GSC.

3. Modal Analysis of Grid-connected DFIG Wind Turbine

3.1. Small-signal Stability Model

By solving the equations of the drive train, DFIG, transmission line and controller models simultaneously, the union model of the grid-connected DFIG wind turbine is obtained as follows:

$$\dot{X} = AX + Bu \tag{12}$$

where $X = \begin{bmatrix} X_{DT} & X_{DFIG} & X_{TL} & X_{DC} & X_r & X_g \end{bmatrix}^T$.

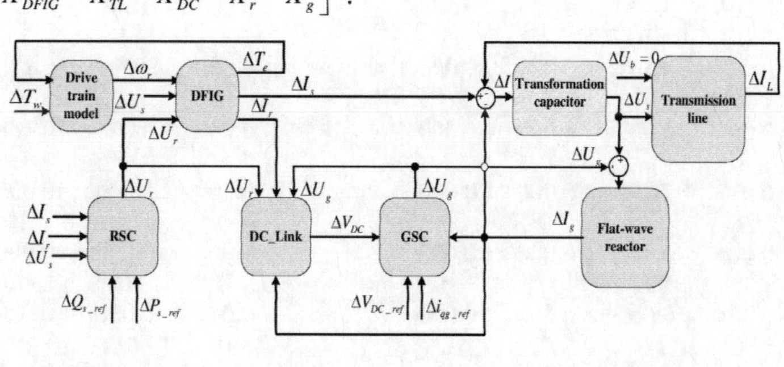

Figure 5. Schematic diagram of the small-signal model

To study shaft torsional vibration issues of DFIG shown in figure 1, it is necessary to build the small-signal stability model of the system including three-mass shaft model, induction generator model, converter, transformer, transmission line model, and the power grid model. Figure 5 below is the schematic diagram of the small-signal model of the DFIG-based wind power system, and presents the interaction among different models. During the process of modelling, to facilitate interaction of the system components, the conversion between state variables, such as that of current and voltage, is indispensable. In this paper, a capacitor is introduced to transform the current deviation to the voltage deviation. In order to avoid accuracy losing resulted from the introduction of capacitor, the value of the capacitor is extremely small.

3.2. Modal Analysis

A 2MW, 690V doubly fed induction generator model is established in MATLAB/SIMULINK based on the models of the aforementioned wind turbine components. The parameters of the drive train, generator and transmission line are the same with those shown in [17], and the parameters of the VSC are listed in the Appendix.

Table 1. Eigenvalues of matrix A

Modes	Eigenvalue	Frequency(Hz)	Damping ratio
1	-15.55 ± 508.40i	80.914	0.031
2	-14.83 ± 125.15i	19.918	0.118
3	-36.05 ± 74.19i	11.807	0.437
4	-0.15 ± 78.08i	12.427	0.0019
5	-0.029 ± 11.09i	1.765	0.0026

Table 2. Participation factors of torsional modes

State variables	Mode 1	Mode 2	Mode 3	Mode 4	Mode 5
$\Delta\theta_1$	0.0000	0.0000	0.0000	0.0002	0.0810
$\Delta\theta_2$	0.0000	0.0000	0.0000	**0.4451**	0.0434
$\Delta\theta_3$	0.0000	0.0000	0.0000	0.0547	**0.3756**
$\Delta\omega_1$	0.0000	0.0000	0.0000	0.0002	0.0810
$\Delta\omega_2$	0.0000	0.0000	0.0000	**0.4451**	0.0434
$\Delta\omega_3$	0.0000	0.0000	0.0000	0.0547	**0.3756**
$\Delta\psi_{qs}$	0.0062	0.0429	**0.1011**	0.0000	0.0000
$\Delta\psi_{ds}$	0.0073	0.0058	0.0371	0.0000	0.0000
$\Delta\psi_{qr}$	0.0004	0.0230	**0.1976**	0.0000	0.0000
$\Delta\psi_{dr}$	0.0013	0.0057	**0.3635**	0.0000	0.0000
Δi_{Lx}	**0.2443**	**0.2294**	0.0556	0.0000	0.0000
Δi_{Ly}	**0.2450**	**0.1929**	0.0440	0.0000	0.0000

State variables	Mode 1	Mode 2	Mode 3	Mode 4	Mode 5
$\Delta u_{sc,x}$	**0.2511**	**0.2206**	0.0203	0.0000	0.0000
$\Delta u_{sc,y}$	**0.2514**	**0.2365**	0.0345	0.0000	0.0000
Δi_{gx}	0.0000	0.0000	0.0007	0.0000	0.0000
Δi_{gy}	0.0000	0.0000	0.0002	0.0000	0.0000
ΔV_{DC}	0.0000	0.0000	0.0009	0.0000	0.0000
Δx_1	0.0003	0.0424	**0.2737**	0.0000	0.0000
Δx_2	0.0000	0.0011	**0.1106**	0.0000	0.0000
Δx_3	0.0005	0.0250	**0.1881**	0.0000	0.0000
Δx_4	0.0000	0.0042	0.0604	0.0000	0.0000
Δx_5	0.0000	0.0000	0.0000	0.0000	0.0000
Δx_6	0.0000	0.0000	0.0002	0.0000	0.0000
Δx_7	0.0000	0.0000	0.0022	0.0000	0.0000

The eigenvalues of the system state matrix A are calculated and shown in table 1 except the non-oscillation modes. As shown, there are five oscillation modes. Since the real parts of the eigenvalues are all negative, the system is stable. The participation factors of corresponding oscillation modes are shown in table 2, which reflect the relative participation of system components. The words marked in bold show that the eigenvalues are highly sensitive to the corresponding variables.

As shown in table 1 and 2, the frequencies of oscillation mode 1 and 2 are 80.914Hz and 19.918Hz. They are highly sensitive to Δi_{Lx}, Δi_{Ly}, $\Delta u_{sc,x}$, $\Delta u_{sc,y}$, in other words, they are greatly influenced by the inductance of the transmission line and series compensation capacitors.

The oscillation mode 3, whose frequency is 11.807Hz, correlates strongly with many state variables. They are mainly affected by the stator flux $\Delta\psi_{qs}$ and rotor flux $\Delta\psi_{dr}, \Delta\psi_{qr}$. The related variables of rotor side controller play a big role besides. The oscillation mode 4 is highly sensitive to the state variables $\Delta\theta_2$, $\Delta\omega_2$, so they are mainly affected by the torsion angle and speed of the low-speed shaft. In addition, it is also affected by the torsion angle and speed of the high-speed shaft to some degree. Accordingly, the oscillation mode 4 is the mechanical oscillation mode, and the oscillation frequency is 12.427 Hz. The oscillation mode 5 is highly sensitive to the state variables $\Delta\theta_3$, $\Delta\omega_3$, and it is mainly affected by the torsion angle and speed of the high-speed shaft. In addition, it is also affected by the torsion angle and speed of the blades and high-speed shaft to some extent. Thus, the oscillation mode 5 is also mechanical oscillation mode, and the oscillation frequency is 1.765 Hz. Since the number of natural frequencies of a multi-mass system are equal to the number of masses minus one [19], the drive train model in this paper has only two natural frequencies, which are 12.427 Hz and 1.765 Hz, and the oscillation mode 4 and 5 are the torsional modes.

4. Effect of Parameters

4.1. Effect of Series Compensation Capacitor

The effect of series compensation capacitor can be described by the series compensation level k_{com} [16], which is expressed as

$$k_{com} = X_C / X_L \tag{13}$$

where, X_C is the capacitive reactance of the transmission line; X_L is the sum of the inductive reactance of the transmission line and the transformer.

Varying the value of k_{com} (0.1~1) while keeping all other parameters at their original values, the frequency and damping ratio variations of the mechanical mode 4 and 5 are shown in table 3.

Table 3. Effect of series compensation level

k_{com}	Mode 4 Frequency(Hz)	Mode 4 Damping ratio	Mode 5 Frequency(Hz)	Mode 5 Damping ratio
0.1	12.43	0.0019	1.765	0.0026
0.2	12.43	0.0019	1.765	0.0026
0.3	12.43	0.0019	1.765	0.0026
0.4	12.43	0.0019	1.765	0.0026
0.5	12.43	0.0019	1.765	0.0026
0.6	12.43	0.0019	1.765	0.0026
0.7	12.43	0.0019	1.765	0.0026
0.8	12.43	0.0019	1.765	0.0026
0.9	12.43	0.0019	1.765	0.0026
1.0	12.43	0.0019	1.765	0.0026

As shown in table 3, the frequency and damping ratio of mode 4 and 5 almost remain unchanged with the value of k_{com} increasing, which can be explained by table 2. The mechanical modes are mainly affected by the angle and speed of the high and low-speed shafts, and the electrical parameters don't cause significant effect on the mechanical modes.

At the same time, vary the value of the flat-wave reactor and the seven PI parameters of the converter controllers respectively. The frequencies and the damping ratios of mode 4 and 5 all remain unchanged, which are the same cases as that of the series compensation capacitor.

4.2. Effect of Speed Controller

As shown in figure 3, the RSC uses the stator active power as the control variable. If the rotor speed is used as the control variable, and apply a PI controller in the speed control scheme to track the reference speed, the association between rotor speed and PI parameters will be strengthened a lot. The block diagram of RSC with speed controller is shown in figure 6, and the speed controller is marked with dotted lines. The speed controller can be expressed as

$$\begin{cases} \dot{x}_8 = \omega_{gen\_ref} - \omega_{gen} \\ P_{s\_ref} = K_{p\omega}\dot{x}_8 + K_{i\omega}x_8 \end{cases} \tag{14}$$

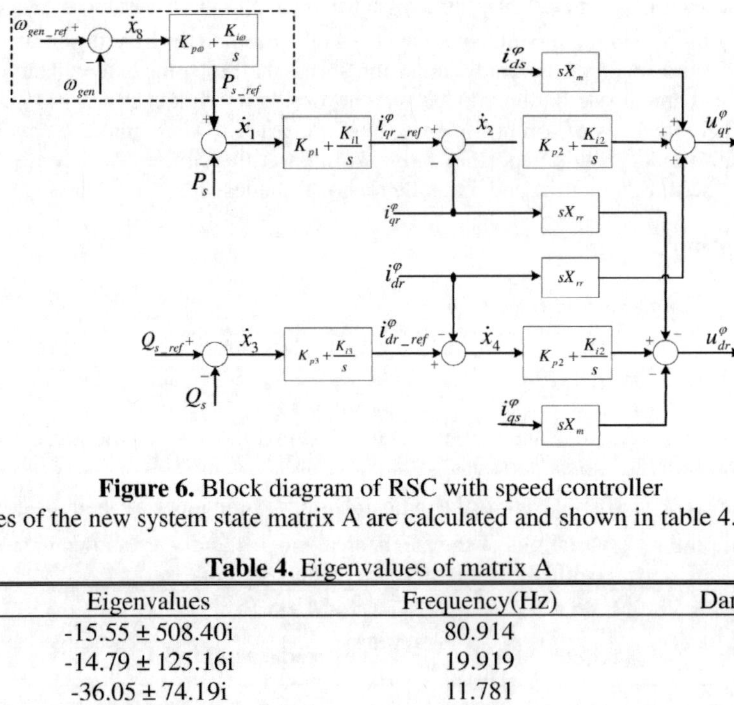

Figure 6. Block diagram of RSC with speed controller

The eigenvalues of the new system state matrix A are calculated and shown in table 4.

Table 4. Eigenvalues of matrix A

Modes	Eigenvalues	Frequency(Hz)	Damping ratio
1	-15.55 ± 508.40i	80.914	0.031
2	-14.79 ± 125.16i	19.919	0.117
3	-36.05 ± 74.19i	11.781	0.437
4	-0.16 ± 78.13i	12.434	0.002
5	-0.27 ± 13.25i	2.109	0.020
6	-0.02 ± 2.55i	0.406	0.009

Compared with table 2, one more oscillation mode with frequency 0.406 Hz and damping ratio 0.009 can be found in table 4. Given that mode 1~3 have few changes and are not mechanical torsional mode, we just take mode 4~6 into consideration, whose corresponding participation factors are shown in table 5.

As shown in table 4, mode 4 has little changes in its frequency and damping ratio, while mode 5 changes with the frequency from 1.765Hz to 2.109Hz and the damping ratio from 0.0026 to 0.02 respectively, which can be explained from table 5. Mode 4 is mainly affected by low-speed shaft, while mode 5 is affected by high-speed shaft. Since the speed controller only enhances effect of PI parameters on rotor speed (of high-speed shaft), the state variable Δx_8 has more significant effect on mode 5 and less on mode 4. In addition, it also can be obtained from table 4 that the speed controller can increase system damping of mode 5.

Table 5. Participation factors of torsional modes

State variables	Mode 4	Mode 5	Mode 6	State variables	Mode 4	Mode 5	Mode 6
$\Delta\theta_1$	0.0002	0.0398	**0.4600**	$\Delta u_{sc,y}$	0.0000	0.0000	0.0000
$\Delta\theta_2$	**0.4438**	0.0495	0.0065	Δi_{gx}	0.0000	0.0000	0.0000
$\Delta\theta_3$	0.0551	**0.2554**	**0.3102**	Δi_{gy}	0.0000	0.0000	0.0000
$\Delta\omega_1$	0.0002	0.0398	**0.4600**	ΔV_{DC}	0.0000	0.0000	0.0000
$\Delta\omega_2$	**0.4439**	0.0495	0.0064	Δx_1	0.0003	0.0048	0.0022
$\Delta\omega_3$	0.0558	**0.4106**	0.0335	Δx_2	0.0001	0.0006	0.0000
$\Delta\psi_{qs}$	0.0001	0.0002	0.0000	Δx_3	0.0001	0.0001	0.0000
$\Delta\psi_{ds}$	0.0001	0.0018	0.0007	Δx_4	0.0000	0.0000	0.0000
$\Delta\psi_{qr}$	0.0001	0.0000	0.0000	Δx_5	0.0000	0.0000	0.0000
$\Delta\psi_{dr}$	0.0004	0.0007	0.0000	Δx_6	0.0000	0.0000	0.0000
Δi_{Lx}	0.0000	0.0000	0.0000	Δx_7	0.0000	0.0000	0.0000
Δi_{Ly}	0.0002	0.0002	0.0000	Δx_8	0.0005	**0.1550**	**0.3437**
$\Delta u_{sc,x}$	0.0000	0.0000	0.0000	—	—	—	—

Mode 6 is mainly affected by the angle and speed of blades and high-speed shaft as well as PI controller, which are concerned with not only the mechanical system but also the electrical system. And the corresponding participation factors are 0.4600, 0.3102 and 0.3437. Therefore mode 6 is the electromagnetic oscillation mode.

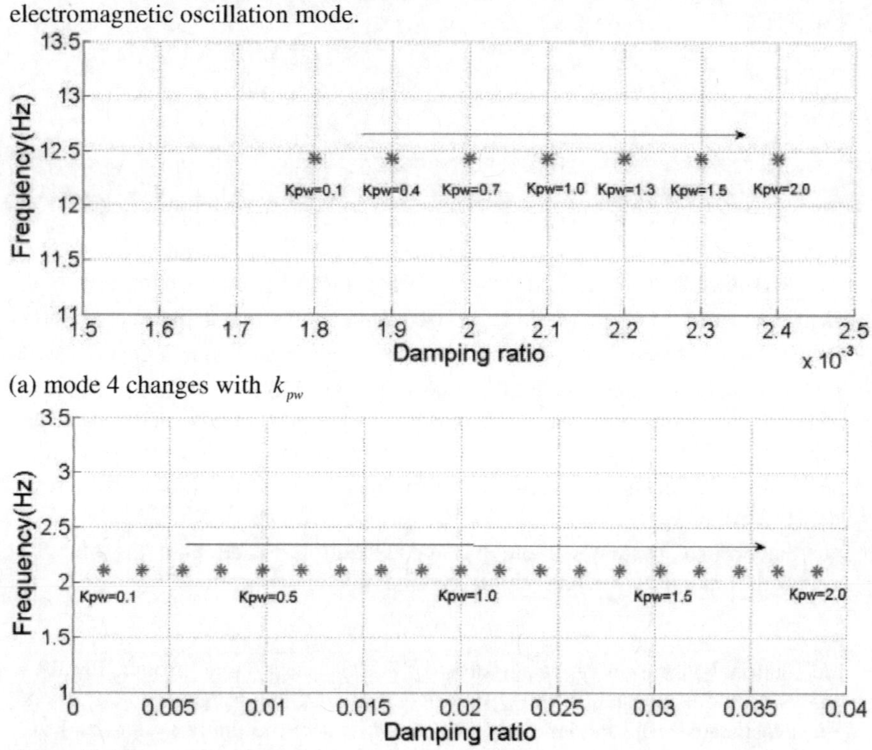

(a) mode 4 changes with k_{pw}

(b) mode 5 changes with k_{pw}

Figure 7. Variation of frequencies and damping ratios of mode 4 and 5 with k_{pw} increasing

As shown in figure 7, with the value of k_{pw} increasing, the frequencies of mode 4 and 5 remain unchanged, while the damping ratios increase significantly. Therefore the increase of k_{pw} can contribute to the torsional vibration suppression, especially the torsional mode affected by high-speed shaft.

Varying k_{iw} of PI controller in the speed control scheme, the variation of frequencies and damping ratios of mode 4 and 5 is shown in figure 8.

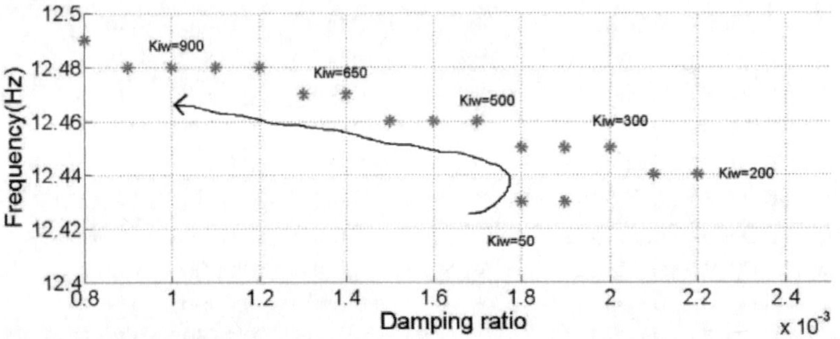

(a) mode 4 changes with k_{iw}

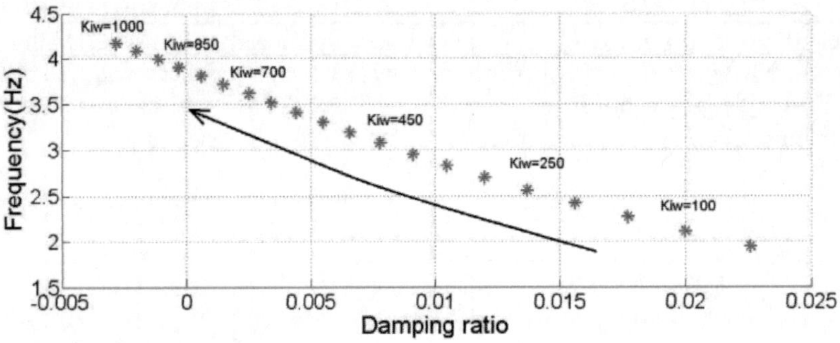

(b) mode 5 changes with k_{iw}

Figure 8. Variation of frequencies and damping ratios of mode 4 and 5 with k_{iw} increasing

As shown in figure 8, with the value of k_{iw} increasing, the frequencies of mode 4 increase slightly, while the damping ratios first increase then decrease at the point $k_{iw} = 200$. And mode 5 changes with the frequencies from 1.941Hz to 4.174Hz and the damping ratios from 0.0226 to -0.0028 respectively. Mode 5 becomes unstable since the damping ratios decrease to negative value, which means the damping is positive. It can be seen that the increase of k_{iw} can cause the damping decrease, even excite the torsional vibration and cause the mechanical failure. In addition, the torsional mode affected by high-speed shaft is highly sensitive to k_{iw}.

In addition, compared to the original system without the speed control, variation of the other PI parameters of converter controller has certain effect on the torsional mode, but limited.

5. Conclusions

This paper describes the torsional vibration characteristics of DFIG-based wind turbine. Firstly a small-signal stability analysis model is established with drive train, DFIG, transmission line and converter controller models associated. The modal analysis is employed in the union system and two oscillation modes are found to be the mechanical mode. Then the effects of several parameters on the system modes are studied, especially that of the speed controller. The conclusions are drawn as follows.

(1) Regarded to the DFIG-based wind turbine without speed controller, the torsional modes are mainly affected by the high-speed and low-speed shaft. Accordingly, system parameters, such as the series compensation capacitor, the flat-wave reactor and each PI parameter of converter controllers, do not have significant effect on the two mechanical modes.
(2)The speed controller strengthens association between the mechanical system and the electrical system, and also produces a low-frequency oscillation mode, which proves to be the electromagnetic oscillation mode. In addition, the speed controller contributes to the torsional damping increase, especially for the mode affected by the high-speed shaft. As well, the torsional damping increases with the increase of the proportion coefficient, but decreases with the increase of the integration coefficient of the speed controller. And other parameters of converter controller have slight effect on the torsional modes.

Acknowledgement
This paper is funded by National Natural Science Foundation of China (51277119), and Surface Project (51477098).

Appendix

Parameters of DFIG converter

DFIG converter	Value	DFIG converter	Value
DC_link capacitor $C_{DC}/(10^{-4}$ pu)	1.337	K_{i3}	256
Grid-side equivalent reactance X_{con}/pu	0.55	K_{pdg}	15
Grid-side equivalent reactance R_{con}/pu	0.006	K_{idg}	65
K_{p1}	0.6	K_{pg}	15.8
K_{i1}	80.6	K_{ig}	145
K_{p2}	0.26	K_{pw}	1.36
K_{i2}	6.0	K_{iw}	256
K_{p3}	1.36	—	—

References

[1] Li S Y, Sun Y, Wu T, Li Q J and Liu H 2010 Analysis of small signal stability of grid-connected doubly fed induction generators *2010 Asia-Pacific Power and Energy Engineering Conf.* 1-4

[2] Luna A, Lima F D A, Santos D, Rodríguez P, Watanabe E H and Arnaltes S 2011 Simplified modeling of a DFIG for transient studies in wind power applications *IEEE Transactions on Industrial Electronics* **58(1)** 9-20

[3] Gautam D, Vittal V and Harbour T 2009 Impact of increased penetration of DFIG-based wind turbine generators on transient and small signal stability of power systems *IEEE Transactions on Power Systems* **24(3)** 1426-34

[4] Badrzadeh B and Salman S K 2009 Investigation of the torsional vibration and transient stability margin for doubly-fed induction generators *2009 IEEE PES/IAS Conf. on Sustainable Alternative Energy* 1-8

[5] Xu L and Wang Y 2007 Dynamic modeling and control of DFIG-based wind turbines under unbalanced network conditions *IEEE Transactions on Power Systems* **22(1)** 314-323

[6] Bejaoui M, Marinescu B, Slama-Belkhodja I and Monmasson E 2014 Control of doubly-fed induction generator for wind energy in network context *IET Renewable Power Generation* **8(2)** 109-118

[7] Yang L, Yang G Y, Xu Z, Dong Z Y, Wong K P and Ma X 2010 Optimal controller design of a doubly-fed induction generator wind turbine system for small signal stability enhancement *IET generation, transmission & distribution* **4(5)** 579-597

[8] Shen B, Mwinyiwiwa B, Zhang Y and Ooi B T 2009 Sensorless maximum power point tracking of wind by DFIG using rotor position phase lock loop (PLL) *IEEE Transactions on Power*

Electronics **24(4)** 942-951

[9] Miao Z, Fan L, Osborn D and Yuvarajan S 2009 Control of DFIG-based wind generation to improve interarea oscillation damping *IEEE Transactions on Energy Conversion* **24(2)** 415-422

[10] Gautam D, Vittal V and Harbour T 2009 Impact of increased penetration of DFIG-based wind turbine generators on transient and small signal stability of power systems *IEEE Transactions on Power Systems* **24(3)** 1426-34

[11] Santoso S and Le H T 2007 Fundamental time–domain wind turbine models for wind power studies *Renewable energy* **32(14)** 2436-52

[12] Salman S K and Teo A L 2003 Windmill modeling consideration and factors influencing the stability of a grid-connected wind power-based embedded generator *IEEE Transactions on Power Systems* **18(2)** 793-802

[13] Li H and Chen Z 2007 Transient stability analysis of wind turbines with induction generators considering blades and shaft flexibility *Industrial Electronics Society, 2007. IECON 2007. 33rd Annual Conf. of the IEEE* 1604-09

[14] Kayikçi M and Milanovic J V 2008 Assessing transient response of DFIG-based wind plants—the influence of model simplifications and parameters *IEEE Transactions on power systems* **23(2)** 545-554

[15] Mei F and Pal B C 2006 Modal analysis of a grid connected doubly-fed induction generator *Power Electronics, Machines and Drives, 2006. The 3rd IET Int. Conf.* 611-615

[16] Fan L, Zhu C, Miao Z and Hu M 2011 Modal analysis of a DFIG-based wind farm interfaced with a series compensated network *IEEE Transactions on Energy Conversion* **26(4)** 1010-20

[17] Zhang Y, Xie D, Feng J and Wang R 2014 Small-signal modeling and modal analysis of wind turbine based on three-mass shaft model *Electric Power Components and Systems* **42(7)** 693-702

[18] Kundur Prabha 1994 *Power system stability and control* (New York: McGraw-hill)

[19] Salman S K and Rida I M 2001 Investigating the impact of embedded generation on relay settings of utilities electrical feeders *IEEE Transactions on Power Delivery* **16(2)** 246-251

IOP Conf. Series: Earth and Environmental Science **52** (2017) 012023 doi:10.1088/1755-1315/52/1/012023

Research on wind and solar energy transmission economic of Asia and Europe under Northern-hemisphere energy connection

Xiaoxia Wei[1], Jian Ding, Jie Liu, Tiezhong Wei

State Grid Energy Research Institute, Guangyi Building A1206, Guangyi Street 5#, Xicheng district, Beijing, China

E-mail: 20583853@qq.com

Abstract. Relying on the northern-hemisphere energy connection, considering the energy implementation of Asia and Europe, carrying out clean energy alternative is mainly to use the clean energy to take place of fossil energy. Under the green development scenario, this research gives the northern-hemisphere energy interconnection development model, makes the Artic as the connection points, gives the Northern hemisphere interconnection model unite the whole world energy. This research also identifies the factors effecting the transmission changes cost, including generation cost, transmission cost and landing cost. And estimate these two continents cost benefit, its economic and variable power-jointed scheme cost competitiveness. It showed that the trans-continent mode had better benefit, and can solve the pollution and energy restriction.

1. Introduction

Energy is an important affection to the development of human civilization [1]. The development of world energy has experienced changes from high to low carbon, from low to high efficiency, from partial equilibrium to large scale configuration [2].

Global energy supply is constrained by resource reserves and rising supply costs. To the amount, fossil energy reserves are limited, non-renewable, large-scale development and utilization will lead to depletion of resources. To the layout, the world's energy resources and consumption are reverse distribution; energy exploitation is centralizing to a few countries [3, 4].

The supply costs of fossil and clean energy is showing a "one rising and one dropping". With the coal, oil, natural gas wide exploration, the marginal cost of fossil energy will be gradually improved. And it will increase the energy cost, while the clean energy cost is gradually declining, but now it is at high level. With the rapid development of clean energy technology, the cost will decrease.

So we really need to exploit the rich energy areas, and transport the energy to the developed areas short of energy. Connect the northern hemisphere continents and balance the supply and demand [5, 6, 7].

2. Northern Hemisphere Energy connection

Northern Hemisphere Energy connection is mainly based on the Arctic power transmission channel. On one hand, it connects the Greenland, Norway, Barents Sea, Kara Hai, Bering Strait, while on the other hand, it is a global energy strategy platform connecting Arctic wind power base to electricity delivery showed in Figure 1.

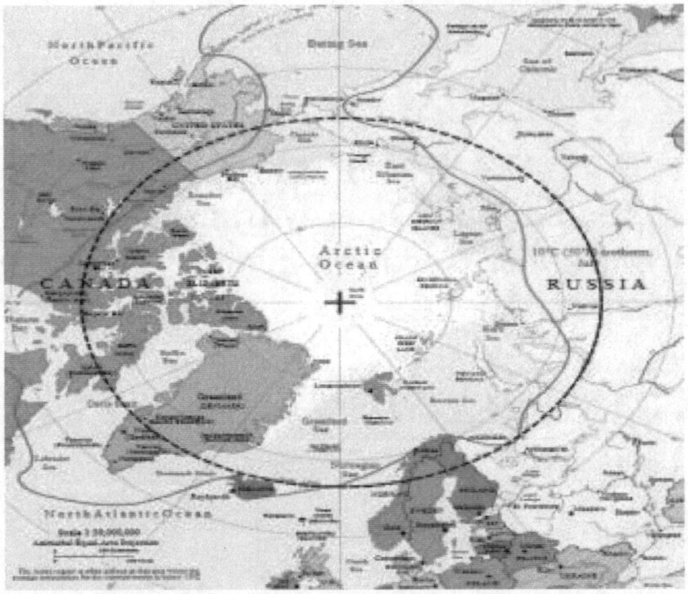

Figure 1. Asia and Europe energy connection

Northern Hemisphere Energy connection not only can solve the problem of wind power transmission in the Arctic, but also can satisfy supply and demand gap of Northeast Asia, Europe and North America, taking the Bering Strait, Greenland, Norwegian Sea and Barents Sea as the fulcrums in realizing the Arctic, Europe and North America connection. This connection can give full play to great grid connection; give many benefits of peak and valley regulation, mutual backup, cross-complementarity. In addition, making use of the time difference between the continents, send the Arctic wind to other continents, meet the day peak load, and improve Arctic wind utilization efficiency showed in Table 1.

Table 1. Global Energy Interconnection Development Model

Connection	Sending and receiving	Line setting	2050 transmission capacity
North Africa to Europe	Solar power base of Morocco, Algeria, Tunisia, Libya, Egypt and other countries of North Africa to the South of Europe	Adopt ± 1100 kV HVDC transmission technology	Transmission capacity is 1.5 trillion kilowatt-hours / year; channel capacity is 300 million kilowatts.
The Middle East to South Asia	Solar power base of the Middle East, Saudi Arabia, Oman, the United Arab Emirates in South Asia to the western of India	Adopt UHV DC submarine cable connecting the Strait of Hormuz to reach Iran, then by land UHVDC from Pakistan to the western India Mumbai load center	Transmission capacity is 2.5 trillion kilowatt-hours / year; channel capacity is 500 million kilowatts.
Australia to Southeast Asia	Australian northern solar power base using high-voltage submarine cable cross-sea landing in Indonesia, and then through a short cross-sea distance through Singapore to Thailand	Adopt ± 1100 kV HVDC transmission technology	Transmission capacity is 1 trillion kilowatt-hours / year; channel capacity is 200 million kilowatts.

3. Wind and Solar Energy Transmission Cost Model of Asia and Europe Connection

Asia is the sending (Russia Siberia, Kazakhstan, Mongolia, Xinjiang, China). Coal, hydropower, wind, solar and other energy resources are abundant. Sending has the ability of establishing of large power base, set energy to the European load centre. The connection of Asia and Europe, the transmission of wind, solar and other clean energy, use cost analysis model to measure the intercontinental transmission economy between the two continents.

3.1. The generation cost of renewable energy

The sending generation cost, is considering renewable energy generation cost and different energy delivery compound mode.

(1) Considering the characteristics of resources in each region, identify the trend of the renewable energy base construction and transportation cost.

(2) Considering the development trend of power generation technology, identify the trend of various power generation internal and external costs.

(3) Identify the costs of different electricity generation, such as hydropower, gas and coal, according to unit cost, coal price, gas price, utilization hours, tax rate, operating period, power consumption rate, annual operation and maintenance cost, and hydropower station submerged compensation.

Wind and solar power, with the variable characteristic renewable energy, bring new challenge to the power system. For the restriction of these two generation technique suffering from wind and solar, they cannot maintain better power supply balance. In the future, system and market will be united, and become the founder of renewable energy large-scale implementation. (1) Improve the net benefit of wind and solar power generation at maximum and using this benefit into the whole system. (2)Improve the management strategy, adopting the advancer renewable energy prediction method and enhancing power plant distribution method. (3) Increase the extra auxiliary work, adding demanding resource, storage, grid foundation equipment and flexible equipment usage.

According to the book of Global Energy Interconnection, in 2015, wind and solar power generation amount accounts 90% to the total new power increased. Between 2008 and 2015, the land wind power's average cost decreased about 35%. While the solar generation cost decreased about 80%. More and more evidence showed that, the nest generation wind and solar power generation technique will be more mature, and much more economy, which can push the wind and solar into the widely usage showed in Figure 2.

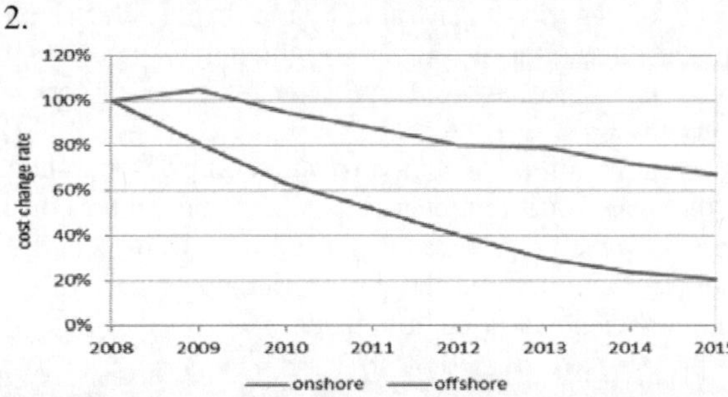

Figure 2. 2008-2015 wind and solar power generation cost curve[7]

Considering the renewable energy resource constraints, the sending side will use more renewable energy to reduce the power cost. The sending renewable energy bases, established in the area of wind and solar energy, facing the energy bundled sending out, its cost will decrease gradually. The solar energy LCOE (levelized cost of electricity) will decrease from 0.13dollar/kw·h of 2015, to 0.055dollar/kw·h of 2025, the decreasing ratio is 59%. The onshore wind cost will drop to 0.053dollar/kw·h of 2025, affected by wind tower, generator, wind farm development and other factors. Consider the renewable energy cost trend, the energy persistence, fast impact the coal, gas and other fossil energy. In the future, the solar and wind power sending costs will have more advantage.

3.2. Transmission cost of channel

Transmission costs of channels depend on how to select transnational power transmission mode. The corresponding project cost and the cost of fluctuations in different countries, line loss and the cost of electricity changes in different situation.

(1) The choice of power transmission type, consider power supply and demand of inter-continent and inter-country, analyse the advantage of choosing transmission mode, analysis characteristics of UHV DC transmission project, transmission channel characteristics, utilization hours and its supporting power supply, including the grid to provide the price of transmission services

(2) Engineering cost, combined with the above project planning to estimate the static cost of construction, project cost fluctuations, the relevant technical costs of the project cost-related.

(3) The line losses cost, should be calculated on the basis of the principle of tariff balance. It is assumed that the cost of landing is the settlement price of the grid companies that provide transmission services and the grid companies in the receiving areas, which is equivalent to the cost of local power purchase.

$$C_{lineloss} = C_{generation} \times K/(1-K) \tag{1}$$

$C_{lineloss}$ is line loss cost, $C_{generation}$ is generation cost, K is line loss.

3.3. The receiving landing cost

The receiving landing cost refers to the sum of the renewable energy, the transmission cost, and the tariff of the line across.

$$C_{landing} = C_{generation} + C_{line}(C_{lineloss}) + T_{tariff} \tag{2}$$

$C_{landing}$ is landing cost, C_{line} is transmission cost, T_{tariff} is tariff.

(1) Landing Cost should be calculated according to the cumulative principle of cost. Power transmission projects across continents, need to consider cross-border tariffs generated by different countries when electricity imports. Each transit country should be a link, based on DC lines, converter stations and supporting exchange projects in the country's investment were calculated transmission costs and line loss costs, the country's landing costs and based on the country's tariff rate The resulting tariffs are added together as the cost of the next country's electricity generation.

(2) There are relay points in some countries. In the landing cost calculation, the DC cost of the before relay point and the cost of electricity generated by the surplus electricity, which are taken as the DC generation cost. It can be used as after the power-weighted average of the respective country.

3.4. The landing cost competitiveness

The landing cost competitiveness refers to analysis and comparison, under the basic and global interconnection scenarios, the use of different energy combinations in the form of energy transmission area, compared with the cost of receiving electricity in the renewable energy power generation. The lower the landing cost, the more competitive the corresponding price will be. It will have the better economical; the transmission efficiency will be obviously.

$$P = \left(1 - C_{landing} / C_{generation}\right) * 100\% \tag{3}$$

P is the landing cost competitiveness.

4. Prediction

The connection of Asia and Europe has make up six intercontinental transmission schemes on two continents, which are estimated according to the investment of DC transmission project.

4.1. The prediction of power generation cost

Based on IEA and GWEC research, Mongolia wind power potential installed capacity is 1-3.1 million kW, East Gobi and Middle Gobi are 0.75-1.0 million kW. The maximum load is projected to increase to 3 million kilowatts by 2030.

Mongolia solar energy scale is about 0.37 million kilowatts, mainly concentrated in the southern Gobi, south Gobi, the East Gobi and Middle Gobi, which are rich of solar. The average radiation per square meter is 1600 kW·h. Total radiation of 570 million kW·h.

Xinjiang wind base is mainly develops Hami, Dabancheng. It predicted the wind capacity will be 20 million kW. We predicted that solar capacity will be reach 400 million kW in 2020.

4.2. The prediction of energy Tariffs

Based on IEA and GWEC research, China's tariff rate ranges from 0% to 35%, Poland ranges from 0% to 15% (average 4.2%), and Germany ranges from 5% to 17%.By the end of 2011, Russia, Belarus, Kazakhstan, the three countries jointly created the Eurasian Economic Commission, on September 1 of 2014, the weighted average tariff rate fell to 7.1%.In this research, it chooses the rate can be seen by Table 2.

Table 2. The sending cost of variable ports [8] (Yuan/kW·h)

	Russia	Mongolia	Kazakhstan	Xinjiang
Coal	0.4442	0.2175	0.2055	0.250
Gas	0.60	-	0.397	-
Hydro	0.627	-	0.627	-
Wind	-	0.3922	-	0.51
Solar	-	0.7945	-	0.8

Consider the proportion of wind and solar capacity, hydropower flood season adjustment, gas power adjustment. It can get different power generation cost showed in Table 3 and 4.

Table 3. The cost analysis of sending side power-jointed [9]

	program design	synthesized generation cost（yuan/kWh）	Synthesized hour（h）
1	Kazakhstan gas / hydro + China Xinjiang wind power	0.4601	6000
2	Kazakhstan gas / hydro + China Xinjiang + Mongolia solar wind power	0.4978	5800
3	China Xinjiang wind power + Russia Siberia hydropower	0.4962	5900

Table 4. Tariffs along the lines cross the different countries

Country	Calculation tariff（%）
China	17
Kazakhstan	7.1
Russia	7.1
Belarus	7.1
Poland	4.2
Germany	11

4.3. The loss rate prediction

Based on Ultra high power line manuscript, ± 800, ± 1100 kilovolt converter station power loss is generally between 0.5% to 1%, this research make the value as 0.75%.

4.4. The transmission cost prediction

According to the transmission scheme and power-jointed program mentioned above, combining with basic usage hours. It can get the relative transmission cost. In Table 5, one, two and three represent the program design. 1, 2 represents different transmission type, which has different line landing port. It can get different transmission cost. These costs can show the comparison and get cost advantage.

Table 5. The corresponding transmission cost(yuan/kWh)

	one/1	one/2	two/1	two/2	three/1	three/2
Transmission cost 1	0.126225	0.140363	0.135270	0.149710	0.185018	0.201080
Transmission cost 2	0.078844	0.081095	0.078844	0.081095	/	/

4.5. The landing cost

According to the relative basic usage hour and its calculation method, the landing cost can show the receiving energy price which has advantage and competitiveness showed in Table 6.

Table 6. The relative landing cost and competitiveness analysis (yuan/kWh、%)

	one/1	one/2	two/1	two/2	three/1	three/2
Landing cost	0.9905	1.0566	0.9121	0.9699	1.0768	1.1694
cost competitiveness	17.46	11.95	24.00	19.17	10.27	2.55

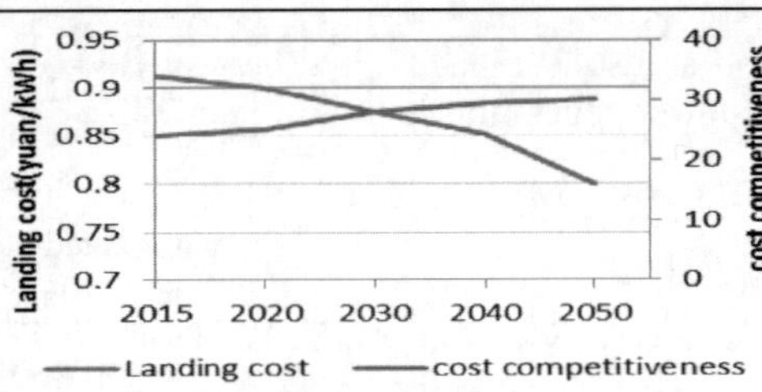

Figure 3. Under scheme two/1 analyse the landing cost and cost competitiveness

In the scheme above, the landing cost of two/1 is lowest, and cost competitiveness is highest. While the landing cost of three/2 is highest, and cost competitiveness is the lowest. This showed that the relative landing cost of scheme two is higher than the scheme one. The relay transmission program's landing cost is lower than the direct transmission program. The power scheme two is higher than one. The loss cost of scheme two is lower than other two schemes.

Considering the tariff of covering countries, China is the DC original point, offset the making Kazakhstan as the original point scheme's transmission loss affection. The landing cost of scheme one and three is higher than scheme two. The tariff of scheme three is obvious high. Cost competitiveness of direct scheme cannot get the same affection with relay scheme.

Figure 3 gives the 2015-2050 landing cost and cost competitiveness. Considering all kinds of factors, as the jointed power basement reset power proportion of the scheme one and two is the same, the after generation power cost of scheme one is higher than scheme two. We can get the result is that; the lower of generation cost, the landing cost affected by generation cost will be great.

5. Conclusion

The global wind energy resources exceeds 1 trillion kilowatt, and solar energy resources exceed 100 trillion kilowatts, surpassing the human society. This research gives the northern-hemisphere energy interconnection development model, connecting Asia and Europe energy together. It also identifies the factors effecting the transmission changes cost, including generation cost, transmission cost (line loss) and landing cost. And estimate these two continents cost benefit and variable power-jointed scheme cost competitiveness. The research showed that under the northern-hemisphere energy interconnection mode, the trans-continent mode had better benefit, and the landing cost is good to be used, can solve the pollution and energy restriction. And in the future, it will add more research focusing on the connection reasonableness.

References

[1] Tian X and Sun Y 2014 *Power System Technology* **39** 663–668
[2] Wei X, Zhang J and Huang H 2010 *Power System Technology* **39** 46–50
[3] Yang H, Wang D and Liu D 2011*Power System Protection and Control* **43** 91–96
[4] Huang Y, Wang Z, Liu J, Zhang L and Li J 2014 *Electric Power Construction* **32** 100–103
[5] Xu J 2013 *North china electric power university* 29
[6] Sun B 2015 *Zhejiang University* 121
[7] IEA Energy balances of OECD countries 2015 112–162
[8] IEA World Energy Outlook 2015 65–78
[9] GWEC Global Wind Report 2015 89-120

Performance analysis of different ORC configurations for thermal energy and LNG cold energy hybrid power generation system

Zhixin Sun[1, 2, 3], Feng Wang[1], Shujia Wang[1], Fuquan Xu[1] and Kui Lin[2]

[1]School of Chemical Engineering, Fuzhou University,No.2 XueYuan Road, University Town, Fuzhou, Fujian, 350108, P R China

[2]Fujian Snowman Co.,Ltd., Dongshan West Road, Minjiangkou Industrial Zone, Fujian, 350200, P R China

E-mail: zxsun@fzu.edu.cn

Abstract. This paper presents a thermal energy and Liquefied natural gas (LNG) cold energy hybrid power generation system. Performances of four different Organic Rankine cycle (ORC) configurations (the basic, the regenerative, the reheat and the regenerative-reheat ORCs) are studied based on the first and the second law of thermodynamics. Dry organic fluid R245fa is selected as the typical working fluid. Parameter analysis is also conducted in this paper. The results show that regeneration could not increase the thermal efficiency of the thermal and cold energy hybrid power generation system. ORC with the reheat process could produce more specific net power output but it may also reduce the system thermal efficiency. The basic and the regenerative ORCs produce higher thermal efficiency while the regenerative-reheat ORC performs best in the exergy efficiency. A preheater is necessary for the thermal and cold energy hybrid power generation system. And due to the presence of the preheater, there will be a step change of the system performance as the turbine inlet pressure rises.

1. Introduction

In recent years, renewable energy utilization and industrial waste heat recovery have gained more and more attentions due to the energy structure and the environmental concerns. Since the temperatures of renewable energy sources and industrial waste heat are relatively lower than conventional fossil fuels, Organic Rankine Cycle (ORC) systems have been considered as a better solution than the water-steam Rankine cycle [1]. Many efforts have been made to improve the performance of ORC. Some researchers point out that reducing the condensing temperature of working fluid, for example, using Liquefied Natural Gas (LNG) as the cooling medium instead of water [2-3], could greatly increase the system thermal efficiency. Similarly, some LNG researchers figure out that increasing the evaporation temperature of working fluid, such as using solar energy or waste heat as the heat sources instead of seawater [4-6], could obviously improve the efficiency of LNG cold energy power generation system. Thermal energy and cold energy are two different types of energy. However, most of the studies focused on the efficiency improvement of one type of energy system achieved by the addition of the other type of energy, assuming the additional energy is offered free of charge. Few researches considered the thermal energy and the cold energy equally as the energy inputs. In this paper, we establish a low temperature thermal energy and LNG cold energy hybrid power generation system.

Low temperature thermal energy is adopted as the heat source of the steam generator (SG) and LNG is taken as the cooling media of the condenser.

The configuration of ORC is critical for achieving higher system efficiency. The basic ORC comprises four key components: the steam generator, the steam turbine or expander, the condenser and the pump. Nowadays, regeneration and reheat have been widely used to improve the Rankine cycle efficiency of large-scale power plants. Hence, some researchers have studied the effects of the addition of regenerator or reheater on the ORC efficiency. The addition of regenerator has been investigated by many researchers and the results show that regenerative ORC has a higher efficiency than the basic ORC [7-11]. Studies on reheat ORC shows that the addition of reheater could increase the net power output obviously but the efficiency improvement is little and even lower than the basic ORC [12-14]. However, most of the studies mentioned above focused on the ORCs utilizing low temperature thermal energy, little attention has been paid to the ORCs utilizing both thermal and cold energy. Moreover, according to the previous studies of the authors [15], a preheater, in which the working fluid is preheated to the nearly environmental temperature, is quite necessary for ORCs utilizing both thermal and cold energy. Thus, the basic ORC utilizing both thermal and cold energy is composed of five key components: the steam generator, the steam turbine or expander, the condenser, the pump and the additional preheater.

The purpose of this paper is to investigate the performance of different configurations of ORCs utilizing both thermal and cold energy. The paper is organized as follows. In Section 2, different configurations of ORCs are described. Section 3 presents the mathematical models and performance criteria. Section 4 is the results and discussion. Conclusions are shown in Section 5.

2. System description

In this paper, four ORC configurations, the basic ORC, the regenerative ORC, the reheat ORC and the regenerative-reheat ORC, were investigated. The schematic and *T-s* diagrams of the four configurations are shown in Figures 1-4 respectively.

2.1. Basic ORC

As observed in Figure 1, there are five processes in the basic ORC:

Process 1-2: a non-isentropic compression process in the pump;
Process 2-3: a constant pressure heat addition process in the preheater;
Process 3-4: a constant pressure heat addition process in the steam generator;
Process 4-5: a non-isentropic expansion process in the turbine;
Process 5-1: a constant pressure heat rejection process in the condenser.

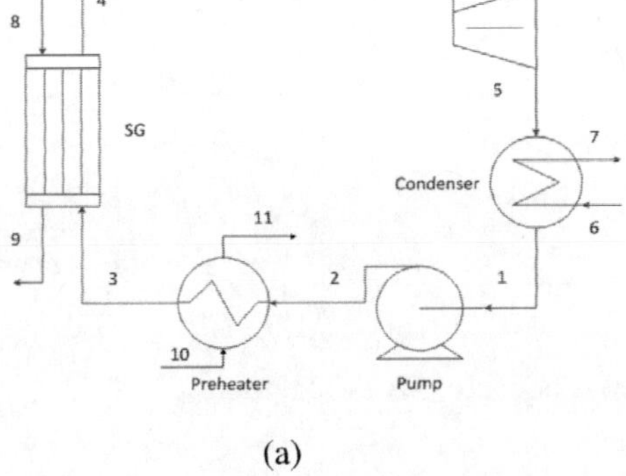

(a)

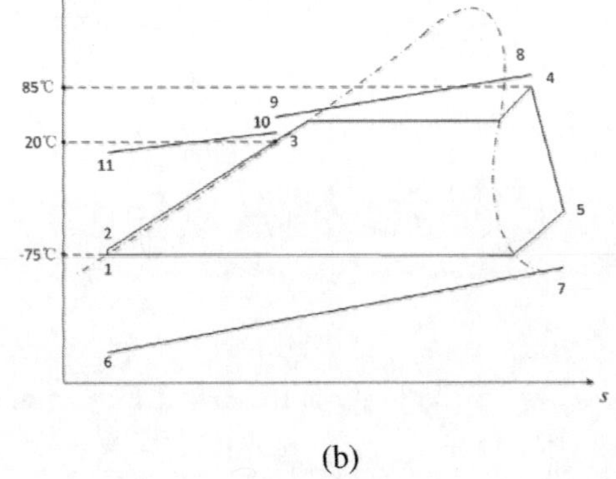

(b)

Figure 1. Basic ORC system: (a) schematic, (b) *T-s* diagrams

A preheater is added in the basic ORC compared to the traditional Rankine cycle. In the preheater, the working fluid is nearly preheated to the environmental temperature by water, which could save the thermal energy in the steam generator.

2.2. Regenerative ORC

For dry fluids, the exhaust vapor exiting the vapor turbine is still in superheated state. Hence, a regenerator is added to recover the remaining energy of the exhaust vapor to preheat the working fluid before it enters the preheater (Figure 2).

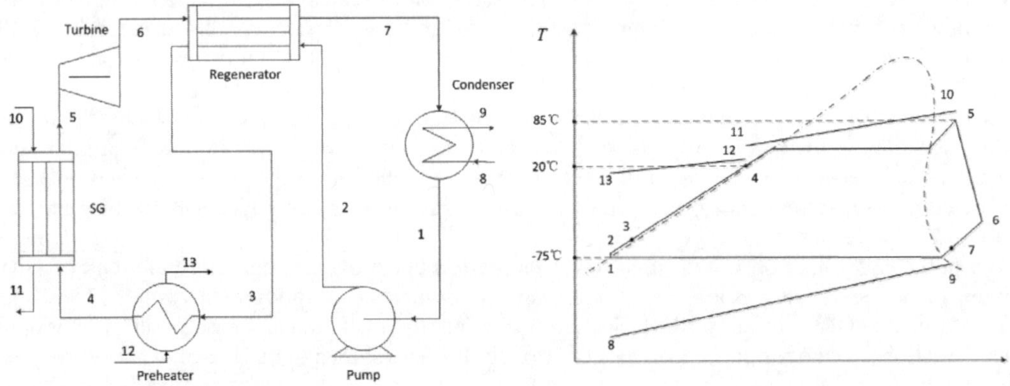

(a) (b)

Figure 2. Regenerative ORC system: (a) schematic, (b) *T-s* diagrams

2.3. Reheat ORC

As shown in Figure 3, the working fluid first expands to an intermediate pressure in turbine I, and then flows into the reheater for temperature boost. The working fluid is heated to the same temperature as the new vapor and then enters turbine II for further expansion.

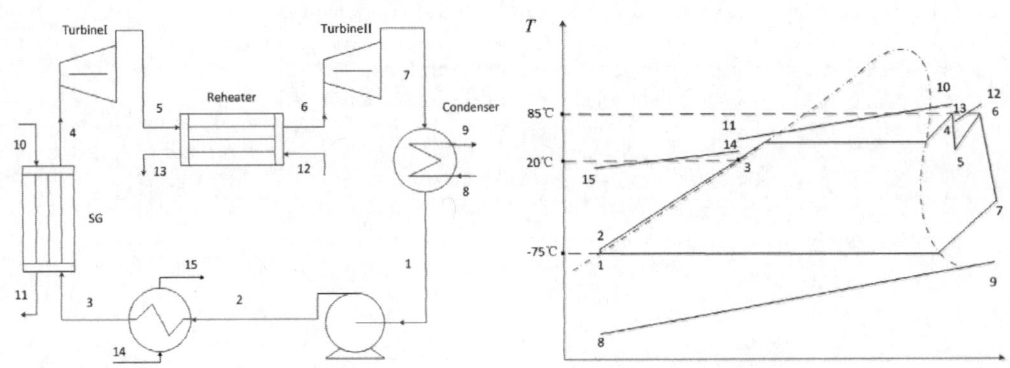

(a) (b)

Figure 3. Reheat ORC system: (a) schematic, (b) *T-s* diagrams

2.4. Regenerative-Reheat ORC

The regenerative-reheat ORC is a combination of the regenerative ORC and the reheat ORC (Figure 4).

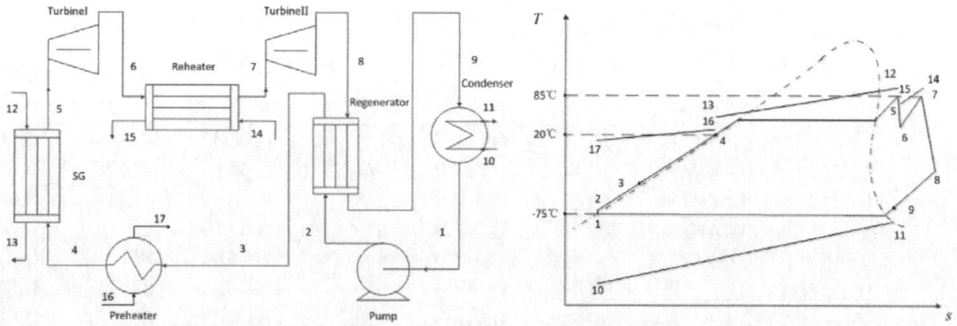

(a) (b)

Figure 4. Regenerative-Reheat ORC system: (a) schematic, (b) *T-s* diagrams

3. Mathematical models and performance criteria

The mathematical models of the devices are established based on the conservation of mass and energy [3]. The thermodynamic properties of working fluid are calculated by the Reference Fluid Thermodynamic and Transport Properties 9.0.

3.1. Performance criteria of the cogeneration system

Both the first and the second law of thermodynamics are employed as the performance criteria. The systems are simulated to compare the performance of the proposed cycle with those of the basic ORC.

The thermal and exergy efficiency can be determined by:

$$\eta_{th} = \left(\frac{\textit{energy in products}}{\textit{total energy inputs}}\right) \tag{1}$$

The exergy efficiency is:

$$\eta_{ex} = \left(\frac{\textit{exergy in products}}{\textit{total exergy inputs}}\right) \tag{2}$$

4. Results and discussion

Table 1. Simulation conditions of the power generation system

Term	Value	Unit	Term	Value	Unit
Environmental temperature	25	℃	LNG pressure	0.6	MPa
Environmental pressure	0.1	MPa	Condensation temperature	-75	℃
Heat source temperature	100	℃	Turbine efficiency	80	%
Heat source pressure	0.5	MPa	Pump isentropic efficiency	80	%
Mass flow rate of heat source	0.8	$Kg\cdot s^{-1}$	Reheater inlet temperature	100	℃
SG inlet temperature	20	℃	Reheater outlet temperature	75	℃
LNG inlet temperature	-161.48	℃	Reheater pressure	0.5	MPa

System simulation was performed in MATLAB. The dry fluid R245fa is usually selected as the working fluid in ORC due to different factors like lower critical temperature almost 154.01℃, critical pressure almost 3.651Mpa, ozone depression potential is 0 and global warming potential is 6300. Then, R245fa was chosen as the typical working fluid. Simulation conditions of the system are displayed in Table 1. Tables 2 show the thermodynamic parameters at each node of the four different configurations respectively.

Table 2. Simulation results of different ORC configurations

State	Basic ORC		Regenerative ORC		Reheat ORC		Regenerative-Reheat ORC	
	t (℃)	*p* (MPa)	*t* (℃)	*p* (MPa)	*t* (℃)	*p* (MPa)	*t* (℃)	*p* (MPa)
1	-75	0.0003	-75	0.0003	-75	0.0003	-75	0.0003
2	-74.87	0.4419	-74.87	0.4419	-74.87	0.4419	-74.87	0.4419
3	20	0.4419	-55.52	0.4419	20	0.4419	-38.06	0.4419
4	85	0.4419	20	0.4419	85	0.4419	20	0.4419
5	-37.15	0.0003	85	0.4419	51.71	0.0884	85	0.4419
6	-161.48	0.6	-37.15	0.0003	85	0.0884	51.71	0.0884
7	-85	0.6	-69.87	0.0003	-9.62	0.0003	85	0.0884
8	100	0.5	-161.48	0.6	-161.48	0.6	-9.62	0.0003
9	62.68	0.5	-85	0.6	-85	0.6	-69.87	0.0003
10	25	0.5	100	0.5	100	0.5	-161.48	0.6
11	15	0.5	62.68	0.5	62.68	0.5	-85	0.6
12	-	-	25	0.5	100	0.5	100	0.5
13	-	-	15	0.5	75	0.5	62.68	0.5
14	-	-	-	-	25	0.5	100	0.5
15	-	-	-	-	15	0.5	75	0.5
16	-	-	-	-	-	-	25	0.5
17	-	-	-	-	-	-	15	0.5

Figure 5 shows the exergy loss distributions of the four ORC configurations. Other exergy losses indicate the sum of the exergy losses in pump, regenerator and reheater. The condenser occupies the largest portion of the total irreversible loss, which is due to the great temperature differences between the working fluid and the LNG. The reheat ORC has the biggest total exergy loss. This can be explained since the reheater boosts the condenser inlet temperature (turbine outlet temperature) of the working fluid from -37.15℃(basic ORC) to -9.62℃, which enlarges the temperature difference between the working fluid and the LNG. As to the regenerative-reheat ORC, the addition of the reheater does not lift the condenser inlet temperature since the regenerator cools down the turbine exhaust steam before it enters the condenser. Furthermore, the regenerator increases the preheater inlet temperature, which results in the decrease of preheater exergy loss. Hence, the regenerative-reheat ORC has the lowest total exergy loss.

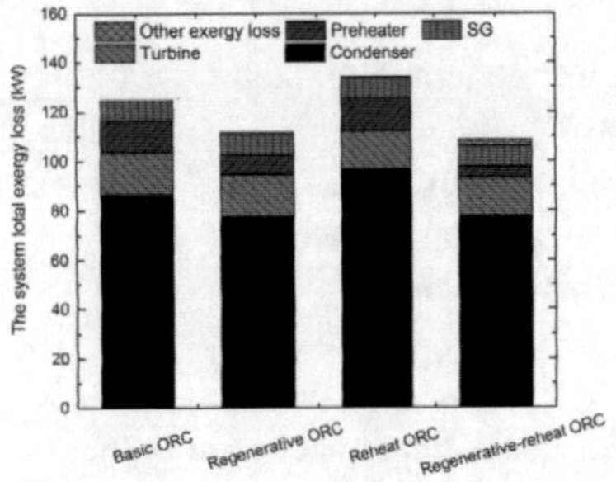

Figure 5. Variation of the exergy loss with the turbine inlet pressure

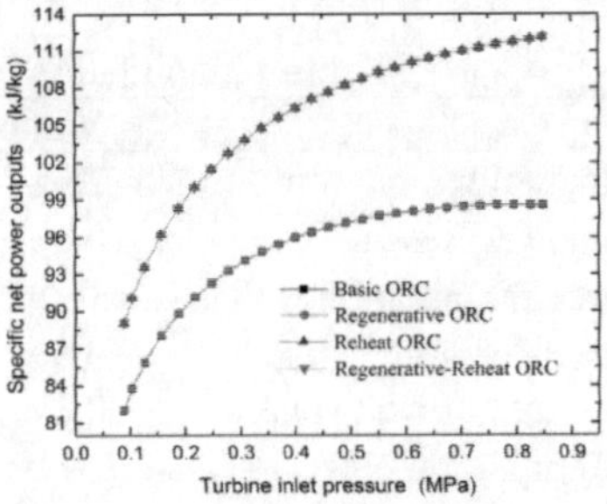

Figure 6. Variation of the specific net power outputs with the turbine inlet pressure

It can be observed from Figure 6 that the specific net power outputs of the basic ORC and the regenerative ORC are identical, which could be explained by the same inlet and outlet parameters of the turbine and the pump as shown in Table 2 .The specific net power outputs of the reheat ORC and the regenerative-reheat ORC are also identical because of the same reason. The specific net power outputs of ORCs with reheat are obviously greater than that of ORCs without reheat since the reheater provides extra energy. It could also be noticed that the increases of net power outputs rise rapidly at first and then become more and more slowly. This is because the increase of turbine power output is becoming more and more smaller but the increase of pump power input is becoming more and more larger as the turbine inlet pressure rises.

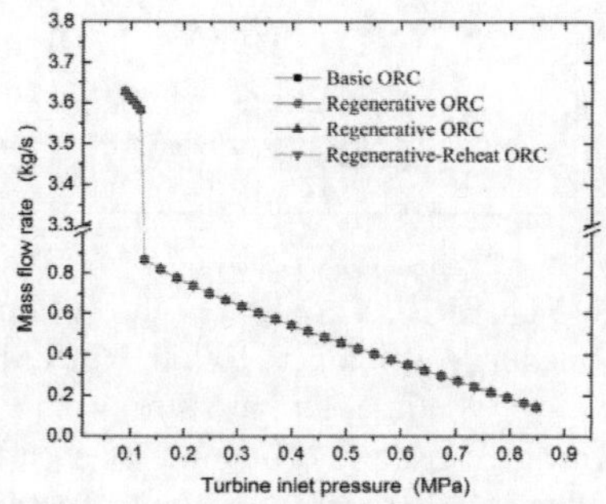

Figure 7. Variation of the mass flow rate with the turbine inlet pressure

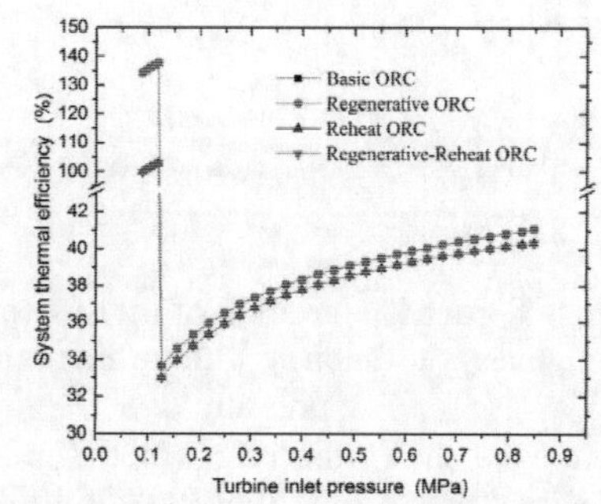

Figure 8. Variation of the system thermal efficiency with the turbine inlet pressure

Figure 7 shows the variation of the mass flow rate with the turbine inlet pressure. There is a step change of the mass flow rate at the turbine inlet pressure of 0.123MPa (the saturation pressure of 20℃ for R245fa). Since the preheater outlet temperature is set to 20℃, when the turbine inlet pressure is lower than 0.123MPa, the working fluid at the preheater outlet is superheated steam, which means the evaporation process of the working fluid is completed in the preheater. However, when the turbine inlet pressure is higher than 0.123MPa, the working fluid at the preheater outlet is subcooled liquid, which means the evaporation process of the working fluid is completed in the steam generator. Hence, the working fluid has a step decrease in the mass flow rate. Another interesting point is that the mass flow rate drops as the turbine inlet pressure rises. This is because the higher the turbine inlet pressure is, the higher the heat source outlet temperature is. That is, less energy provided by the heat source is utilized to heat the working fluid. Hence, the mass flow rate of the working fluid decreases.

Figure 8 illustrates the variation of the thermal efficiency with the turbine inlet pressure. It can be observed that the thermal efficiency is higher than 1 when the evaporation process occurs in the preheater (turbine inlet pressure <0.123MPa) since the cold energy input of LNG is not considered in the calculation of the thermal efficiency. Figure 8 also shows that the addition of regeneration cannot improve the thermal efficiency at all, which is quite different from the results of usual ORC utilizations. The reason for this is that this is a system that using both thermal and cold energy for power generation and a preheater is absolutely necessary as mentioned above. Due to the presence of the preheater, the working fluid will always be preheated to 20℃ by environmental air before it enters the steam generator. Which means the regeneration cannot increase the steam generator inlet temperature since this temperature is kept constant of 20℃ by the preheater. Moreover, it should also be noticed that the thermal efficiencies of ORCs without reheat is greater than that of ORCs with reheat. This is because the reheat process enlarges the specific net power outputs at the cost of more energy input in the reheater.

The variation of the exergy efficiency with the turbine inlet pressure is shown in Figure 9. It can be observed that the regenerative-reheat ORC has the best exergy efficiency since the total exergy loss of this ORC is the smallest as shown in Figure 5. It should also be pointed out that although there is a step decrease of exergy efficiency when the evaporation process is transferred from the preheater to the steam generator, the exergy efficiency after the turning point rises as the turbine inlet pressure increases and finally surpass the previous exergy efficiency peak. The reason for this is that the total exergy loss (especially the loss in condenser) drops more than the total net power output does as the turbine inlet pressure increases.

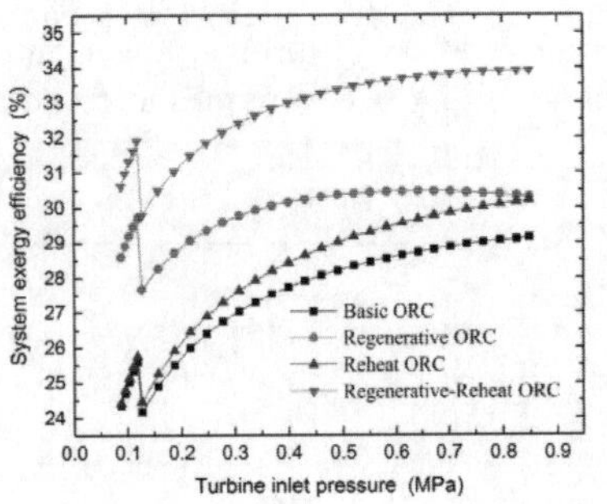

Figure 9. Variation of the system exergy efficiency with turbine inlet pressure

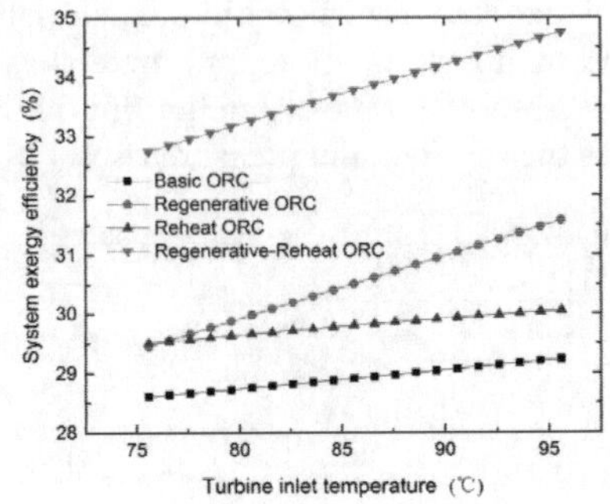

Figure 10. Variation of the system exergy efficiency with the turbine inlet temperature

Figure 10 shows the effect of turbine inlet temperature on system exergy efficiency at a certain turbine inlet pressure (0.706MPa). With the increase of the turbine inlet temperature, the temperature difference between working fluid and the heat source reduces, which results in the decrease of the exergy loss in the steam generator. Hence, the exergy efficiency rises with the increment of turbine inlet temperature for all the configurations.

5. Conclusions

This paper presents the performance analysis of different ORC configurations using dry fluid R245fa as the working fluid based on laws of thermodynamics. The following conclusions may be drawn:

- Reheat could increase the specific net power outputs obviously but it may also reduce the system thermal efficiency.
- As the turbine inlet pressure rises, there is a step change of system performance due to the transfer of the evaporation process from the preheater to the steam generator. The thermal efficiency drop greatly after the turning point while the exergy efficiency rises gradually and finally surpasses the previous peak.
- Condenser accounts for the largest part of the total exergy loss. Utilizing lower condensing temperature could greatly reduce the exergy loss in the condenser.

Acknowledgments

The authors gratefully acknowledge the support of the National Natural Science Foundation Program of China (No.51406037).

References

[1] Liu B T, Chien K H and Wang C C 2004 *Energy* **39** 1207-17
[2] Zhang N and Noam L 2006 *Energy* **31** 1666-79
[3] Song Y, Wang J, Dai Y and Zhou E 2012 *Appl Energy* **92** 194-203
[4] Rao W, Zhao L, Liu C and Zhang M G 2013 *Appl.Therm.Eng* **60** 51-60
[5] V Zare 2015 *Energy.Convers.Manag* **105** 127-138
[6] Bisio G and Tagliafico L 2002 *Exergy.Int.J* **2** 34-50
[7] Gang P, Jing L and Jie J 2010 *Appl Therm.Eng* **30** 998-1004
[8] Roy J P and Ashok M 2012 *Energy* **39** 227-235
[9] Huan X, Li MJ, Xu C and He Y L 2013 *Energy* **58** 473-482
[10] Muhammad I, Muhammad U, Byuny-Sik P and Yang Y 2016 *Energy* **102** 473-490
[11] Feng Y, Zhang Y, Li B, Yang J and Shi Y 2015 *Energy* **82** 664-677
[12] Li G 2016 *Renew.Sust.Energ.Rev* **53** 477-499
[13] Bernardo P, Joaquin N R and Francisco M 2013 *Appl.Therm.Eng* **61** 364-371
[14] Sun Z, Wang J and Dai Y 2012 *Int.J.Hydrogen.Energ* **37** 18731-39

IOP Conf. Series: Earth and Environmental Science **52** (2017) 012025 doi:10.1088/1755-1315/52/1/012025

Computer modeling movement of biomass in the bioreactors with bubbling mixing

L A Kuschev, D Yu Suslov[1], A I Alifanova

Belgorod Shukhov State Technological University, Street Kostyukova, 46, Belgorod, 308012, Russia

E-mail: Suslov1687@mail.ru

Abstract. Recently in the Russian Federation there is an observation of the development of biogas technologies which are used in organic waste conversion of agricultural enterprises, consequently improving the ecological environment. To intensify the process and effective outstanding performance of the acquisition of biogas the application of systems of mixing of bubbling is used. In the case of bubbling mixing of biomass in the bioreactor two-phase portions consisting of biomass and bubbles of gas are formed. The bioreactor computer model with bubble pipeline has been made in a vertical spiral form forming a cone type turned upside down. With the help of computing program of OpenFVM-Flow, an evaluation experiment was conducted to determine the key technological parameters of process of bubbling mixing and to get a visual picture of biomass flows distribution in the bioreactor. For the experimental bioreactor the following equation of V=190 l, speed level, the biomass circulation, and the time of a single cycle of u_{ax}=0,029 m/s; Q_C=0,00087 m^3/s, Δt_{bm}.=159 s. In future, we plan to conduct a series of theoretical and experimental researches into the mixing frequency influence on the biogas acquisition process effectiveness.

1. Introduction

Lately, the biogas stations and equipment have been introduced into the agricultural industry on the territory of the Russian Federation, whose main objective is the recycling of organic waste [1-3]. The recycling takes place due to active life activity of microorganisms which transform proteins, fats and carbohydrates into methane and carbon dioxide. To provide a comfortable environment for reproduction and activity of microorganisms, it is necessary to maintain a certain uniform temperature and distribution of bacteria all over the biomass. For the solution of these problems lately, a combination use of various design devices is applied, currently the systems of mixing of bubbling type have been installed in the biogas reactor [4-6].

An original design of the bioreactor has been developed (fig. 1) [7]. The novelty of the bioreactor construction is the system of mixing device type which represents an evenly perforated pipe created in a form of vertical spirals formed in a turned down cone towards the base. Hence the biogas enters the bubbling pipeline in bubble forms, which when rising under the tangential force, push the biomass in the vertical direction. As a result of the underlying rising bubbles formed there is a biomass suction discharge. Thus, there is an observation of liquid-gas circulation flow and intensive mixing environment in the bioreactor device.

Currently the main method of bubbling hydrodynamics studies remaining is carried out in the laboratory and industrial experiments, of subsequent mathematical handling data have been obtained [5,

6]. However lately the methods of mathematical and computer modeling of two-phase flows of bubble disperse phase have increased [8].

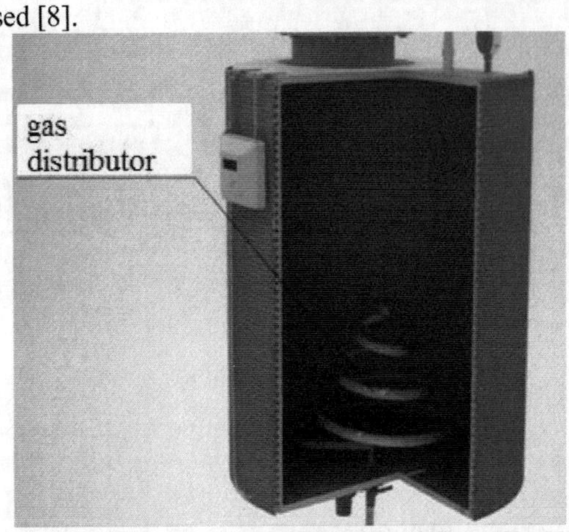

Figure 1. The bioreactor with system of hashing of bubbling type.

2. Methods

2.1. Mathematical model

A mathematical model of the bubbling stirring biomass, involving the biomass and equations of motion of the bubbles of biogas has been developed [9].

Given the high consumption rate of gas bubbling so the description of the motion adopted in the biomass Reynolds equations for average turbulent motion, have been taken [10, 11]:

$$\frac{du_j}{dx_j} = 0 \tag{1}$$

$$\rho\frac{\partial u_i}{\partial t} + \rho u_j \frac{\partial u_i}{\partial x_j} = -\frac{\partial P}{\partial x_i} + \mu\frac{\partial^2 u_i}{\partial x_j \partial x_j} + \frac{\partial}{\partial x_j}\Pi_{ij} + F_i \tag{2}$$

u_i – the velocity vector of biomass, m/s; P – the average pressure value, Pa; μ [mu] – coefficient dynamic viscosity; F_i – strength of interfacial interaction, N; Π_{ij}, – turbulent stress tensor in the biomass, m^2/s^2; ρ [rho] – the density of biomass, kg/m^3.

To describe the motion of bubbles of the biogas a transfer equation of the bubble pulse has been made in a complete quasi environment:

$$\frac{\partial \beta}{\partial t} + \frac{\partial(\beta \upsilon_j)}{\partial x_i} = 0 \tag{3}$$

$$\rho_g \frac{\partial(\beta\upsilon_i)}{\partial t} + \rho_g \upsilon_j \frac{\partial(\beta\upsilon_i)}{\partial x_j} = -\rho_g \beta f_D(\mathrm{Re}_n)\frac{\upsilon_i - u_i}{\tau_P} - \beta(\rho - \rho_g)g_i + \frac{\partial(\beta \Pi_{ij}^P)}{\partial x_j}. \tag{4}$$

β [beta] – The concentration volume of bubbles, %; υ_i [upsilon] – the velocity vector "liquid bubble", m/s; τ_P [tau] - the dynamic relaxation time of bubble, s; f_D - the resistance function; Π_{ij}^P - the turbulent tensor stress in the "liquid bubble", m^2/s^2, ρ_g – [rho] density of the gas fed to the mixing kg/m^3.

2.2. Computer model

The developed mathematical model of bubbling mixing of biomass was made by means of software computing aerodynamics package of OpenFVM-Flow. The volume control method is applied to modeling of the spatial two-phase flow.

The basic data used in the computer modeling of bubbling mixing of biomass are constructive and technological parameters of the experimental bioreactor: the bioreactor height is H_b=1 m, the internal diameter of the bioreactor D_b = 0,5 m, the biomass column height H_{bm}=0,7 m, the density and viscosity of biomass are respectively equal to: ρ=1040 kg/m^3, μ=1,25·10^{-3}, the hole number of the bubbler is n_h=27, the diameter of a hole is d_h=0,001 m, the velocity of the escaping bubbling gas is υ_h=30 m/s, the rotation number of bubbling in the pipeline is n_t=4, the external diameter of the bubbling pipe is d_{bp}=0,017 m, the bubbling gas consumption is Q_g=0,00064 m^3/s, its determined velocity is υ_r=0,0032 m/s, the average diameter of the bubbling gas is d_b=0,005 m.

The computer model of the developed bioreactor is shown in figure 2.

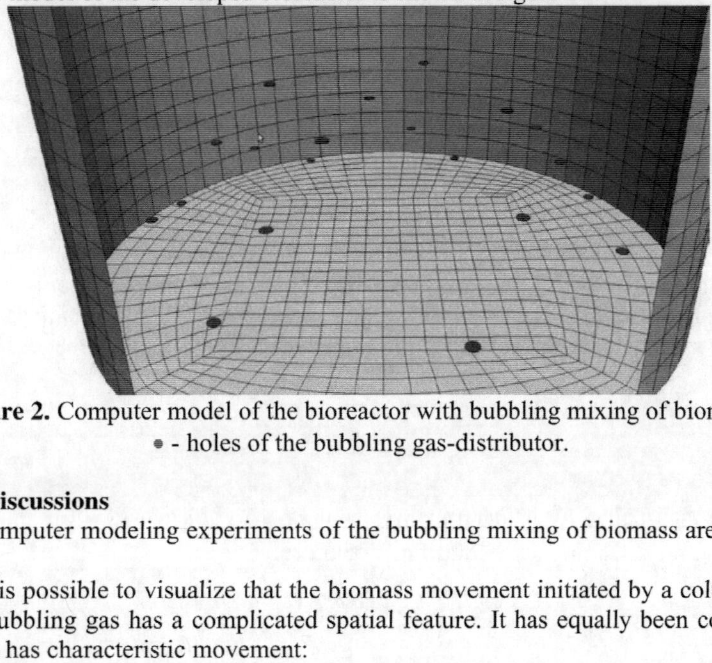

Figure 2. Computer model of the bioreactor with bubbling mixing of biomass:
• - holes of the bubbling gas-distributor.

3. Results and discussions

The results of computer modeling experiments of the bubbling mixing of biomass are given in the figure 3.

In figure 3 it is possible to visualize that the biomass movement initiated by a column of emerging bubbles of the bubbling gas has a complicated spatial feature. It has equally been confirmed that the bubbled biomass has characteristic movement:

– the bubbled movement of the biomass has circular character: the central area of the reactor is occupied by the escaping biomass, reaching the free surface, turning in the opposite direction, forming a circular along the container of the bioreactor;

– the axial component of velocity of the flow of the biomass $u_z(R)$ c with the reduced value of R increases and reaches the maximal value of the reactor axis which is u_{ax};

– the biomass flow velocity occupying the circular area, increases the R in absolute value, reaching the maximum value $R=R_1$, and then reducing to zero against the internal surface of the reactor case.

To preserve the mass of substratum it is necessary that its current complete circulation should be equal to zero:

$$Q = \int_0^{R_b} u_z(R) R \, dR = 0 \tag{5}$$

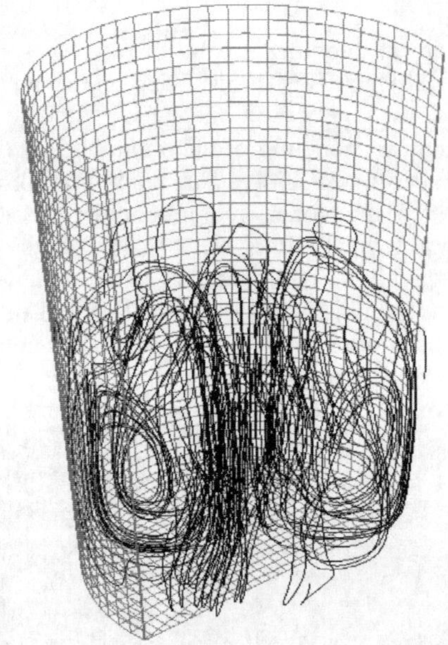

Figure 3. Visual lines of the current of the bubbled biomass.

Under real condition the axial velocity component of biomass depends not only on R, but also on the axial movements of Z. In particular, u_z=0 in the reactor base and the free biomass surface. In this regard $u_z(R)$ is the axial velocity of the biomass averagely distributed on all the layer height:

$$u_z(R) = \frac{\int_0^{H_{bm}} u_z(R,Z)\, dZ}{H_{bm}} \quad (6)$$

The features of distribution of the axial biomass component velocity can approximately be defined along the bioreactor radius in the fourth multifunctional degree:

$$\tilde{u}(x) = Ax^4 + Bx^3 + Cx^2 + Dx + E \quad (7)$$

$\tilde{u}(x) = u_Z / u_r$, $x = R/R_b$.

The coefficients of equation (7) are defined in the system of linear equations which are stated in the following conditions:

$$D = 0, \quad E = \tilde{u}_{ax} \quad (8)$$

$$Ax_0^4 + Bx_0^3 + Cx_0^2 + \tilde{u}_{ax} = 0 \quad (9)$$

$$4Ax_1^2 + 3Bx_1 + 2C = 0 \quad (10)$$

$$A + B + C + \tilde{u}_{ax} = 0 \quad (11)$$

$$A/6 + B/5 + C/4 + \tilde{u}_{ax}/2 = 0 \quad (12)$$

$\tilde{u}_{ax} = u_{ax} / u_r$, $x_0 = R_0 / R_b$, $x_1 = R_1 / R_b$.

The coefficients of A, B, and C are defined in the equations of (9…12):

$$A = -\frac{15x_0^3 - 18x_0^2 + 3}{3x_0^4 - 5x_0^3 - 2x_0^2} \cdot \tilde{u}_{ax} \quad (13)$$

$$B = \frac{5 \cdot (3x_0^4 - 4x_0^2 + 1)}{3x_0^4 - 5x_0^3 - 2x_0^2} \cdot \tilde{u}_{ax} \quad (14)$$

$$C = -\frac{2 \cdot (9x_0^4 - 10x_0^3 + 1)}{3x_0^4 - 5x_0^3 - 2x_0^2} \cdot \tilde{u}_{ax} \quad (15)$$

According to the defined coefficients *A*, *B* and *C* x_1 is determined in relation to the radius on which the velocity of the down biomass flow reaches a maximum value:

$$x_1 = \frac{-3B \pm \sqrt{9B^2 - 16AC}}{8A} \quad (16)$$

The selection of two possible x_1 values depends on the equation obtained out of $x_0 < x_1 < 1$.

To resolve ratio equations of (7), (13…15), that provide the profile determination of the axial biomass velocity it is necessary to calculate the value of $\tilde{u}_{ax}$ and x_0.

A series of computing experiments was conducted to investigate the dependence on the technological parameters which have the greatest impact on the bubbling biomass mixture intensity: at specified velocity of the bubbling gas υ_r ($0{,}0025 \le \upsilon_r \le 0{,}1$ м/с) and the reactor diameter D_b ($0{,}25 \le D_b \le 10$ м).

The processing results of the computing experiment have shown that the parameters of $\tilde{u}_{ax}$ and x_0 can be presented in the form of functions of the Reynolds figures of the bioreactor:

$$\mathrm{Re}_b = \frac{\upsilon_r D_b \rho}{\mu} \quad (17)$$

The method of the smallest squares on depending on the $\tilde{u}_{ax}(\mathrm{Re}_b)$ и $x_0(\mathrm{Re}_b)$ has been also arrived at by the following expressions:

$$\tilde{u}_{ax} = 1{,}06lg^2\,\mathrm{Re}_b - 12{,}43lg\,\mathrm{Re}_b + 37{,}25 \quad (18)$$

$$x_o = 0{,}66 - 0{,}022lg\,\mathrm{Re}_b \quad (19)$$

For the experimental bioreactor the following values have been arrived at by Re_b=1373; $\tilde{u}_{ax}$=8,7; x_0=0,59; x_1=0,83; R_0=0,15 m; $\tilde{u}_{ax}$=0,029 m/s, and the equation (7) describing the axial biomass profile velocity is shown in figure 4.

$$\frac{u}{u_{ax}} = 5{,}63x^4 - 4{,}39x^3 - 2{,}24x^2 + 1 \quad (20)$$

According to the obtained profile from the axial biomass velocity (fig. 4) we were able to determine that the circulation equals to the spatial volume flow:

$$Q_C = 2\pi \int_0^{R_0} u_Z(R)\,R\,dR = 2\pi R_b^2 u_{ax} \int_0^{x_0} \left(5{,}63x^4 - 4{,}39x^3 - 2{,}24x^2 + 1\right) x\,dx \quad (21)$$

For the bioreactor with a biomass volume V_{bm}=0,1372 m^3 the following parameters were received: u_{ax}=0,029 m/s; x_0=0,59; Q_C=0,00087 m^3/s.

As one of the characteristics of intensity of the bubbling mixing, the coefficient of mixing of biomass in the bioreactor which is K_m [12], is similar to the coefficient of air exchange in the ventilated production rooms and equals to the used volume of the current circulating to biomass amount:

$$K_m = \frac{Q_C}{V_{bm}} \tag{22}$$

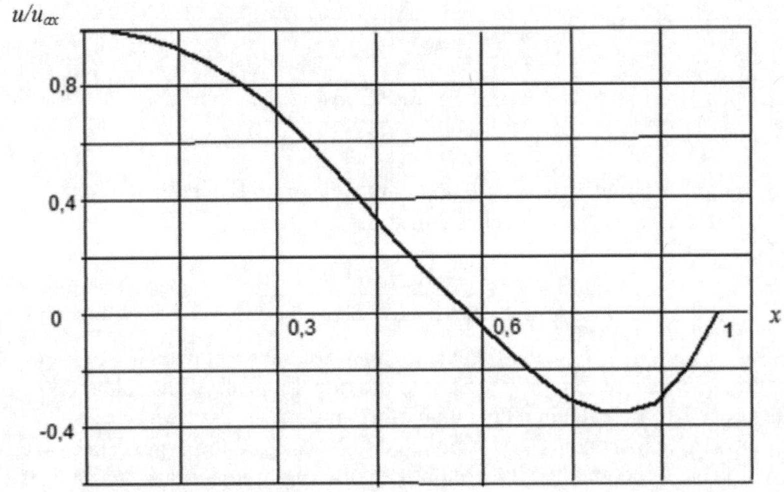

Figure 4. The profile of the average axial biomass velocity.

The size of the reverse coefficient of mixing is equal to the minimum duration of Δt_{bm} one cycle of bubbling mixing of biomass during one cycle of the whole volume:

$$\Delta t_{bm} = \frac{1}{K_m} \tag{23}$$

4. Conclusion

On the basis of the developed mathematical computing model experiment the key parameter processes of the bubbling mixing has been determined and a visual picture of distribution of biomass flows in the bioreactor has been received. As a result of the data processed the experiment of the axial profile velocity of biomass flow was obtained. For the experiment in the bioreactor, the volume of V=190 l, the velocity, the circulation time of the biomass and one cycle time of mixing the following equation has been used u_{ax}=0,029 m/s; Q_C=0,00087 m^3/s, Δt_{bm} =159 s. The results of the entire research were used in the industrial calculation of constructive and technological parameters in industrial bubbling bioreactors. In future, we plan to conduct a series of theoretical and experimental researches into the mixing frequency influence on the biogas acquisition process effectiveness.

Acknowledgments

This work was supported by the Russian President's Scholarship SP - 1716.2015.1.

References

[1] Vasilov R G 2007 Prospects for development of biofuel production in Russia. Message 3: biogas. Bulletin of biotechnology and physical and chemical biology named after Y.A. *Ovchinnikov* **3(3)** 54-61.

[2] Evstuynichev M A and T N Il'ina 2013 The features of raw material base in the Belgorod Region for the biogas production. Bulletin of Belgorod State Technological University named after V.G. *Shukhov* **5** 170-173.

[3] Birkin S M 2009 Perfection of technology and means of disposal of cattle manure: Abstract.

Diss ... cand. tehn. Sciences: 05.20.01. *Birkin Sergei Mikhailovich* Volgograd 23 p.

[4] Karim K, Klasson K T, Hoffmann R, Drescher S R, DePaoli D W and Al-Dahhan M H 2005 Anaerobic digestion of animal waste: Effect of mixing. *Bioresource Technology* **96**(14) 1607-12.

[5] Tur A A 2006 Scientifi c Principles of Calculation of Primary and Auxiliary Bubbling Reactors of Technological Blocks *Auth. Abstr. Dissert. Doct. Techn. Sci* 05.17.08, Angarsk.

[6] Kostromin D V 2010 The anaerobic digestion of organic animal waste in the bioreactor with bubbling stirring: dis ... cand. tehn. sciences. Yoshkar-Ola.

[7] Suslov D Y and Kushchev L A 2014 The device for the production of biogas in anaerobic processing of organic waste, Patent RU No. 144613.

[8] Vavilin V A, Lokshina L Ya, Jokela J P Y, Rintala J A 2004 Modeling solid waste decomposition *Bioresource Technology* **94** 69–81.

[9] Kushchev L.A. and D.Y. Suslov 2016 Modeling of bubble induced mixing of organic substrate in biogas generating bioreactor // Chemical and Petroleum Engineering **52**(**3**) 151-155.

[10] Landau L D and Lifshits E M 2006 Theoretical Physics, Vol. VI, Hydrodynamics, 5th ed.

[11] Loitsyasnskii L G 2003 Liquid and Gas Mechanics: Textbook, Drofa, Moscow, 7th ed.

[12] Dmitrienko, Yu M, Zahoranski R and Fissenko S P 2004 About leveling scalar impurity

Numerical simulation on combustion and exhaust emission of staged combustor

Chengming Yin, Rongchun Zhang, Changjiu Zhang and Weijun Fan[1]

School of Energy and Power Engineering, Beijing University of Aeronautics and Astronautics, Beijing 100191, China

E-mail: fwj1682@126.com

Abstract. Aiming at the goal of low pollution and high efficiency, combustion characteristics and exhaust emissions of radial staged combustor are studied using a numerical simulation method. The result shows that the combustion efficiency of staged combustor is all over 97%. The emission of pollutants, especially the CO and NO_x, is obviously reduced. The advantages of staged combustor cannot be fully exploited, when the airflow is evenly distributed in the double-head. Higher combustion efficiency and lower pollutant emission can be achieved with an appropriate flow distribution ratio.

1. Introduction

With the rapid development of China's aviation career, the emission of pollutants from the aero-engine is increasing, and its' pollution to the atmosphere is becoming more and more serious [1]. At the same time, public awareness of environmental protection and energy saving has been deepening [2]. The main development direction of civil aviation engine has become low pollution and high efficiency [3]· All of countries around the world have carried out extensive research on the low-pollution combustion technology of aero-engines, and developed different low-pollution combustion chambers.

From the 1970s, the United States developed many practical technologies [4, 5], such as the staged combustion, Lean premixed prevaporized (LPP), Variable geometry combustor (VGC), Rich-burn/quick-mix/lean-burn (RQL) [6, 7], et al. In the 1990s, General Electric (GE) Company developed a combustor of double-loop premixed swirler to make the pollution emission be 85% lower than the current standard [7, 8]. In addition, SNECMA in France developed double-head combustion chamber to make the emissions lower. Rolls-Royce in England developed a lean single-stage and axial staged combustion chamber, which can reduce emission of NO_X [1].

China has also conducted research on low-pollution combustion technology, and has made some progress in the low-pollution combustion chamber of gas turbine. Lin Yuzheng [8, 9] et al developed a low-pollution combustion chamber of staged/lean premixed prevaporized to achieve the combination of staged combustion and LPP, which can reduce NO_X emissions. Ma Wenjie [10] et al experimentally investigated the air axially-staged approach's influence on combustion and exhaust emissions.

The main research objectives of advanced aero-engine combustion chamber are reducing the exhaust emissions, especially the NO_X emissions in the future. Therefore, the development of cost-effective low-NO_X emission technologies is important. Based on the radial staged combustion chamber, a double-head staged combustor was designed, and the numerical simulation was carried out to study its combustion characteristics and exhaust emission.

2. The design of double-head staged combustion chamber

When aero-engine combustion chambers are in low-power or high-power operating conditions, emissions of pollutants are different. The staged combustion was proposed to reduce the emissions of pollutants. It divides the combustion chamber into two main combustion zones. As shown in figure 1, each of the main combustion zones has an inlet head, respectively, which is supplied fuel and air alone. Each inlet head is equipped with a swirler, which makes the fuel and air well mixed. Low emissions of pollutants are achieved by controlling the temperature of each main combustion zone.

The double-head staged combustion chamber is an annular combustor. Axial cross-sectional view is shown in figure 1. Flame tube is divided into two primary combustion zone, the length of the main combustion zone is 62mm in First-head, the length of the primary combustion zone is 72mm in Second-head, and total length of flame tube (including double-head) is 127mm. According to the flow distribution, two swirlers are arranged in the head of flame tube, primary holes and dilution holes are provided in the flame tube.

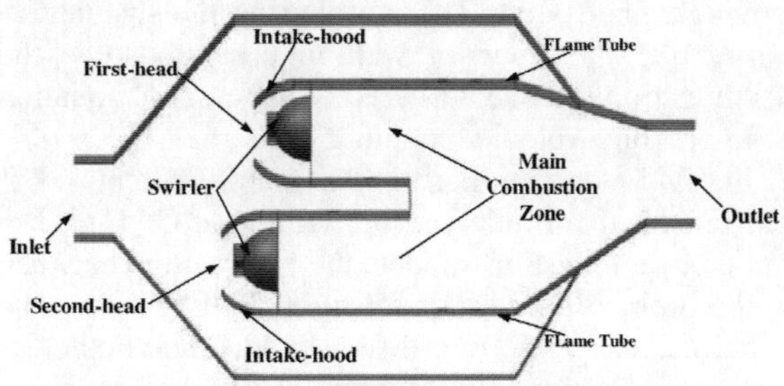

Figure 1. Axial cross-sectional view of the double-head staged combustion chamber.

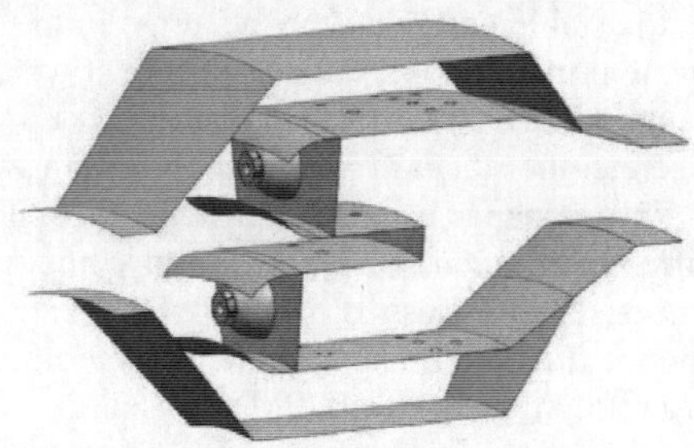

Figure 2. The computational domain.

There are 18 double-head structures evenly distributed in the circumferential direction, so they are periodic symmetrical. Considering the computational resources and calculation precision, a 20-degree fan-shaped region is selected for computational domain. As shown in figure 2, a complete double-head structure, the primary holes and dilution holes, distributed on the downstream surface of flame tube, are in the computational domain.

3. Flow distribution schemes and boundary condition setting

This paper focuses on the impact of air staged proportion on combustion performance and pollutant emissions. Air staged is achieved by different inlet airflow distributions at the double-head sections. As shown in the table 1, two schemes were studied to show the differences in the overall combustion performance resulting from the different inlet airflow distributions at the heads. In the two schemes, the total inlet airflow is the same. The total inlet airflow is evenly divided into two equal portions in Scheme 1, which means the inlet airflow at First-head and Second-head respectively accounts for 50% of total inlet airflow. In the Scheme 2, the inlet airflow at First-head accounts for 45% of total inlet airflow, while the inlet airflow at Second-head accounts for 55%.

The different inlet airflow distributions are achieved by the different area of double-head sections. In order to ensure the same total flow, the total area of double-head sections is designed to be identical. According to different distribution ratio to build the corresponding model. The proportion of the area is exactly the same as the flow rate distribution at each head. The corresponding intake-hood is set outside the head to ensure that the airflow into the flame tube, and improve the static pressure near the head.

Table 1. Schemes of flow distribution.

Scheme	1	2
First-head (%)	50	45
Second-head (%)	50	55

According to the typical working condition of the aero-engine's main combustion chamber, the inlet airflow parameters are selected as follows: the temperature is 850K, inlet Mach number is 0.3, and gas-oil ratio is 0.03. The injector is located behind the swirler.

4. Numerical simulation methods

The numerical simulation is based on 3-dimensional incompressible Navier-Stokes equations. The standard k-epsilon model is a turbulence model, which was proposed by Launder and Spalding. It's a two-equation model and given by the empirical formula. It's used widely in engineering applications and is stable and relatively accurate in high Reynolds number flows. Therefore, using the standard k-epsilon model to calculate turbulent viscosity, selecting the standard wall function to solve the boundary layer flow. The second-order upwind discretization scheme was used in differential equation discrete, the standard SIMPLE algorithm was used in pressure-velocity coupling [11].

In this paper, aviation kerosene was chosen as the fuel. Aviation kerosene is mainly composed of different fractions of hydrocarbon compounds, the chemical formula can be written as $C_{12}H_{23}$. The combustion process is simulated by the non-premixed combustion model, the interaction between chemical reaction and turbulence is simulated by the probability density function (PDF) model. The discrete phase model (DPM) is used to simulate the discrete phase in the flow field. This method is suitable for simulating the problems which must consider turbulent flow and complex chemical reaction mechanism, so it can be well used to calculate the complex combustion in the main combustion chamber. Fuel atomization using Rosin-Rammler distribution, the Santer mean diameter (SMD) is 0.05mm.

The calculate grids adopt unstructured grids. Compared 2.3 million to 3.7 million of grid number, the deviation of combustion efficiency is less than 0.3% in Scheme 1. That means the result of numerical simulation is independent of different grid densities and it is accurate. Therefore, we selected 2.3 million grid numbers to save computing resources and times in this paper.

5. Analysis of numerical results

5.1. Combustion characteristics

Combustion characteristics of aero-engine combustion chamber is mainly represented by the combustion efficiency, higher combustion efficiency represents better combustion performance. In this paper, the combustion characteristics of staged combustion chamber will be studied from three aspects: density, velocity and temperature, which are all closely related to combustion efficiency.

Figure 3 shows the density contour of the combustion chamber along the axial section. The gas is obviously compressed through the swirlers and the holes. The primary combustion zone is in a low-density state, because of the combustion. After the primary combustion zone, the gas gradually expands, and the density of gas is significantly increased near the outlet. Comparison with the schemes, the density changes dramatically at the swirler in Scheme 1, but it is more evenly distributed at the rear of the primary combustion zone.

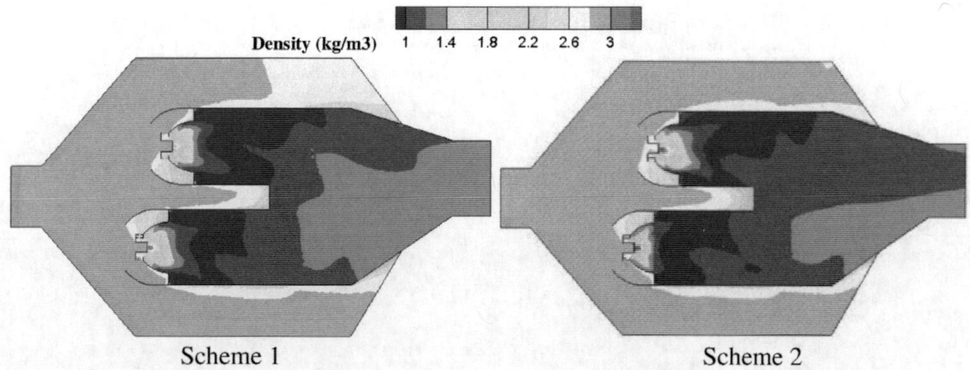

Figure 3. The density contour of the combustion chamber along the axial section.

As shown in figure 4, the zone with the highest speed appears in swirlers, which can make the gas stays for longer periods of time to ensure the gas mixed well with the fuel and burns adequately. The velocity of the primary combustion zones is relatively small, but the changes are quite dramatic. When the air flow to the outlet, the speed is gradually increasing.

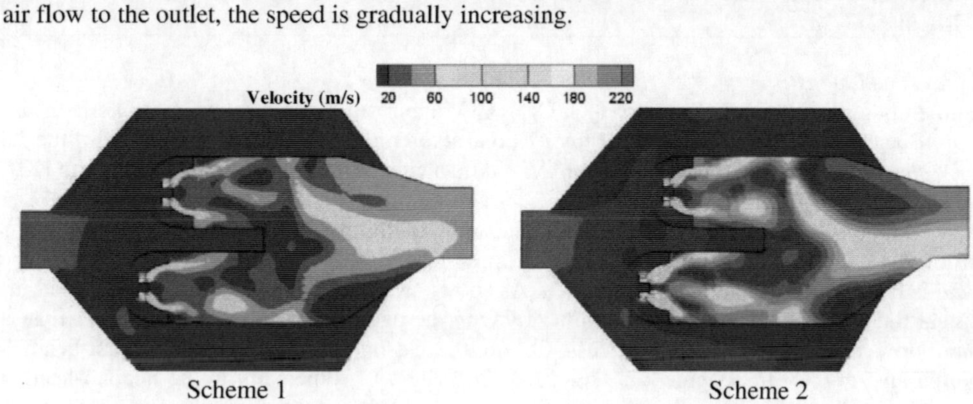

Figure 4. The velocity contour of the combustion chamber along the axial section.

Figure 5 shows the temperature contour of the combustion chamber along the axial section. The primary combustion zone of the Scheme 2 is wider. High-temperature zone concentrated in the main combustion zone, which means the combustion efficiency is pretty well. The combustion efficiency was calculated by the temperature rise method. The formula is:

$$\eta = \Delta T_r / \Delta T_{th} = (\overline{T}_{out-r} - \overline{T}_{in}) / (\overline{T}_{out-th} - \overline{T}_{in}) \quad (1)$$

Character *T* represents the temperature, the subscript of *out* represents the theoretical temperature rise, and the subscript of *r* represents the actual temperature rise.

As shown in table 2, the average velocity at the outlet of Scheme 2 is lower, which means that the air and fuel have more time to mix with each other and burn. It results an increase in the average temperature at the outlet, a decrease in density, and an increase in combustion efficiency. The combustion efficiency is 97.0% in Scheme 1 and 97.6% in Scheme 2. It means that the Scheme 2 has better combustion performance.

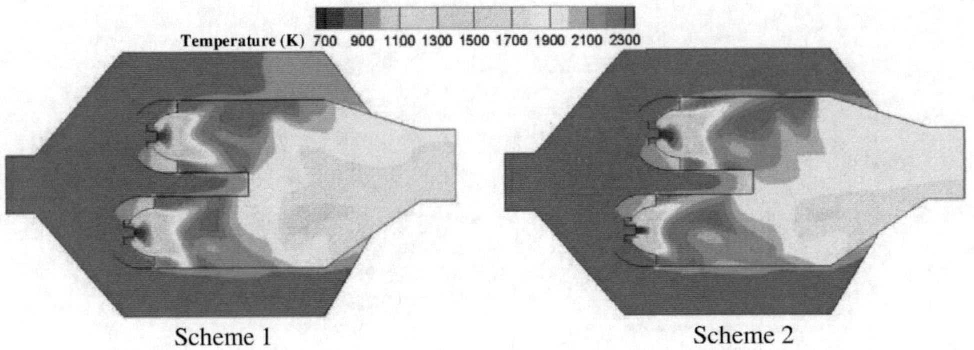

Figure 5. The temperature contour of the combustion chamber along the axial section.

Table 2. Airflow parameters at the outlet and combustion efficiency.

	Density (kg/m^3)	Velocity (m/s)	Temperature (K)	Combustion Efficiency (%)
Scheme 1	1.226	50.0	1791.3	97.0
Scheme 2	1.224	48.8	1796.7	97.6

5.2. Exhaust emissions

The main emissions of aero-engine are: H_2O, CO_2, CO, unburned hydrocarbons (UHC), black smoke and NO_x. Because the combustion efficiency of civil aero-engine combustor is very high, the UHC and black smoke is relatively few and easily oxidized than CO, CH_4 is chosen to represent the UHC in this paper.

Figure 6 shows the distribution of mole fraction of CO, figure 7 shows the distribution of mole fraction of CO_2. CO is a product of incomplete combustion, usually formed in the main combustion zone, and is oxidized to CO_2 in the middle area. As shown in figure 6, CO is mainly distributed in the region near the swirler, and the molar fraction of CO in the two heads is almost the same, because of the same flow distribution. There are less CO in the Second-head than in the First-head. The distribution area of CO_2 in Scheme 2 is larger than in Scheme 1, especially in the Second-head. The reason is that the Second-head of Scheme 2 can get more oxidant to participate in combustion reaction, CO can be quickly oxidized to CO_2 at high temperatures.

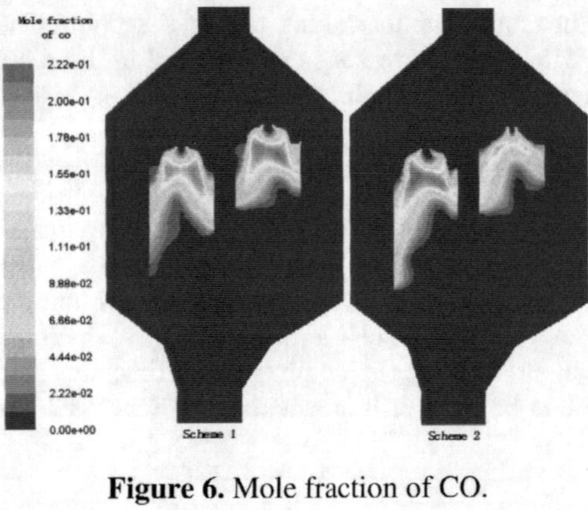

Figure 6. Mole fraction of CO.

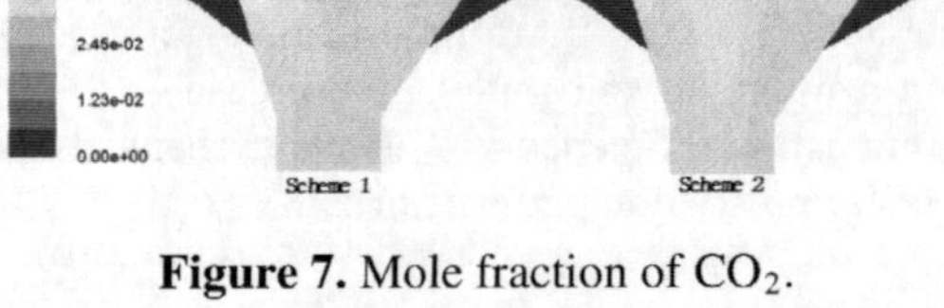

Figure 7. Mole fraction of CO_2.

Figure 8 shows the distribution of mole fraction of NO, figure 9 shows the distribution of mole fraction of N_2O. The molar fraction of NO in the Scheme 1 is obviously higher than in Scheme 2, especially around the swirler. The distribution of N_2O in the 2 Schemes is roughly the same, but the area is larger in the Scheme 2.

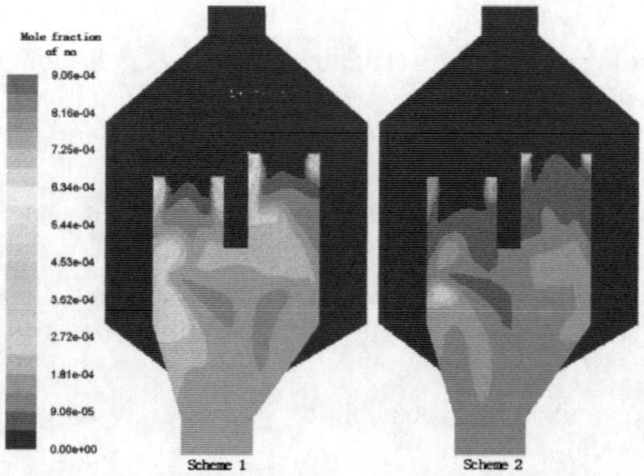

Figure 8. Mole fraction of NO.

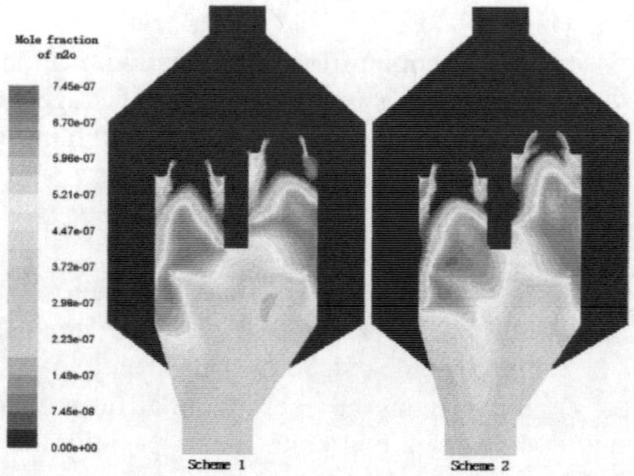

Figure 9. Mole fraction of N_2O.

Figure 10 shows the emission of pollutants at the outlet of the staged combustion chamber. Compared with Scheme 1, the mass fraction of NO at the outlet in Scheme 2 is lower, but the mass fraction of N_2O is a little higher. In terms of the NO_x, the total emissions of Scheme 2 is less, because the mass fraction of N_2O is only one thousandth of NO. And the mass fraction of unburned hydrocarbons is less the 10^{-9}, so it can be ignored. The emission of CO_2 is roughly the same, the CO of Scheme 2 is less.

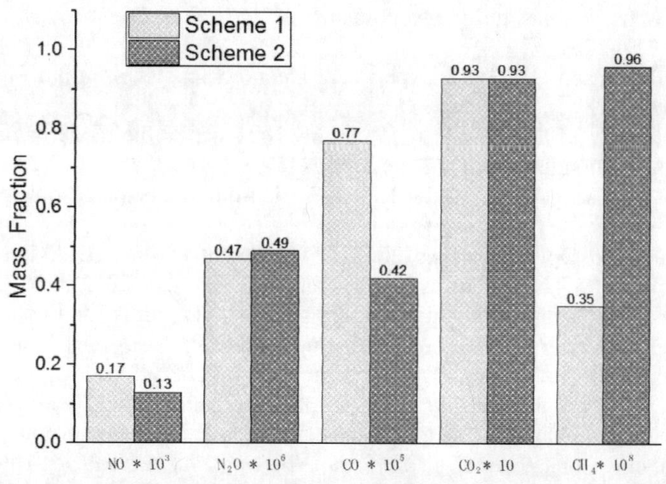

Figure 10. The emission of pollutants at the outlet of the staged combustion chamber.

In general, staged combustion chambers with different flow distributions have less emissions and higher combustion efficiency. There is an optimum flow distribution ratio that will optimize the combustor's combustion performance and pollutant emissions. The most appropriate flow distributions scheme also needs further study in the future work.

6. Conclusion

The paper studied on combustion and exhaust emission of staged combustor in 3 different flow distribution schemes using the numerical simulation method, and obtained some conclusions as following:

- Staged combustion is a new combustion technology with high efficiency and low pollution. The combustion efficiencies of staged combustor are all above 97% under the conditions setting in this paper, and the pollutant emission conforms to the latest civil aviation engine emission standard.
- When the airflow is evenly distributed in the double-head, the advantages of staged combustor cannot be fully exploited.
- By controlling the airflow distribution ratio in the double-head, it was possible to improve the uniformity of the distribution of temperature, density and velocity in the flame tube, and improve the combustion efficiency.
- By selecting an appropriate airflow distribution ratio, it was possible to significantly reduce the emission of pollutants, especially the CO and NO_x.

Acknowledgments

This work was supported by the National Natural Science Foundation of China (No. 51506003).

References

[1] Zhao Jianxing 2008 Pollutant emission and development of low-emission combustion technology for civil aero engine *Journal of Aerospace Power* **23**(6) 986–996

[2] Mongia H C 2003 TAPS-A4th generation propulsion combustor technology for low emissions *AIAA* **2003** 2657

[3] Walker A D, Denman P A and McGuirk J J 2004 Experimental and computational study of hybrid diffusers for gas turbine combustors *ASME J. of Engineering Gas Turbines and Power* **126** 717–725

[4] Bahr W W and Gleason C C 1975 Experiment clean combustor program Phase I: Final report *NACA-CR-134737*

[5] Lew L H, Carl D R and Dezubay E A 1981 Low NO_X heavy fuel combustor concept program Phase I: Combustion technology generation *NASA-CR-165482*

[6] Feitelberg A S, Lacey MA 1997 The GE rich-quench-lean gasturbine combustor *In: ASMEPaper* [C] 97-GT-127

[7] Michael J Foust and Doug Thomsen 2012 Development of the GE Aviation Low Emission TAPS Combustor for Next Generation Aircraft Engines *AIAA* **2012** 0936

[8] Lin Yuzhen and Liu Gaoen 2003 A preliminary study of NO_X emission of staging/premixed and prevaporized lean emission combustor scheme *Journal of Aerospace Power* **18**(4) 492–497

[9] Wang Yansheng and Lin Yuzhen 2016 Experimental Investigation on Ignition Performance of Internally-Staged Combustor *Journal of Propulsion Technology* **37**(1) 98–104

[10] Ma Wenjie, Fan Weijun and Hou Muyu 2004 Influences of air staged proportion on combustion and exhaust emissions *Journal of Combustion Science and Technology* **10**(4) 345–349

[11] Jiang Chunbo, Zhang Yongliang and Ding Zeping 2007 *Computational Fluid Dynamics* (Beijing: China Electric Power Press)

Effect of notch width-ratio of core lobe to bypass lobe on performance of two-dimensional lobed mixer

Jingyi Wu, Weijun Fan, Rongchun Zhang[1], Dawei Zhang

National Key Laboratory of Science and Technology on Aero-Engines Aero-Thermaldynamics, School of Energy and Power Engineering, Beijing University of Aeronautics and Astronautics, Beijing 100191, China

E-mail: zhangrongchun@buaa.edu.cn

Abstract. Geometrical and numerical simulation models of 2-D lobed mixer with differences of b_h/b_c were established to study the effects of b_h/b_c and bypass lobe to thermal mixing efficiency and total pressure recovery coefficient of the mixer. The results show that when the b_h/b_c range from 0.444 to 1.364, the maximum average non-dimensional streamwise vorticity along flow direction rises as b_h/b_c increases. The mixer of which thermal mixing efficiency is maximum varies over stream wise and b_h/b_c of the mixer of which thermal mixing efficiency is maximum increases along the flow direction. Besides, the total pressure recovery coefficient of the lobed mixer ascends as b_h/b_c increases.

Nomenclature

D = diameter of the lobed mixer, 40 mm
d = height of 2-D lobe, 40 mm
x, y, z = coordinates
u, v, w = velocity three components
U_1 = average velocity of the air at the inlet of the lobed mixer, 20 m/s
$\bar{u}$ = average velocity of the air at the inlet of the 2-D lobe mixer, 60 m/s
m = mass flow rare, m/s
T = total temperature, K
ω = non-dimensional streamwise vorticity
η = mixing efficiency
p^* = total pressure
σ = total recovery coefficient
b = notch width
λ = width of computational domain

Subscript

b_h = width of core lobe
b_c = width of bypass lobe
mix = mixed or partly mixed flow

1. Introduction

Lobed mixers, an improved mixing fluid mechanic devices that has corrugated trailing edges, can create large scale streamwise vorticity to realize full mixing in less distance [1-2]. Lobed mixers possess many of advantages such as higher mixing efficiency and excellent stability owing to its special structure. The devices were widely applied to promote aero-engine thrust, reduce infrared radiation of aero-craft, lower jet noise and fuel consumption and improve the performance of reheat combustion chamber [3].

In numerous studies on effective mixing mechanism of lobed mixer since lobed mixer was firstly put forward and applied to aero-engine by Frost in 1966, Paterson [4] proposed the conception of streamwise vorticity and confirmed the existence of streamwise vortices in the flow field behind the lobe edges in the experimental studies. Meanwhile, Paterson expounded the mechanism how streamwise vortices enhance the mixing between core flow and fan flow. Skebe et al [5] further clarified the conformation of streamwise vortices and derived a formula to calculate the circulation at the trailing edge of the lobed mixer which was used to estimate the incremental coefficient of streamwise vortices to the mixing. Elliott et al [6] concluded that three primary contributors concern to the mixing process in mixing flows: the spanwise vortices which occur in any free shear layers; the increased interfacial contact area due to the convoluted trailing edge; the streamwise vortices generated by the radial velocity difference due to special geometry of the lobed mixer.

Besides the theoretical mechanism studies for lobed mixer, many of engineering researches were carried out to design mixer with better performance. Most researches concentrated on the effects of geometry sizes and inlet conditions to the flow field and performance of lobed mixer. Liu et al [7-8] completed a series of quantitative influence researches of geometry on the performance of lobed mixer through numerical simulations based on RANS. However, the studies about the effects of the b_h/b_c on performance of mixer are not comprehensive enough. Besides, the accuracy of numerical simulation still need to be improved.

In this paper, numerical simulation is carried out on the lobe mixer model with different notch width-ratio of core lobe to bypass lobe to find the influences of b_h/b_c on flow field, thermal mixing efficiency and total pressure recovery coefficient of the mixer. The reasons for which the effects were caused are also explored in this paper.

2. Physical model and numerical method

2.1. Computational method verification

The paper established verifying model (Figure 1) and calculated with the methods in the paper referring to the experiments of Hu [9]. Non-dimensional streamwise vorticity in Hu's paper is defined as following:

$$\omega_x = D/U_1\,(\partial w/\partial y - \partial v/\partial z) \qquad (1)$$

Figure 2 shows the non-dimensional streamwise vorticity on x/D=0.5 of numerical simulation results and Hu's experiment. Figure 3 shows the non-dimensional maximum streamwise vorticity on different cross plane from numerical simulation and experimental measurement. According to Figure 2, the flow field of numerical simulation is nearly consistent with experimental results. Similarly, the differences of non-dimensional maximum streamwise vorticity on different cross plane between computational and experimental results are also small, less than 6 percent (Figure 3). It means that the computational method applied in the paper could well reflect the flow changes in lobed mixer.

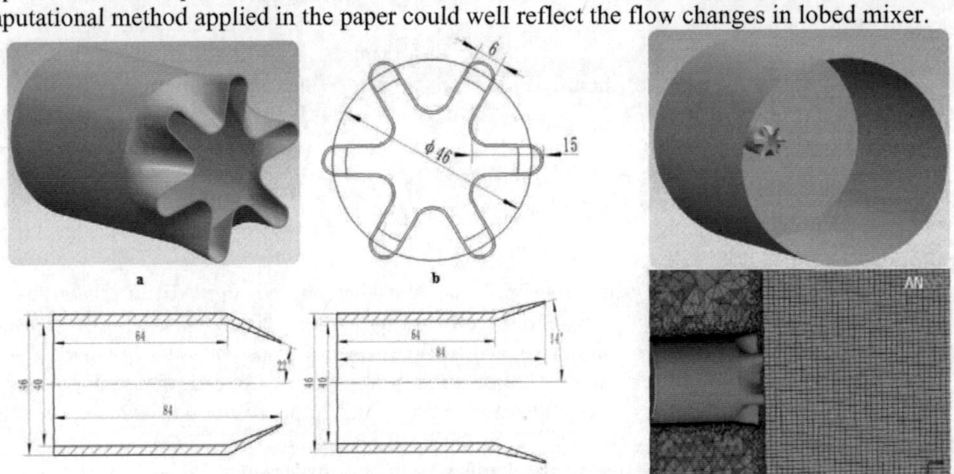

Figure 1. Geometry model and mesh of verification

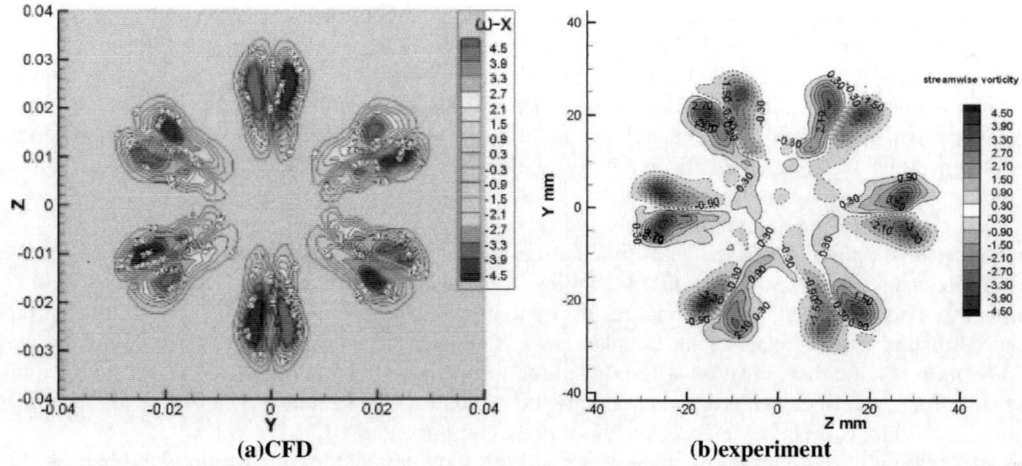

Figure 2. Non-dimensional streamwise vorticity on x/D=0.5

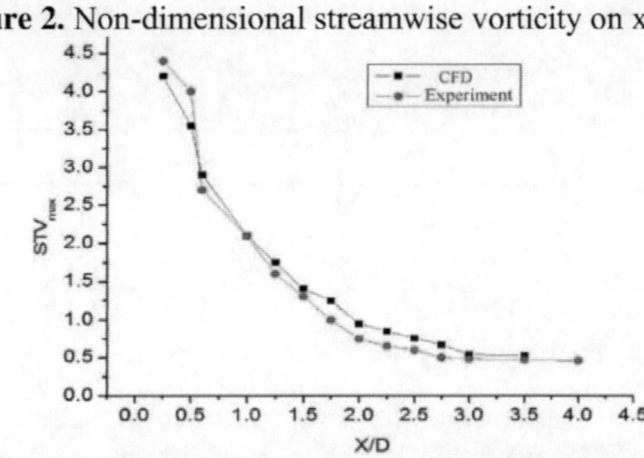

Figure 3. Distribution of maximum non-dimensional streamwise vorticity along streamwise

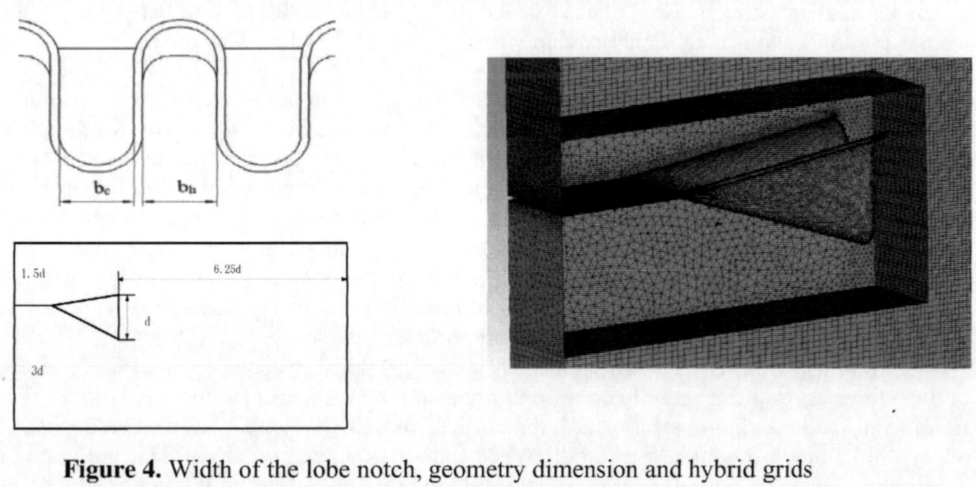

Figure 4. Width of the lobe notch, geometry dimension and hybrid grids

2.2. Geometry modeling and computational mesh

Considering the plane symmetry structure of two-dimensional lobed mixer, a segment (λ = 29mm) containing a whole lobe unit was chosen as computational domain. The geometric dimensions of the domain were shown in Figure 4. Seven computational models with different notch width-ratio (0.444, 0.529, 0.625, 0.733, 0.857, 1.0, 1.167 and 1.364) of core lobe to bypass lobe were established.

Hybrid grid was adopted considering the complicated structure of the lobe. The mesh around the lobes is unstructured tetrahedral cells and the other domain is structured grid as shown in Figure 4. Five-layer prism grid is applied in the walls to adapt to the boundary layer flow. The paper verified the independence of simulation results from mesh and eventually identified the grid number as 1.8 million.

2.3. *Computational method and boundary conditions*

The numerical simulation is based on 3-dimensional incompressible Reynolds Average Naiver-Stokes equation in steady state for the low inlet velocities (less than 0.3 Mach number). Compared with other turbulent models [10], the computational results of Realizable k-ε turbulent model approach best to the measured results in Hu's experiment. Standard wall function was selected to solve the boundary layer flow. A finite volume discretization with convention terms discretized using MUSCL was adopted and pressure term discretized applying second upwind scheme. The standard SIMPLE was chosen as solution algorithm. Convergence accuracy of velocity component is 10^{-6}.

Work medium in two mass flow inlets is air and the state parameters of core flow and bypass flow is shown in table 1. Pressure outlet with atmospheric pressure was selected as outlet boundary condition.

Table 1. Boundary conditions

	v (m/s)	T (K)
Core flow	90	560
Bypass flow	30	300

3. Computation Results and Discussion

3.1. *Analyses of flow field*

When the airflow passes through the lobe, radial velocity of opposite directions appears nearby sidewalls of the lobe which forms the vortex structure consistent with lowing direction. The vortex structure is called "streamwise vortices" which greatly enhance the mixing between the primary and secondary flows. Non-dimensional streamwise vorticity of 2-D lobed mixer is defined as following:

$$\omega_x = d/\bar{u}\,(\partial w/\partial y - \partial v/\partial z) \qquad (2)$$

Figure 5 shows the velocity vector and non-dimensional streamwise vorticity of the lobed mixer model ($b_h/b_c = 1$) in different cross sections. The development of the streamwise vortices in 2-D lobe mixer is generally similar to that of the circular lobe mixer, but the differences still exist. Fig 5(a) shows that the shape of streamwise vortices in x/d=0.2 is consistent with the trailing edge of the lobe in general. The velocity vector distribution in Figure 5(a) reveals the opposite component velocity along x direction nearby sidewalls of the lobe. In vicinity of lobe trough and crest, component velocity along y direction are also existing. However, there is no component velocity along y or x direction far away the lobe, which suggests that the effects of lobe are limited. The shape of streamwise vortices in figure 5(b) are no longer parallel to longitudinal portion of the lobe.

At cross-plane of x/d=1.25, two smaller vortex cores appear in each streamwise vortex and the size of two vortex cores differs greatly because of unequal lobe inclination at top and bottom. When the air flows from x / d = 0.625 to x / d = 1.25, the vortices further developed and the component velocity of the airflow along y direction increased. When the air flow past x / d = 1.875, the vortex begins to decay, the vortex intensity decreases in core and the area influenced by the vortex gradually reduces. At cross-plane of x / d = 6.25, the streamwise vorticity is very small due to the frictional losses from the mixing between hot and cold air.

Figure 6 shows the changes of average non-dimensional streamwise vorticity of lobed mixer models with differentb_h/b_c. When the air flow past the lobe trailing edge, the streamwise vortex begins to develop and the intensity of the vortex increases gradually and reaches a peak on the cross-plane of x / d = 1.875. Subsequently, the vortex dissipates for the mixing of core and bypass airflow. Although the change trends of average non-dimensional streamwise vorticity of different b_h/b_c are similar, the differences of average non-dimensional streamwise vorticity of different b_h/b_c can be observed distinctly in Figure 6. When the b_h/b_c ranges from 0.444 to 1.364, the average non-dimensional streamwise vorticity increases on all cross-planes. However, the augment of the maximum average non-dimensional streamwise vorticity decreases as b_h/b_c increasing. The notch width of core lobe and bypass lobe determines the flow passing through the notch was changed. Only the core and bypass flows passing through the notch are equivalent, the maximum average non-dimensional streamwise vorticity will reach the climax. Thus, it can be inferred that the maximum average non-dimensional streamwise vorticity will reach the climax as b_h/b_c increases to a specific point and will decrease if b_h/b_c further increases.

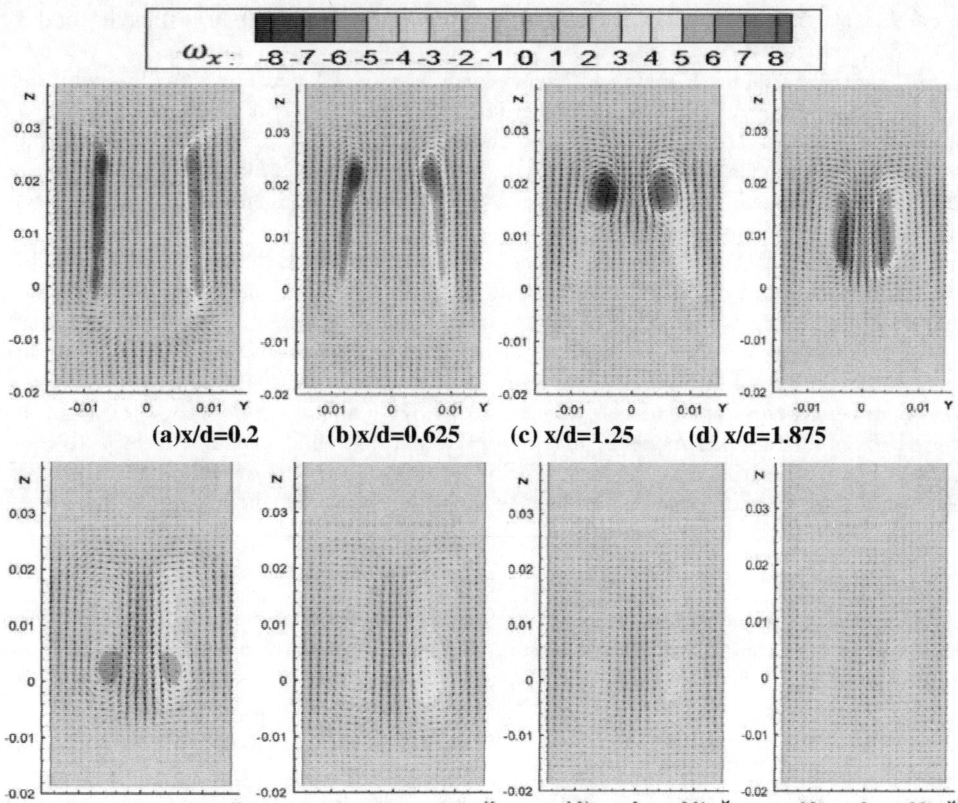

Figure 5. Distribution of non-dimensional streamwise vorticity and velocity vector

3.2. *Thermal mixing efficiency*

Thermal mixing efficiency is adopted to evaluate the mixing degree of lobed mixer. Among numerous definition of thermal mixing efficiency, a revised formula is selected in the paper, which is defined as following [11]:

$$\eta_T = 1 - \frac{\int (T - T_{mix})^2 dm}{m_h T_h^2 + m_c T_c^2 - m_{mix} T_{mix}^2} \tag{3}$$

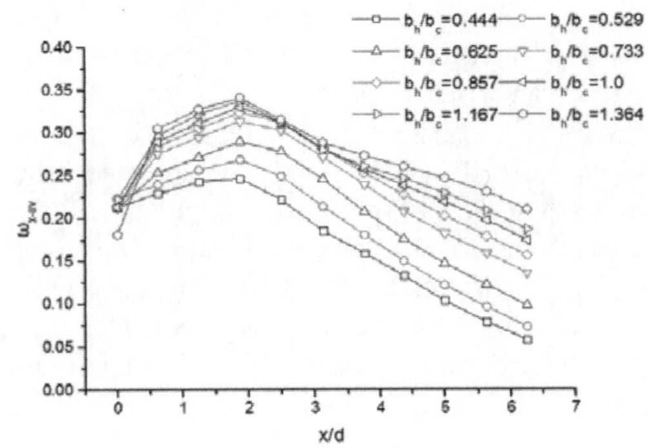

Figure 6 Average non-dimensional streamwise vorticity

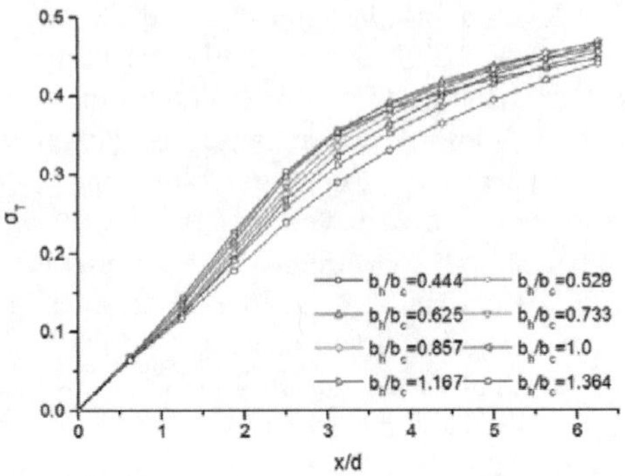

Figure 7 Thermal mixing efficiency of different mixers

Figure 7 shows the thermal mixing efficiency distribution of lobed mixer models with different b_h/b_c along flow direction. Thermal mixing efficiency rises quickly in the front of mixer and the rising rate slow down subsequently. As analysed above, the dissipation of streamwise vortex and large velocity gradient perpendicular to flow direction are the main reasons for mixing. When the mixing advances to a certain extent, the streamwise vorticity and velocity gradient inclined. Then, the mixing slows down.as shown in Figure 7

When x/d<3.125, the thermal mixing efficiency increases slightly as the b_h/b_c decreases. However, when x/d>3.125, the lobed mixer of maximum the thermal mixing efficiency is different in different cross-plane. In order to reveal the effect of the b_h/b_c on thermal mixing efficiency of lobed mixer, the X axis was change to represent b_h/b_c and the distribution of thermal mixing efficiency on different cross-planes was illustrated in Figure 8. The figure shows that the thermal mixing efficiency declines as b_h/b_c increases on cross-plane of x/d=1.25 and x/d=1.825. On cross-plane of x/d=2.5, 3.75, 5.0 and 6.25, the b_h/b_c of lobed mixer model of maximum thermal mixing efficiency are respectively 0.529, 0.625, 0.625 and 0.875. It means that the b_h/b_c of lobed mixer model of maximum thermal mixing efficiency increases as the lobed mixer lengthens, which provides theoretical foundation to the performance optimization of lobed mixer.

3.3. Total pressure loss

For lobed mixer, the mixing of hot and cold air flow is important reason for total pressure loss. The flow loss is often characterized with total pressure recovery coefficient which is defined as following:

$$\sigma = \frac{\int P_{mix}^{*}\,dm/m_{mix}}{\int P_{in}^{*}\,dm/m_{in}} \qquad (4)$$

The change of σ along flow direction is shown in Figure 9. Since the velocity of core flow is larger than the velocity of bypass flow and smaller b_h/b_c means smaller area ratio of core flow and bypass flow, smaller b_h/b_c will expand the velocity gap between core flow and bypass flow which can increase the total pressure loss. Thus, the declination of σ slow down and the total pressure loss reduces as b_h/b_c increases. The total pressure recovery coefficient drops more quickly when x/d<3.125, the reason is that high velocity gradient and streamwise vorticity lead to larger total pressure loss. Correspondingly, the growth rate of thermal mixing efficiency is bigger when x/d<3.125.

4. Conclusion

The paper studied the effect of b_h/b_c (0.444-1.364) on flow field, thermal mixing efficiency and total pressure loss of lobed mixer and obtained some conclusions as following:

1. The mixing process in lobed mixer is dominated by the development of streamwise vortex, which can be influenced byb_h/b_c . The average non-dimensional streamwise vorticity ascends as b_h/b_c increases.

2. The lobed mixer of maximum thermal mixing efficiency varies on different cross-plane, in other words, there is optimal b_h/b_c and the optimal b_h/b_c rises when mixing distance lengthens.
3. Total pressure recovery coefficient decreases as b_h/b_c increases.

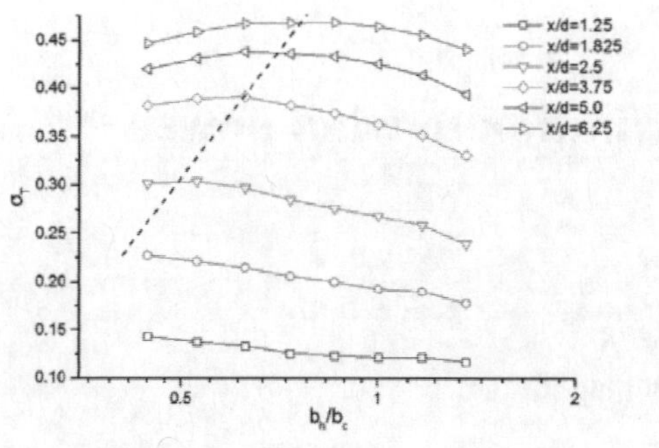

Figure 8. Thermal mixing efficiency of different streamwise location

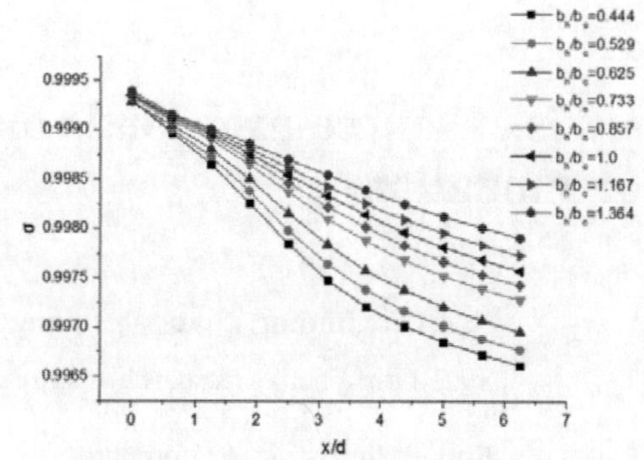

Figure 7. Total pressure recovery coefficient

Acknowledgments

This work Supported by National Natural Science Foundation of China（51506003）.

References

[1] Kuchar A P and Chamberlin R Scale 1980 Model performance test investigation of exhaust system mixers for an energy efficient engine (e3) propulsion system *AIAA* 0229.
[2] Elliott J K, Manning T A, Qiu Y J, Greitzer E M and Tan C S 1992 Computational and experimental studies of flow in multi-lobed forced mixers *AIAA* 3568.
[3] Yang Kai-tian 2001 Aircraft Engine Design Manual *Aviation Industry Press* 50-84.
[4] Paterson R W 1984 Turbofan mixer nozzle flow field-a benchmark experimental study *Journal of Engineering for Gas Turbines and Power* **106** 692-698.
[5] Skebe S A, Paterson R W and Barber T J 1988 Experimental investigation of three-dimensional forced mixer lobe flow fields *AIAA* 385.
[6] Elliott J K, Manning T A, Qiu Y J, Greitzer E M and Tan C S 1992 Computational and experimental studies of flow in multi-lobed forced mixers *AIAA* 3568.
[7] Liu Youhong 2007 Experimental and numerical research on high pumping performance mechanism of lobed exhauster-ejector *International Communication of Heat and Mass Transfer* **34** 197-209
[8] Liu Youhong, Xie Yi 2011 Effects of lobe penetration ratio on performance of mixer exhaust system *Journal of Propulsion Technology* **32**(2) 207-213.
[9] Hu H, Saga T, Kobayashi T 2001 Investigation of the Vortex Structures Downstream of a Lobed Nozzle by Mean of Dual-Plane Stereoscopic PIV *Goettingen: 4th International Symposium on Particle Image Velocimetry*
[10] Xie Yi, Li Teng, Liu You-hong 2011 Numerical Investigation of Vortical Structures in a Jet Mixing Flow of Lobed Mixer *Science Technology and Engineering* **11**(32) 7972-78
[11] Liu Youhong, Guo Nan 2010 Effict of Cut Angles at Trailing Edge of Performance of a Lobed Mixer *Jounal of Aerospce Power* **25**(2) 243-250

International Conference on Energy Engineering and Environmental Protection (EEEP2016) IOP Publishing
IOP Conf. Series: Earth and Environmental Science **52** (2017) 012028 doi:10.1088/1755-1315/52/1/012028

Analysis of source-grid-load contribution to wind power accommodation

Nan Li [1], **Jinfang Zhang, Fuqiang Zhang**

State Grid Energy Research Institute, Beijing, China

E-mail: linanyz@163.com

Abstract. In recent years, China's wind power is developing rapidly, but wind electricity curtailment becomes a serious problem. This paper evaluates wind power penetration limit from peak regulation, analyses relevant technical elements that impact wind power accommodation, coving the source side, the transmission side and the load side. By taking certain measures around the three aspects, for example, reducing the minimum technical output of conventional units, increasing outbound transmission capacity, or increasing the valley load, wind electricity curtailment will be effectively reduced. This paper quantitatively analyses wind power penetration limit under certain conditions and the source-grid-load contribution to wind power accommodation, and explores the important technical factors that affect wind power accommodation.

1. Introduction

Resource shortage, environmental pollution and climatic variation are increasingly severe. The energy development path established on the basis of traditional fossil energy will be difficult to sustain. The low-carbon, renewable clean energy system will alter traditional fossil energy system, which is the core of worldwide energy revolution. At present, China's new energy capacity ranked first in the world, especially the wind power is far ahead, but wind curtailment is very serious. In order to promote the sound development of wind power, earnestly implement the national new energy strategy policy, it is necessary to further study the reason of wind curtailment, come up with reasonable measures to promote wind power accommodation. Combined with domestic [1-3] and international research results [4-7], policies and market factors play a vital role in the development of wind power, but from a technical point of view, the source side, the transmission side and the load side are important factors affecting wind power accommodation [8-9].

2. Technical factors influencing the wind power accommodation

In order to adapt the wind power utilization and accommodation on a large scale, all kinds of power resources need to work in coordination. To implement large-scale wind power development and efficient utilization in our country, we need to analyze the current outstanding problems and solve the problems. About policy and market aspects, please refer to the literature [1] and literature [10]. In the technical level, the solution can be proceeded from the power side, grid side, and load side. We need to improve power supply flexibility in the power supply side, to expand the capacity of the grid side, and to implement demand management and electrical energy alternative in the load side. We need to implement a variety of measures, improving the power system absorption capacity in an all-round way, promoting wind power development better and faster.

Peaking space of the power system is the difference between the maximum output and the minimum technical output. The peaking space need to meet the requirements of peak-valley load difference. When the system contains wind power, it should also meet the peak-valley difference of wind power. From the peak regulation point of view, we can draw the following rules.

The maximum load and power transmission channel capacity decide the operating capacity of the conventional units. Specifically the maximum load plus channel delivering power, minus the output of the wind power at the peak time, determines the conventional unit-operating capacity.

The minimum technical output of operating units decides the wind power accommodation scale. Specifically the valley load minus the minimum technical output of conventional units, determines the wind power accommodation scale.

3. Source-grid-load contribution

From technical perspective, the scale of wind power accommodation is governed mainly by the following factors: the capacity of operating units, the spare capacity, the maximum load, the peak-valley load, the transmission line capacity and regulating ability, and adjustable capacity of different types of units, such as heating units, nuclear power plants, hydroelectric units, thermal power, pumped storage units, captive power plant, etc.

3.1. Wind power accommodation scale calculation

Figure 1 is a schematic diagram about a typical daily load curve and the corresponding wind power accommodation scale. The basic daily load curve has a noon peak, an evening peak and a valley. Noon peak usually appears at about 12:00, the evening peak appears at about 22:00 in summer and 19:00 in winter. Because in winter, the duration of sunshine is shorter than in summer, the lighting loads reach the maximum around 20:00, then the cooling loads are launched successively, so the evening peak will appear at 22:00. The difference between the two peaks is small. The valley usually appears at about 3:00～5:00.

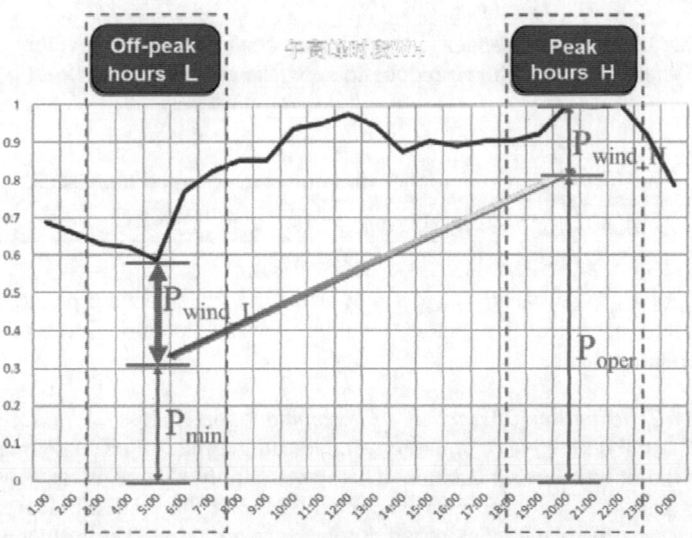

Figure 1. A typical daily load curve and the corresponding wind power accommodation scale

Calculation of wind power accommodation scale will use the following parameters [11-14].

The load side:

P_{load}^{H} : The maximum load, generally accrued in the evening.

P^{L}_{load} : The valley load.

ΔP_{load} : The peak-valley load difference. $\Delta P_{load}=P^{H}_{load}-P^{L}_{load}$.

Power grid side:

P_{trans} : Delivery power by transmission lines at peak load hours. It is usually the capacity of transmission lines.

P^{L}_{trans} : Delivery power by transmission lines at valley load hours.

ΔC_{trans} : The peak-valley difference rate of power transmission. $\Delta C_{trans}=(P_{trans}-P^{L}_{trans})/P_{trans}$.

Power supply side:

P_{spare} : The spare capacity of the system.

P_{oper} : Capacity of conventional operating units.

P_{force} : The forced output of the units which are operating but not adjustable.

C_{tec} : The minimum technical output coefficient of conventional operating units.

Wind power side:

C^{H}_{wind} : Wind power usable coefficient at peak load hours.

C^{L}_{wind} : Wind power usable coefficient at valley load hours.

ΔC_{wind} : The wind power difference between peak and valley coefficients. $\Delta C_{wind}=C^{H}_{wind}-C^{L}_{wind}$.

On the basis of above parameters, we can deduce the accommodation scale of wind power [15]. Assume some wind turbines already exist in the system, so the surplus peaking space at valley load hours is:

$$P_{adjust}=(P^{L}_{load}+P_{trans})-[(P_{oper}-P_{force})\times C_{tec}+P_{force}+P^{L}_{wind}] \quad (1)$$

If we increase the wind power capacity, the surplus peaking space at valley load hours will gradually decrease. When the space almost reduce to zero, the corresponding wind power capacity is the maximum scale.

Hypothesis $P_{adjust}=0$, on account of $P_{oper}=P^{H}_{load}+P_{trans}-P^{H}_{wind}+P_{spare}$, and bring these two parameters into equation(1), we can come to the wind power accommodation scale:

$$S_{wind}=\frac{(P^{H}_{load}+P_{trans}+P_{spare}-P_{force})\times(1-C_{tec})-(\Delta P_{load}+P_{trans}\times\Delta C_{trans}+P_{spare})}{C^{H}_{wind}\times(1-C_{tec})-\Delta C_{wind}} \quad (2)$$

From a technical point of view, many factors will affect the scale of wind power accommodation. The fluctuation of related technical parameters, including source, network, and load, will impact on wind power accommodation scale.

3.2. Source-grid-load Contribution to Wind Power accommodation scale

Through calculating partial derivative of equation (2), we can get the influence degree of source-grid-load parameters to wind power accommodation. The power supply side bringing down the minimum technical output, the grid side adding transmission channels, and the load side increasing the valley load, three measures can effectively promote wind power consumption. The influence degrees are in formulas (3)-(5).

$$\frac{\partial S_{wind}}{\partial P_{tec}}=\frac{\partial S_{wind}}{\partial C_{tec}}\times\frac{\partial C_{tec}}{\partial P_{tec}}=\frac{(-P_{trans}\times\Delta C_{trans}-P_{spare}-\Delta P_{load})\times C^{H}_{wind}+(P^{H}_{load}+P_{trans}+P_{spare}-P_{force})\times\Delta C_{wind}}{[C^{H}_{wind}\times(1-C_{tec})-\Delta C_{wind}]^{2}\times P_{oper}} \quad (3)$$

$$\frac{\partial S_{wind}}{\partial P_{trans}}=\frac{1-C_{tec}-\Delta C_{trans}}{C^{H}_{wind}\times(1-C_{tec})-\Delta C_{wind}} \quad (4)$$

$$\frac{\partial S_{wind}}{\partial P_{load}^{L}}=\frac{1}{C_{wind}^{H}\times(1-C_{tec})-\Delta C_{wind}} \tag{5}$$

Generally, $C_{wind}^{H}<C_{wind}^{L}$, $\Delta C_{wind}<0$, so $C_{wind}^{H}\times(1-C_{tec})-\Delta C_{wind}>0$, we can estimate $\frac{\partial S_{wind}}{\partial P_{tec}}<0$, $\frac{\partial S_{wind}}{\partial P_{load}^{L}}>0$. Therefore, the lower the minimum technical output of conventional units is, the larger the wind power accommodation scale will be. And the bigger the valley load is, the larger the wind power accommodation scale will be.

To grid side, the influence of raising the transmission capacity to wind power accommodation is indeterminate. On one hand, raising the transmission capacity can increase the consumption range of wind power, which is in favour of the wind power accommodation. On the other hand, raising the transmission capacity is equal to increasing the peak load, which will lead to increasing of operating capacity of conventional units and accompanying increasing of the minimum technical output, which is against wind power accommodation. According to formula (4), if $C_{tec}+\Delta C_{trans}<1$, then $\frac{\partial S_{wind}}{\partial P_{trans}}>0$. So if the transmission capacity is raised, we can reduce the peak-valley difference rate of the transmission lines to meet $C_{tec}+\Delta C_{trans}<1$, then the raised transmission lines will promote wind power accommodation .

Under normal circumstances, which can be seen from the formulas, the impact from the power grid side to wind power accommodation is less than the extent of the impact from the load side. Note that this analysis is based on the planning level.

Hypothesis $Q_{wind}=\left|\frac{\partial S_{wind}}{\partial P_{tec}}\right|+\left|\frac{\partial S_{wind}}{\partial P_{trans}}\right|+\left|\frac{\partial S_{wind}}{\partial P_{load}^{L}}\right|$, the contribution of power supply, grid and load side to wind power accommodation, are shown in formulas (6) ~ (8).

$$Q_{tec}=\left|\frac{\partial S_{wind}}{\partial P_{tec}}\right|\Big/Q_{wind} \tag{6}$$

$$Q_{trans}=\left|\frac{\partial S_{wind}}{\partial P_{trans}}\right|\Big/Q_{wind} \tag{7}$$

$$Q_{load}^{L}=\left|\frac{\partial S_{wind}}{\partial P_{load}^{L}}\right|\Big/Q_{wind} \tag{8}$$

4. Examples

Take a typical regional power gird for example, in which the peak load is 888.9MW, and the valley load is 731.7MW. The differences ratio between peak and valley load is 17.7%. In additional, the grid has one power delivering channel, the peak transmission power is 500MW, and the valley transmission power is 350MW. The spinning reserve account for 7% of the maximum load. The force output is about 70 MW. The average minimum technical output of conventional power units is 50%. The simultaneous rate of wind power output in the peak and valley load time are respectively 0.25 and 0.53.

According to formula (2), the maximum wind power accommodation capacity is 600MW. Moreover, based on the proposed method, the contribution degrees of source-grid-load side to wind power consumption could be calculated and shown in the following table.

Table.1 Source-grid-load contribution degree to wind power accommodation

	Contribution degree to wind power accommodation
Reducing minimum technical output in power source side	43.5%
Exaggerating the transmission capacity in grid side	7.4%
Increasing valley load value in load side	49.1%

From table.1, it can be found that demand side has the maximum contribution to wind power accommodation, power supply side takes the second place, and power grid has less impact, which are also shown in figure 2.

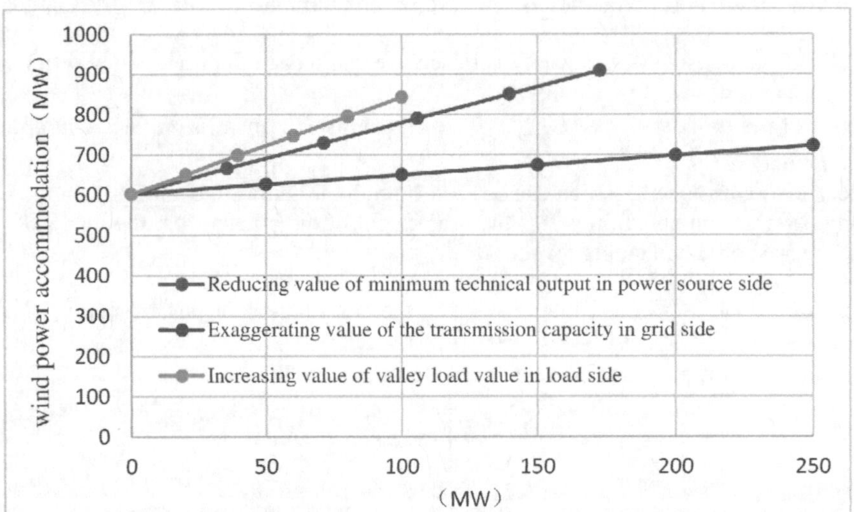

Figure 2. Influence degree of source-grid-load side to wind power accommodation

Based on the three curves, the following conclusions could be figured out. If the valley load increases 100 MW, more 241MW wind power can be consumed. If the average minimum technical output ration reduces to 40%, equal to 173MW, about 307MW extra wind power will be utilized. If the capacity of the transmission line increases 250MW, the scale of wind power consumption will increase 121MW.

5. Conclusion

As a result, this paper has proposed that by improving the operation flexibility in power source side, enlarging the interconnected scale in power gird side and adopting demand management in demand side will guarantee the high efficient utilization of wind power. It is also important that the above mentioned three treasures will play key roles in wind power consumption with different contribution degrees. Finally, the corresponding improvement and construction of three types of measures need to be based on local condition. In order to reduce the wind power curtailment, multidimensional works should be put in practice.

References

[1] Wang Yibo, Cai Guowei, Zheng Cunlong, et al. 2016 Study on the accommodation approach of wind power based on rational wind power casting *Electrical Measurement &*

Instrumentation **53**(11) 45-50.

[2] Liu Xindong, Fang Ke, Chen Huanyuan, et al. 2012 Research on rational wind power casting theory for large-scale wind power integration improvement *Powre System Protection and Contral* **40**(6) 35-39.

[3] Liu Chang, Wu Hao, gao Changzheng, et al. 2014 Study on analysis method of accommodated capacity for wind power *Power System Protection and control* **42**(4) 61-66.

[4] Hugh Sharman 2005 Why wind power works for Denmark *Proceedings of the Institution of Civil Engineers* **158**(2) 66-72.

[5] Evans J, Shawwash Z 2009 Assessing the benefits of wind power curtailment in a hydro-dominated power system *IEEE PES Joint Symposium on Integration of Wide-Scale Renewable Resources Into the Power Delivery System*, Paris; IEEE 1-9.

[6] John Kabouris, Kanellos F D 2010 Impacts of large-scale wind penetration on designing and operation of electric power systems *IEEE Transactions on Sustainable Energy* **1**(2) 107-115.

[7] Callaway D S 2009 Tapping the energy storage potential in electric loads to deliver load following and regulation with application to wind energy *Energy Conversion and Management* **50**(5) 1389-1400.

[8] Yuan Tiejiang, Peng Chaofeng, Mei Shengwei, et al. 2015 A partitioning method to enhance accommodation capacity of power network with balanced region of wind power *Power System Technology* **39**(8) 2129-34.

[9] Niu Dongxiao, Li Jianfeng, Wei Linjun, et al. 2016 Study on technical factors analysis and overall evaluation method regarding wind power integration in tans-provincial powre grid *Power System Technology* **40**(4) 1087-93.

[10] Xia Ye, Kang Chongqing, Chen Tianen, et al. 2015 Day-ahead market mode with power consumers participation in wind power accommodation *Automation of Electric Power Systems* **39**(17) 120-126.

[11] Liu Dewei, Huang Yuehui, Wang Weisheng, et al. 2011 Analysis on provincial system available capability of accommodating wind power considering peak load dispatch and transmission contraints *Automation of Electric Power System* **35**(22) 77-81.

[12] Liu Wenying, Wen Jing, Xie Chang, et al. 2015 Multi-objective optimal method considering wind power accommodation based on source-load coordination **35**(5) 1079-88.

[13] Kang Chongqing, Jia Wenzhao, Xu qianyao, et al. 2013 Capability evaluation of wind power accommodation considering security constraints of powre grid in real-time dispatch *Proceedings of the CSEE* **33**(16) 23-29

[14] Song Yihang, Tan Zhongfu, Li Huanhuan, et al. 2014 An optimization model combining generation side and energy storage system with demand side to promote accommodation of wind power *Power System Technology* **38**(3) 610-615.

[15] Han Xiaoqi, Sun Shouguang, Qi Qingru 2010 Evaluation of wind power penetration limit from peak regulation *Electric powre* **43**(6) 16-19.

Phytotechnological purification of water and bio energy utilization of plant biomass

D I Stom[1, 2, 3, 4], **O V Gruznych** [1], **G O Zhdanova** [1], **S S Timofeeva** [2], **A V Kashevsky** [1], **M N Saksonov** [1], **A E Balayan** [1]

[1] Irkutsk State University (ISU), Irkutsk, Russia
[2] Irkutsk National Research Technical University (INRTU), Irkutsk, Russia
[3] Federal State Budget Institution of Science "Baikal Museum at the Irkutsk Scientific Center of the Siberian Branch of the Russian Academy of Sciences", Listvyanka, Russia

E-mail: stomd@mail.ru

Abstract. The aim of the study was to explore the possibility of using the phytomass of aquatic plants as the substrate in the microbial fuel cells and selection of microorganisms suitable for the generation of electricity on this substrate. The conversion of chemical energy of phytomass of aquatic plants to the electrical energy was carried out in a microbial fuel cells by biochemical transformation. As biological agents in the generation of electricity in the microbial fuel cells was used commercial microbial drugs "Doctor Robic 109K" and "Vostok-EM-1". The results of evaluation of the characteristics of electrogenic (amperage, voltage) and the dynamics of the growth of microorganisms in the microbial fuel cells presents in the experimental part. As a source of electrogenic microorganisms is possible to use drugs "Dr. Robic 109K" and "Vostok-EM-1" was established. The possibility of utilization of excess phytomass of aquatic plants, formed during the implementation of phytotechnological purification of water, in microbial fuel cells, was demonstrated. The principal possibility of creating hybrid phytotechnology (plant-microbe cells), allowing to obtain electricity as a product, which can be used to ensure the operation of the pump equipment and the creation of a full cycle of resource-saving technologies for water treatment, was reviewed.

1. Introduction

Recently, for the protection, treatment and post-treatment of water, and the ecological reconstruction of water bodies, the phytotechnology have been used more frequently. The basis of phytotechnologies is the use aquatic vegetation and its attendant microorganisms. Today the similar research is being conducted in more than 80 countries mainly in North America, Europe, South East Asia [1, 2, 3]. The most active development in recent decades, this trend was in China. It is used for cleaning agricultural and irrigation wastewater [4]. In Russia, research on the development of phytotechnological purification of water is conducted at leading universities for more than 50 years. In particular, we studied the mechanisms and patterns of accumulation and biodegradation of wastewater components of different industries [5, 6]. With regard to the conditions of Siberia design of phytotechnological facilities, technological regulations operation of such facilities, and the options for complex purification schemes was developed [7].

Organic compounds in systems of phytotechnologies are decomposed in aerobic and anaerobic conditions. Oxygen comes to the aerobic zone from the atmosphere as a result of convective-diffusive processes, and through the macrophyte roots. Anaerobic conditions are created in the soil pores. Typical substrates for vegetation are sand, gravel, and the production waste (sawdust, rice and buckwheat husk, ash, coal slag, dead seston, compost).

Also phytotechnology is used in detoxification and removal of the second largest group of pollutants - heavy metals. Plants which involved in the decontamination of heavy metals and radionuclides, must meet certain requirements: 1) to grow rapidly; 2) have a high tolerance to metals; 3) be resistant to diseases and pesticides; 4) have a well-developed root system and shoots; 5) have the ability to synthesize the particular substance in response to toxic external environment; 6) be unattractive to animals in order to avoid the transfer of pollutants to higher trophic levels; 7) not be specific to certain elements, so that there exists the possibility of extracting and other metal detoxification [7].

One of the problems in the operation of phytotechnologies of water purification is the removal and subsequent disposal of excess biomass of aquatic plants. Bearing in mind that plant phytomass can be used as an energy source, in this paper the possibility of its utilization in the biofuel cell was investigated. Complex biotechnological approach will not only clean the waste water, but also to convert chemical energy into electrical energy [8, 9, 10]. Currently, the most significant obstacle to the use of fuel cells is a limited number of strains of microorganisms capable of operating in conditions of microbial fuel cells (MFC). However, there are many commercial microbiological preparations for wastewater treatment, removal of various contaminants. They include a wide range of different microorganisms, utilizing many of the components of sewage [11].

In this regard, the tasks of this article included:

- Assessment of the possibility of using waste phytomass of aquatic plants as a substrate in microbial fuel cells;
- Selection of microorganisms suitable for the generation of electricity on this substrate.

2. Objects and methods of research

We used phytomass higher aquatic plant *Elodea canadensis* Michx. (1803), recommended as the main component of phytocenoses for the extraction of metals from waste water and disposal of organic components of sewage in a sharply continental climate of Eastern Siberia.

We collected plants by trawl in Angara River and the Lake Baikal. Hydrophytes grown in the laboratory with moderate light and temperature 13 ^{0}C. Experiments were carried out in vessels of 5 liters in which the plants were placed at 10 g/l. Dishes filled with solutions studied metals (copper, iron) in a concentration of 1-50 mg/l. Plants exposed in thermoluminostate with illumination 2 thousand Lux and a temperature of 25 C. After certain intervals samples were taken and analyzed the residual content of metals, the cleaning efficiency was determined.

After exposure for 10 days, the plant was used as the organic substrate for microorganisms in MFC. Elodea was used as a crude homogenate and dry material. Phytomass, culture media and model waste water was autoclaved at 1 atmosphere for 30 min.

We used the model sample MFC made of Plexiglas, which consists of two identical chambers of 370 ml separated by a proton exchange membrane Nafion (USA) [12, 13]. Medium filling the anode and cathode space is model waste water (GOST 50595-93). Electrodes for MFC made of electrically silicone carbide ("Podolskogneupor", Russia) (Figure 1).

Figure 1. The microbial fuel cell.

Tests were carried out according to the procedure described in detail in [9]. The measurements of amperage were carried out in the short-circuit mode.

In different series of experiments were added to the anode analyzed chamber substrate – an *Elodea canadensis* homogenate or in dry form. The cathode compartment was filled with a model wastewater.

As biological agents at generating electricity in the MFC was used the commercial microbial preparations "Doctor Robic 109K" and "Vostok-EM-1". Doctor Robic "109K" produced for cleaning of septic tanks and cesspools. It consists of: wheat bran 60%, baking soda 37-38%, 2-3% of non-pathogenic microorganisms containing 3×10^{10} CFU/g: *Bacillus amyloliquefaciens, B. pumilis, B. licheniformis, B. subtilus*. Microbiological preparation "Vostok-EM-1» is used for seed treatment and improve soil quality. It contains lactic acid, photosynthetic bacteria, yeast, fungi (*Lactobacillus plantarum, L. casei, Rhodopseudomonas palustris, Saccharomyces cerevisiae*), and the waste products of microorganisms. Preparations previously centrifuged to separate the microbial biomass, which is introduced into the anode chamber in an amount of MFC ~10^6 CFU/ml.

It is known that a number of strains of *Clostridium* effectively utilize cellulose. In this regard, we evaluated the possibility of intensifying the transformation of Elodea in addition clostridia to biological agents in MFC. We used *Clostridium acetobutylicum* strain VKPM-B-1787. The microorganism is introduced into the anode chamber MFC with Elodea and microbial preparations in an amount ~10^5 CFU/m.

Incubation of microorganisms in the anode chamber of MFC was carried under anaerobic conditions. Cathode chamber was aerated. We used Microsoft Excel software package for statistical analysis of the data. The reliability of the results of the differences was determined by Student's t test. Conclusions are made with probability of faultless prognosis $R\geq0.95$

3. Results and discussion

Table 1 shows the absorption coefficients of biological metals in different groups of plants. Studies have shown that plants belonging to various ecological groups have different storage capacity in relation to heavy metals. Thus, the maximum amount of metal was accumulated in the tissues of the immersed plants. In this connection, preference should be given to the plants submerged primarily in *Elodea canadensis* (Table 1).

Table 1. Coefficients of biological accumulation in plants at pH = 7.2.

Plant	Bioaccumulation factor						
	Fe	Al	Ni	Cu	Cd	Zn	Pb
Acorus calamus	1.1	8.5	2849.0	1.9	9.0	38.0	10.2
Typha angustifolia	1.3	8.5	22.7	9.5	949.0	649.0	409.0
Comarum palustre	0.9	28.8	35.5	4479.0	46.5	9.2	3.5
Potamogeton natans L.	0.6	8.7	5.8	2239.0	18.0	3.1	40.0
Nuphar luteum (L.)	1.1	7.8	34.6	3732.0	236.0	1.0	135.6
Potamogeton lucens L.	3.3	27.8	22.7	21.4	18.0	2.4	40.0
Potamogeton perfoliatus L.	1.1	24.8	311.5	447.0	94.0	974.0	101.5
Potamogeton htctinatus L.	1.2	775.0	39.7	2132.0	13.6	194.0	19.5
Myriophyllum spicatum L.	1.0	484.0	14.0	1947.0	46.5	974.0	18.5
Ceratophyllum demersum L.	1.7	4.5	406.1	173.1	474.0	71.2	44.5
Elodea canadensis L	2.0	8.0	474.0	51.7	157.0	47.7	24.6
Chara vulgaris	1.7	27.9	63.7	741.0	89.0	56.2	10.9
Chara hispida	1.8	78.2	71.3	235.0	134.0	91.1	130.0

Experiments have shown that in MFC generated EMF increases with time when adding a phytomass of Elodea to microbiological preparations (Figure 2). Both drugs had electrogenic activity for disposal Elodea biomass. Thus, after 6 days of exposure in the case of drug Doctor Robic voltage was 407 mV, and an EM-preparation – 304 mV. Prolonged incubation (for 17 days) increased the voltage to 550 - 600 mV (Figure 2).

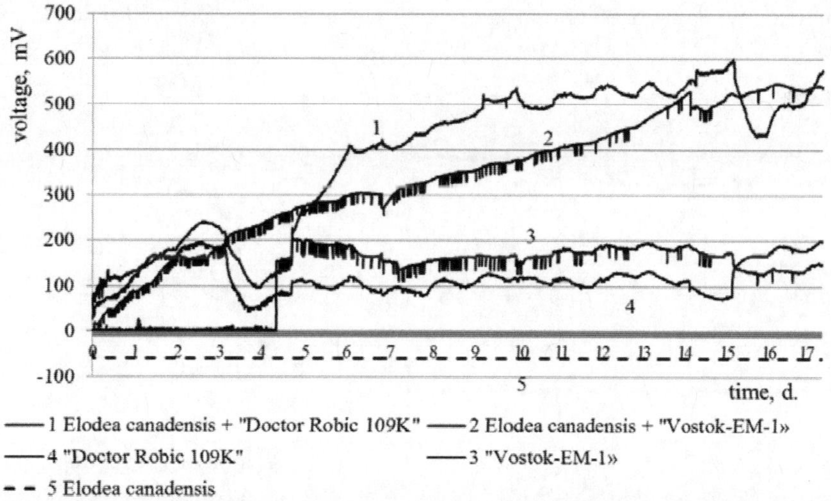

Figure 2. The dynamics voltage in MFC with Elodea when used "Dr. Robic 109K" and "Vostok-EM-1" as electrogenic drugs.

A similar trend was observed in the measurement of the amperage. A gradual increase of this index occurred in the MFC, which containing both microbial drugs and Elodea. On day 17 of the experiment amperage, generated by the drug "Doctor Robic 109K" was 364 mcA, drug "Vostok-EM-1" - 291 mcA. In control cells containing microbial preparations without elodea, amperage does not rise higher than 32 microamps. In the version of the MFC with Elodea, but without the addition of drugs, amperage is not fixed throughout the experiment (Figure 3).

IOP Conf. Series: Earth and Environmental Science **52** (2017) 012029 doi:10.1088/1755-1315/52/1/012029

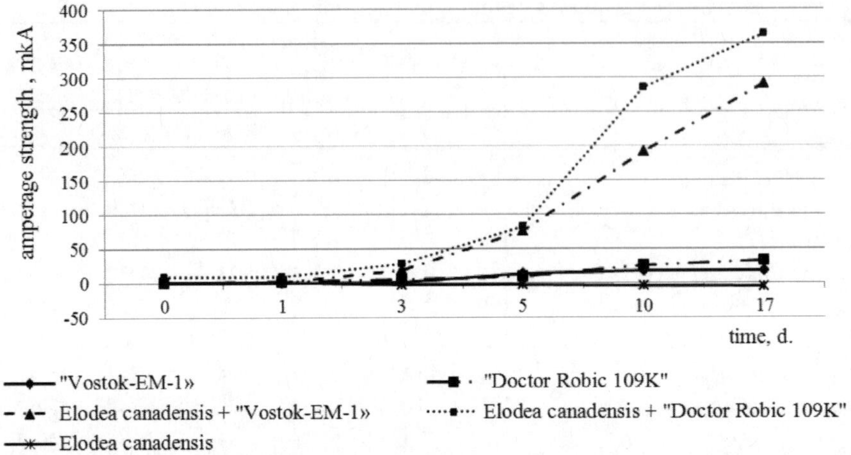

Figure 3. The dynamics of the amperage strength in the MFC with *Elodea* when used as electrogenic drugs "Doctor Robic 109K" and "Vostok-EM-1».

To intensify the process of disposal of plant mass by preparations in MFC additionally introduced the culture of *Cl. acetobutylicum* strain VKPM-B-1787.

Within 7 days of incubation the greatest indicators of the voltage (~500 mV) were shown in the cells containing the drug "Doctor Robic 109 K", Elodea and culture of *Cl. acetobutylicum.* During the same time, the drug "Doctor Robic 109 K" in MFC with no added Clostridium was generated only 300 mV (Figure 4A.). When used as a electrogenic instead of the drug "Doctor Robic 109 K" microbial associations "Vostok-EM-1"watched voltage increase to 300 – 370 mV (Figure 4B).

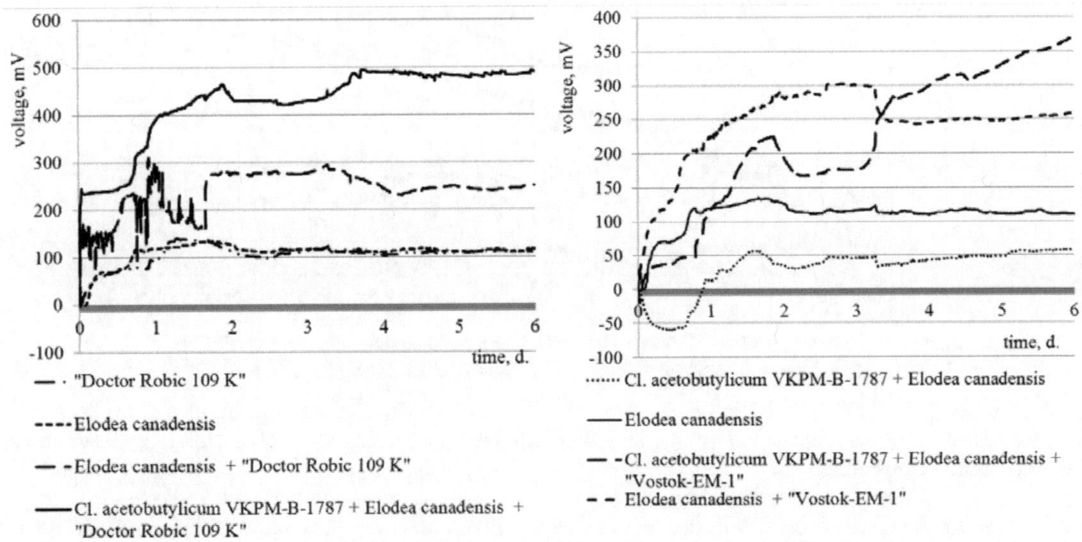

A B

Figure 4. Dynamics of stress in the MFC with Elodea and *Clostridium acetobutylicum* strain VKPM-B-1787 for use as electrogenic drug "Doctor Robic 109K" (A), "Vostok-EM-1" (B).

Parallel analysis of electrogenic characteristics was evaluated the dynamics of the growth of the *Cl. acetobutylicum* strain VKPM-B-1787 and microorganisms of preparations "Doctor Robic 109K" and "Vostok-EM-1".

The most active growth of *Clostridium* cells was observed in cells with Elodea and preparation "Vostok-EM-1". In this case, at 3 days of incubation the number of cells increased from $7.9 \cdot 10^5$ to $3.2 \cdot 10^8$ CFU/ml. The MFC containing Elodea and *Cl. acetobutylicum* without the addition of microbial agents, the number of viable *Clostridium* cells after 3 days of cultivation was slightly below the $1.5 \cdot 10^8$ CFU/ml. In a medium containing *Clostridium* without Elodea culture titer in the specified time interval was $9.2 \cdot 10^7$ CFU/ml. That is, it is actually no different from the same period in the cell with the drug "Doctor Robic 109K", Elodea and *Clostridium* ($1,1 \cdot 10^8$ CFU/ml) (Figure 5).

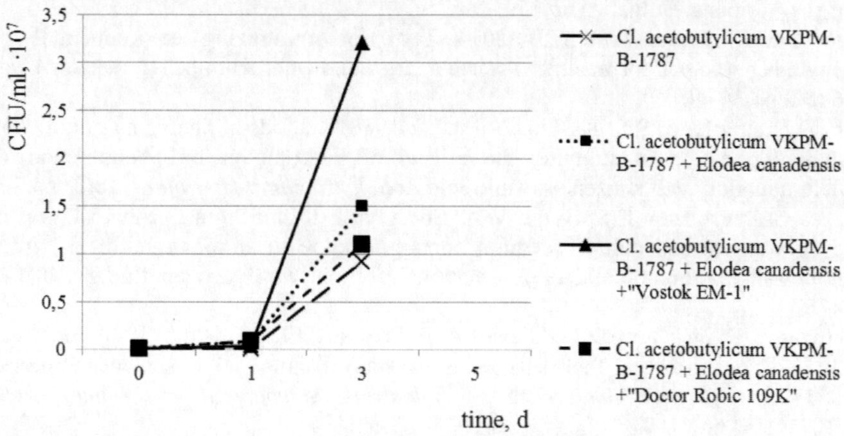

Figure 5. Dynamics of cell culture growth of *Clostridium acetobutylicum* strain VKPM-B-1787 in MFC.

Similar relationships were observed in experiments with the drug "Vostok-EM-1". Over the 5 days of incubation a number of microorganisms was increased from $2.1 \cdot 10^6$ to $8,5 \cdot 10^9$ CFU/ml.

4. Conclusion

Thus, the possibility of disposal of excess biomass of *Elodea* in MFC which generated during the implementation of phytotechnology of water purification was experimentally demonstrated. It was established that as a source of electrogenic microorganisms can be used drugs "Doctor Robic 109K" and "Vostok-EM-1". At the same time the drug "Dr. Robic 109K" has generated somewhat higher emf values than the "Vostok-EM-1."

It should be mentioned that electric indicators of MFC in which biological agents were microbiological drugs "Doctor Robic 109K" and "Vostok-EM-1", were comparable to data given by other specialists were used as biocatalysts other microorganisms [14, 15].

Consequently, there is in principal possibility of creating hybrid phytotechnology (plant-microbe cells), allowing to obtain electricity as a product, which can be used to ensure the operation of the pump equipment and the creation of a full cycle of resource-saving technologies for water treatment.

The authors are grateful to V Ya Severina for providing the drug "Vostok-EM-1", V A Bybin and N S Stepanova for participation in research. The work was supported by the RF Ministry of Education as part of the design of the public tasks in the field of scientific activity (Task № 13.1263.2014/K from 07.11.2014), RFBR and Government of Republic of Buryatia according to the research project № 16-48-030887.

References

[1] Kadlec R H and Wallace S D 2008 *Treatment Wetlands Second edition* (CRC Press, Boca Raton,

Florida)

[2] Mihret D U 2014 Performance evaluation of constructed Wetlands: A review of arid and semiarid climatic region African *J. of Environ. Science and Technol.* **8(2)** pp 99–106

[3] Feng P, Chen Zh and Jing Yu 2002 Review on constructed wetland and – its mechanisms of wastewater treatment *Ecological Science* **21(2)** pp 264–268

[4] Lin Zh 2003 Selenium removal by constructed wetlands: Quantitative importance of biological volatilization in the treatment of selenium laden agricultural drainage water Terry Norman. *Environ. Sci. and Technol.* **37(3)** pp 606–615

[5] Timofeeva S S, Timofeev S S and Ulrich D V 2014 Environmental phytotechnologies in eastern Siberia and south Ural *14th International Multidisciplinary Scientific GeoConference & EXPO* (Albena, Bulgaria) pp 259–266

[6] Timofeeva S S and Ulrich D 2014 Tool for minimizing environmental risks. Using phytotechnologies for treatment waste and groundwater from heavy metals *Water Magazine* **6 (82)** pp 36–40

[7] Stom D I, Timofeeva S S, Kashina N F et al. 1980 Methods of analyzing quinines in water and their application in studying the effects of hydrophytes on phenols. Part 1. Quinone determination with benzene sulfinic acid *Acta hydrochim: Hydrobiol.* **18(3)** pp 203–211

[8] Markov S A, Protasov E S, Bybin V A, Eivazova E R and Stom D I 2015 Using immobilized cyanobacteria and culture medium contaminated with ammonium for H_2 production in a hollow-fiber photobioreactor International Journal of Hydrogen Energy **40(14)** pp 4752–4757

[9] Konovalova E Yu, Stom D I, Balayan A E, Protasov E S, Tolstoy M Yu and Tyutyunin V V 2015 Using microbial fuel cells for utilization of industrial wastewater *Proceedings of the 2014 International Conference on Industrial, Mechanical and Manufacturing Science* (ICIMMS) pp 71–74

[10] Carole Abourached, Marshall J English, Hong Liu 2016 Wastewater treatment by Microbial Fuel Cell (MFC) prior irrigation water reuse *Journal of Cleaner Production* **137** pp 144–149

[11] Puchlik M, Ignatowicz K and Dąbrowski W 2015 Influence of bio-preparation on wastewater purification process in constructed wetlands *Journal of Ecological Engineering* **16(1)** pp 159–163

[12] Lachine A F, Stom D I, Protasov E S, Lapkovsky A A, Saksonov M N, Tolstoy M Yu, Konovalova E Yu, Bybin V A, Ponomareva A L and Mirmanov R J 2014 Bio-electrochemical reactor *RF patent for utility model № 153593 from 27.10.2014*

[13] Konovalova E Yu, Stom D I, Bybin V A, Ponomareva A A, Zhdanova G O and Tolstoy M Yu 2014 Microbial biofuel cells *Patent RF on helpful model № 153691 from 25.12.2014*

[14] Rismani-Yazdi H, Christy A D, Dehority B A, Morrison M, Yu Zh, Tuovinen Olli H 2007 Electricity Generation From Cellulose by Rumen Microorganisms in Microbial Fuel Cells *Biotechnology and Bioengineering* **97(6)** pp 1398–1407

[15] Toczyłowska-Mamińskaa R, Szymonaa K, Madeja H, Wongc W Zhen, Balad A, Brutkowskie W, Krajewskia K, H'ngc P San and Mamiński M 2015 Cellulolytic and electrogenic activity of *Enterobacter cloacae* in mediatorless microbial fuel cell *Applied Energy* **160** pp 88–93

Classification and features of single-well flow units in a carbonate reservoir - Taking the NT oil field at the eastern edge of Pre-Caspian Basin as an example

Zifei Fan, Shuqin Wang[1], Lingli Liu, Jianxin Li, Wenqi Zhao, Meng Sun

PetroChina Research Institute of Petroleum Exploration and Development, Beijing 100083, China

E-mail: wshuqin@petrochina.com.cn

Abstract. Identifying the carbonate reservoirs has always been a challenge to geological exploration, while reasonable classification of flow units is the bottleneck in this exploitation. While taking the NT oil field at the eastern edge of Pre-Caspian Basin as an example, this paper proposes the classification of flow units into five categories based on previous flow-unit classification theory and actual oilfield features by using the pore throat radius at the mercury injection saturation of 35% as the main judging criterion. In this paper, the features of various flow units have also been analyzed. The type-I flow units, mainly found in dolomite and algal reef limestone reservoirs, have the highest production capacity. Given the existence of corrosion and dolomitization, they are mainly fracture-cave composite reservoirs or fracture pore reservoirs. As far as the type-I flow units are concerned, the flow index is > 1.42 for KT-I stratum and > 1.55 for KT-II stratum. The production capacity and reservoir quality of type-II-IV flow units would decline in turn. The type V flow units are argillaceous limestone, with a very low production capacity and a flow index being 0.01-0.05 for KT-I and 0.03-0.05 for KT-II.

1. Introduction

From the perspective of microcosmic origin, these flow units constitute a rock mass with basically similar pore-throat properties determining the fluid flow and are much different from other units among porous media. As the diagenesis of carbonate reservoirs has a great influence on the type and growth characteristics of reservoir space, this study has classified the flow units from the perspective of microcosmic reservoir-space origin through the comprehensive use of research results on core, physical property analysis, capillary pressure, diagenesis, and sedimentary microfacies.

The classification of flow units is based on core analysis data, through which the calculation of R35 and its RQI and FZI can obtain the average FZI reflecting different flow units. When multiple flow units exist, the whole FZI distribution function will be the total contribution of single FZI distribution functions. The identification of single FZI functions (namely flow units) needs to break the whole FZI distribution down into different components. As a result, cluster analysis is needed.

The flow units relate to the distribution of sedimentary microfacies and diagenetic facies, but may not comply with the boundaries of sedimentary microfacies. The most important factor influencing the fluid flow is the geometric feature of pore throat, which is controlled by the mineral composition (type, position, and amount) and structure (size, shape and sorted behavior of granules) of rocks. Different combinations of these properties can lead to totally different fluid flow characteristics. Therefore, a flow unit may contain different types of lithofacies with different sedimentary textures and mineral compositions.

By taking the NT oil field at the eastern edge of Pre-Caspian Basin as an example, this paper describes the method of classifying the flow units of carbonate reservoirs.

2. Regional profile

The NT oil field lies at the eastern edge of Pre-Caspian Basin, including two sets of carbonate strata, namely carboniferous KT-I and KT-II, with a MKT stratum about 300 m thick in between and 6000 m deep at most. The MKT is divided into two sets of strata consisting of 14 substrata, while the KT-I into 9 substrata and the KT-II into 6 substrata. After the sedimentation, the carbonate reservoirs on the target stratum experienced various diagenetic environments, multi-stages of diagenesis, as well as several tectonic activities. Due to the influence of multiple factors such as diagenetic epigenesis, structural fracturing and dissolution on reservoir space, the reservoirs are complexly distributed, strongly anisotropic and badly connected. Therefore, the research on flow units is of great importance to comprehensive reservoir evaluation while providing a basis for the development and adjustment of oilfield flooding. Especially, the correct identification of single-well flow units is the key of overall reasonable assessment of carbonate reservoirs.

3. Classification method of flow units

3.1. Pittman R35 model

R35 was proposed by using the relationships among porosity, air permeability and pore throat radius (at the mercury injection saturation of 35%). A mass of previous findings show that, R35 can reflect better the type of reservoir space and the seepage characteristics of carbonate reservoirs to promote better research on flow units (Amaefule, et al,1993; Ebank,1987; Hearn,et al,1984).

As R35 can reflect not only sedimentary and diagenetic environments but also flow behavior and reservoir performance, it has been widely accepted by geologists and reservoir engineers to calculate R35 through the combination of Winland model with core-related data or by directly using the capillary pressure curves. The flow units of reservoirs can be classified in accordance with the range of R35 variation.

Winland R35 model was only robust to the rocks in the form of pores and pore throats, which are intergranular pores related to rock texture (such as sandstone)(Fan, et al,2014; Liu,et al,2000;Wang,et al ,2010). But for those carbonatites containing not merely intergranular pores, Winland may not correctly evaluate the reservoir qaulity, so the modified Winland R35, namely Pittman model, can be used to calculate the carbonate reservoirs:

$$\mathrm{LogR_{35}} = 0.255 + 0.565 \times \mathrm{Logk} - 0.523 \times \mathrm{Log}\emptyset_e$$

Where: k is air permeability to the core, in mD; $\emptyset_e$ is core porosity, in %.

3.2. Flow unit classification based on both R35 and PTT

By relating R35 values with the size of pore system, pore throat type (PTT) can reflect better the reservoir quality, seepage characteristics and production capacity of different flow units.

Type I (Megaport): R35>10um. This is the best reservoir with the best reserve and seepage and normally the highest output. In such a flow unit, given the consistency between stratum thickness and other conditions, tens of thousands of barrels of medium-gravity crude oil can be output every day.

Type II (Macroport): 2um<R35<=10um. This reservoir has good quality and high output. Given the consistency between stratum thickness and other conditions, this type of flow units can yield thousand of barrels of crude oil every day.

Type III (Mesoport): 0.5um<R35<=2um. This reservoir has average quality and high output. If other conditions are desirable, this type of flow units can yield hundreds of barrels of crude oil every day.

Type Ⅳ (Microport): 0.1<R35<=0.5um. This reservoir has average quality. Although many dense gas reservoirs have such a R35, their production, in most cases, is very low because of small stratum thickness as well as average quality and heavy gravity of crude oil.

Type Ⅴ (Nanoport): R35<0.1um. This reservoir has bad quality and normally very low output (or mostly dry).

By using the 4249 sample point data of 13 wells in the NT oil field for physical property analysis, this paper classifies the flow units into five types (see the Fig.1 and 2, and the Table 1) according to the R35 Pittman model.

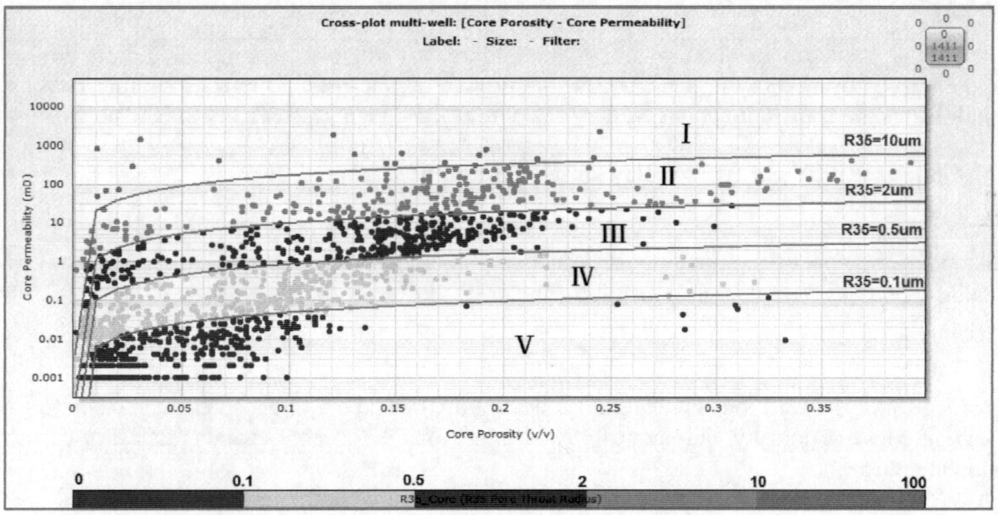

Figure 1. Porosity-permeability crossplot for KT-I with different pore throat radii

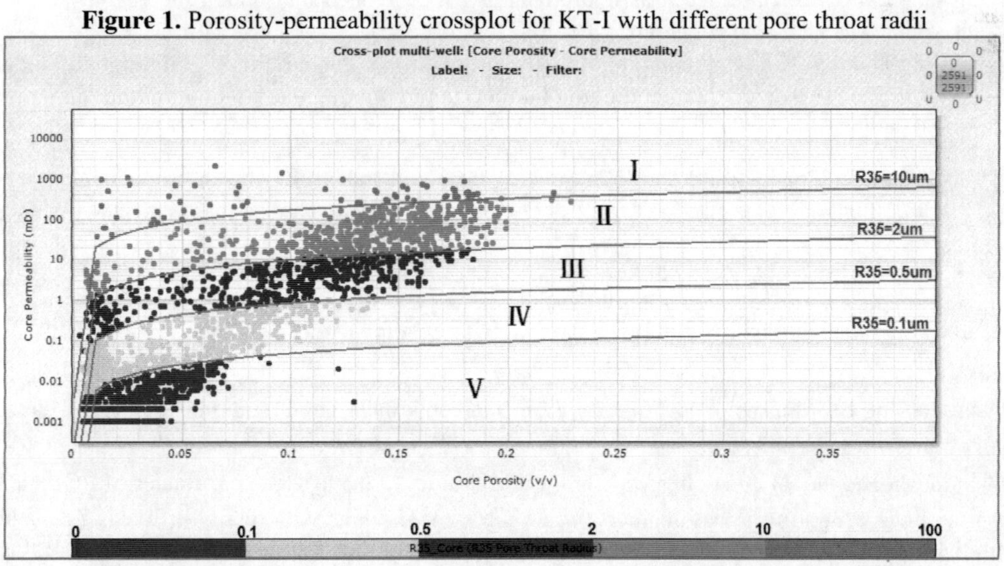

Figure 2. Porosity-permeability crossplot for KT-II with different pore throat radii

Table 1. List of different pore throats for KT1

Stratum	KT-I		KT-I	
Type of pore throat	Number of data points	Percentage (%)	Number of data points	Percentage (%)
I	19	1.35	69	2.46
II	231	16.37	617	21.99
III	377	26.72	535	19.07
IV	431	30.55	609	21.7
V	353	25.02	976	34.78

It is observed from both Fig.1 and 2 that, the growth of fractures in the KT2 stratum is stronger than in KT1 (as there are more points with a <5% porosity and high permeability in KT2).

4. Flow index for different types of flow units in different strata

4.1. Flow index

Simulation of the flow of fluid in porous media by combining Tubular Darcy law with Poiseuille law and replacing porous medium with a series of capillaries showed:

$$k = \frac{r^2}{8}\emptyset_e \ldots\ldots\ldots\ldots\ldots\ldots. \quad (1)$$

Simple as it is, the equation (1) relates pore size (radius) with pore throat shape, demonstrating that the relation between porosity and permeability depends mainly on pore characteristics, in particular pore throat radius.

Therefore, to describe the relation between actual porosity and permeability more relevantly, buckling factor was later introduced as hydrodynamic radius concept to obtain the Kozeny-Carmen equation, which was then converted into:

$$0.0314\sqrt{\frac{k}{\emptyset_e}} = \left(\frac{\emptyset_e}{1-\emptyset_e}\right)\left(\frac{1}{\tau S_{gv}\sqrt{F_s}}\right) \ldots\ldots\ldots\ldots\ldots \quad (2)$$

Through the combination with the definitions of flow unit index FZI and reservoir quality index RQI and the introduction of $\emptyset_e$ as normalized active porosity, the following equation can be obtained:

$$FZI = \frac{RQI}{\emptyset_z} \ldots\ldots\ldots\ldots\ldots\ldots \quad (3)$$

After taking the logarithm on both sides, the following relation can be derived:

$$\log RQI = \log\emptyset_z + \log FZI \ldots\ldots\ldots\ldots\ldots\ldots \quad (4)$$

In the log-log coordinates of FZI and $\emptyset_z$, the intercepts of different straight lines mean the FZI values of different flow units.

In an ideal case, the sample points with similar FZI values in the log-log coordinates of RQI and $\emptyset_z$ are on the lines with an identical intercept, while the sample points with totally different FZI values are on the lines with different intercepts. The sample points on the same line have similar pore throat features, thus forming one flow unit. Every line is a specific flow unit, whose average FZI is expressed by $\emptyset_z$ intercept.

4.2. Determination of FZI for different types of flow units in KT1 and KT2

The parameters input in this study include core porosity $ø_z$ (at the interval of 0.3～1.0 m), core permeability K, and sample analysis data. The corrected core porosity has also been obtained. It can be seen from the equation (4) that, in the log-log coordinates of RQI and $ø_z$, a linear relation is found between RQI and $ø_z$, whose intercept is log (Fig.3 and 4). Thereby, the FZI values for different flow units in different strata, as shown in the Table 2, can be obtained.

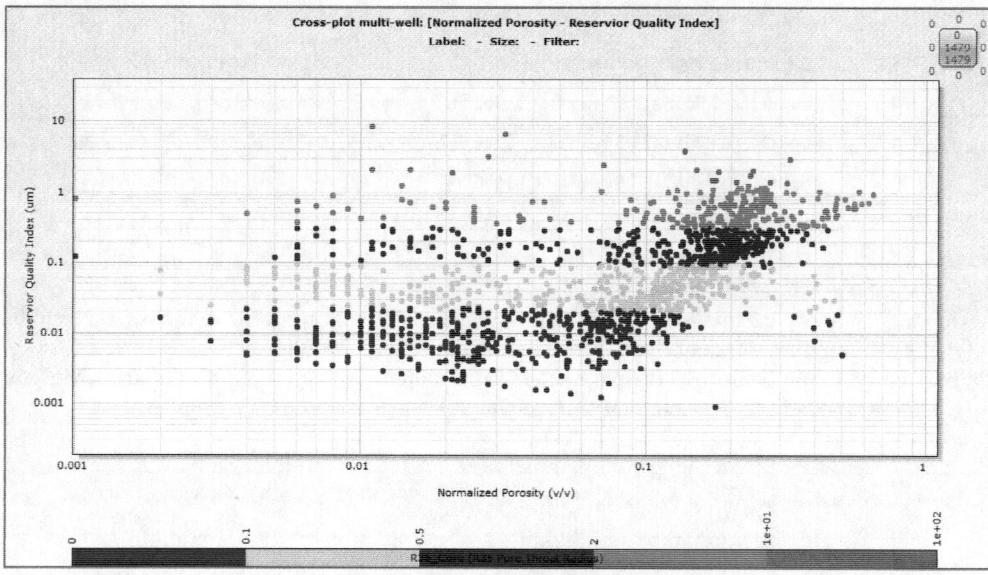

Figure 3. Normalized porosity and RQI for KT1

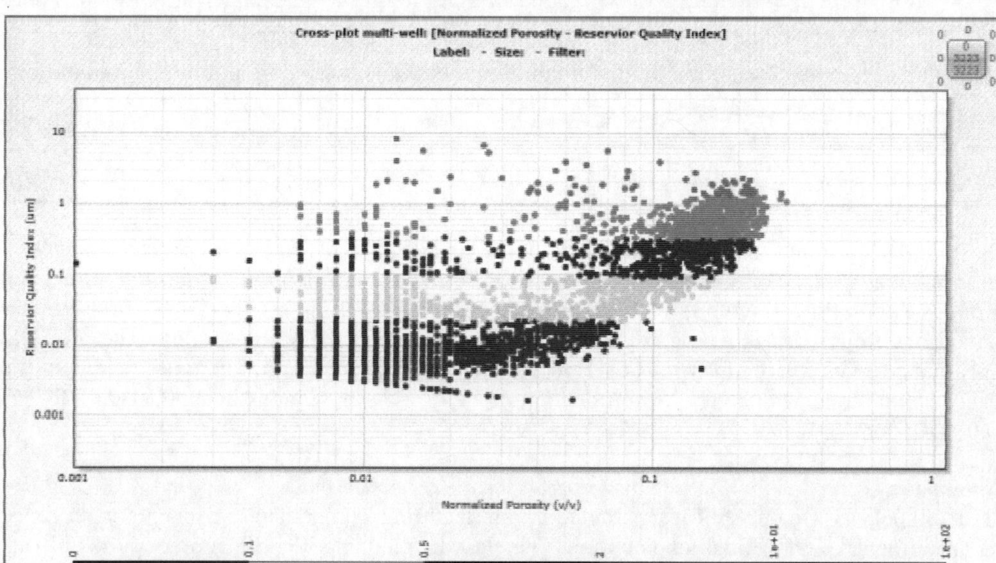

Figure 4. Normalized porosity and RQI for KT2

Table 2. Relation between RQI and ϕ_z for different flow units in KT1 and KT2

	FU Type	Equation	Samples	RMSE	Average FZI (um)
KT1	FU5	$LogRQI = 0.09132852 * Log\Phi_Z - 1.902238$	415	0.26	0.01
	FU4	$LogRQI = 0.03700668 * Log\Phi_Z - 1.304759$	443	0.19	0.05
	FU3	$LogRQI = - 0.01585089 * Log\Phi_Z - 0.786287$	345	0.14	0.16
	FU2	$LogRQI = 0.01013611 * Log\Phi_Z - 0.2473519$	228	0.16	0.57
	FU1	$LogRQI = - 0.1067485 * Log\Phi_Z + 0.1522143$	19	0.14	1.42
KT2	FU5	$LogRQI = 0.2603245 * Log\Phi_Z - 1.557015$	1294	0.18	0.03
	FU4	$LogRQI = 0.05367775 * Log\Phi_Z - 1.26324$	591	0.18	0.05
	FU3	$LogRQI = 0.07655873 * Log\Phi_Z - 0.6546777$	533	0.15	0.22
	FU2	$LogRQI = 0.08281183 * Log\Phi_Z - 0.1570062$	613	0.16	0.70
	FU1	$LogRQI= - 0.05277874 * Log\Phi_Z + 0.1904083$	54	0.09	1.55

4.3. Classification result and distribution pattern of flow units

By considering the above calculation method, the flow units of 13 coring wells in the NT oil field have been classified through discriminatory analysis in accordance with sedimentary microfacies, rock type, reservoir space type, physical property parameters, and mercury-injection curve features. Moreover, the classification results have been comprehensively analyzed, as shown in the Table 3.

Table 3. Features of different flow units in KT1 and KT2

FU Type	R35 (um)	FZI (um)	Main rock type	Main reservoir space type	Main sedimentary microfacies	Main diagenesis	Oil output
FU1	R35>10	FZI>1.42 (KT1) FZI>1.55 (KT2)	Dolomite; algal reef limestone	Fracture-cave type; fracture pore type	Dolomitic flat; algal reef	IA, IB	Highest
FU2	2<R35≤10	0.57<FZI≤1.42 (KT1) 0.7< FZI≤1.55 (KT2)	Limestone; biolithite	Fracture-cave type; fracture pore type	Algal reef; intraplatform shoal	IB, IC	Higher
FU3	0.5<R35≤2	0.16< FZI≤0.57 (KT1) 0.22< FZI≤0.7 (KT2)	Calcarenite; limestone	Fracture pore type; pore type	Intraplatform shoal; grain shoal	IIA, IIB	Average
FU4	0.1<R35≤0.5	0.05< FZI≤0.16 (KT1) 0.05< FZI≤0.22 (KT2)	Limestone	Pore type	Limestone flat; grain shoal	III	Low
FU5	R35<0.1	0.01< FZI≤0.05 (KT1) 0.03<FZI≤0.05 (KT2)	Argillaceous limestone	Pore type	Limestone flat; grain shoal	IV	Very low

The type-I flow units, mainly found in dolomite and algal reef limestone reservoirs, have the highest production capacity. Given the existence of corrosion and dolomitization, they are mainly fracture-cave composite reservoirs or fracture pore reservoirs. As far as the type-I flow units are concerned, the flow index is > 1.42 for KT-I stratum and > 1.55 for KT-II stratum. The production capacity and reservoir quality of type-II-IV flow units would decline in turn. The type-V flow units are argillaceous limestone, with a very low production capacity and a flow index being 0.01-0.05 for KT-I and 0.03-0.05 for KT-II.

5. Conclusion

The NT oil field is a complex carbonatite reservoir featuring complicated sedimentary diagenesis, strong anisotropism, and difficult exploitation. The classification and identification of flow units is of great importance to the exploitation and adjustment of oil fields. The carboniferous reservoir of an oil field has five types of flow units. Among them, the type-I flow units constitute the best stratum, with the biggest flow index but the smallest proportion of only 1.35%. In contrast, the KT-II stratum accounts for 2.46%, and the type-II-V strata for 16.37-34.78% respectively in relative equilibrium. The main contribution to the production capacity of this oil field is made by type-I-III flow units.

References

[1] Amaefule J O, Altunbay M, Tiab D, Kersey D G, Keelan D K 1993 Enhanced reservoir descriptio：Using core and log data to identify hydraulic(flow)units and predict permeability in uncored intervals / wells *SPE* 26436.

[2] Ebank W J 1987 Flow unit concept-integrated approach to reservoir description for engineering projects *AAPG annual meeting AAPG Bulletin* **71(5)** 551-552.

[3] Fan Zifei, Li Kongchou, Li Jianxin 2014 A study on remaining oil distribution in a carbonate oil reservoir based on reservoir flow units *PETROLEUM EXPL0RATION AND DEVELOPMENT* **2014(10)** 578-584.

[4] Hearn C L, Ebanks W J Jr, Tye R S, et al. 1984 Geological factors in fleecing reservoir performance of the Hartzog Dra field, wypming *Petrol Tech* **36** 1335-44.

[5] Liu Jiyu 2000 Advances in the study of reservoir flow unit *Advances in Earth Science* **15(3)** 303-306.

[6] Wang Zhizhang, He Gang 2010 The classification method and application of reservoir flow units *Natural Gas Geoseience* **21(3)** 362-366.

Roadmap of retail electricity market reform in China: assisting in mitigating wind energy curtailment

Dezhao Yu[1], Huadong Qiu[2], Xiang Yuan[3], Yuan Li[3], Changzheng Shao[1], You Lin[1], Yi Ding[1, 4]

[1] School of Electrical Engineering, Zhejiang University, Hangzhou 310027, China;
[2] State Grid Zhejiang Electric Power Corporation, Hangzhou 310007, Zhejiang Province, China;
[3] State Grid Zhejiang Economic Research Institute, Hangzhou 310008, Zhejiang Province, China;

Email:yiding@zju.edu.cn

Abstract. Among the renewable energies, wind energy has gained the rapidest development in China. Moreover wind power generation has been penetrated into power system in a large scale. However, the high level wind curtailment also indicates a low efficiency of wind energy utilization over the last decade in China. One of the primary constraints on the utilization of wind energy is the lack of an electricity market, in which renewable energies can compete equally with traditional fossil fuel generation. Thus the new round electric power industry reform is essential in China. The reform involves implementing new pricing mechanism, introducing retail-side competition, promoting the consumption of renewable energy. The new round reform can be a promising solution for promoting the development and consumption of wind energy generation in China. Based on proposed reform policies of electric power industry, this paper suggests a roadmap for retail electricity market reform of China, which consists of three stages. Barriers to the efficient utilization of wind energy are also analysed. Finally, this paper introduces several efficient measures for mitigating wind curtailment in each stage of reform.

1. Introduction

The development of renewable energies is thriving all over the world. According to REN21's 2016 report [1], worldwide investments in renewable technologies amounted to more than $286 billion in 2015, with countries like China and the United States heavily investing in wind, hydro, solar and biofuels [2]. As a kind of clean energy form, wind energy is relatively abundant and widely used all around the world, such as Europe, Asia, and the United States. Wind power generation has been penetrated into power system in a large scale. As of 2015, more than 100 countries are adopting wind power on a commercial basis [1].

As one of the major contributors of CO_2 emissions, China has made a great effort in developing renewable energies for mitigating air pollution from overuse of fossil fuels. China has constructed many wind power plants, especially in north-western areas, and total installed capacity of wind power generation grows every year. By 2015, the installed capacities of off-shore wind generation in China had reached 145GW, which accounted for 32.5% of the global generation capacities [3,4]. However, the variability and uncertainty of wind power bring a lot of difficulties in power system in term of operation and reliability. Therefore, compared with conventional hydropower and thermal power generation, it is a more serious issue for efficient and effective utilization of wind power resources.

Wind curtailment is one of the severe problems in the utilization of renewable energies in China [5]. In many regions of China, fast increasing installed capacity with low load growth caused widespread overcapacity. Especially in the north-western areas, a lot of wind power generation equipment was forced to deactivate. The curtailment ratio was up to 16.3% in 2015, while the number was only 10.5% in 2014 [3]. One of the primary constraints on the utilization of wind energy is the lack of a competitive electricity market, in which wind power can compete equally with traditional fossil fuel generation [6]. Although the external environment has improved remarkably since the last power industry reform in 2002, the pricing and planning mechanism of China's power industry remain monopoly and opaque [7]. These problems have become increasingly serious and raised public concern. Given on this, the central government has carried out a new round of power system reform in 2015 [8]. The new reform is regarded as the key to solving these problems.

Actually, the new reform serves as a continuation of the last round of reform. A lot of work has been done and achievement has been made since the last reform in 2002 [7]. By now, the generation side has already been relatively competitive although the electricity consumption side (retail side) is still regulated by the government [9]. As a consequence, retail market deregulation is one of the most important task in the new round of reform.

In the following sections, roadmap of retail electricity market reform is proposed and the reasons for wind curtailment in China are revealed. The paper clearly demonstrates that wind curtailment will be mitigated in the process of the reform.

2. Roadmap of retail electricity market reform

The conventional power industry is characterized as a high degree of monopoly. After big users' direct power purchase from generation companies was permitted into electricity market, a competitive market is formed at generation side [10]. However, the monopoly of power industry in retail and consumption side caused an imbalanced situation. Thus the government published a series of reform policies in 2015 to build an efficient pricing mechanism and establish a competitive electricity retail market. In this section, a roadmap of retail electricity market reform is proposed to demonstrate how the new reform will be carried out step by step.

2.1. Power industry reform in China

The power industry of China has been in a process of reforms since the 1980s. In 2002, a reform was launched by the central government to bridge the gap between power supply and demand [7,8]. The reform has made a series of achievements, including the enhancement of operation efficiency of power industry and the fulfilment of power demand all over the country.

However, the reform failed to achieve the predetermined targets. Because there are still a number of problems unsolved, including the twisted pricing mechanism, irrational power system planning, and inefficient utilization of renewable energy and so on. These problems have not only hindered the development of power industry in China, but also brought a negative impact on the business operation of upstream and downstream of corresponding industries.

Therefore, Chinese government has issued "Relative Policies on Deepening the Reform of Power Industry" (No. 9 Document) in 2015 and launched a new round of power industry reform [11]. Policies were carried out to form a new pricing mechanism, new policies about renewable energy generation and power system planning process to solve the problems in the existing power system.

In the policy document "Implementing suggestions of promoting the reform of retail electricity market", it is pointed out that retail electricity business is open to social capital and more options are given to end users [12]. Government tries to bring competition in retail electricity market for improving the quality of services and change the conventional mode of electricity selling, in which the electricity grid is the only permitted electricity seller.

Moreover, China has established some pilots to put the measures proposed in the policy document into practice. By now, China has two retail market reform pilots, including Guangdong Province and Chongqing Municipality and several comprehensive reform pilots [13].

Guangdong province leads the country in implementing the reform. After one-year practice, the achievements of the reform are encouraging in Guangdong. Guangdong has put forward some constructive proposals. For example, as a new market participant, retail companies can participate in the market by trading directly with generation companies on behalf of electricity users. Moreover, Guangdong sets up complete rules for monthly electricity trading based on centralized and transparent auction [14].

2.2. *Roadmap of retail electricity reform*

Following the requirement of NO.9 document [12], this paper analyses the roadmap of electricity retail market reform in China. In the proposed roadmap, the retail market reform process can be divided into three stages, as shown in Fig 1 [11,12]. Each stage is analyzed in the term of market characteristics, customers, electricity sellers and power sale services, as shown in Table 1 [11,12].

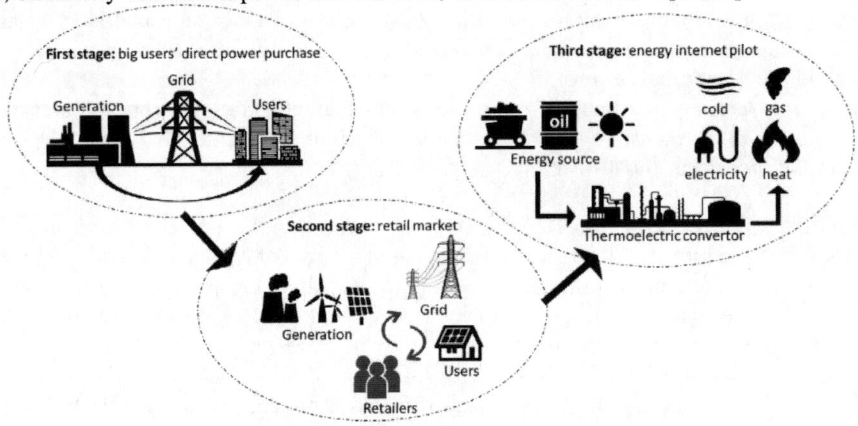

Figure 1. Roadmap of retail electricity market reform

Table 1. Features of each stage in a glimpse

	First stage	Second stage	Third stage
Market characteristics	big users' direct electricity purchase	construction of retail electricity market	power system reform in development of energy internet
Customers	large commercial and industrial users who match the access conditions	other market users	users utilizing different energy forms
Main sellers	power generation group subsidiaries	independent power selling companies	multiple energy integration companies
Power sale services	power supply service based on a long-term contract	customized/personalized services and diversified packages	integrated energy solutions

2.2.1. *First stage* The first stage is characterized by that big users can purchase electricity directly from the generation companies. During the first stage, the permitted market seller is still decided by the government or power industry administrative departments. Among all the electricity sellers, those which are subsidiaries of generation companies have monopolistic superiority in the retail market in term of financial strength, hardware and software conditions, and so on. The main trading types are long-term contract and monthly electricity transactions, which is currently only available to industrial and commercial users who match the access conditions to participate in direct purchase.

Big users' direct power purchase is still the mainstream form of electricity selling during this stage. For generation companies, signing a contract can efficiently protect their benefits when they face the serious problems of excess generation capacity and slow power consumption growth. For big users, they can reduce costs and strengthen economic vitality. It is well known that renewable energies, such as wind and solar power, have low marginal operation cost. However, this advantage cannot work in this trading mode. Actually, the most proper power trading mode is that sellers and users can choose their counterparty freely and negotiate the quantity and price of electricity equally.

2.2.2. Second stage During the second stage, the construction of a competitive electricity retail market will be put a higher priority. During this stage, the government should set different accession standards for different electricity sellers according to their customers' type for encouraging electricity sellers to participate in the market. Moreover, electricity sellers are expected to offer diverse value-added services and personalized services and packages to their consumers.

During this stage, the government should take effort to develop users' market awareness, help them to break the traditional thinking model in electricity consumption and encourage users to participate in the retailer market actively. Besides, users' awareness and ability of participating in the market should be developed. Users, in this way, will be aware of what benefit they can receive in a competitive retailer market, what services they can obtain from electricity sellers and how to choose the most suitable sellers.

2.2.3. Third stage During the third stage, multiple energy transaction is added into retail market due to the development of energy internet. As a consequence, multiple energy integration companies are gradually established and they offer energy supply, demand side management and integrated energy services to energy consumers in energy internet pilots.

During this stage, the government should accelerate the development of energy internet technologies, including information and communication technology, energy storage technology and energy conversion technology, to support the construction of energy internet pilots. Departments concerned should establish a trading platform where sellers can trade with customers by offering integrated energy solutions. Moreover, market mechanisms should be set up to encourage multiple energy sellers to provide customers with choices in different forms of energy utilization which can be switched freely.

3. Mitigation of wind curtailment

The roadmap of China's electricity market reform illustrated the direction of the reform. As the reform moves forward, many existing problems, like wind curtailment, can be solved or mitigated to a certain extent. This section analyses reasons for wind curtailment and explains how to mitigate curtailment during each development stage.

3.1. Why wind curtailment occurs

3.1.1. The existing installed wind power capacity and wind curtailment Wind industry has been expanded year by year since wind was first used for electricity generation in 1980s. The wind power industry set new records in 2014, that is, more than 50GW of new capacity was installed in this year [15]. Another breakthrough occurred in 2015, with 22% annual market growth resulting in the 60 GW mark being passed [1].

To make better use of wind resource and develop wind generation, China has built wind farms with large installed capacity. Wind power installed capacity has boomed in recent year in China since a series of effective support policies were adopted.

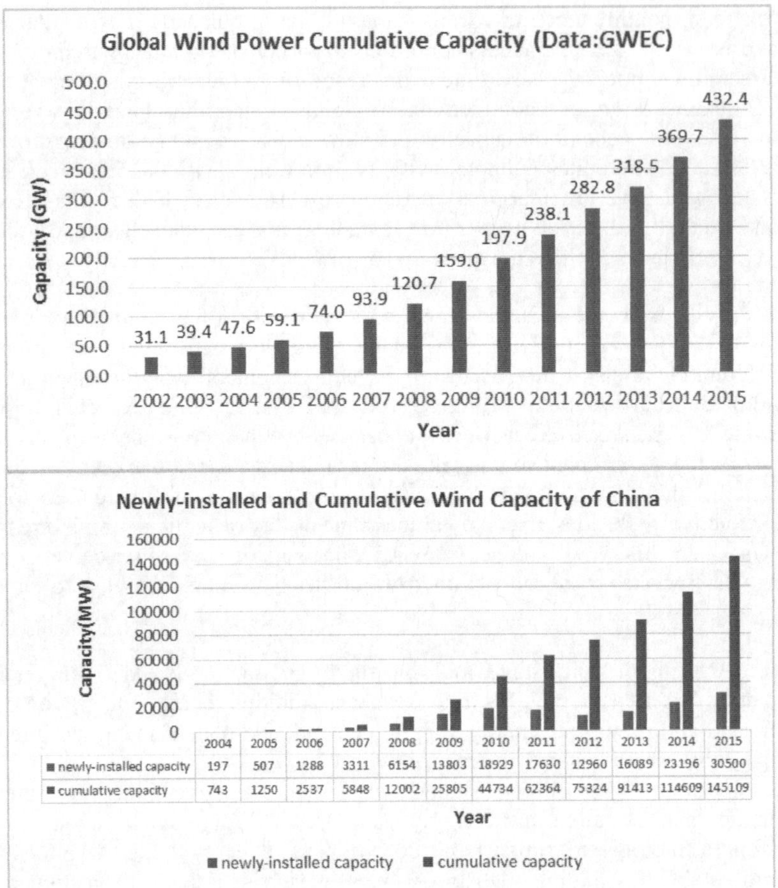

Figure 2. Global and China's installed wind power generation capacity

Figure 2 describes the dramatically increasing trend in installed wind power generation capacity [1,3,15]. Global wind power capacity had expanded rapidly to 432 GW in 2015, and wind energy production was around 5% of total worldwide electricity usage, and kept growing. In 2014, China ranks first in the capacity of installed wind power generation [3].

Table 2. Wind curtailments in 2014 and 2015

Region (province)	The year 2014(%)	The year 2015(%)
Gansu	14.4	37.9
Jilin	27	36.7
Heilongjiang	17.7	22.2
West Neimenggu	14.7	21.7
Xinjiang	13.2	29.1
East Neimenggu	14.5	17.9
Liaoning	16.2	16.4
Hebei	14.3	14.3
Shanxi	1.2	4.2
Ningxia	0	2.58
Yunnan	1.6	1.95
China	**10.5**	**16.3**

However, too much installed capacity of wind power may cause large-scale wind curtailment. Wind curtailments in 2014 and 2015 are shown in Table 2 [3]. For example, in Gansu province, its electricity consumption in 2015 is about 13 million kilowatts while new energy installed capacity is beyond 17 million kilowatts [16]. Therefore, if excess wind power output cannot be sent to other province by trans-province channel, it has to be curtailed.

In fact, in many regions of China, fast increasing installed capacity with low load growth causes widespread overcapacity. Especially in the north-western areas, a lot of wind power generation equipment was forced to deactivate.

3.1.2. Reasons for wind curtailment The most important reasons for curtailment are insufficient transmission, local congestion and excessive supply during periods of lower demand [17-19].

First, inconsistency between the wind power development and the overall grid planning is a crucial issue [18]. The ability to transport electricity between provinces or regions is usually limited by the transmission capacities. China is rich of wind energy resource, which is mainly distributed in north and west areas. However, the power transmission and distribution cabling and grid all over the country have not been comprehensively considered for the large-scale penetration of wind power.

In addition, the peak regulation capacity of thermal power generation is weak [17]. At present, thermal power generation still lacks enough peak regulation capacity for stabilizing the variability and uncertainty of large-scale wind power penetration.

Furthermore, the gird enterprises lack sufficient impetus for purchasing renewable energies in the existing price mechanism since they are usually more expensive. Moreover, conventional generating units involved in the regulation are not given sufficient economic compensation, resulting in the lack of their impetus to participant in providing balancing reserves for accommodating the variable and intermittent renewables.

Therefore, a proper and competitive market mechanism is urgently needed.

3.2. Curtailment mitigation at each stage during the reform

3.2.1. First stage Since the publishment of NO.5 document in 2002, big users' direct power purchase was already a form of electricity trade. However, due to the monopoly of power distribution and selling, this trading mode led to a worse economic result for generation companies. Before the reform in 2015, users can only buy electricity from State Grid if they did not participant in direct power purchase. Thus, generation companies must decrease their price in power purchase agreement to capture market share, which reduced incomes of generation companies. In fact, for generation companies, they would rather sell electricity to transmission companies than big users. Therefore, only when sale market is deregulated can generation companies participate in benefit distribution through big users' direct power purchase.

In competitive market, generation companies can consider offers from sale companies or big users, which provide them with more trading choices. Companies can be more flexible in selling electricity and, hence, they can manage to plan their output and control curtailment.

Moreover, after retailer companies are permitted into market, generation companies can set up their own retailer companies to sell electricity which is generated by their generating units. Through their sales subsidiaries, they can easily sell electricity to users after paying “wheeling cost” to State Grid for transmission. Thus their wind power successfully enters the market, and through proper market strategies they can guarantee their generating capacity, which help mitigate wind curtailment a lot.

3.2.2. Second stage The establishment of spot electricity market is the fundamental solution to promote wind power utilization and mitigate wind curtailment.

In fully competitive markets, generating units which has the lowest marginal cost should be dispatched first. Compared with thermal power plants, wind power generation owns lower marginal cost because no raw energy like coal and oil is required for generating electricity. Therefore, in spot electricity market wind power and other clean energy can be the main electricity resource, while thermal power generation is used to help balance the power system.

Spot price is the basic of spot electricity market, which reflects the supply and demand conditions. That is, when the power output is high, the power price is usually relatively low. Users therefore tend to use more electricity. In this way, wind power can get efficient utilization because the higher power output, the more users will use power. For example, the power price at night is much lower than daytime and hence users will shift part of their production and manufacture to night to decrease their cost. Also, people will charge their electric vehicles and do other energy-using activities at night. The power they use in fact is the wind power which should be abandoned.

3.2.3. Third stage With the development of energy internet, the tools for accommodating the renewables become diversified, which greatly helps mitigate wind curtailment and promotes power utilization.

Firstly, renewable energy output prediction methods using big data break the limitation of the conventional prediction method, which is helpful for scheduling generating units to maintain the balance of supply and demand in the power system. Secondly, the application of multiple energy system imparts flexibility to the operation of power system. In many energy internet pilots, coal -heating is replaced by thermoelectric converters which use excess wind power. According to wind power output and load characteristic in winter, thermoelectric converter starts to work when the load is low, thus excess wind power is converted to the heat we need. Besides, the power price is much lower than heat price when wind power is abundant.

3.3. Cases

Statistics show on-grid wind energy is about 120 billion kWh, while 32.3 billion of which is abandoned in 2016 [20]. The curtailment ratio is beyond 20%. Now the on-grid tariff is 0.47¥/kWh, the income of wind power industry is about ¥56.4 billion. In spot market, wind power is supposed to be dispatched first due to its low marginal cost. If the transaction price is about 0.4¥/kWh (including the subsidy), on-grid 150 kWh can bring ¥60 billion to wind power industry.

That result demonstrates that the existing pricing mechanism hinders the development of wind power and other renewable energy. Construction of spot market can mitigate wind curtailment and increase the profit of wind power industry.

4. Conclusions

Raising the utilization rate of renewable energy is the current trend of power industry all over the world. Among the renewable energies, wind energy has gained the rapidest development, especially in China. As one of the most important contributors of CO_2 emissions, China has made a great effort in developing renewable energies and wind power generation has increased significantly these years. However, the utilization of wind power generation is not optimistic, since the wind curtailment has become an increasing problem year by year. Therefore, China has published a series of policies to boost the new round electric power industry reform, which can be a promising solution for promoting the development and consumption of wind energy generation in China. Based on policies and documents that support China's power industry reform, this paper proposes a roadmap of retail electricity market reform, which consists of three stages. In each stage, different characters and tasks are put forward to solve a series of problems in the existing system, including wind curtailment.

Besides, this paper analyzes factors that affect wind curtailment and clearly reveals the barriers to the efficient utilization of wind energy. Finally, this paper introduces several efficient measures for mitigating wind curtailment in each stage of reform. In conclusion, this paper demonstrates that wind curtailment will be mitigated in the process of retail electricity market reform and provides some approaches to efficient wind energy utilization. While specific pricing mechanism and technical models are not mentioned in this paper, successive research is required.

Acknowledgements
The research is supported by technology projects of State Grid Corporation of China (5211JY160005).

References

[1] REN21, Global Status Report 2016 Retrieved 8th June 2016. [Online]. Available: http://www.ren21.net/wp-content/uploads/2016/06/GSR_2016_FullReport_.pdf
[2] Pazheri F R, Othman M F and Malik N H 2014 A review on global renewable electricity scenario *Renewable & Sustainable Energy Reviews **31***(31) 835-845.
[3] The Globe and Mail, China now the world leader in wind power production. Retrieved 28 February 2016. [Online]. Available: http://www.theglobeandmail.com/report-on-business/industry-news/energy-and-resources/china-now-the-world-leader-in-wind-power-production/article28713509/
[4] Liu Y, Ren L, Li Y and Zhao X G 2015 The industrial performance of wind power industry in china *Renewable & Sustainable Energy Reviews* **43** 644-655.
[5] Cheng Y S, Wong W K and Woo C K 2013 How much have electricity shortages hampered china's gdp growth? *Energy Policy* **55** 369-373.
[6] Holttinen H 2005 Optimal electricity market for wind power *Energy Policy* **33**(16) 2052-63.
[7] Zhao X, Lyon T P and Song C 2012 Lurching towards markets for power: china's electricity policy 1985–2007 *Applied Energy* **94** 148-155.
[8] Zeng M, Yang Y, Wang L and Sun J 2016 The power industry reform in china 2015: policies, evaluations and solutions *Renewable & Sustainable Energy Reviews* **57** 94-110.
[9] DuL, He Y and Yan J 2013 The effects of electricity reforms on productivity and efficiency of china's fossil-fired power plants: an empirical analysis *Energy Economics* **40** 804-812.
[10] Tong M G 2006 Direct power purchase model for large power end-user and its application in our country *Electric Power Technologic Economics*
[11] National Development and Reform Commission, Relative Policies on Deepening the Reform of Power Industry. Retrieved 15th March 2015. [Online]. Available: http://tgs.ndrc.gov.cn/zywj/201601/t20160129_773852.html
[12] National Development and Reform Commission, Implementing Suggestions of Promoting Retail Electricity Market Reform. Retrieved 26th November 2015. [Online]. Available: http://tgs.ndrc.gov.cn/zywj/201601/t20160129_773854.html
[13] National Development and Reform Commission, Reply of Approving Retail Electricity Market Reform in Guangdong Province and Chongqing Municipality. Retrieved 11th November 2015. [Online]. Available: http://www.ndrc.gov.cn/zcfb/zcfbtz/201605/t20160510_801190.html
[14] People's Government of Guangdong Province, Request for Retail Electricity Market Reform in Guangdong Province. Retrieved 9th April 2015. [Online]. Available: http://www.gd.gov.cn/govpub/bmguifan/201505/t20150504_212687.htm
[15] Pazheri F R, Othman M F and Malik N H 2014 A review on global renewable electricity scenario *Renewable & Sustainable Energy Reviews* **31**(31) 835-845.
[16] Li J, Wang N B, Wei Y and Gu Q 2016 The current and solution of wind power consumption for gansu province *Applied Mechanics and Materials **826*** 55-60.
[17] Burke D J and O'Malley M J 2011 Factors influencing wind energy curtailment *IEEE Transactions on Sustainable Energy 2*(2) 185-193.
[18] Rogers J, Fink S and Porter K 2010 Examples of wind energy curtailment practices *National Renewable Energy Laboratory*
[19] Fink S, Mudd C, Porter K and Morgenstern B 2009 Wind energy curtailment case studies: may 2008 - may 2009.
[20] National Energy Administration, Wind Power Integration in the first half of 2016.Retrieved 26th July 2016. [Online]. Available: http://www.nea.gov.cn/2016-07/26/c_135540902.htm

Research on the business QoS and trustworthy and controllable strategies mechanism for energy internet

Geng Zhang[1], Meiling Lu[2, 3], Dahua Zhang[1], Liang Zhou[1] and Likang Li[1]

[1] Information & communication Dept of China Electric Power Research Institute, Beijing, 100192, China

[2] Institute of Electrical and Electronic Engineering, North China Electric Power University, Beijing, 102206, China

E-mail: lumeiling@ncepu.edu.cn

Abstract. Energy internet is a kind of power sharing network, which can realize the bidirectional flow of energy information on the basis of the existing power grid. It puts forward higher requirements for reliability and controllability of information communication, and all kinds of business QoS of the backbone network. So the research of business QoS and trustworthy and controllable strategies mechanism have an important significance for the development of energy internet. This paper mainly studies how to use the software defined network (SDN) to achieve business QoS, and provide QoS support for all kinds of business of the energy internet. Combined with the current development situation of the energy internet in our country, this paper researches the trustworthy and controllable strategies mechanism for energy internet, and proposes the transition scheme of the IPv6 credible network architecture based on SDN. This coordinates the contradiction between the growing demand for energy applications and the limitations of the energy internet technology itself.

1. Introduction

In recent years, with the sustained growth of the world's total energy consumption, energy structure has been adjusted. The global fossil energy resources are facing resource exhaustion and other practical problems. To solve the problem effectively, clean energy is being developed and used in large scale [1].Global energy internet has a global and strategic significance for the coordinated development of the world economy and environment[2].

The normal work of the energy internet needs safe and reliable information and communication network. With the continuous deepening of the construction of energy internet, information and communication infrastructure is becoming more and more important [3]. Compared with the traditional Internet, energy exchange in the energy internet requires more reliability, so energy internet requires more controllability of the information and communication network under the premise of open network. To achieve better controllability and reliability, we can refer to the latest software defined network technology. The data forwarding and controlling is separated through the software defined network (SDN). It not only enhances the controllability, also makes the management more convenient. In addition, SDN technology can also provide more security, flexible security policy, convenience for the dynamic deployment and configuration [4].

With the improvement of the intelligent degree of energy internet, many aspects of energy internet will mostly use IP technology in information and communication area. Intelligent sensor and intelligent terminal has been widely used. With the exhaustion of global IPv4 address resources, IPv4 address resources in China is scarce, is not enough for the needs of the development of energy internet [5]. Large power grid which is as the backbone network of the energy internet is bound to put forward higher requirements for the power quality service level in the future. Also various types of service business of power grid will increase requirements of data integrity, accuracy, timeliness, security and QoS. Limited to QoS capabilities of IPv4 technology, QoS is basically no application in the power grid data communication network of our country. Due to the lack of a reasonable QoS planning, a variety of businesses of different types and different communication performance requirements influence each other with the parallel transmission in the backbone data network.

In [6], an application programmable interface for QoS based on SDN is designed for video conference and streaming media. The application interface has good flexibility and scalability, which provides support for QoS applications. A network performance management system based on SDN is proposed in [7]. And the system has many advantages, such as intelligence, high reliability and security. In [8], the author proposed the trustworthy and controllable network architecture by attaching to the present network architecture with a four lays of trustworthy and controllable logical architecture. In [9], a method to realize IPv4/IPv6 protocol conversion based on SDN is proposed, which separates the protocol conversion function and the message forwarding function.

This paper takes advantage of SDN to achieve business QoS. Business priority module, network monitoring module and strategy formulation module are deployed in application layer of SDN network. It divides and sets the priority of various businesses according to the actual situation of the energy Internet and certain rules. Different priorities of the businesses own different bandwidth resources and transmission flow. The paper chooses queuing mechanisms provided by the OpenFlow protocol to achieve band-width guarantee for different types of services.The implementation architecture of business QoS based on SDN are introduced in detail. Finally, the paper puts forward IPv6 credible network structure transition scheme to help to put the proposed network architecture into practical application as soon as possible.

2. IPv4/IPv6 transition technology

At present, China has only about 0.33 billion IPv4 addresses. It can't meet the rapid growth of application requirements even if a large number of applications of address translation technology to delay the IPv4 address consumption. And it also significantly increased network complexity and difficulty of management, reduced the network and information security level and QoS. Compared with IPv4, IPv6 technology has large amount of address space, and it supports for mobile IP seamless connection. Through new IPv6 flow label field, it is able to identify the same type of data flow and provide efficient QoS support for different businesses. IPv6 technology can provide a good solution to solve the above problems and provide better support for the construction and operation of the energy internet [10].

Table 1. IPv4/IPv6 transition technology comparison

	The technology used	Advantage	Disadvantage
Dual-stack	IPv4/IPv6; dual-stack	simple; mature IPv4/IPv6 coexistence	can't solve the problem of insufficient IPv4 address space
Tunnel	GRE tunnel; 6to4 tunnel ISATAP; 6PE/6VPE DS-Lite; 6RD	Effectively solve interconnection of "isolated island"; no loss of information network; maintenance is simple	increase device complexity; not conducive to large-scale applications
Translation	NAT-PT; NAT64 IVI; NAT444 BIA; SOCK64	network maintenance is simple; mutual access between IPv4 and IPv6	information may be lost; increase the workload of conversion equipment

Promoting the transition from IPv4 to IPv6 evolution of the Internet, and developing the next generation Internet on this basis has become a global consensus [11].The transition from IPv4 to IPv6 is a process which needs for a long-term evolution. There will be many IPv4 and IPv6 islands in the transition process. Due to IPv4 and IPv6 is not compatible, we must introduce IPv6 transition technology to achieve conversion of IPv4 to IPv6 [12].A variety of different approaches have been proposed to solve this problem. The current mainstream transition technology is divided into three types: dual stack technology, tunnel technology and translation technology [13].Table 1 is a comparison of three techniques. In order to develop a reasonable network transition solution, we should combine the use of the three technologies considering the actual network status.

3. Software Defined Network

In the rapidly changing business environment of the energy internet, the high stability and high performance of the network is not enough to meet the business needs, flexibility and agility is more critical. Distributed energy can be accessed to power grid through many forms, the majority of users can also participate in energy transactions through a variety of network access methods, so flexible network access capability is particularly important. Energy Internet supports a variety of control methods and its operating topology is no longer limited to a specific structure. It is able to achieve independent energy allocation, because its decision unit is no longer restricted by the central control node [14]. In order to meet the development needs of energy internet, the Internet architecture based on software defined network has emerged.

Software defined network is a new network innovation architecture proposed by the CleanSlate research group of Stanford University. Its basic idea is to separate complex control logic that determines how packets are forwarded of the current IP network interconnection node from switches or routers or other equipment, realize the hardware control of the data forwarding rules through the software programming, and finally achieve the purpose of free control of the flow [15]. SDN core technology OpenFlow achieves a flexible control of network flow through the separation of the control plane and the data plane of network device. OpenFlow (OF) is a kind of network exchange model of SDN. Openflow network consists of OpenFlow switch, FlowVisor and controller. OpenFlow switch is used for data layer forwarding. FlowVisor is to make virtualization on network. Controller centralizes control of the network, to achieve the function of the control layer [16].

SDN separates the control power from the network device and administered by a centralized controller. SDN separates control plane and forwarding plane, gets rid of the hardware restrictions on network architecture, and realizes better flexibility and agility of the network. In addition, the programmable ability of SDN has brought the unprecedented flexibility, because the control is completely open, users can customize any network routing and transmission rules strategy.

4. Implementation architecture of business QoS based on SDN

4.1. Existing problems

Fast packet forwarding in data plane and route decision of high level in control plane is integrated together in traditional network technology. The development of this kind of tight coupling mainframe limited the emergence of the IP network innovation technology. Traditional IP packet networks all have the problem which can't guarantee the QoS, because it is designed to pay more attention to network accessibility and is based on the destination IP address hop by hop routing， it can't provide QoS bandwidth resources reserved in advance for end-and-end [17].

Large power grid as the energy internet backbone network, QoS is basically no application in its data communication network of our country. Due to the lack of a reasonable QoS planning, many different types of businesses parallel transfer in the backbone data network and influence each other, which has a great influence on the development of the energy Internet data communication network.

4.2. The basic principle of SDN

As the focus of next generation network technology research, SDN defines a new type of network architecture, which simplifies the complexity of packet forwarding rules and separates the control plane from the forwarding plane. The controller in control plane can be centralized management of the switch flow table to achieve the control objective of data forwarding [18]. The decoupling control and forward and centralized control of the network, as well as the controller programming and the flexible control of flow in SDN network provides a good idea for the network QoS planning[19].

OpenFlow is the first interface specification of the open network foundation for the standardization of SDN research system. All versions of the OpenFlow protocol have the description of the support for QoS services. Openflow version 1.3 add meter table the switches, the meter table can't only test packet forwarding rate, but also control packet forwarding rate. Flow table is the only forwarding rule of the data plane in SDN network, and the flow meter is formulated by the controller and centralized management. So, the controller is the core of SDN network routing control [20].

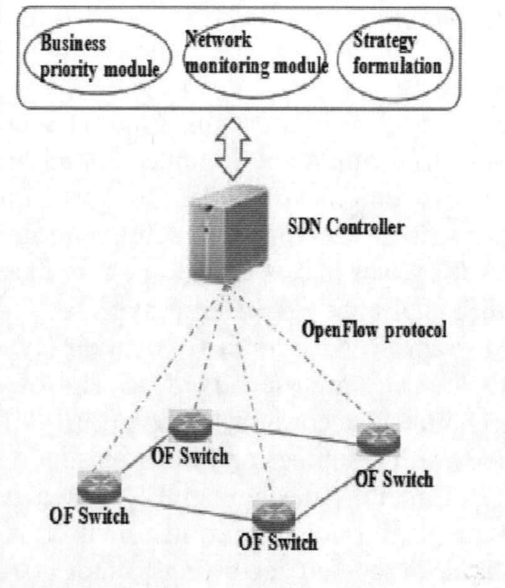

Figure1. Implementation architecture of business QoS based on SDN

4.3. Implementation architecture of business QoS

The implementation architecture of business QoS based on SDN proposed in this paper is shown in Figure 1. Compared to general QoS, business QoS is specifically designed for different business to meet their individual needs. We provide different QoS strategies for different service. In this paper, we mainly consider the business needs of the energy internet.

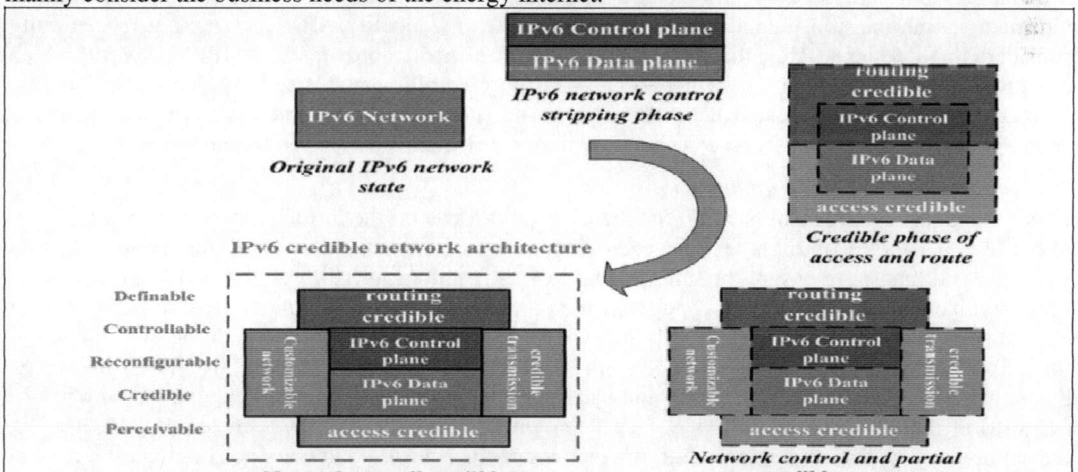

Figure 2. IPv6 credible network structure transition scheme

In the proposed architecture, business priority module, network monitoring module and strategy formulation module are deployed in application layer of SDN network. According to the actual situation of the energy Internet and certain rules, business priority module divides and sets the priority of various businesses. Different priorities of the businesses own different bandwidth resources and transmission flow. Network monitoring module is mainly responsible for monitoring communication network. It can use the controller to check the statistical data on the switch by sending a query message and manage the flow according to the statistical data. Also, it can use OpenFlow switch to collect traffic information, such as sending and receiving data for each port, flow table and queue. Strategy formulation module is mainly responsible for developing the QoS strategy for different business, for example when the queue congestion occurs, the high priority queue should be priority for transmission, or when a communication link fails, how to quickly switch the service.

This paper makes use of queuing mechanisms provided by the OpenFlow protocol to achieve bandwidth guarantee for different types of services. Various services through the OpenFlow access network. The controller distinguishes different types of services by identifying the IP address of the source node, and then the predetermined priority of the services is known. The controller carries on the queuing mechanisms according to the priority of the services. Different priorities of businesses own different bandwidth resources and transmission traffic. OpenFlow switches provide QoS support through such a simple queuing mechanism [21]. Switch port can configure multiple queues. One or more queues can be attached to a port and then be used to forward packets via it. Packets forwarded to a specific queue will be processed according to that queue's configuration (e.g. minimum rate or maximum rate), and control packet forwarding rate of that queue.

5. Research on trustworthy and controllable strategies mechanism

The fundamental purpose of the new network architecture research is to solve the problems encountered in the current energy internet, and to coordinate the contradiction between the growing demand for energy internet applications and the limitations of energy internet technology itself. However, although the design mentality of some new network architecture is good, it can't get practical application because of unable to support incremental deployment. This paper focuses on the research of trustworthy and controllable strategies mechanism for energy internet, and puts forward IPv6 credible network structure transition scheme.

IP network itself tends to be a state of self -control network with high equality. Network itself does not consider the verification of the source address and routing reliability in the basic principle of information transmission, and does not require the service quality. IPv6 network has been fully considered and designed after the long-term development and improvement of IPv4 network, which can provide full support for future network's security, credibility, manageability and controllability. How to realize the IPv6 trusted network needs to be fully considered and designed. IPv6 credible network structure transition scheme is shown in Figure 2, which is divided into four stages.

5.1. IPv6 network control stripping phase

In this stage, the transformation of IPv6 network should focus on the introduction of the control layer to the IPv6 network. Control layer device can be highly customized, and it can design related security guarantee mechanism to ensure the independence of its functions and the compatibility of the original network functions, while ensuring control layer equipment is not hijacked and attacked. Separation of control plane and data plane is the key to the evolution of network, and the basis of the introduction of the definition and reconstruction function in the future. This stage needs to deploy control devices on the network key nodes. Taking into account part network has deployed network equipment which does not support definition function and service deployment and network has been formed, the way of redevelopment can't be implemented, so it is necessary for control equipment not to affect the normal work of the original data transmission service. The new data layer equipment of the key nodes can not only communicate with the control equipment, but also can work with the old equipment, so that the old network can add a control layer and not change the network smoothness.

5.2. Credible phase of access and route

In this stage, the end device access mechanism should be firstly carried on the design and implementation. After the completion of the separation of control and forwarding, the evolution from IPv6 network to IPv6 credible network should start on access credible and routing credible. Identity authentication device should be deployed in the terminal access side to ensure safe and reliable of the access terminal. And then, the key network nodes require the deployment of two-dimensional routing equipment, introducing two-dimensional routing function in the key nodes on the basis of the compatibility of the original destination address routing policy. In addition, the control interface of the identity authentication device and the two-dimensional routing equipment should be retained to set aside the control interface for the future introduction of control equipment. In order to ensure the authenticity of the information, that is, to ensure the authenticity of the identity ID, we should design the true and credible identity ID address distribution system and authenticity verification system for identity ID based on identity authentication technology.

After the terminal equipment is connected to the network, the IPv6 address is obtained through the address allocation system. The back 64 bits of the IPv6 address assigned by the address allocation system carry the true and trusted identity ID information of the user. The address allocation system registers the assigned IPv6 address with the corresponding true trusted identity ID to the two-dimensional information verification system. After the completion of the access and transmission mechanism, the promotion and transformation of the routing function has been gradually carried in the transmission equipment, and the new routing mechanism has been introduced. It takes the innovative IPv6 source address routing method which is compatible with the destination address routing of the original IPv6 network.

5.3. Network control and partial credible stage

Deploy control devices (such as network virtualization controller) in the access layer, and gradually control the function of the key nodes in the network, to achieve the network key nodes terminal can be abstracted, defined, controlled and reconfigurable. The system administrator can log on the control equipment to make the route definition of the data plane equipment, providing the highly customization, flexible and controllable data forwarding, link definition and Qos guarantee. Deploy trusted transfer functions in the export equipment of the subnet or gateway devices, such as address jump communication module, NAT66 module and so on. Through the design of the gateway equipment function, it can realize the security protection mechanism of the data packet routing transmission and ensure the reliability of the transmission after the end device access and verify the reliable identity. The functional equipment in data layer ensures the credible transmission and completes the established security function through the control device configuration and signaling information. In this stage, the control layer of the network basically has the control authority to main node of the network, and the prototype of the credible network is reached, that is, the main network and key nodes can be managed and controlled, access and routing can be trusted and safe. A credible network architecture based on IPv6 has been initially established.

5.4. Network overall credible stage

After the main network and key nodes can be managed and controlled, access and routing can be trusted and safe, with long-term investment, operation and maintenance, the project will gradually replace the equipment which does not support reliable function in turn, until all the equipment can be controlled and can release control signaling through the control plane, and ultimately achieve the credibility of the whole network.

6. Conclusion

Informatization is a public platform and an important means to support the construction of the energy internet. With the improvement of the intelligent degree of energy internet, many aspects of energy internet will mostly use IP technology in information and communication area. Therefore the construction of the energy internet has put forward higher demand for the controllability and reliability of the information communication network and the IP technology. This paper mainly learns how to use SDN to achieve business QoS to provide QoS guarantee for all energy internet services. In addition, this paper focuses on the research of trustworthy and controllable strategies mechanism for energy internet, and proposes the transition scheme of the IPv6 credible network architecture based on SDN. It has important reference value for future design of transition strategy from IPv4 to IPv6 network and the research on trustworthy and controllable mechanism for the energy internet, and it provides a reference for the comprehensive construction of safe and reliable energy internet information communication network as well.

References

[1] Rifkin J 2012 The third industrial revolution, CITIC Press.

[2] Liu Zhenya 2015 Global energy Internet, China Electric Power Press.

[3] Cao Junwei, Yan Mingbo, Zhang Dehua, et al. 2014 Energy internet: aninfrastructure for cyber-energy integration *Southern Power SystemTechnology* **4** 1-10.

[4] Cao Junwei, WangJiye, Ming Yangyang, et al. 2015 Software-defined Information and Communication Technology for Energy Internet *Proceedings of the CSEE* **35** 3649-54.

[5] TianShiming, Luan Wenpeng, Zhang Dongxia, et al. 2015 Technical formsand key technologies on energy internet *Proceedings of the CSEE* **35** 3482-94.

[6] Fu Jian, Shen Subin 2015 A Design Scheme of Application Programmable Interface for QoS Based on SDN *Computer Technology and Develoment* **25** 99-104.

[7] Lin Tang, Chao Wang 2015 The Research on Network Performance Management System Based on SDN Technology *Signal Processing Communications and Computing (ICSPCC).*

[8] Luo JunZhou, Han Zhigeng, Wang Liangmin 2009 Trustworthy and Controllable Network Architecture and Protocol Framework *Chinese Journal of Computers* **32(3)** 391-398.

[9] Huang Zhilan, Fan Yongbing, Chen Nan,et al. 2014 IPv4/IPv6 protocol conversion based on Open Flow SDN *Telecommunications Science* **S2** 61-65.

[10] Zhu Tian, Chen Tao, Ma Di, et al. 2011 Research of IP traceback technology based on transition from IPv4 to IPv6 *Application Research of Computers* **12** 4409-12.

[11] Li Zhenqiang, Zhao Xiaoyu, Ma Yan 2006 IPv6 technology secret People's Posts and Telecommunications Press.

[12] Sun Qiong, Xie Chongfeng, Zhao Huiling,et al. 2014 Design and realization of SDN-based IPv6 transition technology *Telecommunications Science* **4** 15-21.

[13] Tong Yapeng, Li Zhenqiang, Wei Bing 2011 Analysis of IPv6 transition technology *Telecommunications Science* **1** 52-60.

[14] Wang Jiye, Guo Jinghong, Cao Junwei, et al. 2015 Review on Information and Communication Key Technologies of Energy Internet *Smart Grid* **3** 473-485.

[15] Zhang G, Su L, Wang Y 2014 Research on communicationnetwork architecture of energy Internet based on SDN *IEEEWorkshop on Advanced Research and Technology in IndustryApplications (WARTIA)* 316-319.

[16] Tootoonchian A, Ganjali Y, HyperFlow 2010 A distributed control plane for OpenFlow. *Proc. of the 2010 Internet NetworkManagement Workshop/Workshop onResearch on Enterprise Networking (INM/WREN)* . San Jose : USENIX Association http://dl.am.org/citation.cfm?id=1863136

[17] Zhang Shunmiao, Zou Fumin 2013 Survey on software defined network research *Application Research of Computers* **30** 2246-51.

[18] Zhang G, Su L, Wang Y 2014 Research on communication network architecture of energy Internet based on SDN *IEEE Workshop on Advanced Research and Technology in Industry Applications (WARTIA)* 316-319.

[19] Civanlar S, Parlakisik M, Tekalp A M, et al. 2010 A QoS-enabledOpenFlow environment for scalable video streaming *Proc of IEEE GLOBLECOM Workshops* 351 -356.

[20] Open Network Foundation(ONF). OpenFlow switch specificationv1.3. 0 [EB/OL].https: //www. opennetworking. org/images/stories/ downloads / specification /openflow-spec-v 1 .3. 0. pdf.

[21] Yan Jinyao, Zhang Hailong, Shuai Qianjun, et al. 2015 HiQoS:An SDN-Based Multipath QoS Solution *China Communications* **5** 124-132.

Research on heat transfer characteristics and cold trap capacity of a water catcher during vacuum pre-cooling

Enhai Liu[1, 4], Yin Liu[1], Changzhong Wang[2], Shiqi Liu[3]

[1] School of Energy & Environment, Zhongyuan University of Technology, Henan Province, Zhongyuan Road 41, Zhengzhou 450007,China;
[2] Energy Test Center of Industry and Information Technology of Zhengzhou City, Zhengzhou 450007, China;
[3] Department of Building Engineering,College of Information & Business,Zhongyuan University of Technology, Zhengzhou 450007, China.

E-mail: liuenhai1018@126.com

Abstract: Effect of vacuum pre-cooling process on apples was a complex process of heat and mass transfers. The research is based on the physical properties of apples and their heat and mass transfer mechanisms during vacuum pre-cooling. As for the heat transfer characteristics of a water catcher in vacuum pre-cooling, the research studied the heat transfer mechanism and calculated the cold trap capacity by experimental means， and it cold trap capacity were evaluated to supply references for future research into the practical applications of such vacuum pre-cooling techniques. The results provide a theoretical basis for exploring better pre-cooling process conditions and the design of water catchers. The experimental results show that, when the wall temperature of the water catcher is -5°C, the optimal cold trap capacity is about 90.72g and the required cooling capacity is 210.13W in the vacuum pre-cooling of 201.9g of apples.

1. Introduction

Based on the theory that water absorbs the latent heat of vaporization under vacuum, vacuum pre-cooling is designed to improve the storage quality of fruit and vegetables by rapidly cooling them, effectively eliminating field heat and restraining respiratory heat. Simultaneously, owing to heat beings absorbed during the evaporation of water, a cooling effect is achieved without an external heat source in vacuum pre-cooling. Hence, vacuum pre-cooling is characterized by a high cooling rate, uniform temperature distribution, cleanliness, and an absence of pollution. Meanwhile, appropriate pre-cooling over time can maintain the quality of fruit and vegetables to the greatest extent and reduce decay losses [1-7]. In recent years, many scholars have focused on the physical properties of materials after pre-cooling, the physico-chemical changes in vacuum storage, the heat and mass transfer characteristics of fruit and vegetables in storage, etc [8-18]. Many research findings related to vacuum cooling have been published: Wang et al. studied a vacuum model of a meat product using the finite element method in 2002 and they also simulated the cooling process by computational fluid dynamics [19] [20]. Yan et al. (2006)studied the matching of the water catcher and the vacuum pump determined the pre-cooling quality during vacuum cooling processes [21]. Tao et al. (2006) reported the effects of vacuum cooling treatment and storage conditions on lipid oxidation, superoxide anion generation, superoxide dismutase, catalase, peroxidase and polyphenoloxidase in mushrooms [22]. Schmidt et al.

(2010) studied aimed to determine the effects of vacuum cooling on the enzymatic antioxidant system of cherry and inhibition of surface-borne bacteria during storage. By analyzing the literature, it was found that scholars mainly focus on the introduction of vacuum pre-cooling process conditions, the design of pre-cooling equipment, the control of weight-loss ratio of fruit and vegetables in pre-cooling, the physical properties of materials after pre-cooling, etc [23]. However, some scholars paid attention to the structural optimization of the water catcher used in vacuum pre-cooling, and the optimal matching of the vacuum pump, the heat transfer characteristics of water catchers and the theoretical calculation of their cold trap capacity are rarely studied.

Apples collected from Lingbao City apple plantation, Henan Province, China were taken as the study object. According to the requirements of the test, the authors divided and screened the samples. On the basis of the heat transfer characteristics of the water catcher used in vacuum pre-cooling, this research explored the heat transfer mechanism and calculated the cold trap capacity to provide a theoretical basis for the design of the water catcher.

2. The heat transfer characteristics of the water catcher

The water catcher, also called a cold trough or cold trap, is a critical component of any vacuum pre-cooling installation. The air, which contains moisture and is removed by a vacuum pump, is trapped by the water catcher. At 610 Pa and 0 °C, the air volume increases 210,000-fold. Under these conditions, the vacuum pump power consumption becomes significant, especially given its low efficiency. Therefore, the performance of the water catcher is key to the operation of the refrigerating plant.

During vacuum pre-cooling, the water vapor flowing out from the vacuum chamber condenses into liquid water when it meets the condensing surface of the water catcher, and then attaches to the condenser tube. At the same time, the non-condensable gas produced by the system will be removed by the vacuum pump. In the water catcher used for cooling air, there is single phase heat transfer in the air and also a phase change heat transfer because of the change of the moisture content therein. In the heat transfer process, the heat transfer coefficient and heat transfer quantity can be expressed as,

$$\phi = \frac{A_0 \Delta t_m}{\left(\frac{1}{\alpha_i} + \gamma_i\right)\left(\frac{A_0}{A_i}\right) + \left(\frac{\delta}{\lambda}\right)\left(\frac{A_0}{\overline{A}}\right) + \left(\frac{\overline{\delta}}{\lambda_l}\right) + \frac{1}{\varepsilon \varepsilon_e \alpha_d}} \quad (1)$$

where ϕ is the amount of heat transferred, W; α_i and α_d are the heat transfer coefficients of the refrigerant in the tube and of air under dry cooling, $W/(m^2.℃)$; γ_i is the fouling coefficient for the inner side of the tube $0.0004m\cdot$□ / W; A_0, A_i, and $\overline{A}$ respectively represent the heat transfer areas of the outer and inner sides of the tube and the average area, m^2 ; δ and $\overline{\delta}$ denote the thickness of the heat exchange tube and the average water film thickness, m; λ and λ_l are the thermal conductivities of the tubes and water films, $W/(m^2.℃)$; ε_d denotes the reduced convective heat transfer coefficient (air-side)which was set to 0.8; Δt_m is the mean temperature difference, °C; and ε is the dehumidifying coefficient.

$$\varepsilon = 1 + \frac{\gamma_0 + C_{p,v} t_m - C_w t_w}{C_{p,m}} \cdot \frac{d_m - d_w}{1000(t_m - t_w)} \quad (2)$$

where, γ_0 is the latent heat when the water vapor condenses into water at 0 °C; $C_{p,v}$ is the specific heat capacity of water vapor under constant pressure; C_w is the specific heat capacity of water; $C_{p,m}$ is the specific heat capacity of moist air under constant pressure; d_w is the moisture content of saturated wet steam when the wall temperature is , °C; d_m is the average moisture content of a single tube; t_w is the wall temperature, °C; and t_m is the average temperature of the moist air flowing through a single tube, °C.

The heat transfer coefficient in tubes may be calculated using,

$$\alpha_i = Cq^{0.67} M_l^{-0.5} P_r (-\lg P_r)^{-0.55} \tag{3}$$

Where, C is the calculated coefficient; q is the heat flux density, W/m^2; and M_l and P_r are respectively the relative molecular weight and reduced pressure of the liquid.

Condensation heat transfer plays a major role in the water catching process, while convective heat transfer contributes only slightly. Owing to the moist air flow being slow, the convective heat transfer coefficient out of the tubes was replaced with a condensation heat transfer coefficient for these theoretical calculations. The average surface heat transfer coefficient for the surface film condensation of the horizontal circular tube is,

$$h_h = 0.725 \left[\frac{\lambda_1^3 \gamma_0 \rho_1 (\rho_1 - \rho_v) g}{\mu_1 (T_s - T_w) d_0} \right]^{0.25} \tag{4}$$

Where, ρ_1 is the density of water; ρ_v is the density of water vapor; T_s is the saturated temperature corresponding to the prevailing vapor pressure. There into, the characteristic length of the vertical tube is the tube length, represented by L, and the characteristic length of the horizontal circular tube is its outside diameter d_0. Thus, the relationship between the condensation heat transfer coefficients h_v and h_h is,

$$\frac{h_h}{h_v} = 0.77 \left(L/d_0 \right)^{0.25} \tag{5}$$

The operation of the water catcher is a pressure reduction process. The convective heat transfer is less than the condensation heat transfer. Besides, the water film on the outside surface of the tube has a significant influence on the convective heat transfer of moist air. At a 25 °C initial temperature in the vacuum chamber and 60% relative humidity, the partial pressure of saturated water vapor is given by,

$$InP_{q,b} = \frac{c_1}{T_0} + c_2 + c_3 T_0 + c_4 T_0^2 + c_5 T_0^3 + c_6 InT_0 \tag{6}$$

The water vapor pressure is, $P_q = \varphi . P_{q,b}$ (7)

where, $P_{q,b}$ is the partial pressure of saturated water vapor, Pa; P_q is the water vapor pressure, Pa; c_1, c_2, c_3, c_4, c_5, and c_6 are constants and are respectively -5800.2206, 1.3914993, -0.04860239, 0.41764768 × 10^{-4}, -0.14452093 × 10^{-7}, and 6.5459673; φ is set to a value of 0.026; and T_0 is the temperature of the moist air, °C.

3. Experimental work

This research analyzed the physical properties of apples and the heat and mass transfer mechanisms during vacuum pre-cooling. Based on the heat transfer characteristics of the water catcher used in vacuum pre-cooling, the heat transfer mechanism of the water catcher and its cold trap capacity were evaluated to supply references for future research into the practical applications of such vacuum pre-cooling techniques.

3.1. The experimental device

The vacuum pre-cooling testing machine, VFD-2000 (Shanghai Vacuum Preservation Equipment Co., Ltd) was used in this experiment. It contains a refrigeration system, a vacuum system, and a data acquisition and control system (see Fig. 1a). The water catcher, as a key part of the vacuum pre-cooling device, was effectively the evaporator for the circulation of the system and reduced the load on the vacuum pump by condensing the water vapor. In addition, the vacuum system was equipped with a 2XZ-4 rotary vane vacuum pump operating at 4 L/s and at rate of rotation of1,400 rpm. Four thermocouples were used for temperature measurement and a pressure sensor was used to record the state of the applied vacuum. Meanwhile, the data acquisition and control system, comprising a programmable logic controller (PLC), host computer, and electrical system was used to collect, store, and conduct each parametric test in real-time.

a. Test device diagram for refrigeration system. b. Practicality diagram of vacuum chamber.

Figure 1. Experimental table for the vacuum pre-cooling test machine

3.2. Materials and methods

The apples used in this experiment were collected from Lingbao City Science and Technology Pilot Park, Henan Province, China. After being divided and screened, 201.9 g of samples were weighed (on an ACS-3 electronic balance whose maximum load was 3,000 g, with a sensitivity of 0.1 g, and Grade III accuracy) and placed on the shelf in the vacuum chamber. During the experiment, the pressure sensor was placed in the vacuum pump to record the change in pressure in the vacuum pump. The time-span for internal data collection was 30 s and MS-Excel® was used to post-process the raw data. The thermocouples were fixed at the geometrical centre and the margins of the apples as required (see Fig. 1b). Thereafter, the door of the vacuum chamber was locked and the apparatus was turned on and all data recorded.

The compressor and vacuum pump formed the main energy dissipating components in the vacuum pre-cooling process. During the experiment, the author adjusted the load on the compressor and switched the vacuum pump on and off in response to the observed changes in temperature and pressure. After the pre-cooling experiment, the samples were taken out and weighed. The test was then repeated with modified control parameters.

3.3. Analysis of results

By experimental observation, data recording, and analysis, the main factors affecting the cold trap capacity of a water catcher were found to have been: the initial temperature of the vacuum chamber, relative humidity, the temperature and water vapor partial pressure in the water catcher, the wall temperature of the condenser, *etc.* The atmospheric pressure, water vapor partial pressure and relative humidity in the vacuum chamber, and their changes over time, are shown in Figures 2 and 3.

It can be seen from Figure 2 that the gas pressure and dry air pressure in the vacuum chamber gradually decreased during the pre-cooling phase. Figure 3 shows changes in relative humidity in the vacuum chamber over time. Under the initial conditions, the pressure in the vacuum chamber was on the verge of atmospheric pressure. However, the pressure decreased after using the vacuum pump and its rate of change gradually decelerated in the medium-term and stabilized thereafter. At the flash vaporization point (at about 6.2 minutes and 2,338 Pa), the dry air pressure was zero and the volume of water evaporated from the apples gradually decreased. Whereas if the vacuum pump started to work before flash vaporization, almost no water was evaporated from the apples and only a little air in the vacuum chamber will be trapped by the water catcher so that the cold trap capacity becomes the water vapor remaining in the vacuum chamber. At that stage, the pressure in the vacuum chamber gradually decreased and the moisture content of the air decreased rapidly. However, when it reached the flash vaporization point, the moisture content of the apples was practically all evaporated. Figure 2 and Figure 3 show the sharp rising trend in the water vapor pressure and relative humidity in the vacuum chamber. Subsequently, the amount of evaporated water and the water vapor partial pressure were decreased. Therefore, the pre-cooling pressure should be applied on the basis of the different physical properties of different materials to obtain the best pre-cooling results in actual vacuum pre-cooling.

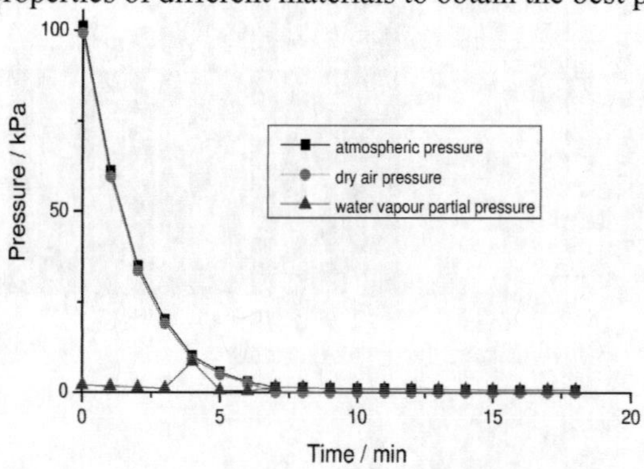

Figure 2. Changes in pressure in the vacuum chamber over time

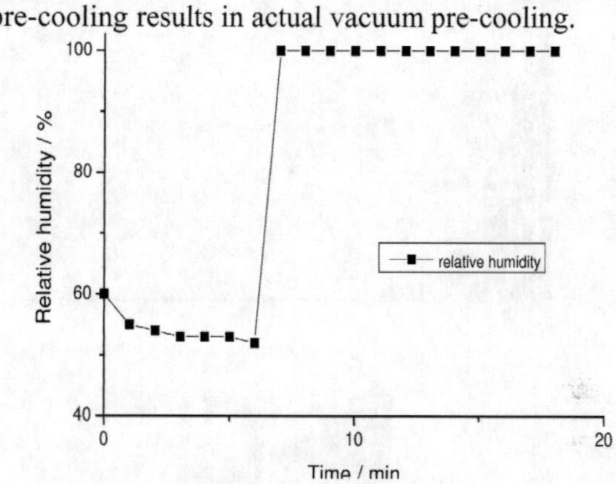

Figure 3. Changes in relative humidity in the vacuum chamber over time

3.4. The calculation of cold trap capacity

The normal operation of a vacuum pre-cooling device mainly depends on the performance of the water catcher. Based on steam condensation theory, and the aforementioned analysis, the cold trap capacity can be calculated by using the formula below. The changes of the cold trap capacity, the quantum of heat transfer from the apples, and the density of the moist air in the vacuum chamber during pre-cooling are shown in Figures 4.The quantities of capture water,

$$G = m_1 + m_2 = \phi / \gamma_0 \quad (8)$$

In formula, G is the quantities of capture water, γ_0 is latent heat, kJ/kg,Where $m_1 = \sum_{i=1}^{5} m_i$ is steam consumption, g, and $m_2 = \sum_{i=6}^{18} m_i$ is water consumption in evaporation, g.

$$q = G\frac{\gamma_0}{t} \quad (9)$$

$$K_{kn} = \frac{kT}{\sqrt{2}\pi L d^2 p} = \frac{M_a}{\mathrm{Re}} \quad (10)$$

where, q is the cooling capacity, W, t is time, min, K_{kn} is Knudsen Number, L is the pipe diameter, m, M_a is Mach number, Re is Reynolds number, γ is fouling factor, $m\cdot$□/ W, k is Coefficient parameters, d is diameter, m, p is Vacuum pressure, Pa, and T is temperature, ℃.

From Figures 4, it can be seen that the density of moist air decreased during pre-cooling. This situation arose because the pressure in the vacuum chamber was close to atmospheric pressure at the start of the pre-cooling phase, and under no influence from the vacuum pump, while the pressure reduced gradually during evacuation of the chamber. At the flash vaporization point, the dry air pressure was zero and both the amount of evaporated water from the apples, and the density, decreased gradually to cause the density of the moist air to continue to decrease while undergoing slight fluctuations. From the aforementioned calculation, at a wall temperature of -5 °C in the water catcher, the cold trap capacity of the vacuum pre-cooling system was 90.72g, and 210.13 W of cooling capacity was required for 201.9g of apples. The computing method and result provided a theoretical basis for the design of such water catchers.

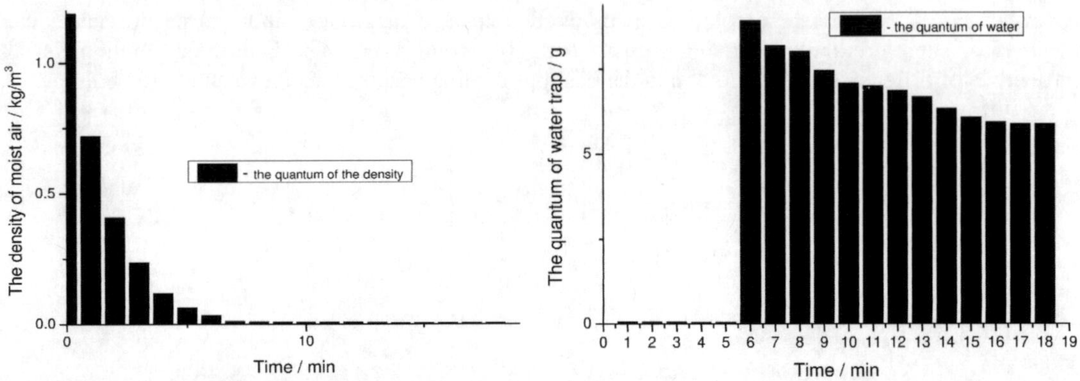

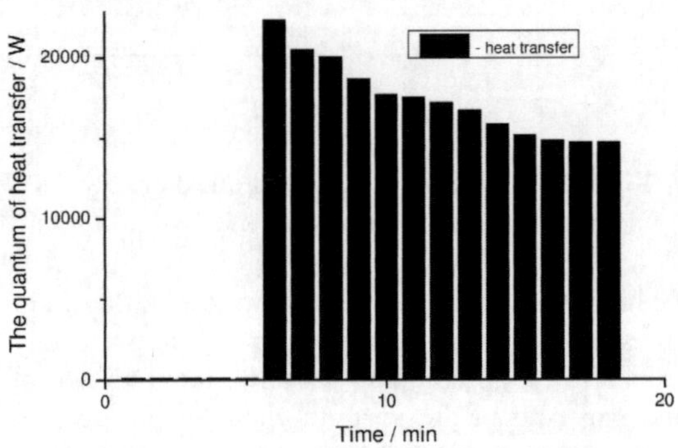

Figure. 4 Changes in the density of moist air, the cold trap capacity and the amount of heat transferred in the vacuum chamber during pre-cooling.

4. Conclusions

In this paper, the heat transfer characteristics of water-catcher used in vacuum pre-cooling system were introduced. Water-catcher was a key component, vacuum extracted with water vapor in the air and subject to capture water, trapping the water down, otherwise it would reduce the pump's performance affects the entire vacuum pre-cooling device performance. The analyses and the results could be useful for operating vacuum pre-cooling.

At first, the gas pressure, dry air pressure, and moist air density in the vacuum chamber decreased during pre-cooling. At the beginning of the pre-cooling phase, the pressure in the vacuum chamber was close to atmospheric pressure. After switching on the vacuum pump, the pressure decreased rapidly and then stabilized (albeit with minor fluctuations therein) in the medium-term, and remained stable thereafter. Besides, the cold trap capacity was related to the surface temperature, evaporating temperature, and cooling capacity of the water catcher. The results suggested that at a wall temperature of -5°C, the optimal cold trap capacity was approximately 90.72gwith 210.13W of cooling capacity required during the vacuum pre-cooling of201.9g of apples.

Acknowledgment
This study supported by the Henan Province Education Department Science and Technology Key Program (14A480005)of China.

References

[1] Brosnan T, Sun D W 2001 Pre-cooling techniques and applications for horticultural products

International Journal of Refrigeration **24** pp154-170

[2] Brosnan T, Sun D W 2003 Influence of modulated vacuum cooling on the cooling rate, mass loss and vase life of cut lily flowers *Biosystems Engineering* **86(1)** pp 45-49

[3] Karl Mc Donald, Da-Wen Sun 2000 Vacuum cooling technology for the food processing industry: a review *Journal of Food Engineering* **45(2)** pp 55-65

[4] Kondjoyan A 2006 A review on surface heat and mass transfer coefficients during air chilling and storage of food products *International Journal of Refrigeration* **29(6)** pp 863-875

[5] Kumano H, Asaoka T, Saito A, Okawa S 2007 Study on latent heat of fusion of ice in aqueous solutions *International Journal of Refrigeration* **30(2)** pp 267-273

[6] Liu Yang, Shen Jiang, Zou Tonghua, et al. 2004 Theoretical research on vapor condenser of vacuum pre-cooling *Vacuum &Cryogenics* **10(4)** pp 230-234

[7] Martens H, Martens M 2001 Multivariate analysis of quality. An introduction. London, UK: Wiley.

[8] He S Y, Zhang G C, Yu G C, et al. 2013 Effects of vacuum cooling on the enzymaticantioxidant system of cherry and inhibitionof surface-borne pathogens *International Journal of Refrigeration* **36** pp 2387-94

[9] Houska M, Landfeld A, Sun D W 2005 Eating quality enhancement of cooked pork and beef by ripening in brine and vacuum cooling *Journal Food Engineering* **68(3)** pp 357-362

[10] Jackman P, Sun D W, Zheng L 2007 Effect of combined vacuum cooling and air blast cooling on processing time and cooling loss of large cooked beef joints *Journal Food Engineering* **81** pp 266-271

[11] LijunWang, Da-Wen Sun 2002 Modelling vacuum cooling process of cooked meat—part 1: analysis of vacuum cooling system *International Journal of Refrigeration* **25** pp 854-861

[12] Lutz J M, Hardenburg R E 1986 The commercial storage of fruits, vegetables, and florist and nursery stocks. In: USDA agriculture handbook (no. 66) Washington, DC: USDA, 55-67

[13] McDonald K, Sun D W 2000 Vacuum cooling technology for the food processing industry *Journal of Food Engineering* **45** pp 55-65

[14] Rennie T J, Raghavan G S V, Vigneault C 2001 Vacuum cooling of lettuce with various rates of pressure reduction *Journal of Electronic Packaging* **44** pp 89-93

[15] Sun D W, Zheng L 2006 Vacuum cooling technology for the agi-food industry: past, present and future *Journal Food Engineering* **77** pp 203-214

[16] Sun H, Qi Z G, Zhang J, Deng D Q 2003 The experimental research of vacuum precooling. In: Cryogenics and Refrigeration- *Proceedings of ICCR. Hongzhou, China*, pp 830-833

[17] Wang L J, Sun D W 2001 Rapid cooling of porous and moisture foods by using vacuum cooling methods *Trends in Food Science and Technology* **12** pp 174-184

[18] Zhang Z H, Drummond Liana, Sun D W 2013 Vacuum cooling in bulk of beef pieces of different sizes and shape-Evaluationand comparison to conventional cooling methods *Journal of Food Engineering* **116** pp 581-587

[19] Wang L J, Sun D W 2002 Modeling vacuum cooling process of cooked meat-part 1: Analysis of vacuum cooling system *Int. J. Refrigeration* **25** pp 854-861

[20] Wang L J, Sun D W 2002 Modeling vacuum cooling process of cooked meat-part 2: mass and heat transfer of cooked meat under vacuum pressure *Int. J. Refrigeration* **25** pp 862-871

[21] Yan Lei, Liu Bin, Zou Tonghua 2006 Study on optimization matching of catcher and vacuum pump in vacuum cooling *Journal of Refrigeration* **27(1)** pp 49-52

[22] Tao F, Zhang M, Hangqing Y 2006 Effects of different storage conditions on chemical and physical properties of white mushrooms after vacuum cooling *Journal Food Engineering* **77(3)** pp 545-549

[23] Schmidt F C, Aragão G M F, Laurindo J B 2010 Integrated cooking and vacuum cooling of chicken breast cuts in a single vessel *Journal of Food Engineering* **100** pp 219-224

The method of a joint intraday security check system based on cloud computing[a]

Wei Dong[1,3], Changyou Feng[2], Caiqi Zhou[2], Zhi Cai[1], Xu Dan[1], Sai Dai[1], Chuancheng Zhang[1]

[1] Power Automation Department, China Electric Power Research Institute , Beijing 100192 China

[2] National Electric Dispatching and Control Center, State Grid Corporation, Beijing 100192 China

E-mail: dongwei3@epri.sgcc.com.cn

Abstract. The intraday security check is the core application in the dispatching control system. The existing security check calculation only uses the dispatch center's local model and data as the functional margin. This paper introduces the design of all-grid intraday joint security check system based on cloud computing and its implementation. To reduce the effect of subarea bad data on the all-grid security check, a new power flow algorithm basing on comparison and adjustment with inter-provincial tie-line plan is presented. And the numerical example illustrated the effectiveness and feasibility of the proposed method.

1. Introduction

The intraday security check (ISC) is one of the four core applications for smart grid dispatching control system. It has been widely used in all levels of the State Grid Corporation dispatching center. With the development of UHV interconnected power grid, the contact among inter-provincial power grid operation is increasingly close [1]. The three level of dispatching centers (national, regional, provincial) are urged to join together and accomplish the main grid operation schedule and security check. The existing ISC can only calculate with the local model and data, lack of synchronous funcation with data and model, and ability of reducing the effect of bad planning data to the power flow in ISC is poor [2-8]. This paper firstly shows a method of joint intraday security check system based on cloud computing. And a new power flow algorithm basing on comparison and adjustment with inter-provincial tie-line plan is presented. It would alleviate the effect of bad planning data coming from subarea on all-grid power flow in the ISC. The numerical example illustrated the effectiveness and feasibility of the proposed method. At last, the paper forecasts the development trend of the key technology in the design and implementation of the ISC system based on cloud technology.

2. Cloud Computing in the Intraday Security Check

Cloud computing technology is mainly applicated on four aspects in the joint intraday security check system.

[a] This paper was supported by Science and Technology Projects of the State Grid Corporation of China

- The Power System Model Cloud: With the virtualization technology, a cloud service of power grid structure and realtime information is supported and managed by the power grid model cloud.
- The Power System Data Cloud: The intraday-plan integrated data is stored in different types of database, such as real-time database, relational database. The integration service is provided by the data cloud.
- The Grid Compute Cloud: The AC power flow and state security analysis using in the Joint ISC system both base on the grid compute cloud service. With the help of its great computational capacity, the all-grid intraday-plan can be handled by the new ISC system much quicker than before.
- The Planning Cloud: The planning cloud proposes the service such as data and parameters preparation, status notification, unified calculation.

The cloud computing platform used in the joint ISC system is composed of the four types of cloud services above. In this paper, three level dispatch center is used to describe the whole power grid manager architecture, which name after Level-A, Level-B, Level-C dispatch center. Level-A generally represents the national dispatch center, Level-B represents the regional dispatch center, and Level-C means the provincial dispatch center. This architecture has been widely established in China power grid Corps. The structure of dispatch system based on cloud computing is shown in Figure 1.

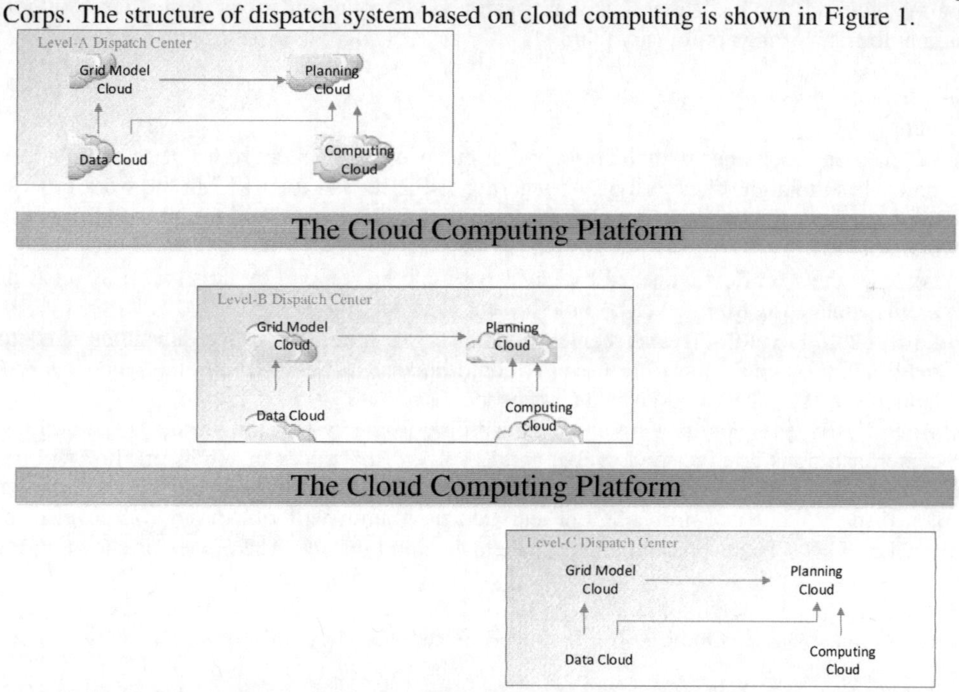

Figure 1. Figure with short caption (caption centred).

3. The Method of A Joint Intraday Security Check Based on Cloud Computing

Based on the multi-level schedule balance management, the function which can automatically and orderly transfer and balance the intraday schedule will be realized by the method. The multi-level plans are calculated and analysed by the ISC. The alarm information is sent to security-constrained economic dispatch(SCED) system where it will be iteratively adjusted until the results meet the safety requirements of power grid.

The general principles of the joint ISC is "top-down", which means the superior dispatch center release new intraday research unit schedule to the subordinate SCED system. The corresponding plans will be automatically programmed and synchronized by the Planning Cloud and the power system data cloud service. The all levels of ISC results are shown to all-level dispatcher to determine whether the schedule data meet the safety operation requirements.

The whole joint ISC system should have the intraday-plan integrated data process, the intraday-plan automatic planning, the security check computation, the check result analysis and the intraday-plan adjustment function.

The Intraday-plan integrated data processing function should possess periodic and artificial trigger control, intraday plan collection and the pretreatment of planning data.

According to the requirements of load balance, regulation, the clean energy accommodation principles, the intraday-plan should be automatically made and adjusted. In this paper, a method based on power balance management authority mode is proposed to solve the complicated problem of the multilevel schedule planning.

3.1. Balance management rights model

a) The local system can trigger the new research plan scheme, the subordinate system would receive the research plan triggered by the higher level and prepare the preliminary subordinate plan according to the planning information received from higher level dispatch center.

b) Using the monitor function, subordinate units can apply recommendations to the higher authorities for changing the data of the preliminary plan.

3.2. The Method Implement

a) The rights:

- Level-A: In accordance with the principle of balance management rights, the Level-A dispatch center has the highest authority for triggering a SCED and joint ISC in the work process. In the meantime, the Level-B, Level-C dispatch center should respond to the workflow initiated by the Level-A.
- Level-B: the workflow launched by Level-B should be respond by Level-C. If Level-A needs it , the application from Level-B could be sent to Level-A too.
- Level-C: If Level-C dispatch center finds that the intraday tie-line schedules need to be modified, the application from Level-C could be sent to Level-B, and the workflow will be launched by Level-B dispatch center if needed.

b) Workflow: Firstly, pre-plan data are created according to the system requirements at the Level A dipatch center, and transferred to the Level-B and Level-C dispatch center. The workflow will trigger their adjustment. Secondly, Level-A dispatch center integrates data collected from local and the Level-B, Level-C dispatch center. Thirdly, SCED and ISC programm will run through all-level dispatch center to make a new schedule synchronized by the data cloud service. The process is shown in Figure 2.

The functions of the security check computation, the check result analysis and the intraday-plan adjustment are mature technologies. Due to space limitations, they are not discussed further in this paper.

4. Equations and mathematics Power Flow Adaptive Adjustment Optimization Algorithm Considering Tie-Line Plan

Compared with real-time calculation, the joint ISC calculation based on cloud platform involves a large number of schedules full of the future forecast data with much uncertainty. In order to improve the accuracy of calculation and avoid bad subarea data influence in the section power flow calculation, this paper proposes a algorithm considering the adaptive optimization adjustment of inter-provincial tie-line for planning power flow. The algorithm makes every part digest the internal unbalance flow of their own according to its proportion by controlling the transmission flow the AC power flow calculation strategy, thereby ensures the accuracy and practicality of the large power grid ISC.

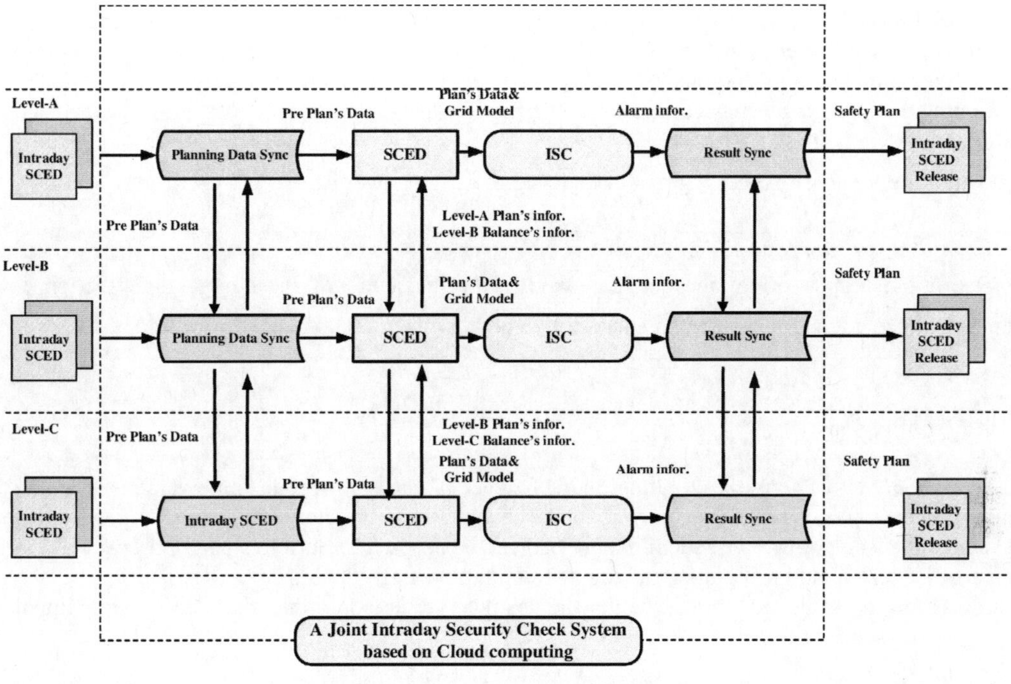

Figure 2. The process of workflow

4.1. Algorithm Steps

4.1.1. Precheck and revise the plan value of tie-line as follows:

a) Determine the initial plan value of the sum Δp by the following formula:

$$\Delta p = \sum_{i=1}^{N} p_i^{sch0} \tag{1}$$

b) If is equal to zero, the initial value of the tie-line plan of each con section is pass for precheck:

$$p_i^{sch} = p_i^{sch0} \tag{2}$$

In (1) and (2), p_i^{sch} is the regional tie-line schedule value for the use of power flow calculation, p_i^{sch0} is the initial regional tie-line schedule value. The large power grid is divided into $N(i \in N)$ area.

c) If Δp is not equal to 0, then the region i 's tie-line power schedule initial value p_i^{sch} will be corrected, including:

$$p_i^{sch} = p_i^{sch0} - \Delta p * \left| p_i^{sch0} \right| / \sum_{i=1}^{N} \left| p_i^{sch0} \right| \tag{3}$$

and then

$$\sum_{i=1}^{N} p_i^{sch} = 0$$

4.1.2. AC Power flow calculation AC power flow calculation is carried out to calculate the unbalanced active power of the nodes and the regions; the power flow calculation is carried out by the future generation planning and bus load forecasting, and the total unbalance active power is concentrated on the balance node. A balance of nodes in the region for , the total power imbalance of active power is Δp_s.

In addition to region , other regional imbalance of active power is:

$$\Delta p_i = p_i^{sch} - \sum_{jk \in L_i} P_{jk} \tag{4}$$

P_{jk} for the branch jk of the active power, L_i for the area i of the contact line section of the member branch collection. The unbalanced active power of regional A is

$$\Delta p_a = \Delta p_s - \sum_{i \in G, i \neq a_i} \Delta p_i \tag{5}$$

G is a set of regions.

4.1.3. The balance of active power Judging the balance of active power absolute value is less than the convergence criteria: if not active power balance absolute value greater than the convergence criteria, and adjust the active power of generator and load can be adjusted, return to step 2; otherwise flow convergence, calculated using the trend the N-1 and fault set calculation.

a) According to the equation, whether the absolute value is less than the active power imbalance convergence criteria is judged

$$|\Delta p_i| < \varepsilon \ (i \in G) \tag{6}$$

b) Convergence accuracy is represent by ε. If the formula (6) is satisfied, the flow convergence is judged, and the result of power flow calculation is used to calculate the N-1 and fault sets;

$$\Delta p_m^{unld} = \frac{p_m^{unld}}{\sum_{j \in R_i} p_j^{unld}} \cdot \Delta p_i (m \in R_i, i \in G) \tag{7}$$

c) Δp_m^{unld} on behalf of the regulatory unit or load to bear the burden of the adjustment of power; p_j^{unld} on behalf of the adjust[[unit or load j to participate in the adjustment of the coefficient; R_i on behalf of the region i involved in the adjustment of the unit or load set. The process of the algorithm is shown in Figure 3.

4.2. Application Examples

In order to verify the effectiveness of the proposed algorithm, the results are compared with actual power grid calculation of the joint ISC.

We select an existing power grid tie-line section to show the effect of the proposed algorithm. The results are shown in Figure 4. It shows that the calculation results of the optimal control algorithm consdering tie-line schedule are consistent with the plan, but the results obtained by the direct calculation are quite different. Using this algorithm, average deviation rate of tie-line calculation values is 0.2%, otherwise it is 6.9%.

A transmission line in the contact line section is selected for showing the algorithm effects. The results are shown in Figure 4. It can be seen from the graph that the difference of calculation results with the algorithm and the actual result is very small, but directly calculation of line flow and the value of actual line flow varies greatly. Using this algorithm, average deviation rate of contact line calculation values is 0.7%, otherwise it is 12%.

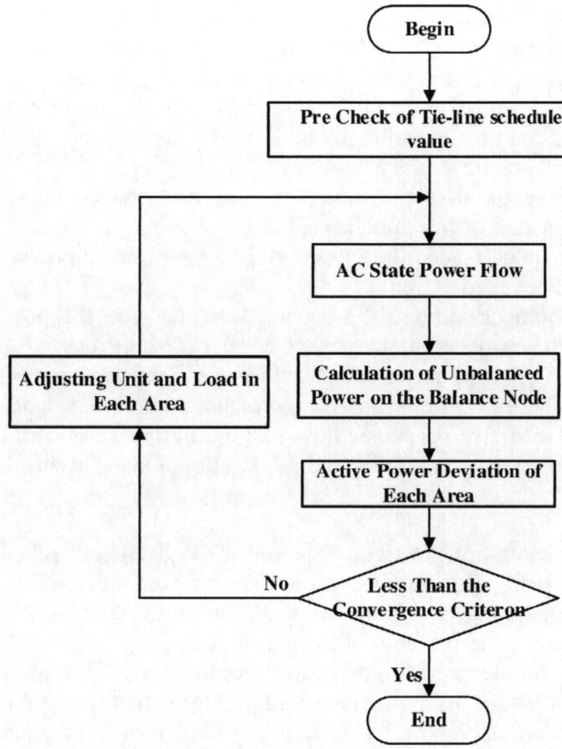

Figure 1. The process of power flow algorithm considering tie-line schedule

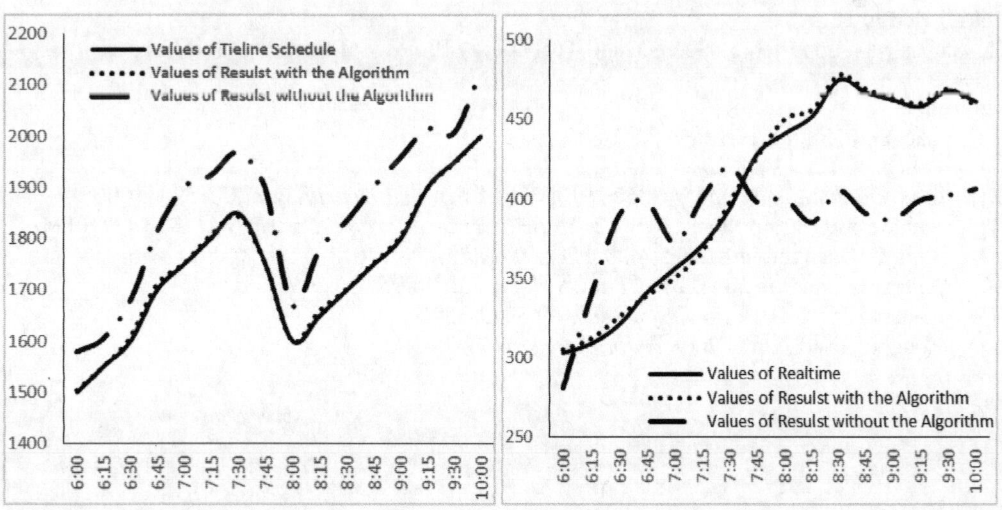

Figure 4. The comparison of a tie-line section and transmission line in tie line section calculation

4.3. The Advantages of the Algorithm

1) The algorithm proposed in this paper according to the plan value control Liaison Section transmission of active power, the partitioned iterated elimination system generation planning and bus load forecast data between mismatch are reasonable recently AC power flow section, on the basis of this section were N-1 and fault set calculation and improve the power system recently static security checking the reliability and practicability.

2) The algorithm can adapt to the different regions reported data quality is inconsistent, the reasonable elimination of active power imbalance.

3) This Algorithm can quickly according to the size of the regional power imbalance, to check the quality of the data reported in each region.

4) It can realize the compatibility of the contact line plan and the power generation plan, and improve the accuracy of the results of static security check before the day.

The algorithm according to the plan value control Liaison Section transmission of active power, the partitioned iterated elimination system generation planning and bus load forecast data between mismatch are reasonable recently AC power flow section, on the basis of this section were N-1 and fault set computing, to improve the reliability and practicality of static security checking.

5. Conclusion

Cloud computing is a research hotspot right now, and it has been researched and developed in both industrial and academic field. This paper proposes the method of a joint intraday security check system based on cloud computing. The power flow algorithm using in joint ISC which considers tie-line plan is proposed to solve the problem occurring on a large scale power system planning section flow calculation. The method described in this paper would be used in multi-level dispatch centers of State Grid Corporation of China. Its application will greatly enhance the ability of all-grid dispatch center to deal with complex large power grid safety operation and the adaptation of power market reform.

Acknowledgments

This paper was supported by Science and Technology Projects of the State Grid Corporation

References

[1] Cao Y 2012 *Chn. Ele. Pow.* **45** 14-17
[2] Popeanga J 2012 *Database Syst.J.* **3** 57–66
[3] Guo Y, Pan M and Fang Y 2012 *IEEE Trans. Parallel Distrib.Syst.* **23** 1593–1606
[4] Samaresh B, Sudip M and Joel J 2015 *IEEE Trans. Parallel Distrib.Syst.* **26** 1477-1494
[5] Ying L, Guangming L 2015 *Auto. Elec. Pow.* **39** 171-176
[6] Xingzhi W and Sheng Y 2011 *Pow. Sys. Pro.* **39** 90-95
[7] Chuancheng Z 2016 *Pow. Sys. Tech.* **40** 3342-3347
[8] Zonghe G and Jian G 2008 *Auto. Elec. Pow* **32** 28-30

Key issues and technical route of cyber physical distribution system

P X Zheng[1], B Chen[2], L J Zheng[2], G L Zhang[2], Y L Fan[2] and T Pei[3, 4]

[1] State Grid Fujian Electric Power Co., LTD., Fuzhou 350003, China
[2] State Grid Fujian Electric Power Research Institute, Fuzhou 350007, China
[3] Electrical Engineering, Zhejiang University, Hangzhou 310027, Zhejiang Province, China

E-mail: welcome_pt@163.com

Abstract. Relying on the National High Technology Research and Development Program, this paper introduced the key issues in Cyber Physical Distribution System (CPDS), mainly includes: composite modelling method and interaction mechanism, system planning method, security defence technology, distributed control theory. Then on this basis, the corresponding technical route is proposed, and a more detailed research framework along with main schemes to be adopted is also presented.

1. Introduction
Cyber physical system integrates computation, networks and physical environments into a multi-dimensional complex system via 3C (Computation, Communication, Control) technology, and enables large-scale project to realize real-time perceiving, dynamic controlling and information services [1]-[2]. As the level of the utilization of information enhances unceasingly, Modern distribution system has developed to be a multi-dimensional cyber physical distribution system (CPDS) which highly relies on information, and follows traditional power system process [1][3]-[4].

Although the traditional distribution system which has a complete "produce-supply-users" ternary coupling mode, has been mature enough, with varies of distributional sources, storages, protections and sensors plugging in, acute problems such as distribution system topology, information interaction, cyber-physical integration come into sight.

Compared to traditional distribution system, CPDS has distinguishing features:

(1) Multi-source isomerism, complex network.

The distributional loads and the intermittent renewable sources changed the feature that power flows in single direction, and the exponential growth in the sensors along with communication devices, also causedthe physical topology to grow enormous and complex, and its dynamic behavior becomes non-linear, uncertain, time-varying, and isomeric.

(2) Frequent interaction, andhence vulnerable

Since smart meters, distributional storages and sources plug in, to realize massive information collecting, transferring, controlling and demand side response, power flow and information flow interacts frequently, thus information security becomes a serious problem. While the communication protocol, which is open and standardized, also bring new hidden dangers to traditional power system field, such as dispatching automation, relay protection and safety equipment automation, load control, user information collection and dispatching automation, relay protection and safety equipment automation, load control, and user information acquisition.

(3) Hierarchical scheduling, difficult to defend

By applying single strategy like centralized control or decentralized control cannot synthetically take investment, control validity, and communication reliability into account. While distributed control can satisfy such system demand. However, power system adopts hierarchical scheduling, hence the system relies on the communication network heavily, and increases the difficulty to control and to defend.

CPDS is a relatively new concept, and research achievements about electric information intelligent synchronization and acquisition, intelligent and digital monitoring equipment have been made. However, how to understand and analyze the inter-action mechanism between power system and cyber system, how to build a mathematical model considering energy flow, information flow and service flow, how to utilize information technology to enhance system security, these questions have not yet been solved.

Therefore, this paper illustrates key issues on CPDS integrated modelling, system organization, security defense and distributed control, then provides research directions and preliminary solutions for other researchers to reference and feasible approaches to consider.

2. Introduction of key issues in CPDS

2.1. Integrated Modelling Method and Interaction Effect Mechanism

CPDS integrated model is the theoretical basis of control strategy, network security and organization. As for the traditional power system analysis and control, the influence made by cyber system is neglected. Generally the assumption that the system information acquired is timely, accurate and reliable. Moreover, power system use dedicated network that has less interfaces to external network, thus the reliability and security can be assured [2][5]-[6]. However, the applying of massive PMU and smart meters, distributed energy devices, and high sample rate sensors, means there will be amazing amount of data existing in the CPDS communication network. Power system's dependency on cyber system has to go deeper. Therefore, when considering built a CPDS model, the assumption above is no longer reasonable. Network security problems such delay, packet loss, and malicious attack will bring influence on power system. Hence in the integrated model, cyber system shall deserve equal attention as power system. But how to quantize and evaluate information, how to organize operation, there is a lack of mature theatrical framework.

To study the interaction between power system and cyber system, firstly the unified modelling method, namely the cyber-physical modeling, shall be determined [3]. Physical system is a continuous system, while cyber system is a discrete system. They have different math basis and simulation method. The primary question to develop a complete modeling theory, is how to eliminate the barrier between power system and cyber system. Up till now, researches on modelling have not yet provide effective modeling theory that can both fully and accurately describe the interaction between cyber system and power system, and the traditional power system modeling architecture has not been broken, thus the influence made by cyber system cannot be accurately depicted.

This paper presents research contents on CPDS integrated modeling and interaction effect mechanism: analyze the sources, load, topology in traditional distribution system and measurement, communication, computation in cyber system, then find out the unified model which can represent the dynamic features of power system and execution order of cyber system: clarify the influence, which is caused by complex distributed cyber system, on power system from both steady and dynamic field, and study how it will affect the stability and reliability of distributed system.

2.2. Planning method

As coverage area, power capacity and the type of equipment grows larger and larger, nowadays distributed system has come close to its limit of operation. So it is vulnerable to varies interference, and then becomes a threat to safe and stable operation for whole system. Besides, with the increase of cross-regional large-scale interconnected power grid, once the effect brought by power failure would be limited in local area, but now the effect will spread out to adjacent power areas, thus expanding the scale of the accident. Therefore security factors shall be fully considered at the stage of power system planning, and narrow the scale of accident, reduce the loss of social economy to the greatest extent.

To those traditional power system which only contains synchronous generator and loads, the main theory to determine power system subareas are: graph theory [7], slow coherency [8][9] and intelligent optimization [10][11]. However, in traditional power system there are few distribution generators, storages and other new power devices such as smart meter, PMU, those methods for subareas above cannot apply to complex power system to answer for cyber security.

This paper presents research contents on CPDS planning method: based on the interaction mechanism among distributed system, communication networkand cyber system, find the composite restrain for both energy transfer property and information demand in CPDS planning, then build a composite model for multi-object CPDS planning; and this paper studies the coordinated planning based on CPDS technology for distributed system, communication network and cyber system; this paper also presents an on-line planning method based on isolated-island, which is used under conditions like serious device failure or malicious communication attack. With the proposal of planning method for CPDS, the corresponding evaluation mechanism will be updated.

2.3. Cyber Security
CPS is a dual heterogeneous composite network, its security problems includes virtual network security for cyber space and real network security for physical space, and along with coupling risk caused by the interaction between two spaces [4][12]-[13]. Since public network and open protocol have been applied, cyberspace security risk degree increased.

At present there are relatively few researches carried on CPDS cyber security, and most researches only focus on process control system, smart meters, communication protocols, wireless communications networks, and other independent areas, lacking complete system. Therefore, to build a complete framework, firstly CPDS property should be clarified, then combined with computer network theory, high risks that threatens power system operation steadily should be analyzed.

This paper presents research contents on CPDS cyber security: the research on cyber space and power space and their risk mapping model, coupling and propagation mechanism. Theillustration on vertical transmission way, namelyhow risks in cyber space spread across space edge and finally cause power system failures, and the back propagation way, namely how power system failures react on cyber system and finally cause cyber security threat. Based on the interaction mechanism between power system and cyber system, propose the analysis method for CPDS cyber security fragility. The research on security flaw in cyber space, and on this basis, carry out research on active defense strategies based on outlier detection and fault diagnosis technology.

2.4. Controlling Strategy
Power balance and voltage/frequency stability is the premise for normal operation of power system. However, with high density, distributed, intermittent, heterogeneity features, renewable energy power generation system and storage system cause traditional primary control and secondary control method cannot satisfy this new situation. Furthermore, increased complexity of communication network and other power devices tightly coupled, this also increased the demand for robustness of system control and interference resistance. How to realize coordinated control under the circumstances of CPDS's time-vary, complex network environment, and realize power balance and voltage/frequency stability control, are important problems needed to resolve.

Traditional control methods for distributed source/storage include centralized control and decentralized control.

Centralized control is one of the most common, and also the simplest kind of control methods. In centralized control mode, every power device communicates with control center (SCADA, EMS) through network lines. The server proceed optimal management, executing information transferring and processing, coordinates the level control system, and give each controlled unit in the system a set operating point. The main corresponding control method for centralized control is Optimal Power Flow [5]-[13]. However, when there are more and more controlled units, it is difficult for centralized control mode to satisfy instantaneityand flexibility. And when some certain communication line or device encounter failure because of attack behaviors (physical failure, malicious data), centralized control mode will face the danger of breakdown, and the corresponding control strategy don't possess the ability of security and stability control.

In contrast to centralized control, decentralized control is a control method that suitable for sources and storages which are decentralized geographically, and have varies features, and this control method enable loads to operate effectively and steadily, meanwhile it can increase system safety margin. The main corresponding control methods for de-centralized control are: droop method [14][15], anti-droop method [16], and maximum power point tracking methods [17][18]. In decentralized control mode, every device connect directly to its monitoring system, and realize the stable control of itself by local control. Hence it possess high flexibility, and strong ability to against network security threats. Decentralized control mode is especially suitable for the remote areas that is difficult for power grid to effectively cover. The effective control among source, storage and load can be achieved simply by local controller, and hidden risks of chain accidents can be also averted. However, because in this control mode, every power device act dedicatedly, they cannot act like "Virtual Power Plant" to support the whole power grid, such as voltage support and frequency stability. Moreover, decentralized control is not suitable for relatively concentrated power area.

With the continuous development of distributed computation, multi-agent and network control theory, distributed energy coordinated control gradually show its steady, safe and reliable control ability. In distributed control mode, every device (source, storage, load) all connect to its adjacent devices. The control method is determined according to the interaction in neighborhood, and in this way the control can control active power and reactive power coordinately, and can also avoid influences brought by cyber network failures, such as communication delay and interruption. Therefore distributed control possess high flexibility, perturbation resistance and the ability to face cyber space risks. Taking the distribution and complexion of CPDS physical environment into account, many researches aimed at CPDS distributed control method have been carried out. The coordinated control strategy using distributed power sources and reactive power compensation equipment is proposed [19]. And there is another coordinated control strategy aiming at solving massive distributed storage devices [20]. However, because of the variety of scale, structure, and control units in distributed system, and different kinds of computation restraints, CPDS distributed control is a quite challenging issue. Among these researches, the main consideration is the effect that communication part brings to control part, and control object is fair simple, moreover, these researches only discuss local optimization control, and there is a lack of the factors about communication and computation.

This paper presents research contents on CPDS control strategy: considering CPDS is a multi-level and distributed coupling system, carry out research on control strategy which contains computation, communication and physical objects; discuss proper distributed source output distribution strategy, and for the micro power grid, to find out the autonomous operation control strategy; aiming at the topology structure already exists, to propose control strategy from aspects such as convergence, robustness and flexibility.

3. Integrated Modeling Method

Further study on CPS such as control algorithm and system evaluation must have an appropriate system model to support. The problem is that cyber system and power system are based on different theoretical basis. More specifically, cyber system is generally driven by information/events, and its theoretical basis is discrete mathematics, its modeling tools are discrete mathematics tools, such as the finite automaton. On the contrary, power system is based on continuous mathematics, and its modeling tools generally are algebraic equations and differential equations. The difference on theoretical basis and modeling tools are the basic reason for the split between power system and cyber system [1]. And up till now, no integrated model effective and accurate enough has been proposed.

Some researchers have obtained achievements on modeling theory. In [21], CPS sub-modules are divided according to its function, after comparing their features in both physical and cyber space, these sub-modules are integrated into a whole CPS model. On this base, in [16], by trial and error process, the C model and P model come to one correspondence, and according to such single cyber-physical model, all C models are used as logical node in connection, so a "C-integration" is obtained. And P model is treated as a function module, and are connected to get a "P-Integration". Finally the two module are integrated into a whole CP model. And by this way, the CP models for photovoltaic, energy storage are deduced, and then proposed a method to simulate on the CP model by dynamic link library. In [17], a CPS computation platform based on large scale distributed computing technology was suggested to realize CPDS controlling. Besides, in [18], the conception of CPS was introduced into relaying of smart distributed system, and some detailed components of this system and their functions, structures were also presented. The paper then compared this system with traditional system, and concluded that the new system could deal with complex algorithm because of its distributed computation ability.

However, the coupling relationship among power system, communication system and cyber system, hasn't been reflected in these modelling methods; for the dynamic property like packet loss and delay, there are not relevant researches, either. More specifically, all the models above trend to the benefits that will come from cyber system, while there is a lack of consideration about the negative effects that cyber system will bring to power system. For instance, when massive monitor data collected suffers cyber space delay, the control center of power system would not react in time, and cyber system would be a burden to power system.

There are two available ways for the co-simulation of CPDS model: simulation with power system expanded, simulation with cyber system expanded, and simulation combined physical system and cyber system.

Simulation with power system expanded means: based on the simulation platform of power system that already exist, integrate simulation model and method of discrete event system. For discrete event system description, a practical way is add continuous system simulation language as packet form into simulation platform, such like Nutaro J et al. [19]. However, because of the lack of necessary elements of discrete event system, like dynamic memory management and dynamic linking, there is difficulty to simulate the dynamic processes in cyber space accurately [20].

Simulation with cyber system expanded means: based on the simulation platform of cyber system that already exists, integrate simulation model and method of continuous system. There are some examples utilized this thought. Such as Agent/Plant and generalized discrete time system specification [21]. The advantage of this thought is high flexibility, convenience for information interaction, and it enables independent subsystem simulation. But since every single discrete event needs computation of all the network, when the continuous system is relatively large scale, the computational work would be quite heavy.

Simulation combined physical system and cyber system means: based on the simulation platforms of physical system and cyber system and the cooperation of both, realize the co-simulation of continuous system and discrete system. For instance, combine physical system platform (Modelica and VTB) with cyber system platform (NS2 and OPNET) to form a new co-simulation platform [22].

These methods described above adopted mature simulation platforms, so the accuracy and reliability of simulation results can be guaranteed; however, the order that physical system follows is time-order, while event-order for cyber system. So the problem about timing synchronization needs further study.

This paper presents a general outline for CPDS integrated modelling: first of all, build appropriate models for computation, communication and sensor devices using theory such as queuing theory, Markov chains, stochastic process, and stochastic differential equations; then, analyze the interface variances between physical and cyber space, such as network delay, message error, to clarify the relation between the two spaces; last, fuse two models by integrate mathematical models, and obtain a complete CPDS model.

As for computation unit (usually made up of processor and cache), queuing theory and stochastic process can be chosen to build models. When a computation task reaches the computation unit, it will queue in data cache. The tasks first come will be first served, so in queue theory, D/G/c/∞ or M/G/c/∞ can be chosen as the model for computationunit. In distributed system, computation tasks include kinds of power system analysis (power flow calculation, on-line security analyze) and control strategy decision (such as calculate the optimal control strategy for distributed sources).

The main concerns of cyber modeling are the properties of data link layer, network layer and transport layer. To simulate network congestion agreement, differential algebraic equations driven by finite automaton can be used to describe three different discrete states (empty queue, full queue and common queue), and their properties will be described by differential algebraic equation, and three states will have interconversion.

Modeling for sensor units needs to know the stochastic process that the data rate of sensor units obeys. Adoptable models include Markov process like Poisson process, and Heavy-tailed process like Pareto process. Considering that sensors may have multiple discrete work states, the model for sensors could be stochastic process driven by Markov.

In conclusion, the continuous changes of state variances (e.g. size of queue, size of congestion window) in CPS can be described by differential algebraic equations; while the discrete state variances can be described by finite automation or Markov Chain. As for uncertainty of system (e.g. data rate of sensors, processor speed), several appropriate stochastic processes can be used to depict. All the models combined with traditional power system model will come to a complete CPDS model. And the time scale conflict, which is caused by continuous time and discrete time, can be solved using ultra-dense time model, which inserts discrete time into continuous time axis, so continuous time and discrete time can be combined. For continuous and discrete transition, a new clock variance should be introduced, and this clock is an infinite discrete time series, which can discretize continuous time by sampling, hence combine continuous transition with discrete transition.

4. Planning method

For modern power system that contains distributed sources, storages, controllable loads and varies communication devices, since the high digitized degree, the planning method for traditional power system cannot satisfy the demand of safe and stable operation. Therefore, scholars home and abroad proposed new planning methods for informationized modern power system.

Arefifar S A et al. [23] proposed a partition strategy for power system which can realize self-sufficiency, with mathematical models of photovoltaic, windmill, energy storage, load and reactive power source, an optimized objective function based on power loss objective functions and power unbalance functions. When the restrictions of power and voltage are satisfied, the optimal partition strategy will be determined by tabu search algorithm, and such strategy can reduce system loss and power unbalance. Atwa Y M et al. [24] proposed a planning strategy for power system by properly configuring different kinds of renewable sources to reduce system loss, and the optimal solution was obtained via a mixed integer nonlinear programming problem whose restrictions are voltage, current, capacity and power. Kleinberg M R et al. [25] proposed a partition strategy based on capacitance reactive power domain, together with a distributed control algorithm based on exhaustive search to reduce voltage fluctuation and capacitance reactive power loss. Mehrjerdi H et al. [26] proposed a scheme for power system partition and secondary voltage control by the means of graph partition and key bus choice, meanwhile after partition the reactive power in every subarea will be controlled by fuzzy controller, so the voltage in every isolated island can keep stable. Arefifar S A et al. [27] proposed a distributed system partition method referring to previous research achievements, and this method considered communication and power in the same time. The optimal partition strategy was determined using tabu search algorithm, graph theory and power flowcalculation based on Newton

Raphson, and it can satisfy communication delay and its cost, power balance, voltage, current, and frequency.

From the analysis above, a conclusion about modern power system (e.g. smart gird) can be drawn: the existing partition method only considers single aspect (e.g. reactive power balance, active power balance, power loss, communication cost), it is hard to satisfy the compositive demand from communication, control, power and security in modern power system; there is a lack of analysis for interactions between physical network and cyber network, and it is difficult to obtain the mathematical description for dynamic properties of the system; and usually the attention is paid to modern power system which contains distributed sources, while for the traditional synchronous generators, the ability to face emergencies should not be neglected.

Power CPS planning method may include three level, namely property analysis for physical device, construction for communication network, and planning design for whole system.

(1) Property analysis for physical device

Physical devices in modern power system trend to be diversified, and there are quite difference in their physical properties, work modes, and control principles, so it is not easy to analysis the whole system. By using probability statistics, power electronic technology and control theory, along with the mathematical models of these devices, the descriptions for power flow and information flow will be preliminarily established, and provide basic elements for CPS integrated model.

(2) Construction for communication network

For physical network that is constructed of complex power devices and communication devices, the heuristic learning method can be used to strengthen security of communication network. Meanwhile, a communication system based on cable communication and wireless communication can be established to increase communication reliability.

(3) Planning design for whole system

First of all, establish appropriate models for computation, communication and sensor devices using proper theory such as queuing theory, Markov chains, stochastic process, and stochastic differential equations; then, analyze the interface variances between physical and cyber space, such as network delay, message error, to clarify the relation between the two spaces and realize fusion of these two models; later, obtain the steady-state operating point of cyber system via the optimal routing problem, and analyze the dynamic property based on "N-1" principle, and then discuss the reliability and security of the system.

To avoid uncontrollable splitting, and increase the ability to face security risk, at present the most efficient method is controllable island based on system partition [28]. The fundamental conception is that, when a risk or failure is detected, the system will be divided into several controllable isolated islands (or micro girds) that are self-sufficient; when the risk or failure is eliminated, these islands can connect with each other by switches, and come back to original state.

Therefore, on the base of CPS model and the analysis of interaction scheme, the partition method that based on slow coherency theory, power domain and distributed optimization can be adopted to divide the power system into controllable isolated islands. And via objective function that reflects information, power and control, the cut set of power system can be determined; when the fault is eliminated, the interconnection switch control strategy will restore the system to original state.

5. Cyber security

Cyber system and physical system have much more interaction than before, cyber security must be reconsidered. For now, three levels of security and stability standards are established for power system. Moreover, relay protection device, stability control device, and split device are applied to ensure safe and stable operation. Since these devices above need to cooperate with computer networks and communication system, hidden risks exist. For instance, the attackers could fake or tamper the message collected, thus cause mistakes in state estimation results, then incorrect operation appears; the attackers could acquire the authority of administrator by sniffing or password cracking, then tamper

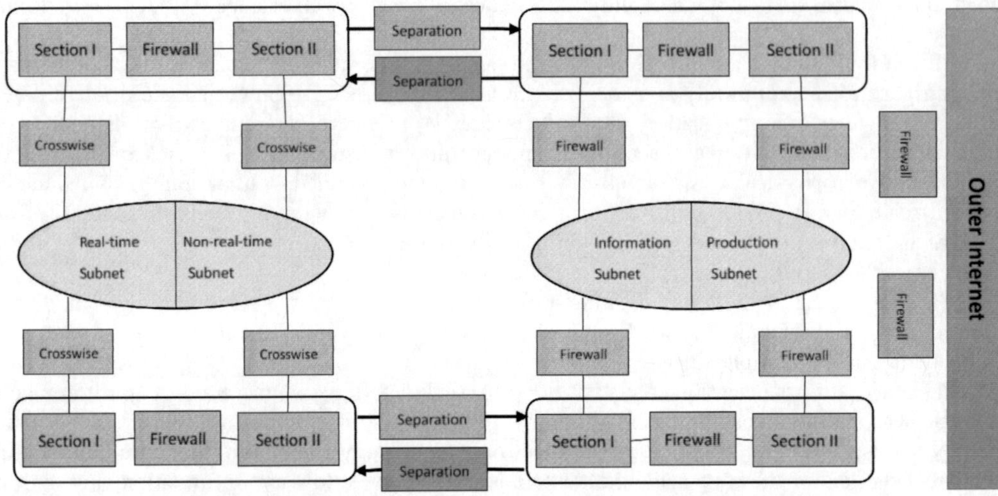

Figure1. Cyber system security framework of power system

the parameters of operation; also, via DoS attack, the cyber system could lose the control of physical system, then physical fault cannot be eliminated and go to further expansion.

The security protectionprinciple of power CPS in our country should basically follow "safe partition, private network, crosswise separation, lengthways identification". To be more specifically, the power enterprises should divide its server system into different safe regions using computers and networks, so the emphasis of system protection can be clear.

As Figure 1 shown, production control region can be divided into control region (safe region I, the core of security protection) and non-control region (safe region II); information management region can be flexibly divided due to different demand from power enterprises, such like production management region (safe region III) and message management (safe region IV). Besides, there are forward and backward physical separation existing between region I, II and region III, IV, added with firewall and encryption certification, the range that cyber risks could influence can be effectively limited, thus the safety of core region can be ensured.

However, the communication network presents the characteristics as "strong high-voltage side, weak low-voltage side", and researches on cyber security mainly concern on generation, transmission and converter processes. Besides, still there are mistakes in daily manual operations: although the encryption scheme is mature, the hidden risks lie in producers and distributors of these encryption equipment cannot be fully eliminated. The same problem troubles severs in region I; the attack incidents against power system that are already known show that, even inside the private network that is separated from the other network, there is still chance for the cyber system to catch virus. Therefore, the traditional safety protection for power system, which mainly focus on physical separation and safe regions, cannot satisfy the demand from CPDS safety protection.

The main attacks against the network of power system are, DoS attack (e.g. SYN flood, smurf, and DDoS), virus (in 2010 nearly 1/5 nuclear power plants in Iran were affected by Stuxnet which entered from USB interface), information collection (e.g. sniffing, eavesdropping), attack against network (e.g. spoofing, replay and man-in-the-middle) and also false data injection, etc. Among them, false data injection attracts more attention, because it is hard to permeate into the rigor power dispatch center, while quite simple if the target is sensors. In [29], attackers could bypass the bad data detection and inject false data into the results of power system state estimation, and bring in errors at will. Kosut O et al. proposed a strategy against malicious data attack, and they utilized graph theory and polynomial time algorithm, deducted the minimum number of sensors that need to be attacked to make the system unobservable [30][31]. There are other attack forms like load re-allocation, and this would change the cost of power system [32]. In the environment of power market, attacker can get economic benefits by injecting false data. The traditional method against false data injection in wireless network includes filteringin the midway and tracking back to source. Filtering in the midway filters false date passively, so the attacking nodes could launch sustained attack and avoid being discovered and separated. While tracking back to source needs plenty of packets from convergent nodes to take effect. Based on these, there are many study on the combination of two methods above to generate a new strategy against false data injection. Despite these ways discussed above, there are other methods, can be considered such as encryption and public opinion report.

The researches on cyber security is quite mature, and specific safety precautions can be applied into CPDS. However, these precautions cannot cover the physical aspect very well. Therefore, this paper presents research contents on cyber security of CPDS: first, classify hidden physical attack and its source, then, develop evaluation scheme for CPDS, and at last propose specific strategies for CPDS security defence. The technical route is shown in Figure 2.

(1) Effect of network attack

Analyse those information nodes and power nodes to find out weakness of them; the emphasis of this work is on the process where the data of power system is collected and transferred. Then find out all possibilities of attack that could be launched to break the weak nodes in CPDS. Based on power system analysis (transient analysis and steady-state analysis) and Bayesian network (conditional probability), the scheme of cross-space failure expansion can be established.

(2) Evaluation for cyber security

Evaluate the influence that communication system may bring on CPDS, and develop a scheme to evaluate weakness in both physical and cyber system: 1) Regard the random attacks as the input of system evaluation, and utilize Fuzzing test to evaluate possible risks according to cross-space failure expansion model of CPDS. 2) Build an attack-and-defence game model, and obtain the next node that the attacker would choose. Then the result of the game will be used as fundamental parameters of the cross-space failure expansion model, thus the evaluation of weakness in CPDS can be carried on.

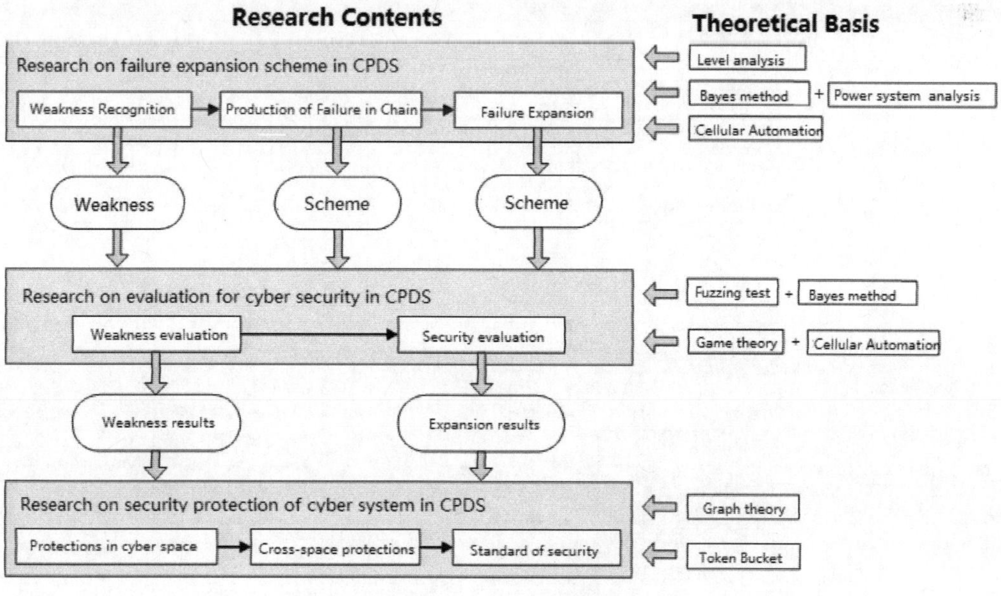

Figure 2. Security framework for CPDS

(3) Protections for cyber security

Since the weak nodes have been found, aiming at false data injection and other attack against network channel, and armed with modern cyber security technology (including encryption, certification, integrity test, routing protocol), now the possibility to be attacked and the possibility the attack succeed can be effectively determined. And the system contains complete message or incomplete message need to be discussed by the use of game theory, and then formulate defensive strategies for CPDS. Or, use graph theory to describe all the proper message flow and control flow: view the whole message set or the operation instruction set as a graph (universal set), next take every piece of message or instruction as a directed edge, then a legal message flow or operation flow can be viewed as a directed route constructed of several directed edges (and this route may be a circle). Then, all the legal message flow and control flow are defined as known route in the graph; shall there appear any unknown route, it must mean illegal message or instruction.

6. Control strategy

A scientific and reasonable system, not only can ensure stable operation in normal situations, but also can strengthen the ability to handle bad situations like communication interference, channel failure, malicious attack and device break down. The stable operation relies on the instructions transferred among dispatching center, distributed sources and user loads. Therefore, CPDS will be a typical networked control system (NCS). NCS means a control system whose sensors, controllers and executionsare located on different nodes, and they must communicate with each other via cyber network [33]. When the number of nodes and the sample rate increased, and there is communication congestion in the system, trans-mission delay and data loss will occur, and even cause instability of the system which may lead to break down. To solve this, here are some available ways: 1) reflect the effects of delay and data loss in the controlling model; 2) compensate for delay and loss, which means at any node in time scale, the control system should work out the present control instructions, and also several instructions for nodes in the future, and send them to execution at the same time.

At present, there are some problems about distributedcontrol for CPDS: 1) the researches now only considered the distribution and complexion of the physical devices in distributed system, and mainly

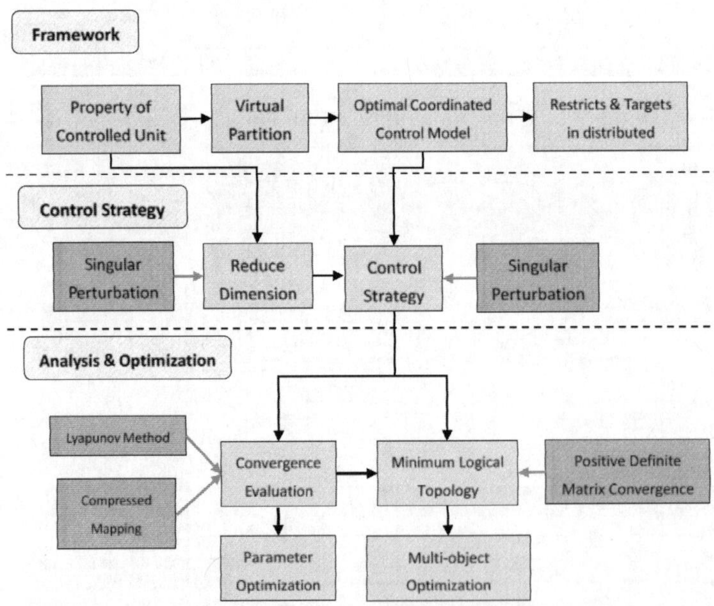

Figure 3. Self-organizing control route for CPDS generation

focus on the controlling problem, but there are not enough attention paid to communication, computation, and the complex coupling of different levels. Thus, those control frameworks proposed cannot fully satisfy the demand of distribution controlling. 2) Although there are researches on coordinated control for distributed system,systemic solutions are still haven't been proposed. For instance, when dealing with feeder current controlling problem, the coupling relationship between physical system and cyber system should be considered more carefully, and a mature algorithm for CPDS coordinated control should be established.

To study on coordinated control strategy for CPDS system based on open communication system, there are three steps, as shown in Figure 3:

The detailed introduction for each step is as follows:

(1) Framework

For the framework of CPDS control strategy, this paper proposed a solution: based on distributed control theory, graph theory and the property of the dynamic mathematic model in power system, build multi-time scale, local information shared distributed control model for different controlled subarea in the distributed system. Taken restraints in power system (e.g. power flow) into consideration, combined with other needs such as frequency, voltage and even the demand of game, the complete optimal coordinated mathematical model can be proposed.

(2) Coordinated distributed control

The core idea of coordinated distributed control is utilizing iteration algorithm while considering the interaction among units in CPDS at the same time.

Aiming at the coordinated distributed control, this paper presents some solutions as below: use singular perturbation and general passivation theory to reduce dimensions, thus the basic needs of coordinated control can be met; and the distributed average filter algorithm can be used to estimate all the real-time demand of the loads in CPDS.

(3) Dynamic property analysis

There are two solutions for dynamic property analysis of control strategy in CPDS. When the Lyapunov function is determined, and the convergence of the strategy is verified, then the corresponding parameter optimization method can be proposed; another solution is that, considering the incidents that could happen in actual operation such as delay and interruption, analyse the convergence and robustness of the algorithm.

(4) Topology integrity and its optimization

Several solutions are presents as below: use the convergence equivalent conditions of matrix sequence to find out the equivalent conditions of minimum logic communication topology; consider the demand of robustness of power system synthetically (e.g. the "N-1" rule) and formulate the optimized model for logical communication network; by the means of optimization theory and its computational methods, find out the optimized communication structure.

7. Conclusion

As the form of energy and electricity become diversified, the distributed system slowly turns into a cyber-physical system, which contains traditional power system, distributed renewable sources, communication devices and varies of loads, and with the cyber system and the physical system being deeply coupled. Based on a "863" project, this project introduced the key issues in developing CPDS, including integrated modelling method, planning method, cyber security and control strategy. Further, the research direction and recent development from home and abroad are introduced, and the existing problems are also pointed out. Last, specific research direction are presented in this paper.

The researches on the key issues of CPDS can promote the integration of the frontier information technology and the physical system, and this also contributes to allow the full potential of distributed devices, increase the absorption of distributed sources, strengthen the cyber security, and hence promote the level of comprehensive utilization, and provide important reference for the development of CPS theory and technology.

References

[1] Zhao J H, Wen F S, Xue Y S, et al. 2010 Cyber physical power systems: architecture, implementation techniques and challenges *Automation of Electric Power Syst* **34** 1-7

[2] Zhao J H, Wen F S, Xue Y S, et al. 2011 Modeling analysis and control research framework of cyber physical power systems *Automation of Electric Power Syst* **35** 1-8

[3] Dong Z Y, Zhao J H, Wen F S, et al. 2014 From smart grid to energy internet: basic concept and research framework *Automation of Electric Power Syst* **38** 1-11

[4] Gao K L 2013 Research on security of cyber physical system *China Information Security* **2** 86-87

[5] Tsikalakis A and Hatziargyriou N 2008 Centralized control for optimizing microgrids operation. *IEEE Trans. Energy Conver* **23** 241-248

[6] Abdelaziz M A, Shaaban M F, Farag H E, et al. 2014 A multistage centralized control scheme for islanded microgrids with PEVs *IEEE Trans. Sustainable Energy* **5** 927-937

[7] Dent C, Ochoa L and Harrison G 2010 Network distributed generation capacity analysis using OPF with voltage step constraints *IEEE Trans. Power Syst.* **25** 296-304

[8] Divshali P H, Alimardani A, Hosseinian S H, et al. 2012 Decentralized cooperative control strategy of microsources for stabilizing autonomous VSC-Based microgrids *IEEE Trans. Power Syst.* **27** 1949-59

[9] Loh P C, Li D, Chai Y K, et al. 2012 Autonomous operation of hybrid microgrid with AC and DC subgrids *IEEE Trans. Power Electro.* **28** 2214-23

[10] Guzman C, Cardenas A and Agbossou K 2014 Load sharing strategy for autonomous AC microgrids based on FPGA implementation of ADALINE&FLL *IEEE Trans. Energy Conver.* **29** 663-672

[11] He L J, Li Y D and Harley R G 2013 Adaptive multi-mode power control of a direct-drive PM wind generation system in a microgrid *IEEE J. Emerging & Selected Topics in Power Electro* **1** 217-225

[12] Abdelsalam A K, Massoud A M, Ahmed S, et al. 2011 High-performance adaptive perturb and observe MPPT technique for photovoltaic-based microgrids *IEEE Trans. Power Electro.* **26** 1010-21

[13] Xin H H, Qu Z H, Seuss J, et al. 2011 A self-organizing strategy for power flow control of photovoltaic generators in a distribution network *IEEE Trans. Power Syst.* **26** 1462-73

[14] Xin H H, Zhang M D, Seuss J, et al. 2013 A real-time power allocation algorithm and its communication optimization for geographically dispersed energy storage systems *IEEE Trans. Power Syst.* **28** 4732-41

[15] DIlic M, Xie L, AKhan U, et al. 2010 Modeling of future cyber-physical energy systems for distributed sensing and control *IEEE Trans. Systems, Man, and Cybernetics-Part A: Systems and Humans* **40** 825-838

[16] Zeng Z Y and Liu D 2013 Study on cyber-physical system modeling on coordinated control of photovoltaic generation and battery energy storage system *Power Syst Tech* **37** 1506-13

[17] Zhao J H, Wen F S, Xue Y S, et al. 2010 Cloud computing: implementing an essential computting platform for future power systems *Automation of Electric Power Syst* **34** 1-8

[18] Zhuang W and Mu D H 2012 Study on cyber-physical relaying protection system of intelligent distribution network *Power Syst Protection & Control* **40** 113-118

[19] Nutaro J, Kuruganti P T, Protopopescu V, et al. 2011 The split system approach to managing time in simulations of hybrid systems having continuous and discrete event components *Simulation* **87** 1-18

[20] Hua L, Sambamoorthy S, Shukla S, et al. 2011 Power system and communicationn network co-simulation for smart grid applications *IEEE PES Conference on Innovative Smart Grid Technologies* 1-6

[21] Nutaro J 2011 Designing power system simulators for the smart grid: combining controls, communications, and electro-mechanical dynamics *Proceedings of IEEE Power and Energy*

Society General Meeting **5** 1-5

[22] Godfrey T, Mullen S, Dugan R C, et al. 2010 Modeling smart grid applications with co-simulation *IEEE International Conference on Smart Grid Communications* 291-296

[23] Arefifar S A, Mohamed Y A and El-Fouly T 2012 Supply-adequacy-based optimal construction of microgridsin smart distribution systems *IEEE Trans. Smart Grid* **3** 1491-1502

[24] Atwa Y M, El-Saadany E F, Salama M and Seethapathy R 2010 Optimal renewable resources mix for distribution system energy loss minimization *IEEE Trans. Power System* **25** 360-370

[25] Kleinberg M R, Miu K, Segal N, Lehmann H and Figura T R 2014 A partitioning method for distributed capacitor control of electric power distribution systems *IEEE Trans. Power System* **29** 637-644

[26] Mehrjerdi H, Lefebvre S, Saad M and Asber D 2013 A decentralized control of partitioned power networks for voltage regulation and prevention against disturbance propagation *IEEE Trans. Power System* **28** 1461-69

[27] Arefifar S A, Mohamed Y and El-Fouly T 2015 Optimized multiple microgrid-based clustering of active distribution systems considering communication and control requirements *IEEE Trans. Industrial Electro* **62** 711-722

[28] Anderson P and LeReverend B K 1996 Industry experience with special protection schemes *IEEE Trans. Power System* **11** 1166-79

[29] Liu Y, Ning P, Reiter M K 2009 False data injection attacks against state estimation in electric power grids *Proceedings of the 16th ACM conference on Computer and communications security* 21-32

[30] Kosut O, Jia L, Thomas R J, et al. 2010 Malicious data attacks on smart grid state estimation: attack strategies and countermeasures *IEEE International Conference on Smart Grid Communications* 220-225

[31] Kosut O, Jia L, Thomas R J, et al. 2011 Malicious data attacks on the smart grid *IEEE Trans. Smart Grid* **2** 645-658

[32] Yuan Y, Li Z and Ren K 2011 Modeling load redistribution attacks in power systems *IEEE Trans. Smart Grid* **2** 382-390

[33] Xue B, Li N, Li S, et al. 2010 Robust model predictive control for networked control systems with quantization *IET Control Theory and Applications* **4** 2896-2906

IOP Conf. Series: Earth and Environmental Science 52 (2017) 012036 doi:10.1088/1755-1315/52/1/012036

A PSFI-based analysis on the energy efficiency potential of China's domestic passenger vehicles

Chuan Chen[1], **Huanhuan Ren and Dongchang Zhao**

China Automotive Technology & Research Center, Tianjin, China

E-mail: chenchuan@catarc.ac.cn

Abstract. In this article, China's domestic passenger vehicles (excluding new energy vehicles) are categorized into two groups: local brand vehicles and vehicles manufactured by joint ventures. Performance-Size-Fuel economy Index (PSFI) will be applied to analyse the speed of technical progress and the future trends of these vehicles. In addition, a forecast on energy efficiency potential of domestic passenger vehicles from 2016 to 2020 will be made based on different Emphasis on Reducing Fuel Consumption (ERFC) scenarios. According to the study, if the process of technical progress continues at its current speed, domestic ICE passenger vehicles will hardly meet Phase IV requirements by 2020 even though companies contribute as much technical progress to fuel consumption reduction as possible.

1. Introduction

On January 1 2013, Fuel Consumption Limits for Passenger Cars (GB19578-2014) and Fuel Consumption Evaluation Methods and Targets for Passenger Cars (GB27999-2014) were officially released, enabling China to enter Phase IV of fuel consumption administration. It is required that the average fuel consumption of passenger vehicles should be reduced to around 5L/100km by 2020 [1][2]. Although new energy vehicles (NEV) enjoy favorable policies on fuel consumption calculation, internal combustion engine (ICE) vehicles will dominate the market in the short term. Therefore, the energy efficiency potential of ICE vehicles is crucial for the target achievement. In this article, a forecast on the fuel consumption reduction potential of domestic passenger vehicles (excluding NEV) will be made based on PSFI analysis and ERFC index scenarios to evaluate the difficulty faced by the domestic companies in achieving the energy efficiency target.

2. PSFI and ERFC index

The progress made in automotive technologies has led to changes in vehicle properties. The improved engine and powertrain efficiency positively impact fuel consumption reduction, but increased curb weight and body size exert negative impacts [3]. Thus, how to evaluate fuel economy accurately has become an industrial focus. In the annual reports on Fuel Economy Trends released by EPA, Ton Mpg (miles per gallon) is adopted as an indicator to describe automotive efficiency [4]. However, the Ton Mpg only indicates vehicle conveying efficiency in regards to weight, not the compound effect of power matching and space change on fuel consumption. Hence it is necessary to introduce a complex indicator with comprehensive consideration of power, economy and space to evaluate the true development level of automotive energy efficiency technology.

Feng A and John D put forward the concept of PSFI for the first time [5], to evaluate U.S. passenger vehicle changes in properties including power, economy and space, and prove the corresponding relationship between PSFI and technical progress in automotive industry. Cheah L W et al. then propose the concept of ERFC index [6], to indicate the degree of fuel economy improvement contributed by technical progress, and forecast fuel economy trends of U.S. passenger vehicles by adopting this index.

The parameters for evaluating passenger vehicle energy efficiency vary between automotive industries in U.S. and in China. China usually use fuel consumption (L/100km) as the unit while U.S. adopt fuel economy unit (MPG). On this account, PSFI and ERFC index proposed by foreign researchers will be restructured in this article to make them applicable to the energy efficiency trend study of Chinese passenger vehicles. See the following formula:

$$PSFI = \frac{P \cdot S}{F} \tag{1}$$

P is engine power rating/curb weight, kW/kg;
S is vehicle length * width * height, m^3;
F is NEDC combined fuel consumption, L/100km.

By calculating average PSFI, the progress speed and development level of vehicles, regarding power, space and fuel consumption for a period can be analyzed. An evaluation of the fuel consumption reduction potential can be gained by a further analysis on PSFI. The fuel consumption reduction potential refers to the resulting fuel consumption after all technical progress is contributed to fuel consumption reduction, assuming the power and space remains the same as the base year. See the formula:

$$FC_{potential} = \frac{(P \cdot S)_{base}}{PSFI_{cal}} \tag{2}$$

$FC_{potential}$ is fuel consumption reduction potential of the applicable year, L/100km;
$(P \cdot S)_{base}$ is the power and space of the base year;
$PFSI_{cal}$ is the PSFI value of the applicable year.

Based on the fuel consumption reduction potential, we can further introduce ERFC index to indicate the degree of consumption made by technical progress. See the following formula [7]:

$$ERFC = \frac{FC_{base} - FC_{cal}}{FC_{base} - FC_{potential}} \tag{3}$$

FC_{base} is the actual fuel consumption of the base year, L/100km;
FC_{cal} is the actual fuel consumption of the applicable year, L/100km;
$FC_{potential}$ is the fuel consumption reduction potential of the applicable year, L/100km.

Based on ERFC index, the degree of technical progress contributed to the fuel consumption reduction can be observed, and the fuel consumption reduction potential can be gained by evaluating future ERFC index.

3. PSFI and ERFC index analysis on domestic passenger vehicles

PSFI trends of domestic passenger vehicles (hereinafter, excluding NEV) can be calculated by the domestic passenger vehicle data including fuel consumption, curb weight, engine power rating and space from 2009 to 2015 (Data Source：MIIT fuel consumption data management system). In figure 1, PSFI uptrends of vehicles manufactured by both joint ventures and local companies could be observed. A good linear relationship could also be found, indicating an upward trend of overall technical level of domestic passenger vehicles. During the period between 2009 and 2015, PSFI value of local passenger vehicles were 19.8% lower than those of vehicles manufactured by joint ventures, but the former grew faster than the latter, indicating a narrowing gap between properties of local vehicles and of joint venture brand vehicles.

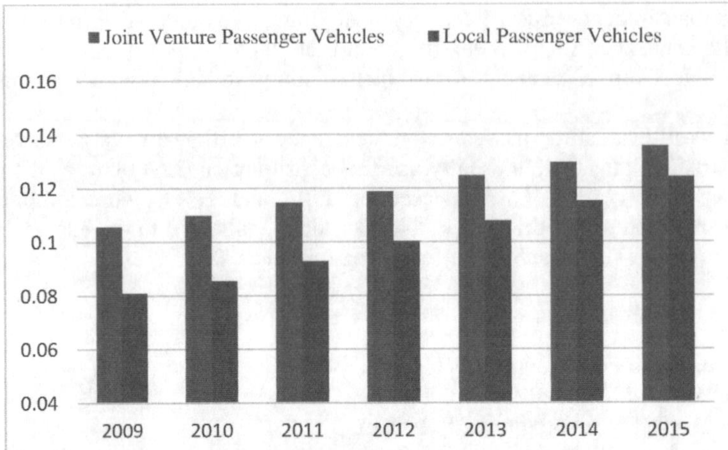

Figure 1. 2009-2015 average PSFI changes of domestic passenger vehicles

Figure 2 indicates ERFC index changes of domestic passenger vehicles. In 2010, ERFC indexes of two vehicle groups were almost the same, showing that only 23% of technical progress were contributed to fuel consumption reduction. However, joint ventures and local companies adopted different attitudes after the release of Fuel Consumption Evaluation Methods and Targets for Passenger Cars (Phase III) in 2011. Joint ventures have attached greater importance to fuel consumption. Nevertheless, local comtablepanies had a late start in the industry. Due to the relative backwardness in power and space, local companies choose to prioritize the power and space development instead of the fuel consumption reduction to ensure their market competitiveness. As a result, the average ERFC index of local passenger vehicles from 2012 to 2015 was only 6.7%, and the fuel consumption even experienced an increase between 2014 and 2015.

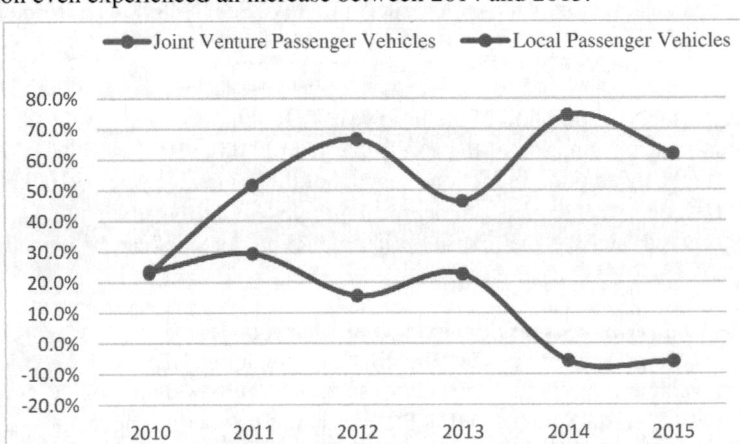

Figure 2. 2010-2015 average ERFC index changes of domestic passenger vehicles

4. An analysis on the fuel consumption reduction potential of Domestic passenger vehicles

As shown in the figure 3, the future fuel consumption reduction potential can be evaluated by observing PSFI and ERFC index development trends of domestic passenger vehicles. PSFI usually presents a good linear relationship according to the historical trends. Accordingly, if we perform linear fitting on past PSFI and extend the trend line, we can forecast PSFI changes of domestic passenger vehicles between 2016 and 2020.

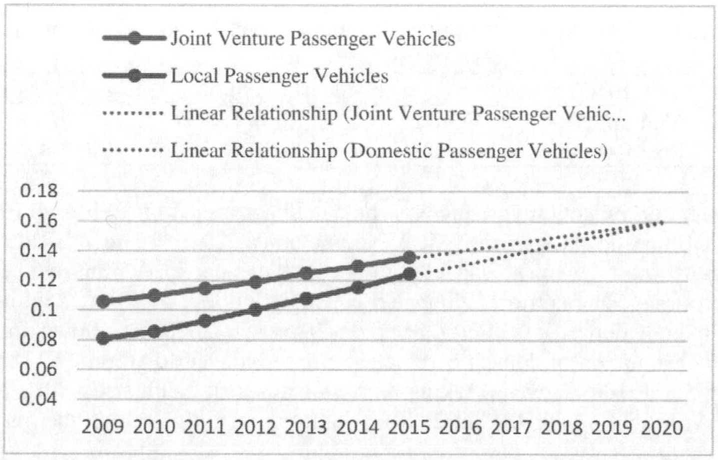

Figure 3. Forecast on average PSFI changes of China's passenger vehicles

As shown in the table 1, it is estimated that in 2020, PSFI of passenger vehicles manufactured by joint ventures will reach 0.1649, at growth rate of 21.4% compared with PSFI in 2015. PSFI of local passenger vehicles will be 0.1667, increased by 34.0% compared with the number in 2015. Therefore, it is expected that the power, economy and space of local passenger vehicles will reach the same development level as joint ventures in 2020.

Table 1. 2016-2020 Forecast on China's passenger vehicle PSFI.

PSFI Forecast	2016	2017	2018	2019	2020
Joint Venture Passenger Vehicles	0.1449	0.1499	0.1549	0.1599	0.1649
Local Passenger Vehicles	0.1375	0.1448	0.1521	0.1594	0.1667

Due to the intensive fluctuations on annual ERFC index and other influencing factors such as policies and market, it is difficult to find a fixed rule to forecast ERFC index. Thus in this article, scenario analysis will be employed by assuming ERFC index based on certain scenario setting principles. Scenario setting principles are shown as follows: 1) From 2016, China has entered the fourth phase of fuel consumption control, and the fuel consumption target is further tightened by 30% compared with phase III, which means in the period from 2016 to 2020, companies need to reach a higher ERFC index level than the average index level between 2012 and 2015. 2) According to the development trend of domestic passenger vehicle market, power and comfortability remain to be top concerns for consumers in the near future. As a result, ERFC index of all passenger vehicles must be kept below 100% to ensure market competitiveness. 3) Compared with passenger vehicles manufactured by joint ventures, local ones should pay more attention to improving the power and space. In addition, a majority of local companies rely on NEV production to meet the fuel consumption target, so ERFC index of local vehicles should be 30% lower than that of joint venture vehicles. Based on these principles, average ERFC indexes of domestic passenger vehicles between 2016 and 2020 are set under two scenarios in table 2. In the low scenario, ERFC indexes of joint venture vehicle brands and local vehicle brands are 75% and 40% respectively; in the high scenario, ERFC indexes are 90% and 60% respectively.

Table 2. 2016-2020 Average ERFC index scenario settings of China's passenger vehicles.

ERFC index	Low Scenario	High Scenario
Joint Venture Passenger Vehicles	75%	90%
Local Passenger Vehicles	40%	60%

The fuel consumption reduction potential can be calculated based on PSFI and ERFC index from 2016 to 2020 and domestic passenger vehicle parameters in 2015. In low ERFC index scenario, companies contribute most technical capacity in the power and space improvement but pay less attention to fuel economy. The corresponding fuel consumption levels of joint venture vehicles, local vehicles and the overall domestic vehicles between 2016 and 2020 are shown in table 3. In this scenario, the fuel consumption of domestic vehicles will be reduced to around 6.11L/100km in 2020, and can only reach 5.53L/100km even if taking NEV into account (calculated with a NEV yield of 1.4 million, 62% of EV and 38% of PHEV in 2020), which is much higher than the 5L/100km target in 2020.

Table 3. 2016-2020 Forecast on the fuel consumption reduction potential of domestic passenger vehicles in low ERFC index scenario (L/100km).

Year	2016	2017	2018	2019	2020
Joint Venture	6.60	6.43	6.28	6.14	6.01
Local	6.75	6.62	6.51	6.40	6.31
Domestic	6.64	6.49	6.36	6.23	6.11

In high ERFC index scenario, companies attach great importance to the fuel consumption and contribute as much technical progress to fuel consumption reduction as possible. The corresponding fuel consumption levels of joint venture vehicles, local vehicles and the overall domestic vehicles from 2016 to 2020 are shown in table 4. It can be observed that if companies make most contribution to reduce fuel consumption, the fuel consumption level of domestic vehicles can only be lowered to 5.87L/100km. Even if taking NEV into account, the number can only be reduced to 5.31L/100km (calculated with a NEV yield of 1.4 million, 62% of EV and 38% of PHEV in 2020), which still fails to reach the 5L/100km target in 2020.

Table 4. 2016-2020 Forecast on the fuel consumption reduction potential of domestic passenger vehicles in high ERFC index scenario (L/100km).

Year	2016	2017	2018	2019	2020
Joint Venture	6.53	6.34	6.15	5.98	5.82
Local	6.62	6.43	6.25	6.10	5.95
Domestic	6.56	6.37	6.19	6.02	5.87

5. Conclusion

In this article, an analysis is performed on PSFI development trends of joint venture passenger vehicles and local passenger vehicles, and a forecast of the fuel consumption reduction potential of domestic passenger vehicles is conducted based on different ERFC index scenario assumptions. The study shows that with current technological speed, even if companies contribute most technical progress to the fuel consumption reduction, the fuel consumption level of domestic passenger vehicles can only be reduced to 5.87L/100km in 2020. Even if new energy vehicles are taken into account, the level will only be lowered to 5.31L/100km, showing a certain gap from the 5L/100km target in 2020. Chinese manufacturers can only achieve the Phase IV target by approaches as follows:

- 1. Increase technology research and development investment and accelerate vehicle property development, which means a higher PSFI growth rate.
- 2. Sacrifice part of the power and comfortability to achieve further reduction of fuel consumption.
- 3. Increase new energy vehicle output to compensate negative impacts caused by the insufficient energy efficiency potential of ICE vehicles.

References

[1] Zhao F, Hao H, Liu Z 2016 Technology strategy to meet China's 5 L/100 km fuel consumption target for passenger vehicles in 2020 *Clean Techn Environ Policy* **18(1)** 7-15.
[2] Zhao W, Jin Y, Wang M, et al. 2010 New fuel consumption standards for Chinese passenger vehicles and their effects on reductions of oil use and CO2, emissions of the Chinese passenger vehicle fleet *Energy Policy* **38(9)** 5242-50.
[3] Greene, David L. 2011 Assessment of Fuel Economy Technologies for Light-Duty Vehicles *Transportation Research Record* 2058.
[4] Hula A, Bunker A, Alson J 2015 Light-Duty Automotive Technology, Carbon Dioxide Emissions, and Fuel Economy Trends: 1975 Through 2015.
[5] Feng A and John D 2007 Trends in technical efficiency trade-offs for the U.S. light vehicle fleet *SAE Technical Paper Series* 2007-01-1325 (SAE International)
[6] Cheah L W, Bandivadekar A P, Bodek K M et al 2008 The trade-off between automobile acceleration performance, weight, and fuel consumption *Sae Int. J. of Fuels & Lubricants* **1(1)** 771-7
[7] Cheah L and Heywood J 2011 Meeting U.S. passenger vehicle fuel economy standards in 2016 and beyond *Energy Policy* **39(1)** 454-66

Research on three-phase unbalanced distribution network reconfiguration strategy

Shuang Hu[1], Ke-Jun Li, Yanshun Xu, Zhijie Liu, Jing Guo, Zhuodi Wang

School of Electrical Engineering, Shandong University, Jinan, China

E-mail:15106994270@163.com

Abstract. With the development of social economy, the loads installed in the distribution network become more and more complex which may cause the three-phase unbalance problems. This paper proposes an optimal reconfiguration approach based on mixed integer quadric programming (MIQP) method to address the three-phase unbalance problem. It aims to minimize the total network losses of the system. By using several square constraints to substitute the circular constraint, the original optimization problem is linearized and converted into a mixed-integer linear programming (MILP) model. Then this MILP problem is solved in general algebraic model system (GAMS) software using CPLEX solver. The additional losses caused by three-phase unbalanced are also considered. An IEEE 34 nodes test system is used to demonstrate the feasibility and effectiveness of the proposed method. The results show that the losses and the voltage violation mitigation in the network can be reduced significantly.

1. Introduction

Distribution system reconfiguration is a classical method which is used to optimize objects such as the reliability operation as well as the power quality by changing the switch status or updating the network structure[1]. In this paper, the main objective is to minimize the losses of active power in transmission lines. With the rapid development of city constructions, the investment and update work of city distribution network is relatively backward. The unreasonable network structure and limited network capacity have difficulty to satisfy the electricity consuming demand of city residents. Reconfiguration of distribution network is one of the viable approaches to address these problems.

Various methods for distribution reconfiguration optimization has been studied, such as traditional mathematical algorithm, the optimal load flow algorithm , branch exchange method , simulated annealing algorithm , genetic algorithm, tabu algorithm , differential evolution method, immune algorithms, and family eugenics algorithm etc [2-7]. In [8], a hybrid particle swarm optimization (PSO) approach composed of the binary PSO and discrete PSO was proposed and successfully applied to distribution network reconfiguration problem. Paper [9] studies a comprehensive optimization model that combined reactive power optimization and network reconfiguration to minimize power losses and eliminate voltage violations. Paper [10] proposes a mixed-integer conic programming formulation for the minimum losses distribution network reconfiguration problem. Even if load balance is taken into consideration in the above studies, the research is usually only measured by the relationship between the apparent power of each branch and its capacity, but little attention is paid to the balance among the three phases in distribution systems.

Although the mentioned methods above have achieved success in distribution configuration, as the consuming demands in distribution network have been more and more complex problems, such as unbalance in the three-phase voltages, over-current in the neural point, the decline in the distribution transformer output and the increased active power losses become more serious than before. Balancing the three-phase loads by reconfiguration becomes more demanding. This paper utilize mixed integer quadric programming (MIQP) model to optimize the distribution network taking comprehensive consideration of three-phase unbalance problem to further raise the efficiency and reducing the computing time. Also to be more practical, the reconfiguration should not only study static optimal network structure but also take consideration of the global optimization under the load changing situation in several continuous time periods.

2. The necessity of three-phase balance in distribution network

With the large increase in power demand, the active losses, voltage drop in system nodes and three-phase unbalance problems become increasingly serious. These problems have already become the serious problems that should be overcome for better distribution management, especially the three-phase unbalance problem. This is because the loads in the low-voltage distribution networks are dominated by single-phase power consuming equipment and there still exist some two-phase power consuming equipment. The unbalance also cause problems including three-phase nodal voltage unbalance, lower output of distribution transformers and increased active losses of distribution transformers, losses in transmission lines. These influence, on the other side, not only causing low quality electricity supply with low reliability but also make it difficult to recover power supply and load transfer [11]. This makes a negative effect on the power system operating security. Besides, unbalance in distribution network would make over-current problems in neutral phase and feeders. These problems need to be solved by vast investment to update the system, but this will bring economic problems. Therefore conducting reconfiguration optimization to balance the three-phase is of necessity.

3. The mathematical model of three-phase unbalance degree

3.1. The objective function

To guarantee the feasibility of the model, this paper chooses the minimum system losses as the target, and penalty terms are introduced to include the three-phase active power unbalance and the three-phase reactive power unbalance. The objective function is as follows:

$$\min \sum_{l=1}^{L} \sum_{ph \in \{a,b,c\}} R\left((P_{ij}^{ph})^2 + (Q_{ij}^{ph})^2\right) / V_1^{\,2} + \sum_{i=1}^{I} \sum_{ph \in \{a,b,c\}} (d_i^{ph} + q_i^{ph}) \tag{1}$$

Where V_1 are 1.0 pu in the objective function; L represents the total number of branches in the system[7], i and j represent the serial number, R represents the line resistance, P_{ij}^{ph} represents the active power in transmission line i-j whereas Q_{ij}^{ph} represents the reactive in line i-j; the term d_i^{ph} is introduced to represent the degree of three-phase unbalanced active power; q_i^{ph} is for the unbalanced reactive power, respectively.

3.2. The constraints

To guarantee the security operation of distribution network, the reconfiguration of distribution network should include the radial operation constraints, the output constraints at root nodes, power balance constraints, the voltage limitation constraints and the transformer capacity constraints, and etc.

1) Radial network constraints

$$g \in G \tag{2}$$

2) Feeder output constraints

$$\underline{P}_g^{ph} \le P_g^{ph} \le \overline{P}_g^{ph}, \forall g, ph \in \{a,b,c\} \tag{3}$$

$$\underline{Q}_g^{ph} \le Q_g^{ph} \le \overline{Q}_g^{ph}, \forall g, ph \in \{a,b,c\} \tag{4}$$

3) Power flow equality constraints:

$$\sum_{i:(k,i)\in W} P_{ki}^{ph} = \sum_{i:(i,j)\in W} P_{ij}^{ph} + P_i, \forall i \in N, ph \in \{a,b,c\} \tag{5}$$

$$\sum_{i:(k,i)\in W} Q_{ki}^{ph} = \sum_{i:(i,j)\in W} Q_{ij}^{ph} + Q_i, \forall i \in N, ph \in \{a,b,c\} \tag{6}$$

4) Branch capacity constraints:

$$(P_{ij}^{ph})^2 + (Q_{ij}^{ph})^2 \le z_{ij} \cdot \overline{S}_{ij}^2, \forall ij \in L \tag{7}$$

5) Voltage limitation constraints:

$$v^{ph} = (V^{ph})^2, \forall t, ph \in \{a,b,c\} \tag{8}$$

$$\underline{V}_i^2 \le v_i^{ph} \le \overline{V}_i^2, ph \in \{a,b,c\} \tag{9}$$

6) Transformer capacity constraints:

$$S_t \le S_{t\max} \tag{10}$$

Where Z_{ij} is a binary variable representing the status of the corresponding branch switch, and note that 1 is represent closed, 0 represent the switch is off; S_t and S_{tmax} are the apparent power and the allowable maximum power of a transformer; g is the current network structure whereas G is the aggregation of all the allowable radial network structure, P_g^{ph}、$\underline{P}_g^{ph}$、$\overline{P}_g^{ph}$ are the real-active power output, the bottom-limit and the upper-limit of active power output; Q_g^{ph}、$\underline{Q}_g^{ph}$、$\overline{Q}_g^{ph}$ are the real-reactive power output, the bottom-limit and the upper-limit of reactive power output; V_i^{ph}、$\underline{V}_i^{ph}$、$\overline{V}_i^{ph}$ are the value and limitations for nodal voltage.

4. The solution of the model forecasting

Heuristic algorithm, mathematical optimization algorithm and artificial intelligence algorithm can be choose for the reconfiguration problems of distribution network [12]. Among them mathematical optimization algorithm has obvious advantages over the others in adaptability and optimizing ability and it is regarded as the most frequently used method. Mixed integer quadratic programming method as the most simple and efficient mathematical optimization method has been developing fast for its fast convergence, stability and high efficiency.

4.1. Three phase transformation of the objective function

The objective function can be rewritten as follow:

$$\begin{aligned} \min \sum_{(i,j)\in W} (r_{ij}^{aa} + r_{ij}^{ab} + r_{ij}^{ac}) \cdot ((P_{ij}^a)^2 + (Q_{ij}^a)^2) / V_1^2 + \sum_{(i,j)\in W} (r_{ij}^{ba} + r_{ij}^{bb} + r_{ij}^{bc}) \cdot ((P_{ij}^b)^2 + (Q_{ij}^b)^2) / V_1^2 + \\ \sum_{(i,j)\in W} (r_{ij}^{ca} + r_{ij}^{cb} + r_{ij}^{cc}) \cdot ((P_{ij}^c)^2 + (Q_{ij}^c)^2) / V_1^2 \end{aligned} \tag{11}$$

Where r_{ij}^{aa} represents the self-impedance of the a-phase between node i and node j, r_{ij}^{ab} represents the mutual impedance between phase a and phase b in transmission line *i-j*, r_{ij}^{ac} represents the mutual impedance between phase a and phase c in transmission line *i-j*; whereas, r_{ij}^{ba} , r_{ij}^{bb} , r_{ij}^{bc} represents the mutual impedance between phase b and three-phase in transmission line i-j respectively; r_{ij}^{bc} , r_{ij}^{cb} , r_{ij}^{cc} represents the mutual impedance between phase c and three-phase in transmission line i-j respectively.

4.2. Linearization of the branch capacity constraints

This paper adopts mixed integer quadratic programming method for distribution configuration. To be convenient, linearizing uses quadratic constraint linearization method [13] which is as shown in Figure 1. This paper uses several square constraints to substitute the circular constraint. For general case, by increasing the number of square constraints, the accuracy will be improved. After linearizing, the problem becomes a mixed integer linear programming problem.

In this paper all constraint equations are linear, except the constraint equation (7), which is called circular constraint [13]. Therefore, two nonlinear constraints equations (12), which is called square constraint [13], are employed to substitute for (7), providing an adequate level of precision for engineering applications.

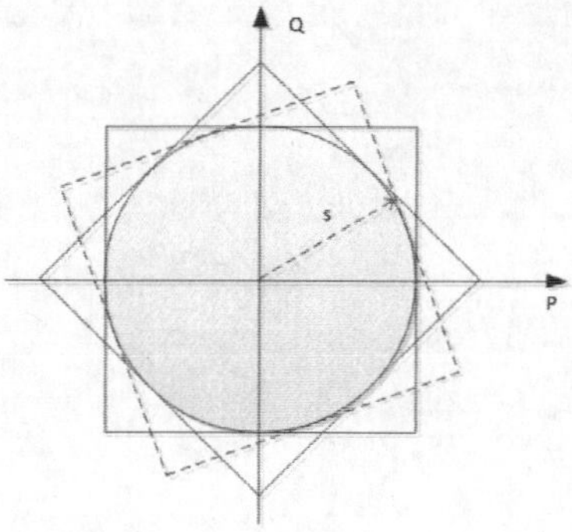

Figure 1. Circular constraint linearization method.

$$\begin{cases} -z_{ij}\bar{S}_{ij} \le P_{ij}^{ph} \le z_{ij}\bar{S}_{ij}, -z_{ij}\bar{S}_{ij} \le Q_{ij}^{ph} \le z_{ij}\bar{S}_{ij} \\ -\sqrt{2}z_{ij}\cdot\bar{S}_{ij} \le P_{ij}^{ph} + Q_{ij}^{ph} \le \sqrt{2}z_{ij}\cdot\bar{S}_{ij}, -\sqrt{2}z_{ij}\cdot\bar{S}_{ij} \le P_{ij}^{ph} - Q_{ij}^{ph} \le \sqrt{2}z_{ij}\cdot\bar{S}_{ij}, \\ \forall(i,j)\in l, ph\in\{a,b,c\} \end{cases} \quad (12)$$

4.3. Linearization of the power balance equations

For the power balance equation, this modification is justified because the squared p.u. resistance and reactance renders this term significantly smaller than the others. The approximation is obtained by replacing (5) and (6) with the following constraints:

$$\begin{cases} v_j^{ph} - v_i^{ph} \le m_{ij} + 2(r_{ij}^{ph}P_{ij}^{ph} + x_{ij}^{ph}Q_{ij}^{ph}),\ \ v_j^{ph} - v_i^{ph} \ge -m_{ij} + 2(r_{ij}^{ph}P_{ij}^{ph} + x_{ij}^{ph}Q_{ij}^{ph}) \\ m_{ij}^{ph} = (1-z_{ij})M,\ \ \forall(i,j)\in l,\ \ ph\in\{a,b,c\} \end{cases} \quad (13)$$

Equation (13) describes the power flow equations, and a big M is introduced to cancel the constraints in disconnected branches. A detailed description is given in [13]. For power balance equations, taking considerations of the three-phase unbalance, each of the three phases must be deduced separately considering the distribution network of three-phase imbalance.

5. The case study

The program is written in GAMS software and the solution is obtained by using CPLEX solver. The case study is the IEEE 34 nodes example. Single line diagram is shown in Figure 2. The original diagram contains 34 nodes. By adding 5 extra transmission branches (numbered by 34, 35, 36, 37 and 38) altogether become 38 branches. At initial state, all the connecting lines are open while the others are closed.

By three-phase load-flow analysis, the three-phase load flow information is derived. The voltage profile is demonstrated in Figure 1, and the unbalance phenomenon is obvious seen from the voltage distributions. The maximum of discrepancy between three-phase voltages for each node is up to 0.054p.u and the minimum nodal voltage is 0.9p.u.

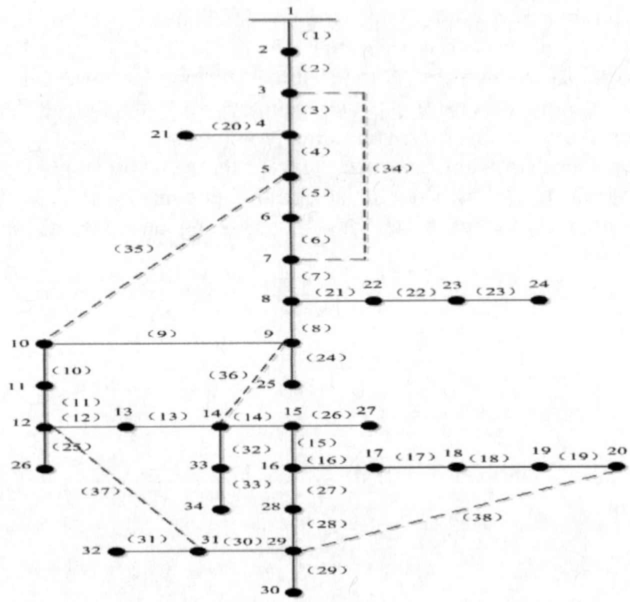

Figure 2. IEEE34 node test feeder

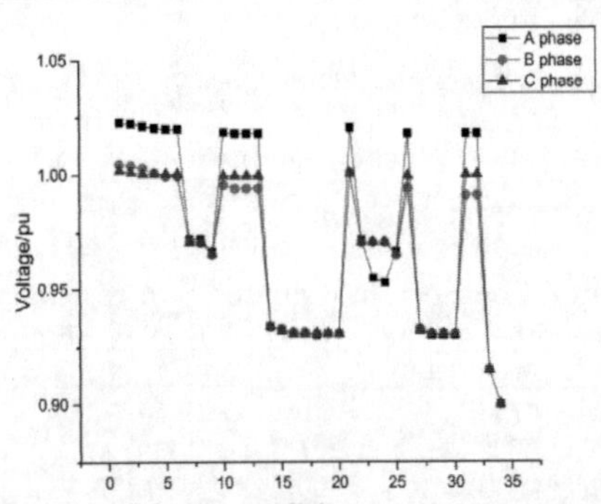

Figure 3. Node voltage before reconfiguration.

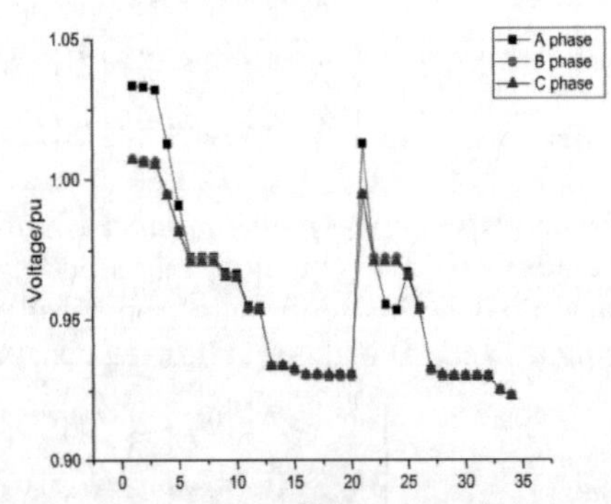

Figure 4. Node voltage after reconfiguration.

Table 1. Comparison between the result before and after reconfiguration.

	Turns on a collection of switches						
Before	2-6	4-9	8-13	11-30	19-28		
After	7-8	8-22	10-11	22-23	12-26	14-15	15-27

Figure 3 and Figure 4 show the node voltage before and after reconfiguration. Using the method proposed in this paper to optimize the three-phase unbalanced system reconstruction, the reconstruction program is shown in Table 1. The switch before reconfiguration has only five tie-switches, after reconfiguration, the contact switches are closed and open the 7 switches as follow: 6-7, 7-21, 9-10, 21-22, 11-25, 13-14, 14-26.

Table 2. Comparison between the result before and after reconfiguration.

	Degree of currents unbalance	Maximum voltage discrepancy(pu)	System losses(kW-hr)
Before	3.135	0.054	205
After	2.517	0.037	180

The distribution of nodal voltages is as shown in Table 2. By comparison, it is concluded that the three-phase currents unbalance is reduced by 19.7% from 3.135 to 2.517 resulting lower level of asymmetry. The maximum voltage discrepancy decreases from 0.054pu to 0.037pu. The objective function is to minimize the system losses, and the losses of the system before reconfiguration is 205 kW-hr while the losses of the system after the reconfiguration optimization is 180 kW-hr.

Table 3. Comparison of nonlinear load flow and linear load flow.

	System losses(kW-hr)	Operation time(ms)
Nonlinear Constraint	198	0.025
Linear Constraint	180	0.016

Table 3 shows the power flow calculation, using nonlinear constraints and linear constraints respectively, in a 34-node system. Considering that the constraint equation (7) is nonlinear, this paper uses equation (12) to substitute for equation (7) in linear constraint power flow calculation, while other constraint equations stay unchanged. The system loss using nonlinear constraint equation is 198 kW-hr, while using linear constraint equation is 180 kW-hr, which is decreased by 9%. The study is performed on a computer with i-3 CPU, 4GB ram, WIN 10 64-bit operating system GAMS version 8.2 and the operation time reduces to 0.016ms from 0.025ms, the calculation time is reduced by 0.051ms.

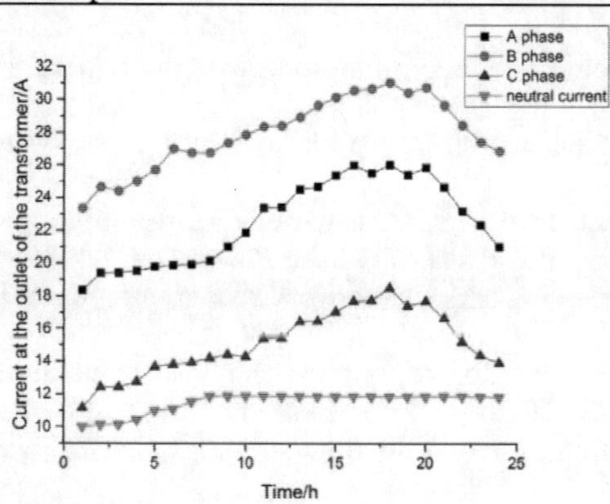

Figure 5. The three-phase currents at the terminal of transformer before reconfiguration.

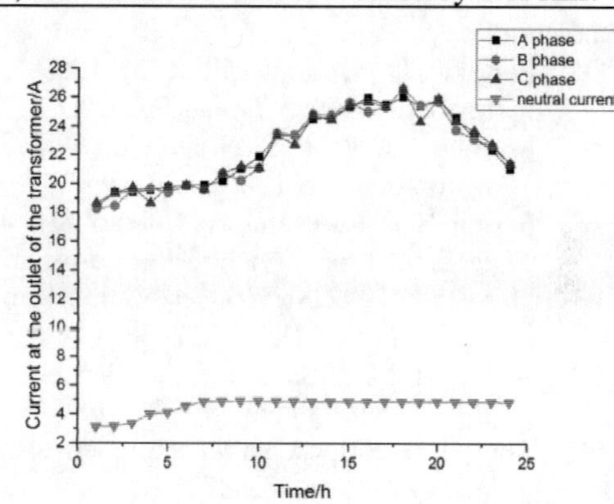

Figure 6. The three-phase currents at the terminal of transformer after reconfiguration.

Figure 5 and Figure 6 show the three-phase current and the current at the neural point at the transformer terminals before and after the reconfiguration optimization respectively. Comparing with the results before and after the reconfiguration, it is obvious that the proposed method can greatly mitigate the unbalance current and nodal voltage.

To demonstrate the advantages of the optimization scheme proposed in this paper, Cases using other methods are simulated including dynamic programming, genetic simulation algorithm, tabu algorithm, immune algorithm and family eugenics algorithm and get the simulation results for comparison which is shown in table 4 below. From the comparison in table 4, this paper can draw the conclusion that the active losses of reconfiguration using hybrid integer second order programing method is only a little more than the losses using hybrid particle swarms method and this method is better than the other four. In this method the minimum nodal voltage is only lower than the result obtained from tabu algorithm and obviously better than the other methods. Theses sufficiently prove the validity of the algorithm.

Table 4. Comparison between the result before and after reconfiguration.

Results	Hybrid integer second order programming	Family eugenics algorithm	Immune algorithm	Fuzzy genetic simulation method	Tabu algorithm	Hybrid particle swarm algorithm
Losses(kW-hr)	180.08	181.01	181.10	180.81	183.86	180.06
The minimum node voltage(pu)	0.923	0.911	0.911	0.913	0.928	0.910

6. Conclusion

A mixed-integer quadratic programming model for the problem of reconfiguration of electric distribution systems is presented in this paper. To address the three-phase unbalance problem, this paper proposes an optimal reconfiguration approach based on mixed integer quadric programming (MIQP) method. It aims to minimize the total network losses of the system. By using several square constraints to substitute the circular constraint, the original optimization problem is linearized and converted into a mixed-integer linear programming (MILP) model. The additional losses caused by three-phase unbalance are also considered in the paper. And then this MILP problem is solved in general algebraic modeling system (GAMS) software using CPLEX solver. The following conclusions can be drawn from this paper： (a) The proposed method can effectively balance the three-phase voltage asymmetry.(b) The method can reduce the network losses and improve the stability of the system.(c) The linearized processing method reduces the computation time.

References

[1] Popovic D S and Ciric R M 1999 A multi-objective algorithm for distribution networks restoration *IEEE Trans. Power Del.* **14** 1134–41

[2] Morelato A L and Monticelli A 1989 Heuristic search approach to distribution system restoration *IEEE Trans. Power Del.* **14** 2235–41

[3] Sarma N D R, Prasad V C, Rao K S P and Sankar V 1994 A new network reconfiguration technique for service restoration in distribution networks *IEEE Trans. Power Del.* **9** 1936–42

[4] Dariush S 1992 Service restoration in distribution networks via network reconfiguration *IEEE Trans. Power Del.* **7** 952–58

[5] Adibi M M 1988 An expert system operational aid for restoration and loss reduction of distribution systems *IEEE Trans. Power Syst.* **3** 619-26

[6] Teng J H 2003 A direct approach for distribution system load flow solutions *IEEE Trans. Power Del.* **18** 882-87

[7] Joshua A T and Franz S H 2012 Convex models of distribution system reconfiguration *IEEE Trans. Power Syst.* **27** 1407-13

[8] Zhenkun L, Xingying C, Kun Yu, Yi S, and Haoming L 2008 Hybrid particle swarm optimization for distribution network reconfiguration *Proceedings of the CSEE* **36** 1-7

[9] Zhuang T, Wenchuan W, Boming Z and Anjan B 2016 Mixed-integer second-order cone programing model for VAR optimisation and network reconfiguration in active distribution networks *IET Generation Transmission & Distribution* **10** 1938–46

[10] Jabr R A, Singh R and Pal B C 2012 Minimum loss network reconfiguration using mixed-integer convex programming *IEEE Trans. Power Syst.* **27** 1106-15

[11] Broadwater R P, Thompson J C, Robert E and Hamidreza M 1991 Computer-aided protection system design with reconfiguration *IEEE Trans. Power Del.* **6** 260–266

[12] Huijuan S, Chunhua P and Tingfang Y 2014 A three-phase equilibrium reconfiguration strategy for distribution network *Power System Technology* **38** 789–94

[13] Xin C, Wenchuan W and Boming Z 2015 Robust restoration method for active distribution networks *IEEE Trans. Power Syst.* **31** 4005–15

Research on power equalization using a low-loss DC-DC chopper for lithium-ion batteries in electric vehicle

Y W Wei[1], G T Liu[2], S N Xiong[1], J Z Cheng[1], and Y H Huang[1]

[1] Hubei Collaborative Innovation Centre for Microgrid of New Energy, China Three Gorges University, Yichang, HB 443002, China

[2] Dongguan Power Supply Bureau, Guangdong Power Grid Corporation, Dongguan, GD 523000, China

E-mail: weiyewen8@126.com

Abstract. In the near future, electric vehicle is entirely possible to replace traditional cars due to its zero pollution, small power consumption and low noise. Lithium-ion battery, which owns lots of advantages such as lighter and larger capacity and longer life, has been widely equipped in different electric cars all over the world. One disadvantage of this energy storage device is state of charge (SOC) difference among these cells in each series branch. If equalization circuit is not allocated for series-connected batteries, its safety and lifetime are declined due to over-charge or over-discharge happened, unavoidably. In this paper, a novel modularized equalization circuit, based on DC-DC chopper, is proposed to supply zero loss in theory. The proposed circuit works as an equalizer when Lithium-ion battery pack is charging or discharging or standing idle. Theoretical analysis and control method have been finished, respectively. Simulation and small scale experiments are applied to verify its real effect.

1. Introduction

The electric car, due to its zero pollution and small power consumption and low noise, is inevitable trend in the development of automobile industry. Recently, lithium-ion batteries have been widely equipped in different electric vehicles (EV) all over the world. Main advantages of this emerging battery include high energy density, no memory effect, and very low self-discharging. These features support the storage cell to obtain more safety, longer lifetime and lighter weight. To achieve the high power requirements, a large number of lithium-ion battery cells are applied and connected to each other in series [1]. Because of the difference on chemical and electrical characteristics, imbalance among these batteries obviously exists and leads to repeated charge and over discharge. It generates bad influence on energy storage capacity and battery lifetime [2-5]. Therefore, an equalization circuit is necessary for lithium-ion battery packs.

Numerous equalization circuits have already been developed, such as individual dc-dc converter, the bidirectional dc-dc converter, the switched capacitor [6-8]. Some of them can solve the imbalance problem. However, these are also suffered from long equalization time, complex circuits, control difficulty, or low efficiency when there are a large number of batteries [9-11]. This paper discusses a concept of modularized equalization for series-connected battery packs by using a low-loss DC-DC chopper. Basic principle can be described as:

- Charging state: Lower SOC batteries are charged and high power ones are bypassed.
- Discharging state: Higher SOC batteries are discharged and low power ones are bypassed.

- Standing idle state: Energy flows from high power batteries to low ones through inductor.

The mentioned modularized equalization circuit, as shown in Figure 1, can be charged by switching SWC or discharged by operating SWD. Equalizer unit includes double switches and an inductor, which works as a storage device. In theory, there is no energy consumption during the process of power regulation. To confirm its performance, theoretical analysis and control method have been developed. Simulation and four battery cells experiments are applied to verify its real effect.

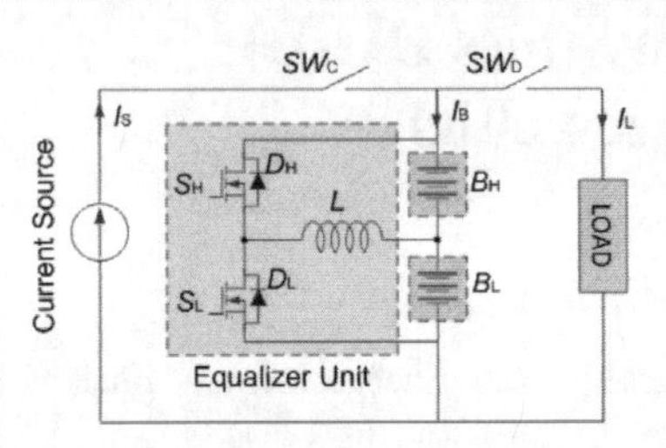

Figure 1. Structure of the researched equalization circuit.

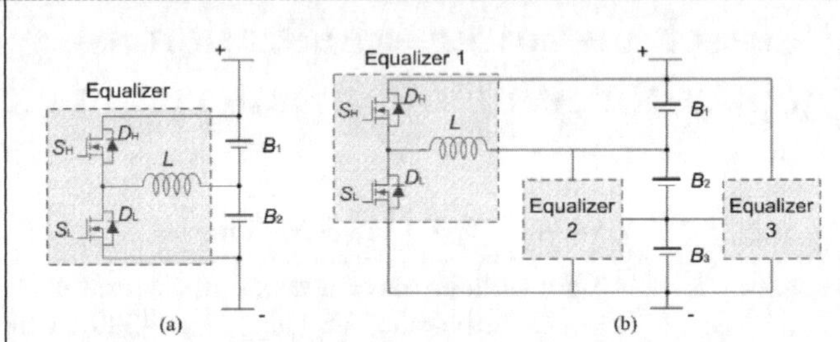

Figure 2. Structure of equalization circuit unit (a) double cells (b) triple cells.

2. Low-loss DC-DC chopper used to realize energy equalization

In practice, running process of DC-DC chopper should be thought about on different states, respectively. This section discusses the circuit structure, operating principle and waveforms.

2.1. *Circuit structure*

In Figure 1, series-connected cells are divided into two parts, BH and BL. The number of cells included in BH or BL can be odd or even number. Smallest unit of these two parts has only one cell. Figure 2 shows two kinds of equalization circuit unit with double cells or triple cells. An expected output voltage, therefore, can be achieved through the combination of these units.

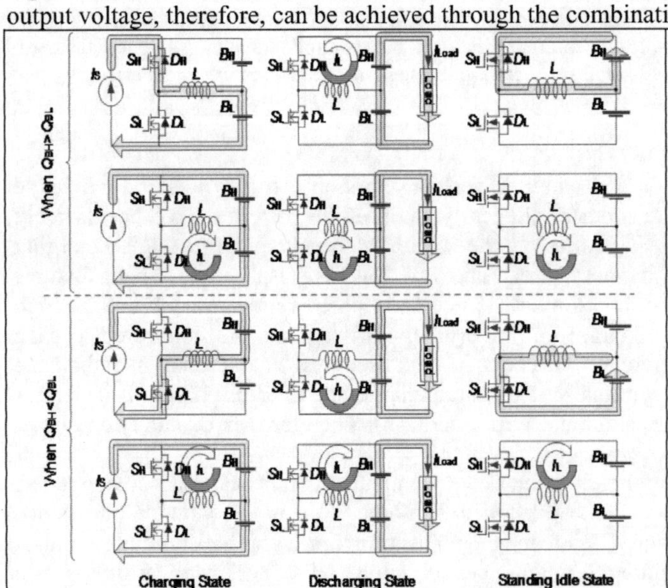

Figure 3. Operating principle of double cells circuit unit.

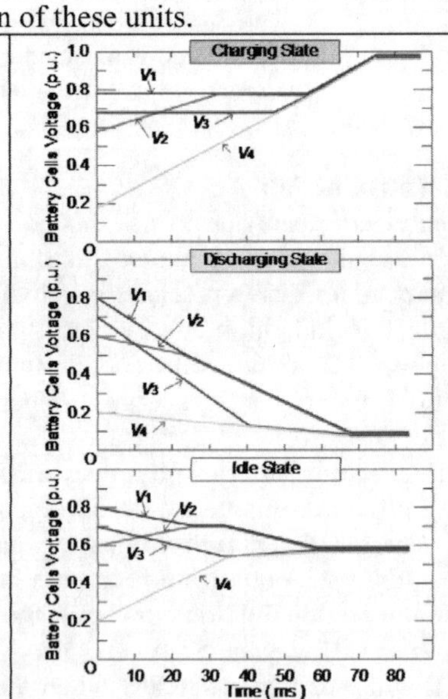

Figure 4. Waveforms for four battery cells energy equalization.

As we see, only one equalizer needed for double cells unit but three ones used in triple one. Inductor or switches in each equalizer is designed to satisfy power transmission requirements and not keep the same parameters. BH or BL includes several circuit units, which these number also maybe odd or even, revealed in Figure 2. Then BH or BL is divided into two parts again just like Figure 1. Additional equalizers can be increased between some circuit units in the adjacent position.

2.2. *Operating principle*

Current path, as Figure 3 shown, is displayed on different states when BH and BL have only one cell. If SOC of BH is larger than BL (QBH>QBL), equalization principle is introduced as follow:

- For charging state, B_H is bypassed when S_H keeps running to reduce energy charged for B_H. In other words, length of charging time is regulated to equalize B_H and B_L by controlling S_H.
- For discharging state, lower battery outputs less energy for load. It means that L is charged during S_H is opened. Inductor L is discharged and supplies a part of load current when S_H is not working. Therefore, the current supplied by B_L is decreased.
- For standing idle state, S_H is turned on to release energy stored in B_H as it higher than B_H. And B_L adopts the power stored in the inductor when S_H is turned off.

Detail process is revealed on the top of Figure 4 and another condition is also thought about. Similar principle is explained in case of $Q_{BH}<Q_{BL}$ with the difference S_L controlled only but not S_H.

For a battery pack with four series-connected cells, equalization process can be described based on the concept discussed above and shown in Figure 4. V1~V4 means voltage of each cell. There are two circuit units and each one consists of double cells and an equalizer. A third equalizer is equipped between the two circuit units. Form the curves in Figure 4, V1 and V2 or V3 and V4 firstly reach the same voltage point before all of these are equalized.

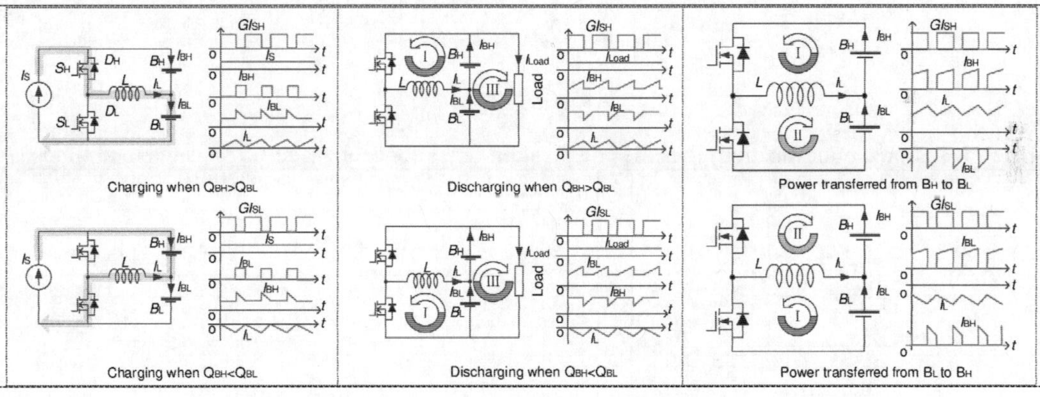

Figure 5. Waveforms of each state for double cells circuit unit.

2.3. *Waveforms and mathematic model*

Generally speaking, terminal voltage is an important factor for SOC evaluation of Lithium-ion batteries. It is also a key standard for energy equalization. Voltage of each cell is defined as Vi (i=1, 2,…, 2n). Then each voltage of circuit unit is Vj + Vj+1 (j=1,…, 2n-1) and voltage of x circuit units is Vk +,…,+ Vk+x (k=1,…, 2n-x). Figure 5 presents the equalization control process of battery packs. In a circuit unit, each cell's current of different states is formulated as follow:

$$\begin{array}{c}\text{Charging. State}\\ Q_{BH}>Q_{BL}\end{array}:\begin{cases} i_{BH}=0 \ \ \& \ \ i_{BL}=I_{CS}, & t\in[0,DT) \\ i_{BL}=i_{BH}+I_{CS}-V_{BL}\cdot(t-DT)/L_1, t\in[DT,T] \end{cases} \tag{1}$$

$$\begin{array}{c}\text{Discharging State}\\ Q_{BH}>Q_{BL}\end{array}:\begin{cases} i_{BH}=i_{BL}+V_{BH}\cdot t/L_1 \ \ \& \ \ i_{BL}=I_{Load}, t\in[0,DT) \\ i_{BL}=i_{BH}-[(V_{BH}+V_{BL})\cdot DT-V_{BL}\cdot t]/L_1, t\in[DT,T] \end{cases} \tag{2}$$

$$\text{Standing Idle State } Q_{BH} > Q_{BL}: \begin{cases} i_{BL} = 0 \quad \& \quad i_{BH} = V_{BH} \cdot t / L_1, & t \in [0, DT) \\ i_{BL} = [(V_{BH} + V_{BL}) \cdot DT - V_{BL} \cdot t] / L_1, t \in [DT, T] \end{cases} \tag{3}$$

Where, L_1 means the inductor in circuit unit, and corresponding equations can be easily deduced in the case of $Q_{BH}<Q_{BL}$. For double circuit units or n double circuit units, all circuit units are divided into two parts just like B_H and B_L. They also satisfy the above equations.

3. Strategy for system equalization control

For the circuits revealed in Figure 2, two kind of strategies, including modularized equalization control (MEC) and energy averaging control (EAC), are developed to realize system equalization control. MEC has the advantage on accuracy but suffer with slower response. To the contrast, EAC has the opposite characteristics.

3.1. Modularized equalization control

Each branch of battery packs can be divided into N double cells and M triple cells circuit units. By using modularized equalization control, every circuit unit is self-regulated at first. Then adjacent circuit units start to transfer electric power when the double cells' SOC reach a same value. Therefore, four circuit units can be equalized after the former two and the latter two finished equalization. In conclusion, MEC is actually hierarchical control and Figure 6 shows its schematic.

In any circuit unit, V_A, V_B and V_C are cells voltage. Output voltage of this unit is $V_A+V_B+V_C$ and each cell's voltage equals to $(V_A+V_B+V_C)$ /3 after it equalized. Double cells unit only have V_A and V_B. Total equalization times can be described as follow:

$$\begin{cases} P(x) = \text{int}[x], e.g.\, P(\frac{10}{3}) = 3 \\ P_i = \text{int}[\frac{M+N}{2^{i-1}}], P_1 = M+N \\ T_{MEC} = \sum_{i=1}^{D} P_i \;\&\; \text{int}[\frac{M+N}{2^{D-1}}] = 1 \end{cases} \tag{4}$$

Where $P(x)$ is an aliquot function, P_i is the equalization times for each stage, T_{EMC} is the total times.

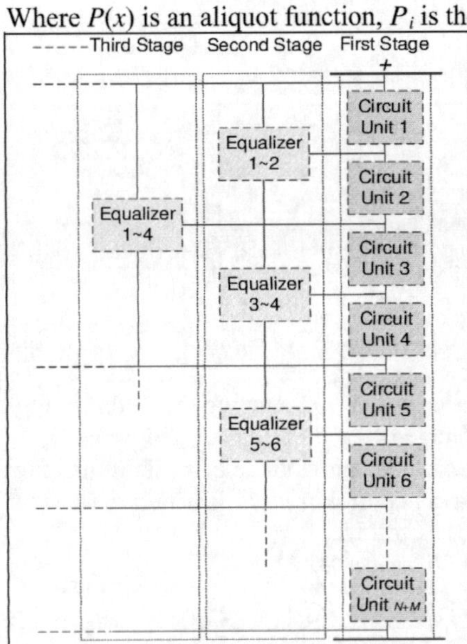

Figure 6. Schematic of MEC.

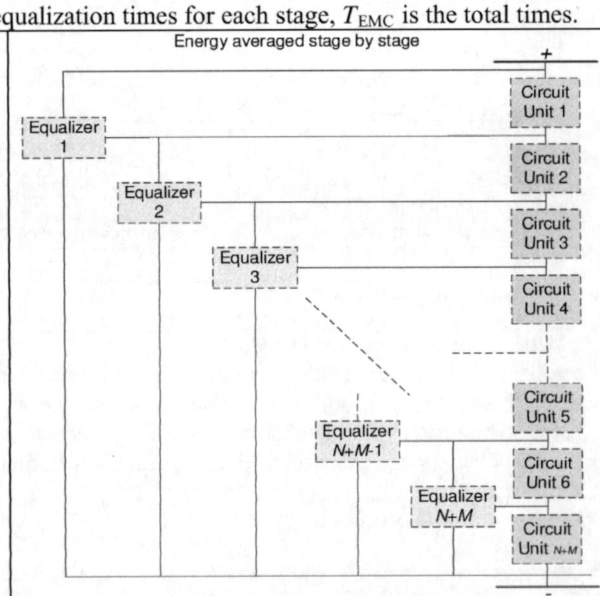

Figure7. Schematic of EAC.

3.2. *Energy averaging control*

For the same branch with $M+N$ circuit units, excess energy stored in unit τ is directly transferred to unit $\tau+1$ to $M+N$ (or unit $\tau-1$ to 1) by using energy averaging control and as shown in Figure 7. The maximum value of total equalization times can be described as follow:

$$T_{\text{EAC-max}} = (M+N) + 0.5(M+N) \tag{5}$$

3.3. *Hybrid equalization control*

Both of MEC and EAC have obviously features on control accuracy and response speed. A hybrid equalization control (HEC) is developed to improve control accuracy and promote response speed simultaneously. In HEC, only stage one and stage two are controlled by MEC. EAC is applied to the rest stages. So the total equalization times can be described as follow:

$$T_{\text{HEC-max}} = (M+N) + 0.5(M+N) + 0.25(M+N) \tag{6}$$

Furthermore, charging and discharging times of a single cell, not only each circuit, is reduced a lot compared to MEC or EAC. Then process of battery packs equalization is improved and it works more safety due to less loss and cooling difficulty.

Table 1. Characteristics of the simplified model

Category	Parameters	Values
Equalizers	MOSFETs, S_H or S_L ×14	20 A, 100 V
	Inductor, L ×7	100 uH
Batteries ×8	Cell's Rated Voltage, V_0	3.2 V
	Cell's Rated Capacity, Q	3000 mAh
Control Variables	Frequency, f	10 kHz
	Reference Error Voltage, ΔV_{REF}	10.0 mV
	Charging Stop Voltage, V_{max}	3.6 V
	Discharging Stop Voltage, V_{min}	2.5 V

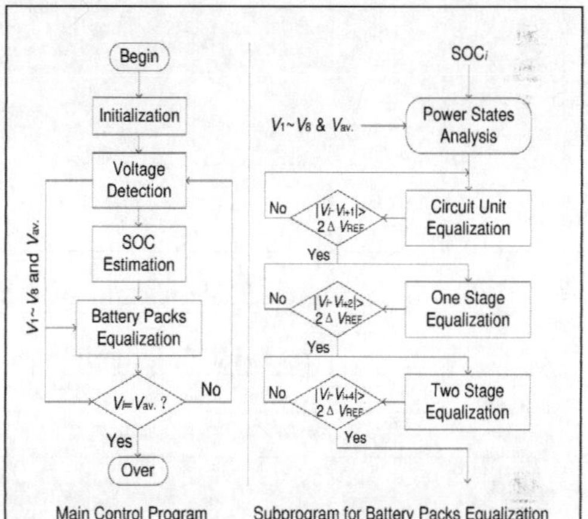

Figure 8. System control program.

4. Verifications

The concept of batteries equalization discussed above is validated using a simplified eight cells circuit in this section. Parameters and control method are designed for the experimental circuit. Then simulation and experiment results have been completed, respectively.

4.1. *Small-scale equalization circuit design*

A small-scale model, including seven equalizers and eight battery cells, has been established and main parameters are displayed on table 1. Applying the HEC strategy, detail running process is revealed in Figure 8. The left side is main program and the right one is subprogram.

4.2. *Performance verification*

Simulation and experimental verifications have been finished, respectively. To any circuit unit in the case of QBH>QBL on different states, current waveforms comparison has been shown in Figure 9 (left part is simulation results, right part is experimental results). These results are obviously consistent with the theory analysis shown in Figure 5. Table 2 reflects the effect of voltage equalization for the eight series-connected battery cells. It also supports the equalization concept proposed in this paper.

Dissipated power loss, compared to traditional circuit which usually connects a resistor, has been discussed during the process of SOC regulation in practical application. Loss types mainly include line loss, loss of semiconductor devices, resistor loss and inductor loss (coil resistance). Comparative analysis, as revealed in Figure 10, has been completed to verify the effect of loss reduction. Obviously, the saving loss has been moved to other battery cells but not consumed by resistor.

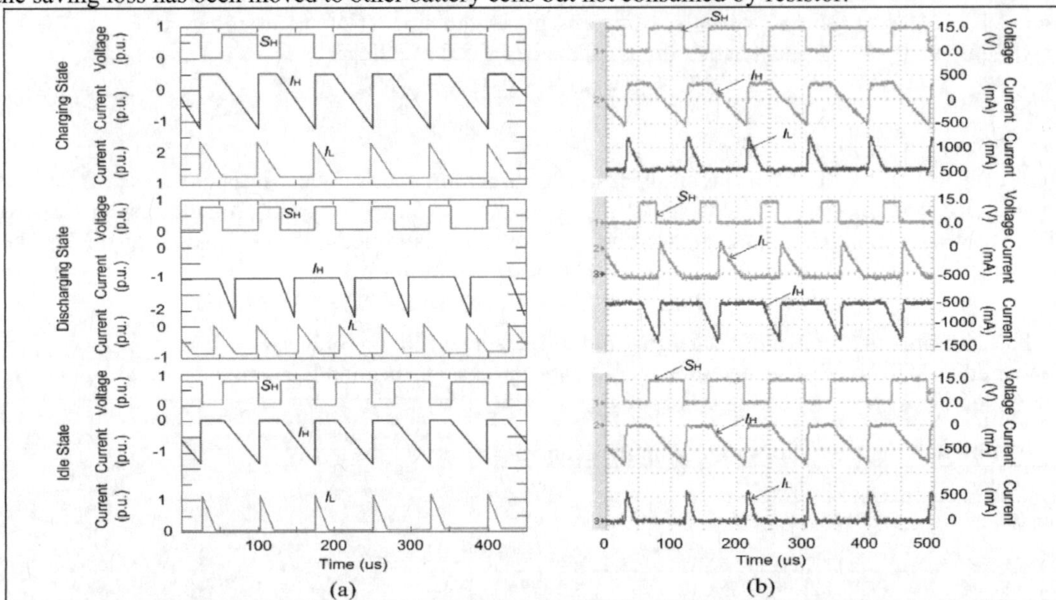

Figure 9. Verification results (a) By simulation (b) By experiments.

Table 2. Experimental results of equalization process.

Battery Cells		Charging State/Time point, ms					Discharging State/Time point, ms					Standing Idle State/Time point, ms				
		0	10	20	30	40	0	10	20	30	40	0	10	20	30	40
Cells' Voltage (V)	B_1	2.90	3.20	3.35	3.50	3.60	2.90	2.80	2.75	2.60	2.50	2.90	2.70	2.70	2.80	2.95
	B_2	2.50	3.10	3.37	3.50	3.60	2.50	2.50	2.50	2.50	2.50	2.50	2.70	2.70	2.80	2.95
	B_3	2.60	3.12	3.36	3.50	3.60	2.60	2.60	2.60	2.55	2.50	2.60	2.70	2.70	2.80	2.95
	B_4	2.80	3.20	3.38	3.49	3.60	2.80	2.79	2.75	2.60	2.50	2.80	2.70	2.70	2.80	2.95
	B_5	3.10	3.30	3.45	3.56	3.60	3.10	3.00	2.80	2.62	2.50	3.10	3.30	3.25	3.10	2.95
	B_6	3.50	3.60	3.70	3.60	3.60	3.50	3.00	2.80	2.62	2.50	3.50	3.30	3.25	3.10	2.95
	B_7	3.00	3.20	3.35	3.58	3.60	3.00	2.90	2.80	2.63	2.50	3.00	3.20	3.25	3.10	2.95
	B_8	3.40	3.50	3.57	3.62	3.60	3.40	3.00	2.80	2.61	2.50	3.40	3.20	3.25	3.10	2.95

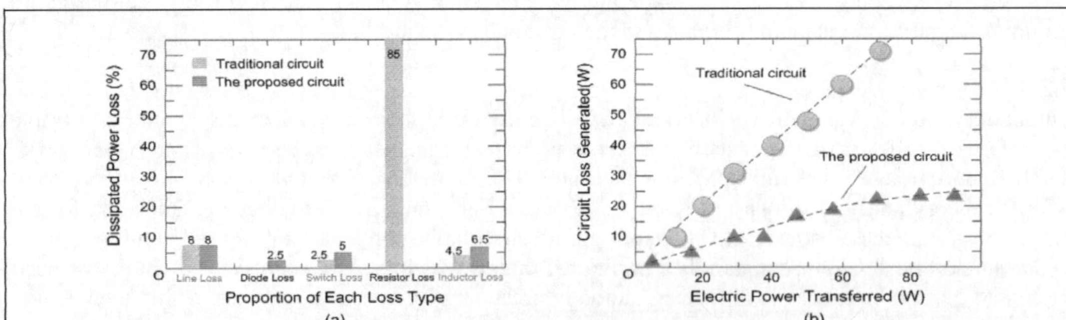

Figure 10. Power loss analysis by comparison (a) Loss proportion (b) Loss reduction effect.

5. Conclusion and future works

This work, applied a DC-DC chopper, plans to decrease the loss of energy equalization for lithium-ion batteries in electric vehicle. Advantages of the researched method include simple circuit structure, high accuracy and less loss. Stability of energy control effect has been verified by simulation and small-scale experiments. SOC estimation technique, which is also important as well as equalization circuit, must be developed in the future.

Acknowledgement

This work is supported by Outstanding Youth Science and Technology Innovation Team Program of Education Department of Hubei Province (T201504).

References

[1] Moo C S , Hsieh Y C and Tsai I S 2003 Apr. Charge equalization for series connected batteries *IEEE Trans. Aerosp. Electron. Syst.* **39(2)** 704-710.

[2] Laldin O, Moshirvaziri M and Trescases O 2013 Aug. Predictive algorithm for optimizing power flow in hybrid ultracapacitor/battery storage systems for light electric vehicles *IEEE Trans. Power Electron* **28(8)** 3882-95.

[3] Baughman A C and Ferdowsi M 2008 Jun. Double-tiered switched-capacitor battery charge equalization technique *IEEE Trans. Ind. Electron.* **55(6)** 2277-85.

[4] Ongaro F, Saggini S and Mattavelli P 2012 Sep. Li-ion battery-supercapacitor hybrid storage system for a long lifetime, photovoltaic-based wireless sensor network *IEEE Trans. Power Electron.* **27(9)** 3944-52.

[5] Einhorn M, Conte F V, Kral C and Fleig J 2013 Mar. Comparison, selection, and parameterization of electrical battery models for automotive applications *IEEE Trans. Power Electron.* **28(3)** 1429-37.

[6] Tan N M L , Abe T and Akagi H 2012 Mar. Design and performance of a bidirectional isolated dc-dc converter for a battery energy storage system *IEEE Trans. Power Electron.* **27(3)** 1237-48.

[7] Lee Y S and Cheng G T 2006 Sep. Quasi-resonant zero-current-switching bidirectional converter for battery equalization applications *IEEE Trans. Power Electron.* **21(5)** 1213-24,.

[8] Yuanmao Y, Cheng K W E and Yeung Y P B 2012 Jul. Zero-current switching switched-capacitor zero-voltage-gap automatic equalization system for series battery string *IEEE Trans. Power Electron.* **27(7)** 3234-42.

[9] Lee Y S and Cheng M W 2005 Oct. Intelligent control battery equalization for series connected lithium-ion battery strings *IEEE Trans. Ind. Electron.* **52(5)** 1297-1307.

[10] Kim C H, Park H S, Kim C E, Moon G W and Lee J H 2009 May Individual charge equalization converter with parallel primary winding of transformer for series connected lithium-ion battery strings in an HEV *J. Power Electron.* **9** 472-480.

[11] Kim J, Shin J, Chun C and Cho B H 2012 Jan.Stable configuration of a Li-ion series battery pack based on a screening process for improved voltage/SOC balancing *IEEE Trans. Power Electron.* **27(1)** 411-424.

Thermal performance analysis of a thermocline thermal energy storage system with FLiNaK molten salt

C Liu[1,2], M S Cheng[1,3], B C Zhao[1,2], Z M Dai[1]

[1] Shanghai Institute of Applied Physics, Chinese Academy of Sciences, Jiading campus, Shanghai 201800, China
[2] University of Chinese Academy of Sciences, Beijing 100049, China

E-mail: chengmaosong@sinap.ac.cn

Abstract. A thermocline thermal storage unit with a heat transfer fluid (HTF) of high-temperature molten salt is considered as one of the most promising methods of thermal storage due to its lower cost and smaller size. The main objective of this work is to analyze the transient behavior of the available molten salt FLiNaK used as the HTF in heat transfer and heat storage in a thermocline thermal energy storage (TES) system. Thermal characteristics including temperature profiles influenced by different inlet velocities of HTF and different void fractions of porous heat storage medium are analyzed. The numerical investigation on the heat storage and heat transfer characteristics of FLiINaK has been carried out. A comparison between two different molten salts, FLiNaK and Hitec, has been explored in this paper with regards to their charging and discharging operations. The results indicate the system with FLiNaK has a greater energy storage capability, a shorter charging time and a higher output power. The numerical investigation reveals heat storage and heat transfer characteristics of the thermocline TES system with FLiNaK, and provide important references for molten salt selection of the TES system in the future.

1. Introduction

Thermal energy storage (TES) is very important in many engineering applications such as solar energy, space heating and waste heat recovery. Generally, in a TES system, the heat transfer fluid (HTF) used to carry heat from heat source to TES and the storage medium capable of retaining heat obtained from HTF for later discharge are required. The HTF, sometimes either used for direct fluid storage must have favourable properties for heat transfer include low freezing temperature and /or high maximum operational temperature, high heat capacity, high thermal conductivity, high density, good stability and low vapor pressure [1, 2]. And it is desirable for the storage medium to have high specific heat capacity, long term stability and compatibility with its containment [3]. Inorganic salt of a molten state has broad application prospects in TES of large-scale renewable energy due to its good performance of wide operation temperature range, low vapor pressure, low viscosity and relatively high density in heat storage and heat transfer [3, 4]. Thermal energy can be stored as sensible, latent, or a combination of both heats by the method of using molten salt as the HTF and storage material. Storage of the sensible heat is achieved by raising the temperature of the storage medium and the release of sensible heat is implemented by lowing the temperature of the storage medium [5]. The sensible heat storage has simpler heat transfer design, more flexible operation, more mature technology and cheaper cost than the latent heat storage [6].

In order to reduce the size and costs, a single tank system full filled with spherical particles which are made of some high thermal capacity material have been proposed [7-9], in which natural thermal stratification driven by buoyancy forces within single storage tank is taken into consideration to isolate hot and cold HTF. The single tank system uses porous medium to replace part of costly molten salt as a storage medium. There is a special region called thermocline characterized by a great temperature gradient between the coldest fluid and the hottest fluid in the storage tank, which separates the hot and cold fluids. The gradient of temperature in the thermocline mainly depends on the characteristics of the storage material and molten salt.

Currently the most widely used molten salt is nitrate-based, and different researches on the thermocline TES system with the nitrate salts have been conducted [10-13]. However, the maximum working temperature of nitrate salts is limited below 600℃ because of its thermal stability, which restricts the improvement of power generation efficiency of the energy storage system. Thus, to find a suitable molten salt so as to raise the working temperature and extend operating temperature range will be of great significance.

The FLiNaK is one of the most common fluoride molten salts possessing a series of advantages include higher heat capacity, higher thermal conductivity, higher operational temperature and good heat transfer performance, which is widely used in the high-temperature hydrogen, nuclear fuel and other energy fields [14, 15].

The present work makes an analysis of the transient behavior and thermal characteristics during the charging and discharging modes of the single tank TES system with FLiNaK molten salt. Thermal characteristics including the temperature profiles influenced by different inlet velocities and different void fractions of the porous medium were analyzed. And Hitec molten salt is also considered as HTF to be compared with FLiNaK.

2. Mathematical model

As the previous studies on the single tank thermolcline TES system, the mathematical model used is developed based on Schumann's [16] original work. The geometry of a cylindrical thermocline single tank is illustrated in the figure 1, where the gray area is regarded as the analysis object. The packed bed is defined by the void fraction of ε and the particles behave as continuous and isotropic porous medium. During the charging period, the hot molten salt flows into the storage tank from the upper inlet in the middle of the top of storage tank and transfer heat to the cold solid particles and molten salt, and then flows out of the tank through the outlet at the bottom of the tank with a lower temperature. In this process, a thermocline is formed between the hot molten salt and the cold one. When the thermocline has exited the storage tank completely and the filler materials have reached a uniform temperature as high as the temperature of inlet molten salt, it means the charging process has been completed. At last, the thermal energy is stored in the filler materials in the tank. During the discharging period, the cooled HTF enters the storage tank from the bottom of the tank, and the hot molten salt leaves the storage tank for applications. When the thermocline begins to get out from the tank, the effective discharging process is finished.

The main objective of this work is to analyze the thermal performance of the available FLiNaK molten salt in heat transfer and heat storage system. The mathematical model is developed with the multi-physics finite element analysis software COMSOL Multiphysics. The heat storage tank is an upstanding cylinder, 2 meter high and 1 meter in diameter. In the calculation and analysis, several reasonable assumptions employed for the simplification of the modeling and corresponding justifications are shown as follows:

(a) The packed-bed region is symmetrical along the axis direction. Distributors are assumed to ensure that the inlet HTF flow is uniformly distributed to create a well-formed thermocline. The characteristics of storage medium in the tank can be considered to be uniform in the radial direction, and the change only occurs in the axial direction;
(b) Physical properties of the porous media are kept constant;
(c) The Biot number of the filler particles is so small that the temperature distribution inside the sphere is regarded as uniform [17];
(d) The filler particles behave as continuous, homogeneous and isotropic porous medium;
(e) There is a complete thermal insulation between storage tank and environment.

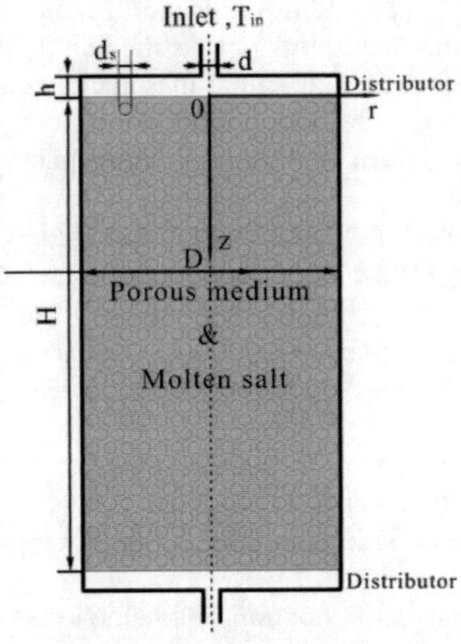

Figure 1. Schematic diagram of the thermocline TES system during charging mode.

The temperatures of both solid and fluid phases are evaluated along the flow direction according to the energy equations (1) and (2) [18].

For the porous medium, the equation is:

$$(1-\varepsilon)\rho_s C_{P,s}\frac{\partial T_s}{\partial t}=(1-\varepsilon)k_s\frac{\partial^2 T_s}{\partial z^2}-h_V(T_s-T_f) \tag{1}$$

and for the HTF, the equation is:

$$\varepsilon\rho_f C_{P,f}\frac{\partial T_f}{\partial t}+\rho_f C_{P,f}u\frac{\partial T_f}{\partial z}=\varepsilon k_f\frac{\partial^2 T_f}{\partial z^2}+h_V(T_s-T_f) \tag{2}$$

where, h_V denotes the volumetric convective heat transfer coefficient between the solid particles and HTF, T_s, T_f, u, $C_{P,s}$, $C_{P,f}$, k_s and k_f respectively represent the temperature of solid material and fluid, the axial apparent velocity of HTF in the storage tank, the heat capacity of solid material and fluid, the thermal conductivities of solid material and fluid.

The h_v is linked to the convective heat transfer coefficient h_s:

$$h_v=\frac{6(1-\varepsilon)}{d_s}h_s \tag{3}$$

where d_s denotes the diameter of solid particles. The heat transfer coefficient h_s can be calculated from [17]:

$$h_s=\frac{Nuk}{d_s} \tag{4}$$

In the Eq.(4), the Nusselt number Nu is a function of Reynolds number and Prandtl number based on particle diameter:

$$Nu = 2 + 1.1\ (\mathrm{Re}^{0.6}\ \mathrm{Pr}^{0.33}) \tag{5}$$

The FLiNaK (LiF-NaF-KF, 46.5-11.5-42.0 mol%) is an eutectic mixture of inorganic salts with melting temperature of 454 ℃ [15] and boiling point of 1570 ℃. The Hitec® (KNO_3-$NaNO_3$-$NaNO_2$, 44-7-49 mol%) is also an eutectic mixture of inorganic salts and in a stable liquid state ranging between 142 ℃ to 600 ℃. The average thermo-physical properties such as specific heat capacity and thermal conductivity of FLiNaK (1882.8 $\mathrm{Jkg^{-1}K^{-1}}$, 0.92 $\mathrm{W\,m^{-1}\,K^{-1}}$) are respectively higher than them of Hitec (1560 $\mathrm{Jkg^{-1}K^{-1}}$, 0.51 $\mathrm{W\,m^{-1}\,K^{-1}}$).

In this study, the thermo-physical properties (heat capacity, thermal conductivity, density and viscosity) of molten salt are functions of temperature [19-22]. The input parameters of porosity and thermal properties of porous medium are tabulated in table 1.

Table 1. Main characteristics and design parameters of the thermocline TES system.

Parameters	FLiNaK	Hitec
Initial temperature of charging mode (K)	773	473
Inlet fluid temperature of charging mode (K)	973	673
Initial temperature of discharging mode (K)	973	673
Inlet fluid temperature of discharging mode (K)	773	473
Diameter of storage tank (m)	1	1
Height of storage tank (m)	2	2
Diameter of spherical particles (m)	0.01	0.01
Void fraction	0.2/0.3/0.4	0.4
Thermal conductivity of porous medium (W $\mathrm{m^{-1}\ K^{-1}}$)	1	1
Specific heat capacity of porous medium (J $\mathrm{kg^{-1}\ K^{-1}}$)	1500	1500
Density of porous medium (kg $\mathrm{m^{-3}}$)	2000	2000
Inlet velocity of charge and discharging mode (m $\mathrm{s^{-1}}$)	0.001/0.005/0.01	0.001
Initial pressure (bar)	1	1

Initial conditions:

$$T_f(z,0) = T_{ini},\ T_s(z,0) = T_{ini},\ (0 \le z \le H) \tag{6}$$

Boundary conditions:

$$T_f(2,t) = T_{in},\ T_s(2,t) = T_{in},\ (t > 0) \tag{7}$$

Where, T_{ini} denotes the initial temperature of porous medium and molten salt, T_{in} represents the inlet temperature of the molten salt.

The reported experimental results of Yang et al. (2012) [13] are used here to validate the numerical model. The diameter and height of the thermal storage tank built for the reported TES system was 263 and 550 mm, respectively. The diameter of the filler particle was 30 mm, whereas the density of the filler material was 2100 $\mathrm{kg/m^3}$. Composite nitrate salt was used as HTF. The numerical results for the temperature profiles of molten salt at position x = 0.25 m and x = 0.5m are compared with the experimental ones in figure 2. The simulation uses the same conditions as described in experiment. Within the experimental uncertainty, the results from the simulations appear to agree well with the experiments. Thus, the numerical method is reliable and the model is suitable for the thermal performance analysis of a TES with molten salts.

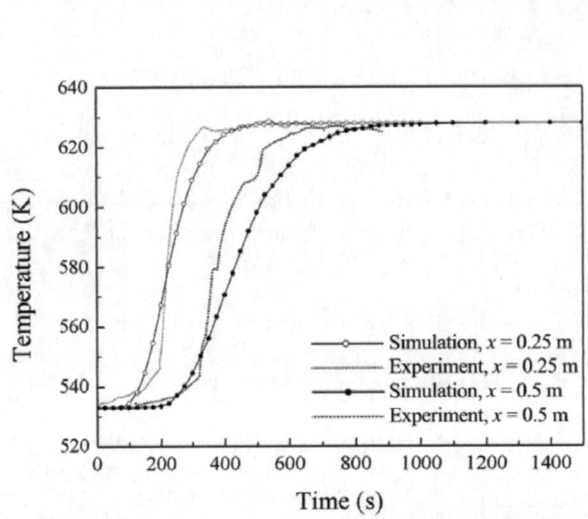

Figure 2. Comparison between the experimental and numerical results for the axial temperature profiles of molten salt at positions of $x = 0.25$ m and $x = 0.5$ m.

3. Results and discussion

3.1. Temperature profiles

The temperature distributions of the FLiNaK molten salt and porous medium during the charging mode are shown in figure 3 (a). The left side of the picture represents the fluid domain and the right side represents the porous medium domain. It is observed that the temperature of the fluid is higher than the porous medium at a same height.

Figure 3 (b) shows the temperature profiles along the HTF flow direction of the FLiNaK and porous medium at a given charging time of 920 sec (to facilitate observation, total charging time is 2230 sec). In this case, the inlet velocity of fluid is 0.001 m/s, the void fraction of porous material is 0.4, and the diameter of filler particle is 0.01 m. This figure indicates that the thickness of thermocline accounts for about 62.5 % of the entire height of storage tank at this moment. A same conclusion with figure 3 (a) can be easily obtained from figure 3 (b) that the temperature of fluid is slightly higher than the temperature of porous medium when compared at a same position, and the temperature difference is mainly occurs in the transition region and peaks in the middle of the thermocline.

3.2. Influence of HTF inlet velocity

The influence of HTF velocity on the thermal performance of the storage system is investigated in this section. figure 4 shows the temperature difference distribution for FLiNaK molten salt along the axial flow direction under various HTF inlet velocities. To facilitate observation, the curves were selected when the flowing distance along axial direction of HTF thermal front reached the half of the tank height. The charging times of each curve are respectively 920, 185, 97 sec, and three inlet velocities 0.001, 0.005, 0.01 m/s are considered respectively. With the increase in the value of inlet velocity, the Reynolds number increases. It is clear that the temperature difference between HTF and porous solid particles occurs in transition region of thermocline becomes larger and the local thermal non-equilibrium becomes more obvious with the rise in the Reynolds number. As a matter of fact, the increase of Reynolds number enhances the effect of heat transfer between HTF and porous medium.

3.3. Influence of void fraction

The temperature distributions at the outlet of storage tank with FLiNaK molten salt during the charging mode for the various void fractions are presented in figure 5. In this case, the inlet velocity of HTF is 0.001 m/s and filler particle diameter is 0.01 m. With the decrease of void fraction, the overall time of charging process becomes shorter. That is because the thermal conductivity of the porous material is larger than the thermal conductivity of FLiNaK molten salt [22]. Reducing the void fraction of porous material results in an increase of the volume fraction of porous medium to the whole storage tank, so that it will enhance the effect of heat transfer between HTF and porous medium and speed up the charging process.

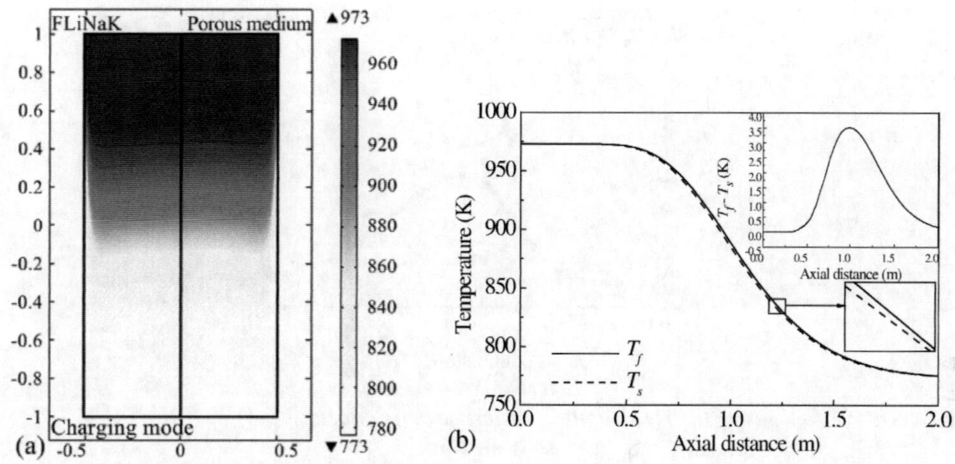

Figure 3. Temperature distributions of the FLiNaK molten salt and porous medium in storage tank during charging mode: (a) 2D temperature distributions, (b) 1D temperature profiles along the axial direction.

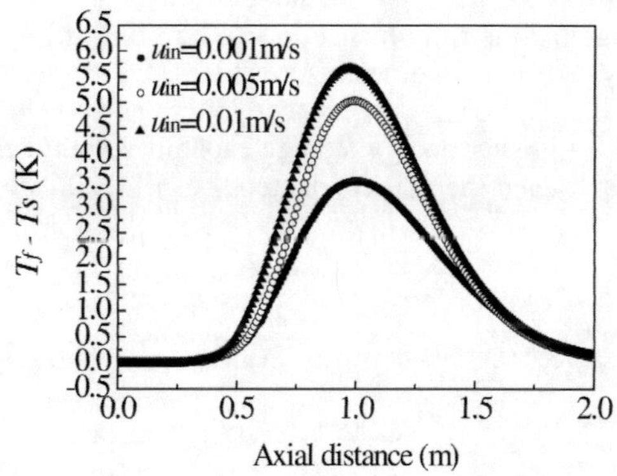

Figure 4. Temperature differences between FLiNaK and porous medium along the axial distance during charging mode at different inlet velocities.

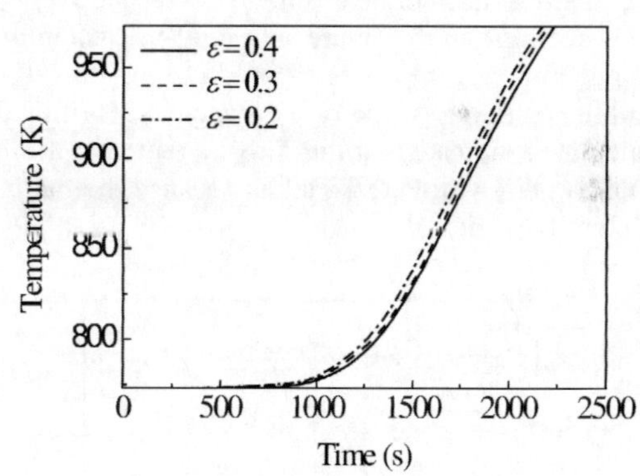

Figure 5. Temperature profiles with time of HTF for various void fractions at the outlet of the storage tank during charging mode.

3.4. Comparison between FLiNaK and Hitec

3.4.1. Influence on the temperature difference Figure 6 shows the temperature differences between the HTF of various salts and the porous medium. In this case, the inlet velocity is 0.001 m/s, the void fraction is 0.4, the diameter of spherical particle is 0.01 m, and the flowing distance along the axial direction was a half of the tank height. It can be clearly demonstrated in the figure 6 that the temperature difference between FLiNaK and porous medium is smaller than that between Hitec and porous medium. The thickness of thermocline in the TES system with FLiNaK is slightly thicker than in the TES system with Hitec. These results are attributed to a fact that the thermal conductivity of FLiNaK is higher than the thermal conductivity of Hitec. Namely, the higher the thermal conductivity of HTF is, the more for the HTF contributes to the temperature uniformity of filler material and the thickness increase of thermocline.

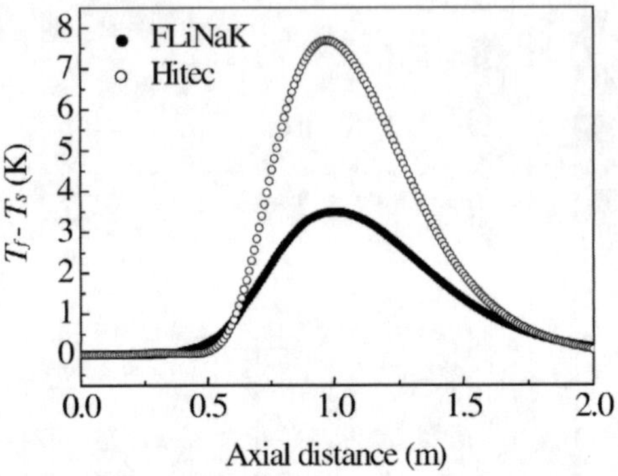

Figure 6. Temperature differences between the HTF of various salts and the porous medium.

3.4.2. Analysis on charging and discharging processes Figure 7 and 8 give the outlet temperature changes during charging and discharging processes for different HTFs. For this case, the void fraction, the inlet velocity, filler particle's diameter and the temperature difference from initial state to the completion status of heat storage are respectively set as 0.4, 0.001 m/s, 0.01 m, and 200 K.

According to the figure 7 (a) and (b), when the storage tank is fully loaded, it can be achieved by calculation that the energy stored in the FLiNaK-TES system is 1.14 times of the Hitec-TES system, and the charging period of the system with FLiNaK is just 95.3% of the system with Hitec. The results calculated indicate that the TES system with FLiNaK has a greater energy storage capability than the TES system with Hitec. This is because the thermal capacity and thermal conductivity of FLiNaK are higher than Hitec.

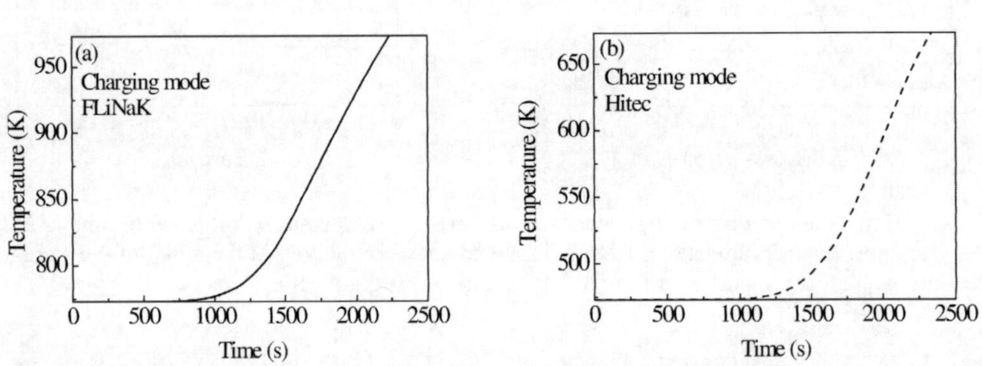

Figure 7. Temperature profiles at the outlet of the storage tank (a) of FLiNaK system during charging mode, (b) of Hitec system during charging mode.

During the discharging process, based on the requirement of a stable thermal output, it is considered that the heat released from the beginning of discharging mode until the moment that the thermocline start to leave the tank as the effective heat storage capacity, and the period of heat releasing as the effective discharging time.

Figure 8 (a) and (b) also indicate that the effective discharging time of the system with FLiNaK is obviously shorter than the system with Hitec. After deduction, the output power for the system with FLiNaK is 1.5 times of the system with Hitec.

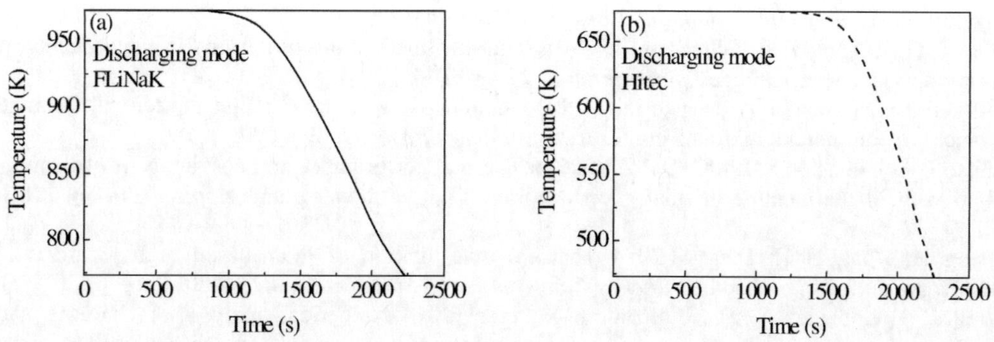

Figure 8. Temperature profiles at the outlet of the storage tank (a) of FLiNaK system during discharging mode, (b) of Hitec system during discharging mode.

4. Conclusion

In this study, a molten salt with higher heat capacity and lower viscosity was brought into the thermocline TES system. Thermal characteristics of the heat transfer and heat storage of the energy storage system with FLiNaK are systematically explored.

During charging process, there is a small temperature difference mainly occurring in the transition region between FLiNaK and porous medium. With the increase of void fraction, the overall time of charging process will become longer. When working in the same condition, the temperature difference between FLiNaK and porous medium will be smaller than that between Hitec and porous medium, and the thickness of thermocline in the system of FLiNaK will be slightly thicker than that in the system with Hitec. Due to the higher heat capacity and thermal conductivity of FLiNaK compared with Hitec, the thermal energy stored by the FLiNaK-TES system is larger and the time spent by the system with FLiNaK is shorter when the storage systems are both fully loaded and the changed temperature from initial state to completion status is set at a constant value.

During the discharging process, the effective discharge time of the system with FLiNaK is obviously shorter than the system with Hitec and the output power for the system with FLiNaK is 1.5 times of the system with Hitec according to numerical results.

The TES system with FLiNaK is a high performance thermal storage system and can be applied to different industries. This study provides important guidelines for designing a storage tank with filler particles for a TES system.

Acknowledgments
This paper is supported by the "Strategic Priority Research Program" of the Chinese Academy of Sciences. (Grant No. XD02001002)

References

[1] Li P W, Lew J V, Karaki W, Chan C L, Stephens J, O'Brien J E 2011 Transient heat transfer and energy transport in packed bed thermal storage systems C. *Developments in Heat Transfert* p 373.

[2] Vignarooban K, Xu X, Arvay A, Hsu K, Kannan A M 2015 Heat transfer fluids for concentrating solar power systems – a review *Applied Energy* **146** p 383

[3] Hasnain S M 1998 Review on sustainable thermal energy storage technologies, part I: heat storage materials and techniques *Energy Convers. Mgmt* **39** p 1127

[4] Herrmann U, Kelly B, Price H 2004 Two-tank molten salt storage for parabolic trough solar power plants *Energy* **29** p 883

[5] Liu M, Tay N.H. S, Bell S, Belusko M, Jacob R, Will G, Saman W, Bruno F 2016 Review on

concentrating solar power plants and new developments in high temperature thermal energy storage technologies *Renewable and Sustainable Energy Reviews* **53** p 1411

[6] Gil A, Medrano M, Martorell I, La´zaro A, Dolado P, Zalba B, Cabeza L F. 2010 State of the art on high temperature thermal energy storage for power generation. Part 1-Concepts, materials and modellization *Renewable and Sustainable Energy Reviews* **14** p 31

[7] Qin F G F, Yang X P, Ding Z, et al 2012 Thermocline stability criterions in single-tanks of molten salt thermal energy storage *Applied Energy* **97** p 816

[8] Angelini G, Lucchini A, Manzolini G 2014 Comparison of thermocline molten salt storage performances to commercial two-tank configuration *Energy Procedia* **49** p 694

[9] Zhao B C, Cheng M S, Liu C, Dai Z M 2016 Thermal performance and cost analysis of a multi-layered solid-PCM thermocline thermal energy storage for CSP tower plants *Applied Energy* **178** p 784

[10] Peng H, Dong H H, Ling X 2014 Thermal investigation of PCM-based high temperature thermal energy storage in packed bed *Energy Conversion and Management* **81** p 420

[11] Bellan S, Gonzalez-Aguilar J, Romero M, et al 2014 Numerical analysis of charging and discharging performance of a thermal energy storage system with encapsulated phase change material *Applied Thermal Engineering* **71** p 481

[12] Zhen Y, Garimella S V 2010 Thermal analysis of solar thermal energy storage in a molten-salt thermocline *Solar Energy* **84** p 974

[13] Yang X P, Yang X X, Ding J, et al 2012 Criteria for performance improvement of a molten salt thermocline storage system *Applied Thermal Engineering* **48** p 24

[14] Delpch S, Cabet C, Slim C, Picard G S 2010 Molten fluorides for nuclear applications *Materials Today* **13** p 34

[15] Zong G Q, Chen B, Zhang L, Sun J H, Dong Q A, Chen W, Xiao J C 2014 Preparation of FLiNaK molten salt *Nuclear Techniques* **37** p 1

[16] Schumann T 1929 Heat transfer: a liquid flowing through a porous prism *Franklin Inst* **208** p 405

[17] Cascetta M, Cau G, Puddu P, et al 2014 Numerical investigation of a packed bed thermal energy storage system with different heat transfer fluids *Energy Procedia* **45** p 598

[18] Yang X P, Yang X X, Ding J, et al 2012 The Thermal Analysis of Porous Medium for High-Temperature Thermal Storage *Journal of Engineering Thermophysics* **33** p 477

[19] Sohal M S, Ebner M A, Saharwall P, et al 2010 Engineering database of liquid salt thermophysical and thermochemical properties R INL/EXT-10-18297

[20] Van Der Meer J P M, Konings R J M 2007 Thermal and physical properties of molten fluorides for nuclear applications *Journal of Nuclear Materials* **360** p 16

[21] Knut T, øye H A 1980 Viscosity of the Eutectic LiF-NaF-KF Melt (FLINAK) *American Chemical Society* **25** p 17

[22] Beneš O, Konings R J M 2009 Thermodynamic properties and phase diagrams of fluoride salts for nuclear applications *Journal of Fluorine Chemistry* **130** p 22

Research on solving the optimal sizing and siting of distributed generation

Bo Wang

Shandong Appraisal Centre for Environment and Engineering, Jinan, 250012, China

E-mail: eia001@126.com

Abstract. In this paper, a distributed network planning model is proposed with the goal of minimizing the sum of distributed power investment cost, network loss and interruption cost .In order to compare the performance of differential evolution algorithm (DE) and genetic algorithm (GA) in solving the optimal sizing and siting of distributed generation in distribution networks, the two algorithms were adopted to optimize the capacities and positions of DGs. Through analysis on a 10-bus test system, the study results show that the proposed model and algorithm can get reasonable planning scheme. And in solving simple optimization problems, both GA and DE Algorithms can get good results, but compare to DE, GA is of slow convergence speed and the convergence process is not quite stable.

1. Introduction

The optimal access problems of the distributed power, as a tool to promote the development of intelligent power grid, are gaining more and more attention from experts and scholars [1-2]. Distributed generation(DG)，a supplement to centralized generation, mainly includes wind power and solar power, small hydropower and miniature gas turbine, etc[3-4] .The combination of distributed generation and centralized generation can leverage their advantages .In fact, if DG is properly placed in a distribution network, it can reduce distributed power investments power losses and interruption cost for network enforcing. But, if it is not correctly applied, it can cause degradation of power quality and reliability , increase in system losses and costs [5] .The problem of DG sizing and siting is of great importance ,for that reason, optimization methods capable of indicating the best solution for a given distribution network have been extensively studied[6-7].

So far, many methods for solving the problem of DG optimization exist in literature. Examples of those methods include Lagrange multiplier [8], genetic algorithm [9], the improved differential evolution algorithm [10], tabu search algorithm [11] and etc. Also, definitions of optimization vary among authors' opinions. The objectives can be the minimization of the system operation cost [8], maximization of the benefit/cost ratio [9], or minimization of network loss [11].

Differential evolution algorithm (DE) and genetic algorithm (GA) are presented in this paper as the optimization techniques for the sizing and siting of DG in distribution network .In order to reduce cost in distribution network, distribution network planning model based on distributed power investment cost, network loss and the interruption cost is developed. Section 4 presents a brief introduction about load flow calculation issues. At last, an IEEE 10-bus application system is used to test our model, and the study results show that the proposed model and algorithms can get reasonable planning scheme, and the advantages and disadvantages of the two algorithms are discussed.

2. Genetic algorithm in DG siting and sizing problem

GA is a search heuristic that mimics the process of natural evolution. In a genetic algorithm, to an optimization problem, a population of strings, which encode candidate solutions, evolves toward better solutions. The evolution usually starts from a population of randomly generated individuals and happens in generations. In each generation, the fitness of every individual in the population is evaluated, multiple individuals are stochastically selected from the current population based on their fitness, and modified to form a new population. The new population is then used in the next iteration of the algorithm. This generational process is repeated until a termination condition has been reached .Then a satisfactory solution has been reached.GA has the features of implied parallelism, global searching capability, simple encoding and implementation, and is not limited by constraint conditions .it is commonly used to solve an optimal problem [12-16]. The flow chart of GA to solve the DG siting and sizing problem is given in figure 1.[17-19]

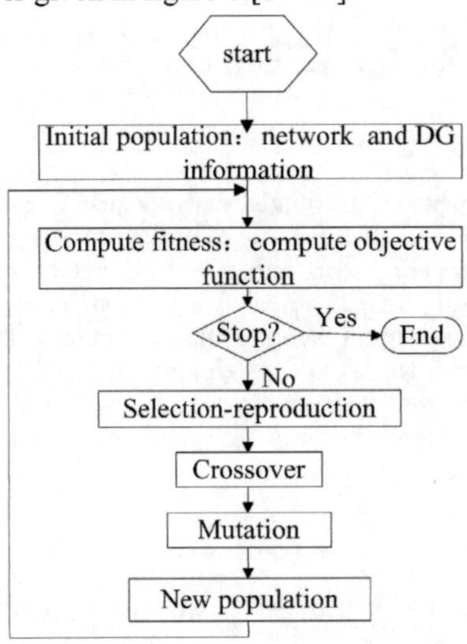

Figure1. Proposed Flow Chart

3. Differential evolution algorithm in DG siting and sizing problem

DE is also a search heuristic that mimics the process of natural evolution. It is a method that optimizes a problem by iteratively trying to improve a candidate solution with regard to a given measure of quality .Also ,DE generates new points that are perturbations of exiting points ,but these deviation are neither reflections ,nor samples from a predefined probability density function , Instead ,DE perturbs vectors with the scaled difference of two randomly selected population vectors. To produce the trial vector, DE adds the scaled, random vector difference to a third randomly selected population vector. In the selection stage, the trial vector competes against the population vector of the same index .the select-and-save step in which the vector with the lower objective function value is marked as a member of the next generation.DE works like GA[20].

DE can be used for problems whose objective functions are non-continuous, non-linear, multi-dimensional, non-differentiable, noisy or flat, etc. Or DE can be used to find approximate solutions to such problems. It has been found that DE is more accurate and efficient than controlled random search and another GA in some problems.

4. Distribution network planning model

The objective function was developed for the DG sizing and siting in distribution network is to decrease the costs of DG investment, the losses of network and the costs of interruption [21-24].

$$\min f = \sum_{i=1}^{n_{DG}} (K_{DGi} C_{DG} S_{DGi} + C_r E_{DGi} r) + C_{pu} \Delta P T_{max} + C_Q \sum_{i=0}^{n_{DG}} \sigma_i T_i \left(P_{loadi} - S_{DGi}\right) \quad (1)$$

Where K_{DGi} presents coefficient of the ith yearly DG investment cost, C_{DG} presents yearly DG investment cost ($/KW), C_r presents yearly DG operating cost ($/KW), S_{DGi} stands for the capacity limit of DG i (MW), E_{DGi} stands for power generated from DG i in a year (MWh), r stands for the discount rate, C_{pu} stands for electricity market price ($/MWh), $\triangle P$ is the network losses ($), T_{max} is maximum load loss hours, C_Q is Economic loss because of unit power lack ($/KWh), σ_i is weight coefficient of the load i, T_i is the island running time of the load, P_{loadi} is the load of node i in network, f is the objective function ($).

The optimization of above objective function is done by considering the following limitations:

(1) The line power flows

$$\begin{cases} P_{DG,j} - P_{L,j} = U_j \sum_{i\in J} U_i (G_{ji} \cos\theta_{ji} + B_{ji} \sin\theta_{ji}) \\ Q_{DG,j} - Q_{L,j} = U_j \sum_{i\in J} U_i (G_{ji} \sin\theta_{ji} + B_{ji} \cos\theta_{ji}) \end{cases} \quad (2)$$

(2) The maximum flow capacity of buses in network

$$0 \le I_i \le I_i^{\max} \qquad i = 1,\ 2, 3. \ldots\ldots n_b \quad (3)$$

(3) The permissible voltage amplitude

$$U_i^{\min} \le U_i \le U_i^{\max} \qquad i = 1,\ 2, 3. \ldots\ldots n_n \quad (4)$$

(4) The permissible DG power

$$0 \le P_{DGi} \le P_{DGi}^{\max} \qquad i = 1,\ 2, 3. \ldots\ldots n_D \quad (5)$$

Where ,$P_{DG,j}$ is the injected DG active power to node j, $Q_{DG,j}$ is the injected DG reactive power to node j, $P_{L,j}$ is the injected active load to node j in network, $Q_{L,j}$ is the injected reactive load to node j in network, G_{ji} is conductivity value, B_{ji} is electricity value, θ_{ji} is voltage phase Angle, I_i^{max} is maximum acceptable flow in path I, n_b is the number of total path, U_i^{min} is minimum voltage amplitude of i in network, U_i^{max} is maximum voltage amplitude of i in network, n_n is the number of total node, P_{DGi}^{max} is maximum acceptable DG capacity, n_D is the number of candidate node.

5. Power flow calculation

According to the radial structure and ring opening operation characteristic of distribution network, backward / forward sweep method for power flower calculation is presented [25].

Let us consider a radial network, as in Figure 2.the backward/forward sweep method for the power-flow computation is an iterative method in which, at each iteration, some computational stages are performed [26]:

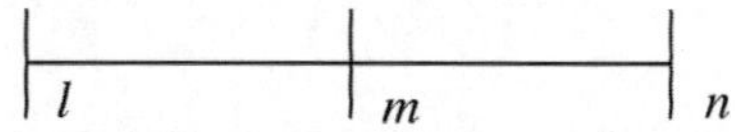

Figure 2. the wiring diagram of feeder

$$\begin{cases} \Delta P_{lm}^{(n)} = \dfrac{r_{lm}[(\sum_{p\in c_m} P_{mp}^{(n)} + P_{L,m})^2 + (\sum_{p\in c_m} Q_{mp}^{(n)} + Q_{L,m})^2]}{(U_m^{(n-1)})^2} \\ \Delta Q_{lm}^{(n)} = \dfrac{x_{lm}[(\sum_{p\in c_m} P_{mp}^{(n)} + P_{L,m})^2 + (\sum_{p\in c_m} Q_{mp}^{(n)} + Q_{L,m})^2]}{(U_m^{(n-1)})^2} \end{cases} \quad (6)$$

$$\begin{cases} P_{lm}^{(n)} = \Delta P_{lm}^{(n)} + P_{L,m} + \sum_{i\in c_i} P_{mp}^{(n)} \\ Q_{lm}^{(n)} = \Delta Q_{lm}^{(n)} + Q_{L,m} + \sum_{i\in c_i} Q_{mp}^{(n)} \end{cases} \tag{7}$$

$$I_{lm}^{(n)} = \frac{P_{lm}^{(n)} - jQ_{lm}^{(n)}}{(U_{lm}^{(n-1)})^*} \tag{8}$$

$$U_m^{(n)} = U_l^{(n)} - I_{lm}^{(n)}(r_{lm} + jx_{lm}) \tag{9}$$

Where, $\triangle P_{lm}{}^{(n)}$ is active network loss, $\triangle Q_{lm}{}^{(n)}$ is reactive network loss, $P_{L,m}$ is active load to node m in network, $Q_{L,m}$ is reactive load to node m in network, $\sum P_{mp}{}^{(n)}$ is active power flow in path *mp*, $\sum Q_{mp}{}^{(n)}$ is reactive power flow in path *mp*, $U_m{}^{(n)}$ is the node voltage, $r_{lm}+jx_{lm}$ is impedance of line *lm*, $I_{lm}{}^{(n)}$ is flow in path *lm*.

6. Example analysis

The studying distribution system in this project is IEEE 10-bus network (figure 3.). It is 10kV voltage grade for distribution network and the system total load is 7.5MW and 5.625 Mvar.

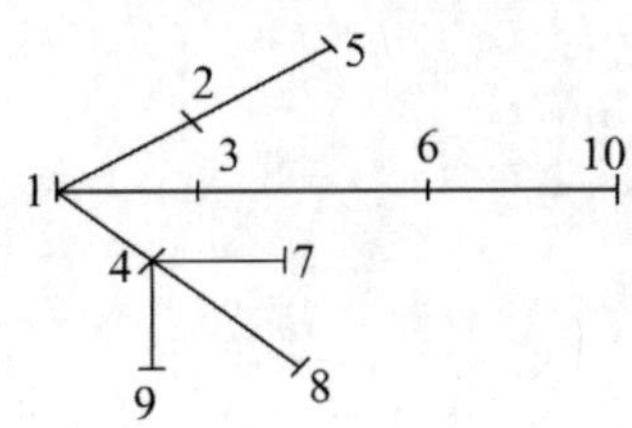

Figure 3. IEEE 10-bus system

In order to test the proposed methodology DG units have been considered, 100 KW .The number of node, to install, is 9. The maximum level of DG penetration admitted for the study is 20% of total load. The cost of Yearly DG investment has been taken as 2039.7$ /KW. The cost of Yearly DG operating has been taken as 0.005$/(KWh).Electricity market price has been assumed 0.078$/(KWh) .Economic loss has been assumed 15.7$/KWh .The period taken into consideration for the planning study is 20 years long.

Based on past experience, as well as the results of the experiment, the size of population has been assumed 50 and the maximum iteration 300.

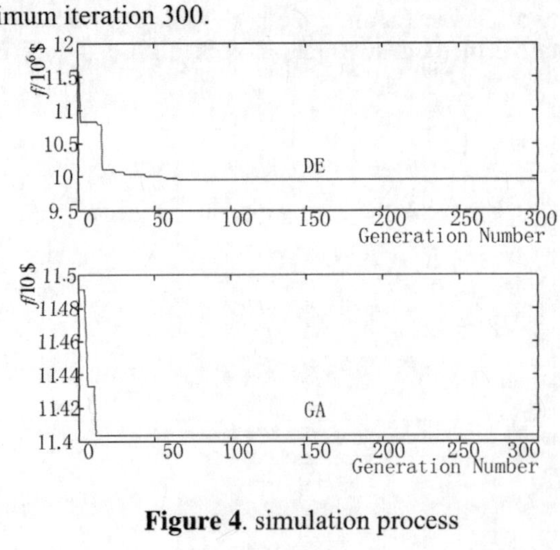

Figure 4. simulation process

Figure 4. shows iterative process of GA and DE methods to reach the optimization answer. By observing the figure, based on the network, GA and DE methods in DG siting and sizing problem in distribution network are efficient in reaching the closely optimization answer. But DE has better speed and is more stable.

Table 1. DG's capacity and position

	capacity and position of DG (kw)
Optimizing by DE Method	400(2),400(3),400(5),200(8)
Optimizing by GA Method	300(2),300(3),200(4),300(5),200(8),200(10)

Table 2. Results by using GA and AC methods

	Optimizing by DE method	Optimizing by GA method	Without DG
Investment cost of DG($)	2,517,987	2,697,843	
Network loss($)	352	347	450
Interruption cost($)	97,057,399	111,344,399	162,683,399
Total($)	99,575,740	114,042,591	

Table 1 shows the capacity and position of DG and table 2 shows the results of investment cost of DG, network loss, and interruption cost ,in this network ,by using DE and GA .It can be found that network loss and interruption cost have decreased after installing DG.

7. Conclusion

DG will play more and more important role in the electric power system. In order to let the application of DG get the best result from DG techniques, researches on an optimization method capable of indicating the best solution for a given distribution network will be extensively conducted. In this paper, distribution network planning model based on distributed power investment cost, network loss and the interruption cost is developed. GA and DE algorithms were proposed to optimize the capacities and positions of DG based on the model. By observing the figure and the table, based on the network, following results obtained:

(1) The distribution network planning model based on distributed power investment cost, Network loss and the interruption cost, in this paper, is very efficient.

(2)Both GA and DE methods in DG siting and sizing problem in distribution network are efficient in reaching the closely optimization answers.

(3)For DG siting and sizing problem , compare to GA,DE has better speed and is more stable to reach the optimization answer

References

[1] Gou J S, Qin C K, Ma F, et al. 2012 Exhaust emissions and electric energy generation in a stationary engine using blends of biogas and natural gas *Energy Education Science and Technology Part A-Energy Science and Research* **29** 193-200

[2] Zuo Y,Wang Y N, Hou Z X, et al. 2012 Simplified adaptive neural network-based maximum wind energy capture control for wind generator system *Energy Education Science and Technology Part A-Energy Science and Research* **29** 129-138

[3] Consortium for Electrical Reliability Technology Solutions "Integration of distributed energy resources the CERTS microgrid concept," April 2002. http://certs.lbl.gov/CERTS_P_DER.html.

[4] EL-KHATTAM W, SALAMA M M A 2003 Distribution system planning using distributed generation *Proceedings of IEEE Canadian Conference on Electrical and Computer Engineering Montreal (Canada)* **1** 579-582.

[5] Wallace A R, Harrison G P 2003 Planning for optimal accommodation of dispersed generation in distributed networks *17 th International conference on Electricity Distribution, CIRED* 2003, Barcelona, Spain.

[6] Zhou Y C, Zhao H M 2011 An object linking and embedding process control-based interconnection method for distributed control system and simulation system in Power station *Energy Education Science and Technology Part A-Energy Science and Research* **28** 55-62

[7] Kazancioglu F S 2012 Assessment of greenhouse production performance of Turkey compared to use of her main energy sources *Energy Education Science and Technology Part A-Energy Science and Research* **26** 851-858.

[8] William R, Nowicki E 2002 Optimal placement of distributedgeneration//14th PSCC . Sevilla, Spain : University of Sevilla

[9] Kim K H, Lee Y J, Rhee S B, et al 2002 Dispersed generator placement using fuzzy-GA in distribution systems *Proe.2002 IEEE Power Engineering Society Aummer Meeting* **3** 1148-53.

[10] Xue Xiaoqiang 2009 Optimal planning for distributed generation in distribution network based on improved differential evolution algorithm *North China Electric Power University* [in Chinese] **10**.

[11] Nara K，Hayashi Y，Ikeda K，et al. 2001 Application of tabu search to optimal placement of distributed generators *IEEE PowerEngineering Society Winter Meeting* Ohio，USA：IEEE

[12] Wang Qian, Zhang Lizi 2011 Multi-objective transmission planning associated with wind farms applying NSGA-II hybrid intelligent algorithm *Proceedings of the CSEE [in Chinese]* **31(19)** 17-31.

[13] Srinivas N, Deb K 1994 Multiobjective optimization using nondominated sorting in genetic algorithm *Evolutionary Compution* **2(3)** 221-248

[14] Akgobek O 2012 Genetic algorithms to determine the critical values of a power energy system for different operating conditions *Energy Education Science and Technology Part A-Energy Science and Research* **29** 971-982.

[15] Akgobek O A 2012 comparative study of genetic algorithm and simulated annealing for solving the operational fixed job scheduling problems *Energy Education Science and Technology Part A-Energy Science and Research* **29** 1417-30.

[16] Gandomkar M, Vakilian M, Ehsan M 2005 A combination of genetic algorithm and simulated annealing for optimal dg allocation in distribution networks CCECE IEEE, Canada, May 2005.

[17] Guvenc U, Altum BE, Duman S 2012 Optimal power flow using genetic algorithm based on similarity *Energy Education Science and Technology Part A-Energy Science and Research* **29** 1-10

[18] Akgobek O 2012 A hybrid approach for improving the accuracy of classification algorithms in data mining *Energy Education Science and Technology Part A-Energy Science and Research* **29** 1039-54

[19] Sulaiman M H, Aliman O, AbdulRahim S R 2010 Optimal allocation of EG in distribution system using genetic algorithm technique *Journal of Energy and Power Engineering* **4(1)** 56-63

[20] Qiu Wei, Zhang Jianhua, Liu Nian 2009 Application of differential evolution to power systems *Modern Electric Power [in Chinese]* **26 (5)** 11-16

[21] Ergonul A H, Kahraman G 2012 Computation of loss of load probability using a Markov Chain with two-day memory *Energy Education Science and Technology Part A-Energy Science and Research* **29** 1115-24

[22] Ozgur M A 2012 Technoeconomic evaluation of solar-wind hybrid power system in Kutahya, Turkey *Energy Education Science and Technology Part A-Energy Science and Research* **28**

785-798

[23] Wang Jing 2010 Study on electrical dirtribution network planning optimization based on reliability of cost-benefit analysis North China *Electric Power University* [in Chinese]

[24] Farnaz Sheidaei, Majid Shadkam, Mahdi Zarei 2008 Optimal distributed generation allocation in distirbution systems employing ant colony to reduce losses *Universities Power Engineering Conference*, UPEC, IEEE, Iran, Sept 2008

[25] Yumurtaci M, Gokmen G, Kilic O, et al. 2012 The application of instantaneous reactive power calculation methods to a real power system *Energy Education Science and Technology Part A-Energy Science and Research* **29** 1083-92

[26] Bompard E, Carpaneto E, Chicco G, Napoli R 2000 Convergence of the backward/forward sweep method for the load-flow analysis of radial distribution systems *Electrical Power and Energy Systems* **22** 521–530

Operational adaptability evaluation index system of pumped storage in UHV receiving-end grids

Bo Yuan[1, 3], **Jin Zong**[2], **Junshu Feng**[1]

[1] State Grid Energy Research Institute, Beijing, China
[2] State Grid Jibei Electric power Co., Ltd Research Institute , Beijing, China

E-mail: yuanbo@sgeri.sgcc.com.cn

Abstract. Pumped storage is an effective solution to deal with the emergency reserve shortage, renewable energy accommodating and peak-shaving problems in ultra-high voltage (UHV) transmission receiving-end grids. However, governments and public opinion in China tend to evaluate the operational effectiveness of pumped storage using annual utilization hour, which may result in unreasonable and unnecessary dispatch of pumped storage. This paper built an operational adaptability evaluation index system for pumped storage in UHV-receiving end grids from three aspects: security insurance, peak-shaving and renewable energy accommodating, which can provide a comprehensive and objective way to evaluate the operational performance of a pumped storage station.

1. Introduction

Recently, the rapid development of renewable energy and large-scale UHV DC transmission have brought significant changes to the basic structure of power grids in China. Peak-shaving, accommodating renewable energy, emergency reserve, frequency control and active power control in UHV receiving-end grids are becoming more and more serious problems. Pumped storage is an effective way to deal with these problems due to its excellent technical and economical characteristics. However, there are no official rules or evaluation systems to evaluate the actual operation of pumped storage in China. Government and public opinions tend to using annual utilization hour to see whether the operation a pumped storage has achieved its full potential. This may result in unnecessary or unreasonable dispatch of pumped storage. For instance, a station that is intended to provide emergency reserve may be frequently dispatched to generating power and would fail to provide enough reserve when needed.

In this paper, the main functions of pumped storage in UHV receiving-end grids are classified into three categories: security insurance, peak-shaving and renewable energy accommodating. An operational adaptability evaluation index system is built considering these functions, which can provide a comprehensive and objective perspective to evaluate the performance of pumped storage.

2. Main Principle and Framework of the Evaluation Index System

2.1. Main Principle

Pumped storage station is intended to serve the entire power system, which means its operational adaptability evaluation index system should also consider the characteristics of the whole power system.

Therefore, the evaluation indexes in this paper are chosen according to the following principles :

- **Scientific principle:** the evaluation index should be built under theoretical basis, which could reflect the connotations of the functions of pumped storage.
- **Feasible principle:** the evaluation index system aims to evaluate the operational performance of pumped storage, therefore the index should be feasible, especially the required data should be easy to get. Indexes that are unable to get should not be included.
- **Representative principle:** the evaluation index should be representative, which means the evaluation system should be able to reflect all the aspects of the operation of pumped storage.
- **Comparable principle:** the evaluation index should be comparable, which means it should be applicable to different pumped storage stations and the result should be comparable.

2.2. Framework

The evaluation index system consist of 4 criteria layers, including overall performance, security insurance, peak-shaving and renewable energy accommodating, and 18 evaluation factors shown as the figure below.

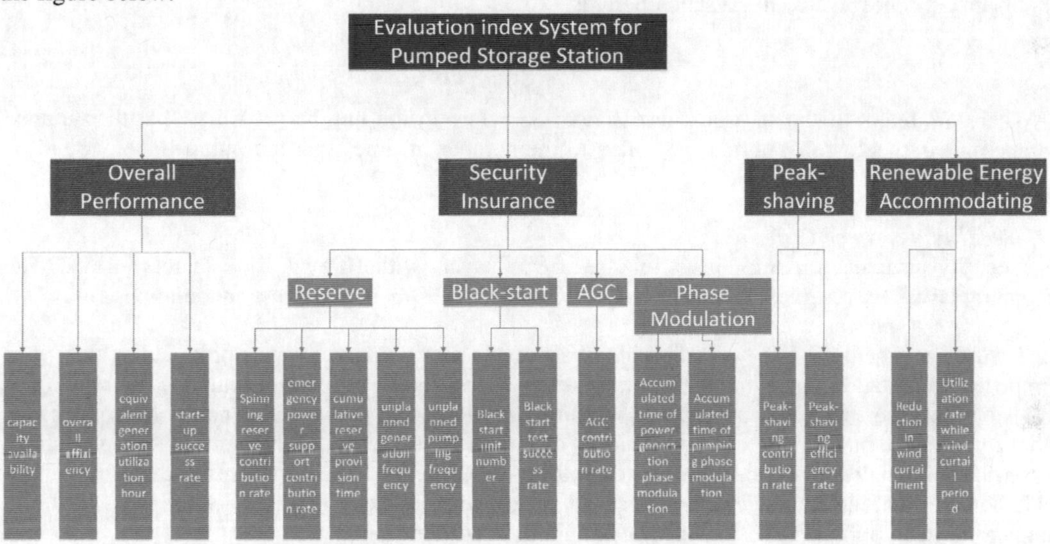

Figure 1. Framework of the evaluation index system

3. The Operational Adaptability Evaluation Index System for Pumped Storage Station

3.1. Overall Performance Criteria

The overall performance criteria layer consist of 4 evaluation factors, including capacity availability, overall efficiency, equivalent generation utilization hour and start-up success rate. The first two factors aims to evaluate the operation and maintenance status of the pumped storage station, the other two factors aims to evaluate how often and how well the station is used by the system operator.

Capacity availability is intended to evaluate the unavailable capacity caused by overhaul and malfunction. It can be calculated by

$$CF = 1 - \frac{CU \times T_U}{CR \times T}$$

Where CF refers to capacity availability; CU refers to unavailable capacity; T_U refers to time of duration of unavailable capacity; CR refers to the rated capacity; T refers to the overall time period.

Overall efficiency is intended to evaluate the losses caused by pumping and generating of a pumped storage station. It is the ratio of the electricity generated by the station to the electricity consumed by the station.

$$OR = \frac{EG}{EP}$$

Where *EG* refers to the electricity generated by the station during the evaluation time period; *EP* refers to the electricity consumed by the station to pump water during the evaluation time period.

Equivalent generation utilization hour provide an intuitional perspective of how often the station is deployed by the system operator.

$$EH = \frac{EG}{CR}$$

Where *EH* refers to the equivalent generation utilization hour.

Start-up success rate represents the operation and maintenance level of the pumped storage station, which is most concerned by the system operator.

$$SR = \frac{SS}{SS + FS}$$

Where *SR* refers to the start-up success rate; *SS* refers to the number of times that the pumped storage unit is successfully started; *FS* refers to the number of times that the pumped storage unit is fail to start.

3.2. *Security Insurance Criteria*

The security insurance criteria consist of 4 secondary layers with 10 evaluation factors in total. The secondary layers are reserve, black-start, AGC frequency adjustment and phase modulation.

3.2.1. Reserve. There are five evaluation indexes to evaluate the reserve service provided by the pumped storage station, including the spinning reserve contribution rate, the cumulative reserve provision time, the emergency power support contribution rate, the unplanned generation frequency and the unplanned pumping frequency. The first one assesses the ability of the pumped storage station to provide spinning reserve capacity and the contribution to the reduction of the reserve capacity of the grid; the latter four indicators evaluate the actual performance of the reserve service provided by the pumped storage station from the standpoint of spinning reserve.

Spinning reserve contribution rate is used to measure the proportion of reserve capacity provided by the pumped storage station to the reserve capacity required by the whole system, which can be calculated by:

$$\alpha = \frac{\int_T (P_R - p(t))dt}{\int_T R(t)dt}$$

Where α is spinning reserve contribution rate; T is the evaluation time period; P_R is the rated power of pumped storage; R is the spinning reserve of the whole system.

The emergency power support contribution rate represents the ratio of the emergency active power the station can provide to the emergency power gap when a large capacity loss occurs.

The cumulative reserve provision time represents the actual cumulative time that the pumped storage provided spinning reserve, which can be drawn from the historical operation statistic data.

The unplanned generation frequency and the unplanned pumping frequency indicate unplanned unit start-up due to emergencies during the evaluation period and are derived from historical operation statistic data.

3.2.2. Black-start. The black start function of the pumped storage power station needs to be certified by the grid dispatching authority in the region. The black start performance of the power plant is evaluated from the following two aspects:

- Black start unit number, that is, number of units certified by the dispatching authority.
- Black start test success rate, that is, the success rate of each black-start unit through the black start test.

3.2.3. AGC frequency adjustment. The AGC function of the pumped storage power station is evaluated by the AGC contribution rate index. AGC contribution rate represents the AGC power consumption of all units in the power station that accounting for the whole network AGC theoretical requirement for the corresponding time, which could evaluate the actual AGC performance of the power station. It can be calculated by:

$$\rho = \frac{\int_{t_1}^{t_2} p_1(\Delta f, t)dt}{\int_{t_1}^{t_2} p_{system}(\Delta f, t)dt}$$

Where ρ is the AGC contribution rate; t_1-t_2 is the AGC responding time, which is usually 60 seconds; $p_1(\Delta f, t)$ is the AGC characteristic of the pumped storage, and $p_{system}(\Delta f, t)$ is the AGC characteristic of the whole system.

3.2.4. Phase Modulation. The evaluation index of phase modulation function of pumped storage power station includes the accumulated time of power generation phase modulation and the accumulated time of pumping phase modulation.

Accumulated time of power generation phase modulation = power generation phase modulation cumulative hours during the evaluation period.

Accumulated time of pumping phase modulation = pumping phase modulation cumulative hours during the evaluation period.

3.3. Peak-shaving Criteria

This paper investigates the peak-shaving criteria of pumped storage power station using two evaluation factors, peak-shaving contribution rate and peak-shaving efficiency rate. The first one is to evaluate the inherent peak-shaving capability of pumped storage; the second evaluation factor is to assess the actual performance of the pumped storage during operation period.

Peak-shaving contribution rate represents how much the construction of pumped storage power station can relieve the peak-shave pressure of the system. It is the ratio of the double installed capacity of the pumped storage to the system peak-valley difference:

$$PC = \frac{2 \times CR}{PV}$$

Where *PC* is the peak-shaving contribution rate; *PV* is the system peak-valley difference.

Peak-shaving efficiency rate is used to evaluate the performance of the pumped storage for peak-shaving during the evaluation period. It is expressed as the absolute value of the correlation coefficient between the power output curve and the load curve:

$$\beta = \left| \frac{\sum (P - \overline{P})(L - \overline{L})}{\sqrt{\sum (P - \overline{P})^2 \sum (L - \overline{L})^2}} \right|$$

where β is the peak-shaving efficiency rate; P is the output curve of the pumped storage; $\overline{P}$ is the average power output of the pumped storage; L is the load curve; $\overline{L}$ is the average load of the load curve.

3.4. *Renewable Energy Accommodating Criteria*

When evaluating the performance of a pumped storage to accommodate renewable energy, this paper uses two evaluation factors, which are the reduction in wind curtailment, and utilization rate while wind curtail period.

Reduction in wind curtailment refers to the reduced amount of wind curtailment due to the construction of the pumped storage. It can be represented by the electricity used to pump water during wind curtailment period.

Utilization rate while wind curtail period represents whether the pumped storage is fully utilized during wind curtailment period. It can be calculated by:

$$\eta = \frac{\sum_{T'} P_{T'}}{\sum_{T'} P_R}$$

Where η is the utilization rate while wind curtail period; T' is wind curtailment period; $P_{T'}$ is the pumping output during wind curtail period; P_R is the rated pumping rate of the pumped storage station.

4. Conclusion

This paper proposed an operational adaptability evaluation index system for pumped storage station in UHV receiving-end grids. The evaluation index system consist of 4 criteria layers, including overall performance, security insurance, peak-shaving and renewable energy accommodating, and 18 evaluation factors, which could comprehensively and scientifically assess the operation of the pumped storage and guide its operation and management.

References

[1] Kusiak A, Zhang Z 2010 Short-Horizon Prediction of Wind Power: A Data-Driven Approach *IEEE Transactions on Energy Conversion* **25(4)** 1112-22.

[2] Heitsch Holger, Romisch Werner 2007 A note on scenario re-duction for two-stage stochastic programs *Operations Research Letters* **35(6)** 731-738.

[3] Pappala V S, Erlich I, Rohrig K 2009 A Stochastic Model for the Optimal Operation of a Wind-Thermal Power System *IEEE Transactions on Power Systems* **24(2)** 940-950.

[4] Po-Hung Chen 2008 Pumped-Storage Scheduling Using Evolu-tionary Particle Swarm Optimization *IEEE Transactions on Energy Conversion* **23(1)** 294-301.

[5] Yuan Xiaohui, Nie Hao, Su Anjun 2009 An improved binary particle swarm optimization for unit commitment prob-lem *Expert Systems with Applications* **36(4)** 8049-55.

[6] Ge Weixun 2010 Ramp Rate Constrained Unit Commitment by Improved Priority List and Enhanced Particle Swarm Optimization *2010 International Conference on Computational Intelligence and Software Engineering (CISE)* 1-8.

[7] Chen Xing 2007 Power system steady state analysis (Third Edi-tion)[M]. Beijing: China Electric Power Press.

[8] Xia Shu 2010 Differential evolution algorithm for solving dy-namic economic dispatch considering wind power pene-tration[D]. North China Power Electric University (Bei-jing).

[9] Fan Mingtian, Zhang Zuping, Yang Shaoyong 2000 Economic dispatch of multi regional power network involving pump storage plants *Power System Technology* **24(8)** 57-61

[10] State Grid Corporation of China 2010 Specification of dis-patching and operating management

for wind power [S].

[11] Pan Wenxia, Fan Yongwei, Zhu Li 2008 The optimal sizing for pumped storage system in wind farm *Transactions of China Electrotechnical Society* **23(3)** 120-124

[12] Liu Deyou, Tan Zhizhong, Wang Feng 2007 Study on com-bined wind power with pump storage hydropower sys-tem *Shanghai Electric Power* **1** 39-42.

[13] Garcia-Gonzalez J, de la Muela R M R, Santos L M 2008 Stochastic Joint Optimization of Wind Generation and Pumped-Storage Units in an Electricity Market *IEEE Transactions on Power Systems* **23(2)** 460-468.

[14] Sun Yuanzhang, Wu Jun, Li Guojie 2009 Dynamic economic dispatch considering wind power penetration based on wind speed forecasting and stochastic programming *Proceedings of the CSEE* **29(4)** 41-47.

[15] Zhou Wei, Peng Yu, Sun Hui 2009 Dynamic economic dis-patch in wind power integrated system *Proceedings of the CSEE* **29(25)** 13-18.

[16] Yuan Tiejiang, Chao Qin, Li Yiyan 2010 Short-term wind power output forecasting model for economic dispatch of power system incorporating large-scale wind farm *Pro-ceedings of the CSEE* **30(13)** 23-27.

[17] State Grid Corporation of China 2010 Specification of wind power forecasting system[S].

[18] Zhang Boming, Wu Wenchuan, Zheng Taiyi 2011 Design of a multi-time scale coordinated active power dispatching system for accommodating large scale wind power pene-tration *Automation of Electric Power Systems* **35(01)** 1-6.

Research on renewable energy power generation complementarity and storage distribution model

Xiaoxia Wei[1], Jinfang Zhang

State Grid Energy Research Institute, Future Science and Technology Park North Area, Beiqijia, Changping, Beijing, 102209, P.R. China

E-mail: 20583853@qq.com

Abstract. This paper mainly studied the equivalent conversion relationships and model of different "quality "energies in process of multi-energy conversion. In energy interconnection system containing wind turbine, photovoltaic cell and energy storage systems, it gives renewable energy and storage distribution development model, considering comprehensive effect of load demand characteristics on energy utilization mode, multi-objective optimization model is established with objectives of both maximized energy utilization ratio and minimized system operation costs. Then, take Chinese one certain area as scenario, and give out "renewable energy utilization", "energy transfer" and "total operating cost" three different analyses, according to the connection model. The result is compared with that for traditional energy utilization model. Feasibility of the proposed model is verified with simulation results.

1. Introduction

Renewable clean energy has become an important part of the current energy structure, and also an important direction of energy development. As a typical representative of renewable energy, wind and solar energy has been large-scale development and wide application. Because of the energy transmission limitation, "energy" has become main form of use. The renewable energy is vulnerable to external environmental factors, showing a strong gap and volatility, large-scale power grid by safe operation of the power system has a significant negative impact. Many studies have proposed the use of micro-grid in the form of a unified integration of various types of energy and load to form independent and interconnected energy systems to reduce the independent operation of the energy supply system to adverse effects, improve the effective utilization of energy by the system [1].

While the single energy utilization mode and the relatively independent operation mode between energy supply systems will result in the problem of insufficient energy consumption and low utilization rate in the system, resulting in excess of renewable energy and not giving full play to the maximum comprehensive benefits of energy. So this research focused on the characteristics of renewable energy interconnection and the interrelationship of energy storage cooperation, and analysed the problems encountered in the process of cooperation, and proposed a model to improve energy efficiency [2].

Solar and wind energy are pollution-free and environment-friendly. However, due to their intermittent and fluctuating characteristics, their utilization and efficiency are greatly limited. In order to improve the utilization efficiency of solar energy, wind energy and other new energy sources, wind and solar power generation system, wind storage system, light storage power generation system, wind solar storage generation system gradually attract people's attention [3].

2. Renewable energy interconnection characteristics

Wind and solar hybrid power systems are generally used in rural and mountainous areas far away from large power grids. Wind and light power supply systems can hardly provide continuous and stable active power output due to the randomness and fluctuation of wind and solar power. Wind power generation system is using the natural complementarity of wind and solar energy, such as sufficient solar energy in the summer, sufficient wind energy in winter, can greatly reduce the amount of battery in the system, thus improving the system reliability and economy.

Wind-solar hybrid power generation and energy storage system is used to store and release electric energy, which can improve the power output characteristic of the whole wind power generation system, reduce the intermittence and randomness of wind power and photovoltaic power generation and power system requirement real-time balance. Reduce its negative impact on the power grid [4].

From the energy conversion, transmission and utilization process, the transformation and transmission of renewable energy and energy storage, are "energy" form. According to the different demand forms of energy on the load side, the energy needs to be converted into multiple energy forms to match the energy demand of the load side, showed in Figure 1. It can be seen that the equivalent transformation between different "energy" can unite the originally independent energy supply systems and finally form a multi-energy interconnection sharing system, showed in Figure 1. This paper will focus on the process of secondary equivalent conversion between different energies, which is the primary transformation energy model [5,6].

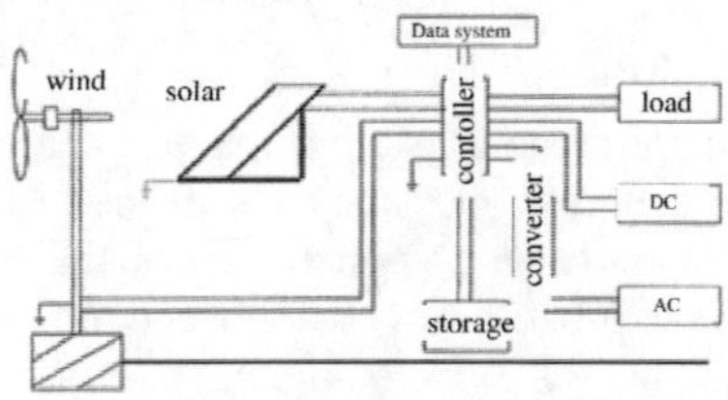

Figure 1. Wind, solar and storage connection system

3. Renewable energy and storage distribution development model

To energy demand, energy demand is treated as a random variable which does not assume any probability distribution. The energy supply, data multi-channel energy centre includes smart grids, storage devices (such as batteries), and new energy sources such as wind energy. The system model is a discrete-time model. From Figure 2, Energy supply and demand model: This paper assumes that energy demand and new energy production are independent of each other a random process[7,8].

At time t, energy demand is D(t), renewable energy production is R(t). The net energy supply amount is $\Delta(t) = R(t) - D(t)$. The excess renewable energy will be stored in the battery until the battery is fully charged; while new energy can not meet the energy demand will be provided by the grid power and battery. G(t) is electricity purchased from the grid in order to meet demand, and satisfy the upper and lower boundary $0 \ll G(t) \ll G_{max}$.

The storage battery model and its parameters include energy storage capacity, rated power, and efficiency. B_{max} is storage capacity refers, which refers to the battery can store the maximum energy.

The rated power refers to the battery charge and discharge during each charge and discharge the maximum charge, using B^c_{max} and B^d_{max} . Battery efficiency includes charge efficiency, discharge efficiency and charge-discharge cycle efficiency. The charging efficiency $\eta_c \in (0,1)$ is the total amount of charging into the battery and the total input energy value. Discharge efficiency $\eta_d \in (0,1)$ is the effective discharge of the battery and the total energy released by the battery quotient. Battery charge and discharge cycle efficiency of the charge efficiency multiplied by the discharge efficiency, $\eta = \eta_c \eta_d$. Although the efficiency of battery charging and discharging will decrease with increasing use time in actual battery management, efficiency will not change much over a certain period of time. Assuming that the battery charge and discharge efficiency is constant, at time t, the battery charging amount is $B(t)$, charging is $B_c(t)$, discharging is $B_d(t)$.

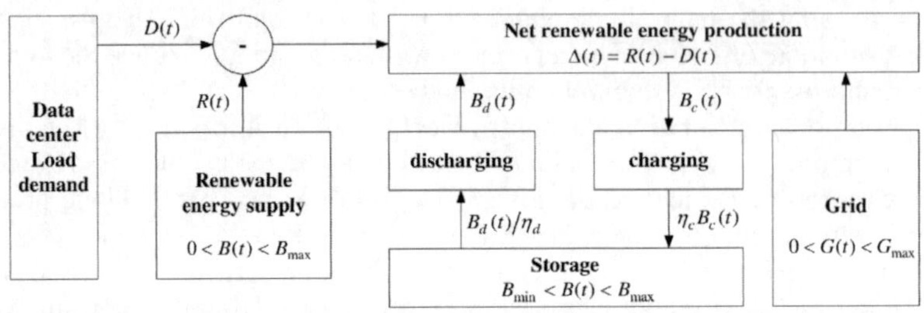

Figure 2. Renewable energy and storage distribution development model

In order to avoid battery over-discharging, the battery must at least reserve energy at each moment to back up power as a data centre. So, the battery power at each moment must be met $0 < B(t) < B_{\max}$. Since the charge and discharge have energy loss, the battery will only charge or discharge at each moment, that is $t = 0$, the dynamic update process of the battery power is:

$$B(t+1) = B(t) + \eta_c B_c(t)\, B_d(t)/\eta_d \tag{1}$$

Among that, $0 \le B_c(t) \le B^c_{\max}$, $0 \le B_d(t) \le B^d_{\max}$.

The use of renewable energy is vulnerable to the impact of the external environment, the energy output has a great deal of passive, generally can not be with the load demand. In the energy transfer process, energy will be "storage" and "release" two stages of the secondary use. Among them, the energy transfer form of energy storage process can be described as:

$$\begin{cases} B(t+1) = B(t)(1-\delta_e) + P_{bt}\eta_c \Delta t \\ B(t+1) = B(t)(1-\delta_e) - P_{bt}\Delta t/\eta_d \end{cases} \tag{2}$$

$B(t)$ is the charging capacity of the energy storage system at time t. $B(t+1)$ is the charging capacity of the energy storage system at time t+1. δ_e is self - discharge rate of power storage system. P_{bt} is the operating power of the power storage system. Positive value is the charging process; the negative value is discharge process. Δt is the gap between t and t+1.

4. The optimal algorithm of investment and operation cost for the cooperative system of renewable energy and storage

The objective function of the optimization design is to reduce the investment and operating costs while satisfying the performance indexes. The reliability cost and the environmental treatment cost can be added in order to minimize the total cost of the system[9,10]. After considering the investment and operating costs and environmental management costs according to the target function, this model can be described as:

$$C = \sum_{i=1}^{2}\sum_{j=1}^{M_j} C_{ij}P_j + \sum_{i=1}^{2} u(P_i) + k_{coe}P_{eens} + e_t\left(\sum_{t=1}^{8760} P_g P_{gridt} + \sum_{k=1}^{n} \beta_k \alpha_k Q_{gridk}\right) \tag{3}$$

Among them, i is the different kinds of energy, i=1,2, which refers to wind and solar. M_j、C_i、P_i refers to number, cost and output power. $u(P_i)$ is operation cost, k_{coe} is compensation factor. P_{eens} is less than system power perdition. $e_i = 0,1$ refer to isolated power system operation and connection to the power line. P_g is the parallel line price. P_{gridt} the grid power supple capacity. α_k、β_k、Q_{gridk} refer to pollution emission factors, pollutant treatment charges and its emission amount.

Considering the boundary condition, power balance limit refers to the wind power; solar power system output power coupled with the power output of the power grid is equal to the user's electricity load. That is:

$$P_{DG} + P_{grid} = P_L \tag{4}$$

Among them, P_L is total power load.

The power output can not larger than the maximum transmission capacity. And at the same time, wind and solar power needs capacity boundary. That is:

$$\begin{cases} P_{grid} \le P_{g\max} \\ P_{DG} \le P_{DG\max} \end{cases} \tag{5}$$

Among them, $P_{g\max}$, $P_{DG\max}$ refer to maximum transmission capacity and maximum distributed power capacity. For the solar and wind power system has certain fault. So the system must have a certain degree of power supply reliability. The out-of-load probability is the ratio of the terminal time to the total supply time in the statistic time, which is usually less than 2%.

5. Analysis of the result of the cooperation of renewable energy and energy storage

Take Chinese one special area as example, to the wind power, photovoltaic power generation system, CCHP system and various types of energy storage system based energy system, the maximum power output of wind power system is 200 kW; the maximum power output of PV system is 200 kW; the maximum power output of CCHP is 200 kW; the maximum cooling power of 300kW; battery maximum charge and discharge power of 100kW; storage system maximum operating power of 200kW.

The mode 1: Considering the CCHP continuous supply system adopts the coupling operation mode of "heat-set" to meet the cooling load demand in the system, the demand of the power load is supplied by other distributed power supply.

The mode2: Renewable energy quota energy storage, the provision of electrical energy supply.

The three operating modes are compared: "renewable energy utilization", "energy transfer" and "total operating cost".

5.1 Renewable energy utilization

In mode 1, renewable energy will be converted into electricity as the form of a single use, so in this mode of operation, the excess energy can not be forced to abandon, resulting in short of energy use. Since the wind and solar energy are renewable energy sources, and the system operation and maintenance costs are low, are 0.11 yuan / (kW · h) and 0.08 yuan / (kW · h) respectively. The renewable energy operating cost of mode 1is 229.89 yuan. Compared with the operating mode 1, the utilization of renewable energy in the system can be increased by 11.5%, and the running cost of the system is increased to 264.45 yuan.

In mode 1, the cooling load demand of the system is all supplied by the CCHP co-supply system in the coupled operation mode of "heat-set", and the same amount of cooling load demand is satisfied with respect to the mode 2. The cost of CCHP energy loss is 2.03 yuan. Resulting in higher operating costs of the system, which will increase the operating cost of 425.38 yuan, the mode 2 operating cost of the system can be reduced by 390.82 yuan , showed in Table 1.

Table 1. Renewable energy utilization type under different modes

	Energy supply(kW·h)		Utilization (kW·h)		Utilization(%)
	solar	wind	Supply	cooling	
Mode1	972.54	1782.73	2355.19	-	85.48
Mode2			2355.19	314.14	96.88

5.2. Energy transfer

In mode 2, with the energy equivalent substitution relationship, according to the load demand system, use the different type of energy to storage, and form into the energy transfer. In the TOU price under the guidance of the flexibility to change their operating conditions, and priority to participate in scheduling, the valley of the energy transfer to the peak hours to use. The renewable energy efficiency will be increased by 3.11%. In order to satisfy the same amount of load demand, the mode 1 is mainly depending on CCHP system (energy loss is 2608.71kW·h), and needing to buy extra power at high price (the total energy amount is 27.5kW·h), which can make the operation cost as 550.43yuan, showed in Table2.

Table 2. Energy transfer structure

Transfer		Energy structure	
Power (kW·h)	Cooling(kW·h)	Renewable energy (kW·h)	Extra grid power (kW·h)
800	1620	85.95	1555.65

5.3. Total operating cost

Mode 1 is short of renewable energy utilization, operating costs are relatively low, showed in Table 3. To the CCHP supply, mode 1 adopts heat decision of electricity, with higher energy consumption and operating costs. In the energy price guidance of mode 2, form the timing of the transfer, and the operation of the system to add a certain operating costs. Mode 1 will be surplus of electricity sent to large power grids, and achieved a certain income, but can not offset the high cost of the system inputs.

Table 3. Costs under two modes

project		Operation cost (yuan)	
		Mode 1	Mode2
wind		152.09	196.10
Photovoltaic		77.80	77.80
CCHP		1676.81	358.54
Energy transfer	Power	-	32
	Cooling	-	32.40
Power connection	Extra sending	693.92	-
	buying	13.92	264.46
synthesize cost		1226.69	961.30

6. Conclusion

This paper analysed different energy equivalent transmission relation. Construct renewable energy and storage distribution development model. And realize the multi-energy connection. Through the comparison with traditional modes, this paper verified the model reasonable and advantage. The main conclusions are:

- Through the energy equivalent, it can enhance system utilization to the renewable energy, and reduce the system operation cost.
- Realize the adjustment jointed, can enhance the renewable energy utilization and reduce the energy exchange of different energy, can increase system operation economic efficiency.
- Through the energy in different forms of energy in the timing of the transfer, to a certain extent, reduce the operating costs of the system.

The example gives three "renewable energy utilization", "energy transfer" and "total operating cost" characteristics, compared with traditional energy utilization model, which can showed that the cooperation of renewable and storage has better efficiency.

Reference

[1] Ji Y,Ai Q and Xie D 2010 P*ower System Technology* vol **34** 15-23
[2] Zhu X, Han X Q and Qin W P 2015 *Renewable and Sustainable Energy Reviews* vol **42** 453-463
[3] Zhang H Y, Li S D 2011 *Research on micro grid.* Beijing, China:Tsinghua University 595-598
[4] Wu X,Yin X G and Song X 2013 *Energy and Power Engineering* **4** 142-149
[5] Xu J,Sui J and Li B *Energy* **35** 4361-67
[6] Cho H, Smith A D and Mago P 2014 *Applied Energy* **136** 168-185
[7] Luo F, Wei G, Xu X and Lu Y 2016 *Journal of Computational and Theoretical Nanoscience* vol **13** 1933-38
[8] Zhang D and Wang J 2016 *Dianwang Jishu/Power System Technology* **40** 451-458
[9] Ming W, Jian S, Liu H, Yang L and Yu J 2014 *POWERCON 2014 - 2014 International Conference on Power System Technology: Towards Green, Efficient and Smart Power System, Proceedings* 3361-65
[10] Zeng Z, Zhao R, Yang H and Tang S 2014 *Renewable and Sustainable Energy Reviews* 701-718

International Conference on Energy Engineering and Environmental Protection (EEEP2016) IOP Publishing
IOP Conf. Series: Earth and Environmental Science **52** (2017) 012043 doi:10.1088/1755-1315/52/1/012043

Research on the function orientation of pumped-storage plant in China

Jun-shu Feng[1,3], Yuan Bo[1], Sheng-yu Wu[1], Fu-qiang Zhang[1] and Wei-hua Hao[2]

[1] State Grid Energy Research Institute, Future Science and Technology Park, Beijing 102209, China

[2] State Grid Shandong Electric Power Corporation Maintenance Company, Jiaodong Converter Station, Shandong 266329, China

E-mail: fengjunshu@sgeri.sgcc.com.cn

Abstract. With the growth of electrical load and development of clean energy, peak regulation of power system has been paid more attention to, in which case pumped-storage plant becomes the most efficient and economical way against peak load and the ensurance of safe and stable system operation. However, the dispatch and operation mode of pumped-storage plant is a little unreasonable at present in our country, due to the insufficient acknowledge of plants' function orientation in different regions. This paper studied the standard of classifying pumped-storage plants, by dividing the plant function into three types security oriented, peak shaving oriented and renewable energy accommodation oriented, based on the local load characteristic, structure of power supply, accommodating capability of renewable energy, nuclear energy, etc. Finally, according to the current and future development situation of five Chinese regions, the function orientation partition of pumped-storage plant is given, which has profound guiding and practical significance for pumped-storage plant dispatch and operation.

1. Introduction

In recent years, power system regulating problem caused by growth of power load and development of clean energy has been increasingly apparent. And limited by resource endowment and no key break-through of most storage technologies, as shown in [1-2], pumped-storage plant will still be the most effective and economic tools in the present and the future to ensure operation safety and stability of power system in China.

Current pumped-storage plants in China lacks of unified rules, and each regions' dispatching institutions about pumped-storage is pretty diverse, which leads to a big difference of actual pumped-storage utilization and its design. It's largely due to the lack of understanding of regional pumped-storage plants' function orientation. In fact, with the rapid energy industry development of China and the huge adjustment of energy structure, power system need to be more safe and reliable, the function orientation of pumped-storage plant has been given new meanings, which are presented in [3-5].

In this paper, standards of classifying pumped-storage plants have been studied, by dividing the plant function into three types security oriented, peak shaving oriented and renewable energy accommodation oriented, based on the local load characteristic, structure of power supply, accommodating capability of renewable energy, nuclear energy, etc.

2. Function orientation of pumped-storage plant

2.1. Overview of pumped-storage plants' function

Pumped-storage plant has two main function characteristics. The first is to converse flexibly between power generation and power consumption, which can be used in power peak cutting and valley filling. Second is start-up quickly and response to the change of the electric load quickly, thus can be used in grid frequency control, phase control, emergency reserve and black-start, also shown in [6-9]. Overall, ordinary function of pumped-storage plants includes peak cutting and valley filling, frequency control, emergency reserve, regulating support for new energy and nuclear power, phase control, black start power supply, etc.

As shown in table 1 which are presented in [10], contrary to coal-fired and gas-fired plants, pumped-storage plant has obvious advantages on regulating ability, start-up rate, rich functions, and so on.

Table 1. Comparison of operation characteristics of different power plants.

		Pumped-storage plant	Coal-fired plant	Gas-fired plant
Unit commitment		peak load	basic load	peak load
Regulating capacity (%)		200	30~50	100
Start-up rate	Start-up to full-load	2~3 min	6~8 h	45 min
	No-load to full-load	30~35 s	2~3% capacity/min	6~8 min
Ramping rate (capacity/min)		50~100%	2~3%	12~15%
Valley filling		√	×	×
Frequency control		√	√	√
Phase control		√	√	√
Spinning reserve		√	√	√
Black start power supply		√	×	√

2.2. Classification of function orientation

Although pumped-storage has various functions, however, because of regional grid differences about source structure, load characteristic, balance between power supply and demand, and the orientation of pumped-storage plants in different region cannot be the same, and should be considered respectively.

Considering the current electric power development in China, the functional requirements for the pumped-storage by power system are mainly three types security oriented, peak shaving oriented and renewable energy accommodation oriented.

2.2.1. Security oriented. It refers to the functions of emergency reserve, black start, frequency control, phase control, which can handle various emergencies appearing in the power system. And this kind of plants are mainly located in the high power reliability requirements of important economic and political centres, or the grid key hub.

2.2.2. Peak shaving oriented. This kind of plants are mainly distributed in the system of load peak and off-peak difference is big and regulating capacity is insufficient. With the rapid development of society, the proportion of residential load keeps rising, and triggers the increase of load peak and off-peak difference. Pumped-storage plant has obvious advantages in Peak cutting and valley filling, especially its unique function in filling valley, which makes it extremely useful in reducing the system load peak and off-peak difference.

2.2.3. Renewable energy accommodation oriented. As discussed in [11-12], this type of pumped-storage plants can cooperate with some certain power plants and provide directional operation service. As for new power plants such as wind power and solar energy, because of its natural characteristics of randomness and intermittent, its effective capacity is low and should have a certain capacity of pumped-storage plants with given proportion can be effectively improved. Besides, to keep safe operation, nuclear power has a very high requirements so should be matching capacity of pumped storage units with appropriate load, energy storage or security, otherwise cannot give full play to the function of design.

2.3. *Standards of function classification*

Considering function orientation given above, this section will introduce the standards of pumped-storage classifying shown in figure 1, based on the local load characteristic, structure of power supply, accommodating capability of renewable energy, nuclear energy, etc.

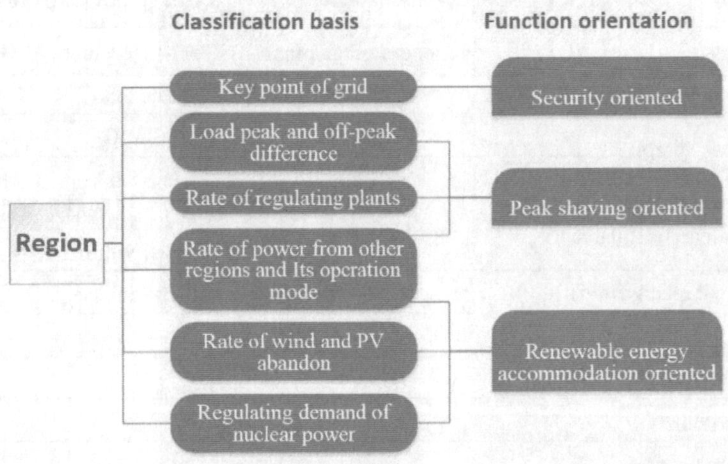

Figure 1. Classification basis for the function orientation of pumped-storage plant.

2.3.1. Key node of grid. It belongs to security oriented. Located in cities having high power supply reliability requirements, such as Beijing, Shanghai and other political and economic centres. Besides, if plants locate in grid key hub node, such as drop point of large-scale UHV DC, pumped-storage can be the effective means of frequency control and active power control.

2.3.2. Load peak and off-peak difference and rate of regulating plants. It belongs to peak-shaving oriented on the situation of large load peak and off-peak difference and low rate of regulating plants.

2.3.3. Rate of power from other regions and its operation mode. It belongs to peak-shaving oriented when pumped-storage located in the area accept lager rate of power from other regions.

2.3.4. Rate of wind and PV abandon. When area's rate of renewable power integration is lager and the rate of power abandon is high because of lacking regulating capacity, the pumped-storage belongs to renewable energy accommodation oriented.

2.3.5. Regulating demand of nuclear power. If there exists nuclear power plant, and needs of regulating and security, the pumped-storage belongs to renewable energy accommodation oriented.

Function orientation of pumped-storage plants is not immutable, instead, it should be adjusted as the grid's characteristics changing. For example, Central China has lots of runoff hydropower, in flood season, pumped-storage plants mainly bear peak shaving function, and when dry season came and the need of regulating reducing, its function turned to be security oriented. Thus, in practice, function orientation of pumped-storage plant should be treated differently in different evaluation periods.

3. Regional function orientation of pumped-storage plant in China

Due to lager geographical area of China and its uneven regional social and economic development level, the load characteristics and power structure also have big difference of different regions, also shown in [13]. This section puts forward on the combination of classification basis, according to the present situation and future development trend of Chinese regional grid, which are presented in [14], and analyzed regional function orientation of pumped-storage plant in China.

3.1. North China grid

In present, the function orientation of pumped-storage plants in north china grid focuses on security and peak-shaving. North china grid mainly depends on coal-fired plants, which accounts for 78.6% of the whole capacity by the end of 2014. Its load peak and off-peak difference is almost 27% no matter in winter or summer. Because of heat supplied plants starts up more in winter, the rate of regulating capacity only gets 10~20%, which needs pumped-storage to filling the valley especially.

In the future, as the increasing incoming power flow scale and new energy bases into operation, the function orientation of pumped-storage plants in north china will focus on security and renewable energy accommodation. During the 13th five-year, as the large-scale wind bases in Hebei province and Inner Mongolia province and UHVDC project from Jalute to Qingzhou and from Shanghaimiao to Linyi building into operation, the capacity of wind power in north china will be 20 million kW and 70 million kW power will come from other regions, thus, corresponding pumped-storage should be constructed for grid security operation and wind power accommodation.

3.2. East China grid

In present, the function orientation of pumped-storage plants in east china grid is peak shaving oriented. East china grid has large load with peak and off-peak difference being 20 million kW, and the grid mainly depends on coal-fired plants, which accounts for 65% of the whole capacity by the end of 2014. East China grid has lots of pressure of regulating, and its existed pumped-storage plants play big role.

To cope with the increasing incoming power flow scale and rising rate of nuclear power, function orientation of pumped-storage plants in north china will be regulating oriented. The ratio of power come from other regions will be over 30% after 2020, and pumped-storage will be the vital plants to provide active power supply.

3.3. Central China grid

In present, the function orientation of pumped-storage plants in central china grid is peak shaving oriented. Central China power grid is the largest hydropower installed capacity, the highest proportion of power grid in 2014, hydropower installed capacity of 130 million kW, the proportion of 46%. The research and practice show that the proper development of pumped-storage plants in the grid with the ratio of the main power of the runoff type hydropower has the comprehensive benefits such as improving the operating condition of the thermal power unit, reducing the running cost of the whole network and rationally utilizing the seasonal electric energy of hydropower. Such as Hubei, Hunan, Henan have different proportions of hydropower, but with the ability to adjust and more than a small proportion of hydropower stations, the dry season can be used for water and electricity peaking, loss of regulation of hydropower during flood season, to use water peak, it can only force to take abandoned water peaking mode. In this kind of power grid, with the pumped storage power station, it can absorb the basic charge in the flood season and convert it into the peak charge, so as to reduce or avoid flooding in the flood season, improve the economic efficiency and improve the hydropower flood season operation condition, of the operating conditions.

The future of central pumped storage power station is mainly located in the support by the end of power to ensure safe and stable operation of power grid. With the Jiuquan-Xiangtan and northern Shaanxi-Wuhan and other UHVDC projects completed, the central China region by the proportion of power into the province by the proportion of electricity will rise to about 20%, part of the delivery of the power flow of 60 million kW, the channel is mainly distributed by transmission and distribution, and it has anti-peaking characteristics. The pumped storage power station in the district will undertake the task of supporting the power supply and regulating the peak of the power network.

3.4. Northeast grid

In present, the function orientation of pumped-storage plants in northeast grid is renewable energy accommodation oriented. Same as the north china grid and east china grid, northeast grid mainly depends on coal-fired plants, and what makes things worse is heat supplied plants accounting for the very great proportion, which will lead to low regulating capacity in winter, and when the wind power blows will makes it even worse.

It will still be renewable energy accommodation oriented to pumped-storage plants in northeast grid in the future. As new wind power bases going into operation in Gansu province, Qinghai province and Ningxia province, more regulating pressure will appear in northeast grid, therefore, more pumped-storage plants should be built to prevent serious wind abandon.

3.5. Northwest grid

Pumped-storage plant hasn't been put into operation in northwest yet, and the planning plants should be oriented to renewable energy accommodation. In northwest grid, hydropower accounts for 17% of the whole capacity, on the other hand, its load peak and off-peak difference was only 20%. However, without regard to the rapid development of renewable power, it turns to be well in grid regulating. According to the national planning, the wind capacity of northwest grid by 2020 will be nearly 50 million kilowatt, which accounts for about 35% of the northwest peak load, and will put pressure on grid regulating. At present, power abandon caused by lack of regulating capacity occurred in Gansu province and Xinjiang province, it will be even harder when the proportion of wind and PV get higher in the future. Thus, to build pumped-storage plants in these areas have become very necessary.

4. Conclusion

Based on the domestic and foreign development experience and the reality of China, this paper puts forward the method of pumped-storage plant's function orientation, and classifies into three types security oriented, peak shaving oriented and renewable energy accommodation oriented. Then, combined with the current and 13th five year plan development in China, five regions' pumped-storage plant's function are given, including North China, East China, Central China, Northeast and Northwest. The research has profound guiding and practical significance for pumped-storage plant dispatch and operation.

References

[1] Yan Zhiyong, Zhai Guoshou 2004 The experienced development course and development prospect of Chinese pumped-storage power stations *Water Power* **30(12)** 73-76.

[2] Ibrahima H, Ilincaa A, Perronb J 2008 Energy storage systems-Characteristics and comparisons *Renewable and Sustainable Energy Reviews* **12(5)** 1221-1250.

[3] Guo Xiaona, Chen Tongfa 2015 Orienting the essential function scientifically to promote the Healthy and rapid development of pumped storages *Water Power* **41(6)**1-4.

[4] Cheng Lu, Bai Jianhua 2013 Role and prospect of pumped storage power stations in China *Electric Power* **46(11)** 155-159.

[5] Zhao Yongsheng 2003 Discussions on the development of pumped-storage power plants in China *Water Power* **29(7)** 69-71.

[6] He Yongxiu, Guan Lei, Cai Qi, et al. 2004 Analysis of securing function and economic benefit

of pumped storage station in power grid *Power System Technology* **28(20)** 54-57,67.

[7] Zhang Shu 2008 The research on strategy of pumped storage power station's development in China [D] Heilongjing, Harbin Institute of Technology

[8] Zhang Dong 2014 Economical benefit study of pumped storage power station based on comprehen-sive evaluation method [D] Beijing, North China Electric Power University

[9] Wu Chin-Chung, Lee Wei-Jen, Cheng Chin-Lung 2008 Role and Value of Pumped Storage Units in an Ancillary Services Market for Isolated Power Systems—Simulation in the Taiwan Power System *IEEE Transactions on Industry Applications* **44(6)** 1924-29.

[10] Guo Chunping, Yu Zhen 2011 Economic operation of pumping-storage power station in China *Hdropower Automation and Dam Monitoring* **35(6)** 1-4.

[11] Denholm P, Ela E, Kirby B, et al. 2010 The Role of energy storage with renewable electricity generation *National Renewable Energy Laboratory* 1-61.

[12] Caralis G, Papantonis D, Zervos A 2012 The role of pumped storage systems towards the large scale wind integration in the Greek power supply system *Renewable and Sustainable Energy Reviews* **16(5)** 2558–65.

[13] Liu Zhenya 2012 Electric power and energy in China *Wiley* 179-193.

[14] Zhang Yunzhou, Cheng Lu 2015 Research on major issues of the 13th five-year plan development of power industry *Electric Power* **48(1)** 1-5.

Measurement and control system design for the assembly of stator can

H H Liu and W Jiang[1]

School of Mechanical Engineering, Dalian University of Technology, NO. 2, Linggong road, GangjingziDistrict,Dalian, P.R. China

E-mail:liuhh1987@126.com;jiangwei@dlut.edu.cn

Abstract. Stator can assembly is one of difficult issues in the manufacturing of AP1000 nuclear reactor coolant pumps. A new measurement and control system is developed to ensure precise assembly of the ultrathin stator can with large length to thickness ratio on the stator core. The system includes three parts: hydraulic pressure measurement and control, deformation measurement of the can and displacement measurement of valve core. Since deformation of the can is hard to measure, a new method by measuring the bulge displacement of the can is proposed. With the help of the system, the loading path can be accurately controlled and the deformation can be precisely measured. The system provides possibility for the detailed study of assembly process. Finally, the feasibility of the system in the experiment rig of a reduced scale is verified.

1. Introduction

The AP1000 reactor coolant pump (RCP) is one of the key components in the primary loop of the third generation nuclear plant, which is a single-stage, hermetically sealed, high-inertia, centrifugal, canned motor pump. It provides power for the coolant to circulate in the nuclear island core, RCPs, heat leg main pipes, cold leg main pipes and steam generator[1, 2]. In order to avoid coil corrosion, the stator and rotor are encased in corrosion-resistant cans (made by Hastelloy C-276) [3, 4]. The clearance between the stator can and the rotor can is 4.83mm after assembly, which is the unique room for the coolant to pass. Length of stator core is over 2m with the inner diameter more than 500mm and the outer diameter more than 1175mm. Length of stator can is more than 3m with the thickness less than 0.5mm and the diameter over 500mm. The stator can is ultrathin can with large length to thickness ratio. There is a clearance between the stator can and the stator core after fitting them together, and methods should be used to eliminate this clearance.

Assembly methods currently available include: vacuum technique[5]and hydrostatic test[6]. For the first one, uneven deformation of the can will appear along the axial direction of the stator core and the load generated is too small to make the can deform to the expected profile. As for the second one, since the hydraulic pressure path corresponding to the time cannot be controlled step by step, we still cannot obtain the expected effects. Therefore, a new method based on the theory of tube hydroforming was presented by authors[7]. Pioneer work has been done [8, 9]. It is shown that the proposed method is feasible for the assembly of the stator can.

In this paper, an assembly experiment rig with reduced size was setup to study the assembly process at lower cost. Measurement and control methods were proposed, and the system were analyzed and designed. The stator assembly, hydraulic system and measurement and control system were included in the experiment rig. Deformation of the stator can, hydraulic pressure and displacement of the overflow valve core are measured. At the same time, the location of relief valve core should be adjusted in order to precisely control the loading path. The effects of the system, such as, pressure-time curves, displacement-time curves, were verified by experiments. Finally, the assembly error of the displacement measurement equipment was discussed.

2. Measure & control methods for the assembly process

The experiment rig is shown in figure1.Three components are included: hydraulic system, measurement & control system, and stator assembly. Parameters needed to be measured in the assembly process are the hydraulic pressure, displacement of the valve core and the radial displacement of the stator can.

Figure 1. The experiment rigand measured points arrangement

2.1. Hydraulic pressure measurement

Figure 2(a) shows cross section profile of the stator assembly. Figure 2(b) shows the actual stator assembly. Pressure measurement between the inner wall of stator can and the outer wall of supportive tube is hard to realize. A method by mounting a pressure transducer on the input pipe to measure the pressure indirectly is proposed, as shown in figure 1.

2.2. Deformation measurement of stator can

For proper installation of the ultrathin stator can, the can is required to deform uniformly and attached to stator core closely. Deformation measurement of the stator can is utterly important. Experiments combined with FEA method [8] are applied to study the assembly process. To simplify the structure, a reduced size is used in the axial direction, the coils and the slot groove are neglected in the radial direction, and only the components above the slot wedges are considered. Such simplification saves cost without reducing the assembly accuracy.

Several steps are involved in study of the assembly process: **Step 1**, numerical simulations have been conducted for the assembly process to obtain an ideal assembly process, and the strain and stress of the stator can were analyzed[7]. **Step 2**, assembly experiments have been conducted according to the loading path determined in **Step 1**, and the deformation and stress are measured by the designed measurement & control system. **Step 3**,FEA model established described in[10] is revised based on analyzing the difference between experiment data and simulation data, then numerical simulation was conducted again. A reliable FEA model and the best assembly process will be obtained by conducting the above steps repeatedly. Eventually, assembly processes of full size stator can deduced by numerical simulations will be reliable.

The stress, strain, thickness of the stator can need to be known in the assembly process study. However, the stator can is enclosed in the cavity formed by the supportive tube and the stator: the room between the stator can and the supportive tube is filled with high pressure oil. The other room between the stator can and the stator core is needed to be eliminated. After assembling, the stator can is attached to inside of the stator core closely. It is difficult to measure the deformation of the stator can. If strain gauges are mounted on the inner wall which is filled with high pressure oil, the strain gauges are hard to be fixed and the wire is not easy to be connected to the outside equipment. On the other hand, if strain gauges are mounted on the outer wall which will be attached on stator core, the gauges are easily to be damaged.

Based on the difficulties discussed above, a method for measuring the stator can' deformation by measuring bulge amount of the stator can in the radial direction is proposed. The new method is shown in figure2(a) in detail. Four holes are bored in the stator core around the circumferential direction and the displacement sensors are mounted in the holes to measure the deformation in real-time. Other deformations can be deduced from the bulging amounts. The curves of pressure-time, displacement-time can be obtained through the post-processing of the computer.

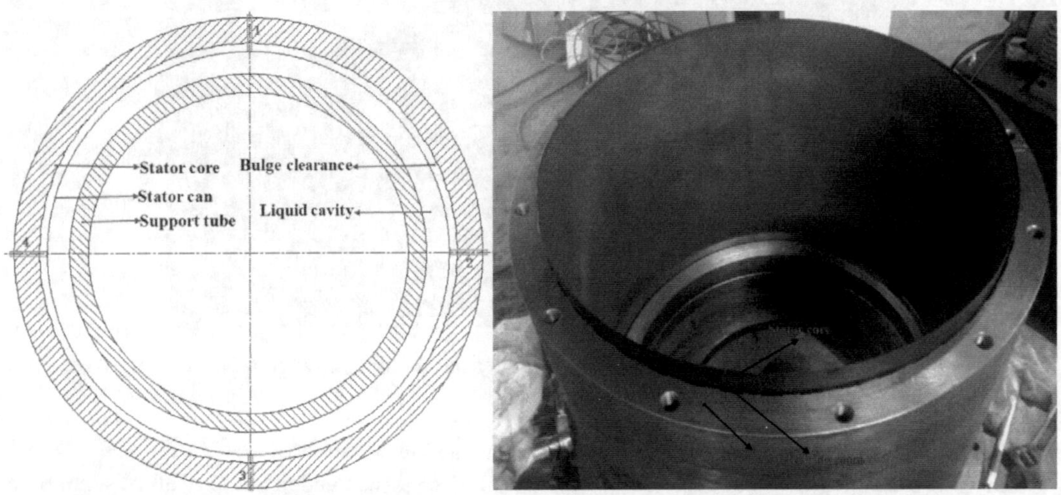

(a)(b)

Figure 2. The stator assembly and the illustrations of the control & measurement methods

2.3. Displacement measurement for proportional relief valve

In the assembly process, the proportional relief valve is used to increase the pressure increasing rate. Given a signal by electromagnet, the valve core will move and adjust the opening size of the nozzle, which will change the pressure increasing rate of the valve. It is thus necessary to measure and control the displacement of the valve core accurately so as to precisely control the loading path. The pressure increasing rate can be calculated by analyzing the displacement of the valve core. The movement of the valve core can be controlled by the computer through comparing the pressure measured with the input. The displacement sensor is used to measure the displacement. The feedback control of the pressure is realized by using the digital module of the PLC.

3. Design of the measure & control system

3.1. Functional requirements of the system

The requirements of the designed system are as follows:(1) Functional requirements. Displacement of the stator can in radial direction and pressure in the stator core should be measured in precision real-time tracking. Location of the valve core should be measured precisely. The curves of pressure-time, displacement-time and displacement-pressure can be shown with the post-processing progress. Six variables are needed to be measured. Five displacement sensors and one pressure transducer are required. It is important that the values are collected synchronously to avoid errors.(2) Precision requirements. Clearance between the stator can and surface of slot wedge of the stator core is 2mm, which is the maximum displacement of the stator can in the assembly process. The maximum allowable pressure of hydraulic system is 25MPa. The requirements for the sensors are shown in table 1.

Table 1. The precision requirements

	Types	Precision requirements
Displacement measurement	Test precision	The sensitivity is 0.01mm
	Resolution	0.001mm
	Dynamic frequency	200HZ
	Measuring range	0-3mm
Pressure measurement	Test precision	The sensitivity is 0.01MPa
	Measuring range	0-30MPa

3.2. System design

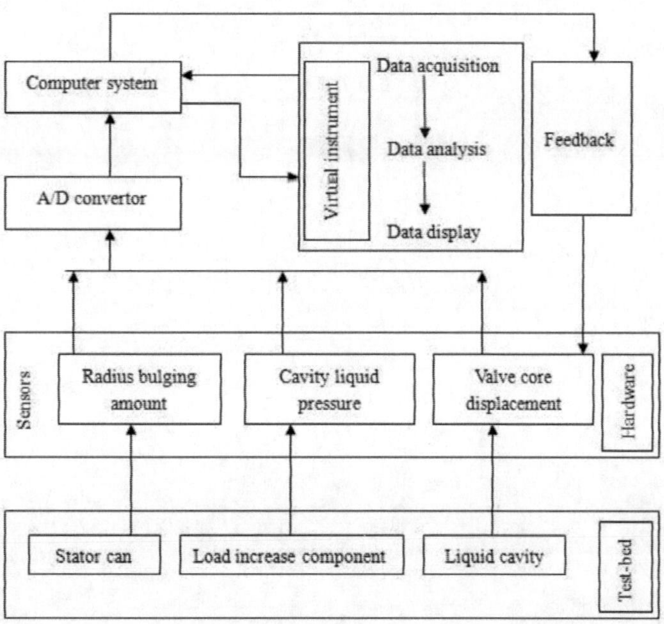

Figure3. The design of the system

Figure 3 shows the structure of the system. Two measurement subsystems and one measuring-feedback control subsystem are included. The experiment rig is shown at the bottom. The hardware for collecting information (namely sensors) which is used to convert the non-electrical variables into electrical signals is shown in the middle. The controlling and processing system of the computer which is used to convert the analog quantity into digital quantity is shown on the top. Adding the designed control-measurement system on the hydraulic system (previous work) is shown in figure 4. Every component is composed of signal generator, signal detection and conditioning circuit[11, 12], data acquisition card[13], network card. In the post-processing phase, the data processing software collects data and exports the relationships of displacement-time, liquid pressure-time in the form of curves and tabular[14].

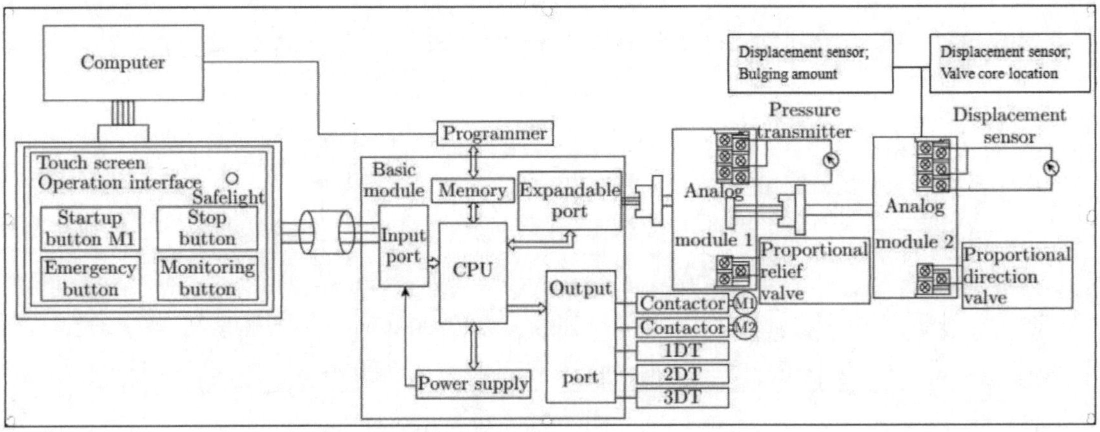

Figure 4. The hardware design of the system

4. Test result analysis

4.1. Analysis of pressure test results

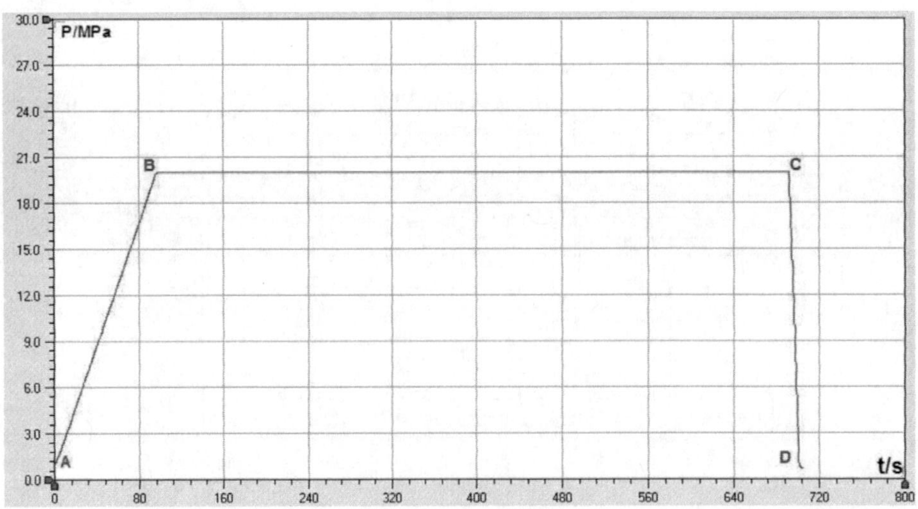

Figure 5. The testing pressure curve

The maximum pressure of the hydraulic system was firstly set 20MPa with the holding time of 600s and the pressure increasing rate is 0.2MPa/s. Then it is unloaded. The testing pressure curve is shown in figure5. The points A, B, C, D represent 0.87MPa (t=0.1s), 20.00MPa (t=98.3s), 20.2MPa (t=691.8s), 0.74MPa (t=702.8s), respectively. Analysis of the points and every phase' errors (loading phase, holding phase, unloading phase) are shown in table 2: (1) Loading phase error is defined as the percentage ratio of the absolute difference of the pressure increasing rate between the input and test to the input. Then the percentage is 2.6% and considered to be within the allowable error range. (2)Holding phase error is defined as the percentage ratio of the absolute difference of the holding time between the input and test to the input. Then the percent is 1.08%, which is also considered to be within the allowable error range. (3) Unloading phase error is defined as the percentage ratio of the absolute difference of the pressure decreasing rate between the input and the test to the input. Then the percentage is 54.1% and the deviancy is significant. Many reasons account for this, i.e. large setting for the unloading rate as 4MPa/s, the low sensitivity of the system. As the system becomes complicated, the more lag as the rate increases. Compared to the loading rate of 0.2MPa/s, it can be concluded that the precision of the system increases with the decrease of loading rate.

From figure 5 and table 1, it could be observed that measurement-control system meets our precision control requirements in certain rate range. Under the circumstance, the loading and unloading velocity will be linearly increased and decreased steadily. It is found that the start point is not at the origin and this is attributed to the sensitivity and error of the system. But it will be back to the point which has the same pressure value as the start point after unloading. The start point will be considered as the origin point. Then it indicates that the designed system is steady.

Table2. Testing errors analysis

	Loading phase		Holding phase		Unloading phase	
	A	B	B	C	C	D
Testing results	0.1s, 0.87MPa	98.3s, 20.00MPa	98.3s, 20.00MPa	691.8s, 20.02MPa	691.8s, 20.02MPa	702.8s, 0.74MPa
Theory results	0.0s, 0.00MPa	100.0s, 20.00MPa	100.0s, 20.00MPa	700.0s, 20.00MPa	700.0s, 20.00MPa	705.0s, 0.00MPa
Error analysis	$\varepsilon_{loading}=\left\|v_{test}-v_{input}\right\|/v_{input}\times 100\%=2.6\%$		$\varepsilon_{hold}=\left\|T_{test}-T_{input}\right\|/T_{input}\times 100\%=1.08\%$		$\varepsilon_{unload}=\left\|v_{test}-v_{input}\right\|/v_{input}\times 100\%=54.1\%$	

4.2. Assembly results

Figure 6 shows the assembly results obtained by the designed system under maximum pressure of 20MPa, the pressure increasing rate of 0.2MPa/s and the holding time of 600s. It is found that the stator can was already attached inside of the stator core closely. It is indicated that the system can meet the requirements of precise measuring and controlling of the assembly process.

Figure 6. Assembly results

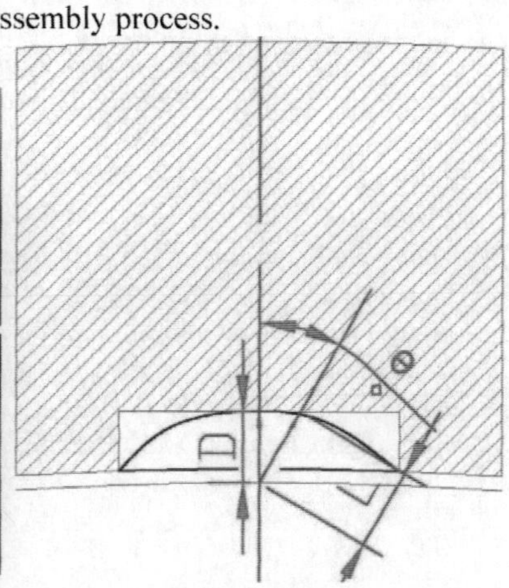

Figure 7. Error analysis for measuring

5. Precision analysis of the deformation measurement

Figure7 is the detailed enlarging picture at the position of displacement sensor 1 shown in figure 3. The sensor is mounted in the hole which is drilled in the stator wall and the sensor head is contacted with the outer wall of the stator can. The bulging of the stator can push the sensor bar which will cause the change of the electric current. The errors of drilling the hole and mounting the sensors may occur which are shown with the type of skew (axis of the sensor bar isn't vertical to the surface of the stator can). There may be a skew angle θ between two axis of the sensors caused by manufacturing error or assembly error, shown in figure7.Displacementmeasuredwill be not accurate due to this angle. The error increases with the increase of the deflection angle and the displacement measured is larger than the theoretical one. The error is analyzed by equation $L = D/\sin\theta$ and $\varepsilon = D(1-\sin\theta)/\sin\theta$.

Measures must be taken to eliminate testing errors. The finish machining of the hole is necessary. As to reducing the mounting errors, several sensors can be mounted around the circumferential direction of the stator core and the displacement curve values can be obtained by calculating the average values of sensors. The measurement error will be minimized by these improved methods.

6. Conclusion

A new measurement and control system used in the precision assembly of the stator can is proposed based on the presented assembly method. Due to the difficulties in the measurement of the stator can, an indirect method by measuring the bulge amount in the radial direction of the stator can is proposed. The reliability of the designed system was verified by testing results: the loading path can be controlled precisely and the loading rate can be set in a broad range. The loading rate and unloading rate must be set at a small rate in order to obtain an accurate loading path. Also, the reasons of the displacement measuring errors are discussed and the improving methods are presented. In the future work, the experiments will be performed to study on the best assembly process of the stator can, and the deformation mechanism of the stator can effected by the pressure will also be investigated.

Acknowledgement

Financial support for this work has been provided by the National Key Basic Research Program of China (Grant No. 2009CB724307).

References

[1] Ma X P 2007*Electrical Equip*. **8**20-25.
[2] Schulz T L 2006 *Nucl Eng Des*. **236**1547-57.
[3] Scott C 2005 *World Pumps*. **462**22-24.
[4] Wed M 1996 *World Pumps*.30-34.
[5] Feng Y H 2011 Mechanical analysis of the cans in AP1000 reactor coolant pump *Master thesis*(Shanghai: Shanghai Jiaotong University) .
[6] Guan R and Gao Y 2008 *China Nucl Power*. **1**49-50.
[7] Liu H, Jiang W, and Hao M, 2013 *Nucl Eng Des*. **256**256-263.
[8] Liu H and Jiang W 2012 *Appl Mech Mater*. **190-191** 381-384.
[9] Hao M and Jiang W 2013 *J. of Shanghai Jiaotong Univ*. **18** 1-7.
[10] Garrido O C, Cizelj L, and Sirnonovski I 2012 *Nucl Eng Des*. **246**115-122.
[11] Yang H, Zhang S, and Li Z 2000 *Appl of Electron Tech*. **2**64-68.
[12] Yin Q and Han S 2003 *Electron Des& Appl*. **3** 81-82.
[13] ADVANTECH Device Driver V2.0 for Windows User Manual.
[14] An P and Cheng X 2002 Precise measurement method of high-speed dynamic object based on IRIG-B

Optimal power allocation based on sum-throughput maximization for energy harvesting cognitive radio networks

Zhenwei Xie[1] and Qi Zhu[2, 3]

[1] The Key Wireless Laboratory of Jiangsu Province, Nanjing University of Posts and Telecommunications, Nanjing 210003, China
[2] Key Laboratory on Wideband Wireless Communications and Sensor Network Technology of Ministry of Education, Jiangsu Nanjing 210003, China

E-mail: zhuqi@njupt.edu.cn

Abstract. In this study, an optimal power allocation algorithm by maximizing the sum-throughput in energy harvesting cognitive radio networks is proposed. Under the causality constraints of the harvested energy by solar radiation, electromagnetic waves and so on in the two secondary users (SUs), and the interference constraint in the primary user (PU), the sum-throughput maximization problem is formulated. The algorithm decomposes the interference threshold constraint to the power upper bounds of the two SUs. Then, the power allocation problems of the two SUs can be solved by a directional water-filling algorithm (DWA) with the power upper bounds, respectively. The paper gives the algorithm steps and simulation results, and the simulation results verify that the proposed algorithm has obvious advantages over the other two algorithms.

1. Introduction

Energy harvesting technology can continuously gather energy from the surrounding environment, such as solar radiation, electromagnetic waves and so on, which greatly extends the lifecycle of equipment, reduces the use of traditional energy sources, protects environment and improves the performance of wireless networks [1], [2]. In addition, the cognitive radio technology can find and rationally use spectrum holes [3], which improves the spectral resource utilization rate of the system. So it is very significant to study the resource allocation problems in energy harvesting cognitive radio networks.

The power allocation problems in energy harvesting networks have been studied extensively. In [4], the authors assume that the source can utilize conventional energy and harvested energy, and harvested energy spends less than conventional energy. The problem of the total energy cost minimization is solved by the theory of linear programming limited by a minimum capacity and harvested energy. In [5], the problem to maximize throughput is studied by Lagrangian function and KKT conditions in a two-node energy harvesting network, and a DWA scheme is proposed. In [6], the authors assume that data packets and harvested energy gradually arrive in the process of the data transmission, and then the throughput maximization problem is decoupled to a power allocation problem and a data transmission problem, which can be solved by alternating maximization method.

In [7], [8], both cognitive radio technology and energy harvesting technology are considered. In [7], the authors deduce the upper bound of the throughput that the SU can achieve under a collision constraint. In [8], the authors propose a geometric water-filling algorithm, in which the interference constraint is converted into the power upper bound for the SU.

The main contributions of this paper are as follows:

- This paper considers a new energy harvesting wireless network scenario in which two SUs can share a common spectrum of the PU [9], so we must jointly optimize the power allocations of the two SUs. But both [7], [8] consider that a single SU occupies the spectrum of the PU, so the interference constraint is easily converted into the power upper bound of the SU.
- The harvested energy can come from solar radiation, the wind, electromagnetic waves and so on, so the network can utilize a variety of new energy, reduce the use of traditional energy, and protect environment.
- In [6], the constraints of harvested energy and data are separable, and the original problem can be decoupled directly. But this paper need to decompose the interference threshold constraint to the power upper bounds of the two SUs. Then an iteration water-filling algorithm is proposed for the power upper bounds of the two SUs. In addition, we prove that the equivalence problem is a convex optimization problem.

2. Energy harvesting cognitive radio networks model and problem formulation

This paper considers an energy harvesting cognitive radio network, which consists of a PU, two SUs and a base station (BS), and the two SUs are both energy harvesting nodes, as shown in figure 1. Assuming that the timeslot is the minimum transmission unit, and there are N equal length timeslots. For convenience, we set the timeslot length is 1s, which can be extended to arbitrary timeslot length. Assuming that harvested energy arrives before the data transmission of each timeslot starts. We denote the channel fading coefficients from the two SUs to the BS as $h_{1,i}$ and $h_{2,i}$, and the channel fading coefficients from the two SUs to the PU as g_i and f_i in the i-th timeslot. All channels are additive white Gaussian noise channels with the noise power spectral density N_0. The bandwidth of the PU is W. Let $E_{1,i}$ and $E_{2,i}$ denote the harvested energy amounts of the two SUs in the i-th timeslot.

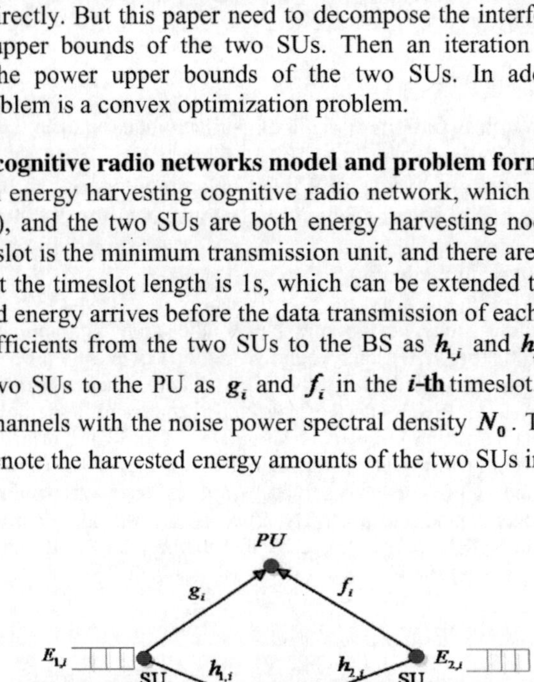

Figure 1. Energy harvesting cognitive radio networks model.

Assuming that the two SUs work simultaneously in each timeslot with half the bandwidth of the PU, that is, FDMA (frequency division multiple access) mode. The two SUs work with power $P_{1,i}$ and $P_{2,i}$ in the i-th timeslot. Meanwhile, the sum of the interference in the PU is required to be less than the interference threshold I [9], so the interference threshold constraint is

$$P_{1,i}g_i + P_{2,i}f_i \le I \tag{1}$$

At the same time the causality constraints of the harvested energy in the two SUs are

$$\sum_{i=1}^{n} P_{1,i} \le \sum_{i=1}^{n} E_{1,i} \tag{2}$$

$$\sum_{i=1}^{n} P_{2,i} \le \sum_{i=1}^{n} E_{2,i} \tag{3}$$

The sum-throughput within N timeslots can be expressed as

$$C=\sum_{i=1}^{N}\left[\frac{1}{2}W\log_2(1+\frac{h_{1,i}P_{1,i}}{WN_0/2})+\frac{1}{2}W\log_2(1+\frac{h_{2,i}P_{2,i}}{WN_0/2})\right] \tag{4}$$

where we denote $\bar{h}_{1,i}=\frac{h_{1,i}}{WN_0/2}$ and $\bar{h}_{2,i}=\frac{h_{2,i}}{WN_0/2}$ to simplify the derivation process since $WN_0/2$ is a constant.

Hence, under the criterion of the sum-throughput maximization, we can formulate the problem as

$$\begin{aligned} \max_{\{P_{1,i},P_{2,i}\}} \quad & \sum_{i=1}^{N}\left[\log_2(1+\bar{h}_{1,i}P_{1,i})+\log_2(1+\bar{h}_{2,i}P_{2,i})\right] \\ s.t. \quad & P_{1,i}g_i+P_{2,i}f_i\le I,\quad \forall i \\ & \sum_{i=1}^{n}P_{1,i}\le\sum_{i=1}^{n}E_{1,i},\quad \forall n \\ & \sum_{i=1}^{n}P_{2,i}\le\sum_{i=1}^{n}E_{2,i},\quad \forall n \\ & 0\le P_{1,i},\quad \forall i \\ & 0\le P_{2,i},\quad \forall i \end{aligned} \tag{5}$$

where we omit the $\frac{1}{2}W$ in the objective function, which does not affect the optimal power allocation result.

Since the objective function in equation (5) is concave and the constraint conditions are linear, equation (5) is a convex optimization problem with a unique optimal value.

3. Proposed power allocation algorithm

This paper proposes a new iteration water-filling algorithm for the power upper bounds of the two SUs through converting an inequality constraint $P_{1,i}g_i+P_{2,i}f_i\le I$ to an equality constraint $P_{1,i}^I g_i+P_{2,i}^I f_i=I$.

As shown in figure 2, we denote the power upper bounds of the two SUs as $\{P_{1,i}^I\}$ and $\left\{P_{2,i}^I=\frac{I-P_{1,i}^I g_i}{f_i}\right\}$, and then we must have $P_{1,i}\le P_{1,i}^I$, $P_{2,i}\le P_{2,i}^I$, $\forall i$ for figure 2(a) and (b).

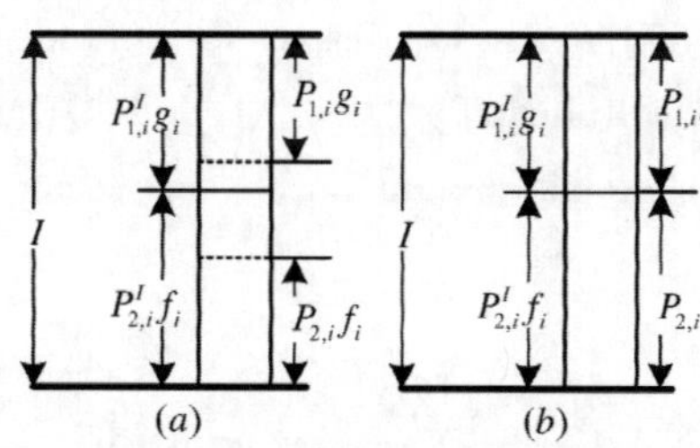

Figure 2. Schematic diagram of converting an inequality constraint to an equality constraint.

Assuming that the power upper bounds of the two SUs are known, and $\{P_{1,i}^I\}$ is denoted as vector $\mathbf{P}_1^I$. The equation (5) can be decoupled into the following two sub-problems:

$$\begin{aligned} g(\mathbf{P}_1^{\mathbf{I}}) = \max_{\{P_{1,i}\}} \quad & \sum_{i=1}^{N} \log_2(1+\bar{h}_{1,i}P_{1,i}) \\ s.t. \quad & P_{1,i} \le P_{1,i}^{I}, \quad \forall i \\ & \sum_{i=1}^{n} P_{1,i} \le \sum_{i=1}^{n} E_{1,i}, \quad \forall n \\ & 0 \le P_{1,i}, \quad \forall i \end{aligned} \tag{6}$$

where $g(\mathbf{P}_1^{\mathbf{I}})$ denotes the maximum value of equation (6) in the case of $\mathbf{P}_1^{\mathbf{I}}$ is determined.

$$\begin{aligned} f(\mathbf{P}_1^{\mathbf{I}}) = \max_{\{P_{2,i}\}} \quad & \sum_{i=1}^{N} \log_2(1+\bar{h}_{2,i}P_{2,i}) \\ s.t. \quad & P_{2,i} \le \frac{I - P_{1,i}^{I} g_i}{f_i}, \quad \forall i \\ & \sum_{i=1}^{n} P_{2,i} \le \sum_{i=1}^{n} E_{2,i}, \quad \forall n \\ & 0 \le P_{2,i}, \quad \forall i \end{aligned} \tag{7}$$

where $f(\mathbf{P}_1^{\mathbf{I}})$ denotes the maximum value of equation (7) in the case of $\mathbf{P}_1^{\mathbf{I}}$ is determined. Since the objective function in equation (6) is concave for $\{P_{1,i}\}$, and all the constraint conditions are linear, equation (6) is a convex optimization problem. Similarly, equation (7) is also a convex optimization problem for $\{P_{2,i}\}$. Then, equations (6) and (7) can be solved by a DWA with the power upper bounds [6], respectively.

Now, the causality constraints of the harvested energy in the two SUs and the interference constraint in the PU both are satisfied, and equation (5) is converted into optimizing $\{P_{1,i}^{I}\}$ to maximize the sum-throughput of equations (6) and (7). The equation (5) is equivalent to

$$\begin{aligned} \max_{\{P_{1,i}^{I}\}} \quad & g(\mathbf{P}_1^{\mathbf{I}}) + f(\mathbf{P}_1^{\mathbf{I}}) \\ s.t. \quad & P_{1,i}^{I} g_i \le I, \forall i \end{aligned} \tag{8}$$

Then we will prove that equation (8) is a convex optimization problem for $\mathbf{P}_1^{\mathbf{I}}$.

Proof: First, we prove equations (6) and (7) both are concave for $\mathbf{P}_1^{\mathbf{I}}$.

For equation (7), let $0 \le \xi \le 1$, and $\mathbf{Q}_1^{\mathbf{I}}$, $\mathbf{Q}_2^{\mathbf{I}}$ denote two power upper bound vectors to substitute $\mathbf{P}_1^{\mathbf{I}}$. Then, $\mathbf{Q}_1$ denotes the solution of $f(\mathbf{Q}_1^{\mathbf{I}})$, and $\mathbf{Q}_2$ denotes the solution of $f(\mathbf{Q}_2^{\mathbf{I}})$, respectively. Since equation (7) is a convex optimization problem for $\{P_{2,i}\}$, $\mathbf{Q}_1$ and $\mathbf{Q}_2$ are existing and unique because of convexity. Since the constraint conditions are linear, the vector $\xi\mathbf{Q}_1$+(1-ξ)$\mathbf{Q}_2$ is feasible for $f(\xi\mathbf{Q}_1^{\mathbf{I}}+(1-\xi)\mathbf{Q}_2^{\mathbf{I}})$. So

$$\begin{aligned} f(\xi\mathbf{Q}_1^{\mathbf{I}}+(1-\xi)\mathbf{Q}_2^{\mathbf{I}}) & \overset{(a)}{\ge} \sum_{i=1}^{N} \log_2(1+\xi Q_{1,i} + (1-\xi)Q_{2,i}) \\ & \overset{(b)}{\ge} \sum_{i=1}^{N} \xi \log_2(1+Q_{1,i}) + \sum_{i=1}^{N} (1-\xi)\log_2(1+Q_{2,i}) \\ & \overset{(c)}{=} \xi f(\mathbf{Q}_1^{\mathbf{I}}) + (1-\xi)f(\mathbf{Q}_2^{\mathbf{I}}) \end{aligned} \tag{9}$$

where the reason of (a) is that $f(\xi\mathbf{Q}_1^{\mathbf{I}}+(1-\xi)\mathbf{Q}_2^{\mathbf{I}})$ denotes the maximum value of equation (7) in the case of $\mathbf{P}_1^{\mathbf{I}}$ is substituted by $\xi\mathbf{Q}_1^{\mathbf{I}}+(1-\xi)\mathbf{Q}_2^{\mathbf{I}}$, but $\sum_{i=1}^{N} \log_2(1+\xi Q_{1,i} + (1-\xi)Q_{2,i})$ denotes the sum-throughput which is calculated by the feasible point $\xi\mathbf{Q}_1+(1-\xi)\mathbf{Q}_2$, and then we must have that the former is greater than the latter. The reason of (b) is that log is a concave function. The reason of (c) is that $\mathbf{Q}_1$ denotes the solution of $f(\mathbf{Q}_1^{\mathbf{I}})$, and $\mathbf{Q}_2$ denotes the solution of $f(\mathbf{Q}_2^{\mathbf{I}})$, respectively.

Similarly, we can prove that equation (6) is concave for $\mathbf{P}_1^{\mathbf{I}}$.

Then, we can obtain that $g(\mathbf{P_1^I})+f(\mathbf{P_1^I})$ is concave for $\mathbf{P_1^I}$ because all the independent variables in log are linear. Meanwhile, the constraint condition $P_{1,i}^I g_i \le I$ is linear, so equation (8) is a convex optimization problem and it can be solved by the iteration water-filling algorithm in table 1.

In table 1, note that the prerequisite for the next iteration is that the sum-throughput increases, if we repeat iteration many times, the objective function will increase monotonically until it converges to the optimal value. Since each iteration increases the sum-throughput, and equation (8) is a convex optimization problem with a unique optimal value, the convergence of the problem is guaranteed. At the same time, the number of iterations depends on the step size to flow energy from $P_{1,i}^I$ to $P_{2,i}^I$ or from $P_{2,i}^I$ to $P_{1,i}^I$. However, $\mathbf{P_1^I}$ corresponding to the optimal value is not unique in table 1. Assuming that the optimal power allocations of the two SUs are respectively $P_{1,i}^*$ and $P_{2,i}^*$ in the i-th timeslot, which satisfy $P_{1,i}^* g_i + P_{2,i}^* f_i < I$, and then $P_{1,i}^I$ can be arbitrary value within the limit of $P_{1,i}^* \le P_{1,i}^I \le \dfrac{I-P_{2,i}^* f_i}{g_i}$.

Table 1. Proposed iteration water-filling algorithm.

1. Initialize $P_{1,i}^I = \dfrac{I}{g_i}$, $P_{2,i}^I = \dfrac{I-P_{1,i}^I g_i}{f_i}$, $1 \le i \le N$, $temp = 0$, the minimum step size s;
2. Solve equations (6) and (7) by the DWA with the power upper bounds;
3. Compute $C = \sum_{i=1}^{N} \frac{1}{2} W \left[\log_2(1+\bar{h}_{1,i} P_{1,i}) + \log_2(1+\bar{h}_{2,i} P_{2,i}) \right]$;
4. **while** $C > temp$ **do**
5. $\quad temp = C$;
6. $\quad$ **for** $i = 1:N$ **do**
7. $\qquad$ Let $P_{1,i}^I = P_{1,i}^I - s$, $P_{2,i}^I = P_{2,i}^I + s g_i / f_i$, repeat steps 2 and 3;
8. $\qquad$ As long as the sum-throughput C increases, repeat step 7, otherwise, let $P_{1,i}^I = P_{1,i}^I + s$, $P_{2,i}^I = P_{2,i}^I - s g_i / f_i$;
9. $\qquad$ Let $P_{1,i}^I = P_{1,i}^I + s$, $P_{2,i}^I = P_{2,i}^I - s g_i / f_i$, repeat steps 2 and 3;
10. $\qquad$ As long as the sum-throughput C increases, repeat step 9, otherwise, let $P_{1,i}^I = P_{1,i}^I - s$, $P_{2,i}^I = P_{2,i}^I + s g_i / f_i$;
11. $\quad$ **end for**
12. **end while**

4. Simulation and analysis

The simulation parameters of this paper mainly refer to [10]. The distance between the PU and the BS is 1000m. The Two SUs are uniformly distributed in a circle with a radius of 1000m centered on the mid-point of the PU and the BS. Assuming that the bandwidth of the PU is $W = 2\text{MHz}$, so the bandwidth occupied by the two SUs is $\frac{1}{2}W = 1\text{MHz}$, respectively. The large-scale fading with distance d is approximately $(-31.5 - 30\log_{10} d)$ dB, and the small-scale fading is modeled as Rayleigh fading. The noise power spectral density is $N_0 = -174$ dBm/Hz.

Let $N = 20$. Then, the proposed algorithm is compared with the maximum capacity transmission algorithm and the optimal power allocation (OPA) of a single SU. The single SU can occupy all the bandwidth of the PU, but its capacity of energy harvesting is only equivalent to the SU_1. The basic steps of the maximum capacity transmission algorithm are as follows: First, the harvested energy of the two SUs in each timeslot is respectively allocated to the current timeslot, i.e., $P_{1,i} = E_{1,i}, P_{2,i} = E_{2,i}, 1 \le i \le N$. Then, the allocated power of the two SUs is reduced proportionally for the timeslots in which the total interference of the two SUs exceeds the interference threshold, until the total interference is equal to the interference threshold. Suppose that $\overline{P}_{1,i}, \overline{P}_{2,i}$ are the reduced power of the two SUs, then $\overline{P}_{1,i} g_i = \overline{P}_{2,i} f_i$ is satisfied.

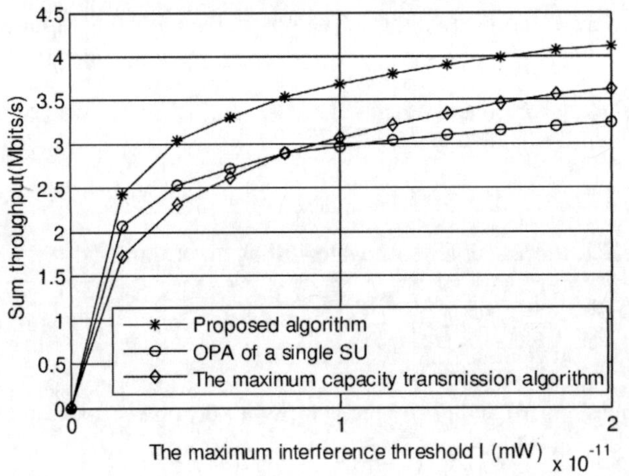

Figure 3. The comparison of the sum-throughput under different I.

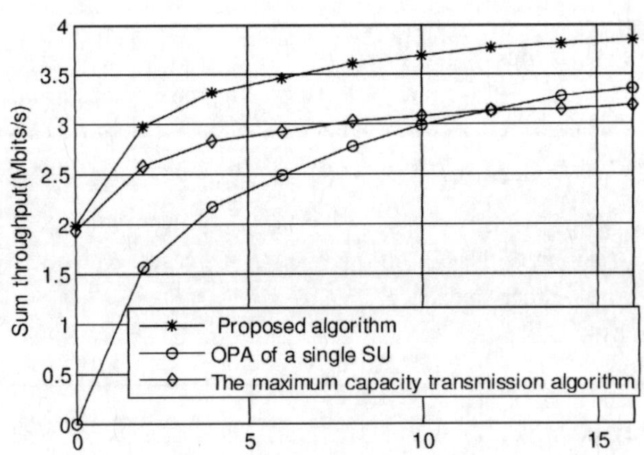

Figure 4. The comparison of the sum-throughput under different $E_1^{\max}$.

Figure 3 shows the comparison of the sum-throughput under different interference thresholds I. Assuming that the harvested energy of the two SUs in each timeslot is uniformly distributed between $[0, E_1^{\max}]$ and $[0, E_2^{\max}]$, where $E_1^{\max} = E_2^{\max} = 10\text{mJ}$. From the figure we can see that the sum-throughputs of the three algorithms will gradually increase with the increase of the interference threshold I . After about $I = 0.8 \times 10^{-11}$ mW , the sum-throughput of the maximum capacity transmission algorithm will exceed the OPA of a single SU, because the harvested energy amounts of the two SUs are more than the single SU. Finally, the sum-throughputs of the three algorithms will converge due to the constraint of harvested energy.

Figure 4 shows the comparison of the sum-throughput under different $E_1^{\max}$. Assuming that $I = 1 \times 10^{-11}$ mW , and the harvested energy of the two SUs in each timeslot is uniformly distributed between $[0, E_1^{\max}]$ and $[0, E_2^{\max}]$, where $E_2^{\max}$=10mJ . From the figure we can see that with the $E_1^{\max}$ gradually increases, the sum-throughputs of the three algorithms will gradually increase, too. Note that after $E_1^{\max} = 12$ mJ , the sum-throughput of the OPA of a single SU will exceed the maximum capacity transmission algorithm, because the interference threshold I becomes the main limiting factor for the sum-throughput of the system, and the OPA of a single SU obtains the optimal power allocation, but the maximum capacity transmission algorithm uses the harvested energy to the current timeslot.

5. Conclusions

In this paper, an optimal power allocation algorithm to maximize the sum-throughput in energy harvesting cognitive radio networks is proposed. The algorithm decomposes the interference threshold constraint to the power upper bounds of the two SUs. The power allocation problem within the two SUs can be solved by a DWA with the power upper bounds, respectively. The power allocation problem between the two SUs can be solved by the iteration water-filling between the power upper bounds. Finally, the simulation results verify that the proposed algorithm has obvious advantages.

Acknowledgements

This work is supported by National Natural Science Foundation of China (NO. 61571234, 61631020.), and National Basic Research Program of China (NO. 2013CB329005.)

References

[1] Ulukus S, Yener A, Erkip E, Simeone O and Huang K 2015 Energy harvesting wireless communications: A review of recent advances *IEEE J. Sel. Areas Commun.* **33** 360-81

[2] Ku M L, Li W, Chen Y and Liu K J R 2016 Advances in Energy Harvesting Communications: Past, Present, and Future Challenges *IEEE Commun. Surveys Tuts.* **18** 1384-1412

[3] Duy T T, Li W and Kong H Y 2012 Exact outage probability of cognitive two-way relaying scheme with opportunistic relay selection under interference constraint *IET Commun.* **6** 2750-59

[4] Kang X, Chia Y, Ho C K and Sun S 2014 Cost Minimization for Fading Channels With Energy Harvesting and Conventional Energy *IEEE Trans. Wireless Commun.* **13** 4586-98

[5] Ozel O, Tutuncuoglu K, Yang J and Yener A 2011 Transmission with Energy Harvesting Nodes in Fading Wireless Channels: Optimal Policies *IEEE J. Sel. Areas in Commun.* **29** 1732-43

[6] Varan B and Yener A 2016 Delay Constrained Energy Harvesting Networks with Limited Energy and Data Storage *IEEE J. Sel. Areas in Commun.* **34** 1550-64

[7] Park S and Hong D 2014 Achievable Throughput of Energy Harvesting Cognitive Radio Networks *IEEE Trans. Wireless Commun.* **13** 1010-22

[8] He P and Zhao L 2015 Optimal power control for energy harvesting cognitive radio networks *IEEE ICC* 92-97

[9] Liang J and Chen J 2013 Resource Allocation in Cognitive Radio Relay Networks *IEEE J. Sel. Areas in Commun.* **31** 476-88

[10] Zhou X, Ho C K and Zhang R 2016 Wireless Power Meets Energy Harvesting: A Joint Energy Allocation Approach in OFDM-Based System *IEEE Trans. Wireless Commun.* **15** 3481-91

Homogenization methods for the Sea Surface Temperature Data over the South China Seas

G S Wang[1], M Hou[2], Y Li[1,3], H Wang[1], W J Fan[1], K X Liu[1], J Gao[1], C Li[1]

[1] National Marine Data and Information Service, Tianjin 300171, China;
[2] Meteorological Bureau of Binhai New Area, Tianjin 300457, China

E-mail address: ok8200@sina.com

Abstract. Based on the metadata, Monthly Sea Surface Temperature (SST) series from nine marine stations over the South China Sea (SCS) are homogeneity detection and correction by Penalized Maximum T Test (PMT) method. The reference stations are developed using surrounding meteorological stations. Correction results show that: (1) The homogeneity detection and correction of marine observation stations should be based on the metadata, meanwhile, fully consider the influence of regional climate change factors. (2) Correlation analysis found that, the air temperature series from the surrounding meteorological stations is currently the optimal reference series. (3) The marine stations has 1~2 change points average, among them, changes of instrumentation and changes of location environment has great impact on the discontinuities. (4) The trend of the SST over SCS have a more pronounced warming trend during the past 52 years. Correction results indicate that the homogenization research in the SCS has an important meaning for the study of the SCS coast SST changes and climate change.

1. Introduction

The South China Sea (SCS) is the most typical monsoon region in the world, and it is influenced by the local Hadley circulation [1-2]. The inter-annual variability of SST over the SCS has important significance for the climate change research [3-5]. The quality of climate data is of extreme relevance for environment analysis, geographic information and remote sensing. Due to the changes of observation environment, measuring instrumentation, observation method or the human habit , the observed climate data sequences often includes many non-natural irregularities, which could lead to errors and bias in the results of environment analysis. For instance, the conclusion of coastal environment change research may has certain uncertainty caused by the influence of inhomogeneity [6]. As a result, the SCS marine observation data need to be detection and correction of irregularities.

Many countries take the climatological data homogeneity as an important task, and has develop many homogeneity datasets, which as the primary reference for the IPCC report [7]. Numerous studies has found [8-9], climatological data homogeneity detection must use direct and indirect homogenisation methods in both, meanwhile, absolute and relative homogenization methods are also very popular homogenization methods for climatic variables. The application of metadata is very important in homogeneity detection [10]. The Standard Normal Homogeneity Test (SNHT), Multiple Analysis of Series for Homogenization (MASH), Penalised Maximal F-test (PMF) and Penalised Maximal T-test (PMT) have been widespread used for assessing the significance trend in hydrological time series. Stephenson [10] was systematic evaluation of the homogeneity of daily SST for the Caribbean and neighbouring regions by used RHtest software. Zhai Panmao [11] firstly analysed gross errors and biases in China's historical radio sonde datasets. Li Qingxiang [12] combination statistical methods with station metadata, designed several tests and tested for inhomogeneity in all Chinese historical surface air temperature series from 1951 to 2001.

In summary, many domestic and scholars have made various researches on homogeneity detection and correction. However, previous studies focused on the meteorological data, and less for the SST data over the SCS. This paper will based on the metadata, surrounding meteorological stations as reference stations, and climate rationality analysis was carried out on the detection of change points and correction, monthly SST series of nine marine stations in coastal zone of the SCS are from 1960 to 2001 would be homogeneity detection and correction by PMT method.

2. Data and Methods

2.1. Data

The SST series form marine stations over the SCS used in this paper, which are obtained from the National Marine Data and Information Service (NMDIS), has been quality control checked , includes checks for physically unreasonable values, unreasonably long consecutive occurrences of the same values , illegal code, consistency inspection , and so on. The missing data has been interpolated by multiple linear regression method (Qinglan, Dongshan and Nanao marine station exit some missing data). SST series time range is from 1960 to 2011, the position distribution is shown in Figure 1.

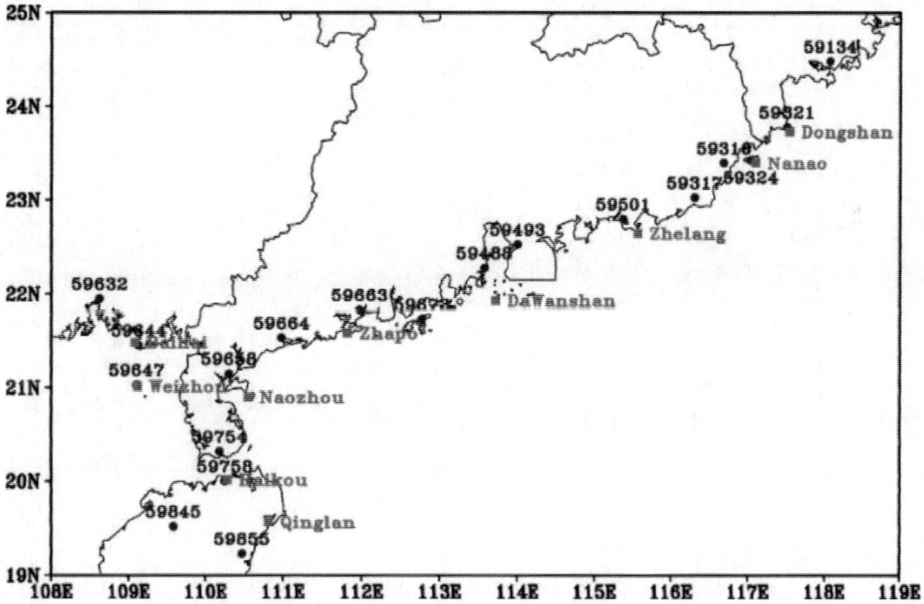

Figure 1. Locations of marine stations (block) and meteorological stations (circle) around the SCS.

The metadata used in this paper are obtained from the NMDIS, including relocation of monitoring stations, changes of measuring instrumentation, observing practices or environment change, and so on. In addition, the monthly mean surface air temperature (SAT) from meteorological observation stations over the SCS are obtained from National Meteorological Information Centre (NMIC). Surface air temperature series from 1960 to2011, has been quality controlled and procedures to detect and correct inconsistencies have been applied to homogenize the data. The position distribution is also shown in Figure 1.

The SST reanalysis data used in this paper includes: The Hadley Centre Global Sea Ice and Sea Surface Temperature (HadISST v1.1) data (1°×1°, 1961-2011) is provided by UK Met Office Hadley Centre. OISST (Optimum Interpolation sea Surface temperature, 1°×1°, 1981-2011,) monthly mean reanalysis data [13] and higher resolution daily average SST datasets (OISST.v2, 1/4°×1/4°, 1981-2011) [14] are both provided by US NOAA (National Oceanic and Atmospheric Administration) .(HadISST v1.1 data can be download from https://climatedataguide.ucar.edu/ climate-data/sst-data-hadisst-v11, OISST data can be download from http://www.ncdc.noaa.gov/oisst/data-access, OISST.v2 data can be download from ftp://eclipse.ncdc.noaa.gov/pub/OI-daily-v2/)

2.2. Methods

We use the RHtest version 4 software package to detect and adjust inhomogeneity in this study. This package includes the PMTred algorithm, which is based on the penalized maximum t (PMT) test. This algorithm deal with multiple change points using a recursive testing algorithm and account for the first order autocorrelation. The PMT test is greatly depend on the reference series. Since the much lower density of the SCS marine stations, it is difficult to choose the SST time series from the nearby marine stations to make the reference series. Compared to other homogenization methodology, PMT method can eliminate local trend signal from climate change, obtain better correction and results, but this method requires complete and high-quality reference comparisons. Rayner [14] adopted HadISST1 datasets (1°×1°) as a reference sequence to homogenise the candidate stations and has achieved good results. Since the SCS stations location of sparse, it is difficult to choose the right surrounding stations (reference stations) to homogenise the candidate station.

Table 1. The correlation coefficients of SSTA. The most significant correlation coefficients at confidence 0.01 level are in bold.

Marine Stations (1960-2011)	Meteorological Stations (1960-2011)			HadISST1 (1960-2011) 1°×1°	OISST (1981-2011) 1°×1°	OISST.v2 (1981-2011) 1/4°×1/4°
Baihai	59644	59647	59632	0.395	0.335	0.023
	0.907	**0.931**	**0.856**			
Dongshan	59324	59134	59321	-0.124	0.058	-0.134
	0.790	**0.746**	**0.833**			
Da Wanshan	59488	59673	59493	0.022	-0.088	-0.120
	0.568	**0.569**	**0.544**			
Haikou	59758	59754	59855	0.420	**0.492**	**0.613**
	0.776	**0.788**	**0.765**			
Naozhou	59658	59754	59664	0.370	**0.558**	**0.704**
	0.84	**0.81**	**0.83**			
Qinglan	59855	59845	59758	-0.022	-0.341	-0.381
	0.856	**0.836**	**0.784**			
Weizhou	59647	59644	59754	**0.528**	**0.471**	**0.654**
	0.819	**0.761**	**0.772**			
Nanao	59324	59321	59316	-0.130	-0.043	-0.186
	0.768	**0.810**	**0.720**			
Zhelang	59493	59317	59501	**0.681**	**0.506**	**0.582**
	0.674	**0.748**	**0.759**			
Zhapo	59663	59664	59673	0.378	0.409	**0.665**
	0.876	**0.890**	**0.898**			
Average		0.784		0.252	0.225	0.347

Studies have found that [15], the correlations must be greater than 0.70, which is indicate that at least 50% of the correction sequence variability is captured by the reference series. Therefore, this article compared the SST anomaly of candidate station correlation between the SAT data (distance criterion), OISST and HadISST, respectively. The HadISST1 data sets largely span 1960-2011, while the OISST and OISST.v2 data span 1981-2011. Due to the different lengths of the SST time series from different sources, we calculate the correlation coefficient between SST time series from candidate stations and the SAT from meteorological stations and SST time series from the International gridded data sets on different period which is based on the common data set lengths, respectively. In our study, when n=52, $r\alpha=0.05=0.27$, $r\alpha=0.01=0.35$; when n=31, $r\alpha=0.05=0.35$, $r\alpha=0.01=0.45$, here α denotes the confidence level. The results show that the average correlation coefficient of the SAT data was 0.784, which is much higher than OISST, OISST.v2 and HadISST1 datasets (Table 1).

Based on the above analysis, three surrounding meteorological stations, which is the correlation coefficient is higher than 0.70, is selected as reference stations for each candidate station. Because of lack of metadata information, and the correlation coefficients is lower than 0.70, the Da Wanshan station is temporarily revised. The Zhelang station uses only two meteorological stations as a reference stations due to the low correlation coefficient with No.59493 meteorological station. For each tested station, the arithmetic average of the three or two SAT time series from the reference stations is defined as a reference series. Using the PMT method as well as the reference series, the SST time series of the candidate stations are detected and adjusted. Consequently, the statistical break points will be obtained.

3. Detection of change points over the SCS

Tan Jun [1] pointed out that the SST anomaly over SCS has close relation to the ENSO phenomenon. Wang Weiqiang [4] found that the SST over SCS has inverted relationship with the Southern Oscillation Index. Based on the preliminary analysis of the marine station SST series over the SCS (Figure 2), the interannual variability of SST over the SCS has closely related with the climate change. In the process of station data correction, the inhomogeneity may be superimposed on a true climate trend [16], so the breakpoints are generally more difficult to detect. Several statistical methods have been proposed for detection of "undocumented change points" [7]. It is need to be fully consider the impact of local climate factors in the detection of "undocumented change points" of SST over the SCS. In this paper, the SST over the SCS are conducted by using PMT test methods [17]. Change points that are significant at the 95% level are significant change points for further investigation. Except for Qinglan station and Nanao station, the remaining seven candidate stations have full and detail metadata. Here in Dongshan station and Nanao station stand two typical example, briefly explain the process of detection of "undocumented change points" and detection of "documented change points". Then, the identified change points are adjusted using the quantile-matching (QM) adjustment method which is also included in the RHtest-v4 software.

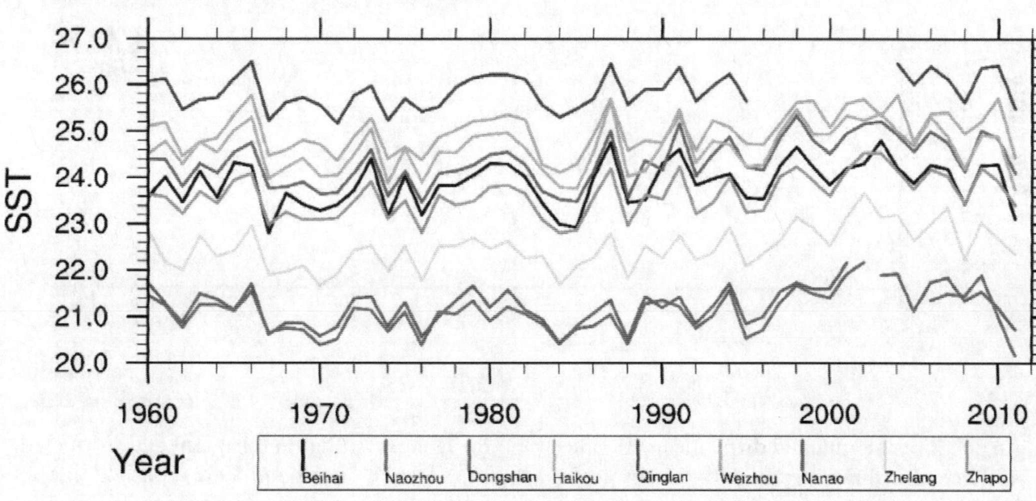

Figure 2. SST change along the coastal region of SCS from 1960 to 2011

3.1. Detection of documented change points

Based on metadata information of Dongshan station can be found, the temperature and salinity measuring site of Dongshan station was change in January 1969, the new station site of about 1km far away the old site. Since in September 1959, the temperature measurement instrument had changed many times (Table 2). Early adopt the SWL1-1 type SST thermometer measurements. The SST observation was cancelled in July 2002, and restore observation in April 2006. In January 2007 started using YSI-600-1 SST automation system for SST observation. According to the above metadata, Dongshan station has experienced a transformation from manual observation to automated observation in 2006-2007 period, observation instrument accuracy also exists a substantial change.

Figure 3 shows plots of the anomalies of the difference series between SST and SAT. The plots indicate that, the SST anomaly time series has a significant shifts in the 1969-70 period compare with SAT anomaly time series. This is the new location evident for the 1969-70 period for Dongshan station. Meanwhile, the SST anomaly time series has a significant shifts in the 2006-07 period compare with SAT anomaly time series, which is the new measuring instrumentation and new observing procedures evident for the 2006-07 period for Dongshan station.

Synthetic analysis indicates that, the SST anomaly time series for Dongshan station has been detection of change points following variance comparisons with SAT anomaly time series. This approach is considered to be proper and reasonable because the homogeneous SAT series works to detection of change points of SST anomaly time series.

Table 2. Changes of measuring instrumentation.

No.	Observation period	Measuring instrumentation	Accuracy of observation	Specification
1	1959.9.1-1959.9.30	SWL1-1	0.2°C	GB/T 14914-94
2	1959.10.1-2002.6.30	SWL1-1	0.2°C	GB/T 14914-94
3	2006.4.1-2006.12	SWL1-1	0.2°C	GB/T 14914-94 GB/T 14914-2006
4	2007.1—now	YSI-600-1	0.05°C	GB/T 14914-94 GB/T 14914-2006

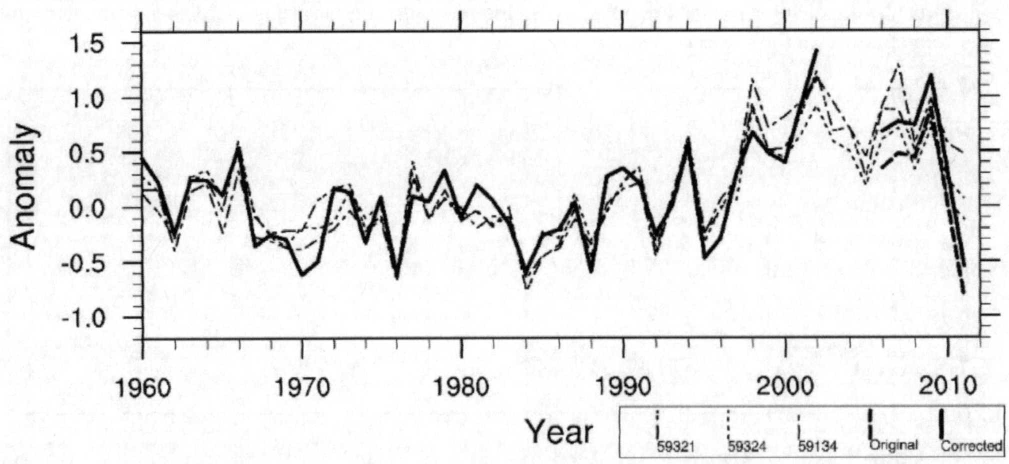

Figure 3. Plots of annually difference anomalies (°C) for Dongshan station (Original and Corrected, bold line) and three surrounding meteorological stations (59324, 59134 and 59321, dashed line).

Dongshan station, for example, shows the process of detection of documented change points. First, select three surrounding meteorological stations (59324, 59134 and 59321, the anomaly correlation coefficient are over 0.7) as reference stations to homogenise the Dongshan station. Second, with in significance at the 95% level , the homogeneity tests show that, only three breakpoints (January 1969 , April 2006 and January 2007) is consistent with the metadata (Figure 3). Third, eventually determine the three breakpoints and corrected.

3.2. Detection of undocumented change points

By analysing the marine station metadata has found, in addition to changes of location, changes of instrumentation, the effects of urbanization, land use of the surroundings, and other anthropogenic emissions of heat also caused SST data inhomogeneity. Nanao Station from October 1959 to begin SST observations, relying on temperature and salinity pier reef construction, conditions are suitable for SST observations. Because of the refrigeration factory is completed in 2007, SST measurement was crashed many times by the fishing boat. As a result of the fishing boat drainage, heat removal etc., can affect the accuracy of observation. Due to construction, strengthening pier, put the rocky drift into measurement area, causing the measurement area becomes shallow (Figure 4). Because the metadata not clear, based on the research of Peterson [18], in case of the metadata is unknown, after a significant test, and requires further climate rationality analysis of discontinuities. In this paper, in view of the stations are greatly influenced by local environmental change, based on the ENSO decadal distribution [19-20] and significant feature to detection of undocumented change points.

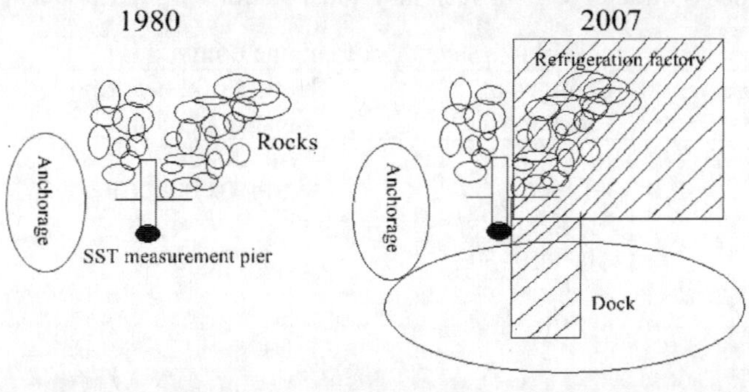

Figure 4. Changes of local environmental in Nanao station

In Nanao station as an example, explain the process of undocumented change points. Firstly, select three surrounding meteorological stations (59324, 59321 and 59316, the anomaly correlation coefficient are over 0.7) as reference stations to homogenise the Nanao station. Second, use PMT methods to detection the undocumented change points, then base on the metadata, there are six undocumented change points in 2005-2011 period (Table 3). On the basis of decadal distribution of ENSO, found that in August 2006-January 2007 and June 2009-April 2010 are weak and moderate warming events respectively, on June 2010-March 2011 is medium strength cold events. Significance test show that, there are three discontinuous points are the significant change points, that is, November 2006, April 2007 and October 2009.

Table 3. Change points of Nanao station.

No.	Change points	Significant test	Significant test range	ENSO	Significant change points	Rationality analysis
1	20050200	3.2442	(-2.8389-3.2816)	/	NO	Unreasonable
2	20060500	2.4962	(-2.5917-2.9981)	/	NO	Unreasonable
3	20061100	3.4407	(-2.2759-2.6289)	El Nino	YES	Unreasonable
4	20070400	3.8755	(-2.8369-3.2811)	/	YES	Reasonable
5	20091000	3.7217	(-2.8469-3.2935)	El Nino	YES	Unreasonable
6	20110100	2.3412	(-2.8172-3.2536)	La Nina	NO	Unreasonable

According to Peterson [18] and T.S.Stephenson [10] correction methods, climate rationality analysis of the above change points, at the same time satisfy the significant change points and the non-ENSO events both, established the change point of April 2007 of Nanao station and corrected.

3.3. statistics of change points

The statistics of nine marine stations change points in the SCS found (Table 4), the homogeneity tests produced a total of 14 change points. There are 5, 5 and 4 number of change points ,which is caused by changes of measuring instrumentation, changes of location or changes of local environmental of the surroundings, respectively, and the average absolute value of the corrected was 0.321 °C , 0.414 °C and 0.272 °C.

Correction of the four marine stations which is changes of location is about 1 time, moving station distance between 2 km and 8 km away. Six change points of instrument changes occurred after 2002, which is the observation to transmission and mechanisation processes.

Time distribution from the change points can be found, anthropogenic observation to automated observation is the main reasons for SST inhomogeneity in the nearly two decades. Shallow measuring area, port congestion, human emissions has a greater influence on the marine stations SST data. Time distribution of change points is scattered, did not focus in a certain time period (except for 2002), indicating that the process of detection between individual stations are independent.

Table 4. Statistics of change points.

Marine Stations	Change points	Changes of location	Changes of measuring instrumentation	Changes of local environmental
Beihai	1	/	200505	/
Dongshan	4	196901,	200604,200701	/
Haikou	1	196601	/	/
Naozhou	3	199001,199606	200206	/
Qinglan	1	/	/	198307
Weizhou	1	200301	/	/
Nanao	1	/	/	200704
Zhelang	1	/	200201	/
Zhapo	2	/	/	197912,198803

4. Result analysis

To verify the correction of PMT methods, the OISST datasets and HadISST datasets were calculated the correlation coefficient respectively with the SST anomalies after correction (Table 5). After the corrected, the average correlation coefficient were 0.492, 0.482 and 0.594, respectively. Compared with the original, correlation coefficient increased about more than doubled, which Dongshan station, Nanao station and Qinglan station improvement is most evident. The linear trend of annual mean SST over the SCS can be found (Table 6), after the correction, the linear warming trend of the nine marine stations average annual SST over the SCS has obvious increase, this is consistent with the previous research [6]. The correlation coefficient of the homogenous SST series have significantly improved.

Table 5. The correlation coefficient of the homogenous SST series and the candidate SST time series. The most significant correlation coefficients at confidence 0.01 level are in bold.

	Beihai	Dongshan	Haikou	Naozhou	Qinglan	Weizhou	Nanao	Zhelang	Zhapo	Average
HadISST1	**0.400**	**0.608**	**0.421**	**0.366**	**0.458**	**0.531**	**0.616**	**0.652**	**0.373**	**0.492**
OISST	0.303	0.432	**0.507**	**0.575**	**0.632**	**0.461**	**0.496**	**0.521**	0.413	**0.482**
OISST.V2	0.031	**0.732**	**0.631**	**0.713**	**0.527**	**0.670**	**0.761**	**0.602**	**0.682**	**0.594**

Table 6. The linear trend of annual mean SST over the SCS.

Marine	Linear trend (°C/century)		Significantly test	
Stations	Original	Corrected	Original	Corrected
Beihai	0.94	1.45	0.45	0.67
Dongshan	0.81	1.27	0.56	0.72
Haikou	0.96	1.44	0.44	0.77
Naozhou	1.70	1.60	0.38	0.59
Qinglan	0.78	1.62	0.41	0.64
Weizhou	0.84	1.01	0.48	0.78
Nanao	0.85	1.14	0.51	0.65
Zhelang	1.43	1.53	0.32	0.72
Zhapo	1.28	1.57	0.52	0.69
Average	1.04	1.37	0.45	0.69

Figure 5 shows the corrected (dashed line) and original (solid line) mean annual SST curves for the all SST series around the SCS. This result shows that the inhomogeneous time series appears to underestimates the warming trend during the past 52 years. The average warming rate based on the homogenous SST series over the SCS is about 1.37 °C/century, and increased by a total of 0.71 °C in the last 52 years (1960- 2011). The warming trend in the central SCS is most prominent, and coastal SST changes in Guangdong province is the largest

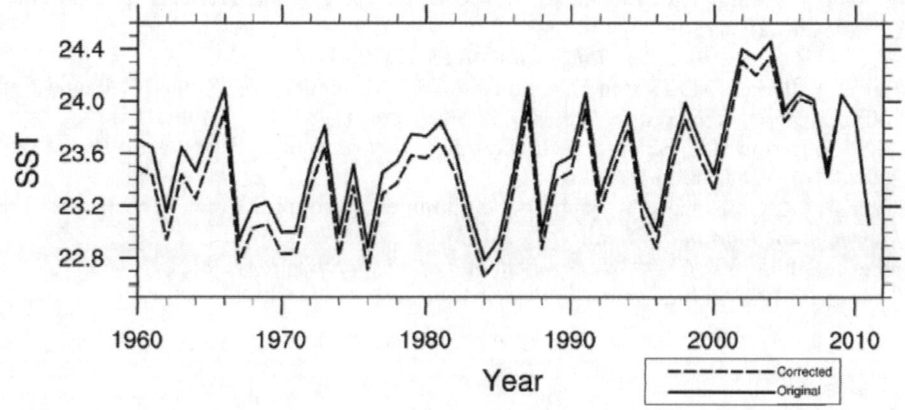

Figure 5. Corrected (dashed line) and original (solid line) mean annual SST curves for the all SST series around the SCS.

5. Conclusion

The SST series over the SCS is inhomogeneity, which has influence on the research of climate change and marine. Based on the metadata, the SST series of nine marine stations over the SCS are homogeneity detection and correction by PMT method. Select the surrounding meteorological stations as reference stations to homogenise the candidate station, and climate rationality analysis was carried out on the detection of change points and correction.

Correction results show that: (1) The homogeneity detection and correction of marine observation stations should be based on the metadata, meanwhile, fully consider the influence of regional climate change factors. (2) Correlation analysis found that, the surrounding meteorological station temperature data is the optimal reference data. (3) The average marine stations has 1~2 change points, among them, changes of instrumentation and changes of location environment is the most common. The anthropogenic observation change to automated observation is the main reasons of SST inhomogeneity in the last two decades. (4) The trend of the SST over SCS have a more pronounced warming trend, increased by an average 0.0137 °C each year. Correction results indicate that the homogenization research in the SCS has an important meaning for the study of the SCS coast SST changes and climate change.

References

[1] Tan J, Zhou F X and Hu D X 1995 *Oceanology Et Limnology Sinica* **26(4)** 377-82
[2] Liu X and Lin X P. 2013 *Periodical Of Ocean University Of China* **43(12)** 001-6
[3] Yu S Y, Zhou F X, Fu G, Wang D X. 1994 *Oceanologia Et Limnologia Sinica* 25(5) 546-51
[4] Wang W Q, Wang D X, Qi Y Q. 2000 *Acta Oceanologica Sinica*. 22(4) 8-16
[5] Arguez, A., Huang, B., Mcmahon, J. R., Menne, M. J., and Zhang, H. 2015 *Science* 348(6242) 1469-72
[6] Li Q X, Li W. 2007 Acta Meteorologica *Sinica* **65(2)**:293-300
[7] Cao L J, Ju X H, and Liu X N 2010 *Meteorological Monthly* **10** 52-6
[8] Ren G Y, Guo J and Xu M Z 2005 *Acta Meteorologica Sinica* **63(6)** 948-52
[9] Li Q X, Liu X N, Zhang H Z, Tu Q Q 2003 *Meteorological Science And Technology* **31(1)** 3-10
[10] Stephenson T S, Goodess C M and Haylock M R 2008 *Journal of Geophysical Research Atmospheres* (1984–2012) **113(D21)** 23-31
[11] Zhai P M 1997 *Acta Meteorologica Sinica* **55(5)** 563-72
[12] Li Q, Liu X, Zhang H 2004 *Advances in Atmospheric Sciences* **21(2)** 260-8
[13] Reynolds R W, Rayner N A, Smith T M and Wang W 2002 *Journal of climate* **15(13)** 1609-25
[14] Reynolds R W, Smith T M, Liu, C, Chelton D B, Casey K S, and Schlax M G 2007 *Journal of Climate* **20(22)** 5473-96
[15] Lund R and Reeves J 2002 *Journal of Climate* **15(17)** 2547-54
[16] Rayner N A, Parker D E, Horton E B, Folland C K, Alexander L V, Rowell D P, and Kaplan A 2003 *Journal of Geophysical Research* Atmospheres (1984–2012) **108(D14)**
[17] Chapin F S, Berlow E L, Bloomfield J, Dirzo R, Hubersanwald E, Huenneke L F, and Wall D H 2000 *Science* **287(5459)** 1770-74
[18] Easterling D R, Groisman P Y, Nicholls N, Plummer N, Torok S, Auer I and Parker D E 1998 *International Journal of Climatology* **18(13)** 1493-517
[19] Malcher J, Schönwiese C 1987 *Theoretical and applied climatology* **38(3)** 157-66
[20] Schönwiese C 1987 *Beiträge zur Physik der Atmosphäre* **60(1)** 49-64

IOP Conf. Series: Earth and Environmental Science **52** (2017) 012047 doi:10.1088/1755-1315/52/1/012047

Assessing spatial distribution of soil erosion in a karst region in southwestern China: A case study in Jinfo Mountains

H Y Zhou[1, 2], X Y Pan[1, 3] and W Z Zhou[1, 4]

[1] School of Geographical Sciences, Southwest University, Chongqing 400715, China
[2] No.1 Experimental Middle School Attached to Sichuan Normal University, Sichuan 61000, China
[3] Yonye Saiboneng Land Surveying and Planning Company, Wuhan 40061, China

E-mail: zhouwz@swu.edu.cn

Abstract. Soil erosion is serious with rocky desertification areas appearing in mountainous Karst regions in southwest China due to a conspicuous contradiction between man and the land resource. Land use and land cover play significant roles in regional soil erosion by water. This paper aimed to quantify regional soil erosion and to explore relationships between soil erosion and land use/land cover in order to locate high risk areas requiring soil conservation. Based on GIS, the Revised Universal Soil Loss Equation (RUSLE) was employed for erosion assessment for a typical Karst region, Jinfo Mountain region in southwest China, using local parameters. Spatial distribution of topsoil erosion was analyzed and relationships between soil erosion and land use/land cover changes (LULC) were statistically explored and discussed for regional erosion control. The overall values were under 25 t.ha.a, with the medium erosion areas accounting for 12.7% and the intense and very intense erosion areas totalled about 6%. The relations between soil erosion and LULC are complicated in this Karst mountainous region. Generally, the amount of ground cover, soil conservation measures, and cultivation disturbance have played critical roles in topsoil loss in the Jinfo mountain region. The reduced ground cover levels accompanying greater cultivation disturbance lead to higher erosion intensity in each landscape, and vice versa.

1. Introduction

Soil erosion is one of major global environmental problems causing negative impacts that limit the sustainable development of land resources [1]. It is a cumulative process that depends on factors such as land use, vegetation cover, climate, topography, and on soil physical properties such as texture, organic matter, structure and porosity. Among these, land use and land cover are two of the most important factors that influence the occurrence and intensity of surface runoff and soil erosion [2, 3]. Land cover can improve some physical–chemical properties of the soil through both physical protection of the soil surface, the contribution of organic matter from the plant canopy and the root system [4], and by the increased soil structural stability due to increasing soil organic carbon and aggregation [5]. Land uses which change the type and coverage of vegetation contribute not only to the changes of the runoff and both the physical and chemical properties of soil, but also the development of the soil erosion. Soil erosion causing the loss of fertile soil constrains the structure and distribution of the land use which intensify the contradiction between human and land resource [6]. Rational land use /land cover changes can greatly improve soil properties [7] and reduce soil erosion [8, 9]. Scholars have explored the relationship between LULC and soil erosion from different perspectives using different methods[10-13], such as analysis of driving factors in the process of soil erosion by observation stations data[14], effect of LULC on soil erosion and sediment transport by the decision support system and GIS[15], regional land use change contributing to reducing soil erosion using models[16], and spatial relationships between natural and socio-economic indicators and the index of soil erosion risk[17].

Nevertheless, studies on the relationship between soil erosion and LULC in Karst regions are fewer than in other areas. The studies on soil erosion in Karst region mainly focused on the characteristics [18], cause [19], process [20], and erosion risk assessment[21]. Studies on soil erosion of limestone hill slopes in semi-arid or arid areas have been carried out from the perspective of the hydrology based on statistical and parametric models [22-24]. New methods have been introduced in studies of soil erosion in Karst area, such as ^{137}Cs tracer [18] and fuzzy modeling [21]. Human activities on land use and management have been considered as one of main influential factors in Karst regions [19].

The Karst region of southwestern China is the largest Karst region in the world as well as one of the most fragile ecotones [25]. With high and steep mountains, low vegetation coverage, high precipitation intensity and variability, conspicuous contradiction between human activities and the land resource, soil erosion is serious and even karst rocky desertification areas appear in the southwestern China. As an ultimate state of soil erosion, rocky desertification means collapse of the ecology and environment, and it is hard to restore because of its low rate of soil formation, thin solum, low soil erosion tolerance and fragile ecosystem. Improper land use may cause severe soil erosion and Karst rocky desertification. So it is important to quantify regional soil erosion and to explore relationships between soil erosion and LULC in order to locate the critical areas for conservation in this region. The objectives of this paper were to quantify the characteristics and spatial differentiation of soil erosion in the Jinfo mountain region and to explore the relationships between soil erosion and LULC in Southwestern China.

2. Materials and methods

2.1. *Study area*

The Jinfo Mountain region which is one of typical Karst areas in southwestern China covers about 1260.58 km^2,lies between 106° 54′E~107° 27′E and 28° 46′N~29° 30′ in Nanchuan district of Chongqing municipality, the southeast boundary of Sichuan Basin with elevation from 500m~1000m. The topography in this area is complicated with hills and slots composed of karst monadnocks and terraces in the north, with folded mountains with high steep cliffs, valleys and karst caves in the south. The subtropical humid monsoon climate in this area gives it not only abundant heat but also wide variation in its annual and seasonal rainfall patterns. Typically, mean annual temperature is about 9.5℃ and the mean precipitation is about 1390 mm annually with most rain occurring from May to October when the temperature is relatively high. Rain with heat at the same period has accelerated karst development. The main soil types are yellow brown soil, yellow soil, limestone soil, purple soil and paddy soil. The main land use types are forestlands, hill dryland, and hill paddy land. The geography in the Jinfo Mountain is a typical pattern of the Karst region in southwestern China.

2.2. *Data sources*

The Landsat-5 TM (2010-08-19) remote sensing image was geometrical precise adjusted by reference to the DEM with a resolution of 1:50,000. The NDVI was derived from the remote sensing image by the index calculating in the spectral enhancement tool. The local land use map with 1:25,000 scale was obtained from the Bureau of Land and Resource of Nanchuan County and classified into 8 types, i.e. dryland, paddy field, woodland, garden-plot, grassland, bare land, construction land, and water area. Slope map of study area was produced by the surface analysis module tool in ArcGIS 9.3 software according to its Digital Elevation Model (DEM). Precipitation data recorded by nearby weather observation stations was interpolated into raster precipitation data and adjusted according to the DEM. Soil texture and organic matter data come from the second Chinese soil survey data.

2.3. Soil Erosion Model

The Revised Universal Soil Loss Equation (RUSLE) [26], formula 1, explains how climate, soil, topography, and land use affect rill and inter rill soil erosion caused by raindrop impact and surface runoff. It has been extensively used to estimate soil erosion loss and to assess soil erosion risk.

$$A = R\times K\times L\times S\times C\times P \quad (1)$$

where A is the annual soil erosion module ($t.ha^{-2}.a^{-1}$) , R the rainfall erodibility factor ($MJ\bullet mm\bullet ha^{-2}\bullet h^{-1}\bullet a^{-1}$), K the soil erodibility factor (m.t.ha....) , L the slope length (m), S the slope gradient(%), C the crop management factor, and P the erosion control practice factor. Factors C and P are dimensionless.

The grid map of soil erosion in 2010 was produced according to the RUSLE equation using grid calculator in ArcGIS 9.3 software. Then the soil erosion class map and soil erosion intensity classification were obtained according to SL 190-2007 standards [27] and the spatial differences in annual soil erosion were analyzed. The relationships between LULC and soil erosion were explored in the study area.

2.4. Soil erosion intensity

Researchers demonstrated that the index of soil erosion intensity can reflect the soil erosion intensity on different land use [28, 29]. The index of soil erosion intensity is calculated by using equation 2.

$$E_j = 100\times\sum_{i=1}^{n} C_i \times A_i / S_j \quad (2)$$

where E_j refers to soil erosion intensity index of the j kind of land use, C_i refers to the *i* grade value of the soil erosion intensity, A_i refers to the area of *i* grade soil erosion intensity in *j* kind of land use, S_j refers to the area of the *j* kind land use, *n* refers the total grade of soil erosion intensity. The value of E_j was multiplied 100 times for statistical purposes. The soil erosion intensity was classified into six grades as very low, mild, medium, intense, very intense and severe which were assigned with the values of 1, 2, 3, 4, 5, 6 respectively in which the larger the value, the more intense the soil erosion. Then the response relationship between soil erosion and land use was analyzed quantitatively according to the soil erosion intensity index.

2.5. Validation

To check the performance of the RUSLE model, we set up four 10x5m observing plots, two for dryland and two for grassland, in the study area. Each plot had been set a dam to deposit sediment and to let out water. The field observations were from January to December in 2010 and the sediment was sampled, air-dried and weighed at the end of each season. The annual soil erosion of every plot was totaled from seasonal sediment to test the simulated site data.

2.6. Statistics of soil erosion characteristics

The soil erosion module, the area and percentage of different soil erosion intensity in different land use were obtained with statistic according the land use map and the soil erosion map of study area. Two special land covers, bare rocks and water areas had no soil erosion due to lacking soil on their surface.

3. Results and discussion

3.1. The spatial distribution of soil erosion

The annual soil erosion was classfied into very low, mild, medium, intense, very intense and severe erosion grades while soil erosion rates were less than 5, 5-25, 25-50, 50-80, 80-150 and more than 150 t.ha.a respectively according to the SL 190-2007 standards (Table 1). The soil erosion intensity in the study areas are mainly very low, accounting for 61.9% of the study area. Secondly, mild and medium erosion areas accounted for 19.3% and 12.7% respectively. The intense and very intense erosion areas were small comparatively and summed to about 6%. The severe erosion areas were few and covered only 2.3 ha in the study area.

Table 1. The statistics of annual soil erosion of study area

erosion grade	standard for erosion/t.ha.a	erosion area/ ha	percentage of erosion area
very low	<5	78057.1	61.9
mild	5-25	24322.1	19.3
medium	25-50	15983.4	12.7
intense	50-80	6901.2	5.5
very intense	80–150	792	0.6
severe	>150	2.3	0.0

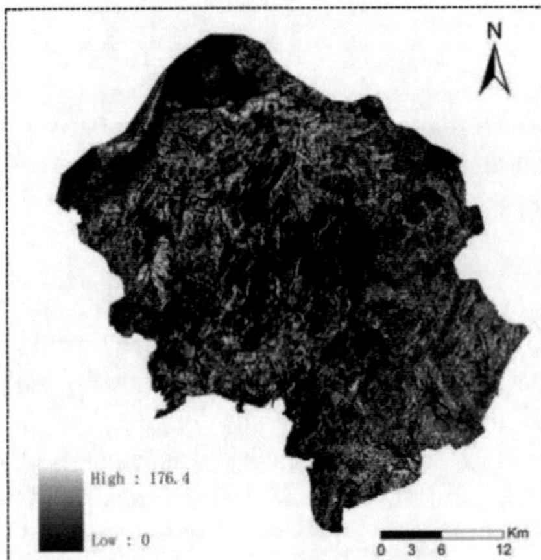

Figure 1. The soil erosion map

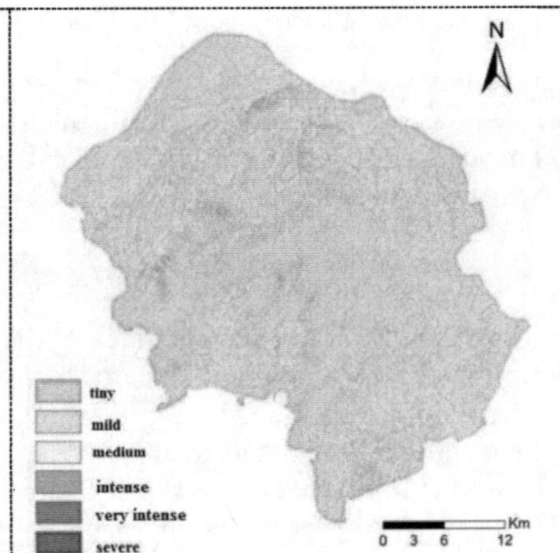

Figure 2. The erosion intensity grade map

The soil erosion (Figure 1) and erosion intensity grading (Figure 2) maps demonstrate that the distribution of soil erosion was different in space in terms of geographical environment and land use. According to statistics, the annual average erosion rate is about 11.3 t/ha, which is a little more than the tolerance level set for the southwest China by the Ministry of Water Resources of China. The total annual erosion quantity is about 142.4 x10^4 tons in the study area. The mass values were under 25 t.ha.a and the maximum erosion rate amounted to 176.4 t.ha.a. Horizontally, the erosion intensity increases from the middle to the surroundings in the study area. Vertically, it increased from the peak of Jinfo Mountain not only to the west but also northeast and southeast. In general, the soil erosion intensity in the east and southeast mountains are mainly mild and medium which are higher than other areas. Conversely, most of the soil erosion intensity in the west and middle region is very low and lower than other areas.

3.2. The analysis of relationship between soil erosion and LULC

Table 2 shows the statistics of annual soil erosion characteristics of different land use. From the soil erosion map, grassland had the highest estimated erosion rate, amounting to 15.7 t.ha.a, followed by the woodland and garden plot which were 12.9 and 12.4 t.ha.a respectively and had fairly high soil erosion rates. Grassland was usually distributed in areas with relatively high slope and low cover levels where the topsoil is easily eroded by runoff. The woods in woodlands are usually mixed conifers and broadleaf trees which grow in the subtropics. With low ground cover, the topsoil in those woodlands was eroded by raindrops from broadleaf on which produce influx. On the other hand, some of woodlands were returned from slope farmland and planted some young trees with low vegetation cover since projects of returning farmland for forest and grass was enfored by the goverment in cultured land areas with slope which were more than 25 degree recent years. The returned woodland area amounted to 11775 hm^2 from 2000 to 2010 in the study area. As for the garden plots, beside the raindrop conflux erosion from broad leaves of trees, frequent disturbance by cultivation cause the relatively high erosion. This may be the result of human activity such as weeding and fertilization which not only improve the growth of the fruit tree or the tea tree but also decrease the vegetation cover which may cause soil erosion over topsoil. The annual soil erosion module of dryland was 10.3 t·ha·a and lower than that of grassland, woodland and garden plot due to low slope and preferable erosion control practice. The erosion rate of dryland in the study area was different from those studied results that dryland had the most erosion rates in Loess Plateau of China [8, 30]. The erosion modules of paddy field and construction land, where the erosion protection control is best in those land use and land cover, is 5.7 and 3.3 t.ha.a respectively. .

The index of soil erosion intensity *E*j showed the impacts of soil erosion by water on different land uses/land cover in the study area. Like the mean soil erosion module, the index of the soil erosion intensity on construction land was 120.5, which was smaller than those on other land uses. The paddy field had a middle erosion intensity index that was 144.6. While the indexes of grassland, garden plot, dryland and woodland were fairly high and amount to 183.9, 183.9, 168.5, and 167.7 respectively. Although having different mean erosion moduls, the woodland and dryland had approximate indexes of soil erosion intensity.

Table. 2 The soil erosion characteristics of different land use

		dryland	**paddy field**	**woodland**	**garden plot**	**grassland**	**construction land**
erosion rate (t.ha.a)		10.3	5.7	12.9	12.4	15.7	3.3
E_j		168.5	144.6	167.7	174.0	183.9	120.5
very low	area（hm^2）	7645.9	8848.8	51495.6	1019.9	3060.9	4496.1
mild	area（hm^2）	9912.1	5335.1	7969.6	608.8	780.3	558.4
medium	area（hm^2）	1389.4	397.6	12230.3	390.6	1139.2	168.3
intense	area（hm^2）	107.8	127.4	5937.7	43.1	451.1	52.3
very intense	area（hm^2）	9.3	13.3	696.3	2.4	46.2	8. 0
severe	area（hm^2）	0	0.1	1.9	0	0	0.1

In the study area, the erosion intensity of very low and mild accounted for the main area and the intense, very intense and severe intensity were small in the six kinds of land use and land cover. But the proportion of erosion intensity area varies for one land use/ land cover. The very low intensity accounted for 40.1%, 60.1%, 65.7%, 49.4%, 55.9% and 85.1% respectively in dryland, paddy field, woodland, garden plot, grassland and construction land. The mild intensity had 52.0%, 36.2%, 10.2%, 29.5%, 14.2% and 10.6% respectively in those types of land use and land cover. Nevertheless, what was noteworthy was that more than half of the dryland had mild erosion intensity. Although the average erosion rate is relatively low in the woodland, garden plot and grassland, the medium intensity took up 15.6%, 18.9% and 20.8% respectively, and the intense class did 7.6%, 2.1% and 8.2% respectively. For those land uses and land cover, the very intense erosion is no more than 1% and the severe intensity areas were hardly any in the study area. For the grassland and woodland, the terrain with escarpment and poor ground cover had led to medium and intense intensity. As for the garden plot, low cover rate and frequent disturbance by cultivation had contributed to the topsoil loss. So those areas in woodland, garden plot and grassland would be the critical ones to prevent soil from water erosion in the study area.

4. Conclusion

The spatial distribution of anual soil erosion by water varied due to the terrain, soil properties, frequency and intensity of rainfall, land cover, land use and its management in the Jinfo Karst Mountain region. Although the annual average erosion is about 11.3 t.ha.a, the maximum erosion rate amounted to 176.4 t.ha.a. The mass values were under 25 t.ha.a , including very low intensity which accounted for 61.9% and mild intensity which accounted for 19.3% of the study area. Medium erosion areas accounted for 12.7%, and intense and very intense erosion areas for about 6%.

The relations between soil erosion and LULC are complicated in the Karst mountainous region. Generally, the ground cover rate, protective measures, and cultivation disturbance had played critical roles in topsoil loss in Jinfo Karst Mountain region. The lower the ground cover rate and the greater the cultivation disturbance, the higher the erosion intensity became in each landscape, and vice versa. In terms of annual erosion rates and indexes of erosion intensity of each land use/land cover, the grassland had the maximum erosion rate and highest index of soil erosion intensity; while there are medium erosion and erosion intensity index for garden plot, woodland and dryland, and low values for paddy field and construction land. It was critical that 52.0% of the dryland was under mild erosion intensity, with 7.3% under medium intensity. What was worth paying more attention to was that there were 15.6%, 18.9% and 20.8% of areas with medium erosion intensity, and 7.6%, 2.1% and 8.2% of intense erosion for woodland, garden land and grassland respectively, although the topsoil erosion rates in most areas with different land use/land cover was under the tolerance level set for the southwestern China by the government. Those areas with relatively high erosion would be the key ones to focus on for prevention of soil loss from water erosion in the study area.

Acknowledgement
The work was financially supported by the Fundamental Research Funds for the Central Universities in China (no. XDJK2014C013). We also thank the two anonymous reviewers for their suggestions.

References

[1] Marques M J, Bienes R, Perez-Rodriguez R, et al. 2008 Soil degradation in Central Spain due to sheet water erosion by low-intensity rainfall events *Earth Surf. Proc. Land* **33 (3)** 414 - 423.
[2] Morgan P C 1995 *Soil Erosion and Conservation* Longman, London.
[3] Cerdan O, Le Bissonnais Y, Couturier A, et al. 2002 Modelling interrill erosion in small cultivated catchments *Hydrological Processes* **16** 3215–26.
[4] Bronick C J, Lal R 2005 Soil structure and management: a review *Geoderma* **124** 3–22.
[5] Six J, Elliott E T, Paustian K, Doran J W 1998 Aggregation and soil organic matter accumulation in cultivated and native grassland soils *Soil Science Society of America Journal* **62** 1367–77.
[6] Li R 2011 The research on the process and adjustment of soil erosion in the main water eroded region of China *Bulletin of soil and water conservation* **31(5)** 1-6. In Chinese with English abstract.
[7] Kosmas C, Gerontidis S, Marathianou M 2000 The effect of land use change on soils and vegetation over various lithological formations on Lesvos (Greece) *Catena* **40** 51–68.
[8] Fu B J, Wang Y F, Lu Y H, et al. 2009 The effects of land-use combinations on soil erosion: a case study in the Loess Plateau of China *Progress in Physical Geography* **33** 793–804.
[9] Zhang B, Yang Y S, Zen H 2004 Effect of vegetation restoration on soil and water erosion and nutrient losses of a severely eroded clayey Plinthudult in southeastern China *Catena* **57** 77–90.
[10] Islam K R, Weil R R 2000 Land use effects on soil quality in a tropical forest ecosystem of Bangladesh *Agriculture ecosystems & Environment* **1(79)** 9-16.
[11] Adelia N, Nunes, Antonio C, De Almeida, Celeste O A 2011 Impacts of land use and cover type on runoff and soil erosion in a marginal area of Portugal *Applied Geography* **31** 687-699.

[12] Henry A, Mabit L, Jaramillo R E 2013 Land use effects on erosion and carbon storage of the Rio Chimbo watershed, Ecuador *Plant and Soil* **1-2(367)** 477-491.

[13] Ziadat F M, Taimeh A Y 2013 Effect of rainfall intensity, slope, land use and antecedent soil moisture on soil erosion in an arid environment *Land degradation & Development* **6(24)** 582-590.

[14] Ries J B 1995 Geomorphodynamic processes in the eastern central Himalaya - Consequences of land use and soil erosion for the lowlands *Mitteilungen der osterreichischen geographischen gesellschaft* **137** 187-202.

[15] Rodda, Stroud M J, Shankar U 2001 A GIS based approach to modelling the effects of land-use change on soil erosion in New Zealand *Soil use and Management* **1(17)** 30-40.

[16] Panagos Christos, Karydas Cristiano, et al. 2014 Seasonal monitoring of soil erosion at regional scale: An application of the G2 model in Crete focusing on agricultural land uses *International Journal of Applied Earth Observation and Geoinformation* **4 (27)** 147-155.

[17] Sabbi A., Salvati L. 2014 Searching for a downward spiral soil erosion risk, agro-forest landscape and socioeconomic conditions in Italian local communities *Land Use Policy* **41** 388–396.

[18] Fu W L, Zhang Z W, Zhang H, et al. 2007 Study on characteristics of soil erosion on Karst Hillslope J. *Soil Water Conserve*. **21** 38–41. In Chinese with English abstract.

[19] Drew D P 1983 Accelerated soil erosion in a karst area: the Burren *Western Ireland. Journal of Hydrology* **61** 113–124.

[20] Jian P, Yue Q X, Ren Z, et al. 2013 Soil erosion monitoring and its implication in a limestone land suffering from rocky desertification in the Huajiang Canyon, Guizhou, Southwest China *Environ Earth Sci* **69** 831–841.

[21] Yang Q Y, Xie Y Q, Li W J, et al 2014 Assessing soil erosion risk in karst area using fuzzy modeling and method of the analytical hierarchy process *Environ Earth Sci* **71** 287–292

[22] Calvo-Cases A, Boix-Fayos C, Imeson A C 2003 Runoff generation, sediment movement and soil water behaviour on calcareous (limestone) slopes of some Mediterranean environments in southeast Spain *Geomorphology* **50** 269–291.

[23] Imeson A C, Lavee H, Calvo A, Cerdà A 1998 The erosional response of calcareous soils along a climatological gradient in Southeast Spain *Geomorphology* **24** 3–16.

[24] Kheir R B, Abdallah C, Khawlie A 2008 Assessing soil erosion in Mediterranean karst landscapes of Lebanon using remote sensing and GIS *Engineering Geology* **99** 239–254.

[25] Xiao H, Weng Q 2009 The impact of land use and land cover changes on land surface temperature in Karst area of China *Journal of Environmental Management* **15** 115-121.

[26] Wischmeier W H and Smith D D 1978 *Predicting Rainfall Erosion Losses* USDA Agricultural Research Services handbook 537, Washington, DC.

[27] Ministry of Water Resources of China 2008 Standards for Classification and Gradation of Soil Erosion. Beijing. In Chinese

[28] Wang S Y, Wang G Q, Chen Z X 2005 Relationship between land use and soil erosion in Yellow River Basin *Journal of Natural Disasters* **14(1)** 32-37.

[29] Yao H R, Cui B S 2006 The effect of Land use and its change on soil erosion of the Lancang River watershed in Yunnan Province *Acta Scientiae Circumstantiates* **26(8)** 1362 -71.

[30] Li Y K, Ni J, Yang Q K, Li R 2006 Human impacts on soil erosion identified using land-use changes: a case study from the Loess Plateau, China. *Physical Geography* **27(2)** 109 – 126.

Effects of pruning intensity on jujube transpiration and soil moisture of plantation in the Loess Plateau

Zhenyi Nie[1], Xing Wang[2, 5], Youke Wang[2], Jianpeng Ma[2], Xinguang Wei[3] and Dianyu Chen[4]

[1] Institute of Soil and Water Conservation, Northwest Agriculture and Forestry University, Yangling, 712100, China.
[2] Institute of Soil and Water Conservation, Chinese Academy of Sciences and Ministry of Water Resources, Yangling, 712100, China.
[3] College of Water Conservancy, Shenyang Agricultural University, Shenyang, 110866, China.
[4] College of Water Resources and Architecture Engineering, Northwest Agriculture and Forestry University, Yangling, 712100, China.

E-mail: gjzwyk@vip.sina.com

Abstract. In order to ease soil desiccation and prevent ecological deterioration in the Loess Plateau, where jujube (Zizyphus jujube MIll) is widely cultivated as a drought tolerant plant, four pruning intensities (PI), from PI-1 (light) to PI-4 (heavy) were set up based on total length of secondary branches to study the effects of pruning on transpiration and soil moisture in jujube plantations. Furthermore, growth indexes were regularly monitored to estimate jujubes biomass. Sap flow, meteorological and soil moisture conditions were monitored using thermal dissipation probes (TDP), weather station (RR-9100) and the combination of time domain transmission (TDT) technology and neutron moisture gauges (CNC503B), respectively. The results showed that daily actual transpiration of jujube was positively correlated with leaf biomass. Compared with PI-1, jujube transpiration during growth period under PI-2, PI-3, and PI-4 dropped by 11.1%, 29.2%, and 47.9%, respectively. On the contrary, annual water storage under PI-2, PI-3, and PI-4 increased by 6.29 mm, 25.78 mm and 34.74 mm while water use efficiency increased by 5.1%, 15.7% and 24.2%, respectively. Overall, increase in pruning intensity could significantly reduce water consumption of jujube and improve soil moisture in jujube plantations.

1. Introduction

Drought and soil erosion are critical limitations for ecological restoration and agricultural development in the Loess Plateau of China [1]. Soil and water loss in the Loess Plateau have significantly improved since the passing of the policy to reconvert farmlands to forests by the Chinese Government in 1999. However, due to the farmland-to-forest conversion policy, increase in vegetation cover has increased demand for soil water. Meanwhile, the increasing vegetation cover has also increased soil water deficit on the Loess Plateau. Large areas of dry soil layers have developed with deregulation of forest soils, a condition that has affected normal vegetation growth and has resulted in degradation and decline in the community and ecosystem [2-4]. Preventing the severe soil desiccation and the deterioration of ecological environment in the semiarid hilly region is a major ecological issue in the study area.

Jujube (*Zizyphus jujube* MIll) is a major economic forest in the Loess Plateau. The planting area of jujube in Shanxi province exceeds 100 acres, mostly planted in the mountains where dry farming is the main mode of cultivation due the severe scarcity of water for irrigation [5]. In the effort to protect the fragile local ecological environment, various water-saving technologies have been proposed, including rainwater harvesting for micro-irrigation, interception ditches and mulching to hold back soil moisture, all of which aim at increasing rainfall infiltration, reducing surface evaporation, increasing water use efficiency and increasing yield of jujube [6-9]. But as water consumption by jujube plant increases with increasing plant age [10], soil desiccation has worsened [4] and traditional water-saving technologies in the region have become ineffective.

Pruning not only regulates fruit tree canopy by increasing light conditions, tree growth and fruit quality [11, 12], but also reduces plant transpiration and improves soil moisture conditions [13-15]. Zhao argued that pruning of jujube plantations can reduce plant water consumption, which can contribute towards water-saving. The study further noted that pruning maintains target yield of plantations under natural rainfall and soil moisture conditions, also prevents soil desiccation and ecological degradation [16]. In recent years, a series of studies have been conducted on jujube water consumption mechanism [5, 10, 17-19], biomass modeling [20] and transpiration response to main branch pruning [21].

Due to limitations in observations and analysis methods, the complete effect of pruning on plant characteristics remain unclear. Issues such as insufficiency and imprecision of pruning index are still not conclusive. Thus in this study, jujube plantations in semiarid hilly region of the Loess Plateau were investigated as total length of secondary branches was used to establish pruning index. The established index was in turn used to determine the correlations between tree size and water consumption of jujube plantation. The study would provide a theoretical basis for soil desiccation prevention and cure, water-saving agriculture and perfect water-saving pruning method for jujube plantations in semiarid hilly regions.

2. Materials and Methods

2.1 Study site

This study was conducted for the period from 1st January to 31st December 2015 at the Mizhi Experimental Station (38°11'N, 109°28'E) of Northwest Agriculture and Forestry University. The study area belongs to the Loess Plateau region, where topographical characteristics are typically loess/gullies with semi-arid temperate climate. Mean annual rainfall in the station is 451.6 mm (which mostly occurs in July through September), with mean annual temperature, sunshine duration, percent sunshine, total radiation and mean elevation of 8.4 ℃, 2761 h, 62%, 580.5 kJ/cm^2, 1049 m, respectively. The average depth to water table is over 50 m with loess soil of uniform texture and moderate permeability. The mean bulk density of soil (depth of 1.0 m) is 1.29 g/cm^3, with field capacity of 23% and wilting moisture content of 5.16% (percent mass).

2.2 Pruning design

Jujube trees of 9-year-old tested (Lizao jujube variety) were planted in row and stand spacing of 3 m × 2 m on horizontal bed sloping eastwards at 25° and bed width of 2 m. A total of 16 jujube trees were observed on 4 horizontal beds (Figure 1), each treated as an independent test plot with 4 trees measured and others in a plot pruned at the same intensity.

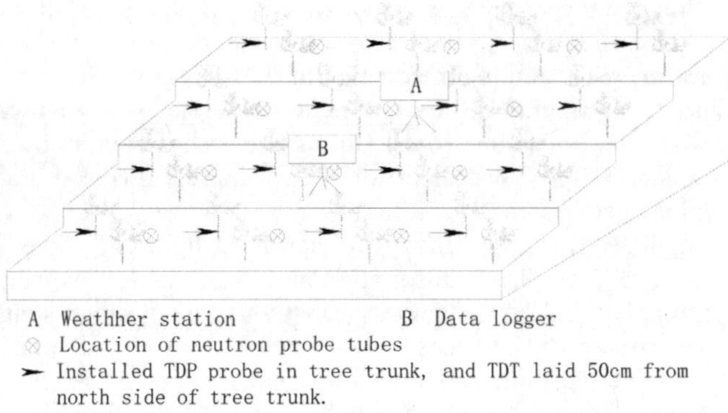

Figure 1. A diagram of the experimental layout used in the study in 2015.

Shoots with fruits and leaves were mainly secondary branches of the jujube trees and therefore the mainly fruit bearing parts. Besides, Wei proved that number of main branches of jujube affect its transpiration mainly due to size of trees after pruning, to regulate size of trees more accurately and study effects of tree size on transpiration, total length of secondary branches was used as pruning intensity (PI) index and four PIs (from PI-1 to PI-4) were set for four plots (Table 1).The control treatment was the jujube tree of PI-1, was the traditional PI with maximum yield. Pruning was taken every 5 days from 1st May (before sprouting) to 15st October to maintain the tree size. For jujube under each pruning intensity (PI), if the number or length of secondary branches on a main branch was not enough for the design, the number or length of secondary branches on other main branches was increased to make up for the deficiency. Thus, PI was decided by total length of secondary branches.

Table 1. A table detailing the pruning design of each pruning intensity (PI) used in the study.

PI	Number. of main branches	Number. of secondary branches	Length of secondary branches (cm)	Total length of secondary branches (cm)
PI-1	3	27	33.3	800±20
PI-2	3	24	28.6	600±15
PI-3	2	14	28.6	400±12
PI-4	1	6	33.3	200±10

2.3 Measured and calculated indices

2.3.1 Jujube growth indices. Using measuring tools such as the vernier caliper and measuring tape, the growth index of jujube was measured every 7 days. The indices measured included the number of main and secondary branches, the length and diameter of main and secondary branches, the number, length and diameter of jujube branches, the number of fruits and leaves, and the transverse and longitudinal diameter of fruits and leaves. Jujube plant biomass of each part was calculated using the specialized jujube biomass model [20] as follows:

$$W_{branch} = 0.002 \times D^{1.564} \times H^{1.016} \quad (1)$$

$$W_{\text{fruit shoot}} = 0.005 \times D^{1.02} \times H^{1.078} \quad (2)$$

$$W_{leaf} = 4.568 \times 10^{-5} \times Z^{1.374} \times T^{0.901} \quad (3)$$

$$W_{fruit} = 0.631 \times D_1^{3.601\times10^{-8}} \times D_2^{0.999} \quad (4)$$

where W is biomass of each jujube part (g); D is diameter of jujube branch and shoot (mm); H is length of jujube branch and shoot (mm); Z is longitudinal length of leaf (mm); T is transverse length of leaf (mm); D_1 is longitudinal diameter of fruit (mm); and D_2 is transverse diameter of fruit (mm).

2.3.2 Jujube sap flow and transpiration. Thermal diffusion method was used to monitor jujube sap flow and the data recorded from April 30 through October 15 of 2015. To eliminate detection errors due to position and height, thermal diffuse probes (TDP, Dynamax Co., USA) were installed to the north of jujube trunks at about 20 cm above the land surface. Silver membranes of 30 cm width were used to package the probes in order to reduce the impact of external environment [22]. The CR1000 data logger (Campbell, Co., USA) was used to record sap flow every 10 min. Sap flow density was calculated [23] as follows:

$$Js = 119\left(\frac{\Delta Tm-\Delta T}{\Delta T}\right)^{1.231} \tag{5}$$

where Js is sap flow density (g/m^2s); ΔT is the temperature difference between heated and unheated probe (°C); and ΔTm is the temperature difference at zero sap flow (°C).

Therefore, daily actual transpiraition of jujube was calculated as follows:

$$AT = \sum_{i=1}^{144}(Jsi \times As \times 10^{-5}) \tag{6}$$

where AT is daily actual transpiraition (mm/d); As is area of sap wood (cm^2); and Jsi is sap flow density at 10*i min. To determine sapwood area of tested jujube, a regression equation between As and the diameter at breast height (DBH) was obtained through the investigation of jujubes as: As = 0.8249 x DBH + 1.5634, R^2 = 0.8901, where As is sapwood area (cm^2) and DBH is breast-height diameter of jujube (cm). Transpiration water use efficiency of Jujube was calculated as follows:

$$WUE = Y \div AT' \tag{7}$$

where WUE is transpiration water efficiency of jujube (kg/m^3); Y is yield of jujube (kg/hm^2); and AT' is total transpiration during jujube growth (m^3/hm^{-2}).

2.3.3 Soil moisture. Roots of the jujube trees older than 10 years can reach the depth of 10 m [24]. Based on preliminary investigations, roots jujube trees can reach the depth of 7 m in the study area. Thus the soil moisture monitoring neutron tubes were set to the depth of 10 m. In the study area, however, it has been noted that change in soil moisture mainly occurs in the 0–2.6 m depth soil layer, and inter-annual variations in soil moisture are low [19]. In order to determine the effects of pruning intensity on soil moisture of jujube plantation, 3 neutron probes were set 3 m apart in each plot. The neutron probes (CNC503B, China) were used to collect soil moisture data at every 20 cm depth of 0-300cm soil layer every 10 days. Tree plant water demand is generally affected by available moisture in the soil layer with fine root network [25]. In this study, fine roots (mainly water-absorbing tissues) of jujube trees were mainly concentrated in the 0–1 m depth soil layer [24]. To get detailed soil moisture data for the 0–100 cm depth soil layer to analyze the effects of soil moisture on transpiraition, TDT soil moisture probes (Acclima, USA) were set at 20 cm, 40 cm, 60 cm, 80 cm and 100 cm depths, respectively at about 50 cm north of the jujube branches. The probes were connected with the CR1000 data logger and data collected every 1 h. Soil moisture condition was classified based on relative effective water (REW) as follows[23]:

$$REW = \frac{SWC-SWC_{min}}{SWC_{max}-SWC_{min}} \tag{8}$$

where SWC is soil moisture (%); and SWC_{max} and SWC_{min} are the maximum and minmum soil moisture (%) for the period of study.

The soil water storage was calculated as follows:

$$W = \theta_V \times h \tag{9}$$

where W is soil water storage (mm); θ_v is percent soil moisture by volume (%); and h is soil depth (cm).

2.3.4 Meteorological factors. Weather station (RR-9100, UK) was in the study area and the weather station data collected every 10 min. Meteorological indices monitored included rainfall (mm), total radiation (W/m^2), net radiation (W/m^2), photosynthetically active radiation ($\mu mol/m^2s$), wind speed (m/s), temperature (°C) and relative humidity (%). Vapor pressure difference (VPD) was calculated from temperature and relative humidity [26] as follows:

$$VPD = 0.6108 \times (1 - RH) \times e^{12.27T/(T+273.3)} \quad (10)$$

where *VPD* is vapor pressure difference (kPa); *RH* is relative air humidity (%); and *T* is temperature of air (°C).

Generally, vapor pressure difference accounts for over two-thirds and solar radiation accounts for most of the remained one-third of total transp iration [5, 27-29]. Thus variable transpiration [5, 27-29], which is an integrated index, was used to classify meteorological condition and *VPD* and radition (*Rs*) calculated as follows:

$$\mathrm{VT} = \mathrm{VPD} \times Rs^{1/2} \quad (11)$$

where *VT* is variable transpiration [kPa(W/m^2)$^{1/2}$]; *VPD* is deficit saturated vapor pressure (kpa); and *Rs* is total radiation (W/m^2).

2.4 Data analysis

Excel 2012, PASW Statistics 18.0 and Sigmaplot 12.5 were used to analyze the data and draw relevant figures. The variables measured with their equipments, time interval and start & finish date could be seen in Table 2.

Table 2. A table detailing variables measured in the study.

Measured variable	Equipment	Time interval	Start & Finish date
Jujube growth index	Vernier caliper, etc	7 days	2015/4/30-2015/10/15
Jujube sap flow	TDP probes	10 minutes	2015/4/30-2015/10/15
Soil moisture	TDT probes	1 hour	2015/1/1-2015/12/31
Soil moisture	Neutron probes	10 days	2015/1/1-2015/12/31
Meteorological factors	Weather station	10 minutes	2015/1/1-2015/12/31

3. Results

3.1 Pruning intensity and transpiration

3.1.1 Pruning and sap flow. Following different PIs, jujube sap flow was monitored under different meteorological conditions (sunny, cloudy and rainy days) in September 2015 to determine the effect of pruning on sap flow. Vegetative growth of jujube trees basically ended before September and therefore the amount of branches and leaves was stable. The measured sap flow in jujube trees is plotted under different PIs in Figure 2. Note that plate (a, d) is for sunny day on September 6, 2015; plate (b, e) is for cloudy day on September 8, 2015; and plate (c, f) is for rainy day on September 9, 2015.

Overall, sap flow in jujube trees increased in the morning with increasing VT, reached peak amount in the afternoon and then dropped before finally stopping at about 20:00 hours local time. Also sap flow was significantly correlated with VT and was highest on sunny days, followed by cloudy days and then lowest on rainy days under the same PI. Although sap flow under different PIs had similar trends under the same meteorological condition, it decreased with increasing PI. The sap flow curve had a double peak on September 6 (Figure 2d) might due to the "noon break" phenomenon of plant transpiration. Irrespective of the effect of meteorological factors on peak sap flow, pruning limited sap flow in jujube partly due to pruning-driven physiological changes in the plants.

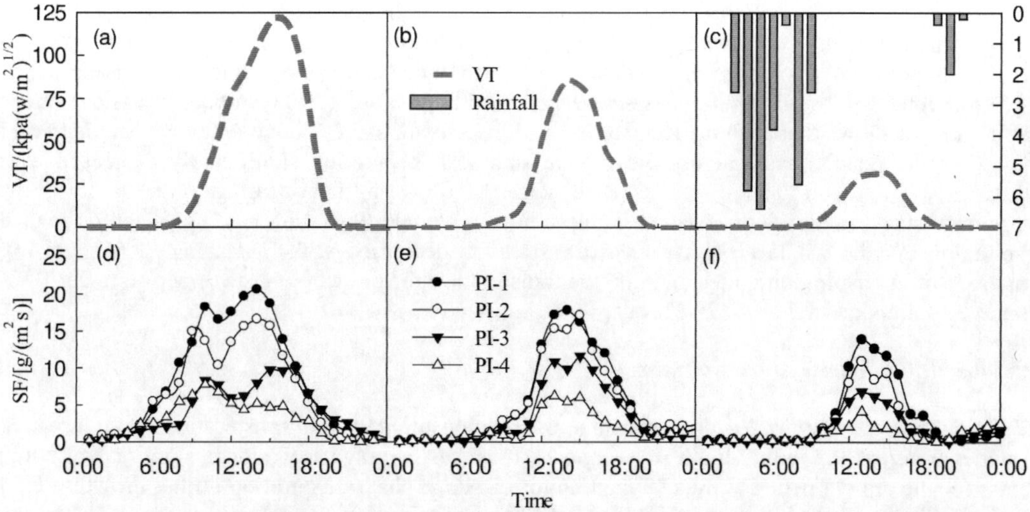

Figure 2. Plots of daily dynamics of variable transpiration (VT, a, b, c), Rainfall (c) and sap flow (SF, d, e, f) under different pruning intensities on typical sunny, cloudy and rainy days.

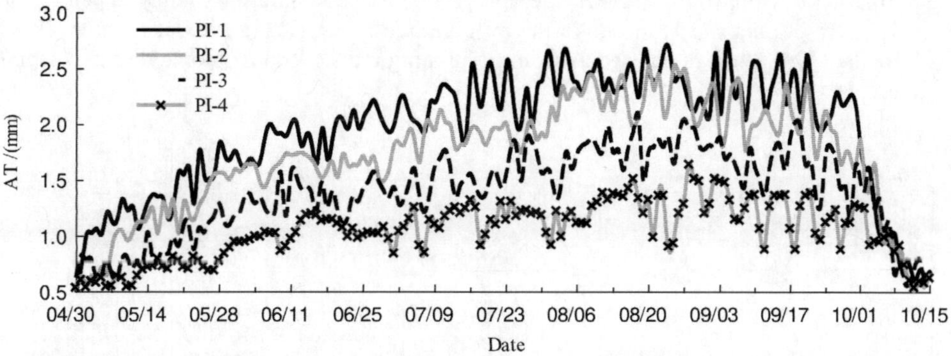

Figure 3. Dynamics of daily actual transpiration (AT) under different pruning intensities during jujube growing season in 2015.

3.1.2 Pruning intensity and daily transpiration. Figure 3 depicts daily actual transpiration for the whole growth period under different PIs. Although the trends were similar, value of the transpiration varied (within a small range) with weather conditions, soil moisture and other environmental factors. With dormancy break in May, the range and difference variations in daily actual transpiration gradually increased. The range of daily actual transpiration was driven by PI, generally decreasing with increasing PI. Peak daily actual transpiration under PI-1, PI-2, PI-3 and PI-4 was 2.64 mm, 2.51 mm, 2.10 mm and 1.65 mm, respectively. There was no obvious difference between daily actual transpiration under PI-1 and PI-2 in August. This was attributed to the arrival of the rainy season during which time mild pruning has no significant effect on water use of jujube. By the end of September, daily actual transpiration under different PIs started declining and become the same. With continuous defoliation in October, leaf transpiration deduced and the difference in daily actual transpiration among the different PIs gradually weakened during the period, jujube trees went dormant.

3.1.3 Pruning intensity, transpiration and growth stage. The growth period of jujube trees lasted for 164 days in 2015. Based on observed germination, flowering and other physiological processes during the growth period of jujube, growth period was divided into four stages. The period from May 5 to June 12 was germination-frondescence stage, June 13 to July 15 was flowering-fruit-bearing stage, July 16 to September 16 was fruit-enlargement stage and September 17 to October 15 was deciduous stage. Figure 4a shows that pruning affected each of the growth stages above, although the degree of the effects were not the same across the growth stages. Transpiration significantly decreased with increasing PI from germination-frondescence stage to flowering-fruit-bearing stage. From fruit-enlargement stage to deciduous stage, increase in PI beyond PI-2 did not significantly reduce transpiration. For the whole growth period, transpiration decreased with increasing PI (Figure 4b). Compared with transpiration under PI-1, jujube transpiration for the whole growth period under PI-2, PI-3 and PI-4 dropped by 11.1%, 29.2% and 47.9%, respectively.

3.2 Jujube growth index and transpiration

3.2.1 Tree size and transpiration. To determine the relationship between transpiration and jujube tree size under different PIs, total length of secondary branches, average leaf biomass and average tree biomass for the growth period were analyzed. Figure 5 shows that transpiration during growth period increased with increasing tree size. After the total length of secondary branches exceeded 600 cm, the rate of increase in transpiration decreased. Compared with average tree biomass, the relationship between transpiration and average leaf biomass was much closer to that between transpiration and total length of secondary branches. This was because jujube tree transpiration mainly depended on the proportions of leaves, fruits and branches in the average tree biomass. Figure 5 also shows that total length of secondary branches can be used as a proxy (pruning index) for leaf biomass and transpiration in jujube trees.

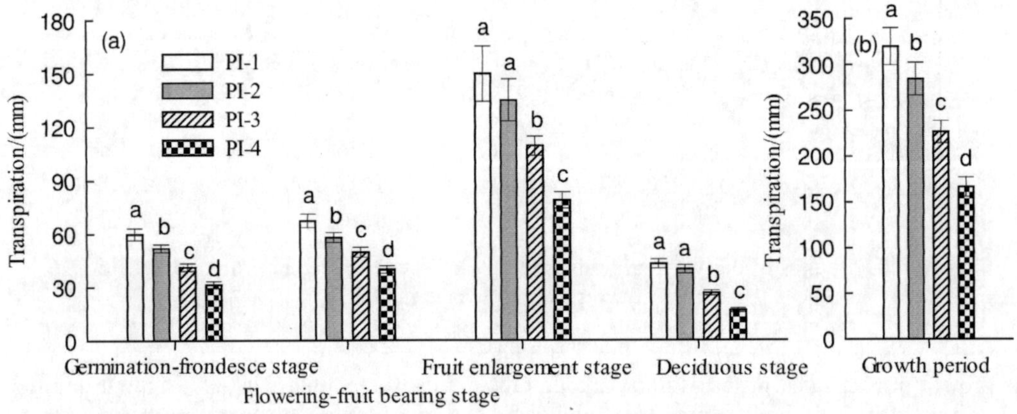

Figure 4. Plots of transpiration at different growth stages (a) and growth periods (b) of jujube. Note that means with the same letter at the same stage or period are not significantly different at the 0.05 level (Duncan's test).

3.2.2 Leaf biomass and daily actual transpiration. To further explore the relationship between transpiration and leaf biomass, the plant growth model $BL = a / (1+e^{b \times DOY + c})$ was used to fit leaf biomass (BL) under different PIs and days of year (DOY) during growth period, where a, b and c were the model constants. For DOY 140–273 (i.e., May 20 to September 30), jujube growth indices measured every 7 days were used to calculate jujube leaf biomass (see Table 3 for the fitted results). Taking into account the high error in TDP probes for vapor pressure differences (VPD) less than 0.6 kPa [27, 28], sap flow data for VPD was less than 0.6 kPa was removed. Then the fitted leaf biomass

(BL) and daily actual transpiration (AT, mm) under different PIs was expressed as AT = 2.1710 x BL + 1.0602 (R^2 = 0.6918, n = 468). This indicated that a significant positive correlation existed between daily actual transpiration and jujube leaf biomass.

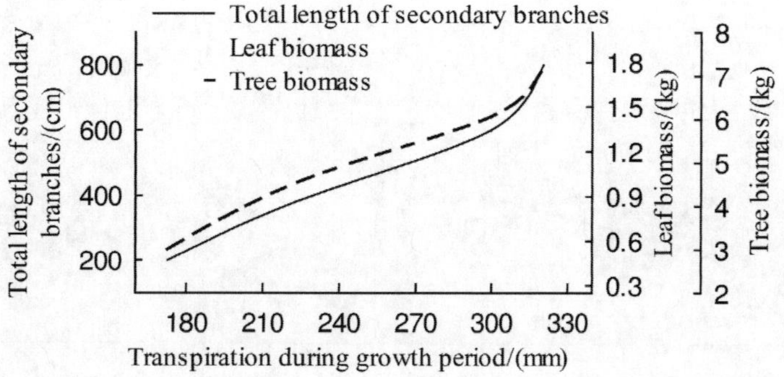

Figure 5. Relationships between size of jujube tree and transpiration during growth period.

Table 3. A table detailing the results of statistical fitting between leaf biomass (BL) and day of year (DOY) under different pruning intensities (PIs) in the study area.

PI	a	b	c	R^2	n	*p*
PI-1	2.584	-0.083	14.117	0.995	17	0.001**
PI-2	2.468	-0.080	14.002	0.991	17	0.018*
PI-3	1.484	-0.071	12.547	0.994	17	0.004**
PI-4	0.762	-0.063	10.868	0.975	17	0.004**

Note that a, b and c were parameters in the growth model BL= a / $(1+e^{b \times DOY}$ + c), taken to fit leaf biomass (BL) and day of year (DOY); R^2 is coefficient of determination; n is sample number; and P is significant difference. The superscript of P "*" expresses that $p < 0.01$ and the difference is highly significant; then the superscript of P "**" expresses that $0.01 < p < 0.05$ and the difference is significant.

Transpiration was not only closely related to tree size, but it was also affected by soil moisture, weather conditions and other environmental factors. For higher correlations between BL and AT, soil moisture and meteorological conditions were classified into two groups. During the whole jujube growing period, average daily relative effective water (REW) of 0-100 cm soil layer was 0–0.98 and that of VT was 5.78–40.90 kPa $(W/m^2)^{1/2}$. Thus the soil moisture was classified into REW < 0.5 and REW > 0.5 and then meteorological conditions classified into VT < 20 kPa $(W/m^2)^{1/2}$ and VT > 20 kPa $(W/m^2)^{1/2}$ — see Figure 6 for the fitted results.

Little difference existed between the correlations under the two soil moisture conditions, suggesting that soil moisture in the study area was as much driven by leaf biomass and daily actual transpiration. Conversely, significant difference existed in terms of the correlation between the two meteorological conditions. The coefficients of the correlations increased to up to 0.8, suggesting a significant impact of daily actual transpiration on meteorological factors. The correlation between leaf biomass and daily actual transpiration was even more significant under the classified meteorological conditions.

As REW < 0.5, soil moisture of 0-100 cm soil layer was between 5.27%-6.37%; while REW > 0.5, soil moisture of 0-100 cm soil layer was between 5.63%-7.99%. Jujube trees with the same leaf biomass consume more water in higher soil moisture condition. However, the correlations under the two soil moisture conditions were not affected much while the range of soil moisture was only 2.73% in growth period. Besides, as VT < 20 kPa$(W/m^2)^{1/2}$, daily mean temperature was 8.06-28.46 °C, relative humidity was 28.77%-72.89%, total radiation was 87.16-286.20 W/m^2 and vapor pressure difference was 0.60-1.63 kPa; while VT > 20 kPa$(W/m^2)^{1/2}$, daily mean temperature was 15.39-

31.19 °C, relative humidity was 18.38%-57.08%, total radiation was 182.05-314.0 W/m^2 and vapor pressure difference was 1.23-2.44 kPa.

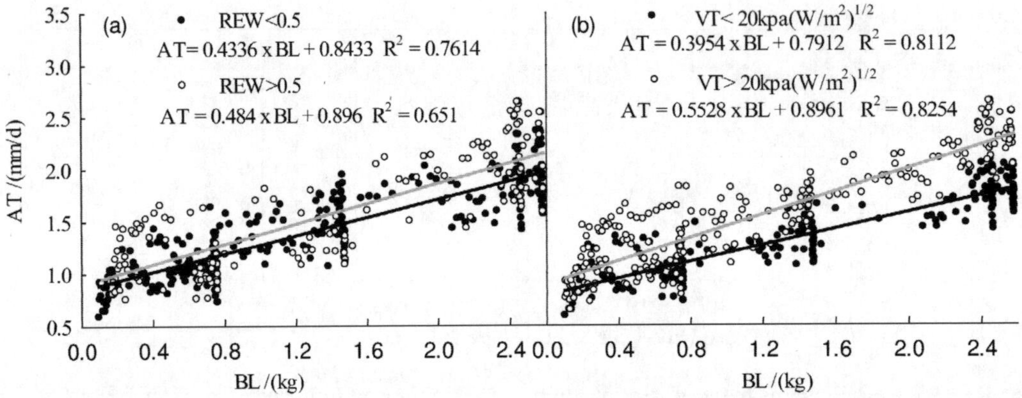

Figure 6. Correlation plots for biomass of leaves (BL) and daily actual transpirations (AT) under two soil moisture (a) and two meteorological (b) conditions.

The coefficient of determination, intercept and slope for VT > 20 kPa(W/m^2)$^{1/2}$ and REW > 0.5 were much higher than those for VT < 20 kPa (W/m^2)$^{1/2}$ and REW < 0.5, respectively. This implied that better correlation existed between leaf biomass and daily actual transpiration for VT > 20 kPa(W/m^2)$^{1/2}$ and REW > 0.5, in other words, transpiration of jujube was more sensitive to leaf biomass under higher soil moisture and higher temperature, radiation, vapor pressure difference, and low relative humidity. Presumably, jujube transpiration consume more water in good environment, but the correlation between transpiration and tree size would became more reliable because bad environment obstructed transpiration to some extent.

3.2.3 Yield and transpiration. Table 4 shows reductions in jujube yield under PI-2, PI-3 and PI-4, compared with that under PI-1. No significant difference existed between yield under PI-1 and PI-2, possibly due to the lack of significant difference between transpiration under PI-1 and PI-2 in the late growth stage of jujube (Figure 4a). Although yield and transpiration under PI-3 and PI-4 were significantly smaller than under PI-1, water use efficiency significantly increased with increasing PI. Compared with PI-1, water use efficiency under PI-2, PI-3 and PI-4 increased by 5.1%, 15.7% and 24.2%, respectively.

Table 4. A table listing yield, transpiration and water use efficiency of jujube tree under different pruning intensities (PIs) in the study area.

PI	Yield（kg/hm^2）	Transpiration（m^3/hm^2）	Water use efficiency (kg/m^3)
PI-1	11908.94^a	3200.70^a	3.70^a
PI-2	11534.66^a	2836.44^b	3.89^{ab}
PI-3	9701.04^b	2265.81^c	4.28^{bc}
PI-4	7662.98^c	1667.15^d	4.60^c

Note that means with the same letter in the same column are not significantly different at the 0.05 level (Duncan's test). Also, water use efficiency was calculated from transpiration and yield of jujube.

3.3 Pruning intensity and soil moisture

Figure 7 depicts rainfall and dynamic changes in soil moisture at the 3 m depth soil layer under different PIs. Rainfall was only 46.2 mm before jujube growth period, soil moisture mainly driven by rainfall and evaporation and the differences in soil moisture between different PIs not obvious.

Rainfall increased to 246.4 mm during jujube growth period and the difference in soil moisture under different PIs increased with tree pruning. After August, soil moisture under different PIs had obvious differences, with higher PI resulting in higher soil moisture. Compared with PI-1, volumetric soil moisture under PI-2, PI-3 and PI-4 increased by 3.2%, 8.5% and 14.2%, respectively. During this period, even increase in rainfall could not significantly improve soil moisture content because of high water demand of jujube under PI-1 and PI-2. As rainfall was only 88.2 mm after jujube growth period in 2015 and soil moisture no longer affected by vegetative water consumption, rainfall, as much as pruning, greatly improved soil moisture. For the investigated period, water storage of 0-300cm soil layer increased significantly with increasing PI. Compared with 1st January 2015, water storage under PI-1, PI-2, PI-3 and PI-4 increased by 0.76 mm, 7.05 mm, 26.54 mm and 35.50 mm, respectively on 31st December 2015.

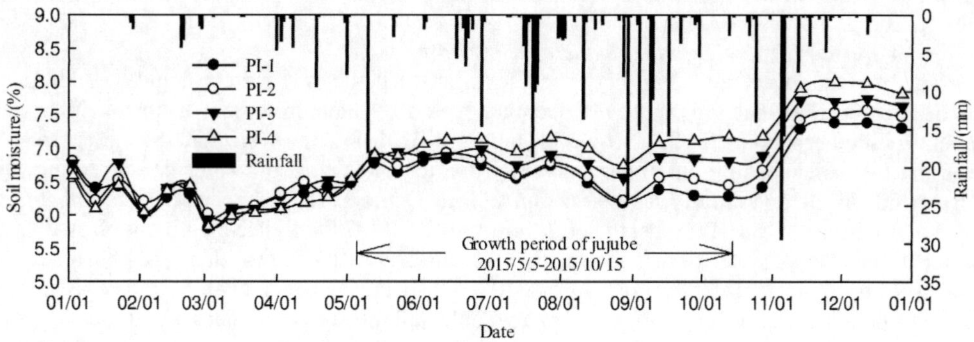

Figure 7. Dynamics of rainfall and soil moisture (percent volume) of jujube plantation under different pruning intensities in 2015.

4. Discussions

4.1 Transpiration and pruning intensity

In addition to the effect of environmental factors, transpiration of a tree plant is generally correlated with leaf area index [29, 30]. This could be the reason why transpiration of jujube trees decreased with increasing PI. Namirembe noted that pruning of *Cassia spectabilis* narrowed trunk xylem vessel diameter and decreased water conductivity in environmental conditions with limited water resources, which suppressed canopy transpiration rate and reduced soil moisture consumption [31]. Also Zhao observed that tylosis originated from catheter of grape branches after pruning, which reduced branch water transportation and sap flow rate maximally by 21.10% [32]. This study showed that sap flow in jujube trees decreased with increasing PI (Figure 2), partly due to physiological changes induced by pruning. There was no significant difference between daily actual transpiration of jujube under PI-1 and PI-2 in August (Figure 3). Also there was no significant difference between jujube transpiration under PI-1 and PI-2 at fruit enlargement and deciduous stages (Figure 4). This was attributed to mild pruning which effectively regulated canopy water demand in jujube plantations. It suggested then that it was only at certain PIs that pruning effectively regulated tree water demand. This observation was in agreement with previous reports on the effects of pruning on *Grevillea robusta* agroforestry system on hill slopes [33].

4.2 Total secondary branch length and pruning

As leaf stomata are the main transpiration organs of plants, plenty research has been conducted on the relationship between transpiration and leaf area. Generally, transpiration has positive significant correlation with leaf area. The rate of transpiration increases with increasing leaf area or it even stops completely unless leaf area reaches a certain value [18, 30, 34]. Tree leaf area is usually calculated by multiplying leaf area index with projected area. However, as crown volume of dwarf jujube under close planting is small and irregular in shape, high errors persist in measuring leaf area index and projected area. The correlation between transpiration and biomass was similar to that between transpiration and leaf area. This implied that total length of secondary branches can be used as a proxy (index) to estimate the range of pruning more accurately for the efficient regulation of transpiration. Furthermore, jujube biomass can be calculated using the high correlation biomass model [20] after monitoring specific growth indices. This facilitates ready analysis of water use efficiency and biomass-transpiration correlations. Theoretically, PI is the factor used to control tree structure and volume and thereby control water consumption. As the practical definition of pruning remains largely unclear, total length of secondary branches could be a reliable pruning index to improve water-saving pruning method. Although further studies are required in this direction, this study has successfully demonstrated the use of this index in precision pruning.

4.3 Soil moisture, meteorological factors and transpiration

Significant difference were noted in the correlations between the two meteorological conditions for Jujube trees, reflecting that sap flow was deeply affected by meteorological factors, which was in agreement with other studies [5, 35]. However, only a little difference were found in the correlations between the two soil moisture conditions, possibly due to the severe soil water deficit during jujube growth period. In 2015, rainfall was 292.6 mm before October 15, which made the soil dry with average soil moisture of the 0–1 m depth soil layer of only 5.27–7.99%. The classification standards of soil water deficit shows water deficit for relative soil moisture (θ_e) of less than 0.4 [5, 36]. Relative soil moisture for the 0–1 m depth soil layer was 0.01–0.14 during the growth period. This indicated severe water deficit under different PIs, lack of available soil moisture and narrow soil moisture range, which weakened the difference between the correlations of the two soil moisture conditions. The severe water deficit also affected the correlation between daily actual transpiration and leaf biomass. Daily actual transpiration for similar jujube treatments exceeded 3.5 mm [21], with the highest value (2.64 mm) under the different PIs. Under this soil moisture condition, simple linear correlation was found between daily actual transpiration and leaf biomass. Daily actual transpiration might be limited due to severe water deficit.

4.4 Pruning effect on soil moisture

In recent years, a series of research has been done on the effects of pruning on soil moisture under fruit plantations. Li observed that compared with traditional pruning method, soil moisture in the 2.4 m depth soil layer greatly improved under repeated pruning at higher PI [15]. Wei also found that maximum PI (of one main branch per plant) increased soil water storage by 40.5 mm after 2 years [21]. The results of this study agreed with the findings of the above reports. With annual rainfall of 380.8 mm, 2015 was, far less than the average local annual rainfall (451.6 mm), even slight pruning could increase the soil storage water under tree plantations. This suggested that pruning can effectively improve soil moisture in 0-300 cm soil layer in jujube plantations. The long-term effects of pruning on deep soil moisture (of over 3 m deep) remains unclear, and therefore require further research.

5. Conclusions

(1) Pruning can significantly reduce transpiration water demand of jujube plant at different periods of growth. Compared with PI-1 (the control treatment) jujube plant transpiration for the whole growing season under PI-2, PI-3 and PI-4 dropped by 11.1%, 29.2%, 47.9%, respectively. At daily scales, the range of daily actual transpiration decreased with increasing pruning intensity (PI). Peak daily actual transpiration under PI-1, PI-2, PI-3 and PI-4 were 2.64 mm, 2.51 mm, 2.10 mm and 1.65 mm, respectively. Sap flow in jujube trees was similar in trend under different PIs, while sap flow rate decreased with increasing pruning intensity (PI). Jujube transpiration was easily influenced by meteorological factors at small time scale (day and hour).

(2) Jujube transpiration during the growing season increased with increasing plant size. The rate of increase in transpiration slowed after the tree attained a certain size (when the total length of secondary branches exceeded 600 cm). Otherwise while increase of pruning intensity (PI) reduced yield of tree, it reduced water consumption that in turn significantly increased water use efficiency of fruits.

(3) The correlation between daily actual transpiration and leaf biomass of jujube tree was positive and significant, but with a significant difference between the correlations for the two sets of meteorological conditions. When VT < 20 kPa $(W/m^2)^{1/2}$, AT = 0.7908 x BL + 0.7912, $R^2 = 0.8112$, n = 220. Then when VT > 20 kPa $(W/m^2)^{1/2}$, AT = 1.3379 x BL + 1.0843, $R^2 = 0.8254$, n = 248.

(4) Increased pruning intensity (PI) could effectively prevent soil desiccation and improve soil moisture under jujube plantations. During dormancy period, the difference in soil moisture due to pruning was significant after rainfall. Annual change in water storage in jujube plantations under PI-1, PI-2, PI-3 and PI-4 increased by 0.76 mm, 7.05 mm, 26.54 mm and 35.50 mm, respectively.

Acknowledgments

We are grateful for the grants from China National Science & Technology Supporting Plan (No. 2015BAC01B03) and Forestry Industry Research Special Funds for Public Welfare Projects (No. 201404709). We also acknowledge the support of the staff of Experimental Station of Northwest Agriculture and Forestry University.

References

[1] Hui S and Mingan S 2000 Soil and water loss from the Loess Plateau in China *Journal of Arid Environments* **45(1)** pp 9-20

[2] Hongsong C, Mingan S and Yuyuan L 2008 Soil desiccation in the Loess Plateau of China *Geoderma* **143(1-2)** pp 91–100

[3] Li W, Quanjiu W, Sanping W, Mingan S and Yi L 2008 Soil desiccation for Loess soils on natural and regrown areas *Forest Ecology & Management* **255(7)** pp 2467-77

[4] Yiping C, Kaibo W, Yishan L, Weiyu S, Yi S and Xinhua H 2015 Balancing green and grain trade *Nature Geoscience* **8(10)** pp 739-741

[5] Dianyu C, Youke W, Shouyang L, Xinguang W and Xing W 2014 Response of relative sap flow to meteorological factors under different soil moisture conditions in rainfed jujube (Ziziphus jujuba Mill.) plantations in semiarid Northwest China *Agricultural Water Management* **136(2014)** pp 23-33

[6] Jun L, Youke W, Xindong W, Sen X and Xue Z 2013 Water impounding characteristics of bamboo-shaped rainwater harvesting ditch in the hilly loess region *Chinese Journal of Applied Ecology* **24(12)** pp 3373-80

[7] Youke W, Fuli X and Xiaogui X 2008 Application of micro-irrigation technology in the production of Jujube plantations in the north of Shanxi *Bulletin of Soil and Water Conservation* **28(4)** pp 198-200+203-204

[8] Pute W, Youke W, Xiaogui X and Delan Z 2008 Integration and demonstration of the date micro-irrigation technology in the hilly of Shanbei *Agricultural Research in the Arid Areas* **26(4)** pp 1-6+12

[9] Xuefeng Y, Pute W, Youke W and Fuli X 2006 Soil moisture conserving irrigation under straw mulch with no-tillage *Transactions of the Chinese Society of Agricultural Engineering* **22(7)** pp 22-26

[10] Xinguang W, Tieliang W, Shouyang L, Bo L, Zhenyi N and Youke Wang 2015 Effect of Planting Years on Jujube Transpiration in Semi-arid Hilly Areas of Loess Plateau *Transactions of the Chinese Society for Agricultural Machinery* **47(7)** pp 171-180

[11] Asrey R, Patel V B, Barman K and Pal R K 2013 Pruning affects fruit yield and postharvest quality in mango (Mangifera indica L.) cv. Amrapali *Fruits* **68(5)** pp 367-381

[12] Kai S, Qinping W, Yuling Y, Xiaowei W and Jixiang Z 2010 Effects of different pruning modes on the light distribution characters and fruit yield and quality in densely planted `RedFuji' apple orchard *Chinese Journal of Applied Ecology* **21(5)** pp 1224-30

[13] David I F, John J C, Christopher L B, Charles R W and Thomas G B 2012 Effect of thinning, pruning and nitrogen fertiliser application on transpiration, photosynthesis and water-use efficiency in a young Eucalyptus nitens plantation *Forest Ecology & Management* **266(2012)** pp 286-300

[14] Kuo T L, Alan N L, Richard P and Terence R 2003 Summer pruning reduces whole-canopy carbon fixation and transpiration in apple trees *Journal of Horticultural Science & Biotechnology* **78(6)** pp 749-754

[15] Mingxia L, Sheni D, Gangshuang B and Guijun G 2012 Effects of renewal pruning on soil moisture and growth of apple tree *Journal of Zhejiang University (Afric. & Life Sci.)* **38(4)** pp 467-476

[16] Xia Z, Youke W, Shouyang L, Xinguang W and Penghong L 2012 Effects of two new dryland farming technologies on growth and fruit bearing of pear jujube trees in hilly regions *Agricultural Research in the Arid Areas* **30(4)** pp 157-160+171

[17] Shouyang L, Youke W, Xindong W, Xinguang W, Wang X and Linlin Z 2013 Measured and estimated evapotranspiration of Jujube (Ziziphus jujuba) forests in the Loess Plateau, China *International Journal of Agriculture & Biology* **15(5)** pp 811-819

[18] Xinguang W, Chenyu C, Xing W, Youke W and Xindong W 2014 Time scale effect on main factors that influence jujube transpiration in hillside jujube orchard *Transactions of the Chinese Society of Agricultural Engineering* **30(17)** pp 149-156

[19] Xinguang W, Zhenyi N, Shouyang L, Tan S, Jianpeng M and Youke W 2015 Soil Moisture Characteristics and Its Influence on Jujube Tree Transpiration in Loess Hilly Region *Transactions of the Chinese Society for Agricultural Machinery* **46(6)** pp 130-140

[20] Tan S, Youke W, Zhiyong G, Xinguang W and Jianguo D 2015 Biomass Models for Jujube in Loess Hilly Mountain of Northern Shanxi Province *Bulletin of Soil and Water Conservation* **35(3)** pp 311-316

[21] Xinguang W, Dianyu C, Shouyang L, Xing W, Zhiyong G and Youke W 2014 Effect of Trim on Jujube Transpiration in Loess Hilly Region *Transactions of the Chinese Society for Agricultural Machinery* **45(12)** pp 194-202+315

[22] Ping L, Laurent U and Ping Z 2004 Granier's thermal dissipation probe (TDP) method for measuring sap flow in trees: Theory and practice *Acta Botanica Sinica* **46(6)** pp 631-646

[23] Granier A 1987 Evaluation of transpiration in a Douglas-fir stand by means of sap flow measurements *Tree Physiology* **3(4)** pp 309-320

[24] Lihui M, Pute W and Youke W 2012 Spatial distribution of roots in a dense jujube plantation in the semiarid hilly region of the Chinese Loess Plateau *Plant Soil* 354 pp 57-68

[25] Tomo'Omi K, Makiko T, Takanori S and Kyoichi O 2008 Transpiration and canopy conductance at two slope positions in a Japanese cedar forest watershed *Agricultural & Forest Meteorology* **148(10)** pp 1444-55

[26] Norman J M 1998 An Introduction to Environmental Biophysics (Springer-Verlag, New York New York Springer-Verlag)

[27] Ewers B E and Oren R 2000 Analyses of assumptions and errors in the calculation of stomatal conductance from sap flux measurements *Tree Physiology* **20(9)** pp 579-589

[28] O'Brien J J, Oberbauer S F and Clark D B 2004 Whole tree xylem sap flow responses to multiple environmental variables in a wet tropical forest *Plant Cell & Environment* **27(5)** pp 551-567

[29] Yuqiang S, Fang G, Jinsong Z, Ping M and Jun G 2009 Variation in Transpiration of Populus bolleana and the Influence Factors in Maowusu Sandy Area *Scientia Silvae Sinicae* **45(9)** pp 66-71

[30] Yongzhe Y, Jianming L, Rong Z, Sanjie S and Kaili C 2012 Greenhouse tomato transpiration and its affecting factors: Correlation analysis and model simulation *Chinese Journal of Applied Ecology* **23(7)** pp 1869-74

[31] Sara N, Robert M B and Chin K O 2009 Manipulating phenology and water relations in Senna spectabilis in a water limited environment in Kenya *Agroforestry Systems* **75(3)** pp 197-210

[32] Xianhua Z 2013 Pruning Effects on Sap Flow and Transporation Assignment of Photosynthates in Grapevine Shanxi Place Northwest Agriculture and Forestry University

[33] Jackson N A, Wallace J S and Ong C K 2000 Tree pruning as a means of controlling water use in an agroforestry system in Kenya *Forest Ecology & Management* **126(2)** pp 133-148

[34] Zhaoquan G, Xianchuan Z and Xiaowei W 2006 Mathematical simulation of canopy transpiration rate of peach tree canopy *Acta Ecologica Sinica* **26(2)** pp 489-495

[35] Sheng D, Yilong W, Tomonori K, Jianguo Z and Kyoichi O 2011 Sapflow characteristics and climatic responses in three forest species in the semiarid Loess Plateau region of China *Agricultural & Forest Meteorology* **151(1)** pp 1-10

[36] Black T A 1979 Evapotranspiration from Douglas fir stands exposed to soil water deficits *Water Resources Research* **15(15)** pp 164-170

Variations in dust contributions to air quality impairment in a temperate grassland of Inner Mongolia, China

Baolin Zhang[1], Chunzhu Guan, Junling Zhao and Jiannan Li

College of Chemistry and Environmental Sciences, Inner Mongolia Normal University, Hohhot 010020, P.R. China

E-mail:nmzhangbaolin@hotmail.com

Abstract. Dust associated visibility impairment is the most noticeable air pollution phenomena, has important implications regarding air quality in developing cities. We integrated the commonly reported visibility with remote sensing, found that dust emission tended to decrease over Xilingol grassland. The temporal coherency between dust event and visibility reduction demonstrated dust was an important factor responsible for air quality impairment, but the differences in seasonal patterns and long-term trends among stations manifested some drops in visibility couldn't be solely attributed to dust activity. It is urgent to conclude the causes of visibility reduction in developing cities susceptible to dust impact in recent years.

1. Introduction

Visibility refers to the clarity or transparency of the atmosphere and the ability to see distant objects [1]. Global visibility had decreased substantially during 1973-2007 over land except in Europe [2]. Visibility reduction is one of the most noticeable effects of air pollution [3]. Visibility has been used as a surrogate of air quality [4]. Besides anthropogenic aerosols, sand and dust storms also contribute significantly to visibility reduction [5]. Before the advent of satellite remote sensing, visibility was the dominant variable used in mapping dust activity [6]. Visibility has been used as a substitute for aeolian dust source [6] and activity [7], because dust loading impairs visibility [8, 9]. In recent years, remote sensing has been proven to be very suitable for monitoring dust events. MODIS (Moderate Resolution Imaging Spectroradiometer) boarded on the Terra and Aqua satellites operated by the NASA (National Aeronautics and Space Administration) provides us with data on diurnal aerosol variations [10]. MODIS provides us with more detailed information on dust emission, transport and deposition [11-13].

Recent studies have been directed to visibility impairment in cities, and the contribution of aeolian dust as well [14]. As one of the most abundant aerosols in the atmosphere, dust has important implications regarding air quality [15]. In this study, we demonstrated the simultaneity of dust events and visibility reductions in 2000-2013 over a temperate grassland of Inner Mongolia, China, integrating satellite remote sensing with traditional visibility observations. The objective of the study was to clarify the contribution of aeolian dust to air quality impairment, to unveil other potential contributors, and to consider measures to improve air quality in the future development of cities.

2. Data and methods

2.1. Study area

Xilingol is located at the center of Inner Mongolia (figure 1), has high-quality natural pastures, used traditionally for nomadic farming. Rapid economic development and population growth since the 1980s increased the demand for agricultural production, resulting in higher stock densities and a shift in agricultural policy from nomadism to stationary farming in the late 1980s [16]. Xilingol has a continental climate, with a long, cold winter and a warm summer with little rain [17].

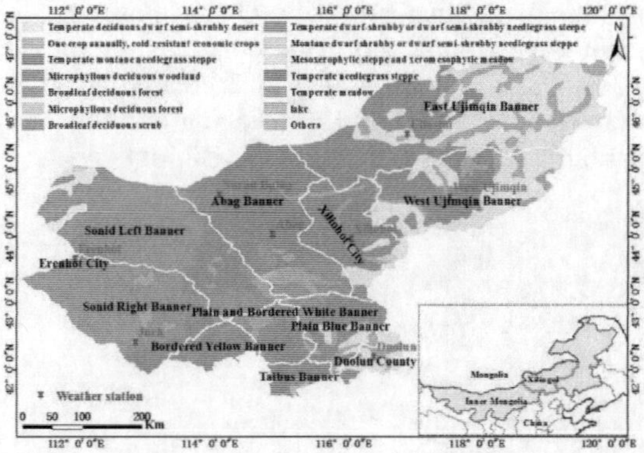

Figure 1. Landscape and observation station distribution in the study area.

2.2. Data and methods

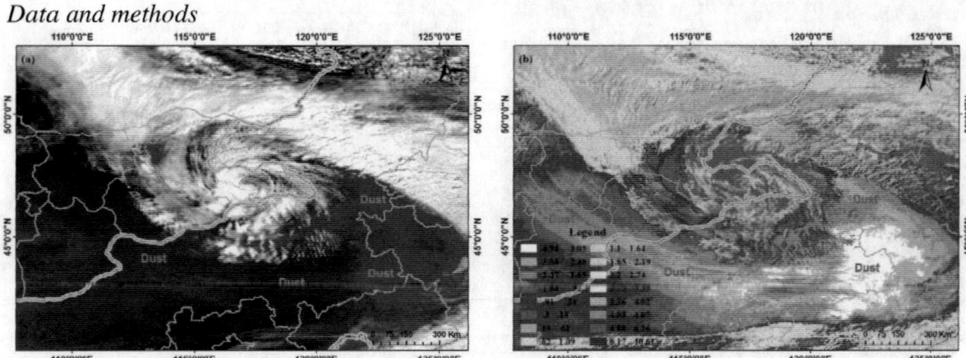

Figure 2. Dust identified by MODIS true-color composite image (a) and BTD (k) algorithm (b).

Daily mean visibility data were downloaded from NCDC (NOAA's National Climatic Data Center) climate data online. Data are available at eight weather stations in the study area (figure 1). In view of data availability, continuity and change of visibility specification in China [18], we analyzed visibility data in 2000-2013. MODIS L1B data from LAADS (the Level 1 and Atmosphere Archive and Distribution System) were used in this study. We analyzed 88 dust events observed from 2000 to 2013, including all the dust emitted from Xilingol Grassland [12, 13].

Dust can be identified using aerosol optical thickness algorithms [19], absorbing aerosol index [20], (thermal) infrared dust index [21] and other indices [22]. We compared dust discrimination methods [13], concluded that MODIS true color image (figure 2a) captured the emission and transport of dust, and that brightness temperature difference (BTD) (figure 2b) was the most consistently reliable technique for dust identification [13, 23]. Therefore, we used MODIS true color images and BTD algorithm to capture dust events. Contribution of dust to visibility impairment was measured by temporal consistency between satellite observed dust events and visibility reduction observations: (visibility reduction (days)/dust-induced visibility reduction (days)) × 100%.

3. Results

3.1. Dust event and visibility impairment: a case study

On May 11, 2011, MODIS captured dust blew out of the Gobi Desert and across the Mongolia-China border. The natural-color image presented that a counter-clockwise arc of dust swept across the border into China, then backed over the border into Mongolia. Clouds hovered over the dust plume, and some fringed the plume's northern edge (figure 3a). This dust event significantly impaired visibility in Xilingol grassland (figure 3b). The visibility on May 11 decreased greatly to <15 km at eight synoptic weather stations, demonstrating the strength and influencing extent of the dust event. The temporal agreement verified the contribution of dust to mean daily visibility decline.

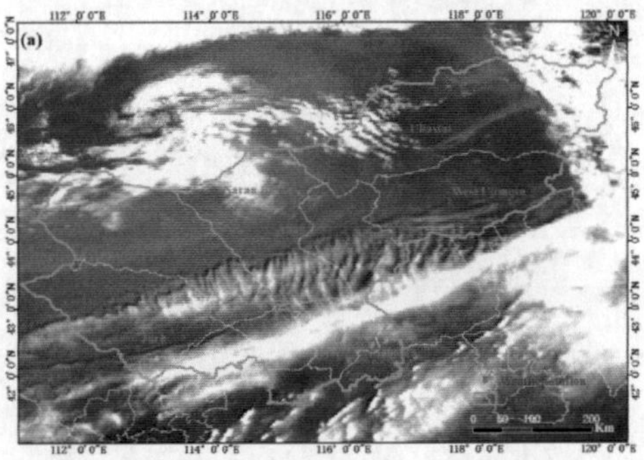

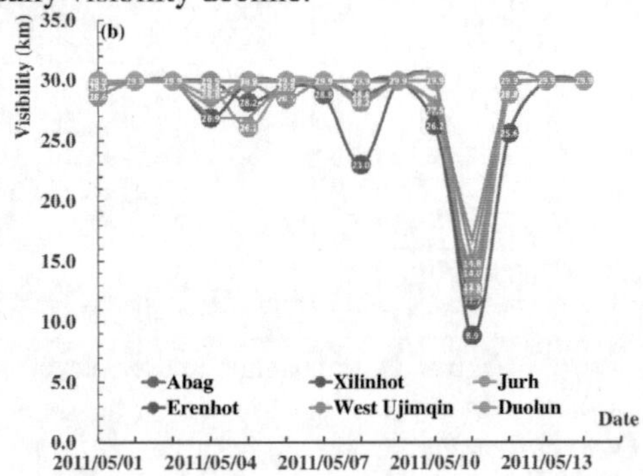

Figure 3. Dust event observed by MODIS on May 11, 2011 (a) and its impact on visibility (b).

3.2. Dust activities in 2000-2013

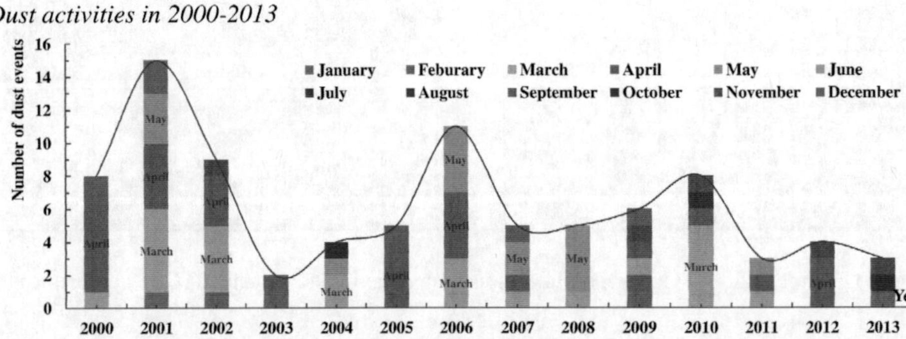

Figure 4. Dust events observed by MODIS in 2000-2013.

According to MODIS observations in 2000-2013, dust activities (figure 4) in Xilingol grassland were detected by analysing true colour and BTD images. Dust emissions changed much in the last 14 years. More dusty days occurred in 2000-2002 and 2004-2008, but fewer in 2003, and after 2010. About 75 percent dust events were observed during March to May.

3.3. Temporal and spatial coherency between dust events and visibility impairments in 2000-2013

Visibility impairments and their temporal associations with dust activities at eight stations over Xilingol grassland demonstrated the simultaneous occurrences of dust emission and visibility decline (figure 5). The dates of dust events captured by MODIS were coherent with the occurrences of visibility impairment. Concurrent with visibility impairment phenomena, dust events indicate its impacts on air quality. Visibility reduction demonstrated dust activity. Annual visibility reductions showed similar patterns to those of dust emissions at eight stations, verifying aeolian dust was an important factor leading to visibility impairment. Visibilities decreased in 2000-2002, 2005-2008, especially 2001-2002, 2006-2008, coincided with dustier spring in the study area. Seasonal visibility changes varied among stations, stations in the north (Uliastai (figure 5a), Erenhot (figure 5b), Naran Bulag (figure 5c), Abag (figure 5d) and West Ujimqin (figure 5f)) showed stronger annual change patterns than those in the south (Jurh (figure 5d), Xilinhot (figure 5g) and Duolun (figure 5h)). Besides seasonal patterns, differences among stations still included long-term trends. Jurh experienced more dusty days in 2000-2009; seven stations except Xilinhot city appeared to suffer less from visibility reductions in recent years, the frequencies of visibility impairments tended to be declining, especially at Jurh (figure 5e). In addition, some dust events associated visibility decreases were local phenomena, not all stations showing visibility declines; some were of great strength and extent, e. g., dust on May 27 2008 (figure 2), dust veiled the whole Xilingol grassland. Visibility impairment differed among stations, showing the differences in the strength and extent of dust events.

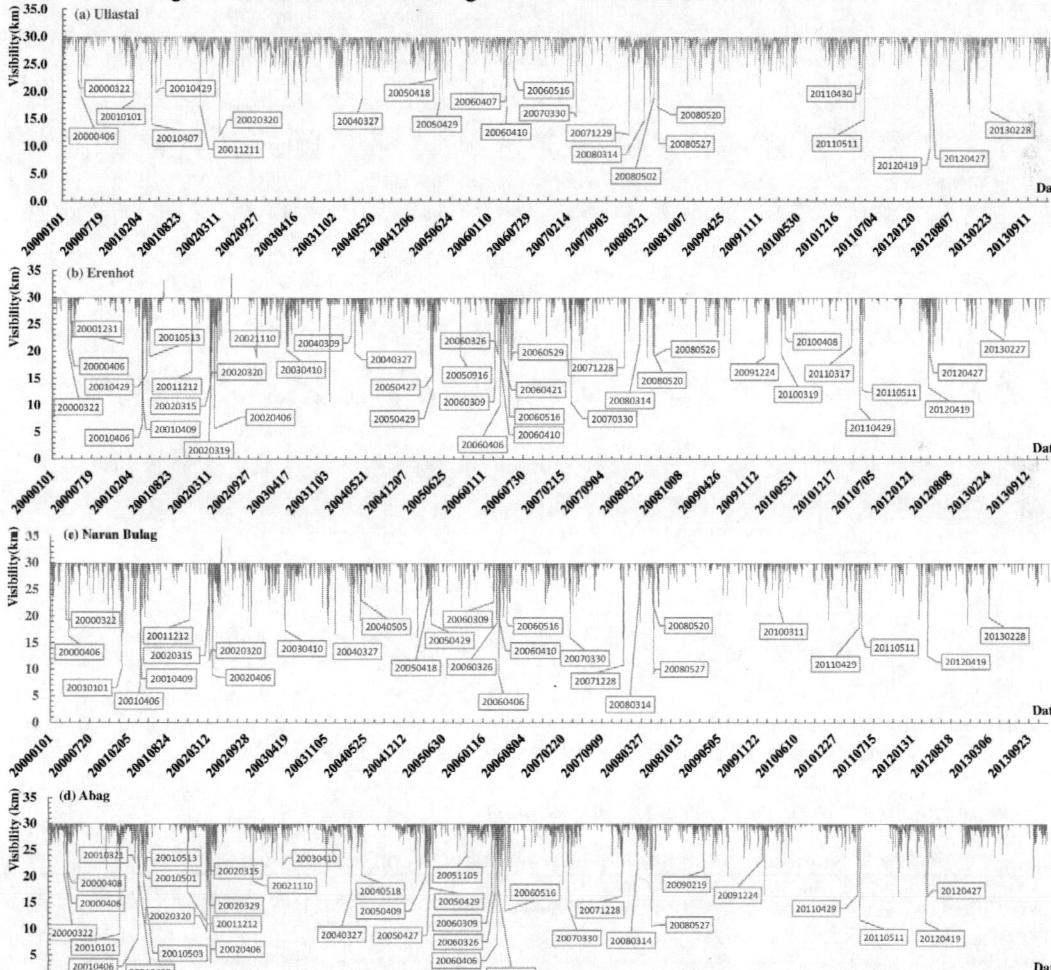

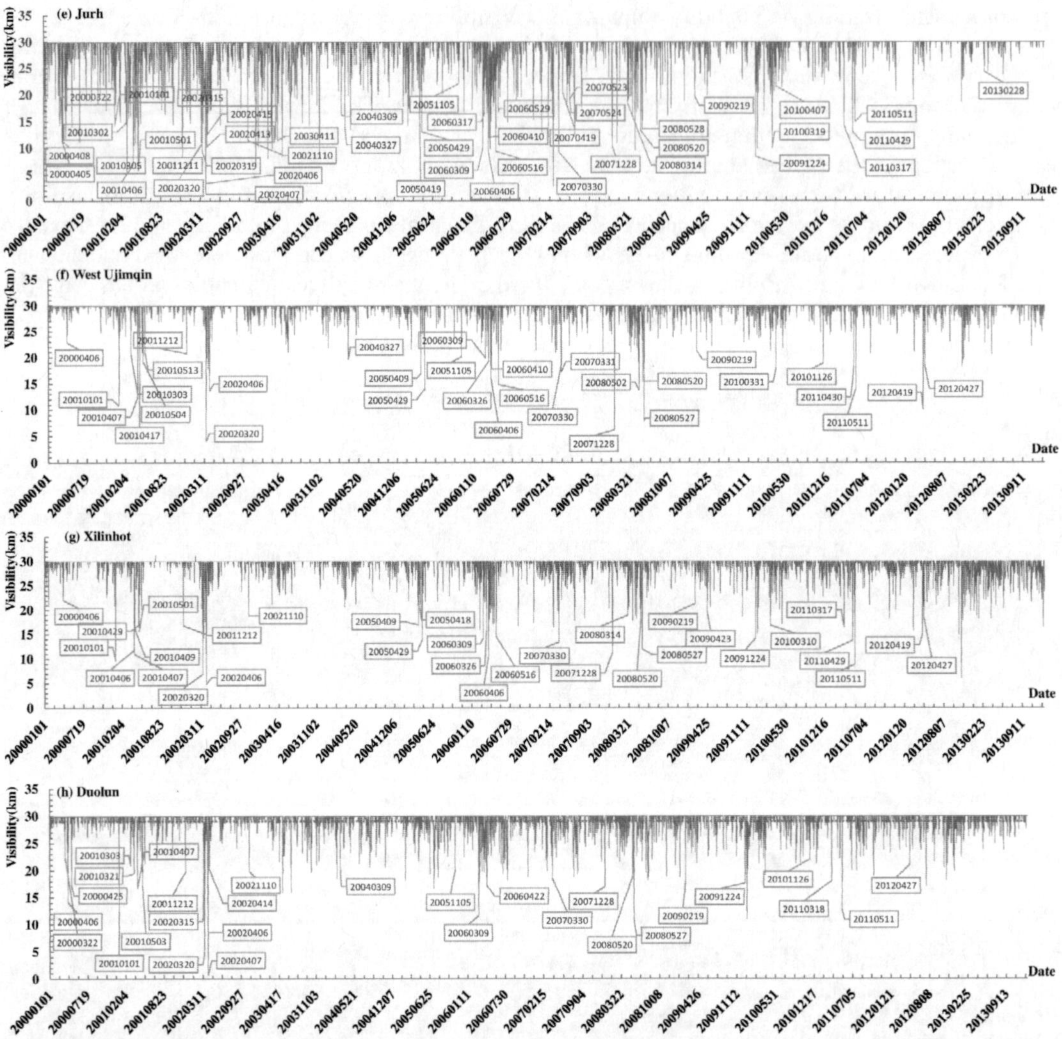

Figure 5. Visibility reductions and associated dust events in 2000-2013. Dates shown in the figures demonstrated the concurrences of dust events and visibility reductions.

3.4. Contributions of dust events to visibility impairments

Table 1. Number of visibility impairments (days) at different thresholds (km).

Station	<21	<20	<19	<18	<17	<16	<15	<14	<13	<12	<11	<10
Uliastai	43	36	27	20	12	10	8	4	4	3	2	1
Erenhot	49	42	30	22	20	17	14	13	9	8	8	5
Naran Bulag	49	43	30	20	18	12	10	10	9	7	6	4
Abag	46	37	30	24	18	12	12	11	9	9	7	6
Jurh	245	198	162	131	108	88	83	58	52	43	31	23
West Ujimqin	47	36	28	23	18	16	13	9	7	6	6	5
Xilinhot	94	71	63	47	37	26	21	15	12	12	9	7
Duolun	73	60	42	22	16	12	10	9	9	7	5	3

Table 2. Contributions (%) of dust events to visibility impairment at different thresholds (km).

Station	<21	<20	<19	<18	<17	<16	<15	<14	<13	<12	<11	<10
Uliastai	62.8	61.1	70.4	70.0	83.3	90.0	87.5	75.0	75.0	66.7	100.0	100.0
Erenhot	75.5	81.0	80.0	86.4	85.0	82.4	85.7	84.6	77.8	75.0	75.0	80.0
Naran Bulag	57.1	58.1	66.7	65.0	72.2	91.7	90.0	90.0	88.9	85.7	100.0	100.0
Abag	71.7	75.7	73.3	83.3	77.8	91.7	91.7	90.9	88.9	88.9	85.7	83.3
Jurh	43.3	46.0	49.4	52.7	56.5	56.8	57.8	58.6	59.6	65.1	64.5	69.6
West Ujimqin	72.3	80.6	85.7	95.7	94.4	93.8	92.3	88.9	85.7	83.3	83.3	80.0
Xilinhot	52.1	56.3	60.3	61.7	70.3	73.1	76.2	80.0	75.0	75.0	66.7	57.1
Duolun	52.1	55.0	57.1	72.7	68.8	58.3	70.0	66.7	66.7	57.1	40.0	66.7

In 2000-2013, the days of visibility below 21 km in Xilingol Grassland varied much, with the numbers at Jurh 2-5 times higher than those at other stations. Except Jurh, Xilinhot and Duolun, other five stations experienced similar low visibility days (figure 4, table 1). The contribution of dust activity to visibility impairment was measured by temporal consistency between MODIS captured dust events and observed visibility reduction phenomena (table 2). The contributions of dust to visibility reduction were different at eight stations, lower contributions were found at Jurh, Xilinhot and Duolun in Xilingol grassland.

4. Discussion and conclusions

Dust events significantly impaired visibility over Xilingol grassland. Dust emissions observed by MODIS tended to decrease over the time, as was proved by visibility records. Dust was more active in 2000-2002 and 2005-2009, these higher frequencies have been verified by earlier works [13].

The coherency demonstrated dust was an important factor leading to air quality impairment, but the differences in seasonal patterns and long-term trends among stations manifested that some drops in visibility could not be solely attributed to dust activity. Other factors might also be the contributors, such as meteorological variables [24], fog, haze [25], and others.

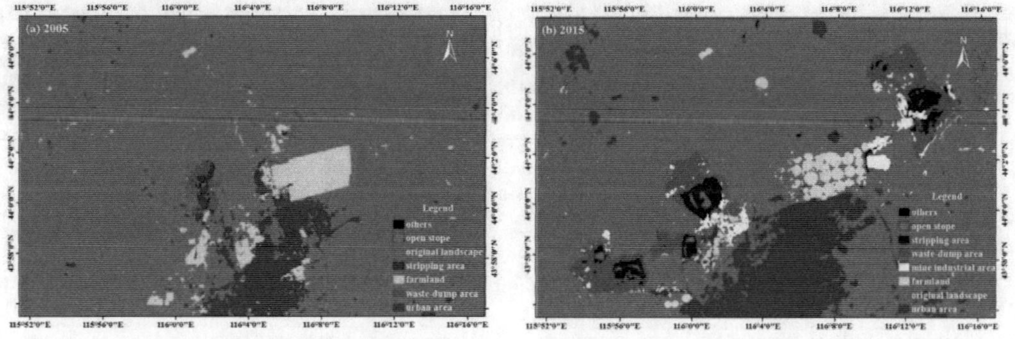

Figure 6. Land uses around Xilinhot in 2005 (a, Landsat 5 TM) and 2015 (b, Landsat 8 OLI).

Many publications discussed visibility changes in China [25, 26], but few gave its details in a temperate grassland. Taking Xilinhot as an example, supervised classification, with the support from GPS data, was used to divide land uses around Xilinhot into seven types (figure 6). The mining area increased in 2005-2015, e. g., waste-dump area grew by 3183.48 hm^2. With the growth of mining and urban areas, grassland decreased by 8661.15 hm^2. The markedly lower visibilities in Chinese megacities suggested that elevated aerosol loadings were responsible for the visibility degradation [26], due to the increased emission of pollutants from urbanization, economic and industrial developments [27]. Xilinhot ranked the second in days of low visibility, besides dust activity, mining and urbanization should also be important contributors to visibility reduction. Therefore, it is urgent to clarify the causes of visibility reduction in developing cities in China in recent years.

Acknowledgements
The authors gratefully acknowledge financial support from the National Natural Science Foundation of China (Grant Nos. 41261048 and 40961014).

References
[1] Hyslop N P 2009 Impaired visibility: the air pollution people see *Atmospheric Environment* **43(1)** 182-195
[2] Wang K, Dickinson R E, Liang S 2009 Clear Sky Visibility Has Decreased over Land Globally from 1973 to 2007 *Science* **323(5920)** 1468-70
[3] Craig C D, Faulkenberry G D 1979 The application of ridit analysis to detect trends in visibility *Atmospheric Environment (1967)* **13(12)** 1617-22
[4] Huang W, Tan J, Kan H, Zhao N, Song W, Song G, et al. 2009 Visibility, air quality and daily mortality in Shanghai, China *Science of The Total Environment* **407(10)** 3295-3300
[5] Mahowald N M, Ballantine J A, Feddema J, Ramankutty N 2007 Global trends in visibility: implications for dust sources *Atmospheric Chemistry And Physics* **7** 3309-39
[6] Baddock M C, Strong C L, Leys J F, Heidenreich S K, Tews E K, McTainsh G H 2014 A visibility and total suspended dust relationship *Atmospheric Environment* **89** 329-336
[7] Orgill M M, Sehmel G A 1976 Frequency and diurnal variation of dust storms in the contiguous U.S.A. *Atmospheric Environment* **10(10)** 813-825
[8] Darmenova K, Sokolik I N, Darmenov A 2005 Characterization of east Asian dust outbreaks in the spring of 2001 using ground-based and satellite data *Journal of Geophysical Research: Atmospheres* **110(D2)** D02204
[9] Goudie A S, Middleton N J 1992 The changing frequency of dust storms through time *Climatic Change* **20(3)** 197-225
[10] Chu D A, Kaufman Y J, Zibordi G, Chern J D, Mao J, Li C, et al. 2003 Global monitoring of air pollution over land from the Earth Observing System-Terra Moderate Resolution Imaging Spectroradiometer (MODIS) *Journal of Geophysical Research: Atmospheres* **108(D21)** 4661
[11] Kaufman Y J, Koren I, Remer L A, Tanré D, Ginoux P, Fan S 2005 Dust transport and deposition observed from the Terra-Moderate Resolution Imaging Spectroradiometer (MODIS) spacecraft over the Atlantic Ocean *Journal of Geophysical Research: Atmospheres* **110(D10)** D10S12
[12] Zhang B L, Tsunekawa A, Tsubo M 2008 Contributions of sandy lands and stony deserts to long-distance dust emission in China and Mongolia during 2000-2006 *Global and Planetary Change* **60(3-4)** 487-504
[13] Zhang B L, Tsunekawa A, Tsubo M 2015 Identification of dust hot spots from multi-resolution remotely sensed data in eastern China and Mongolia *Water, Air, & Soil Pollution* **226(4)** 1-18
[14] Li Z Q, Gu X, Wang L, Li D, Li K, Dubovik O, et al. 2013 Aerosol physical and chemical properties retrieved from ground-based remote sensing measurements during heavy haze days in Beijing winter *Atmospheric Chemistry and Physics Discussions* **13(2)** 5091-5122
[15] Prospero J M 1999 Long-term measurements of the transport of African mineral dust to the southeastern United States: implications for regional air quality *Journal of Geophysical Research: Atmospheres* **104(D13)** 15917-27
[16] Chen J, Hori Y, Yamamura Y, Shiyomi M, Huang D 2008 Spatial heterogeneity and diversity analysis of macrovegetation in the Xilingol region, Inner Mongolia, China, using the beta distribution *Journal of Arid Environments* **72(6)** 1110-19
[17] Gao T, Yang X, Jin Y, Ma H, Li J, Yu H, et al. 2013 Spatio-temporal variation in vegetation biomass and its relationships with climate factors in the Xilingol grasslands, Northern China *PLoS One* **8(12)** e83824
[18] Li X 2010 Effect of changing of ground weather monitoring specification in 1980 on visibility dat continuity (in Chinese with English abstract) *Meteorological Monthly* **36(3)** 117-122

[19] Hsu N C, Chee T S, King M D, Herman J R 2004 Aerosol properties over bright-reflecting source regions *IEEE Transactions on Geoscience and Remote Sensing* **42(3)** 557-569

[20] Prospero J M, Ginoux P, Torres O, Nicholson S E, Gill T E 2002 Environmental characterization of global sources of atmospheric soil dust identified with the NIMBUS 7 Total Ozone Mapping Spectrometer (TOMS) absorbing aerosol products *Reviews of Geophysics* **40(1)** 1002

[21] Ackerman S A 1997 Remote sensing aerosols using satellite infrared observations *Journal of Geophysical Research: Atmospheres* **102(D14)** 17069-79

[22] Han L, Tsunekawa A, Tsubo M, Zhou W 2013 An enhanced dust index for Asian dust detection with MODIS images *International Journal of Remote Sensing* **34(19)** 6484-95

[23] Karimi N, Moridnejad A, Golian S, Samani J M V, Karimi D, Javadi S 2012 Comparison of dust source identification techniques over land in the Middle East region using MODIS data *Canadian Journal of Remote Sensing* **38(5)** 586-599

[24] Tsai Y I 2005 Atmospheric visibility trends in an urban area in Taiwan 1961-2003 *Atmospheric Environment* **39(30)** 5555-67

[25] Ding Y, Liu Y 2014 Analysis of long-term variations of fog and haze in China in recent 50 years and their relations with atmospheric humidity *Science China: Earth Sciences* **57(1)** 36-46

[26] Che H, Zhang X, Li Y, Zhou Z, Qu J J 2007 Horizontal visibility trends in China 1981-2005 *Geophysical Research Letters* **34(24)** L24706

[27] Lin M, Tao J, Chan C, Cao J, Zhang Z, Zhu L, et al. 2012 Regression analyses between recent air quality and visibility changes in megacities at four haze regions in China *Aerosol and Air Quality Research* **12(6)** 1049-61

IOP Conf. Series: Earth and Environmental Science **52** (2017) 0130 doi:10.1088/1755-1315/52/1/0120

Comparison of metals levels in two mangrove species (*Rhizophora stylosa* and *Sonneratia hainanensis*) from Hainan Island, South China

Y W Qiu[1] and H L Qiu[2]

[1] State Key Laboratory of Tropical Oceanography, South China Sea Institute of Oceanology, Chinese Academy of Sciences, Guangzhou 510301, China

[2] School of Materials Science and Engineering, Hubei University, Wuhan 430062, China

E-mail: yqiu@scsio.ac.cn; zoe_cn_q@hotmail.com

Abstract. Trace metals in mangrove tissues (leaf, branch and root) of two species (*Rhizophora stylosa* and *Sonneratia hainanensis*) from Dongzhai Harbor and Sanya Bay of Hainan Island were studied. The total average concentrations of Cu, Pb, Zn, Cd, Cr, Hg and As in the two mamgrove species were 2.4±1.3, 1.1±0.7, 7.8±8.0, 0.03±0.05, 1.4±1.6, 0.03±0.01 and 0.2±0.2 μg g-1 dw, respectively. Metals concentrations among different tissues of mangroves showed different pattern. In general, Zn, Cd and Hg were slightly enriched in leaf, Cu, Pb and As was enriched in root, and Cr were enriched in branch. Metals levels in *R. stylosa* and *S. hainanensis* from both Dongzhai Harbor and Sanya Bay were compared, which suggested that different mangrove species have their unique mechanism to bioaccumulate metals and TOC in the mangrove sediment could be one of the important factors for regulating metals in mangrove tissues. The biota-sediment accumulation factors (BSAF) of metals in mangrove tissues were calculated. The distribution of metals concentrations in mangrove tissues against metals levels in sediment demonstrated that mangrove leaves could be employed as a bio-indicator for some metals (Cu, Zn, Cd and Hg) with temporal monitoring.

1. Introduction

Mangroves are specific intertidal ecosystems, covering between 160,000 and 200,000 km^2 along more than 70% of tropical and subtropical coastlines [1]. This ecosystem is highly rich in biodiversity, and acts as a habitat for many species. It is estimated that almost 80% of global fish catches are directly or indirectly dependant on mangroves [2]. Mangroves can protect inland human communities from damage caused by coastal erosion and storms, provide critical habitat for a variety of terrestrial, estuarine and marine species, and serve as both a source and sink for nutrients/contaminants and sediment for other inshore marine habitats including seagrass bed and coral reef [3-5]. Mangroves can sequester up to 25.5 million tonnes of carbon per year, and provide more than 10% of essential organic carbon to the global oceans [6]. Mangrove forest is an efficient biogeochemical barrier to the transport of heavy metals to coastal areas [7]. However, mangrove areas are declining rapidly in the past several decades as they are cleared for coastal development and aquaculture and logged for timber and fuel production, and eleven of the 70 mangrove species (16%) are at elevated threat of extinction [5]. Currently, mangroves are also increasingly threatened due to anthropogenic chemicals sourced from uncontrolled agricultural runoff, urban and industrial effluent and wastewaters, as well as with urbanization and population growth [8-10]. Further, trace metals taken up by mangroves and concentrated in exported leaf detritus could be reintroduced to the adjacent deeper waters [11]. The biomass, productivity and litter of mangrove forests varied with the age of mangrove, dominant species, and locality. The bioaccumulation and/or fixation of trace metals in the mangrove tissues may prevent release of metals to the water column and limit entry of these contaminants to the food chain. The knowledge of bioaccumulation of trace metals in different mangrove species is important to understand the fates of these pollutants and, can alert coastal managers of possible impacts upon the detritus driven food web which can potentially lead to the bioaccumulation of contaminants in organisms.

Hainan Island, the second largest Island of China, has mangrove areas of 4772 hm^2, accounting for one third of the total mangrove forest areas of China. There are 26 mangrove plant species distributing in Hainan while 27 species in China [12]. Hainan Dongzhai harbor mangrove protection zone, as the first national mangrove reserve of China, was established in 1986 and was listed as the International Important Wetland in 1992. While Hainan Sanya mangrove protection zone, as the southmost mangrove reserve of China with the typical tropical features, was established in 1990. Since 1978 when China started the reform and opening up policy, the development of coastal zones throughout Hainan Island has put immense pressures on biological communities including mangrove ecosystem. More than 60% mangrove forest has disappeared since 1950s [13], and many mangrove forest areas were developed as culturing ponds and constructing sites. As an international important wetland, a few information is available on the status of trace metals in mangrove ecosystem of Hainan Island [9, 14], though many studies on trace metal pollution in mangrove wetlands around the world have been reported [15-16]. Further, there are very few data on mercury and arsenic concentrations in mangrove tissues around the world so far. Based on the concentrations of trace metals (Cu, Pb, Zn, Cd, Cr, Hg and As) in two mangrove species (including in root, branch and leaf) and sediments from two typical mangrove areas of Hainan Island, the present work aimed to compare metals levels in *R. stylosa* and *S. hainanensis* between Dongzhai Harbor and Sanya Bay; and to assess the biotic response of metals contamination in mangrove ecosystem, e.g., the relationship of metal concentrations between mangrove tissues and sediment.

2. Materials and methods

2.1. Study areas

Dongzhai Harbor, located in the northern Hainan Island, is the most well-preserved mangrove forest in China and suffers agricultural stress; Sanya Bay, situated in the center of Sanya city of the southern Hainan Island, is subjected to both sewage discharge and agricultural pollution. The two studied areas were predominantly colonized by mangrove species including *Rhizophora stylosa*, *Rhizophora apiculata*, *Bruguiera sexangula*, *Ceriops tagal*, *Aegiceras corniculatum*, *Avicennia marina*, and *Sonneratia hainanensis* [14, 17].

2.2. Sample collection and processing

Mangrove tissues at six sites of Dongzhai Harbor (DZ1, DZ2, DZ3, DZ4, DZ5 and DZ6) and three sites of Saya Bay (SY1, SY2 and SY3) (Figure 1), were collected in October 2008 using pre-determined GPS co-ordinates to accurately locate each station. Sampling sites were selected based on the spatial and the mangrove species community distribution and can be accessible. Mangrove tissues, including root, perennial branch (diameter about 5-8 cm) and leaf of two species (*R. stylosa* and *S. hainanensis*), were simultaneously collected. Mangrove samples were washed by deionized water in laboratory to remove the possible dust, and naturally dried, and were then sampled with a thin stainless steel blade.

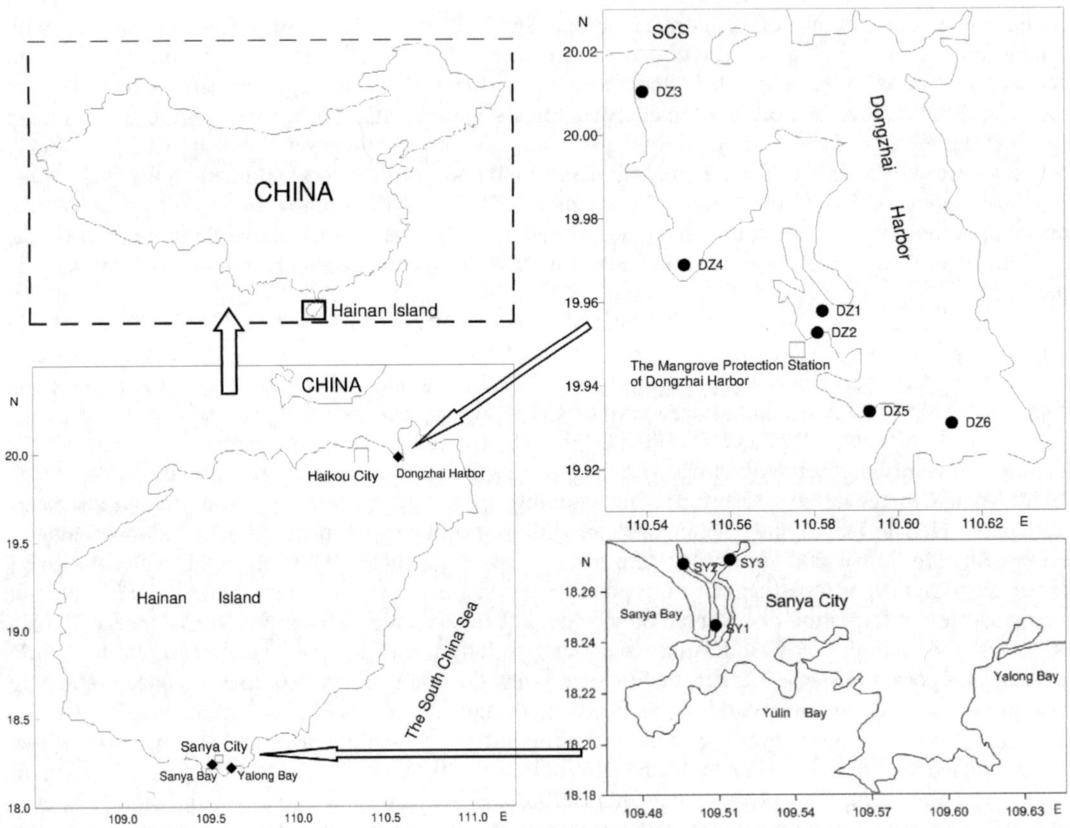

Figure 1. Sampling locations in two mangrove wetlands (Dongzhai Harbor and Sanya Bay) of Hainan Island, South China.

2.3. *Chemical analysis*

Measurements of trace metals followed the procedures described previously [14, 18]. 100.0 mg mangrove tissue sample was placed into Teflon pot pre-cleaned with high purity nitric acid; then 2.0 ml concentrated nitric acid (HNO_3) was added into the pot lasting for 2 h; and then heated in an electric board at 150 °C for 24 h. After cooling, the solution was diluted by high pure water to 40 g and the concentrations of Cu, Pb, Zn, Cd, Cr, Hg and As were measured using ICP-MS (model DRC II, Perkin Elmer, USA). Measurements of trace metals in sediment followed the procedures in Qiu et al [14].

2.4. *Quality assurance and quality control*

All glass- and plastic-ware were soaked in 10% nitric acid overnight and rinsed thoroughly with deionized water before use. For quality control, reagent blanks, the Chinese national standard samples of GBW08571, GBW08517 and GBW010016 (reference materials for mussel, kelp, tea, respectively) were used to monitor the analytical quality. The results were consistent with the reference values with relative differences within 10% (mostly within 5%). Blank determinations were carried out for each set of analysis. The determination limits of Cu, Pb, Zn, Cd, Cr, Hg and As in mangrove tissues were 0.02, 0.002, 0.02, 0.02, 0.02, 0.002 and 0.02 μg g^{-1} dry wt., respectively. Statistical analysis was conducted using SPSS 18.0.

3. Results and discussion

3.1. Metal levels in mangrove tissues and sediment

The total average concentrations of Cu, Pb, Zn, Cd, Cr, Hg and As both in *R. stylosa* and *S. hainanensis* (n=23) from Dongzhai Harbor and Sanya Bay of Hainan Island were 2.4±1.3, 1.1±0.7, 7.8±8.0, 0.03±0.05, 1.4±1.6, 0.03±0.01 and 0.2±0.2 μg g^{-1} dw, while in sediment were 12.2±8.4, 23.9±10.6, 55.6±32.8, 0.14±0.06, 20.3±11.4, 0.07±0.04 and 9.2±4.4 μg g^{-1} dw, respectively. Previous study [19] showed that the metal concentrations in trunk of *Rhizophora apiculata* from Leizhou Peninsula, south China were 10.89, 0.51, 14.73, 0.15 and 26.64 μg g^{-1} for Cu, Pb, Zn, Cd and Cr, respectively, which all the target metals except for Pb were higher than those in the present study. The concentrations of Cu, Pb, Zn, Cd and Cr in roots of three mangrove species *(Acanthus ilicifolius, Aegicerus corniculatum* and *Kandelia candel*) in Mai Po, Hong Kong were 32.6, 34.4, 137.5, 0.4 and 3.5 μg g^{-1}, respectively [20], which were much higher than the present results. The concentrations of Cu, Pb, Zn, Cd and Cr in tissues (leaf, branch and root) of *Rhizophora mangle* in Natal, Brazil [7] were also higher than those in mangrove (*Avicennia marina*) leaves were 3.2, 1.7 and 14.3 μg g^{-1} [21], which were slightly higher than those in Hainan Island. Nath et al [22] recently reported the concentrations of Cu, Pb, Zn, Cd, Cr and As in *Avicennia marina* leaves in Sydney Estuary (Australia) of 10, 0.74, 14, 0.01, 0.49 and 0.48 μg g^{-1}, respectively, which were comparable to those in Hainan Island. Mercury concentrations in the present nine mangrove species (0.03±0.01μg g^{-1}) was lower than those in eight mangrove species from Tamil Nadu of India (0.06±0.03 μg g^{-1}; [23]). All above results suggested that Hainan mangrove was in a relatively unpolluted status.

In the present study, trace metals concentrations among different tissues of two mangrove species were not very significantly different, as shown in Figure 2. Comparatively, Zn, Cd and Hg were slightly enriched in leaf, Cu, Pb and As was enriched in root, and Cr were enriched in branch. The highest average concentrations for Cu (3.3 μg g^{-1}), Pb (1.3 μg g^{-1}), and As (0.34 μg g^{-1}) were detected in the belowground roots. It is well known that metal-rich deposits on roots of aquatic plants may moderate the uptake of potentially toxic metals [7, 24]. In the present study, the highest average concentrations for Zn (11.4 μg g^{-1}), Cd (0.05 μg g^{-1}), and Hg (0.04 μg g^{-1}) were in the leaf, while the branch presented the highest average concentrations for Cr (2.0μg g^{-1}). As the essential elements for the growth of plants, Zn tended to accumulate in leaf, while Hg, a semi-volatile metal, was deposited and accumulated in leaf. It was reported that *A. marina* mangroves of Sydney estuary selectively excluded non-essential elements (As, Cd, Co, Cr and Pb), while regulating essential elements (Cu, Ni, Mn and Zn) and limiting toxicity to plants [22]. Except for the pollution level in the local mangrove ecosystem, several factors (e.g., time of collection, tissue age and tissue type) may also influence the concentrations of tissue accumulated chemicals and their order of accumulation.

3.2. Comparision of trace metals levels in different mangrove species

Trace metals levels in *R. stylosa* and *S. hainanensis* (leaf, branch and root) from Dongzhai Harbor and Sanya Bay were shown in Figure 2. Generally, targeted metals levels in mangrove tissues of *R. stylosa* were lower than those in *S. hainanensis* from Dongzhai Harbor and Sanya Bay. According to the annual National Marine Pollution Bullet in China, pollution status went worse only in the recent decades in Hainan Island. As *R. stylosa* grows much slower than *S. hainanensis*, we may expect that *R. stylosa* grows in relative low level of metals for the longer period and will accumulate less metals, while *S. hainanensis* grows in the recent polluted years and will accumulate more metals. Many studies have showed that concentrations for the same trace metal and tissue can differ within a population of the same species and between different species [7, 15]. Interspecific differences have also been commonly reported and attributed to differences in physiology [23]. This is mainly due to physiological differences, variable translocation and excretion mechanisms of different plant organs. The present results suggested that different mangrove species have their unique mechanism to bioaccumulate metals.

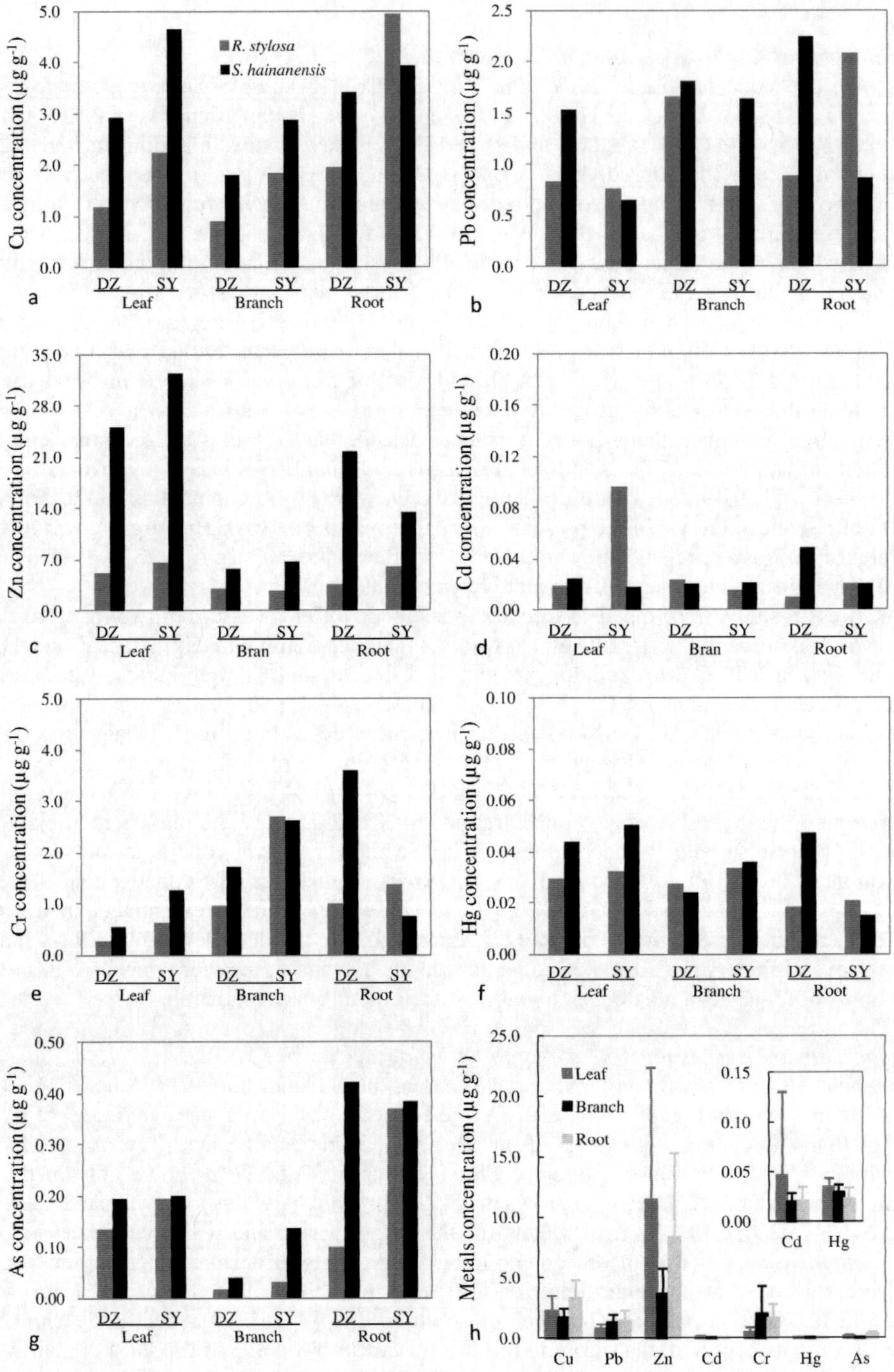

Figure 2. Comparision of metals levels in two mangrove species (*R. stylosa* and *S. hainanensis*) from Dongzhai Harbor (DZ) and Sanya Bay (SY) of Hainan Island.

From the Figure 2, we also can see that targeted metals levels in mangrove tissues of *R. stylosa* and *S. hainanensis* were generally lower from Dongzhai Harbor than those from Sanya Bay. Our previous results showed that the vertical profiles of trace metals concentrations in mangrove sediment cores of Dongzhai Harbor were elevated compared to those in Sanya Bay, but TOC contents in mangrove sediment cores had different patterns, with the higher TOC contents in sediment from *R. stylosa* zones in Dongzhai Harbor than those in Sanya Bay, and the similar TOC contents in sediment from *S. hainanensis* zones in Dongzhai Harbor and Sanya Bay [14]. Positive correlations between TOCs in mangrove sediments and the levels of metals in Hainan Island were also observed [14]. The profiles of TOC in Dongzahi Harbor appear not be different from the profiles of trace metals comparing with Sanya Bay. The present studies showed that TOC in the mangrove sediment could be one of the important factors for regulating metals in mangrove tissues. There were reports that organic matter from leaf fall contributes to the depletion of dissolved oxygen in the sediment with a consequent influence on the mobility of heavy metals [7, 25]. Organic matter is also known to chelate with trace metals, therefore affecting their mobility and bioavailability [16]. The present results further confirm the hypothesis that bioaccumulation of metal in organism depends on the biological availability of metal and not the total metal concentration in the natural environment.

Apart from the metals levels in the media of mangroves, the uptake of metals is primarily influenced by mangrove plant metabolic requirements (e. g., Cu and Zn, which are essential micro-nutrients) are more mobile than non-essential elements (such as Pb) resulting in variable metal accumulation in plant tissues [8, 22, 26-27]. Oxidation of the upper rhizosphere leads to precipitation of Fe-oxyhydroxides and formation of iron plaques around root surfaces and ultimately results in significant accumulation of trace metals in these zones [10, 28-29]. The formation of Fe plaque on the root surface not only reduced the metal bioavailability toward roots and plant metal uptake, it may also play a key role in the removal and detoxification of pollutants and protect plants against pollutants in constructed wetlands [30]. Recent reports showed that metal-tolerant mangrove species often exhibited a thick exodermis with high lignification and suberization which could delay the entry of metals into the roots, and thereby could contribute to a higher tolerance to heavy metals [31].

3.3. The biota-sediment accumulation factors (BSAF) of metals

Trace metals may enter the mangrove through transport by local rivers, marine waters intrusion or through atmospheric deposition. Plants could uptake contaminants from a polluted atmosphere by the deposition of the particle-phase contaminants on the waxy leaf cuticle or by uptake in the gas phase through stomata, plants could also take up contaminants from contaminated sediments or soils by roots, transferred to shoots and leaves [32]. Aquatic organism can absorb contaminants from sediment, and its ability can be expressed by the biota-sediment accumulation factors (BSAF). In the present study, the BSAFs (defined as the ratio of metal concentration in mangrove tissues to metal concentration in underlying sediment) based on all trace metals data both in sediment and mangroves were calculated and shown in Figure 3. Target metals with descending values of BSAFs (in parentheses) for leaves of all studied mangrove species in Hainan Island were all less than 0.5 (<1 unit) and followed the sequence of Hg (0.49) > Cu (0.26) > Cd (0.21) > Zn (0.17)> Cr (0.10) > Pb (0.06) > As (0.02). Mercury exhibited the highest BSAFs values because of its physical property, a semi-volatile metal. The essential metals such as Cu and Zn showed a greater mobility than the non-essential metals such as Pb, Cr and As. As for cadmium, which atomic structure being similar to Ca, also exhibited a high BSAF value. This may be explained by its similar chemical behavior as Cu and Zn [33], but the exact reasons still need further study. Our results were generally similar to those in Sydney Estuary of Australia, with BSAFs of Cu (0.38), Zn (0.25), As (0.09), Cr (0.05), Cd (0.04) and Pb (0.02) in leaves of *Avicennia marina* [22].

The distribution of metals concentrations in mangrove tissues (leaf, branch and root) against metals levels in sediment was shown in Figure 3. Overall, the data did not show a clear positive linear trend in targeted metals accumulations in mangrove tissues (branch and root) with increasing sediment concentrations. However, the slightly positive relationships for trace metals (Cu, Zn, Cd and Hg) concentrations between leaf tissues and sediment were observed, with the relative coefficients (R) of 0.51, 0.32, 0.40 and 0.24, respectively. The results suggested that mangrove leaves could be employed as a bio-indicator for these metals with temporal monitoring. Previous study also showed a restricted mobility of Cu and Zn and a strong exclusion mechanism of Pb in *A.marina* leaf tissues and concluded that *A. marina* leaves could be employed as a bio-indicator for Zn [27].

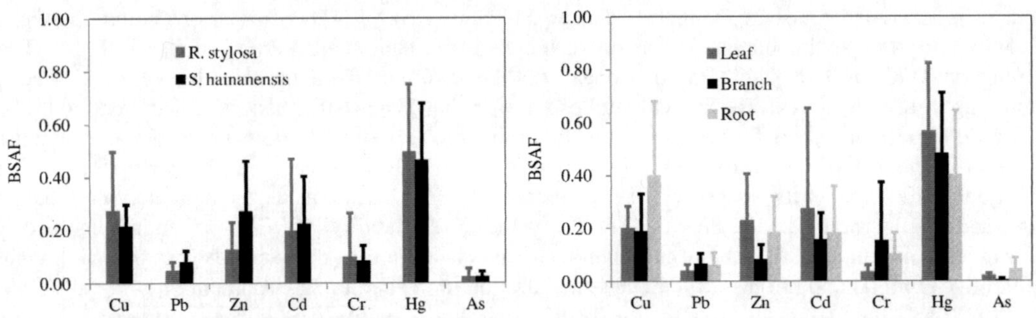

Figure 3. The average BSAF of metals in mangrove tissues (leaf, branch and root) from Dongzhai Harbor and Sanya Bay of Hainan Island.

4. Conclusions

Concentrations of trace metals (Cu, Pb, Zn, Cd, Cr, Hg and As) in mangrove tissues (leaf, branch and root) in Hainan Island, were analyzed. Trace metals levels in *R. stylosa* and *S. hainanensis* from Dongzhai Harbor and Sanya Bay were compared and the results showed that metals concentrations among different tissues of mangroves were different and different mangrove species have their unique mechanism to bioaccumulate metals. TOC in the mangrove sediment could be one of the important factors for regulating metals in mangrove tissues. Mangrove leaves could be employed as a bio-indicator for some metals (Cu, Zn, Cd and Hg) with temporal monitoring.

Acknowledgments

The present study was supported by National Natural Science Foundation of China (No. 41376091) and National Key Basic Research Program of China (2013CB956101) to Y.-W.Q.

References

[1] FAO 2003 Status and Trends in Mangrove Area Extent Worldwide. Forest Resources Division, FAO, Paris
[2] Ellison A M 2008 *J. Sea Res.* **59** 2–15
[3] Peters E C, Gassman N J, Firman J C, et al. 1997 *Enviro Toxico Chem.* **16** 12–40
[4] Tam N F Y, Wong Y S 2000 *Environ. Pollu.* **110** 195–205
[5] Polidoro B A, Carpenter K E, Collins L, et al. 2010 *PLoS One* **5** e10095
[6] Dittmar T, Hertkorn N, Kattner G, et al. 2006 *Global Biogeochem. Cycles* **20** GB1012, doi:10.1029/2005GB002570.
[7] Silva C A R, Silva A P, Oliveira S R 2006 *Mar. Chem.* **99** 2–11
[8] MacFarlane G R, Koller C E, Blomberg S P 2007 *Chemosphere* **69** 1454–1464
[9] Vane C H, Harrison I, Kim A W, et al. 2009 *Mar. Pollut. Bull.* **58** 134–144
[10] Chaudhuri P, Nath B, Birch G 2014 *Mar. Pollut. Bull.* **79** 284–292
[11] Odum W E, Johannes R E 1975 In E.J. Ferguson Wood and R.E. Johannes, eds., Tropical Marine Pollution, Elsevier, New York, NY, USA, pp. 52–62
[12] Lin P 2001 *J. Xiamen U.* **40**(2) 592–603
[13] Mo Y N, Geng Z Z, Su W B 1999 *Tropical Forestry* **27**(1) 19–22
[14] Qiu Y W, Yu K F, Zhang G, Wang W X 2011 *J. Hazard. Mater.* **190** 631–638

[15] Lewis M, Pryor R, Wilking L 2011 *Environ. Pollu.* **159** 2328–2346
[16] Bayen S 2012 *Environ. Int.* **48** 84–101
[17] Lin P 1997 Mangrove Ecosystem in China. Beijing: Science Press, 184–212
[18] Qiu Y W 2015 *Estuar. Coast Shelf Sci.* **163** 7–14
[19] Yu K F, Kamber B S, Lawrence M G, et al. 2007 *Nucl. Instrum. Meth. B* **255** 399–408
[20] Ong Che R G 1999 *Mar Pollut. Bull.* **39** 269–279
[21] MacFarlane G R, Burchett M D 2002 *Mar. Environ Res.* **54** 65–84
[22] Nath B, Chaudhuri P, Birch G 2014 *Ecotox. Environ. Safe.* **107** 284–290
[23] Agoramoorthy G, Chen F A, Hou M J 2008 *Environ. Pollut.* **155** 320–326
[24] Silva C A R, Lacerda L D, Rezende C E 1990 *Biotropica* **22** 339– 345
[25] Aragon G T, Ovalle A R C, Carmouze J-P 1999 *Mangrove Salt Marsh* **3** 85– 93
[26] Baker A J M 1981 *J. Plant Nutr.* **3** 643–56
[27] MacFarlane G R, Pulkownik A, Burchett M D 2003 *Environ. Pollu.* **123** 139–151
[28] Otte M L, Kearns C C, Doyle M O 1995 *Bull. Environ. Contam. Toxicol.* **55** 154–161
[29] Zhou Y W, Peng Y S, Li X L, Chen G Z 2011 *Environ. Earth Sci.* **64** 799–807.
[30] Pi N, Tam N F Y, Wong M H 2011 *Mari. Pollut. Bull.* **63** 402–411
[31] Cheng H, Jiang Z Y, Liu Y, et al. 2014 *Tree Physiol.* **34** 646–656
[32] Simonich S L, Hites R A 1995 *Environ. Sci. Technol.* **29** 2905–2914
[33] Diao S Y, Zhang L Z, Yuan H 2005 *Prog. Veterinary Med.* **26**(5) 49–51.

Spatial and temporal variation of soil temperature of *Taxodium Distichum* Shelterbelts in south China

Lu Zhang[1]**, Ding Ma**

College of Forestry and Landscape Architecture, South China Agricultural University, Guangzhou, 510642, China

Email: zhanglu@scau.edu.cn

Abstract. Start your abstract here With the global environmental change, more attention was paid to the effect of farmland shelterbelt. In this study, spatial and temporal patterns of soil temperature in shelterbelts of Taxodium distichum in South China was monitored using gradient observed method through observation plots placed in Taxodium distichum shelterbelts, sugarcane field and blank field (open site) in January, April, July, and October, respectively. It was found that soil temperature changed with time in a day with a single peak in all the four phases among three sites. However, compared with blank field, the vertical patterns of soil temperature of Taxodium distichum shelterbelts and sugarcane field showed special characteristic. One-way ANOVA and Tukey's HSD showed soil temperature was significantly different with annual and seasons among Taxodium distichum shelterbelts, sugarcane field and blank field along a seasonal dynamic ($P < 0.05$). Effective measures should be taken to regulate the soil temperature in order to satisfy the needs of growth of sugarcane.

1. Introduction

FARMLAND shelterbelt are believed to provide a number of ecosystem services and environmental benefits [1], which is well recognized for reducing soil erosion [2], increasing diversity [3] and has gained attention because of its perceived potential to sustainably enhance agricultural production [4]. Three-Norths Forest Shelterbelt program was carried out in 1970s to combat desertification and control dust storms in China, and various shelterbelts have been constructed in China today. Studies were mainly concentrated on air temperature and relative humidity [5], soil temperature [6], photosynthetically active radiation [7], soil carbon dioxide flux. There was a remarkable difference in the soil temperature between a broadleaved - Korean pine mixed forest and its adjacent open site in Northeast China, and the forest had a lower soil temperature than the open site when the soil temperature was above 0 ℃, but was in adverse when the soil temperature was below 0 ℃ [8]. The soil temperature of forest is one of the important environmental factors of forest ecological system. However, as a managed ecosystem, soil temperature in farmland shelterbelt has seldom been reported.

In this paper, we investigated the temporal variation of soil temperature and the site difference at farm shelterbelts of *Taxodium distichum* in South China. The objectives of our study were: (1) to examine the temporal variation and vertical patterns of soil temperature among *Taxodium distichum* shelterbelt, sugarcane field and blank field, and (2) to compare the annual and seasonal differences of soil temperature among three sites, and discuss the reason if the difference was significant.

2. The study areas

The study area is located at the Lile town of Jiangmen (112° 47′ 13″ - 113° 15′ 24″ E, 22° 05′ 43″ - 22° 48′ 24″ N) in the southern part of Guangdong province, South China. The main zonal soil is Lateritic red soil. The area is characterized by a subtropical continental monsoon climate, which is hot in the summer. The mean annual temperature is 21.8 ℃, and the mean annual precipitation is 1784.6 mm. Three sampling sites (*Taxodium distichum* shelterbelt, sugarcane field, and blank field were located at the Lile town. The dominant species of the study area is *Taxodium distichum*, *Taxodium ascendens*, *Glyptostrobus pensilis*, and *Livistona chinensis*.

3. Methodology

Using gradient observation methods, soil temperature of the shelterbelt microclimate on both horizontal and vertical scales were measured with surface thermometer (WQG-15) and angle geothermometer (WQG-16) through observation plots placed in *Taxodium distichum* shelterbelts, sugarcane field, and blank field, respectively. Our field monitoring was conducted in four phases, i.e. January, April, July, and October. The soil surface temperature, the soil temperature under soil layer 5 cm, the soil temperature under soil layer 10 cm, the soil temperature under soil layer 15 cm, and the soil temperature under soil layer 20 cm were measured in the four phases. Each phases choose typical weather continuous monitoring for three days, 12 monitoring days in total. Statistical software program was employed for One-way ANOVA to compare temporal differences of soil temperature among *Taxodium distichum* shelterbelts, sugarcane field, and blank field.

4. Results and discussion

4.1. Temporal patterns of soil temperature

Soil temperature changed with time in a day with a single peak in all the four phases, i.e. January, April, July, and October, which was higher in the afternoon and lower in the early morning and evening (Figure 1). The peak temperature of three kinds are all at the time of 14:00 in January. However, in July the peak temperature of blank field is much easier than the other two. Soil temperature of blank field was the highest, while the soil temperature of *Taxodium distichum* shelterbelts and sugarcane field were the lower in all the four phases. Moreover, the soil temperature of sugarcane field was higher than the soil temperature of *Taxodium distichum* shelterbelts in July and October, while the soil temperature of *Taxodium distichum* shelterbelts was higher than the soil temperature of sugarcane field in January. The mean daily soil temperature was 24.1 ℃, which was 0.9 ℃ and 5.6 ℃ lower than that of sugarcane field and blank field. Patterns of daily soil temperature change in *Taxodium distichum* shelterbelt in January were similar to that in sugarcane field, whose diurnal ranges were lower than that of blank field. As compared to sugarcane field and blank field, *Taxodium distichum* shelterbelts had a significantly greater cooling effective in July and October, while the soil temperature changes with time not so dramatically in January.

Hydrothermal conditions of the forest is the result of nutrient and energy exchange between shelterbelts and the environment. Our findings showed that *Taxodium distichum* shelterbelts mitigated soil temperature in July. Results show that the shelterbelt can increase the soil temperature of the protected farmland as compared with no shelterbelt zone at regional scale based on MODIS data [9]. Our Study revealed that soil temperature of different monitoring time of the day of *Taxodium distichum* shelterbelts was lowest, while blank field's was highest in July. For example, average soil temperature of *Taxodium distichum* shelterbelts in July was 29.9 ℃, which 11.0 ℃ lower than the same period of blank field. For all year, in the forest the soil temperature were lower than on the blank field. On the one hand, it is available to plant seed germination and survival, on the other hand lower soil temperature provides a stable external environment for soil microbial, and conducive to soil organic matter decomposition and nutrient enrichment10。 From this standpoint, shelterbelts not only play its ecological barrier function, but also had great significance for the local plants grow, regulating local climate and optimizing the environment.

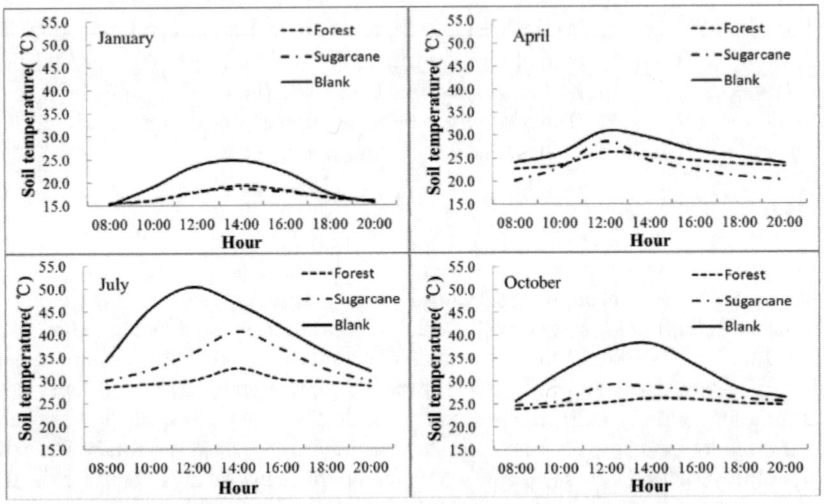

Figure 1. Temporal patterns of soil temperature change

4.2. Vertical patterns of soil temperature

Vertical patterns of soil temperature of *Taxodium distichum* shelterbelt, sugarcane field and blank field exhibited a inconsistent pattern. Compared with soil temperature of blank field at the ground > soil temperature at 5 cm depth> soil temperature at 10 cm depth > soil temperature at 15 cm depth > soil temperature at 20 cm depth, the vertical patterns of soil temperature of *Taxodium distichum* shelterbelts and sugarcane field showed special characteristic (Figure 2). Soil temperature of *Taxodium distichum* shelterbelts at 5 cm depth was the lowest, with the increase of depth, soil temperature increased in January and October. That is because the soil of *Taxodium distichum* shelterbelts at 0 cm - 5 cm depth had already from exotherm to endotherm, while the soil at 10 cm - 20 cm depth had remain exotherm at that time. Soil temperature of *Taxodium distichum* shelterbelts at the ground > soil temperature at 5 cm depth> soil temperature at 10 cm depth > soil temperature at 15 cm depth > soil temperature at 20 cm depth in April and July.

Meanwhile, the vertical patterns of soil temperature of sugarcane field was similar to *Taxodium distichum* shelterbelts in January and October, except for soil temperature in different depths were lower than *Taxodium distichum* shelterbelts. This may be associated with higher groundwater level of sugarcane field. Soil temperature of sugarcane field at the ground > soil temperature at 5 cm depth> soil temperature at 10 cm depth > soil temperature at 15 cm depth > soil temperature at 20 cm depth in July. It's worth pointing out that soil temperature of sugarcane field at the ground < soil temperature at 10 cm depth < soil temperature at 15 cm depth < soil temperature at 20 cm depth < soil temperature at 5 cm depth. This may mainly by the artificial factors, such as tillage, fertilization of sugarcane field in April, which may be related to tillage, fertilization and so on of sugarcane field in Spring. Meanwhile, vertical soil temperature of sugarcane field lower than that of *Taxodium distichum* shelterbelt, which be connected with underground water level and artificial irrigation. In addition to the influence of climate, artificial measures had greater effect on the vertical change of soil temperature.

4.3. Annual and seasonal analysis of soil temperature

One-way ANOVA show that annual patterns of soil temperature were significantly different between *Taxodium distichum* shelterbelt, sugarcane field and blank field ($P < 0.05$). Tukey's HSD test revealed that mean annual soil temperature were significant differences between *Taxodium distichum* shelterbelts and blank field. But, mean annual soil temperature had no significant difference in sugarcane field and in blank field, and soil temperature had no significant difference in sugarcane field and in *Taxodium distichum* shelterbelts too (Figure 3).

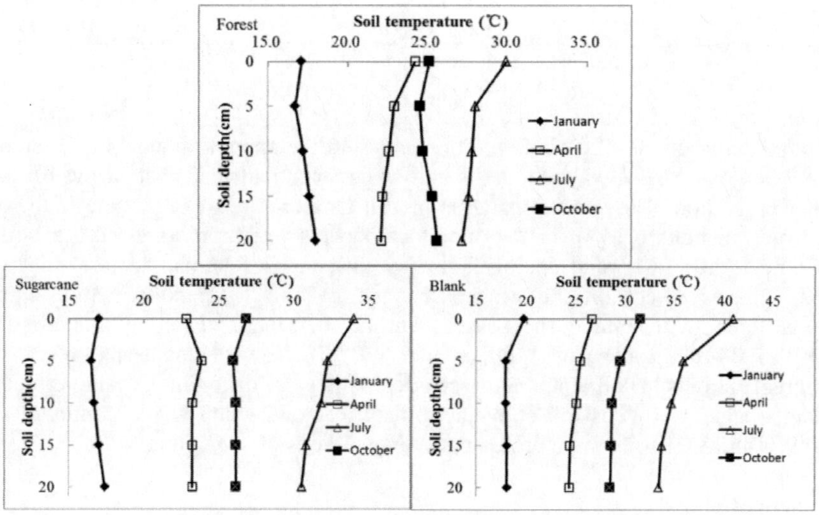

Figure 2. Vertical variation of soil temperature

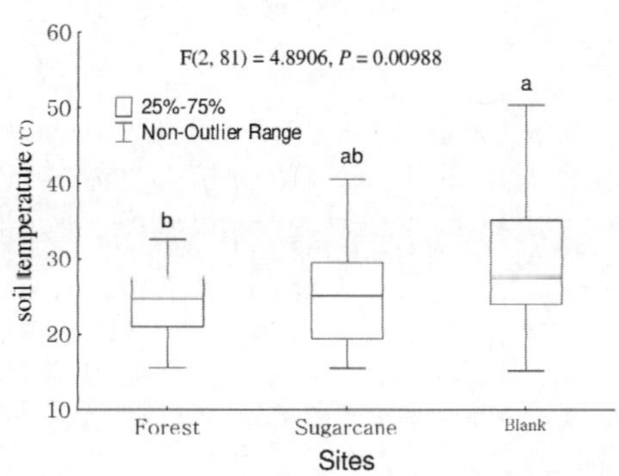

Figure 3. Annual analysis of soil temperature

Soil temperature of *Taxodium distichum* shelterbelt, sugarcane field and blank field exhibited a same pattern: soil temperature of July > soil temperature of October > soil temperature of April > soil temperature of January (Table 1). However, seasonal patterns of soil temperature were different among *Taxodium distichum* shelterbelt, sugarcane field and blank field. One-way ANOVA show that seasonal patterns of soil temperature of *Taxodium distichum* shelterbelt, sugarcane field and blank field had no significantly different between any two of the three sampling sites in January ($P > 0.05$). Seasonal patterns of soil temperature were significantly different between any two of the three sampling sites in April ($P < 0.05$). Tukey's HSD test revealed that there were significant differences between sugarcane field and blank field, however, there was no significant difference between *Taxodium distichum* shelterbelt and sugarcane field, and between *Taxodium distichum* shelterbelt and blank field. However, seasonal patterns of soil temperature were no significant difference between *Taxodium distichum* shelterbelt and sugarcane field in July and October.

Table 1. Seasonal analysis of soil temperature

	January	April	July	October
Forest	17.1±0.5^{a}	24.2±0.5^{ab}	29.9±0.5^{b}	25.1±0.3^{b}
Sugarcane	17.0±0.5^{a}	22.8±1.1^{b}	34.0±1.5^{b}	26.8±0.6^{b}
Blank	19.7±1.5^{a}	26.6±1.0^{a}	40.9±2.6^{a}	31.5±1.9^{a}

Mean ± S.E. followed by the same letter is not significant at 0.05

5. Conclusion

Our study found that at the 0 – 20 cm depth the daily and the annual ranges of soil temperature were lower in the *Taxodium distichum* shelterbelts and sugarcane field than that of the blank field which outside of the vegetation. The vertical patterns of soil temperature of sugarcane field was similar to *Taxodium distichum* shelterbelts in January and October. It's worth pointing out that soil temperature of sugarcane field show a special characteristic by influence of human factors in April. Artificial measures had greater effect on the vertical change of soil temperature. Soil temperature was significantly different with annual and seasons among *Taxodium distichum* shelterbelts, sugarcane field and blank field along a seasonal dynamic ($P < 0.05$). However, patterns of seasonal variation between different habitats in soil temperature characteristic were different. Cooling effect of *Taxodium distichum* shelterbelts is obvious in South Subtropical Region, which was a dominant factor driving the temporal variability of soil temperature in growing seasons of sugarcane.

Acknowledgement

In this paper, the project supported by the Science and Technology Planning Program of Guangdong Province (Project No. 2013B020305009) and Guangdong Natural Science Foundation (Project No. 2015A030313403).

References

[1] Zhang L H, Chen Y N, Zhao R F and Li W 2012 *Eur. J. Soil. Biol.* **48** 24-31
[2] Brandle J R, Hodges L and Zhou X H 2004 *Agroforest. Syst.* **61** 65-78
[3] Gámez-Virués S, Gurr G, Raman A, Salle J L and Nicol H 2009 *Biocontrol.* **54** 211-218
[4] Ding S H and Su P X 2010 *Agroforest. Syst.* **80** 117-129
[5] Zhang L, Su Z Y and Wang X L 2013 *Adv. Mater. Res.* **610-613** 1150-1154
[6] Sindhu S S, Suneja S, Goel A K, Parmar N and Dadarwal KR 2002 *Appl. Soil Ecol.* **19** 57-64
[7] Mõttus M, Ross J and Sulev M 2001 *Agr. Forest Meteoro.* **109** 161-170
[8] Sun J W, Wu J B, Guan D X, Wang A Z and Yuan F H 2011 *Chin. J. Ecol.* **30** 2685-269
[9] Deng R X, Li Y and Zhang S W 2011 *J. Forest. Res.* **22** 65-70

Temporal and spatial changes of land use and landscape in a coal mining area in Xilingol grassland

Chunzhu Guan, Baolin Zhang[1], Jiannan Li and Junling Zhao

College of Chemistry and Environmental Sciences, Inner Mongolia Normal University, Hohhot 010020, P.R. China

E-mail:nmzhangbaolin@hotmail.com

Abstract. Coal mining, particularly surface mining, inevitably disturbs land. According to Landsat images acquired over Xilingol grassland in 2005, 2009 and 2015, land uses were divided into seven classes, i. e., open stope, stripping area, waste-dump area, mine industrial area, farmland, urban area and the original landscape (grassland), using supervised classification and human-computer interactive interpretation. The overall classification accuracies were 97.72 %, 98.43 % and 96.73 %, respectively; the Kappa coefficients were 0.95, 0.97 and 0.95, respectively. Analysis on LUCC (Land Use and Cover Change) showed that surface coal mining disturbed grassland ecosystem: grassland decreased by 8661.15 hm2 in 2005-2015. The area and proportion of mining operation areas (open stope, stripping area, waste-dump area, mine industrial field) increased, but those of grassland decreased continuously. Transfer matrix of land use changes showed that waste-dump had the largest impacts in mining disturbance, and that effective reclamation of waste-dump areas would mitigate eco-environment destruction, as would be of great significance to protect fragile grassland eco system. Six landscape index showed that landscape fragmentation increased, and the influences of human activity on landscape was mainly reflected in the expansion of mining area and urban area. Remote sensing monitoring of coal surface mining in grassland would accurately demonstrate the dynamics and trend of LUCC, providing scientific supports for ecological reconstruction in surface mining area.

1. Introduction

Coal is a major energy resource in China, and the related industries cause a series of environmental issues in the mining, burning and use of its products. The environmental impacts include problems such as coal mine encroaching on forest, farmland and grassland [1], geological disaster, air and water pollution [2-4], vegetation destruction [5], soil pollution and land degradation [6] and occupation [7], etc. These influences may seriously affect the ecological processes, human health [8] and the coordinated development of social economy.

The consequences of terrestrial ecosystem destruction in surface mining are the most obvious and direct. LUCC in surface coal mining areas therefore attracted widely concerns. Studies, such as surface mining monitoring [9], environmental impact indicators [10, 11], vegetation succession [12], relationships between land use change and landscape structure [13], and reclamation of abandoned mines [14], have been carried out in Turkey, Brazil, Australia, Germany, the United States and other countries. In China, extensive studies have also been conducted in surface mining areas, including the extraction of land disturbance features [15], the changes of landscape pattern [16], the evolution of land use pattern [17], the influence of mine subsidence [18], land reclamation and land structure change [19], etc. Satellite remote sensing has the characteristics of wide coverage and good temporal and spatial continuities, providing us a reliable means for studying large-scale environmental change. Remote sensing technique has been widely applied to mine investigation [20], coal mine monitoring [9], mining supervision [21], mine environment monitoring [22] and ecological recovery [23].

Since the 1990s, LUCC has become the essential component and the primary cause of global environmental change, attracted worldwide concerns [24]. Large-scale exploitation of mineral resources in Xilingol grassland (Inner Mongolia, China) brought great pressure on ecological environment, leaving the local in the dilemma of economic development and ecological protection. Despite of numerous reports on coal mining monitoring and ecological environmental consequences [22, 23], few are about coal mining in Xilingol grassland. Analysis on LUCC and landscape index of coal mining areas in Xilingol grassland will help us to understand the change of land use and landscape patterns, providing a guide for land use planning of surface coal mining in arid and semi-arid grassland, and a scientific basis for land reclamation and ecological compensation.

2. Data and methods

2.1. *Study area*

Xilingol grassland is on the edge of the east Asian monsoon, located in the northeast China sample belt of IGBP (International Geosphere-Biosphere Program). Coal mines investigated are located in Xilingol grassland (figure 1), the main body of the natural grassland of Inner Mongolia. The study area is located in the mid-latitude temperate monsoon zone of semi-arid and arid continental climate. Soil are mainly chestnut, suffering wind erosion and hydraulic erosion. The grassland vegetation is dominated by *Stipa grandis*, *Leymus chinensis* and *S. krylovii*, etc., with a coverage of 50%.

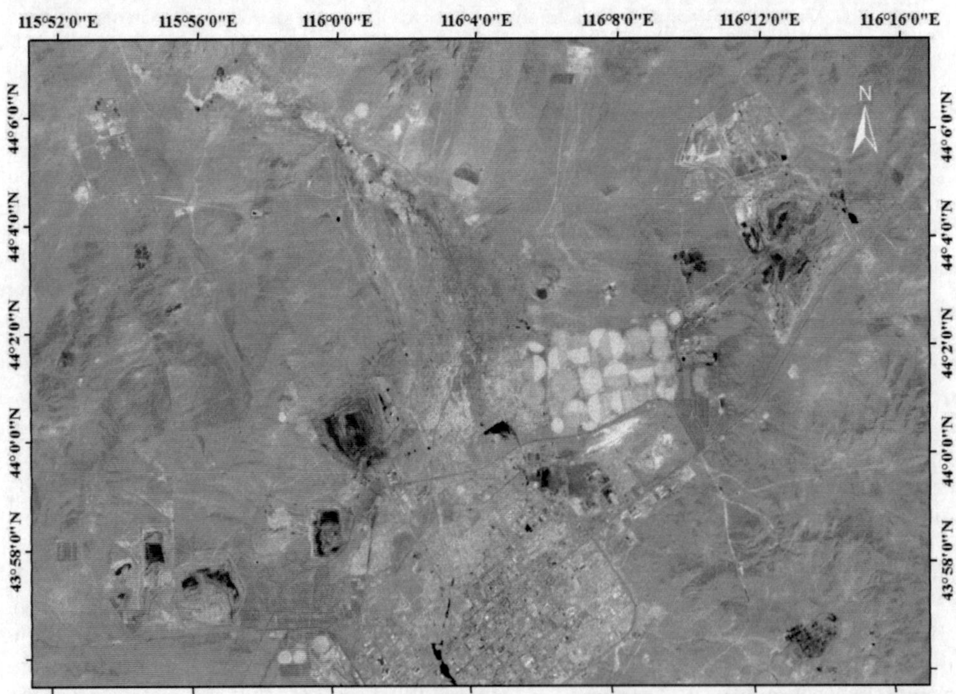

Figure 1. Landsat 8 OLI false colour composite image of the study area.

2.2. Data and method

Landsat series data used in the study are from the USGS (the United States Geological Survey), acquired on September 2, 2005 (Landsat 5 TM (Thematic Mapper)), August 12, 2009 (Landsat 5 TM) and July 12, 2015 (Landsat 8 OLI (Operational Land Imager)), respectively.

Data processing includes radiometric calibration, atmospheric correction and image enhancement, etc. Using field survey data, supervised classification and man-computer interactive interpretation, land uses were divided into seven categories: open stope, stripping area, waste-dump area, mine industrial area, farmland, urban area and the original landscape (grassland). Post-classification change detection was employed to determine LUCC over the time. Transfer matrix were used to depict the temporal and spatial changes (speed and direction) of land uses. Average annual change rate was used to describe land use changes (in area and speed) and their differences:

$$K = \frac{U_b - U_a}{U_a} \times \frac{1}{T} \times 100\%$$

where K is the average annual change rate of land use; U_a, U_b, the starting and the end of the study period, respectively; T, time (year) [25].

Spatial patterns of LUCC were characterized by landscape metrics calculated using FRAGSTATS. Land use/cover maps derived from classification images were used to quantify the areal extent and spatial configuration of patches within a landscape. We selected Total Area (TA), Percentage of Landscape (PLAND), Number of Patches (NP), Patch Density (PD), Largest Patch Index (LPI) and Euclidean Nearest Neighbor Distance (ENN) to represent the landscape structure and configuration.

3. Results

3.1. Land use change

Table 1. Land use changes in area and proportion in 2005-2015.

land use/cover	2005		2009		2015	
	area (hm^2)	proportion (%)	area (hm^2)	proportion (%)	area (hm^2)	proportion (%)
open stope	22.32	0.03	130.50	0.17	341.73	0.45
stripping area	187.17	0.25	756.18	1.00	1474.29	1.95
waste-dump area	461.34	0.60	1860.84	2.45	3644.82	4.82
mine industrial area	0.00	0.00	981.00	1.29	1261.35	1.67
farmland	3036.24	3.98	2060.55	2.71	1851.48	2.45
urban area	3904.92	5.11	4843.89	6.37	7362.36	9.73
original landscape	68418.27	89.59	65059.11	85.57	59757.12	78.91

Table 2. Periodical variations in average annual change index of different land uses.

land use/cover	2005-2009		2009-2015		2005-2015	
	change (hm^2)	annual gradient (%)	change (hm^2)	annual gradient(%)	change (hm^2)	annual gradient(%)
open stope	108.18	96.94	211.23	23.12	319.41	130.10
stripping area	569.01	60.80	718.11	13.57	1287.12	62.52
waste-dump area	1399.5	60.67	1783.98	13.70	3183.48	62.73
mine industrial area	981.00	---	280.35	4.08	1261.35	---
farmland	-975.69	-6.43	-209.07	-1.45	-1184.76	-3.55
urban area	938.97	4.81	2518.47	7.43	3457.44	8.05
original landscape	-3359.16	-0.98	-5301.99	-1.16	-8661.15	-1.15

With the support from field survey GPS data, supervised classification was used in the study. Land uses were divided into seven types, i. e., open stope, stripping area, waste-dump area, mine industrial area, farmland, urban area and the original landscape (grassland), using supervised classification and human-computer interactive interpretation. The overall classification accuracies were 97.72 %, 98.43 % and 96.73 %, respectively; the Kappa coefficients were 0.95, 0.97 and 0.95, respectively. The areas and proportions of land uses in the mining area, i.e., open stope, stripping area, waste-dump area and mine industrial area, increased in the 11 years (table 1-2, figure 2). Analysis on changes of waste-dump area showed that the area accounted for 0.60 % in 2005, 2.45 % in 2009 and 4.82 % in 2015, showing a faster increase by 3183.48 hm^2. Open stope increased from 0.03 % in 2005 to 0.45 % in 2015, the growth rate was relatively low. Urban area accounted for 5.11 % in 2005, 6.37 % in 2009 and 9.73 % in 2015. Xilinhot expanded and grew fast, urban area increased by 3457.44 hm^2. With the growth of mining and urban areas, farmland and original landscape decreased continuously. Farmland reduced by 1184.76 hm^2, dropping from 3.98% in 2005 to 2.45 % in 2015. Original landscape decreased by 8661.15 hm^2, area ratio decreased from 89.59 % in 2005 to 78.95 % in 2015. Average annual change rates showed original landscape presented an accelerated decline. In addition, according to patch changes (figure 2) and field survey, agricultural intensification and urban expansion could be found: large irrigation systems appeared and the city grew substantially in the east and the west.

According to land use maps in 2005 and 2015 (figure 2) and land use change matrix (table 3), the conversions of land uses showed that original landscape transformed to urban and waste-dump areas by 3637.89 hm^2 and 3479.04 hm^2, respectively. Conversions from original landscape to stripping area and mine industrial site were also over 1000 hm^2. Therefore, original landscape mainly changed into mining area and urban constructions.

3.2. Changes of landscape metrics

The original landscape dominated the study area, with proportions of over 75 %, but the area and proportion appeared to decline (figure 3a and 3b), showing original landscape were destroyed in the 11 years. NP and PD of the mining area increased during 2005-2015 (figure 3c and 3d), demonstrating that fragmentation and ecological risk increased. The effects of mining activity on the landscape patterns can be reflected. LPI of urban and mining operation areas tended to grow (figure 3e), the growth rate of the city was higher. LPI of farmland was decreasing from 2.21 % in 2005 to 0.95 % in 2009, and then rising to 1.49 % in 2015. ENN generally showed a decreasing trend, as showing that the nearest neighbour distance between patches decreased (figure 3f), leading to the increase of the aggregation of patches and the decrease of elongation of various land uses. In the 11 years from 2005 to 2015, landscape changed significantly, the number of patches and patch density increased, and the degree of landscape fragmentation increased. The areas under high fragmentation are located close to coal mines, and the high fragmentation areas increased in areas which were previously under low fragmentation (figure 2).

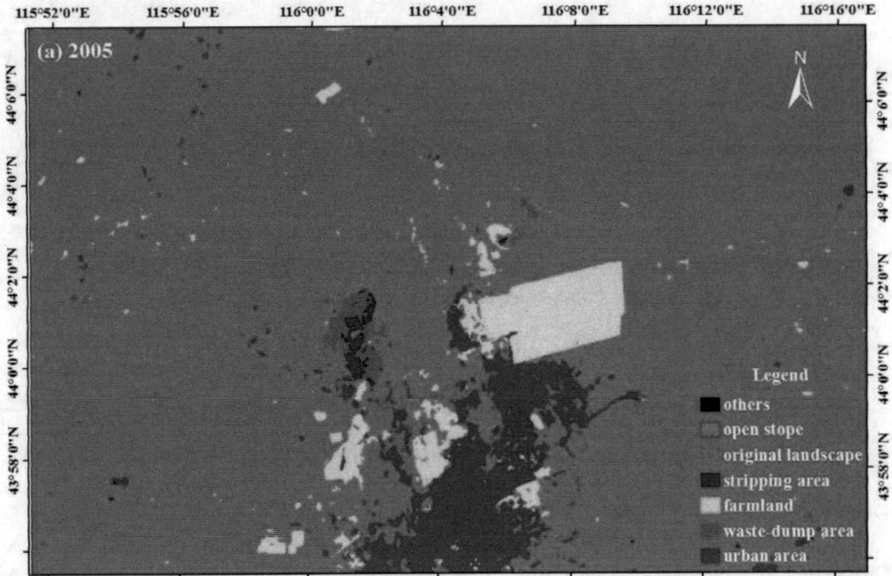

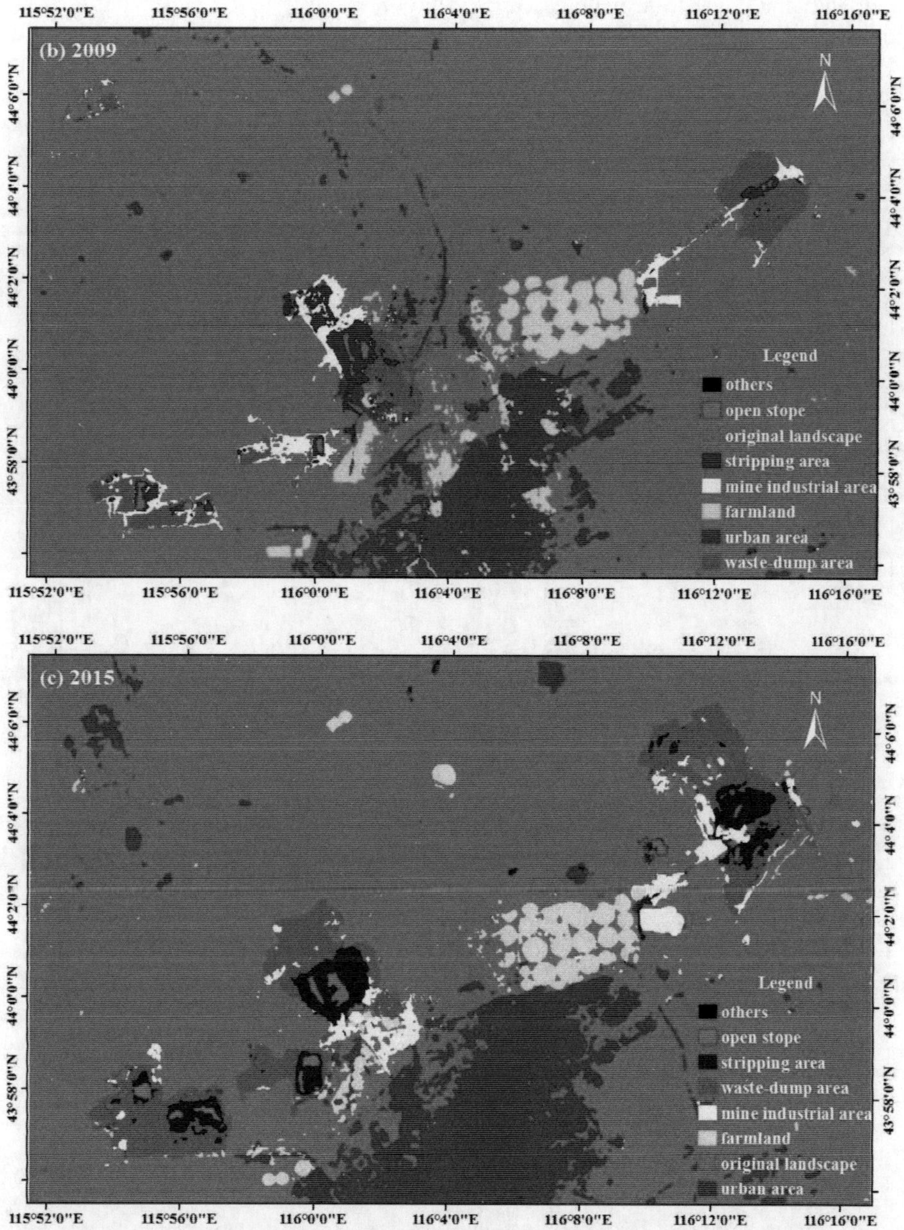

Figure 2. Land use changes in 2005-2015: a 2005, b 2009 and c 2015).

4. Conclusions and discussion

Surface coal mining is a disturbance to the grassland ecosystem. In 2005-2015, mine areas in the study area increased, while the original grassland landscape showed an accelerated decrease. Damages of surface mining to grassland were obvious and direct. Studies showed that the ecological and environmental loss caused by coal mining was far more than its economic benefits [26], sustainable mining requires continuous monitoring of LUCC to identify the long-term impacts of mining on environment and land cover to provide essential safety measures [9]. To understand and predict the status and temporal changes of land uses, studies on land use changes in surface mining area have important significances for rational planning, long-term use and sustainable development of eco-environment.

Table 3. Transfer matrix of land use changes in 2005-2015.

land use/change (2005)	land use/cover (2015)							
	open stope	stripping area	waste-dump	industrial area	farmland	original landscape	urban area	total
open stope	---	8.55	8.55	2.52	---	2.70	---	22.32
stripping area	4.41	75.24	37.44	22.50	0.90	39.42	6.21	186.12
waste-dump	3.24	94.86	83.79	18.63	2.07	189.00	69.21	460.80
farmland	---	1.35	9.09	31.95	1380.78	1268.64	341.37	3033.18
original landscape	328.14	1280.07	3479.04	1121.76	449.91	57662.10	3637.89	67958.91
urban area	5.94	14.22	26.10	62.64	17.46	428.67	3307.68	3862.71
total	341.73	1474.29	3644.01	1260.00	1851.12	59590.53	7362.36	75524.04

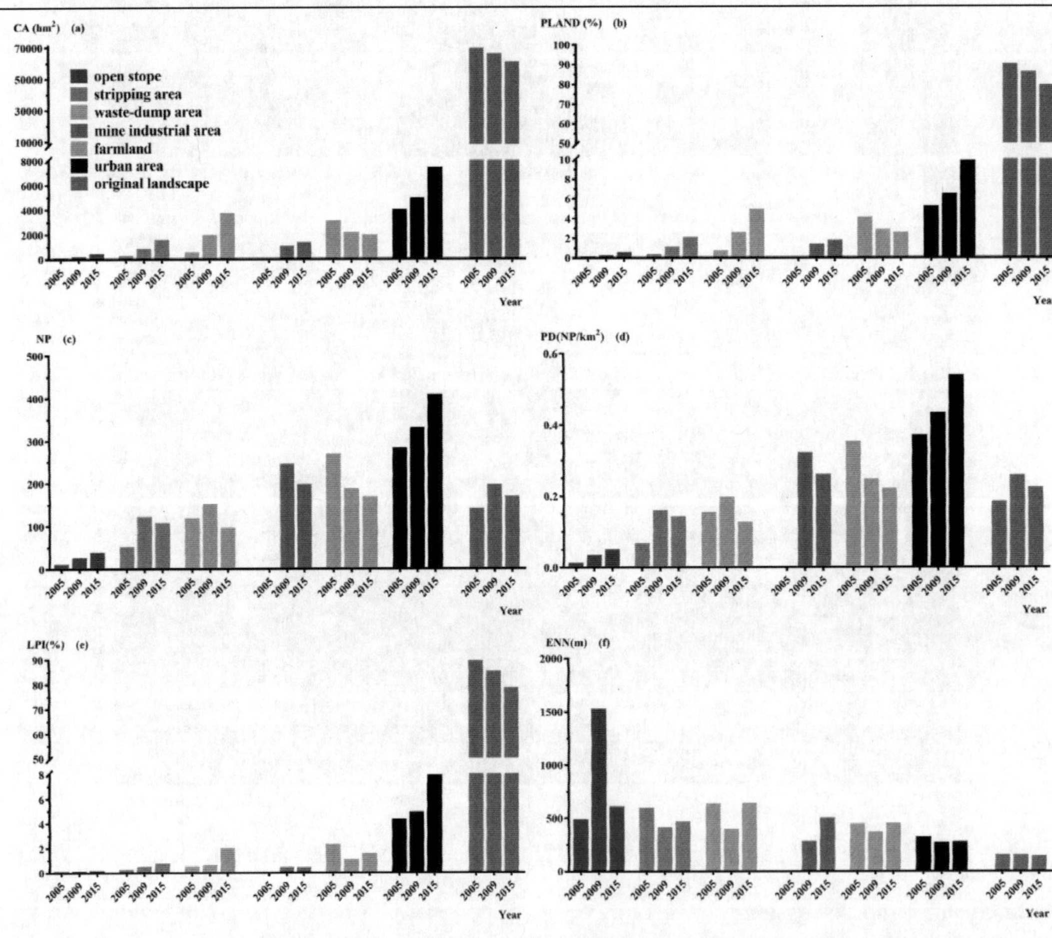

Figure 3. Landscape metrics changes in 2005-2015.

Waste-dump had great impacts on land use in surface coal mining area. According to land use dynamics in 2005-2015, the main changes were the conversions of original landscape into artificial features, such as mining and urban areas. Researchers argued that mining and mineral-processing wastes are one of the world's largest chronic waste concerns [27] , environmental monitoring and assessment should never be neglected. In this study, the influence of waste-dump field on land use was the largest in a surface coal mining area. In European countries, more than 50 % of previously mined lands are reclaimed as forest or grass lands; in China, more than 70 % for agricultural purposes [4]; however, no mined lands in the study area were observed to be reclaimed. Effective reclamation of waste-dump field therefore is very important; this will mitigate eco-environmental degradation, showing great significance to protect grassland ecosystem.

Landscape fragmented in the mining area. The areas under high fragmentation, which were the areas at risk [5], are located close to coal mines. The high fragmentation areas increased in areas which were previously under low fragmentation, but the fragmentation impact on terrestrial environment have been largely overlooked in mining [28]. If eco-environment of grassland was destroyed, ecological restoration would be difficult, high-costed and time-consumed, as is the common issue in the development of coal resources; improper handling will affect the sustainable development of regional economy.

Acknowledgements
The authors gratefully acknowledge financial support from the National Natural Science Foundation of China (Grant Nos. 41261048 and 40961014).

References

[1] Zhai M, Xu X, Jiang D, Jiang X 2012 Remote sensing monitoring of the ecological environment in Wuhai mining area since 1979 (in Chinese with English abstract) *Remote Sensing Technology and Application* **27(6)** 933-940

[2] Lindberg T T, Bernhardt E S, Bier R, Helton A M, Merola R B, A Vengosh, et al. 2011 Cumulative impacts of mountaintop mining on an Appalachian watershed *Proceedings of the National Academy of Sciences of the United States of America* **108(52)** 20929-34

[3] Bernhardt E S, Palmer M A 2011The environmental costs of mountaintop mining valley fill operations for aquatic ecosystems of the Central Appalachians *Annals of the New York Academy of Sciences* **1223(1)** 39-57

[4] Bian Z, Inyang H I, Daniels J L, Otto F, Struthers S 2010 Environmental issues from coal mining and their solutions *Mining Science and Technology (China)* **20(2)** 215-223

[5] Sarma K 2005 Impact of coal mining on vegetation a case study in Jaintia Hills District of Meghalaya, India *The International Institute for Geo-information Science and Earth Observation*, The Netherlands

[6] Qian T, Bagan H, Kinoshita T, Yamagata Y 2014 Spatial-temporal analyses of surface coal mining dominated land degradation in Holingol, Inner Mongolia *IEEE Journal of Selected Topics in Applied Earth Observations and Remote Sensing* **7(5)** 1675-87

[7] Bian Z, Miao X, Lei S, Chen S, Wang W, Struthers S 2012 The challenges of reusing mining and mineral-processing wastes *Science* **337(6095)** 702 -703

[8] Palmer M A, Bernhardt E S, Schlesinger W H, Eshleman K N, Foufoula-Georgiou E, Hendryx M S, et al. 2010 Mountaintop mining consequences *Science* **327(5962)** 148 -149

[9] Demirel N, Emil M K, Duzgun H S 2011 Surface coal mine area monitoring using multi-temporal high-resolution satellite imagery *International Journal of Coal Geology* **86(1)** 3-11

[10] Santo E, Sánchez L 2002 GIS applied to determine environmental impact indicators made by sand mining in a floodplain in southeastern Brazil *Environmental Geology* **41(6)** 628-637

[11] Kong J L, Xian T, Yang J, Chen L, Yang X T 2016 Monitoring soil moisture in a coal mining area with multi-phase LANDSAT images *The International Archives of the Photogrammetry, Remote Sensing and Spatial Information Sciences* **XLI-B7** 537-542

[12] Norman M A, Koch J M, Grant C D, Morald T K, Ward S C 2006 Vegetation succession after bauxite mining in Western Australia *Restoration Ecology* **14(2)** 278-288

[13] Baessler C, Klotz S 2006 Effects of changes in agricultural land-use on landscape structure and arable weed vegetation over the last 50 years *Agriculture, Ecosystems & Environment* **115(1-4)** 43-50

[14] Gorokhovich Y, Reid M, Mignone E, Voros A 2003 Prioritizing abandoned coal mine reclamation projects within the contiguous United States using geographic information system extrapolation

Environmental Management **32(4)** 527-34

[15] Bi R, Bai Z 2007 Land characteristic information and classification in opencast coal mine based on remote sensing images (in Chinese with English abstract) *Transactions of the CSAE* **23(2)** 77-82, 291

[16] Bian Z, Zhang Y 2006 Land use changes in Xuzhou coal mining area (in Chinese with English abstract) *Acta Geographica Sinica* **61(4)** 349-358

[17] Zhang Q, Bai Z, Hao J, Fan J, Zhao J 2006 Pattern succession analysis of agricultural land converted from large opencast mine in loess areas (in Chinese with English abstract) *Transactions of the CSAE* **22(11)** 98-103

[18] Bai Z, Duan Y, Yang H, Fu H, Lv C, Ma R 2006 Forecast of influence of coal-mining subsidence on soil erosion and land use (in Chinese with English abstract) *Transactions of the CSAE* **22(6)** 67-70

[19] Hu Z, Xie H 2005 Study on land use/cover change of coal mining area based on remote sensing images (in Chinese with English abstract) *Journal of China Coal Society* **30(1)** 44-48

[20] Li C, Sun S, Wang X, Zhang G, Niu J 2010 Application of remote sensing technology in coal resource investigation and evaluation in Zhaotong,Yunnan (in Chinese with English abstract) *Coal Geology of China* **22(10)** 17-21

[21] Kang G, Lu Z, Li S, Jin M 2008 Application of remote sensing technology in coal resource exploitation monitoring and management (in Chinese with English abstract) *Coal Geology of China* **20(1)** 13-16

[22] Quan Z, Cheng H, Yu Y, Zou X 2006 Assessment of subsidence impact on vegetation landscape in coal mining area-a case study of Dongda mine in Jincheng city, Shanxi Province (in Chinese with English abstract) *Journal of Plant Ecology* **30(3)** 414-420

[23] Liu J 2008 Remote sensing dynamic monitoring and landscape ecological reconstruction of mineral resources development in Cangshan County (in Chinese with English abstract) [D] *Shandong Normal University*

[24] Chen Y, Yang P 2001 Recent progresses of international study on land use and land cover change(LUCC) (in Chinese with English abstract) *Economic Geography* **21(1)** 95-100

[25] Long H, Li X 2001 Land use pattern in transect of the Yangtse River and its influential factors (in Chinese with English abstract) *Acta Geographica Sinica* **56(4)** 417-425

[26] Li F, Liu X, Zhao D, Wang B, Jin J, Hu D 2011 Evaluating and modeling ecosystem service loss of coal mining A case study of Mentougou district of Beijing, China *Ecological Complexity* **8(2)** 139-143

[27] Bian Z, Miao X, Lei S, Chen S, Wang W, Struthers S 2012 The challenges of reusing mining and mineral-processing wastes *Science* **337(6095)** 702

[28] Wickham J, Wood P B, Nicholson M C, Jenkins W, Druckenbrod D, Suter G W, et al. 2013 The overlooked terrestrial impacts of mountaintop mining *BioScience* **63(5)** 335-348

Multidimensional analysis of particle size fractal characteristics in a farmland soil profile

Y C Du[1,2,5], J C Han[2], S W Zhang[3], Y F Huang[4], H Y Wang[2], L T Luo[2] and W H Zhang[2]

[1]Department of Resources and Environment, Northwest A&F University, Yangling, 712100, China
[2]Key Laboratory of Degraded and Unused Land Consolidation Engineering, Ministry of Land and Resources; Shaanxi Provincial Land Engineering Construction Group Co., Ltd.,Xi'an 710075, China
[3]College of Earth and Environmental Sciences, Anhui University of Science and Technology, Huainan 232001, China;
[4]College of Resources and Environment, China Agricultural University, Beijing 100193, China

E-mail: linghuxzj@sina.com

Abstract: Soil particle size distribution (PSD) is a basic soil property, closely related to soil erosion, soil moisture, and so on. The multidimensional fractal method has been proved to be a better description of soil PSD. By choosing the four fractal parameters of Δa, $\Delta a_R/\Delta a_L$, D_1/D_0 and, $\Delta f(a)_R/\Delta f(a)_L$, and using the multifractal spectrum, this paper analyzed one- and two-dimensional fractal characteristics of the soil profile (A, B, and C layers from top to bottom) PSD and their relationship on a field scale. The results showed that the PSD characteristics of different soil profiles were more heterogeneous in the east and west (WE) than the north and south (NS) in the study area, and the A layer soil PSD was the most uneven, with a larger degree of dispersion. With the soil layer depth, the soil PSD tended to be uniform. Regardless of one-dimensional or two-dimensional characteristics, the proportion of the low value of soil particle mass percent was relatively larger. The results of this study will not only provide a reference method for analyses of the fractal characteristics of soil PSD on a field scale, but also provide guidance in further understanding the soil structure and soil forming process.

1. Introduction

Soil is composed of particles with irregular shapes and self-similar structures, which form a porous medium. Soil particle size distribution (PSD) has fractal properties[1-5]. PSD is closely related to soil erosion, organic matter content, and soil moisture content. A comprehensive and accurate understanding of the soil PSD fractal characteristics is crucial to understanding soil structures and the process of soil formation. Fractal theory and analysis can effectively describe soil texture [6]. Multifractal analysis can effectively capture the essence of change in the measured values [7]. The effectiveness of utilizing fractal parameters to estimate and analyze soil texture exceeds that of the traditional soil texture triangle model [8-11]. Soil scientists have already utilized fractal principles and methods to analyze geological characteristics, which cannot be easily obtained using a traditional geology method[12-13]. PSD analysis requires a large amount of sampled data. If the overall PSD can be reconstructed from a determined percentage of the soil particle mass, it will have a significant impact on the mathematical modeling of soil moisture migration, soil erosion, and the soil formation process. Thus, it is necessary to achieve a clear understanding of the soil PSD structure, including the spatial variability in a given direction and the cross-sectional soil structure. Multifractal methods based on soil PSD studies already exist, such as the study of PSD characteristics under different soil management strategies [14-16], and the influence of vegetation on PSD[17-18]. However, there are a number of limitations in the current research methods, including the following: (1) Single dimension studies and concentratedresearchcontent. Most studies only utilized the two-dimensional surface area information to analyze different influential conditions (such as land use, soil texture, and vegetation) on soil PSD. It is difficult to fully explore the different PSD fractal characteristics given these limitations. (2) The fractal parameters employed are simple and lack representation. Most studies used a fractal dimension method to analyze soil PSD fractal characteristics. Multifractal parameters have rarely been applied; therefore, it is difficult to give a multidimensional description of the soil PSD characteristics. There have been very few reports of multidimensional, multidirectional, and multifractal analyses of farmland soil profiles. Thus, this area could benefit from further exploration.

A multifractal program based on the MATLAB platform was utilized to calculate four fractal parameters, Δa, $\Delta a_R/\Delta a_L$, D_1/D_0, and $\Delta f(a)_R/\Delta f(a)_L$, and to carry out multidimensional profiling. This paper describes an in-depth study of the farmland soil PSD's one-dimensional and two-dimensional fractal characteristics and their relationship with different dimensions. Therefore, this paper proposes an analytical method for determining farmland soil PSD characteristics. In addition, this study also serves as a guide for further understanding soil structure and the soil formation process.

2. Materials and Methods

2.1. Description of sites and soil sampling

The study area was located in Dongbeiwangxiang, Haidian District, Beijing. It was on anpiedmontalluvial plain and the soil type was meadow cinnamon soil. The 60 m × 55 m study area was divided into 156 grids based on sample units of 5 m × 5 m. Sampling was carried out with a 1 m soil auger in the selected 100 grid points (Figure 1). In accordance with the soil profile layers, sample aliquots were extracted from each layer and the depth of each layer was measured and recorded. A total of 300 soil samples were taken. Soil particle composition was determined using thetraditonal pipette method

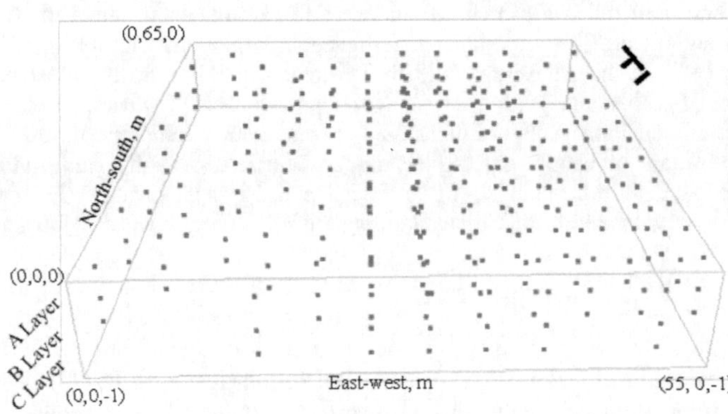

Figure 1. The distribution map of sampling points.

2.2. Data preparation and processing

In order to ensure that one-dimensional north-south (NS) and west-east (WE) data were comparable and representative, the original sampling point data was processed as follows: The average value of soil particles mass percent was obtained in the direction of west to east from samples from rows 1-4 on the same vertical axis. The obtained value was denoted as the new NS sample point 1. Next, a new NS sample point 1 was obtained. The same method as outlined for rows 1-4 was repeated for samples from rows 2-5 to obtain the new NS sample point 2. This process was repeated until nine new NS sample points were obtained as required for one-dimensional analysis of the soil PSD. Likewise, from south to north, samples from points 1-5 on the same horizontal axis were used to obtain the average value of soil particles mass percent. The obtained value was denoted as the new sample value, which allowed for the establishment of a new WE sample point 1. The mean value from rows 2-6 was used as the new WE sample point 2. The same method was ued to obtain nine new WE sample points for analysisingPSD fractal characteristics of the WE soil. MATLAB-based software was employed to calculate the multifractal parameters of the different soil particle compositions in the A, B, and C layers, while accounting for the WE and NS direction of the particles in each layer. The subsequent multifractal spectrum was also drawn using MATLAB.

2.3. Theory and methods

The multifractal method was used to describe multidimensional soil PSD fractal characteristics and their correlations. The most commonly applied multifractal parameters are multifractal spectrum width (Δa), multifractal spectrum left-side width (Δa_R), multifractal spectrum right-side width (Δa_L), asymmetry index ($\Delta a_R/\Delta a_L$), generalized fractal dimension ($\Delta D(q)$), dispersion index (D_1/D_0), right-side branch length ($\Delta f(a)_R$), left-side branch length ($\Delta f(a)_L$), and the ratio of the right-side to the left-side branch length ($\Delta f(a)_R/\Delta f(a)_L$). In order to select representative parameters that could comprehensively describe multidimensional PSD fractal characteristics, the fractal parameters of soil sand particles from different dimensionswere used as analytic target, and the Pearson correlation analysis was carried out to determine the correlation between the different multifractal parameters.

Table 1. Multifractal parameter correlation analysis of clay in layers A,B, and C.

Parameters	Δa	Δa_R	Δa_L	$\Delta a_R/\Delta a_L$	$\Delta D(q)$	D_1/D_0	$\Delta f(a)_R$	$\Delta f(a)_L$	$\Delta f(a)_R/\Delta f(a)_L$
Δa	1	0.968**	0.803**	-0.053	0.932**	-0.983**	0.802**	0.452**	-0.109
Δa_R		1	0.627**	0.099	0.235	-0.930**	0.877**	0.238	0.038
Δa_L			1	-0.396**	0.125	-0.839**	0.408**	0.832**	-0.426**
$\Delta a_R/\Delta a_L$				1	0.098	0.102	0.249	-0.513**	0.663**
$\Delta D(q)$					1	0.643**	0.723**	0.403**	-0.093
D_1/D_0						1	-0.726**	-0.474**	0.159
$\Delta f(a)_R$							1	0.018	0.289*
$\Delta f(a)_L$								1	-0.624**
$\Delta f(a)_R/\Delta f(a)_L$									1

* Correlation is significant at the 0.05 level (2-tailed).
** Correlation is significant at the 0.01 level (2-tailed).

The results indicate that in two dimensions, there was a significant negative correlation between Δa and D_1/D_0(Table 1).Parameters Δa and D_1/D_0 were not correlated with$\Delta a_R/\Delta a_L$. However, Δa_R, Δa_L,$\Delta D(q)$, $\Delta f(a)_R$, and $\Delta f(a)_L$ all had a significantly positive correlation with Δa. One-dimensional particle analysis showed the same correlation characteristics. The properties of different parameters, which were the represented PSD characteristics, were taken into consideration and combined with the results shown in Table 1. This study has selected the parameters Δa, $\Delta a_R/\Delta a_L$, D_1/D_0, and $\Delta f(a)_R/\Delta f(a)_L$

and utilized a multifractal spectrum to conduct a multidirectional, multidimensional analysis of the fractal characteristics of the farmland soil PSD.

The research area was assumed to be a square with length d. In the study area, there were M sampling points, and soil particle composition was measured at each sampling point. An observation scale of ε = 1/n was set, where n > 1, and the entire research area was divided into n rows and n columns, which formed a total of N = n2 small grids, where length = d/n. The average of the measured value in the i-th grid ai (ε) was calculated, and a summation carried out for the entire study area. The probability of soil particle composition in the i-th grid can be determined as follows:

$$p_i(\varepsilon)=\frac{a_i(\varepsilon)}{\sum_{i=1}^{N}a_i(\varepsilon)} \tag{1}$$

wherepi(ε) is the probability of soil composition in the i-th grid. Given real number q, the definition of a partition function is as follows:

$$\chi_i(q,\varepsilon)=\sum_{i=1}^{N}p_i(\varepsilon)^q \tag{2}$$

where$\chi_i(q,\varepsilon)$ is the q order of probability in the i-th grid.

$$D_q=\frac{ln\,\chi_i(q,\varepsilon)}{(q-1)ln\,\varepsilon} \tag{3}$$

$$a(q)=\lim_{\varepsilon\to 0}\frac{\sum_{i=1}^{N}\chi_i(q,\varepsilon)lg\,p_i(\varepsilon)}{lg\,\varepsilon} \tag{4}$$

When q = 0 and 1, the corresponding D_0 andD_1 denote the capacity dimension and dimension entropy, respectively. Dimension entropy, D_1,measures the soil unit aggregation abundance and degree of aggregation. For a particular distribution, D_0 is the maximum value of D_1. When $D_1= D_0$, it becomes a single fractal, with minimum distribution heterogeneity. This implies that D_1/D_0can be used as an indicator for soil dispersion.

$$f(a(q))=\lim_{\varepsilon\to 0}\frac{\sum_{i=1}^{N}\chi_i(q,\varepsilon)lg\,p_i(\varepsilon)}{lg\,\varepsilon} \tag{5}$$

After obtaining the values mentioned above, multifractal parameters of soil PSD would be obtained,. For more information regarding multifractalmodeling, one may refer to the relevant publications [19-23].And then,amultifractal program was developed based on the MATLAB platform which was used to calculate soil PSD fractal parameters D_1/D_0, $\Delta a_R/\Delta a_L$, and $\Delta f(a)_R/\Delta f(a)_L$, and to draw the multifractal spectrum.

3. Results and Discussions

3.1. Soil PSD one-dimensional fractal characteristics

The one-dimensional soil particle mass percentand the multifractal parameters Δa, D_1/D_0, and $\Delta a_R/\Delta a_L$ were compared between WE and NS direction (Figure 2).

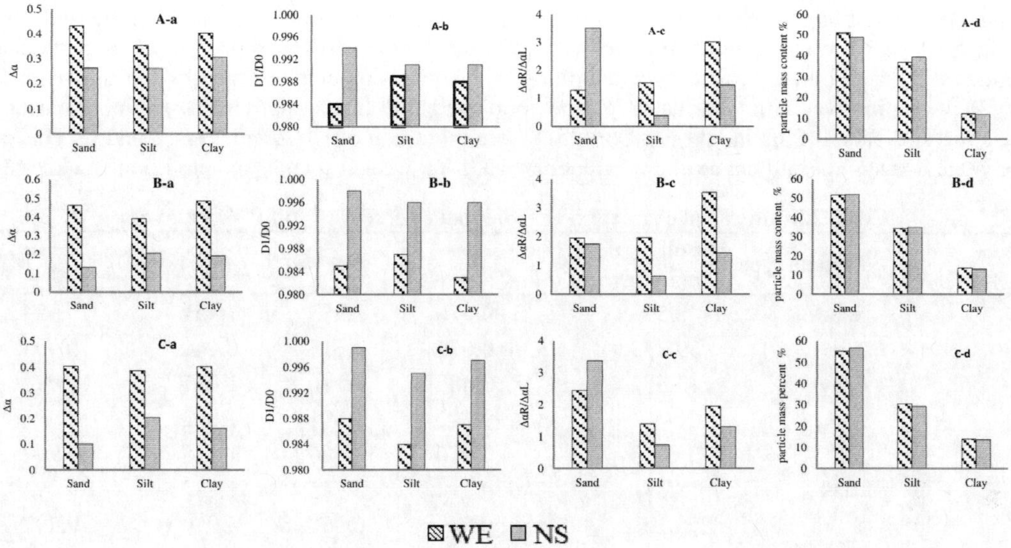

Figure 2. Comparison of one-dimensional multifractal parameters of PSD between WE and NS directions. Four icos as a, b, c and d respectively represented Δa、D_1/D_0、$\Delta a_R/\Delta a_L$and particle mass percent

Figure 2 shows that regardless of the WE or NS direction, the difference between the same types of soil particle percentages were relatively small. However, there was a significant difference between the multifractal parameters Δa, D_1/D_0,and $\Delta a_R/\Delta a_L$.In addition, the same fractal parameter in different layers tended to have similar characteristics. In each layer, the sand, silt, and clay Δa values were greater in the WE direction than in the NS. Whereas $D_1/D_0$was the opposite, suggesting that the soil PSD in the WE direction of the research area showed greater heterogeneity and uneven distribution. Δa_R and Δa_L represent the probability of a high value (or a large value, $q>>1$) and low value (or a small value, $q<<1$), respectively. $\Delta a_R/\Delta a_L$ is the ratio of the right-side spectral width to the left-side spectral width. It is used to measure the symmetry (or asymmetry) of the left and the right sides of the f(a) multispectral spectrum. The closer the value is to 1, the closer the ratio of the larger value to the small value[16],[21]. For the soil profile of each layer, the $\Delta a_R/\Delta a_L$ values of WE were greater than 1. For the sand and clay particles in the NS direction, the $\Delta a_R/\Delta a_L$ values were also greater than 1. So, it indicates that the proportion of the low value of soil particle mass percent was relatively large.

3.2. Soil PSD two-dimensional fractal characteristics

Equations (1) through (5) are used to calculate the multifractal parameters of the two-dimensional soil particle percentage content in different layers. The parameters calculated includeΔa, D_1/D_0, $\Delta a_R/\Delta a_L$,and $\Delta f(a)_R/\Delta f(a)_L$ (Table 2).

Table 2 shows that the two-dimensional sand particle Δa value decreased with each layer (from top to bottom). Silt and clay particles had a minimal Δa value in layer C, which indicated the soil profile PSD heterogeneity wasthe weakest in the bottom layer. In accordance with the U.S. soil texture classification system and soil particles distribution at each layer, layer C was sandy loam soil ,layers A and B were loam soil. This was consistent with the finding that the coarsersoil texture, the smaller the fractal dimension[24-26]. D_0 is the capacity dimension or box dimension, which can be used to measuresoil particle abundance. D_1 is the dimension entropy or information entropy, which can be used to measure dispersion [27]. D_1/D_0 can be used to measure the degree of uniformity. When D1/D0 is close to 1, this indicates a more uniform distribution[16]. The D_1/D_0 value of the sand, silt, and clay inthe soil profile showed that the D_1/D_0 value was the smallest in layer A (Table 2). The corresponding particle type had a larger value in the B and C layers, and a D_1/D_0 value closer to 1. In the study area, the layer A PSD was the most uneven, with a larger degree of dispersion. This was particularly prominent for sand particles, which also confirmed the conclusion demonstrated by the Δa value. The sampling location was farmland that might be concluded the human activity as the main reason to cause the uneven soil PSD in layer A. Soil PSD were more homogenous in layers B and C. This was likely due to a stronger influence of the parent material, and weaker human and environmental impacts.

Table 2. Multifractal characteristics and mass percent of soil PSD in layers.

Soil profile-layer	Particle type	Soil particle mass percent (%)	Δa	$\Delta a_R/\Delta a_L$	$\Delta f(a)_R/\Delta f(a)_L$	D_1/D_0
	Sand	48.59	0.876	2.983	1.451	0.994
A	Silt	39.393	0.676	0.615	0.641	0.993
	Clay	12.017	0.683	1.116	1.017	0.992
	Sand	51.313	0.642	1.591	1.022	0.997
B	Silt	35.356	0.876	1.686	1.135	0.994
	Clay	13.331	0.714	2.445	1.432	0.994
	Sand	55.303	0.539	6.358	4.01	0.997
C	Silt	30.666	0.578	0.599	0.678	0.993
	Clay	14.031	0.44	1.053	0.841	0.995

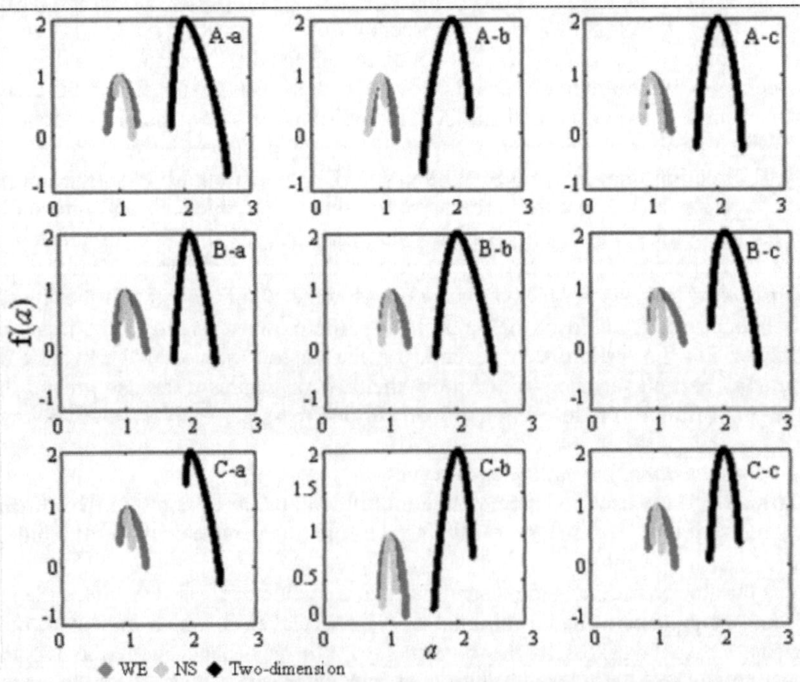

Figure 3. Multifractal spectra of soil PSD for different layers (A, B, and C, respectively, represent the different soil; a, b, and c, respectively, represent sand, silt, and clay).

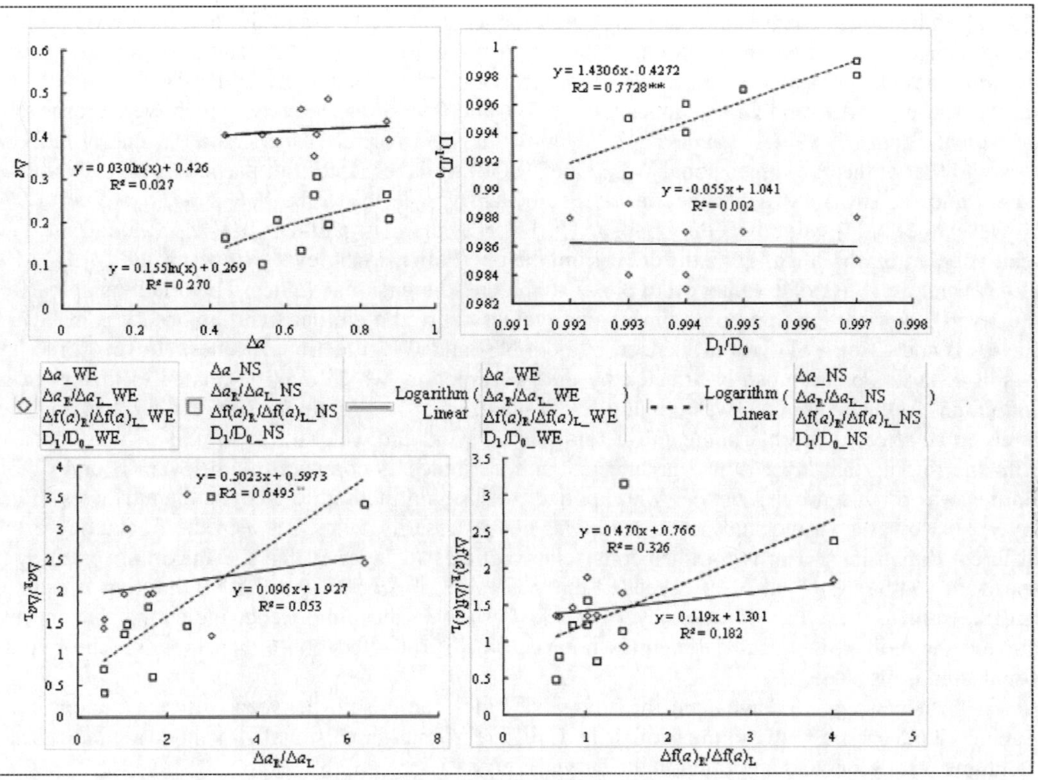

Figure 4. The scatter diagram of different fractal dimensions for soil PSD.

The f(a) value reflects the shape of the curvature and the degree of the symmetry (Figure 3). The shape and the symmetry of the multifractal spectrum are critical in the understanding of soil PSD heterogeneity[11]. The ΔaR/ΔaL value (ratio of right to left spectral width), is the parameter that describes the symmetry of the multifractal spectrum. Figure 3 shows that in each layer, there was a stronger degree of spectral width asymmetry observed in the sand particles. The width of the left-side branch ($q > 0$) was significantly smaller than the right side ($q < 0$). In the terms of different layers, the difference between the left spectral width (ΔaL = 0.218) and the right spectral width (ΔaR = 0.652) of the layer A sand particles was maximum (0.434). Likewise, in layer C sand particles, the difference between the left and the right spectral width was maximum that ΔaR (0.446) was 6.358 times larger than ΔaL (0.073). In layer B sand particles, the right spectral width was 1.591 times the width of the left. The clay particles in layers A and C showed a relatively symmetrical spectral width. In layer C clay particles, the left and the right spectral widths were very similar (ΔaL= 0.226 and ΔaR= 0.214). In layer A clay particles, the left and the right spectral widths were also very similar (ΔaL= 0.323 and ΔaR= 0.360). With the exception of silt particles in layers A and C, the multifractal spectrum of all the different types of particles in the soil profile showed a consistent trend in terms of spectral width. The right spectral widths were greater than the left, which indicates the proportion of low value of soil particle mass percent was relatively large.

3.3. Multifractal characteristics of different dimensions

Figure 3 and Table 2 show that in layer A, the two-dimensional sand particle $\Delta a_R/\Delta a_L$ value was 2.983, and the right-side spectral width was large,indicating that the portion of low value of sand particle mass percent was relatively large. The one-dimensional NS $\Delta a_R/\Delta a_L$ value was 3.512, which was closer to two-dimensional than the WE $\Delta a_R/\Delta a_L$ value (1.289). Sand particle $\Delta f(a)_R/\Delta f(a)_L$ was > 1

(1.451), indicating the right spectral length was longer than the left, and the NS value was closer to two-dimensional. In layer A, the silt and the clay particle D_1/D_0, $\Delta a_R/\Delta a_L$, and $\Delta f(a)_R/\Delta f(a)_L$ showed that the one-dimensional NS value was closer to two-dimensional. In layer B, the two-dimensional sand particle $\Delta a_R/\Delta a_L$ and D_1/D_0 values were 1.591 and 0.997, respectively, which were closer to the one-dimensional NS values. However, in layer B, the sand particle $\Delta f(a)_R/\Delta f(a)_L$ value (1.022) was closer to that of the one-dimensional WE value (1.350). In layer B, the silt particle $\Delta a_R/\Delta a_L$ value was 1.686, and the $\Delta f(a)_R/\Delta f(a)_L$ value was 1.135, which was closer to the WE $\Delta a_R/\Delta a_L$ value (1.971). However the D_1/D_0 value was the opposite. In layer B, the clay particle $\Delta a_R/\Delta a_L$ value (2.445) was relatively large, which was quite different from the one-dimensional level. However, the $\Delta f(a)_R/\Delta f(a)_L$ (1.432) and D_1/D_0 (0.994) values were closer to the one-dimensional values. The relationship between the layer C soil particles and one-dimensional values showed a similar trend as observed in layer B. Layers B and C soil PSD was influenced by both NS- and WE-directional values. However, the layer A soil PSD was predominantly affected by the NS direction, which may be related to factors of soil formation. As the study area was on an alluvial plain, layer A was situated on the surface, and more likely to be affected by environmental factors such as wind and water movement. This was consistent with the finding that layer A had greater azimuth attributes as compared with layers B and C. This finding was produced by Gao*et al.*, who applied the Kriging method to analyze the attributes of each layer[28. However, in the study area, there was not necessarily a link between the fractal degrees for different dimensions. The two-dimensional clay particles of layer B had the maximum multifractal spectrum width of Δa= 0.876. It exceeded the WE value of Δa = 0.391, the NS value of Δa= 0.206, and the sum of these two values. Layers A and C showed the same trend, indicating that the two-dimensional soil PSD fractal strength is not necessarily reflected one-dimensionally. The opposite conclusion is also true.

To further explore linkages and the strength of the relationship between different dimensions of fractal characteristics, two-dimensional and different one-dimensional fractal parameter scatter diagrams were generated and fitted to the curve (Figure 4).

Figure 4 illustrates that, D_1/D_0 and $\Delta a_R/\Delta a_L$ had a significant linear relationship with the two-dimensional level ($p < 0.01$), indicating the NS direction was closely related to the two-dimensional level..However, no significant relationship was observed in the WE direction. Instead, the D_1/D_0 value was negatively correlated with the two-dimensional level. This also confirmed that at the two-dimensional level, NS had some degree of impact on the different particles in the different layers. Figure 4 Δa reflects that there was not necessarily a link between the fractal degrees for different dimensions. So, in order to reveal the relationship between the fractal characteristics of soil PSD and soil formers, it is necessary to carry out multidirectional and multidimensional analyses.

4. Conclusions

This paper utilized a multifractal method to analyze one- and two-dimensional fractal characteristics of the soil profile (A, B, C layers, from top to bottom) PSD and their relationship on a field scale. The following conclusions can be drawn:

(1) In the study area, the multifractal parameters of the soil profile from different layers and particle types indicated that the WE-directional Δa and D_1/D_0 values were always smaller than those of the NS direction. The soil PSD in the WE direction of the research area showed greater heterogeneity and an uneven distribution. In general, the one-dimensional WE and NS $\Delta a_R/\Delta a_L$ values were greater than 1. This indicates that the proportion of low value (small value) of soil particle percentage content was relatively large.

(2) The A layer two-dimensional soil PSD was the most heterogonous, and with an increase in depth, the soil PSD became more uniform. The soil particle multifractal spectrum showed that the right-side spectral width was generally greater than the left. The lower value (small value) of the soil PSD content was more prominent.

(3) The A layer two-dimensional PSD was influenced by the NS-directional effect. In layers B and C, the soil PSD was influenced by a combination of the NS and WE effects. Overall, the different

parameters between different fractal dimensions were not necessarily linked. Therefore, the soil fractal strengths were not necessarily related. In order to reveal the relationship between the fractal characteristics of soil PSD and soil factors, it is necessary to carry out multidirectional and multidimensional analyses.

(4) The results have shown that the use of multifractal parameters and a multifractal spectrum to conduct multidirectional, multidimensional analyses of the soil PSD results in a comprehensive analysis of farmland soil particle characteristics. This paper has provided a methodological guidance for future research of this type. As the study area was relatively small, located on an alluvial plain, and mostly used for agricultural purposes, the difference observed between the influences of topography, parent material, human activities, and the external environment was relatively small. Therefore, it is difficult to carry out quantitative comparisons. In order to reveal a more comprehensive system of the regional soil formation process, future studies should consider using soil formation factors as fractal parameters in subsequent quantitative analyses.

References

[1] Taguas F J, MartíN M A and Perfect E1999 Simulation and testing of self-similar structures for soil particle-size distributions using iterated function systems. *Geoderma***88(00)**191-203

[2] Tyler S W and Wheatcraft S W 1992Fractal Scaling of Soil Particle-Size Distributions: Analysis and Limitations*Soil Science Society of America Journal***56(2)**362-369.

[3] Montero E2005 Rényi dimensions analysis of soil particle-size distributions2005*Ecological Modelling***10(3-4)**305-315

[4] Horton R2011 Single and Joint Multifractal Analysis of Soil Particle Size Distributions. *Pedosphere***21(1)**75-83

[5] Arias M, Gumiel P and Izard A M2012Multifractal analysis of geochemical anomalies: A tool for assessing prospectivity at the SE border of the Ossa Morena Zone, Variscan Massif (Spain)*Journal of Geochemical Exploration***122**101-112

[6] Filgueira R R, Fournier L L, Cerisola C I,Gelati P and Garcia M G 2006Particle-size distribution in soils: A critical study of the fractal model validation*Geoderma***134(s3-4)**327-334

[7] Falconer K1990Fracal geometry - mathematical foundations and applications1990*Biometrics***46(4)**499

[8] Martin M A 1998Fractal modelling, characterization and simulation of particle-size distributions in soil*Proceedings Mathematical Physical&Engineering Science***454(1973)**1457-68.

[9] And P, Gimenez D, Bittelli M, Cmp V and Flury M 2001 Multifractal Characterization of Soil Particle-Size Distributions*Soil science Society of America Journal***65(5)**1361-67

[10] Martín M A and Montero E2002Laser diffraction and multifractal analysis for the characterization of dry soil volume–size distributions*Soil and Tillage Research***64(1-2)**113-123

[11] Miranda J G V, Montero E, Alves M C, Gonzǎlez A P and Vǎzquez E V2006Multifractal characterization of saprolite particle-size distributions after topsoil removal*Geoderma***134(3)**373-385.

[12] Holtz F, Lenné S, Ventura G, Vetere F and Wolf P 2004 Non-linear deformation and break up of enclaves in a rhyolitic magma: A case study from Lipari Island (southern Italy)*Geophysical Research Letters***31(24)**357-370

[13] Prosperini N and Perugini D2008 Particle size distributions of some soils from the Umbria Region (Italy): Fractal analysis and numerical modelling *Geoderma***145(3-4)**185-195

[14] Filgueira R R, Fournier L L, Sarli G O, Aragón A and Rawls WJ 1999Sensitivity of fractal parameters of soil aggregates to different management practices in a Phaeozem in central Argentina*Soil & Tillage Research***52(s3–4)**217-222

[15] Wang D, Fu B, Zhao W, Hu H and Wang Y 2008 Multifractal characteristics of soil particle

size distribution under different land-use types on the Loess Plateau, China*Catena*72(1)29-36.

[16] Paz-Ferreiro J, Vázquez E V and Miranda J G V 2010Assessing soil particle-size distribution on experimental plots with similar texture under different management systems using multifractal parameters*Geoderma***160(1)**47-56

[17] Liu X, Zhang G, Heathman G C, Wang Y and Huang C H 2009 Fractal features of soil particle-size distribution as affected by plant communities in the forested region of Mountain Yimeng, China*Geoderma***154(s1-2)** 123-130.

[18] Xu G, Li Z and Li P2013Fractal features of soil particle-size distribution and total soil nitrogen distribution in a typical watershed in the source area of the middle Dan River, China*Catena***101(2)**17–23

[19] Agterberg F P,Cheng Q,Brown A and Good A 1996 Multifractalmodeling of fractures in the Lac Du Bonnet batholith,Manitoba *Computers and Geosciences***22(5)**497-507

[20] Evertsz C J G and Mandelbrot BB 1992Multifractal measures (Appendix B). In: Peitgen HO,JurensH,SaupeD,Eds. Chaos and Fractals 1992*New York: Springer Verlag*922-953

[21] Agterberg F P2001 Multifractal simulation of geochemical map patterns.In : Merriam D.F.,Davis J.C.,eds. Geologic Modeling and Simulation: Sedimentary Systems*New York: Kluwer*327-346

[22] Cheng Q and Agterberg F P 1995Multifractalmodeling and spatial statistics 1995*Mathematical Geology***28(1)**1-16

[23] Chen G, Gao R , Zhang S W, Zhang L, Ye H C and Huang Y F 2015 Spatial prediction of soil nutrients based on multidimensional fractal methods*Transactions of the Chinese Society for Agricultural***46(8)** 159-168 (in Chinese)

[24] Huang G H, Zhan W H2002 Fractal property of soil particle size distribution and its application*ActaPedologicaSinica***39(04)**490-497 (in Chinese)

[25] Zeng X Q, Liu H P, Lu B J, Wang X Y and Yang W 2008 Fractal dimension fo soil particle-size distribution characteristic in the Beijing mountain*Journal of Mountain Science***26(1)**65-70 (in Chinese)

[26] Zhang S W, Zhang L P, Yuan J, Shen C Y, Chen X Y, Ye H C and Huang Y F 2014Characterizing Variation of Topsoil Particle Size Distribution Based on Fractal Theory and Geostatistics*ScientiaAgriculturaSinica***13**2591-2601(in Chinese)

[27] Caniego J, Ibáñez J J and Martínez F S J2006 Selfsimilarity of pedotaxa distributions at the planetary scale: A multifractal approach*Geoderma***134(3-4)**306-317

[28] Gao J, Huang Y F, Li B G, Zhang S R AND Zhang F 2003Spatial variability analysis of soil particle compositions and soil profile horizon in field*Journal of Plant Nutrition and Fertilizer***9(2)** 151-157 (in Chinese)

Polychlorinated biphenyls in fly ashes collected from five coal-fired power plants in North China

Zhiyong Li[1, 2], Xu Li[1], Huiqiao Ma[1] and Lin Fan[1]

School of Environmental Science and Engineering, North China Electric Power University, 071000 Baoding, China

E-mail: lzy6566@126.com

Abstract. The 5 coal fly ash samples (CFA) were systematically collected from 5 coal-fired power plants (CFPPs) in North China for analysis of 86 PCB congeners. The predominant PCB congeners were PCB-6, -4/10, -28, -18, -19 and -16/32, which belonged to lighter molecular weight (LMW) congeners. The Σ_{86}PCBs for 5 CFPPs ranged from 10.93 to 32.06 ng/g with the mean value as 16.01 ng/g. The PCBs in CFA were dominated by LMW-PCBs with 2-, 3- and 4-Cl PCBs contributed 34.80%, 39.18% and 9.21% to the Σ_{86}PCBs. The TEQ concentrations for 5 CFPPs was 42.54 pg/g higher than 4 pg/g designed by Canada for soil quality, indicated the using of CFA as soil amendment should be cautioned.

1. Introduction

China is the largest coal consumer in world and consumed 50.2% of world coal consumption in 2012 [1]. The CFPPs have the 78% of total power capacity in China. The large amounts of coal fly ashes (CFA) were emitted from the coal-fired power plants (CFPPs), which would be indentified as pollutants if without proper disposed [2, 3]. Some researchers reviewed the utilization methods of CFA such as used as a soil amendment, used in glass and ceramics manufacturing, zeolites and mesoporous materials productions, geopolymers, synthesis, catalysts and catalyst supports [4, 5] Although some recycling disposal methods of CFA were adopted recently, only 45% of CFA was utilized, the rest was often stockpiled before landfill and storing in lagoons [6]. Coal is likely to be the dominant fuel for power generation in energy intensive country such as China in the future [7]. The CFA generation in China has been increased from 2001 to 2015, and has reached 580, 000, 000 tons in 2015 [8]. Clearly more and more CFA need to be disposed in future China as the highest CFA generation in the world. At the same time, large amounts of unintentionally pollutants such as heavy metals, organic pollutants entered the soil, atmosphere and water, had threat to human health [9]. These pollutants in CFA significantly make the utilization of CFA limited. As a result, the investigation of unintentionally organic pollutants in CFA is imperative before it was environmental friendly utilized. For example, the organic toxic compounds and trace elements in CFA restrict the using of CFA as soil remediation agent.

Polychlorinated biphenyls (PCBs) are environmental persistent organic pollutants (POPs) have been studied extensively in recent years characterized by high lipophilicity, extreme toxicity and high resistance to environmental degradation process [10, 11]. The CFA has been identified as the important source of PCBs to atmosphere, water and soil [10]. So the study on emission profiles of PCBs in CFA is very imperative work. Sahu et al. (2009) studied the PAHs and PCBs in CFA from

Indian five CFPPs and indicated the restriction of using of CFA as soil amendment [9]. But the data of PCBs in CFA from CFPPs were not available in China.

Along with the rapid development of economy and the strict requirement for environment protection in China, the CFPPs with higher individual block power capacity (IBPC) are becoming the mainstream power generation units in China. In present China, IBPC as 600 MW is the mainly representative generation unit and accounted 39% of all the coal-fired power generation units [12]. The CFA samples from 5 CFPPs with IBPC as 600 MW were collected during March to November, 2015 for the analysis of 86 PCB congeners.

The main aims of this study were to investigate: 1) the concentrations and homolog distribution of PCBs in CFA from 5 CFPPs in North China with individual block power capacity (IBPC) as 600 MW; 2) the toxic equivalency (TEQ) concentration of PCBs in CFA to evaluate the using of CFA as soil amendment.

2. Experiment Methods

2.1. CFA Sample collection and preparation before extraction

The CFA samples were collected systematically from 5 large-scaled CFPPs in North China during March to November, 2015. These 5 CFPPs were all with the individual block of power capacity as 600 MW, which was the mainstream generation unit in China. The systematic sampling procedure lasting a month was adopted to obtain the representative samples. Every sample were collected for two days interval and the representative sample was consisted these samples. The samples were collected using a stainless spoon from the hoppers of the electrostatic precipitators used to clean the emissions [9]. The collected samples were stored immediately in sealed polyethylene bags to prevent contamination and weathering. The 1kg of each CFA sample was transported to the laboratory in solvent rinsed amber glass jars with Teflon lined aluminium caps. Before analysis, each sample was homogenized using coning and quartering. Each sample was weighed after being dried in a vacuum freeze dryer to remove the moisture.

2.2. Sample extraction for PCBs analysis

Three groups of PCB standards with the catalog numbers as C-IADN-01, 02 and 03 (containing 84 congeners) and PCB-157 and -189 (Accustandard, Inc.) were used in this study. PCB-14, 65, 166 were used as surrogate standards, 2,4,5,6-tetrachloro-m-xylene and PCB-209 were used as internal standards. Other solvents including n-hexane and acetone (pesticide grade, Fisher Scientific, Inc. Fair Lawn, New Jersey) were also used to the analysis of PCBs.

The detailed analyzed method was shown in references [11]. The 60 g of CFA sample was Soxhlet extracted with acetone: n-hexane. The extracts were concentrated by rotary vacuum evaporation and purified by a silica gel column. When the extracts passed through this column, PCBs fractions were eluted with n-hexane. The collected eluents were concentrated by a rotary evaporator and then reduced to 0.5 mL under a gentle N_2 stream for analysis. The extracts were analyzed using a HP 6890GC/5973i MS system equipped with a DB-5 fused silica capillary column (60 m×0.25 mm i.d.). The GC condition adopted was same as references [10] and [11]. GC/MS system was processed using selected ion monitoring mode with m/z as 222, 224 for di-PCBs, 256, 258 for tri-PCBs, 290, 292 for tetra-PCBs, 326, 328 for penta-PCBs, 360, 362 for hexa-PCBs, 394, 396 for hepta-PCBs, 428, 430 for octa-PCBs, 462, 464 for nona-PCBs and 498 for deca-PCBs, respectively.

2.3. Quality control and quality assurance

The recoveries for surrogate standard of PCB-14, 65, and 166 were 93±15%, 89±10 and 88±13%, respectively. The recoveries for 86 PCB congeners in 4 matrix-spiked samples were all within 90% to 118% and the standard deviation was lower than 8%. The standard deviations for 4 pairs of repeated samples were all lower than 10%. The method detection limits for 86 PCBs ranged from 2 to 33 pg/g (mean: 7.5 pg/g)

3. Results and Discussion

3.1. Individual PCB congener in CFA from 5 CFPPs

As shown in Table 1, 47 PCB congeners among 86 PCBs were detectable in this study. The predominant PCBs in CFA for 5 CFPPs in north China were PCB-6, -4/10, -28, -18, -19 and -16/32, which accounted for 43.80% to 80.82% (61.91±15.32) of total PCBs and correlated well with total PCBs (R^2=0.96, p<0.005).

Table 1. Statistic values of individual PCB congener in CFA from 5 CFPPs (pg/g)

PCBs	Min	Max	Mean±SD[a]	PCBs	Min	Max	Mean±SD[a]
PCB4/10	400.21	7089.47	2922.53±1930.51	PCB83	n.d.	n.d.	n.d.
PCB7/9	n.d.	1729.33	435.74±747.08	PCB97	n.d.	n.d.	n.d.
PCB6	800.01	6810.26	3942.81±1113.20	PCB81	30.24	90.25	60.36±20.44
PCB5/8	n.d.	n.d.	n.d.	PCB87	23.54	139.41	51.98±36.66
PCB19	501.24	3289.66	1078.22±565.15	PCB85	n.d.	n.d.	n.d.
PCB12/13	n.d.	1464.86	462.87±321.45	PCB77	n.d.	n.d.	n.d.
PCB18	182.83	5689.45	2581.33±1895.12	PCB110	n.d.	n.d.	n.d.
PCB15	n.d.	n.d.	n.d.	PCB144/135	n.d.	n.d.	n.d.
PCB17	n.d.	325.45	86.40±89.12	PCB123	9.82	46.71	66.17±43.57
PCB16/32	600.00	2695.93	1056.92±942.63	PCB149	n.d.	n.d.	n.d.
PCB26	371.60	700.56	491.56±132.09	PCB118	304.02	460.01	375.94±58.04
PCB28	1160.21	5423.15	2559.39±1651.08	PCB114	n.d.	197.95	103.49±193.94
PCB31	n.d.	n.d.	n.d.	PCB131	n.d.	n.d.	n.d.
PCB33/53	n.d.	300.01	119.02±152.36	PCB153/132	n.d.	n.d.	n.d.
PCB22	100.01	1489.40	491.56±371.83	PCB105	96.58	375.46	204.83±46.54
PCB45	n.d.	172.25	37.65±44.24	PCB138/163	n.d.	n.d.	n.d.
PCB52	800.41	2502.30	1128.42±826.18	PCB126	198.25	300.84	220.01±33.56
PCB49	n.d.	164.59	36.12±71.82	PCB128	n.d.	n.d.	n.d.
PCB47+48	n.d.	18.73	8.95±5.47	PCB167	73.55	100.25	79.20±11.63
PCB44	n.d.	n.d.	n.d.	PCB174	n.d.	n.d.	n.d.
PCB37+42	300.14	540.12	404.70±108.65	PCB171/202	n.d.	n.d.	n.d.
PCB41+71+64	n.d.	n.d.	n.d.	PCB156	n.d.	11.13	4.23±3.86
PCB100	n.d.	n.d.	n.d.	PCB157	319.77	721.21	474.20+167.71
PCB74	n.d.	n.d.	n.d.	PCB172	n.d.	n.d.	n.d.
PCB76+70	n.d.	n.d.	n.d.	PCB180	n.d.	n.d.	n.d.
PCB66+95	n.d.	n.d.	n.d.	PCB201	n.d.	n.d.	n.d.
PCB91	n.d.	n.d.	n.d.	PCB169	550.99	910.66	681.86±141.27
PCB56+60	n.d.	n.d.	n.d.	PCB170/190	n.d.	85.00	44.40±40.61
PCB92	n.d.	154.77	54.54±61.13	PCB199	n.d.	48.09	10.42±21.06
PCB84	60.12	153.98	94.59±37.23	PCB189	125.00	412.21	249.60±124.94
PCB89	22.65	488.79	222.29±167.61	PCB207	n.d.	88.18	44.78±27.76
PCB101	n.d.	n.d.	n.d.	PCB194	n.d.	360.03	72.81±111.38
PCB99	n.d.	82.94	18.99±35.75	PCB205	n.d.	n.d.	n.d.
PCB119	n.d.	n.d.	n.d.	PCB206	n.d.	n.d.	n.d.

[a]SD: Standard deviation

3.2. PCBs levels and homolog compositions for five CFPPs

As shown in Figure 1(a), the Σ_{86}PCBs for 5 CFPPs ranged from 10.93 ng/g to 32.06 ng/g with the mean value as 16.01 ng/g. The PCBs congeners were classified into two categories including low molecular weight (LMW) PCBs (2- to 4-Cl substituted PCB congeners) and high molecular weight (HMW) PCBs (the chlorine atoms higher than 5). As shown in Figure 1(b), the PCBs in CFA were dominated by LMW-PCBs with 2-, 3- and 4-Cl substituted PCBs contributed 34.80%, 39.18% and 9.21% to the Σ_{86}PCBs.

Seven indicator PCB congeners (IPCBs) including PCB-28, 52, 101, 118, 138, 153, and 180 were commonly measured and often compared in PCBs analysis [10]. Among seven IPCBs, PCB-101, -138,

-153 and -180 were all not detected in 5 CFA samples. The sum of 7 IPCBs was 4.06±2.32 ng/g and not correlated with total PCBs. Li et al. (2011) also reported the 7 IPCBs were not correlated with total PCBs in surface soils of Tianjin due to the emission of PCBs from industrial sources [10]. The 7 IPCBs were mainly existed in conventional PCB products, the emissions from industrial stacks resulted this lower correlation coefficient.

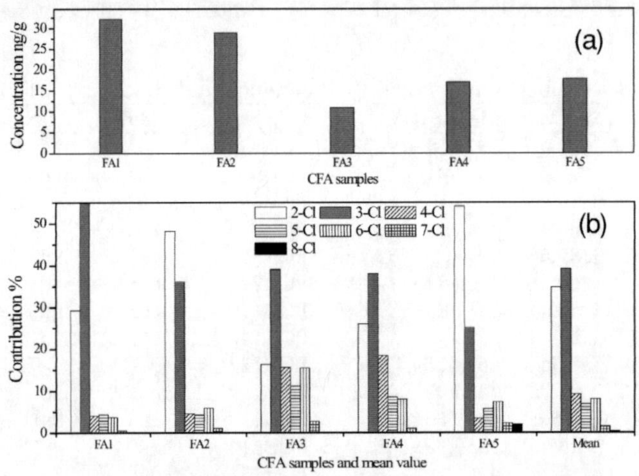

Figure1. a) PCBs concentrations in CFA from 5 CFPPs and b) the contribution of PCB homologs with different chlorine atoms to total PCBs

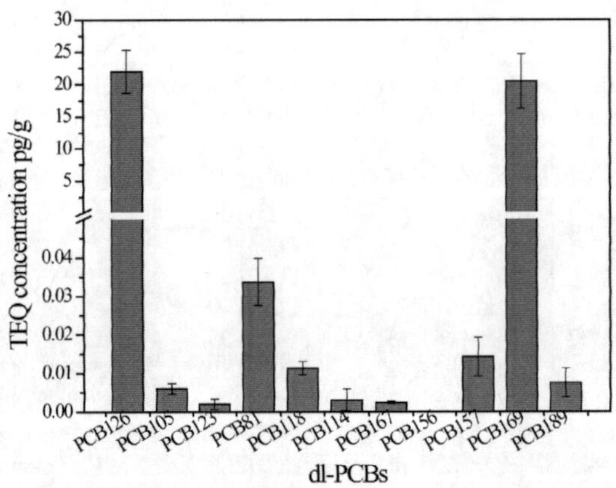

Figure 2. Mean TEQ concentration for each dl-PCBs for 5 CFPPs

3.3. Toxic equivalency concentrations of PCBs in CFA from five CFPPs

The 11 dioxin-like PCBs (dl-PCBs) except PCB-77 such as PCB-81, -105, -114, -118, -123, -126, -156, -157, -167, -169 and -189 were all detected in this study, and the TEQ was calculated for each dl-PCB. The TEQ concentration for each dl-PCB was calculated by multiplying its concentration and WHO 2005 toxic equivalent factor (TEF) [10, 11].

TEQ=(mass concentration of PCB)×TEF

The TEF values for 12 dl-PCBs were 0.0001, 0.0003, 0.00003, 0.00003, 0.00003, 0.00003, 0.1, 0.00003, 0.00003, 0.00003, 0.03 and 0.00003 for PCB-77, -81, -105, -114, -118, -123, -126, -156, -

157, -167, -169 and -189, respectively. The sum of 11 dl-PCBs for 5 CFPPs ranged from 2.24 to 2.92 ng/g with the mean value as 2.57 ng/g. The value of 2.57 ng/g was higher than 1.65 ng/g for 15 Chinese large-scale municipal solid waste incinerators (MSWIs) and 0.5-2.3 ng/g for 4 Chinese MSWIs [13, 14], much lower than 8.972 ng/g for a MSWI in Medellin, Colombia [15]. These differences possibly resulted from the different combustion condition, feed fuel among different industrial stacks. The TEQ concentrations for 5 CFPPs ranged from 37.60 pg/g to 46.59 pg/g with the mean as 42.54 pg/g. PCB-126 and -169 have the higher TEQ concentrations due to their higher TEF values as 0.1 and 0.03. The TEQ for PCB-126 and -169 were 22.00±3.36 and 20.45±4.24 pg/g (see Figure 2). The TEQ concentrations for 18 CFPPs in the range of 38.71-78.55 pg/g (mean:52.63 pg/g), significantly higher than 4 pg/g designated by Canada for soil quality [16], which did not supports their suitability for soil amendment.

4. Conclusions

The 5 CFA samples were collected from 5 CFPPs in North China for analysis of 86 PCB congeners. The predominant PCB congeners were PCB-6, -4/10, -28, -18, -19 and -16/32. The mean value of Σ_{86}PCBs for 5 CFPPs was 16.01 ng/g. The PCBs in CFA were dominated by LMW-PCBs with 2-, 3- and 4-Cl PCBs. The mean value of sum of 12 dioxin-like PCBs for 5 CFPPs was 2.57 ng/g. The TEQ concentrations for 5 CFPPs was higher than that designed by Canada for soil quality, indicated the using of CFA as soil amendment should be cautioned.

Acknowledgment

This study was supported by National Natural Science Foundation of China (21407048).

References

[1] Yao Z T, Ji X S, Sarker P K, Tang J H, Ge L Q, Xia M S and Xi Y Q 2015 *Earth-Sci. Rev.* **141** 105-21.
[2] Skousen J, Yang J E, Lee J S and Ziemkiewicz P 2013 *Geosyst. Eng.* **16** 249-56.
[3] Ram L C and Masto R E 2014 *Earth-Sci. Rev.* **128** 52-74.
[4] Blissett R S and Rowson N A 2012 *Fuel* **97** 1-23.
[5] Shaheen S M, Hooda P S and Tsadilas C D 2014 *J. Environ. Manage.* **145** 249-67.
[6] Pandey V C and Singh N 2010 *Agr. Ecosyst. Environ* **136** 16-27.
[7] Lior N 2010 *Energy* **35** 3976-94.
[8] Yao Z T, Xia M S, Sarker P K and Chen T 2014 *Fuel* **120** 74-85.
[9] Sahu S K, Bhangare R C, Ajmal P Y, Sharma S, Pandit G G and Puranik V D 2009 *Microchem. J.* **92** 92-6.
[10] Li Z Y, Kong S F, Chen L, Bai Z P, Ji Y Q, Liu J W, Lu B, Han B and Wang Q W 2011 *Chemosphere* **85** 494-501.
[11] Li Z Y, Zhang Y L and Chen L 2015 *Aerosol Air Qual. Res.* **15** 1059-71.
[12] Zhu Z F, Wang L P, Qi H, Zhang C, Zhao Y C, Chen G, Qiu J H and Zhang J Y 2014 *J. Fuel Chem. Technol.* **42** 323-28. (in Chinese)
[13] Pan Y, Yang L B, Zhou J Z, Liu J Y, Qian G R, Ohtsuka N, Motegi M, Oh K and Hosono S 2013 *Chemosphere* **92** 765-71.
[14] Wu S, Zhou J, Pan Y, Zhang J and Zhang L 2016 *Waste Manag.* **50** 113-20.
[15] Cobo M, Gálvez A, Conesa JA and de Correa C M *J. Hazard. Mater.* **168** 1223-32.
[16] CCME 2007 *Canadian Soil Quality Guidelines for the Protection of Environmental and Human Health, Winnipeg*.

Homogenization of Sea Surface Temperature at Xiao Changshan marine station in the east of the Bohai Sea using the PMT method

Yan Li [1], Guosong Wang [1], Xue Han[2], Huan Li [1, 3], Wenjing Fan[1], Kexiu Liu[1] , Hui Wang[1]

[1] National Marine Data and Information Service, Tianjin, China;
[2] National Marine Environmental Forecasting Center, Beijing, China

E-mail: usher02@126.com

Abstract: Using the PMTred algorithm (penalized maximum T test) and the detailed metadata archive, the monthly mean sea surface temperature (SST) datasets from 1960 to 2011 have been detected and adjusted. Results show that the SST time series has serious problems of inhomogeneity. The changes in observation instruments and observation system are the main causes of the discontinuity. For the monthly SST time series, the negative adjustments have the high proportion, which may be greatly due to the SST decreasing after automation. It is found that the annual mean SST trends have changed obviously before and after adjusted. The increasing trend of annual mean SST after adjustment is higher than before, by up to 0.252 °C/10 yr.

1. Introduction

Long time observation series is fundamental and essential for the describing climate change. However, these observation series without effect of non-climatic factors are rare [1]. These non-climatic factors including station relocation, instrument change, environment changes and so on may introduce artificial discontinuities or inhomogeneities, in the time series [2, 3]. Such discontinuities can lead to misinterpretations of the studied climate. In order to avoid errors and obtain homogeneous climate time series, non-natural irregularities in climate data series must be detected and removed [2, 3, 4]. In recent years, methods considering effects of inhomogeneity on meteorological elements have been well developed in China, e.g., surface air temperature (SAT), precipitation, pressure and so on. Homogenized meteorological elements datasets have been constructed [5-9]. But, like many long-term climate data series, the oceanographic element time series along coast of China Seas also contains discontinuities or breakpoints caused by relocations or instruments changes. There are few scientists focusing on the homogenization of oceanographic climate series along coast of China. It may dues to the lower density of the marine stations, the complex geography of the coast and the limited metadata, etc [10].

Sea surface temperature (SST) is one of the most important variables used to quantify climate change and elements for the oceanic environment as well [11-15]. In this study, we take the SST time series of Xiao Changshan marine station in coast of the Baihai Sea as an example to examine the homogeneity in this station and then adjust it in detail.

The rest of the paper is organized as follows. Section 2 briefly introduces the data sources. Section 3 describes the methods used in our study. Section 4 analyses the detected change points and the related causes. Section 5 presents the observed trends from the raw time series and the homogenized time series. Finally, a summary is given in section 6 to complete this article.

2. Data sources

The SST data in Xiao Changshan marine station from January 1960 to December 2011 is collected and processed by the National Marine Data and Information Service (NMDIS) of the State Marine Administration (SOA) of China. NMDIS employed a variety of quality control (QC) procedures on the data including checks for physically unreasonable values, unreasonably long consecutive occurrences of the same values, daily maximum temperature less than minimum temperatures, and so on. In addition, few monthly missing data are reconstructed by regression analysis method.

Xiao Changshan marine station is located in the Xiao Changshan Island of Dalian city (Figure 1). This island is dominated by gravels, no large coastal projection. The observational environment conforms to the specification for offshore observations in China [16]. The station history data file (i.e. metadata) which used to verify the veracity of statistically detected change points is checked and stored by NMDIS, including the time of relocation, environment change, algorithms for calculating a daily mean, observing procedures and so on [16]. This station experienced a transformation from artificial observation system to automatic observation system in 2001 and totally four temperature instruments have been used from 1960 to 2011. The station location and environment are almost no significant change. Thus, in the paper, we mainly consider the effect of the observational system and instrument change on the homogeneity of the SST series. In addition, the homogenized monthly mean SAT time series from National Meteorological Information Center (NMIC) are used to construct reference time series.

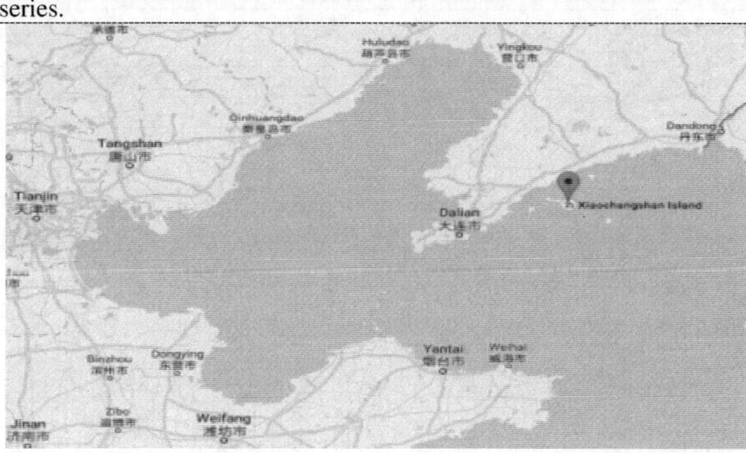

Figure 1. Geographic position of Xiao Changshan marine station, source: Google Maps.

3. Change point detection method of SST series

After comparatively analysing the performance of several homogenization methods for time series [17], the RHtests V4 software package [18] is used in our study. This package includes the PMTred algorithm, which is based on the penalized maximum t (PMT) test. The PMTred algorithm deal with multiple change points using a recursive testing algorithm and account for the first order autocorrelation, and thus have the word "red" in their names [19, 20]. The PMT test is a relative homogeneity test which relies on the reference series in a greater degree. In this condition, it is very important and necessary to create a homogeneous climatological reference series in the PMT test. Since the sparsity of the marine stations along the coast of the Bohai Sea, it is difficult to develop a SST reference series. In this study, we use homogeneous SAT series from neighbouring meteorological stations to construct a reference series for each marine station. This approach is considered to be proper and reasonable due to the fact that the variability of SST in the coastal zone of Circum Bohai Sea is the only inner sea of China and effect mainly by land climate. Meanwhile, SST series always reflect the similar regional-scale climate variability in the study region as the SAT series [21]. In particular, the interannual variation trend of the candidate and reference series is removed in order to obtain the anomaly series that more directly reflects the climate fluctuations in the homogeneity assessment.

The reference meteorological stations are chosen by the following conditions:

(1) The horizontal distance between candidate marine station and reference meteorological stations should be within 350km;
(2) The correlation coefficient between monthly SST time series and the SAT time series should be 0.7 or higher;
(3) The urban heat island effect of the meteorological station is minor or insignificant.

The correlation coefficients of the SST time series of Xiao Changshan marine station with each SAT time series of the neighbouring meteorological stations are calculated. SAT time series of Zhuanghe, Wa Fangdian and Changhai meteorological stations (Table 1, Figure 2) which highly correlated with the candidate SST time series are chosen (the correlation coefficients are 0.77, 0.74, 0.79, respectively, exceeding 99% confidence level). In addition, there are also similar evolution characters between the annual mean SST time series and selected SAT time series (Figure 3).

Table 1. The code and coordinate of the reference meteorological station.

Name	Code	Latitude	Longitude
Zhuanghe	54584	39°72′	122°95′
Wa Fangdian	54563	39°63′	122°02′
Changhai	54579	39°27′	122°58′

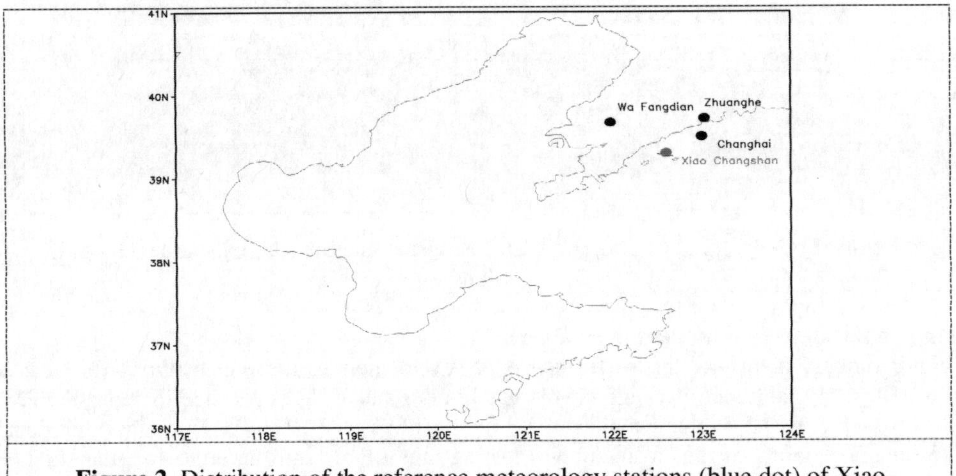

Figure 2. Distribution of the reference meteorology stations (blue dot) of Xiao Changshan ocean station (red dot).

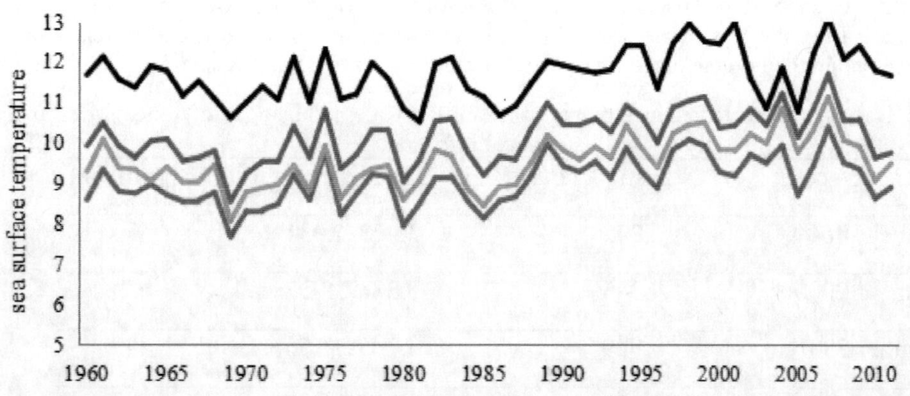

Figure 3. The time series of annual mean SST before adjustment at Xiao Changshan station (black solid line) and the time series of annual mean surface air temperature at Meteorological Observation Stations Wa Fangdian (green solid line), Changhai (blue solid line) and Zhuanghe (purple solid line) from 1960 to 2011.

The homogenous nearby meteorological stations with high correlation coefficients with the SST time series are chosen to construct reference series by correlation coefficient weighted average method. The formula is shown as follows:

$$\overline{y_i} = \frac{\sum_{j=1}^{n} \rho_j^2 \times y_{ji}}{\sum_{j=1}^{n} \rho_j^2} \tag{1}$$

where i denotes the time, j represents the number of the reference meteorological stations, is the cocorrelation coefficients between SST series of Xiao Changshan and the SAT series of the reference meteorological stations station, y is the monthly mean SAT time series of each reference station and means the calculated reference SAT series.

Combined with the detailed metadata of Xiao Changshan station, the resulting change points from PMT method are synthesized and verified. In general, we retain one type of change points for adjustment: change points which are supported by metadata and are identified to be a significant documented change point in the monthly mean series. We consider a detected change point to be a documented one when metadata indicate a documented change within 1 year before or after the detected change point. In this case, the documented time of change is used to replace the estimated time of change when they are not identical (due to estimation error).

4. Statistics of Detected Change Points and the Related Causes

The result of PMT test on SST time series at Xiao Changshan marine observational station shows that there are 5 tested suspicious discontinuities in the raw SST time series, that is, October 1964, September 2000, December 2001, January 2005 and February 2009 (Table 2). Consequently, the station historical records are used to verify the rationality of these tested discontinuities. According to metadata (Table 3), 3 tested discontinuities (October 1964, December 2001, January 2005) are supported by metadata and identified to be significant documented change point in the monthly mean series.

SST time series which is identified to contain significant change points is adjusted to the latest segment of the data series, using the quantile-matching (QM) algorithm from RHtests V4 [18]. Table 4 lists the three detected change points and the adjustment value in different periods. This result shows that the negative correction values have the high proportion, about -0.67 °C and -0.52°C. The latter is linked tightly to the SST decreasing after automation [21]. The homogenization adjusts the warming bias resulting from the less observation times in the earlier time using artificial observation system. Finally, we obtain the homogenized time series of monthly mean SST in Xiao Changshan station.

Table 2. Results of detected change points based on PMT method using a reference series and related statistical information.

Tested change point	Test statistic PTmax	95% confidence interval	significant
October 1964	6.245	3.473-4.034	Yes
September 2000	4.345	3.381-3.937	Yes
December 2001	5.929	3.245-3.756	Yes
January 2005	6.092	3.279-3.802	Yes
February 2009	4.109	3.378-3.920	Yes

Table 3. The observational instruments changes from the metadata of Xiao Changshan ocean observation station.

No	time	Instrument	Observational system
1	Jan. 1959-Dec.1964	Analytik Jena 16 Mercury Thermometer	artificial system
2	Jan. 1965-Dec. 2001	SWY1-1 Thermometer	artificial system
3	Jan. 2002-Jun.2004	YZY4-1 Themohaline Sensor	automatic system
4	Jul. 2004 till now	SWL1-1 Thermometer	automatic system

Table 4. Results of detected break points and the adjustment range

No	Break points	Adjustment range
1	Nov. 1965	-0.67°C
2	Feb. 2002	-0.52°C
3	Jul. 2004	0.1°C

5. SST variation before and after homogeneous adjustment

The annual mean SST time series at Xiao Changshan marine observational station before and after adjustment are shown in Figure 4. The linear SST trend from 1960 to 2011 is estimated by linear regression method. The linear trend of unadjusted annual mean SST time series is 0.157 °C /10 yr. Compared to unadjusted time series, the warming rates of the homogenized SST time series are notably larger, about 0.252 °C/10 yr. This result shows that the inhomogeneous time series appears to underestimates the warming trend during the past 52 years. Furthermore, the correlation coefficients between the adjusted annual mean SST series and annual mean SAT series of Zhuanghe, Wa Fangdian and Changhai are much higher than before, by up to 0.86, 0.83 and 0.86.

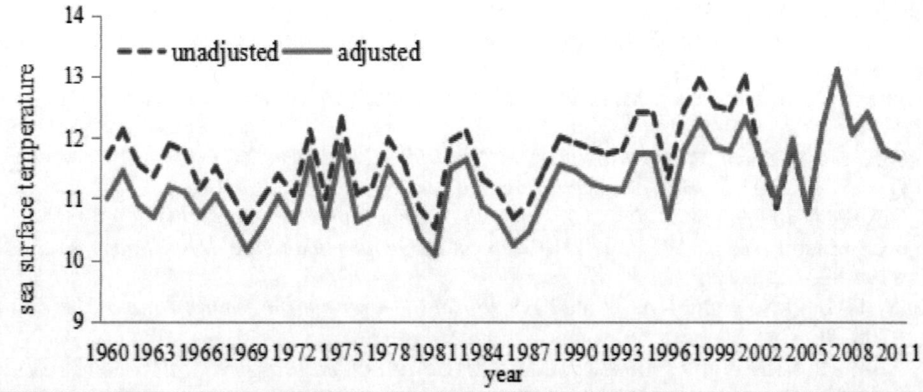

Figure 4. The time series of annual mean SST at Xiao Changshan station before (black dashed curve) and after (red solid curve) adjustment.

6. Conclusion

In this study, we homogenized the SST time series of Xiao Changshan marine observational station over the period from 1960 to 2011, using PMTred algorithm to detect change points, and using the QM adjustment method conducted with SAT reference series to adjust the SST time series. The non-climatic factors causing the inhomogeneity of the SST time series have been analyzed. The changes of instruments and observational system are the main causes for the identified inhomogeneities. The change of observational system causes a significant decrease in temperature in the new observational system.

We also assess the warming trend of the SST time series before and after adjustment. The correlation coefficient between the adjusted SST time series and time is much higher which means that the homogenized SST time series is improved well .The warming rate of SST series after adjustment is much larger than before, up to 0.252 °C/10 yr implying that the inhomogeneous series underestimates the warming trends during the past 52 years.

The density of marine observational stations in China is very sparse. SST reference series are difficult to develop by the traditional method. Therefore, in the present study, we try to use SAT time series and PMTred algorithms to detect the inhomogenities in the SST time series of Xiao Changshan marine station. Some meaningful results have been obtained. However, we have to point out that due to the limited metadata, biases located in the reference time series, biases in homogenization methods and the subjective judgements, etc., there must be some uncertainty in the homogenized time series. In other words, although homogenized data always reflects actual climate changes better than the original data does, it still cannot reflect the "real" climate changes. In order to eliminate the uncertainty and more closer to the "real" one, there is still much work should to be done, including collection of more high quality metadata, using multiple detected methods and then deciding the result by averaging and so on [22]. Our work is only the beginning and much hard work is needed for the oceanographic element homogenization.

Acknowledgements

This study was jointly supported by the program of the National Natural Science Foundation of China (Nos. 41306007, and 41376014), the Natural Science Foundation of Tianjin (No. 14ZCZDSF00012), the Project on Basic Research of the National Department of Science and Technology (No. 2014BAB12B02).

References

[1] Alexandersson H and Moberg A 1997 *International Journal Climatology* **17** 25-34.

[2] Karl T R and Williams C N 1987 *Journal Climate Appl. Meteorolo.* **26** 1744-63.
[3] Peterson T C, Easterling D R, Karl T R, et al. 1998 *International Journal of Climatology* **18** 1493-1517.
[4] Stephenson T S, Goodess C M, Haylock M R, et al. 2008 *Journal of Geophysical Research* **113** 1-17.
[5] Li Q X, Liu X, Zhang H 2004 *Advances in Atmospheric Sciences* **21**(2) 260-8.
[6] Li Q X, W Li 2007 *Journal of Meteorological Research* **65**(2) 293-300.
[7] Li Q X, Peng J D and Shen Yan 2012 *Journal of Geographic Science* **22**(4) 579-593.
[8] Cao L J, Zhao P, Yan Z W, et al. 2013 *Journal of Geophysical Research: Atmospheres* **118** (15) 8197-8207.
[9] Xu W H, Li Q X, Wang X L, et al. 2013 *Journal Geophysical Research: Atmospheres* **118**(17) 9708-20.
[10] Li Yan, Mu Lin, Wang Guosong, et al. 2016 *Acta Oceanologica Sinica* **38** (3) 27-39.
[11] Yeh S Wand Kim C H 2010 *Continental Shelf Research* 1428-34.
[12] Xie S P, Deser C, Vecchi G A and et al. 2010 *Journal of Climate* **23** 966-986.
[13] Jin Q H, Wang H 2011 *Acta Oceanologica Sinica* **30**(4) 14-23.
[14] Liu Q Y and Zhang Q 2013 *Journal of Ocean University China* **12**(2) 295-300.
[15] Bao B and Ren G Y 2014 *Continental Shelf Research* **77** 96-106.
[16] State Oceanic Administration of China 2006 GB/T14914-2006 60-61.
[17] Cao L J and Yan Z W 2011 *Advance Climate Change Research* **7** 129-135.
[18] Wang X L and Feng Y 2013 RHtestV4 User Mannual, UserManual.doc. Available online at http://cccma.seos.uvic.ca/ETCCDMI/software.shtml.
[19] Wang X L 2008 *Journal Apply Meteorological Climatology* **47** 2423-44.
[20] Cao L J, Zhao P, Yan Z W, et al. 2013 *Journal of Geophysical Research: Atmospheres* **118** 8197-8207.
[21] Sun X P 2006 Beijing, China Ocean Press 201-228.
[22] Li Q X, Menne M J, Williams J C N, et al. 2010 *Climate Environment Research* **10** 736-742.

Climatic characteristics and regionalization of fogs in China

T Y Hao[1, 5], S C Chen[2], S Q Han[3], X L Shan[4], L H Meng[3]

[1] Tianjin Environmental Meteorological Centre, Tianjin 300074, China
[2] Tianjin Meteorological Observation Centre, Tianjin 300061, China
[3] Tianjin Institute of Meteorological Science, Tianjin 300074, China
[4] Tianjin Binhai New Area Meteorological Bureau, Tianjin 300457, China

E-mail: Tianyi_hao@126.com

Abstract. Using trend coefficient method, vector analysis method, and monitored meteorological data across China, climatic characteristics and spatial pattern of fogs in China were investigated. The results show that most fogs occur in southeastern China. Thin fogs usually occur in fog-rare regions and dense fogs take place in fog-prone regions. The number of annual fog days in most regions of China exhibits a decreasing trend from 1980 to 2010. It also found that the regions with more fog days correspond to the lower concentration degree of fogs, and vice versa. In terms of the national scale, the concentration periods of fogs are mainly in November, December, and January in China. We further classified the occurrence frequencies of fogs into five spatial distribution patterns over a single year according to the spatial distribution characteristics of fogs occurrence frequencies of 36 dekads, namely, a whole year can be correspondingly divided into five phases. Based on this, multi-year average fog-prone regions in the five phases are obtained. Our results also identify the high incidence periods of fogs in different fog-prone regions.

1. Introduction

Fogs have strong influence on transportation, industrial and agricultural production, and human health. The understanding to the climate characteristics of fog could help greatly to reduce and avoid the disasters caused by fogs. A fog is defined as a collection of suspended water droplets or ice crystals near the Earth's surface that lead to a reduction of horizontal visibility below 1 km [1]. It has been reported that more fogs occurred in Southeast China and less in Northeast China based on statistics of fogs from 1971 to 2000 [2]. The number of heavy fog days in winter half year (October to March) was greater considerably than that in the summer half year (April to September) [3]. The fogs usually begin at 20 pm (Beijing time) to 8 am in the next day with highest rate of occurrence at 6 am ~7 am, ceasing at 8~12 am. The life time of fogs extended usually from 1 to 10h, and mostly about 3h[4,5]. Lin et al. [4] also characterized fog patterns based on low level atmospheric circulation patterns and revealed that uniform pressure and fore-front pattern were most popular fog patterns in China. Wang et al. [6] identified six major fog regions across China based on the analysis of average annual fog days. Niu et al.[7] revealed that the frequencies of fog events during the wintertime over Eastern-central China doubled over the past three decades (1976-2007), confirming with the results from Chen et al. [5]. They linked this considerable increase of fog events with the weakening of the Eastern Asian monsoon circulation and the increasing aerosol loadings.

This study aims at better understanding the climatic characteristics of fogs and helping the prediction to fogs occurrence in China. The objectives are (1) to elucidate the spatial and temporal distribution characteristics and trends of fog days for different fog grades, (2) to investigate the concentration degree of fogs occurrence and major concentration periods of fogs occurrence, (3) to distinguish the fog-prone areas and its high-incidence periods.

2. Material and methods

2.1. Data sources and processing
Data at 581 observation stations across China were collected. They include atmospheric horizontal visibility (4 times per day), daily precipitation, and daily average relative humidity from January 1980 to December 2010. A fog day is defined if the atmospheric horizontal visibility is less than 1km and daily average relative humidity is greater than 90%, except for rainy days and dusty days.

2.2. Methods
The fog concentration degree (FCD) and the fog concentration period (FCP) are defined based on vector analysis [8]. The amount of fog days in each dekad can be regarded as the length of a vector whereas the corresponding dekad is viewed as the direction of the vector.

Using this approach, 36 vectors representing each dekad's fog days are synthesized as a new vector. The FCD can then be defined as the ratio of the modulus of the new synthetic vector to the total number of the annual fog days.

$$FCD = \frac{\sqrt{R_x^2 + R_y^2}}{R} \tag{1}$$

Where $R_x = \sum_{i=1}^{36} r_i \sin\theta_i$ and $R_y = \sum_{i=1}^{36} r_i \cos\theta_i$. R is the amount of annual fog days. The r_i is the amount of fog days in each dekad. Hence the FCD reflects whether the fogs occurrences are concentrate within a year.

The FCP measures the direction angle of the new synthetic vector.

$$FCP = \arctan(\frac{R_x}{R_y}) \tag{2}$$

The FCP is used to find the dekad of a year during which the maximal numbers of fog days are accounted for.

3. Results and discussion

3.1. Spatial distribution of average annual fog days
The spatial distribution of average annual fog days in China averaged over 1980 through 2010 is shown in Figure 1. The highest fog days can be observed in the east of Sichuan, Chongqing, the north of Hunan, the east of Guizhou, most areas of Fujian, and part of Yunnan where there were at least 30 fog days on an annual basis. Those areas filled by green color in Figure 1 also observed relatively high fog days with annual fog days greater than 10 days. The average annual fog days in the rest of China were less than 10 days.

To facilitate analysis, the fogs are divided into three classes according to the atmospheric horizontal visibility. The first class is termed dense fog when atmospheric horizontal visibility is less than 0.2km. The second class is termed thick fog when atmospheric horizontal visibility is less than 0.5km and greater than 0.2km, and the third is termed thin fog when atmospheric horizontal visibility is less than 1km and greater than 0.5km. Figure 2 marks fogs in different classes at 581 stations across China. The results suggest that thin fogs occur in those regions where fogs seldom occur whereas dense fogs usually take place in those regions with high occurrence frequency of fogs.

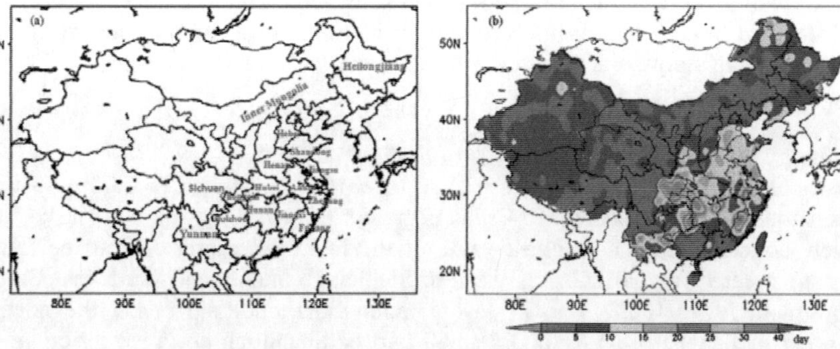

Figure 1. (a) Distribution of provinces and cities of China (b) The spatial distribution of average annual fog days in China from 1980 to 2010.

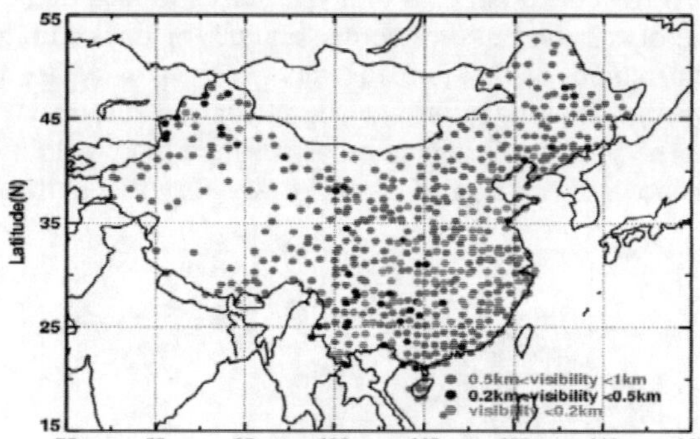

Figure 2. Atmospheric horizontal visibility at all 581 stations under condition of fog occurrence.

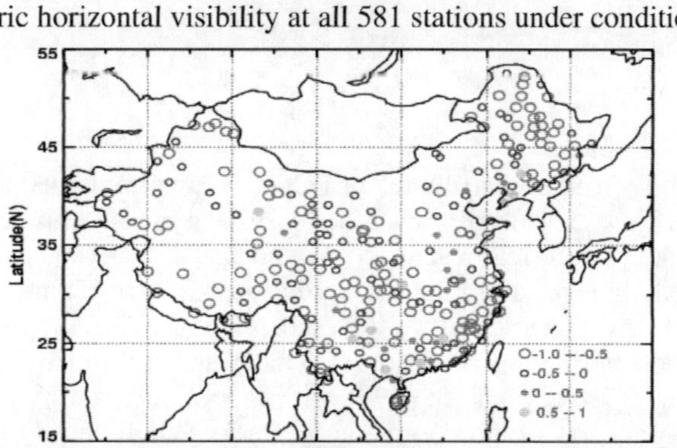

Figure 3. Spatial distribution of trend coefficients from 1980 to 2010. Hollow points represent negative trend and solid points represent positive trend, respectively.

3.2. Trend of fog days

The trend of annual fog days of 1980 through 2010 is presented in Figure 3. As shown, the fog days exhibit a negative trend in most areas of China. In particular, a strong decreasing trend can be observed in most places in Northeast China, Sichuan Basin, and the southeast Wuyi Mountain. On the other hand, there was an increasing trend of fog days at a certain number of stations in southern and

Eastern China, as well as in Bohai Rim region. Our results agree with previous works from Liu et al. [2], Lin et al.[4], and Liu et al. [9]. While the spatial distribution of the long-term trend of fog days was not fully understood, aerosols activity [7, 10, 11] may play a very important role in the long-term trend of fog days.

3.3. Concentration degree and concentration period of fog days

The spatial distribution of FCD in China from 1980 to 2010 is illustrated in Figure 4. It can be seen from the figure 4 that FCD in most areas of China is greater than 0.5, suggesting that the occurrence of fogs is relatively concentrated. The highest value of the FCD exceeding 0.8 can be found in Inner Mongolia, Gansu, eastern Qinghai, Xizang, western Sichuan, Yunnan, and southwest Guangxi. There are two regions with lowest FCD, one in east Yunnan-Guizhou Plateau and the north of Hunan province, the other in Jiangsu province and a small part of Shandong province. In contrast, the FCD values in North China Plain, the middle and lower reaches of the Yangtze River Plain, Loess Plateau, the east of Yunnan-Guizhou Plateau, and Wuyi Mountains are lower than that in the rest of China. This suggests that fogs occurring in the 36 dekads of a year decentralized in these regions.

Combined with the spatial distribution of the average annual fog days and the spatial distribution of FCD, we can found that the regions with more fog days corresponded to the lower concentration degree of fogs, and vice versa. This suggests that fogs occurred dispersedly in fog-prone regions within a year. Exception can be seen in the Sichuan basin where there exhibited the most annual fog days, while its FCD is not the lowest. But its FCD is still lower than that in those regions with less fog.

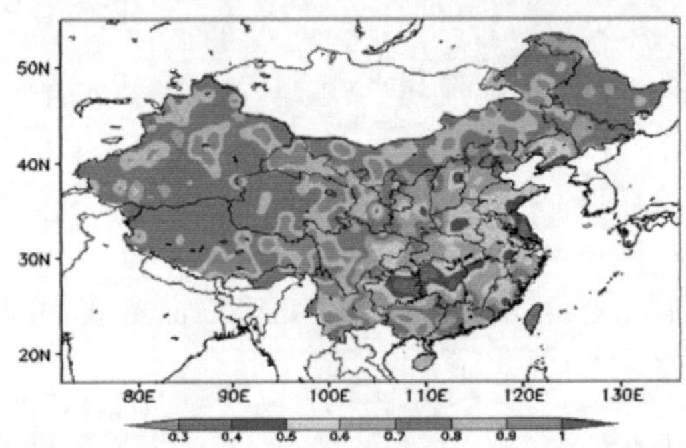

Figure 4. The spatial distribution of FCD in China from 1980 to 2010.

It can be seen from Figure 5 that the values of FCP have an obvious regional pattern. On the country scale, the FCP mainly occurred in January, November, and December. The regions with FCP values in January were mainly located in Inner Mongolia and Western China, where fogs seldom took place and the FCD was also higher than that of other regions. The area where FCP occurred in November was primarily the northern part of North China Plain. In Sichuan Basin and the middle and lower reaches of the Yangtze River Plain, the FCP was in December. Besides, in the southern China, the FCP mainly started from early October to the end of next March, mostly in January. In Northern and Northeastern China, the FCP included July and August. In the coastal area of Bohai Rim, closing to Yellow Sea, the FCP included June, July, and August.

3.4. Fog-prone regions and periods

We have analyzed the spatial distributions of fogs occurrence frequencies in 36 dekads from 1980 to 2010, respectively, in order to identify the high fog incidence areas for each dekad, and to reveal the high fog incidence periods for fog-prone regions in China. To do so, the spatial distributions of fogs occurrence frequencies for 36 dekads have been classified into 5 patterns. This is equivalent to separate 36 dekads into 5 phases, as shown in Figure 6. The duration of the each pattern is shown in Table 1.

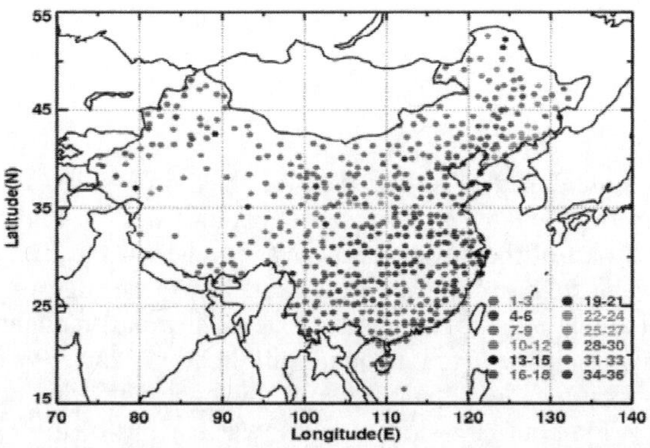

Figure 5. The spatial distribution of FCP in China from 1980 to 2010. (1-3 represents dekad1-dekad3, and so on)

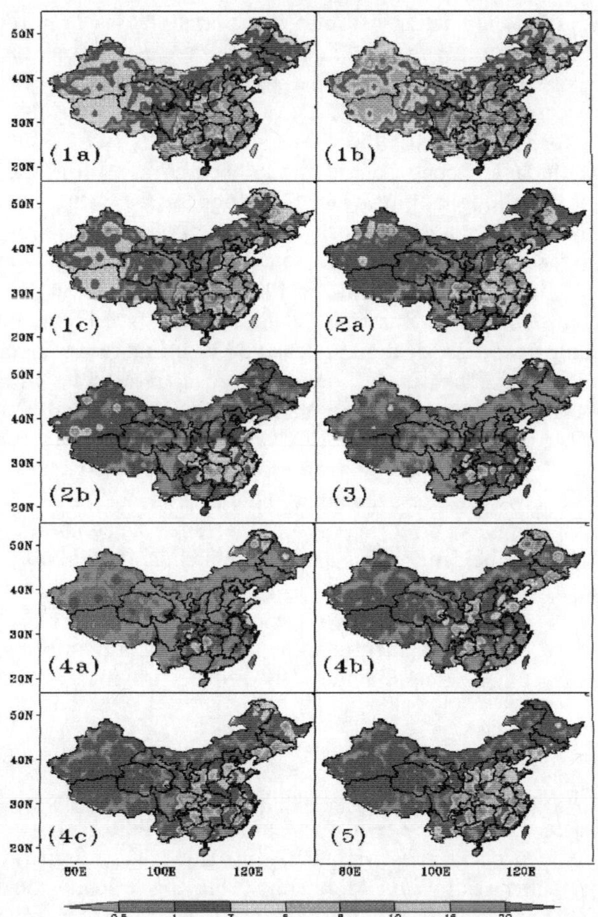

Figure 6. Spatial distribution of fogs occurrence frequency (%) in dekad 31 (1a), dekad 33 (1b), dekad 3 (1c), dekad 5 (2a), dekad 10 (2b), dekad 15 (3), dekad 17 (4a), dekad 24 (4b), dekad 24 (4c), dekad 29 (5), respectively.

Table 1. The duration of the spatial distribution for 5 types of fogs occurrence frequencies.

Phase1(Pattern1)	Phase2(Pattern2)	Phase3(Pattern3)	Phase4(Pattern4)	Phase5(Pattern5)
Dekad31-Dekad3	Dekad4-Dekad12	Dekad13-Dekad16	Dekad17-Dekad27	Dekad28-Dekad30

In the first phase, spanning dekad 31 to dekad 3 (1a, 1b, 1c in Figure 6), the highest fog incidence was found in Sichuan Basin and southwest corner of Yunnan-Guizhou Plateau, followed by western North China Plain, the west part of the middle and lower reaches of the Yangtze River Plain, and the east of Wuyi Mountains. The fogs occurrence frequencies in these regions were greater than 10%. This phase is the main period that fogs occurred most extensively and frequently.

The fogs occurrence frequencies reduced gradually till the 5th dekad over these areas with the high fogs incidence. The frequency was less than 10%. In this case, the fogs occurrence frequencies in Sichuan Basin, North China Plain and the middle and lower reaches of the Yangtze River Plain were greater than other parts of China at the values between 3% and 10%. This can be regarded as the second spatial distribution pattern of fogs occurrence frequencies, spanning dekad 4 to dekad 12 (2a, 2b in Figure 6).

From dekad 13 to dekad 16, the spatial distribution patterns of fogs occurrence frequencies were analogous (As is shown in Figure 6 (3)). From Figure 6 (3) we can also see that the fogs occurrence frequencies in the 15th dekad were low across China, except for small areas. From the beginning the 13th dekad, fogs seldom occurred throughout China.

In the fourth phase, the fogs in China still did not occur frequently (see Figure 6 (4a)). In this phase the fog-prone areas mainly include Chengshantou and its vicinity as well as the northernmost border between Inner Mongolia and Heilongjiang province. The fogs in Chengshantou vanished in dekad 24 whereas in the northernmost border area between Inner Mongolia and Heilongjiang province fogs were featured by an occurrence-dissipation process from dekad 17 to dekad 27. From the beginning of dekad 24, the fogs started to increase in southeastern Jilin province, Liaoning province, eastern Loess Plateau, Sichuan Basin, and southeastern Yunnan-Guizhou Plateau. (see Figure 6 (4b) (4c)).

The fifth phase, including dekad 28, dekad 29, dekad 30 (see Figure 6 (5)), can be regarded as the prophase of the first phase. At this phase, the fogs occurrence frequencies in Sichuan Basin, North China Plain, the middle and lower reaches of the Yangtze River Plain, the east of Wuyi Mountains, and part of Liaoning province was more than 3%. However, the occurrence frequency of fogs in above mentioned regions was lower than that in fog-prone regions in the first phase. And the occurrence range of fogs during this phase was also less than that in the first phase.

In summary, a wide range fogs occurred mainly in November, December, and January (dekad31-dekad3) in China. The high fogs incidence areas are mainly Sichuan Basin, northern part of the North China Plain, the southeast of Wuyi Mountains, the east of the middle and lower reaches of the Yangtze River Plain, and the westeast of Yunnan-Guizhou Plateau. Between April and July (dekad10-dekad22), the high fogs incidence areas shifted to Chengshantou and its surrounding regions. The fogs in the northernmost border between Inner Mongolia and Heilongjiang province occurred between July and September.

4. Conclusions

In this study extensive analyses were conducted to identify spatial and temporal characteristics of fogs using the meteorological data. The primary results are summarized below.

(1) The regions where fogs occur most frequently in China were Sichuan Basin, the Yunnan-Guizhou Plateau, and the southeast of Wuyi Mountains. Thin fogs usually took place in the regions where fogs seldom occurred whereas dense fogs usually occurred in the fog-prone regions.

(2) The number of fog days in most regions of China has been decreasing in the past 31 years.

(3) Given that the fogs concentration degree (FCD) was associated inversely with fog days. the FCD in fog-prone regions is relatively low. This suggests that fogs in fog-prone regions occurred

dispersedly. On a national scale, the fogs concentration periods (FCP) mainly were January, November, December, namely the wintertime.

(4) Seven major fog-prone regions were identified, and its high incidence periods were also clear. This maybe help to fog prediction and regional research.

Acknowledgements

The work was supported by the public service industrial special research of National Environmental Protection Administration under Grant (201409001), the science and technology support program of the ministry of science and technology of China (2014BAC23B00), the Tianjin science and technology projects (14JCYBJC22200) and the Tianjin important science and technology specific project (14ZCDGSF00027).

References

[1] National Oceanic and Atmospheric Administration 1995 Surface weather observations and reports *Federal Meteorological Handbook* **1** 94

[2] Liu X, Zhang H Z, Li Q X, Zhu Y J 2005 Preliminary research on the climatic characteristics and change of fog in China *Journal of Applied Meteorological Science* **16(2)** 220-229

[3] Cao Z Q, Wu D, Wu X J 2008 Climatic Characteristics of Heavy Fog Weather in China from 1961 to 2005 *Meteorological Science and Technology* **36(5)** 556-560

[4] Lin J, Yang G M, Mao D Y 2008 Spatial and Temporal Characteristics of Fog in China and Associated Circulation Patterns *Climatic and Environmental Research* **13(2)** 171-181

[5] Chen X X, Guo P W, Luo Y 2008 Climate Characteristics of Fog Days of Various Categories in China *Advances in Climate Change Research* **4(2)** 106-110

[6] Wang L, Chen S, Dong A 2005 The distribution and seasonal variations of fog in China *Acta Geographica Sinica-Chineses Edition* **60(4)** 689

[7] Niu F, Li Z, Li C, Lee K H, Wang M 2010 Increase of wintertime fog in China Potential impacts of weakening of the Eastern Asian monsoon circulation and increasing aerosol loading. *Journal of Geophysical Research Atmospheres* (1984–2012) 115(D7)

[8] Zhang L, Qian Y 2003 Annual distribution features of precipitation in China and their interannual variations *Acta Meteorologica Sinica-English Edition* **17(2)** 146-163

[9] Liu H N, Luo Y 2009 Projection of future fog in China for 2001-2050 *Advances in Climate Change Research* **5(5)** 266-270

[10] Fu C B, Wu J 2011 The different characteristics of sunny visibility over southwest china in recent 50 years. *Procedia Environmental Sciendes* 10 247-254

Deng X, Tie X, Wu D, Zhou X, Bi X, Tan H, Jiang C 2008 Long-term trend of visibility and its characterizations in the Pearl River Delta (PRD) region, China *Atmospheric Environment* **42(7)** 1424-35

Analysing the temporal water quality dynamics of Lake Basaka, Central Rift Valley of Ethiopia

Megersa Olumana Dinka

Department of Civil Engineering Sciences, University of Johannesburg APK Campus 2006, South Africa.

E-mail: mdinka@uj.ac.za

Abstract. This study presents the general water quality status and temporal quality dynamics of Lake Basaka water in the past about 5 decades. Water samples were collected and analysed for important physico-chemical quality parameters following standard procedures. The result showed that Lake Basaka water is highly saline and alkaline and experiencing a general reducing trends in ionic concentrations of quality parameters due to the dilution effect. About 10-fold reduction of total ionic concentration occurred in the Lake over the period of 2 decades (1960-1980). There was a sharp and fast decline in EC, Cl, SO_4, Na, and K ions from early 1960s up to the late 1980s, and then became relatively stable. Some ions (eg. Na, Ca, Mg, Cl, SO_4) are showing increment in recent years. This characteristics of the lake water is terrible in relation to its potential to inundate the nearby areas in the near future. The expansion of such quality water has negative effects on the water resources of the region, especially soil quality, drainage and groundwater, in terms of salinity, sodicity and specific ion toxicity. The regimes of soil moisture, solute and groundwater could be affected, concurrently affecting the productivity and sustainability of the sugar estate. Thus, there is an urgent need to identify the potential sources of water and chemicals to the lake and devise an appropriate mitigation and/or remedial measures.

1. Introduction

Lake Basaka, located adjacent to Matahara sugar estate (MSE), Matahara town and Fantalle village (Ethiopia), has been expanding at a very fast and dramatic rate over the last about 5 decades [1-9]). Surface area of the lake was increased from 3 km^2 in 1960s to about 48.5 km^2 in 2010 [8]. Interestingly, the other Ethiopian Rift Valley Lakes are shrinking, except Lake Hawassa [1-2, 6-8]. The expansion of the lake is attracting the attention of the government and other stakeholders, mostly due to its quality. The lake water is saline (EC~6.3 dS/m), sodic (SAR ~300) and alkaline (pH~9.6) and not usable for irrigation and domestic purposes [6, 8]. The expansion of the lake with its poor water quality is expected to create various negative (socio-economic and environmental) impacts to the region, particularly the nearby sugarcane plantationr. The expansion of the lake was started after the establishment of Matahara Sugar Estate (MSE) in the 1960s. Before the establishment of the sugar plantation, the lake was like a small surface pond created during rainy season and used as grazing area [6-7].

A quite different number of studies were conducted on the quality of Lake Basaka for different purposes. The general lake limonology (morphology and chemistry) (including Awassa, Hora Hado, and Shala Lakes) are summarised by different studies [3, 6, 10-13]. This study attempted to present the general water quality status and temporal dynamics of Lake Basaka in the past about 5 decades.

2. Methodology

2.1. *Brief Description of the Study Area*

Lake Basaka (8°51.5' N, 39°51.5' E) is a terminal lake located in the Fentalle Woreda, East Showa Zone of Oromiya regional state at about 200 km southeast of the capital city of Ethiopia, Addis Ababa (Figure 1). It is volcanically dammed, endoheric lake located in the northern part of the Main

Ethiopian Rift (MER), at the near distance to the Afar triangle [3, 6]. According to the catchment delineation made by Dinka [6]. The total surface water catchment of the lake is about 500 km^2. The lake catchment has variable elevation; ranging from 950 m at the lake to over 1700m (m.a.s.l) at Fentalle Crater (Volcanic Mountain) [6-8]. As the area is situated in the upper most part of MER, central rift valley region of Ethiopia, Lake Basaka is vulnerable to the occurrences of different tectonic and volcanic activities. Topographically, the lake is surrounded by various plateus, which are the extensions of Haraghe Highlands. Geologically, the area is characterized with features of past and recent volcanic events. It is bordered by older volcanoes and a rift margin in the eastern side, by young Quaternary complexes of Fantalle crater in the northern part, and by Kone in the western sides [3, 8].

Matahara plain area has semi-arid climate, characterized by bimodal and erratic rainfall distribution. The major rainy season occurs from July to September and the minor/occasional rain occurring from March to April [3, 6-8]. The long-term average annual rainfall, temperature and evaporation of the area are 543.7 mm, 26.5 °C and 2485 mm, respectively [6, 8]. The Lake Basaka catchment have different land use/cover (LUC) (Figure 2) and soil units [6-7]. The soil units include Lepthosols, Cambisols, Podzoluvisols, Luvisols, Podzol, Fluvisols and Solnchaks. The LUC change analysis made by Dinka [7] indicated that Lake Basaka catchment had experienced a drastic change in its LUC conditions over the last 4–5 decades because of the rapid increase in human settlement, deforestation, and the establishment of irrigation schemes and Awash National Park. About 18 924 ha of forest and 4730 ha of grazing lands were devastated from 1973 to 2008; reducing the forest coverage from 42% in 1970s to only 6% in 2000s [7]. The detail description of Matahara Plain and Lake Basaka Catchment can be obtained from different research reports [1-9, 13-14].

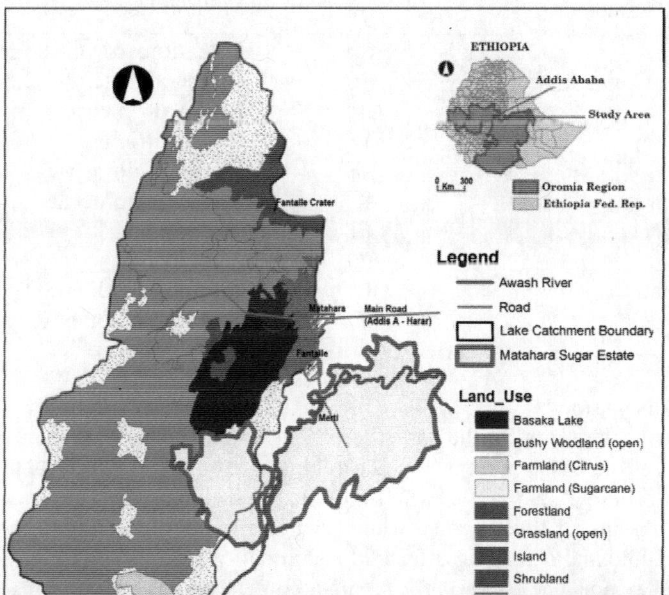

Figure 1. Map showing the location of the study area and the land use-cover for Lake Basaka Catchment and Matahara Sugar Estate (Source: Dinka [6])

2.2. Water sampling and analysis

Water samples were collected in 2015 from Lake Basaka using clean polyethylene bottles (0.5 L). The water pH, EC and temperature were measured actually at field using portable pH and EC meters (Figure 2). All water samples were collected on the same day (morning) and immediately taken to the Matahara Breeding Station Laboratory for important physico-chemical analysis: soluble cations and anions. The water sampling and analysis were done as per the APHA [15] standard test guidelines [13] (Table 1). The previous water quality condition of the lake was taken from previous studies [3, 6-8, 10-11,16, etc].

Figure 2. Water sampling and measurement of pH, EC and Temperature from Lake Basaka

Table 1. Methods adopted for water quality analysis (Source: Dinka et al. [13]).

Quality Parameter	Symbol	Method Used
pH	pH	Potentiometric (1:2.5 H_2O v/v)
Electrical Conductivity	EC	Conductometery (1:2.5 H_2O v/v)
Calcium	Ca^{2+}	EDTA titrimetric
Magnesium	Mg^{2+}	EDTA titrimetric
Sodium	Na^{+}	Flame photometric
Potassium	K^{+}	Flame photometric
Chloride	Cl^{-}	Titration
Carbonate	CO_3	Titration
Bicarbonate	HCO_3	Titration (with H_2SO_4)
Sulphate	SO_4	Spectro Photometric

3. Results and Discussion

The analytical analysis results for the considered physico-chemical parameters are presented in Table 2 (last two rows). Table 2 also presents the general historic limonology (morphology and chemistry) of Lake Basaka by different studies [3, 6-8, 10-13, 15] and this study. Based on its high salinity and alkalinity levels, the use of Lake Basaka water for irrigation and drinking purpose is impossible. It is easy to deem the potential damaging effects of such quality Lake water on soil structure, soil fertility and other soil behaviors. The excessively high Na content of Lake Basaka has a potential to destroy soil structure and aggregate stability, which can lead to reduced organic matter/carbon content and other problems. There is a tendency of Na salts to exceed above their solubility limit in lake water, which lead to precipitation of Ca salts as $CaCO_3$.

As shown in Table 2 and Figure 3 the salinity (EC) concentrations are exhibiting the reduction trends since 1960s. The reduction in EC is very dramatic until the end of 1970s. Since then, the reduction is gradual and very slow. Salinity level measured by Talling & Talling [19] was about 74.17 dS/m [6] and hence, decreased by about 12-fold (~6.15 dS/m in 2015) in the past about a half-century (1960-2015). This changing property of the lake water quality is of special interest as far as its drastic expansion is concerned. Unfortunately, its decrement in the past 30 years (since 1980s) is not significant in relation to its potential to inundate the nearby areas in the near future [6, 8].

Table 2. Physico-chemical parameters for Lake Basaka water in different periods (1961 -2015)

Year	pH	EC	Na^{+}	K^{+}	Ca^{+2}	Mg^{+2}	Cl^{-}	SO_4^{-2}	HCO_3^{-2}	CO_3^{-2}
[a] 1961(May)	9.94	74170	17800	406	<3	<7.5	5480	4680	580	-
[b]1971		79230								
[b]1973		25000								
[b]1974		25410								
[c]1978		5620	-	-	26	5.0	-	-	-	-
[d]1991 (Mar)	9.4	7440	1810	67	2.2	0.49	450	600	46	
[e]1993 (Feb)	9.55	-	1900	20	1.0	0.40	572	540	-	-
[b]1996	9.70	6385	882	67.6	4.2	3.7	863	-	515	622
[f]2000	9.71	-	1740	64	3.15	0.70	542	494	-	-
[g]2003 (Jan)	9.54	-	2160	80	2.18	0.38	421	620	-	-
[b]2002 (Jun)	9.6	7335	2449	38	18.2	6.1	1062	1504	2177	385
[b]2003 (Apr)	9.8	7804	1831	15	8.4	12	996	1080	1932	624
[g]2006	9.7	6730	1805	63.8	3.02	1.2	571	540	-	-
[b]2007 (May)	9.57	6280	2583	61.4	7.0	1.7	927	1020	1213	350
[b]2009 (May)	9.52	6170	2579	61.8	4.5	1.1	915	986	1140	340
[h]2010 (May)	9.52	6168	2582	62.1	4.5	1.5	922	987	1174	322
[h]2015 (May)	9.51	6150	2587	62.0	4.2	1.2	937	990	1180	320

All are in units of mg L^{-1}, except EC ($\mu S\ cm^{-1}$)
Sources: [a] Talling & Talling [16], [b]Dinka [8], [c]Halcrow [17], [d]Kebede et al. [18], [e] Gizaw [19], [f]Ayenew [11]; [g] Klemperer & Cash [12]; [h] This study
Note that the values provided for the years 2010 and 2015 are average values of measurements made at different sites of the lake. Sampling sites are from the North, West and South.

It is evident from Table 2 that Lake Basaka is experiencing a reducing trend in water quality due to the dilution effect. Interestingly, about 10-fold reduction of total ionic concentration occurred in the Lake Basaka over the period of 2 decades (1960-1980) (Table 2). The other RVLs, except Abiyata with increasing trend [18], have relatively stable quality (with no or little change) [1-2, 6]. Actually, the reduction of ionic concentrations with the increment in lake volumes was expected. The volumetric increment of Lake Basaka has resulted in reduction of ionic concentration and concomitant shift in phytoplankton community [18].

The ionic reduction in most of the ionic concentrations are not significant in recent years (post 2000) and remain almost stable. Even, some parameters (eg. Na, Ca, Mg, Cl, SO_4) are showing increment in recent years. With the incremental trends of lake volume [8-9], the ionic concentrations should have been reduced further [6]. The relatively stable or increasing ionic concentration of some parameters in the past 30 years may indicate there are chemical sources to Lake basaka by natural factors and/or anthropogenic effects. Natural factors could be weathering processes, soil erosion, sediment loading, deposition of animal and plant debris, and solution of minerals in the basin. Geologic condition of Lake Basaka catchment and its high evaporation can be another natural factor. Human activities can accelerate some of the natural factors. Tropical areas are characterized by high rainfall intensity, which can easily wash away debris, top soil and various forms of contaminants through surface runoff [20]. This indicates there are anthropogenic sources of chemicals in to Lake Basaka. As suggested by Dinka et al. [13], the local anthropogenic processes could be discharges from factory, domestic sewage and farming activities. Agricultural activities introduces ions and metals from agro-chemicals such as fertilizers, herbicides, fungicides, etc.

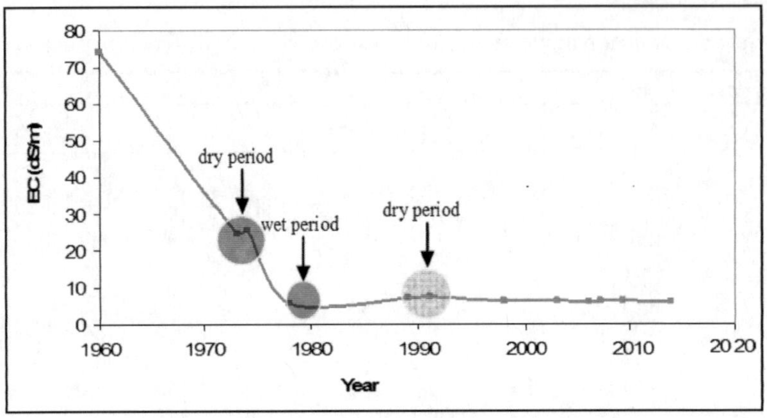

Figure 3. Temporal trend of the salinity (EC) of Lake Basak (data source: from others and this study).

The ionic ratios of ((Ca+Mg)/Na) obtained in this study is about 0.0022. Halcrow [17] reported an ionic ratio for Lake Basaka as 0.028 in 1971 (which was similar to that of hot springs) and 0.01 in 1978 [6, 21]. Hence, the ionic ratios are reduced by approximately 12-fold since 1970s. There was a sharp and fast decline in EC, Cl, SO_4, Na, and K ions from early 1960s up to the late 1980s, and then became relatively stable (Table 2). The sharp decline pre-late 1970s was probably due to the excessively high amount of drainage water discharged into the Lake from flood irrigation from Abadir farm in the period 1964-1978, which was then changed to furrow irrigation. The excess drainage discharge to the Lake was further accelerated by malfunctioning of the gate in Abadir-E side in the period between 1976 and 1978, which was then stopped based on the recommendation made by Halcrow [17].

4. Conclusion

This study result indicated that the use of Lake Basaka for irrigation and drinking purpose is hardly possible. The lake water is saline, alkaline and sodic. It is easy to deem the potential damaging effects of such quality lake water on water resources and soil quality of the region groundwater quality, soil structure, soil fertility and other soil behaviours. The regimes of soil moisture, soil structure and groundwater quality could be affected, concurrently affecting the productivity and sustainability of the nearby sugar estate. The excessively high Na content of Lake Basaka has a potential to destroy soil structure and aggregate stability, which can lead to reduced organic matter/carbon content and other problems. There is a tendency of Na salts to exceed above their solubility limit in lake water, which lead to precipitation of Ca salts as $CaCO_3$. Further, the predominance of Na, K, Cl and HCO_3 in Lake Basaka indicates the recharging water from limestone, sandstone and sand aquifers.

The study result further indicated that Lake Basaka has experienced a reducing trend in ionic concentration in the past about 5 decades due to dilution effect. About 10-fold reduction of total ionic concentration occurred in the period of 2 decades (1960-1980). However, some parameters (eg. Na, Ca, Mg, Cl, SO_4) are showing increment in recent years. The relatively stable or increasing ionic concentration of some parameters in the past 30 years may indicate the potential causes of pollution due to natural and anthropogenic sources. This indicates that there are natural and anthropogenic sources of chemicals in to Lake Basaka. Natural sources could be weathering of rocks, soil erosion, sediment loading, deposition of animal and plant debris, and solution of minerals in the basin, etc. The local anthropogenic processes could be discharges from factory, domestic sewage, and farming activities which introduce ions and metals from agro-chemicals (fertilizers, herbicides, fungicides, etc).

Finally, the author would like to suggest a further detailed study on the potential sources of water and chemicals to Lake Basaka. Identification of the potential sources of water and chemicals plays a great role for understanding the hydrochemical facis and evolution of the lake water, and hence significantly contribute for the sustainable lake water management program in order to limit the further expansion of the lake and its potential damaging effects on the region.

References

[1] Ayenew T 2004 Environmental implications of changes in the levels of lakes in the Ethiopian rift since 1970 *Regional Environ. Change* **4** pp 192–204.

[2] Alemayehu T, Ayenew T and Kebede S 2006 Hydrogeochemical and lake level changes in the Ethiopian Rift *J. Hydrology* **316** (1-4) pp 290-300.

[3] Belay E A 2009 Growing Lake with Growing Problems: Integrated Hydrological Investigation on Lake Basaka, Ethiopia. PhD Dissertation, online publication. Available form URL: http://hss.ulb.uni-bonn.de/diss_online. Accessed 15 January 2010.

[4] Goerner A, Jolie E and Gloaguen R 2009 Non-climatic growth of the saline Lake Beseka, Main Ethiopian Rift *J Arid Environ.* **73** pp 287–95.

[5] Dinka M O, Loiskandl W and Furst J 2009 The expansion of highly saline Basaka Lake and its effects on the sustainability of Matahara Sugar Estate. Proceedings of the 34th–WEDC International Conference Refereed Paper (296), pp 571–9, Addis Ababa, Ethiopia.

[6] Dinka M O 2010 Analyzing the extents of Basaka Lake Expansion and Soil and Water Quality Status of Matahara Irrigation Scheme, Awash Basin (Ethiopia) PhD Thesis submitted to BOKU University (Vienna) Austria.

[7] Dinka M O 2012a Analysing decadal land use/cover dynamics of the Lake Basaka catchment (Main Ethiopian Rift) using LANDSAT imagery and GIS *Lake Reserv. Res. Manag.* **17** pp 11-24.

[8] Dinka M O 2012b Analysing the extents (size and shape) of Lake Basaka Expansion (Main Ethiopian Rift Valley) using Remote Sensing and GIS *Lake Reserv. Res. Manag.* **17** pp 131-141.

[9] Dinka M O, Loiskandl W, Ndambuki J M 2014 Hydrochemical modelling for Lake Basaka: development and application of a conceptual water budget model *Environ. Monit. Assess.* **186** pp 5365–5379.

[10] Baxter R M 2002 Lake morphology and chemistry. In: Tudorancea, C. and Taylor, W.D. (Eds). Ethiopia Rift Valley Lakes. Backhuys, Leiden pp 45-60.

[11] Ayenew T 2007 Water management problems in the Ethiopian rift: Challenges for development. *Afr. J. Earth Sci.* **48** pp 222–36.

[12] Klemperer S L and Cash M D 2007 Temporal geochemical variation in Ethiopian Lakes Shala, Arenguade, Awasa, and Beseka: possible environmental impacts from underwater and borehole detonations *J. African Earth Sci.* **48** pp 174–98.

[13] Dinka M O, Loiskandl W, Ndambuki J M 2015 Hydrochemical characterization of various surface water and groundwater resources available in Matahara areas, Fantalle Woreda of Oromiya Region *Journal of Hydrology: Regional Studies* **3** pp 444-456.

[14] Dinka, M O 2016. Quality Composition and Irrigation Suitability of Various Surface Water and Groundwater Sources at Matahara Plain *Water Resources* **43 (4)** pp 677–689.

[15] APHA 2005 Standard Methods for the Examination of Water and Wastewater, APHA, AWWA, WPCF, 21st Centennial Edition, Washington DC.

[16] Talling J F and Talling I B 1965 The chemical composition of African lake waters *Int. revue. ges. Hydrobiol* **50** pp 421 – 463.

[17] Halcrow W 1978 The study of Beseka Lake Levels. Report for the Government of Ethiopia. Awash Valley Development Agency, Addis Ababa. p 83.

[18] Kebede E, Gebramariam Z and Ahlgren A 1994 The Ethiopian rift valley lakes. Chemical characteristics along a salinity-alkalinity series *Hydrobiologia* **288** pp 1–2.

[19] Gizaw B 1996 The origin of high carbonate and fluoride concentrations in waters of the MER Valley, East African Rift system *Journal of African Earth Science* **22 (4)** pp 391 – 402.

[20] Yillia P T 2009 Linking land use to stream pollutant dynamics and management implications. PhD Thesis submitted to Vienna University of Technology, Austria.

[21] MoWR (Minstry of Water Resources) 1998 Lake Basaka Study and Design Project. Inception Report, Federal Democratic Republic of Ethiopia, MoWR, Addis Ababa.

Climate response and spatial-temporal model on the inter-annual change of winter temperature-salinity in the East China Sea

Jin-kun Yang, Qing-sheng Miao[1], Yang Yang, Shan-shan Xu

National Marine Data and Information Service, Tianjin 300171, China

E-mail: mqs1986@sina.com

Abstract: Spatial distributions and time variation characteristics were analyzed using Rotated Empirical Orthogonal Function (REOF) and spectrum analysis methods using surface and bottom temperature and salinity data in February of 1976-2013 along 30°N section in the East China Sea. Result showed that temperature trends can be divided into western part and east part, salinity trend divided into western, middle and eastern part. The first mode of surface temperature presented a quasi-equilibrium trend and the range was higher in the near-shore than the offshores, first mode of bottom temperature presented a decreasing trend; surface salinity had a decreasing trend and the extent was higher in the near-shore than the offshores, the bottom salinity showed a decreasing trend in recent years. The temperature inter-annual variability related to El Niño closely; short-term shocks of salinity related to El Niño, and long-term changes had something to do with PDO.

1. Introduction

Located to the east of China, East China Sea (ECS) is a marginal sea of the western Pacific. Surrounded by the Chinese mainland, Taiwan Island, Korean Peninsula, Kyushu and Ryukyu Island, its average depth is 349m and the maximum depth is 2719m. With a vast continental shelf in the west and a trough in the east, ESC has both the deep and shallow sea characteristics. The west bank, which is consisted of China's Fujian, Zhejiang and Taiwan west coast, has numerous ports and bays, the largest of which is Hangzhou Bay. While in the east, from Kyushu to Okinawa and east Taiwan there are many straits, waterways communicating with the Pacific Ocean. In the past century, the Earth's climate is characterized by a period of significant changes of global warming, which means a warming of both the atmosphere and sea. Temperature rise resulted lack of oxygen which is essential for creatures such as fish, threatened the marine life in the currently and future [1]. Rise of temperature reduces the number of marine phytoplankton plants in the form of microorganisms, while the fall increases the number of marine plants [2]. The accurate study of the ocean thermohaline helps to improve our understanding of marine ecosystem change. The equatorial Pacific Ocean plays an important role and as an indicator in the global climate system. Kuroshio and its branches, which come from the North Equatorial Current northward branch, transported the warm equatorial waters to the ECS and exert a severe impact on the water temperature there [3,4]. Sea surface temperature anomaly in ECS has a close relationship with Nino3.4 Index. Therefore, variations of the East China Sea thermohaline is an important part in the regional response to global climate change. ECS marine environment factors had been extensively studied during the past decades. CODAS SST data from 1951 to 1996 was used to study the East China Sea continental margin inter-annual temperature and decadal variation, indicating that phase transformation occurs of the region SST on decadal scales and noting that the phase transformation related with western equatorial Pacific [5]. Thermohaline distribution of 32°N section was analyzed and found that this region exist an obvious thermocline in spring. In autumn the bottom cold water has a trend moving to the east. In the east part of the section, a high salt water intrusion presents at the bottom and the isohaline has a vertical distribution [6]. Interannual variation of SST of ECS was studied using AVHRR data and discussed its relationship with East Asian monsoon interannual variation [7]. The possible mechanism of seasonal distribution and change of SST in East China Sea was analyzed based on satellite remote sensing SST data [8-11].

The research on the climate response and spatial-temporal distribution on the inter-annual change of temperature-salinity in the last four decades will help to understand the local spatial-temporal structure of wither time ECS and further find out the impact of other biochemical factors.

2. Data and Methods

2.1. Data source

11 Monitoring stations of 30° N section located in the ECS continental slope, substantially equal-spaced (Figure 1). The westernmost section depth is approximately 60 meters and easternmost 400 meters. From 1976 to 2013, the State Oceanic Administration (SOA) monitored the temperature and salinity at the surface and bottom of this section in the winter (February). An Ocean Station-Dajishan, attached to SOA, located near the east end of the section, has collected hydrometeorology data from 1978.

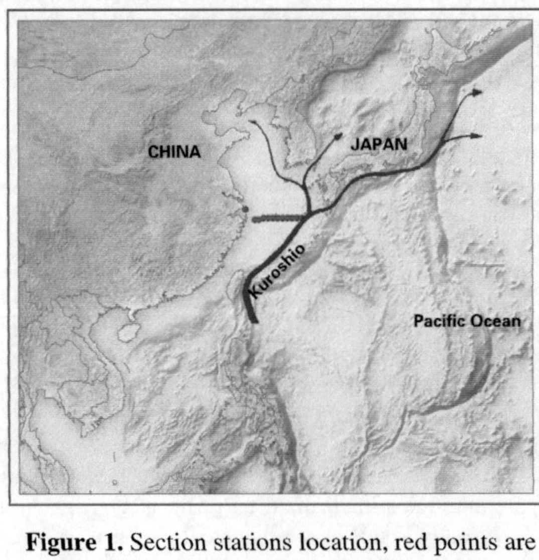

Figure 1. Section stations location, red points are section stations, green point is ocean station Dajishan

2.2. Methods

2.2.1. Rotated empirical orthogonal function. Rotated empirical orthogonal function (REOF) method is an improvement of empirical orthogonal function (EOF). EOF is a method to analyze data structure of the matrix and extract the main spatial and temporal characteristic [12, 13], which can reflect the overall spatial-temporal change of the region. When it comes to analyze the S-T structure of regional variables, REOF is a suitable method.

2.2.2. Jump analysis method Signal to noise ratio (SNR) formula was used to determine the intensity and time when time jump occurred within a time-series data[14], SNR formula is as follows:

$$J = \frac{|M_1 - M_2|}{\sigma_1 + \sigma_2}$$

Where M_1, M_2 and σ_1, σ_2 are mean and standard deviations before and after the reference time, respectively. When the local maximum ratio appears more than 1.0, the jump phenomenon occurs. In this study, the time scale n is set at 15 years.

2.2.3. Normal Distribution Test and Nonlinear Correlation Coefficient When the sample sequence follows a normal distribution, the sample mean could be used as the statistical characteristics of a sample sequence. In this paper, unless special stated, the calculated means of temperature and salinity all have passed normal distribution kurtosis and skewness coefficient criterion test.

Nonlinear correlation coefficient is defined as the ratio of explainable variation to total variation. It ranges from 0 to 1. The explainable variation is calculated using the second-order polynomial fitting. If the highest order of polynomial is reduced from 2 to 1, the nonlinear correlation coefficient degenerates into linear correlation coefficient.

3. Results and discussion

3.1. Annual trends of climatic factors and SST & salinity

The long-term trend is analyzed by averaging air temperature in February and mean value of zonal and meridional wind stress in January of Dajishan Ocean Station of ECS, using Five Gaussian Low-pass Filter method. And the average air temperature in February has a significant linear increase trend (Figure 2); average meridional pseudo-wind stress in January has a significant linear weakening trend (Figure 3), average zonal pseudo-wind stress in January has a non-significant linear weakening trend (Figure 4). The climatic trend of increasing winter temperature and weakening intensity of northwest monsoon will affect the interannual changes of temperature and salinity structure in the ECS.

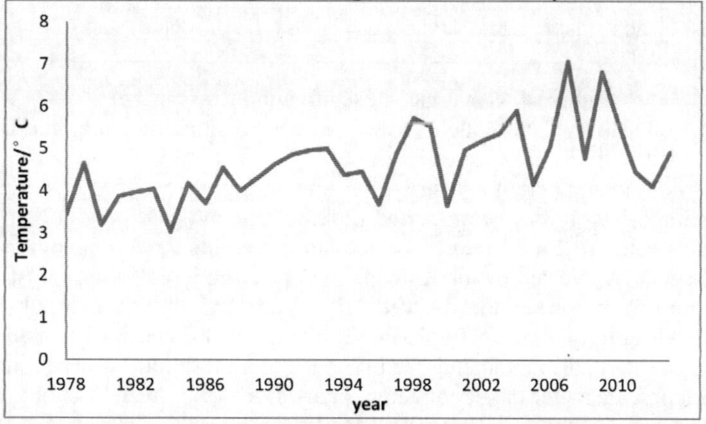

Figure 2. 5 Gaussian low-passes filtering of temperature in Feb of DJS

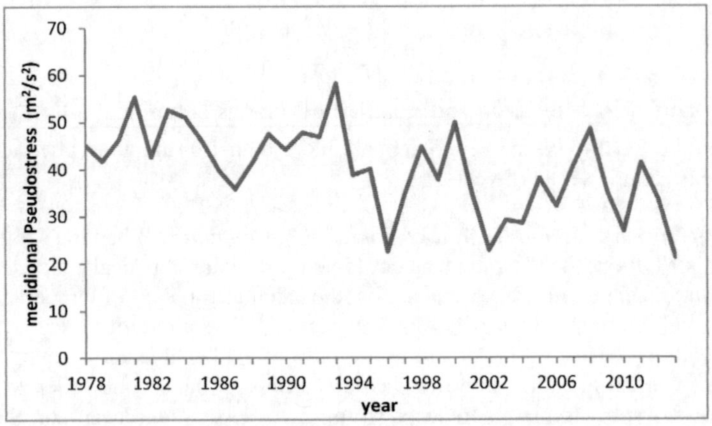

Figure 3. Meridional mean Pseudo wind stress in Jan

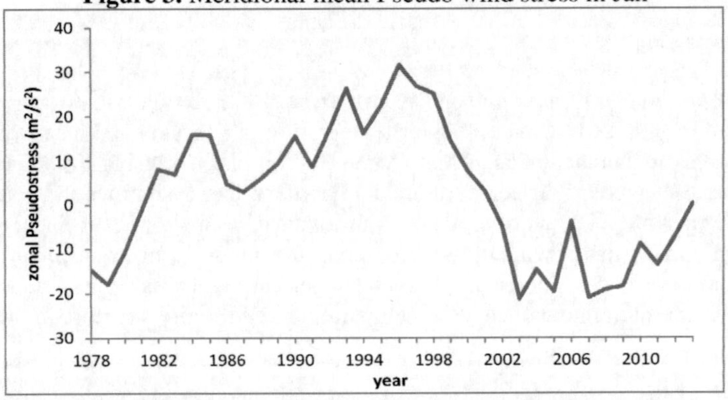

Figure 4. Zonal mean Pseudo wind stress in Jan

The temperature and salinity data sequences were smoothed along the temporal direction to reduce short-wave noise.

As surface and bottom temperature distribution shown(Figure 5), the temperature experienced changes of cold period (1976-1985), warm period (2005-1986) and cold period(2006-2013), in which 2005 was the warmest and 2012 the coldest,. Temperature presents a trend of low in the west section and high in east. Especially, affected by the Kuroshio, temperature in the easternmost is highest.

For salinity (Figure 5), it appears that between 1976 and 1985 there was an obvious low salinity period and in 2002 the salinity was highest. Low salinity water located in the middle of section and highest salinity in the eastern, also indicating the influence of high salinity water brought by Kuroshio.

The average variation character of section temperature from 1976 to 2013 is obtained by obtaining average surface temperature anomaly (figure 6). Meanwhile, to reduce the impact of extreme events on long-term trend, anomalies were averaged using 5-point sliding method. In the past 4 decades, the variation of winter SST anomalies is very intense, varied from -2 ~ 2.4° C. The minimum SST occurred in 1977 while the maximum in 1984. Prior to 1997, there were negative and positive anomalies both, but after 1997 there were basically all positive anomalies. The strongest negative anomalies occurred in 1977, 1982 and 1986. The most significant positive anomalies occurred in 1979, 1984, 1999 and 2009, respectively. These two periods correspond to the strong El Niño and La Nina years.

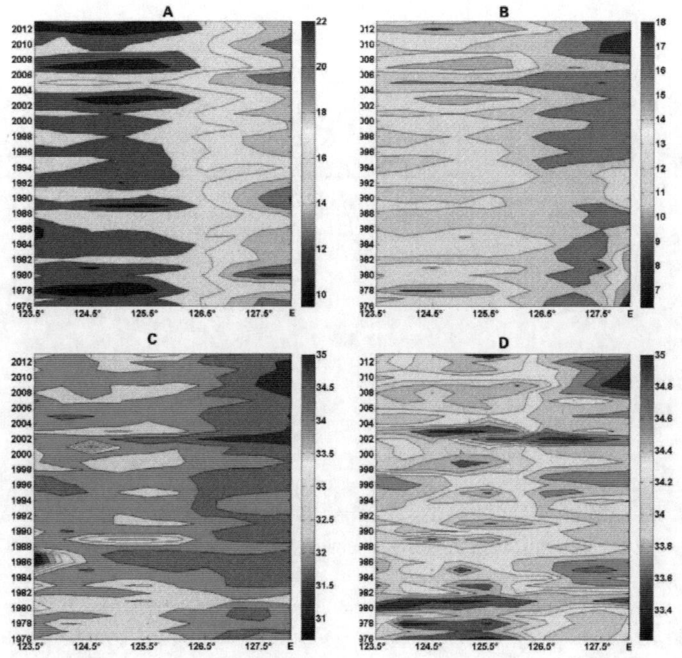

Figure 5. Temperature and salinity distributions along 30°N, 1976-2013. A is surface temperature, B is bottom temperature, C is surface salinity, D is bottom salinity

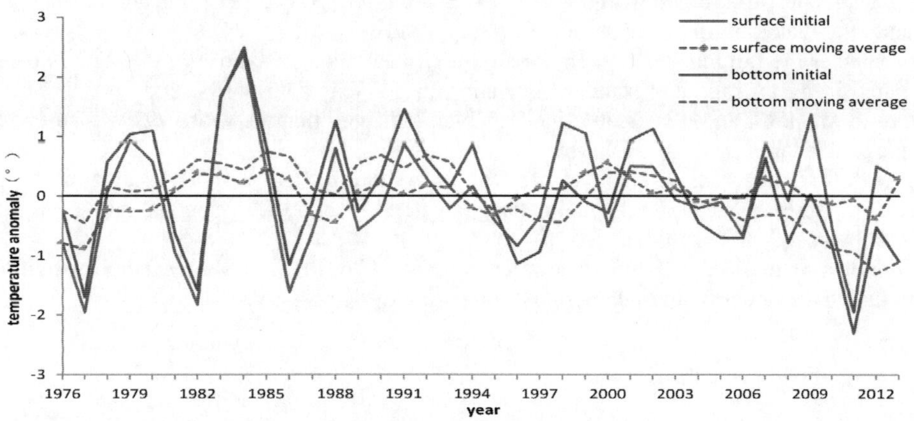

Figure 6. Inter-annual variation of surface and bottom temperature anomaly (blue and red solid line for surface and bottom temperature initial anomaly respectively, blue and red dotted line for surface and bottom temperature moving average anomaly respectively)

Similar averaging is implemented to salinity, (figure 7) resulting surface salinity anomaly varies from -0.58 to 0.53, with the maximum positive anomaly occurred in 1987 while the largest negative anomaly in 2008. 1998 is the dividing line between positive and negative anomalies, i.e. a positive anomaly before and a negative anomaly after.

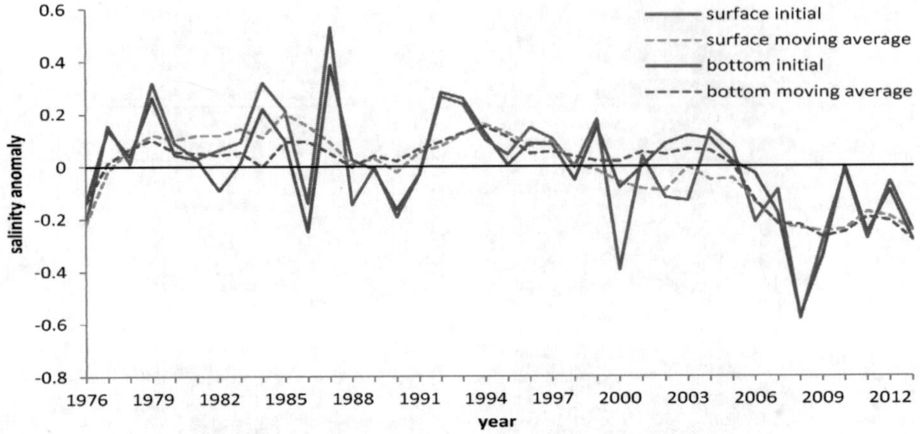

Figure 7. Inter-annual variation of surface and bottom salinity anomaly（blue and red solid line for surface and bottom salinity initial anomaly respectively, blue and red dotted line for surface and bottom salinity moving average anomaly respectively）

3.2. Temperature interannual temporal and spatial variation of surface and bottom layer

REOF was adopted to analyze temperature anomaly sequence data of surface and bottom layer(figure 8-9), the first 2 terms are taken out to do rotation transformation, and Gaussian low-pass filtering was used to eliminate short-term disturbances for time component.

The REOF first mode (REOF1) accounted for 58% of the proportion, which is the main form of temperature anomalies. Combined with the time series, we infer prior to 1997, the anomalies is mainly negative and after that is positive and show a warming trend.

To test whether the turning point reach the abrupt climate change standards, the signal-noise ratio was calculated using average and standard deviation of 15 year data before and after 1997. Result shows the ratio was 1.17, and it is greater than 1, which indicated that the year 1997 was surely abrupt climate change standards.

Magnitude of Temperature anomaly wanes from nearshore to open sea. By analyzing the power spectrum of time series, we can get the main change cycle: 4.8 years, 5.9 years and 3.6 years. And 4.8 year is most obvious.

REOF1 indicated that SST shows a warming trend in 1976-2013, and the offshore magnitude is larger than that in the opens, with main period of changing cycle of 4.8 years.

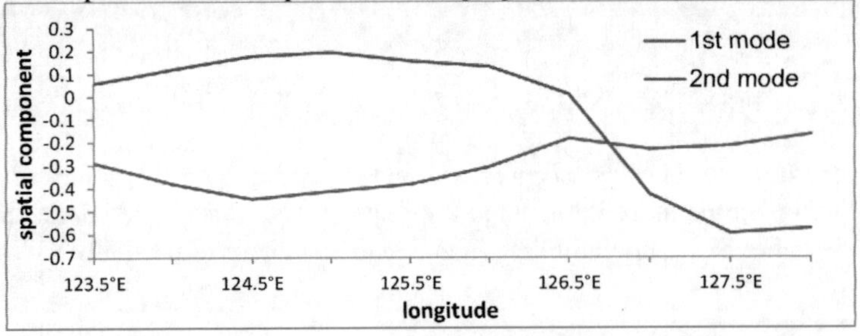

Figure 8. Spatial mode of surface temperature anomaly

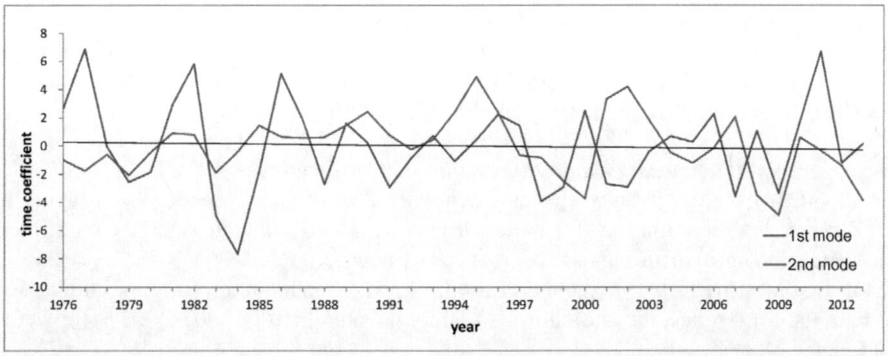

Figure 9. Time-varying coefficients of the first two modes of surface temperature

3.3. Climate response of SST

Delay-lag analysis was made between DJS ocean station air temperature in February and the first and second time mode of section SST (figure 10). A significantly positively correlated response is found, each mode has three peaks. Delay / advance at the peak may associate with the Kuroshio.

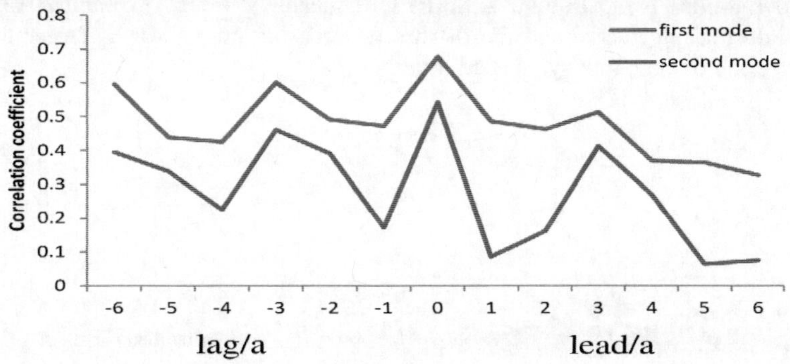

Figure 10. Correlation coefficient of air temperature in Feb of Dajishan and sea surface temperature modes

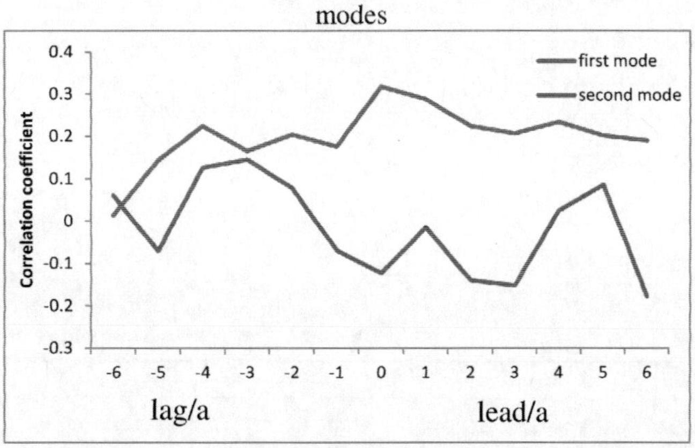

Figure 11. Correlation coefficient of meridional pseudo wind stress in Jan of Dajishan and sea surface temperature modes

Delay-lag analysis was made between DJS ocean station Pseudo wind stress in January and the first and second time mode of section SST (figure 11). Result shows a non-significantly positively correlated response with the first mode, but didn't relate with the second mode. That means the first mode of SST was affect by wind and the signal is 1 month delay.

3.4. Salinity interannual temporal and spatial variation of surface layer

REOF was adopted to analyze salinity anomaly sequence data of surface and bottom layer, the first 2 terms are taken out to do rotation transformation (figure 12-13). Then Gaussian low-pass filtering was used to eliminate short-term disturbances for time component.

The REOF first mode (REOF1) accounted for 45.3%, which is the main form of salinity anomalies. Combined with the time series, the anomalies is mainly negative, but in 2002-2003 there were two big positive and then shift to negative,.

Therefore, prior to 1997, the anomalies are mainly negative and after that turned to positive and showed a warming trend. That is to say that salinity has went through a decreasing trend.

Away from the shore, salinity anomaly showing negative-positive trend-0; The REOF second mode (REOF2) take up 27.1%. Positive salinity anomaly decreases when getting offshore. Time factor shows a roughly positive-to-negative trend.

By analyzing the power spectrum of time series, two main change cycles of 15.4 years and 34.1 years show. And 15.4 year is most obvious. So REOF1 means between 1976 and 2013 salinity of the surface shows a decreasing trends, and the offshore decreasing magnitude is larger than that in the opens, the main period of changing cycle is 15.4 years.

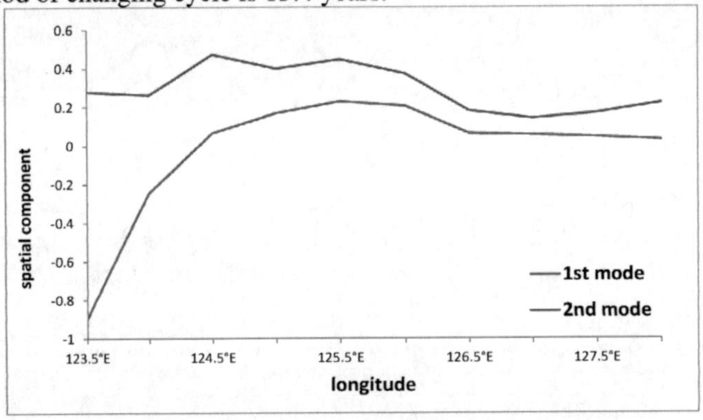

Figure 12. Spatial mode of surface salinity anomaly

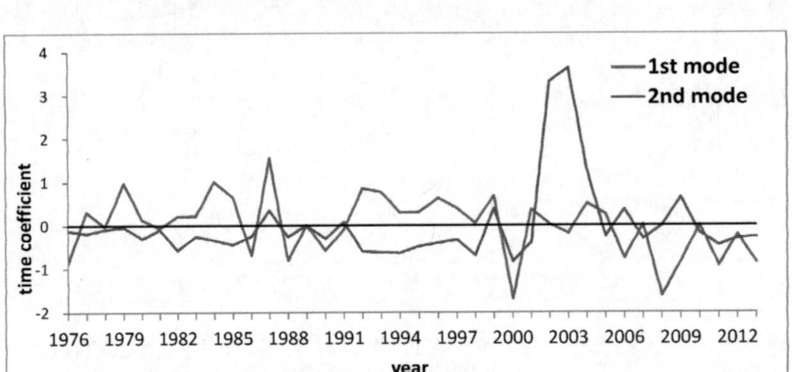

Figure 13. Time-varying coefficients of the first two modes of surface salinity

3.5. Climate response of sea surface salinity

Salinity, especially surface salinity, has a direct relationship with precipitation, and ENSO has an effect on precipitation, so ENSO will affect salinity. Generally speaking, rainfall amount in the North area of China will decrease, if we compare surface salinity with Nino 3.4 index after standardization (figure 14), a significantly positively correlated response is found, the correlation coefficient is 0.66, which means surface salinity increase in the El Niño year and is consistent with decrease of rainfall. However, the correlation is not good after 3 year filtering, which means short term shock of salinity may related with El Niño. Then when comparing salinity after 3 year filtering with PDO index(figure 15), we found that there is a significantly positively correlated response with correlation coefficient is 0.74, which means the outer-annual scale change has a relationship with signal from out seas.

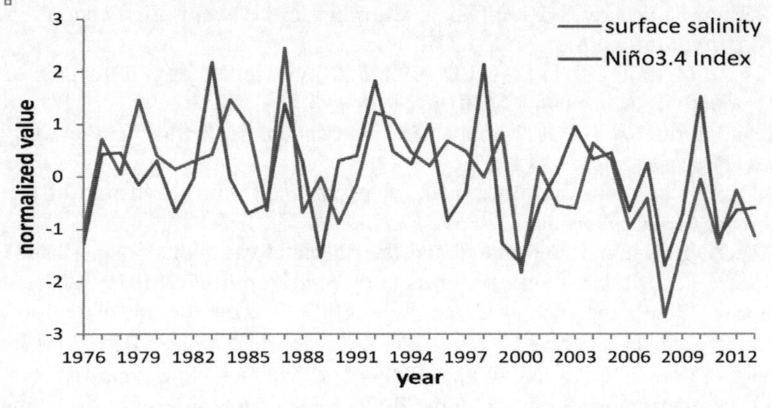

Figure 14. Inter-annual variation of surface average salinity and Niño3.4 Index

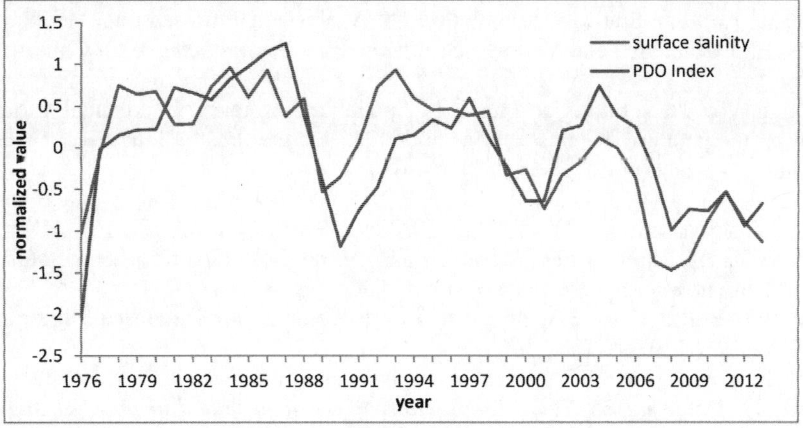

Figure 15. Inter-annual variation of surface average salinity and PDO Index

4. Conclusions

The temperature experienced changes of cold period (1976-1985), warm period (2005-1986) and cold period (2006-2013), in which 2005 was the warmest and 2012 the coldest, Temperature presents a trend of low in the west section and high in east. Especially, affected by the Kuroshio, temperature in the easternmost is highest.

For salinity, it appears that between 1976 and 1985 there was an obvious low salinity period and in 2002 the salinity was highest. Low salinity water located in the middle of the section and highest salinity in the eastern.

We infer prior to 1997, the anomalies are mainly negative and after that are positive and show a warming trend, and the offshore magnitude is larger than that in the opens. The main change cycle of the first mode is 4.8 years, 5.9 years and 3.6 years. SST has a significantly positive response with air temperature with ocean station nearby, also it was affect by wind and the signal is 1 month delay.

The salinity has gone through a decreasing trend. Away from the shore, salinity anomaly shows negative-positive trend-0. The main change cycle of the first mode is 15.4 years and 34.1 years. Short term change has a relationship with ENSO and long term change has a significantly positively correlated response with PDO index.

Reference

[1] Han sen J, Saoto M, Ruedy R 2000 Global warming in the twenty-first century: an alternative scenario *Proc Natl Acad Sci* **97(18)** 9875-80.

[2] Behrenfeld M J, O Malley R T, Siegel D A, et al. 2006 Climate driven trends in contemporary ocean productivity *Nature* **444(7120)** 752-755

[3] Mizuno K and White W B 1983 Annual and interannual variability in the Kuroshio Current system *J. Phys. Oceanogr.* **13** 1847-67.

[4] Kawabe M 1995 Variations of current path, velocity, and volume transport of the Kuroshio in relation with the large meander *J. Phys. Oceanogr.* **25** 3103-17.

[5] Park W S And Oh I S 2000 Inter-annual and inter-decadal variations of sea surface temperature in the East Asian Marginal Seas *Progress in Oceanography* **47(2/4)** 191-204.

[6] Zhang Hui, Shi Xiao-Yong, Zhang Chuan-Song 2009 Distribution and variation trends of the environmental factors in the 32° N section. *Acta Oceanol Sinica* **31(3)** 121-126.

[7] Zhang Song, Yu Fei, Diao Xin-Yuan 2009 The characteristic analysis on sea surface temperature inter-annual variation in the Bohai Sea, Yellow Sea and East China Sea *Marine Sciences* (in Chinese) **33(8)** 76-81.

[8] Zeng Guang-En, Lian Shu-Min et al. 2006 EOF Analysis of Intra-seasonal Variabilities of SST in the East China Sea and Yellow Sea *Advances in Marine Science* (in Chinese) **24(2)** 146-154.

[9] Li Jia-Xing, Wei Hao, Luo Xiao-Fan 2012 Spatial and temporal interannual variations of sea surface temperature in the continental shelf of the East China Sea *Oceanologia Et Limnologia Sinica*(in Chinese) **43(3)** 624-634.

[10] Wu Yu-Mei, Xu Zhao-Li, Fan Wei, Cui Xue-Sen 2009 Change of sea surface temperature in East China Sea during 1985-2005 *Acta Oceanol. Sin*(in Chinese) **33(8)** 76-81.

[11] Yu Fei, Xu Yi 2003 Study of long-term variational trend of sea surface temperature in the East China Sea. *Advances in Marine Science* (in Chinese) **21(4)** 477-481.

[12] North G R T, Bell R C 1982 Sample errors in the estimation of empirical orthogonal function *Mon Wea Rev* **110** 699-706.

[13] Keiner L E, Yan X H 1997 Empirical orthogonal function analysis of sea surface temperature patterns in Delaware Bay *IEEE Transaction on Geoscience and Remote Sensing* **35(5)** 1299-1305.

[14] Wei Feng-Ying 2007 Modern Weather Forecasting Techniques and Statistical Diagnosis. China Meteorological Press, Beijing, ISBN: 978-7-5029-4299-1.

Predicting hydrological response to forest changes by simple statistical models: the selection of the best indicator of forest changes with a hydrological perspective

D Ning[1], M Zhang[1, 2, 3], S Ren[1], Y Hou[1], L Yu[1] and Z Meng[1]

[1] School of Resources and Environment, University of Electronic Science and Technology of China, Chengdu, China 611731
[2] Research Institute of Forest Ecology, Environment and Protection, Chinese Academy of Forestry, Beijing, China 100091

E-mail: mingfangzhang@uestc.edu.cn

Abstract. Forest plays an important role in hydrological cycle, and forest changes will inevitably affect runoff across multiple spatial scales. The selection of a suitable indicator for forest changes is essential for predicting forest-related hydrological response. This study used the Meijiang River, one of the headwaters of the Poyang Lake as an example to identify the best indicator of forest changes for predicting forest change-induced hydrological responses. Correlation analysis was conducted first to detect the relationships between monthly runoff and its predictive variables including antecedent monthly precipitation and indicators for forest changes (forest coverage, vegetation indices including EVI, NDVI, and NDWI), and by use of the identified predictive variables that were most correlated with monthly runoff, multiple linear regression models were then developed. The model with best performance identified in this study included two independent variables -antecedent monthly precipitation and NDWI. It indicates that NDWI is the best indicator of forest change in hydrological prediction while forest coverage, the most commonly used indicator of forest change is insignificantly related to monthly runoff. This highlights the use of vegetation index such as NDWI to indicate forest changes in hydrological studies. This study will provide us with an efficient way to quantify the hydrological impact of large-scale forest changes in the Meijiang River watershed, which is crucial for downstream water resource management and ecological protection in the Poyang Lake basin.

1. Introduction

The relationship between forest change and streamflow is an important research subject for a century [1]. To predict hydrological response to forest changes, the first thing is to quantify forest changes. Forest coverage is often used as an indicator to express forest changes simply because it is easy to be obtained [2-8]. However, forest coverage only serves as a basic indicator without differentiating forest species and forest change types, and fails to express the spatial pattern of forest changes and subsequent forest recovery processes. Equivalent roaded area (ERA) and equivalent clear-cut area (ECA) are believed to be better indicators than forest coverage because they can account for dynamic vegetation conditions or changes following forest disturbances. ERA was originally developed in the early 1980s by Region 5 of the USDA Forest Service to evaluate channel destabilization [9]. It works for assessing sediment and erosion yield and is not spatially explicit and the impacts of an activity cannot be tested against its location in a watershed [10, 11]. A similar index developed by the USDA Forest Service is ECA, which is often used to assess the cumulative effects of forest harvesting on annual runoff. The ECA concept has also been widely used in Canada, particularly in British Columbia (BC) and Alberta. Roads, clear-cuts, burned areas, and partial cuts can all be expressed as "equivalent clear-cut area." There are various revised versions of ECA calculation procedures, but the core concepts are similar [12-15]. In a revised version developed by the BC Ministry of Forests, ECA is defined as the area that has been clear-cut, with a reduction factor to account for the hydrological recovery due to forest regeneration [14]. Although it was originally designed for clear-cut areas, ECA can be applied to wildfire-killed areas, roads, and other open spaces. Research has established the relationships between vegetation growth (ages or tree heights) and hydrological recovery rates following logging so that ECA can be derived spatially and temporally in a watershed [16-18]. Although the ECA is believed to be the best indicator for assessing forest change effects on hydrology in large forest-dominated watersheds, its application is limited mainly due to the fact that the ECA calculation for a watershed is time-consuming, and requires detailed historical data of over millions of harvested, burned, and infested blocks. Moreover, professional judgments are always needed in determining the hydrological recovery rates of different tree species for each disturbance type in different watersheds. In China, there is a lack of continuous forest coverage data since forest resources inventory is conducted every five years. In most watersheds, detailed historical records on forests including species, disturbance type, and disturbed area are deficient. This calls for a need to find a new indicator of forest change for predicting forest change-induced hydrological response in China.

Remote sensing sensor systems detect reflected or emitted radiation from features on the Earth's surface. New techniques have been developed for future extraction where Enhanced Vegetation Index (EVI) [19], Normalized Difference Vegetation Index (NDVI) [20, 21] and Normalized Difference Water Index (NDWI) [22] are most widely used for satellite image processing in recent decades. The emergence of these vegetation indices (NDVI, EVI, NDWI) derived from remote sensing data makes the assessment of historical forest changes feasible at a larger spatial scale. The remote sensing vegetation indices-based approach requires fewer fieldworks to collect detailed vegetation information from stand-level to watershed-level as compared to their counterparts. Moreover, it is an integrated index that can express all types of forest changes including both forest loss (deforestation due to logging, pest infestation, fire, urbanization, landslides) and forest gain (afforestation and reforestation). However, a question arising here then is which vegetation index has best performance when assessing hydrological response to forest changes in a large forested watershed. In order to answer this question, this study used the Meijiang River, one of the headwaters of the Poyang Lake as an example to identify the best indicator of forest changes to predict forest change-induced hydrological responses. Non-parametric correlation analysis was conducted first to detect the relationships between monthly runoff and its predictive variables including antecedent monthly precipitation and indicators for forest changes (forest coverage, vegetation indices including EVI, NDVI, and NDWI), and by use of the identified predictive variables that were most correlated with monthly runoff, multiple linear regression models were then developed. The indicator of forest change included in the model with best performance can be viewed as the best indicator to predict forest change-induced hydrological impact.

2. Study Area

The Meijiang watershed, situated in the upper of the Ganjiang River, the Poyang Lake basin of China (Figure 1). The majority of its tributaries start from Ningdu and Shicheng. The Meijiang River originates from the northeast of Ganzhou and flows southwest, and then merges into the Gongjiang River near Yudu. The Gongjiang River and Zhangjiang River converge in the territory of Ganzhou Region and make up the Ganjiang River. Hence the Meijiang River is called the source of the Ganjiang River. The drainage network of the Meijiang River watershed is very dense and the drainage area is approximately 6310 km^2.

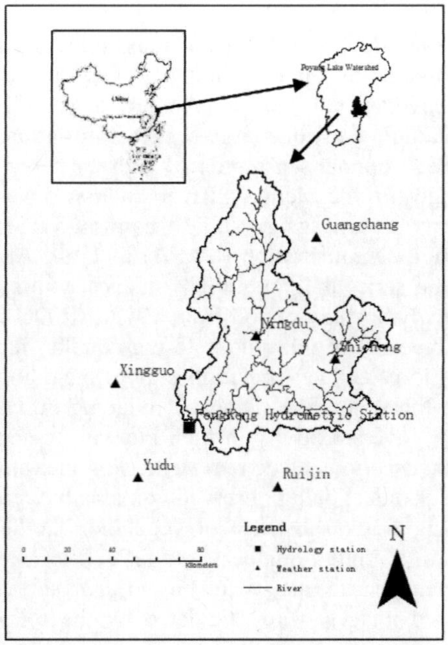

Figure 1. Location of the Meijiang watershed.

The average elevation for the Meijiang watershed is 379 m and most of it is from 300 to 500 m. Soil types mainly include mountainous red soil, purple soil and yellow soil, with light loam and sandy loam in texture. The watershed is rich in forest resources with diverse tree species. Forest coverage is about 71% in 2006. Main vegetation types in this watershed are subtropical evergreen broad-leaved and coniferous forests. The dominant tree species include Pinus massoniana Lamb., Castanopsis fabri, Castanopsis fissa, Cunninghamia Lanceolata and moso bamboo. Most of the forests distribute on the north, southeast and southwest of the Meijiang River watershed.

The climate in the Meijiang watershed belongs to humid subtropical monsoon featured with hot summer and long rainy season. Average annual mean temperature is 18.7 ℃. The maximum temperature is observed in July (monthly average 28.5 ℃) with the minimum temperature in January (monthly average 7.5 ℃). The rainy season lasts from April to June and the dry season is from September to November. Average annual precipitation is 1804.1 mm from 1989 to 2006. From the 1960s to the early 1980s, the watershed underwent many disturbances, especially, soil erosion due to early deforestation. After the large-scale reforestation and protection programs in the mid-1980s, the forest coverage increased from 41% in 1989 to 71% in 2006 in the watershed [23].

3. Data description

3.1. Hydrological and climate data

Daily flow data from 1989 to 2006 were collected from Fenkeng hydrometric station (Figure 1), situated at the outlet of the Meijiang watershed, which were used to calculate monthly runoff. There are six active national weather stations available within and around the Meijiang watershed (Figure 1). In this study, the historical climate data including precipitation and temperature was used. Given the high spatial variability of precipitation in the mountainous region, precipitation data from one station fails to capture the detailed spatial variation in precipitation input for the whole watershed. Thus, we use the ANUSPLIN model to generate spatially interpolated precipitation data. This approach is widely used to generate spatial interpolation of hydrometeorological variables [24].

3.2. Vegetation data

There are two types of vegetation data: forest coverage and remote sensing-based vegetation indices. Forest coverage data were provided by the Ganzhou Regional Forestry Bureaus of Jiangxi Province [23]. According to the forest coverage data, forest coverage rapidly increased from 41% in 1989 to 65% in 1994 due to the implementation of various large-scale reforestation programs, and steadily rose to 71.0% in 2006 in this watershed. Remote sensing data Landsat Thematic Mapper (TM) and Enhanced Thematic Mapper (ETM) images for the Meijiang River watershed were downloaded from the United States Geological Survey (http://glovis.usgs.gov/). The Landsat-5 TM satellite was launched by The National Aeronautics and Space Administration (NASA) in 1984. Although more advanced satellite images are available, only time series of Landsat TM images with cloud cover less than 10% were selected in order to cover the whole study period from 1989 to 2006. Several ETM images were also used when TM images were missing. In this study, 48 high quality images with a resolution of 30 m were processed and extracted to calculate vegetation indices including EVI, NDVI, and NDWI. The EVI is developed to enhance the vegetation signal by reducing influences form the atmosphere and canopy background and to improve sensitivity in high biomass regions [19]. The NDVI is a simple numerical indicator that can be used to analyze remote sensing measurements from a remote platform and assess whether the target or object being observed contains live green vegetation or not [25]. The NDWI is utilized for assessing water substance of vegetation canopy [22]. The processing of the images involved the conversion of digital numbers into reflectance values and a relative atmospheric correction to normalize remotely sensed images for further analysis. Then, the values of EVI, NDVI, and NDWI from the processed images were calculated by the following equations for the whole watershed (Figure 2).

$$EVI = 2.5\times(\rho_{nir} - \rho_{red}) / (1 + \rho_{nir} + 6\rho_{red} - 7.5\rho_{blue}) \tag{1}$$

$$NDVI = (\rho_{nir} - \rho_{red}) / (\rho_{nir} + \rho_{red}) \tag{2}$$

$$NDWI = (\rho_{nir} - \rho_{swir}) / (\rho_{nir} + \rho_{swir}) \tag{3}$$

where ρ_{nir}, ρ_{red}, ρ_{blue} and ρ_{swir} are the reflectance of the near-infrared, the reflectance of the red, the reflectance of the blue, and the reflectance of the short wave infrared, respectively.

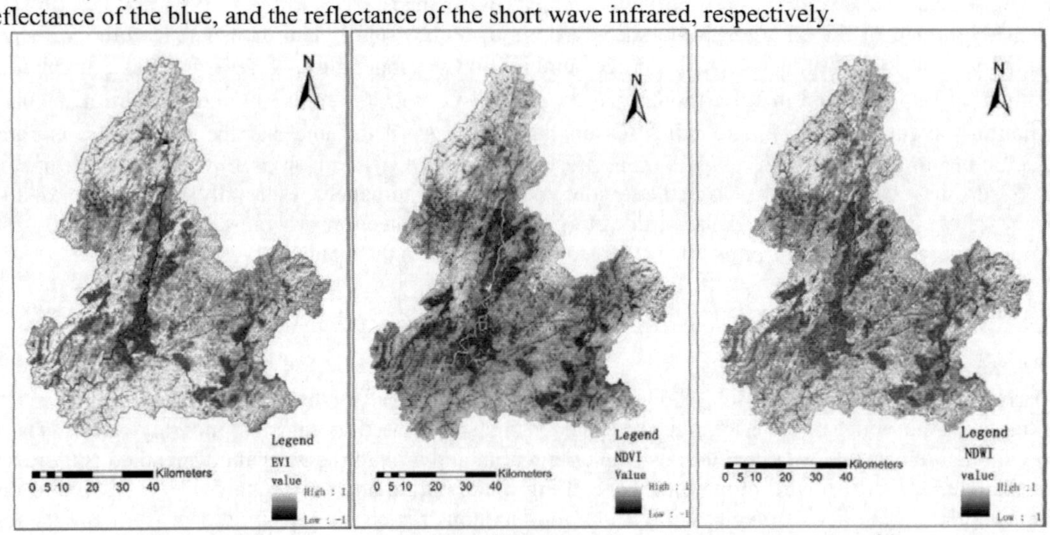

(a) (b) (c)

Figure 2. (a, b, c) The values of EVI, NDVI and NDWI distributed in the Meijiang watershed on December 21, 2006.

4. Method

In this study, statistical analysis including non-parametric tests (Mann-Kendall and Spearman rank tests) and stepwise liner regression were utilized to identify the best indicator of forest change in hydrological prediction.

4.1. Correlation analysis

The correlation analysis is found to be an effective approach to investigate the relationships among environmental variables, particularly in identifying the causal relationships from possible explanatory variables [26, 27]. In this study, non-parametric correlation tests including Mann-Kendall and Spearman rank tests were firstly performed to detect the relationships between monthly hydrological variables (monthly runoff (Q(t)), one-month antecedent monthly runoff (Q(t-1)), one-month and two-month subsequent monthly runoff (Q(t+1) and Q(t+2))) and their predictive variables including precipitation, forest coverage and vegetation indices (Table 1). In the prediction of monthly runoff, climatic variables such as antecedent monthly precipitation (P(t-1)) is believed to a powerful predictor. Thus, antecedent monthly precipitation was also adopted in this analysis. Vegetation indicators used included antecedent monthly forest coverage (FC(t-1)), monthly EVI(t), NDVI(t), and NDWI(t). The predictive variables that were most correlated with hydrological variables were further adopted in the development of multiple linear regression model for the prediction of hydrological response to forest changes.

4.2. Stepwise linear model

Multiple linear regression is a popular alternative technique for hydrological modelling with limited data requirements on watershed conditions. In this study, the choice of predictive variables is carried out by the stepwise selection that automatically keeps the significant independent variables in the multiple linear regression model. The generated multiple linear regression models were then used to predict monthly runoff response to forest changes. The model performance was evaluated by the statistics coefficient of determination (R^2) and Nash-Sutcliffe model efficiency coefficient (E), which are widely used in hydrological modelling and are proved to be excellent in measuring the effectiveness of a model in terms of its prediction ability [28, 29]. Value of R^2 and E can be calculated by Equation 4 and 5. Traditionally, hydrological models were calibrated and validated by split-sample tests where available data were simply divided into two sets by time, which may failed to capture dynamic interactions between runoff and its drivers. To overcome the shortcomings of the traditional split-sample test, a random selection approach was adopted where the data were randomly selected and then grouped into calibrated and validated periods. In this study, 36 sets of data were used in the calibration while 12 sets of data were applied in the validation.

$$R^2 = [\sum_{i=1}^{n}(Q_{o,i}-\bar{Q}_o)(Q_{s,i}-\bar{Q}_s)]^2 / [\sum_{i=1}^{n}(Q_{o,i}-\bar{Q}_o)^2 \sum_{i=1}^{n}(Q_{s,i}-\bar{Q}_s)^2] \tag{4}$$

$$E = 1-\sum_{i=1}^{n}(Q_{o,i}-Q_{s,i})^2 / \sum_{i=1}^{n}(Q_{o,i}-\bar{Q}_o)^2 \tag{5}$$

where $Q_{s,i}$is the simulated monthly runoff for the ith month; $Q_{o,i}$ the observed monthly runoff for the ith month;$\bar{Q}_s$ is the average simulated monthly runoff for the whole period;$\bar{Q}_o$the average observed monthly runoff for the whole study period and n the number of months for the whole study period.

5. Results

5.1. Correlation analysis

As suggested by both Mann-kendall and Spearman rank tests, significant correlations between antecedent monthly runoff (Q(t-1)) and antecedent monthly precipitation (P(t-1))/NDVI(t) / NDWI(t) were detected at α =0.01(Table 1). Thus, these variables were kept as input candidates in stepwise linear models with antecedent monthly runoff (Q(t-1)) as a dependent variable. However, in order to compare the performance of different vegetation indicators, we also built the model including FC (t-1) and EVI(t).

Table 1. Results of non-parametric correlation tests.

Parameter	Test	Q (t-1)	Q(t)	Q(t+1)	Q(t+2)
P(t-1)	Kendall's correlation coefficient	**0.627***	**0.439***	**0.444***	0.259
	Spearman correlation coefficient	**0.807***	**0.611***	**0.642***	0.377
FC(t-1)	Kendall's correlation coefficient	0.021	0.179	0.229	0.174
	Spearman correlation coefficient	0.044	0.244	0.327	0.290
EVI	Kendall's correlation coefficient	0.166	0.173	-0.030	-0.182
	Spearman correlation coefficient	0.229	0.245	-0.050	-0.283
NDVI	Kendall's correlation coefficient	**0.368***	0.173	0.034	-0.272
	Spearman correlation coefficient	**0.507***	0.289	0.045	-0.410
NDWI	Kendall's correlation coefficient	**0.375***	0.281	0.081	-0.252
	Spearman correlation coefficient	**0.538***	0.428	0.122	-0.380

*Significant at α =0.01

We took the time of remote sensing images (t) for benchmarks

5.2. Model selection

Table 2 provides the descriptions of five multiple linear regression models developed using selected input variables. Table 3 shows the performance of fitted models in both calibration and validation periods. According to the values of R^2 and E in both calibration and validation periods, model RM3 with P(t-1) and NDWI(t) as predictors showed best performance as compared to RM1, RM2, RM4 and RM5. Thus, the RM3 model was eventually used to predict the monthly runoff change due to forest changes (Figure 3) and NDWI was identified as the best indicator of forest changes in hydrological prediction in the Meijiang watershed.

6. Discussion

6.1. Relationship between vegetation change and monthly runoff

As suggested by the correlation analysis, monthly runoff can significantly grow with increasing vegetation indices due to continuous reforestation or afforestation. It indicates that more forests will lead to more monthly runoff in the study watershed. This could be related to the fact that planted tree species was mainly Pinus massoniana Lamb after native forests were logged. Unlike previous native forests, most planted forests in the Meijiang watershed are featured with low canopy closure and very poor understory vegetation. After plantation, the watershed may thus have lower transpiration by newly planted trees and understories than before. Therefore, more runoff is expected after large-scale tree plantation. This is in accordance with some studies in Russia and China, as well as in some cloud forests [1]. However, this finding is different from many small watershed studies where forest increase has reduced monthly runoff due to the fact that forest growth can increase interception and evapotranspiration, resulting in less water being available for runoff generation [30]. Thus, a definite conclusion the relationship between forest and water remains inconsistent.

Table 2. The list of model candidates.

Models	Dependent	Independent	Model Structure
RM1	Q(t-1)	P(t-1)	Q(t-1)=15.98+0.43×P(t-1)
RM2	Q(t-1)	P(t-1),NDVI	Q(t-1)=-45.55+0.41×P(t-1)+107.41×NDVI
RM3	**Q(t-1)**	**P(t-1),NDWI**	**Q(t-1)=-30.23+0.40×P(t-1)+240.17×NDWI**
RM4	Q(t-1)	P(t-1), FC	Q(t-1)=69.84+0.43×P(t-1)+77.45×FC
RM5	Q(t-1)	P(t-1), EVI	Q(t-1)=0.85+0.41×P(t-1)+60.10×EVI

Table 3. The performance of five models for both the calibration and validation periods.

Period	Evaluation statistics	RM1	RM2	**RM3**	RM4	RM5
Calibration period (training)	R^2	0.74	0.78	**0.84**	0.75	0.76
	E	0.74	0.78	**0.83**	0.75	0.76
Validation period (testing)	R^2	0.82	0.85	**0.87**	0.81	0.74
	E	0.73	0.79	**0.77**	0.73	0.67

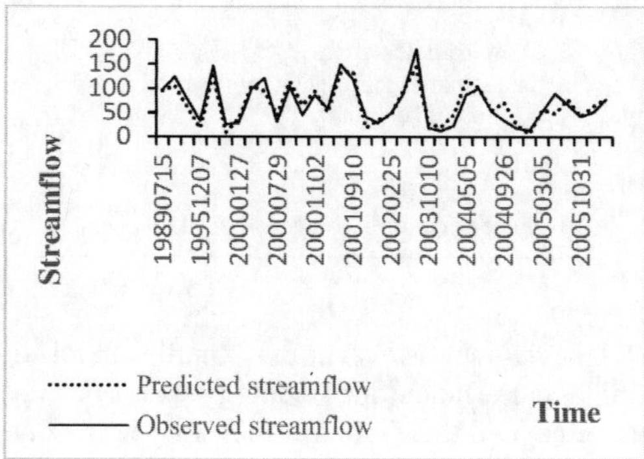

(a)

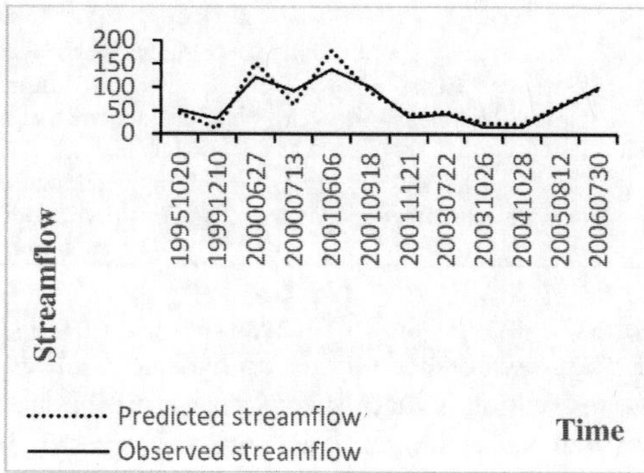

(b)

Figure 3. Predicted and observed streamflow in the calibration period (a) and the validation period (b)

6.2. *Quantification of forest changes*

Forest change is often expressed by forest coverage simply because it is easy to be obtained [7, 8]. However, forest coverage only serves as a basic indicator without differentiating forest species and forest disturbance types, and fails to express the spatial pattern of forest changes and subsequent forest recovery processes. Equivalent roaded area (ERA) and equivalent clear-cut area (ECA) are believed to be better indicators than forest coverage because they can account for dynamic vegetation conditions or change following disturbance, but their application is limited mainly due to the fact that the ERA or ECA calculation for a watershed is time-consuming, and requires detailed historical data of over millions of harvested, burned, and infested blocks. In China, there is a lack of continuous forest coverage data since forest resources inventory is conducted every five years. In most watersheds, detailed historical records on forests including species, disturbance type, and disturbed area are deficient or with poor control of data quality, especially in the remote mountainous region. This study suggests that vegetation indices such as NDVI and NDWI derived from remote sensing data can be effective indicators of forest changes when assessing hydrological response to forest changes. The NDWI was found to be the best indicator of forest changes in the Meijiang watershed as indicated by model performance of five multiple linear regression models. This is may be due to the great sensitivity of NDWI to water substance of vegetation canopy. The quantification of forest changes by NDWI requires fewer fieldworks to collect detailed vegetation information from stand-level to watershed-level as compared to other indicators of forest changes. Moreover, it is an integrated index that can express all types of forest changes including both forest loss (deforestation due to logging, pest infestation, fire, urbanization, landslides) and forest gain (afforestation and reforestation) over time (since 1980s). Therefore, NDWI can be widely applied in forested watersheds in China to quantify forest changes as well as to be used for the prediction of forest change-induced runoff changes. A brief comparison of pros and cons between these indicators is presented in Table 4.

Table 4. Methods for quantifying forest disturbances

Name	Advantage	Disadvantage
Forest coverage	Simple calculation	Only available for single disturbance No consideration of hydrological recovery
ERA (Equivalent roaded area/acre)	Accounts for various types of disturbance: assesses erosion risk and sediment yield	Complex calculation No consideration of hydrological recovery Lacks spatial representation(such as position of harvest)
ECA (Equivalent clear-cut area)	Accounts for various types of disturbance, Considers disturbance severity and hydrological recovery	Complex calculation Lacks spatial representation (such as position of harvest) [31]
NDWI (Normalized Difference Water Index)	Simple calculation, objectivity, Considers water substance of vegetation canopy	Poor accuracy No consideration of hydrological recovery

6.3. A new approach for forest-runoff modelling
The large watershed studies on hydrological response to forest changes are fewer than small watershed studies, which is mainly constrained by the lack of a suitable and efficient methodology, the lack of a comprehensive indicator for forest changes over space and time, and the availability of long-term data on hydrology, climate, and forest. In this study, the multiple linear regression model where the antecedent monthly runoff served as a dependent variable, and the antecedent monthly precipitation and NDWI served as independent variables was found as the best fitted model to predict monthly runoff response to forest changes. In comparison to expensive experimental watershed approach and time-consuming and data-intensive hydrological modelling, this statistical model appears to be a very efficient tool for assessing hydrological response to forest changes. It can be applied to predict historical or future runoff response due to forest changes in both small and large watersheds where remote sensing, hydrological and climate data are available. It can also be used to predict runoff with precipitation and NDWI data in ungauged forested watersheds. This can be particularly true given that in the remote area of Southwest China, for example, covered by dense native forests with only short-term or without hydrological data, we can use this simple approach to estimate runoff by use of radar-precipitation and remote sensing data. This prediction can also be very useful in water resource management and forest protection. For example, the West Route of South–to-North Water Transfer Project involves drawing water from the Upper Yangtze River in the Southwest China and supplying water to the Yellow River. The large-scale project will cause negative effects on forests in water source region. Thus, we can use this model to predict the impact of the project-associated forest changes on runoff of watersheds in Southwest China in an efficient way.

7. Conclusions
This study used the Meijiang watershed as an example to identify the best indicator of forest changes to predict forest change-induced hydrological responses and to develop a statistical model to provide an efficient assessment on hydrological response to forest changes. According to this study, NDWI is the indicator of forest change that is most related to monthly runoff in the Meijiang watershed. It indicates that NDWI is the best indicator of forest change in hydrological prediction while forest coverage, the most commonly used indicator of forest change is insignificantly related to monthly runoff. It also showed that a multiple regression model two independent variables -antecedent monthly precipitation and NDWI can be a very efficient way to predict monthly runoff, as well as to quantify the hydrological impact of large-scale forest changes in the Meijiang River watershed or other similar watersheds, which is crucial for downstream water resource management and ecological protection in the Poyang Lake basin. However, it is important to note that the success of this statistical model is heavily dependent on the quality of remote sensing images, for example, images with low cloud cover

10%. And hydrological processes cannot be fully explained in this model. Therefore, this approach as a substitute for hydrological models can be a very useful tool for forest resources and water resources management in watersheds with limited data.

Acknowledgements
This research was jointly supported by Special Research Program for Public-welfare Forestry (N0. 201404201) ,China National Science Foundation (No.31290223 and No.31400489), China Postdoctoral Science Foundation Grant (No.2014M560137), and the Fundamental Research Funds for the Central Universities (No.ZYGX2014J093) .

References
[1] Andréassian V 2004 Waters and forests; from historical controversy to scientific debate *Journal of Hydrology* **291** 1-27
[2] Bates C G and Henry A J 1928 Forest and streamflow experiment at Wagon Wheel Gap, Colorado *Monthly Weather Review Supplement* **30** 1–79
[3] Swank, W T, Vose, J M and Elliott, K J 2001 Long-term hydrologic and water quality responses following commercial clearcutting of mixed hardwoods on a southern Appalachian catchment *Forest Ecology and Management* **143** 163–178
[4] Hornbeck J W, Martin C W and Eagar C 1997 Summary of water yield experiments at Hubbard Brook Experimental Forest, New Hampshire *Canadian Journal of Forest Research* **27** 2043-52
[5] Troendle C A and King R M 1985 The effect of timber harvest on the Fool Creek watershed, 30 years later *Water Resources Research* **21** 1915–22
[6] Lavabre J, Sempere Torres D and Cernesson F 1993 Changes in the hydrological response of a small Mediterranean basin a year after a wildfire *Journal of Hydrology* **142** 273–299
[7] Fritsch J M 1990 Les effets du de´frichement de la foreˆtamazonienne et de la mise en culture sur l'hydrologie des petitsbassins versants. PhD Thesis *Universite´ des Sciences etTechniques du Languedoc, Montpellier* p 392
[8] Burton T A 1997 Effects of basin-scale timber harvest on water yield and peak streamflow *Journal of the American Water Resources Association* **33** 1187–96.
[9] McGurk B J and Fong D R 1995 Equivalent roaded area as a measure of cumulative effect of logging *Environ Manage* **19** 609-621
[10] Menning K, Erman D, Johnson K and Sessions J 1996 Modeling aquatic and riparian systems, assessing cumulative watershed effects, and limiting watershed disturbance. Addendum, chapter 2, *Sierra Nevada Aquatic Research Lab*(USA: Mammoth Lakes) 33-51
[11] Cobourn J 1989 Application of Cumulative Watershed Effects (CWE) Analysis on the Eldorado National Forest in California. In: the symposium on headwaters hydrology *American Water Resources Association*(Bethesda Maryland) 449-460
[12] USDA Forest Service 1974 Part II: Hydrologic effects of vegetation manipulation *Forest hydrology* (USA: Missoula, Montana) 229
[13] King J 1989 Streamflow responses to road building and harvesting: a comparison with the equivalent clearcut area procedure. *Intermountain Research Station* (Ogden: USDA Forest Service) 401
[14] BC Ministry of Forests and Rangeland 1999 Interior Watershed Assessment Guidebook (Appendix 8) 2nd edition (Canada: Victoria, British Columbia)
[15] Silins U 2003 An integrated forest-watershed planning and assessing model: ECA-Alberta. *http://www.cefm.rr.ualberta.ca/Research_Notes/Research%20Note%2007-03.pdf* (Accessed 11 Nov 2008)
[16] Hudson R 2000 Snowpack recovery in regenerating coastal British Columbia clearcuts *Canadian Journal of Forest Research* **30** 548-556
[17] Talbot J and Plamondon A P 2002 The diminution of snowmelt rate with forest regrowth as an

index of peak flow hydrologic recovery, Montmorency Forest, Quebec *Proceedings of the 59th Eastern Snow Conference* (USA: Stowe, Vermont)

[18] Winkler R, Spittlehouse D L and Golding D L 2005 Measured differences in snow accumulation and melt among clearcut, juvenile, and mature forests in southern British Columbia *Hydrological Processes* **19** 51–62

[19] Huete A, Didan K, Miura T, Rodriguez E P, Gao X and Ferreira L G 2002 Overview of the radiometric and biophysical performance of the MODIS vegetation indices *Remote Sensing of Environment* **83** 195−213

[20] Goward S N, Markham B, Dye D G, Dulaney W and Yang J 1991 Normalized difference vegetation index measurements from the Advanced Very High Resolution Radiometer *Remote Sens Environ* **35** 257–277

[21] Van de Griend A A and Owe M 1993 On the relationship between thermal emissivity and the normalized difference vegetation index for natural surfaces *International Journal of Remote Sensing* **14** 1119–31

[22] Gao B C 1996 NDWI—a normalized difference water index for remote sensing of vegetation liquid water from space *Remote Sens Environ* **58** 257–266

[23] Liu W, Wei X, Liu S，Liu Y, Fan H, Zhang M, Yin J and Zhan M 2015 How do climate and forest changes affect long-term streamflow dynamics?: a case study in the upper reach of Poyang River basin *Ecohydrology* **8** 46-57

[24] Mcvicar T, Van Niel T G, Li L, Hutchinson M F, Mu X and Liu Z 2007 Spatially distributing monthly reference evapotranspiration and pan evaporation considering topographic influences *Journal of Hydrology* **338** 196–220

[25] Santos P and Negri A J 1996 A comparison of the normalized difference vegetation index and rainfall for the amazon and northeastern brazil *Journal of Applied Meteorology and Climatology* **36** 958–965

[26] Dawson C and Wilby R 1998 An artificial neural network approach to rainfall-runoff modelling *Hydrological Sciences Journal* **43** 47–66

[27] Dawson C and Wilby R 2001 Hydrological modelling using artificial neural networks *Progress in Physical Geography* **25** 80–108

[28] Cannon A J, Whitfield P H 2002 Downscaling recent streamflow conditions in British Columbia, Canada using ensemble neural network models *Journal of Hydrology* **259** 136–151

[29] Jain A, Srinivasulu S 2004 Development of effective and efficient rainfall-runoff models using integration of deterministic, real-coded genetic algorithms and artificial neural network techniques *Water Resources Research* **40** W04302. DOI: 10.1029/2003WR002355

[30] Zhang M, Wei X, Sun P, Liu S 2012 The effect of forest harvesting and climatic variability on runoff in a large watershed: the case study in the Upper Minjiang River of Yangtze River basin *Journal of Hydrology* **464-465** 1-11

[31] Pike R G, Redding T, Wilford D, Moore R D, Ice G, Reiter M and Toews D 2007 Chapter 14 – Detecting and predicting changes in watersheds [Draft]. In: Compendium of Forest Hydrology and Geomorphology in British Columbia [In Prep.] Pike, R.G. et al. (Eds). *Land Management Handbook (TBD). B.C. Ministry of Forests and Range Research Branch* (Canada: Victoria, B.C. and FORREX Forest Research Extension Partnership, Kamloops)

Application of reconstructed background in assessing heavy metal pollution of soils near Jilin Section of the Songhua River, China

Libo Hao, Mi Tian[1], Xinyun Zhao, Tong Wang

Department of Geochemistry, College of GeoExploration Sciences and Technology, Jilin University, Changchun 130026, China

E-mail: tianmi62080608@126.com

Abstract. The distinguishing and evaluation of natural and anthropogenic sources of heavy metals in soils have always been the important issues in the field of environmental science. In this research, natural sources of heavy metals in soils are calculated according to the relationships between major oxides and heavy metals. The individual sample has independent backgrounds of heavy metals, which is more reasonable. If the single environmental background is used as baseline, anomalies in regions with low background might be decreased or hided and those in regions with high background will be overstated. In this study, enrichment factor is defined as the ratio of measured concentration and reconstructed background of heavy metals, which is of specific significance in practical applications with no assumptions. The agricultural soils near Jilin section of the Songhua River are polluted by Hg seriously with more than half samples polluted greatly. And the soils are polluted by Cd moderately. A few samples are polluted by Zn moderately. The other heavy metals in soils are almost in the level of background with little human disturbance.

1. Introduction

Heavy metals in soils mainly originate from natural sources and anthropogenic sources. Soil parent materials and soil-forming processes may influence the concentrations of natural heavy metals. Industry, mining, agriculture and transportation, etc are the major anthropogenic sources. The two sources of heavy metals usually form two kinds of anomalies, natural anomaly and anthropogenic anomaly. In regions that disturbed by human, the anomalies often exist together. So it is very important to separate natural and anthropogenic anomalies when evaluating the environment pollution. In recent years, people began to attach importance to the separation and determination of natural and anthropogenic anomalies and put forward different methods. For example, some identify the background and anomalies with the method of robust geostatistics [1], discriminate anthropogenic anomalies on the basis of the environmental behaviour and geochemical characteristics of elements [2], [3] apply the accumulation index [4] to discriminate anthropogenic heavy metal pollutions. Nevertheless, most of the methods are relatively complicated in practical application. And some use sparse samples referring to single environmental background to determine the heavy metal pollution [5], actually pollutions evaluated by such a method contains the mixture of anthropogenic and natural sources, through which we cannot effectively distinguish and track the sources of pollutants.

Soil includes residual soil and transported soil [6]. Residual soil is formed from the materials accumulated in situ after rock weathering which inherits the properties of the underlying rocks, and the transported soil is formed from the materials transported remotely by wind and water after rock weathering which inherits the properties of the rocks in the distance. Formed from different soil parent materials, soils may have different concentrations of rock-forming elements, therefore natural sources of heavy metals are correlated with major oxides. Based on this relationship, we put forward the method of quantifying the natural sources using reconstructed background regressed by soil major oxides. Through the ratios of measured and background concentrations of heavy metals, the degree of anthropogenic pollution can be obtained. The procedures could be applied to other similar environments.

2. Materials & method

2.1. Study Area

We conducted our research along the downstream of the Songhua River (figure 1). The Songhua River is located in Jilin, China. Regions along the river are important grain production bases in China. In 1960s-1970s, this river was seriously polluted by wasted water from petrochemical factories in Jilin City.

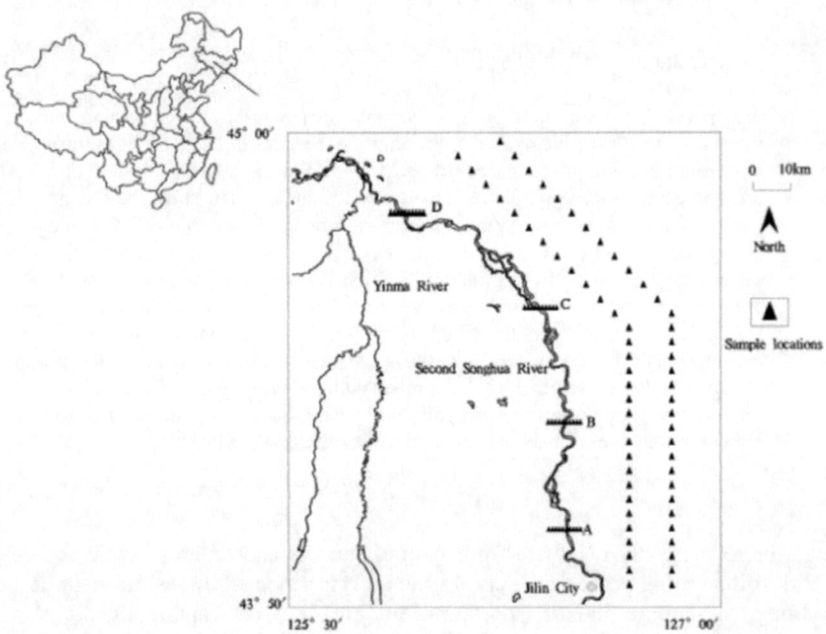

Figure 1. Soil sample locations in the middle-down stream of the Songhua River.

2.2. Sampling and analysis

We set four sections in the irrigation area across the Songhua River and took 10 soil samples every 250 meters in each section. A total of 300 surface soil samples (0-20cm) were collected for studying their contamination levels. Then two sections were set away from the river irrigation area, and a total of 63 samples were collected which are deemed as unpolluted samples. Each sample was composed of four sub-samples collecting randomly within each grid. Each sample weighed more than 1 kg. After natural drying and the removal of gravel and plant roots, samples were sieved using a 20-mesh screen and then sent to a laboratory to be further processed to 200 mesh.

All analyses were carried out in the Ministry of Land and Resources Supervision and Inspection Center, Changchun. The concentrations of As and Hg were analyzed by atomic fluorescence spectrometry (AFS) and the detection limits were 0.9 $mgkg^{-1}$ and 0.003$mgkg^{-1}$, respectively. Samples were prepared by mixing 0.5g of soil with 5.0 ml of nitric and hydrochloric acid in a 25 ml colorimetric tube and shaken well. Samples were then dissolved in boiling water for 2 h, during which they were shaken every 15 min. Next, samples were cooled and then diluted to scale. Finally, the solution was filtered through dry filtration and the filtrate was used for the determination of As and Hg.

Concentrations of Cd were determined using graphite furnace atomic absorption spectrometry (GF-AAS) with a detection limit 0.02 $mgkg^{-1}$. Samples were prepared by placing 0.3 g of soil into a 50 ml polytetrafluoroethylene crucible, wetting with water, and then adding 5 ml of hydrochloric acid. Samples were then decomposed on an electric heating plate under a fume hood on low heat. Following evaporation to ~2–3 ml, samples were removed and cooled, before 5 ml of nitric acid, 4 ml of hydrofluoric acid, and 2 ml of perchloric acid were added. Samples were then returned to the electric heating plate at high temperature for 1 h. After 1 h, the cover was opened and heating was continued in order to remove Si. According to the condition of digestion, 2 ml of nitric acid, 2 ml of hydrofluoric acid, and 1 ml of perchloric acid were added and the above process was repeated. The mixture was then dissolved in 1 ml of nitric acid and transferred to a 25 ml flask, where 3 ml of diammonium phosphate solution was added. Finally, after diluting the solution with water, it was cooled and well shaken for analysis.

The contents of Cu and Pb were measured using X-ray fluorescence spectrometry (XRF) with a detection limit 1 $mgkg^{-1}$ for both. After thorough mixing, soil samples were formed into pressed pellets for analysis. In order to guarantee the accuracy of the method, all analyses were tested against the national primary standard material (GBW).

2.3. Principles

Researches showed that concentrations of trace elements in different types of rocks differ greatly [7], [8].SiO_2 is of the highest contents in rocks which can reflect the basic property of rocks. There are always significant correlations between the concentrations of trace elements and SiO_2 in magmatic rocks. Soil is the product of rock weathering, and thus inherits its chemical composition [9], [10]. And elements in soil migrate, disperse and enrich, etc in a certain degree during the pedogenesis. According to researches on natural soil, correlations between As、Cr、Cu、Mn、Ni、Pb、Zn and SiO_2, Al_2O_3, TFeO or MgO are significant and that of Cd and Hg are relatively weak. So only by eliminating the natural geochemical variability of heavy metals, can we better identify the natural and anthropogenic sources.

The following multivariate linear equation is used to describe the relationships of heavy metals and major oxides:

$$C_{h_i} = \alpha + \sum \beta_j C_{oj} + \varepsilon \tag{1}$$

Where C_{hi} is the concentration of heavy metal i and C_{oj} is the content of oxide j from unpolluted samples, α and β_j are undetermined coefficients, β_j represents the concentrations of heavy metals which change with soil parent materials, ε is random error. Using a multivariate linear regression program, the undetermined coefficients can be calculated. Consequently, we think that the relation of trace elements and oxides of study samples is exactly that of unpolluted samples. And we can use the concentrations of oxides of study samples to calculate the backgrounds of heavy metals for individual sample. Measured concentrations that are higher than background are thought to be polluted by anthropogenic activities, and the ratios of the measured and background concentrations are defined as EFs. The measured concentrations subtract background then divided by measured concentrations can estimate the percentage of anthropogenic sources.

3. Results & discussion

3.1. Natural concentrations of heavy metals

Correlation analysis on basis data shows that there are significant correlations between all heavy metals and major oxides in different degrees (table 1). Therefore multiple regression analysis was carried to obtain the regression equations and then the equations were applied to the data for study to reconstruct the backgrounds of heavy metals, namely the natural sources.

Table 1. Correlations of heavy metals and major oxides.

		As	Cd	Cr	Cu	Mn	Ni	Pb	Zn	Hg
SiO_2	Pearson	-.424	-.098	-.488	-.681	-.369	-.726	-.293	-.594	.005
	Significance	.000	.122	.000	.000	.000	.000	.000	.000	.940
Al_2O_3	Pearson	.353	.115	.730	.641	.205	.567	.370	.556	-.010
	Significance	.000	.070	.000	.000	.001	.000	.000	.000	.881
MgO	Pearson	.528	.136	.735	.833	.372	.794	.436	.679	-.022
	Significance	.000	.031	.000	.000	.000	.000	.000	.000	.735
TFeO	Pearson	.564	.070	.815	.807	.392	.656	.464	.802	.249
	Significance	.000	.269	.000	.000	.000	.000	.000	.000	.000

Graphs illustrating the measured and background concentrations of heavy metals are shown in figure 2. In general, measured concentrations of As are 4-12 mg kg^{-1} which are close to the backgrounds of 2-14 mg kg^{-1}. If the single environmental background is used, most of the measured values are less than it showing that As is not disturbed by human. However, there are specific samples (arrow 1 and 2) with the measured concentrations higher than the single environmental background, but less than the reconstructed background. In fact, natural source of As is relatively high due to the influence of soil parent materials. So the measured values are higher than the single environmental background, but this is not anthropogenic pollution. Measured concentrations of some samples (arrow 3) are less than single environmental background, but are higher than the reconstructed concentrations, which suggests that the samples are disturbed by human activities. This suits for the other elements.

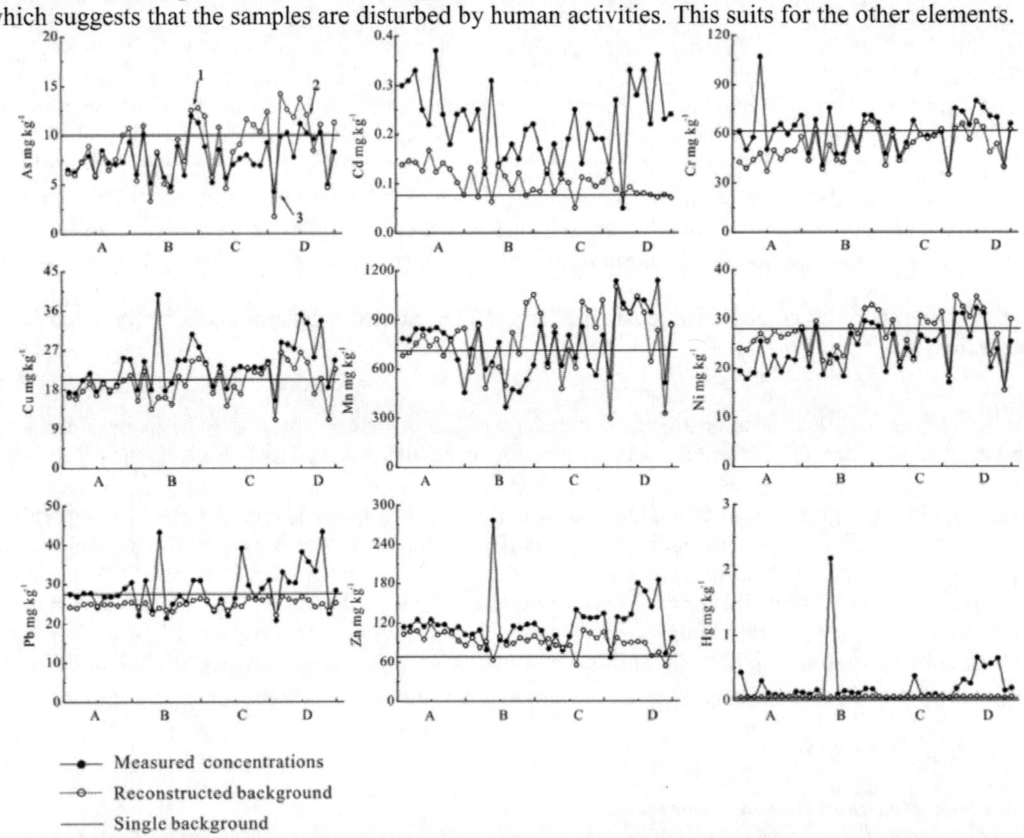

Figure 2. Measured concentrations and reconstructed backgrounds of heavy metals.

Measured concentrations of Cd are 0.05 to 0.37 mg kg^{-1}, which range greatly in different sections. Reconstructed backgrounds of Cd range from 0.05 to 0.17 mg kg^{-1}, which are higher than single environmental background suggesting that soil parent materials contribute much to the natural sources of Cd in study area. Measured concentrations of Cr, Cu, Mn, Ni are 40-110 mg kg^{-1}, 10-25 mg kg^{-1}, 300-1150 mg kg^{-1}, 17-33 mg kg^{-1} and reconstructed backgrounds are 35-70 mg kg^{-1}, 17-40 mg kg^{-1}, 300-1100 mg kg^{-1} and 15-35 mg kg^{-1}, respectively. Reconstructed backgrounds of the above heavy metals fluctuate near the single environmental background. Measured concentrations of Pb are 20-43 mg kg^{-1}, and the reconstructed backgrounds are 23-27 mg kg^{-1}, which are less than single environmental background showing that natural sources of Pb are relatively small. Measured concentrations of Zinc are 70-270 mg kg^{-1} and the reconstructed background are 60-125 mg kg^{-1}. The reconstructed background are higher than single environmental background overall showing that the soil parent materials may contribute largely to the natural sources of Zn. Measured concentrations of Hg are 0.05 to 2.2 mg kg^{-1}, and reconstructed background concentrations are about 0.05 mg kg^{-1}, which are nearly equal to the single environmental background.

Thus it can be seen that the reconstructed backgrounds of most heavy metals differ from the single backgrounds greatly. Anomalies in regions with low backgrounds may be decreased or hided and that in regions with high background will be overstated if the single environmental background is used.

3.2. Enrichment levels of heavy metals

The enrichment factor is usually used to evaluate the enrichment degree of heavy metals [11], [12]. Enrichment factor is expressed as EF:

$$EF = \frac{(C_x / C_n)_s}{(C_x / C_n)_b}$$

Where C_x is the element under consideration, C_n is the chosen reference element and the subscript "s" or "b" indicates which medium the concentration refers to. Sutherland [13] classified the EF into five grades to judge the enrichment degree of heavy metals. When EF is approximately equal to 1, heavy metals are depleted or not polluted by human activities; $1 < EF < 2$ is suggestive of minimal pollution; $2 < EF < 5$, moderate enrichment, suggestive of moderate pollution; $5 < EF < 20$, significant enrichment, suggestive of a significant pollution signal; $20 < EF < 40$, very highly enriched, indicating a very strong pollution signal; $EF > 40$, extremely enriched, indicating an extreme pollution signal.

Reimann [14] pointed out that the normalizer must be one or more important constituents of carriers of heavy metals reflecting the change of granularity in sediments. Previous studies usually take Al, Ni, Ti, V, Y, Li or Rb as normalizers. Three assumptions fundamental to the EF concept are (1) the low variability in concentration for the reference elements relative to the elements of interest, (2) the similarity in regional distribution patterns for various reference elements, and (3) the conservation of crustal (regional) element ratios through various compartments of the ecosystem. Obviously, the above three assumptions in practical application are hard to meet. Many studies use the elements directly or indiscriminately as normalizers ignoring the inherent assumptions. Therefore, when the assumptions are not met, high EFs may not indicate human disturbance.

In the previous researches, average crustal abundance was usually chosen as reference baselines [15-18] proposed the regional geological background should be used as the reference baselines. However, as the above elaboration, for the heavy metals that influenced strongly by soil parent materials, single environmental background cannot be used as the baseline. Owing to the disadvantages of traditional EF method, the ratio of measured concentration and reconstructed background is defined as enrichmenfiguret factor of heavy metal. This method has no assumptions. As shown in figure 3, only one sample in section D is polluted by As weakly, and the other samples are not affected by human disturbance. Half of the samples are polluted by Cd moderately and four samples are polluted by Cd weakly. One sample in section A is polluted by Cr weakly, one sample in section B and two samples in section D are polluted by Cu weakly. Two samples in section D are polluted by Mn weakly. Ni in all samples are in the level of background with no human disturbance.

Individual samples in section B, C, D are polluted by Pb weakly. Individual samples in section B, C, D are polluted by Zinc weakly to moderately. One sample in section B is polluted by Hg extremely and some samples in section A, C and D are polluted by Hg moderately to significantly.

The agricultural soils near Jilin section of the Songhua River are polluted by Hg seriously with more than half samples of over moderate pollutions. The polluted heavy metal is Cd with half samples of moderate pollutions. And individual samples are polluted by Zn moderately. The other heavy metals are almost in the level of background with little human disturbance.

For the heavy metals that are polluted by human obviously, anthropogenic source of Hg accounts for 39-98%, and the percentage of anthropogenic Cd are 37-80% and that of Zn are 47-77%.

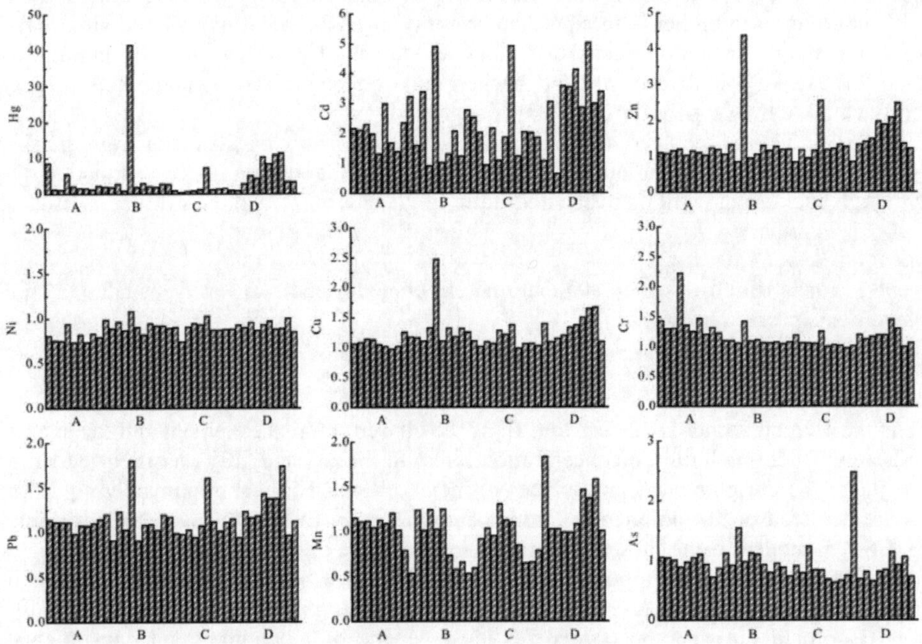

Figure 3. EFs of heavy metals

4. Conclusions

The soil geochemical background traditionally refers to concentration of trace element formed naturally. The influence of soil parent materials should be taken into account to determine the background of heavy metals. For heavy metals that have significant correlations with soil parent materials, natural sources may differ obviously. If the single environmental background is used the anomalies in regions with low background may be hided and that in regions with high background may be overstated and thus affecting the effect of environmental assessment.

In this study, EF is defined as the ratio of measured concentration and reconstructed background. This method has no assumptions and one sample with individual background. EF can represent the enrichment degree of anthropogenic sources relative to the natural sources, which is of significance in practical application relative to traditional method.

Our study shows that the soils near Jilin section of Songhua River are polluted by Hg and Cd, to a lesser extent, Zn. Plants and humans are facing great threat, therefore, more efforts should be made to prevent and control pollution.

References

[1] Matschullat J, Ottenstein R and Rmannei C 2000 Geochemical background-can we calculate it? *Environmental Geology* **39(9)** 990-1000.

[2] Covelli S and Fontolan G 1997 Application of a normalization procedure in determining regional geochemical baselines *Environ Geol* **30(1–2)** 34–45.

[3] Forstner U, et al. 1990 Sediment criteria development-contributions from environmental geochemistry to water quality management 311-338.

[4] Wei B and Yang L 2010 A review of heavy metal contaminations in urban soils, urban road dusts and agricultural soils from China *Microchemical Journal* **94(2)** 99-107.

[5] Sundaray S K 2011 Geochemical speciation and risk assessment of heavy metals in the river estuarine sediments—a case study: Mahanadi basin, India *Journal of hazardous materials* **186(2)** 1837-46.

[6] W Xueqiu and X Xuejin 2000 Exploration Geochemistry of Gold -Principles and Methodology, Strategy and Tactics. (Ji'nan: Shandong Science &Technology Press)

[7] C Zhizhong and X Xuejin 2006 Influence of variation in element background values in rocks on metallogenic prognosis in geochemical maps *Geology in China* **33(2)** 411- 417.

[8] Y Mingcai and C Qinghua 1997 The Chemical Composition of the Eastern China Continental Crust and Rocks. (Beijing: Science Press)

[9] H Libo, M Li and Z Haibin 2004 Elemental homogenization during weathering and pedogenesis of volcanic rocks from North Da Hinggan Ling *Geochimica* **33(2)** 131-138.

[10] H Libo, L Jilong and M Li 2005 Relation between the chemical compositions of residual soils and bedrocks in shallow overburden areas and its significance--a case study of the northern Da Hinggan Mountains *Geology in China* **32(3)** 477- 482.

[11] Reimann C and de Caritat P 2000 Intrinsic flaws of element enrichment factors (EFs) in environmental geochemistry *Environ Sci Technol* **34** 5084–91.

[12] Glasby G P, et al. 2004 Heavy-metal pollution of sediments from Szczecin Lagoon and the Gdansk Basin, Poland *Sci Total Environ* **330** 249–269.

[13] Sutherland R A 2000 Bed sediment-associated trace metals in an urban stream, Oahu, Hawaii *Environmental Geology* **39(6)** 611-627.

[14] Reimann C and de Caritat P 2005 Distinguishing between natural and anthropogenic sources for elements in the environment: regional geochemical surveys versus enrichment factors *Sci Total Environ* **337** 91–107.

[15] Halstead M J R, Cunninghame R G and Hunter K A 2000 Wet deposition of trace metals to a remote site in Fjordland, New Zealand *Atmospheric Environment* **34** 665–676.

[16] Loska K, et al. 1997 Use of enrichment and contamination factors together with geoaccumulation indexes to evaluate the content of Cd, Cu, and Ni in the Rybnik water reservoir in Poland *Water, Air, and Soil Pollution* **93** 347– 365.

[17] Varrica D, Aiuppa A and Dongarra G 2000 Volcanic and anthropogenic contribution to heavy metal content in lichens from Mt. Etna and Vulcano island (Sicily) *Environmental Pollution* **108** 153– 162.

[18] Blaser P, et al. 2000 Critical examination of trace element enrichments and depletions in soils: As, Cr, Cu, Ni, Pb, and Zn in Swiss forest soils *Science of the Total Environment* **249** 257–280.

The interannual trend and preliminary quantitative estimation of the oceans condition in the Bohai Sea area

Yang Yang [1], Jin-kun Yang [1, 3], Qing-sheng Miao [1], Xiu-min Gao [2]

[1] National Marine Data and Information Service, Tianjin 300171, China;
[2] First Institute of Oceanography, SOA, Qingdao 266061, China

E-mail: yang03034101@126.com

Abstract. After different frequency observed temperature data of Bohai sea marine observation stations was analyzed, results showed that daily average sea surface temperature value obtained from 3-hour-observations (08h, 14h, 20h) which was slightly lower than that from 24-hour-observations. The general gap was within 0.10℃, while the value was 0.05℃ for the monthly mean temperature. Daily average sea surface temperature values of 3-hour-observations have a little effect both in statistical properties and the accuracy of statistical data, which does not affect the representativeness of the data. It can be used in studying long time series problems. Analyzing the trend of the SST data changes in 1960-2012 and the SAT data changes in 1965-2012 by using the linear tendency estimate and cumulative distance square method, it can prove that SST annual variation rate was 0.010℃/a, with a total increase of 0.53℃ in last 53 years; and SAT annual variation rate was 0.043℃/a, with a total increase of 2.06℃ in last 48 years. Although the long-term trend of these two factors is significantly increased, but there was a significant mutation around 1987. From 1960 to 1987, it had a downward trend, after 1987 it began to grow, the upward trend has not diminished until 2009.

1. Introduction

Bohai Sea is a continental enclosed sea which lies in north of China, its mean depth is 18 meters and the deepest area is located in Laotieshan channel with a depth of 86 meters [1]. The seawater exchange between the Bohai figure Sea and the Yellow Sea through the Bohai Strait, and it is more difficult to exchange with the deep sea outsides. At the same time, the coastal areas of Bohai is densely populated and within industrial development, and it is also a very active area with human economic activities. The status of its marine hydrology has become a concern [2].

In this paper, based on the representation of sea surface temperature (SST), the long-term trend of the basic hydrometeorological elements (SST and sea surface air temperature(SAT)) in the Bohai sea were analyzed by using recent 50 years of long-term continuous observation data of the oceanic stations. Calculated the elevating trend of annual variation rate and diagnosed roughly time of its mutations, which is helpful to deepen understanding of hydrological and meteorological conditions changes, is beneficial to the evaluation and analysis of the marine environment in Bohai Sea.

2. Basic Information

The observation data of oceanic stations is widely used because of the characteristics of long-term, continuity and near-shore. The practical value and application significance are obvious. When long time series data are needed, the importance and valuable information of oceanic station data are more obvious. The study of marine hydrometeorological elements (such as SST, SAT, etc.) has always been a traditional subject in marine field and has been chosen as a decisive factor by other disciplines [3-13]. These elements have long been the basic observation items of oceanic stations for a long time. Therefore, it is very important to research and analyze the hydrometeorological data of oceanic station, which is very important for marine development and ocean engineering.

In this paper, we have selected more than 50 years, observation data of eight oceanic stations along the Bohai Sea (stations A-G, Figure 1). Analysed SST and SAT in 1960-2012(Table.1) has the most closed relationship with the ecological environment. The SST data is limited because of the freezing sea water in winter, so there is no observational data. Therefore, only the months without frost are recorded for calculating annual mean. For example, the annual average value of station C is the average value of 9 monthly mean average values (from Apr to Dec). The annual average values of the other 3 stations(D、E and F) are the average values of 10 monthly mean average values(from Mar to Dec). The annual average value of station A is the average value of 7 monthly average values (from May to Nov). Although the average annual value of SST is higher than the real situation, the results of this study have a little effect for the long-term trend of the study [4].

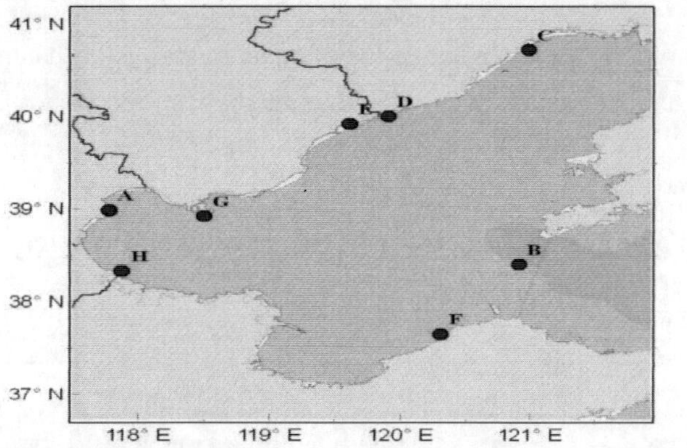

Figure 1. Locations of the Bohai Sea marine observation stations

Table 1. Record length of the obtained data of the Bohai Sea marine observation stations

Stations	SST(Timing)		SST(Hourly)		SAT(Timing)	
	years（a）	Data length	years（a）	Data length	years（a）	Data length
A	1960-2012	53	2002-2012	11	1965-2012	48
B	1960-2012	53	none	none	1965-1982、1984-2012	47
C	1963-2012	50	2010-2012	3	1965-2012	48
D	1960-2012	53	2008-2012	5	1965-1982、1984-2012	47
E	1960-2012	53	2011-2012	2	1965-2012	48
F	1960-2012	53	2011-2012	2	1965-2012	48
G	2011-2012	2	2011-2012	2	2008-2012	5
H	2011-2012	2	2011-2012	2	2008-2012	5

3. The influence of observational frequency on SST statistical characteristics continuity

In the 1950s, due to observation conditions, SST was achieved artificially in a long time. The "Beach Observing Standard" (GB/T 14914-1994) [13] was implemented in 1994, stipulated for the timing observation clearly: "Sea surface temperature should be observed at 08h, 14h, 20h every day." These three times represent the SST of the morning, noon and night respectively. Although they are representative, they are lacking continuity and cannot realize the daily continuity of SST of the oceanic stations. The daily mean values of the SST can be calculated by three times of observations, which may lead to the low or high daily mean value due to the maximum or minimum water temperature. Therefore, in the revised "Beach Observing Standard" (GB/T 14914-2006) [14] in 2006, it is required to keep the original SST timing observation, and to equip with automatic observation equipment to make the SST continuous observations.

Since 1950s, the oceanic station has been using SWY1-2 type sea surface temperature measuring instrument(made by National Ocean Technology Center) to measure the SST, the measurement accuracy is ± 0.1℃.For the continuous, only the last 10 years of data have the hourly SST data. So in some research of long time series, it can only be studied depends on the timing of observation data. However, due to the frequency of observation is less, statistical characteristics of the daily average SST, which calculating with 3 moments SST, should be demonstrated and analyzed.

Using two consecutive years (2011-2012) timing and hourly SST data of 7 oceanic stations(station A, stations C to H) in the Bohai sea area, we calculate the 3 moments daily average SST ($\overline{T}_4$) and 24 moments daily average SST($\overline{T}_{24}$) respectively. And then calculating the monthly average SST $\overline{M}_4$ and $\overline{M}_{24}$ by $\overline{T}_4$ and $\overline{T}_{24}$ respectively, we do the comparative analysis. Specially, the calculation of the $\overline{T}_4$, SST on the 08h should be count twice. Such as follow [15]:

$$\overline{T}_4 = (2t_8 + t_{14} + t_{20})\ /4$$

In the course of the calculation, we select the complete data of the natural day to measure the statistics. The natural day is counted as an effective number of days, when $\overline{T}_4$ and $\overline{T}_{24}$ both have statistical results. The effective number of days is up to 731 days of 2011-2012. Specific comparison results are in Table 2 and Table 3. ($\Delta = \overline{T}_4 - \overline{T}_{24}$)

Table 2. Statistical results of daily average SST difference

Station	Effective Days(d)	$\Delta_1 > 0$ Days(d)	$\Delta_1 > 0$ Percentage (%)	$\Delta_1 < 0$ Days(d)	$\Delta_1 < 0$ Percentage (%)	$\Delta_1 = 0$ Days(d)	$\Delta_1 = 0$ Percentage (%)
A	731	202	27.63	407	55.68	122	16.69
C	696	342	49.14	328	47.13	26	3.74
D	731	156	21.34	504	68.95	71	9.71
H	484	130	26.86	297	61.36	57	11.78
E	587	131	22.32	408	69.51	48	8.18
F	725	162	22.34	501	69.10	62	8.55
G	477	194	40.67	255	53.46	28	5.87
Total	4431	1317	29.72	2700	**60.93**	414	9.34

Table 3. Statistical results of daily average SST difference 2

Station	Effective days(d)	average (℃)	$\|\Delta_1\| < 0.05$ days(d)	$\|\Delta_1\| < 0.05$ Percentage (%)	$0.05 \le \|\Delta_1\| < 0.10$ days(d)	$0.05 \le \|\Delta_1\| < 0.10$ Percentage (%)	$\|\Delta_1\| \ge 0.10$ days(d)	$\|\Delta_1\| \ge 0.10$ Percentage (%)
A	731	-0.014	635	86.87	80	10.94	16	2.19
C	696	0.001	246	37.19	156	47.75	294	42.24
D	731	-0.045	445	60.88	141	19.29	145	19.84
H	484	-0.018	372	76.86	78	16.12	34	7.02
E	587	-0.038	369	62.86	124	21.12	94	16.01
F	725	-0.031	464	64.00	162	22.34	99	13.66
G	477	-0.023	221	46.33	92	19.29	164	34.38
Total	4431	-0.024	2752	**62.11**	833	18.80	846	19.09

The number of effective days is a lot in stations A, C, D and F. The data is integrated. While the number of effective days is relatively less in stations H, E and G, this is still exceeding 65% of the total sample. The results have a certain degree of credibility. From Table 2 and Table 3, it is not difficult to see that, $\overline{T}_4$ is slightly smaller than $\overline{T}_{24}$, the difference value Δ between them is very small. 62.11% of the difference value is within 0.05℃, the difference value within 0.10℃ accounts for 80.91%.

On the basis of this, three stations (station A, D and F) with the highest number of effective days were selected and the monthly average SST is continued to be calculated. From Table 4, the results showed that the difference between $\overline{M}_4$ and $\overline{M}_{24}$ was within 0.05℃, the correlation coefficients were as high as more than 99.99%.

Table 4. Statistical results of monthly average water temperature difference

Year	Month	A			D			F		
		$\overline{M}_4$ (℃)	$\overline{M}_{24}$ (℃)	Δ_2 (℃)	$\overline{M}_4$ (℃)	$\overline{M}_{24}$ (℃)	Δ_2 (℃)	$\overline{M}_4$ (℃)	$\overline{M}_{24}$ (℃)	Δ_2 (℃)
2011	1	-0.90	-0.89	-0.01	-1.50	-1.49	-0.01	-1.00	-0.98	-0.02
	2	-0.90	-0.90	0.00	-1.00	-0.98	-0.02	-0.60	-0.56	-0.04
	3	2.30	2.31	-0.01	2.70	2.73	-0.03	3.50	3.53	-0.03
	4	8.50	8.51	-0.01	9.30	9.35	-0.05	9.10	9.14	-0.04
	5	14.70	14.70	0.00	15.00	15.05	-0.05	15.70	15.75	-0.05
	6	21.20	21.20	0.00	20.10	20.13	-0.03	21.10	21.13	-0.03
	7	25.80	25.81	-0.01	23.70	23.75	-0.05	24.70	24.74	-0.04
	8	27.20	27.22	-0.02	25.50	25.59	-0.09	26.30	26.32	-0.02
	9	24.60	24.62	-0.02	22.30	22.40	-0.10	23.00	23.03	-0.03
	10	19.10	19.12	-0.02	17.00	17.06	-0.06	17.60	17.61	-0.01
	11	13.80	13.81	-0.01	9.60	9.63	-0.03	12.90	12.93	-0.03
	12	5.70	5.72	-0.02	1.20	1.23	-0.03	5.20	5.22	-0.02
2012	1	0.80	0.81	-0.01	-0.80	-0.80	0.00	1.00	1.01	-0.01
	2	-0.70	-0.69	-0.01	-0.80	-0.80	0.00	-0.30	-0.28	-0.02
	3	1.80	1.81	-0.01	2.70	2.73	-0.03	2.40	2.43	-0.03
	4	8.30	8.30	0.00	9.20	9.23	-0.03	9.10	9.15	-0.05
	5	16.10	16.13	-0.03	16.50	16.55	-0.05	17.10	17.15	-0.05
	6	21.80	21.82	-0.02	20.80	20.85	-0.05	21.80	21.83	-0.03
	7	25.90	25.92	-0.02	24.80	24.85	-0.05	25.30	25.33	-0.03
	8	26.00	26.03	-0.03	24.10	24.15	-0.05	25.90	25.93	-0.03
	9	24.60	24.61	-0.01	21.90	21.93	-0.03	23.80	23.83	-0.03
	10	19.50	19.52	-0.02	16.40	16.45	-0.05	18.70	18.73	-0.03
	11	12.10	12.12	-0.02	7.70	7.75	-0.05	10.90	10.94	-0.04
	12	4.00	4.02	-0.02	-0.20	-0.17	-0.03	3.60	3.63	-0.03

In sum, $\overline{T}_4$ is slightly less than $\overline{T}_{24}$, the difference between them is not too much. The influence on data statistics and statistical data accuracy is very small, which does not affect the data representation. In the study of the long time series, the two results can be replaced with each other.

4. Interannual variation trend and preliminary quantitative estimation of SST and SAT

In order to reveal the long-term trend of SST and SAT in the Bohai sea area, and to give a preliminary quantitative estimation of the annual change of them, this paper adopts the linear trend estimation method to study SST and SAT[16]. X_i is used to represent a hydrometeorological variable with sample size n, and t_i is the time corresponding to X_i. The linear regression equation between X_i and t_i is established: $X_i = a + bt_i$ (i = 1, 2, ... , N), this formula can be seen as a special, and the simplest form of linear regression.The meaning of it is to express the relationship between X and time t with a reasonable straight line. In this method, B is the regression coefficient. Its sign indicates the trend of climate variables X, and its size reflects the rate of X rise or fall, which indicates the change rate of the research variables.

The annual average values of SST and SAT in Bohai sea area, the regression lines and equations are shown in Figure 2- Figure 3. The red curve represents the average value of 11-point slide, which reflects the interannual variation. It shows clearly in Figure 2 that the SST annual change of the 6 oceanic stations is slightly different in the past 53 years, but the upward trend is obvious. The trend of change is highly consistent with the interdecadal variation trend. The average SST annual variation of these 6 stations can be roughly calculated as an increase of 0.53℃ for 53 years with an average annual increase of 0.010℃.

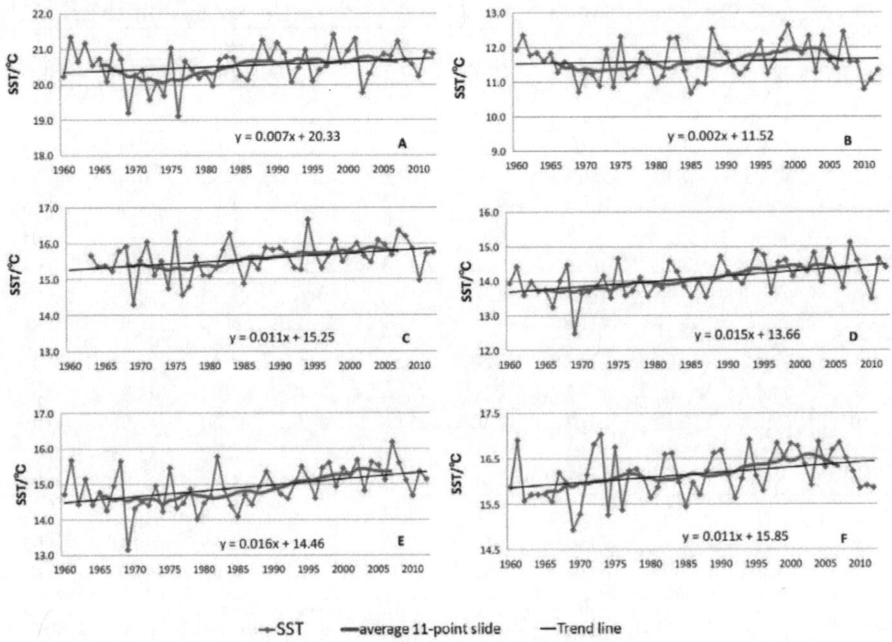

Figure 2. Yearly mean average and regression line and equation of SST in Bohai seas

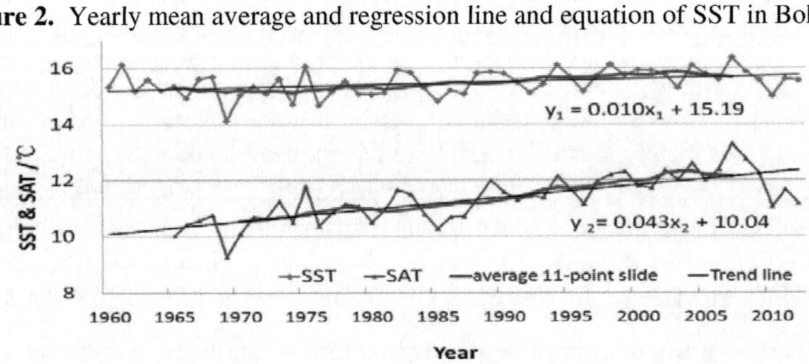

Figure 3. Yearly mean average and regression line and equation of SST and SAT in Bohai Sea

Figure 3 shows the annual average of SST and SAT (mean SST and SAT averaged over 6 stations) in the Bohai Sea and the regression lines. It can be seen from the Figure 3 that SAT, like SST, has an increasing trend in the past 50 years. The annual rate of SAT (0.043℃/a) is much larger than that of SST (0.010℃/a). The former is 4.3 times of the latter. It indicates that the meteorological elements are more sensitive to the changes of environmental conditions than the hydrological elements. The phenomenon of atmospheric warming in the Bohai Sea is very significant.

In order to furtherly understand the long-term significant trends and persistent changes of SST and SAT, this paper uses the method of cumulative anomaly curve to study and analyze. Cumulative anomaly is a commonly used method to determine the trend of change from the curve. Figure 4 is the cumulative anomaly curves of the 6 oceanic stations SST in the Bohai sea area. Although the curves of each station have obvious ups and downs, it can be clearly diagnosed that the trend of the SST of each station occurred a significant mutation around 1987.

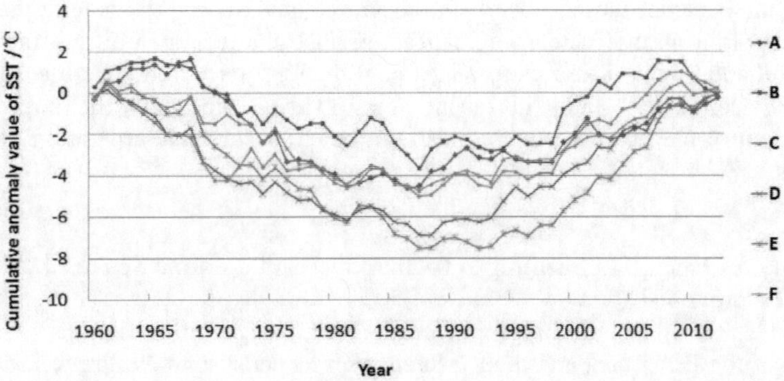

Figure 4. Cumulative anomaly curve of SST in Bohai Sea

The cumulative anomaly values of SST and SAT in the Bohai sea area are plotted as shown in Figure 5. Although the cumulative anomaly values of SST are negative, most of the curves show that the SST of Bohai sea experienced a significant fluctuation in the past 53 years. From 1960 to 1987, SST had a downward trend, and began to increase after 1987, the upward trend has not been reduced until 2009. The trend of the cumulative anomaly curve of SAT is consistent with SST.

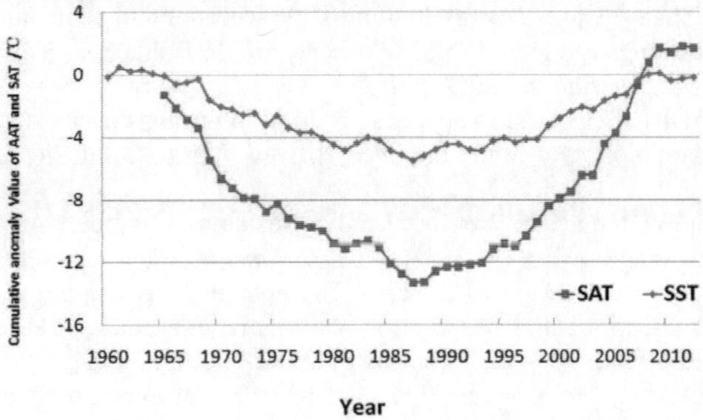

Figure 5. Cumulative anomaly curve of SST and SAT in Bohai Sea

5. Conclusions

- The results of the measured SST data at different observing frequency of the oceanic station in the Bohai sea area show that 3 moments daily average SST is slightly lower than 24 moments daily average SST. There is a little difference between them. The gap is generally less than 0.10℃. The monthly average SST difference is in 0.05℃ or less. The influence of 3 moments daily average SST of data statistics and statistical data accuracy is very small, which does not affect the data representation. It can be used to study the long time series problem.
- In the past 50 years, the marine environmental conditions in the Bohai sea area have changed significantly. The sea surface temperature and sea surface air temperature have increased significantly. The sea surface air temperature is more obvious. The phenomenon of atmospheric warming is significant. The annual variation rate of SAT was 0.043℃/a, which increased by 2.06℃ in the past 48 years. The annual variation rate of SST was 0.010℃/a,

which increased by 0.53℃ in the past 53 years. Although the long-term trend of these two factors is significantly increased, but there was a significant mutation around 1987. From 1960 to 1987, it had a downward trend, after 1987 it began to grow, the upward trend has not diminished until 2009.

- By using the oceanic stations observation data, the quantitative estimate of the long-term trend of the hydrological and meteorological factors in Bohai sea area during the past 50 years is very useful and important for us to understand the long-term trend and mechanism of Chinese coastal environmental status in recent years. This is not only the key of the scientific development of their own needs, but also the key to sustainable utilization of Chinese marine resources.

References

[1] Feng Shizuo, Li Fengqi, Li Shaojing 2003 Introduction to Marine Science [M]. Beijing: *Higher education press* **434**.
[2] Lin Chuanlan, Su Jilan, Xu Bingrong, et al. 2001 Long-term variations of temperature and salinity of the Bohai Sea and their influence on its ecosystem *Progress in Oceanography* **49** 7-19.
[3] Fang Guohong, Wang Kai, Guo Fengyi, et al. 2002 Long-term changes and inter relations of annual variations of the hydrographical and meteorological parameters of the Bohai Sea during recent 30 years *Oceanologia et Limnologia Sinica* **33**(5) 515-525.
[4] Bai Hong, Hu Dunxin, Chen yongli, et al. 2004 Statistic characteristics of thermal structure in the southern Yellow Sea in summer *Chinese Journal of Oceanology and Limnology* **22**(3) 237-243.
[5] Huo Wenfeng, Zhang Aijun, Zhang Shudong 1996 Decadal and pentadal distribution of number of stations for temperature and salinity observation in Bohai Sea, Yellow Sea and East China Sea *Marine Science Bulletin* **15**(1) 67-77.
[6] Intergovernmental Panel on Climate Change 2014 Working Group II. Climate Change 2014: Impacts, Adaptation, and Vulnerability[M]. New York: *Cambridge University Press* 411-484.
[7] Ji Shijian, Zhou Weifeng, Fan Wei, et al. 2016 Sea surface temperature anomaly's interannual variability in pelagic fishing grounds of China *Marine Sciences*, **49**(1) 85-93.
[8] Ma Chao, Ju Xia, Wu Dexing, et al. 2010 The spatial distribution and temporal evolution of salinity at the sections and observation stations in the Yellow Sea and Bohai Sea *Marine Sciences* **34**(9) 70-81.
[9] Ma Chao, Wu Dexing, Lin Xiaopei 2006 The characters of interannual and long-term variations of salinity in the Bohai Sea and Yellow Seas *Periodical of Ocean University of China* **36**(6) 7-12.
[10] Wu Dexing, Mu Lin, Li Qiang, et al. 2004 Long-term varitation characteristics and possible dominant factors of salinity in Bohai Sea *Progress in natural science* **14**(2) 191-195.
[11] Xia Huayong, Gu Wancai 2008 Statistical analysis of mean sea temperatures at tidal gauge station along Guangxi coast *Marine Science Bulletin* **19**(4) 15-21.
[12] Diao Xuexian 1996 Statistical analysis of mean sea surface temperatures in Shijiu port *Marine Science Bulletin* **15**(6) 15-19.
[13] Wang Lina, Chen Xiaohong, Li Yuean, et al. 2009 Heuristic segmentation method for change-point analysis of hydrological time series *Yangtze River* **40**(9) 15-17.
[14] State Technical Supervision Bureau of the people's Republic of China. GB/T 14914-1994. The specification for offshore observations[S]. Beijing: *Standards Press of China*, 1994: 6.
[15] State Technical Supervision Bureau of the people's Republic of China. GB/T 14914-2006. The specification for offshore observations[S]. Beijing: *Standards Press of China*,2006: 6, 79.
[16] Wei Fengying 1999 Statistical diagnosis and prediction technology of modern climate[M]. Beijing: *China Meteorological Press* 82-122.

The summer Bohai Sea hydrological environment analysis

Xu Shanshan[1], Wu Shuangquan[1], Du Ling[2], Li Huan[1, 3], Li Cheng[1]

[1] National Marine Data and Information Service, Tianjin, People's Republic of China, 300171
[2] Ocean University of China, Qingdao, People's Republic of China, 266100

E-mail: usher02@126.com

Abstract. The Bohai Sea is a nearly closed shallow sea which goes deep into china mainland, it has a supper shallow water shelf properties. The research about the Bohai Sea environment must consider the wind, tide, temperature and salinity. The fixed point observed temperature, salinity and current data were analysed. The results show that the Bohai Sea is regular half-day tidal current trend, and the tidal current trend is the substantially reciprocating flow properties. The reciprocating flow direction within the bay is trend towards the bay's direction; the rest is trend towards the sea shore. During the observation, the current direction is stable and the current rate is small at all fixed points, with a good correlation with wind. Thereinto the Liaodong Bay has a clockwise circulation and the Laizhou Bay flows substantially clockwise either, which consistent with the historical trend observation. One fixed point in the Bohai Bay flows different trend compared with the historical observation. The Bohai Sea surface temperature is slightly lower than that in August of the calendar year; salinity has a decreased trend compared to historical statistics.

1. Foreword

Bohai Sea is a nearly closed shallow sea which penetrate deeply into China mainland [1], it belongs to ultra-shallow continental shelf properties, it's response forcing factor is relatively short, the circulation dynamics studies about the Bohai Sea must consider the combined effects of wind、tide、heat and salt [2]. In September 2003, State Oceanic Administration organized and implemented a marine synthesized survey and evaluation [3]. Its investigation ranges includes the Bohai Sea. This paper use this survey data to analyze the summer Bohai Sea hydrological environment.

2. Data Source

Data this article refer to are derived from marine synthesized survey and evaluation, summer physical oceanographic survey. Survey area is the Bohai Sea west of the Bohai Strait.

Observation includes section observation and fixed moorings observation. Section observation contain 188 stations, time ranges from July 2006 to August 2006, and observation elements include station information、 water depth, temperature, salinity and turbidity. The interlayer spacing is 1meter. There are 4 fixed moorings stations, time ranges from July to September 2006, observation elements mainly are currents, and the interlayer spacing is 1 meter. Figure 1 is observation station distribution image.

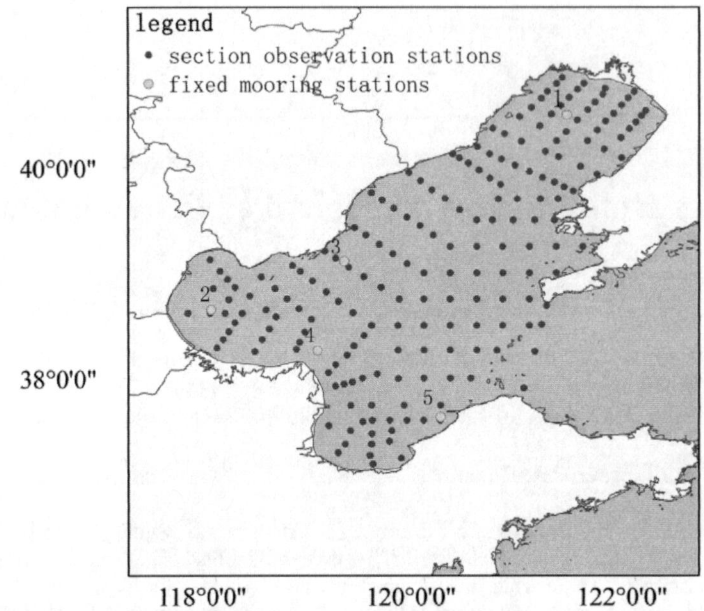

Figure 1. Observation station distribution image

3. Data Processing

Quality control the temperature and salinity data for use, quality control methods include reasonable inspection, range inspection, latitude and longitude reasonable inspection, spike inspection, landing inspection, gradient test, speed test, climatology inspection, constant test, stability test, level inverted repeat inspection. Test out erroneous data and annotate QC mark.

Quality control the current data for use, mainly for range test, peak test, continuity inspection, gradient inspection and any more. Test out erroneous data and annotate QC mark.

4. Current Characteristics

Fixed mooring station referred to as No.1-No.5 stations.

4.1. Primary current direction frequency in fixed mooring station

The following table (Table 1) shows the frequency primary current direction appears at fixed mooring stations, including primary rising trend and frequency; primary falling trend and frequency. Wherein the primary rising trend of No.1 station point to the Liaodong Bay seashore, the primary falling trend point to estuary as shown in Figure 2. No.2 station is similar to No.1 station, rising trend point to the Bohai Bay seashore while falling trend point to estuary (see Figure 3). As shown in Figure 4, 5, and 6, the primary rising and falling trend of No.3, No.4, No.5 station all parallel the seashore.

Table 1. Primary current direction frequency

Station number	Primary rising trend	Frequency	Primary falling trend	Frequency
1	northeast and east- northeast	40-47%	southwest and southwest-west	30-39%
2	west-southwest and west	30-37%	east and east-southeast	36.5-43.5%
3	southwest and west-southwest	39-48.8%	northeast and east-northeast	25.8-35%
4	west-northwest and northwest	35-40.5%	east-southeast and southeast	37.8-39.6%
5	south-southwest and southwest	37.5-47.8%	north-northeast and northeast	22.3-29.4%

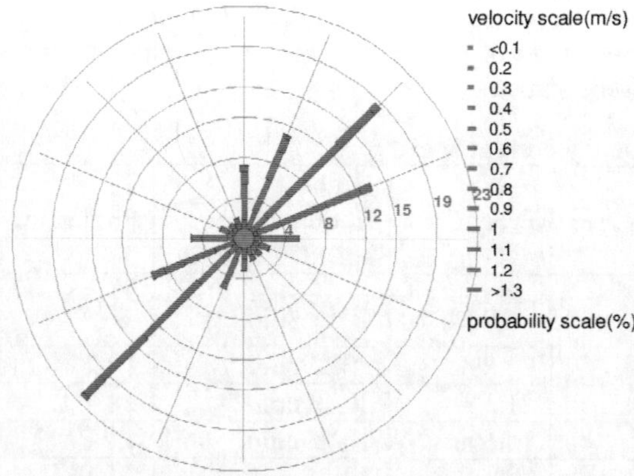

Figure 2. No.1 station 3 meter current speed and direction rose diagram.

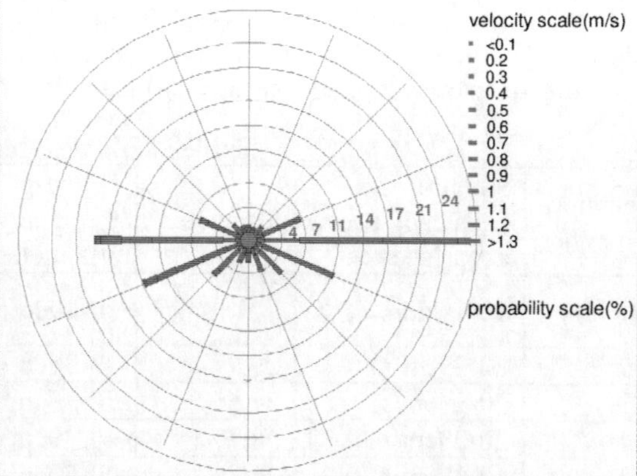

Figure 3. No.2 station 2.5 meter current speed and direction rose diagram.

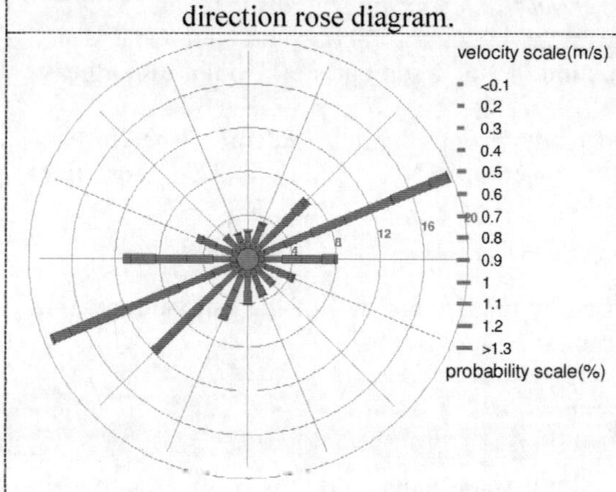

Figure 4. No.3 station 3 meter current speed and direction rose diagram.

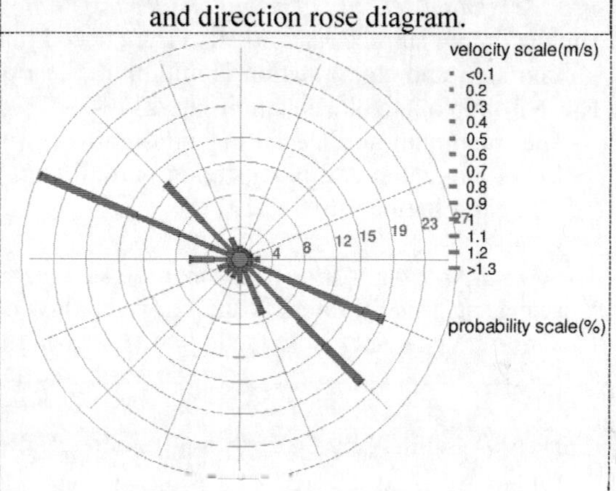

Figure 5. No.4 station 2.5 meter current speed and direction rose diagram.

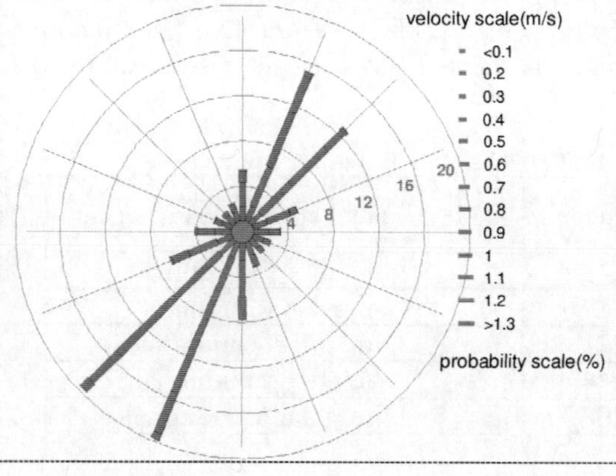

Figure 6. No.5 station 2.5 meter current speed and direction rose diagram.

4.2. Frequency of each current speed levels in fixed mooring station

The following table (Table 2) lists the highest frequency current speed in every fixed mooring station. From the table we can know the current speed dimension is not large. The highest frequency current speed at deeper layers is larger than surface layer at some stations.

Table 2. Highest frequency current speed

Station number	Highest frequency speed at surface	Frequency	Highest frequency speed at 5 meter	Frequency	Highest frequency speed at bottom	Frequency
1	10-19cm/s	30%	10-19cm/s	27.6%	10-19cm/s	31%
2	20-29cm/s	29.9%	30-39cm/s	23.8%	20-29cm/s	19.2%.
3	20-29cm/s	21.5%	30-39cm/s	21.3%	30-39cm/s	28.3%
4	20-29cm/s	18.7%	40-49cm/s	15.5%	10-19cm/s	14%
5	10-19cm/s	35.8%	20-29cm/s	28%	20-29cm/s	24.1%

4.3. Maximum velocity of rising and falling tidal current at fixed mooring station

The maximum surface rising tidal current speed ranges from 45 cm/s to 73cm/s, thereinto No.2 station is maximum and No. 5 station is minimum. The direction in No. 1 station is 50°, point to northeast, These from No.2 to No. 5 stations are 217°-295°.

The maximum surface falling tidal current speed ranges from 44cm/s to 82cm/s, wherein No.4 station is maximum and No.5 station is minimum. The direction in No. 1 station is 227°, those from No.2 to No.5 stations are 34°-127°.

4.4. Fixed mooring station tidal current character

Tidal current usually refers to the seawater flow caused by the rise and fall of the astronomical tide. Tidal current character is divided by the following formula:

$$K = \frac{W_{K_1} + W_{O_1}}{W_{M_2}}$$

Where W_{K_1}、W_{O_1}、W_{M_2} respectively are tidal ellipse semi-major axis of K_1、O_1 and M_2 partial tides. If $K \leq 0.5$ for regular semidiurnal tidal current; if $0.5 < K \leq 2.0$ for irregular semidiurnal tidal current; If $K > 2.0$ as the case may date as regular diurnal tidal current or irregular diurnal tidal current.

According to this formula, tidal current character criterion K of every depth of five fixed mooring stations is calculated. K value is shown in the following table (Table 3). These five fixed mooring station in addition to 5th station is irregular semidiurnal tidal current character, the rest are all regular semidiurnal tidal current properties.

Table 3. Tidal current character

Station number	Water depth rang (m)	K value range	Tidal current character
1	3-15	0.34-0.46	regular semidiurnal
2	3.5-10.5	0.34-0.39	regular semidiurnal
3	3-10	0.67-0.71	irregular semidiurnal
4	2.5-16.5	0.35-0.44	regular semidiurnal
5	2.5-11.5	0.80-1.08	irregular semidiurnal

4.5. Fixed mooring station tidal current trend form

Tidal current trend form depends on the ellipse elements of main partial tides in the sea. By the tidal current character analysis showed above we know that the trend of five fixed mooring stations are semidiurnal tidal current character, so the trend form of main semidiurnal tide (M_2 and S_2) could represent the tidal current trend form of this sea.

The parameter reflecting the trend form of tidal current is rotating rate (also known as ellipticity), its value is the ratio of minor axis and major axis of the ellipse of this partial tide, which sign has the difference of "+", "-", positive sign denotes that the partial tidal current rotates counterclockwise, negative sign means clockwise rotation.

As shown in Figure 7 and Figure 8, No.1 station and No.2 station all layers are counterclockwise rotation, No.3 station and No.5 station all layers substantially rotate clockwise, No.4 station surface layer is counterclockwise rotation, deeper clockwise.

Rotation ellipse of partial tide M_2 and S_2 manifest strong reciprocating current properties, major axis of these two partial tides are consistent. Wherein the rising tidal current direction of No 1 station point to seashore, falling tidal current direction point to Liaodong Bay mouth; Rising tidal current direction of No.2 station also point to seashore and falling tidal current direction point to Bohai Bay mouth; Since No.3、4 and 5 stations are not located within the Bay, and close to the shore, so the direction of the reciprocating current parallel the seashore. Compare to section 4.1 current direction frequency, semidiurnal partial tide rotation ellipse represent the same direction with primary current direction. This shows that tidal current plays important role in current in Bohai Sea.

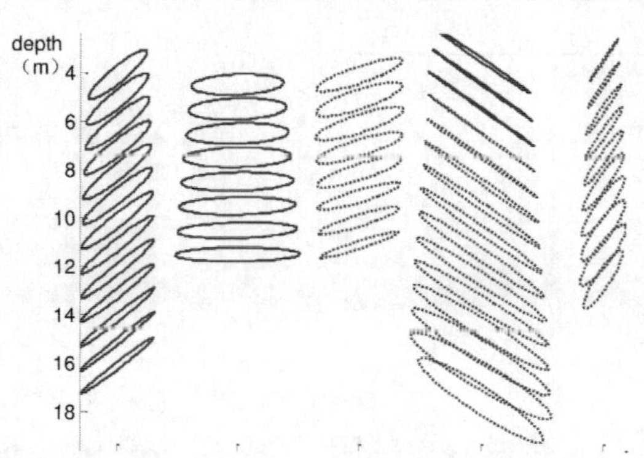

Figure 7. M_2 partial tide ellipse of every fixed mooring station (from left to right is No.1 station to No.5 station).

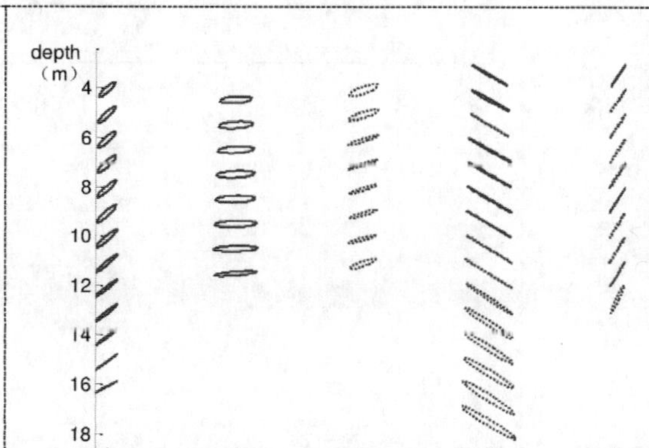

Figure 8. S_2 partial tide ellipse of every fixed mooring station (from left to right is No.1 station to No.5 station).

4.6. Residual current analysis in observation period

The residual current this article refers to is the sum of the various flow after deducting the tidal current from the measured flow, The following figure (Figure 9) shows residual current of every layer of every fixed mooring station during the investigation. Seen from the figure, during the observation No.1 station all layers are flowing northeast except 15 meter layer, wherein the 12 m-15 m layer flow rate are greater than 7cm/s, the flow rate of the other layers is smaller. No2 station all layers substantially are flowing southward, flow rate is less than 5cm/s. No 3 station all layers are flowing southwest, flow rates are small in addition to 10 meter layer. No.4 station several surface layers are flowing southward, deeper layers turn to the westward start from 10 meter, all layers' flow rate are small in addition to bottom layer. No 5 station all layers are flowing southwest, the velocity of the bottom layer is larger and the rest layers is small. On the whole, during the observation period, in addition to No.4 station the residual current have a turn at 10 meter layer, that of the rest stations are stable and the flow rate is smallish. A point worth noting is that residual current at the bottom layer are larger than at the upper layer of each station. Which agree with Jiang WSH[4] measured the C11 station in Bohai Bay.

Compared with previous research results, the surface residual current within the Liaodong Bay (No.1 station) corroborates the viewpoint of Zhao BR[5], who considered that it is clockwise residual current in Liaodong Bay.

Jiang WSH[4,6] launched artificial jellyfish in Bohai Bay and the whole Bohai Sea in summer 1993 and summer 2000, the analysis concluded that the bottom residual current in Bohai is basically less than 5cm/s, the results of this paper show that the bottom residual current in Bohai Sea is mostly about 5cm/s or larger.

Jiang WSH considered that there was an anticlockwise circulation in north Liaodong Bay, and there was a clockwise circulation in Bohai Bay. However, further analysis revealed that the migration of bottom artificial jellyfish didn't fully reflect the direction of circulation, even reversed the trend in some stations. The result of this paper is different with his result, residual current shows up clockwise trend in Liaodong Bay, and there exists an anticlockwise residual current in Bohai Bay.

The residual current direction within Bohai Bay (No.2 station) is anticlockwise, which is different with one clockwise trend of the double current loop structure analysed by Zhao BR[5], we infer it is maybe the influence of wind.

Laizhou Bay (No.5 station) residual current flows substantially in a clockwise trend, which supports the viewpoint of Zhao BR[5].

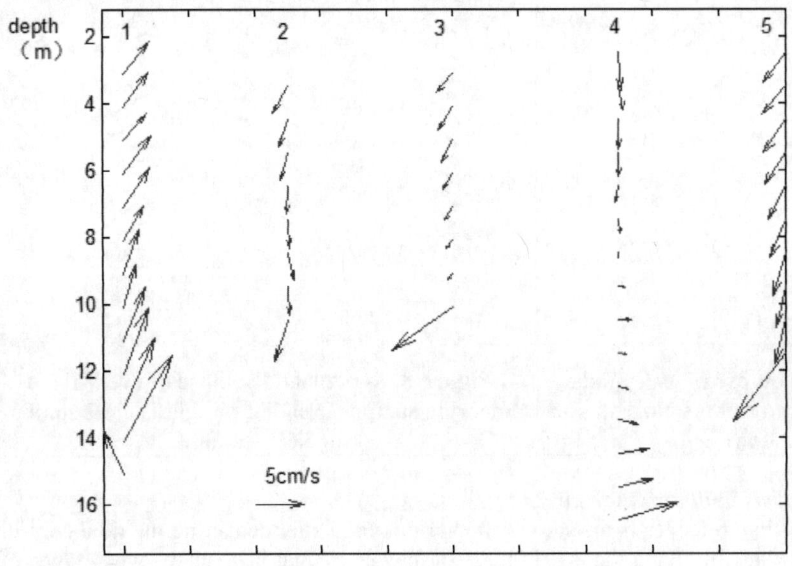

Figure 9. Residual current of each layer of each station (from left to right is No.1 station to No.5 station)

5. Thermohaline characteristics analysis

5.1. Seawater temperature plane distribution characteristic

As shown in Figure 10, the surface temperature distribution situation of the Bohai Sea in summer is that temperature is high along the coast, there present a low-temperature region near the Bohai Strait, and it extends to the direction of the Bohai Bay. With the depth increase, temperature decrease, and the scope of the cold water zone gradually expand. The average surface temperature is 23.58℃.

IOP Conf. Series: Earth and Environmental Science **52** (2017) 012062 doi:10.1088/1755-1315/52/1/012062

Sea surface temperature is lower consider to August plane temperature distribution over the years analysed by He XM[7].

The temperature distribution at 5 meter layer 、 10 meter layer (see Figure 11) and bottom layer is consistent with the temperature distribution measured in summer 2006 which is analysed by Song WP[8].

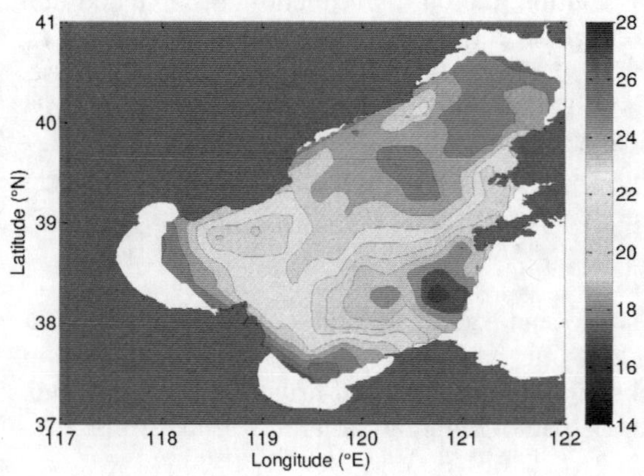

Figure 10. Temperature plane distribution in summer at 0 meter layer (unit: ℃).

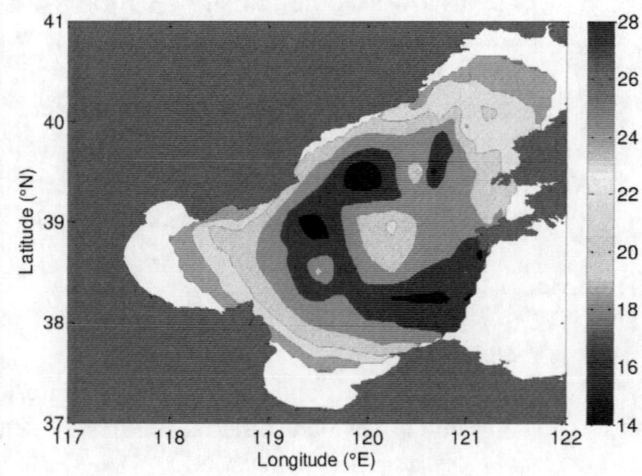

Figure 11. Temperature plane distribution in summer at 10 meter layer (unit: ℃).

5.2. Salinity plane distribution characteristic

The surface salinity in summer as shown in Figure 12 has three high salinity areas and three low salinity areas, they cross each others, the Bohai Bay, middle of Liaodong Bay and the Bohai Strait are high salinity areas, the estuary of Liaodong Bay and middle of Bohai sea are low salinity areas. The high salinity areas under 10 meter (see Figure 13) represent herringbone. The salinity characteristic is consonant with that analyzed by Song WP [8].

The mean surface salinity is 30.93, lower than the mean surface salinity after 90 age analyzed by Bi CC [9].

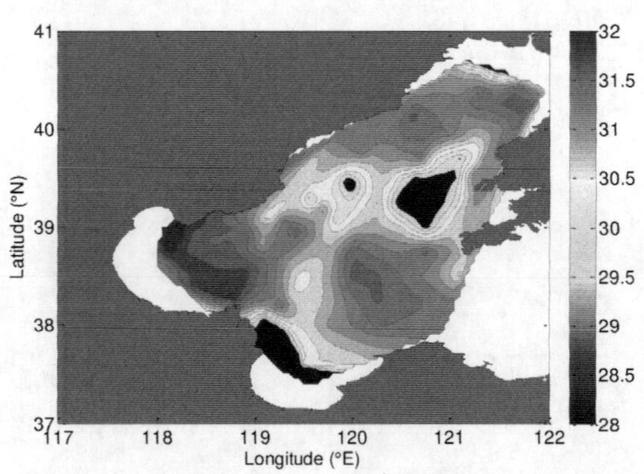

Figure 12. Salinity plane distribution in summer at 0 meter layer.

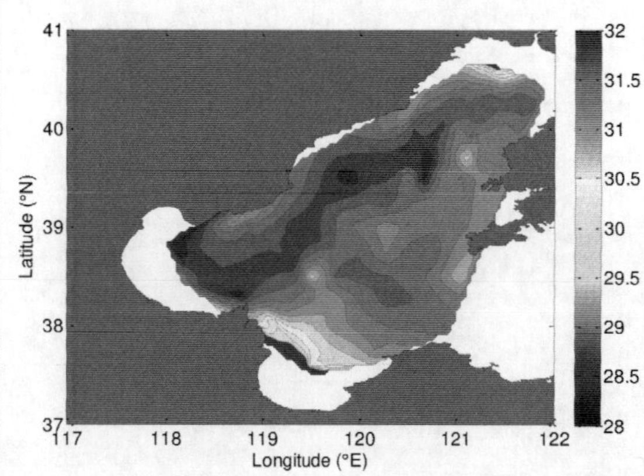

Figure 13. Salinity plane distribution in summer at 10 meter layer.

6. Conclusion

The fixed point observed temperature and salinity, and current data which were quality controlled were analyzed. The results show that:

- The Bohai Sea is the regular half-day tidal current trend, and the tidal current trend is the substantially reciprocating flow properties.
- The reciprocating flow direction within the Bay is trend towards the Bay' s direction, the rest is trend towards the sea shore. During the observation, the current direction is stable and the current rate is small all fixed points, with a good correlation with wind. Thereinto the Liaodong Bay has a clockwise circulation, and the Laizhou Bay flows substantially clockwise either, which consistent with the historical trend observation. One fixed point in the Bohai Bay flows different trend compare with the historical observation.
- The Bohai Sea surface temperature is slightly lower than the in August of the calendar year, salinity has a decreased trend compared to historical statistics.

References

[1] Feng S Z, Li F Q and Li S J 1999 Introduction to ocean science *Higher Education Press* 434-435.

[2] Li G S, Wang H L and Li B L 2005 Numerical simulation on the wind-driven and tide-induced lagrangian residual circulation and its seasonal spatial-temporal variations in the Bohai Sea *Geographical Research* **24**(3) 359-370.

[3] Xu H F, Jiang B, Zhao S H M, Wu H, et al. 2011 Analysis of hydrological environment of Tianjin coast *Ocean Technology* **30**(2) 63-68.

[4] Jiang W S H, Wang J Y, Zhao J Z 1997 An observation of current in BoHai Gulf and its analysis *Journal of Ocean University of QingDao* **27**(1) 23-32.

[5] Zhao B R, Zhuang G W, Cao D M 1995 Circulation, tidal residual currents and their effects on the sedimentations in the Bohai Sea *Oceanologia Et Limnologia Sinica* **26**(5) 466-473.

[6] Jiang W S H, Su J, Yang H, Zhang Y, et al. 2002 Therelation ship between SPM concentration and hydrodynamic condition in the Bohai Sea *Acta Oceanologica Sinica* **24**(1) 311-316.

[7] He X M and Zhang Y K 1990 The temperature distribution characteristic and interannual variability of south central Bohai sea in spring, summer and autumn *Marine Forecasts* **7**(3) 44-50.

[8] Song W P 2009 The Analysis of the structure of T-S and the current characteristics in Bohai Sea during winter and summer, QingDao *Ocean University of China*.

[9] Bi C C, Bao X W, Wan K 2015 The effect of decadal salinity variation on circulation in the Bohai Sea *Periodical of Ocean University of China* **45**(1) 1-8.

Upper ocean response to typhoon Megi (2010)

Yiding Wang[1], Qiang Liu[2, 3] and Yongshuai Ma[2]

[1] Naval Institute of Hydrographic Surveying and Charting, Tianjin 300061, China
[2] Department of Hydrography and Cartography, Dalian Naval Academy, Dalian 116018, China

E-mail: liuqiang3931@163.com.

Abstract. In this paper, the response of the upper ocean to typhoon Megi is analyzed and studied by using the sea surface temperature (SST), 500hPa height field and wind field of NCEP reanalysis data and Argo buoy data. The results show that sea surface temperature (SST) will decrease significantly after typhoon, and on the right side of typhoon path cooling is more obvious than the left; the temperature changes dramatically on the right side of Megi path, and correspondingly the depth variation in mixed layer is also intense, and the distance between buoy and the center of the typhoon will al so affect the severity of the temperature variavtion, the salinity may increase or decrease after typhoon, it depends on the evaporation and precipitation or other relevant factors in this area.

1. Introduction

The tropical cyclone (TC) is a strong low pressure phenomenon that occurs in the tropical or subtropical oceans and is a very strong and wide range of tropical weather system, its energy is mainly provided through the upper ocean [1]. Not only its own rotation, but also moves forward at a fast pace, TC may have great effects on the surrounding sea or land. When the TC passes by, the sea surface wind will increase rapidly, which makes the cold water at the bottom of mixed layer upwelling, mixing with the upper warm water, and further leads to the temperature of mixed layer decreasing and the depth of mixed layer increasing at the same time [2]. Liu Zenghong and Xu Jianping use the data of Argo to analyze the depth variation of mixed layer after TC, and obtain preliminary results about upper ocean response to tropical cyclones in the Pacific Northwest, they consider that the depth of mixed layer increases in more than 50% observed profile after TC passes by, and about 78% mixed layer temperature decreases, the decreasing maximum value reaches 5 degrees, and there is a close relationship between the decreasing temperature and the TC path; in addition, they also find that more than half of the mixed layer salinity slope will decrease [3]. Prie (1981) points out that the most significant phenomenon that the response of the upper ocean to typhoon is the decreasing SST, and the research shows that it is cooler on the right of the typhoon path than the left, SST is an important factor that cause s typhoon and impacts the intensity variation of typhoon [4]; Emanuel (1983) points out that typhoon intensity has a strong correlation with SST [5]. In this paper, we attempt to use the information of SST and sea surface temperature anomaly, NCEP reanalysis data and Argo buoy data to analyze the response of the upper ocean to typhoon Megi on the basis of previous studies, and look for the variation regulars of oceanic elements in the upper ocean after typhoon.

2. Data selection and methods

In this paper, we will study the variation of the SST, salinity and the depth of the mixed layer in the upper ocean due to the impact of typhoon. The SST and temperature anomaly are the daily average value which are from the United States Earth System Research Laboratory (ESRL), the grid resolution is 0.25°x 0.25°; The Argo buoy data is from the U.S. Global Ocean Data Assimilation Experiment Center (USGODAE), combining with the information of typhoon Megi that China typhoon Meteorological Bureau provides, the typhoon center in every moment is the origin, and 4 buoys are screened within the range of longitude, then processing the buoy data selected to draw corresponding change graphs about SST, salinity, the depth of mixed layer when typhoon passes by, to find out the main factors in the variation.

2.1. *NCEP reanalysis data*

The NCEP reanalysis data is exploited to map out the SST daily changes around typhoon Megi path. By comparing the variation to find out the relationship between SST and the typhoon path, and from the graph of daily temperature anomalies can see more intuitive and precise about variation values and the specific range of the SST [6]. By contrasting the upper wind field data and the height field data, we can find reasons for typhoon steering.

2.2. *Argo buoy data*

All Argo buoy data used is selected in a specific latitude and longitude range of the sea area after typhoon, due to the large scope of the typhoon path, so we consider several buoy data within a certain longitude range as the same record. The selected buoy should satisfy: (1) the distance between the location of the typhoon center and the position of buoy should be in four longitude range; (2) buoy should record information for 5-10 days before and after the typhoon comes; (3) the selected Argo buoy data should have been processed [7,8]. In the definition about the depth of the mixed layer, one definition: the temperature in the lowest depth of the mixed layer is 0.5 ℃ colder than the SST, $|T(z)-T(0)| < 0.5$, we consider z as the depth of mixed layer (Sprintall and Tomczak ,1992) [9]; the other definition: mixed layer is defined the part above the pycnocline, so in this paper we select the density gradient is not less than $0.015 kg/m^{-4}$ as the depth of mixed layer (The specification for oceanographic survey data processing, 1992) [10]. The Argo buoy data has a high accuracy, and the thermocline also has certain thickness, in order to reduce the error and ensure the calculated depth of mixed layer more accurate, we select the meridional gradient is less than 0.1 ℃ • *dbar* as the depth of mixed layer that buoys survey.

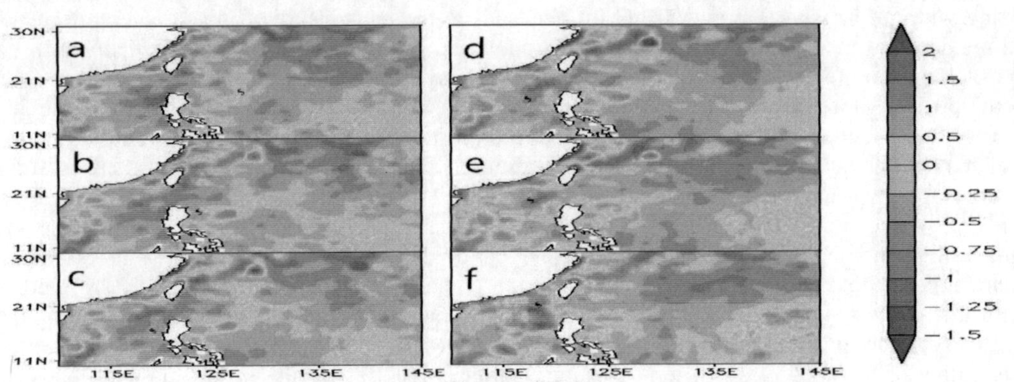

Figure1. Graph a to f respectively show the daily temperature anomalies of typhoon Megi from October 17 to October 22, 2010.

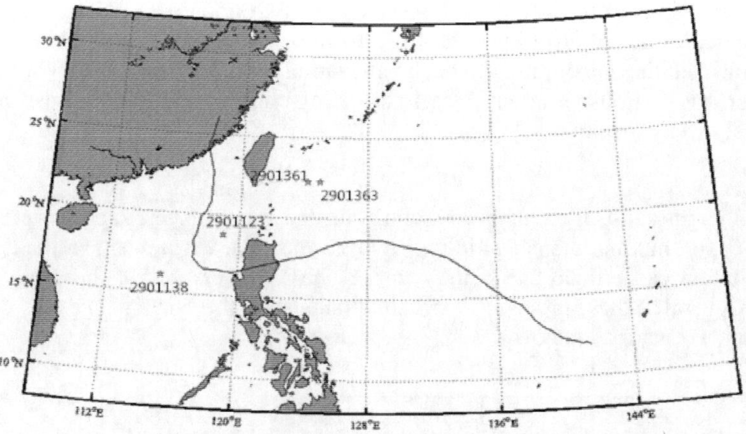

Figure 2. The chart of typhoon Megi path and buoy position

3. Upper ocean response to typhoon Megi

3.1. The response of SST to Megi

Synthetically comparing from graph a to f in figure 1, it can be seen SST where the typhoon passes by is obviously lower than other sea area, the lowest value is about 2 ℃, therefore, we can infer that the typhoon can reduce SST. Meanwhile, further observation we can see it has a significant cooling area on the southwest side of the typhoon before turning from chart d in figure 1.

3.2. The variation of mixed layer depth

The locations of Argo floats and the typhoon path are drawn in figure 2, we can intuitively see the position of buoy 2901138 is close to the turning point of typhoon Mgei, so we can have a good analysis about information of turning point by using data buoy 2901138 records; on the other hand, buoy 2901123 is on the left side of the typhoon, so the changes of response factors at both sides of typhoon Megi can be compared, while the positions of buoy 2901363 and 2901361 are relatively far from the typhoon, so the change range of response factors are relatively small.

Table 1. The depth variation of mixed layer that buoy 2901123 and 2901138 record.

Float label	Recording date	Latitude/°	Longitude/°	Depth variation /m
2901123_159	2010. 10. 13	18. 887	118. 923	50. 10
2901123_160	2010. 10. 17	18. 786	119. 025	39. 90
2901123_161	2010. 10. 21	18. 682	118. 803	90. 60
2901123_162	2010. 10. 25	18. 723	119. 012	70. 50
2901138_138	2010. 10. 13	16. 104	115. 243	79. 80
2901138_139	2010. 10. 17	16. 099	115. 357	50. 50
2901138_140	2010. 10. 21	16. 045	115. 667	60. 50
2901138_141	2010. 10. 25	15. 978	115. 245	59. 20

We know buoy 2901123 is on the right side of the typhoon path, so the change of its temperature amplitude is greater than buoy 2901138, which also shows that SST at the right side of typhoon path has a more obviously cooling than the left around the sea area of typhoon path. We can observe from table 1 the depth variation of mixed layer buoy 2901123 records is particularly intense, the difference between the minimum and the maximum of the depth variations is 50.9m, meanwhile, we have analyzed that SST also has obvious variation from figure 1, by contrast, the depth surveyed by buoy 2901138 has a relatively weak variation.

3.3. The variation of salinity

There are many factors that cause the change of salinity in the mixed layer, such as: precipitation, evaporation of water vapor, mixing intensifies in mixed layer and seawater upwells in the thermocline etc., only the precipitation can reduce the salinity in the above factors, while the other three will increase the salinity. Generally, these processes exist simultaneously; the final change also depends on the strength of the impact of various factors.

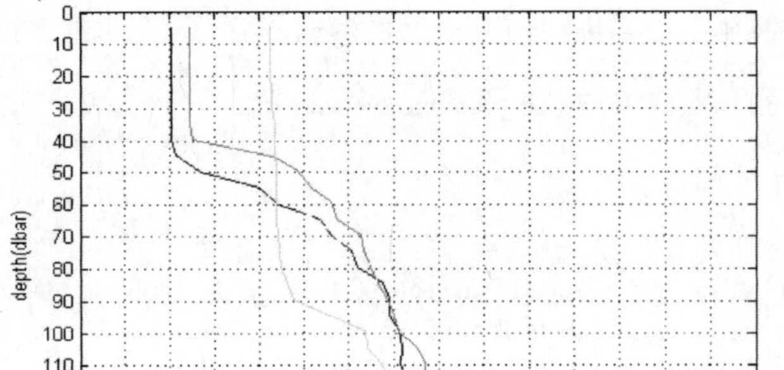

Figure 3. The above two graphs show the variation of salinity with the depth by buoy records, the left is buoy 2901123 and the right is buoy 2901138. Different color curves represent the variation of salinity in different days before and after typhoon comes, black line is the 4 days before the typhoon, red is during typhoon comes , the green is the 4 day after typhoon, yellow is the 8 day after typhoon.

According to the data buoy 2901123 records, we can compare the salinity in the same depth before and after typhoon comes from figure 3(left), we can see the salinity firstly increases after typhoon, the increasing maximum value reaches 0.45 in four days after typhoon by contrasting black line to green line in figure 3(left), and then it comes down, but it is still higher than the salinity before typhoon. While the salinity is firstly down after typhoon according to buoy 2901138 in figure 3(right), the decreasing maximum value is 0.2 by contrasting black line to green line in figure 3(right).Because of the salinity data that buoy 2901138 records is close to the turning point of typhoon, this shows that the typhoon causes the precipitation strengthening in the area when turning, which is more than any other influence of several factors, so the salinity decreases.

4. Conclusions

In this article, we draw the SST profile, the depth map of mixed layer and salinity maps by using the data of SST, sea surface temperature anomalies and reanalysis data from NCEP, and combining with information collected by Argo floats, we analyze and compare the variations of the depth in mixed layer, the SST and salinity at left and right sides of typhoon Megi path before and after typhoon comes. We draw the following conclusions:

The SST will significantly decrease after typhoon, and it is cooler on the right side of typhoon path than the left;

It is noteworthy that there will be a significant "cold spot" on the west or southwest side of the typhoon center before turning , which may be helpful to judge when typhoon will turn;

The temperature changes dramatically on the right side of Megi path, and correspondingly the depth variation in mixed layer is also intense. Also the distance that buoy is from the center of the typhoon will also affect the severity of the temperature changes.

The salinity may increase or decrease after typhoon, it depends on the evaporation and precipitation or other relevant factors in this area.

Reference

[1] BenderM A, Ginis I and KuriharaY 1993 Numerical simulations of tropical cyclone ocean interaction with a high-resolution coupled model *JOURNAL OF GEOPHYSICAL RESEARCH* **23** 245-262.

[2] Ginis I 2002 Atmosphere-Ocean Interactions *Advances in Fluid Mechanics Series* **33** 83-114.

[3] Zenghong Liu and Jianping 2006 Xu Upper ocean response to tropical cyclones in northwestern Pacific during 2001-2004 by Argo data *Journal of Tropical Oceanography* **25(1)** 1-8.

[4] Price J F 1981 Upper ocean response to a hurricane *J. Phys. Oceanogr* **11** 153-175.

[5] EmanuelK A 1986 An air-sea interaction theory for tropical cyclones *J. Atmos. Sci.* **43** 585-604.

[6] Yang Xiaoxia and Tang Danling 2010 Location of sea surface temperature cooling induced by typhoon in the South China *Journal of Tropical Oceanography* **29(4)** 26-31.

[7] Xu Dongfeng, Liu Zenghong and Xu Xiaohua 2005 The influence of typhoon on the sea surface salinity in the warm pool of the western Pacific *ACTA OCEANOLOGICA SINICA* **27(6)** 10-15.

[8] Tong Mingrong, Liu Zenghong and Sun Chaohui 2003 Analysis of the quality control process of the ARGO profiling float data *Marine Technology* **22(4)** 79-84.

[9] Janet Sprintall and Matthias Tomczak 1992 Evidence of the barrier layer in the surface layer of the tropics *J.G .R* **97(C5)** 7305-16.

[10] Xu Yinglong 2011 Forecast analysis on the abrupt northward recurvature of super typhoon Megi (1013) *Meteorological Monthly* **07** 821-826.

Major influencing factors of water flooding in abnormally high-pressure carbonate reservoir

Hou Qingying[1, 2, 3], **Chen Kaiyuan**[1], **Fan Zifei**[2], **Fu Libing**[2], **Chen Yefei**[2]

[1] School of Energy Resources, China University of Geosciences, Beijing 100083, China

[2] Research Institute of Petroleum Exploration & Development, PetroChina, Beijing 100083, China

E-mail: houqingying@petrochina.com.cn

Abstract: The higher pressure coefficient is the major characteristics of the abnormal high pressure carbonate reservoirs, which the pressure coefficient generally exceeds 1.2 and the initial formation pressure is higher than normal sandstone reservoirs. Due to the large pressure difference between initial formation and saturated pressure, oil wells are capable to production with high flow rate by the natural energy at early production stage. When the formation pressure drops to the saturation pressure, the water or gas is usually injected to stabilize the well productivity and sustain the formation pressure. Based on the characteristics of Kenkiak oilfield, a typical abnormal high pressure carbonate reservoir, a well group model is designed to simulate and analyze the influence factors on water flooding. The conclusion is that permeability, interlayer difference and reserve abundance are the main three factors on the water flooding development in these reservoirs.

1. Introduction

Kenkiak pre-salt reservoir is located in the southern of the eastern border zone of the Peri-Caspian depression. Main productive reservoir was formed in the Carboniferous age [1]. It is characterized by abnormally high-pressure (80 MPa, pressure coefficient of 1.84) with the bury depth about 4300 meters and serious heterogeneity [2]. The difference between initial formation pressure and saturated pressure is 46.2MPa. The density of the crude oil is 0.816g/cm^3. This reservoir is classified as low permeability, high difference between initial formation pressure and saturated pressure, and weak volatile oil [3]. Reservoir distribution is controlled by sediment faces and vuggy effects. Based on the statistics of fractures in cores, the matrix porosity averages 9.35% with permeability of 0.44 md, while the fracture porosity ranges from 0.001 to 1.5% with fracture permeability from 0.025 to 800md[4].

Oil well production in this field is controlled by fractures and vuggy development. Analyzed on the stratum structures, sediment phase, fractures, reservoir development features and production performance distribution, the study area can be divided into high permeability zone and low permeability zone, as showed in Figure 1.

High permeability zone, located below salt-dome in the central-western area, is a fractured-porosity reservoir. Mini-fractures are developed and the permeability in dual-porosity media is over 5md. Low permeability zone is located in the south part of the reservoir. Mini-fractures are underdeveloped and permeability in dual-porosity ranges from 0.5 to 5md.

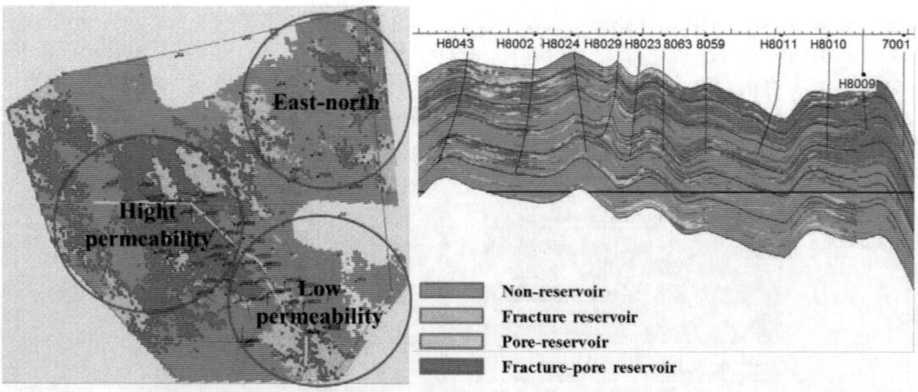

Figure 1. Classification of Carboniferous Reservoir under Salt-Dome (Left: Plane; Right: cross)

2. Production performance

Kenkiak oilfield is an abnormally high pressure carbonate reservoir with large pressure difference between initial formation pressure and saturated pressure which brings enough natural energy in the initial stage of development. However, oil daily production in different zones is significant due to the permeability. In the high permeability zones full of fractures, the horizontal and vertical connectivity greatly increased and the reserve controlled by single well, the fracture plays an important role of communication in the reservoir development. The initial average daily production can reach to 250t/d per well. While the low permeability zone is lack of fractures and the communications both in vertical and horizontal directions are poor resulting in relatively lower initial well productivity of 130t/d per well, which is lesser than that in high permeability zone[5].

In this kind of oilfield, the initial well productivity can be achieved at high level. However, oil rate will decrease rapidly due to formation pressure depleted fast. After more than one decade of development, formation pressure dropped to the level of 39% to 56% of the initial formation pressure, especially the formation pressure in high-permeability zones will be close to the saturation pressure. Then gas will come out of the crude oil with rising producing GOR. In order to sustain formation pressure, water flooding pilot project was initiated in 2009. However, pressure surged near the injector and the water injection rate was too low accordingly. After acidizing, injected volume rose slightly, and well injectivity did not improve much. Energy deficit in the reservoir is so serious that adjustment is an urgent need [6-7].

3. Analysis of major influencing factors in water flooding

In most cases, large pressure drawdown is characterized in the initial development stage of carbonate reservoir with abnormal overpressure. The formation pressure gradually decreases with production time, which leads to the increase of effective stress. The fractures and caves gradually close with the increased effective stress, which finally results in the decrease of reservoir permeability and porosity [8]. Therefore, to recover pressure loss and to maintain formation pressure, water flooding should be implemented when the formation pressure closely drops to saturation pressure. The primary target of water flooding is to keep fractures and pores in open status, which could provide flow channels for oil and improve development performance [9-10]. Based on the geological and physical properties of Carboniferous pre-salt reservoir in Kenkyak Oilfield, a well group model is established and simulated with numerical simulation software to analyse the effects of geological factors on water flooding performance and also provide an optimization criterion for water flooding [11-12].

3.1. Model Parameters and Well Pattern

Geologically, the Carboniferous formation in Kenkyak field can be classified as 9 zones: oil zone - barrier - oil zone - barrier - oil zone - barrier - oil zone-barrier-aquifer (Figure 2). The corresponding parameters of model are listed in Table 1. The porosity and permeability of each zone are determined by geological statistics of pre-salt reservoir in Kenkyak Oilfield. The pre-salt oil reservoir in Kenkyak Oilfield can be divided into high-permeability and low-permeability regions. In calculations, the parameters of C2bI1, C2bI2, C2bI3, and C1sII1 respectively correspond to four oil zones. C1sII2 and C1sII3 zones are almost below oil-water-contact, which can be treated as a water zone [13-14].

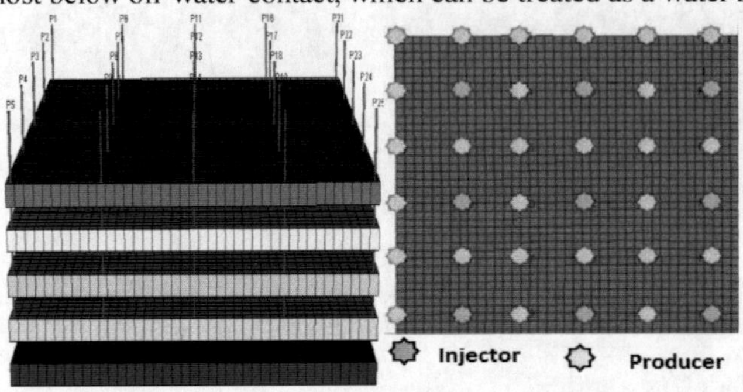

Figure 2. Theoretical model in Water flooding

Table 1. Parameters of geological model

Parameters	Low permeability	High permeability
Cell numbers	2050×2050×180	2100×2050×180
Well pattern	inverted nine-spot	inverted nine-spot
Oil-water interface(m)	4226	4226
Pi(MPa)	78.38	78.38
Pb(MPa)	33.80	33.80

The permeability contrast ranges of high-permeability and low-permeability regions can be obtained according to the statistics of field geological model (Table 2). The matrix permeability in high-permeability region is 0.05-20 md, fracture permeability ranges from 0.05-50 md, K_f/K_m is 1-100/200 and matrix porosity is 6-20%. The matrix permeability in low-permeability region is 0.05-5 md, fracture permeability is 0.05-5 md, K_f/K_m is 1-100 and the matrix porosity is 6%-18%.

Table 2 Parameters of High-Permeability and Low-Permeability zones

Simulation Zone		Oil zone 1	Oil zone 2	Oil zone 3	Oil zone 4	Aquifer	
Layer		C2bI1	C2bI2	C2bI3	C1sII1	C1sII2	C1sII3
High-permeability zones	Netpay（m）	10.93	14.34	12.9	12.56	6.15	2.5
	Por（%）	9.78	8.86	8.45	8.24	7.31	6.89
	Kx（$10^{-3}\mu m^2$）	4.11	2.95	3.17	4.22	1.3	0.78
	Ky（$10^{-3}\mu m^2$）	4.31	3.14	3.36	4.38	1.41	0.85
	Kz（$10^{-3}\mu m^2$）	1.17	0.68	0.73	1.19	0.27	0.17
Low-permeabilty zones	Netpay（m）	7.94	7.72	7.48	8.24	9.42	5.28
	Por（%）	8.45	7.62	7.61	7.48	7.61	6.9
	Kx（$10^{-3}\mu m^2$）	0.33	0.29	0.36	0.2	0.26	0.076
	Ky（$10^{-3}\mu m^2$）	0.35	0.32	0.41	0.22	0.29	0.079
	Kz（$10^{-3}\mu m^2$）	0.18	0.15	0.15	0.1	0.12	0.062

Considering the field test pressures and formation breakdown pressure, the bottom-hole flowing pressure upper limits of water injector and gas injector are set 78MPa and 50MPa respectively, and the lower limit of bottom-hole flowing pressure for oil producer is 25MPa. Formation pressure should be maintained above the bubble point pressure. According to the analysis of actual well production performances in the high-permeability and low-permeability regions of pre-salt reservoir in Kenkyak Oilfield, the allocated initial well productions for low-permeability and high-permeability regions are $50m^3/d$ and $100m^3/d$ respectively.

An inverted nine-spot well pattern is both deployed in the high-permeability and low-permeability regions in this reservoir. The corresponding well spacing is 500m and 350m respectively in high-permeability and low-permeability regions (Figure 2). According to the image principle, the corner-well and edge-well can be treated as a quarter and a half of a whole well respectively. After producer-injector conversion, 12 producers and 4 injectors are designed in the high-permeability region and 25 producers and 9 injectors are in the low-permeability region.

3.2 Impacts of geological factors on water flooding

The effects of multiple geological factors on the water flooding performance of Carboniferous pre-salt reservoir in Kenkyak Oilfield are investigated by numerical simulation, mainly including permeability, inter-zone permeability contrast, and reserve abundance. First of all, permeability is given within the range of reservoir permeability. Inter-zone permeability variation is utilized to compare the heterogeneity of vertical zones. Barriers are used for the study of lateral dimensions and the extent of development. Reserve abundance is gained on the basis of average reserve within the reservoir.

3.2.1. Impact of Permeability. The injection capacity is greatly dependent on reservoir permeability in water flooding development, which directly affects the water flooding performance. Therefore, the effects of reservoir permeability on the water flooding performances of high-permeability and low-permeability regions in the Carboniferous formation of Kenkyak Oilfield are analysed and the corresponding permeability contrast values used in simulation are listed in Table 3.

Table 3. Model Parameters for Permeability Impact

Zone	K_{max}/K_{min}	K_{max}
Low Permeability Zone	1	0.2, 0.5, 1, 5, 10
High Permeability Zone	1	1, 5, 10, 20, 70

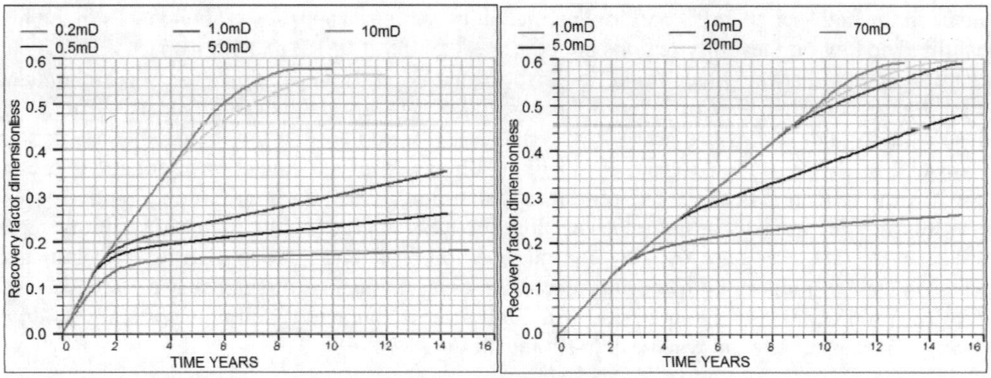

Figure 3. Impact of permeability on recovery factor to date in water flooding (Left: low-permeability zone; Right: high-permeability zone)

Figure 3 illustrates that the water flooding performance is greatly dependent on reservoir permeability. The recovery degree increases with the higher reservoir permeability. However, with the increase of permeability, the recovery degree of its impact will be reduced.

3.2.2. Impact of Inter-zones Heterogeneity. The inter-zone permeability contrast should be defined before analyzing the effect of inter-zone heterogeneity on water flooding performance. A reservoir will be divided into multiple subzones in the longitudinal direction, which have maximum permeability (K_{max}) and minimum permeability (K_{min}). Therefore, K_{max} / K_{min} can be used to the inter-zone permeability contrast.

Pre-salt reservoir of carboniferous formation in Kenkyak Oilfield is mainly dominated by reverse rhythm. The maximum permeability appears in the uppermost zone. Therefore, assuming the permeability of the uppermost zone is fixed, by altering inter-zone permeability contrast is to analyze the effect of inter-zone heterogeneity contrast on water flooding performance. The permeability of other zones is determined by geometric progression. The injected water tends to break through along high-permeability zone in case of relatively high inter-zone permeability contrast, which will lead to poor water flooding performance. Figure 4 indicates that the water flooding performance is also greatly dominated by inter-zone permeability contrast. The simulated recovery degree decreases with the higher inter-zone permeability contrast.

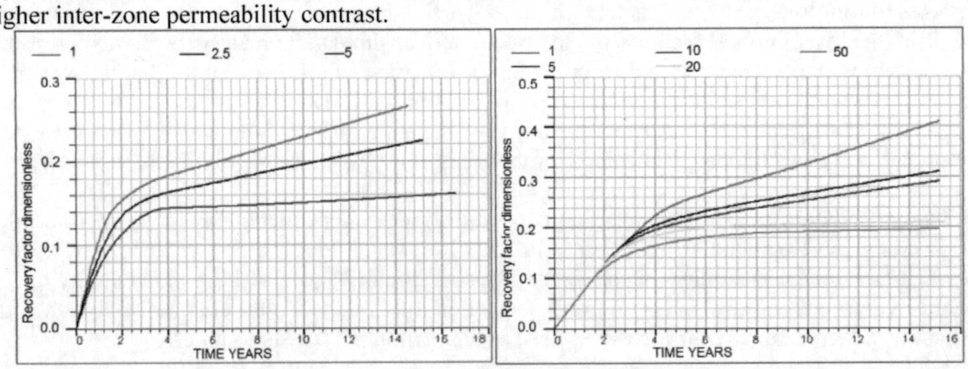

Figure 4. Impact of inter-zones permeability variation on recovery factor (Left: low-permeability zone; Right: high-permeability zone)

3.2.3. Impact of reserves abundance. The influence of reserve abundance on the water flooding performance of Carboniferous pre-salt reservoir is inquired by slightly altering the average reserve abundance in the high-permeability and low permeability regions. The reserve abundances in the high-permeability and low-permeability regions are all set as 0.5, 0.8, 1.0, 1.2 and 1.5 respectively.

Figure 5 reveals that the water flooding performance is partly dependent on reserve abundance, which is slightly less than that of reservoir permeability and inter-zone heterogeneity. The reserve abundance and formation pressure in the low-permeability region are too low to achieve a production with constant recovery rate. Actual production follows constant injection-production pressure drawdown with production decline. The producer is shut down when daily oil production decreases to less than $8m^3/d$, which results in the relatively low recovery degree. In the high-permeability region, constant recovery rate can be achieved by constant well oil production due to high reserve abundance. The recovery degree reduces with the increase of reserve abundance under equal oil recovery rate. However, the production lasts a long period with high reserve abundance and no significant difference in recovery degree can be found when water-cut reaches to 98%.

4. Conclusions

Permeability, inter-zone permeability variation, and reserve abundance are the major factors to influence water flooding performance in both porosity-type and fractured-porosity reservoirs. Permeability depending on fractures has the most pronounced impact. The higher permeability is, the higher the oil recovery degree is. The difference between layers may result in reducing the water drive effect. The reserve abundance also has influence on the water flooding development.

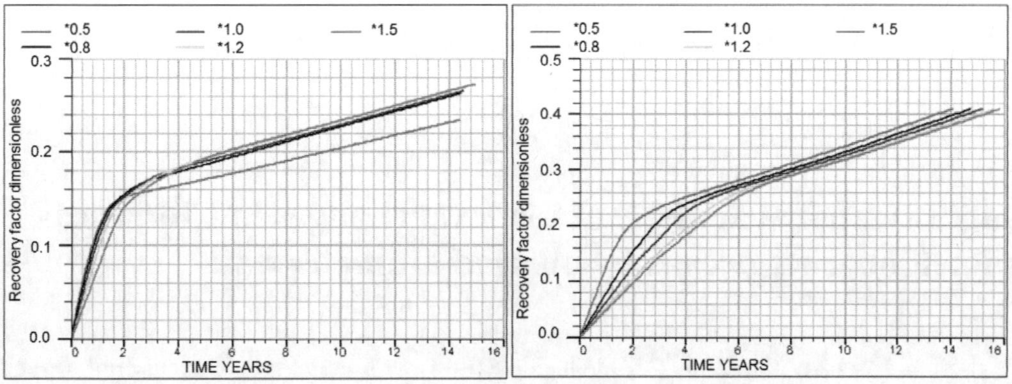

Figure 5. Impact of reserves abundance on recovery factor (Left: low-permeability zone; Right: high-permeability zone)

References

[1] Miao Qianyou, Li Yunjuan 2005 Stress-sensitive reservoir is a reasonable system of work *Daqing Petroleum Institute* **29**(6) 49-51.

[2] Bian Dezhi, Zhao Lun, Chen Yefei, et al. 2011 Characteristics and formation mechanism of fractures in abnormally high - pressure carbonate reservoirs: A case study of Kenjiak oil field in Kazakhstan *Petroleum Exploration and Developmen* **38**(4) 394-399.

[3] Zhaolun, Chen Yefei, et al. 2013 Abnormal high-pressure carbonate reservoir stress-sensitivity evaluation *Petroleum Exploration and Development* **40** 194-199.

[4] Bai Guoping 2006 Characteristics of the distribution of large carbonate reservoirs in the world *Journal of Palaeogeography* **8**(2) 241-250.

[5] Song Heng, Fu Xiujuan, Fan Hailiang, et al. 2009 Characteristics and technical policies of fractured carbonate reservoir with gas cap *Petroleum Exploration and Development* **36**(6) 756-761.

[6] Li Chuanliang 2008 Stress sensitivity and productivity characteristics of fractured reservoirs *Xinjiang Petroleum Geology* **29**(1) 72-75.

[7] Liu Luofu, Zhu Yixiu, Hu Aimei, et al. 2002 Oil and Gas Geological Characteristics of the Lower Cretaceous in the Caspian Basin *Southwest Petroleum Institute* **24**(3) 11-15.

[8] Wilkinson J R, Teletzke G F, King K C 2006 Opportunities and challenges for enhanced recovery in the Middle East *SPE* 101679

[9] Zhao Lun, Fan Zifei, Song Heng, et al. 2009 Improving technology of utilization of low permeability carbonate reservoir *Petroleum Exploration and Development* **36**(4) 508-512.

[10] Wang Yuying, Wang Xiaodong, Wang Yifei, He Ru et al. 2005 Analysis of reasonable production pressure difference and productivity of oil reservoirs in deformation media *Journal of Daqing Petroleum Institute* **29**(4) 51-54.

[11] Li Chuanliang 2005 Reservoir Engineering Principles *Beijing: Petroleum Industry Press*

[12] Zhang Jianguo, Lei Guanglun, Zhang Yanyu 1998 Fluid seepage mechanics *Dongying: Petroleum University Press*

[13] Ge Jiali 1998 the application of complex flow systems and experimental fluid mechanics *Beijing University of Petroleum Press* **27**.

[14] Li Road 2006 Efficient and low-permeability reservoirs *Low permeability oil and gas fields* **11**(3).

Spatial-temporal analysis of sea level changes in China seas and neighboring oceans by merged altimeter data

Yao Xu[1], Bin Zhou[1, 3], Zhifeng Yu[1], Hui Lei[1], Jiamin Sun[1], Xingrui Zhu[1], Congjin Liu[2]

[1] Institute of Remote Sensing and Earth Sciences, Hangzhou Normal University, Hangzhou, Zhejiang, 311121, P.R.China
[2] Hangzhou Institute of Service Engineering, Hangzhou Normal University, Hangzhou, Zhejiang, 311121, P.R.China

E-mail: zhoubin@hznu.edu.cn

Abstract. The knowledge of sea level changes is critical important for social, economic and scientific development in coastal areas. Satellite altimeter makes it possible to observe long term and large scale dynamic changes in the ocean, contiguous shelf seas and coastal zone. In this paper, 1993-2015 altimeter data of Topex/Poseidon and its follow-on missions is used to get a time serious of continuous and homogeneous sea level anomaly gridding product. The sea level rising rate is 0.39 cm/yr in China Seas and the neighboring oceans, 0.37 cm/yr in the Bo and Yellow Sea, 0.29 cm/yr in the East China Sea and 0.40 cm/yr in the South China Sea. The mean sea level and its rising rate are spatial-temporal non-homogeneous. The mean sea level shows opposite characteristics in coastal seas versus open oceans. The Bo and Yellow Sea has the most significant seasonal variability. The results are consistent with in situ data observation by the Nation Ocean Agency of China. The coefficient of variability model is introduced to describe the spatial-temporal variability. Results show that the variability in coastal seas is stronger than that in open oceans, especially the seas off the entrance area of the river, indicating that the validation of altimeter data is less reasonable in these seas.

1. Instruction

As a result of global warming and glacial melting, global sea levels have an obviously rising trend[1]. The rising of sea levels would lead to more flooding. It's a disaster to human long-term development, especially for the coastal areas [2, 3]. Knowledge of rate and acceleration of sea level changes is critical important for a variety of social, economic and scientific development. And it is also important for planning the response of coastal and island population to rising sea levels [4].

During past hundreds of years, the tide gauges were used to study sea level changing [5], while it has some uncertain problems in analysis. The references of different places are not the same, so corrections must be applied individually [6, 7]. And it's difficult to get a full map of sea level changes worldwide because most tide gauges are at coastal places, not open seas [8, 9].

People began to use radar altimeter on satellite to monitor the sea level in late 1960s. In the 1970s and 1980s, several satellites were sent and laid the foundations for a new generation of ocean satellites. Since the launch of Topex/Poseidon (T/P) in 1992, with precise orthography and location system, the accuracy of the satellite positioning began to enable monitoring of sea surface height variations due to low-amplitude ocean dynamical processes. Besides of T/P, ERS-1/2, Envisat, Spot, Jason-1/2/3 and HY-2A were sent in the last 20 years [10]. Researchers began to study sea level changes in large scales with satellite altimeter data, which was difficult with tide gauges [11, 12]. National Center for Space Studies (CNES) of France calculated the mean sea surface level worldwide with a 19 year satellite dateset [13]. Pascuel et al [14] analyzed the ocean mesoscale variability with merged data from 4 satellites and compared the result from tide gauge, showing that satellite data had a better description. Stammer et al [15] studied the variability and spatial structure of world sea surface height with data from T/P. Some others studied the rising rate of sea level with a single or merged missions' data, showing that the global rising rate is around 0.32 cm/yr during last 20 years [16-18].

In China, half of the population and product are in the coastal area. According to the Bulletin of China Sea Level by the National Ocean Agency (NOA) of China, the coastal sea level rises at 0.3 cm/yr in last 35 years, which is higher than the global average [19]. Researches also find out that the sea level changes are spatially no-homogeneous and vary in different seasons [20-22]. Most people's researching focuses are on the South China Sea (5° N-23° N, 105° E-123° E) as it' s a half-open ocean connecting with the Pacific Ocean and has a deep basin bordered by two broad shelf regions [22]. The sea level variability in the South China Sea is proved to have deep correlation with global climatology and oceanography [23]. Guo et al [24] studied sea level rising characteristics on all China seas and the neighboring oceans, the altimeter data was updated to 2012. Results show that the most significant period of sea changing over seas is 1 year, in addition to changing cycle of 9 year in the South China Sea.

In this paper, we continue updating the altimeter data to the latest year 2015, make several improvements in data editing and merging, analyze the seasonal signals and introduce a spatial-temporal analysis method to estimate the level of spatial-temporal variability. To get a homogeneous along track data, the latest mean sea surface model CLS2011 [13] is applied to each dataset instead of its default model. We merge the along track data from different missions before gridding in order to reduce the systematical errors between missions. A modified Shepard algorithm is used for mapping, which can solve the uneven distribution and slope of samples [25]. When analyzing spatial-temporal characteristics, a coefficient of variability (CV) model is applied to divide the China seas into several levels respecting the variability.

2. Data and method

Among previous and current missions, Jason-1 (2001-2013) and Jason-2 (2008-now) are the follow-on missions to Topex/Poseidon (1992-2006), as a cooperation between NASA and CNES. These missions have the same reference ellipsoid orbit parameters [26-28]. So they are chosen to map the sea level in focus seas to get a continuous and homogeneous result. The Geophysical Data Record (GDR) of the missions are processed to monthly gridded map with following steps:

2.1. Quality control and geophysical corrections

For each parameter in the Geophysical Data Record (GDR) file, it has a flag indicating whether the data is reliable. The bad data is a result of instrument conditions or processing error. The bad data must be removed from the dataset first. And editing criteria is a further way to improve the quality. For a homogeneous standard and result, the latest mean sea surface height model CLS2011 [13] is applied to all GDR data. Geophysical corrections are applied individually to get the sea level anomaly (SLA).

2.2. Along track gradient and crosscover adjustment

In theory, the repeating pass of different cycles should be in the same ground track. Because of various factors such as the earth gravity field model and the solar radiation, the satellite ground track is usually drift by ± 1 km. The problem is solved by interpolating samples to a given orbit [27]. Also, SLA at the crosscover point of different passes should be the same while the difference is a result of systematical bias between two passes. An optimal adjustment model is designed to analyze and make adjustments to different month-mean passes to minimal the difference at crosscover point [16].

2.3. Data merging

There is a period of time that T/P and Jason-1, Jason-1 and Jason-2 are on board together. The validation and calibration between two missions were finished during the period [29]. There is a significant systematical bias between two missions. At China seas, different researchers give a bias around -8.6cm between T/P and Jason-1[30], and around 10cm between Jason-1 and Jason-2 [30]. In this paper, we compare the month by month data, and get a bias of -13.27 cm between T/P and Jason-1, and 10.45 cm between Jason-1 and Jason-2. The absolute bias in this paper is bigger than earlier research is a result of changing the mean sea level model. The default model of T/P is based on limited data while CLS2011 is derived from multiple-satellites as long as 19 years. CLS2011 is also the follow-on version of CLS2001, which is the default model of Jason-1.

2.4. Gridding

The along track scatter data is spatial interpolated regular maps to show the continuously spatial variance. As the satellite track points's distribution is uneven in longitude and latitude, it must be considered to balance the weight of different directions. Some scholars give a proper weight to different points before interpolation, while a modified Shepard method take the weight problem and slope of the area in account[30]. Modified Shepard method is improved from inverse-distance algorithm and makes several improvements to consider the uneven distributions of samples, remove samples with too-short distance and add a proper weight on specific direction if samples are on the slope.

3. Result and discussion

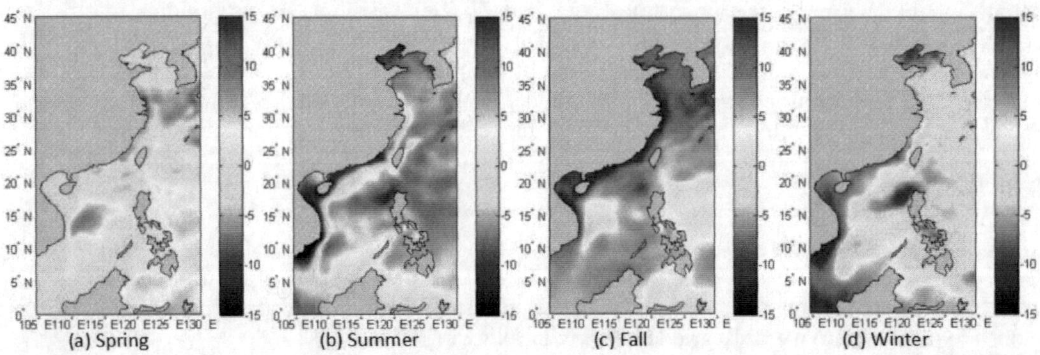

Figure 1. 1993 to 2015 season mean of SLA (cm). The season mean is a simple averaging of month mean.

From the month mean product, we calculate the season (Spring: Mar to May, Summer: Jun to Aug, Fall: Sep to Nov, Winter: Dec to Feb) mean from 1993 to 2015. From Figure 1, the sea level in coastal seas has a opposite characteristics to that in open oceans, especially in Fall. In Spring and Summer, the mean sea level in coastal seas is lower than that in open oceans while in Fall and Winter, the sea level in coastal seas is higher than that in opens oceans. Compared with in situ data in coastal Guangxi and Xisha, which is in the middle of the South China Sea, from NOA'S Bulletin of China Sea Level[19] in past 4 years, sea levels at coastal Guangxi are higher than the reference in 3 years, while those at Xisha are lower than the reference. In general, sea levels in Winter and Spring are 5 to 10 cm lower than those in Summer and Fall in the same year, the result is consistent with in situ observations [19].

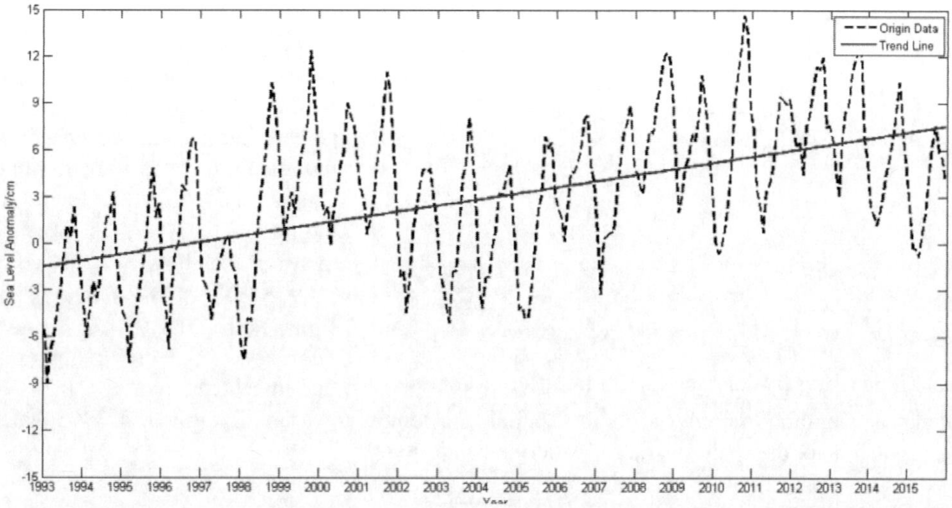

Figure 2. The sea level rising rate in China seas and neighboring oceans. The trend line is $y = 0.39x - 785.65$ .The model is significant different from zero at 95% confident. Statistical calculations are based on the student's t-test and the degrees of freedom are estimated by considering the autocorrelation of data [24, 31]

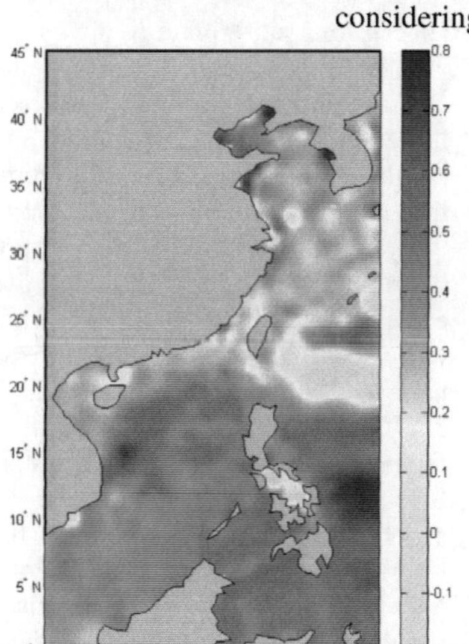

Figure 3. Spatial distributions of 1993-2015 rising rate (cm/yr). Each rising rate is calculated from the 23-year time series data. 99% rate falls into the domain [-0.2 0.8].

To analyze the rising rate of sea level, least-square method is used to solve the linear model $y = ax + b$ to get the rising rate 0.39 cm/yr in China seas and the neighboring oceans. It is 0.08 cm/yr lower than Guo's [24] result 0.47 cm/yr, which is derived from the 1993-2012 altimeter data of the same missions. From Figure 2, we can find out that in the last three years (2013-2015), the sea level has a descending trend. If we chose the same period as Guo's research, the rising rate is 0.48 cm/yr. The 3-year descending trend and the 1999-2004 descending trend are also found in inter-decadal cycles of the SLA temporal variability [32].

The 0.39 cm/yr rising rate is higher than 0.30 cm/yr (from 1980 to 2015) given by NOA [19]. The Bulletin of China Sea Level' s focus area is coastal seas, and from Figure 3, coastal seas rising rates in the Southeast coastal seas are 0.1-0.2 cm/yr lower than those in the open oceans. Another reason is that the rising rate in last two decades are high than ever before [3, 19, 33]. Figure 3 also shows the spatial distributions of rising rate, it' s the highest in the South China Sea (5° N—23° N, 105° E—125° E) and lowest in the East China Sea (23° N—32° N, 117° E—130° E). In most cases, the rising rate in coastal seas are lower than those in open oceans.

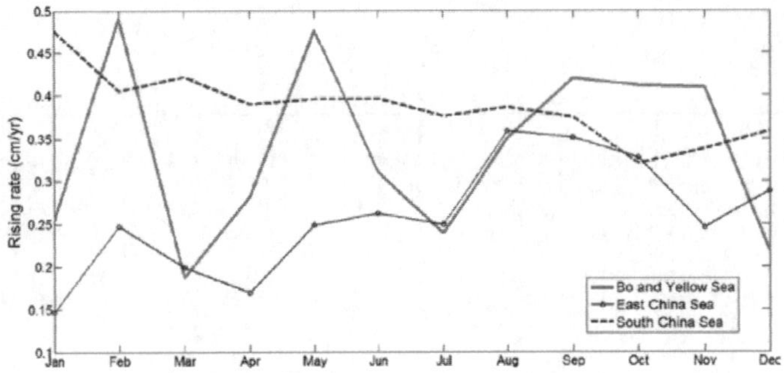

Figure 4. The rising rate of each month in different regions.

From Figure 4, we confirm the temporal variability in different regions, the rising rate in the South China Sea is the highest in 7 months in the year while in the East China Sea, it' s the lowest in 7 months in the year. The variability in Bo and Yellow Sea (32° N—45° N, 115° E—127° E) is the most significant, the rising rate in February is 0.3 cm/yr higher than that in March.

To make a quantitative analysis on the spatial and temporal variability, given a 3×3 window around a pix in a map, the spatial variance is defined as follows:

$$CV_s = \frac{s_{window}}{\mu_{window}} \quad (1)$$

s_{window}, μ_{window} is the standard deviation and average value of data in the 3×3 window. The spatial-temporal coefficient of variance $CV_{s,t}$ is the time average of CV_s. In this paper, month mean SLA data is standardized to domain [0,1], and the results are given in Figure 5 and Figure 6.

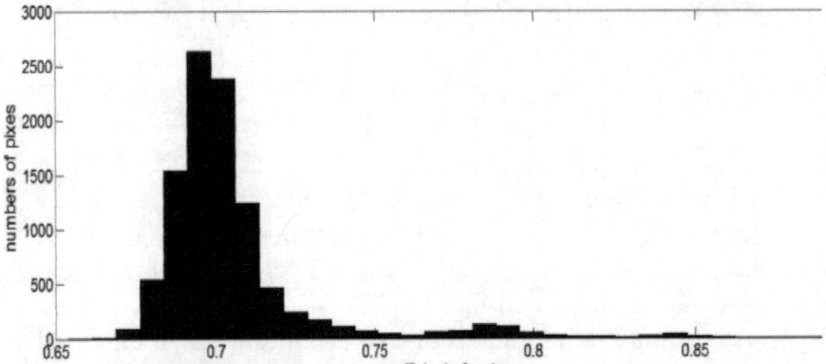

Figure 5. The distribution of CV in China seas and neighboring oceans. X-axis stands for the coefficient of variability and Y-axis stands for the number of pix with the specific CV. 99% of the CV falls into the domain [0.65 0.9]

The coefficient describes the spatial-temporal variability of sea level changes in China seas and neighboring oceans. From Figure 5, the coefficient between 0.67 and 0.75 follows the normalized distribution while there is 5% extreme data higher than 0.75. From Figure 6, the coastal seas are the areas with the extremely high coefficient. It means the spatial-temporal variance in coastal sea is 0.05 higher than that in the open sea, especially at the entrance of the river. The Yangtze River, the Yellow River and the Pearl River are the three biggest rivers in China and the Mekong River is the biggest river in Southeast Asia. There are 8 sea gates in the entrance of Pearl River and 8 sea gates in the entrance of the Mekong River. So the seas off these areas show a significant high variability. The middle of the South China Sea is in the low variability level while most area of the Bo and Yellow Sea and the East China Sea are in the Medium level.

The map of CV can be applied to determine the position of offshore platform of altimeter absolute calibration. In absolute calibration of HY-2A radar altimeter, two positions in the Bo Sea and the South China Sea were chosen, whose seasonal signals of sea surface height were weak[3, 19, 33]. The method of determining the position could be more precise and reasonable with the CV model.

4. Conclusion

We process the GDR data to a merged month mean gridding product. The method has less systematical errors in theory and be consistent with others' results and in situ observation. From the season mean product, we find out that the sea level in China seas has significant difference among seasons. In general, the sea level in Summer and Fall is higher than that in Winter and Spring. Among all regions, the Bo and Yellow Sea has the most significant seasonal variability, the mean sea level in Summer is 15 cm higher than that in Winter. The rising rate from 1993-2015 altimeter data is 0.39 cm/yr, which is higher than the global average and NOA's observation but lower than the rate from 1993-2012 altimeter data because of the descending trend in last 3 years. The rising rate of the South China Sea is the highest in 7 months in the year while the rising rate of the East China Sea is the lowest in 7 months in the year. The temporal variance of the Bo and Yellow Sea is strictly stronger than the others. The result from altimeter data is consistent with the Bulletin of China Sea Level given by NOA. The CV model shows that the spatial-temporal variance in coastal seas is extremely higher than that in open oceans, especially the seas off the entrance area of rivers. The CV model can be applied to improve the method of determining the position in absolute calibration.

Figure 6. The coefficient of spatial-temporal variance of sea level changes. The CV lower than 0.7 is believed to be in the low level, CV between 0.70 and 0.75 in the medium level and that higher than 0.75 in the high level.

Acknowledgement

We are very grateful to AVISO for providing the data of T/P, Jason-1 and Jason-2. The Research is supported by the State High-Tech Development Plan of China (2014AA12330), the Public Benefit Technique Research and Social Development Key Program of the Science Technology Department of Zhejiang Province, China (2014A23002) and the Summer Internship Program for College Student 2016 of the State Ocean Administration of China.

References

[1] Lambeck K and Chappell J 2001 Sea level change through the last glacial cycle *SCIENCE* **292** 679-86

[2] Nicholls R J and Cazenave A 2010 Sea-level rise and its impact on coastal zones *SCIENCE* **328** 1517-20

[3] Houghton J T, Jenkins G J and Ephraums J J 1990 Climate change: the IPCC scientific assessment *American Scientist;(United States)* **80**

[4] Dasgupta S, Laplante B, Meisner C M, Wheeler D and Jianping Yan D 2007 The impact of sea level rise on developing countries: a comparative analysis *World Bank policy research working paper*

[5] Church J A, Clark P U, Cazenave A, Gregory J M, Jevrejeva S, Levermann A, Merrifield M A, Milne G A, Nerem R S and Nunn P D 2013 Sea level change.: PM Cambridge University Press)

[6] Trupin A and Wahr J 1990 Spectroscopic analysis of global tide gauge sea level data

GEOPHYS J INT **100** 441-53

[7] Holgate S J and Woodworth P L 2004 Evidence for enhanced coastal sea level rise during the 1990s *GEOPHYS RES LETT* **31**

[8] Church J A, White N J, Coleman R, Lambeck K and Mitrovica J X 2004 Estimates of the regional distribution of sea level rise over the 1950-2000 period *J CLIMATE* **17** 2609-25

[9] Church J A and White N J 2006 A 20th century acceleration in global sea-level rise *GEOPHYS RES LETT* **33** n/a-n/a

[10] Xu K, Jiang J and Liu H 2013 HY-2A radar altimeter design and in flight preliminary results. In: *2013 IEEE International Geoscience and Remote Sensing Symposium-IGARSS*: IEEE) pp 1680-3

[11] Imawaki S, Uchida H, Ichikawa H, Fukasawa M and Umatani S I 2001 Satellite altimeter monitoring the Kuroshio transport south of Japan *GEOPHYS RES LETT* **28** 17-20

[12] Thomas R, Rignot E, Casassa G, Kanagaratnam P, Acuña C, Akins T, Brecher H, Frederick E, Gogineni P and Krabill W 2004 Accelerated sea-level rise from West Antarctica *SCIENCE* **306** 255-8

[13] Schaeffer P, Faugere Y, Legeais J F and Picot N 2011 The CNES CLS 2011 global mean sea surface *OST-ST, San-Diego*

[14] Pascual A, Faugère Y, Larnicol G and Le Traon P Y 2006 Improved description of the ocean mesoscale variability by combining four satellite altimeters *GEOPHYS RES LETT* **33**

[15] Stammer D 1997 Global characteristics of ocean variability estimated from regional TOPEX/POSEIDON altimeter measurements *J PHYS OCEANOGR* **27** 1743-69

[16] Traon P L and Dibarboure G 1999 Mesoscale mapping capabilities of multiple-satellite altimeter missions *J ATMOS OCEAN TECH* **16** 1208-23

[17] Berry P, Garlick J D, Freeman J A and Mathers E L 2005 Global inland water monitoring from multi - mission altimetry *GEOPHYS RES LETT* **32**

[18] Le Traon P Y, Nadal F and Ducet N 1998 An improved mapping method of multisatellite altimeter data *J ATMOS OCEAN TECH* **15** 522-34

[19] NOA 2016 Bulletin of China Sea Level.

[20] Li L, Jindian X and Rongshuo C 2002 Trends of sea level rise in the South China Sea during the 1990s: An altimetry result *CHINESE SCI BULL* **47** 582-5

[21] Ho C R, Zheng Q, Soong Y S, Kuo N J and Hu J H 2000 Seasonal variability of sea surface height in the South China Sea observed with TOPEX/Poseidon altimeter data *Journal of Geophysical Research: Oceans* **105** 13981-90

[22] Shaw P, Chao S and Fu L 1999 Sea surface height variations in the South China Sea from satellite altimetry *Oceanologica Acta* **22** 1-17

[23] Ho C, Kuo N, Zheng Q and Soong Y S 2000 Dynamically active areas in the South China Sea detected from TOPEX/POSEIDON satellite altimeter data *REMOTE SENS ENVIRON* **71** 320-8

[24] Guo J, Hu Z, Wang J, Chang X and Li G 2015 Sea Level Changes of China Seas and Neighboring Ocean Based on Satellite Altimetry Missions from 1993 to 2012 *J COASTAL RES* **73** 17-21

[25] Yang C, Kao S, Lee F and Hung P 2004 Twelve different interpolation methods: A case study of Surfer 8.0. In: *Proceedings of the XXth ISPRS Congress*, pp 778-85

[26] Benada R J 1997 Merged GDR (TOPEX/POSEIDON). Generation B Users Handbook, Version 2.0 Physical Oceanography Distributed Active Archive Center (PODAAC), Jet Propulsion Laboratory, Pasadena, JPL D-11007 **131**

[27] Zanife O and Dumont J S T. Guinle (2004), SSALTO Products Specifications-Volume 1: Jason-1 User Products, Issue 3.1.: SMM-ST-M-EA-10879-CN, CLS/CNES, Toulouse, France)

[28] Dumont J P, Rosmorduc V, Picot N, Desai S, Bonekamp H, Figa J, Lillibridge J and Scharroo R

2009 OSTM/Jason-2 products handbook *CNES: SALP-MU-M-OP-15815-CN, EUMETSAT: EUM/OPS-JAS/MAN/08/0041, JPL: OSTM-29-1237, NOAA/NESDIS: Polar Series/OSTM J* **400**

[29] Leuliette E W, Nerem R S and Mitchum G T 2004 Calibration of TOPEX/Poseidon and Jason altimeter data to construct a continuous record of mean sea level change *MAR GEOD* **27** 79-94

[30] Sun W, Wang Q B and Zhu Z D 2013 Sea level anomaly series in China Sea and its vicinity based on multi-generation satellite altimetric data *Acta Geodaetica et Cartographica Sinica* **4** 493-500

[31] Sea level trends, interannual and decadal variability in the Pacific Ocean.

[32] Bretherton C S, Widmann M, Dymnikov V P, Wallace J M and Bladé I 1999 The Effective Number of Spatial Degrees of Freedom of a Time-Varying Field. *J CLIMATE*

[33] Cabanes C, Cazenave A and Le Provost C 2001 Sea level rise during past 40 years determined from satellite and in situ observations *SCIENCE* **294** 840-2

[34] Liu Y 2014 Calibration Technology for HY-2A radar altimter Sea Surface Height. (Tsingdao: Ocean University of China)

Regional atmospheric visibility characteristics and its effect on radiation in China

S C Chen[1,4], T Y Hao[2], J Shi[1], X L Shan[3]

[1] Tianjin Meteorological Observation Centre, Tianjin 300061, China
[2] Tianjin Environmental Meteorological Centre, Tianjin 300074, China
[3] Tianjin Binhai New Area Meteorological Bureau, Tianjin 300457, China

E-mail: csc_zhuce@126.com

Abstract. Spatial and temporal characteristics of atmospheric visibility in China were investigated by using monitored meteorological data. The results show that The atmospheric lowest visibility (<10km) occurrence frequency in China can be observed in the east of Sichuan, Chongqing, and the east of Guizhou. The visibility in most regions of China exhibits a decreasing trend from 1980 to 2010. It is also found that fog is major weather phenomenon affecting visibility in China. The second is precipitation except the Northwest and Inner Mongolia, where sand and dust is the major weather phenomenon affecting visibility. Our results also identify the total radiation under the condition of low visibility is less than that under the condition of normal, meanwhile, the atmospheric scattered radiation increase. Net radiation in the condition of low visibility is less than that in the normal condition in most cases over China.

1. Introduction

Atmospheric visibility is an indicator to reflect the degree of atmospheric transparency. So it is an obvious qualitative indicator of air quality. Atmospheric visibility in the daytime is defined as the distance at which an average observer can just barely distinguish a black object silhouetted against the horizontal skyline. In an uncontaminated atmosphere, visibility normally ranges from 145 to 225 km [1]. Atmospheric visibility has not only attracted the attention of the public, but also provides a useful way for investigating variations in the atmosphere [2]. Atmospheric visibility is well received by the impact of meteorological factors. Visibility increases with temperature and wind speed, while declines with relative humidity and atmospheric pressure [3, 4]. Visibility is also affected by human activities. Many studies have researched long term visibility trends and the impacts of dominant air pollutants [5-8]. Important studies have been undertaken to examine visibility characteristics at regional, continental, and global scales [9-11].

This study aims at better understanding the climatic characteristics of atmospheric visibility and its impact. So we conducted research on atmospheric visibility in China by regional division. In view of the different regions, regional characteristics and influences of atmospheric visibility were analyzed.

2. Material and methods

Data at 593 observation stations (excluding those mountainous stations) across China were collected. They include atmospheric horizontal visibility (4 times per day), wind speed (4 times per day), and daily average relative humidity from January 1980 to December 2010, as well as observed radiation data at 122 observation stations among 593 from January 1980 to December 2000.

The trend of annual average visibility in different regions was analyzed. Further efforts have been also made to explore temporal distribution and duration of low visibility occurrence frequencies in different regions. Finally, the influence of low visibility to radiation has been examined.

3. Results and discussion

3.1. Spatial distribution of atmospheric low visibility occurrence frequency

The spatial distribution of atmospheric low visibility (<10km) occurrence frequency (occurrence times/total observation times) in China averaged over 1980 through 2010 is shown in Figure 1. The highest atmospheric low visibility (<10km) occurrence frequency can be observed in the east of Sichuan, Chongqing, and the east of Guizhou where there were at least 0.5% on an annual basis in winter (Figure 1d). Those areas filled by green color in Figure 1a and Figure 1c, namely in spring and autumn, including Jiangxi, Hunan, Hubei, Fujian, and Guangdong provinces also observed relatively high occurrence frequency with annual fog days greater than 0.2%. The average annual fog days in the rest of China were less than 0.2%. In terms of nationwide, the visibility is best in summer.

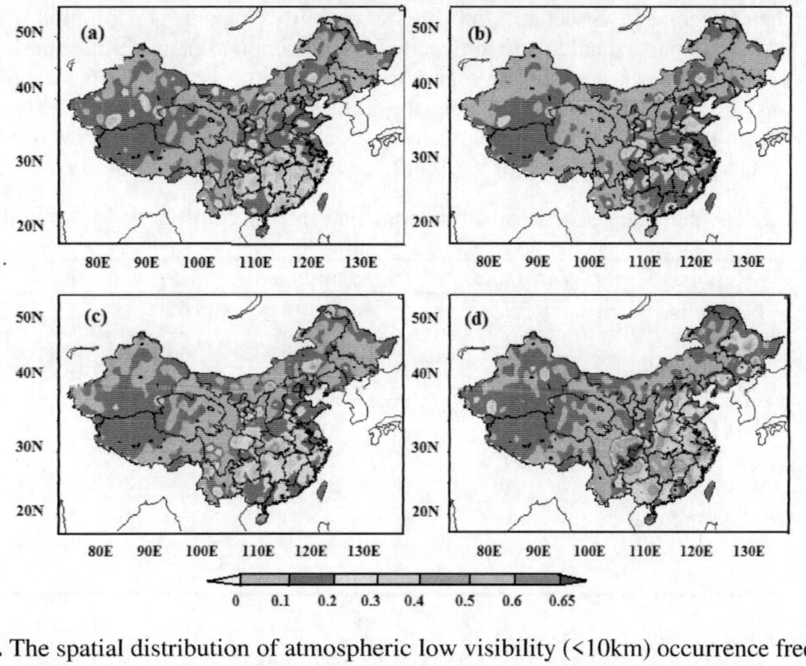

Figure 1. The spatial distribution of atmospheric low visibility (<10km) occurrence frequency in China from 1980 to 2010. (a) Spring (b) Summer (c) Autumn (d) Winter

3.2. Trend of atmospheric visibility

The variation of 10-years average visibility is presented in Figure 2. According to the administrative division, China is divided into nine regions. As shown, the visibility variation in total 30 years exhibit a decreasing trend in most areas of China, including Northwest, Central China, North China, East China, Northeast, and South China(according to the administrative division of China). But in Tibet, Inner Mongolia, and Southwest, the visibility first increased and then decreased subsequently. In particular, a strong decreasing trend can be observed in all regions in China.

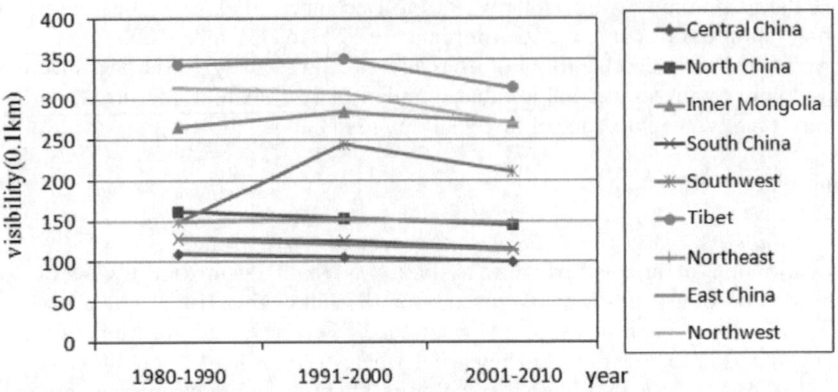

Figure 2. The interdecadal variation of atmospheric visibility (4 times) in 9 regions of China

3.3. The proportion of weather phenomena affecting visibility

The main weather phenomena, affecting atmospheric visibility, include precipitation, dust weather, fog and haze. In order to understand the major weather phenomenon affecting atmospheric visibility in different regions of China, we analyze the major weather phenomenon under the condition of low visibility. The result is shown in Table 1. From the Table 1, we can get that fog is major weather phenomenon affecting visibility in China. The second is precipitation except the Northwest and Inner Mongolia, where sand and dust is the major weather phenomenon affecting visibility.

Table 1. The percentages of weather phenomena causing atmospheric low visibility.

Region	Precipitation	Sand and dust	Fog	Haze
Northeast	25	3	67	3
North China	20	2	71	2
Northwest	26	36	32	0
Central China	31	0	61	2
East China	27	0	63	3
South China	29	0	55	10
Southwest	33	2	55	2
Inner Mongolia	14	40	40	0
Tibet	37	24	25	0

3.4. Duration of atmospheric low visibility

Figure 3 shows that the duration of atmospheric low visibility in 9 regions of China. The proportion which low visibility lasted one day is the highest in vast majority areas of China. Secondly, the proportion which low visibility lasted two days is the higher. Meanwhile, the proportion of sustained two days in North China and East China is the highest than that in the other regions. Overall, the duration of atmospheric low visibility is not very long, more than 70% lasted for one day. But East China is an exception, this shows that the duration of low visibility in East China is longer compared to other regions.

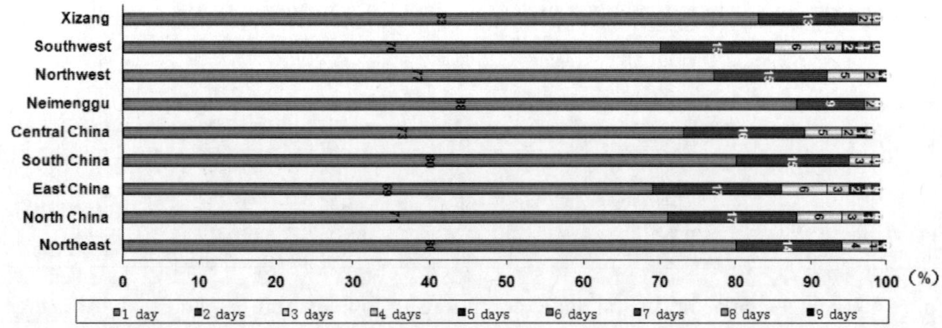

Figure 3. Duration of atmospheric low visibility in 9 regions of China

3.5. The influence of atmospheric visibility to radiation

Atmospheric visibility is an indicator to reflect the degree of atmospheric transparency. The atmospheric aerosols have an important impact on the solar radiation. In order to understand the low visibility on the influence of solar radiation, we compared and analyzed the variation of total radiation, scattered radiation, and net radiation under the condition of low visibility and normal.

3.5.1. Total radiation. The variation of total radiation under the conditions of low visibility and normal is shown in Table 2. Total radiation in the condition of low visibility is less than that in the normal condition in four seasons over China. If the reduction amount to counting, the largest of the three regions are Northwest, Southwest, and Tibet. The reduced maximum value of radiation appeared in different seasons for different regions. The reason could be the weather phenomena influencing the regional visibility is different. Overall, the effect of low visibility is to reduce the total radiation.

Table 2. The average total radiation under conditions of low visibility and normal ($10^4 J/m^2$)

	Winter		Spring		Summer		Autumn	
Region	Low visibility	Normal	Low visibility	Normal	Low visibility	Normal	Low visibility	Normal
Northeast	808	894	1884	2065	2343	2565	1171	1339
North China	889	1079	2043	2221	2370	2576	1272	1471
Northwest	870	1089	2017	2340	2487	2719	1329	1553
East China	1106	1322	2113	2309	2380	2545	1480	1653
Central China	1033	1157	2122	2169	2345	2461	1457	1554
South China	1411	1570	2125	2285	2235	2396	1697	1863
Southwest	1253	1724	1985	2272	2421	2733	1791	1867
Tibet	946	1434	1941	2364	2024	2827	1202	1778
Inner Mongolia	719	1003	1982	2235	2489	2772	1051	1461

3.5.2. Scattered radiation. The variation of scattered radiation under the conditions of low visibility and normal is shown in Table 3. Scattered radiation in the condition of low visibility is more than that in the normal condition in four seasons over China. If the increase amount to counting, the largest regions is Northwest. The reduced maximum value of radiation appeared in spring and summer of Northeast. Overall, the effect of low visibility is to increase the scattered radiation.

3.5.3. Net radiation. The variation of net radiation under the conditions of low visibility and normal is shown in Table 4. Net radiation in the condition of low visibility is less than that in the normal condition in most cases over China. The influence of visibility on the net radiation is more complex. When visibility is low, the total radiation is less than that under the condition of normal, meanwhile, the atmospheric counter radiation and scattered radiation increase. But seen from the Table 4, net radiation under the condition of low visibility is less than that under the condition of normal. In Tibet and North China, the net radiation under the condition of low visibility is more than that under the condition of normal.

Table 3. The average scattered radiation under conditions of low visibility and normal ($10^4 J/m^2$)

Region	Winter		Spring		Summer		Autumn	
	Low visibility	Normal	Low visibility	Normal	Low visibility	Normal	Low visibility	Normal
Northeast	305	266	562	512	666	565	322	301
North China	362	313	713	539	780	584	441	362
Northwest	366	324	973	597	986	558	569	390
East China	419	423	747	710	767	620	475	477
Central China	479	407	857	727	826	709	529	482
South China	492	422	759	699	793	592	555	459
Southwest	511	319	816	567	780	620	538	319
Tibet	366	210	611	428	658	319	388	209
Inner Mongolia	351	233	659	446	577	404	478	255

Table 4. The average net radiation under conditions of low visibility and normal ($10^4 J/m^2$)

Region	Winter		Spring		Summer		Autumn	
	Low visibility	Normal	Low visibility	Normal	Low visibility	Normal	Low visibility	Normal
Northeast	-87	-154	614	651	1217	1298	146	206
North China	44	62	793	681	1276	1180	408	362
Northwest	-13	-54	484	719	542	1014	216	318
East China	158	197	872	881	1166	1293	463	498
Central China	178	224	883	883	1308	1265	557	526
South China	403	407	964	1100	1297	1370	650	680
Southwest	213	708	748	775	1433	--	548	583
Tibet	120	29	582	629	1208	752	390	182
Inner Mongolia	-76	-41	583	600	746	878	--	188

4. Conclusions

In this study extensive analyses were conducted to identify spatial and temporal characteristics of visibility and its effect on the radiation using the meteorological data from 593 observational stations across China from 1980 to 2010. The primary results are summarized below.

- The highest atmospheric low visibility (<10km) occurrence frequency can be observed in the east of Sichuan, Chongqing , and the east of Guizhou where there were at least 0.5% on an annual basis in winter.
- The visibility variation in total 30 years exhibit a decreasing trend in most areas of China. In Tibet, Inner Mongolia, and Southwest, the visibility first increased and then decreased subsequently.
- Fog is major weather phenomenon affecting visibility in China, The second is precipitation except the Northwest and Inner Mongolia, where sand and dust is the major weather phenomenon affecting visibility.
- The proportion which low visibility lasted one day is the highest in vast majority areas of China. The duration of low visibility in East China compared to other regions is longer.
- When visibility is low, the total radiation is less than that under the condition of normal, meanwhile, the atmospheric scattered radiation increase. Net radiation in the condition of low visibility is less than that in the normal condition in most cases over China.

Acknowledgements
The work was supported by the project of Tianjin Meteorological Bureau (201727ybxm19).

References

[1] Environmental Protection Agency 2001 Visibility in mandatory federal class I areas (1994 - 1998), report, Off. Of Air Qual. Plan. and Stand., Research Triangle Park, N.C.
[2] Watson J G 2002 Visibility, science and regulation *Journal of the Air and Waste Management Association* **52** 628–713
[3] Tsai Y I 2005 Atmospheric visibility trends in an urban area in Taiwan 1961–2003 *Atmos. Environ* **39** 5555–67
[4] Tang I, Wong W, Munkelwitz H 1981 The relative importance of atmospheric sulfate and nitrates in visibility reduction *Atmos. Environ.* **15** 2463-71

[5] Schichtel B A, Husar R B, Falke S R, Wilson W E 2001 Haze trends over the United States, 1980–1995. *Atmos. Environ.* **35**, 5205–5210
[6] Doyle M, Dorling S 2002 Visibility trends in the UK 1950–1997 *Atmos. Environ.* **36** 3161–72
[7] Molnár A, Mészáros E, Imre K, Rüll A 2008 Trends in visibility over Hungary between 1996 and 2002 *Atmos. Environ.* **42** 2621–29
[8] Chang D, Song Y, Liu B 2009 Visibility trends in six megacities in China 1973–2007 *Atmospheric Research* **94(2)** 161-167
[9] Mahowald N M, Ballantine J A, Feddema J, Ramankutty N 2007 Global trends in visibility: implications for dust sources *Atmos. Chem. Phys.* **7** 3309–39
[10] Park R J, Jacob D J, Kumar N, Yantosca R M 2006 Regional visibility statistics in the United States: Natural and transboundary pollution influences, and implications for the Regional Haze Rule *Atmospheric Environment* **40(28)** 5405-23
[11] Cheng M T, Tsai Y I 2000 Characterization of visibility and atmospheric aerosols in urban, suburban, and remote areas. *Science of the total environment* **263(1)**, 101-114

The salinity gradient power generating system integrated into the seawater desalination system

Yongqiang Zhu, Wanjun Wang[1], Bingqian Cai, Jiacheng Hao and Ruihua Xia

North China Electric Power University, Beijing 102206, China

E-mai: wwj19931021@163.com

Abstract. Seawater desalination is an important way to solve the problem of fresh water shortage. Low energy efficiency and high cost are disadvantages existing in seawater desalination. With huge reserve and the highest energy density among different types of marine energy, salinity gradient energy has a bright application prospect. The promotion of traditional salinity gradient power generating systems is hindered by its low efficiency and specific requirements on site selection. This paper proposes a salinity gradient power generating system integrated into the seawater desalination system which combines the salinity gradient power generating system and the seawater desalination system aiming to remedy the aforementioned deficiency and could serve as references for future seawater desalination and salinity gradient energy exploitation. The paper elaborates on the operating principles of the system, analyzes the detailed working process, and estimates the energy output and consumption of the system. It is proved that with appropriate design, the energy output of the salinity gradient power generating system can satisfy the demand of the seawater desalination system.

1. Introduction

Energy consumption is increasing rapidly along with technology development across the world, therefore exploitation of new energy resources and reductions of energy loss have become a task of top priority. On the other hand, fresh water shortage has become an indisputable fact among the world. Huge marine energy and potential to be desalinated to fresh water makes ocean a research hotspot. Marine researches follow two major routes including marine energy generation and seawater desalination. Seawater desalination is one of the major approaches to solve the problem of fresh water shortage. According to [1], current frequently applied methods of seawater desalination in the world include multi-stage flash evaporation (MSF), low temperature multi-effect distillation and reverse osmosis. Statistics at the end of year 2003 by International Desalination Association shows that market share of MSF accounts for 66.3% of the aforementioned three major methods of seawater desalination. However, the essence of seawater desalination is to exchange electricity for fresh water. Roughly, one third of cost in seawater desalination is energy consumption. It has always been the goal of researchers and one of the major repetition and development during the past several decades to cut the energy consumption during seawater desalination in order to lower the cost of seawater desalination. Consequently, reducing the energy consumption of seawater desalination is topic of great importance faced by researchers.

Salinity gradient energy is one of many ocean energy sources. It is estimated by experts that the energy capacity of salinity gradient energy is approximately 10 GkW [2]. Furthermore, salinity gradient energy has the largest energy density in comparison with other types of marine energy sources, which is of great value in real projects. Salinity gradient energy which is also called concentration gradient energy is a kind of electrochemical potential energy generated by the gradient difference between seawater and fresh water or seawater of different salinity gradient. According to [3], major salinity gradient energy exploitation method includes pressure retarded osmotic; reverse electrodialysis; vapour pressure differences. At present, the application of salinity gradient energy is limited to estuaries. The efficiency of salinity gradient energy is relatively low because of the limited concentration gradient between seawater and fresh water, therefore applications of salinity gradient energy don't apply to areas that are short of fresh water. Rivers in China are severely polluted, which makes a relatively high pre-processing cost, therefore traditional power-generating method based on salinity gradient energy does not meet conditions for large-scale applications. Salinity gradient energy power generating system broadens integrated in seawater desalination system takes seawater of high concentration outputted by seawater desalination as draw solution and seawater as raw material solution, which broadens application range of salinity gradient energy, adds to osmotic pressure which will boost power density, reduces energy cost of seawater desalination system, and adds to economic value.

Based on principles of osmotic pressure energy method, this paper proposes a salinity gradient energy power generating system integrated in seawater desalination system which combines salinity gradient energy power generating system and seawater desalination system, which will reduce the energy cost during the process of seawater desalination, and boost the economic value of the system.

2. Scheme Design and Principle Explanation

2.1. System Structure and Working Principle

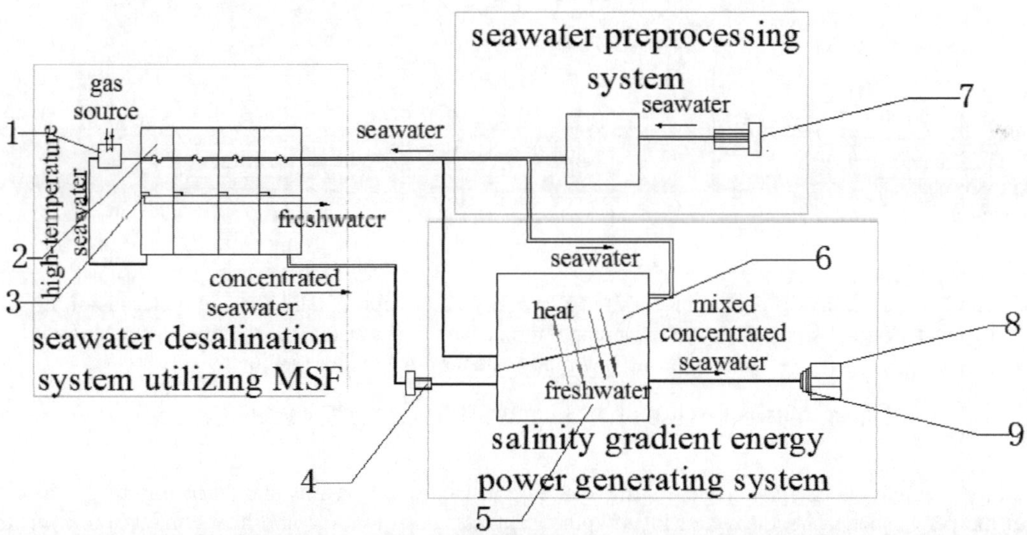

1. seawater heater 2. MSF room 3. gas-water separator 4. pump 1 5. tank for draw solution 6. tank for raw material solution 7. pump 1 8. generator 9.water turbine.

Figure 1. Structure chart of the seawater desalination system combined with the salinity gradient power generation system.

As shown in Figure 1, the system consists of seawater preprocessing system, seawater desalination system utilizing MSF and salinity gradient energy power generating system. Water pump 1 draws ordinary seawater into seawater preprocessing system where regular preprocessing including sterilization, sediment and deoxidize is conducted; then, seawater after preprocessing is fed into cooling water pipe at the upper section of the last stage of MSF. The seawater flowing into the MSF during operation of the device is preheated, and serves as cooling water at the same time. The preheated seawater is discharged from upper section of the first stage of MSF, and turns into high-temperature seawater. Then, high-temperature seawater is fed into the bottom of first flashing stage of MSF. Pressure inside the flash evaporation room is lower than the pressure of saturated vapor corresponded to the high-temperature seawater, therefore as superheated vapor, high-temperature seawater is evaporated rapidly. Tiny droplet of seawater brought by evaporated vapor is eliminated by gas-water separator at the middle section of the flash evaporation room. Separated vapor ascends to the upper section of the flash evaporation room, and condensates fresh water when coming across pipes in condensation zone. Drops of fresh water drips into plates of fresh water, and is led to water tank outside the MSF room for storage after converge. Desalinated seawater discharged after desalination flows into tank of draw solution of salinity gradient energy power generating system through water pump 2, and ordinary seawater drawn by water pump 1 flows into tank of raw material solution after preprocessing. Following the principles of osmotic pressure, fresh water in ordinary seawater is drawn by desalinated seawater. The mixed desalinated seawater drives the water turbine through water pipe. Then, generator is driven by water turbine. The ordinary seawater discharged by salinity gradient energy system is fed into MSF seawater desalination system again through recycle water pipe for cyclic utilization.

2.2. Salinity Gradient Power Generation System

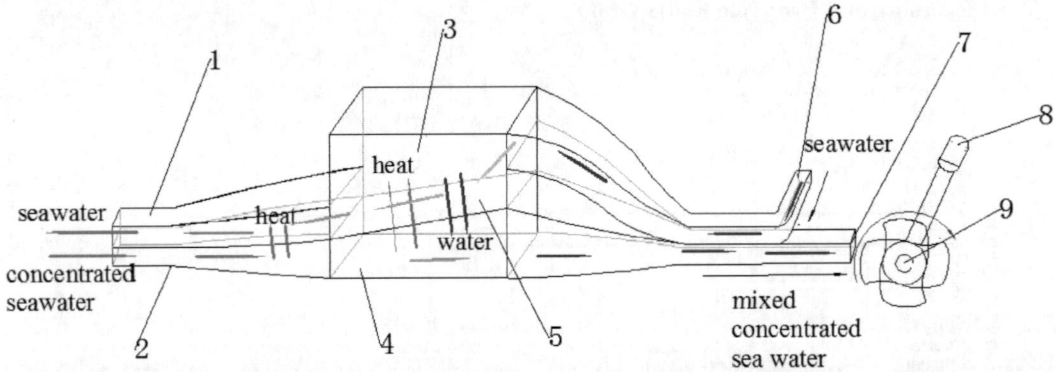

1. seawater outflow pipe 2. concentrated seawater inflow pipe 3. tank for raw material solution 4. tank for draw solution 5. semipermeable membrane 6. seawater inflow pipe 7. concentrated seawater outflow pipe 8. generator 9.water turbine.

Figure 2. Structure chart of the salinity gradient power generation system.

Salinity gradient energy power generating system consists of 4 parts, as shown in Figure 2. The first part includes concentrated seawater inflow pipe, seawater outflow pipe and heat conduction layer; the second part includes tank for draw solution, tank for raw material solution and semipermeable membrane; the third part includes seawater inflow pipe, concentrated seawater outflow pipe and heat conduction layer; the fourth part includes water turbine and generator. Desalinated seawater discharged from seawater desalination system flows into tank for draw solution through desalinated seawater inflow pipe. Seawater is pumped into tank for raw material solution from upper section through seawater inflow pipe, and is discharged through seawater outflow pipe at the bottom of the tank for raw material solution. Then, seawater discharged from tank for raw material solution is fed into MSF seawater desalination system through circulating pipe for cyclic utilization. According to the principle of osmotic pressure, water of ordinary seawater in tank for raw material solution passes through osmotic membranes and flows into tank for draw solution. The mixed desalinated seawater discharged from tank for draw solution of which the hydraulic pressure is relatively high outflows through desalinated seawater outflow pipe, and drives the water turbine. Then, the generator is driven by water turbine and electric power is generated. Circulating pipes of desalinated seawater and circulating pipes of seawater in the whole system is closely fitted, which contributes to the adequate heat exchange with desalinated seawater during the whole circulating process of the seawater, and reduces the energy cost of the whole system.

3. Capacity and Energy Consumption Analysis

Reviewed references of seawater desalination and salinity gradient energy power generation, this assumes operation mode and parameters of the system based actual situation, and analyzes energy cost, power generating efficiency of salinity gradient energy and status of auxiliary facilities.

3.1. Energy Consumption of MSF Seawater Desalination System

Assume that a MSF seawater desalination system which can produce 1t fresh water per hour is to be constructed in a location that suffers from electricity deficit and fresh water shortage. Flash evaporation rate of MSF is generally between 0.4% and 12%, and 5% is chosen as flash evaporation rate in this paper. With flash evaporation rate of 5%, seawater that should be fed into the system per hour is 20t. Energy cost of seawater desalination is calculated in [4], ratio energy cost of MSF is 17.62kWh/m^3. Assume that the density of fresh water is 1000kg/m^3, then energy cost of MSF seawater desalination system per hour is 17.62kW.

3.2. Energy Consumption of Seawater Pretreatment System

Preprocessing technology of seawater desalination is one of the key factors to ensure the stable, reliable and long-term operation of the seawater desalination system. State-of-the-art preprocessing methods for seawater desalination mainly includes regular preprocessing, membrane preprocessing and Actiflo high efficiency sedimentation tank [5]. Seawater desalination methods varies according to technique requirements differences. Requirement of preprocessing of seawater for MSF seawater desalination is relatively simple, and energy cost of cyclic process is low. Energy cost by seawater preprocessing can be ignored, because only regular preprocessing is needed.

According to the assumption, 20t seawater should be fed into the system. Assume that m1t seawater pumped by pump 1 into the seawater preheating system per hour, and m2t of seawater is fed into the tank for raw material solution. As is known that quality of seawater fed into the system by water pump is equal to the quality of water discharged from the system. According to the calculation of water flow of the osmotic membranes, osmotic water flow is 0.3t when 1t of fresh water is discharged. Therefore, quality mixed desalinated seawater discharged from salinity gradient energy power generating system is 19.3t. m1+m2=20.3t/h is fed into the system by pump 1. Pump 10m head is chosen for calculation in this section, and power of the pump is calculated as （1）:

$$W = \frac{Vgh}{3600\eta} \tag{1}$$

Where V is seawater flow, h is the head of the pump, n is the efficiency of the pump, and is generally chosen as 0.6.

Substitute v=20.3t/h into equation (1), power cost by pump 1 is calculated as w1=0.92kW.

3.3. Capacity of the Salinity Gradient Power Generation System

Osmotic pressure of sweater with different concentration can be calculated according to following formulas (temperature is chosen as 25℃) [6]:

$$\pi 1 = \frac{2RT}{1000}\int_0^I \frac{1}{V_1}\left\{1 + 2bI + 3bI^2 - \frac{S(I)^{1/2}}{2[1+1.5(I)^{1/2}]^2}\right\} dI \tag{2}$$

$$\pi 2 = \frac{RTM_1}{1000V_1}\sum \phi_i v_i m_i \tag{3}$$

According to the commonly used marine table [7], salinity(S‰) and osmotic pressure of seawater (MPa) at 25℃ can be found. As known that concentration of seawater is 34.448‰, osmotic pressure can be deduced by interpolation technique. As shown in table 1.

Table 1. The osmotic pressure of seawater at different concentrations at 25 ℃(MPa).

concentration (ppm)	π1 (MPa)	π2 (MPa)	π1 (MPa)
6000	0.424	0.428	0.420
7000	0.493	0.496	0.488
8000	0.563	0.566	0.558
9000	0.632	0.636	0.628
10000	0.697	0.707	0.698
20000	1.401	1.422	1.408
34448	2.466	2.464	2.465
40000	--	2.876	2.884
50000	3.676	3.643	--
75000	5.815	5.664	--

Ordinary seawater of 34448ppm and desalinated seawater of 50000ppm are chosen as example for analysis. As shown in table 1, osmotic pressure of seawater of 24448ppm is 2.464MPa and 2.466MPa according to equation (2) and (3). The average value 2.465MPa is chosen in this paper. Osmotic pressure of desalinated seawater of 50000ppm is 3.676MPa and 3.643MPa according to equation (2) and (3). The average value 3.65MPa is chosen in this paper. Osmotic pressure between two sides of the semipermeable membrane is $\Delta\pi = 3.65 - 2.465 = 1.185\text{MPa}$.

A preliminary overall estimation of the power of the semipermeable membrane is performed in this paper regardless of the energy lost and the operation efficiency of parts. Formula calculating the water flux of water molecules passing through the semipermeable membrane is :

$$J_W = A \times (\Delta\pi - \Delta P) \quad (4)$$

Where J_W is the water flux of semipermeable membrane, A is the permeability coefficient of semipermeable membrane, $\Delta\pi$ is the osmotic pressure of the system, ΔP is the pressure from external.

Water flux through the semipermeable membrane depends on many factors, and the most important factor is the characteristic of the semipermeable membrane. Efficiency of the salinity gradient energy devices is influenced by water flow directly. Cellulose acetate membrane produced by Hydration Technologies Inc（HTI）is the only semipermeable membrane for business use until now.

Table 2. Water flux of CTA-W forward-osmosis membrane in different conditions.

temperature (℃)	raw material solution (mol/L)	draw solution (mol/L)	active layer orientation	water flux (LMH)
20	0.05NaCl	1.5 NaCl	DS	31.78[8]
20	0.5 NaCl	1.5 NaCl	FS	5.77[8]
20±1	0.05NaCl	4.0NaCl	FS	27[9]
20±1	0.05NaCl	4.0NaCl	DS	37.8[9]
23	10m NaCl	1.0 NaCl	DS	26.8[10]
23	10m NaCl	1.0 NaCl	FS	15.8[10]
23	10m NaCl	0.5 NaCl	FS	4.4[11]
25	DIwater	0.87 NaCl	FS	12.07[12]
50	0.05 NaCl	6NH4HCO3	FS	36.5[13]

As shown in table 2, water flux is influenced by factors such as temperature、raw material solution、draw solution、active layer orientation. When temperature is 20℃, raw material solution is 0.05 mol/L NaCl, draw solution is 1.5 mol/L NaCl, active layer orientation is DS, the water flux is 5.77 $L/m^2 \cdot h$. When temperature is 20℃, raw material solution is 0.5 mol/L NaCl, draw solution is 1.5 mol/L NaCl, active layer orientation is DS, the water flux is $31.78L/m^2 \cdot h$. By comparison, we can know that when the working environment is higher than 20℃, the active layer orientation is DS, the concentration difference between the absorbing liquid and the raw material is higher than 1 mol/L, the water flux should be much higher than 5.77 $L/m^2 \cdot h$. Operation temperature of the salinity gradient energy power generating system in this paper is higher that 25℃. 5.77 $L/m^2 \cdot h$ is chosen as water flow with the consideration for convenience of analysis of the following article. It is assumed in this paper that the area of semipermeable membrane in salinity gradient energy power generating devices is $50m^2$, therefore volume of water passing through the semipermeable membrane every half an hour is 300L.

As known that desalinated seawater delivered by pump 2 is 19t/h. As long as desalinated seawater can flow into the tank for draw solution, and the concentration of tank for draw solution remains constant, the whole cycle can be completed. Therefore, requirements on the head of the pump is not very difficult to fulfill.

The calculation of this part choices the pump with 8m head, the power consumption of pump 2 is $W_2 = 0.69\text{kW}$.

The hydraulic turbine power calculation formula is:

$$P = gHQ = 9.81HQ \tag{5}$$

Where H is the water head（m）, Q is the water flow（m^3/s）.

According to calculation performed, osmotic pressure is $\Delta\pi = 1.185\text{MPa}$. 1Mpa is equal to water head of 100m, therefore water head H=118.5m. Water flow passing through osmotic membranes and flowing into the system is $Q_0 = 0.003 \times 50 = 0.15\text{m}^3$. Mixed desalinated seawater discharged per hour is 19.3t. Assume that the density of mixed desalinated seawater is $\rho = 1025\text{kg/m}^3$, water flow per hour will be $Q = 18.83\text{m}^3$. Substituting aforementioned data into (5), power capacity of semipermeable membrane per unit area can be calculated

$$P = 9.81 \times 118.5 \times 18.83 \div 3600 \div 50 = 0.122\text{kW}$$

As is known that the efficiency of the water turbine is 0.7, the power capacity of per unit area of membrane is

$$P = 0.7 \times 0.122 = 0.085\text{kW}$$

Overall power cost of the system can be concluded as table 3.

Table 3. Table of the energy consumption of the system.

energy-consuming system	Energy consumption（kW）
multi-stage flash seawater desalination system	17.62
seawater pretreatment system	0.92
salinity gradient power generation system	0.69
Total energy consumption	19.23

Therefore, area of osmotic membranes required for MSF seawater desalination system that can produce 1t fresh water per hour is:

$$S = \frac{W}{P} = \frac{19.23}{0.085} = 226.24\text{m}^2$$

Salinity gradient energy power generating system can generate electricity that can compensate the power cost by the whole system if it is equipped with osmotic membranes with area of $226.24m^2$.

4. System Characteristic Analysis

The salinity gradient energy power generating system integrated in the seawater desalination system proposed by this paper which combines salinity gradient energy power generating system and seawater desalination system is superior to either salinity gradient energy power generating system or seawater desalination system in the following aspects: electricity required by the system can be generated by the system itself; site selection limitation is broken, size of devices is reduced, flexibility is increased; differential osmotic pressure is increased, and overall power generating efficiency is increased. The proposed system will be introduced from following three aspects including energy cost reduction, site selection and electricity generation cost.

(1) From the perspective of energy cost consumption, salinity gradient energy power generating system integrated in seawater desalination system combines salinity gradient energy power generating system and seawater desalination system. Taking discharge (desalinated seawater) from seawater desalination system as draw solution and ordinary seawater as raw material solution, and applying principles of osmotic pressure, the system converted salinity gradient pressure into the electric power we need. In reasonable scope of design, while producing fresh water, the system can generate electricity utilizing salinity gradient energy power generating system which could compensate the electricity cost of the whole system. Therefore, self-balance of electricity consumption is reached, cost of seawater desalination is reduced and development of seawater desalination is promoted.

(2) From the perspective of site selection, traditional salinity gradient energy power generating system takes ordinary seawater as draw solution and fresh water as raw material solution, which requires sites of salinity gradient energy power generating system must be chosen as estuaries where both fresh water and ordinary seawater can be found. Therefore, salinity gradient energy power generation is difficult to develop in areas short of fresh water. The seawater desalination system powered by salinity gradient energy power generating system proposed by this paper increases the possibility of the construction of seawater desalination plant in areas lack of energy sources, which contributes to mitigates the problem of seawater shortage to some extent.

(3) From the perspective of economics of power generation, traditional salinity gradient energy power generation system takes ordinary seawater as draw solution and fresh water as raw material solution for electricity production. For fresh water in rivers in China is severely polluted, preprocessing cost of fresh water is relatively high. Seawater preprocessing in the proposed system needs only simple and regular treatment, which cuts the cost to a certain degree. At present, efficiency of salinity gradient energy plant is relatively low that generating capacity of semipermeable membrane is 1.3W per square meter at a latest established salinity gradient energy plant in the Netherlands. The generating capacity of semipermeable membrane of the salinity gradient energy power generating system proposed in this paper is increased to 8.5W per square meter.

The use of sections to divide the text of the paper is optional and left as a decision for the author. Where the author wishes to divide the paper into sections the formatting shown in table 2 should be used.

5. Conclusion

This paper proposes a salinity gradient power generating system integrated in seawater desalination system which includes a seawater preprocessing system, a seawater desalination system based on multi-stage flash desalination and a salinity gradient power generating system. Salinity gradient energy power generating system together with seawater desalination system based on MSF reduces energy consumption required during seawater desalination, which enables energy balance of the whole system during fresh water production. The proposed integrated seawater desalination system breaks the site selection limitation of salinity gradient energy power plant, increases power generation efficiency and economic value of the overall system, adds to the possibility of establishing seawater desalination plant and salinity gradient plant in areas with few resources, and boosts the exploitation of marine energy. Based on principles of osmotic pressure and quantitative calculation of energy consumption and salinity gradient energy power generation, this paper proves that the proposed system can realize energy self-balance and produce fresh water in reasonable range. On the basis of conducted research, future study will focus on the feasibility analysis of the system in real projects, which can offer comprehensive references for development and construction of future seawater desalination system and salinity gradient energy system.

References

[1] Li Changhai, Zhang Yaxiao 2011 Discussion on seawater desalination technology and its application *Electric power technology and environmental protection* **01** 48-51.
[2] China industrial information network 2013 Depth analysis and strategic investment advisory report of china ocean energy industry during 2013-2018 *Intelligence research consulting group*
[3] Liu Boyu, Li Shaohong, Wang Gang 2010 Progress in extracting power trom salinity gradient *Renewable energy resources* **02** 141-144.
[4] Darwish M A 1995 Desalination processes: A technical comparision 95' IDA World Congress on Desalination and Water Sciences Abu Dhabi 149-173.
[5] Chang L F, Li X M 2010 Comparison and selection of pretreatment system for seawater desalination *Cfhi Technology* **04** 7-12.
[6] Yin Qi, Zhu Huajie 1989 Discussion for osmotic pressure of sea salt solution *Technology of Water Treatment* **06** 30-35.
[7] Zhu Bofu 1958 Ocean common table Science Press 138.
[8] McCutcheon J R, Elimelech M 2006 Influence of concentrative and dilutive internal concentration polarization on flux behavior inforward osmosis *Journal of Membrane Science* **284** 237-247.
[9] Mi B, Elimelech M. 2008 Chemical and physical aspects of organic fouling of forward osmosis membranes *Journal of Membrane Science* **320** 292-302.
[10] Wei J, Qiu C, Y C Y, et al. 2011 Synthesis and Characterization of Flat-sheet Thin Film Composite Forward Osmosis Membranes *Journal of Membrane Science* **372** 292-302.
[11] Zou S, Gu Y S, Xiao D Z, et al. 2011 The role of physical and chemical parameters on forward osmosis membrane fouling during algae separation *Journal of Membrane Science* **336** 356-362.
[12] Achilli A, Cath T Y, Childress A E 2010 Selection of inorganic-based draw solutions for forward osmosis applications *Journal of Membrane Science* **364** 233-241.
[13] McCutcheon J R, McGinnis R L, Elimelech M 2006 Desalination by ammonia-carbon dioxide forward osmosis: Influence of draw and feed solution concentrations on process performance *Journal of Membrane Science* **278** 114-123.

The application of environmental numerical simulation in pollutants migration into river basin

Z R Liang[1], Y J Qin[1], J G Zhou[1], H Y Nan[2, 3]

[1] Chongqing Environment Engineering Appraisal Center, Honghu East Road No. 9, Yubei District, Chongqing city 401121, China
[2] Faculty of urban construction and environmental engineering, Chongqing University, Sand Road No. 83, Shapingba District, Chongqing city, 400045, China

E-mail: 1161454533@qq.com

Abstract: For the sustainable development of the social economy, it is very important that the water environment quality is analyzed, evaluated and predicted rationally, so that it could be planned, harnessed and managed effectively. To analyze and predict the water environment, the hydrodynamics behaviors and water pollution situations of the water body must be analyzed first based on hydrodynamics and water quality models. The objective of this work is to introduce how to establish river water dynamics and water quality model through the environment in the basin pollutant migration transformation rule, to simulate the rules and dynamics of river water pollution situation, to introduce the water quality model with geographic information system (GIS), and to apply the combination of environment numerical simulation to basin pollutant migration problem.

1. Instruction

Environment is the main living and developing condition of human being. Environmental pollution, especially the water environment pollution, has been one of the primary issues that threaten the survival of mankind. Migration motion of water as well as attenuation and transformation motion of pollutant will gradually dilute and diffuse the pollutants into the water: It's the important "self-purification of water body" [1]. However, once the pollutant level is over the self-purification, water pollution will endanger the health and life of human being.

Some laboratories have tried to adopt smaller model systems to study the migration and transformation law of pollutants under the effects of various factors, but it is rather difficult to do with the current technology. At the same time, numerical model can avoid the similarity criterion of physical model, and have the advantages of small invest and short period. This paper summarizes the theory of the pollutants' transition and translation in water body, and sums up the research progress from the numerical simulation angle and its application prospect combined with GIS, which provide an important theoretical significance and engineering practical value for water pollution research.

2. Summary of environment numerical simulation

It is of great significance for numerical simulation of environment to analyze, evaluate and predict the river water quality. Moreover, it is also of great importance to apply the diffusion rules of pollutants in water to establish the numerical model, to predict the long time influence pollutants exert on water and development tendency, and finally to take measures to control and prevent the pollutants.

In general, there are two water environment simulations. One is a physical model, which, according to the principle of similitude, simulates the fluid motion or changes of pollutants. The other is a numerical simulation, which, according to hydromechanics and migration and transformation law of pollutants, establishes numerical model to analyze the fluid motion and changes of pollutants in a certain boundary conditions and initial conditions [2].

Theoretical analysis, experimental analysis and numerical simulation are three basic approaches in the study of river hydraulic dynamics. But because of the complexity of river water environment, it will require a lot more manpower and material resources in some hydraulic model tests such as field or flume, while the similarity between physical model and real natural water environment cannot be guaranteed. Therefore, under the most conditions, numerical simulation becomes economical and practical research techniques.

According to assumption and simplification, we can establish mechanical model to find the math expression to equivalent models and create a set of mathematical equations, and finally form a mathematical model with a fixed solution,combined with initial conditions and boundary conditions. Thus, within the margin of error, adopting numerical algebra, numerical approximation, numerical differentiation and integration, ordinary differential equation and partial differential equation can lead to numerical solutions conforming to engineering application. These solutions can not only be numerical points, but also array or matrix form [3].

Nowadays, river water environment model does not just to contain diffusion in spatial and temporal distribution which can calculate several contamination index. With the rapid development in computer technique and the maturity of various theories in environment hydraulics, the discharge of pollutants and its dilution and dispersion mechanism into water can be studied. For example, radiant flux theory near zone and diffusion theory of water quality far zone, and the studies and applications of contemporary control theory, systematics and inverse problem theory all have been developed in environment hydraulics. These strike the traditional hydraulics violently, and urge water environment model nowadays to develop in the direction of integration, complication and globalization.

3. Motion characteristics of pollutants in water

The purification process of pollutants in water usually includes physical, chemical and biological process. The most fundamental process is mixing and transporting, of which temporal mean flow transport pollutants and turbulent fluctuation diffuses pollutants. Besides, attenuation and conversion further reduce pollutants concentration [4]. The chemical and biological processes decide the final destination of pollutants. These processes are influenced by temporal mean flow, turbulent fluctuation, hydrology-weather condition, chemical, biological actions and more, which makes the migration and transformation rule of pollutants in water more difficult to investigate [5].

According to the characteristics of water in nature, different types of diffusion can be classified, such as river, estuary, lake, and bay. Here, the type of diffusion in river will be introduced.

(1) flow and migration

Flow and migration generate flow action. It can only change the site of pollutants in water, but not the concentration. Under the flow and migration, transformation flux of pollutants can be calculated as follows:

$$f_x=u_x c, fy=u_y c, fz=u_z c \qquad (1\text{-}1)$$

where, f_x, f_y and f_z represent the transformation flux in x, y and z direction, respectively; u_x, u_y and u_z represent the water velocity in x, y and z direction; c represents the concentration of pollutants in water.

(2) dispersion effect

Dispersion effect of pollutants in water includes molecular diffusion, turbulent diffusion and dispersion.

When deciding the dispersion effect of pollutants, it is important to assume that dynamic characteristics between pollutants particles and water are in agreement. The assumption satisfies the majority of dissolved pollutants in colloidal state. Molecular diffusion generates the random vibration of molecules. Molecular diffusion meets the Fick's first law [6,7], namely that mass flux is in direct proportion to the concentration gradient of diffusate. That is,

$$I_x' = -E_M \frac{\partial c}{\partial x}, I_y' = -E_M \frac{\partial c}{\partial y}, I_z' = -E_M \frac{\partial c}{\partial z} \quad (1\text{-}2)$$

Where E_M and c represent molecular diffusivity coefficient and pollutant concentration transferred through molecular diffusion, respectively.

Molecular diffusion is an isotropy, so the negative sign above represents migration of particles pointing to the negative gradient direction. Turbulent diffusion is a dispersion phenomenon where the instantaneous value of particles in various states fluctuates randomly relative to its mean value in the turbulent flow-field in water. When the instantaneous fluctuating velocity of particles in water becomes a steady random variable, turbulent diffusion rule can be expressed with the Fick's first law, namely,

$$I_x^2 = -E_x \frac{\bar{c}}{x}, I_y^2 = -E_y \frac{\bar{c}}{y}, I_x^2 = -E_z \frac{\bar{c}}{z} \quad (1\text{-}3)$$

Where E_x, E_y and E_z represent turbulent diffusivity in x, y and z direction.

Turbulent diffusivity follows the anisotropy due to the characteristics of turbulent flow. Turbulent diffusivity generates the calculation of time mean value. There won't be turbulent diffusion term with instantaneous value.

Dispersion arises from heterogeneous velocity distribution in the section cross, so when describing the practical movement using mean velocity at section cross, an additional action arising from heterogeneous velocity, namely dispersion, must be considered. Mass flux generated by dispersion effect can also be described with the Fick's first law:

$$I_x^3 = -D_x \frac{\partial \bar{c}}{\partial x}, I_y^3 = -D_y \frac{\overline{\partial c}}{\partial y}, I_x^3 = -D_z \frac{\partial \bar{c}}{\partial z} \quad (1\text{-}4)$$

Where D_x, D_y and D_z represent the dispersion coefficient in x, y and z direction.

Given that it is usual to use time mean value in practical calculation in turbulent flow, a turbulent diffusivity must be introduced. In general, molecular diffusivity value is between 10^{-5}-$10^{-4}m^2/s$, so turbulent diffusivity value is larger. In river, its order is 10^{-2}-$10^{0}m^2/s$. Dispersion effect usually starts when time means value spatial mean value is used, so the effect always appears in river. In general, the value of dispersion effect in river is 10^{1}-$10^{4}m^2/s$.

In fact, the molecular diffusion is caused by the random motion. Rivers, atmospheric problems always can be ignored on the study of the actual molecular diffusion. This is because in the flow field, the effect scale of molecular diffusion is very small, compared with turbulent diffusion or diffuses diffusion effect. But you should consider molecular diffusion when it refers to the problem of lakes and reservoirs, because in the static water environment, the molecular diffusion may become major diffusion effect.

Turbulent diffusion is caused by the random fluctuation of the various states of instantaneous value relative to its time average value of turbulent flow field in the molecules. So when the medium is in the flow field with a range of velocity, if the time average value is used to describe the various states of particles, it should introduce the turbulent diffusion, and it also should consider turbulent diffusion about the study of deep sea pollution emissions, air quality simulation and the pollutant diffusion of some rivers.

Dispersion effect is caused by uneven cross-sectional flow velocity .So the dispersion effect must be considered when it is described by using cross-sectional average flow velocity.

(3) attenuation and transformation of pollutants

Pollutants into water can be divided into two groups: conservative and non-conservative substances.

In water environment, conservative substance continuously changes its spatial position with the water flow movement. At the same time, it instantly diffuses around and decreases in concentration without changes in its total quantity due to the dispersion effect. As for non-conservative substance, its position changes with water flow and its concentration decreases with dispersion effect in water. Besides, the concentration will also decrease quickly with the attenuation of pollutant itself.

Both experimental and practical data prove that the attenuation of pollutant in water basically conforms to the first order reaction kinetics, namely,

$$\frac{dc}{dt} = -K_c \tag{1-5}$$

Where c, t and K represent concentration of pollutant, reaction time and rate constant.

Flow and migration of river water and diffusion and attenuation of pollutant can be described using Fig 1-1.Assuming x is at x_0, quantum of pollutants into water is A and presents in straight square, and all pollutants go through x_0 at Δt (Figure 1). After a time, pollutants shift to x_1, and its quantum changes to a. If only flow, then a is equal to A, and the distribution shape of pollutants at x_1 is equal to at x_0. While if containing flow and diffusion, then a is still equal to A, but the distribution of these two is different, and transit time is also longer. If flow, diffusion and attenuation all exist, then distribution shape changes, and a is smaller than A.

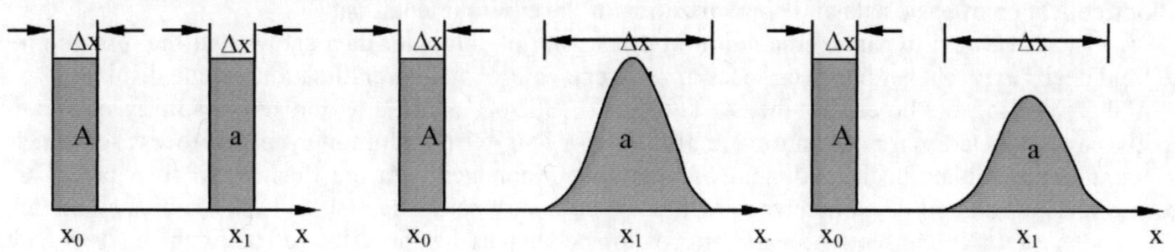

A. flow and migration $a=A, \Delta x_1=\Delta x_0$

B. A and diffusion $a=A, \Delta x_1>\Delta x_0$

C. B and attenuation $a<A, \Delta x_1>\Delta x_0$

Figure 1. flow and migration, diffusion and attenuation

4. The combination of water quality model and GIS [8]

Even if standard weakly compressible SPH model has been employed for the numerical simulations [9], GIS begins to play a more and more important role in water environment numerical simulation. For GIS, it takes spatial data of geo-location as research object and spatial database as core, and uses spatial analysis and modeling method for providing multiple spatial and dynamic resource and environment information timely. It is an interpretation of geographic data characterizing geographic feature and geographic phenomenon. These data include spatial location, attributive character and time domain feature. Spatial location data describe the location of surface features, attribute data belongs to certain surface features and describe their characteristics qualitatively and quantitatively. Time domain feature is the time or moment when geographic data is collected or geographic phenomenon happens (at WPCP, such as, relevant index of source of pollution, monitoring index of fracture surface, emergency and more). Spatial location, attribute and time are three basic elements for geospatial analysis.

Because it can analyze the space features for catchment area for GIS, A. Goonetilleke[10] thinks that it has a great influence on urban hydrology, for the availability of spatial distribution database for catchment area can eliminate the attenuation of research quality due to simplification assumption. On the other hand, it also indicates the importance of accurate explanation of space-time in land use of urban catchment area. William Dixon [11] uses matrix form to combine GIS and metal forming theory to describe the structure of river network when he optimizes the sampling location of river network. Based on GIS technology and numerical simulation equation of river water environment, studying the metastatic spread rule of pollutants in water, realizing the visual representation of pollutants diffusion and discussing new method of pollution monitoring in water environment all have great practical significance in effective treatment of pollution[12]. As a kind of space, dynamic of more sources, and multidisciplinary information science and technology GIS will be a watershed hydrological system simulation research was one of the key technologies [13].

Combined water quality model with GIS, Xia Bingxue realized the unified storage and management for mass data, made water quality simulation results more clear and visual to display and record. Meanwhile, they analyzed two-dimensional water quality into library wanzhou Yangtze river before and after the period of two-dimensional damage zone with the same method, which provided powerful technical support to relevant departments [14].

5. Conclusion and prospect

Water environment is a complex system influenced by many factors. Thus, it is unrealistic to simulate it in detail with limited time, manpower, material and financial resources, which leads to water environment model only play a role of demonstrative reference and cannot display its potential effect.

Fundamental data collection: considering that it gradually varies in the environment, it is not easy for people to observe the influence in short time. Finding the law of environment change requires tedious collection of basic data and popularization of the environmental data.

For visual research: dynamic simulation in processing of computer data shows that the research in the field needs to go deeper into the aspect of real-time control and diversification of side display.

With increasing public environmental awareness, elusion, chronicity and irreversibility of water quality have attracted more and more attention. It is a long-term and uninterrupted process for water pollutant to accumulate in the sediment of river, and it changes with the flushing of river bed. The process mechanism and its parameter selection are both very sophisticated, and the established model is still not perfect. In consequence, it's crucial for understanding the effect of pollutant in river and establishing rational equation of numerical simulation.

GIS can establish complex spatial model, manage spatial data, call water environment modules, and use computation based on water environment model and spatial data to do querying. It can be helpful to analyze spatial distribution pattern and offer decision support for river environment management. Because of parataxis of GIS and water environment model in function, even if now combined Water quality model with the GIS has a certain limitation[15], in future the combination of them may be a hot field in environment research.

Acknowledgement

Center leaders and colleagues provided the help in the essay writing process.

References

[1] Chen Y X 2001 Environment ology *China Environment Science Press* (in Chinese)
[2] Li Y S, Liu W J, et al. 2006 A review of progress in research on water quality models in America *Water Resource and Hydropower Engineering* **37(2)** pp 68-73 (in Chinese)
[3] Yang D D, Tu G Y 1998 Numerical analysis *Chongqing University Press*(first edition) (in Chinese)
[4] Li W 1999 Advances in environment hydraulics *Wuhan Hydraulic and Hydropower Engineering University Press* (in Chinese)
[5] Gong Y J 2008 study of pollutant diffusion numerical simulation and environment capacity in Weihai Bay: *Ocean University of China* pp **8-9** (in Chinese)
[6] Zhao Z S 2009 Environment hydraulics *Peking University Press* pp **65** (in Chinese)
[7] Fick A E 1885 Poggendorffs Annelen der Physic **94** pp 1–59
[8] Liu M 2006 study of pollution diffusion numerical simulation and total quantity control in Liaohe estuary *Dalian Maritime University* (in Chinese)
[9] Aristodemo F, et al. 2010 Two-phase SPH modeling of advective diffusion processes *Environmental Fluid Mechanics* **10(4)** pp 451-470.
[10] Goonetilleke A, et al. 1999 The role of geographical information systems in urban hydrological modeling. Lecturers, school of civil engineering *Queensland university of technology* **13(3)**

pp 200-206

[11] William Dixon, et al. 1999 Optimized selection of river sampling sites *Water research* **33(4)** pp 971-978

[12] Guo J. S., Li H. S., Long T. R. 2002 Study and progress of water quality model and its application *Journal of Chongqing Jianzhu University 4* **24(2)** pp 100-115(in Chinese)

[13] Kovar K Naehtrebel 1993 Application of geographic information system in hydrology and water resources management *International Association of Hydrological Sciences*

[14] Xia Bingxue 2007 The three gorges reservoir water quality simulation based on GIS *Chongqing University* (in Chinese)

[15] Sui Dz, Maggio RC. 1999 Integrating GIS with hydrological modeling: practice, problem and prospect *Computer, Environment and Urban System* **23** pp 33-51.

Optimization of biosurfactant production from *Vibrio sp.* BSM-30 isolated in tropical waters

Zengjian Su[1, 2], Min Li[3] and Yuxiu Zhang[1, 4]

[1] China University of Mining and Technology (Beijng), School of Chemical and Environmental Engineering, Beijng 100083, China;
[2] Environment and Plant Protection College, Hainan University, Haikou, Hainan Province 570228, China
[3] CATAS, Environment and Plant Protection Research Institute, Haikou, Hainan Province 571737, China

E-mail: zhangyuxiu@cumtb.edu.cn

Abstract. The strain BSM-30 (*Vibrio sp.*), isolated from Chinese tropical waters, could be a biosurfactant producing bacteria according with results obtained by the oil spreading method. The culture conditions for biosurfactant production were tested respectively such as inoculation (2%,6%,10%,14% as setting), shaking speed(120 r/min,150 r/min,180 r/min as setting), temperature (25°C,30°C,35°C as setting), pH (7,8,9 as setting), salinity (1.5%, 2.5%, 3.0%, 4.5%, 5.5% as setting), which results showed that the best culture conditions for BS production were 10% inoculation quantity, 180 r/min, 25°C, pH 8, and 3.5% salinity. The optimization of carbon sources (20g/ of glucose, 20g/L of starch, 20g/L of paraffin oil 20g/L of diesel, 20g/L of oil as setting) and nitrogen sources (6g/L of $NaNO_3$,7.1g/L of KNO_3,5.6g/L of NH_4NO_3,9.3g/L of $(NH_4)_2SO_4$, 4.2g/L of $CO(NH_2)_2$ as setting) were also tested, which results showed that the best nitrogen source and carbon source were $(NH_4)_2SO_4$ and soluble starch.

1. Introduction

Biosurfactant (BS) are metabolites in the microorganism metabolism which reduce surface tension. Other properties associated with BS are increase of solubility, emulsifying, foaming, wetting, and dispersing. At present, most BS producing bacteria come from the oil polluted environment, and they can also be applied to petroleum hydrocarbon pollutant remediation in environment due to BS's properties such as non-toxic or low toxicity, readily biodegradable, environmentally friendly. [1, 2] Some scholars studies [3,4] showed that BS producing bacteria made very good effect of oil removing in oil sludge by washing directly with its bacterial liquid. The focus of this research is screening BS producing bacteria and the optimization of strain living conditions to prepare for later repair test of petroleum hydrocarbon pollutants.

2. Materials and method

2.1. Materials

Sampling

In May 2, 2015, seawater samples were obtained from Baishamen offshore, Haikou, China. Samples were below 0 ℃ during transportation.

Culture medium

Rich culture medium (RCM)[5]: Beef Extract 0.5%, Peptone 1%, NaCl 0.5%, Agar 1.5-2%, stale seawater 1L, pH 7.5, sterilized 20min under 121℃.

Filtration culture medium: Rich culture medium containing 10ml oil for screening BS producing bacteria.

Bacterial fermentation medium (BFM) [6]: glucose 10g, NH_4Cl 1.5g, $K_2HPO_4 \cdot 12H_2O$ 3g, KH_2PO_4 0.5g, $CaCl_2$ 0.01g, $FeSO_4 \cdot 7H_2O$ 0.015g, $MnSO_4$ 0.002g, $CuSO_4 \cdot 5H_2O$ 0.002g， stale seawater 1L, pH 7.5 , sterilized 20min under 121℃.

Nitrogen deficient medium: Bacterial fermentation medium without nitrogen compounds Carbon deficient medium: Bacterial fermentation medium without carbon compounds.

2.2. *Method*

2.2.1. Degradation bacterium enrichment. The seawater samples, which were diluted in gradient 10^{-1}, 10^{-2}, 10^{-3}, 10^{-4}, were spread (1 ml) on RCM. After culture, colonies were streaked for purification. All of single bacteria isolated were inoculated BFM (150r/min, 30℃, 4-5d), [4] and from of them, the dominant surfactant bacterium were screened for BS production by the oil spreading method. [7~10] The bacteria BSM-30 selected in this research is a strain of Vibrio sp. by the molecular identification (RNAr 16r Sequencing).

2.2.2. Culture conditions optimization for BSM-30. The orthogonal method was used to optimize BSM-30 culture conditions with inoculum, temperature, pH, speed as factors. Results obtained are measured as bacterial growth (OD_{600}) and BS production (clearing zone diameter). Studied conditions for orthogonal design are shown in Table1 and Table2.[11,12]

Table1. The orthogonal test and the selection of level

Level	Factor			
	Inoculum /%	pH	T/℃	Speed r/min
1	2	7	25	120
2	6	8	30	150
3	10	9	35	180

Table2. The optimization of orthogonal experimental design

Sequence	**Inoculum /%**	**pH**	**T/℃**	**Speed r/min**
1	2	7	25	120
2	2	8	30	150
3	2	9	35	180
4	6	7	30	180
5	6	8	35	120
6	6	9	25	150
7	10	7	35	150
8	10	8	25	180
9	10	9	30	120

2.2.3. Nitrogen source optimization experiment. $NaNO_3$, KNO_3, NH_4NO_3, $(NH_4)_2SO_4$, $CO(NH_2)_2$ were used as nitrogen source of nitrogen deficient medium respectively(concentration of nitrogenous species are same as BFM), in which BSM-30 was continuously cultured (25 ℃, 180 r/min, 108 h), and then, the OD_{600} incremental and the clearing zone diameter of BSM-30 fermentation liquid were measured.

2.2.4. Carbon source optimization experiment. Glucose, Starch, Paraffin oil, Diesel oil, Oil (from Fushan oilfield Hainan province, China) were used as carbon source of carbon deficient medium respectively (concentration of carbon species are same as BFM), in which BSM-30 was continuously cultured (25 ℃, 180 r/min, 108 h), and then, the OD_{600} incremental and the degreasing cycle of BSM-30 fermentation liquid were measured.

2.2.5. NaCl optimization experiment. Different NaCl concentrations (such as 1.5%, 2.5%, 3.0%, 4.5%, 5.5%) were given to bacterial fermentation medium respectively, in which BSM-30 was continuously cultured (25 ℃, 180 r/min, 108 h), and then, the OD_{600} incremental and the clearing zone diameter of BSM-30 fermentation liquid were measured.[13]

3. Results and Discussion

3.1. The growth curve and degreasing cycle

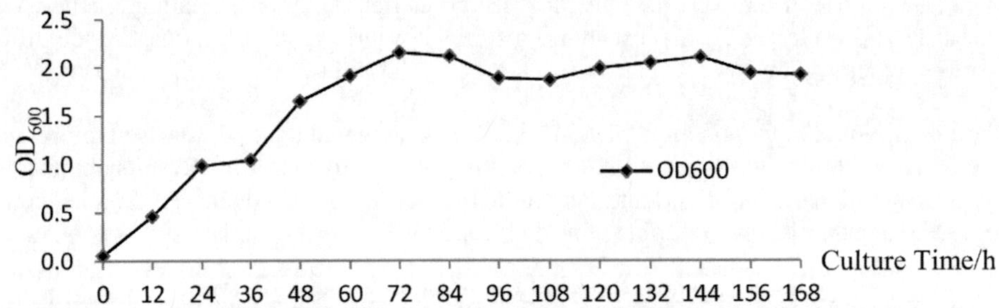

Figure 1. The growth curve with fermentation time

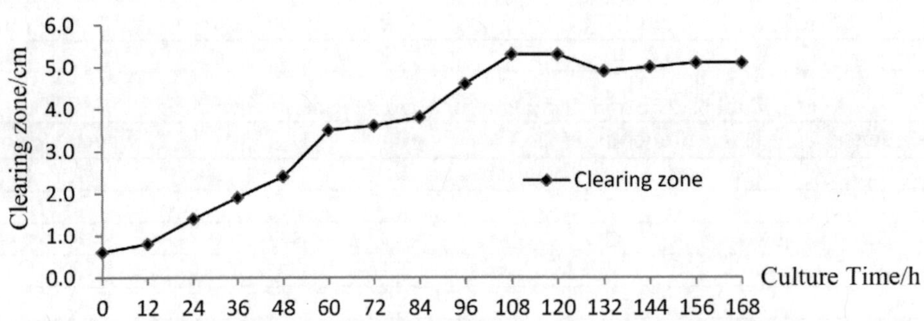

Figure 2. Clearing zone changes with fermentation time

The strain BSM-30 was grown rapidly till 72 *h* after inoculation, when its grown entered the stationary phase (Figure 1). But BS production of BSM-30 has a certain lag than bacterial growth, the clearing zone diameter of BSM-30 grown supernatant increased with time till to max 5.3 cm, which time is about 108 *h*, and then remained unchanged (Figure 2). So the 108 *h* is the optimum culture time when the aim is to obtain the biosurfactant of BSM-30 secretions.

Table 3. The result of orthogonal experimental optimization

Sequence	Inoculum /%	pH	T/℃	Speed r/min	Clearing zone /cm
1	2	7	25	120	3.9
2	2	8	30	150	3.9
3	2	9	35	180	5.5
4	6	7	30	180	6.0
5	6	8	35	120	1.8
6	6	9	25	150	2.9
7	10	7	35	150	2.8
8	10	8	25	180	6.3
9	10	9	30	120	5.0
T1	13.3	13.0	13.2	10.7	
T2	11.0	12.1	15.2	9.6	
T3	14.2	13.4	10.1	18.2	
t1	4.4	4.3	4.4	3.6	
t2	3.7	4.0	5.1	3.2	
t3	4.7	4.5	3.4	6.1	
R	1.0	0.5	1.7	2.9	

3.2. Culture conditions optimization

The Temperature and Speed have greater influence on clearing zone diameter of BSM-30 than other factors (the grey parts in table 3). The optimal culture conditions should be 10% inoculation quantity, pH 8, 25 ℃, 180 r/min, while the clearing zone diameter can grow to 6.3 cm which increase 20.8 % more than the original one.

3.3. Nitrogen source optimization experiment

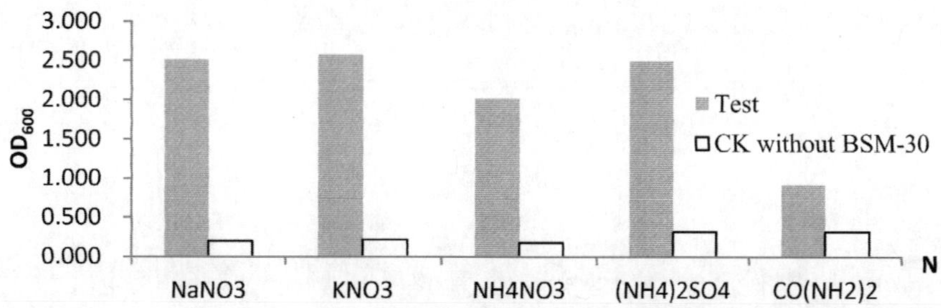

Figure 3. N optimization (OD)

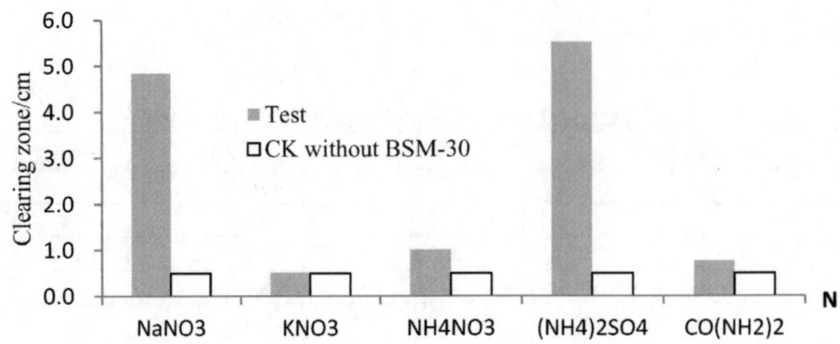

Figure 4. N optimization (Clearing zone)

The strain BSM-30 can get better growth whatever $NaNO_3$, KNO_3, NH_4NO_3, $(NH_4)\ _2SO_4$ were respectively used as nitrogen source (Figure 3), while the strain BSM-30 not only grow well but also could produce more biosurfactant when $(NH_4)\ _2SO_4$ was used as nitrogen source than others, which clearing zone diameter was big as 5.5 cm (Figure 4).

3.4. Carbon source optimization experiment

The test results showed that Glucose and Starch were more suitable for strain BSM-30 culture as carbon source (Figure 5), while although the strain BSM-30 showed moderate levels of proliferation when the petroleum hydrocarbons were used as carbon source, but their fermentation liquid had more biosurfactant than others for their clearing zone diameter even bigger (Figure 6).

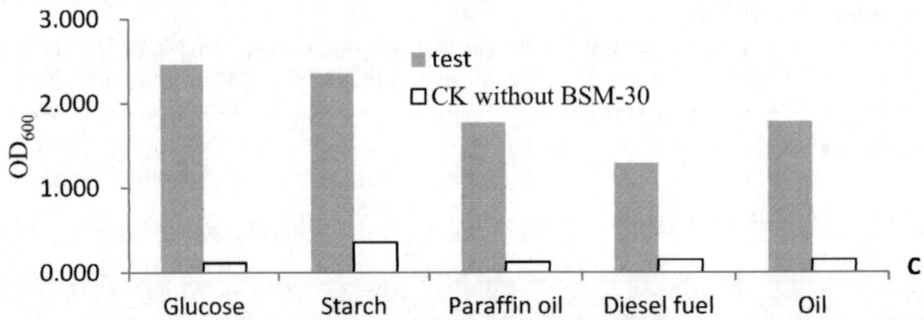

Figure 5. C optimization (OD)

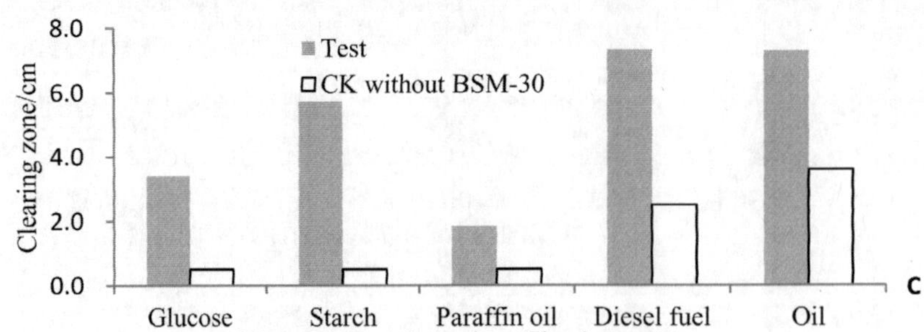

Figure 6. C optimization (Clearing zone)

3.5. NaCl optimization experiment

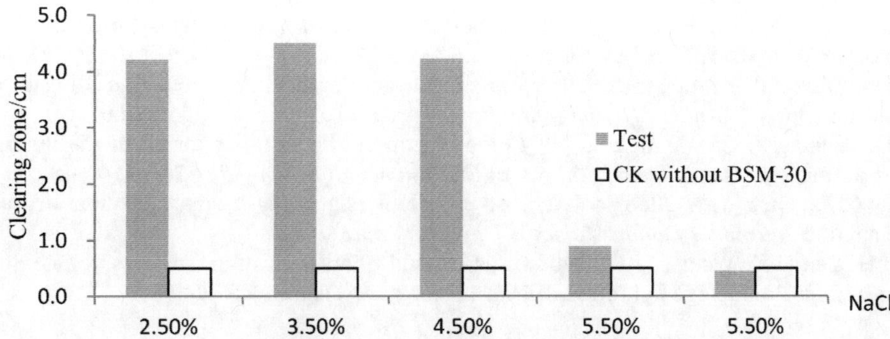

Figure 7. NaCl optimization (Clearing zone)

The salinity of about 3.5% was more suitable for the growth of strains, which is consistent with the mid latitude ocean salinity and the seawater salinity of strains screening (Figure 7), and the test results also showed that the salinity was suitable for biosurfactant secretion activity of the strain BSM-30.

4. Discussion

The results of this study showed that the optimum culture conditions of strain BSM-30 were 10% inoculation quantity, pH 8, 25 ℃, 180 r/min, $(NH_4)_2SO_4$ as nitrogen source, soluble starch as carbon source and 3.5% salinity. Under these conditions, the proliferative activity of BSM-30 peaked at around 72 hours and surfactant secretion peaked at 108 hours, which described that BS production is a non-associated to bacterial growth., the restriction of those nutrition factors may be the reason of that. Although the strain BSM-30 showed moderate levels of proliferation when the petroleum hydrocarbons were used as carbon source, but their fermentation liquid supernatant produce more biosurfactant than others from their clearing zone diameter, which will be conducive to its application to assist petroleum hydrocarbon degrading strains in environment polluted by petroleum hydrocarbon. In addition, the growth and surfactant secretion of BSM-30 were not affected by the salinity of 3.5%, which indicated that it does not exist adaptable problems for strain BSM-30 to the marine environment, being these features consistent with our research purpose.

References

[1] Shavandi M, Mohebali G, Haddadi A, et. al. 2011 Emulsification potential of a newly isolated biosurfactant-producing bacterium, *Rhodococcus* sp. strain TA6 *Colloids and Surfaces B: Biointerfaces* **82(2)** 477-482.

[2] Reddy M S, NA R ESH B, LEELA T, et. al. 2010 Biodegradation of phenanthrene with biosurfactant production by a new strain of Brevibacillus sp *Bioresource Technology* **101(20)** 7980-83.

[3] Chen Z X, Guo S H, Liu G M, et al. 2007 Isolation and characteristics of the degreasing biosurfactant bacteria *Journal of Harbin Institute of Technology* **39(4)** 586-588

[4] Niu M F, Li F M, Han X R, et al. 2005 Isolation of biosurfactant-producing microorganisms and their stability *Chinese Journal of Ecology* **24(6)** 631-634

[5] Lu L J, Huang X F, Liu J, et al. 2008 Isolation, identification and characterization of a highly efficient biosurfactant producing strain *Industrial Microbiology* **05** 34-39.

[6] Christopher W K, Christopher L K 2008 Bacterial succession in a Petroleum land treatment unit *Applied and Environmental Microbiology* **70(3)** 1777-86

[7] Zhang F, She Y H 2005 Quantitative and qualitative analysis of biosurfactant by oil spreading

method *Chemical Engineer* **112(1)** 14-15,38

[8] Kang H, Zhao S J, Guo A L, et al. 2012 Isolation and identification of biosurfactant producing microorganisms *Journal of Natural Science of Heilongjiang University* **02** 239-242.

[9] Wang J, Chen Y 2009 Categories and evaluating methods of biosurfactants *Chemical Research and Application* **21(2)** 137-140

[10] Bi S N, Wang Y J, Zuo Y H 2009 Improvement and Application of Oil Spreading to Detect Biosurfactant *Journal of Heilongjiang Bayi Agricultural University* **06** 58-60.

[11] Gao X P, Jiang Z, Gao X M, et al. 2013 Optimization of conditions for oil-degrading bacteria producing surfactant *Chinese Journal of Environmental Engineering* **7(8)** 3244-48

[12] Wang C Y, Ding Y S 2008 Isolation of crude oil degrading strains and their degradation capability *Journal of Dalian Maritime University* **34(3)** 9-12

[13] Zhao H, Yan H X, Yang T, et al. 2008 Creening and Characterization Analysis of One Strain of Bacterium Producing High-effect Biosurfactant *Bio-technology* **5** 34-39.

Study on the influence of the decoking agent on the activity of limestone in wet flue gas desulfurization

Qianjun Li[1], Dongyang Xu and Yunxia Wu and Jin Yu

School of Environmental Engineering, Nanjing Institute of Technology, Nanjing, 211167, China

E-mail: hjliqianjun@njit.edu.cn

Abstract: Influence of the main components of decoking agent (magnesium nitrate, aluminum nitrate, copper nitrate, ammonium nitrate and actual decoking agent) on the activity of limestone is studied in laboratory by MET method. Results show that magnesium nitrate, ammonium nitrate and copper nitrate almost has no effect on the activity of limestone. With the concentration increasing, aluminum nitrate has an increasing inhibition on the dissolution of limestone. Fly ash has inhibition on dissolution of limestone due to the blockage of limestone pore by fly ash. The actual decoking agent has almost no effect on the limestone.

1. Introduction

Thermal power plants are important power production enterprises in China. The main fuel is coal in thermal power station [1]. The boiler will be influenced by coal quality, combustion condition and other factors. Coal coking phenomenon will occur in the combustion process when low quality coal is burned [2]. Decoking agent can prevent slagging and improve the thermal efficiency of the boiler [3]. As the world's largest coal producing and consuming country, SO_2 emissions from coal combustion account for about 85%-90% of the total SO_2 emissions in China [4]. Effects of magnesium carbonate, pH value, temperature, particle size on the activity of limestone have been studied systematically [5-7]. However, the effect of the coke removal agent on the activity of limestone has not been reported. The main components of decoking agent are magnesium nitrate, aluminum nitrate, copper nitrate and ammonium nitrate. The effects of these components on the limestone activity will be researched in this article. On the base of above experiments, real decoking agent will be used to study the effect on the activity of limestone.

2. Experiment

2.1. Experimental method

Activity of limestone is measured by MET method. MET method is to measure the relationship between the pH value and the reaction time under the condition of the flow of sulfuric acid. Adding acid, there will be a platform when the pH value reducing. It can be used to compare the reaction activity of different limestone. The desulfurization reaction activity of different limestone can be compared. According to the criteria provided by the United States MET company, the higher the pH retention platform, the better the limestone activity [6]. Experimental device is shown as figure 1.

Limestone samples as 1 grams (±0.01) $CaCO_3$ alkalinity is weighed and added in a 500ml beaker. 100ml deionized water is added in the same beaker. Then the beaker is placed on a hot plate mixer.

Setting mixing rate 600 rpm and inserting thermometer and pH meter electrode, the beaker is heated to 60℃(± 1°C). The sulfuric acid solution is 0.72mol/l. The titration of calcium carbonate solution is carried out at the rate of 0.75ml/min. Within the first 10 minutes, the slurry pH value is recorded every 2 minutes interval and once every 4 minutes in the next 15 minutes. The slurry pH value should be continuously recorded for 25 minutes. The operating time will be available in 25 minutes to add an excess of acid to the limestone solution to neutralize the equivalent of 1 g $CaCO_3$. Taking the time t as the horizontal coordinate, the pH value is the vertical coordinate, makes the pH-t curve. That is the characteristic curve of the limestone activity.

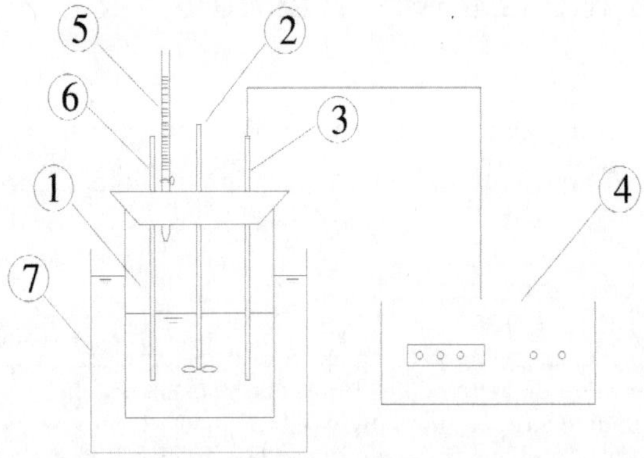

Figure 1. Experimental device
① Glass tank reactor; ②Digital electric mixer; ③ pH electrode;④ Digital pH meter; ⑤ Acid burette; ⑥ Thermometer; ⑦ Super thermostatic water bath

Table1. Experimental cases

Case number	Drugs	Concentration(mol/l)	Dosage(ml)
Case 1	ammonium nitrate	0.1	2
Case 2	ammonium nitrate	0.1	5
case3	ammonium nitrate	0.1	8
case4	aluminum nitrate	0.1	2
case5	aluminum nitrate	0.1	5
case6	aluminum nitrate	0.1	10
case7	magnesium nitrate	0.1	2
case8	magnesium nitrate	0.1	6
case9	magnesium nitrate	0.1	10
case10	copper nitrate	0.1	4
case11	copper nitrate	0.1	6
case12	copper nitrate	0.1	10
case13	Decoking agent	/	5
case14	Fly ash	/	0.2g
case15	Decoking and fly ash	/	5ml+0.2g

2.2. Experimental cases
The main components of decoking agent are ammonium nitrate, aluminum nitrate, magnesium nitrate, copper nitrate. Not only effects of single component, but also effects of a certain brand of decoking agent and fly ash on the activity of limestone are carried out in this paper. Experimental cases are shown in table 1.

3. Results and discussion

3.1. Effect of ammonium nitrate
Figure 2 shows the results of case1-3. Dropping 2 ml and 5ml ammonium nitrate solution, the pH value curves coincide with the curve of pure limestone. Stable platform of pH value is approximately 5.5. When dropping 8ml ammonium nitrate solution, the stable platform of pH value slightly decrease and the value is about 5.2. But within 5min-7min, limestone activity curves of different concentrations of ammonium nitrate nearly coincident with that of pure limestone. During the reaction process, three curves are close to overlap. So the effects of ammonium nitrate on limestone activity are not obvious.When ammonium nitrate concentration is high enough, stable pH platform slightly decreases. This cannot explain that ammonium nitrate has a slight adverse effect on the reactivity of limestone. This may be due to the hydrolysis of ammonium ions to produce partial hydrogen ions.

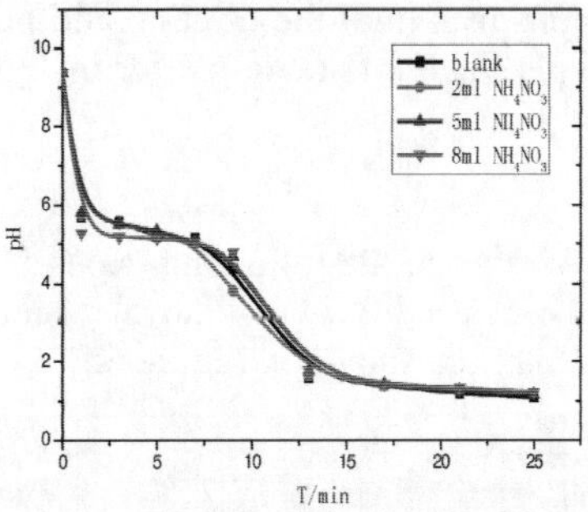

Figure 2. Effect of ammonium nitrate on the activity of limestone

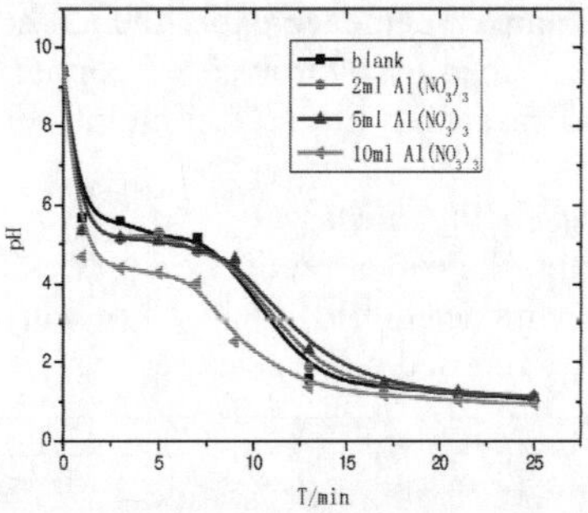

Figure 3. Effect of aluminum nitrate on the activity of limestone

3.2. Effect of aluminum nitrate
Effects of Aluminum nitrate (case 4-6) on the activity of limestone are shown in Figure 3. Dropping 2 ml and 5ml Al $(NO_3)_3$ solutions, the curve is basically coincident. The pH value of stable platform is slightly lower than the blank. Which shows that low concentration of Al $(NO_3)_3$ has little influence on limestone activity. But when dropping 10ml Al $(NO_3)_3$ solutions, pH value of the stable platform decreased to 4.4 and the pH value curve is also below that of pure limestone during the whole process.On the one hand, hydrolysis of aluminum ions to produce hydrogen ions, on the other hand, aluminum hydroxide formed by hydrolysis has flocculation which inhibits the dissolution of limestone particles. So with increasing of Al $(NO_3)_3$ concentrations, it has more obvious inhibition on limestone dissolution.

3.3. Effect of magnesium nitrate
Results of case7-9 are shown in Figure 4. Compared to the blank, the four curves in the whole process of titration basically coincide, only in the late period of reaction, 14 minutes to the end point of titration, the pH value decreases slightly compared with the blank condition. So we can judge that the magnesium nitrate has no distinct influence on limestone activity and can ignore its effect on wet desulfurization effect.

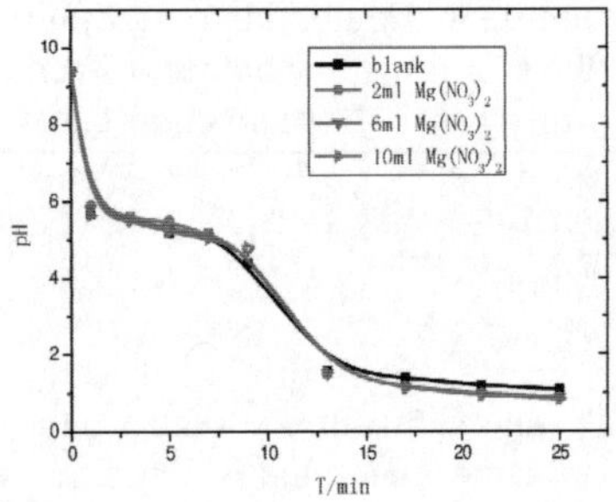

Figure 4. effect of magnesium nitrate on activity of limestone

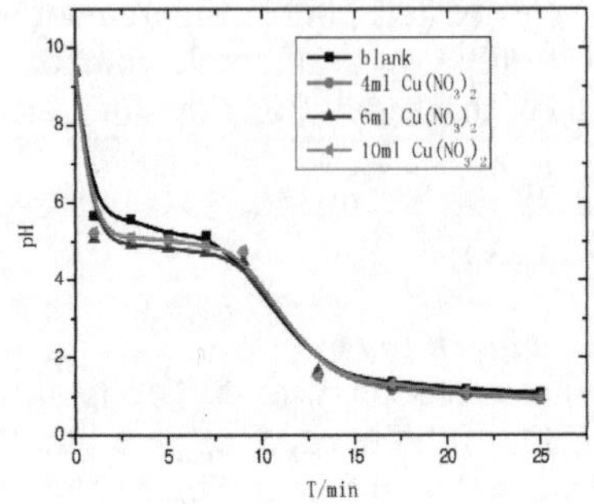

Figure 5. Effect of copper on limestone activity

3.4. Effect of cupric nitrate

Experimental results of case10-12 are shown in Figure 5. According to Figure 5, adding copper nitrate, pH value drops quickly in first 1-2 minutes and the pH value of stable platform drops from 5.5 to 5.1 compared to the blank. Then each experimental curve basically coincides until the titration end point. Hydrogen ions produced by hydrolysis of copper ions make the pH value of stable platform slightly decreased. Cupric nitrate has little effect on limestone dissolution.

3.5. Effect of the decoking agent

Taking into account the actual production process, the existence of fly ash, the following experiments also consider the impact of fly ash. Comprehensive study of the effect of coke removal agent on limestone activity is carried out. The experimental results of case13-15 are shown in Figure 6.

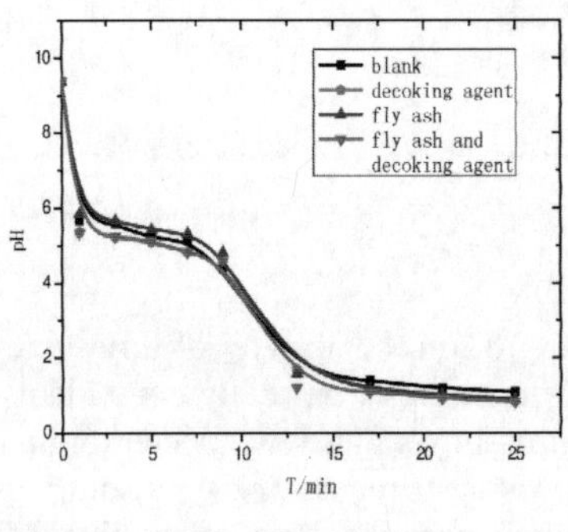

Figure 6. Effect of fly ash and decoking agent on the activity of limestone

From Figure 6 it can be found that with fly ash the pH value of stable platform increases slightly than that of the blank. This is not the dissolution rate of limestone becoming faster, but the fly ash shows alkalescent. In the late reaction, pH value curve of fly ash is below that of the blank sample. The reason maybe the fly ash clogs micro-porous of limestone particles and inhibits the dissolution of limestone. The curve of case13 coincides with that of case15. The pH value of stable platform is slightly lower than that of the blank until the end of the titration. This may be due to the hydrolysis of nitrate in decoking agent. There is no obvious evidence that the decoking agent has an inhibitory effect on dissolution of limestone.

4. Conclusions

Effects of decoking agent component on the limestone dissolution are different. Even if the increase of concentration, the effects of ammonium nitrate, nitrate magnesium and cupric nitrate on limestone dissolution are very small; Aluminum nitrate is inhibited to limestone dissolution, and the inhibition becomes more pronounced with the concentration; Fly ash has inhibition to the dissolution of limestone due to blockage of limestone pore by fly ash; Practical decoking agent almost has no inhibition on the dissolution of limestone.

Acknowledgements

Financial supports from the Fund of Nanjing Institute of technology (CKJB201217).

Reference

[1] Wang Haitao 2014 *Electronics World* **14** 311-312.
[2] Zhang Yubo 2015 *Technology Innovation and Application* **18** 127.
[3] Zhao Junbin, Tian Jian, Zhou Feng 2012 *Electric power safety technology* **09** 64-65.
[4] Li Chunyu, Dong Yue, Qi Yong 2014 *Environmental Engineering* **S1** 513-518
[5] Xu Shengrong 2007 *Journal of Nanjing Normal University(Engineering and Technology Edition)* **03** 44-47.
[6] Gu Yinna 2006*Experimental study on limestone activity in wet flue gas desulfurization* (Nanjing: Nanjing University of Science and Technology)25
[7] Xu Junde 2013 *basic science* **3** 71-79

Study on feasible technical potential of coal to electricity in china

Dexiang Jia[1]**, Xiandong Tan**

State Grid Energy Research Institute, "State Grid Corporation of China" Area, Future Science and Technology City, North Qijia, Changping District, Beijing, 102209, PRC

E-mail: jiadexiang71@163.com, jiadexiang@sgeri.sgcc.com.cn

Abstract. The control of bulk coal is one of the important work of air pollution control in China's future. Existing research mainly focuses on the adaptability, economy, construction and renovation plan, and operation optimization of specific energy substitution utilization, and lacks the strategy research of long-term layout of energy substitution utilization in large area. This paper puts forward a technical potential prediction method of coal to electricity based on the thermal equivalent method, which is based on the characteristics of regional coal consumption, and combined with the trend of adaptability and economy of energy substitution utilization. Also, the paper calculates the comprehensive benefit of coal to electricity according to the varieties of energy consumption and pollutant emission level of unit energy consumption in China's future. The research result shows that the development technical potential of coal to electricity in China is huge, about 1.8 trillion kWh, including distributed electric heating, heat pump and electric heating boiler, mainly located in North China, East China, and Northeast China. The implementation of coal to electricity has remarkable comprehensive benefits in energy conservation and emission reduction, and improvement of energy consumption safety level. Case study shows the rationality of the proposed method.

1. Introduction

Large-scale exploitation and utilization of fossil energy results in many global problems such as shortage of resources and energy security. Development of human society is facing increasingly serious fossil energy predicament. Greenhouse gas emissions and air pollution problems that result from fossil energy production and consumption are increasingly arousing people's concern. According to scholars study, one of the main sources of air pollution problem is burning bulk coal. Coal to electricity directly consumes electricity instead of coal in the energy consumption, which can increase the proportion of electricity in terminal energy consumption, and fundamentally solves the problems that restrict sustainable development of human society such as energy environment and climate change. In paper [1], the author establishes an equivalent substitution model between different energy based on the analysis of the equivalent conversion relationship of a variety of energy in the process of energy conversion and utilization. The author also considers the influence of price factor on energy utilization type, and establishes a multi-objective optimization model with the goal of maximum energy efficiency and minimum system operation cost, and verifies the rationality and superiority of the proposed model through the comparative analysis with the traditional energy utilization way. In paper [2], the author studies the difference of pollution emission rate between power industry and other industries by exploring the internal relation between energy, economy and environment. The author also uses system dynamics theory to build benefit simulation system of coal to electricity, and establishes a benefit analysis model of coal to electricity emission reduction, and verifies the implementation effect of coal to electricity using Beijing as an example. Paper [3] studies the market equilibrium problem of multi-energy systems with electric-gas equipment, and models the multi-energy systems with electric-gas equipment based on the energy center modeling method, and describes the market equilibrium problem of multi-energy systems as a game problem that all energy centers simultaneously participate in. The optimal response equations of the Nikaido-Isoda function are used to solve the iteration problem in the multi-energy market. In paper[4], a cost model and operation constraint are established for ground source heat pump system with the heating and cooling function, and for heating supply system which supplies cool with cooling water generation unit in summer, and heat with heat storage electric boiler in winter. An optimization model with direct solution of the boundary conditions is established, taking the minimum system cost deviation of the above two types as objective, which is used to determine the applicable areas of the two systems. In paper [5], several typical scenarios of energy substitution in rural areas of developing countries are studied. In paper [6], the ground source heat pump system with both heating and cooling functions is modeled, and the artificial neural network (ANN) is used to predict the EER of the system. In paper [7], the electric boiler and ground source heat pump technology are integrated. The author analyzes the operation strategy of the electric boiler and the heat pump in the joint use of different performance curves based on the energy consumption and economic cost. The integrated solution of using ground source heat pump as the main heating facility and electric boiler as auxiliary heating facility is put forward. Based on the consumption situation of coal, natural gas and electricity, the author in paper [8] analyzes the situation and potentiality of energy substitution in Tianjin, and estimates the economic efficiency of several energy substitution utilization. The existing researches mainly focus on the adaptability, economy, construction and operation optimization of specific energy substitution technologies, and lack strategic research on long-term planning of energy substitution in large area.

In view of the next 5-10 years, one of the focus of China's air pollution control is to control bulk coal, this paper puts forward a coal to electricity technical potential prediction method based on thermal equivalent method, which is according to the characteristics of coal consumption in various regions, and combined with development trends of energy substitution utilization adaptability and economy. Then the paper calculates the comprehensive benefits of coal to electricity. Case studies show that the proposed method is reasonable.

2. Analysis Model of Technical Potential for Coal to Electricity

Calculation method: The national and local air pollution control action plans and implementation plans, energy consumption and coal consumption control objectives, as well as the city, industry and other development planning of the corresponding energy substitution areas are all considered. China's energy substitution utilization technology level, its application status and development trend, energy consumption of diffident species, the economy and comparability trends of energy substitution, the manufacturing capacity of main kinds of electricity substitution equipment and other factors are also taken into account to make a comprehensive speculation. Taking the energy efficiency into consideration, the fossil energy consumption which has feasible energy substitution technology is converted to the technical potential of coal to electricity according to the thermal equivalent method, which is called electricity substitution potential of coal to electricity.

Formula: Technical potential of coal to electricity

$$p = \sum_{i=1}^{N} \frac{c_i \times u_i}{e_i} \tag{1}$$

p represents the technical potential of coal to electricity.

c_i represents coal consumption of the i-th industry which has feasible coal to electricity technology.

u_i represents coal utilization efficiency of the i-th industry which has feasible coal to electricity technology.

e_i represents electricity utilization efficiency of the i-th industry which has feasible coal to electricity technology.

N represents the quantity of industry.

There are more than 50 kinds of coal to electricity technologies, such as distributed electric heating, heat pump, electric boiler heating, building material kilns, electric cookers and medium frequency induction furnaces. The costs of migrating from coal to these technologies include their construction cost, fuel cost, electricity consumption cost and other operation cost. As for a consumer, the total cost of electricity utilization technology should be comparable to that of coal utilization technology.

Calculation range: First, bulk coal is all replaced, and the coal used for centralized heat supply in the area that is not suitable for combined heat and power cogeneration is also been replaced, steam coal (excluding thermal coal and industrial furnace coke) is partially replaced according to the characteristics of coal consumption. Second, the substitute power generation of the captive power plant's thermal coal is the incremental power generation by increasing the power generation efficiency through the high-efficiency generating unit considering generation rights transactions. Third, other coal consumption shall be considered as bulk coal, centralized heating coal, steam coal and chemical raw coal according to the proportion of all kinds of coal consumption, and the corresponding substitute electric energy shall be calculated.

Main calculation approach: First, coal consumption in different industry is counted with its utilization technology. Second, whether there is a feasible coal to electricity technology is checked for each main industry. Third, whether the total cost of electricity utilization technology is comparable to that of coal utilization technology is checked for each main coal consumer. Fourth, if the above two steps are yes, the technical potential of the main coal consumer is calculated according to formula (1). Fifth, coal consumption in the i-th industry is added up according to step fourth. Sixth, if various national and local planning has more coal consumption control target than that to be replaced in the i-th industry, and there is sufficient manufacturing capacity of corresponding coal to electricity equipment, and the technical potential in the i-th industry is modified according to formula (1). Seventh, technical potential of coal to electricity in each industry is added up. Eighth, according to the coal consumption to be replaced and its original pollutant emissions per unit of coal consumption, the comprehensive benefits of coal to electricity are estimated.

3. Case Study

According to the data of China coal consumption survey directed by the research group, and that from National Bureau of Statistics of the PRC and National Energy Administration, the national primary energy consumption in 2015 is 4.40 billion tons of standard coal equivalent, the total coal consumption reaches 3.965 billion tons, 3.7% down as compared to the previous year, but still accounts for half of the world's coal consumption. Among them, the power industry consumes 1.839 billion tons of coal, while 627 million tons of coal is consumed in iron and steel industry, 525 million tons in building materials industry, and 253 million tons in chemical industry. The proportion of coal in China's primary energy consumption structure is about 64%, far higher than the average world coal consumption ratio level [9], which is 30%. Much input data can be referred to in report [9].

Considering the development of major types of coal to electricity technologies and their application scale in past 10 years, such as electric boilers, electric kilns [9], using the above forecast method to estimate the technical potential of China's future coal to electricity, coal to electricity technical development potential is huge, which is about 1.8 trillion kWh, equivalent to reduction of 700 million tons of bulk coal. The types of technology for coal to electricity mainly include distributed electric heating, electric boiler heating, heat pumps, industrial boilers, industrial kilns and other coal to electricity areas. Among them, distributed electric heating, heat pump and electric boiler heating technical potentials are 500 billions, 500 billions and 300 billions kWh respectively, accounting for 28%, 28% and 17% of the technical potential for coal to electricity, which is more than 70% in total. The technical potential for coal to electricity of building material kilns, electric cookers and medium frequency induction furnaces are 190 billions, 150 billions and 85 billions kWh respectively, accounting for 11%, 8% and 5% of the technical potential of coal to electricity. Other coal to electricity technology has less potential, only 3% of the total, as shown in Figure 1.

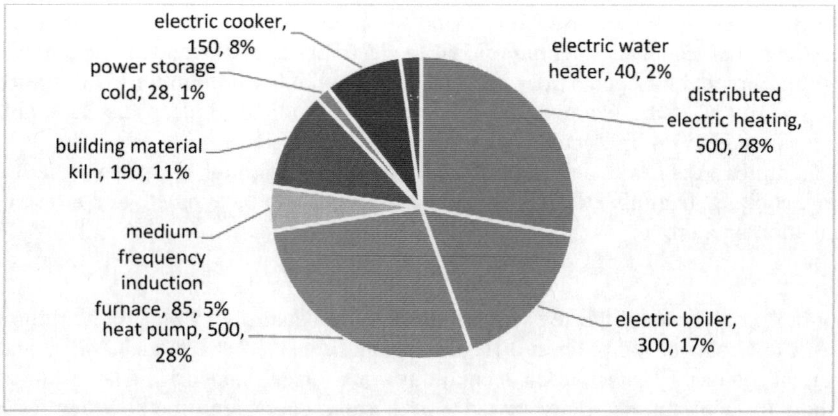

Figure 1. Distribution of technical potential for different types of coal to electricity (billion kWh, ratio)

The technical development potential of coal to electricity is mainly distributed in North China, East China, Northeast China, which is 655.6 billions, 319.7 billions, 221.8 billions kWh respectively, accounting for 37%, 18%, 12% of the technical potential for coal to electricity, which is about two-thirds in total. The technical potential for coal to electricity in the Central China, South China and Northwest China is 186.6 billions, 135 billions and 133.8 billions kWh respectively, accounting for 10%, 8% and 7% of the total. Other areas have less technical potential for coal to electricity. The technical potential for coal to electricity in the West Inner Mongolia and Southwest China is 111.4 and 29.4 billion kWh, less than 10% of the total, as shown in Figure 2.

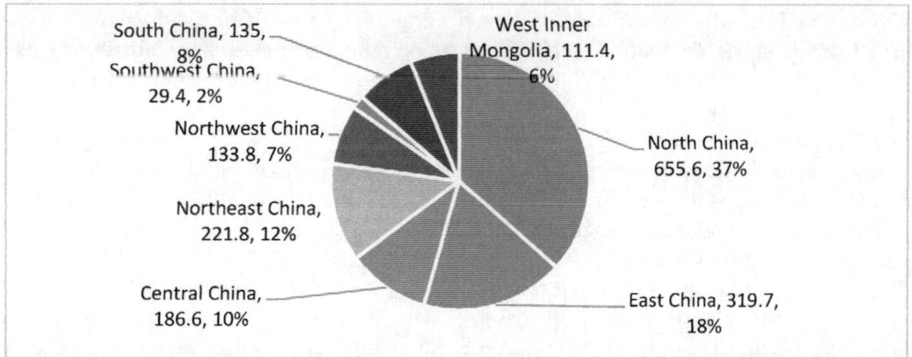

Figure 2. Area distribution of technical potential of coal to electricity (billion kWh, ratio).

The implementation of coal to electricity has significant comprehensive benefits, such as energy-saving, emission reduction, improving the level of energy utilization safety and others.

First, it can improve energy utilization efficiency, reduce energy waste. China's waste of energy due to inefficient utilization of bulk coal is more than 114 million tons of standard coal equivalent, accounting for about 4% of the total terminal energy consumption. Most of the industrial coal-fired boiler's actual operating efficiency is between 60-65%, and some are between 30-40%, 10-20% lower than the average level of advanced countries. Electric boiler's operating efficiency is about 85- 90%. Industrial kiln efficiency is about 40%, while electric kiln efficiency is up to 70-80%. Energy utilization efficiency of resident's bulk coal burning is generally lower than 20%, while the efficiency of electric cookers is up to 50-80%. Implementation of coal to electricity can effectively reduce the waste of distributed inefficient coal consumption and improve the overall energy efficiency level.

Second, it can significantly reduce air pollution and greenhouse gas emissions. According to China's future energy consumption of different kinds and their pollutant emissions per unit of energy consumption, if the total technical potential of coal to electricity is implemented, the annual emissions reduction of sulfur dioxide, nitrogen oxides and PM2.5 are about 8.4 million tons, 2.8 million tons, 0.3 million tons respectively. If the increment electric energy from coal to electricity is supplied by the clean energy, it can reduce 1.4 billion tons of carbon dioxide emission.

Third, it can improve the health and safety level of residents. Through the promotion of distributed electric heating, cooking, bathing, etc., gas poisoning incidents can be avoided, and the safety level of energy consumption is improved.

4. Conclusion

The control of bulk coal is one of the key work of air pollution control in China in the future. Based on the characteristics of coal consumption in different regions, considering the adaptability and economy development trend of energy substitution technology, this paper puts forward a technical potential forecast method of coal to electricity based on thermal equivalent method and estimates the comprehensive benefits of coal to electricity. The results show that China's coal-to-electricity technical development potential is huge, about 1.8 trillion kilowatt-hours, which is equivalent to utilization reduction of 700 million tons of bulk coal. Distributed electric heating, heat pump and electric boiler heating technical potential accounts for 73% of coal to electricity technical potential. North China, East China, Northeast China's coal to electricity technical potential is about 2/3 of the total. The implementation of coal to electricity has significant comprehensive benefits, such as energy saving, improving energy utilization safety levels and other. The case study shows the rationality of the proposed approach.

References

[1] Yan Z X, Liu J Y, Wei Z B, et al. 2016 Investigation on Multi-Energy Equivalent Substitution Mode and Transition Benefit Model *Power System Technology* **40**(6) 1620-26.

[2] Zhang L H, Xiong J, Yu Xiaobao 2015 Analysis on reduction benefits of substituting electricity for coal based on system dynamics *Journal of Central South University (Science and Technology)* **46**(4) 1527-33.

[3] Wang Y L, Zhao J H, Wen F S, et al. 2015 Market Equilibrium of Multi-energy System with Power-to-gas Functions *Automation of Electric Power Systems* **39**(21) 1-10,65.

[4] Qu Z Q, Xin J Q, Wu L, et al. 2016 Argument on boundary conditions of selection electric energy substitution technologies with heating for commercial customers *Automation of Electric Power Systems* **40**(13) 48-54.

[5] B J van Ruijven, J Schers, D P van Vuuren. 2012 Model based scenarios for rural electrification in developing countries *Energy* **38**(1) 386-397.

[6] Sun W, Hu P F, Lei F, et al. 2015 Case study of performance evaluation of ground source heat pump system based on ANN and ANFIS models *Applied Thermal Engineering* **87**(5) 586-594.

[7] Wu Z, Li Q, Xu H T 2015 Efficiency evaluation of alternative energy in heat supply system *Journal of Zhejiang University of Technology* **43**(5) 508-511.

[8] Yuan X R, Wu L, Zhang J, et al. 2015 Situation of energy substitution and economic analysis of electric heating in Tianjin *POWER DSM* **17**(5) 24-29.

[9] Research on the optimum road map for coal to electricity based on control of atmospheric pollution ［R］ State Grid Energy Research Institute, 2016.

Regional sulfur dioxide emissions: shall we achieve the goal?

X Tan[1,2], L Shi[2,3], M Wang[2] and JY Wang[2]

[1] State Grid Energy Research Institute, Beijing 102209, PRC
[2] Renmin University of China, Beijing 100872, PRC

E-mail: shil@ruc.edu.cn

Abstract. Although economic growth is slowing down in the new normal period, air pollution is still a very serious problem in China. The 15% binding goal of sulfur dioxide emission reduction from 2016 to 2020, as stipulated in the 13th Five-Year Plan, has been an ambitious target for the Chinese government. This paper studies the synthetic evaluation and forecasting analysis of sulfur dioxide in China by means of a "grey model" approach combined with the grey relational analysis methods, with the panel data of 31 provinces from 2005 to 2015. Grey analysis used to analyse a system with imperfect information, such that a variety of available solutions is reviewed, and the optimal solution is identified. Some encouraging results show that national emissions and a majority of provinces will achieve the target. Over time, the gap of regional differences is rapidly closing. According to the results of grey relational analysis, we find industrial structure and energy consumption have a more significant impact on sulfur dioxide emissions than GDP. Atmospheric treatment investment and environmental protection manpower play a more important role in emissions variation. Based on the findings, we should distinguish different factors and take different measures to protect the environment.

1. Introduction

China has maintained a rapid economic growth since its economic reform and opening-up in 1978. According to the data from National Bureau of Statistics of China, in the 12th Five-Year Plan period, an average of about 8% annual GDP growth, much higher than the world average, has made China one of the fastest growing economies in the world. The dark side of that achievement is a heavy environmental cost that China has paid. The overall environmental quality becomes worse and worse, and air pollution, such as smog, has become a hot topic in the society. Even though in recent years, sulfur dioxide emissions have declined slowly, that could not mitigate the problems brought by sustained and rapid growth of emissions in the past few decades. After 2011, the beginning of the 11th Five-Year Plan, a mandatory emissions control was implemented. The 15% reduction goal of sulfur dioxide emission has been an ambitious but constraining target to both central and local government. But, can that goal be achieved?

To examine whether China can meet that goal or not, this paper forecasts the sulfur dioxide emissions trends in China. Many factors influence the emissions and most information is inadequate or difficult to obtain. All of these factors can be reflected in the time trend of sulfur dioxide emission. So, emission of sulfur dioxide is both a grey system and a fuzzy dynamic system. As an interdisciplinary scientific area, grey system theory is the very method of analysis and prediction using small data sets and inadequate information to achieve an effective forecast of future trends optimally.

In China, early in 1980s, Professor Deng brought forward the grey system theory[1], an applicable method to analyze an unknown situation and predict trends of certain terms with poor information or limited observed data[2]. Soon afterwards, many scholars developed the theory. It expanded from management information system to social system, economic system, natural system, etc[3,4]. So far, it is applied extensively in several fields to forecast trends in economy, industry, environment, etc. Many scholars have been using a grey model approach to find solutions, adjust data [5,6], or improve models[7,8]. Among the methods in grey theory system, the method referred to as GM (1, 1) is used frequently by scholars as a prediction tool. Many modeling practices indicate that the GM (1, 1) model has a much higher accuracy than the exponential curve regression model[9,10]. The grey relational analysis method provides a quantitative tool to analyze a two-factor system development trend, indicating the characterization of numerical referred to as the grey relation [11,12]. It has been applied to various fields to measure the relationship between different variables, such as economics, chemistry, biology, environment and so on [13].

Admittedly, the GM (1, 1) model can forecast the regional emissions of sulfur dioxide in 2020, but it cannot tell the reasons behind the regional difference. So grey relational analysis method is conjointly used in our analysis, to mine the reasons behind sulfur dioxide emissions' variation. This article starts with a discussion on the GM (1, 1) model applied in environmental prediction and management. We study and forecast the sulfur dioxide emissions in China using the provincial data from 2005 to 2015. Furthermore, we apply grey relational analysis twice to search the relationship between emissions and social economy. This article intends to not only advance the prediction of the sulfur dioxide emissions during *the 13th Five-Year Plan for Economic and Social Development of the People's Republic of China* (the 13th Five-Year Plan) period, but also explain the reasons of emissions' variation with factors of economic development, industrial structure, and environmental protection efforts.

2. Data and methods

2.1. Data selection

Based on the availability and validity of the data, in this article, data from 2005 to 2015 are selected from national and regional *Statistical Yearbook*, and 2015's data are mined from provincial *National Economy and Society Developed Statistical Bulletin*, including sulfur dioxide emissions, gross domestic product (GDP), per capita GDP, proportion of secondary industry, total energy consumption. Other data are selected from *Annual Statistic Report on Environment in China* from 2005 to 2015, including per capita environmental protection investment, every ten thousand people with staff in environmental protection systems, and the proportion of atmospheric treatment in environmental protection investment.

2.2. Methods description

Since some sectors' emissions data are not collected, the data of sulfur dioxide is a grey system. We construct the traditional grey GM (1, 1) model and briefly look at the relevant basics of analysis of both parameter a and b. Assume $X^{(0)}$ is the original sequence of data, $X^{(0)}=\{x^{(0)}(1), x^{(0)}(2), \cdots, x^{(0)}(n)\}$,which stands for the non-negative original historical time series data. Then indicate new generate series of data as $X^{(1)}$, and $X^{(1)}=\{x^{(1)}(1), x^{(1)}(2), \cdots, x^{(1)}(n)\}$,. If there is a relationship between $x^{(0)}$and $x^{(1)}$ as $x^{(1)}(k)=\Sigma^{k}_{i=1} x^{(0)}(i)^2$, it is called the first order accumulated generating operation , marked as 1-AGO, and $X^{(1)}$ is called the first order accumulating generation sequence of $X^{(0)}$.The sequence $Z^{(1)}=\{z^{(1)}(2), z^{(1)}(3), \cdots, z^{(1)}(n)\}$ demonstrates the mean consecutive neighbor's generation operator for $X^{(1)}$,and $Z^{(1)}(k)= 0.5[x^{(1)}(k) + x^{(1)}(k\text{-}1)]$. The following differential equation represents the approximation of $x^{(0)}(k)$ data:

$$\frac{dx^{(1)}}{dt}+ax^{(1)}=b \tag{1}$$

Where a and b represent development coefficient and grey effect coefficient, both are estimated parameter in equation (1). Solving Eq. (1), we can obtain the discretization of the Eq. (2):

$$\Delta^{(1)}(x^{(1)}(k+1))+az^{(1)}(x(k+1))=b \tag{2}$$

It can be simplified:

$$x^{(0)}(k+1)=a[-\frac{1}{2}(x^{(1)}(k)+x^{(1)}(k+1))]+b \tag{3}$$

The matrix form of Eq. (1) will be described as follows:

$$\begin{bmatrix} x^{(0)}(2) \\ x^{(0)}(3) \\ \vdots \\ x^{(0)}(n) \end{bmatrix} = \begin{bmatrix} -\frac{1}{2}(x^{(1)}(1)+x^{(1)}(2)) & 1 \\ -\frac{1}{2}(x^{(1)}(2)+x^{(1)}(3)) & 1 \\ \vdots & \vdots \\ -\frac{1}{2}(x^{(1)}(n-1)+x^{(1)}(n)) & 1 \end{bmatrix} \tag{4}$$

The parameter matrixes are $\hat{a}=[\hat{a},\hat{b}]^T=(B^TB)^{-1}B^TY$,

$$Y=\begin{bmatrix} x^{(0)}(2) \\ x^{(0)}(3) \\ \vdots \\ x^{(0)}(n) \end{bmatrix} \text{ and } B=\begin{bmatrix} -\frac{1}{2}(x^{(1)}(1)+x^{(1)}(2)) & 1 \\ -\frac{1}{2}(x^{(1)}(2)+x^{(1)}(3)) & 1 \\ \vdots & \vdots \\ -\frac{1}{2}(x^{(1)}(n-1)+x^{(1)}(n)) & 1 \end{bmatrix} \tag{5}$$

If we plug the estimated result into Eq. (2), the time respond function of the grey differential equation is:

$$\hat{x}^{(1)}(k+1)=[x^{(1)}(1)-\frac{\hat{b}}{\hat{a}}]e^{-\hat{a}k}+\frac{\hat{b}}{\hat{a}} \tag{6}$$

We can obtain the formula with regard to original series data:

$$\hat{x}^{(0)}(k+1)=\hat{x}^{(1)}(k+1)-\hat{x}^{(1)}(k)=(1-e^{\hat{a}})[x^{(1)}(1)-\frac{\hat{b}}{\hat{a}}]e^{-\hat{a}k} \tag{7}$$

When the generation sequence approximately conforms to a linearity power function law and the sequence of data varies approximately according with the law of exponential function, GM (1, 1) method has a better prediction effect. The data of sulfur dioxide emissions of China from 2005 to 2015 and the accumulation generation sequence correspond to rules as above.

Based on the results of grey forecasting, combining the economic level and environmental management of different regions, we attempt to mine the reasons behind sulfur dioxide emissions growth or decline by grey relational analysis.

Assume X_0 is the consensus sequence of data, $X_0=(x_0(1),x_0(2),\cdots x_0(n))$, which indicates the characteristics of the system behavior. The sequence of data $X_i=(x_i(1),x_i(2),\cdots x_i(n))$ is a comparative sequence, which represents the factors that affect the system. ρ is the identification coefficient, $0<\rho<1$. The relation coefficient of two sequences in moment k can be described as follows.

$$\xi_i(k)=\frac{\min|X_0(k)-X_i(k)|+\rho\max|X_0(k)-X_i(k)|}{|X_0(k)-X_i(k)|+\rho\max|X_0(k)-X_i(k)|} \tag{8}$$

It is customary to use the mean of results to describe the relation between consensus sequence and comparative sequence, since results in all moments are too myriad to compare, as follows.

$$\gamma(X_0,X_k)=\frac{1}{n}\xi_i(k) \tag{9}$$

Since 0< $\zeta_i(k)$<1, 0< γ<1. When grey relation γ tends to 1, it shows that the development trend of the two sequences is consistent. According to the degree of the relation between different elements, we can quantify and judge the factors' influence.

This paper measures the influence of different factors on the emissions and emissions reductions of sulfur dioxide via two approaches. Firstly, we estimate economy, industrial structure and energy consumption's impact on sulfur dioxide emissions, trying to find reasons for the emissions variations. Secondly, we calculate the grey relation between environmental protection investment, atmospheric governance investment, and labor and emission variation.

3. Result Analysis and Discussion

3.1. Grey model forecasting for the nation

We obtain the value of estimated parameter $\hat{a}$ and $\hat{b}$. $\hat{a}=0.0323$, $\hat{b}=2655.858$. Backward error-detection is necessary to test the model accuracy. Two indicators used in this paper are C (variance ratio) and P (small error probability), obtained by calculation of the formulas as below, where S_x^2 means variances of original sequence and S_ε^2 means residual error sequence.

$$S_x^2=\frac{1}{n-1}\sum_{k=1}^{n}(x^{(0)}(k)-\bar{x}^{(0)})^2 \tag{10}$$

$$S_\varepsilon^2=\frac{1}{n-1}\sum_{k=2}^{n}(\varepsilon^{(0)}(k)-\bar{\varepsilon}^{(0)})^2 \tag{11}$$

$$C=S_\varepsilon^2/S_x^2\ ,\quad P=P\{0.6745S_x>\left|\varepsilon^{(0)}(k)-\bar{\varepsilon}^{(0)}\right|\} \tag{12}$$

Table 1. Accuracy grade of grey model.

Grade	R	C	P
excellent	$r\le 0.01$	$C\le 0.35$	$0.95\le P$
well	$0.01<r\le 0.05$	$0.35<C\le 0.50$	$0.80\le P<0.95$
qualified	$0.05<r\le 0.1$	$0.50<C\le 0.65$	$0.70\le P<0.80$
unqualified	$0.1<r$	$0.65<C$	$P<0.70$

We obtain the values of C and P, and test results by calculating as following:

C=0.0400, P=1.6990

From the grade in Table 1, it is obvious that this model can be applied to forecast with an excellent precision. The results of GM(1,1) are shown in Table 2. The maximum relative error of prediction value is -3.78%; the minimum one is 0.67%, and its mean relative error is -1.90%.

Table 2. GM (1, 1) analysis results of national data[1].

Year	sulfur dioxide emission	value of simulation	Residue (r)	relative error (%)	Year	sulfur dioxide emission	value of simulation	Residue (r)	relative error (%)
2005	2549.40	2549.40	-	-					
2006	2588.80	2532.24	56.56	2.18%	2011	2217.91	2154.17	63.74	2.87%
2007	2468.10	2451.66	16.44	0.67%	2012	2117.60	2085.61	31.99	1.51%
2008	2321.20	2373.64	-52.44	-2.26%	2013	2043.00	2019.24	23.76	1.16%
2009	2214.40	2298.10	-83.70	-3.78%	2014	1974.40	1954.98	19.42	0.98%
2010	2185.15	2224.97	-39.82	-1.82%	2015	1859.10	1892.77	-33.67	-1.81%

[1] Units of sulfur dioxide emission is 10000 tons, similarly hereinafter.

According to the results, sulfur dioxide emissions in 2020 will be 1610.17 ten thousand tons, which is 14.93% less than that in 2015. This number is close to the emission reduction targets of the 13th Five-Year Plan (-15%), which looks encouraging on a national basis.

3.2. Regional sulfur dioxide emissions prediction

Figure 1 shows provincial sulfur dioxide emissions in 2015 and 2020 based on the results of GM (1, 1). In light of the accuracy grade and Table 4, most provincial models have good precision, such that the maximum relative error of provincial prediction value is 8.49% (Guangxi) and the minimum one is 0.64% (Anhui).

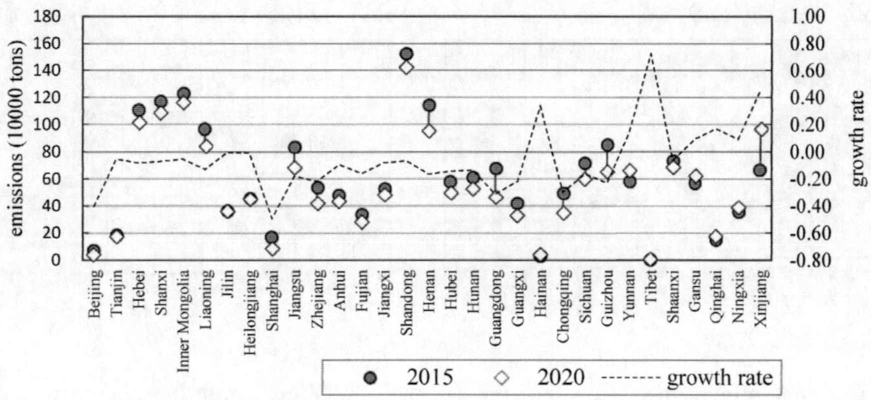

Figure 1. Provincial sulfur dioxide emissions in 2015 and 2020.

Compared with 2015, shown in Figure 1, in the light of its own system trends and present environmental management, 7 provinces' sulfur dioxide emissions will increase in 2020, most of which are located in western inland areas, such as Tibet, Xinjiang, Qinghai, and Gansu. In all, 11 provinces have better emission reduction rates than the national average and they achieve the reduction target of -15%, most of which are eastern developed regions, such as Shanghai, Beijing, and Guangdong.

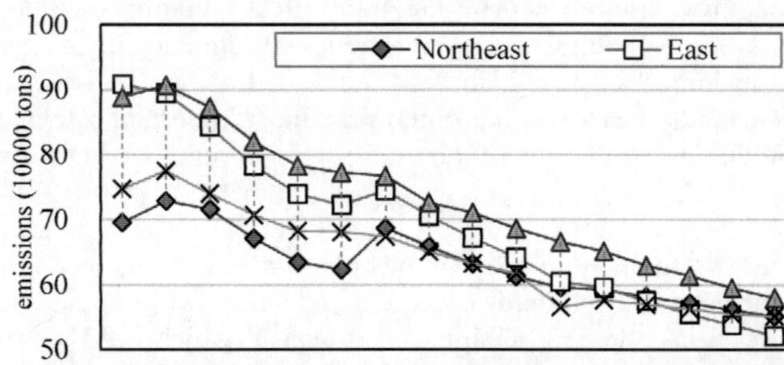

Figure 2. Sulfur dioxide emissions trend of the Four Areas).

According to the forecasting results of 31 provinces, over time, middle areas' average sulfur dioxide emissions still are higher than northeast and west areas, but Figure 2 shows that the gap nationwide is rapidly closing. Among the four areas, the eastern area has the biggest emission reduction, and this area has a more developed economy generally. We find that economic development brings large amounts of emissions but also substantial funding for emissions reductions. That is to say, the advanced economy is the main driver of emissions reduction in these provinces as large injections of funds are available to improve the technology of environmental production and coal-fired power plants.

3.3. Influencing factor analysis

In the factor analysis, we first estimate economy, industrial structure and energy consumption's impact on sulfur dioxide emissions, with indicators of "per capita GDP (p_GDP)", "total energy consumption (Energy_consump)", "proportion of secondary industry (Prop_S-industry)", mentioned above. The results shown in Figure 3 help explain the factors which influence the regional emissions, with wide difference among regions.

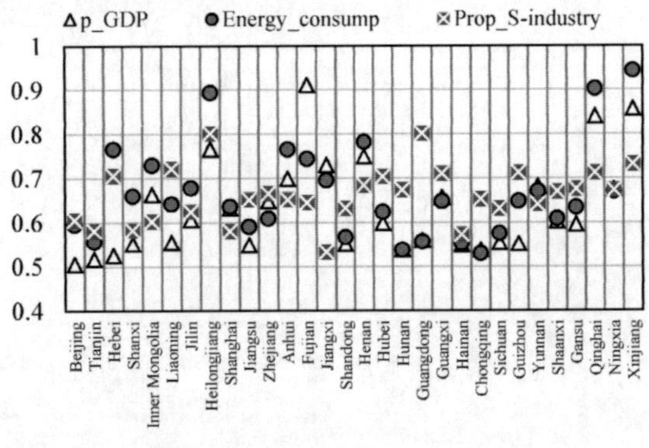

Figure 3. Influencing factors on sulfur dioxide emissions.

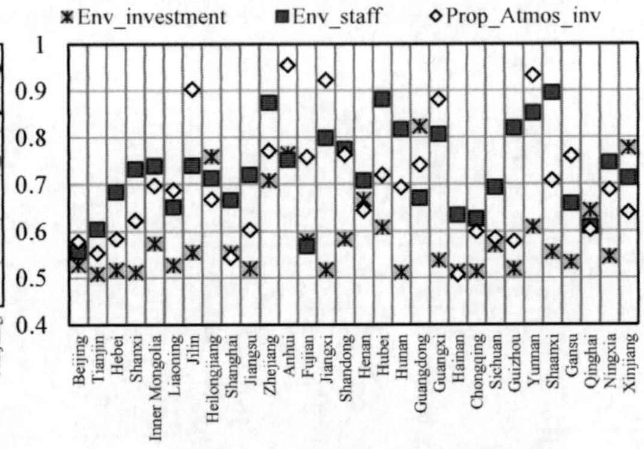

Figure 4. Influencing factors on sulfur dioxide emissions variation.

Secondly, we estimate the grey relation among environmental protection investment, atmospheric governance investment, and labor and emission variation, with indicators of "per capita environmental protection investment (Env_investment)", "every ten thousand people with staff in environmental protection system (Env_staff)", "proportion of atmospheric treatment investment in environmental protection investment (Prop_Atmos_inv)".

Figure 4 shows the results of the grey relational analysis, with features of wide regional differences. Generally, environmental protection investment has little effect on sulfur dioxide emission variation. On the contrary, proportion of atmospheric treatment investment in environmental protection investment has significant impact on it, which is reasonable. In some regions, to name only a few, Jinlin, Anhui, Jiangxi, Guangxi, manpower input plays the most important role in emissions variation, which can partly explain that the magnitude of the environmental regulation is very important.

4. Discussion

It's gratifying that China will be likely to achieve the emission reduction targets of the 13th Five-Year Plan. Most regions will achieve sulfur dioxide emissions reduction, while some western inland regions will go to the contrary, such as Xinjiang, Qinghai, and Gansu, which need to strengthen prevention and treatment.

In this paper, the grey relational analysis method is applied to mine the reasons for sulfur dioxide emissions and emissions variation. We find industrial structure and energy consumption have more significant impact on sulfur dioxide emissions than GDP. In view of emissions variation, atmospheric treatment investment level and environmental protection manpower play a more important role in most regions. Overall, a critical element is the different levels of resources applied to this problem, including political will, economic investment, and technological resources and so on. The advanced regions, to only name a few, Beijing, Shanghai, have more resources to determine their own development path and structure, and to choose better methods of pollution control, as opposed to less advanced regions that assume more passive roles. It is difficult to consider the political factor into the grey relational analysis model in this paper, and that is a topic for a future discussion.

Based on the findings, different regions should adjust measures to local conditions. For advanced regions, such as Beijing, Tianjin, Shanghai, since their forecasted emissions and grey relations are generally smaller, it's feasible to maintain the existing economic growth pace and environmental protection efforts. The regions with increasing forecasted emissions should make more efforts on the points where their relational values are high. Especially for the western inland regions, just name only a few, Xinjiang, Qinghai, Gansu, not only influence factors of reduction but also of emissions should be considered. Those regions should explore solution from the essential development patterns, energy conservation and emissions reduction, for instance, more clean energy instead of fossil energy consumption to optimize the energy consumption structure.

5. Conclusions

We have forecasted regional sulfur dioxide emission growth of 2015 to 2020, and proposed a potential relationship among emission reduction, economic development and environmental protection efforts with an approach of grey relational analysis. Economic development brings large amounts of emissions but also substantial of funding for emissions reductions. The strength of the current environmental protection is generally inadequate in the western inland regions. For different regions, we should distinguish different influence factors and take different measures to promote emissions reduction and environmental protection.

Acknowledgement

We gratefully acknowledge the valuable support by State Grid science and technology project "Method Research on Smog Formation Mechanism's Influence on Power Flow Planning", and Asian Development Bank project "BTH Regional Air Pollution Control".

References

[1] Deng J L 1982 Control problems of grey system *Systems & Control Letters* **1** 288–294

[2] Lin Y, Liu S 2004 A historical introduction to grey systems theory *Proceedings of IEEE international conference on systems, man and cybernetics* **1** 2403–08.

[3] Bo Z, Chuan L 2016 Forecasting the natural gas demand in China using a self-adapting intelligent grey model *Energy* **112** 810-825

[4] Liu S F 2007 The current developing status on Grey system theory *Journal of Grey System* **19** 111–123

[5] Kayacan E, Ulutas B, Kaynak O 2010 Grey system theory-based models in time series prediction *Expert Systems with Applications* **37** 1784–89

[6] Chen J, Xu C X 2005 An improvement of grey forecasting model *Journal of Liaoning Normal University (Natural Science Edition)* **3** 262-264

[7] Xu L, Tan T, Tu M, Qi L 2011 Improvement of grey models by least squares *Expert Systems with Applications* **38** 13961–66

[8] Wang Y N 2003 An extended step by step optimum direct modeling method of GM (1, 1) *Systems Engineering-Theory &Practice* **2** 120–124.

[9] Chen C I, Huang S J 2013 The necessary and sufficient condition for GM (1,1) grey prediction model *Applied Mathematics and Computation* **219** 6152–62

[10]Hsu L C, Wang C H 2009 Forecasting integrated circuit output using multivariate grey model and grey relational analysis *Expert Systems with Applications* **36** 1403–09

[11]Rajesh R, Ravi V 2015 Supplier selection in resilient supply chains: a grey relational analysis approach *Journal of Cleaner Production* **86** 343-359

[12]Chou C H 2001 A variable structure controller based on the grey prediction technology *Proceedings of the American Control Conference* 1505-06.

[13]Ying W, Chang Z, Gaopeng J 2016 Priority-sequence of mineral resources' development and utilization based on grey relational analysis method 2016 *International Journal of Mining Science and Technology* **26**(3) 395-400.

Seasonal variability of heavy metal content and its chemical speciation in excess sludge by vermi-stabilization

H Mu[1], Y X Zhao[1], Y Li[1], X D Zhang[1], D L Hua[1] and C H Zhao[2, 3]

[1] Energy Research Institute of Shandong Academy of Sciences, Key Laboratory for Biomass Gasification Technology of Shandong Province, Jinan 250014, China
[2] School of Resources & Environment, University of Jinan, Jinan 250022, China

E-mail: stu_zhaoch@ujn.edu.cn

Abstract. Vermifiltration system has been reported to enhance sludge degradation and relieve the potential risks of heavy metals, whereas few studies focus on the seasonal variability of heavy metal content and its chemical speciation in excess sludge by vermi-stabilization. The results of this study showed that total contents of Zn, Pb, Cr and the treatment efficiency of vermifilter were higher in summer and lower in winter due to the temperature effect. While Cu was observed to not be dependent on season. Further investigation using a five-step fractionation procedure to evaluate the main chemical speciations of metals in the sludge showed that there were similar variations in the five fractions (F1-F5) between each season, which indicated that the chemical speciation of heavy metals were most affected by vermifiltration technique than seasonal factor. The transformations among the five fractions suggested that the vermifiltration significantly reduced the mobility and bioavailability of Zn and Pb due to their increased stable fractions and decreased unstable fractions. The higher stable fraction of Cu led to the stable morphology in sludge, which favoured the insignificant variations with seasons.

1. Introduction

Larger amount of excess sludge has been produced in municipal wastewater treatment plants (WWTPs) in China due to the development of urbanization during the last few decades [1]. Accordingly, a mixture of heavy metals, organic compounds, non-essential trace elements, and microorganisms included in excess sludge would have negative effects on water, soil and floral, thus the sludge needed to be treated before being discharged to the environment [2, 3]. However, the high heavy metal content in excess sludge likely caused a serious risk to soil when the treated sludge was used preferentially as the fertilizer. This will become a more serious problem if the toxic metals are taken up by plants or transported through water after heavy metal migration and transformation.

Vermifiltration system, a low-cost and eco-friendly treatment technique for sludge by the interaction of earthworms and microorganisms in conventional filtration systems, has been reported to enhance sludge degradation and relieve the potential risks of heavy metals [4-8]. Nevertheless, some uncertainties like a sudden leakage accident of metals, rainstorm and seasonal temperature change often affect the properties of sludge. Especially for the seasonal factor, the different sludge properties would lead to the changes of heavy metal contents and chemical speciations in sludge after vermi-stabilization. The purpose of this study is to investigate the variations of heavy metal content and its chemical speciation in excess sludge after vermi-stabilization at different seasons.

As we known, only using the measure of total heavy metal contents in sludge as the evaluation index of pollution does not give a comprehensive overview. Because it cannot evaluate the potential migration and transformation of the metals under field conditions. Therefore, the mobility and bioavailability of heavy metals are also major limitations of land application of excess sludge. The bioavailability of heavy metals in sludge is strongly dependent on its specific chemical speciation or ways of binding [9,10]. Its specific chemical speciation is fractionized into exchangeable fraction (F1), carbonate precipitated fraction (F2), Fe–Mn oxide binding fraction (F3), organic and sulfide binding fraction (F4), and the residual fraction (F5) [11]. F5 is often immobilized by insoluble salts such as phosphate and humic acid-like organic matters, and is regarded as the mostly stable fraction and no toxic to environment [12,13]. Many studies have reported the changes of heavy metal bioavailability due to the earthworm activity [14-16]. For example, Devliegher et al. thought that the decreasing bioavailability of heavy metals by earthworms was due to their redistributing of metal speciations [17]. Other researchers reported that the vermicomposting technology also played a positive role in stabilizing heavy metals in the treatment of waste [18-20].

Under field conditions, the dynamic variations of total content and chemical speciations of heavy metals were found to be dependent on the individual metal and sludge characteristic and the treatment processes involved. So far, few studies focus on the seasonal variability of heavy metal content and its chemical speciation in excess sludge by vermi-stabilization. The aim of this study is to investigate the impact of vermi-stabilization on the total content of selected heavy metals (Zn, Pb, Cr and Cu) at different seasons. And the variations of their corresponding chemical speciation (F1-F5) transformations with seasonal variability were further investigated.

2. Materials and methods

2.1. Raw excess sludge and vermifiltration treatment

Raw excess sludge (RES) used in this study was taken from the secondary sedimentation tank of Guang Da Municipal WWTP in Jinan, China monthly from 2015 to 2016. The earthworms, purchased from a farm in Jinan, were cultured in the vermifilter (diameter of 30 cm and height of 90 cm, made of Teflon, showed in Figure 1) at an initial earthworm density of approximating 32 g/L (fresh weigh basis). A filter material (ceramsite) height of 80 cm was placed in the vermifilter to entrap the sludge and support the earthworms. The hydraulic load of the vermifilter was 3 $m^3/m^2 \cdot d$, and the volatile solid (VS) concentration of diluted RES was 292-320 mg/L.

During a 12-month steady operation (earthworms were healthy), the vermifilter treated sludge (VTS) was collected three times a month to determine the total contents and speciation of heavy metals (Zn, Pb, Cr and Cu). The concentrated VTS was firstly centrifuged at 9000 rpm for 15 min after natural sedimentation for 2 h; and then the residua were freeze-dried for 48 h. The samples were passed through 0.15 mm mesh prior to further analysis. The RES was sampled and treated using the same process described above on the same days.

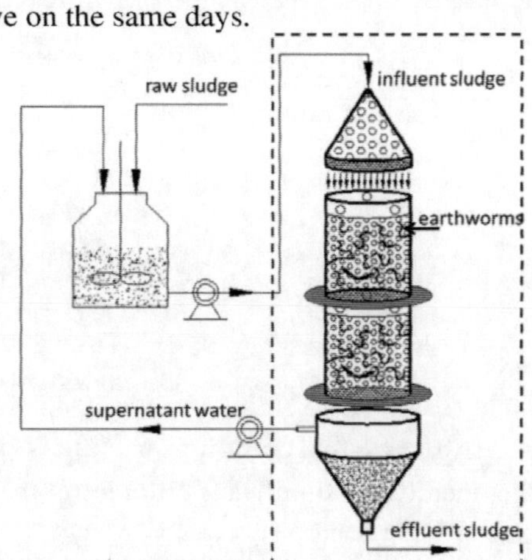

Figure 1. Schematic diagram of the vermifilter (with earthworms in the filter bed).

The important properties of sludge pertaining to the study, such as pH, the ratio of VS to total suspended solid (TS) (VS/TS), total N and C were determined following the reference [21]. The C/N ratio and total organic C were measured using an elemental analyzer Vario EL III (German).

2.2. Determination of heavy metals

The sludge samples needed to be digested prior to heavy metal contents being determined based on the reference [22]. These samples were digested in a microwave (Berghof, Germany) after being acidified with a mixture of 6 ml HNO_3, 2 ml HCl and 2 ml HF. The procedure went as follows: a 10 min gradual temperature increase to 200 °C (1000 W, 106 Pa), a 15 min step of 200 °C and a ventilated cooling stage. Then the digested samples were heated on a hot plate (120 °C) until they became near dry for inductively coupled plasma optical emission spectrometry (ICP-OES, PerkinElmer Optima 2100 DV, USA) analysis. The heavy metal chemical speciations of five fractions (F1, F2, F3, F4, and F5) were determined by a five-step sequential extraction procedure proposed by Tessier et al. [8,11].

2.3. Statistical analysis

Samples of RES and vermifilter treatment sludge (VTS) were respectively collected from wastewater treatment and vermifilter at the beginning, middle, and end of each month. All assays were conducted in triplicate, and the results were expressed as mean ± standard deviation. An analysis of variance (ANOVA) was used to test the significance of results, and $p < 0.05$ was considered to be statistically significant.

3. Results and discussion

3.1. Performance of sludge stabilization by vermifilter

The treatment performance of vermifilter was firstly investigated, and the degree of the sludge stabilization at each month was assessed by the VS/TS (Figure 2). In this study, seasons were divided as follows: spring was March 2016 to May 2016; summer was June 2016 to August 2016; autumn was September 2015 to November 2015; winter was December 2015 to February 2016. Figure 2 showed that the variation trend of VS/TS in the RES was similar to that in the VTS during the whole year, and the VS/TS value was higher in winter and spring, and lower in summer and autumn, suggesting a relative strong relationship with temperature. The VS/TS value of the RES and VTS were 70.2-79.1% and 57.9%-68.6% during the whole year, and their comparisons indicated that the vermifiltration greatly improved the degradation of the organic matter in the RES, which was consistent with the report of Zhao et al. [1], who thought that the sludge stabilization in vermifilter was achieved by earthworms digestion and earthworm-microorganism interactions.

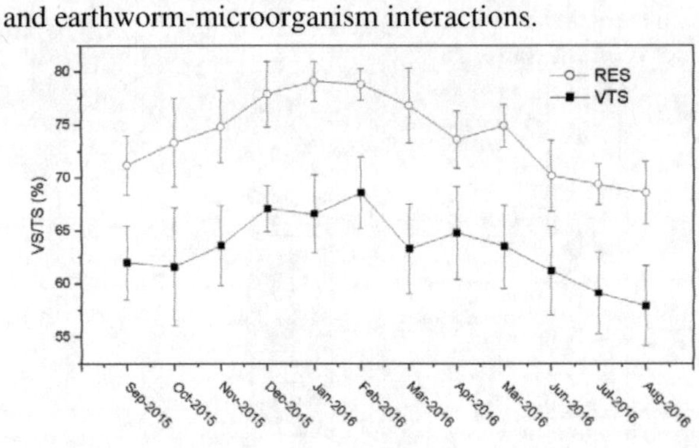

Figure 2. Variations of VS/TS in the RES and VTS during the whole year.

By the vermifiltration of sludge, there were significant different in the physic-chemical properties between the RES and VTS. For example, average value of pH was decreased from 7.2 in the RES to 6.5 in the VTS, average values of total C and N were decreased from 310.6 and 37.7 mg/kg-VS in the RES to 211.3 and 25.3 mg/kg-VS in the VTS, respectively. For their specific seasonal variations, they were also observed to be interconnected with temperature. The values of pH, total organic C and N were higher in winter and spring than those in summer and autumn. The different physic-chemical characteristics of sludge would result in the variations of the total content and the chemical speciation of heavy metals. Thus, the comparisons of heavy metal contents between the sludge before (RES) and after (VTS) vermifiltration during the whole year were then investigated in the following text.

3.2. Seasonal variations of heavy metal contents by vermifiltration

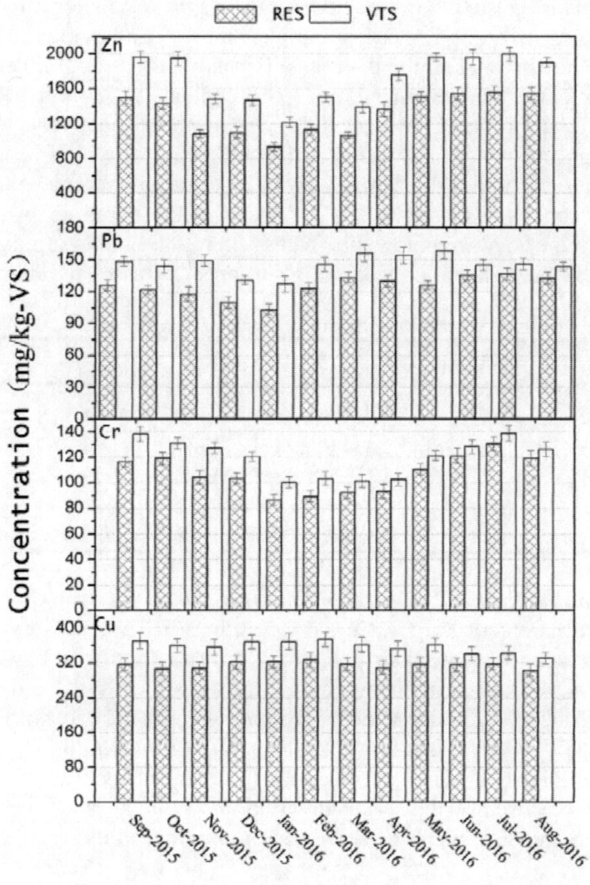

Figure 3. Comparisons of total contents of Zn, Pb, Cr, and Cu between the RES and VTS during the whole year.

Figure 3 shows the monthly variations of selected heavy metal (Zn, Pb, Cr, Cu) contents in the RES and VTS at different months from Sep. 2015 to Aug. 2016. It can be seen that total contents of these metals were increased in the VTS compared with the RES, especially for Zn, which seemed to heighten its potential risk to the environment; however, all heavy metal contents obtained in the VTS were well below the threshold value required in China for land application [23]. The increases of the total contents of Zn, Pb and Cr were mainly due to the fact that the organic matters in the sludge were biodegraded by the earthworms and the interactions of earthworm-microorganisms in the vermifilter. A similar phenomenon was found that total contents of heavy metals might be increased after sludge stabilization by composting, liquefaction and phytostabilization [24-26], and all these were due to the degradation of organic matters. While some studies pointed that total contents of heavy metal were decreased during composting process, with the reason being that metals were released from the decomposed organic matter [26,27]. In this study, the detected soluble metal ions in the effluent were negligible. Therefore, the higher total contents of Zn, Pb and Cr in the VTS than those of RES mainly resulted from the organic matter degradation by vermifiltration. The reason for the inconspicuous variation of Cu during the whole year was likely due to its stable characteristic.

Heavy metal composition of excess sludge can be extremely variable depending on the weather and human activities. Furthermore, the average value of heavy metal total contents in the RES and VTS at each season was included in Table 1. It can be observed that total contents of Zn, Pb and Cr in both RES and VTS were higher in summer and lower in winter, and the similar phenomenon was also found by Garcia [28]. This suggested that the temperature was a key factor affecting total contents of heavy metals in excess sludge. Taking Zn as an example, since of Zn having the highest percentage (> 70%) among the four metals, its total contents in the RES and VTS were 1950 and 1548 mg/kg-VS in summer. When the lower temperature (winter) came, total contents of Zn in the RES and VTS were all decreased, and its concentration was 1394 and 1052 mg/kg-VS, respectively. We suspected that the activity of the microorganisms was easily affected by temperature, which resulted in more organic matter was degraded under the higher microbial activity condition in summer. Since heavy metals are non-biodegradable, their concentrations in the VTS tended to be increased. All total contents of the selected heavy metals in the RES were below the values reported by other authors [24-26], which were probably related to the low heavy metal concentrations of area characterized domestic sewage in Jinan.

Table 1. Total contents of heavy metals (Zn, Pb, Cr and Cu) in the RES and VTS at each season.

	Autumn		Winter		Spring		Summer	
	RES	VTS	RES	VTS	RES	VTS	RES	VTS
Zn	1337±183	1796±223	1052±86	1394±129	1312±183	1701±237	1548±6	1950±37
Pb	122±3.5	147±2.5	112±8.3	135±7.8	130±2.9	157±1.8	135±1.8	145±1.0
Cr	113±6.5	132±4.6	93±7.4	108±8.9	98.6±8.3	108±9.1	124±4.9	131±5.5
Cu	310±4.4	361±6.5	324±2.1	369±2.6	314±3.8	358±4.5	311±1.9	388±5.1

3.3. Variations of heavy metal chemical speciations by vermifiltration

The specific chemical speciations of Zn, Pb, Cr and Cu in the VES and VTS were further investigated (Figure 4) to comprehensively evaluate their potential risk. The data from Figure 4 were determined based on the sludge samples collected in the beginning, middle and end of April in 2016. In this study, the proportions of heavy metal fractions at other months were also analyzed (data not shown), and their variation trends were found to be in correspondence with the results showed in Figure 4. The reason being that the variations of heavy metal speciation caused by vermifiltration technique was much higher than the effect caused by the seasonal variation. Thus, the transformations among the five fractions between the RES and VTS were the key discussion in this study.

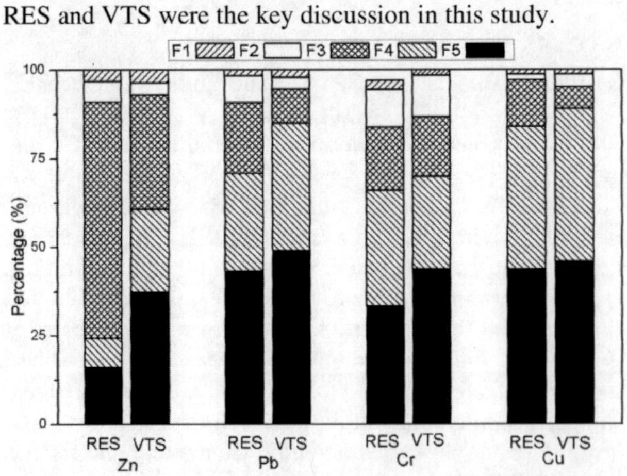

Figure 4. Proportions of Zn, Pb, Cr, Cu fractions (F1, F2, F3, F4, and F5) in the RES and VTS at April 2016.

According to the reference [29], F1 represents the readily bioavailable proportion of metal and it can be defined as toxic. As seen from Figure 4, F1 percentage of Zn, Pb, Cr and Cu in the RES were low to 3%, suggesting its faint and direct toxicity to the environment [25,30]. After vermifiltration, F1 percentages of Zn and Pb were observed to be increased ($p<0.05$), while its variations in Cr and Cu were negligible. F2 and F3 are sensitive to pH and can be easily transformed into F1 under weak acid condition [31,32]. F2 proportion of Zn and Pb were found to be significantly different ($p<0.05$) before and after vermifiltration, and their portions decreased from 5.81% and 7.54% to 3.32% and 3.33%, respectively. F3 was the largest percentage among the five fractions of Zn in both the RES and VTS, which were 66.71% and 32.37%, respectively. This phenomenon was consistent with the report of Peruzzi [24]. The similar decreasing F3 in Pb and Cu were also observed after vermifiltration, while its variation for Cr was inconspicuous. F4 of Zn, Pb and Cu in the VTS were clearly higher than those in the RES, which indicated that the vermifiltration technique could promote Zn, Pb and Cu bound to organics and sulfide and decrease their bioavailability. F5 is immobilized by insoluble salts such as phosphate and humic acid-like organic matters, and is regarded as the mostly stable fraction and no toxic to environment [12,13]. F5 was the main chemical speciation of Pb, Cr and Cu, of the proportion were 43.16%, 33.43% and 43.62% in the RES, and 48.95%, 43.72% and 45.80% in the VTS, respectively, indicating an insignificant variation ($p>0.05$). As for Zn, its F5 fraction was increased from 16.14% to 37.34% after vermifiltration ($p<0.05$). Due to the multivalent characteristic of Cr, it can be existed stably in the anionic speciations of Cr_2O_3, chromate (CrO_4^{2-}) or dichromate ($Cr_2O_7^{2-}$) in the vermifiltration system, which accordingly made its faint complexation with the negative charged sludge and resulted in the weaker transformations among the five fractions. The sum of F4 and F5 fractions of Cu was greater than 83.54%, which was the highest among the four heavy metals and the most stable. The higher stable fraction of Cu was consistent with its balanced total content variation at different months (Figure 3).

4. Conclusion

The marked organic degradation in excess sludge by vermifiltration technique indeed enhanced total contents of selected heavy metals (Zn, Pb, Cr and Cu). Total contents of Zn, Pb and Cr in excess sludge before and after vermifiltration were found to change with the seasons due to the temperature effect; while this phenomenon does not apply to Cu. Further investigation of the chemical speciation transformations showed that there were similar variations in the five fractions (F1-F5) between each season, which indicated that the chemical speciations of heavy metals were most affected by vermifiltration technique than seasonal factor. Further analyses of transformations among the five fractions suggested that the vermifiltration significantly reduced the mobility and bioavailability of Zn and Pb due to their increased stable fractions and decreased unstable fractions. The higher stable fraction of Cu led to the stable morphology in sludge, which favoured the insignificant variations with seasons.

Acknowledgments

This study was supported by National Natural Science Foundation of China (51608314), Shandong Province Natural Science Foundation (ZR2014EEQ023), and Special Fund for Agroscientific Researchin the Public Interest (201503135-04).

References

[1] Zhao L M, Wang Y Y, Yang J, Xing M Y, Li X W, Yi D H and Deng D H 2010 *Water Res.* **44** 2572-82
[2] Wang X J, Chen L, Xia S Q and Zhao J F 2008 *J. Environ. Sci.-China* **20** 156-160
[3] Lasheen M R and Ammar N S 2009 *J. Hazard. Mater.* **164** 740-749
[4] Wang L M, Guo F H, Zheng Z, Luo X Z and Zhang J B 2011 *Bioresource Technology* **102** 9462-70
[5] Xing M Y, Yang J A, Wang Y Y, Liu J and Yu F 2011 *J. Hazard. Mater.* **185** 881-888
[6] Xing M Y, Zhao L M, Yang J, Huang Z D and Xu Z 2011 *Environ. Eng. Sci.* **28** 619-626

[7] Jiang L H, Liu Y G, Hu X J, Zeng G M, Wang H, Zhou L, Tan X F, Huang B Y, Liu S B and Liu S M 2016 *Bioresour. Technol.* **200** 999-1007
[8] Yang J, Zhao C H, Xing M Y and Lin Y A 2013 *Bioresour. Technol.* **146** 649-655
[9] Jin C W, Zheng S J, He Y F, Di Zhou G and Zhou Z X 2005 *Chemosphere* **59** 1151-59
[10] Fuentes A, Llorens M, Saez M J, Aguilar I, Perez-Marin A B, Ortuno J F and Meseguer V F 2006 *Environ. Poll.* **143** 355-360
[11] Tessier A, Campbell P G C and Bisson M 1979 *Analyt. Chem.* **51** 844-851
[12] Pare T, Dinel H and Schnitzer M 1999 *Biol. Fertility Soils* **29** 31-37
[13] Walker D J, Clemente R, Roig A and Bernal M P 2003 *Environ. Poll.* **122** 303-312
[14] Amir S, Hafidi M, Merlina G and Revel J C 2005 *Chemosphere* **59** 801-810
[15] Wang D D, Li H X, Wei Z G, Wang X and Hu F 2006 *Biol. Fertility Soils* **43** 120-123
[16] Udovic M and Lestan D 2007 *Environ. Poll.* **148** 663-668
[17] Devliegher W and Verstraete W 1996 *Soil Biol. Biochem.* **28** 489-496
[18] Song X, Liu M, Wu D, Qi L, Ye C, Jiao J and Hu F 2014 *Waste Manage.* **34** 1977-83
[19] Singh J and Kalamdhad A S 2013 *Ecol. Eng.* **60** 214-223
[20] Lv B, Xing M and Yang J 2016 *Bioresour. Technol.* **209** 397-401
[21] Bossio D A and Scow K M 1998 *Microbial Ecol.* **35** 265-278
[22] Nemati K, Abu Bakar N K, Bin Abas M R, Sobhanzadeh E and Low K H 2010 *J. Hazard. Mater.* **182** 453-459
[23] State Environmental Protection Administration of China 2002 Discharge standard of pollutants for municipal wastewater (GB 18918–2002). China Environmental Science Press, Beijing, China (in Chinese).
[24] Peruzzi E, Masciandaro G, Macci C, Doni S, Ravelo S G M, Peruzzi P and Ceccanti B 2011 *Ecol. Engin.* **37** 771-778
[25] Yuan X Z, Huang H J, Zeng G M, Li H, Wang J Y, Zhou C F, Zhu H N, Pei X K, Liu Z F and Liu Z T 2011 *Bioresour. Technol.* **102** 4104-10
[26] Jordao C P, Fialho L L, Neves J C L, Cecon P R, Mendonca E S and Fontes R L F 2007 *Bioresour. Technol.* **98** 2800-13
[27] Haroun M, Idris A and Omar S 2009 *J. Hazard. Mater.* **165** 111-119
[28] Garcia-Delgado M, Rodriguez-Cruz M S, Lorenzo L F, Arienzo M and Sanchez-Martin M J 2007 *Sci. Total Environ.***382** 82-92
[29] Hare L, Tessier A and Borgmann U 2003 *Human and Ecol. Risk Assess.***9** 779-793
[30] Chen M, Li X M, Yang Q, Zeng G M, Zhang Y, Liao D X, Liu J J, Hu J M and Guo L 2008 *J. Hazard. Mater.* **160** 324-329
[31] Li L X Y, Wu J Y, Tian G M and Xu Z L 2009 *J. Hazard. Mater.* **167** 634-640
[32] Tandy S, Healey J R, Nason M A, Williamson J C and Jones D L 2009 *Bioresour. Technol.* **100** 4220-26

IOP Conf. Series: Earth and Environmental Science (52) 012074 doi:10.1088/1755-1315/52/1

Foreign trade and pollution: the case of South China water quality

Hui Zuo[1], Danxiang Ai[2, 4] and Yuling Lin[3]

[1] School of Economics and Commerce, Guangdong University of Technology, Longdong, Yinglong Road, Tianhe District, Guangzhou City, Guangdong Province, China

[2] School of Management, Guangdong University of Technology, Longdong, Yinglong Road, Tianhe District, Guangzhou City, Guangdong Province, China

[3] Counrty Garden Co. LTD, Country Garden Avenue, Shunde District, Foshan City, Guangdong Province, China

E-mail: aidx@gdut.edu.cn

Abstract. With rapid economic growth, South China has to face the most serious water pollution. However, whether or not such kind of water pollution is mainly caused by foreign trade is questionable. And, how the trade mode will be changed by pollution and corresponding regulation is also uncertain. In this paper, a fully endogenous model, which integrate economic growth, energy use and pollution, is designed to interpret the interrelation among these key variables in South China. Through this model, a new possibility of water environment Kuznets curve change has been investigated. Attribute to mixed two stage feasible general least square estimation method, we conclude that foreign trade has strong influence on environment change rate and the turning point. It can make the virtuous circle of between economic growth and environment improvement come early or later in different circumstances. Export and import play different role in such process and have counter effects on environment.

1. Introduction

Since 1978, the reform and opening policy has brought a tremendous economic growth of the South China Area. Known as "World Factory", this area has experienced international trade booming of about 30% growth annually. Unfortunately, the spring up of manufactures heavily polluted the air, water and soil and ecology system of this place. Environmental deterioration and resource exploitation hinder the economy from sustainable development. Nearly 40% cultivated land in the South China is heavily metal polluted and more than 10% is in fatal condition. Some people suffered ptomaine poisoning because of eating larminaria product of Guangzhou and Zhuhai City. The 18th Chinese Communist Party congress asserted that constructing eco-civilization is the long term strategy for people and pledged to enhance protection for environment and ecosystem. The central and local government are programming important trade transformation and upgrade in the South China region. However, can such trade policy direction helps the region to escape from heavy pollution? Will trade structure adjustments work in improving interrelationship between environment and economic growth of this pollution intensive export dependent region?

Environment Kuznets Curve explains relationship of the economic growth and environment. EKC hypothesis usually shows that pollution increases firstly and decreases afterwards, shaped as an inverted U-shaped curve (Panayotou, 1993) [1]. This point is also given by Grossman and Krueger (1993)[2] in their study on economy and environment in NAFTA. In the beginning, pollution grows rapidly, which actually happened in most developed countries. As countries begin to pay more attention to ecosystem and health. Environment will improve gradually with economic growth. Selden and Song (1995) [3] introduced a time contiguous model and concluded that the capital distribution between economic growth and environment protection is important in EKC.

It is occasionally mentioned in literature that international trade, as important mobilization of factors and products, will also influence EKC dynamics. Some researchers regarded that liberalization of international trade is beneficial to environment because it promotes income so as to secure financial and technology support for environment (Egli and Steger, 2007) [4]. This is the channel of technology and preference. Cole (2004)[5] studied the interrelationship between international trade and environment. He saw liberalization of international trade as an instrument to help environmental resources global distributed optimally so as to facilitate production in the most efficient way. Suri and Chapman (1998) [6] found that liberalization of international trade is helpful to remove trade distortion produced by subsidy, custom and other trade barriers, which are proved to be deleterious to environment. Lopez (1994)[7] regarded the scale expansion will both increased return of environmental friendly technology, and increased marginal cost of pollution.

Opposite to viewpoints above, some researchers thought that liberalization of international trade is destructive to environment. The Pollution Heaven, a famous hypothesis, is supported widely by normative and positive studies. The Pollution Heaven hypothesis predicts that trade liberalization will cause production of pollution intensive goods to transfer to countries with relatively weak environmental policy of those strictly environmental regulated countries. Copeland and Taylor (1995) [8] developed a model with endogenous environmental policy. With assumptions that demand for better environment quality increases with income and that government implements environmental policy considering the preferences of residents, they predicted that South (developing countries) has a comparative advantage in producing pollution intensive good for trading. Liberalization of international trade transfers the pollution intensive industries from North (developed countries) to South.

However, some researchers considered that pollution intensive industry transfer from North to South is not inevitable. Antweiler et al (2001) [9] and Copeland and Taylor (1995) [8] introduced models for empirical test. With the assumption that pollution intensive industry is also capital intensive, North's abundance of capital also makes them comparative advantageous to produce pollution intensive products. If the effect of stringent pollution control policy is less important than the effect of capital abundance, trade liberalization will make pollution intensive industry transfer from South to North even the former is more comparative advantageous in environment. Furthermore, Busse (2004)[10] regarded that labor cost, technology, political system, market scale and infrastructure also influence trade. Pollution intensive industry transfer mode doesn't only depend on environment comparative advantage or factor endowment.

According to Race to Bottom hypothesis, posed by Dua and Esty (1997) [11], if South is abundant of environment, it has a comparative advantage in producing and export pollution intensive good to increase its balance of payment surplus and gross domestic product. Meanwhile, if North has abundant capital and technology, they may also have comparative advantage in similar industry (Westerlund, 2007)[12]. Therefore, South has incentive to loosen control on pollution so as to improve competitiveness. If both North and South do the same to secure their competitiveness and export, global environment will be damaged. In case that developed country doesn't race to bottom because of its stringent legal system, vicious competition in environmental regulation will almost happen in developing countries.

Since the factor endowment and trade mode decided influences an economy in Race to Bottom and Pollution Heaven, arbitrary judgment on trade liberalization's positive or negative effects on environment without considering economic development stage is less persuasive. Environment of an economy suffered from trade liberalization in relative low income stage may benefit from the same trade mode in higher income stage. It is hard to say whether or not the opening policy which leads to some pollution currently is optimum when considering the future environmental benefits. Therefore, the key is the trade influence in the interrelationship of environment and economic growth rather than environment itself. Lots of literatures discussed that international trade will be beneficial or deleterious to environment, but few of them thoroughly investigated its effect on dynamics of EKC, the turning point, the deterioration or improvement speed, the elasticity and subsequent channel after crossing the pollution ceiling. Will structure and scale alteration of trade influence environment-economy dynamics? Will export of some products slow down the deterioration at a certain income level? Dose pollution control has persistent effects on environment or just transitory? We will investigate the trade influence in EKC as well as answering these questions. In the next section, we introduce the model, variables and data. In section three, feasible general least square estimation method and two stages least square estimation method are adopted to estimate the simultaneous equation group. Section four is the result discussion and section five is conclusion.

2. Model, variables and data

Our models are endogenously constructed by three equations. The first one describes the relationship between the pollution P and the income Y and foreign trade. We set squared income Y^2 to capture the inverted U change of the relationship between pollution and income. Both export X and import M are considers. Also, we take technology progress A into consideration.

$$P = b_{11}Y + b_{12}Y^2 + b_{13}X + b_{14}M + b_{15}A \tag{1}$$

The second equation describe the relationship between energy E and income. We suppose that energy demand is decided by three factors. The first one is the income. Higher income suggests higher productions and higher energy use at given level of technology. The second one is the stock of capitals K, which includes two effects on energy demand. One effect is the scale effect, which regards that capital accumulation is accompanied by the enlargement of industry scale and needs more energy input. The other effects is the structural effects, which says capital accumulation is beneficial to energy saving due to its contribution to structure alteration from energy intensive industry to technology intensive industry. The third variable influencing the energy use is the technology. It includes the technology effects on the energy. Advancement of technology can improve the producing process and reduce the undesired energy loss.

$$E = b_{21}Y + b_{22}K + b_{23}A \tag{2}$$

The third equation capture the effects on income. This equations enable the model process the relationship among income, environment, energy and human resources fully endogenously. We consider all variables at per capita level. Thus the population variable can be removed from production function. Totally, we consider four factors for production, i.e. capital stock, environment, energy, technology and human resources H.

$$Y = b_{31}A + b_{32}K + b_{33}P + b_{34}E + b_{35}H \tag{3}$$

And corresponding production function could be Cobb-Douglas form like this:

$$Y = AK^{\alpha}K^{\beta}E^{\gamma}H^{\delta} \tag{4}$$

A is also the total factor productivity (TFP). And we suppose that the product has constant scale return, so we have:

$$\alpha + \beta + \gamma + \delta = 1 \tag{5}$$

The last two equations describe the endogenous relationship between international trade and income. For simplicity, we suppose that import is affected only by income and the export is affected only by capital, so we have:

$$M = b_{41}Y \tag{6}$$

$$X = b_{51}K \tag{7}$$

All equations include constant variables and all endogenous and exogenous variables are selected by tests of Akaike information criterion (AIC) and Bayesian information criterion (BIC).

Income data are collected from the six province statistic yearbooks 2004 till 2015. They are Guangdong, Guangxi, Yunnan, Fujian, Guizhou and Hainan province. Real GDP per capita is chosen so as to remove price fluctuation impacts. Water quality score includes two part, the integral part is defined by national environmental quality standards for surface water GB3838-2002, which sets water quality to six basic grade, i.e., 1-5 and w5. Grade 1 is the best quality and grade w5, which we set to 6, is the worst level. Considering the aquatic environment is closely related to biochemical oxygen demand (BOD), the decimal part of the water quality score is defined and sorted by BOD. The highest BOD of each grade of water quality in our sample is defined to grade number plus 0.9 and the lowest BOD is 0.0. Water quality data are collected from water quality year report 2004 till 2015 from Geographic Information System of Environmental Protection Department Data Centre and water quality monitoring station in six provinces.

As to import M and export X, we use their ratio to GDP per capita. Capital stock K is calculated by perpetual inventory method. The base year is 1990 and the accumulated part is calculated by investment of each year. All data are adjusted by price and the base year is also 1990. Human resource H is defined by educated years per capita and calculated by the ratio of sum of educated years in population. All of above data are collected from statistic books from 1990-2015 of each province.

3. Estimation results

3.1. Model set

Three steps are applied to set this endogenous model. Firstly, Breusch LM Test is applied to determine whether or not panel estimation should be used. Secondly, we use Hausman Test to choose between individual fixed-effect model and random-effect model. Thirdly, inter-class heteroscedasticiy, inter-class correlation and intra-class correlation are tested by Wald Test and Frees Test. The results of B LM support that panel data estimation is better than mixed regression. Hausman Test supports the model to adopt individual fixed-effect estimation. Adjusted Wald test and Frees Test suggest that inter-class heteroscedasticiy and cross correlation exist in all equations significantly.

3.2. Simultaneous equations estimation

We combine two stage least square estimation method (2SLS) and feasible general least square estimation method (FGLS) to 2SFGLS to estimate the coefficients. 2SLS is used to avoid bias and inconsistency of estimators. FGLS can avoid heteroscedasticity and auto-correlation. Ordinary panel regression results are also reported for comparison. The results are listed in Table 1 and Table 2.

Table 1. Simultaneous Equations Estimation for Pollution, Energy and Foreign Trade[a]

	Y	Y^2	X	M	K	A	Con	WaldX^2
P	2.3300	-0.1050	-1.2332	1.9990	--	-1.9422	1.8773	115.60
2SFGLS	10.38***	-4.20***	-4.05***	10.00***		-3.03***		***
P	2.8563	-0.1295	-1.4835	1.5974	--	-1.4490	2.0403	90.90***
Ordinary	11.98***	-8.01***	-2.45**	1.08		-1.30		
E	0.4376	--	--	--	2.8778	2.7735	0.2889	87.72***
2SFGLS	10.22***				4.89***	1.18		
E	0.9990	--	--	--	3.1556	2.4330	0.6440	62.88***
Ordinary	4.54***				6.43***	1.05		
M	0.1173	--	--	--	--	--	--	--
2SFGLS	4.48***							
M	0.1240	--	--	--	--	--	--	--
Ordinary	2.01**							
X	--	--	--	--	0.0145	--	--	--
2SFGLS					3.44***			
X	--	--	--	--	0.0090	--	--	--
Ordinary					2.00**			

aAs to FGLS +2SLS, data below coefficients are Z statistics and to Non-FGLS, t statistics. WaldX2 shows the joint significance. ***, ** and * denotes 1%, 5% and 10% critical level.

Table 2. Simultaneous Equations Estimation for Income[a]

	P	E	H	K	A	Con	WaldX^2
Y	0.3340	0.2886	2.0008	0.2098	0.1190	1.0004	109.34***
2SFGLS	6.77***	3.87***	10.98***	10.44***	3.56***		
Y	0.1445	0.4545	2.2020	0.1299	0.0998	1.0080	109.93***
Ordinary	1.05	2.13**	7.66***	2.01**	6.77***		

aAs to FGLS +2SLS, data below coefficients are Z statistics and to Non-FGLS, t statistics. WaldX2 shows the joint significance. ***, ** and * denotes 1%, 5% and 10% critical level.

4. Discussion

We focus on the impact on environment by the international trade. We find that smaller export ratio X or larger import ratio X will both shift the turning point of EKC to the right, i.e. higher income level and worse water quality. And either smaller export or larger import will make the EKC slope smaller, i.e. plainer on the left of the turning point and steeper on the right.

First, we find that with the increase in export ratio or decrease in import ratio, EKC slope will decline at the identical income level. Just as illustrated above, plainer on the left and steeper on the right. When a region strengthens technology intensive and less pollution intensive export or control such kind of import, it will slow down their pace on pollution deterioration before the EKC turning point and speed up environmental improvement after it. In another word, reaching the same EKC pollution degree will need more time before the turning point and less time after it. This EKC change effected by international trade is different from the change by consumer preference. Consumer preference is an important explanation to the EKC. However, preference change tends to reshape the EKC more like A, i.e. 1st order derivatives is relatively larger on the left of the turning point and smaller on the right, because in the earlier stage of economic growth, preference alteration favours the low-end manufacturers of cost competition. With the upgrade of the economy, consumers demand more on comfort, experience and environment, thus, environment will deteriorate faster in early stage as well as improve faster in later stage and the EKC looks like more convex.

Quite different from the effects of preference, effect from international trade makes the left part of EKC plainer for this region. With the economic growth, a region may apply such strict environmental regulations to manufacture that only environmental or pollution free industry can survive in competition domestically. Pollution intensive industries need to cut down their output as well as emissions, just as many enterprises need to do in the near future in some cities of the South China, according to their thirteenth five-year development planning. At the same time, increasing cost doesn't allow labour intensive industry to remain. Therefore, pollution intensive and less technology intensive production transports to lower growth stage region or country. With the process, scale effects are somewhat offset by structural effects and the move to the pollution peak slows down. After the turning point, technology advancement has more and more positive influences on environment and works with which, trade structural change even has more effects on pollution control.

Change of export and import not only changes the slope of EKC, but also the turning point of this region. It can be observed that the contour line moves to the far end of the income growth with the increase in import or decrease in the export. When cities in the South China tend to export more intensive pollution goods for earning foreign exchange, or import less, the situation is worse for their environment. This is just what they had done since 1978.

Second, the result supports the adjustment to trade structure as a balanced development choice and the way of scale firstly and structure secondly is not optimal. Slope of EKC only measures the change rate of environment without considering effects other than income, e.g. international trade. Tangent plain in space measures the aggregate change rate of two factors, one is scale effect by growth and the other is structure effect by trade. Path on the surface perpendicular to the contour illustrates the nearest way to the pollution peak, supposing that this pollution level is inflexible. So, we regard that the usual way adopted by developing regions, which is scale firstly and structure secondly, is a bad choice because there will be a longer period of their suffering from pollution. Unfortunately, this is just the industry history of the South China since opening and reform. One of the most beautiful city of scenery in the South China, even in South China, Zhaoqing City, is polluted severely with their ceramic and stone material factories. Many cities in the region is going to make amendments in their next development strategies. Simultaneous progress of economy and international trade mode is better for its smaller cost, e.g. on health, species diversity, etc. Considering that some damage by pollution is irreversible after all, balanced development is more advantageous.

Third, we find the influence in the South China environment by import and export are flexible and different in marginal percentage. We give the marginal growth rate of pollution regarding export and import ratio in Figure 1. It looks like somewhat linear, but both are convex curves essentially. Positive slope curve illustrates the elasticity of water quality to import ratio and negative slope curve is for elasticity to export ratio, given that human resource level H and capital stock K fixed at 2004 level.

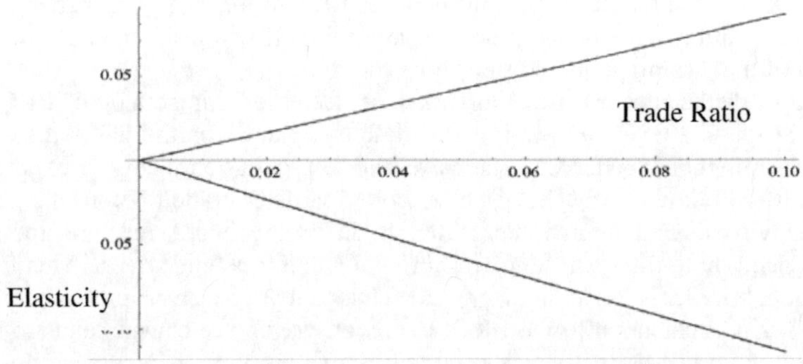

Figure 1. Environmental Elasticity to Trade

Just as we mentioned above, export ratio of technology intensive and pollution free good is beneficial to environmental improvement as shown by the negative slope line in Figure 1. With the increase in the export, pollution decreases and the marginal percentage of the decrease rise along with the export ratio growth. Although we mentioned a relative close economy will perform better in EKC turning point, this result shows that the marginal return in environment is increasing along with such beneficial export growing. On the other side, import ratio of technology intensive and pollution free good hinder environment from improving and the elasticity is also increasing with ratio raise. It suggests that influence on environment by such trade mode is increasing with higher proportion of trade to GDP, including both export and import. There are three reasons can explain this result. First, higher trade dependence often corresponds to advanced stage of economic growth and technology effects in this stage is usually stronger than early stages. Second, rise of trade ratio also brings greater structural change to industry and the favoured region has less possibility in becoming Pollution Heaven. Third, scale enlargement has decreased marginal pollution, and demand for better environment is increasing. We compare the influences among three groups, the South China, non-Pearl-River-Delta region in Guangdong province and the Changjiang River Delta. The estimation result shows that the elasticity of environment to export and import in each group is 0.029 and -0.045, 0.005 and -0.013, 0.020 and -0.037 respectively on average at 5% critical level, corresponding to their openness in such industry. So our perspective to the environmental change due to trade adjustments is more optimistic in the South China than the region of control group.

In our data, the South China has higher export ratio as well as higher import ratio than other region in the same province, so the aggregate water pollution elasticity to trade structure is not so far away as the elasticity to either export or import alone from other regions in Guangdong province. That is why a high trade dependence region can seldom find the trade scale enlargement improves environment significantly. Balanced trade in technology intensive and environmental good has less effects on pollution because the structural effect is offset and only technology effect remains.

5. Conclusions

In most discussion about EKC, international trade is only considered as a subordinate factor influencing the relationship between income and environment. The model in this paper reconsiders the mixed effect of income and trade and offers some important influences made by trade. From the estimation, we conclude that interrelationship between the South China economic growth and water pollution is inverted-U-shaped EKC. The pollution will possibly reach an undesirable level far beyond the point that the region ever experienced. Less opening economy may be beneficial to environment and opening modes also influence the environmental performance in this region. As to environmental performance, the South China suffered more from international trade than North China, which is less opening or structural different. The economic growth of the South China hasn't passed the turning point and the temporary pollution change is not the evidence of long run water quality improvement. The most interesting implication is that balanced trade growth may not result in EKC change. If openness is neutral to environment, balanced pollution intensive trade also has no significant influence on EKC. It suggests that China needs to direct trade to environmental friendly export or pollution intensive import. Will it be a Race to Top rather than Race to Bottom when all countries do the same thing? And will it be a good choice for developing countries in trade liberalization negotiations? It should be answered with more caution after investigation.

Acknowledgements

This work is financially supported by the National Natural Science Foundation of China (71203037), the Foundation for Distinguished Young Talents in Higher Education of Guangdong (2012WYM-0049) and the Natural Science Foundation of Guangdong Province (2015A030313499).

References

[1] Panayotou T 1993 Empirical tests and policy analysis of environmental degradation at different stages of economic development Geneva, Technology and Employment Programme
[2] Grossman G M and Krueger A B 1993 Environmental impacts of the North American Free Trade Agreement *The U.S. –Mexico Free Trade Agreement* (Cambridge: MIT Press) pp 13–56
[3] Selden T and Song D 1995 Neoclassical growth, the J Curve for abatement, and the inverted-U Curve for pollution *Journal of Environmental Economics and Management* **29** pp 162–168
[4] Egli H and Steger T M 2007 A dynamic model of the environmental Kuznets curve: turning point and public policy *Environmental and Resource Economics* **36 (1)** pp 15–34
[5] Cole M A 2004 Trade, the pollution haven hypothesis and environmental kuznets curve: examining the linkages *Ecological Economics* **48** pp 71–81
[6] Suri V and Chapman D 1998 Economic growth, trade and the environment: implications for the environmental Kuznets curve *Ecological Economics* **25** pp 195–208
[7] Lopez R 1994 The environment as a factor of production: the effects of economic growth and trade liberalization *Journal of Environmental Economics and management* 27 pp 163–184
[8] Copeland B R and Taylor M S 1995 Trade and environment: a partial synthesis *American Journal of Agricultural Economics* **77** pp 765–71
[9] Antweiler W, Copeland B R and Taylor M S 2001 Is free trade good for the environment? *American Economic Review* **91(4)** pp 877–908
[10] Busse M 2004 Trade Environmental regulations and the world trade organization--new empirical evidence (Copeland: World Bank Policy Research working paper 3361)
[11] Dua and Esty 1997 *Sustaining the Asia Pacific miracle* Washington DC, Institute for international economics
[12] Westerlund 2007 Testing for error correction in panel data *Oxford Bulletin of Economics and Statistics* **69** pp 709–48

The application of DEA model in enterprise environmental performance auditing

F Li, L Y Zhu, J D Zhang[1], C Y Liu, Z G Qu and M S Xiao

Research Center for Environment and Health, Zhongnan University of Economics and Law, Wuhan 430073, China

E-mail: jdzhang@zuel.edu.cn

Abstract. As a part of society, enterprises have an inescapable responsibility for environmental protection and governance. This article discusses the feasibility and necessity of enterprises environmental performance auditing and uses DEA model calculate the environmental performance of Haier for example. The most of reference data are selected and sorted from Haier's environmental reportspublished in 2008, 2009, 2011 and 2015, and some of the data from some published articles and fieldwork. All the calculation results are calculated by DEAP software andhave a high credibility. The analysis results of this article can give corporate managements an idea about using environmental performance auditing to adjust their corporate environmental investments capital quota and change their company's environmental strategies.

1. Introduction

In recent years, China's environmental problems are increasing, especially in aspects of air pollution and water pollution[1].As a part of society, enterprises have an inescapable responsibility for environmental protection and governance [2].With the improvement of environmental management, and consumer awareness of the environment, more and more enterprises are willing to assume the responsibility of environmental protection, and realize environmental protection and management will become the important factors affecting the sustainable development of enterprises [3].

With the development of the concept of environmental tax, so many scholars begin to pay close attention to the enterprise environmental performance auditing [4]. The Data Envelopment Analysis model (DEA) is a simple and effective analysis method used to calculate Input-output efficiency and widely used in management science field.

The objectives of this study were to (1) introduce the model of DEA and adjust some parameters to suit the actual situation; (2) use this model in calculating Haier's environmental performance and analyst the calculation results; (3) judge the feasibility and necessity of enterprises environmental performance auditing through Haier's environmental performance calculation analysis results.

2. Methods and materials

2.1. Data source

Haier is the leader of China's appliance brand and focus on leading china's appliance industry becoming international and environmentally friendly. Since 2005, Haier has began to publish corporate environmental report and aimed at convert environmental management into the sustained driving force for enterprise development. Haier is the pathfinder and brave practitioner of environmental information disclosure, and also is one of the policymakers and positive participants of "the Guidelines for the Establishment of Corporate Environmental report" (HJ617-2011) and "the Guidelines for the Establishment of Corporate Environmental report in Shandong Province" (DB37/T1086-2008).

Most of the data used in this article are selected from the corporate environmental reports published in 2008, 2009, 2011 and 2015, and some of the data from some published articles and fieldworks.

2.2. The data envelopment analysis model

The Data Envelopment Analysis model (DEA) is a simple and effective analysis method used to calculate Input-output efficiency and widely used in management science field [5].C^2R is a basic DEA model used to study on multi-inputs and multi-outputs. Through the analysis of input and output data, C^2R can calculate the comprehensive efficiency indexes of each Decision Making Units (DMUs), select the effective and order them. For DMUs being evaluated as inefficient, C^2R can find the ineffective reasons, and address an improvement procedure to enable them to be efficient [6].

2.3. The selection of decision making units

There are three main selection principles of DMUs selection. Firstly, each DMU needs to perform the same working task and has the same working aim. Secondly, each DMU needs to operate in the same marketing environment [7]. Thirdly, each DMU needs to select the same input-factors and output-factors, and DMUs' performance is determined by the intensity or magnitude of each factor [8].

Comparing the level of environmental information disclosure and the development of enterprise scale, Haier is the only company can be selected. This article selects Haier's environmental performances in different years as DUMs in order to compare different rates of return on investment in environment protection.

2.4. The operation of the DEA model

This article selects 2 kinds of input-output index system.

Table 1. Annual environmental performance input-output index data in 2005-2011

			2005	2006	2007	2008	2009	2010	2011
Input	R M[a]	10^4 ton	59.0	49.9	64.2	60.4	83.3	113	125
	Electricity	10^4kW·h	1.0	2.4	2.5	2.5	3.8	4.3	4.5
	Coal	10^3 ton	1.9	2.2	1.7	1.8	1.7	2.0	2.4
	Water	ton	104	224	221	214	188	240	249
	E M[b]	yuan	285	957	1746	368	1016	1560	1657
Output	CO_2	kg/10^4yuan	6.4	6.1	5.9	5.9	4.6	4.2	4.1
	SO_2	kg/10^4yuan	34.9	24.4	17.3	16.9	15.7	15.2	14.8
	Waste water	10^2 $m^3/10^4$yuan	5.4	4.1	3.8	3.7	3.5	3.4	3.1
	COD	g/10^4yuan	20.3	19.1	18.1	17.5	16.8	15.5	12.9
	Solid waste	10^4 ton	1.1	1.0	0.5	0.5	0.6	0.7	0.7
	Energy use	kg/10^4yuan	15.2	14.5	13.9	12.6	13.1	12.6	12.2
	Water use	10^2 $m^3/10^4$yuan	6.7	5.1	4.7	4.6	4.4	4.2	3.8

[a]Raw materials is abbreviated as R M in this table.
[b]Environmental management is abbreviated as E M in this table.

The first input-output indexes system selects the environmental achievements in 2005-2011 as DMUs. Each DMU has 5 input indexes, such as the total weight of raw materials (like copper, steel, aluminum, plastic and zinc plated sheet), the consumption amount of electric energy consumption, the consumption amount of coal, the consumption amount of water and the investment in environmental management, abbreviated as INT1-5 respectively. Each DMU has 7 output indexes, such as the emission amount of CO2, the emission amount of SO2, the emission amount of water waste, the emission amount of COD, the emission amount of solid waste (slag, phosphate slag, waste mineral oil, oil, cotton yarn, etc.), and the usage of energy and water, abbreviated as ONT1-7 respectively. The input and output indexes data per yearof this system are listed in Table 1.

The second input-output indexes system selects the environmental achievements in 2008, 2009, 2011 as DMUs. The input indexes only consider the investment in environment protection, including the cost of air pollution control, the cost of water pollution control, the cost of daily monitoring, the cost of process equipment renewal and maintenance, the cost of staff environmental awareness training and the cost of environmental protection publicity, abbreviated as NINT1-6 respectively. The output indexes only consider the emission of waste gas, the emission of wastewater, the emission of solid waste (slag, phosphate slag, waste mineral oil, oil, cotton yarn, etc.), the consumption of energy, and the consumption of water, abbreviated as NOUT1-5 respectively.The input and output indexes data per year of this system are listed in Table 2-3.

Table 2. Annual environmental performance input index data in 2008-2015

	Environmental protection investments (10^4yuan)					
	Air	Water	Monitor	Equipment	Training	Publicity
2008	45	90	5	202	10.7	13
2009	80	30	10	853	30	13
2011	120	850	20	510	22	135
2015	2000	500	25	2410	55	95

Table 3. Annual environmental performance output index data in 2008-2015

	Waste gas ($kg/10^4$yuan)	Waste water ($m^3/10^4$yuan)	Solid waste 10^4 ton	Energy cost (kg coal/10^4yuan)	Water cost (10^2 $m^3/10^4$yuan)
2008	34.7	0.46	0.5	13.5	0.46
2009	32.6	0.44	6.6	13.1	0.44
2011	28.9	0.42	16.3	12.1	0.42
2015	11.89	0.215	3.6	10.27	0.215

2.5. The construction of input-output indicators

According to the characteristics of the multi-output indexes system of Haier, this article selects the C^2R model to evaluate its environmental performance.

C^2R model is a simple form model, which has a long history and a perfect system in its theories [9]. The thought of model building is clear, and has a strong operability. C^2R model assuming that there are n comparable DMUs, each DMU has m kinds of input indexes and s kinds of output indexes [10]. The input indexes are the smaller the better, and conversely, the output indexes are the bigger the better [11]. C^2R model assuming that n has a comparable decision unit, each decision unit has a M type S type input and output, the input and output of the understanding is, the smaller the better output input, the bigger the better[12]. The general enterprise to maximize the benefit but also the pursuit of such input-output ratio, but when the evaluation object is the Haier's environmental performance, the situation is not the same, in the input stage, we still pursue the minimization of raw materials, energy, environmental management costs, but in the output stage, emissions and energy consumption of water we selected pollutants consumption as the evaluation index, so the index is as small as possible. Therefore, we need to modify the data of the environmental performance indicators, that is, to take its reciprocal, so that it is in line with the DEA performance evaluation of the output indicators, the greater the better [13]. Normally, enterprises use this principle to maximize their benefits, however, when the evaluation object comes to environmental performance of Haier, the situation is totally different. In the input stage, the evaluation object also needs smaller indexes of raw materials costs, energy costs, and environmental management costs, but in the output stage, the output indexes of pollution emission are the smaller the better. Therefore, the output indexes of environmental performance need to be modified (take its reciprocal) to suit the real needs of calculation. The modified data are listed in Table 4-7.

The next step is to take all the input indexes and output indexes to the DEAP software and write commands to calculate the order of the years' environmental performances[14]. Then, neaten all the effective data to the form of table or graph.

Table 4. Fixeddata of annual environmental performance input index data in 2005-2011

	INT1	INT2	INT3	INT4	INT5
2005	59.0	1.0	1.9	104.0	285.0
2006	49.9	2.4	2.2	224.0	957.0
2007	64.2	2.5	1.7	221.0	1746.0
2008	60.4	2.5	1.8	214.0	368.0
2009	83.3	3.8	1.7	188.0	1016.0
2010	113.0	4.3	2.0	240.0	1560.0
2011	125.0	4.5	2.4	249.0	1657.0

Table 5. Fixeddata of annual environmental performance output index data in 2005-2011

	OUT1	OUT2	OUT3	OUT4	OUT5	OUT6	OUT7
2005	0.156	0.029	0.185	0.049	0.909	0.066	0.149
2006	0.164	0.041	0.244	0.052	1.000	0.069	0.196
2007	0.169	0.058	0.263	0.055	2.000	0.072	0.213
2008	0.169	0.059	0.270	0.057	2.000	0.079	0.217
2009	0.217	0.064	0.286	0.060	1.667	0.076	0.227
2010	0.238	0.066	0.294	0.065	1.429	0.079	0.238
2011	0.244	0.068	0.323	0.078	1.429	0.082	0.263

Table 6. Fixeddata of annual environmental performance input index data in 2005-2011

	NINT1	NINT2	NINT3	NINT4	NINT5	NINT6
2008	45	90	5	202	10.7	13
2009	80	30	10	853	30	13
2011	120	850	20	510	22	135
2015	2000	500	25	2410	55	95

Table 7. Fixeddata of annual environmental performance output index data in 2005-2011

	NOUT1	NOUT2	NOUT3	NOUT4	NOUT5
2008	0.029	2.174	2.000	0.074	2.174
2009	0.031	2.273	0.152	0.076	2.273
2011	0.034	2.381	0.061	0.083	2.381
2015	0.084	4.651	0.278	0.097	4.651

3. Results and discussion

The Table 8 shows that the scores of 2005, 2006, 2007, 2008 and 2009 are all equal to 1, and all of their slacks of input and output are equal to 0. That means the environmental performance in 2005-2009 of Haier all efficient on DEA efficiency and their input/output ratios are optimal.

The scores of 2010 and 2011 are less than 1 that means the environmental performance of these two years are DEA inefficient. In the aspect of the factor inputs, there are 159.38 thousand tons of raw materials, 23.158 tons of water, and 3885.76 thousand yuan of environmental management costs have not been fully utilized in 2010. In 2011, there are 181.18 thousand tons of raw materials, 4910 thousand yuan of environmental management costs have not been fully utilized. In the aspect of output

IOP Conf. Series: Earth and Environmental Science 52(2017) 012075 doi:10.1088/1755-1315/52/1/0120

factors, all of the indexes have exceeded the aim values except the emission of CO_2 in unit output value in 2010. In 2011, all of the indexes have exceeded the aim values except the emission of COD in unit output value.

The Table 9 shows that the scores of 2008 and 2009 are equal to 1, and their slacks of input and output are equal to 0. That means the environmental performance in 2008 and 2009 of Haier all efficient on DEA efficiency and their input/output ratios are optimal.

The scores of 2011 and 2015 are less than 1 that means the environmental performance of these two years are DEA inefficient. In the aspect of the factor inputs, there are 156.68 thousand yuan of air pollution control costs, 3791.69 thousand yuan of water pollution control costs, 55.42 thousand of daily detective costs, and 539.84 thousand of staff to participate in environmental protection technology and management training costs have not been fully utilized in 2011. In 2015, there are 10282.76 thousand yuan of air pollution control costs, 289.96 thousand yuan of water pollution control costs, 8110.34 thousand yuan of staff to participate in environmental protection technology and management training costs and 8.69 thousand yuan of environmental protection related publicity costs have not been fully utilized. In the aspect of output factors, only the emission of waste gas hasn't exceeded the aim values both in 2011 and in 2015.

Table 8. Calculation results of environmental performance in 2005-2011

	2005	2006	2007	2008	2009	2010	2011
θ	1.000	1.000	1.000	1.000	1.000	0.949	0.972
S1-	0.000	0.000	0.000	0.000	0.000	15.938	18.118
S2-	0.000	0.000	0.000	0.000	0.000	0.000	0.000
S3-	0.000	0.000	0.000	0.000	0.000	0.000	0.000
S4-	0.000	0.000	0.000	0.000	0.000	23.158	0.000
S5-	0.000	0.000	0.000	0.000	0.000	388.576	491.454
S1+	0.000	0.000	0.000	0.000	0.000	0.000	0.028
S2+	0.000	0.000	0.000	0.000	0.000	0.003	0.011
S3+	0.000	0.000	0.000	0.000	0.000	0.019	0.040
S4+	0.000	0.000	0.000	0.000	0.000	0.001	0.000
S5+	0.000	0.000	0.000	0.000	0.000	0.385	0.733
S6+	0.000	0.000	0.000	0.000	0.000	0.005	0.019
S7+	0.000	0.000	0.000	0.000	0.000	0.010	0.026

Table 9. Calculation results of environmental performance in 2008-2015

	2008	2009	2011	2015
θ	1.000	1.000	0.570	0.579
S1-	0.000	0.000	15.668	1028.276
S2-	0.000	0.000	379.169	28.966
S3-	0.000	0.000	5.542	0.000
S4-	0.000	0.000	53.984	811.034
S5-	0.000	0.000	0.000	0.869
S6-	0.000	0.000	61.738	17.379
S1+	0.000	0.000	0.000	0.000
S2+	0.000	0.000	0.168	1.646
S3+	0.000	0.000	2.284	5.515
S4+	0.000	0.000	0.004	0.117
S5+	0.000	0.000	0.168	1.646

4. Conclusion

The above calculation results illustrate that Haier have a effective performance in 2005-2011 when the input indexes select the cost of raw materials, electric energy, coal, water and environmental management, and the output indexes select the mission of CO_2, SO_2, waste water, COD, solid waste and the cost of energy and water. The wastes only exist in the use of raw materials and the cost of environmental management. The possible reason is with the development of economy, the problem of resource shortage and environmental pollution gets worse, the costs of raw materials and environmental management dramatically increased which cannot have longitudinal comparison with the past years' data. Another input-output indexes system only have a effective performance in 2008 and 2009 when the input indexes select the cost of waste gas control, waste water control, daily monitor, process equipment renew and maintain, employee environmental knowledge training and environmental protection campaign, and the output indexes select the emission of waste gas, waste water, solid waste, and the cost of energy and water. The waste gas control costs include waste gas pollution discharge fee. The process renew costs include the costs of ensuring the sewage treatment facilities operating well, the costs of ensuring the reclaimed water reuse operating well and the costs of improving industrial equipment and production efficiency. The costs of training consist of the costs of environmental management, inspection technology, and other organizations' environmental protection training participation. The costs of propaganda consist of the costs of environmental protection campaign and energy saving campaign.

Half of the DMUs are DEA ineffective and have a large amount of waste in investment capital. Most of the capital wastes exist in air pollution control, water pollution control, equipment renew and maintain, and environmental protection propaganda. Only in 2015, there are 10282.76 thousand yuan have not been fully utilized. However, consult the environmental report in 2015 found that air pollution control costs in 2015 is mainly used for improved SO_2 removal equipment, so SO_2 emissions for this year equal to 0. Due to the large emissions of CO_2 emissions, the total waste gas emissions decrease is not significant. Similarly, the benefits of environmental protection propaganda and process renew and maintain are long-term and sustainable, which cannot see the effect immediately. However, the emission of wastewater has already arrived optimal, the over costs of water pollution control are useless. Therefore, Haier is suggested to reduce the cost of the related items in the next year.

To summarize, through the calculation and analysis of Haier's environmental performance in 2005-2015, corporate internal environmental performance auditing is considered to be necessary and feasible, and DEA model is considered to be a effective metering method which has a high reliability [15]. The evaluation results of input and output efficiency can helps corporate' policymakers to analyst weather their investments in each aspect are effective, and adjust their corporate' capital quota in time.

References

[1] Li F, Huang J H, Zeng G M, Yuan X Z, Li X D, Liang J, Wang X Y, Tang X J and Bai B 2013 *J. Geochem. Explpr.* **132** 75-83
[2] Li F, Zhang J D, Huang J H, Huang D W, Yang J, Song Y W, Zeng G M 2016 *Environ. Sci. Pollut. R.* **23** 13100-13
[3] Li F, Zhang J D, Yang J, Liu C Y and Zeng G M 2016 *Hum. Ecol. Risk Assess.* **22** 1224-41
[4] Huang J H, Li F, Zeng G M, Huang X, Liu W, Wu H and Li X 2016 *Sci. Total Environ.* **541** 969-76
[5] Chen S Z, Lai B C and Chen X H 2005 *Systems Eng.* **6** 99-104
[6] He Q 2013 *China Secur. Future* **4** 330-1
[7] Wang Z, Wu Q and Hu L 2013 *J. China Geosci. U.* **4** 106-111
[8] Bian Y W 2009 *Bus. J.* **6** 64-5
[9] Hu J, Li X Y and Sun J H 2009 *Sci. R. Manag.* **2** 150-6
[10] Zhang L and Qi Z Y 2009 *J. Harbin Eng. U.* **4** 456-60
[11] Pi C G, Dong L D and Hou G M 2013 *Sci. Technol. Prog. Policy* **21** 122-7
[12] Zhang S R, Sun H J and Wang S J 2014 *Friends of Accounting* **24** 41-5
[13] Zhang Y M, Lu J and Yuan L X 2014 *Bus. Accounting* **18** 36-7
[14] Ye W L, Wen Y L, Guo M and Wu Y Y 2014 *Environ. Pollut. Control* **10** 102-10
[15] He P L, Shi Y D and Li T 2012 *Accounting R.* **2** 11-7

Assessment method for the prevention effectiveness of $PM_{2.5}$ based on the optimization development of coal-fired power generation

Kuan Zheng[1,3], Jun Liu[1], Jin-fang Zhang[1], Weihua Hao[2]

[1] State Grid Energy Research Institute, Beijing, 102209, China
[2] State Grid Shandong Electric Power Corporation Maintenance Company, Jinan, 250118, China

E-mail: Kzheng302@126.com

Abstract. A large number of combustion of coal is easy to lead to the haze weather which has brought a lot of inconveniences and threat to people's living and health in E&C China, as the dominant power source of China, the coal-fired power generation is one of the main sources to the haze. In this paper, the contribution of the combustion of coal and development of coal-fired power generation to the $PM_{2.5}$ emissions is summarized based on the analysis of the present situation, the mechanism and the emission source of $PM_{2.5}$. Considering the peak of carbon emissions and the constraints of atmospheric environment, the quantitative assessment method of $PM_{2.5}$ by optimizing the development of coal-fired power generation is present. By the computation analysis for different scenarios, it indicates that the optimization scenario, which means the main new-installed coal-fired power generation is distributed in western and northern China, can prevent the $PM_{2.5}$ effectively for both the load center and coal base regions of China. The results of this paper not only have reference value for the optimized layout of coal-fired power generation in the "13rd fifth-year" power planning, also is of great significance to deal with problems that the atmospheric pollution and climate warming in the future .

1. Introduction

In recent years, the haze weather occurs quite often in eastern and central China (E&C China), which brings a lot of inconveniences and threat to people's living and health. $PM_{2.5}$ is the most important and harmful component of the haze weather, thus how to reduce the $PM_{2.5}$ emission become a major challenge of the air pollution. Coal is the dominant energy in the China, and a large number of utilization of coal is easy to lead to the $PM_{2.5}$. According to the statistic of China's National Bureau of Statistics, the total amount of coal consumption in 2013 amounted to 2.475 billion tons, which accounts for 59.4% of total primary energy consumption and this proportion is far higher than the world average level (30%). A large amount of dust, carbon dioxide (CO_2) and nitric oxide (NO_x) are produced by the combustion of coal, which become one of the main sources of $PM_{2.5}$. Therefore, it is of great significance to control the $PM_{2.5}$ during the coal utilization to realize the prevention and improvement of air pollution.

Coal-fired power generation (CFPG) is the main consumer of coal in China, although the technologies such as dust removal, desulfurization and denitrification have been promotion a lot in power industry, but considering the large coal utilization in the power industry, there is still a large space of emission reduction with different methods. In the meanwhile, the coal combustion in the other industries is also contributes a lot to the air pollution, such as the coal burning scattered. Therefore, to control the $PM_{2.5}$ pollution, it is not only need to reduce the emission for CFPG, but also need to control the pollutants emission in other industries.

In this paper, we summarized the contribution of the coal combustion to the $PM_{2.5}$ emissions based on the analysis of the present situation, the mechanism and the emission source of $PM_{2.5}$. Considering the peak of carbon emissions and the constraints of atmospheric environment we present a optimization model of the CFPG development and give the quantitative assessment method of the $PM_{2.5}$ emission of CFPG. In view of the new trend and characteristics of the CFPG development, we designed different developing scenarios based on the optimization model of the CFPG, and assessed the $PM_{2.5}$ emissions of CFPG quantitatively for each scenario, then give the suggestions for the development of the CFPG in the future. The results of this paper have reference value for the optimized layout of CFPG in the "13rd fifth-year" power planning, also is of great significance to deal with problems that the atmospheric pollution and climate warming in the future.

2. Analysis on the formation mechanism and emission source of $PM_{2.5}$

2.1. Present situation of $PM_{2.5}$ pollution

According to the data published by the Ministry of Environmental Protection of China, there is only 3 cities (Lhasa, Haikou, Zhoushan) achieve the secondary standard of the air quality on the list of the 74 cities which have been implemented the new air quality standards in 2013 (74 cities). During the whole 2013, the air quality of the 74 cities could achieve the standard only within 221 days averagely, which means the standard-reaching rate is only 60.5%, and the yearly average $PM_{2.5}$ concentrations is $72\mu g/m^3$, which is 2.1 times of the limit value for the secondly regions. The ratio of primary pollutants and the average concentrations of $PM_{2.5}$ of the 74 cities in 2013 are shown in Figure 1.

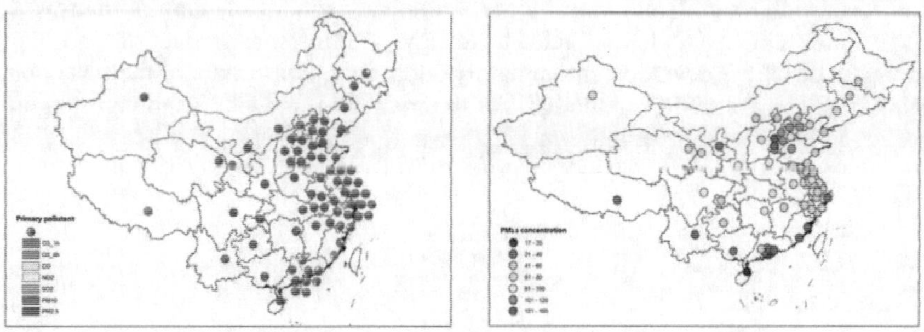

Figure 1. The ratio of primary pollutants and the average concentrations of $PM_{2.5}$ of the 74 cities in 2013 (g/m^3).

It indicates from Figure 1 that the concentration of $PM_{2.5}$ changes greatly in different geographical position, and usually the concentration is higher in the urban, and the regional distribution of $PM_{2.5}$ present much stronger in some megalopolis and big cities as the center. In the Beijing-Tianjin-Hebei region, there are 11 cities listed in the top 20 of the 74 cities mentioned above, and 7 cities in the top 10. For the Beijing-Tianjin-Hebei region, the Yangtze River Delta region and the Pearl River Delta region, which are the most developed regions in China, the population accounts for 24% of China, the consumption of coal accounts for 43% of China, but the land only accounts for 8% of China, so the unit area pollutant emissions is nearly five times of the average value.

2.2. Formation mechanism and emission source of $PM_{2.5}$

The emission source of $PM_{2.5}$ is very complicated, in terms of $PM_{2.5}$ produce process, one source is directly by the pollutant emission, which is called primary particulate matter (PPM). The main component of PPM are carbon, organic carbon and soil dust, etc., which arises mainly from the burning of fossil fuels and biomass fuel, and also from the industrial manufacture, the mineral processing, the construction, and farming, etc.; the other source is the complex which is formed by the condensation of gaseous pollutants discharged from various sources or by heterogeneous chemical reaction, which is called secondary particulate matter (SPM). The main component of SPM are semi volatile organic compounds and ammonium sulfate, nitrate, etc [1]. Among them, the SPM have greater contribution to the formation of $PM_{2.5}$ [2].

For the emission source of $PM_{2.5}$, there are some controversies in the current academic circle. Take Beijing as an example, the Atmospheric Physics of Chinese Academy of Sciences thought the combustion of coal, motor vehicles and industrial production are the main sources of $PM_{2.5}$ pollution [3]. The Environmental Department of Peking University thought the contribution of $PM_{2.5}$ was 35.8% by the motor vehicle emissions and 19% by the coal burning [4]. The Carnegie Center for Global Policy thought the contribution of $PM_{2.5}$ by the motor vehicle emissions is more than 1/4 [5].

3. Optimization model of CFPG development and its assessment method of $PM_{2.5}$

3.1. Improved optimization model of CFPG development

Traditional optimization model of CFPG development is based on principal that the total cost of power supply is lowest during the whole planning period, there the total cost includes the investment and operation cost of the power source and grid, and make the external environment cost as internalization. The optimization analysis not only consider the factors such as the power flow and target market of the CFPG in the future, but also take the following factors as the constraint conditions: power balance, balance of coal supply, supply ability of water, environmental carrying capacity, transport capacity of coal and power market space, by means of the mixed integer programming to get the optimal solution of the install capacity of the CFPG and its delivery power flow [6]. With the action of the air pollution prevention has been gradually strict in China, the optimization of the traditional fossil energy consumption must be the significant step of the whole adjustment of China's energy structure, as development of the CFPG is closely related to $PM_{2.5}$ emissions reduction, so the improved optimization model of CFPG development is mainly adding the constraints of the $PM_{2.5}$ emission and the peak installed capacity of CFPG on the basis of the traditional model, as shown in Figure 2.

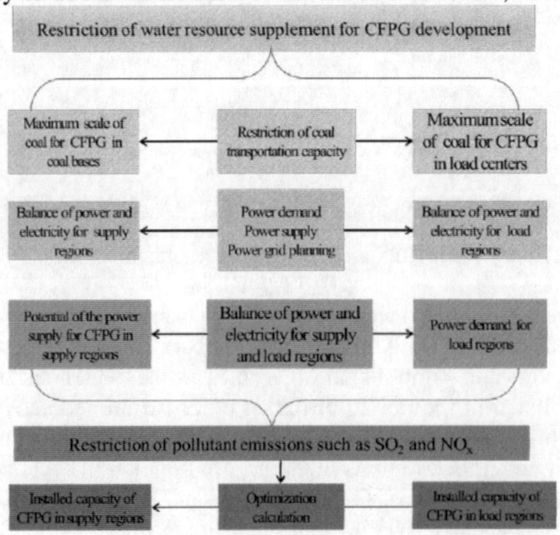

Figure 2. Improved optimization model of CFPG development

According to the action plan to combat air pollution released by State Council of China in September 2013, the $PM_{2.5}$ concentration will reduce by 25%, 20% and 15% respectively in the Beijing-Tianjin-Hebei region, the Yangtze River Delta region and the Pearl River Delta region in 2017. If this target accomplished, the $PM_{2.5}$ concentration will reduce by more than 30% for the three regions above in 2020.

At present, the per capita electricity consumption is still low in China, the demand of electricity will continue to increase for a long time though the present growth rate has slowed down. Considering the double constraints of the conventional air pollutant and the CO_2 emission, the installed-capacity peak of the CFPG is expected around 1200GW in 2030. According to the action plan mentioned above, the new added CFPG will be the main built in western and northern China (W&N China), and the electricity will be supplied through the interregional transmission to the load centers such as the Beijing-Tianjin-Hebei region, the Yangtze River Delta region and the Pearl River Delta region.

3.2. Assessment method of $PM_{2.5}$

As a large coal consumer, the power industry is undoubtedly one of the main sources of $PM_{2.5}$ pollution. Take CFPG as example, the contribution to the $PM_{2.5}$ is from 2 parts: one is the direct emissions (called primary $PM_{2.5}$), the other part is the sulfate and nitrate which are generated from the emissions of SO_2 and NO_x in the atmosphere (called secondary $PM_{2.5}$). According to the related research results [7, 8], 10% of the $PM_{2.5}$ emission of the CFPG comes from the primary particle emitted directly, 50% comes from secondary particle converted from the SO_2, and 40% comes from secondary particle converted from the NO_x, as shown in Figure 3.

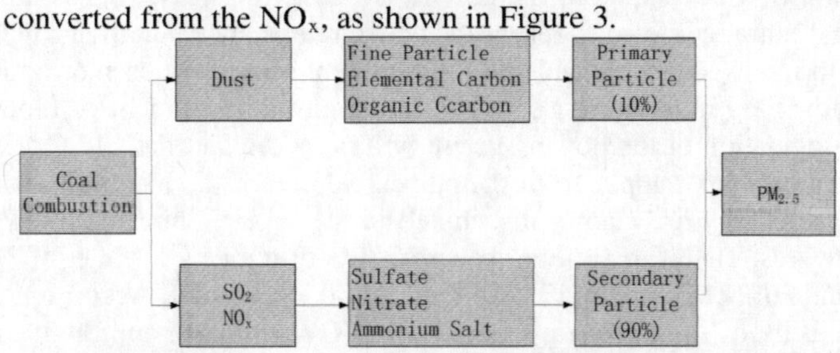

Figure 3. Component analysis of $PM_{2.5}$ for CFPG

The emission of $PM_{2.5}$ can be calculated from the equation (1) [2]:

$$E_{PM_{2.5}} = E_{PM_{2.5}\square\text{Primary}} + E_{PM_{2.5}\square\text{Secondary}} \tag{1}$$

there the $E_{PM_{2.5}\square\text{Primary}}$ is generated from the coal combustion directly, which is dependent on the quality of the coal which mainly includes the ash proportion, the fly-ash ratio and the efficiency of dust removal, it can be calculated by equation (2):

$$E_{PM_{2.5}\square\text{Primary}} = C_{coal} \times \eta_{ash} \times \eta_{FAR} \times \eta_{AP} \times (1-\eta_{AR}) \tag{2}$$

there C_{coal} is the consumption of the coal, η_{ash} is the ash proportion of the coal, η_{FAR} is the fly ash ratio, η_{AP} is the proportion of $PM_{2.5}$ to the ash, η_{AR} is the dust removal efficiency.

The SO_2、SO_3、NO_x emitted from the CFPG are the main sources to form the secondary $PM_{2.5}$, which can be calculated by equation (3-5).

$$E_{PM_{2.5}\square SO_2} = C_{coal} \times \eta_{SO_2} \times \eta_{PM_{2.5}\square SO_2} \tag{3}$$

$$E_{PM_{2.5}\square SO_3} = C_{coal} \times \eta_{SO_3} \times \eta_{PM_{2.5}\square SO_3} \tag{4}$$

$$E_{PM_{2.5}\square NOx} = C_{coal} \times \eta_{NOx} \times \eta_{PM_{2.5}\square NOx} \tag{5}$$

The secondary $PM_{2.5}$ generated from the CFPG is the summation of the three results above:

$$E_{PM_{2.5}\lceil Secondary} = E_{PM_{2.5}\lceil SO_2} + E_{PM_{2.5}\lceil SO_3} + E_{PM_{2.5}\lceil NOx} \quad (6)$$

4. Assessment of the prevention effectiveness of $PM_{2.5}$ considering different development scenarios of CFPG

4.1. Scenarios design

By the end of 2013, the total installed capacity of all power generations in China is 1247 GW, and the installed capacity of CFPG is 862 GW, accounted for 69.2%. The total electricity consumption is 5.32 million GWh, CFPG contributes 3.95 million GWh which accounted for 74.2%. According to the equations presented in Section 3.2, the total emissions of SO_2, NO_x and $PM_{2.5}$ of all CFPG in China in 2013 are 5.8 million tons, 8.10 million tons and 6.4 million tons respectively. According to the action plan mentioned in Section 3, the proportion of CFPG in China will remain around 60% by 2020. Considering the different roadmaps, two development scenarios of CFPG are designed in this paper.

Scenario 1: the uniform growth of the CFPG. During the "twelfth five-year" period, the average annual growth of the CFPG in China is 5.9%, the growth of the coal bases in W&N China such as Shanxi, Shaanxi, Gansu, Inner Mongolia, Ningxia and Xinjiang is 7.2%, and the growth in E&C China is 5.6%. According to this calculation, the total installed capacity of the CFPG in China is 1200GW in 2020, and 25.3% is in W&N China while 46.4% in E&C China.

Scenario 2: optimization development of the CFPG. Although the CFPG is still the main power supply in China in the future, while considering the new circumstance that the growth of the economic and power demand has slowed down, and the environment restriction is more and more strict, the utilization time of the CFPG must decrease. With the prompt development of renewable energy, the pressure of the peak regulation of the power system will increase, and the CFPG will be changed into the main source of the capacity supply from the current source of the electricity supply. Considering the coal and renewable energy are mainly distributed in the W&N China, the CFPG will develop to westwards and northwards. Therefore, in this scenario, 80% of the new installed CFPG will be located in the W&N China and 20% is located in eastern and central load center. According to this calculation, the total installed capacity of the CFPG in China is 1197GW in 2020, and 39.3% is in northern and western China while 32.8% in E&C China.

Considering the proportion of the unit composition of CFPG in different regions, and the emission performance of the dust, SO_2 and NO_x for different units, also some other factors such as the coal quality and conditions of pollutant diffusion conditions, the emissions of $PM_{2.5}$ for CFPG in W&N China and E&C China with Scenario 1 in 2020 and with Scenario 2 in 2020 are calculated by the assessment method mentioned in Section 3.2. The results are shown in Table1.

Table 1. Emissions of PM2.5 of CFPG in different regions of China

Scenario analysis		Installed capacity of CFPG (GW)	SO_2 (10^4ton)	NO_x (10^4ton)	Primary $PM_{2.5}$ (10^4ton)	Secondary $PM_{2.5}$ (10^4ton)	Total $PM_{2.5}$ (10^4ton)
Present situation	E&C China	356.57	262	369	28	260	289
	W&N China	217.01	251	293	17	210	227
Scenario 1 in 2020	E&C China	568.98	358	486	35	351	386
	W&N China	405.89	422	493	29	353	382
Scenario 2 in 2020	E&C China	392.62	371	424	25	311	337
	W&N China	470.42	489	572	30	368	398

4.2. Comparative analysis

According to the calculated result above, it shows that the emissions of $PM_{2.5}$ are increasing since 2013 to 2020 in all the regions because the installed capacity of the CFPG still increases during this period, while the growth speeds are different obviously in the different scenarios. As shown in Figure 4, the emission of $PM_{2.5}$ with Scenario 2 is 490,000 tons less than that with Scenario 1 in the E&C

China, which means the prevention of $PM_{2.5}$ in the E&C China is effective by means of moving the layout of CFPG westward and northward; while for the W&N China, although the new added installed capacity of the CFPG with Scenario 2 is much more than that with Scenario 1 because of the layout of CFPG moving westward and northward, the emissions of $PM_{2.5}$ for both scenarios are almost the same. The reasons for this phenomenon are that the adding units of CFPG in W&N China are the more-developed units with lower pollutant emissions, and the emissions have to meet the increasingly stringent policies of environment protection.

Also, from the point of view that increasing the peak regulation capability of power system to analyze this phenomenon, the increasing amount of the CFPG unit in the W&N China could guarantee the system accepting more clean power generated from the renewable energy. Which is a big bonus not only for the consumption of the renewable energy in the W&N China, but also for the clean alternative effect to reduce the $PM_{2.5}$ emission in the E&C China. All things considered, the prevention effectiveness of $PM_{2.5}$ for China with Scenario 2 is much better, which also means the optimization layout of the CFPG is of great significance for the prevention of $PM_{2.5}$.

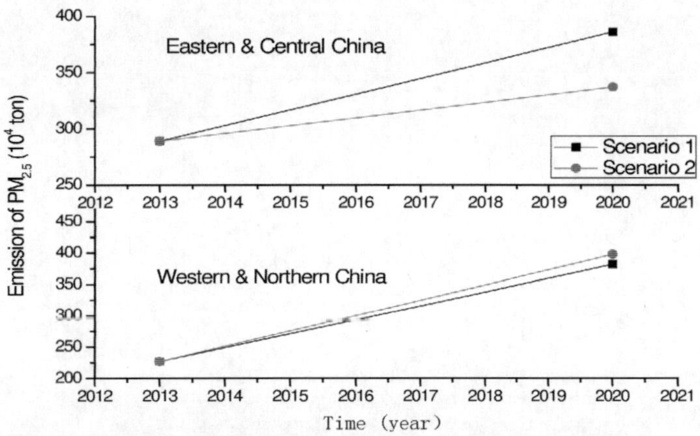

Figure 4. Comparison of $PM_{2.5}$ emissions of CFPG with different scenarios.

5. Conclusion

In this paper, the quantitative assessment method of the $PM_{2.5}$ emissions for CFPG is proposed, the contribution to form the $PM_{2.5}$ of CFPG is analyzed. By the computation analysis for different scenarios, it shows that the optimization development of CFPG with 80% adding CFPG in W&N China and 20% in E&C China contribute a lot for the prevention effectiveness of $PM_{2.5}$ for China. Consequently, as one of the important step of the power industry reform, the optimization development of CFPG can prevent the $PM_{2.5}$ for China and it is of great significance to deal with problems that the atmospheric pollution and climate warming in the future.

References

[1] Xu Lan, et al. 2013 Sources and research status of the air fine particulate matter (PM2.5) *Environmental Protection and Technology* **19(3)** 5-10.

[2] Mo Hua, et al. 2013 Contribution to PM2.5 of Atmospheric Pollutant Emission from Thermal Power Sector and Emission Reduction Countermeasures *Electric Power* **46(8)** 1-6.

[3] Wang Yuesi, et al. 2013 Formation of haze pollution in Beijing-Tianjin-Hebei region and their control strategies *Bulletin of Chinese Academy of Sciences* **3** 353-363

[4] He lingyan, et al. 2005 Determination of organic molecular tracers in PM2.5 in the atmosphere of Beijing *Acta Scientiae Circumstantiae* **25(1)** 23-29.

[5] Wang Tao The contribution of motor vehicle to haze may be undervalued in Beijing.

http://carnegietsinghua.org/2014/02/28/.

[6] Bai Jianhua, et al. 2007 Optimizing research on power transmission scale and target market ofmain coal-power bases *Electric Power* **40(12)** 10-13.

[7] Yue Yong, et al. 2005 Experimental study on characteristics of particulate matter size distribution and trace elements enrichment in emissions from a pulverized coal-fired boiler *Proceeding of the CSEE* **25(18)** 74-79.

[8] Wang Sheng, et al. 2011 Fine particle emission characteristics from coal -fired power plants based on field tests *Acta Scientiae Circumstantiae* **31(3)** 630-635.

Removal of organic matter and heavy metals of low concentration from wastewater via micellar-enhanced ultrafiltration: an overview

F Li[1,4], X Li[2,3], J D Zhang[1], L Peng[2] and C Y Liu[1]

[1] Research Center for Environment and Health, Zhongnan University of Economics and Law, Wuhan 430073, China

[2] Department of Biological and Environmental Engineering, Changsha University, Changsha 410003, China

E-mail: [3]gina0131@163.com; [4]lifei@zuel.edu.cn

Abstract. As a new and effective means of wastewater treatment, the micellar-enhanced ultrafiltration (MEUF) has been extensively studied. In this paper, MEUF was introduced from the aspects of theory basis, ultrafiltration membranes, and surfactants. Additionally, the latest research achievements in removing organic matter and heavy ions, its application in actual wastewater, and the characterization parameters of MEUF were introduced and summarized. Then, influences and mechanisms of the primary operation parameters, including surfactant concentration, pH, electrolytes, and transmembrane pressure on the performance of the MEUF process were analyzed. Finally, existing problems in the MEUF process were identified and developmental trends were predicted.

1. Introduction

Micellar-enhanced ultrafiltration (MEUF), which was first developed by Dunn *et al.* in 1985 for the treatment of wastewater, combines the use of surfactants and ultrafiltration membranes for the removal and recycle of organic matter and heavy metal ions of low concentration in wastewater [1].Its beneficial characteristics includestrong separation capability, easy extension, andlow energy consumption [2]. In virtue of efforts over the past two decades, MEUF has achieved great progresses and its industrial applications have been observed.

1.1. Principles of MEUF

In cases the surfactant concentration is higher than the critical micelle concentration (CMC),the surfactant forms an aggregate with a hydrophobic domain inside and hydrophilic region outside, i.e. micelle. Based on the like dissolve like theory, organics have increased solubility inside the micelle while charged groups on thesurface attract counter ions in the solution through electrostatic interaction. Therefore, aggregates of organics-micelle or metal ions-micelle in the solution can be blocked by an ultrafiltration membrane with a pore size smaller than the diameter of the aggregates.

With a deeper understanding of MEUF, studies in recent yearshave focused on reverse micelles [3]. Compared to normal micelles, reverse micelles are surfactant aggregates in apolar solvents with hydrophilic groups inside and hydrophobic groups pointing toward an apolar phase. The hydrophilic inner-core of reverse micelles can dissolve hydrophilic materialinside and achieve successful removal or recycle through ultrafiltration.

1.2. Ultrafiltration membranes

Defined as the maximum molecular weight of substances of which 90% are retained, the molecular weight cutoff (MWCO) of a membrane is an important factor that influences the treatment efficiency of MEUF. A membrane with a MWCO larger than the micelle molecular weight is often adopted in the MEUF process. Usually, the permeation flux is higher for a membrane with a larger MWCO, although the removal efficiency for contaminants is lower. The hydrophobicity of a membrane also has a certain impact on MEUF. If a hydrophilic membrane is used to remove hydrophobic organics, hydrogen bonds between the membrane surface and water molecules attract water molecules, thus reducing attraction and contamination to the membrane.Commonly used hydrophilic ultrafiltration membranes include polyacrylonitrile hollow fiber membranes, cellulose acetate membranes, polyamide membranes, and polyethersulfone membranes; hydrophobic ultrafiltration membranes include polysulfonemembrances, PESF membranes, polyphenylene ether membranes, and ceramic membranes. Advances on ultrafiltration membranesaim to reduce membrane contamination and improve removal efficiency. Ngang*et al.* improved hydrophilicity using ultrafiltration membranes doped withTiO_2 nanoparticles, which increased the permeation flux and reduced contaminationof the membrane [2]. Furthermore, an external magnetic field was applied to the membrane to reduce the concentration polarization phenomenon during removal of charged ions with MEUF, hence reducing membrane contamination,increasing permeation flux, and improving removal efficiency [4, 5].

1.3. Surfactants

Surfactants can be classified into anionic, cationic, and nonionic surfactants. Common anionic surfactants include sodium dodecyl sulfate (SDS), sodium dodecylbenzenesulfonate (SDBS), and linear alkylbenzenesulfonate (LAS) [6,7]; common cationic surfactants include cetyltrimethylammonium bromide (CTAB), n-octyltrimethylammonium bromide (OTAB), and cetylpyridinium chloride (CPC) [8]; nonionic surfactants include polyoxyethyleneoctyl phenyl ether (TX-100), oxyethylated coconut fatty acid methyl esters (OMC), and polyoxyethyleneglycol dodecyl ether (Brij-35) [9,10]. In recent years, research trends have included the pursuit of surfactants to remove certain contaminants. EI-Abbassi*et al.* discovered that SDS efficiently removes polyphenols in MEUF [11]. Fang *et al.* reported that CPC is superior to OTAB and CTAB in removing phenols [8]. Usually, anionic surfactants are chosen for the removal of metal ions, while cationic surfactants are preferred for anions and organics [12].

Mixed surfactant systems are attractingincreasing attention and have been extensively studied in recent years. The advantages of using mixed surfactant systems includegreatly reduced amountsof surfactant being used, cost, surfactant concentrations of permeated solution, and secondary pollution. Nonionic surfactants and ionic surfactants are often combined. This principle is based on the concept that nonionic surfactants act as a grid on the anionic surfactants, reducing the repulsion between anionic surfactants and thus promoting the formation of micelles. Beislak*et al.* adopted a mixture of SDS and OMC-10 for the removal of methylene blue (MB) [9]. The CMC of the surfactant decreased from 8.3×10^{-3} M to 2.0×10^{-3} Mwhen the ratio of SDS to OMC-10 was 4:1, and using a mixed surfactant showed no impact on the removal of MB. Paiboon*et al.* mixed SDS with nonionic surfactant of TX-100 and NP12 in MEUF [13]. This methodimproved the removal efficiency of Pb^{2+} and the surfactant, which solved the problems of low SDS removal efficiency when using only SDS, and low Pb^{2+} removal efficiency when using only anionic surfactants. Huang *et al.* used a mixture of SDS and TX-100or Brij-35 in MEUF for the removal of Cd^{2+} ions and discovered that although the CMC was effectively reduced by adding nonionic surfactants to the solution of anionic surfactants, excess nonionic surfactants reduced the micelle adsorption of Cd^{2+} ions [10]. This was because the charge density was reduced with the addition of nonionic surfactants,which in turn reduced the electrostatic adsorption. Consequently, there is an optimum ratio when using multiple surfactants.

The Gemini surfactant is a new and effective surfactant that is different from traditional surfactants in structure and propertyin that it contains two hydrophobic chains and two hydrophilic head groups. The CMC of the Gemini surfactant is quite low and its adsorption of charges of different polarity is stronger due to its stronger charge compared to traditional surfactants. Zhang *et al.* studied cationic Gemini surfactants in MEUF for the removal of phenols and obtained excellent removal efficiency, whichshowed great application potential [12].

Along with the development of MEUF, requirements for surfactants are also advancing. Using biodegradable surfactants for reducing secondary pollution to the environment and the subsequent treatment expenditures are trends in MEUF development. Monem*et al.* adopted arhamnolipid bio-surfactant for the removal of heavy metal ions, copper, zinc, nickel, lead, and cadmium, and demonstrated the ability to lower the concentrations to below federal Canadian regulations [14].

2. Application of MEUF

2.1. Removal of organics with MEUF

Great efforts have been made to investigate the mechanism of organic removal by MEUF. The removals of methylene blue and various reactive dyes such as Reactive Black 5 and Reactive Orange 16 have been demonstrated (>99%) [15]. Xu *et al.* studied the interaction between phenols and CPC and found that phenols were adsorbed at the interface between water and micelle due to the electrostatic interaction between$C_6H_60$and $C_5H_5N^+$or increased solubility in hydrophilic domain of CPC micelles at a low concentration of phenolswhile at a higher concentration of phenols, phenols at the surface of micelles and in the hydrophilic domains saturated and entered the palisade layer of micelles but were still closed to the hydrophilic regions of CPC [16]. Deriszadeh*et al.* removed naphthenic acids by MEUF and discovered that the amphiphilic nature of the acids reduced the CMC value of the CPC surfactants, hence reducing membrane contamination and CPC concentration in the permeated solution and improving the removal efficiency [17]. To date, the application of MEUF in organics-containing wastewater treatment has not gained sufficient significance. Vinder*et al.* applied MEUF in the treatment of adsorbable organic halogens (AOX) in industrial wastewater and found that the removal efficiency was mainly determined by the characteristic of the surfactant and electrolyte concentration. In the presence of nonionic surfactants, when the solution conductivity is 4.0 mS/cm, the removal efficiency can exceed 90%, and the concentrations of anionic surfactants, SDS,and SDBS are lower than their CMC. El-Abbassi*et al.* treated wastewater from an olive mill, which contained polyphenols using MEUF [11].The removal efficiency of polyphenols was 74%, and a faintly-colored permeate (>87%, less dark)was achieved at the transmembrane pressure of 4 bars and SDS concentration of 10CMC under pH 2. Furthermore, MEUF has been tested in the removal of sulfonamide antibiotics of low concentrations in pharmaceutical wastewater with a removal efficiency that exceeded 70% [18]. In addition to the removal of contaminants, MEUF applications for recycling of valuable low-concentration organics in solutions have also been explored. For example, Schwarze*et al.* reported recycling an expensive catalyst in solutions using MEUF [19]. Geanta*et al.* reported recycling lactic and citrate acids in solutions using MEUF [20].

2.2. Removal of metal ions by MEUF

Many studies have been conducted on using MEUF for the removal of low-concentration heavy metal ions including arsenic, zinc, cadmium, copper, nickel, and lead [13-14]. In recent years, the removal of heavy metal ions has advanced from removing only one kind of heavy metal ion to removing various heavy metal ions simultaneously. Fang *et al.* found that in the systems of one metal ion and mixed metal ions under the same experimental conditions, the removal efficiency of Cd^{2+}, Cu^{2+}, and Ni^{2+} ions was similar (all higher than 98%), indicating that the removal mechanism of MEUF was based on the electrostatic adsorption of heavy metals onto the micelles, and was dependent only on the valence number of heavy metals and independent of the kind of heavy metals [21]. Using MEUF for removing heavy metal ions is progressing towards treating actual wastewater. Aguirre *et al.* removed heavy metalsfrom phosphorous-rich drainage waters (pH of 3.2 and SDS concentration of 75.6 mM) of a fertilizer company using MEUF; the removal efficiency of cadmium and copper was 84.3% and 75.0%, respectively [22]. At the same time, MEUF is making its way into the recycling of low-concentration heavy metals in solutions. Kedari*et al.* utilized MEUF for the recovery of low-concentration americium (III), and achieved a recovery efficiency of 100% with SDS as the surfactant and a solution of pH level of ≥ 3 [23].

2.3. Simultaneous removal of organics and heavy metals using MEUF

With environmental concerns growing and contaminant compositions in the environment becoming more complex, the simultaneous removal of organics and heavy metals is gradually becoming a necessity; this need has attracted growing attention and extensive investigations. Tung *et al.* used a mixture of SDS and Triton X-100 to simultaneously remove Cu^{2+} and phenols, and discovered that the removal efficiency of the Cu^{2+} was higher than that of Cu^{2+}alone, while the removal efficiency of phenols was lower than that of phenolsalone [24]. Witek*et al.* proposed that the presence of Cr^{3+}did not influence the removal of phenols with a surfactant system of SDS or CTAB [25]. Misra*et al.* demonstrated experimentally that in a SDS surfactant system, the presence of UO_2^{2+}did not affect the removal of di-butyl phosphate (DBP) [26]. Zeng*et al.* found during the simultaneous removal of Cd2+ and MB that the presence of MB increased the removal efficiency of Cd^{2+} when the SDS concentration was lower than 1 CMC [27]. The removal efficiency of Cd^{2+} reached 98.8% when the concentration of MB and SDS was 4 mg/L and 1 CMC, respectively. However, the MB removal efficiency reached 96.5% in the presence of Cd^{2+}, which was a decrease of 3.4% compared to that of MB alone.

2.4. Technical parameters of MEUF

Common technical parameters of MEUF include removal efficiency and permeation flux. Removal efficiency is defined as the ability to retain or reject contaminants with the use of a membrane under certain conditions, and is often expressed as follows:

$$R=\frac{C_R\ C_P}{C_R}=1-\frac{C_R}{C_P} \tag{1}$$

where R is the rejection rate or removal efficiency of the contaminant; C is the concentration of the contaminant; subscripts R and P denote concentrated solution and permeated solution, respectively.

Permeation flux is defined as the volume of permeated solution passing through the membrane per unit time and per unit area. It is often used to indicate wastewater treatment efficiency of the membrane and the contamination level. Permeation flux is usually expressed by the following equation:

$$J=\frac{Q}{A} \tag{2}$$

where J ($L\cdot m^{-2}\cdot h^{-1}$) is the permeation flux of the membrane; Q ($L\cdot h^{-1}$) is the feed solution flow passing through the membrane; and A (m^2) is the effective area of the membrane.

In addition to removal efficiency and permeation flux, new parameters have been deduced in succession and adopted in current studies; examples include the distribution coefficient (D), adsorption amount of micelle per unit (L), and relative adhesion coefficient of contaminant (K). D represents the distribution of contaminants between the concentrated and permeated solutions. A higher D value indicates a higher degree of enrichment of contaminants and better removal efficiency. It can be expressed by the following equation [28, 29]:

$$D=\frac{C_R}{C_P} \tag{3}$$

Adsorption amount of micelle per unit (L) characterizes the amount of contaminant adsorbed by a micelle unit, and can be expressed by the following equation [27, 28]:

$$L=\frac{C_R\ C_P}{S_R\ S_P} \tag{4}$$

where S is the concentration of surfactant. In the case of a mixed surfactant system, SR is the sum of the concentrations of all surfactants in a solution.

The relative adhesion coefficient of contaminant (K) represents the relative adsorption amount of contaminant on the micelle, and can be expressed by the following equation [28, 29]:

$$K=\frac{C_M}{C_W \quad S} \tag{5}$$

where subscripts M and W denote micelle and water phases, respectively; and S is the concentration of surfactant in the micelle phase.

With the introduction of new parameters, the principle of the enhanced ultrafiltration of the micelle is understood more profoundly. Paiboon*et al.* used distribution coefficient (D) and relative adhesion coefficient of contaminant (K) to describe the removal of lead ions in wastewater using MEUF based on mixed surfactants, and demonstrated that an increased surfactant concentration simultaneously improved the removal efficiencies of lead ions and surfactants as more surfactant monomers formed micelles, which strengthened the micelles' adsorption of lead ions [29]. In a study of the simultaneous removal of nickel and cobalt using MEUF, Karate *et al.* manifested the highest efficiency of heavy metals removal at the early stage of ultrafiltration through a comparison of the changes of removal efficiency and distribution coefficient under different ultrafiltration volumes [28]. The results were also verified by the adsorption amount of micelle per unit (L) and relative adhesion coefficient of contaminant (K).

3. Factors that influence the treatment efficiency of MEUF

3.1. Effects of surfactant concentrations

Micelles are only observed when the surfactant concentration in the solution is no less than 1 CMC; however, the concentration polarization effect resulting from the membrane filtration in MEUF causes abundant surfactants to aggregate near the membrane surface, which leads to the formation of micelles and a much higher concentration of CMC. Therefore, with MEUF, contaminants can be removed even if the surfactant concentration is lower than 1 CMC [6]. Samper *et al.* used SDS and low-concentration LAS to remove heavy metal ions in solutions; more than 90% of the heavy metal ions were removed when the SDS concentration was lower than 1 CMC, and almost 100% were removed when the LAS concentration was lower than 1 CMC. In a certain range, the removal efficiency of contaminants increases with an increasing surfactant concentration with MEUF. However, excess surfactants can result in compact micelles and a reduced hydraulic radius, which causes the solubilization of organics by the micelles to be reduced. This results in a simultaneous change in the morphology of micelles from ball-shape to rod-shape. In such a case, both the solubilization of organics and adsorption of heavy metals by micelles are significantly reduced, and they penetrate the membrane into the permeated solution easily, which leads to a reduced treatment efficiency [30, 31]. Huang *et al.* increased SDS concentration from 8 mM to 80 mM in MEUF and the resultant MB removal efficiency decreased from 99.34% to 87.14% [31]. Moreover, it was discovered that excess micelles readily aggregated on the membrane surface and formed a gel layer, which lead to a reduced permeation flux.

3.2. Effects of pH value

The heavy metal ions removal efficiency of MEUF increases gradually with increasing pH levels. This can be attributed to the fact that H^+ ions in the solution are adsorbed by charges of different polarity on the micelle surface and compete with metal ions, which reduces the available adsorption sites of the micelle surface for the metal ions [23]. Therefore, it is not suitable to use MEUF for heavy metal removal under acidic conditions. Aguirre *et al.* found that the optimum pH was 3.2 for the removal of copper and cadmium from phosphorous-rich drainage waters of a fertilizer company using MEUF[22].

The influence of pH on organics removal using MEUF is less obvious than that on heavy metals, which relates to the property of organics to some extent. Studies have shown that pH change does not influence the removal of dyes, e.g. MB and Safranin T; the removal efficiency of both is above 99% [31, 32]. Based on a study of phenols removal using CTAB as the surfactant in MEUF, Sabatino *et al.* found that the removal mechanisms were affected by the pH value, although the pH changes did not significantly affect the removal efficiency [33]. Phenols are adsorbed on the micelles surface by ionic adsorption under neutral conditions (pH=6.7) while under basic conditions (pH=9.94), phenols are removed due to increased solubility in the palisade layer of micelles. Venkataganesh *et al.* studied the removal of naphthenic acid (NA) using MEUF and found that under acidic conditions, carboxylic groups of NA were ionized to neutral and NA molecules became nonionic, which resulted in its increased solubility in micelles; under basic conditions, the NA decomposed and its water solubility increased [4]. At the same time, the CMC of SDS micelles decreased at low pH levels while the CMC remained unchanged at higher pH levels (>5). The NA removal efficiency decreased from 82% to 68% when the pH level increased from 3 to 9.

3.3. *Effects of electrolytes*

Due to the presence of various electrolytes in wastewater, it is necessary to understand the influence of electrolytes in the MEUF process. Electrolytes reduce the CMC of surfactants. This is primarily realized by the salting-in and salting-out effects from hydrophobic groups of surfactants for nonionic surfactants, by the neutralization of charges in micelles, and the reduction of electrostatic adsorption for ionic surfactants. The addition of concentrated electrolytes significantly reduces the metal ions removal efficiency because cations in the electrolytes compete with metal ions in the solution for the adsorption sites on the micelles [21]. Since the number of micelles increases and micelle volume decreases after the addition of electrolytes, hydrocarbons soluble in the micelle core increase while polar molecules soluble in the palisade layer of micelle decrease. Therefore, organics removal effects of electrolytes are related to the property of organics. Organics, like dyes, usually have increased solubility in the micelle core and palisade layer; thus, electrolytes have no obvious effect on organics removal efficiency [34, 35].

Previous studies showed that the permeation flux decreases with the addition of electrolytes because the repulsion between surfactants in the solution decreases, the number of micelles increases, and the concentration polarization phenomenon becomes severe, resulting in increasing deposition of micelles on the membrane surface [31]. Simultaneously, electrostatic adsorption causes the electrolyte ions to be adsorbed onto the charged micelle surface, which reduces the repulsion between micelles, densifies the gel layer on the membrane surface, and increases the resistance on the membrane surface [4]. Furthermore, viscosity of the solution and hydraulic resistance increase with the addition of electrolytes.

3.4. *Effects of trans-membrane pressure*

The effects of trans-membrane pressure on MEUF can be discussed based on contaminant removal efficiency, permeation flux, and flux reduction volume. Usually the change in trans-membrane pressure does not influence the removal of organics and heavy metals as contaminant removal is largely achieved by its combination with micelles [35, 36]. However, micelles are squeezed when the trans-membrane pressure is over-high, resulting in reduce dorganics solubility. At the same time, overlarge pressure increases the possibility of contaminant permeation through the membrane [31]. A change in trans-membrane pressure significantly affects permeation flux. Increasing the trans-membrane pressure overcomes permeation pressure and membrane resistance, and increases permeation flux [31]. However, higher trans-membrane pressure accelerates the growth of the gel layer, which increases the thickness and flux reduction amount [4].

4. Conclusions and future research needs

Traditional ultrafiltration membranes are limited by contamination issues, flux reduction and rapid membrane consumption. Hence, a novel lanti-pollution ultrafiltration membrane and efficient membrane cleaning method are needed. Since a surfactant of about 1 CMC remains in a permeated solution, it is necessary to have a biodegradable surfactant with a low CMC. The permeated solution can be simultaneously treated again at a low cost, and currently used methods include foam fractionation and activated carbon adsorption [37, 38]. There are usually various organics and heavy metal ions in actual wastewater; therefore, in order to realize industrialization and based on the treatment of uncomplicated wastewater, studies regarding MEUF should focus on the treatment of wastewater with multi-components. Besides, there are concentrated surfactants, organics, and heavy metals in concentrated solutions, and it is an inevitable trend of MEUF development to efficiently recycle surfactants, valuable organics, and heavy metals from such solutions.

Acknowledgements
This study was supported by the National Natural Science Foundation of China (51308076) and the Open Fund from Institute of Wuhan Studies (IWHS2016106).

References

[1] Dunn R O, Scamehorn J F, and Christian S D 1985 Sep. *Sci. Technol.* **20** 257–284
[2] Ngang H P, Ahmad A L, and Low S C et al. 2012 *Desalination* **293** 7–20
[3] Nguyen L A T, Minding M, and Schwarze M et al. 2013 *J. Membrane Sci.* **433** 80–87
[4] Venkataganesh B, Maiti A, and Bhattacharjee S et al. 2012 Sep. *Purif. Technol.* **98** 36–45
[5] Sarkar B, DasGupta S, and Sirshendu D 2009 Sep. *Purif. Technol.* **66** 263–272
[6] Samper E, Rodríguez M, and La Rubia M A et al. 2009 Sep. *Purif. Technol.* **65** 337–342
[7] Vinder A and Simonič M 2012 *Desalination* **289** 51–57
[8] Luo F, Zeng G M, and Huang J H et al. 2010 *J. Hazard. Mater.* **173** 455–461
[9] Bielska M and Szymanowski J 2006 *Water Res.* **40** 1027–1033
[10] Huang J H, Zeng G M, Fang Y Y et al. 2009 *J. Membrane Sci.* **326** 303–309
[11] El-Abbassi A, Khayet M, and Hafidi A 2011 *Water Res* **45** 4522–4530
[12] Zhang W X, Huang G H, and Wei J 2013 *Desalination* **311** 31–36
[13] Yenphan P, Chanachai A, and Jiraratananon R 2010 *Desalination* **253** 30–37
[14] Monem El Zeftawy M A and Mulligan C N 2011 Sep. *Purif. Technol.* **77** 120–127
[15] Ahmad A L, Puasa S W and Zulkali M M D 2006 *Desalination* **191** 153–161
[16] Xu K, Ren H Q, and Zeng G M et al. 2010 *Colloid. Surface. A* **356** 150–155
[17] Deriszadeh A, Harding T G, and Husein M M 2009 *J. Membrane Sci.* **326** 161–167
[18] Exall K, Balakrishnan V K, and Toito J et al. 2013 *Sci Total Environ* **461–462** 371–376
[19] Schwarze M, Rost A, Weigel T et al. 2009 *Chem. Eng. Process.* **48** 356–363
[20] Geanta R M, Ruiz M O, and Escudero I 2013 *J. Membrane Sci.* **430** 11–23
[21] Fang Y Y, Zeng G M, Huang J H 2011 *Zhongnan University Academic Journal* **42** 266–271
[22] Aguirre J L, Pongrácz E, and Sarpola A 2012 Sep. *Purif. Technol.* **88** 130–137
[23] Kedari C S, Pandit S S, and Tripathi S C 2009 *J. Membrane Sci.* **341** 122–130
[24] Tung C C, Yang Y M, and Chang C H 2002 *Waste Manag.* **22** 695–701
[25] Witek A, Kołtuniewicz A, and Kurczewski B 2006 *Desalination* **191** 111–116
[26] Misra S K, Mahatele A K, and Tripathi S C 2009 *Hydrometallurgy* **96** 47–51
[27] Zeng G M, Li X, Huang J H et al. 2011 *J. Hazard. Mater.* **185** 1304–1310
[28] Karate V Dand Marathe K V 2008 *J. Hazard. Mater.* **157** 464–471
[29] Yenphan P, Chanachai A, and Jiraratananon R 2010 *Desalination* **253** 30–37
[30] Fang Y Y, Zeng G M, and Huang J H 2008 *J. Membrane Sci.* **320** 514–519
[31] Huang J H, Zhou C F, and Zeng G M 2010 *J. Membrane Sci.* **365** 138–144
[32] Zaghbani N, Hafiane A, and Dhahbi M 2008 *Desalination* **222** 348–356
[33] Sabatino P, Szczygiel A, Sinnaeve D et al. 2010 *Colloid. Surface.* A **370** 42–48
[34] Huang J H, Zhou C F, Zeng G M et al. 2012 Sep. *Purif. Technol.* **98** 497–502
[35] Zaghbani N, Hafiane A, and Dhahbi M 2009 *J. Hazard. Mater.* **168** 1417–1421
[36] Rahmanian B, Pakizeh M, and Maskooki A 2010 *J. Hazard. Mater.* **184** 261–267
[37] Qu Y H, Zeng G M, and Huang J H et al. 2008 *J. Hazard. Mater.* **155** 32-38
[38] Channarong B, Lee S H and Bade R 2010 *Desalination* **262** 221–227

An objective decision model of power grid environmental protection based on environmental influence index and energy-saving and emission-reducing index

Jun-shu Feng[1, 3], Yan-ming Jin[1] and Wei-hua Hao[2]

[1] State Grid Energy Research Institute, Future Science and Technology Park, Beijing 102209, China

[2] State Grid Shandong Electric Power Corporation Maintenance Company, Jiaodong Converter Station, Shandong 266329, China

E-mail: fengjunshu@sgeri.sgcc.com.cn

Abstract. Based on modelling the environmental influence index of power transmission and transformation project and energy-saving and emission-reducing index of source-grid-load of power system, this paper establishes an objective decision model of power grid environmental protection, with constraints of power grid environmental protection objectives being legal and economical, and considering both positive and negative influences of grid on the environmental in all-life grid cycle. This model can be used to guide the programming work of power grid environmental protection. A numerical simulation of Jiangsu province's power grid environmental protection objective decision model has been operated, and the results shows that the maximum goal of energy-saving and emission-reducing benefits would be reached firstly as investment increasing, and then the minimum goal of environmental influence.

1. Introduction

Along with economy swift development and living standards yearly enhancement, the requirement of environmental protection is increasing in many countries. Under state environmental protection strategy, any behavior of power grid enterprises should follow the demands of low carbon economy and can perform higher than legal environmental standards, as well as provide leading technology actively for regulatory agencies and industry, to establish market barriers or the first-mover advantage of the enterprise. Driven by external environmental pressure and internal clean development, power grid environmental protection has gradually become an important content of grid enterprise development strategies, and a top priority of grid environmental protection is its objective, which is the general strategy and roadmap of grid development, also is a fundamental basis of compiling power grid plan.

Environmental protection was seldom considered in the existing grid planning works or researches which are presented in [1-5], and influence of some environmental protection factors, such as electromagnetic environment, noise, waste and landscape, to decide environmental protection objective also seldom involved, and the objective optimizing decision of power grid environmental protection. At present, some grid environmental policies and standards has been enacted in china. However, only environmental limits are given in all these standards, the staged quantitative indicators can be used to guide the grid environmental protection work haven't get involved. Scholars have made some exploration in grid environment protection planning, such as, Wang discusses some ideas about planning environmental impact assessment in [6], MAO establishes the regional grid environmental protection planning method and framework in [7], ZHU quantifies some indexes to establish a grid planning model related to environmental protection in [8]. Generally, grid environmental protection planning just started, researches of grid environmental protection objectives optimizing decision should be strengthened.

In this paper, an objective decision model of power grid environmental protection is established to guide power grid enterprises planning their grid environmental protection strategies, based on modeling the environmental influence index of power transmission and transformation project and energy-saving and emission-reducing index of source-grid-load of power system, and considering both positive and negative influences of grid on the environmental in all-life grid cycle.

2. Mosel design

Power grid environmental protection has the characteristics of wide range and big significance. For example, equipment like transformers and transmission lines has different environmental standards in their different life cycle, and as the connection of generation side and demand side, power grid will affect environmental protection work of the whole electric power industry indirectly.

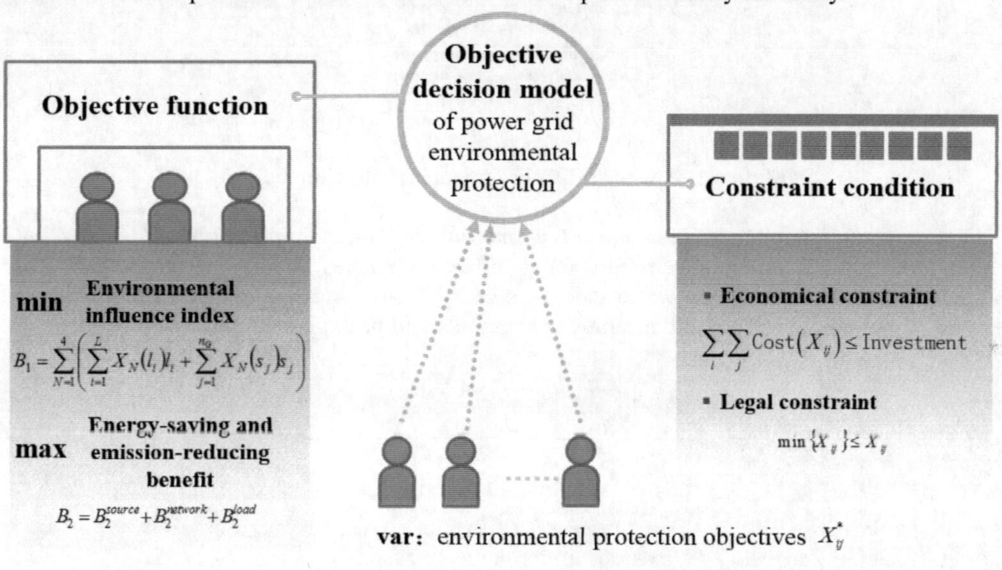

Figure 1. Objective decision model of power grid environmental protection.

Comprehensively considering both the positive role and negative influence of grid to the environment in its whole life cycle, including construction, operation and retirement, the concept of grid environmental protection factor is given. The positive role of grid mainly reflects in promoting its upstream and downstream, which is power generation side and demand side, to conserve energy and reduce emissions. The negative impact of grid mainly reflects in producing waste, landscape effect, electromagnetic environment and noise in its life cycle process. Therefore, the objective decision model of power grid environmental protection will be established as the following steps.

Step 1. Define grid environmental protection factor, including electromagnetic, noise, landscape and waste, and model all the factors specifically in their all-life grid cycle, including construction, operation and retirement.

Step 2. Study and model the each benefit of energy-saving and emission-reducing from power generation side, grid side and demand side, caused by environmental protection strategies of grid enterprises.

Step 3. Establish objective decision model of power grid environmental protection, with minimal grid environmental protection factor and maximum benefits of energy-saving and emission-reducing.

3. Power grid environmental protection factor modelling

To structure the power grid environmental protection factor, we mainly considers environmental influence indexes of power transmission and transformation project in the processes of construction, operation and retirement, including electromagnetic index, noise index, landscape index and waste index. And then modelling the energy-saving and emission-reducing index of source-grid-load of power system, by used of analyzing the benefits of grid promoting its upstream and downstream, which is power generation side and demand side, to conserve energy and reduce emissions.

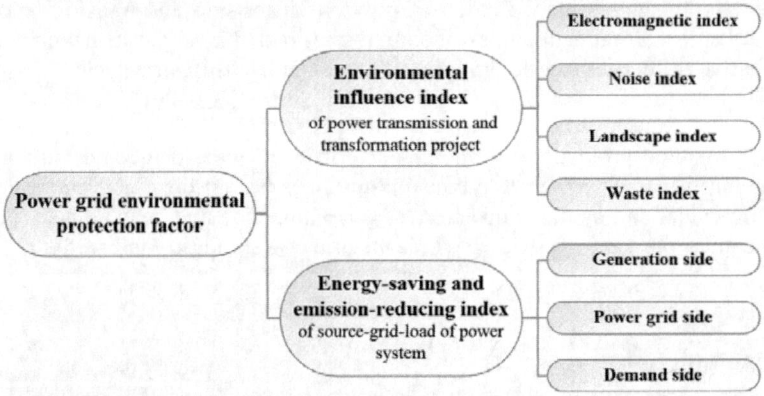

Figure 2. Power grid environmental protection factor.

3.1. *Environmental influence index of power transmission and transformation project*

Refer to some models in [9-11], environmental influence indexes includes electromagnetic index, noise index, landscape index and waste index. Power grid environmental influence index in all-life grid cycle is showed in figure 3, and meanings of symbols will be explained in the following.

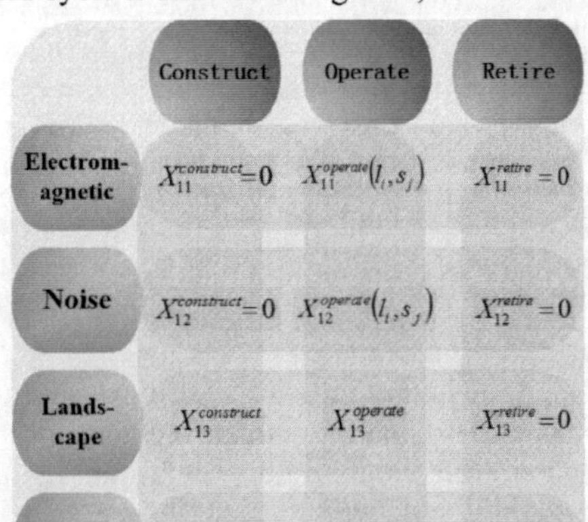

Figure 3. Power grid environmental influence index in all-life grid cycle.

3.1.1. Electromagnetic index. In the process of operation, $X_1^{operate}(l_i)$ is the electromagnetic influence of transmission line l_i. In equation (1), n is the segments of line l_i, p_i is the population density of the segment i region, L_i is the length of segment i, W_L is the distance of transmission line's side and the central line, $E_L(x)$ is the function of electromagnetic influence depends on the distance from the central line which can be calculated by the method recommended by International Conference on Power Grid.

$$X_1^{operate}(l_i)=\sum_{i=1}^{n}\left(p_iL_i\int_{-W_L}^{W_L}E_L(x)dx\right) \tag{1}$$

In the process of operation, $X_1^{operate}(s_j)$ is the environmental influence of substation s_j. In equation (2), p_s is the population density of the substation, W_s is the distance from the substation, $E_s(x)$ is the function of electromagnetic influence depends on the distance from the substation.

$$X_1^{operate}(s_j) = p_s \int_0^{W_s} (E_s(x) 2\pi x) dx \tag{2}$$

3.1.2. Noise index. In the process of operation, $X_2^{operate}(s_j)$ is the noise influence of transmission line l_i. In equation (3), $N_L(x)$ is the function of noise influence depends on the distance from the central line which also can be calculated by the method recommended by International Conference on Power Grid.

$$X_2^{operate}(l_i) = \sum_{i=1}^{n} \left(p_i L_i \int_{-W_L}^{W_L} N_L(x) dx \right) \tag{3}$$

In the process of operation, $X_2^{operate}(s_j)$ is the noise influence of substation s_j. In equation (4), $N_s(x)$ is the function of noise influence depends on the distance from the substation.

$$X_2^{operate}(s_j) = p_s \int_0^{W_s} (N_s(x) 2\pi x) dx \tag{4}$$

3.1.3. Landscape index. In equation (5), X_3 is the landscape index, and S_d is the distance of a line or a substation away from the affected point, S_p is the affected population, which includes floating population and fixed population active around the affected point, S_t is the possibility of lines or substations appearing by the viewer around the affected point, which can be calculated by the ratio of lines' or substations' area being viewing and the whole affected point area. Landscape index in the process of construction and operation can be expressed as $X_2^{construct}$ and $X_2^{operate}$.

$$X_3 = S_d S_p S_t \tag{5}$$

3.1.4. Waste index. In the process of operation, $X_4^{operate}$ is the waste influence, including liquid waste and solid waste. In equation (6), $|\alpha pD|$ is the normalized index of liquid waste emissions, which α is the waste reduction factor, p is the resident person-time during the construction process, D is engineering water quota. $|W_{muck}|$ is the normalized index of solid waste emissions, which W_{muck} includes concrete waste of telegraph poles, insulator waste, and so on.

$$X_4^{construct} = |\alpha pD| + |W_{muck}| \tag{6}$$

In the process of operation, $X_4^{operate}$ is the waste index, including liquid waste, gaseous waste and solid waste. In equation (7), $|W_{oil}|$ is the normalized index of substation oil waste emissions, $|W_{SF6}|$ is the normalized index of SF_6 emissions.

$$X_4^{operate} = |\alpha pD| + |W_{muck}| + |W_{oil}| + |W_{SF_6}| \tag{7}$$

In the process of retirement, X_4^{retire} is the waste index, including oil waste, gaseous waste and battery waste.

$$X_4^{retire} = |W_{muck}| + |W_{oil}| + |W_{SF_6}| \tag{8}$$

3.2. Energy-saving and emission-reducing index of source-grid-load of power system

The mechanism of power grid environmental protection to promoting the whole power system of energy conservation and emissions reduction mechanism shows in figure 4. Respectively considering power grid enterprise environmental protection strategies to promote the benefits of the generation side, grid side and user side of energy conservation and emissions reduction. B_2 is the energy-saving and emission-reducing index of source-grid-load of power system in equation (9), and B_2^{source}, $B_2^{network}$ and B2load represent the benefits of generation side, grid side and user side, respectively.

$$B_2 = B_2^{source} + B_2^{network} + B_2^{load} \quad (9)$$

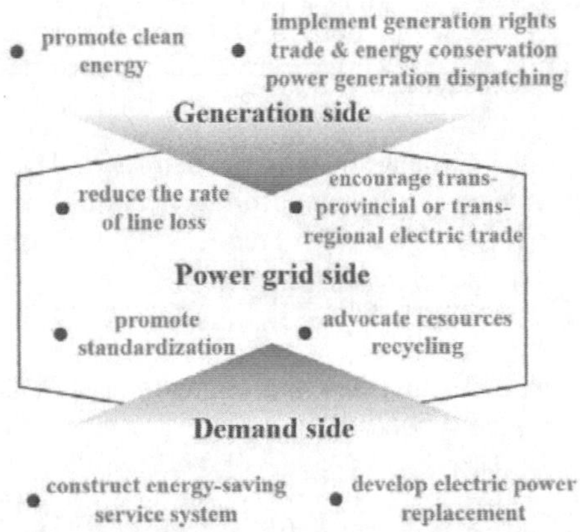

Figure 4. Process of energy-saving and emission-reducing of source-grid-load of power system.

3.2.1. Generation side. Mainly considered promoting clean energy and implementing generation rights trade or energy conservation power generation dispatching. In equation (10), X_{21}^{source} is the goal of clean energy development, X_{22}^{source} is the goal of implementing generation rights trade or energy conservation power generation dispatching, and B_{21}^{source} and B_{22}^{source} are their energy-saving and emission-reducing benefits, respectively.

$$B_2^{source} = B_{21}^{source}\left(X_{21}^{source}\right) + B_{22}^{source}\left(X_{22}^{source}\right) \quad (10)$$

3.2.2. Power grid side. Mainly considered reducing the rate of line loss, encouraging trans-provincial or trans-regional electric trade, promoting standardization and advocating resources recycling. In equation (11), $X_{2i}^{network}$ (i=1,2,3,4) are the goals of reducing rate of line loss, encouraging trans-provincial or trans-regional electric trade, promoting standardization and advocating resources recycling, respectively. And $B_{2i}^{network}$ (i=1,2,3,4) are their benefits.

$$B_2^{network} = \sum_{i=1}^{4} B_{2i}^{network}\left(X_{2i}^{network}\right) \quad (11)$$

3.2.3. Demand side. Mainly considered constructing energy-saving service system and developing electric power replacement. In equation (12), X_{21}^{load} and X_{22}^{load} are the goals of constructing energy-saving service system and developing electric power replacement, and B_{21}^{load} and B_{22}^{load} are their benefits.

$$B_2^{load} = B_{21}^{load}\left(X_{21}^{load}\right) + B_{22}^{load}\left(X_{22}^{load}\right) \quad (12)$$

4. Objective decision model of power grid environmental protection

After defining and modeling every environmental protection objectives, objective decision model of power grid environmental protection is established. In the following model, *L* is the total transmission line length during the planning period and *S* is the total amount of substations during the same period.

4.1. Objective function

Objective function considers both the positive role and negative influence of grid to the environment in its whole life cycle caused by the environmental influence index of power transmission and transformation project and energy-saving and emission-reducing index of source-grid-load of power system.

$$\min\ B_1 = \sum_{N=1}^{4}\left(\sum_{i=1}^{L} X_N l_i + \sum_{j=1}^{S} X_N s_j\right)$$
$$\max\ B_2 = B_2^{source} + B_2^{network} + B_2^{load} \quad (13)$$

In equation (13), B_1 is the environmental influence objective, and every environmental protection influence objective function X_N includes construct objective $X^{construct}$, operate objective $X^{operate}$ and retire objective X^{retire}, which can be calculated by $X = X^{construct} + X^{operate} + X^{retire}$.

4.2. Constraint condition

Constraint conditions includes environmental protection objective being legal and environmental protection cost being under limit.

Equation (14) gives the legal environmental protection objective constraint.

$$\min\{X_{ij}^*\} \le X_{ij}^* \quad (14)$$

In equation (14), $\min\{X^*_{ij}\}$ is the minimum requirements to environmental protection objective of state policies, laws or regulations。

Equation (15) gives the environmental protection cost under limit constraint.

$$\sum_i \sum_j \sum_* C_{ij}^*\left(X_{ij}^*\right) \le C_{\max} \quad (15)$$

In equation (15), $C^*_{ij}(X^*_{ij})$ is the cost of achieving the environmental protection objective X^*_{ij}, and C_{max} is the maximum grid environmental protection investment of grid enterprises.

5. Example

Taking Jiangsu province as a sample example, to calculate the objective decision model of power grid environmental protection, and figure 5 shows the feasible solutions and optimal solutions under different grid investment. The results shows that, the maximum goal of energy-saving and emission-reducing benefits and the minimum goal of environmental influence will be both considered when the grid environmental protection investment being small, the maximum goal of energy-saving and emission-reducing benefits will be reached firstly as investment increasing, and when the investment being big enough, both goal will be achieved.

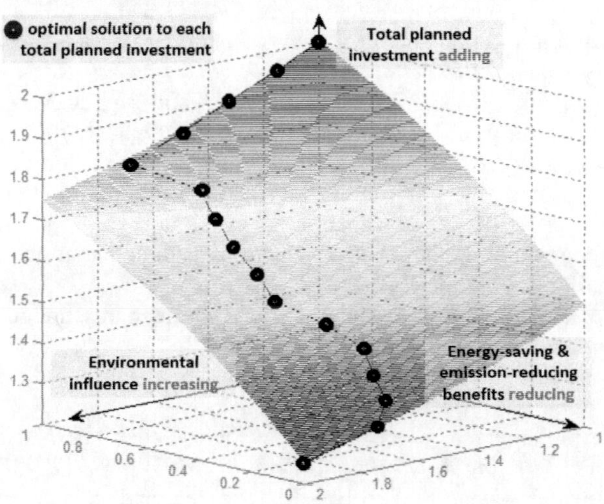

Figure 5. Environmental protection objective decision under variable investment.

6. Conclusion

In this paper, an objective decision model of power grid environmental protection is constructed, based on the constraints of grid environmental protection factors and grid environmental protection investment. Under the premise of electric power enterprises carrying out environmental protection policy actively, to optimize and control several kinds of grid environmental protection index can minimize the negative impact of grid construction and operation on the environment, also maximize the positive impact to the environment, and eventually to achieve the coordinated development of grid and environment. A sample example of Jiangsu province is given, and the results shows that the maximum goal of energy-saving and emission-reducing benefits would be reached firstly as investment increasing, and then the minimum goal of environmental influence.

Based on this method, future grid environmental protection planning objectives can be helped to determine, and then the quantified objectives can be calculated by used of data from the power grid enterprises' environmental information platform. Such as, to realize the coordinated development of grid construction or operation and ecological environment, indications like rates of environmental impact assessment and environmental protection acceptance check, rates of sulfur hexafluoride gas recycling and utilization, etc., to reach the standards, to complete state's or enterprise's energy conservation and emissions reduction requirements, to improve environmental protection organization system and keep the grid environmental protection supervised and managed in the whole process, to reach the international leading level of grid environmental protection science and technology, and so on.

References

[1] Mo Hua, Wang Meng 2007 Analysis & suggestion on some existing issues about power transmission and transformation EIA *Electric Power Environmental Protection* **4** 4-7.

[2] Yao Jian 2009 Environmental planning and management *Beijing Chemical Industry Press*

[3] Ma Rui 2002 A Novel Bi-objective Fuzzy Optimal Model of Short-term Trade Planning Considering Environmental Protection and Economic Profit in Deregulated Power System *Proceedings of the CSEE* **22(4)** 104-106.

[4] Liang Dong, et al. 2011 Discussion on strengthening environmental protection planning and design for power grid projects *Electric Power Construction* **32(2)** 35-38.

[5] Xiao Liang, et al. 2015 Grid Dispatching of Energy Saving and Environmental Protection Based on Thermodynamic System Model with Flow Dimension *Proceedings of the Second International Conference on Mechatronics and Automatic Control* **334** 333-341.

[6] Wang Zhibiao, et al. 2011 The perception and practice on system of plan eia of transmission electricity power grid *Energy Environmental Protection* **25(2)** 39-45.

[7] Mao Guozhu, Zhang Fengrong, Li Jing, et al. 2015 Framework and application of environmental planning for regional power grid *Electric Power Construction* **36(11)** 70-76.

[8] Zhu Jianqiang, Cheng Haozhong, Wu Peng, et al. 2009 A preliminary method for urban transmission network planning required by environmental protection *Electric Power Construction* **33(16)** 30-34.

[9] Tan Jianjun 2003 Study on environmental pollution by electromagnetic radiation for transformers and transmission lines[D]. *Chongqing University*

[10] Xu Lu-wen, et al. 2010 Audible Noise Survey and Analysis of Environment in Chongqing Electric Grid *Environmental Monitoring in China* **26(5)** 29-31.

[11] Zhu Min, et al. 2016 Current condition of environmental waste treatment and disposal policy in power grid enterprise *ShanXi Electric Power* **44(4)** 71-74.

[12] Qun Wang, et al. 2005 Analysis & Assessment of Electromagnetic Field Under 110kV High Voltage Transmission Lines *Journal of Beijing Polytechnic University* **31(3)** 308-312.

[13] Zhu Fahua, et al. 2015 Environmental impacts and benefits of regional power grid interconnections for China *Energy Policy* **33(14)** 1797-1805.

Analysis on spatial transfer model of energy development layout and the ecological footprint affection

Xiaoxia Wei[1], Jinfang Zhang

State Grid Energy Research Institute, Future Science and Technology Park North Area, Beiqijia, Changping, Beijing, 102209, P.R. ChinaGuangyi Building A1206, Guangyi Street 5#, Xicheng district, Beijing, China

E-mail: 20583853@qq.com

Abstract. Consider the global energy interconnection, the global is concentrating on carrying out clean energy alternative, which is mainly focusing on using the clean energy to take place of fossil energy, and change the global energy layout and ecological atmosphere condition. This research gives the energy spatial transfer model of energy development layout to analyse the global energy development layout condition and ecological affection. And it is a fast and direct method to analyse its energy usage process and environmental affection. The paper also gives out a system dynamics model of energy spatial transfer shows , which electric power transmission is better than original energy usage and transportation. It also gives the comparison of different parameters. The energy spatial transfer can affect the environment directly. Consider its three environmental factors, including energy saving, climate changing and conventional pollutant emission reduction, synthetic combine with the spatial transfer model, it can get the environmental change parameters, which showed that with the clean energy wide usage, the ecological footprint affection will be affected significantly.

1. Introduction

The energy interconnection can be equivalent to the spatial distribution transfer of energy development layout. On the one hand, the global energy spatial transfer, can drive economic and social transformation; on the other hand, this spatial transfer layout, greatly improving global environmental pollution and reduce global environmental stress[1].

Energy transfer, spatial layout changes and ecological environment, can be affected by these three factors showed in Figure 1. Energy transferring combined with energy extraction, processing and transportation and other series of energy activities, will inevitably damage the ecological environment. Take the coal as example, coal transportation by railway, affecting on the ecological environment mainly includes the impact of coal mining, the occupation and destruction of land resources, water consumption and water pollution, air pollution, noise pollution, mining ecosystem damage and coal transportation process, long-distance transport dust and coal storage from dust[2,3].

The ecological environment provides the space and carrier for human beings. Due to the backwardness of economic development and the lack of environmental awareness, energy output is often excessive development and use of energy in order to seek rapid economic growth in the short term, leading to serious waste of energy and environmental pollution. Because of the non-renewable characteristics of fossil fuels, extensive energy consumption behaviour will lead to energy depletion.

Adopting the global energy interconnection, promote the development and utilization of renewable energy, use renewable energy to replace a large number of fossil energy consumption, reduce large amounts of pollutants and greenhouse gas emissions, fossil energy development and utilization to avoid the consumption of water and other resources and other resources damage to ecosystems. The global energy interconnection, mainly consider energy and resource conservation, climate change, conventional pollutants emission reduction[4-6].

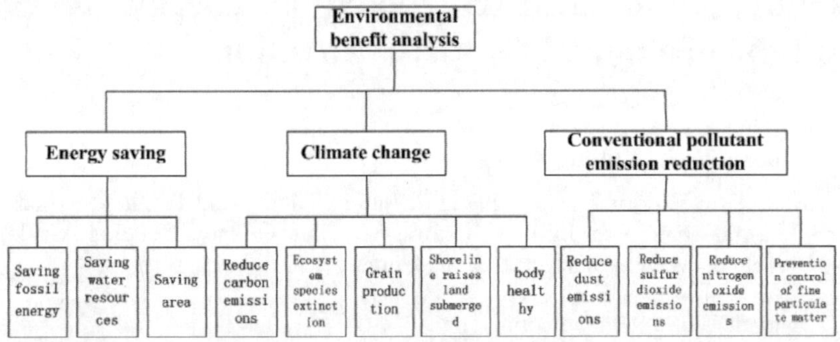

Figure 1. The global energy interconnection impaction on the environment

2. The relationship of energy transfer and ecological environment stress transfer

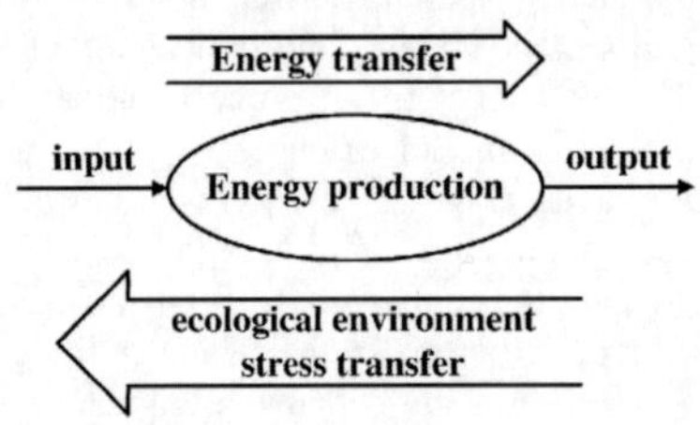

Figure 2. Sources of the Ecological Environment Stress of Energy Resource Export

Figure 2 showed, take one area as the energy production, energy production is the input direction, the receiving area is the energy consumption direction, the positive direction is the energy transfer direction, while the opposite is the ecological environmental stress direction. The model is the relation of energy transfer and ecological environmental stress[7].

Energy transfer, changing by the "power alternative", becomes an important way to transfer the ecological environment pressure. Energy input and consumption, energy production areas, the use of local energy, improve energy efficiency, the extensive use of direct energy consumption into clean, convenient intensive power consumption.

Ecological environment pressure: The ecological environment pressure of one region is composed of two parts, one part comes from the local, and another part comes from the field. Local ecological stress refers to the destruction of the ecological environment caused by resource exploitation, production activities and pollutant discharge to meet local living, production and socio-economic activities. The input of ecological environment refers to the destruction of the local ecological environment caused by the activities of resource exploitation and production which are carried out locally to meet the needs of other regions.

Thus, to the energy production areas, the output of resources means the input the ecological environment of the pressure generated by damage. Considering the large amount of wind energy and solar energy in the energy production area, on the one hand, to solve the local environmental pressure, on the other hand, considering the minimum cost of wind power and photovoltaic power generation, the two kinds of energy are bundled to solve the fossil energy substitution in energy production areas and environmental pressures in the energy consuming regions.

3. System dynamics model of energy spatial transfer

From the perspective of the interaction between regional development and energy transmission, the impact of energy transmission on energy output includes both background variables, such as economic development and environmental pollution, as well as energy policy planning. It is a complex process involving multiple levels. Only by building a scientific and targeted dynamic analysis model, can the relationship between the various variables in the system and the results of joint research conducted in-depth study. In the development model of global energy interconnection, the energy transfer model of energy transfer is identified as shown in the following figure.

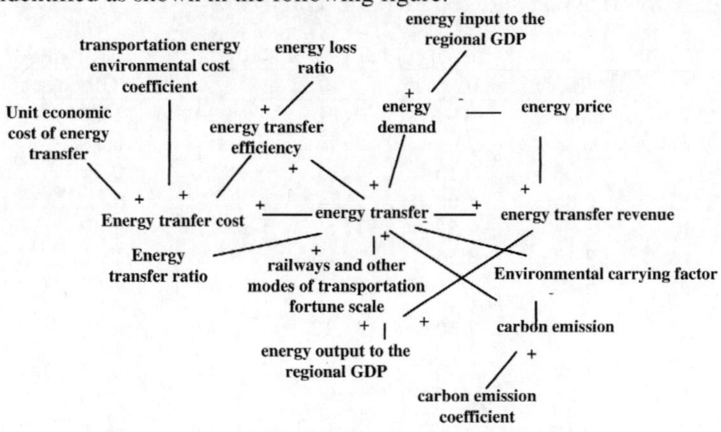

Figure 3. Dynamics model of energy spatial transfer

Figure 3 shows that, from the cost system flow of the energy transfer, consider the energy transfer process relative economic parameters, energy parameters, efficiency parameters and environmental parameters. Analysing the causal relationship between model variables is the initial stage of the system dynamics model, and it is convenient to describe the system structure non technically and macroscopically. In the process of establishing the causal relationship between the model variables, the development of energy transfer system can be the main line to determine the variables involved in the model.

The system dynamics model of energy transmission impacts includes the following variables:

(1) Economic variable: GDP, the cost of energy transfer, the income of energy transfer, energy price, unit economic cost of energy transmission, rail transportation scale under different modes of transport.

(2)Energy variable: energy quantity demand, energy transmission quantity, the ratio of energy transmission.

(3)Efficiency variable: energy transmission efficiency, the energy loss ratio of transportation.

(4)Environment variable: carbon emission, carbon emission coefficient, transportation energy environmental cost coefficient, environmental carrying factor.

In the casual loop graph, there are two main feedback loops:

The first loop 1, energy transmission growth—+energy production income growth—+economic level increasing—+energy transport channel scale enlarge—+ energy transmission growth. This loop circuit is the positive loop, which is the energy transmission and economic development between the causal feedback mechanisms.

The second loop 2, energy transfer growth—+carbon emission growth—+environmental carrying capacity reduction—- energy transfer loss. This loop is the opposite loop, which shows energy

IOP Conf. Series: Earth and Environmental Science **52** (2017) 012079 doi:10.1088/1755-1315/52/1/012079

transmission and ecological environment interaction mechanism, environmental constraints is to adjust the energy transmission mode of the important factors.

4. The parameters of energy spatial transfer

Considering the affection of energy demand of energy consumption, transmission of fossil energy and power are both declining year by year, while the power transmission volume is increasing year by year. Fossil energy production in energy-producing areas has been declining year by year, and the amount of clean energy power generation has been increasing year by year.

Under the basic scenario, every continent power supply and demand is self-balance, while under the green scenario, and consider every continent get power connection. The structure of the power source showed above showed in Table 1[8,9,10].

Table 1. The structure of global power source (trillion kilowatts)

	Basic scenario				Green scenario			
	2020	2030	2040	2050	2020	2030	2040	2050
Coal	7.27	6.13	5.22	4.56	9.07	8.78	6.44	2.13
Oil	0.87	0.36	0.15	0.15	0.73	0.21	0.00	0.00
Gas	6.16	6.86	8.60	8.54	8.34	10.51	11.97	4.98
Hydro	4.79	8.27	9.57	11.53	4.96	7.40	9.28	10.08
Nuclear	3.14	3.12	3.36	3.85	3.14	2.62	2.26	2.10
Wind	1.91	5.80	12.07	18.23	0.81	4.55	12.62	22.37
Photovoltaic	1.86	4.22	7.10	11.12	0.30	2.50	7.00	16.72
Heat	0.92	1.80	5.55	9.19	0.04	0.52	2.55	9.32
Biomass	1.30	2.66	4.24	4.95	0.84	2.13	3.74	4.42
Ocean	0.01	0.18	0.62	1.01	0.01	0.18	0.62	1.01
Total	28.23	39.40	56.48	73.14	28.23	39.40	56.48	73.14

From the parameters above, can see that coal power transmission ration is declining, while the wind and photovoltaic power transmission ratio is growing. So it shows that the coal transmission amount is declining significantly, while the power transmission amount will grow. The change is calculated by (green scenario – basic scenario)/basic scenario *100% showed in Figure 4.

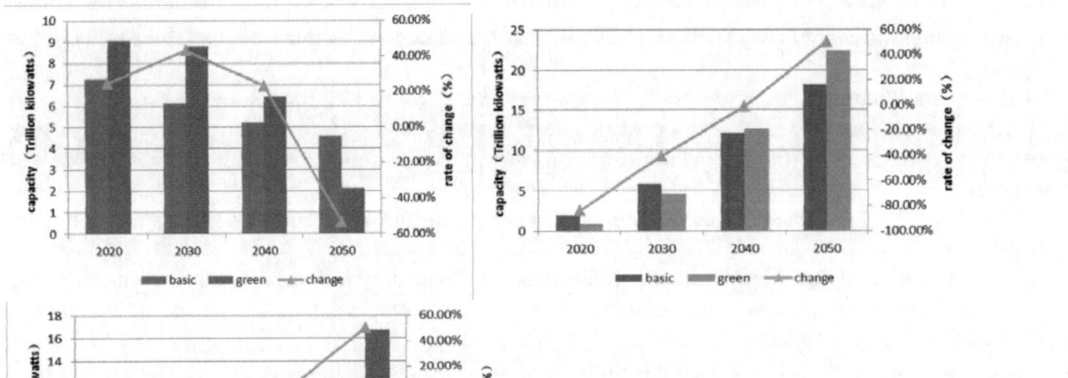

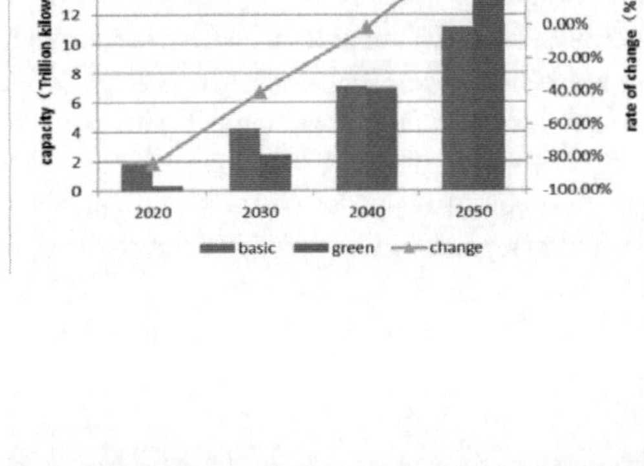

Figure 4. Coal, wind and photovoltaic power transfer changes

To the energy transmission efficiency, the transmission efficiency of the two energy type is not very different. This is mainly due to coal transmission loss rate. Power transmission loss rate is basically the same, so change the proportion of coal and electricity transmission ratio will not affect the energy transmission efficiency. Figure 5 gives out the energy transmission rate, according to the Table 1, and used the same rate to predict the energy transmission rate connected with the energy exploitation.

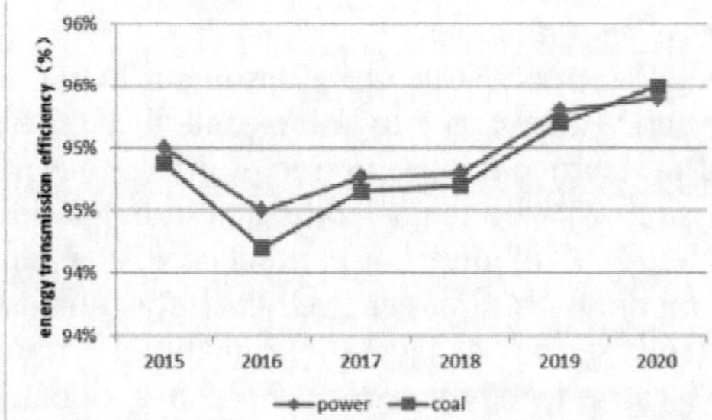

Figure 5. Energy transmission rate change

To the perspective of carbon intensity, because of the increasing of power transmission, coal-fired power generation produces more carbon dioxide, so the carbon emission intensity of the second coal transportation is significantly higher than the electricity power transmission.

To the energy transmission cost, power transmission is more economical than coal transportation, but the energy transmission cost of the second energy transmission mode is higher because of the more serious environmental pollution caused by power transmission to the energy output, and the environmental cost is increased higher.

To the energy transmission income, the two energy transmission incomes are not much difference, but over time, the income gap gradually expanded.

To sum up, to reduce the transmission of fossil fuels such as coal, increase the amount of power transmission, energy output to improve energy transmission revenue, but energy transmission costs and carbon intensity will be increased accordingly. To the energy input, the input power is more cost-effective than the input of coal, but also greatly reduces the environmental pollution caused by thermal power generation. However, to the energy output, the coal-fired power generation will cause land use, water consumption, pollution, gas emissions and other issues, increased environmental pollution. Energy transfer of space should be borne by the energy input to the ecological and environmental problems transferred to the energy output.

5. Energy spatial transfer affection to the environment

5.1. Energy source saving

Energy spatial transfer can accelerate the development and utilization of renewable energy, alternative to traditional fossil fuels, fossil energy conservation development and utilization. Global interconnection can use more resource-rich areas of energy, out of the economically developed areas of their own energy supply constraints, saving fossil energy development and utilization. In some regions, economic development needs a lot of energy support, but is limited by local resources. Considering the ecological environment capacity, resource endowment, geographical traffic conditions and energy demand pattern, the energy spatial transfer will form a more scientific and rational energy distribution pattern. Some resource-rich areas, its energy transcontinental transport, can better protect the economic stability of the affluent areas.

Consider the global energy spatial transfer, with a large amount of development and utilization of renewable energy. According to the forecast, in the global energy interconnection to accelerate the development of scenarios, in 2050, non-fossil energy power generation will reach 66 trillion kilowatts, an increase of nearly 60 trillion kilowatts, accounting for 90% of the total generating capacity. If use the WEO 2014 in the proportion of clean electricity, the 2050 global energy green scenarios, the clean energy electrical quantity will be 29 trillion kilowatts • hour. The equivalent carbon dioxide, sulfur dioxide, nitrogen oxides and soot emissions will reach 25 billion tons / year, 53.7 million tons / year, 56.4 million tons / year, and the annual output of carbon dioxide, Year, 9.4 million tons / year, saving water 70 billion tons / year.

5.2. Conventional pollutant emission reduction

Energy spatial transfer, changing the supply structure of resources in the two regions, the development and utilization of renewable energy alternative to a large number of fossil fuels, coal-fired power generation is no longer leading to improve the proportion of direct consumption of coal, direct impact on conventional pollutant emissions. Of the major pollutants in the atmosphere, about 80% of sulfur dioxide, 60% of nitrogen oxides, 50% of the fine particulate matter comes from coal combustion, while nearly half of the coal emissions from direct-fired coal, this unreasonable energy consumption structure and consumption patterns, serious damage to air quality. By optimizing the allocation of the global energy supply structure, can effectively reduce the conventional pollutant emissions. To the industrial SO_2, NO_x, the largest contributor to the total thermal power industry, accounts for 45%, 65%, 23%.

According to the development progress of the energy base, it is estimated that the electric power output scale of "one pole and one pole" of 2030, 2040 and 2050 will be 0.9 trillion, 4.2 trillion and 12 trillion kilowatt. If calculated by the same amount of coal equivalent power generation, can reduce the consumption of 300 million standard coal, 1.3 billion, 3.8 billion tons, equivalent to reduce carbon dioxide emissions by about 800 million, 3.7 billion, 10.5 billion tons / year. Reduce annual emissions of sulfur dioxide 1.8 million, 7.9 million, 22.3 million tons, reduce nitrogen oxide emissions 1.9 million, 8.3 million, 23.4 million tons, reduce soot emissions of about 300,000, 1.4 million, 3.9 million tons / year, save water use of about 20 billion, 100 billion, 29 billion cubic meter.

5.3. Affection to the climate change

Energy spatial transfer change the global power industry and energy industry development pattern, the development of clean energy development can reduce greenhouse gas emissions. The power industry in the energy conversion process emissions of carbon dioxide emissions, accounted for about 50% of the total global emissions of carbon dioxide, which is one of the key industries. As the global total installed capacity and the rapid growth of power generation, power sector energy consumption growth is the overall energy consumption and emissions growth of the main driving factors. However, with the rapid development of various clean energy technologies and the rapid maturation of new technologies, the future development of electricity will greatly affect the emission of greenhouse gases.

According to forecasts, relying on the global energy Interconnection, by 2050, clean energy can replace the equivalent of 24 billion tons of standard coal fossil energy, abatement of carbon dioxide 67 billion tons, sulfur dioxide 580 million tons. It is assumed that the global energy carbon emissions 11.5 billion tons, only 33% in 2013 and 2009, about 50%. According to the IPCC report, To achieve the "United Nations Framework Convention on Climate Change," proposed in 2050 will be the global average temperature increase within 2 ℃ in control "target, a fundamental solution to glacier melting, sustainable development.

The energy spatial transfer can realize the temperature control. Taking into account the "Paris agreement" clearly put the global temperature control within 2 ℃ target, while temperature control within 1.5 ℃ at effort ; as soon as possible to achieve greenhouse gas emissions peak, the second half of this century to achieve zero net greenhouse gas emissions. From 2011 to 2011, the case of clean energy spatial transfer development scenarios, the cumulative global emissions of greenhouse gas emissions of about 1.4 trillion tons; to reach the peak carbon emissions by 2020, about 55 billion tons of carbon dioxide equivalents; in 2060, energy and related industries will achieve net zero emissions.

6. Conclusion

This research gives the energy spatial transfer model of energy development layout to analyse the global energy development layout condition and ecological affection. And it is a fast and direct method to analyse its energy usage process and environmental affection. It gives out a system dynamics model of energy spatial transfer showed, which electric power transmission is better than original energy usage and transportation. The energy spatial transfer can affect the environment directly. Energy saving, climate changing and conventional pollutant emission reduction, synthetic combine with the spatial transfer model, it can get the environmental change parameters, which showed that with the clean energy wide usage, the ecological footprint affection will be affected significantly.

Reference

[1] Zhi J, Gao J X 2009 *Green Economy* **8** 80-83.
[2] Yan Z X, Liu J Y, Wei Z B, Wu Y, Xu L X and Zhang L 2016 *Power System Technology* **40** 1621-26.
[3] Yang H, Wang D and Liu D 2012 *Power System Protection and Control* **43** 1621-1626.
[4] Han Q 2013 *North China Electric Power University* 162-166.
[5] Xu J 2013 *North china electric power university* 18-24.
[6] Sun B 2015 *Zhejiang University* 28-35.
[7] Liu J P 2013 *North China Electric Power University* 62-76.
[8] IEA Energy balances of OECD countries 2015 112–162.
[9] IEA World Energy Outlook 2015 65–78.
[10] GWEC Global Wind Report 2015 89-120.

Measures to restore metallurgical mine wasteland using ecological restoration technologies: A case study at Longnan Rare Earth Mine

Yunzhang Rao [1, 2, 3], Ruizhi Gu [1], Ruikai Guo[1] and Xueyan Zhang[1]

[1]School of Resources and Environmental Engineering, Jiangxi University of Science and Technology, Gan zhou, Jiangxi 341000, China
[2]Jiangxi Key Laboratory of Mining Engineering, Jiangxi University of Science and Technology, Ganzhou, Jiangxi 341000, China

Email: raoyunzhang@sohu.com

Abstract: Whereas mining activities produce the raw materials that are crucial to economic growth, such activities leave extensive scarring on the land, contributing to the waste of valuable land resources and upsetting the ecological environment. The aim of this study is therefore to investigate various ecological technologies to restore metallurgical mine wastelands. These technologies include measures such as soil amelioration, vegetation restoration, different vegetation planting patterns, and engineering technologies. The Longnan Rare Earth Mine in the Jiangxi Province of China is used as the case study. The ecological restoration process provides a favourable reference for the restoration of a metallurgical mine wasteland.

1. Introduction

The abandoned mine land in China totaled 3 x 106hm^2 at the end of twentieth century, with no recovery measures being in place for 80% of this wasteland [1-2].Furthermore, the scale of the effects of abandoned mines expands rapidly as the demand for resources increases. Moreover, the damage to the environment is exacerbated by associated activities such as road building, strip mining, and constructing of waste-rock yards and tailings ponds, as these occupy large tracts of land and lead to the destruction of vegetation. Toxic heavy metals such as Pb, Zn, As, Cr, Cu, and others that remain in the waste-rock yards and tailings ponds of sulfide mines can potentially pollute the surface water and groundwater through surface runoff and leakage. The destruction of normal vegetation detrimentally affects soil fertility, leading to mud-rock flows, soil erosion, and landslides down mountains. In addition, abandoned mining areas pose other geologic hazards, for instance, exposed mining pits that are dangerous to humans and animals. In view of the extensive damage to the environment in the mining areas of China, this study discusses several ecological restoration methods. Furthermore, by employing the area surrounding the Longnan Rare Earth Mine as case study, the general process of restoration is explored, with a focus on the restoration of the ecosystem. Finally, concerns for the future are discussed.

2. Repair Methods

2.1. Principle of adjusting restoration measures to suit local conditions
Mining activities can have disastrous effects on the surrounding geochemical environment and ecosystem, such as soil disturbance, surface collapse, acid mine drainage, heavy metal pollution of the soil, atmospheric pollution, upsetting of the ecological environmental balance, and others. The source of such problems has to be determined and special technologies have to be employed to solve them. In addition, the local climate and soil conditions have to be considered, when selecting appropriate plants for vegetation remediation.

2.2. Methods of synchronization
Ideally, the ecological restoration plan should be applied throughout the entire period of the mining activity, rather than simply focusing on the recovery of the land after the end of the exploitation. According to Dai [3], the different development stages of mining areas can have crucial effects on the ecological remediation of such sites. Therefore, the reclamation of the site has to be considered at the start of the exploitation; for instance, in the construction and management of tailings ponds and the slopes of waste-rock yards and steps. After closing of the mining activities, the retention, or the dismantlement of the remaining structures mainly depends on an assessment of their value. In addition, the reclamation of each mining tunnel should be done in consideration of its characteristic in order that ecological restoration of each site can be effected.

2.3. Combination with landscape study
Another method, based on the original landform, is available for appropriately planning the reclamation efforts for mine environments. This method combines the knowledge of the geography of the sites, landscape study, ecology, and the aesthetics of the particular sites in the remediation efforts. Since the 1990s, landscape design practices have been utilized often to transform the mine environments[4]. A case in point is Nanhu Park, the former site of the Kailuan Coal mine, which is part of the heavily industrialized Tangshan City, a coal mining in Hebei Province. The environment was restored by utilizing the principles of bionics in the design of wetlands to purify the area, and, subsequently, wetland ecosystems have formed that contain aquatic and terrestrial plants[5].

3. Remediation Measures

3.1. Restoration and improvement of the soil matrix

3.1.1. Soil replacement method. The crucial problem in restoring the soil matrix is the lack of soil; consequently, searching for replacement soil is of vital importance in remediating abandoned mine lands. However, a new layer of soil can be formed artificially to solve the problem. As regards topsoil contaminated with heavy metals, a layer of impervious paving is essential to separate the heavy metals from the soil layer. So for mine wastelands, the main problems in applying this measure are finding a suitable soil source and confirming the thickness of the soil layer[6].

3.1.2. Improvement of soil matrix nutrients. This measure includes two main actions, namely, decreasing the toxicity, and enhancing the nutrients of the soil. As an example, toxic metals from sulfide mining can affect the groundwater and soil adversely by surface runoff and infiltration, causing environmental pollution, such as acid mine drainage and the like. In such an instance, the toxicity should first be reduced by utilizing the chemical reaction technique, i.e., mainly adding sulfate, sulfur, and other acidic modifiers to alkaline mining areas, and lime, fly ash, and the like to acidic mining areas. This technique is often employed to modify the pH level of the soil in order to reduce the toxicity of the metals. Generally, low pH conditions can potentially increase the solubility of heavy metals, which increases toxicity, whereas high pH conditions can potentially reduce solubility and reduce toxicity. This is because heavy metals precipitate as metal hydroxide or exist in carbonate and phosphate forms, which, accordingly, reduces toxicity[7]. Basically, pH can control the bioavailability of Zn and Cd[8]. The biological method is utilized mainly to enable vegetation to adsorb metals from the soil, which facilitates soil remediation while being free from secondary pollution, but leaves a long life cycle of plants.

On the other hand, the selection of appropriate fertilizer is crucial for soil that lacks nutrients in order to encourage vegetation cover. Moreover, the role of soil organisms, e.g., earthworms and microbial fungi in maintaining soil fertility should not be overlooked. Various studies have tested the inoculation efficiency of the arbuscular mycorrhizal fungus (AMF) on plants growing in mine areas. The results indicate thatthe fungi could promote the growth of the root system, enhance the absorption ability of the roots, and increase the soil organic content. Furthermore, the inoculated fungi accelerate the degree of aging of the matrix, which enhances the nitrogen fixation ability and the absorption of the available potassium[9]. Furthermore, the action of earthworms increases the speed of soil turn over, accelerates the ripening process of humus, and encourages the aeration and hydration of the soil.One study has pointed out that the waste of organisms, such as the compost sludge of resources-based cities, could be used for soil modification, as the nutrients contained in the sludge, such as nitrogen, phosphorus, and potassium, could supplement the soil nutrients[10]. Although this technique has various advantages, the sewage sludge could contain pathogenic microorganisms, parasitic ova, and trace metal elements, which are harmful to plant growth. Consequently, using sewage sludge has to be considered carefully[11].

3.2. *Phytoremediation technology*

Phytoremediation has recently emerged as a restoration technology because of its slight secondary pollution and low cost[12]. The purpose of using phytoremediation is to decrease the toxic metals content and the degree of toxicity through adsorption, transformation, and the deposition of heavy metals in the soil by using hyperaccumulator vegetation. This technique includes three main mechanisms, namely, phytoextraction, phytostabilization, and phytovolatilization[7]. It is interesting to note that there is no definite criterion for every hyperaccumulator plants; different plants have different criteria. For instance, as regards Zn, it could be appropriate to regard Zn levels exceeding 3000 μg/g as remarkable,and the particular plant deserving of being described as a hyperaccumulator of Zn[13].

It has been reported that Pb metal hyperaccumulator plants, such as *Sedum Tatarinowii*, *Ailanthus Desf.*, rosebush, and *Gold Vein Forsythia* could accumulate levels of Pb metals as high as 11.87–164.502 mg/kg in their aboveground biomass [14].Plants such as *Sedum spectabile Brilliant* have been proven to remediate mine areas highly contaminated (up to 3 000 mg/kg)by Zn^{2+}. As *Sedum spectabile Brilliant* has high tolerance for $ZnCl_2$, as well as phytostability, it should be the primary choice as hyperaccumulator plant[15].

3.3. *Compound planting patterns*

A technique that combines various planting patterns could increase the nutrients in soils, improve the soil physicochemical character, and accelerate the speed of remediation. In addition, employing this technique could establish a stable ecological structure. Chen[16] conducted a comparison between six planting patterns and unplanted (bare) land, finding that after planting the soil moisture capacity and the soil organic matter content had increased, the soil structure had become more stable, and the soil physical and chemical properties were superior in comparison with those of the bare land. Peng et al.[17] have investigated the effects of different vegetation restoration patterns on the physicochemical properties of and the heavy metals in soil. For this purpose, they employed six different vegetation patterns in the abandoned lands containing tailings of the Fujian Gold Mine, comparing the land under a natural vegetation community and the bare tailings without vegetation restoration. Compared with the soil properties and soil bulk density of the bare tailings, the results showed the soil porosity and water retention capacity significantly increased, and the soil organic matter increased 12.23–27.74 times.

3.4. Engineering technologies

The loose soils on the slopes of mine dumps could cause geologic hazards such as landslides, mud-rock flows, and collapse; therefore, it is essential to reduce the slope gradient.

Steep slopes have little ability to retain soil and water. Combining an adequate drainage system with a technique to convert the slope into ladder-type topography,or to lower the gradient of the slope is crucial to enhancing such ability. The technique for blocking stonewall in mine slopes is building a stonewall in the root of the mine areas. The principle is to rely on the self-weight capacity of the wall to resist the lateral thrust of the soil mass in order to maintain the stability of the soil and to prevent the slope from falling down. At the same time, drainage ditches should be arranged sensibly to ensure that the water flows away easily in order to decrease the possibility of landslides and debris flow disasters and to reduce the surface water erosion to the soil of the slope.

Subsequently, ecological repair measures can be employed, such as integrating engineering technologies with the planting of vegetation to stabilize the slope and retain soil and water. Additional methods include:

- The vegetation carpet restoration technique, which is a carpet of mixed straw and wheat straw, a water retention agent, nutrient soil layer, adhesive, and water together to made it be a whole. Then put it on the mines slope. This technique is suitable for stable mines with a slope percentage lower than 1:1.5[18].
- The ecological vegetation bags method. The vegetation bags are made of special degradable fiber material and have nutrient soils containing proper seeds. The bags are placed on the slope, and are watered regularly and maintained. Over time, the bags degrade completely and the vegetation and soil layer could cover the entire area[18].
- The hydraulic seeding method (hydroseeding), in which anchor stock is used to fix wire netting on the slope, after which a high-voltage machine is used to attach the mixture of soil, grass seeds, adhesive, and a water retention agent to the rock face of the slope[4]. It is important to employ the principle of adjusting the measures to suit the local climatic conditions in the different mining areas. The most important aspect is the selection of plants, as choosing local plant varieties increases the possibility of the plants growing successfully.

4. Ecological restoration of the Longnan Rare Earth Mine

The Longnan Rare Earth Mine is located in Ganzhou city, Jiangxi Province, and has been mined for the past 40 years. Since 1994–1995[19], in conformance with government requirements, the mine started using the in-situ leaching method. However, this has caused various environmental problems, such as vegetation destruction (Figure 1) and soil erosion (Figure 2). In addition, pollution of the water and soil has occurred, including tailings effluent containing Pb up to 14 mg/L and Cd up to 0.024 mg/L. This pollution has affected the drinking water of more than 30 000 people, with more than 400 acres (161.874 ha) of farmland being destroyed[20].

In consideration of these conditions, the Longnan Rare Earth Mine has implemented various remediation schemes.

4.1. Environmental and geological exploration of the Longnan Mine

Remote Sensing (RS) technology was employed to survey the situation of the mine rare earth resources in the vegetation pollution and destruction of land [21].Subsequently, with the help of the Ganzhou government, access to relevant information was gained and accurate topographic maps were drafted. The physical and chemical properties of water and soil samples were determined to establish the mechanical stability of the rock and the weathering crust in the mining area, providing the basic data required for the subsequent activities[21].

Figure 1. Vegetation destruction

Figure 2. Soil destructio

4.2. Cutting slope and smoothing the mining area

In the rare earth mining area, in accordance with the principle of local conditions and, therefore, based on the terrain and landforms, measures should be employed to convert the abandoned mine land into small plain or artificial terraces. At the same time, for flat mining areas, the first step of restoration is sourcing soil, subsequently adding organic fertilizer, and sprinkling grass seeds directly in the matrix. In the instance of a larger slope, other suitable engineering measures should be employed, such as laying down an ecological vegetation blanket or ecological vegetation bags, and the like. Subsequently, suitable ditches and drainage systems can be constructed to ensure that each part of the regional drainage system is able to remove the water without difficulty.

4.3. Selection of vegetation species

The soil of rare earth mining waste mainly comprises loose rocks and hardened tailings, which have poor water conservation ability, and contain low levels of organic matter, phosphorus, and potassium, but high levels of ammonium ions. Therefore, it is preferable to choose tall plants for vegetation restoration that have a developed root system and rapid growth ability[22].In the restoration of the Longnan Rare Earth Mine, in accordance with the soil and climatic conditions, suitable biological measures and various planting patterns were employed. Elephant grass, economic forestry, and sericulture were established in three planting bases at the Zudong mining area. In this area, the planting layout is pine trees at the top of the mountain, grass on the slopes, mulberry trees on the terrace, and bamboo in the valley[20]. Plants such as *paspalum notatum* and *green bristlegrass* are used in the tailings area to reduce surface runoff erosion. Cash crops, such as pine and fir trees, and bamboo, and fruit trees, such as pear and navel orange (Figure 3) complete the restoration project, with considerable economic benefits (Figure 4).The so-called "Longnan model" successfully established a large area of green coverage, improved the soil, and facilitated water conservation. Moreover, it established the unity of the economy, society, and the environment.

Figure 3. Restoration at the early stage

Figure 4. Restoration at a later stage

4.4. *Strengthening policy management*

A sewage disposal plant for mine wastewater has been set up in the downstream of the river that runs through the Longnan Mine area and the water quality has been improved to discharge standard. At the same time, the government has established a rare earth ceramic factory to make full use of the mine tailings. In addition, the government of Longnan County has implemented various legal measures, such as requiring the mine to take environmental responsibility for the mine and paying environmental governance fees[20].

Although the "Longnan model" has achieved a certain degree of success, and has increased the vegetation reclamation rate, achieving complete success is difficult because of the poor natural conditions. Nonetheless, this model is worth employing[19].

5. Conclusion

Based on the knowledge acquired from the Longnan project,more research is needed on the following aspects:

- The selection of plants and the application of compound planting patterns.
- Enhancing the exploration of the survival environment and the limiting factors of hyperaccumulator plants, and increasing the research on new materials, especially a new soil matrix. For example, adding biological bacteria that are suitable to the soil environment and are able to live in mine areas in an effort to increase the fertility of the soil.
- Finding new technologies suitable to the mine environment, for example, the application of nanotechnology and transgenic technology in mine areas could be worth the effort and cost of extensive research. The nanotechnology technique is often applied for metal remediation in wastewater treatment because of its special adsorption and mechanical properties and high chemical stability. However, there has been little progress in the study on reducing metal contamination in the soil and research on improving soil fertility [23]. However, Groza[24] has used ionic phosphate nanoparticles (vivianite) to immobilize Pb^{2+} in soils by in-situ remediation. The results of the study indicated that the nanoparticles could reduce the leachability and bioaccessibility of Pb^{2+} of soils effectively. In conclusion, more attention should be paid to the application of nanotechnology in mine areas. At the same time, transgenic technology can be employed in plant cultivation to improve the absorption rate of plants and the extraction rate of heavy metals of the plants.
- Enhancing government supervision and management. Mines should accept responsibility for the "who pollutes, who controls" mining model, whereas the government should introduce appropriate policies to regulate mining models and mining methods, including post mining management.

Acknowledgements

Fund project: The National High Technology Research and Development Program of China (863 Program). (Number: 2012AA061901)

References

[1] Shu W S, Zhang Z Q and Lan C Y 2000 The countermeasure of reclaiming of Chinese Mining Industry abandoned land *Ecologic Science* **19** 24-29 (in Chinese)
[2] Lin G C S and Ho S P S 2003 China's resources and land-use change: insights from the 1996 land survey *Land Use Policy* **20** 87-107
[3] Dai H W 2010 The techniques of ecological remediation and rehabilitation for derelict mine la-nd *Chinese Mining magazine* **19** 58-61 (in Chinese)
[4] Liu G 2008 Ecological rapid recovery technology of abandoned mine *SHANGHAI LAND&RESOURCE* 16-19 (in Chinese)
[5] Xu J R, Wu J M, Dong R Y and Xu X S 2016 A Primary Research of Mining Wasteland Landscape Regeneration in the View of Bionic Design---A Case Study of Huolinhe

Opencast Coal Mine *Jour of Fujian Forestry Sci and Tech* **43** 182-188 (in Chinese)

[6] Wu W Y 2004 Ecological Restoration Research Progress of Mining Area *Guangdong Chemical Industry* **41** 82-83 (in Chinese)

[7] Guo S C and Yang W Q 2015 Research Progress of Phytoremediation Technology to Remediate Heavy Metal Contaminated Soil *Journal of Northweat Forestry University* **30** 81-87 (in Chinese)

[8] Farrag K, Senesi N, Rovira P S and Brunetti G 2012 Effects of selected soil properties on phytoremediation applicability for heavy-metal-contaminated soils in the Apulia region Southern Italy *Environmental Monitoring & Assessment* **184** 6593-606

[9] Qian K M, Wang L P, Li J, Zhang J and Jia L F 2010 Inoculation Effect of AMF in Reclamation of Abandoned Mining Area *Environmental Science and Technology* **23** 5-9 (in Chinese)

[10] Zhong T, Shen G, Wang W and Yang H J 2015 Prospect analysis of resource based on urban compost sludge for ecological restoration of abandoned mine in mine *Resources Economization & Environmental Protection* 147-48 (in Chinese)

[11] Zhao M H 2008 The Research on Amelioration of Soil in the Mining Wasteland *Chinese Agricultural Science Bulletin* **24** 128-31 (in Chinese)

[12] Guo B, Li W D, Ding N F, Fu Q L, Liu C and Li N Y 2013 Problems and Countermeasures of phytoremediation of heavy metal contaminated soil *Acta Agriculturae Zhejiangensis* **25** 852-57 (in Chinese)

[13] Ent A V D, Baker A J M, Reeves R D, Pollard A J and Schat H 2013 Hyperaccumulators of metal and metalloid trace elements:facts and fiction *Plant and Soil* **362** 319-34

[14] Gao H Z, Guo W Z and Bi J 2014 Study on the Accumulation and Tolerance of 20 Plants to Heavy Metal Pb in soil *Chinese Agricultural Science Bulletin* **30** 19-24 (in Chinese)

[15] Wang C, Bi J, Song X L and You H Z 2014 Study on the Accumulation and Tolerance of 6 Plants to Heavy Metal ZN in soil *Environmental Science & Technology* **37** 62-65 (in Chinese)

[16] Chen J H 2016 Effect of Mine Wastelands Vegetation on Soil Properties *METAL MINE* **45** 147-50 (in Chinese)

[17] Peng D H, Hou X L,He Z M，Liu L Q, Cai L P, Lin J W, Jiang R R, Huang F C and Zhong J H 2015 Effects of Different Vegetation Restoration Patterns on Soil Physical and Chemical Properties in Wasteland of Gold Tailings *Journal of Soil and Water Conservation* **29** 137-42 (in Chinese)

[18] Zhao F T and Sun B P 2009 Technique of mine ecological vegetation restoration (Beijing: China Forestry Publishing House) pp 80-93 (in Chinese)

[19] Liu F 2013 Eco-environment Status and its Comprehensive Countermeasures of Longnan Ionic Rare-earth Mine *METAL MINE* **42** 135-38 (in Chinese)

[20] Chen J G and Li Z M 2010 Rare earth mine environmental governance and land reclamation—Gannan "Longnan model" as an example Proceedings of the annual conference of the 2010 China Environmental Science Society (Shanghai,China) vol **4** (in Chinese)

[21] Lei G J, Liu S F and Cheng S Y 2006 Application of RS in the extraction of vegetation on pollution information in rare earth mining area *Nonferrous Metals Science and Engineering* **20** 1-5 (in Chinese)

[22] Rao Y Z, Deng Y W, Li G Q and Hu W etc 2015 Study on comprehensive treatment technology of waste rare earth mine environment in Ganzhou City (Changsha: central South University Press) pp **46**,29-30 (in Chinese)

[23] Singh A and Prasad S M 2015 Remediation of heavy metal contaminated ecosystem: an overview on technology advancement *International Journal of Environmental Science&Technology* **12** 353-66

[24] Groza N 2008 Ecological Solutions of Contaminated Environment Remediation from Uranium Mining Activities in Romania// Uranium, Mining and Hydrogeology (Springer Berlin Heidelberg) pp 333-334.

Synthesis and photocatalytic properties of Co- and Cu-doped $Bi_2Sn_2O_7$

Jing Zhuang[1, 2], Chaohao Hu[1, 2, 3], Binqing Zhu[1, 2], Yan Zhong[1, 2], and Huaiying Zhou[1, 2]

[1] Guangxi Key Laboratory of Information Materials, Guilin University of Electronic Technology, Guilin 541004, P.R. China
[2] School of Materials Science and Engineering, Guilin University of Electronic Technology, Guilin 541004, P.R. China

E-mail: chaohao.hu@guet.edu.cn

Abstract. $Bi_2Sn_2O_7$ photocatalysts doped by Co and Cu ions were successfully synthesized by using the hydrothermal process and impregnation method. The products were characterized using X-ray diffraction analysis (XRD), Field emission scanning electron microscopy (FESEM), Energy dispersive X-ray detector (EDS), infrared spectroscopy (IR), the photoluminescence (PL) spectra and UV-visible diffuse reflectance spectroscopy (DRS). The photocatalytic properties were further evaluated by degrading rhodamine B (RhB) as a model pollution under visible-light irradiation. The results indicated that Co- and Cu-doped $Bi_2Sn_2O_7$ photocatalysts have a cubic pyrochlore phase with the hybrid metals. The metal-loaded photocatalysts show the enhanced photocatalytic efficiency for degradation of RhB under visible-light (λ> 420 nm). The mechanism of improved photocatalytic activity is also discussed in detail.

1. Introduction

Environmental pollution has become a serious problem in the past few decades with the rapid development of industrial technology and economy. Among a great number of the advanced oxidation processes, semiconductor photocatalytic oxidation is regarded as a green technology to decompose the soluble dyes in wastewater and eliminate toxic chemicals in the environment [1-6]. So far TiO_2 is the most studied semiconductor photocatalyst due to the outstanding oxidative power, low cost, non-toxicity and chemical stability. However, TiO_2 is only active under ultraviolet (UV) light irradiation because of its large band gap (about 3.2 eV for anatase). It is important to develop new photocatalysts responsive to visible-light irradiation with the enhanced photocatalytic activity in order to make full use of solar energy [7-9].

Currently, $Bi_2Sn_2O_7$ has received considerable attention because of its wide application in catalysis and gas sensors in recent years [10,11]. $Bi_2Sn_2O_7$ belongs to the pyrochlore family with the general formula $A_2B_2O_7$, where A-cations are eight coordinated and B-cations are six coordinated [12]. As shown in Figure 1, the crystal structure of $Bi_2Sn_2O_7$ is constructed of SnO_6 octahedra and BiO_8 dodecahedra in which SnO_6 and BiO_8 connect to each other by sharing vertexes and edges, respectively. Recently, Tian et al. [13] have investigated the photooxidation of As(III) under visible-light irradiation of $Bi_2Sn_2O_7$. Wu et al. [14] have synthesized nano-$Bi_2Sn_2O_7$ by a hydrothermal method and studied its photocatalytic activity in the degradation of methyl orange. However, the photocatalytic activity of individual $Bi_2Sn_2O_7$ is relatively low. The photocatalytic performance of $Bi_2Sn_2O_7$ should be further improved for practical use. Recently, Liu et al. [15] have synthesized $Bi_2Sn_2O_7$/reduced graphene oxide nanocomposite for efficient photocatalytic degradation of organic contaminants. Mohamed et al. [16] have prepared Pd-doped-Bi_2O_3/$Bi_2Sn_2O_7$ hybrid nanocomposites for photocatalytic fluorene oxidation.

In this study, metal-doped catalysts Cu-$Bi_2Sn_2O_7$ and Co-$Bi_2Sn_2O_7$ were successfully synthesized by using the hydrothermal process and wet-impregnation method. The crystal structure, morphologies, particle sizes and optical properties were investigated by various techniques. The photocatalytic activities of the composite samples were evaluated by the photocatalytic discoloration of rhodamine B (RhB) under visible light irradiation.

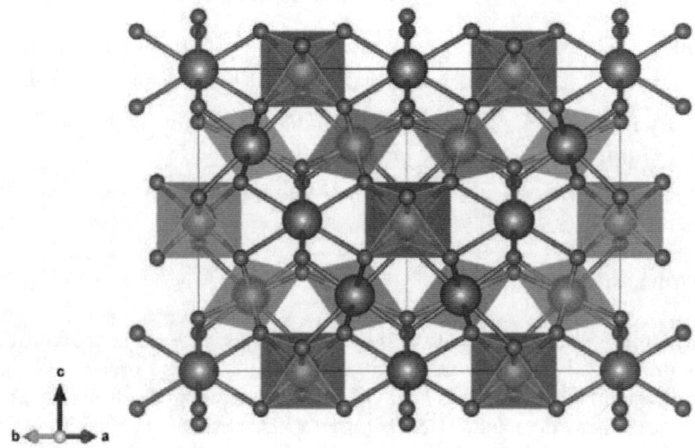

Figure 1. Crystal structure of $Bi_2Sn_2O_7$. The purple, gray, and red spheres are for Bi, Sn, and O atoms, respectively.

2. Experimental procedures

2.1. Synthesis

All chemicals were reagent grade and used without further purification. The cubic pyrochlore $Bi_2Sn_2O_7$ powders were synthesized in a typical hydrothermal processing. First, 0.01 mol $Bi(NO_3)_3 \cdot 5H_2O$ and 0.01 mol $SnCl_4 \cdot 5H_2O$ were dissolved in 30 ml deionized water and stirred for 20 min at room temperature. Then the pH value of the mixture was adjusted to 12 by using 4 mol/L NaOH. The mixed solution was stirred continuously at room temperature for 30 min, then transferred into Teflon-lined autoclave (100 mL capacity, 80% full) and kept at 200 ℃ for 16 h in an oven. After the reaction, The as-prepared products were collected after washing with deionized water and anhydrous ethanol several times, respectively. At last, the products were dried at 60℃ for 12 h in an oven.

M-$Bi_2Sn_2O_7$ (M = Cu, Co) catalysts were prepared by the impregnation method from the aqueous solutions of $Cu(NO_3)_2$ and $Co(NO_3)_2$. The mass ratio of $Cu(NO_3)_2$ and $Co(NO_3)_2$ to $Bi_2Sn_2O_7$ is 1:20. The as-prepared $Bi_2Sn_2O_7$ powder was added to anhydrous ethanol (10 mL) containing an certain amount of $Cu(NO_3)_2$ and $Co(NO_3)_2$ in a ceramic dish. The suspension was stirred using a glass rod during the irradiation of infrared light until the evaporation of anhydrous ethanol. Finally, the dried powders were calcined in alumina crucible at 600℃ for 4 h.

2.2. Characterization
Crystal structures of $Bi_2Sn_2O_7$ and M-$Bi_2Sn_2O_7$ catalysts were characterized by powder X-ray diffraction (XRD) (Bruker D8-2-Advanced XRD, Germany) using Cu Kα radiation (λ = 0.1541 nm) from 10° to 80°, with a scan speed of 5° min^{-1}. The morphologies and microstructures of the as-prepared catalysts were observed using the field emission scanning electron microscopy (FESEM, Quanta 450 FEG, USA). The chemical composition of the samples was examined using the energy dispersive X-ray detector (EDS) (Thermo Noran VANTAG-ESI, USA).The photoluminescence (PL) spectra were measured using a fluorescence spectrophotometer at room temperature (Varian, USA) with an excitation wavelength of 380 nm. Fourier transform infrared (FT-IR) spectra were recorded on FTIR spectrometer (Thermo Fisher Nicolet 6700) using the standard KBr disk method. The UV-vis absorption spectra of the photocatalysts were obtained by a UV-vis spectrometer (Puxi TU-1901, China) in the range of 200-700 nm, using $BaSO_4$ as the reference sample.

2.3. Photocatalytic activity test
The photocatalytic efficiency of all of the prepared samples was evaluated by degrading RhB under visible-light irradiation at room temperature. The 0.10 g powder of photocatalyst was immersed in a quartz beaker containing 100 mL RhB solution with a concentration of 5 mg/L. In order to ensure the adsorption/desorption equilibrium between RhB and photocatalyst, the suspension was stirred in the dark for 30 min. Xeon arc lamp with 300 W through UV cut-off filters placed at about 10 cm from the reactor completely removes any radiation below 420 nm and offer visible light source. After the photocatalytic reactions, 5 mL of suspensions was taken out at given time intervals. At last, a centrifuge was used to remove the powder and the supernatant was taken for analysis. The absorbance of centrifugal solution was determined at 553 nm by UV-vis spectrophotometer.

3. Results and Discussion

3.1. Structural and morphological characterization
The measured XRD shown in Figure 2 indicates that the diffraction peaks of all of the samples were indexed to the standard card of pure cubic phase of $Bi_2Sn_2O_7$ (JCPDS No. 87-0284). The characteristic peaks of the three samples at 2θ = 28.8°, 33.3°, 47.9°, 56.9° are obvious and no other impurity peaks can be observed, indicating high purity of the final products. The doped Co and Cu do not change the crystal structure of $Bi_2Sn_2O_7$, implying that Cu and Co are loaded onto the surface of the $Bi_2Sn_2O_7$. According to Debye-Scherrer equation [17] with the shape factor $K = 0.89$, the calculated crystallite size of three samples is about 21.1 nm based on the main diffraction peak (222). In addition, the diffraction peaks of Co and Cu cannot be observed probably due to the low amount of the hybrid metals.

The IR spectra of the pure $Bi_2Sn_2O_7$ and M-$Bi_2Sn_2O_7$ samples are shown in Figure 3. The bands centered at about 3420 and 1640 cm^{-1} are assigned to O-H stretching and bending modes of water, respectively [18]. The band at about 621 cm^{-1} can be assigned to the B-O stretching vibration in the BO_6 octahedron, which is comparable with the A-O stretching vibration of pyrochlore $A_2B_2O_7$ centered at about 517 cm^{-1} reported by Subramanian and co-authors [19]. The weak bands at about 422 cm^{-1} are ascribed to the Bi-O-Bi bending vibrations, which is similar to the previous experimental results presented by Xu and co-authors [20]. Compared with the pure $Bi_2Sn_2O_7$ sample, the peak at about 621 cm^{-1} for M-$Bi_2Sn_2O_7$ is sharper and has a little blue shift.

Figure 4 shows the SEM images of pure and metal doped $Bi_2Sn_2O_7$. Our results indicate that the size of particles mainly ranges from 2 to 4 μm. It can be seen that the micro morphology of pure $Bi_2Sn_2O_7$ catalysts are composed of layered pompons and particles. The pompons in M-$Bi_2Sn_2O_7$ are packed closer to each other, which may be caused by a heterojunction between the metal particles and $Bi_2Sn_2O_7$ after loading the metals. The morphology and microstructure of Cu-$Bi_2Sn_2O_7$ is rose-shaped and that in Co-$Bi_2Sn_2O_7$ is consisted of more regular layered pompon. The EDS analysis was carried out in order to determine the elemental composition of M-$Bi_2Sn_2O_7$. Bi, Sn, O, Co and Cu elements can be found from the EDS energy spectra as presented in Figure 5(a-b).

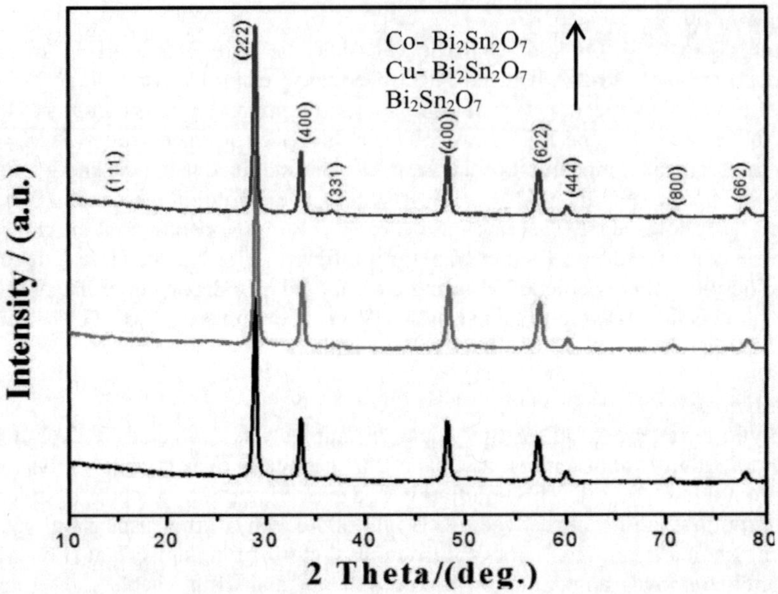

Figure 2. XRD patterns of pure $Bi_2Sn_2O_7$ and M-$Bi_2Sn_2O_7$

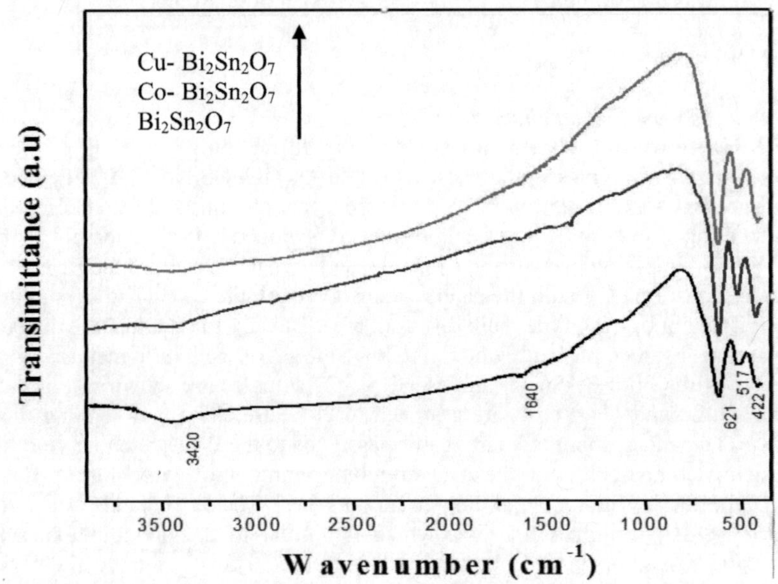

Figure 3. IR spectra of the as-prepared pure $Bi_2Sn_2O_7$ and M-$Bi_2Sn_2O_7$ powders

Figure 4. SEM images of pure $Bi_2Sn_2O_7$ and M-$Bi_2Sn_2O_7$: (a) $Bi_2Sn_2O_7$, (b-c) Cu-$Bi_2Sn_2O_7$, (d) Co-$Bi_2Sn_2O_7$

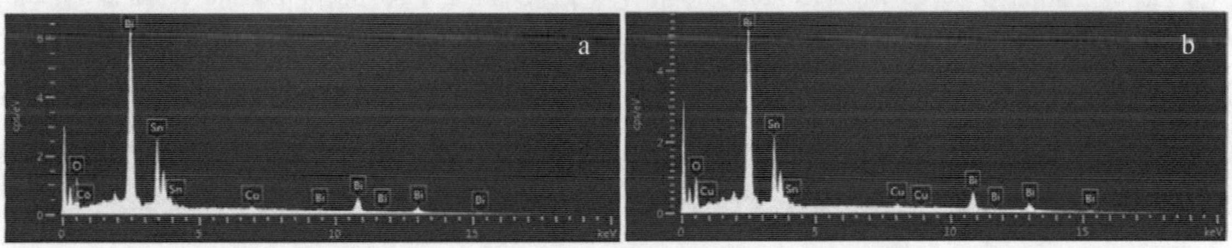

Figure 5. EDS analysis of M-$Bi_2Sn_2O_7$ powders: (a) Co-$Bi_2Sn_2O_7$, (b) Cu-$Bi_2Sn_2O_7$

3.2. UV-visible absorption spectra

The optical absorption properties of semiconductor materials are essential to determine their photocatalytic activities. Diffuse reflection spectra of the pure $Bi_2Sn_2O_7$ and M-$Bi_2Sn_2O_7$ catalysts are shown in Figure 6(a). It can be found that all of the samples show very good absorption in the visible region. An obvious red shift in spectra can be observed for the Cu- and Co-doped $Bi_2Sn_2O_7$ compared to that of pure $Bi_2Sn_2O_7$. The considerable absorption band in the visible light region would make full use of sunlight for the degradation of the pollutants. For a crystalline semiconductor, the band gap energy (E_g, in eV) of catalyst can be calculated according to the following equation [21]:

$$(\alpha h\nu)^n = A\,(h\nu - E_g)$$

Where α, $h\nu$, E_g and A are an absorption coefficient, the absorption energy, band gap, and a constant, respectively. Here n depends on the characteristic of the transition in a crystalline semiconductor [22].

For $Bi_2Sn_2O_7$, the value of n is 1/2 for the indirect transition [23]. The E_g of $Bi_2Sn_2O_7$ can be estimated from a plot of $(\alpha hv)^2$ versus the photon energy hv. The intercept of the tangent to the x axis gives rise to a good approximation of the E_g value for as-prepared $Bi_2Sn_2O_7$ photocatalysts. As depicted in Figure 6(b), the band gap of pure $Bi_2Sn_2O_7$ is about 2.75 eV, which is similar to the previously reported values [24]. The band gap of Cu-$Bi_2Sn_2O_7$ is 2.72 eV and that of Co-doped $Bi_2Sn_2O_7$ is about 2.68 eV. The measured results have suggested the photocatalytic activity of metal loaded catalysts would be better than that of pure $Bi_2Sn_2O_7$ catalyst with response to visible light irradiation.

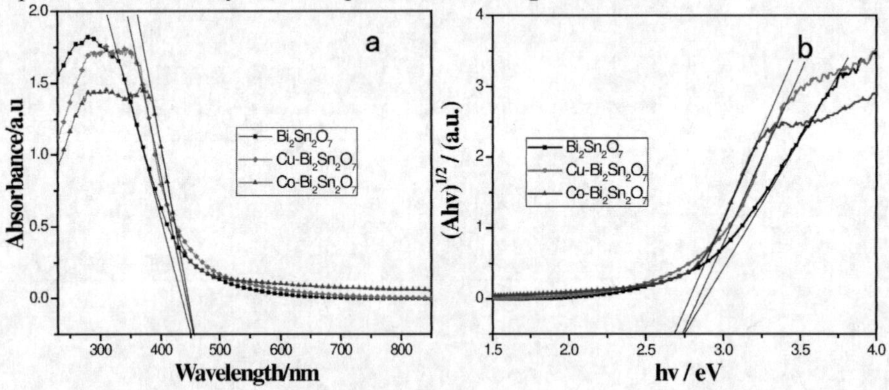

Figure 6. (a) UV-vis diffuse absorption spectra and (b) estimated band gap of pure $Bi_2Sn_2O_7$ and M-$Bi_2Sn_2O_7$ photocatalysts.

3.3. *PL emission spectra*

In order to establish the above analysis, the photoluminescence (PL) spectra of pure $Bi_2Sn_2O_7$ and M-$Bi_2Sn_2O_7$ photocatalysts were further measured because it can disclose the migration, transfer, and recombination processes of the photo-generated electron-hole pairs. As shown in Figure 7, an obvious peak at 520 nm in the PL spectrum for pure $Bi_2Sn_2O_7$ can be found, which is attributed to the recombination of holes and electrons in the valence band and conduction band. Apparently, the M-$Bi_2Sn_2O_7$ composite exhibits a much lower emission intensity than pure $Bi_2Sn_2O_7$, implying that the recombination of photo-generated electrons and holes is inhibited.

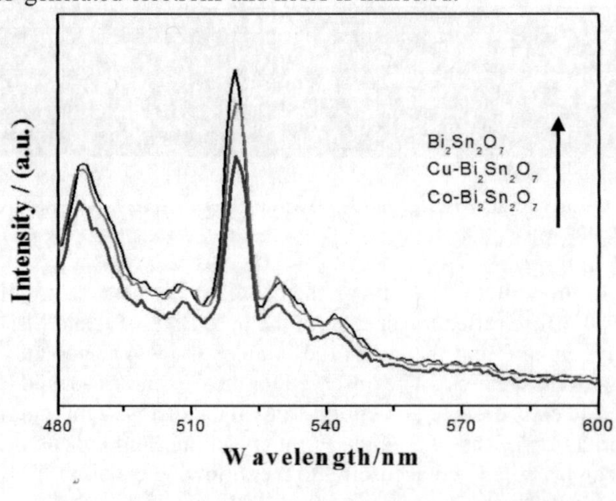

Figure 7. PL spectra of pure $Bi_2Sn_2O_7$ and metal-loaded $Bi_2Sn_2O_7$ samples

3.4. Photocatalytic activity

The photocatalytic performance of all the samples were studied by the decomposition of RhB under visible-light irradiation and presented in Figure 8, where C and C_0 represent the time-dependent concentration and the initial concentration of pollutants solution, respectively. As shown in Figure 8, the degradation efficiency of RhB of pure $Bi_2Sn_2O_7$ is about 59.2% after 250 min under visible-light irradiation. However, the degradation efficiency of metal-loaded $Bi_2Sn_2O_7$ catalysts shows a significant improvement compared to pure $Bi_2Sn_2O_7$. It is obvious that the about 83.6% RhB can be degraded by Cu-doped $Bi_2Sn_2O_7$ catalyst and the Co-$Bi_2Sn_2O_7$ catalyst shows the best performance of 89.5% degradation efficiency within the same period. The enhancement in photocatalytic activity of M-$Bi_2Sn_2O_7$ is probably ascribed to the generation of more active electrons and holes and the improvement of the separation of the photogenerated electron-hole pairs on the $Bi_2Sn_2O_7$ surface due to the introduction of metal particles.

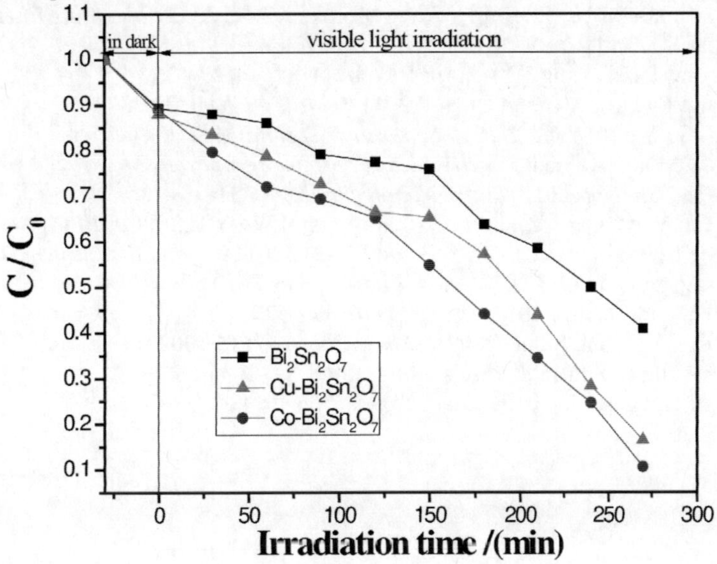

Figure 8. Photodegradation of RhB with pure $Bi_2Sn_2O_7$ and metal-loaded $Bi_2Sn_2O_7$ powders under visible-light irradiation

4. Conclusion

In summary, M-$Bi_2Sn_2O_7$ (M = Cu, Co) catalysts have been successfully synthesized by using the hydrothermal process and impregnation method. All of the as-prepared samples have been characterized by XRD, FESEM, UV-vis, IR, and PL. The degradation experiments have revealed that M-$Bi_2Sn_2O_7$ (M = Cu, Co) catalysts exhibit significant photocatalytic activity and the photocatalytic efficiency in degradation of RhB of Cu-$Bi_2Sn_2O_7$ and Co-$Bi_2Sn_2O_7$ is about 1.41 and 1.51 times compared to the pure $Bi_2Sn_2O_7$. The enhancement in photocatalytic efficiency for M-$Bi_2Sn_2O_7$ (M = Cu, Co) catalysts mainly result from the lower band-gap energies and weak fluorescence effect, which can response in visible light and lead to the more effective charge separation.

Acknowledgement

This research was supported by the National Basic Research Program of China (Grant No. 2014CB643703), the National Natural Science Foundation of China (Grant No. 11464008 and No. 51401060), the Guangxi Natural Science Foundation (Grant No. 2014GXNSFGA118001), Guangxi Key Laboratory of Information Materials (Grant No. 1210908-215-Z), and the Innovation Project of Guilin Graduate Education (Grant No. 2016YJCX20).

References

[1] Kudo A, Omori K and Kato H 1999 *J. Am. Chem. Soc.* **121** 11459
[2] Liu Z, Bai H and Sun D 2011 *Appl. Catal. B: Environ.* **104** 234
[3] Chaiwichian S, Inceesungvorn B, Wetchakun K, Phanichphant S, Kangwansupamonkon W and Wetchakun N 2014 *Mater. Res. Bull.* **54** 28
[4] Li Z, Chen X and Xue Z 2013 *J. Colloid Interface Sci.* **394** 69
[5] Chai S Y, Kim Y J and Jung M H 2009 *J. Catal.* **262** 144
[6] Xing Y, Que W and Liu X 2014 *RSC Adv.* **4** 49900
[7] Anpo M and Takeuchi M 2003 *J. Catal.* **216** 505
[8] Kikugawa N, Yang L Q, Matsumoto T and Ye J H 2010 *J. Mater. Res.* **25** 177
[9] Zhang H, Wang W, Shang M and Yin W 2010 *Catal. Commun.* **11** 982
[10] Kim H W, Shim S H, Lee J W, Park J Y and Kim S S 2008 *Chem. Phys. Lett.* **456** 193
[11] Udod L V, Sitnikov M N, Aplesnin S S and Molokeev M S 2013 *Solid State Phenomena* **215** 503
[12] Bala I, Barbar S K and Roy M 2012 *Physica B* **407** 3939
[13] Tian Q, Zhuang J and Wang J 2012 *Appl. Catal. A: Gen.* **425-426** 74
[14] Wu J J, Huang F Q, Lu X G and Chen P 2011 *J. Mater. Chem.* **21** 3872
[15] Liu H, Jin Z, Su Y and Wang Y 2015 *Sep. Purif. Technol.* **142** 25
[16] Mohameda M M and Ahmed S A 2015 *Microporous Mesoporous Mater.* **204** 62
[17] Holzwarth U and Gibson N 2011 *Nature Nanotechnology* **6** 534
[18] Long M C, Cai W M, Cai J, Zhou B X, Chai X Y and Wu Y H 2006 *J. Phys. Chem. B* **110** 20211
[19] Sun W T, Xie M Z, Jing L Q, Luan Y B and Fu H G 2011 *J. Solid State Chem.* **184** 3050
[20] Xu W C, Liu Z and Fang J Z 2013 *Int. J. Photoenergy* **2013** 394079
[21] Bian Z, Zhu Y and Zhang J 2014 *Chemosphere* **117** 527
[22] Yu K, Yang S, He H and Sun C 2009 *J. Phys. Chem. A* **113** 10024
[23] Gao E P and Wang W Z 2015 *J. Inorg. Mater.* **30** 87
[24] Xu W, Fang J and Chen Y 2015 *Mater. Chem. Phys.* **154** 30

Solar-energy mobile water aerators are efficient for restoring eutrophic water

Y Y Wang[1,2], Z X Xu[3,4]

[1] Chengdu Academy of Environmental Sciences, Chengdu, China, 610041
[2] Chengdu Environment Engineering Appraisal Center, Shaocheng mansion, Chengdu, China, 610041
[3] Institute of Water Environment Rehabilitation, College of Environmental Science and Engineering, Tongji University, Shanghai, China, 200092

E-mail: wyy1042@aliyun.com

Abstract. Surface water eutrophication has become a worldwide social issue. large amounts of secondhand energy, high capital investment are required, and most ecosystem disturbances will arise in the conventional eutrophication restoration measures. However, mobile solar-energy water aerator has the better oxygen transfer rate, hydrodynamic condition and can be used in the large waterbody for its cruising character. Second, the device is low carbon and sustainable for the solar photovoltaic system applications. So the device can be widely used in the eutrophication restoration.

1. Eutrophication and measures

The rapid increase in human activity has substantially accelerated the eutrophication process, which become a worldwide water quality issue. For example, the potential annual value losses in recreational water usage and water-front real estate were approximately $2.2 billion as a result of eutrophication in U.S. freshwaters [1]. A recent study of over 14000 lakes <1 ha size in Great Britain found that 51% are likely to take measures to meet the Water Framework Directive to achieve 'good status' by the year 2015 [2]. However, most freshwater lakes in China had faced serious challenge in the eutrophic problems[3], and nearly 2*104 ha aquaculture ponds in China which exports 60% aquatic products to the world have the same trouble too[4]. Currently, many countries and districts have taken measures against eutrophication, such as the "Water Framework Directive" in Europe and the "Action Plan for Prevention and Treatment of Water Pollution (i.e., ten key water pollution control plans)" announced in 2015 in China. Except the reduction of discharge, in-site restoration and ectopic treatment are important measures against eutrophication. Sediment dredging, water transfer are major actions to against aquatic eutrophication[5]. However many scientists think sediment dredging measures should be discreetly implemented as it's big cost and ecological risk of huge volume sludge ,also the lake ecosystem disturbance[6]. Transferring water needs diversion project and a lot of energy consumption, but it can only maintain the 'good status' in a short period after one time pour into [7].

2. Aeration and new demands

Artificial aeration which includes the fixed aerator and aeration vessel was considered as the cheapest and best preventive action against eutrophication [8]. Vertical-pump-sprayer, propeller-aspirator-pump, paddle-wheel, and diffused-air-system are the most widely used aerators, all of which must be fixed in a site, so that energy was focused and cannot be distributed evenly in the water. Thus many installation sites are needed and large quantity of energy dissipation are arising in the fixed aerators. In addition, aeration vessels need high cost and manual operation, making the systems cannot be popularized in the eutrophic restoration. Research interests in mobile aerator have grown in the past 10 years, and several devices were published [9-12]. However their defects of the low mobile capacity, the small working area, the energy-extensive consumption, the dependence of secondary energy and the low overall performance in practice induce the low market application in the aquatic ecosystems rehabilitation. To this end, there is a need to develop novel devices that are able to perform on the larger water area with high efficiency and low-carbon footprint in operation. Emerging solar-energy mobile water aerator (SEMWA) will play a key role in the aquatic ecosystems restoration of low cost application, good performance, and without secondary energy.

3. Device advantages

3.1 Water exchange and Hydrodynamic force

The major difficulty of surface water (lake, reservoir, pond and river) eutrophic restoration is the large area, where the traditional restoration devices are not suitable to put into use. Compared with the small service area (usually less than 0.05 ha) of fixed aerators [13], SEMWA can cruise constantly with the help of time-switch or horizontal location-switch electronic program, and reaches every corner of the water. Hypoxic/anoxic water can be air-lift from the bottom to the surface for oxygenation and photo-chemical treatment, which can control taste and odor problems as well as nuisance algal blooms [14], together its running characteristic water can be maintained fresh as "running water is never stale". Hence, we only need to spend little cost input than the fixed aerator for the less devices to use.

3.2 Clean energy and sustainability

Solar photovoltaic system is one of the potential low-carbon measures, as its energy payback time is only 1~4.1 years, and system lifetime reaches 30 years [15]. Solar photovoltaic panels and conversion system help SEMWAs obtain the power, which can save 36.8 TWh yr-1 electricity if them were applied in the eutrophic lakes and aquaculture ponds of China sufficiently(lake allocation (photovoltaic area: surface water area) = 2:104; aquaculture pond allocation=4:104). Meanwhile, we can avoid project construction troubles and decrease the current-leakage possibility of traditional electricity power deliver system. Therefore it is one of the best management practices (BMPs) to restore eutrophic water.

4. Function evidences

Aerators had better oxygen transfer rate in the state of mobility, for example, suspended chain moving aeration has lots of practical applications in the BIOLAK process of wastewater treatment plant [16] . In these years, so many aeration vessels have had good applications in the eutrophic lakes and rivers, such as Seine river [17], which the outstanding mobile characters were favored by many customers, "Pioneer One" aeration vessels will play important role in Taiwan fresh water eutrophic restoration [18]. SEMWAs also have been applied to some lakes and ponds (figure 1), which gets many approvals from scientists and customers for their performance of enhancing the water dissolved oxygen effectively, promoting the nutrients circulate and transform, had won the award of "2014 ten most innovative programs in China's aquaculture industry" and "2015 environmental innovation and entrepreneurship competition in China"[19].

Figure 1. Application of solar-energy mobile water aerator to a freshwater lake, China

For rivers with narrow shape, SEMWAs have advantages of cruising back and forth while avoiding oxygenation asymmetrical and shipping interference, however the range limit is also needed as they would disappear without control. Alike rivers, eutrophic lakes usually need intensifying aeration in target area through positioning mode. Different power and mobile velocity of SEMWA can be set to satisfy different water area. Additionally, power of traditional aerators usually is below 3kW h-1 due to the energy transfer limits, thus the manufacture cost of the latter is too high.

There are many cases of solar photovoltaic system applications in natural water restoration, for example mechanical water mixer of SolarBee that uses a solar powered pump to create water circulation [20]. Besides, there are also the solar-powered impeller system and the solar powered aeration system [21]. In the water source of Dongguan Songmu mountain reservoir in China, 16 hectares solar energy restoring areas were construct for decreasing the chlorophyll-a concentration and optimizing algae community structure[22]. Extensive applications of solar energy in eutrophic water restoration in the future not only need high electricity transform efficiency of photovoltaic technology, but also depend on reduction of the system cost. In fact, restoration performance will be better when the sunlight is stronger, and energy transform efficiency is the highest when the sunlight is used for the eutrophic restoration directly.

5. Bright future

Although the comprehensive assessments of resource cost savings, device stability and flexibility in the field use still need to be addressed, we envisage that the real-time restoration of eutrophic waters will soon be realized using sustainable methods. From this perspective, SEMWA is one of effective and efficient mobile aerators due to its high efficiency, low carbon, convenience in operation and maintenance, which can enhance the water dissolved oxygen effectively, remit the state of water unfavorable appearance, unpleasant odors and algal blooms as well.

Acknowledgments

We thank Gao Hong, Haiyuan Liu, Guoqiang Luo, Jingyan Xia for their support. We thank Hailong Yin, Sheng Wang, Penghui Li for editing the manuscript.

References

[1] Dodds W K, Bouska W W, Eitzmann J L, Pilger T J, Pitts K L, Riley A J, Schloesser J T, Thornbrugh D J 2009 Eutrophication of U.S. freshwaters: analysis of potential economic damages *ENVIRON SCI TECHNOL.* **43** 12.

[2] Carvalho L, Maberly S, May L, Reynolds C, Hughes M, Brazier R 2005 Risk assessment methodology for determining nutrient impacts in surface freshwater bodies *Report to the environment agency* 220.

[3] Qin B, Gao G, Zhu G, Zhang Y, Song Y, Tang X, Xu H, Deng J 2013 Lake eutrophication and its ecosystem response *CHINESE SCI BULL.* **58** 961.

[4] Zhang Y, Bleeker A, Liu J 2015 Nutrient discharge from China's aquaculture industry and associated environmental impacts *ENVIRON RES LETT.* **10** 1.

[5] Cooke G D, Welch E B, Peterson S, Nichols S A 2005 *Restoration and management of lakes and reservoirs, third ed* (*Boca Raton, Florida: CRC Press*)

[6] Manap N, Voulvoulis N 2015 Environmental management for dredging sediments - The requirement of developing nations *J ENVIRON MANAGE.* **147** 338.

[7] Hu L, Hu W, Zhai S, Wu H 2010 Effects on water quality following water transfer in Lake Taihu, China *ECOL ENG.* **36** 471.

[8] Pastorok R A, Ginn T C, Lorenzen M W 1981 Evaluation of aeration/circulation as a lake restoration technique. *U.S, Environmental Protection Agency, Office of Research & Development*.

[9] Sack E 2012 Solar-powered mobile water aerator useful for water-air swirling, comprises floating body exhibiting square-, rectangular-, triangular-, rhombic- or round- and oval shape, DE 201012008952-U1 (p).

[10] Fang S 2012 Mobile microporous aeration aerator, i.e. ship aerator, comprises ship body, oxygen increasing pump, gas pipe, front diffuser, aeration disc and control valve, CN 202643445-U(p).

[11] Cheng G, Liu X, Shi X, Wang X, Wu Z, Zhang Z 2013 Pond mobile aerator device includes solar power assembly, upper layer water body extraction assembly, floating body, connecting rod, mounting bracket, solar power supply, guide bracket, solar energy power generating plate and lantern ring. CN103588309-A(p).

[12] Al-anzi B S 2012 Mobile buoyant aerator, useful as free floating device for aerating the water to facilitate the breakdown of organic contaminants, comprises buoyant capture tray, fluid entrainment tube, fluid delivery device, and drain passage, US20140110323-A1 (p).

[13] Abdelwahed M, Dabaghi F, Ouazar D 2002 A virtual numerical simulator for aeration effects in Lake Eutrophication *INT J COMPUT FLUID D* **16** 119.

[14] Gerling A B, Browne R G, Gantzer P A, Mobley M H, Little J C, Carey C C 2014 First report of the successful operation of a side stream supersaturation hypolimnetic oxygenation system in a eutrophic, shallow reservoir *WATER RES.* **67** 129.

[15] Bhandari K P, Collier J M, Ellingson R J, Apul D S 2015 Energy payback time (EPBT) and energy return on energy invested (EROI) of solar photovoltaic systems: A systematic review and meta-analysis *RENEW SUST ENERG REV.* **47** 133.

[16] Yu K, Wei S 2009 Application of BIOLAK process to wastewater treatment plant in Baoying County *CHINA WATER & WASTE WATER*. **25** 44.

[17] Thibodeaux L, Poulin M, Even S 1994 A model for enhanced aeration of streams by motor vessels with application to the Seine river *J HAZARD MATER* **37** 459.

[18] Daily C 2010 Taiwan multifunction aeration vessel is applied in environment protection (DB/OL). *http://www.chinanews.com/tw/2010/11-05/2636252.shtml.*

[19] BBS T 2015 The award list of ten innovation programs of China aquaculture industry in 2014 (DB/OL). *http://bbs.tongwei.cn/thread-77802-1-1.html*

[20] Barkoh A, Begley D C, Smith D G, Kurten G L, Fries L L, Schlechte J W 2011 Can solar powered circulation control Prymnesium parvum blooms and toxicity in fish hatchery ponds? *HARMFUL ALGAE* **10** 173.

[21] Prasetyaningsari I, Setiawan A, Setiawan A A 2013 Design optimization of solar powered aeration system for fish pond in sleman Regency, Yogyakarta by HOMER Software *Energy Procedia* **32** 90.

[22] Li W, Yuanqi B, Yongjian S, Guoqin L, Wei L 2012 The performance study of solar-energy water ecology restoration to the micropollution water source *Advanced water treatment seminar of water industry meeting-CSCE 2010 (Huaian: Jiangsu)* p 512.

Selective recovery of copper and zinc from mine dump waters of mining enterprises in precipitates

N N Orekhova[1], G A Tarybaeva, D S Muravev

Mining Engineering and Transport Institute, Nosov Magnitogorsk State Technical University, Pr. Lenina, 38, Magnitogorsk, Russia

E-mail: n_orehova@mail.ru

Abstract. The acid mine dump waters at mining companies that mining the copper and copper-zinc sulphide ore have the high concentrates of metals and classified in our view as the raw materials for selective recovery of metals in precipitates comparable in quality with concentrates of ores enrichment and suitable for further metallurgical processing. Authors have implemented three two-stage schemes for sequential extraction of copper and zinc: cementation-sulphide precipitation, galvanocoagulation- sulphide precipitation, precipitation-precipitation. Moreover, parameters of processes and quality of the obtained precipitates have described. The achieved copper recovery is 89% and 94% respectively with the application of cementation and galvanocoagulation, in sulphide precipitate exceeded 75%. Furthermore, the copper recovery from decoppering in precipitates amounted to more than 65%. Zinc-containing precipitates because of coprecipitation of iron, magnesium and calcium contain zinc from 14% to 28%, in two to three times less than the quality of conditioned zinc concentrates. The content of precipitates allows to apply them for additional charging to concentrates of enrichment or for the production of metals in metallurgical treatment. As a result of the studying the effect of reduced total salinity (S) on mass fraction of zinc in precipitate (β) with the constant concentration of zinc (C_{Zn}), the changes in concentration with constant salinity and reduction in total salinity (S) with constant ratio S/C_{Zn}, the following dependencies have obtained: (S): $(\frac{\partial\beta}{\partial C_{Zn}})_S > 0$, $(\frac{\partial\beta}{\partial S})_{C_{Zn}} < 0$, in the range of an index S from 4.5 to 90.0 *g/L* $(\frac{\partial\beta}{\partial S})_{\frac{C_{Zn}}{S}} < 0$, where $(\frac{\partial x}{\partial y})_z$ is a partial derivative of x to y, whereas the value z is fixed.

1. Introduction

The acidification of mine and mine dump waters happens over time with increased concentrations of metals at mining plants that produce and process copper and copper-zinc sulphide ores. The key factor of forming the ionic composition of waters is the sulphide oxidation rate, the most moving of which are the minerals of copper, zinc and iron [1]. The concentrations of chalcophylic elements reach several grammes per litre in the mine dump waters of the Shemur and the Tarnier copper-zinc pyrite deposits that characterise the non-stationary period of sulphide oxidation. Characteristics of mine dump waters in 2012-13 are presented in Table 1.

These waters together with pit sump are guided to the treatment facilities. A treatment is carried out by reagent precipitation with the application of lime and sodium sulphide. The obtained precipitates the low content of copper (to 2%) and zinc (to 7%) and cannot be used in pyrometallurgical extraction.

Table 1. Characteristics of mine dump waters in 2012-2013.

Name of the ore deposit (type of water)	Average water flow (m^3/h)	Concentration of ions (mg/L)				pH
		Cu^{2+}	Zn^{2+}	Fe_{total}	Fe^{3+}	
Tarnier (Copper-zinc mine's dump water)	6.02	402.5 – 2 082.0	2 156.0 – 8 680.0	4 075.0 – 9 334.0	1 652.1 – 2 952.2	2.2–2.9
Shemur (Copper- mine's dump water)	0.45	626.4 – 2 940.0	43.1 – 300.0	3 126.0 – 16512.0	358.7 – 958.7	1.9–2.4

To precipitate the copper from multicomponent solutions it is most often applied electrochemical methods, such as electrolysis [2,3] with the use of carbon as the cathode, cementation [4-6] and galvanacoagulation as its derivative [7]. Copper recovery reaches from 92 to 99% by these methods. It is proposed to use hydrolytic [8] and sulphide [9] precipitation with controlled pH range for the selective recovery of copper and zinc from acidic drainage waters. In most studies [2-4,7-9], the main controlled output parameter is the residual concentration of the polluting component. Moreover, little attention is given to the mass fraction of metal in precipitation and their phase content.

Recently, the technological schemes of copper and zinc recovery from the water have developed with the application of hydrolytic precipitation methods [10], sulphide precipitation [11], cementation [6,12] and ionic flotation [13]. In accordance with the existing practice of the treatment of acidic mine dump waters [14] and theoretical research [1-4] in the area of copper and zinc recovery from acidic sulfate mine water [2-13], we have chosen such well-known processes as galvanocoagulation, cementation and sulphide precipitation for copper recovery and sulphide precipitation was chosen for zinc recovery.

According to the authors' opinion, the main issue in this paper is to determine the possibility of consistent recovery of copper and zinc-containing precipitates with the high content of metal at the realisation of three two-stage schemes: cementation-precipitation, galvanocoagulation-precipitation and precipitation-precipitation. Also, the searching for simple solutions is carried out to increase the content of zinc in a precipitate.

2. Experimental design

The experiments were carried out in a two-stage sequence. The first stage involved the precipitation of copper as a sulphide, cemented or ferrite precipitate. The second stage involved the precipitation of zinc as a sulphide.

2.1. Cementation method

The cementation is carried out in the flowing mode with the application of laboratory, cementation, the drum-type unit with a diameter of 20mm in rotary washing screen providing the continuous mixing of mine dump water with iron turnings. The ratio of the drum's length and width is 3:1 and the optimum rotation speed is calculated according to the following formulas [6]:

$$A\,min = 10.1/\sqrt{D} A\,max = 19.7 \cdot \sqrt{D}\,, \quad (1)$$

where D is the diameter of the drum and equal to 35.7-69 rpm.

The actual rotation speed of the drum under load is 36 and 64 rpm, depending on the position of the transmission gears in the gearbox.

2.2. Galvanocoagulation method

The galvanocoagulation is carried out with the application of laboratory type drum galvanocoagulator with the drum. The size is 110x350 mm, the mass load of ferrum: carbon = 3:1, with the coarseness 5-10mm is 1200 g. The parameters of galvanocoagulation adopted from our experience on mine dump waters at mining enterprises of South Ural [15].

The experiments were conducted under dynamic conditions. Furthermore, the liquid phase and precipitates have been analysed, as well as charge before and after the contact with a solution.

To prevent accidental errors before each series of experiments, the galvanocoagulator was loaded with a new portion of the deoil charge of a known mass.

2.3. Precipitation of the metals with sodium hydrosulphide

The precipitation of the metals with sodium hydrosulphide was carried out in accordance with the results of studies described in [16]. The consumption of the sulphide reagent was taken as a percentage of the value of the stoichiometrical ratio for the reduction of Fe III to Fe (II) according to the reaction: $Fe_2(SO_4)_3 + NaHS = 2FeSO_4 + NaHSO_4 + S$, and the metal's full precipitation according to the reaction: $MeSO_4 + NaHS = MeS + NaHSO4$, where Me stands for Cu^{2+}or Zn^{2+}. The conditioning with the reagent was performed in plastic containers with a working volume of 4 dm3. The contents were stirred for 5-15 minutes. Moreover, the treatment solution of sulphide reagent injected into the water mass closer to the reactor's bottom with constant mixing. The technical sodium hydrosulfide with the concentration of 31% was used as the precipitating reagent. The working solution with the concentration of 5% NaHS is produced by dilution of the 31% of concentration solution.

The main research methods of products have been selected as follows, chemical analysis for solutions, X-ray phase analysis and chemical analysis for dispersing sediments. The samples of sediments have been analysed at the Shimadzu XRD-6000 diffractometer, with a Cu anode and a graphite –monochromator. The sampling results were calculated with the Peter-Kalman method [17] with the application of coefficients taken from tabular data with maximum diffraction consistency. To determine the quantity of the X-ray amorphous component the addition technique was used. Three X-ray photograms were taken in the range of 24-30 degrees, then the average ratio of the reflection intensity for silicon (3,136 Å) and the intensity of maximum diffraction reflection on the X-rayogram of the sample is calculated. After that, the minor amount of silicon was again added to the sample and the whole photographing procedure was repeated.

3. Discussion

The results of experiments showed that the specific consumption of iron during the cementation of water from the «Tarnierskoe» deposit (4,7g of iron per gram of copper) is higher than the average consumption of iron according to the data of practical application of copper cementation (1.32-3.00 gFe/gCu) [7], which is most probably caused by the high content of Fe (III) in the mine dump water [4,11].

To reduce the iron consumption at acidifying the effluents to pH 2 for reduction of the precipitation process of hydroxides and basic iron sulphates. When the copper content in the initial mine water rises from 920 to 1250 mg/L (all other things being equal) we have seen a gradual increase in the rate of copper recovery from 88.6 to 93.1% which are linked to the shift in the concentration ratio of copper and zinc in favor of copper content. Mass fraction of copper in precipitate ranged from 17% to 23%.

The specific iron consumption at cementation of the copper from mine dump water of «Shemurskoe» deposit basically similar to the known range of iron consumption based on application practice data of copper cementation. When the copper content rises from 876.24 to 2 940.00 mg/L in the initial water with the relative concentrations of Cu^{2+}, Zn^{2+} and Fe^{3+} being equal, we have seen the decrease in the rate of recovery from 94.5% to 85.77%. The parameters of the cementation recovery of copper are given in Table 3. When you change the rotational speed of the drum, the iron consumption and copper recovery has hardly changed.

The ferrite precipitate obtained on the scheme of galvanocoagulation- precipitation from mine dump waters at the «Tarnierskoe» and «Shemurskoe» deposits, satisfying five components such as, weight content, recovery rate, selectivity, low iron consumption and low iron residual concentration in the drainage obtained at the rate of galvanocoagulator drum rotation of 12 rpm and the duration of Valvano coagulation to 6 minutes. The zinc-containing precipitate has obtained at the charge of NaHS, i.e. 125% on stoichiometry to precipitation of zinc. The values of the galvanocoagulation recovery are given in Table 3. Specific consumption of iron is 4-5 times lower than with cementation (Table 3) and equal to 0.73-0.67 gFe/gCu.

The results obtained for mine waters with concentrations (mg/dm3) at «Tarnierskoe» deposit Cu^{2+}= 950.13, Zn^{2+}= 6 960.50 and Fe^{3+} = 8 899.67 mg/L, and Cu^{2+} = 2 091.0, Zn^{2+} = 300.0 and «Shemurskoe» deposit Fe^{3+} =9 800.83 are presented in Tables 2-4.

Copper precipitation by sulphides is performed with acidification of the water before pH-rate of the selective extraction of 1.7–2.0. The consumption of concentrated sulfuric acid for waters with an initial pH range of 2.7 is 0.5-0.6 mL/L.

In figure 1 the results of copper recovery from the dump water of a copper and zinc deposit with a $NaHSO_4$ consumption in the range from 95% to 135% of the corresponding stoichiometrical ratio value have shown that it is not possible entirely to separate copper from zinc. More than 5.49% of zinc is precipitated with a copper recovery of 65%. The predominance of zinc several times over copper in mine dump water necessitates the mass fraction of zinc in precipitate at the level of the mass fraction of copper. The increase in consumption of the sodium sulphide precipitator to over 115% of the stoichiometrical ratio one leads to the precipitation of Fe (II) and zinc in quantities exceeding that of copper. The optimal consumption was taken to be 115% of the relevant level of stoichiometrical ratio. The parameters of recovery of Cu^{2+} from /mine dump water are presented in Tables 2-4.

Table 2. The results of copper recovery by cementation and the recovery of zinc by the sulphide precipitation.

Process	Specific weight in precipitate, g/L	Mass fraction of metal in precipitate, (%)			Recovery rate (%)		Specific consumption of iron	
		Cu^{2+}	Zn^{2+}	Fe_{total}	Cu^{2+}	Zn	*(g/L)*	*(g_{Fe}/g_{Cu})*
		Copper-zinc mine dump water						
Cementation	4.53	18.64	1.05	7.89	88.84	0.68	3.90	4.70
Zinc precipitation	30.80	0.34	14.32	15.00	10.62	63.03	-	-
		Copper mine dump water						
Cementation	8.26	24.12	0.49	13.53	95.63	13.55	4.30	2.10

Table 3. The results of copper recovery by galvanocoagulation and zinc by the sulphide precipitation.

Process	Specific weight in precipitate, g/L	Mass fraction of metal in precipitate, (%)			Recovery rate (%)			Specific consumption of iron	
		Cu	Zn	Fe	Cu^{2+}	Zn^{2+}	Fe_{total}	*(g/L)*	*(g_{Fe}/g_{Cu})*
		Copper-zinc mine dump water							
Galvanocoagulation	6.69	16.73	0.83	14.42	94.40	0.90	9.71	0.75	0.67
Zinc precipitation	17.22	0.37	21.75	12.98	4.90	63.10	24.8	-	-
		Copper mine dump water							
Galvanocoagulation	15.84	17.67	0.56	18.42	95.20	29.33	29.78	0.61	0.73

The results of zinc recovery from decoppered mine water with a consumption of $NaHSO_4$ in the stoichiometrical ratio in the range of 115-200% which are presented in Figure 2, showed that an increase in the consumption more than 135% leads to an explosive precipitation of iron and the decrease in zinc weight content in the sediment. According to the results of experiments, the optimum consumption of the precipitator adopted to be 125% of the stoichiometrical ratio value. The obtained values of selective sulphide precipitation of Zn^{2+} are presented in Table 4.

Table 4. The results of copper and zinc recovery by the sulphide precipitation

Process	Specific weight in precipitate, g/L	Mass fraction of metal in precipitate, (%)			Recovery rate (%)		
		Cu	Zn	Fe	Cu^{2+}	Zn^{2+}	Fe_{total}
Copper-zinc mine dump water							
Precipitation of copper	5.01	14.25	7.23	7.50	75.15	5.82	4.43
Precipitation of zinc	14.50	1.60	28.29	16.70	24.41	65.95	28.51
Copper mine dump water							
Precipitation of copper	6.20	28.67	0.46	4.57	86.39	9.75	2.94

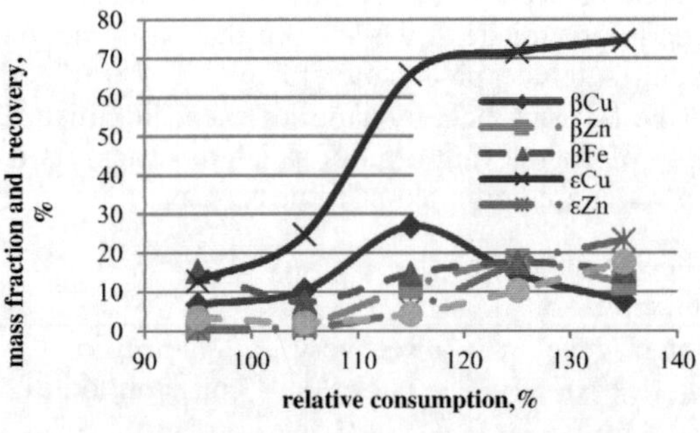

Figure 1. Influence of NaHS consumption on the parameters of recovery of Cu^{2+}, Zn^{2+} and Fe_{total} from mine dump water (**β** is a mass fraction of metal in the precipitate; ε is a rate of recovery of metal in precipitate).

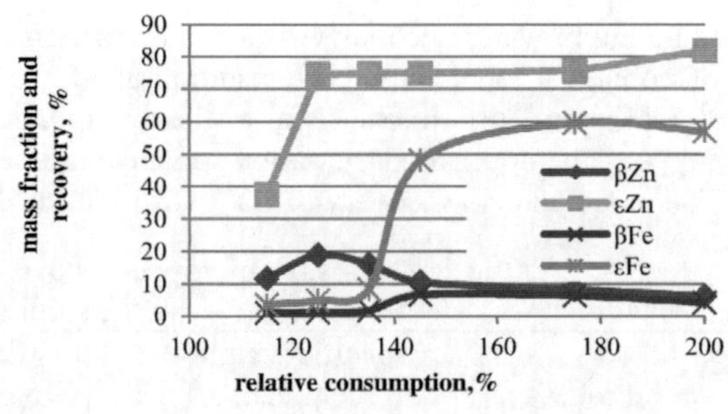

Figure 2. Influence of NaHS consumption on the parameters of recovery of Zn^{2+} and Fe_{total} from decoppered mine dump water (**β** is a mass fraction of metal in the precipitate; ε is a rate of recovery of metal in precipitate).

Precipitates obtained by sulphide precipitation contain from 14% to 28% of zinc, that two, three times less than the quality of certified enrichment concentrates. Therefore, we have analyzed the influence of the following factors on mass fraction of zinc in precipitate (β): the decrease in salt content (S) with constant concentration of zinc (C_{Zn}), the change of zinc concertation with constant salinity and decrease in total salinity (S) with constant ratio of S/CZn. These experiments were carried out with simulated mine dump water of copper-zinc deposit and the following concntrations of components g/L: Ca^{2+} - 4.26, Mg^{2+} -7.00, Na^{+} - 0.0081, K^{+} - 0.00096, Al^{3+} - 1.754, Mn^{2+} - 0.348, Fe^{2+} - 7.86, Cu^{2+} - 0,0152, Zn^{2+} - 6.7, NH^{4+} - 0.0105, Cl^{-} - 0.425, SO_4^{2-} - 102.48, NO^{2-} - 0.00122, NO^{3-} - 0.00295, pH - 2.4.

It was not able to achieve in the simulated water the same calcium concentration as in the mine dump water. The concentration of calcium in the water was limited by ion product of calcium sulphate Ca^{2+} - 0,466 g/L. The mass fraction in precipitates of the base experiment with the NaHS flocculent's consumption of 100% amounted to 34.2%, as well as extraction amounted to 68.27%.

In the first case, the mass fraction of zinc in precipitate gradually grows with the decrease of the indicator S. The extraction of zinc from solution and mass fraction of zinc precipitate increases with decreased total salinity. The mass fraction of zinc increases by one and half times and accounts for 49% with the reduction of indicator S in 10 times. The extraction increases by 11.2%.

The mass fraction of zinc in the precipitate is decreased with fixed total salinity and decrease in concentration of zinc. The decrease in the mass fraction of zinc in precipitate from 34,2 to 17,6% is observed with the decrease in concentration of zinc in 4 times. The mass fraction of zinc in the precipitate is increased before dilution by 20 times with precipitation of zinc from dilute multicomponent model solutions. Further dilution leads to the decrease in the mass fraction of zinc in a precipitate. For example, the mass fraction of zinc is 42,2 % with the dilution by 20 times (S =4,5 g/L), however, the mass fraction of zinc is 30,1 % with the dilution by 30 times (S =3,0 g/L). The obtained patterns may be represented in the following way): $(\frac{\partial \beta}{\partial C_{Zn}})_S > 0$, $(\frac{\partial \beta}{\partial S})_{C_{Zn}} < 0$, in the range of an index S from 4.5 to 90.0 *g/L* $(\frac{\partial \beta}{\partial S})_{\frac{C_{Zn}}{S}} < 0$, where $(\frac{\partial x}{\partial y})_z$ is a partial derivative of x to y, whereas the value z is fixed.

The above results proved that quality indicators of zinc-containing products obtained by metal extraction from mine dump waters with sulphide precipitation may be increased by diluting the initial solution. As shown by the X-ray phase analysis, precipitates formed in the process of sulphide precipitation contain sulphide and sulphate phases of copper and zinc, elemental sulphur, and sulphates of calcium and magnesium that are presented in Table 5.

The precipitation of calcium and magnesium sulphates in the range of copper and zinc-selective recovery parameters resulted from the high concentration calcium and magnesium in the water and its saturation with sulphur derivatives as a result of sulphide injection.

Table 5. Semi-quantitative phase analysis of precipitates

Phase with mass fraction more than 1%		Copper precipitate	Zin precipitate
Name	Formula	Content,%	
		Cu=14,25; Zn=7,23; Fe=7,50; S=35,4	Cu=1,6; Zn=28,2; Fe=16,7; S=20,45
Sulfur	S^0	17	5,5
Poitevinite	$(Cu,Fe)SO_4 \cdot H_2O$; Cu:Fe≈1:1	4	-
Covelline	CuS	11	
Gunningite	$Zn_{0.75}Mn^{2+}{}_{0.25}(SO_4) \cdot (H_2O)$	26	15
Rozenite	$FeSO_4 \cdot 4H_2O$	14	14
Greigite	Fe_3S_4	4	8
Kieserite	$Mg[SO_4] \cdot H_2O$	11	-
Gypsum	$CaSO_4 \cdot 2H_2O$	5	7
Blinde	ZnS	-	14
Bianchite	$FeZn_2[SO_4]_3 \cdot 18H_2O$	-	3
Boyleite	$(Zn,Mg)SO_4 \cdot 4H_2O$	-	2
Unrecristallization phase		5	20

4. Conclusion

The processing stage of acid mine dump waters for selective recovery of zinc and copper from them in precipitates is possible with the use of any of three two-step schemes: cementation-precipitation, galvanocoagulation-precipitation, precipitation-precipitation.

Obtaining the coppery precipitates with high copper content is not a difficult task. Selectivity of copper recovery from the solution is provided by the position of copper in electrochemical series and it's the fact that ion product for copper sulphide is less than for zinc or ferric sulphides. The most difficult task is to obtain zinc-containing precipitate with the high content of zinc in the process of sulphide precipitation. The higher content of zinc in precipitates (28 – 29 %) is obtained by sulphide precipitation with reagent consumption equal to 125% of the stoichiometrical ratio after preliminary copper precipitation from the water by hydrosulphide with consumption equal to 115% of the stoichiometrical ratio.

The phase composition of the precipitate indicates coprecipifation of calcium, magnesium and aluminium with obtaining of substances that can crystallise forming a joint crystalline grid, while mixed crystals are obtained. The reduction of total salt content by dilution of decoppered mine dump water happening because of the decrease in coprecipifation and increase in the content of zinc in precipitates. Moreover, for mine dump waters of copper-zinc deposits, the obtained dependence of the zinc content in precipitates on the dilution is exponential. The highest content of zinc has obtained with decreased total salinity to 4.5 g/L.

Acknowledgements
This research was financially supported by the Ministry of Science and Education of Russian Federation (State assignment № 2014/80).

References

[1] Blowes D W, Ptacek C J, Jambor J L, Weisner C G 2003 The geochemistry of acid mine drainage *Treatise on geochemistry* **9** 149-04

[2] Hatfield T L, Kleven T L, Pierce D T 1996 Electrochemical remediation of metal-bearing wastewaters Part I: Copper removal from simulated mine drainage waters *Journal of Applied Electrochemistry* **26 (6)** 567–74

[3] Gorgievski M, Božić D, Stanković V, Bogdanović G 2009 Copper electrowinning from acid mine drainage: A case study from the closed mine "Cerovo" *Journal of Hazardous Materials* **170 (2)** 716-21

[4] Chanturia V A, Shadrunova I V, Medyanik N L, Mishurina O A 2010 Electric flotation extraction of manganese from hydromineral wastes at yellow copper deposits in the South Ural *Journal of Mining Science* **46 (3)** 311-16 (in Russian)

[5] Alkatsev M I 1981 *Cementation processes in nonferrous metallurgy* (Moscow: Metallurgy) (in Russian)

[6] Halezov B D, Vatolin N A, Makurin Yu N and Bykov N A 2005 Research of extraction of copper in drum cementator *Russian Journal Min. Inform. Anal. Bull* **3** 302-11 (in Russian)

[7] Solozhenkin P et al 2001 The effect of microgalvano pairs in waste-water treatment. New Developments in Mineral Processing *Proc. of the 9th Balkan Mineral Processing Congress* (Istanbul: Turkey) pp 11-13

[8] Balintova M, Petrilakova A 2011 Study of pH influence on selective precipitation of heavy metals from acid mine drainage *Chemical Engineering Transactions* **25** 1-6

[9] Sampaio R M, Timmers R A, Xu Y, Keesman K J, Lens P N 2009 Selective precipitation of Cu from Zn in a pS controlled continuously stirred tank reactor *Journal of Hazardous Materials* **165 (1-3)** 256-65

[10] Shidlovskaya I P, Maltsev G I and Naboichenko S S 2005 Determination of optimal conditions for the deposition of metal hydroxide impurities in wastewater treatment *Russian Journal Non-Ferrous Metals* **6** 14-17 (in Russian)

[11] Khalezov B D 2009 *Research and Development of a Process for Heap Leaching of Copper and Copper-Zinc*, Dissertation, Institute of Metallurgy, Ural Division, Russian Academy of Sciences of Yekaterinburg (in Russian)

[12] Medianik N, Chanturia V, Shadrunova I 2012 Quantum chemical method of the selection of a collecting agent and its use in the process of the floatation extraction of cationic zinc and copper (II) from mine water *Russian Journal of Mining Science* **1** 155-65 (in Russian)

[13] Global Acid Rock Drainage Guide (GARD Guide) INAP: The International Network for Acid Prevention. http://www.gardguide.com/index.php?title=Main_Page. Accessed 28 September 2013

[14] Chanturia V, Shadrunova I, Orekhova N and Chalkova N 2011 Process for recovery of zinc from mine and mine dump water *Russian Journal of Ore Beneficiation* **1** 35-9 (in Russian)

[15] Shadrunova I V, Orekhova N N 2015 A process for advanced recycling of water originating from mining operations, with metal recovery *Mine Water Environent Journal* **34 (4)** 478–84

[16] Khalezov B D, Nezhivykh V A, Ovchinnikova L A 2005 Semiindustrial trials of zinc recovery by a sodium hydrosulphide from a liquors of heap leaching *Russian Journal of Min. Inform. Anal. Bull.* **4** 278-79

[17] Peter E, Kalman A 1964 Quantitative X-ray analysis of crystalline multicomponent systems *Acta Chimica Hungarica* **41** 413–22

Level and ecological risk of four common metals in surface water along the Qinhuangdao coastal areas, China

Liping Wang[1], Kun Lei, Yanzhen Qiao and Chenlin Hao

State Key Laboratory of Environmental Criteria and Risk Assessment, Chinese Research Academy of Environmental Sciences, Beijing, China

E-mail: wanglp@craes.org.cn

Abstract: Heavy metals have been a widespread environmental contamination. Due to their associated ecological risk, the presence in water environment has attracted broad attention to public. Here 4 most common metals including copper (Cu), lead (Pb), cadmium (Cd) and zinc (Zn) were determined in surface water along the Qinhuangdao coastal areas, China. And their ecological risk was assessed using species sensitivity distribution (SSD) method. Total 12 stations were designed near the main estuary in the study area. The results showed that the concentrations of Cu, Pb, Cd and Zn of surface water were in the range of 847.81-1602.81μg/L, 0.42-1.59μg/L, 1.82-7.99μg/L and 26.9 -59.36μg/L, respectively. According to the National Seawater Quality Standard of China (GB3097-1997), Cu concentration in each station was much higher than the standard value of Ⅳ level (50μg/L), thus Cu could not even met the lowest level of water quality. In contrast, Pb met the Ⅰ or Ⅱ level of water quality, Cd and Zn met the Ⅱ or Ⅲ level. The HC_5 (hazardous concentration for 5% of species) of each metal was obtained from their corresponding SSD curve. In case of Cu and Zn, the concentration at all sites exceeded their HC_5 values, suggesting both of them had adverse effect on the aquatic organism, especially Cu. While Pb concentration at all sites was much lower than its HC_5 value, thus Pb had no negative effect on aquatic life. In case of Cd, the concentration at 5 sites was higher than its HC_5 value, and the other 7 sites was lower than its HC_5 value, suggesting that adverse effect only occur at partial region in the study area. The RQ (risk quotient) value of Cu varied between 1355.41 and 2621.28, far larger than 1, indicating that 100% of sites had a much higher risk. The RQ of Zn varied between 6.06 and 13.88 (>1) indicating that Zn had a high risk in the study area. In case of Cd, the RQ ranged from 0.94 to 4.41 and about 92% of sites were larger than 1, suggesting that Cd had a high risk in these regions, while 8% of sites (0.1<RQ<1)had a medium risk. In case of Pb, the RQ ranged from 0.04 to 0.13 and about 92% of sites (<0.1) had a low risk in these regions and about 8% of sites (0.1<RQ<1) had a medium risk. The results indicated that the ecological risk was in the following: Cu>Zn>Cd>Pb. The Cu pollution level extremely high during this investigation, and a much higher ecological risk could already occur, thus Cu should be paid more attentions. Meanwhile, both Zn and Cd also were important contaminations in the study area.

1. Introduction

Heavy metal has been one of main environmental pollutants, and aroused much attention because of their potential toxicity, persistence, bioaccumulation and biomagnifications through food chains [1-5]. It is well-known that organisms in different trophic level have been suffering from metal toxicities [6]. Accumulation of heavy metals in organisms shows different tolerant capabilities among aquatic food chains [7] and human health is under threats from exposure to heavy metals through seafood intake [8].

Coastal and estuarine areas are often the final repository of anthropogenic pollutants. Therefore heavy metal pollution in coastal and estuarine ecosystems has long been widely concerned and has been studied by many researchers worldwide [9-11]. Pan&Wang (2012) [12] reported that high metal contents can be detected across the coasts in China, which is closely associated with accelerated economic growth in the past decades.

Liaodong Bay locates in the north of Bohai Sea, and belongs to a semi-enclosed bay with a long residence time. Qinhuangdao coastal area locates at the south-western corner of Liaodong Bay, and this region is a shallow embayment. There are multiple rivers into sea including Tang River, Xin River, Dai River, Yang River, Man-made River and Sha River along Qinhuangdao city. Here the main objective is to reveal the polluted status of 4 common metals including Cu, Cd, Pb, Zn, and assess their ecological risk near estuary along the Qinhuangdao coastal area, China.

2. Materials and methods

2.1. Study area and sample collection

Twelve stations were selected in the Qinhuangdao coastal area where several rivers flowed into the Liaodong Bay (Figure 1). All the sites located near the estuary and coastal tourist resort. The field sampling was carried out on May 10-11, 2016. The sea water samples were collected in clean polyethylene sampling bottles, and then were filtered through the 0.45μm synthetic fabric membranes. The nitric acid solution (2 mol/L) was added to adjust the pH of water samples to be <2. All the samples were transported to the laboratory in an icebox, and stored in dark at 4 ℃ for metal analysis.

The temperature (T), pH, dissolved oxygen (DO) in surface water were measured in situ using a Multi-parameter water quality meter (HACH HQ30d, USA). Salinity was measured on-site by a Master Refractometer (ATAGO, Japan).

2.2. Sample analysis and risk assessment

Metal mixed standards of liquid state (GSB04-1767-2004) were purchased from National Standard Material Center (China). The Cu, Cd, Pb and Zn concentrations were analyzed using ELAN6100 DRC ICP-MS (Perkin-Elmer, USA). All analytical procedures were strictly monitored through quality assurance and quality control. Quantification was performed by the external standard method using metal reference material mixture, with correlation coefficients for calibration curves all higher than 0.999. The range of recovery rate was 95.5~107.61%. The sample determination results were corrected by results of blank controls and recovery rates.

The National Seawater Quality Standard of China (GB3097-1997) was used to assess the seawater quality. And the species sensitivity distribution (SSD) was used for ecological risk analysis.

The hazardous concentration for 5% of species (HC_5) is the estimated 5th percentage of the distribution, i.e., the concentration makes the 95% of the species in an ecosystem protected [13]. The HC_5 value was obtained from the corresponding SSD curve. And then the PNEC (predicted no effect concentration) is calculated by the following equation:

$$\mathrm{PNEC} = \frac{HC_5}{\mathrm{AF}} \tag{1}$$

AF is an assessment factor in the range of 1 and 5, reflecting the uncertainties of data [14]. In this study, the value of AF is set as 3 according to the number of species tested, quantity and quality of toxicity data and model goodness of fit. The 'risk quotient' (RQ) was used to determine the ecological risk [15]:

$$\mathrm{RQ} = \frac{MEC}{\mathrm{PNEC}} \tag{2}$$

Here MEC means measured environmental concentration. A risk level can be ranked as low, medium and high if the RQ is <0.1, 0.1~1.0 and ≥1.0, respectively.

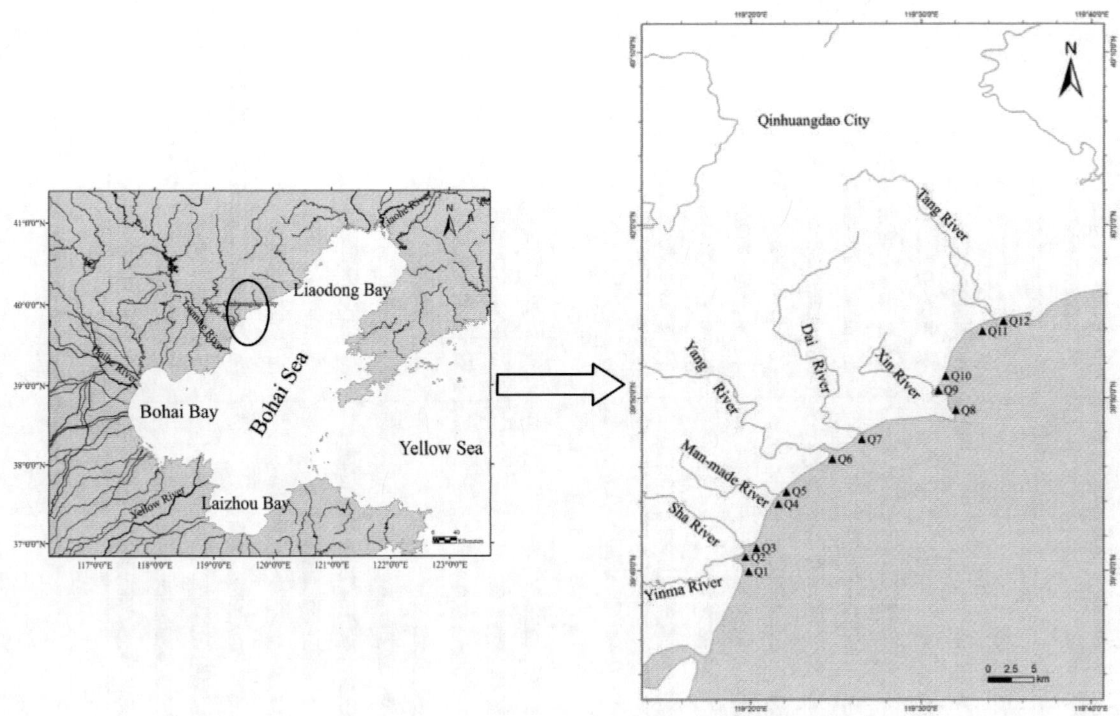

Figure 1. Study region and the sampling sites

3. Results and discussion

3.1. The primary environmental conditions at the sampling time

The results showed that the T, pH, DO and salinity ranged from 18.8 to 21.7 ℃, from 8.92 to 9.44, from 8.17 to 13.45mg/L and from 26 to 36‰, respectively.

3.2. Pollution levels and water quality assessment

The concentrations of Cu, Pb, Cd and Zn in surface water were in the range of 847.81-1602.81μg/L, 0.42-1.59μg/L, 1.82-7.99μg/L and 26.9-59.36μg/L, respectively. The comparative levels of four metals in the different stations were showed in Figure 2. It showed the highest concentration of Cu was found at site Q12 and the second was at site Q6, while the lowest one was found at site Q10 following by Q11. The highest concentration of Pb was at Q9 and the second was at site Q12, while the lowest was at site Q4 following by Q3. The highest concentration of Cd was at Q12 and the second at site Q6, while the lowest was at site Q5 following by Q1. The highest concentration of Zn was at site Q6 and the second at site Q12, while the lowest one at site Q4 following by Q5.

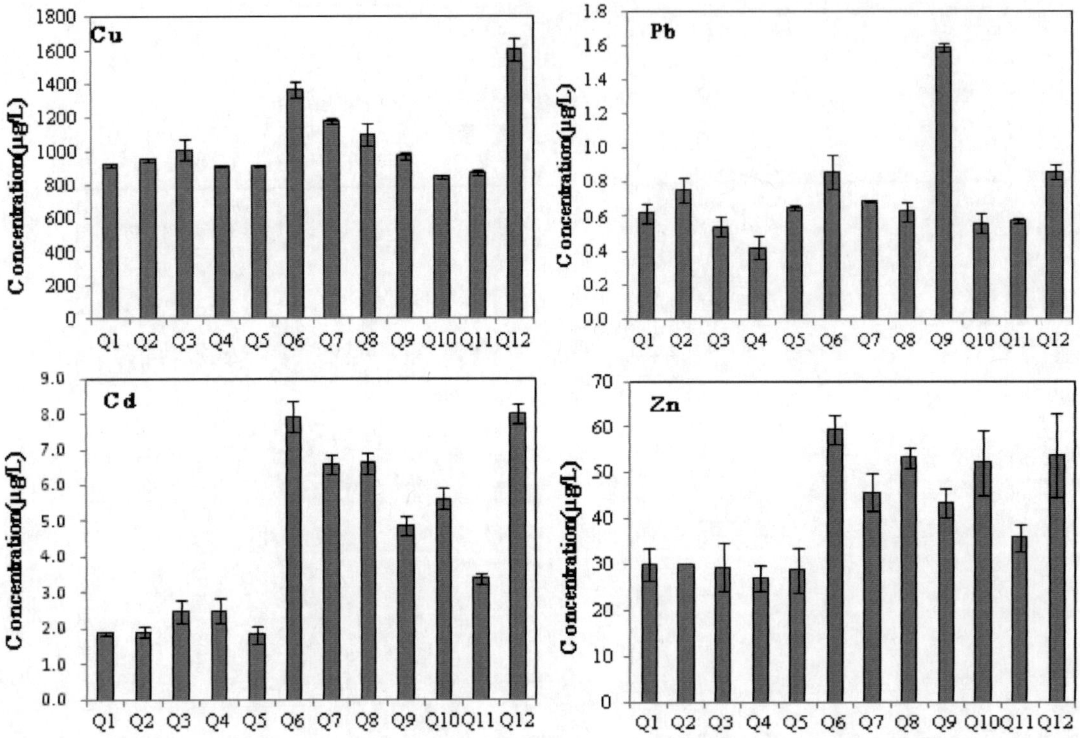

Figure 2. The concentration level in twelve stations in study area

3.3. Risk assessment of four heavy metals in surface water

According to the National Seawater Quality Standard of China (GB3097-1997), the quality of seawater is divided into four levels. Based on the site-specific concentration, Cu concentration at all stations were much higher than the standard value for Ⅳ level (50μg/L). The Cu levels were 16.96-32.06 times in 12 stations as high as 50μg/L, indicating that the study area suffered from heavy Cu pollution. For Pb, water quality at 11 stations attained Ⅰ level (<1μg/L), while only 1 stations meet Ⅱ level (1-5μg/L). In case of Cd, water quality at 7 stations attained Ⅱ level(1-5μg/L), and the other 5 stations meet Ⅲ level (5-10μg/L). In case of Zn, water quality at 8 stations attained Ⅱ level (20-50μg/L), the other 4 stations meet Ⅲ level (50-100μg/L). The results suggested that Cu pollution was particularly serious.

The ecotoxicological data of the representative species from the main taxa of marine aquatic organisms are used to develop the SSD curves. Durán and Beiras (2013) [16] have established the SSD curves of Cu, Pb, Cd and Zn according to the above method, and the curves were used as reference in the present study. The four heavy metals shows different toxicity to the tested organisms, and Cu is the most toxic metals for most of the species. The HC_5 value of each metal was obtained from its SSD curve, and they were 1.90μg/L for Cu, 36.90μg/L for Pb, 5.31μg/L for Cd and 12.36μg/L for Zn, respectively [16]. And then the HC_5 of these four metals was used for the following analysis. The concentration levels of both Cu and Zn in all stations (Figure 2) were much higher than their corresponding HC_5 value, suggesting that the aquatic organism had been affected adversely by Cu and Zn pollution. In contrast, the Pb levels in all stations were much lower than its HC_5 value, suggesting that Pb pollution was negligible and no adverse effect happened to the local organism. In case of Cd, there were 5 stations (Q6,Q7,Q8,Q10 and Q12) where the concentration of Cd was higher than its HC_5 value, while the one in the other 7 stations was lower than its HC_5 value, suggesting that Cd at Q6,Q7,Q8,Q10 and Q12 stations had negative effect on the aquatic life.

Based on the HC_5 values, the RQ values of each metal in all the stations were calculated using Eq2 (Figure 3). The results showed that the RQ values of Cu varied between 1355.41 and 2621.28, 100% of sites have a very high ecological risk with RQ values much larger than 1.0. In contrast, the RQ values of Pb ranged from 0.04 to 0.13, about 92% of sites (11 stations) showed a low ecological risk and 8% (only 1 station) showed a medium ecological risk with a RQ value between 0.1 and 1.0. In case of Zn, the RQ values ranged from 6.06 to 13.88, 100% of sites showed a high ecological risk with RQ >1. In case of Cd, the RQ values varied between 0.94 and 4.41, about 92% of sites (11 stations) showed a high ecological risk with a RQ>1.0 and 8% (1 station) showed a medium ecological risk with a RQ value between 0.1 and 1.0. According to the RQ, the ecological risk caused by metal pollution in the study area decline in order: Cu>Zn>Cd>Pb. Comparing with the Cu concentration of the surface water in other coastal area [2,6,8,17], the pollution was extremely high along the Qinhuangdao coastal area. It should arouse much attention of relevant departments, and the pollution source of Cu needs further investigation.

4. Conclusions

The concentrations of Cu, Pb, Cd and Zn in surface water along the Qinhuangdao coastal area varied between 847.81-1602.81μg/L, 0.42-1.59μg/L, 1.82-7.99μg/L and 26.9-59.36μg/L, respectively. According to the National Seawater Quality Standard of China (GB3097-1997), Cu at all the stations did not attained Ⅳ level (>50μg/L); Pb attained Ⅰ level (<1μg/L) or Ⅱ level (1-5μg/L); For Cd and Zn, water quality meet Ⅱ level (1-5μg/L for Cd, 20-50μg/L for Zn) or Ⅲ level (5-10μg/L for Cd, 50-100μg/L for Zn). Based on SSD curves, the HC_5 were 1.90μg/L for Cu, 36.90μg/L for Pb, 5.31μg/L for Cd and 12.36μg/L for Zn, respectively. The RQ values ranged from 1355.41 to 2621.28 for Cu, 0.04 to 0.13 for Pb, 0.94 to 4.41 for Cd and 6.06 to 13.88 for Zn, respectively. The results indicated that Cu and Zn at 100% sites had a high ecological risk, while Cd at 92% sites had a high ecological risk. In contrast, the ecological risk of Pb in the study area was negligible. Therefore, the ecological risk decline in order: Cu>Zn>Cd>Pb. Comparing to other coastal area, the Cu pollution was extremely high and should be paid more attention. Furthermore, the people living in this region try to eat less the seafood that is easily enriched metals such as shellfish.

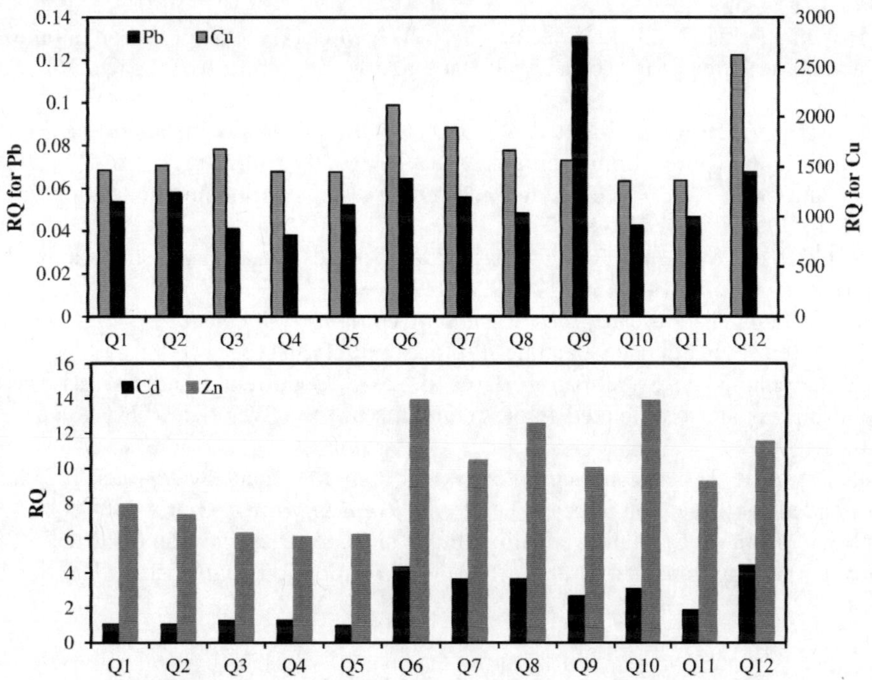

Figure 3. The RQ values of each metal in different stations

Acknowledgements
This research was financially supported by the central basic scientific research project in the Public Welfare for the scientific research institutes (gyk5091301) and the basic special project of science and technology (2013FY111100).

References
[1] Álvaro N V, Neto A I, Couto R P, et al. 2016 Crabs tell the difference-Relating trace metal content with land use and landscape attributes *Chemosphere* **144** 1377-83.
[2] EI-Sorogy A S and Attiah A 2015 Assessment of metal contamination in coastal sediments, seawaters and bivalves of the Mediterranean Sea coast Egypt *Mar Poll Bull* **101** 867-871.
[3] Deycard V N, Schäfer J, Blanc G, et al. 2014 Contributions and potential impacts of seven priority substances (As, Cd, Cu, Ni, Pb, Zn) to a major European Estuary (Gironde Estuary, France) from urban wastewater *Mar Chem* **167** 123-134.
[4] Shi Y J, Wang R S, Lu Y L et al. 2016 Regional multi-compartment ecological risk assessment: Establishing cadmium pollution risk in the northern Bohai Rim, China *Environ Int* **94** 283-291.
[5] Zhu Z M, Xue J H, Deng Y Z, et al. 2016 Trace metal contamination in surface sediments of intertidal zone from Qinhuangdao, China, revealed by geochemical and magnetic approaches: Distribution, sources, and health risk assessment *Mar Poll Bull* **105** 422-429.
[6] Wang S L, Xu X R, Sun Y X, et al. 2013 Heavy metal pollution in coastal areas of South China: A review *Mar Poll Bull* **76** 7-15.
[7] Youssef M 2015 Heavy metals contamination and distribution of benthic foraminifera from the Red Sea coastal area, Jeddah, Saudi Arabia *Oceanologia* **57** 236-250.
[8] Mendoza-Carranza M, Sepúlveda-Lozada A, Dias-Ferreira C, et a. 2016 Distribution and bioconcentration of heavy metals in a tropical aquatic food web: A case study of a tropical estuarine lagoon in SE Mexico *Environ Poll* **210** 155-165.
[9] Lim D, Choi J W, Shin H H, et al. 2013 Toxicological impact assessment of heavy metal contamination on macrobenthic communities in southern coastal sediments of Korea *Mar Poll Bull* **73** 362-368.
[10] Kibria G, Hossain M M, Mallick D, et al. 2016 Trace/heavy metal pollution monitoring in estuary and coastal area of Bay of Bengal, Bangladesh and implicated impacts *Mar Poll Bull* **105** 393-402.
[11] Rashdi S A, Arbi A A, Howari F M, et al. 2015 Distribution of heavy metals in the coastal area of Abu Dhabi in the United Arab Emirates *Mar Poll Bull* **97** 494-498.
[12] Pan K, and Wang W X 2012 Trace metal contamination in estuarine and coastal environments in China *Sci Total Environ* **421-422** 3-16.
[13] Wheeler J R, Grist E P M, Leung K M Y, et al. 2002 Species sensitivity distributions: data and model choice *Mar Pollut Bull* **45(1-12)** 192-202.
[14] ECB(European Chemicals Bureau) 2003 *Technical Guidance Document on Risk Assessment-* Part II. Institute for Health and Consumer Protection, Italy, Ispra.
[15] Cristale J, Katsoyiannis A, Sweetman A J, et al. 2013 Occurrence and risk assessment of organophosphorus and brominated flame retardants in the River Aire (UK) *Environ Pollut* **179** 194-200.
[16] Durán I and Beiras R 2013 Ecotoxicologically based marine acute water quality criteria for metals intended for protection of coastal areas *Sci Total Environ* **463-464** 446-453.
[17] La Colla N S, Negrin V L, Marcovecchio J E, et al 2015 Dissolved and particulate metals dynamics in a human impacted estuary from the SW Atlantic *Estuar Coast Shelf Sci* **166** 45-55.

Study of the effect of bacteria on the disappearance and transformation of CO in the sealed fire zone of coal mine

Xiaowei Zhai[1, 2, 5], Shibo Wu[1, 2], Jun Deng[1, 2], Yifan Yang[3], Hua Jiang[1, 4] and Kai Wang[1, 2]

[1] College of Safety Science and Engineering, Xi'an University of Science & Technology, Xi'an 710054 China

[2] Key Laboratory of Western Mine Exploitation & Hazard Prevention Ministry of Education, Xi'an 710054 China

[3] Environmental Service Center of Clean Development Mechanism in Ningxia Province, Yinchuan 750000 China

[4] Department of Mining Engineering, West Virginia University, Morgantown, WV, 26505

E-mail: langran221@163.com

Abstract. When the underground coal mine gob area has been sealed due to the coal spontaneous combustion, under the low oxygen and potentially high temperature environment, the CO concentration could drop sharply and disappear quickly. But it could rise rapidly after re-opening. These indicate that the disappearance is the only index for coal burnt out. In order to find a way how let CO disappear, experiments have been conducted using the newly developed experiment setup for three samples, raw, watered and bacteria-free coal sample. The CO and CO_2 concentration have been monitored and analyzed. The results show the bacteria in the coal do consume CO and increase the chance of CO transfer to CO_2. These results reveal how let CO disappear in a sealed zone from a new aspect. And the accuracy was improved when used gas index to determine combustion status for coal spontaneous combustion.

1. Introduction

Underground coal fire is one of the most common disasters that affects the safe production of coal mines [1]. According to statistics, there are 72.86% of coal mines that has ever occurred coal spontaneous combustion phenomenon in China [2]. A phenomenon was found from several coal mine fire accidents (such as the fire accident of Duan Poqiao coal mine of Hunan province in 2008 and the coal mine fire accident of Liaoning province in 2005 [3]). After a gob fire being sealed, the CO concentration inside the gob can drop as low as 0 ppm. However, as soon as the gob was re-opened, the CO concentration can rise rapidly in a short period (1-3 days), which may cause the 2nd seal of the gob. Therefore, it can be found that not only the extinction of coal fire but also other factors can lead to the disappearance of CO. Wang [4] has measured the adsorbing capability of coal sample to CO at the different temperature by high-pressure volumetric method to study the variation of CO concentration in sealed gob area. They concluded that the adsorption of coal seam to CO was not accord with the monomolecular layer adsorption model by theoretical analysis. The coke produced by coal combustion can react with CO_2 to form CO under high temperature condition [5], while a large amount of CO can be generated during coal oxidization [6]. However, CO may react with O_2 to produce CO_2 in the process of coal combustion [7]. The generation rate of CO_2 produced by oxygen oxidation of CO is very small in the atmosphere air [8].

There are a diversity of aerogenes in the coal seam. According to the three-stage theory of anaerobic fermentation, the aerogenes can be divided into the hydrolysis and fermentation bacteria, hydrogen-producing aerogenes and methanogenus [9, 10]. Orem et al. [11, 12] confirmed that the microbes obtained from coal seam or coal seam water can transfer coal into methane via experiment. Wang [13] obtained stable gas-producing bacteria group by enrichment from Lignite coal. Conrad & Seiler [14] believed that CO can be oxidized by certain organisms, these organisms can achieve benefits which can only obtain from the oxidation process of CO (such as obtain the energy required for life activities). Some chemoautotrophic bacteria can oxidize a big amount of CO to form CO_2 under the common temperature range (5℃-30℃) and the oxidation rate of CO gradually increased with the increasing of temperature [15]. Carbon source and water are important factors that affect the growth of bacteria [16]. Therefore, it can be deduced that bacteria have a positive effect on the disappearance and transformation of CO in gob. Experimental apparatus and technology for studying the effects of bacteria on the disappearance of CO has been designed and set up to start a new perspective in this paper. Experiments have been conducted for raw coal, raw coal with water and bacteria free raw coal under same and different pressure in an airtight vessel, the variation of concentration of CO and CO_2 has been monitored and analysed. These results can improve the accuracy when using the CO index to determine combustion status in sealed gob.

2. Sample preparation and experiment Process

2.1. Sample preparation

2 kg coal sample were selected in the size range from 0.9mm to 3mm and stored in a nitrogen environment for 24 hours. Then the experimenters made three equal weight (370g) sample labelled as A, B and C. For sample B, 20 mL pure water has been added into the vessel; for sample C, with UV light irradiation for 4 hours. Then the experimenters packed three coal samples into three sterile vessels and sealed them. All these processes were operated under bacteria-free environment. The three vessels were sterilized by high-pressure steam and were dried by vacuum wind drum. The sterilized gas was used during this experiment was sterilized by air sterilization filter (model DR-LT-2S).

2.2. Experiment Procedure

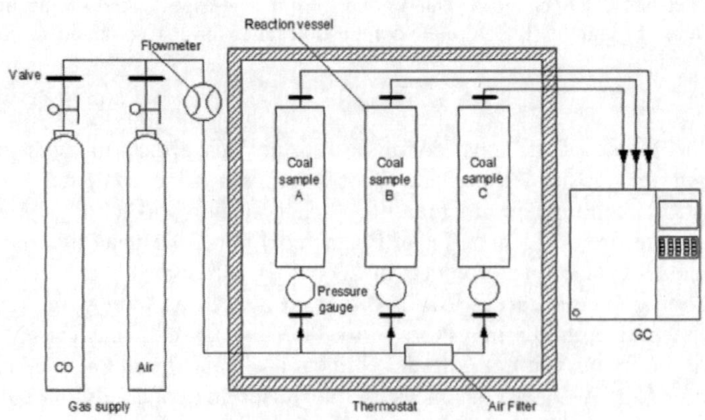

Figure 1. Experiment setup and schematic.

The experiment schematic was shown in Figure 1. Raw coal sample A and watered sample B have been pumped 300 mL atmosphere air and 150 mL CO respectively. Bacteria free sample C has been pumped the same amount of sterilized gases. Then three vessels are sealed and have been put into the thermostat. According to Zhang's [15] study, the appropriate temperature for this experiment is 30℃. The gas sample has been collected every 12 hours and analysed by gas chromatograph after the pressure in the vessels gets re-balanced. Another two sets of experiments with different CO injection amount, 75 mL and 40 mL, have been conducted as well.

3. Results and discussion

3.1. CO concentration comparison under same amount of gas injection

As the generation rate of CO_2 is very small in the atmosphere air [17], this paper isn't considered the effect of oxygen oxidation of CO on experimental results. And the total volume of gas is not changed in the experimental process because of the sealing vessels. After processing the reduction rate of CO concentration and the rise rate of CO_2 concentration by two cumulative method, Figure 2, 3 and 4 have been plotted, respectively. In the figures, A CO, B CO, C CO refer the reduction rate of CO concentration in sample A, B and C, respectively. A CO_2, B CO_2, C CO_2 indicate the rise rate of CO_2 concentration in sample A, B and C, respectively. The effects of bacteria on the disappearance of CO has been analysed as follow.

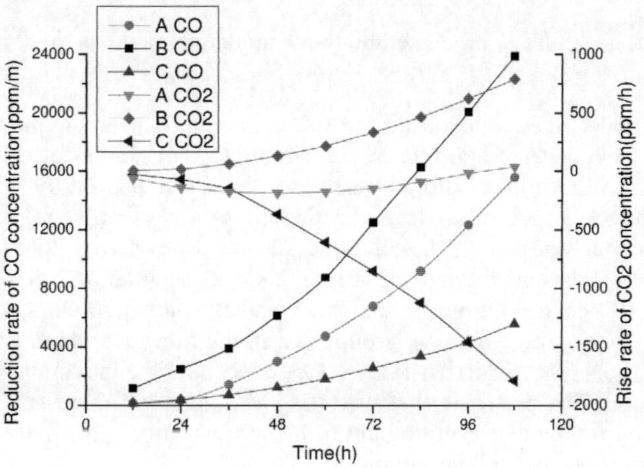

Figure 2. Rate tendency processed by two cumulative methods in 150 mL CO gas.

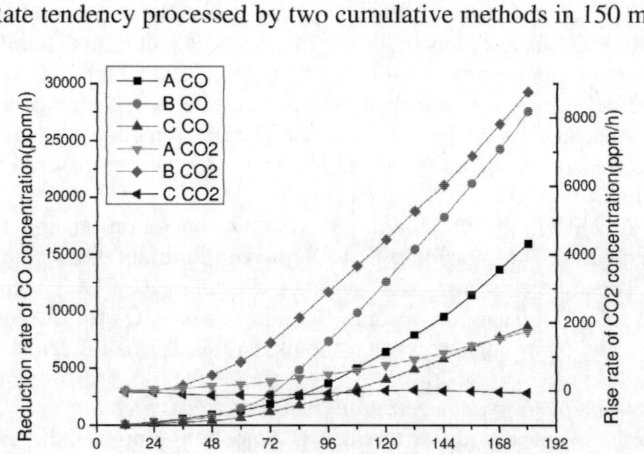

Figure 3. Rate tendency processed by two cumulative methods in 75 mL CO gas.

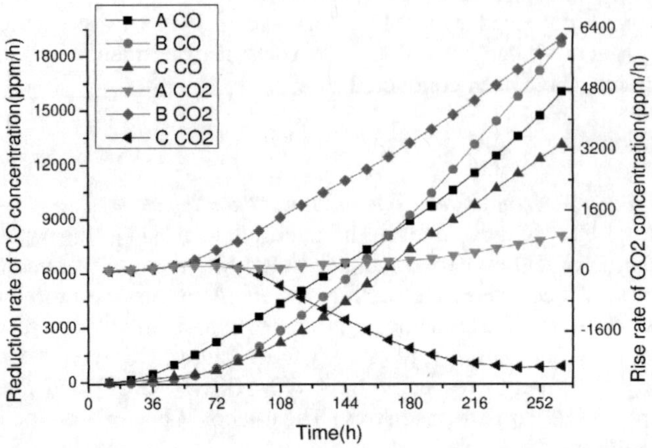

Figure 4. Rate tendency processed by two cumulative methods in 40 mL CO gas.

From Figure 2, 3 and 4, the reduction rate of CO concentration for sample C is positive number, which indicates that CO concentration reduces in sample C. The reason of this phenomenon is that coal samples can adsorb CO in the sealing vessels, which can be proved by Wang's [4] study. The production of CO and CO_2 are much less and limited as a result of coal oxidation in the low-temperature oxidation stage of coal [17], so that aerial oxidation of coal doesn't affect experimental results in the paper. It's clear that the reduction rate of CO concentration has an increasing trend and the reduction rate of CO concentration for sample A and B is greater than sample C. This indicates that some kinds of CO consuming bacteria group exist in the raw coal which accelerate the reduction rate of CO concentration. The reduction rate of CO concentration in sample B is the greatest and sample A is the second. The reason is that water in the sample promotes the rapid propagation of bacteria in the coal and resulted a larger amount of bacteria in sample B. So the reduction rate of CO concentration in sample B is bigger than sample A.

In Figure 2, 3 and 4, the rise rate of CO_2 concentration for sample A and B is positive number, which indicates that CO_2 concentration rises in sample A and B. But the rise rate of CO_2 concentration for sample C is negative number, which indicates that CO_2 concentration reduces in sample C. The reason is that coal samples can adsorb CO_2 in the sealing vessels, which can be proved by [18]. The rise rate of CO_2 concentration has a rising trend in both samples A and B, especially in sample B. But the rise rate of CO_2 concentration drops in sample C. The phenomenon turns out that the bacteria in the coal promote the generation of CO_2 and this effect becomes more obvious when there more bacteria exist in the coal.

In both samples A and B, the reduction rate of CO concentration and the rise rate of CO_2 concentration have increased. But in sample C, CO concentration and CO_2 concentration decrease at the same time. Thus, it can be concluded that the bacteria can cause CO concentration to reduce and CO_2 concentration to rise. The bacteria consume CO and release CO_2 to form organic matter to keep vital activity and reproduction. So the process promotes the transformation from CO to CO_2.

3.2. CO concentration comparison under different amount of gas injection

For each of the three samples with different CO injection amount, the rate tendency of CO concentration has been plotted in Figure 5, 6 and 7 by two cumulative methods, respectively. In the figures, 150mL CO, 75mL CO, 40mL CO refer the reduction rate of CO concentration in sample with different CO injection amount, 150mL, 75 mL and 40 mL, respectively.

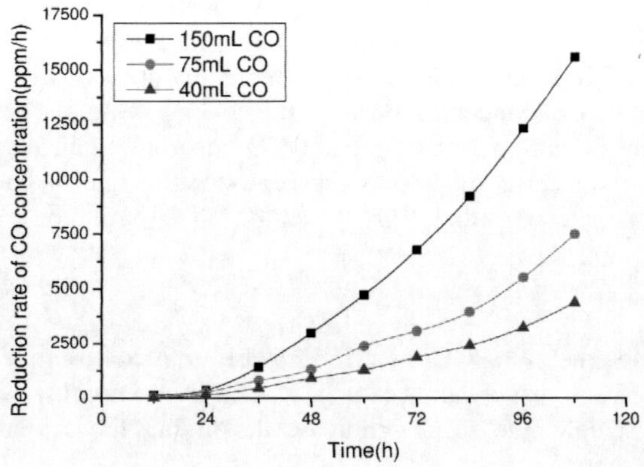

Figure 5. Rate tendency processed by two cumulative methods in sample A.

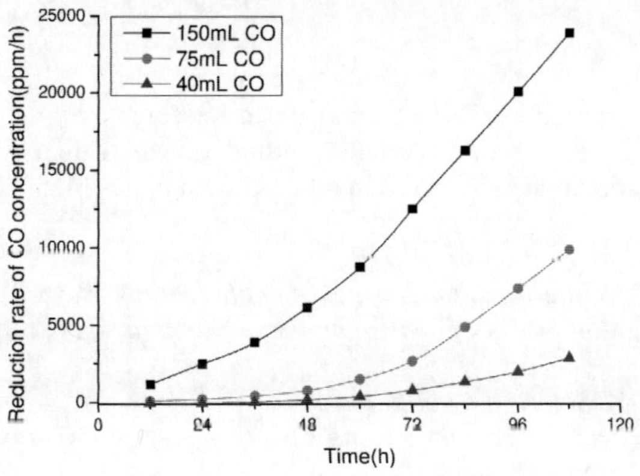

Figure 6. Rate tendency processed by two cumulative methods in sample B.

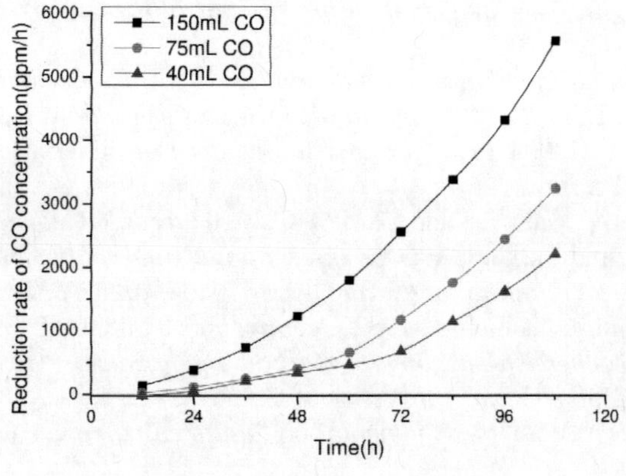

Figure 7. Rate tendency processed by two cumulative methods in sample C.

Different amount of CO injected to the same coal sample can cause different pressure and CO concentration in the sealing vessels. Figure 5, 6 and 7 show for the same sample the reduction rate of CO concentration is the greatest for 150 mL CO injection and the smallest for 40 mL injection. These indicates the reduction rate of CO concentration is also affected by pressure. The higher the pressure is, the greater the reduction rate of CO concentration is in the sealing vessels. There are two reasons for this phenomenon. One reason is that coal has a greater CO adsorption capacity under high pressure. And another one is that CO concentration affects the reproduction rate of bacteria. The higher the reproduction rate of bacteria is under high CO concentration environment, the faster the rate of bacteria consuming CO is.

4. Conclusions

From the comparison of same amount of CO injection for different coal samples, it can be concluded that bacteria have accelerated the disappearance of CO. And by comparing with different injection amount for the same coal sample, the results indicate the higher the pressure is, the greater the reduction rate of CO concentration is.

Under normal temperature, bacteria take a positive effect on the disappearance of CO and generation of CO_2. The reason is that chemoautotrophic bacteria consume CO to form organic matter and release CO_2 during vital activity. In general, the existence of bacteria accelerates the transformation of CO to CO_2.

Acknowledgments

This work was financially supported by The National Basic Research Program (973 Program) (Grant Nos. 2014CB260404), the National Natural Science Foundation of China (Grant Nos. 51404195), Project Funded by China Postdoctoral Science Foundation (Grant Nos. 2015M572582).

References

[1] Gray J and Colaizzi 2004 *International Journal of coal Geology* **59** 75-81
[2] Xu J C 2001 *Determination theory of coal spontaneous combustion* (China: China Coal Industry Publishing House)
[3] Yang X Q and Li N 2010 *Safety in Coal Mines* **9** 4-7
[4] Wang Y H 2006 *Research on regulation and effect factors of CO absorption by coal seams* (China: Hebei Polytechnic University)
[5] Yu Z 2011 *Chinese Journal of Inorganic Analytical Chemistry* **1** 77-80
[6] Zhu L Q, Zhou X Q, Xie J G and Zhu Y C 2008 *Journal of Mining & Safety Engineering* **25** 440-4
[7] He Z G 2004 *The engineering and application of controlling air pollution* (China: Chemical Industry Press)
[8] Wu Z B 2002 *The technology of controlling air pollution* (China: Chemical Industry Press)
[9] Gong W J, Li W Z and Liu J Y 2007 *Journal of Northeast Agricultural University* **3** 838-841
[10] Ren N Q and Wang A J 2004 *Principle and application of anaerobic biotechnology* (China: Chemical Industry Press)
[11] Orem W H, Voytek M A, Jones E J and Clark A C 2010 *Organic Geochemistry* **41** 997-1000
[12] Harris S H, Smith R L and Barker C E 2008 *International Journal of Coal Geology* **76** 46-51
[13] Wang A K and Qin Y 2011 *Journal of China University of Mining & Technology* **6** 888-893
[14] Conrad R, Seiler W, Bunse G, Giehl H 1982 *Geophys-Res* **87** 8839-52
[15] Zhang Y 2008 *Biogeochemical cycling of carbon monoxide in the St-Lawrence estuarine system(Canada)* (China: Ocean University of China)
[16] Ji X H and Zhang J J 2001 *Medical immunology and medicine microbiology* (China: Science Press)
[17] Guo X Y, Wang D M and Li J S 2011 *Coal Engineering* **5** 102–4
[18] Jian X, Guan P and Zhang W 2012 *Sci China Earth Sci* **42** 492–504

IOP Conf. Series: Earth and Environmental Science **52** (2017) 012086 doi:10.1088/1755-1315/52/1/012086

Simulation study on the impact of air distribution on formaldehyde pollutant distribution in room

Jingtao Wu[1, 3], Jun Wang[1] and Zhu Cheng[2]

[1] School of Aeronautic Science and Engineering, Beijing University of Aeronautics and Astronautics, Beijing 100191, China

[2] China Aircraft Strength Research Institute, Xi'an 710065, China

E-mail: wuxiaodan324@163.com

Abstract. In this paper, physical and mathematical model of a room was established based on the Airpak software. The velocity distribution, air age distribution, formaldehyde concentration distribution and Predicted Mean Vote(PMV), Predicted Percentage Dissatisfied(PPD) distribution in the ward of a hospital were simulated. In addition, the air volume was doubled, the change of indoor pollutant concentration distribution was simulated. And further, the change of air age was simulated. Through the simulation, it can help arrange the position of the air supply port, so it is very necessary to increase the comfort of the staff in the room. Finally, through the simulation of pollutant concentration distribution, it can be seen that when concentration of indoor pollutants was high, the supply air flow rate should be increased appropriately. Indoor pollutant will be discharged as soon as possible, which is very beneficial to human body health.

1. Introduction

Computational fluid dynamics (CFD) is based on the heat transfer, mass transfer and momentum transfer theory. In addition, it combines with a number of significant technologies, such as multiphase flow technique, combustion process and chemical reaction etc. It is applied very commonly in many fields, for example environmental engineering field, mechanical field, thermal energy and power field, aerospace field and chemical field. What's more, the field of heating, ventilation and air conditioning is one of the important application field. We can use CFD technology by establishing mathematical model and setting reasonable boundary conditions and related parameters to simulated airflow distribution in certain space [1].

However, the software Airpak is further developed on the basis of the FLUENT software. Compared with Fluent software Gambit in modeling, Airpak software is very convenient for modeling and it provides wall, staff, tuyere and tables and chairs and other modules. Different modules can also be set the corresponding parameters according to the actual situation [2].

In the grid division for Airpak software, it is also based on the Fluent software processor and the grid contains also a variety of shapes. For the two-dimensional flow model, it can be set for both rectangular and triangular form grid. For three-dimensional flow model, it can be set into a triangular prism, tetrahedron and hexahedron mesh. At the same time, it also provides encrypted setting function of the model. The important modules can be targeted encrypted in order to make the results more accurate. Equations can be selected based on the flow state correspondingly. For the turbulent flow model, zero equation model, indoor zero equation model, two equation model can be chosen to calculate. The calculation result of Airpak can also be used with Tecplot postprocessing software for further processing. Flow chart of the Airpak calculation process is shown in Figure 1.

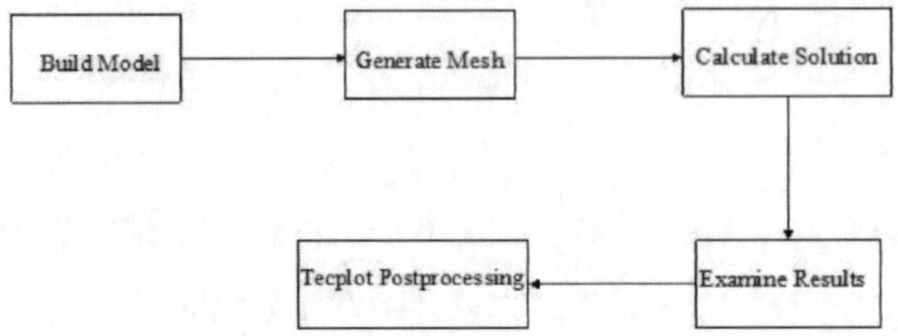

Figure 1. Airpak calculation flow chart.

Through the detection of different types of indoor pollutant concentration, indoor air pollutants have a great effect on human comfort. Indoor air quality becomes more and more poor and it is high time to take certain measures to improve the indoor air quality for our health.

Indoor air quality (IAQ) is defined in a specific environment, some elements of people's life and work suitability in the air. It is a concept which reflects the specific requirements of people. How to evaluate the quality of the indoor environment is one of the main problems that people concerned [3]. It includes indoor air temperature, humidity, air freshness and cleanliness and so on.

In recent years, due to the popularity of air conditioning system in modern architecture and improvement in the airtight construction, the indoor harmful gas can't be discharged. The carbon dioxide emissions increased, resulting in air conditioning disease, Sick Building Syndrome (SBS) and other sub-health states [4]. Now, the research of indoor air purification has become one of the most concerned problems in the world [5]. Further research and discussion on the influence of indoor air quality on human health is very important.

Formaldehyde, a kind of popular indoor air pollutant, is an endogenous chemical. It could cause human asthma and syndrome like asthma. Formaldehyde is the main pollutant after the room decoration [6].

In this paper, taking one of the room for an example, three dimensional model was established for the ward. The computational domain was determined for the numerical model and boundary conditions were set. And then, turbulence model was selected and the division of grid was done. In the end, the indoor airflow distribution in the ward can be simulated by calculating with computer. And further by importing formaldehyde pollutant diffusion model, the pollutants distribution in the ward was simulated. Through the simulation study, it provides references for ventilation scheme selection and ventilation system design in order to improve indoor air quality in the room.

2. Air flow simulation in the room

2.1. The establishment of the numerical model

Because various boundary conditions can be set, when the numerical simulation model was built, a representative room can be selected as the calculation model instead of calculating the whole building. Thereby, on the one hand, it reduces the complexity of modeling work. On the other hand, it also improves the accuracy of the model calculation and computing speed [7].

Numerical model was set up by using Airpak software, the detail parameters are listed as follows:

(1) The model is a standard room which contains two beds and a small bathroom in the room. Length, width and height of the ward were 5.3 meters, 3.6 meters, 3.6 meters respectively. The length, width and height of toilet were 2.1 meters, 1.5 meters, 1.5 meters respectively. The thickness of the outer wall was 0.24 meter and the thickness of the inner wall was 0.09 meter. The length, width and height of the coffer were 0.5 meter, 0.4 meter, 0.4 meter and the bed 2 meters, 1 meter, 0.63 meter.

(2) To establish a coordinate system: the width of the room acts as the X axis right direction, the length of the room being the Y axis right direction and the height of the room being the Z axis right direction.

(3) Both the air supply outlet and exhaust outlet adopt opening model. The entrance of air supply outlet is set to speed type and the exhaust outlet being pressure type. The air supply outlet sends the air from the roof and exhaust outlet also exhausts the air from the roof.

(4) It was assumed that the wall of the room was adiabatic.

(5) The indoor zero equation (the indoor zero equation) model was chosen so as to calculate more accurately. In this paper, calculation domain of a typical room was shown in Figure 2.

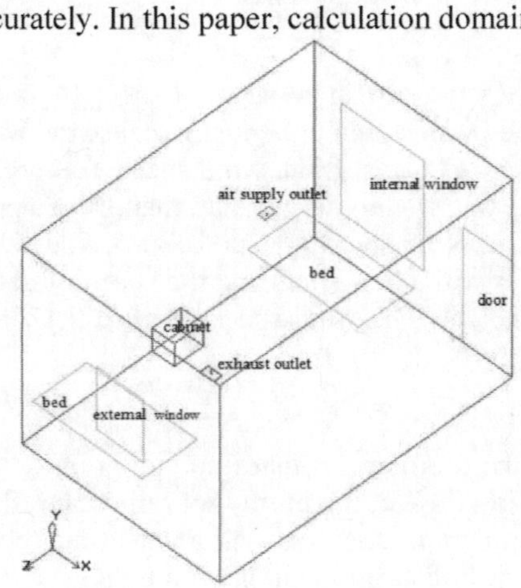

Figure 2. Model of the typical ward.

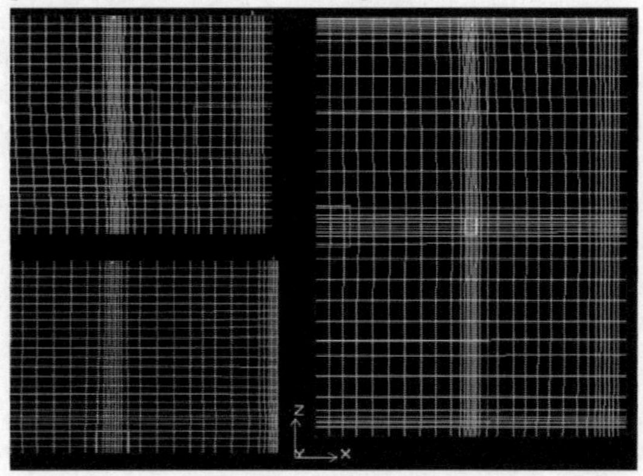

Figure 3. Grid of model.

2.2. Calculation of the numerical model

In order to make the computing grid more homogeneous and the calculation result more accurate, the parameters were set as follows:The minimum number of calculation grid for solid boundary was set to 2. Maximum size of the X axis direction is 0.18. The maximum size of the Y direction is 0.18. Maximum size of the Z axis is 0.265. At the same time, the air supply outlet and exhaust outlet were encrypted processing. Through the calculation of Airpak, the grid can be gotten as shown in Figure 3.

3. Analysis of the numerical simulation calculation results

3.1. The velocity distribution

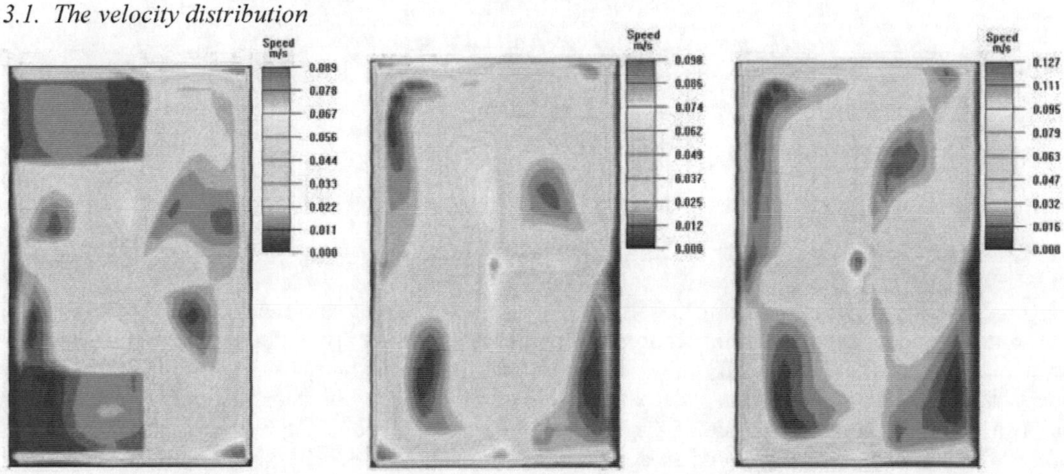

Y=0.65m Y=1.2m Y=1.5m

Figure 4. Velocity distribution cloud map of different cross sections.

Because the height of the bed is 0.63 m, when the patient laid in bed, breathing area is around 0.65 m high from ground in the Y right direction. Therefore, researching on the distribution of the velocity field can be more accurately analyzing personnel comfort [8]. In addition, when people stand, the height of person's breathing zone was between 0.8m and 1.5m. The analysis and calculation of velocity field in this height can be very necessary. So in this article take Y = 1.2 m and Y = 1.5 m to analyze the velocity field. Therefore, take Y=0.65m, Y=1.2m, and Y=1.5m three cross section separately to analyze velocity field and the results were shown in Figure 4.

From the Figure 4, in the Y=0.65m section, wind speed was lower than 0.089m/s. Especially in the area in the upper part of the both sides of the bed, whose speed was within 0.03 m/s. Because the wind speed was small, people won't feel a sense of blowing. In the Y=1.2m section, wind speed was within 0.098 m/s. In the corridor between the two beds, air velocity was slightly larger than that of the upper part of the bed and the maximum speed was 0.098m/s. Because the speed was not big, people won't feel a sense of blowing. In the Y=1.5m section, the speed was within 0.127m/s and the speed of under the air diffuser was larger than the other area in the same height. The wind speed reached 0.127m/s and the researchers also won't feel the sense of blowing [9].

3.2. The air age distribution in room

Indoor air age refers to the time when air particles enters the room and reaches certain point of the room, which is the time of old air being replaced by new air. Air age, therefore, not only can reflect the freshness of the indoor air but also can response the ventilation effect of each point in the room. Age of indoor air can be used to evaluate the pros and cons of air distribution in the room [10]. Air age has a direct relationship with the air quality of breathing. In order to be able to form bright contrast, we selected three different cross sections to calculate. They were Z = 0.7 m, Z=2.65m and Z=4.6m respectively. The Z=0.7m section is near the door on the side of the center of the bed, Z=2.65 m is the center of the room and Z= 4.6m section is far away from the door on the side of the center of the bed. The air age distribution in room was shown in Figure 5.

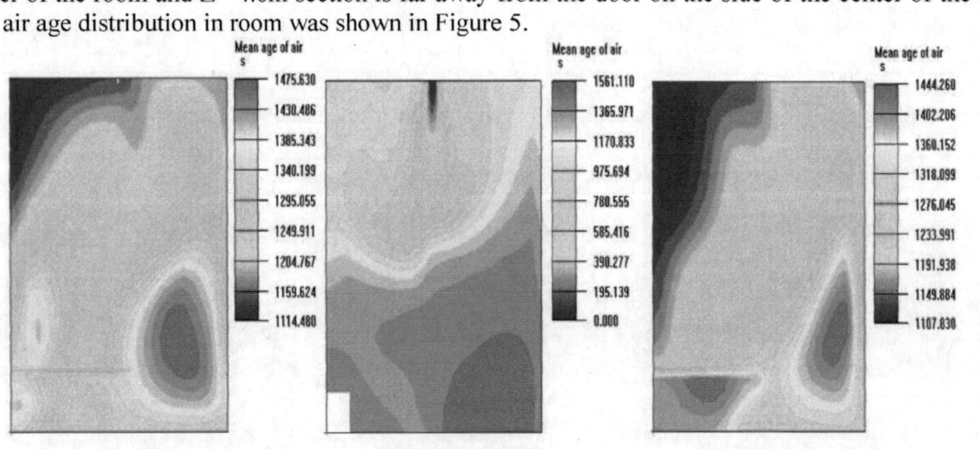

Z=0.7m Z=2.65m Z=4.6m

Figure 5. The cloud map of indoor air age distribution in different sections.

From the Figure 5, in Z=2.65m section, air age near the air supply outlet was minimum and the value was 0. The air at here was fresh, which points out that the reliability of the result [11]. As can be seen from the diagram, the air age value on the side of the bed near the door was greater than that of far away from the door on the side of the bed. Age of air at the end of the bed was maximum in the area. At Z=2.65m section, maximum air age was 1561.110s. It showed that the pollutant concentration was larger here and phenomenon of pollutants gathering was more serious. Therefore, people should not stay in the area for a long time.

3.3. *The formaldehyde pollutant distribution in room*

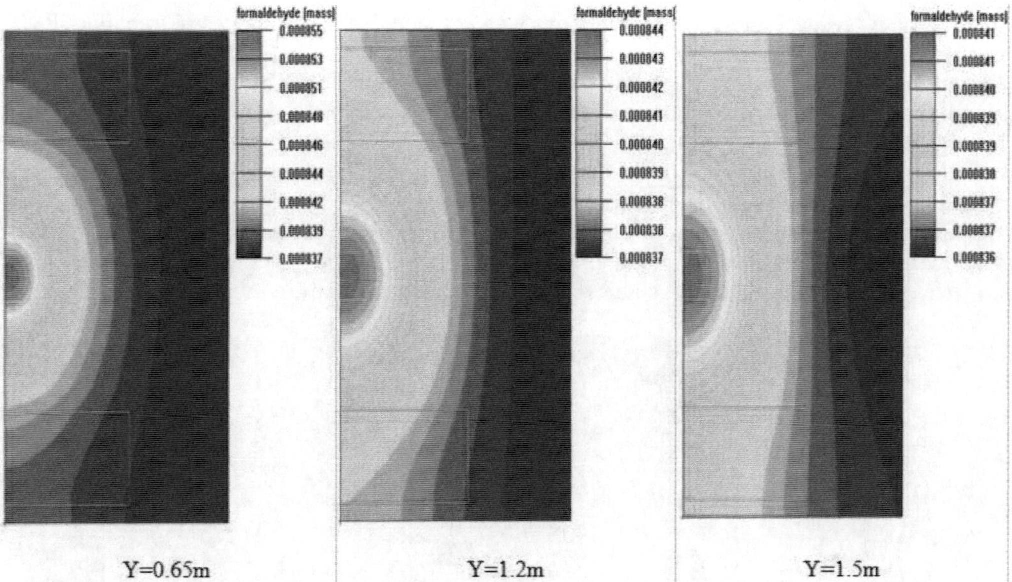

Figure 6. Formaldehyde concentration distribution cloud map in different cross sections.

From the Figure 6, along the height direction of the room, pollutants concentration continuously reduced. Pollutants spread from the diffusion source to outward and the concentration also gradually reduced. Pollutant dispersion origination concentration reached 0.855mg/m3. Due to the less time of air supply volume, the lowest concentration of pollutants concentration in the whole room also reached as high as 0.836 mg/m3. Therefore, it needs to take a breath by ventilation for a long time in order to realize the aim to eliminate pollutants and reach the limited values of indoor pollutant concentration [12]. As to the patients, it is suggested that the farther the pollutant source, the better for the health.

3.4. *Distribution of Predicted Mean Vote (PMV) and Predicted Percentage Dissatisfied(PPD) in room*

Thermal comfort index PMV and PPD are the subjective feeling of thermal environment. PMV index represents the cold and hot feeling of the majority of people in the the same environment. So PMV evaluation index can evaluate the human body's comfort or not. But usually as a result of great difference in human physiological and psychological aspects, the PMV index only is not comprehensive, which does not reflect the extent of each person's hot and cold feeling and comfort. Therefore Predicted Percentage of Dissatisfied (PPD) was put forward to indicate the percentage of dissatisfied with thermal environment.

PMV index adopted 7 level degrees, as shown in Table 1.

Table 1. PMV index of thermal sensation.

thermal sensation	hot	warm	slight warm	moderate	slight cool	cool	cold
PMV value	+3	+2	+1	+0	-1	-2	-3

According to IS07730 standard, PMV, PPD indexes were given the recommended value, namely the PMV value should be between 0.5~+0.5. But it was allowed to have 10% of people still feeling not satisfied with comfort of indoor thermal environment. By calculating, the distribution of PMV and PPD were shown in Figure 7 and Figure 8.

From the Figure 7 and Figure 8, in Y=0.65m, Y=1.2m, and Y=1.5m three cross sections, PMV index is in the range of -0.5~ +0.5, so under this condition, thermal sensation of the indoor personnel was between moderate and micro warm and comfort level is higher. But by comparison, when the patient lay in bed, it was closer to moderate level. At this time personnel comfort was better than that of the personnel standing. It is suggested that the patient should lay in bed for the health.

PPD index was less than 10% in different cross sections. It proved that the environment temperature and wind speed were suitable for people's life [13]. With the increase of height, personnel dissatisfaction percentage also increased gradually. When the patient lay in bed, the dissatisfaction percentage was lower than that of standing. It also proves that the PMV and PPD have some relationship in evaluating the personnel comfort. And it further verifies the correctness of result.

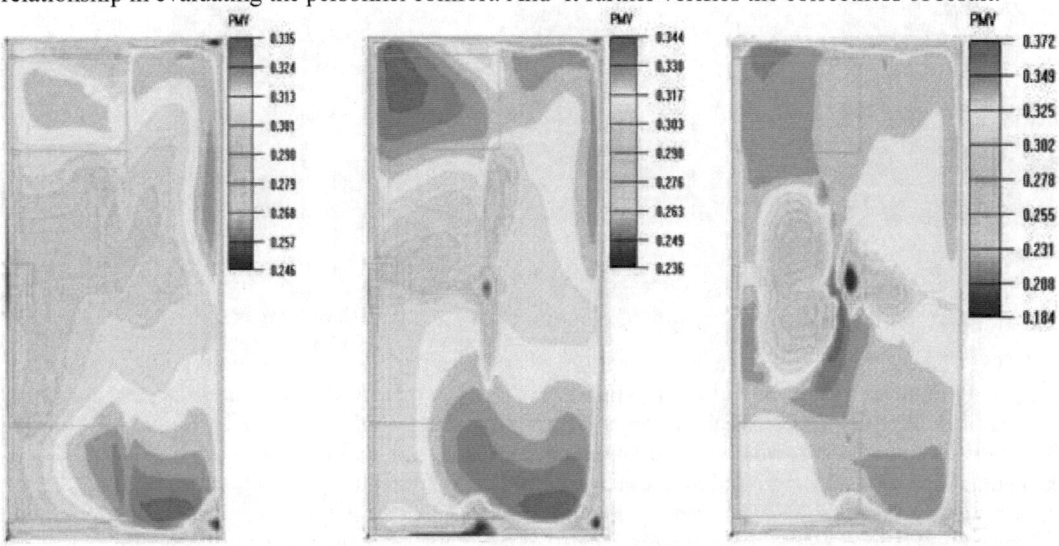

Y=0.65m Y=1.2m Y=1.5m

Figure 7. PMV distribution cloud map in different cross-sections.

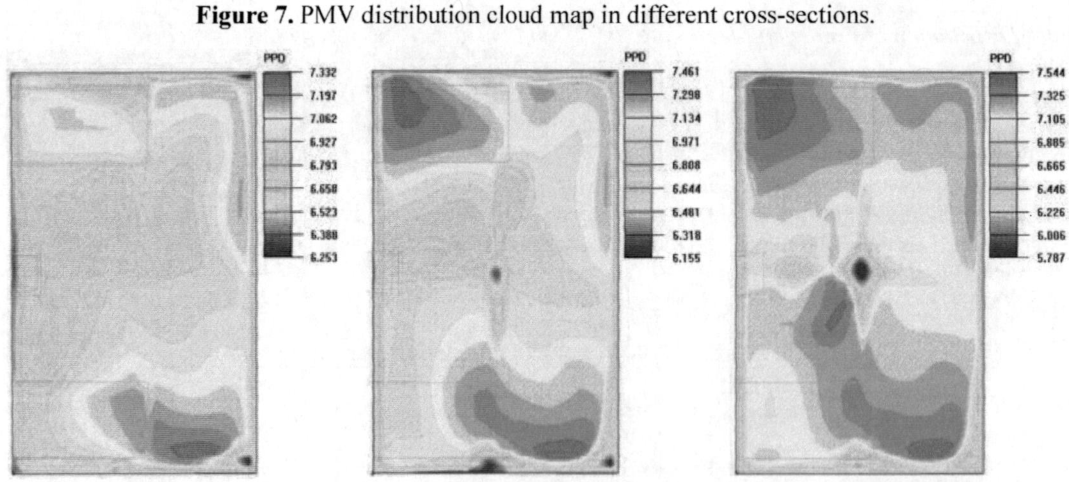

Y=0.65m Y=1.2m Y=1.5m

Figure 8. PPD distribution cloud map in different cross sections.

4. Summary

In this paper, physical and mathematical model of a room was established with the the Airpak software. The velocity distribution, air age distribution, formaldehyde concentration distribution and PMV, PPD distribution in the ward of a hospital were simulated. Through the simulation, it can help arrange the position of the air supply port, so it is very necessary to increase the comfort of the staff in the room. Finally, through the simulation of pollutant concentration distribution, it can be seen that when concentration of indoor pollutants was high, the supply air flow rate should be increased appropriately. Indoor pollutant will be discharged as soon as possible, which is very beneficial to human body health.

References

[1] Zheng Ping 2006 Study on the influence of natural ventilation on indoor air quality *Master thesis of Chongqing University*
[2] Spengler J D, Chen Q 2000 Indoor air quality factors in designing a healthy building Annual Review of Energy and the Environment **25** 567-601
[3] Min J K, Richard D B, et al. 2015 Indoor air quality control for improving passenger health in subway platforms using an outdoor air quality dependent ventilation system *Building and Environment* **92** 407-417
[4] Tom B D, Michael S W 2016 Impact of natural versus mechanical ventilation on simulated indoor air quality and energy consumption in offices in fourteen U.S. cities *Building and Environment* **104** 320-336
[5] Jun Wang, Yong Chen 2016 Concentration characteristics of ozone and product for indooroccupant surface chemical reaction under displacement ventilation *Building and Environment* **130** 378-387
[6] Walid A H, Nesreen G, et al. 2016 Optimized performance of displacement ventilation aided with chair fans for comfort and indoor air quality *Building and Environment* **127** 907-919
[7] Niu Jianlei 2004 Some significant environmental issues in high-rise residential building design in urban areas *Energy and Buildings* **36(12)** 1259-63
[8] Wu Di 2006 Airpak simulation of indoor decoration formaldehyde emission *Master thesis of Chongqing University*
[9] Chen Jing, Zhu Neng, Lu Nan, et al. 2012 Detection and evaluation of air quality of two kinds of public places in Tianjin area *Journal of Beijing University of Technology* **38(2)** 295-299
[10] Li Heping, Zhong Ke 2007 Effect of fresh air volume and air distribution on removal of indoor pollutants *Clean and air conditioning technology* 18-21
[11] Jiao Junjun, Gong Guangcai, Shao Chunsheng 2008 Numerical simulation of the influence of air distribution on indoor air quality *National HVAC Proceedings of the 2008 academic annual conference*
[12] Mats Sandberg 2004 Building ventilation theory measure *Environmental Science and Technology* **37(5)** 472-473
[13] Yang X, Srebric J, Li X, et al. 2004 Performance of three air distribution systems in VOC removal from an area source *Building and Environment* **39(11)** 1289-99

IOP Conf. Series: Earth and Environmental Science **52** (2017) 012087 doi:10.1088/1755-1315/52/1/012087

Thermodynamic evaluation of the impact of strongly swelling polymer hydrogels with ionic silver on the water retention capacity of sandy substrate

A V Smagin

Institute of Forestry, Russian Academy of Sciences, ul. Sovetskaya 21, Uspenskoe, Moscow oblast, 143030 Russia; Moscow State University, Leninskie gory 1, Moscow, 119991 Russia

Email: smagin@list.ru

Abstract. The impact of two types of strongly swelling polymer hydrogel (SSPH) on the water retention capacity of quartz sand in pure water and Ag^+ solutions (10-100 mg/l) has been studied by using a centrifugation method in a wide range of thermodynamic water potential (Gibbs energy) from 0 to 3030 J/kg. The experimental data for the water retention curves (WRC) were estimated by the van Genuchten model. Both hydrogels – the Aquasorb preparation (Germany) with hydrophilic properties and high degree of swelling in pure water (700–1000 g H_2O/g) and the new Russian amphiphilic SSPH with a peat filler (degree of swelling 500-700 g H_2O/g) were very effective as water adsorbing soil conditioners in relatively small doses from 0.05 to 0.3% per mass of dry (105°C) soil substrate. The water retention capacity of sandy substrate increases under the influence of SSPH with 2-3 times up to the level of native loamy sands and loams. Adding Ag^+ to the water solution results just for the highest concentration of SSPH (0.3%) and iconic silver (100 mg/l) in a significant decrease of the water retention in the soil-gel compositions.

1. Introduction

A new generation of synthetic materials – strongly swelling polymer hydrogels (SSPH) belong to a class of super water absorbers and are used in agricultural melioration practices, gardening and landscaping to improve soil water-holding capacity and structure [1-8]. The high water retention capacity of the SSPH is used, which is characterized by an equilibrium free-state swelling up to 700-1000 g H_2O per gram of soil substance. However, the soil as biologically and chemically active porous media has negative impact on the SSPH due to biodegradation, osmotic stress, limited pore space, envelope pressure, etc. These factors can strongly decrease the SSPH efficiency in the soil that requires the development of more suitable and resistant media materials. Incorporating the monogenic groups into the polymer matrix causes a blockade of the osmotic mechanism of the gel's swelling decrease (destruction of the ion-electrostatic barrier), according to [1]. In addition, we suggest to fill the polymer matrix by natural amphiphilic components like humates or peat, that increases the resistance of SSPH to microbial destruction in the soil on one side and improves their quality as structural agent and effective lock of microelements and crop protector in the root zone on the other [9]. This research compares a new type of acryl SSPH filled by peat with traditional hydrophilic preparation (trade mark "Aquasorb") in pure water and Ag^+ solution as microbial inhibitor and crop protection agent.

It is clear that an increase in water-holding capacity by SSPH of one soil layer is insufficient for optimization of its moisture regime, since the latter is dictated by water movement in the entire profile, together with the water influx and discharge (evapotranspiration) rates. Consequently, for proper evaluation of the efficacy of the SSPH, as well as any other soil conditioner, a system analysis of of water and polymer behaviour in a given soil-bioclimatic condition is required. Such analysis is conveniently achieved using a computer model of water transfer in soils affected by SSPH, which should account for the properties and dynamics of the polymer as well [4]. A central part of the computer models (see for example HYDRUS-1D [10]) is a hydraulic block, which requires input data on the changing soil water-retention curve (WRC) or the dependences between soil moisture and thermodynamic potential of soil water (specific Gibbs energy) affected by SSPH. So in the current research, special attention is paid to instrumental evaluation of WRC in the soil substrate and its alteration affected by SSPH. The objective was, besides collecting experimental results of WRC as a function of the doses (concentration) of SSPH in soil substrate, to estimate the WRC by the standard van Genuchten model using in contemporary computer media for the modelling of energy and mass transfer in soils [11]. Thus determination of WRC as a basic thermodynamic characteristic of the "soil-gel" system allows not only comparison of different types of SSPH but also obtains important information for the future technological modelling the water accumulative soil construction based on SSPH [4].

2. Materials and methods

The tested materials included two types of SSPH: the first sample (№1) is well-known in floriculture and gardening, namely acryl hydrophilic Aquasorb preparation (Germany) with high degree of swelling in pure water (700–1000 g H_2O/g) and the second sample (№2) is the new Russian amphiphilic acryl SSPH with a peat filler and medium degree of swelling (500-700 g H_2O/g) [12]. The new product (sample №2) along with the co-polymer of acrylamide and sodium acrylate contains as filler (23.5%) finely divided (less than 0.25 mm) peat; it has not characteristic for the SSPH black colour. Soil substrate in experimental studies served as the sample monomineral medium-grained quartz sand (so-called white "glass sand") was used. The SSPH doses 0.05; 0.1; 0.2; 0.3% per dry mass of mineral soil substrate were used. The experiments were conducted in pure deionized water and in water solutions of 10-100 mg/l Ag^+ in nitrate form. It is assumed that incorporation of Ag^+ can protect SSPH as well as the root zone in case of SSPH application from adverse microbial activity and pathogenic microorganisms. The samples were prepared by placing a calculated amount of dry SSPH in a measuring cup, which then was filled with distilled water or Ag^+ solution; the amount required to saturate a given soil substrate sample weight to a a state of full capacity (water saturation). After swelling the gel substance was mixed with mineral soil substrate, allowing it to achieve a uniform distribution in the host material. Otherwise, (in the case of mixing dry gel) it is virtually impossible because of the low doses of SSPH, not exceeding 0.3 g per 100 g of the mineral mass. All compositions were studied in 3-fold replicates.

Thermodynamic analysis of water-retention capacity of soil samples and their compositions with hydrogels was carried out by equilibrium centrifugation with author's modification [13] with a high-speed laboratory centrifuge Hettich Universal 320 (Germany) with a water-retention energy range (soil water potential or equivalent soil water pressure) from 0 to 3030 J/kg (kPa). With the centrifugation method water was removed from the sample under the effect of centrifugal force. The developed pressure on the liquid phase (P) or equivalent of soil water potential in first approximation (small or so-called undistributed samples) can be determined from the equation [13]:

$$P = \rho_\ell \omega^2 R h \quad (1)$$

where ω is the rotation speed, ρ_ℓ is the liquid's density, R is the distance from the rotation axis to the mass centre of the sample. The construction of the high-speed Hettich centrifuge involves an inclined position of the sample with respect to the rotation axis; therefore, a correction factor cos(α), where α is the angle between the horizontal and the central symmetry axis of the sample, should be included in Eq. (1). For low speeds regimes of the centrifuge, the gravity component of the pressure

(P_g = g·*h*·sin(α), where *h* is the sample's height, and g is the acceleration of gravity) should be taken into consideration. The rotation speed (ω) is calculated from the number of centrifuge revolutions per minute (*N*) using the equation $\omega = \pi N/30$, where [ω] = s^{-1}. Therefore, the final equation for determining the absolute values of the soil water pressure (or equal potential *ψ*) depending on the centrifugation parameters for an undistributed sample has the following form [13]:

$$|P|\ [\text{kPa}] = |\psi|\ [\text{J/kg}] = -\ \{0.011{\cdot}n^2 R{\cdot}\cos(\alpha) + g{\cdot}h{\cdot}\sin(\alpha)\}{\cdot}h, \quad (2)$$

where [*N*] = rpm, [*R*] = m, and [*h*] = m, |*P*|; |*ψ*| are absolute values as the water pressure and potential in porous media have negative sign by definition.

Experimental WRC data were approximated using the van Genuchten [11] model, which relates the variable values of pressure (P) and soil moisture (W) by the following nonlinear relationship:

$$W = W_r + \frac{(W_s - W_r)}{\{(1+(\alpha P)^n\}^m}, \quad (3)$$

where W_r - residual moisture, corresponding to almost a fixed tightly bound moisture, W_s - moisture in a state of saturation of the soil (full water capacity), *α, n, m* - empirical constants, and $m = 1 - 1/n$. To estimate the parameters W_r, W_s, *α*, *n* according to experimental data for WRC as a sequence of *W(P)* was used the program S-Plot, version 7, which allows to solve the problem of non-linear regression by iteration method through the built-in least squares algorithm (Regression Wizard).

3. Results of study

The WRC of sandy soil substrate had a characteristic shape, as was shown by the analysis of experimental results (Figures 1; 2, control) and known literature data [11, 14, 15]. Weak energy of retention in coarse media with large pores caused an almost complete drainage in a very narrow range of water suction (about 10 kPa). At higher suctions, the angle of the WRC increases steeply, since the water remaining in few fine pores at the bridges between coarse particles and sorbed water are retained by mush stronger forces. A field water capacity (FWC) parameter, which is determined, according [14] as cross point between the WRC and the line with equation |P| [kPa]=$10^{(1.17+W\%/100)}$ (trace line in Figs. 1-2), is not exceeding 5% in this material. It means that the main amount of water did not retain in the soil due to gravitation losses of water as the differences between Ws and FWC are equal to 22% or more than 80% of the porosity. Hence, because of the wide spread especially in sandy soils in arid environments, there is a need for an improvement of the water retention and SSPH can be perspective agents for this purpose. The comparative analysis of the WRC of a sandy substrate under the effect of two different SSPH showed that all the studied samples usually increased the water retention capacity proportionally to the preparation rate in a whole range of absolute soil water potentials from 0 to 3030 J/kg (Figures 1; 2). The FWC value in the presence of the SSPH (0.2–0.3 %) exceeded 18–25% compared to 5% in the control. Thus, the water retention at these rates increased with 3–5 times and corresponded to the parameter changing from the region of the sandy substrate to that of the loamy substrate at the addition of organic substance in small amounts (0.2-0.3%). Small doses SSPH (0.05-0.1%) were also effective and they lead to an increase in water retention up to the level of loamy sandy soils with FWC=12-15% and more (Figures 1; 2). Both types of SSPH show the same results despite on the differences in degree of free swelling. This is easy to explain if we take into account that SSPH never reached their maximal swelling capacity in pore space [9]. At water saturation level of the sand-gel composition (Ws = 40-50%) the SSPH degree of swelling as ratio "Ws/dose" will be equal to 100-500 gH_2O/g or less than the maximal degree of free swelling, which can reach 700-1000 gH_2O/g for Aquasorb super absorber. This circumstance allows to produce agriculture products less expensive and more resistant to unfavourable soil factors like new Russian preparations filled by humates or peat and mineral components with medium degree of swelling [9]. Inclusion of ionic groups of electrolytes in the acrylates and humates form in the structure of a polymer matrix allows to solve the problem of reducing the SSPH swelling as affected by osmotic stress [1]. In this case, embedded cations can exchange ions and serve as a source of elements of plants mineral nutrition. As shows the experiment with SSPH, swelling in a solution of ionic silver, the composition based on it did not worsen much the water retention properties, which remained at the frames of standard deviations of studied WRC. Only high dose of Aquasorb SSPH (0.3%) at maximal concentration of Ag^+ (100 mg/l) significantly lost swelling capacity at whole diapason of water retention energy (Figure 1-C). The SSPH at the more effective, from the economical point of view, doses 0.05-0.1% were resistant to osmotic stress. This allows us to recommend a simple technological way of injecting water-soluble plant protection products and eliciting admixtures directly into the matrix solution while preparing gel compositions from dry preparations before these compositions enter the soil. We can assume that as well for the hydrophobic agents, being used in crop production as fungicides and antimicrobials, such mechanical embedding in gel compositions on basis of peat and their following strengthening will be successful owing to amphiphilic polymer structure.

Table 1. Parameters of WRC approximation by van Genuchten model (3)

SSPH rates and solutions:	Model parameters:				Statistics:	
	W_s	n	α	W_r	R^2	s
Monomineral quartz sand (control)						
0%	22.7±0.7	2.25±0.20	0.32±0.05	1.33±0.30	0.996	0.68
Hydrophilic Aquasorb SSPH (Germany).						
0.05% pure water	32.9±1.8	2.10±0.31	0.18±0.04	3.97±0.84	0.981	1.74
0.1% pure water	38.7±1.7	1.68±0.16	0.14±0.04	4.84±1.28	0.998	1.66
0.1%+Ag^+10mg/l	36.4±2.0	1.98±0.90	0.15±0.04	4.89±1.10	0.973	2.07
0.1%+Ag^+100mg/l	38.6 ±1.1	1.87±0.13	0.16±0.02	4.23±0.66	0.994	1.13
0.2% pure water	49.2±2.5	1.40±0.13	0.21±0.10	3.79±2.05	0.984	2.36
0.2%+Ag^+10mg/l	45.7±2.0	1.50±0.13	0.21±0.07	5.52±2.15	0.988	1.87
0.2%+Ag^+100mg/l	44.3±2.6	1.38±0.14	0.25±0.16	-	0.973	2.36
0.3% pure water	47.3±1.8	1.37±0.10	0.27±0.10	4.33±3.03	0.990	1.65
0.3%+Ag^+10mg/l	48.6±2.4	1.30±0.13	0.18±0.10	1.42±0.83	0.992	2.25
0.3%+Ag^+100mg/l	48.6±0.6	1.39±0.03	0.42±0.05	3.07±0.79	0.999	0.50
Amphiphilic SSPH with a peat filler (Russia)						
0.05% pure water	33.4±0.93	1.77±0.13	0.06±0.01	2.23±0.89	0.995	1.08
0.1% pure water	37.1±2.3	1.81±0.30	0.18±0.07	5.84±1.37	0.973	2.22
0.1%+Ag^+10mg/l	35.3±2.2	1.83±0.28	0.15±0.05	5.04±1.33	0.976	2.16
0.1%+Ag^+100mg/l	37.3±2.0	1.82±0.24	0.14±0.04	4.78±1.29	0.982	2.04
0.2% pure water	40.7±2.3	1.67±0.22	0.13±0.05	5.60±1.93	0.979	2.33
0.2%+Ag^+10mg/l	39.6±2.0	1.50±0.17	0.19±0.08	5.77±2.30	0.973	1.96
0.2%+Ag^+100mg/l	38.6±1.5	2.00±0.22	0.11±0.02	7.00±0.83	0.990	1.52
0.3% pure water	51.9±2.56	1.49±0.16	0.21±0.09	7.74±2.88	0.983	2.43
0.3%+Ag^+10mg/l	48.1±2.4	1.31±0.13	0.23±0.12	7.77±3.79	0.984	2.19
0.3%+Ag^+100mg/l	48.9±2.8	1.28±0.14	0.20±0.12	-	0.989	2.60

- no significant differences from zero

At the end of the paper, we briefly touch on the mathematical simulation of WRC for soil substrate under the impact of SSPH. Almost all the WRC was adequately described by the van-Genuchten function with the nonlinear regression coefficients $R^2 \geq 0.97$-0.99, estimation parameters α and m (n) statistically significant at the probability level of $p \leq 0.001$, and standard errors not exceeding confidence intervals of variation of the $P(W)$ dependences studied (Table 1). All the curves in Figures 1; 2 represent the results of experimental data approximation by the van-Genuchten equation (3). Reliable approximation results by the van-Genuchten equation allow us to use the computer models of HYDRUS-type [10] in the future where it is the standard function to forecast the dynamics of water consumption and plants productivity under the SSPH effect.

Summarizing the results of assessing the water retention capacity we can conclude that a significant increase in the water retention capacity was revealed for the mineral light-textured soil substrate, which corresponded to the heaving of the soil texture by 1–2 gradation levels at SSPH rates of 0.05–0.3%. The highest effect is observed for the nonsaline samples (pure water), however incorporation 10-100 mg/l ionic silver into the soil solution leads to not significant decreasing of water retention of the 0.05-0.1% SSPH compositions. It allows us to inject soil-gel compositions with various substances, including stimulants and inhibitors of the biological activity without substantial damage to the main water absorption function, and thus to regulate the intricate set of biophysical processes inherent to the soil.

Acknowledgments

This work was supported by the Russian Scientific Foundation, project no. 16-16-04014.

References

[1] Kazanskii K S, Rakova G V, and Enikolopov N S 1986 Strongly swelling polymer hydrogels as water-retaining soil amendments, in Natural Resources of Deserts and Their Development, Ashkhabad, Ylum 147-148 [in Russian]

[2] Al-Darby A M, Al-Asfoor S I, and El-Shafei Y Z 2002 Effect of Soil Gel-Conditioner on the Hydrophysical Properties Sandy Soil *J. Saudi Soc. for Agric.Sci.* **1**(1) 14-40.

[3] Smagin A V and Sadovnikova N B 1994 Impact of Strongly Swelling Hydrogels on Water-Holding Capacity of Light-Textured Soils *Eur. Soil Sci.* **27** (12) 26–34

[4] Smagin A V and Sadovnikova N B 2015 Creation of Soil-Like Constructions *Eur. Soil Sci.* **48** (9) 981–990 doi10.1134/S1064229315090100

[5] Raju M P and Raju K M 2001 Design and synthesis of superabsorbent polymers *J. of Appl. Polymer Sci.* **80**(14) 2635–39

[6] Wu L and Liu M 2008 Preparation and characterization of cellulose acetate-coated compound fertilizer with controlled-release and water-retention *Polymers for Advanced Technologies* **19**(7) 785–792

[7] Demitri C, Scalera F, Madaghiele M, Sannino A, and Maffezzoli A 2013 Potential of Cellulose-Based Superabsorbent Hydrogels as Water Reservoir in Agriculture *International J. of Polymer Science* **2013** Article ID 435073, 6 pages, http//dx.doi.org/10.1155/2013/435073

[8] Narjary B, Aggarwal P, Kumar S, and Meena M D 2013 Significance of Hydrogel and its application in agriculture *Indian Farming* **62**(10) 15-17

[9] Smagin A V, Sadovnikova N B, and Nikolaeva E I 2014 Thermodynamic Analysis of the Effect of Strongly Swelling Polymer Hydrogels on the Physical State of Soil and Sediment Samples *Eur. Soil Sci.* **47**(2) 78–88. doi 10.1134/S1064229314020100

[10] Šimůnek J, van Genuchten M T, and Šejna M 2006 The HYDRUS Software Package for Simulating Two- and Three-Dimensional Movement of Water, Heat, and Multiple Solutes in Variably-Saturated Media. Technical Manual, Version 1.0, PC Progress, Prague, Czech Republic, 241

[11] van Genuchten M T 1980 A closed form equation for predicting the hydraulic conductivity of unsaturated soils *Soil Sci Soc. Am. J.* **44** 892–898

[12] Budnikov V I, Fedchenko V N, Drobinin D V, Kus'mitsky G E, Sinkin V V, Lokotkov A N, Smagin A V, and Nazarov V B 2014 The composite water-holding material and its production method *Patent RU* №2536509, data 27.12.2014
[13] Smagin A V, Sadovnikova N B and Mizuri Maauia Ben Ali 1998 The Determination of the Primary Hydrophysical Function of Soil by the Centrifuge Method *Eur. Soil Sci.* **31**(11) 1237-44.
[14] Voronin A D 1990 Energy Concept of the Physical State of Soils *Eur. Soil Sci.* **23**(5) 7-19
[15] Dexter A R 2004 Soil physical quality part I. Theory, effects of soil texture, density, and organic matter, and effects on root growth *Geoderma* **120** 201–214.

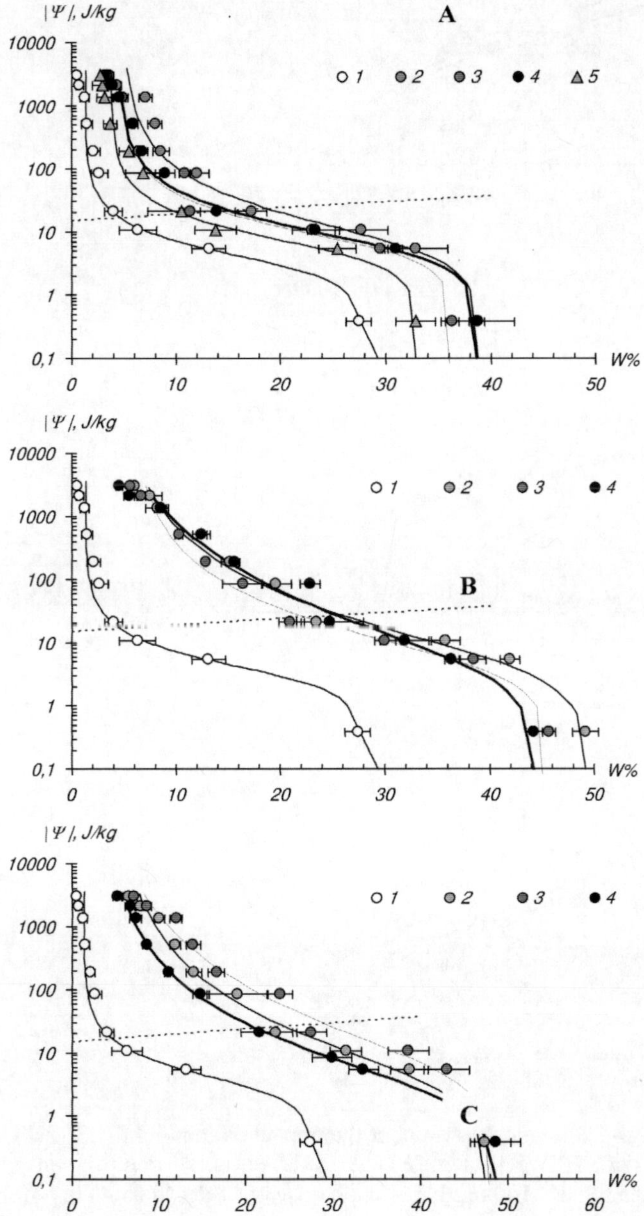

Figure 1. WRC of mineral-gel compositions under the impact of hydrophilic Aquasorb SSPH (Germany). Doses of SSPH: (A) 0.1%; (B) 0.2%; (C) 0.3%; 1 – control; concentration of Ag+: 2 – 0 mg/l; 3 – 10 mg/l; 4 – 100 mg/l; 5 – small dose of SSPH (0.05%) in pure water.

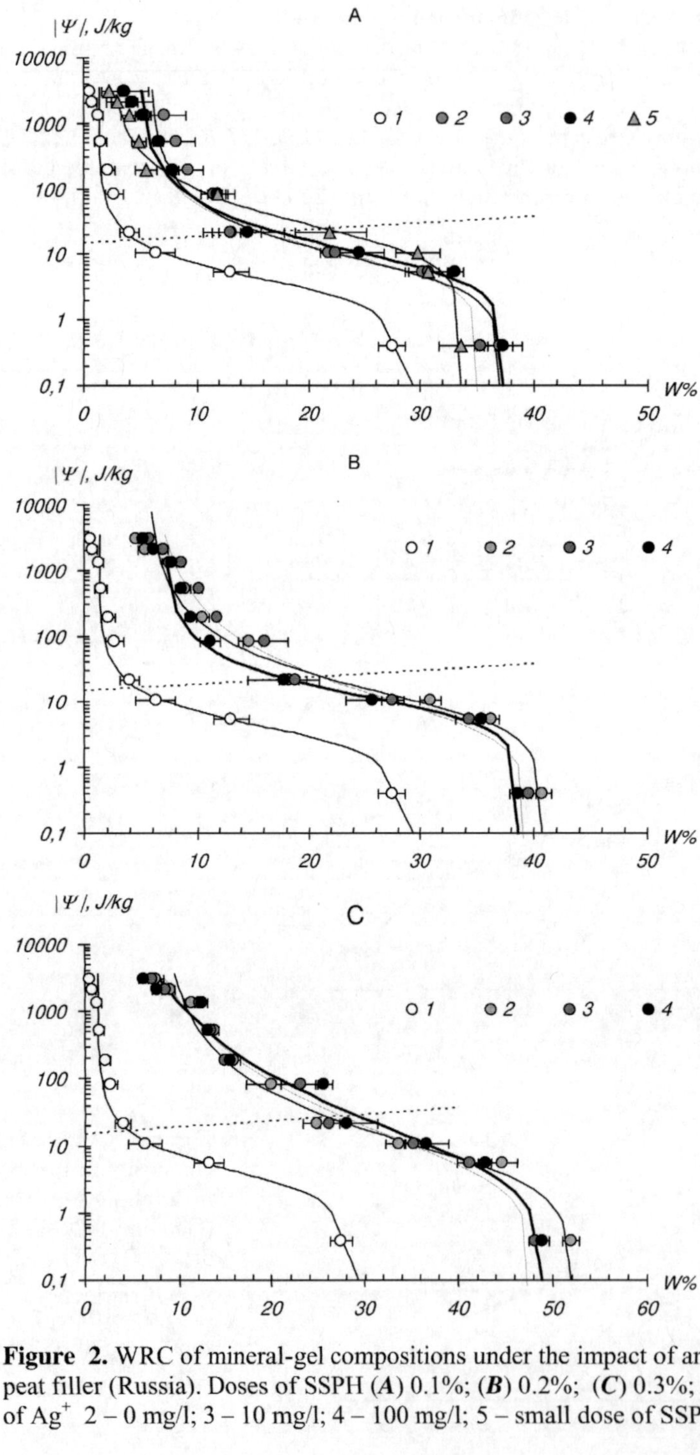

Figure 2. WRC of mineral-gel compositions under the impact of amphiphilic SSPH with a peat filler (Russia). Doses of SSPH (***A***) 0.1%; (***B***) 0.2%; (***C***) 0.3%; 1 – control; concentration of Ag^{+} 2 – 0 mg/l; 3 – 10 mg/l; 4 – 100 mg/l; 5 – small dose of SSPH (0.05%) in pure water.

Long-term remote sensing monitoring coal mining activity in resource-based cities: a case study of Qitaihe City, Northeastern China

L Liu and J S Zhou[1]

School of Humanities and Economic Management, China University of Geosciences; No 29, Xueyuan Road, Haidian District, Beijing 100083, China.

E-mail: 13601005388@163.com

Abstract: Mining activity has strongly impacted the sustainable socioeconomic development of resource-based cities. The systematic monitoring of the change in mining activity can provide evidence for the transition and future development of resource-based cities. This paper chose Qitaihe, one of the four coal mining cities in northeastern China as the study area. Remote sensing and Geographical Information Systems (GIS) technique, as well as methods on landscape pattern analysis were used to study the evolution of mining activity from 4 different periods over 58 years' time. Results showed that the area of land used in mining increased by about six times during the study period with cultivated land the main type that contributed to this increase. Mining activity showed an eastward trend, developing from one concentration circle to four circles, from a disordered system to a relatively integrated system. It was also suggested that differentiated policies should be adopted in different mining circles. This study also provides a framework for future city planning and sustainable development.

1. Introduction

Mining activity has made a significant contribution to regional and national economics, creating new job opportunities and revenue flows. Despite the contribution of the resource sector to economic growth, questions are frequently raised about the concomitant negative social, economic and environmental impacts [1]. Over these years, these questions have become a worrying subject in resource-based cities all around the world. Thus, it is of great significance to monitor the change of mining activity and its impact on the city and the country as a whole.

Remote sensing data provides an objective, reliable, safe, and cost-effective way to monitor mining activities and their environmental and societal impacts [2]. In recent times, remote sensing monitoring mainly focused on the land use and cover change caused by mining activities [3], or specific environmental problems such as land subsidence as a result of mining activity [4]. Few studies focused on the dynamic change of mining activity and its spatial variation. Focusing on Qitaihe city in northeastern China's Heilongjiang province; this paper tracks the changes mining activity has brought to the city over the years. These changes can be seen as representative of the wider impact to mining on northeastern China, specifically the four coal cities which comprise Longmei mining holding group. Remote sensing conducted in this area included landscape pattern change from 2000-2010 [5], vegetation coverage change from 1993-2011[6], and landscape ecological security monitoring from 1984-2014 [7]. However, most of these studies concentrated mainly on land cover change of the overall area; no studies have focused on the evolution of mining activity or on a relatively long period high temporal frequency and spatial resolution.

In this paper, we combine a topographic map of the year 1957 and 1979 with Landsat images from 1985, 2000 and 2015 to study the temporal and spatial process of mining activities in Qitaihe city over 58 years. Methods including dynamic indexes, transfer matrix, landscape pattern indexes and centroid algorithm are used to present the evolution of mining activity in the city.

2. Study area

Located in the east of Heilongjiang province, Qitaihe city is one of the four famous coal cities together with Jixi city to the south, Shuangyashan city and Hegang city to the north (see the Figure. 1). The city is also in the belt of northeast economic zone with an easy access to Russia in the north. Geographically, it is located between longitude 130°06′-131°58′E, latitude 45°16′-46°37′N, and lies in the narrow convergence of Zhangguangcai range and Wanda range. Qitaihe has a semi-humid continental climate, and the average annual temperature is 4.0℃.

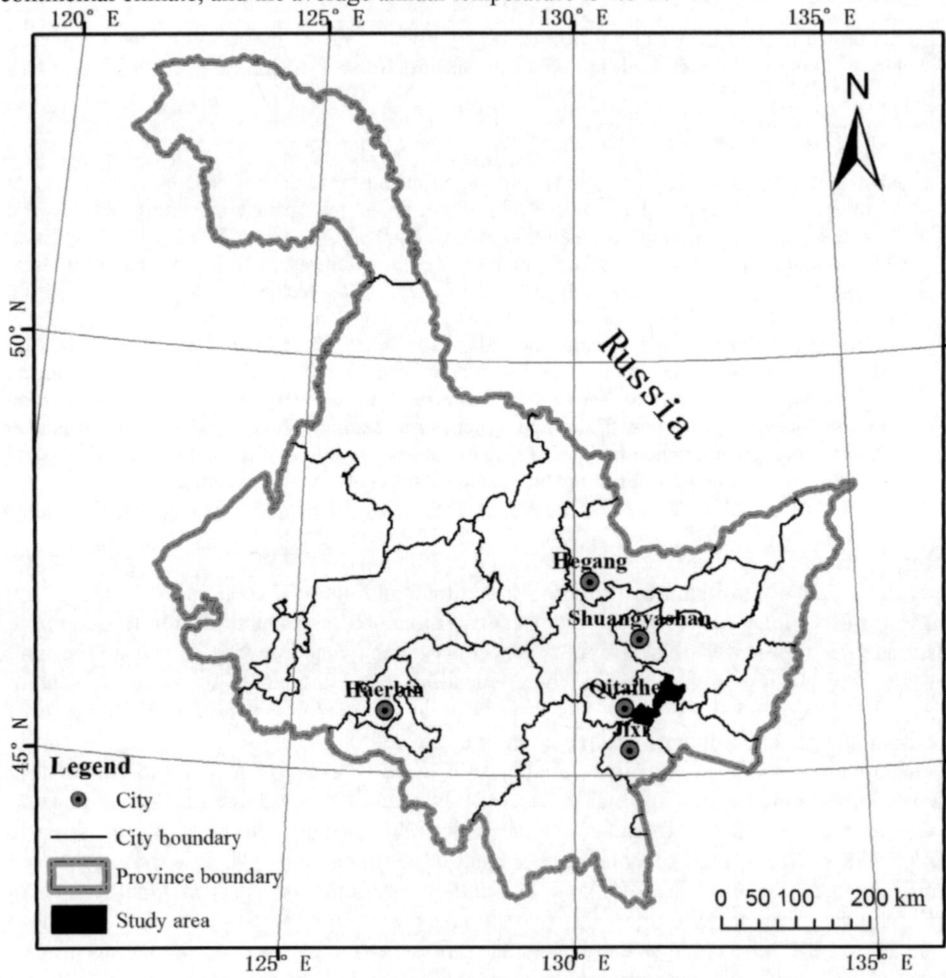

Figure 1. Location of the study area

The city is rich in coal with the proven reserves exceeding 1.7 billion tons. Complete types and high quality make it one of the three rare coal fields that need protective mining in China and the largest coking coal production base in northeastern China. Mining started in 1958 and expanded at a rapid pace. Over the past few decades, mining activity caused changes in the population, economy, society and landscape of the area and also contributed to a series of geological disasters and environmental problems which hindered regional development. Covering an area of 6223 km^2, Qitaihe city is divided administratively into 3 districts and 1 county, this paper makes the central urban area of Qitaihe city as the area of interest.

3. Materials and methods

3.1. Data sources

The remote sensing data used to derive the temporal and spatial evolution of the mining activity and land cover type resulting from the mining activity comprised Landsat TM, and OLI images with a 30m spatial resolution. We acquired TM images for the year 1985 and 2000, and OLI images for the year 2015. In addition, the topographic map with 1:50.000 resolution of the year 1957/1979 as well as land use map with 1:100.000 of the year 2014 were used for geometry correction and verification. The details of data used were given in Table 1.

Table 1. Data used for monitoring of coal mining in Qitaihe city

Data	Path/row	Data of acquisition	Scale or Resolution	Number
Topographic map		1957/1979	1:50000	2
Land use map		2014	1:100000	1
Landsat5	150/28	1985-10-14	30	1
Landsat5	150/28	2000-9-1	30	1
Landsat8	150/28	2015-9-15	30	1

3.2. Visual interpretation

In this study, a classification system for the land use/cover in the region has been introduced according to national land use classification standard and previous studies on this field [8] as well as the research need (Table 2).

Table 2. Land use/cover classification system used in the study

Type	Description
Cultivated land	Dry farmland, irrigable land
Forest	Trees, shrubs, sparse vegetation, and other forested lands
Grass land	High coverage grassland, medium coverage grass land, low coverage grass land
Water area	River, lake, reservoir or pond,
Construction land	Urban area, rural residential area, other built-up area
Mining land	Open-pit area, overburden-dump area
Unused land	Sandy land, saline land, bare land, bare rock, other unused land, swamp

Before interpretation, false-color composite images were used according to the characteristics of different bands and statistical characteristics, band 8,5, and 3 were used for Landsat OLI 8 in 2015; band 7,4,2 were used for Landsat TM 5 in 1985 and 2000.

The geometry rectification of the topographic map of 1957 and 1979 were conducted using Erdas 2014. Then the rectified topographic map of 1979 was used to check the Landsat 5 TM data of 1985.

More than 10 control points were selected with the mean location error less than 1 pixel. The resolution with 30m was adopted to ensure the matching of the spatial data. Finally, the rectified TM image of the year 1985 was used to check other Landsat TM images.

Visual interpretation was based on image characteristics such as size, shape, tone, texture, site association and pattern [9]. Visual interpretation work was done using GIS 10.X. During interpretation, the following principles were adopted: (1) the minimum mapping patch was 6*6 pixels; (2) indoor validation work was done to the interpretation results using the land use map of 2014; (3) the overall accuracy interpretation was greater than 90%.

3.3. Models

3.3.1. Dynamic index of mining land The dynamism of a land use class represents change in quantity of a certain land use class in a unit time [9], so is a key index for evaluating spatial change of a land class. The dynamism of mining land can be calculated according to the following formula:

$$R=\frac{U_a-U_b}{U_a}\times\frac{1}{T}\times 100\%$$

Where U_a and U_b denote areas of a certain land use class at time a and time b respectively; T denotes the length of time from time a to time b. When T is in a unit of year, then R is the annual rate of change in area for this land use class.

3.3.2. Transfer matrix of mining land Transfer matrix can reflect the interconversion relationship between different land use types in the same area at different time, including unchanged part, transfer part, transfer direction, newly increased part and so on [10]. Cross tabulation is often used in transfer matrix to calculate quantities of conversions from a particular land cover to another land cover category at a later date. It is defined as follows:

$$S=\begin{pmatrix} S_{11} & \cdots & S_{1n} \\ \vdots & \ddots & \vdots \\ S_{n1} & \cdots & S_{nn} \end{pmatrix}$$

S_{1n} refers to the transfer area of land type 1 to land type n; S_{n1} refers to the transfer area of land type n to land type 1. The transfer variation of mining land area and other land types can be calculated by transfer matrix. All the area can be calculated through ArcGIS 10.X.

3.3.3. Landscape pattern indexes Landscape indexes has become one of the main methods to quantify the landscape patterns. Some indices are with the same ecological meaning, while others have the opposite and even some have no specific one [11]. In this study, landscape indexes PD, MPS, LPI and DIVISION were chosen to quantify the dominance and fragmentation degree of mining land and the diversity variation change of the landscape under the disturbance of mining activity [12-14] (Table 3).

Table 3. Landscape indexes and their descriptions

Landscape index	Description
Percentage of Landscape (PD)	The percentage the landscape comprised of the corresponding type.
Mean Patch Size (MPS)	The total area of a certain land cover type divides by the total number of patches.
Largest Patch Index (LPI)	Defines the percentage of the largest patch area accounting for the total area and reflects the dominance of the patch level
DIVISION	The length of the landscape boundary per unit area. Reflects the exchange potential and the influence strength of the material, energy and species between patches.

3.3.4. Centroid algorithm The spatial change of land use can be reflected by the change of gravity centre [15]. Similarly, the centroid algorithm was employed to study the rule of spatial change of mining land. It is defined as follows:

$$X^t = \sum_{i=1}^{n}(S_{ti} \times X_i)/\sum_{i=1}^{n} S_{ti} \qquad Y^t = \sum_{i=1}^{n}(S_{ti} \times Y_i)/\sum_{i=1}^{n} S_{ti}$$

X^t and Y^t refers to the centroid coordinate of mining land in time t. S_{ti} refers to the area of patch i in mining land. X_i and Y_i is the centroid coordinate of patch i. All the centroid coordinate and its area can be calculated by ArcGIS 10.X.

4. Results

4.1. Area change of mining land

4.1.1. Land use and cover change The land use and cover change in the study area from 1957 to 2015 were shown in Figure 2. In 1957, 1398.31km^2 being 80.29% of the total area was covered with forest; the study area remained undeveloped. From 1957 to 1985, cultivated land increased quickly, reached 649.15 km^2 being 37.27% of the total area, and became the dominant land type thereafter (866.92km^2, 49.78% in 2000; 862.51 km^2, 49.53% in 2015). At the same time, forest kept declining, from 1398.31km^2 in 1957 to 615.54km^2 in 2015. Likewise, the unused land declined from 276.32km^2 in 1957 to 14.99km^2 in 2015 (see Table 4).

In general, except for grass land and water area, which remained basically unchanged, cultivated land, construction land and mining land increased largely, which implied that human activity played an important role. With the expansion of mining activity, population flowed in and became the main driving force of the total region.

Table 4. Land use/cover change in the study area from 1957 to 2015

Land use/cover type	1957		1985		2000		2015	
	Area km^2	AreaPercentage %	Area km^2	AreaPercentage %	Area km^2	AreaPercentage %	Area km^2	AreaPercentage %
Cultivated land	3.41	0.20%	649.15	37.27%	866.92	49.78%	862.51	49.53%
Forest	1398.31	80.29%	849.85	48.80%	627.98	36.06%	615.54	35.35%
Grass land	36.91	2.12%	92.07	5.29%	80.16	4.60%	30.11	1.73%
Water area	10.12	0.58%	1.05	0.06%	18.53	1.06%	28.91	1.66%
Construction land	7.6	0.44%	84.15	4.83%	82.02	4.71%	138.56	7.96%
Mining land	8.92	0.51%	19.33	1.11%	36.94	2.12%	50.87	2.92%
Unused land	276.32	15.87%	46	2.64%	28.81	1.65%	14.99	0.86%

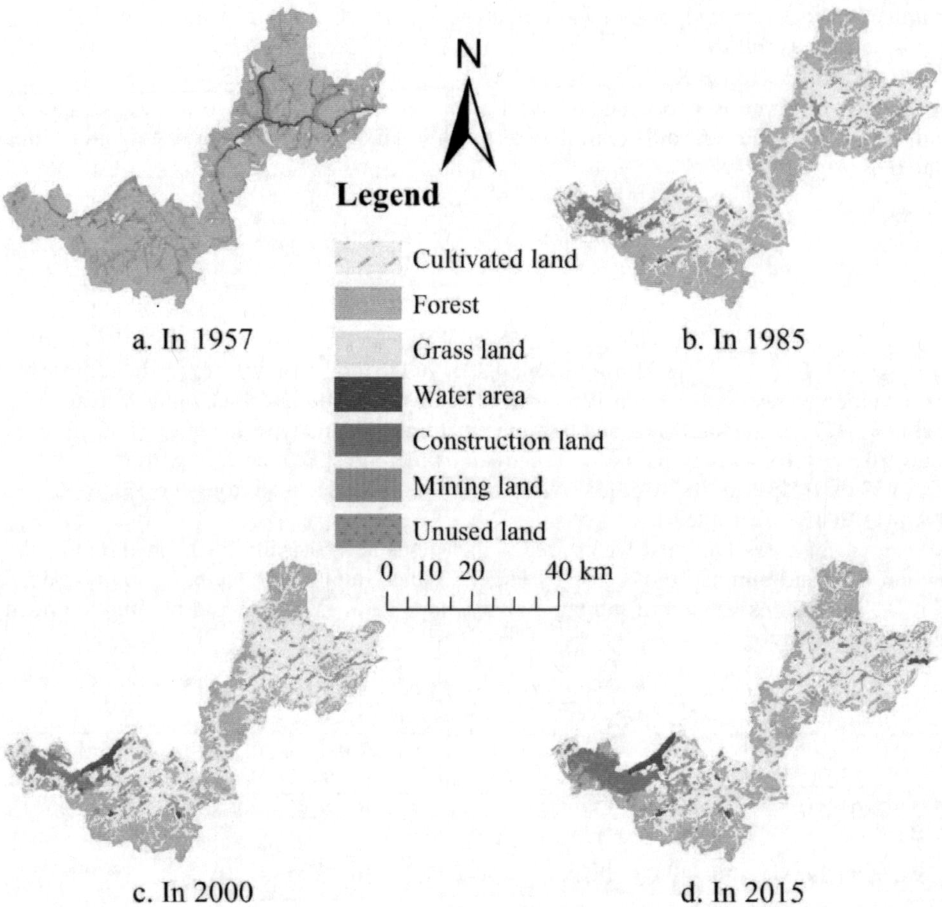

Figure 2. Land cover map in 1957,1985, 2000 and 2015 in the study area

4.1.2. Speed of mining land change The dynamic degree of land use can quantitatively describe the speed of land use change, and play an important role in predicting the future trend of land use change [16]. The mining land area, its proportion in the built-up area in the corresponding year and its dynamic index are shown in Table 5. It was obvious that the area of mining land kept increasing during the study area (8.92 km^2 in 1957; 19.33 km^2 in 1985; 36.94 km^2 in 2000; 50.87 km^2 in 2015). In 1957, mining land took up 54% of the built-up area, which proved that it was a city built and developed by coal mining activity. With the large expansion of mining activity in 1958, a lot of people flowed in and the built-up area expanded dramatically. In 1985, the population reached 309,912, about 10 times as that in 1957. The proportion of mining land in built-up area declined to 18.68% as the living area was built near mining deposits. From 1985-2000, after the implementation of reform and opening-up policy, mining activity entered into its golden period, representing a dynamic index of 6.15% which was the highest among the three indexes. Meanwhile, due to low access threshold, a lot of small and irregular deposits occurred, causing a series of problems such as surface subsidence and tailing stacking. From 2000-2015, after entering into the new century, the increase rate declined, representing a 2.51% dynamic index. Problems as environmental pollution, resource depletion were the main causes.

Table 5. The dynamic change of mining land from 1957 to 2015 in Qitaihe city

Item	1957	1985	2000	2015
Mining land area (km^2)	8.92	19.33	36.94	50.87
% of total built-up area	54%	18.68%	31.05%	26.85%
Increased area of miningland (km^2)	/	10.46	17.89	14.06
R(%)	/	4.19	6.15	2.51

4.1.3. Type of mining land change Mining activity can cause the change of other types of land cover. In a mining city, different types of land cover have different ability to protect the ecological environment. Thus, it is significant to monitor the land cover changes caused by expansion of mining activity. Table 3 presented the main types of land cover transferring to mining land during 58 years. Here, we define a ratio, equal to the size of each land type transferring to mining land divided by the total size of mining regions in corresponding period, as a quantitative parameter shown in Table 2.

As was shown in Table 6, one can see that from 1957-1985, forest was found to be the most contribution to the mining land (18.17km^2, 94%), which indicated that mining activity expanded at the cost of deforestation at the early stage of Qitaihe's development. From 1985-2000, the gain in mining land area was largely from cultivated land (12.36 km^2, 33.16%), then urban land (12.28 km^2, 32.95%) and mining land itself (7.96 km^2, 21.36%). Cultivated land was severely spoiled by expansion of mining activity during this period. From 2000-2015, cultivated land (23.21 km^2, 45.21%) continued playing a dominant role in the increase of mining land, then mining land (14.55, 28.35%) and forest land (7.93km^2, 15.45). So the conclusion could be drawn that the increase in mining land was largely taken from cultivated land, especially in the last 30 years. Forest land and construction land were affected by the mining expansion too. Besides, the rate of mining land transferring to mining land kept increasing (3.97% from 1957-1985, 21.36% from 1985-2000, and 28.35% from 2000-2015), indicating that the mining activity expanded on the basis of its own scale.

Table 6. Transfer matrix of the mining region from the six land cover types in 58 years.

Type of land cover	1957-1985		1985-2000		2000-2015	
	Area (km^2)	Rate (%)	Area (km^2)	Rate (%)	Area (km^2)	Rate (%)
Cultivated land	0	0	12.36	33.16	23.21	45.21
Forest	18.17	94	3.77	10.12	7.93	15.45
Grass land	0.23	1.19	0.80	2.15	0.56	1.09
Construction land	0.04	0.21	12.28	32.95	5.07	9.88
Mining land	0.77	3.97	7.96	21.36	14.55	28.35
Unused land	0.17	0.88	0.1	0.27	0	0
Water area	0	0	0	0	0	0

4.2. Landscape pattern change of mining land

Landscape indexes including PD, MPS, LPI and DIVISION were calculated using Fragstats 4.2 (see Table 7) to analyze the dominance, fragmentation and division degree of mining land.

Largest patch index (LPI) was a measure of dominance. The rising curve indicated that mining land became more and more important in the study area. The rising trend was more obvious from 1957 to 1985. Patch density (PD) and Mean patch size (MPS) were indicators for fragmentation and heterogeneity. Fragmentation means breaking up of habitat, ecosystem or land cover types into smaller parcels [17]. Figure 3 showed that PD increased from 1957 to 1985 and then declined from 1985 to 2015, while MPS showed an opposite trend. This implied that in 1985, the mining land was the most dispersed distributed. After 1985, some related mining regulations were enacted to limit the small mines and regular the large ones, so the fragmentation degree declined and the mining land became more integrated than that in 1985.

DIVISION (Landscape division index) referred to the dispersed degree of spatial distribution for different land types. The division degree kept declining through the study period. The highest decline rate was seen from 1985 to 2000, indicating that mining land become more integrated during this period.

So the conclusion can be drawn that the large expansion of mining activity made the mining land more and more important in the study area. The year 1985 had seen the largest fragmentation degree, which showed that mining activity was conducted disorderly at the first stage of exploitation, but then the situation improved in the following 30 years. As mining land become more integrated and industrialized, more policies should be constituted to promote its further development.

Table 7. Landscape pattern indexes of mining land

	1957	1985	2000	2015
LPI	0.13	0.22	0.28	0.33
PD	3.36	5.84	3.6	2.77
MPS	29.73	17.26	27.77	36.08
DIVISION	1.28	1.15	0.65	0.49

4.3. Spatial change of mining land

The evolution of mining activity can be also found in spatial distribution. Centroid algorithm was employed to better present the change trend (see Figure. 3).

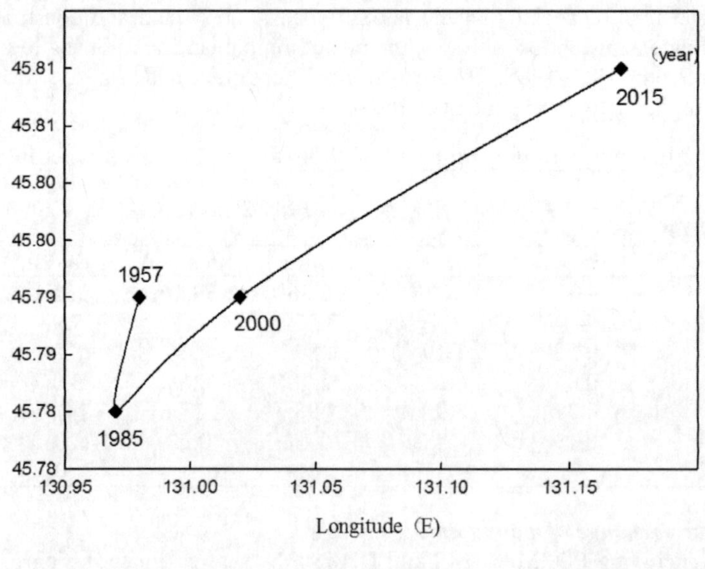

Figure 3. The centroid change of mining land during 1957-2015.

In general, as was shown in Figure 3, centroids of mining land showed an eastward trend as time went by. But centroids shifted 0.01° westward from 1957-1985, showing that mining activity expanded at its own scale during this period. Mining activities started to shift eastwards after 1985, and the trend became more obvious during 2000-2015. The whole process transferred about 0.18° eastward, and 0.02° northward. To better present the spatial distribution of mining activity, four concentration circles were drawn using ArcGIS 10.X (see Figure.4). Circle 1 was the earliest concentration point that was formed in the city. In 1957, there was some small and scattered mining sites, and then the mining activity expanded quickly at its own scale, but until 1985, all the mining activity in the city gathered in circle 1, the west part of the city. Five large deposits were included in circle 1, including Xinjian, Xinxing, Taoshan, Xinli, and Dongfeng deposit. These five deposits are the earliest to be explored, and had the longest development period. Circle 1 was also where the former central urban district-Xinxing district located, but then the central district moved due to serious subsidence caused by large scale of mining activity.

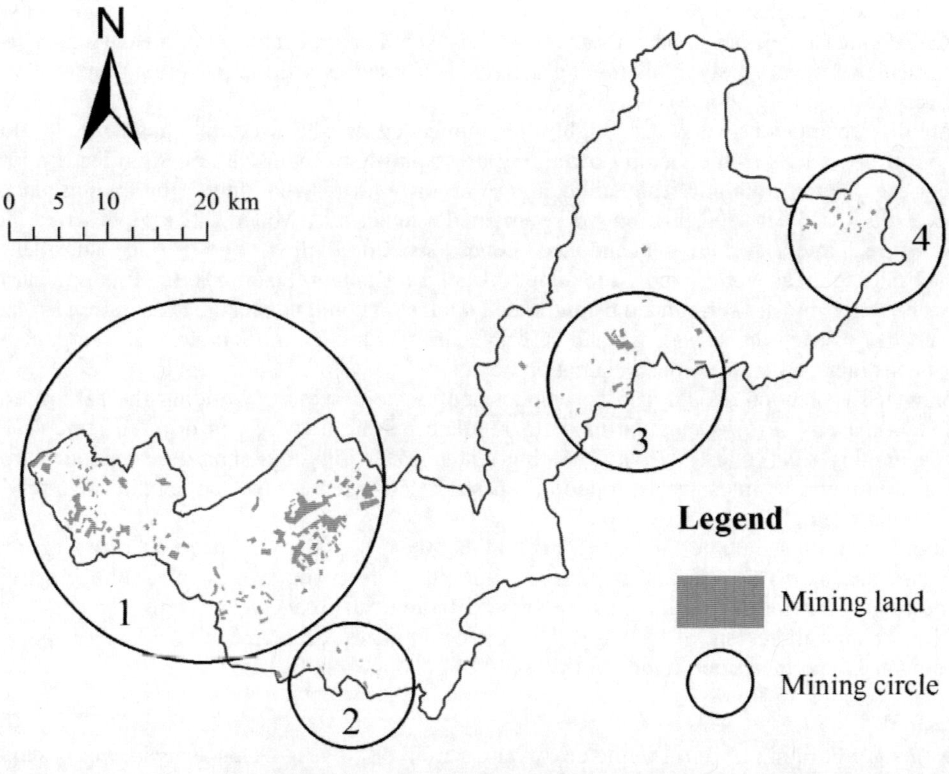

Figure 4. Four mining circles in Qitaihe city

Mining activity moved eastward with the founding of new resources. After 1985, Circle 2, 3 & 4 occurred constantly. Circle 2 included Tiedong, Fuqiang and Longhu deposit. These deposits had the relatively long development period only second to the deposits in circle 1, and higher development level as well. On the contrary, the eastern circle 3&4 were not well developed yet, including Lushan coal deposits and Beixing farm mining sites. These deposits were the main object of future mining activity.

The spatial transfer of mining activities was closely linked with the condition of coal resources, especially the occurrence condition of coal seams. The coal field in the city showed an arched structure, with more stable and simpler structure in the west flank than in the east [18], so the deposits in Circle 1, as located in the west flank, was the easiest to explore and had the longest development period.

The second point that matters was the problem caused by coal mining. Mining subsidence was the important factor that largely hindered the development of the city. The expansion of mining activity accelerated the surface collapse. In 1985, the collapse appeared in circle 1, and then in Circle 2 in the year 2000. The collapse area accounted for 10.3% of the total coal field area at the beginning of new century. In 2011, 30.7% of the main urban area was mine-out and subsidence areas. Large area of subsidence in Xinxing district caused damages to roads and houses as well as city's infrastructure, disturbing resident's daily life, so the central urban district removed twice, once from Xinxing to Beishan district, and finally to Taonan district [19]. The frequent removal of the main central area also caused a lot of resource waste to the city.

5. Discussion

Since the city of Qitaihe was listed as the transition pilot city in 2009, a series of polices have been enacted to promote its future development, among these, the coordination between economic development and environmental protection were highly emphasized. However, as the coal deposits in the city are differing in their development stages, policies were not applicable to all areas. Suggestions were given that an integrated layout of mining activity is needed, and different mining areas should adopt different development policies.

First at all, reclamation work was highly recommended in the western subsidence region. Reclamation of these sites in the vicinity of mining areas plays an important and significant role in regenerating ecosystem health and regional ecological security [20]. According to the mining plan of Qitaihe city (2009-2015), the reclamation work remained a tough one. About 3.63 km2 deserted land needed to be reclaimed, including abandoned stones, stacking tailings and surface subsidence. Different subsidence areas were supposed to adopt different reclamation methods. In Xinxing district, the subsidence area is mostly surrounded by cultivated land, so it could be reclaimed as cultivated land, while In Taoshan district, the subsidence had an easy access to the main urban area, so reorganizing the land for construction should be the reclamation object.

Environmental protection should also be emphasized in the process of building the eastern coal mining base in the city. Efforts must be made to regulate mining activity and develop large-scaled mines while limiting small ones. To avoid wasting resources, living areas should be separated from mining areas. Furthermore, intensive exploitation was supposed to replace the former extensive way in the future mining work.

In addition, attention should be paid to the fact that that the city is also characterized by a large area of forestry, as it has the world's largest artificial Korean pine forest. The expansion of mining activity has also destroyed the forest, resulting in the loss of biological diversity and a rise in ecological vulnerability. Though the extent of the damage to the ecological processes due to mining activity is not well estimated from this paper, future work should be conducted in this area.

6. Conclusions

This study analyzed mining land in Qitaihe city over a period of 58 years. This analysis was based on remote sensing data, as well as their dynamic spatial and temporal distribution variation. Results showed that mining activity expanded at a fast pace. The area of mining land increased by about 6 times its original area with the fastest increase happening between 1985 and 2000. Cultivated land, as the crucial land cover type for northeastern China, contributed most to the increase of mining regions.

The large increase in mining activity made the mining land more and more important in the study area. Mining activity was largely carried out in a disorganized fashion before 1985, but then become more integrated and industrialized in the following 30 years with the implementation of related regulations. Moreover, the spatial location of mining activity transferred eastward and developed from one concentration circle to four circles. Mining activity also directly contributed to a series of socioeconomic problems, such as subsidence and stacking of tailings, the relocation of urban centers and a direct and detrimental effect on urban development. The monitored results provide evidence for managers to evaluate the effects of mining activity in order to constitute new planning for the continued sustainable development of mining activity.

The study further showed that dynamic monitoring of the mining activity by means of multi-temporal remote sensing images was a highly efficient and feasible technique, which provides scientific criteria for reasonable planning and decision-making in urban management.

Despite the promising results, there are also limitations on the accuracy of the remote sensing data. The main constraints are the long time interval and a lack of experienced visual interpreters. We believed that if more images are used at a shorter time interval, and more quantitative methods are employed in the analytical part, monitoring of mining activity will become more objective and effective.

Acknowledgments
This research was supported by China Geological Survey Bureau (No. 1212011220851).

References

[1] Windle J, Rolfe J 2014 Assessing the trade-offs of increased mining activity in the Surat Basin, Queensland: preferences of Brisbane residents using nonmarket valuation techniques *Aust J Agri Res Eco* **58** 111-129
[2] Chaussard E, KeroskyS 2016 Characterization of black sand mining activities and their environmental impacts in the Philippines using remote sensing *RemoSensi* **8** 100-106
[3] Hou H, Zhang S, Yan Y, et al. 2009 Study on ecological landscape restoration in mining areas based on RS and GIS: a case of Xuzhou northern coal mining area *China Land Sci* **23** 62-27
[4] Manna A, Maiti R 2016 Alteration of surface water hydrology by opencast mining in the Raniganj coalfield area *India Mine Water Environ* **35** 156-167
[5] Wang T, Hou S, Tang J, Zhu Y 2012 Analysis on dynamics of landscape types in Qitaihe City *Res Soil Water* **19** 90-93
[6] Zhang Y 2012 Study on vegetation coverage in Qitaihe City based on remote sensing image *Heilongjiang Agri Sci* **6** 139-141
[7] Yu X, Wu K, Yun W, et al. 2016 Analysis on temporal and spatial variation of landscape ecological security in modern agricultural area *Trans Chin Soc Agri Engi* **32** 253-259
[8] Li B, Gu H, Ji Y 2010 Effects of land use change on values of ecosystem services of Xuzhou Based on RS &GIS: Taking Jiuli mining area for example *Res Soil water cons* **17** 123-128
[9] Li N, Yan C, Xie J 2015 Remote sensing monitoring recent rapid increase of coal mining activity of an important energy base in northern China, a case study of Mu Us Sandy Land *Res Conserv Recy* **94** 129-135
[10] Liu R, Zhu D 2010 Methods for detecting land use changes based on the land use transition matrix *Res Sci* **32** 1544-50
[11] Schumaker N H 2010 Using landscape indices to predict habitat connectivity *Ecology* **77** 1210-25
[12] Mcgarigal K, Marks B J 1994 Spatial pattern analysis program for quantifying landscape structure *Dolo Po Box* 1-134
[13] Chen L, Fu B 1996 Analysis of impact of human activity on landscape structure in Yellow river delta-a case study of Dongying region *Acta Ecolo Sini* **16** 337-344
[14] Mariano MDLH, Saco P M, Willgoose G R et al 2011 Assessing landscape structure and pattern fragmentation in semiarid ecosystems using patch-size distributions *Eco Appl Pub Ecolo Soci Ame* **21** 2793-2805
[15] Bai J, Ouyang H, Cui B, et al. 2008 Changes in landscape pattern of alpine wetlands on the Zoige Plateau in the past four decades *Acta Eco Sinica* **5** 2245-52
[16] Niu X, Ou M 2007 On dynamic change of regional land use and its structure forecast: A case study in Nantong City *J Nanjing Agric Univ* **7** 64-70
[17] Forman, RTT 1995 Land Mosaics The Ecology of Landscapes and Regions Cambridge University Press, Cambridge, UK
[18] Yin X, Wang Y, Hua L, et al. 2003 Current situation, cause, harm and prevention measure on surface subsidence in Qitaihe mining area *Energy Environ Prote* **17** 33-36
[19] Gao Y, Chu Y 2014 Remote sensing monitoring geological disaster in Qitaihe city *Heilongjiang Sci Tech Info* **26** 41-42
[20] Huang X, Yuan H, Yu F, et al. 2014 Spatial-temporal succession of the vegetation in Xishuangbanna, China during 1976-2010: A case study based on RS technology and implications for eco-restoration *Ecolo Engi* **70** 255-262

Design and implementation air quality monitoring robot

Yuanhua Chen[1, 3], **Jie Li**[1], **Chunxue Qi**[2]

[1] CECEP Liuhe and Talroad Environmental Technology Co., Ltd, Beijing, 100082, China.

[2] ChinatechTalroad(Beijing) Technology Co., Ltd, Beijing, 100085, China.

E-mail: yuanhua.chen@talroad.com.cn

Abstract. Robot applied in environmental protection can break through the limitations in working environment, scope and mode of the existing environmental monitoring and pollution abatement equipments, which undertake the innovation and improvement in the basin, atmosphere, emergency and pollution treatment facilities. Actually, the relevant technology is backward with limited research and investment. Though the device companies have achieved some results in the study on the water quality monitoring, pipeline monitoring and sewage disposal, this technological progress on the whole is still much slow, and the mature product has not been formed. As a result, the market urges a demand of a new type of device which is more suitable for environmental protection on the basis of robot successfully applied in other fields. This paper designs and realizes a tracked mobile robot of air quality monitoring, which can be used to monitor air quality for the pollution accident in industrial parks and regular management.

1. Introduction

With the pace of the reform of traditional industries by Internet plus,the robot brings industry change.According to a series policies released in 2015, such as the "Made in China 2025" plan [1] and the robot industry development plan (2016-2020) [2],the development of service robot industry has been top-level designed, realizing the intelligent equipment andautomation of manufacturing process.

Currently, the frontier technology research on environmental protection equipment industry in China is not enough[3], and the market share of the advanced environmental protection technology and equipment is still under 10% [4], causing high investment and operation cost in environmental engineering projects under unstable operation. Therefore, the environmental protection equipments call for intelligent upgrade, and that has certain market capacity.

Some research has been carried out, an international team of scientists in Germany's Max Planck institute recently developed a miniature robot to quickly remove pollutants and heavy metals in industrial waste water[5], researchers at the MIT developed a group of robots called "Seaswarm" to adsorb and process oil pollution[6], and researchers at the university of Bristol developed sewage disposal robot using microbes in the stomach to purify waste water and generate electricity[7].

Through the analysis of market size and demand urgency of the segment field of the environmental protection equipment, the development direction of environmental protection robot mainly contains: 1) emergency robots, 2) water and air pollution control robot, 3) solid waste treatment robot.

Furthermore, with the rapid development of social economy, especially in recent years, a lot of high-rise buildings, underground buildings and large petrochemical enterprises continue to spring up in China, leading the number of potential accidents of hazardous chemicals and the radioactive material leakage and combustion, explosion, collapse increasing. Due to the particularity of the construction and enterprise's production, serious environment pollution will be accompanied by the accident. In addition, emergency workers cannot make a quick and efficient disposal decision under the high temperature, dark, toxic and smoke circumstance, unless confirm the kind of pollutant, pollutant emissions and concentration. Only then can they reduce the environmental damage and casualties. It follows that, in the face of increasingly complex environment pollution accidents in industrial park, emergency monitoring robot appears increasingly important to solve the mentioned problem better. Therefore, this paper presents a tracked mobile robot to monitor air quality in cases of emergency.

2. Function Design

Instead of on-site personnel into the complex and hazardous atmosphere, the emergency monitoring robot, as a special disposal method, must have the following features:

2.1. Mobile function.

Adopting the caterpillar way, the intelligent family moving platform must have a certain cross-country ability and adaptive ability to judge alone and cross barriers in view of the situation of the pollution in situ.

Table1. The external sensors

No.	measured parameter	effective range	resolution	data output method	volume
1	CO	0~1000ppm	0.05ppm	UART(3V)、DAC(0.4~2V)、PWM	57.23*18*38.61mm
2	NO2	0~20ppm	0.1ppm	UART(3V)、DAC(0.4~2V)、PWM	57.23*18*38.61mm
3	O3	0~20ppm	0.05ppm	UART(3V)、DAC(0.4~2V)、PWM	57.23*18*38.61mm
4	SO2	0~100ppm	0.05ppm	UART(3V)、DAC(0.4~2V)、PWM	57.23*18*38.61mm
5	PM2.5	0~999ug/m3	1ug/m3	digital	50*38*21mm
6	PM10	0~999ug/m3	1ug/m3	digital	50*38*21mm
7	CHOH	0~5mg/m3	0.01mg/m3		27.16*24.9*9.2mm
8	HCL	0～20ppm	0.1ppm	analog voltage /UART	ϕ23.5*30.5mm
9	H2S	0~100ppm	0.1ppm	analog voltage /UART	ϕ23.5*30.5mm
10	HCN	0~10ppm		DAC/UART	ϕ22*31.2mm
11	VOC	0~6000ppm	100ppb	analog voltage	ϕ20*21.1mm
12	wind speed	0~70m/s		RS485/RS232	180*180*160mm
13	wind direction	0~360º		RS485/RS232	230*190*70mm
14	anemometer	0~60m/s	0.01m/s	RS485/RS232	Φ148×158mm

2.2. Monitoring function.

Carrying air quality monitoring sensors on the mobile platform of robot to perceptpollution scene, whileusing cameras to check the real-time pollution field condition through the communication network transmission for system processing.The sensorsare mainly divided into two kinds:1) the internal sensors, used to detect robot working parameters in the scene of the accident and feedback to the control system for the reasonable path planning, make the robot work orderly by reasonable arrangements according to the scene of the accident situation. 2) the external sensors, shown in the table 1, provide original monitoring data to the processing system, such as temperature, wind speed, wind direction, oxygen content and concentrations of toxicor flammable gases,and details as attached list.

2.3. Control and communication function.

It is composed of robot internal processing system, communication network, and the assistant decision system.

2.4. Anti-corrosion and explosion-proof function.

Because of some extreme environment at the scene of the pollution accident, the electronic components and machinery materials of the robot must be anti-corrosion and explosion-proof processed to improve its corrosion, high temperature, dust, smoke, and other severe environment tolerance, which will ensure the stable working stateto achieve their missions.

3. Robot implement

On the basis of function design, the appearance design was done as shown in figure 1-2, and the robot prototype was finally manufactured in accordance with the design scheme, as shown in figure 3.

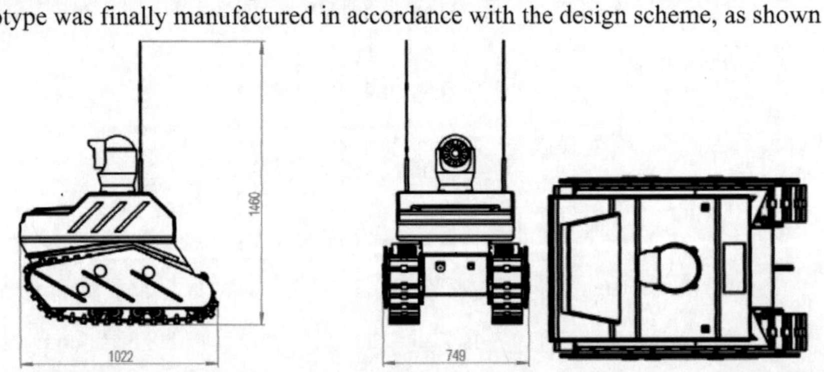

(a) lateral view (b) front view (c) vertical view

Figure 1. Size of the robot.

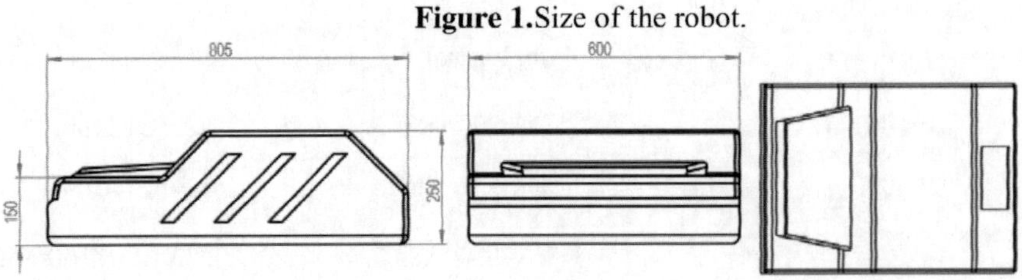

(a) lateral view (b) front view (c) vertical view

Figure 2. Size of Electric control panel.

(a) lateral view (b) back view

Figure 3. Product picture.

4. Conclusion

As the intelligent equipment, robot has become the focus of researches and application at home and abroad in recent decades. It can be operated by human command,preset programs and the principle of artificial intelligence technology,and can work instead of human beings at where humans cannot reach. Therefore, the environmental protection robot has become an inevitable tendency with the development of environmental intelligent equipment.This paper designed an advanced air quality monitoring robot, which not only promotes the combination of robot and the environmental protection industry, but also gathers experiences for relevant intelligent equipment's research and development.

References

[1] http://www.mof.gov.cn/zhengwuxinxi/zhengcefabu/201505/t20150519_1233751.html.
[2] http://www.miit.gov.cn/n1146285/n1146352/n3054355/n3057585/n3057590/c4746240/content.html.
[3] http://www.miit.gov.cn/n1146285/n1146352/n3054355/n3057542/n3057544/c3651058/content.html.
[4] http://www.miit.gov.cn/n1146285/n1146352/n3054355/n3057542/n3057544/c3651058/content.html.
[5] http://www.wokeji.com/kbjh/zxbd_10031/201604/t20160413_2414519.shtml.
[6] http://edition.cnn.com/2010/TECH/innovation/08/26/mit.oil.robot/index.html.
[7] http://www.cnbeta.com/articles/448103.htm.

Delineating the decadal expansion of Lake Basaka (Ethiopia) using various image interpretation techniques

Megersa Olumana Dinka

Department of Civil Engineering Sciences, University of Johannesburg APK Campus 2006, South Africa.

Email: mdinka@uj.ac.za

Abstract. Lake Basaka is expanding at a significant rate. Before the establishment of Matahara Sugar Plantation, the lake was like a small surface pond created during rainy season and used as grazing area. The lake expansion has certain negative consequences to the region. Thus, appropriate method of quantification of the Lake expansion is extremely important. In this particular study, the areal expansion of the Lake Basaka and sugarcane plantation was analysed. Four LandSat images (1973, 1986, 2000 and 2008) were taken for the cloud-free period and processed in ERDAS Imagine and ArcGIS softwares. Three techniques were employed in the delineation of the areas of the lake and plantation: visual interpretation of FCC in GIS, enhancements and advanced classification in ERDAS Imagine. The study result shows that Lake Basaka expansion is very significant and the increment is geometric rather than linear. Overall, the finding indicates that visual image interpretation gives a fast and accurate indication of the Lake Basaka expansion compared to image enhancement and classification techniques.

1. Introduction

Lake Basaka is expanding at a significant rate [1-2] as opposed to the other Ethiopian rift valley lakes, which are shrinking [3-4]. The total surface area of the lake is estimated to be about 50 km^2 in the year 2015 [5], which was about 3 km^2 in the 1960's [3-6]. Before the establishment of the Matahara Sugar Estate (MSE) in the 1970s, according to the information from elders of the indigenous (Karaayyuu) people, the Lake was like a small surface pond created during the rainy season and used as grazing area and watering for their livestock during the dry season.

The lake expansion has certain negative consequences to the region. The lake expansion affects the surface- and ground-water dynamics and soil properties of the region and the condition is specifically dangerous for the sustainability of MSE and Matahara town in particular and the Awash basin irrigation development in general [2]. An appropriate method of quantification of the Lake expansion is extremely important.

This study was initiated with the objective to determine the expansion extent (spatio-temporal) of Lake Basaka in the past about 40 years using different image interpretation techniques. Four LandSat images (1973, 1986, 2000 and 2008) were taken and processed in ERDAS Imagine and ArcGIS softwares. Satellite image processing is a digital techniques consisting of a number of operations and modules as expedients to elucidate the information contents of images to the utmost [4] .Various satellite image interpretation methods and techniques have been in use. The three most widely used image interpretation techniques are visual interpretation, enhancement and image classification. Visual satellite image classification techniques are robust, effective and efficient methods, but time consuming [7].

2. Methodology

2.1. Brief Description of the Study Area

Lake Basaka is located in the Middle Awash Basin, central rift-valley of Ethiopia at about 200 km SE of the capital city, Addis Ababa. Matahara area is situated on a nearly levelled plain (slope < 3%) surrounded by mountain chains of variable elevation in the south, south east, south west and north west (Figure 1). The area is situated in the central rift valley region and hence vulnerable to the occurrences of different tectonic and volcanic activities. As a result, the area is characterized with features of past & recent volcanic events [1-2, 4, 8]. Matahara area is characterized by bimodal rainfall distribution pattern with mean annual rainfall of about 543.7 mm. The long-term mean maximum and minimum temperatures are 32.9 and 17.5 °C, respectively. The climate of the area, in general, is classified to be semi-arid.

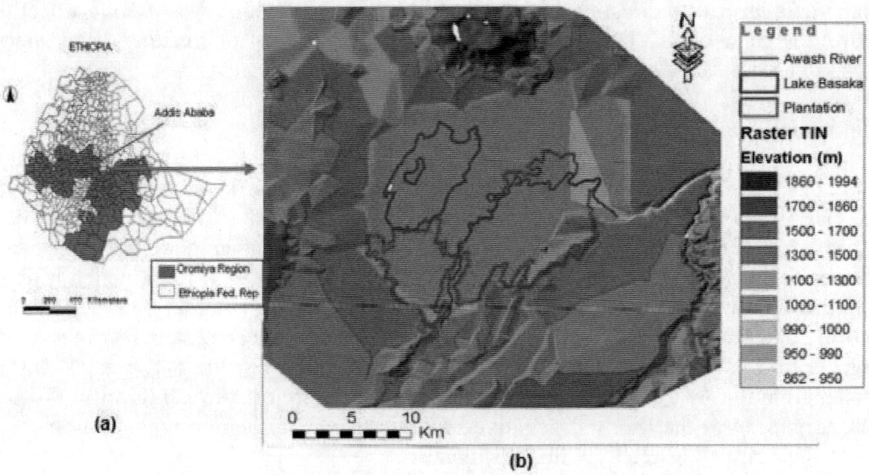

Figure 1. The Study Area: (a) Location; (b) TIN view of the study area and the surrounding escarpments as created from the DEM

2.2. Satellite Image Acquisition, Processing and Interpretation

Four series of Landsat images (1973-MSS, 1986-TM, 2000-ETM+ & 2008-ETM+) were acquired from the FREE Global Orthorectified Landsat Data via FTP (http://glovis.usgs.gov). The selected images were all cloud free and covered both Matahara & Wonji Sugar Estates. The Digital Elevation Model (DEM) was downloaded from the NASA Shuttle Radar Topography Mission (SRTM) through the National Map Seamless Data Distribution System site (http://www2.jpl.nasa.gov/srtm/) and processed/extracted in ArcGIS (9.2) [9] for the study area and the surrounding features.

Then the satellite images with good resolution (panchromatic) was georeferenced with the base map (topomap and plantation map) of the project area to the required accuracy (<30m) in GIS and then the other images were registered with the already georeferenced image (image-to image registration) in ERDAS Imagine by matching some of the easily identifiable features (like crossing of roads, rivers, railways, irrigation/drainage canals, bridges, etc) on both the map and images. The images were imported to ERDAS Imagine [10] and ILWIS softwares for identification of important features using the different image interpretation techniques. The image geometric and radiometric corrections were carried out in ERDAS Imagine. Then the images were converted to false colour images (FCC) in ERDAS and imported to GIS for delineation based on the different interpretation (visual) elements such as tone, shape, hue and association/pattern in addition to colour.

After testing the different image enhancement techniques, the filtration, histogram equalization, brightness inversion, stretching and indices were adopted depending upon the Landsat type. High pass filtering (high frequency kernel) with user defined Laplacian (non-directional) edge filter, which has the effect of increasing spatial frequency [11] was used in order to bring edge between homogeneous groups of pixels. This technique is especially useful for some areas of the image where the agricultural plots and some line features such as roads are not clearly visible. A strectch (histogram equalization) and brightenss inversion operations were carried out as a radiometric enhancement after making a haze and noise reduction for the TM images. Spectral enhancement techniqes using the application of specific band ratios (indices) such as Normalised Difference Vegetation Index (NDVI),Time Composited Normalised Difference Vegetation Index (TNDVI), Salinity Index (SI), Normalised Difference Salinity Index (NDSI), Normalised Difference Water Index (NDWI) were employed for the different Landsat types following standard algorithms [1, 4, 12-14].

Furthermore, image classifications (supervised and unsupervised) were done for each the images in ERDAS. First the unsupervised classification was carried out, which are useful for selecting training areas for the supervised classification system. The training samples for the supervised classification (maximum likelihood) were identified using user-defined polygon in the area of interest (AOI). The class from a thematic data layer of the results of the unsupervised classification was used to select the parametric signature for each class. These training samples were used to create the parametric signature of each class based on field observations and image data file values.

3. Results and Discussion

The shapes and areas of Lake and sugarcane plantation were delineated by the three image interpretation techniques and the results are summarized in Figure 2. Based on average values, the surface area of the lake increased from 1028 ha in 1973 to 4337 ha in 2008. Similarly, the sugarcane plantation area increased from 6105 ha to 13151 ha in the same period. The average of the three image interpretation indicates that Lake Basaka expanded by about 2017 ha from 1973–1986, which is about 150 ha/year considering linear increment. From 1986–2000, the expansion is about 875 ha, which is 63 ha/yr. In the period 2000–2008 the total expansion is about 417 ha, which is 52 ha /yr.

As evident from Figure 2, the expansion extent of the lake is very terrible and creates a great developmental challenge in the Awash basin. The lake has a potential to join Awash River in the near future [1-2, 4]. The expansion of the lake has negative socio-economic and environmental impacts in the region, especially the nearby sugarcane farm and villages.

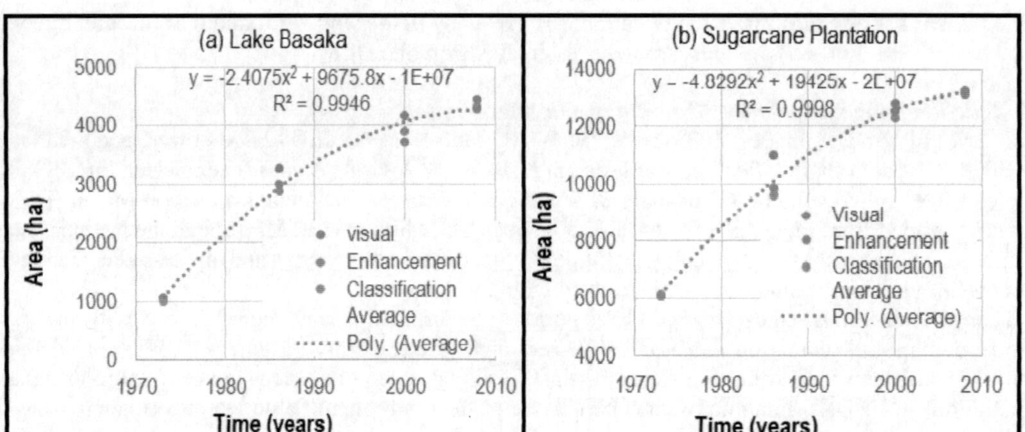

Figure 2. Area of Lake Basak: comparison of the three image interpretation techniques. Note that the fitted equation is the 2nd degree polynomial.

3.1. Visual Interpretation Techniques

Satellite Images can be interpreted using the different visual interpretation elements such as structure, tone, texture or shape variations in individual bands. However, the precise identification of some features is very difficult from a single band. The use of false colour composite images provides an aid for better and faster visual interpretation. Thus, the images for the different periods were converted to FCC (RGB) (where the red (R) indicates vegetation, the green (G) indicates the bare soil and the blue (B) represents the presence of water) so that it is easily interpretable to human naked eye.

Different band combinations were adopted depending upon the sensor type: 4-3-2 for MSS images and 6-4-1 for TM and ETM+ images. Monoscopic visual interpretation of the standard FCC were carried out using the combination of different visual interpretation elements such as tone, shape, hue, pattern, texture and association. The colour of Lake Basaka is not blue, rather deep dark to dark blue and observed to be black in the FCC (see Fig. 2). Since the vertisols (volcanic soils with inverting property) near the lake are also showing almost similar colour, the texture was used an important visual element to differentiate the tonnage. At the interface between the two (Lake and soil), the DN (Digital Number) values are changing abruptly and used as demarcation.

The 1973 MSS image, even after FCC, is difficult to identify the shape of the lake by naked eye and makes the visual interpretation difficult and unrealistic. Agricultural plots are easily identified and visually observe as the area is in semi-arid climatic zone and the images were taken during dry period (Dec-Feb). In the area, only sugar farms are available during these periods and hence, the hue, size, shape, and texture of the plantation areas are identifiable, which can also be easily delineated from the base map of the area.

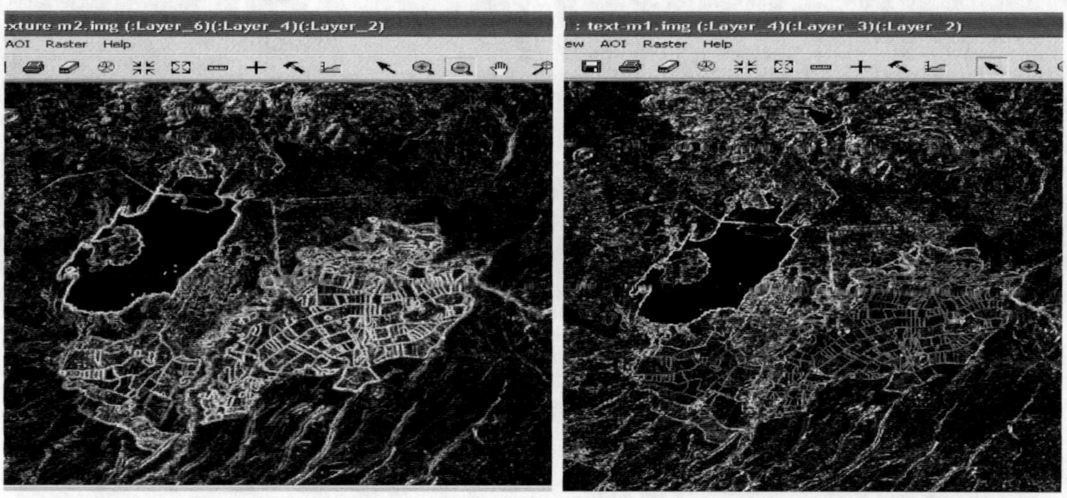

Figure 3. Textural analysis of the 1986 TM image (Left: 6-4-2 FCC, and Right: 4-3-2 FCC)

3.2. Image Enhancement Techniques

The three image enhancement techniques (spatial, radiometric and spectral) were applied depending upon the sensor type. After checking the different spatial enhancement techniques, the textural analysis and non-directional edge filter were used. Figure shows the results of the texture analysis of layer-stacked image for the 1986 image. The left image is the texture analysis for layer combination 6-4-2 and the right image indicates the 4-3-2 combination. From the two images, the boundaries of the two images can be delineated easily.

Figure 4 (left) illustrates the results of user defined Laplacian (non-directional) edge filtering operation and Fig. 3 (right) shows the brightness inversion as radiometric enhancement technique for the 2000 ETM+ image. It can be clearly observed from the Figure 3 that the brightness inversion is better than the non-directional edge filtering since it gives a clear demarcation between the Lake and the vertisol (volcanic soil) in the north side of the lake. Also it gives a clear distinction between soil (bare/ploughed) and vegetation cover.

The use of spectral enhancement (using indices) is more useful in extracting features as compared to the other spatial and radiometric enhancement techniques. After checking the different spectral enhancement techniques, the indices including NDWI, NDSI and TNDVI were used. Therefore, the different indices were selected for the different images and used for the preparation of the map of the lake. For the 1973 MSS image, NDSI gave satisfactory results and for the 1986 TM image, TNDVI was adopted. For the 2000 and 2008 ETM+ images, NDWI was adopted to produce the final map (shape and expansion) of the lake.

Figure 5 indicates the results of spectral enhancement using TNDVI and NDWI. The results of NDWI (Figure 5, right) indicate that the sugarcane (matured) has the highest value and water has the lowest value and therefore, the Lake and the plantation area can easily be delineated. Therefore, the salinity index (SI) of the Lake Basaka is as high as (1/0.42=2.38). The SI of some of the crop field is as low as (1/2.17=0.46). This value indicates that the Lake Basaka has at least SI as high as five times that of the plantation soil, which is also confirmed from their measured EC values.

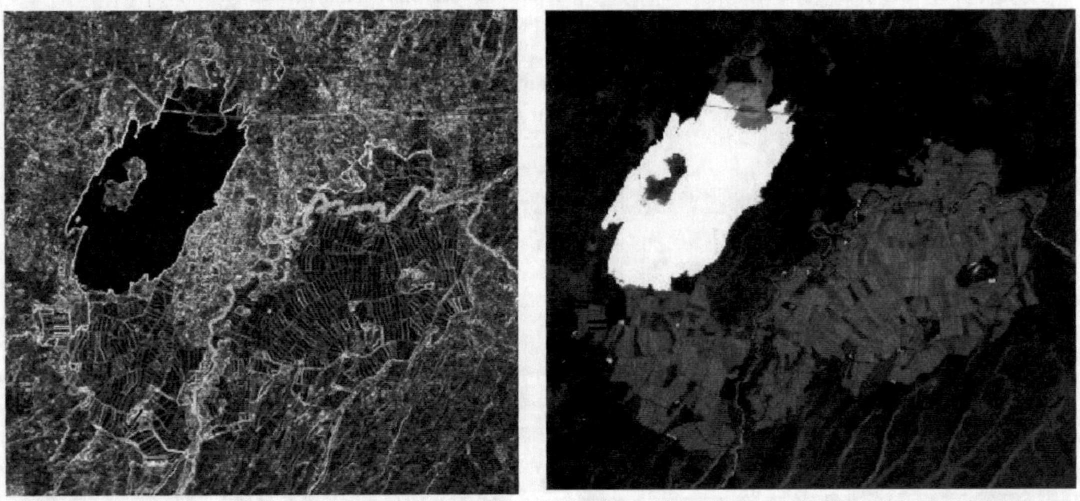

Figure 4. Non-directional edge filter spatial enhancement (Left) and brightness inversion radiometric enhancement (Light) for the 2000 ETM+ image

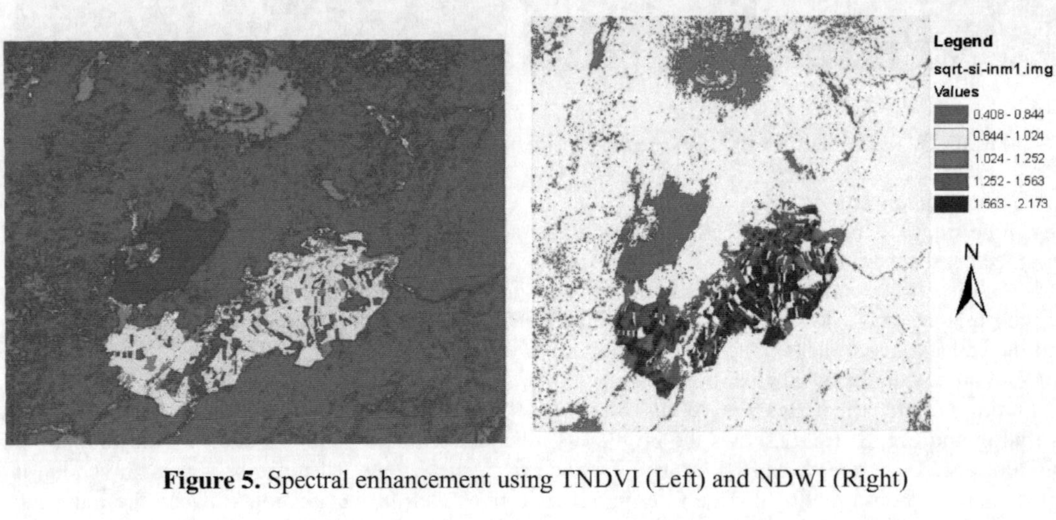

Figure 5. Spectral enhancement using TNDVI (Left) and NDWI (Right)

3.3. Image Classification

Image classification forms an important tool for examination of the digital images produces a map like image as the final product of the analysis [14]. Here the Landsat images were classified in ERDAS Imagine [9] in order to delineate the Lake and plantation areas. The accuracy values were checked and found to be within the acceptable range. The good accuracy obtained was due to the fact that the classification signatures used were obtained from unsupervised classification followed by training samples. The results of supervised classification (maximum likelihood) are shown in Figure 2.

3.4. Quantitative Evaluation of the three Methods

As observed in Figure 2, the results obtained from the three image interpretation techniques are almost similar and follows the same general pattern. The lake increment is geometric rather than linear. In general, the visual interpretation gives an accurate result almost comparable with the image enhancement and classification techniques. Thus, it is possible to suggest that the visual satellite image classification techniques are relatively fast and effective methods for the delineation of the surface area of the lake and plantation. Moreover, it overestimates the surface area of the lake. In general, the average of the three techniques gives a relatively better estimation for the area of the lake and plantation. The average value is fitted with a 2nd degree polynomial function as shown in Figure 2.

4. Conclusion and Recommendation

The surface area of Lake Basaka and sugarcane plantation has been delineated using three image interpretation techniques. The analysis result indicates that visual image interpretation gives a fast and efficient ways for the delineation of the Lake Basaka expansion compared to image enhancement and classification techniques. However, the average of the three image interpretation techniques usually gives a better result. The average of the three image interpretation indicates that Lake Basaka expanded by about 2017 ha from 1973–1986, which is about 150 ha/year considering linear increment. In the period between 1986 to2000, the expansion is about 875 ha, which is 63 ha/yr. In the period 2000 to 2008 the total expansion is about 417 ha, which is 52 ha /yr. The expansion of the lake has already started affecting Matahar Sugar Estate (MSE) in terms of groundwater dynamics, soil salinization, yield decline, etc and it should be considered seriously before it brings total devastation. If the past expansion trend continues, the sustainability of the plantation itself is under great risk. The expansion extent of the lake is very terrible and creates a great developmental challenge in the Awash basin. The lake has a potential to join Awash River in the near future, thereby impacting all downstream irrigation developments in the Awash Basin, and affecting the livelihood of the people depending on the water resources of this basin. All the beneficiaries of the basin, concerned institutions and the decision-makers of the country should consider the condition seriously and adopt mitigation measures before it brings irreversible damage to the region.

References

[1] Dinka M O 2012a Analysing decadal land use/cover dynamics of the Lake Basaka catchment (Main Ethiopian Rift) using LANDSAT imagery and GIS. *Lake Reserv. Res. Manag.* **17** pp 11-24.

[2] Dinka M O 2012b Analysing the extents (size and shape) of Lake Basaka Expansion (Main Ethiopian Rift Valley) using Remote Sensing and GIS. *Lake Reserv. Res. Manag.* **17** pp 131-141.

[3] Alemayehu T, Ayenew T and Kebede S 2006 Hydrogeochemical and lake level changes in the Ethiopian Rift. *J. Hydrology* **316** (1-4) pp 290-300.

[4] Dinka M O 2010 Analyzing the extents of Basaka Lake Expansion and Soil and Water Quality Status of Matahara Irrigation Scheme, Awash Basin (Ethiopia) PhD Thesis submitted to BOKU University (Vienna) Austria.

[5] Dinka (under review). Lake Basaka Expansion: Challenges for the Sustainability of Matahara

Irrigation Scheme, Awash River Basin (Ethiopia).

[6] Abebe G 2000 Feasibility study on the proposed remedial measures of the Lake Beseka level rise MSc Thesis submitted to the Graduate School of Alemaya University Ethiopia.

[7] Mor, P A 1971 The Geology of Ethiopia. University College of Addis Ababa Press, Ethiopia.

[8] ERDAS (Earth Resources Data Analysis System) (2006). ERDAS Field Guide, 5th edn. ERDAS, Inc, Atlanta, Georgia. pp. 671.

[9] ESRI 2006 ArcGIS Desktop Help 9.2. Environmental Systems Research Institute, Inc. Available from URL: http://webhelp.esri.com/arcgisdesktop/9.2. Accessed: 15 Oct 2009.

[10] Jensen J R 2005 Introductory Digital Image Processing: A remote Sensing Perspective. Upper Saddle River, Prentice Hall.

[11] Gao B C 1996 NDWI - A normalized difference water index for remote sensing of vegetation liquid water from space. *Remote Sensing of Environment* **58** pp 257-266.

[12] McFeeters S K 1996 The use of the Normalized Difference Water Index (NDWI) in the delineation of open water features *Int. J. Remote Sensing* **17** pp 1425 -1432.

[13] Albert D P, Gesler W M, Levergood B 2000 Spatial Analysis, GIS, and Remote Sensing Applications in the Health Sciences Ann. Arbor. Press p 217.

[14] Campbell J B 2007 Introduction to Remote Sensing (4th Ed.). The Guilford Press.

IOP Conf. Series: Earth and Environmental Science 52(2017) 012091 doi:10.1088/1755-1315/52/1/012(

An assessment of environmental and toxicological risk to pesticide exposure based on a case-based approach to computing

Cristina Coelho[1], Henrique Vicente[1, 2], M. Rosário Martins[3], Nelson Lima[4], Mariana Neves[5] and José Neves[2, 6]

[1] Departamento de Química, Escola de Ciências e Tecnologia, Universidade de Évora, Évora, Portugal
[2] Centro Algoritmi, Universidade do Minho, Braga, Portugal
[3] Departamento de Química, Laboratório HERCULES, Escola de Ciências e Tecnologia, Universidade de Évora, Évora, Portugal
[4] Centro de Engenharia Biológica, Micoteca da Universidade do Minho, Universidade do Minho, Braga, Portugal
[5] Deloitte, London, United Kingdom

E-mail: jneves@di.uminho.pt

Abstract. Pesticide environmental fate and toxicity depends on its physical and chemical features, the soil composition, soil adsorption, as well as residues that may be found in different soil slots. Indeed, pesticide degradation in soil may be influenced by either biotic or abiotic factors. In addition, the toxicity of pesticides for living organisms depends on their adsorption, distribution, biotransformation, dissemination of metabolites together with interaction with cellular macromolecules and excretion. Biotransformation may result in the formation of less toxic and/or more toxic metabolites, while other processes determine the balance between toxic and a nontoxic upcoming. Aggregate exposure and risk assessment involve multiple pathways and routes, including the potential for pesticide residues in food and drinking water, in addition to residues from pesticide use in residential and non-occupational environments. Therefore, this work will focus on the development of a decision support system to assess the environmental and toxicological risk to pesticide exposure, built on top of a *Logic Programming* approach to *Knowledge Representation and Reasoning,* complemented with a *Case Based* attitude to computing. The proposed solution is unique in itself, once it caters for the explicit treatment of incomplete, unknown, or even self-contradictory information, either in terms of a qualitative or quantitative setting.

1. Introduction

The use of agricultural pesticides is mounting nowadays, aiming at the control of insect pests and diseases. They are a diverse group of inorganic and organic chemicals that include herbicides, insecticides, nematicides, fungicides and soil fumigants. Pesticide sprays may directly hit non-target vegetation, or may drift or volatilize from the treated area and contaminate air, surface, ground water, soil, and non-target plants. They can be moved from soil by runoff and leaching, thereby constituting a potential problem for the supply of drinking water to the inhabitants [1]. Pesticide environmental fate and toxicity depends on the physical and chemical characteristics of pesticide, the soil composition, and soil adsorption. Pesticide degradation in soil may be influenced by both biotic and abiotic factors. Microbial activity has been considered to be the most influential and significant cause of degradation of organic pesticides in soil and surface waters. Nevertheless, other physic-chemical factors such as light, volatility, pH, temperature, soil moisture and organic carbon content, adsorption to sediment particles and lixiviation can influence the degradation of a pesticide and the exposure pathway [2].

In addition, pesticides may be toxic to non-target organisms that may range from beneficial soil bacteria, to insects, plants, fish, or birds. Insecticides are generally the most acutely toxic class of pesticides, but herbicides may also pose risks to non-target organisms. The excessive and frequent application of pesticides may also result in high level of pesticides residues accumulated on vegetables, which poses a potential health risk to consumers [3]. On the one hand, the human hazard is determined by the pesticide properties, exposure time and the individual's susceptibility, affecting the magnitude of these processes and the final fate of pesticide. On the other hand, the toxicity of pesticides for living organisms depends on their adsorption, distribution, biotransformation, distribution of metabolites or the interaction with cellular macromolecules and excretion [4]. Indeed, biotransformation may result in the formation of less toxic and/or more toxic metabolites, while the various other processes determine the balance between toxic and a nontoxic upcoming [5].

Human exposure may be dietary recreational and/or occupational, and toxicity might be acute or chronic [6], i.e., to ensure the safety of the food supply for human consumption, *Maximum Contaminant Levels* (*MCLs*) set the legal limits for the amount of pesticides allowed in food and drinking water, which is correlated with *Acceptable Daily Intake* (*ADI*), defined as the amount of a chemical that can be intake on a daily base [7-10]. More recently some works established pesticide impact and toxicity based on chemical properties, environmental fate and exposure considerations [8-10].

Therefore, this work will focus on the development of a decision support system to assess the environmental and toxicological risk to pesticide exposure, built on top of a Logic Programming approach to Knowledge Representation and Reasoning, complemented with a Case Based attitude to computing. The proposed solution is unique in itself, once it caters for the explicit treatment of incomplete, unknown, or even self-contradictory information, either in terms of a qualitative or quantitative setting.

2. Background

Many approaches to Knowledge Representation and Reasoning have been proposed using the *Logic Programming* (*LP*) epitome, namely in the area of *Model Theory* [11, 12] and *Proof Theory* [13, 14]. In the present work the *Proof Theoretical* approach in terms of an extension to the *LP* language is followed. An *Extended Logic Program* is a finite set of clauses, given in the form:

$$\{ \quad \neg\, p \leftarrow not\ p, not\ exception_p$$

$$p \leftarrow p_1, \cdots, p_n, not\ q_1, \cdots, not\ q_m$$

$$?\,(p_1, \cdots, p_n, not\ q_1, \cdots, not\ q_m) \ \ (n, m \geq 0)$$

$$exception_{p_1} \quad \ldots \quad exception_{p_j} \ \ (j \leq m, n) \ \ being\ k\ an\ integer\ number \quad \} :: scoring_{value}$$

where the first clause stand for predicate's closure, "," denotes "*logical and*", "*?*" is a domain atom denoting "*falsity*", "::" stands for "*where*", the p_i, q_j, and p (e.g., consider *tera* (((0.02, 0.02)(1, 1)), …, (((1, 1)(1, 1))) :: 1 :: 0.85 as it is depicted at the end of section 4) are "*classical ground literals*", i.e., either positive atoms or atoms preceded by the classical negation sign "$\neg$" [13]. Indeed, "$\neg$" stands for a strong declaration that speaks for itself, and *not* denotes *negation-by-failure*, or in other words, a flop in proving a given statement, once it was not declared explicitly. Under symbols' theory, every program is associated with a set of "*abducibles*" [11, 12], given here in the form of exceptions to the extensions of the predicates that make the program, i.e., clauses of the form:

$$exception_{p_1}, \cdots, exception_{p_j} \ (0 \leq j \leq k), \ \ being\ k\ an\ integer$$

that stand for data, information or knowledge that cannot be ruled out. On the other hand, clauses of the type:

$$?\,(p_1, \cdots, p_n, not\ q_1, \cdots, not\ q_m) \ (n, m \geq 0)$$

also named *invariants*, allows one to set the context under which the universe of discourse has to be understood. The term $scoring_{value}$ stands for the relative weight of the extension of a specific predicate with respect to the extensions of peers ones that make the inclusive or global program.

2.1. *Knowledge Representation and Reasoning – Quantitative Knowledge*

In order to set one's approach to knowledge representation and reasoning, two metrics will be set, namely the Quality-of-Information (*QoI*) of a logic program that will be understood as a mathematical function that will return a truth-value ranging between 0 and 1 [15, 16], once it is fed with the extension of a given predicate. Indeed, $QoI_i = 1$ when the information is *known* (*positive*) or *false* (*negative*) and $QoI_i = 0$ if the information is *unknown*. For situations where the extensions of the predicates that make the program also include *abducible* sets, its terms (or clauses) present a $QoI_i \in]0, 1[$, in the form:

$$QoI_i = {}^{1}/_{Card} \tag{1}$$

if the *abducible* set for *predicates i* and *j* satisfy the *invariant*:

$$?\left(\left(exception_{p_i};\ exception_{p_j}\right), \neg\left(exception_{p_i};\ exception_{p_j}\right)\right)$$

where ";" denotes " *logical or*" and "*Card*" stands for set cardinality, being $i \neq j$ and $i, j \geq 1$. A pictorial view of this process is given in Figure 1(a), as a pie chart.

On the other hand, the clauses cardinality (*K*) will be given by $C_1^{Card} + \cdots + C_{Card}^{Card}$, if there is no constraint on the possible combinations among the abducible clauses, being the *QoI* acknowledged as:

$$QoI_{i_{1 \leq i \leq Card}} = {}^{1}/_{C_1^{Card}}, \cdots, {}^{1}/_{C_{Card}^{Card}} \tag{2}$$

where C_{Card}^{Card} is a card-combination subset, with *Card* elements. A pictorial view of this process is given in Figure 1(b), as a pie chart.

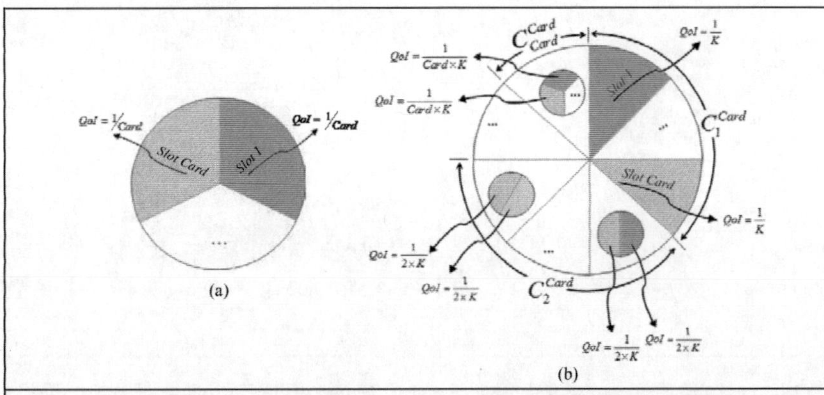

Figure 1. *QoI's* values for the abducible set for $predicate_i$ given in terms of an hatched area (a), i.e., considering only the abducible clauses, or with a circle (b), i.e., considering the possible combinations among the abducible clauses.

However, a term's QoI also depends on their attribute's QoI_s. In order to evaluate this metric, look to Figure 2, where the segment with bounds 0 and 1 stands for every attribute domain, i.e., all the attributes range in the interval [0, 1]. [*A*, *B*] denotes the range where the unknown attributes values for a given predicate may occur (Figure 2):

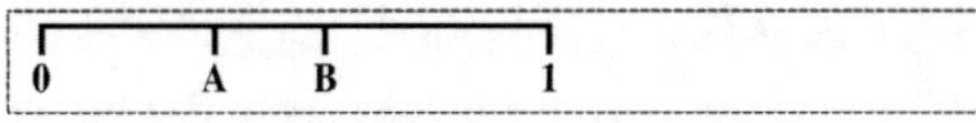

Figure 2. Setting the QoI_s for each attribute's clause.

$$QoI_{attribute_i} = 1 - \|A - B\| \quad (3)$$

where ||A–B|| stands for the modulus of the arithmetic difference between *A* and *B*, i.e., taking the absolute value. It must be also stated that unsharp (e.g., *fuzzy* or *probabilistic*) or linguistic attribute values (e.g., *good*, *bad*, …) may be transferred into an arithmetic difference as it is shown below (Figure 3 and subsection 2.2). Indeed, this generalized conception of observable enables a consistent notion of *unsharp reality* and with it an adequate concept of *joint properties*.

Under this setting, another metric has to be considered, which will be denoted as *DoC* (*Degree-of--Confidence*), that stands for one's confidence that the argument values or attributes of the terms that make the extension of a given predicate, having into consideration their domains (which were set to the interval [0, 1], are in a given interval [17]. Therefore, the *DoC* is figured as $DoC = \sqrt{1 - \Delta l^2}$, where Δl stands for $\|A - B\|$ (Figure 4).

Thus, the universe of discourse is engendered according to the information presented in the extensions of such predicates, according to productions of the type:

$$predicate_i - \bigcup_{1 \leq j \leq m} clause_j\left(\left((A_{x_1}, B_{x_1})(QoI_{x_1}, DoC_{x_1})\right), \cdots, \left((A_{x_l}, B_{x_l})(QoI_{x_l}, DoC_{x_l})\right)\right) :: QoI_j :: DoC_j \quad (4)$$

where *U*, *m* and *l* stand, respectively, for *set union*, the *cardinality* of the extension of $predicate_i$ and the number of attributes of each clause [17]. On the other hand, either the subscripts of the QoI_s and the DoC_s, or those of the pairs (A_s, B_s), i.e., $x_1, \ldots, x_l$, stand for the attributes' clauses values ranges.

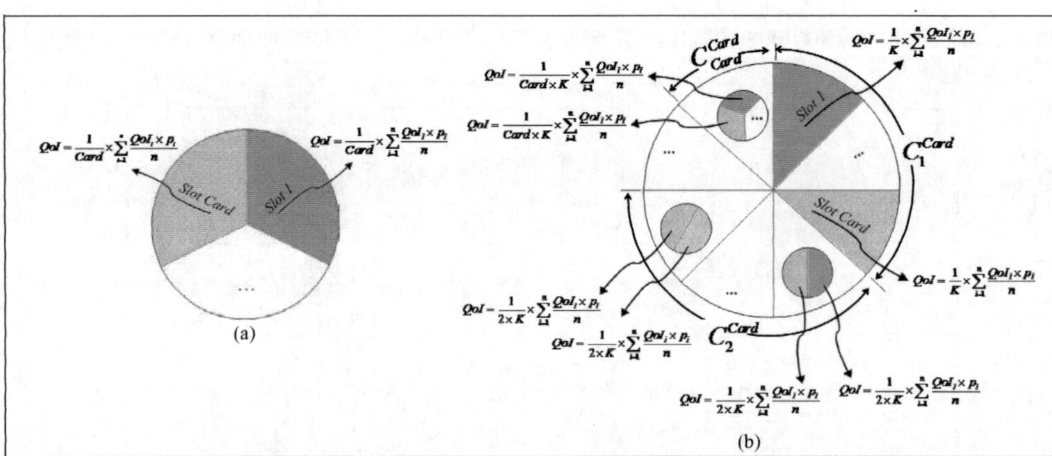

Figure 3. *QoI's* values for the abducible set for $predicate_i$ given in terms of an hatched area (a) considering only the abducible clauses, or with a circle (b), considering the possible combinations among the abducible clauses. $\sum_{i=1}^{n}(QoI_i \times p_i)/n$ denotes the *QoI's* average of the attributes of each clause that sets the extension of the predicate under analysis. *n* and p_i stand for, respectively, for the attribute's cardinality and the relative weight of attribute p_i with respect to its peers ($\sum_{i=1}^{n} p_i = 1$).

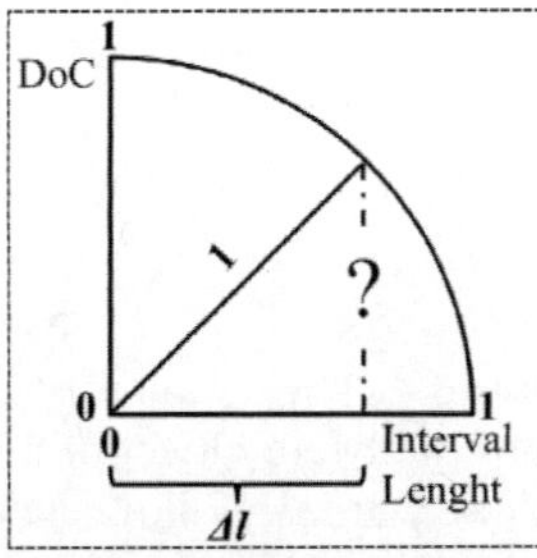

Figure 4. Evaluation of the Degree of Confidence.

2.2. Knowledge Representation and Reasoning – Qualitative Knowledge

In present study both qualitative and quantitative data/knowledge are present. Aiming at the quantification of the qualitative part and in order to make easy the understanding of the process, it will be presented in a graphical form. Taking as an example, consider a set of n issues regarding a particular subject, where the values of the k criteria, understood as linguistic ones, are *none*, *low*, ..., *high* and *very high*. Now, enumerating a unitary area circle split into n slices (Figure 5), the marks in the axis resemble each of the possible criteria` values. If the answer to issue 1 is *high* the corresponding area is $\pi \times \left(\sqrt{(k-1)/k \times \pi}\right)^2/n$, i.e., $(k-1)/(k \times n)$ (Figure 5(a)). Assuming that in the issue 2 are chosen the alternatives *high* and *very high*, the resultant area ranges in the interval $\left[\pi \times \left(\sqrt{(k-1)/k \times \pi}\right)^2/n,\ \pi \times \left(\sqrt{k/k \times \pi}\right)^2/n\right]$, i.e., $[(k-1)/(k \times n),\ k/(k \times n)]$ (Figure 5(b)). Finally, in issue n if no alternative is ticked, all the hypotheses should be considered and the area varies in the interval $\left[0,\ \pi \times \left(\sqrt{k/k \times \pi}\right)^2/n\right]$, i.e., $[0,\ k/(k \times n)]$ (Figure 5(c)). Thus, the total area is the sum of the partial ones (Figure 5(d)), i.e., $[(2k-2)/(k \times n),\ (3k-1)/(k \times n)]$.

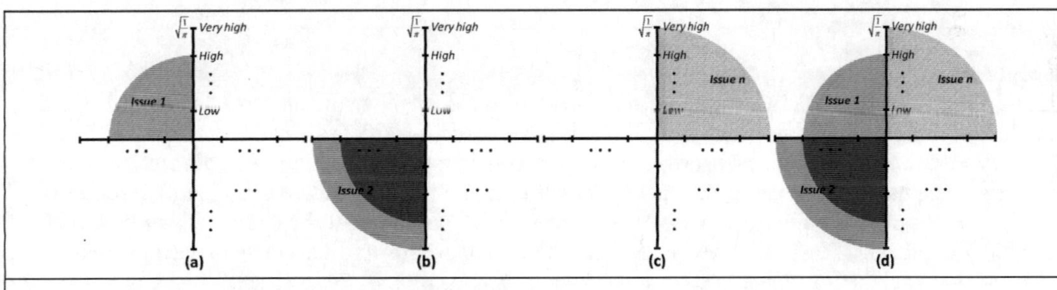

Figure 5. A view of a qualitative data/knowledge processing.

3. Case Based Reasoning

The *Case Based* (*CB*) approach to computing stands for an act of finding and justifying a solution to a given problem based on solutions of past ones [18, 19]. In *CB* the *Cases* are stored in a *Case Base*, and those cases that are similar (or close) to a new one are used in the problem solving process (see section 5). The typical *CB* cycle presents the mechanism that must be followed, where the former stage entails an initial description of the problem. The new case is used to retrieve one or more cases from the *Case Base* [17], a process that will be explained below.

Undeniably, despite promising results, the current *CB* systems are neither complete nor adaptable for all domains. Moreover, in real problems, the access to the necessary information is not always possible, since existent *CB* systems have limitations related to the capability of dealing, explicitly, with unknown, incomplete, and even self-contradictory information. To make a change, a different *CB* cycle was induced (Figure 6). It takes into consideration the case's *QoI* and *DoC* metrics. It also contemplates a cases optimization process in the *Case Base*, whenever the retrieved cases do not comply with the terms under which a given problem as to be addressed (e.g., the expected *DoC* on a prediction was not attained). This process that uses either *Artificial Neural Networks* [20, 21], *Particle Swarm Optimization* [22] or *Genetic Algorithms* [14], just to name a few, generates a set of new cases from the retrieved ones which must be in conformity with the invariant:

$$\bigcap_{i=1}^{n}(\boldsymbol{B}_i, \boldsymbol{E}_i) \neq \emptyset \tag{5}$$

i.e., it denotes that the intersection of the attribute's values ranges for the cases' set that make the *Case Base* or their optimized counterparts ($\boldsymbol{B}_i$) (being n its cardinality), and the ones that were object of a process of optimization ($\boldsymbol{E}_i$), cannot be empty (Figure 6), where the cases in the *Case Base* follow the pattern $Case = \{< Raw_{data}, Normalized_{data} >\}$.

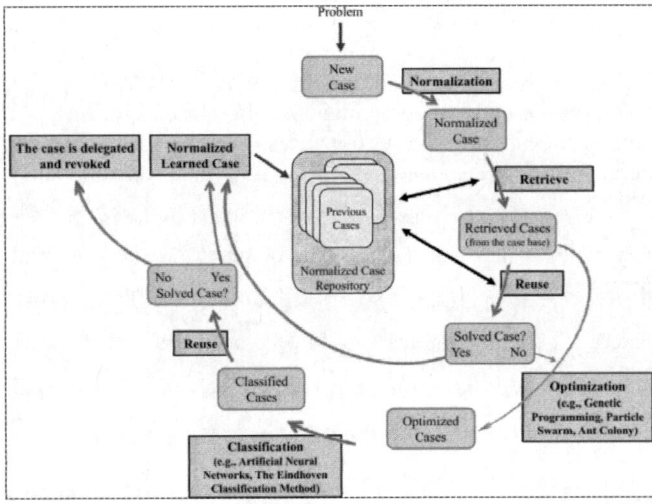

Figure 6. The CB cycle updated's view.

4. Methods

Aiming to develop a predictive model to assess the pesticides environmental and toxicological risk a knowledge database was set, and built around the pesticides records of the *National Pesticide Information Center* [23]. For each pesticide it was considered information regarding environmental fate, human exposure and toxicity (i.e., acute and chronic) both in qualitative and quantitative terms.

It is now possible to build up a knowledge database given in terms of the extensions of the relations (or tables) depicted in Figure 7, which denote a situation where one has to manage information in order to evaluate the *Pesticide Environmental and Toxicological Risk*. Under this scenario some incomplete and/or default data is present. For instance, in the former case the *ADI* is unknown (depicted by the symbol $\perp$), while the *Acute Toxicity* for *Mice/Rats* is not conclusive (*Slightly/Moderate*). The *Human Exposure* table is populated with 0 (zero) that stands for absence, 1 (one) that denotes only food or drinking water (in the *Dietary Exposure* (*DE*) column), and only dermal or inhalation exposure (in the *Occupational Exposure* (*OE*) column), and 2 (two) that stands for simultaneous exposition. The issues presented in *Environmental Fate* (*EF*) table are populated with *absence*, *low*, *medium*, *high* and *very high*, while the issues present in columns of *Acute Toxicity* (*AT*) and *Chronic Toxicity* (*CT*) tables are qualified as *absence*, *slightly*, *medium*, *high* and *very high*. In order to measure the information present in these tables the procedures already described in section 2.2 were followed.

Applying the algorithm presented in [17] to the table or relation's fields that make the knowledge base for *Pesticide Environmental and Toxicological Risk Assessment* (Figure 7), and looking to the DoC_s values obtained as described in [17], it is possible to set the arguments of the predicate ***t****oxicological and* ***e****nvironmental* ***r****isk* ***a****ssessment* (***tera***) referred to below, whose extension also denotes the objective function with respect to the problem under analyze:

$$tera: A_{cceptable}D_{aily}I_{ntake}, M_{aximum}C_{ontamination}L_{evel}, E_{nvironmental}F_{ate}, A_{cute}$$
$$T_{oxicity}, C_{hronic}T_{oxicity}, D_{ietary}E_{xposure}, O_{ccupational}E_{xposure} \rightarrow \{0,1\}$$

where 0 (zero) and 1 (one) denote, respectively, the truth values *false* and *true*.

Indeed, in this work it is presented a construction method for interval-valued incomplete preference relations (or predicates) from an incomplete preference relation (or predicate) and the representation of the lack of knowledge or ignorance that experts suffer when they define the membership values of the elements of such incomplete preference relation (or predicate). We also prove that, with this construction method, we obtain membership intervals for an element which domain's length is equal to the ignorance associated with that element. Undeniably, the algorithm presented in [17] encompasses these different phases. In the former one the clauses or terms that make extension of the predicate under study are established. In the subsequent stage the arguments of each clause are set as continuous intervals. In a third step the boundaries of the attributes intervals are set in the interval [0, 1] according to a normalization process given by the expression $(Y - Y_{min})/(Y_{max} - Y_{min})$, where the Y_s stand for themselves. Finally, the *DoC* is evaluated as described in section 2.1.

Environmental Fate (EF)

#	*Soil Adsorption*	*Soil Mobility*	*Soil Persistence*	*Microbial Degradation*	*Groundwater*	*Superficial Waters*	*Air*
1	Medium	High	Very High	Moderate/High	Very High	High	Absence
2	Low/Medium	Low	Low	Very High	Low	Medium	⊥
...	...	...	...	...	...	...	...
148	High	Low	Medium/High	Low	⊥	Moderate	Low

Human Exposure

#	*Dietary Exposure (DE)*	*Occupational Exposure (OE)*
1	2	2
2	2	2
...	...	...
148	2	2

Pesticide Information

#	*Name*	*Chemical Class*	*Pesticide Class*	*ADI*	*MCL*
1	Atrazine	Triazine	Herbicide	⊥	0.003
2	Alachlor	Acetanilide	Herbicide	⊥	0.003
...	...	...	...	...	...
148	Tebuconazole	Azole	Fungicide	[0, 0.03]	⊥

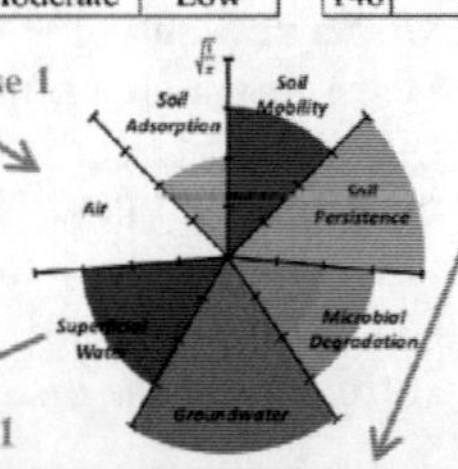

Case 1

Case 1

Toxicological Risk Assessment

#	*Name*	*ADI*	*MCL*	*EF*	*AT*	*CT*	*DE*	*OE*
1	Atrazine	⊥	0.003	[0.64, 0.68]	[0.3, 0.55]	[0.08, 0.58]	2	2
2	Alachlor	⊥	0.003	[0.32, 0.5]	[0.25, 0.5]	[0.08, 0.58]	2	2
...	...	...	...	...	...	...	...	...
148	Tebuconazole	[0, 0.03]	⊥	[0.36, 0.54]	0.3	[0.25, 0.42]	2	2

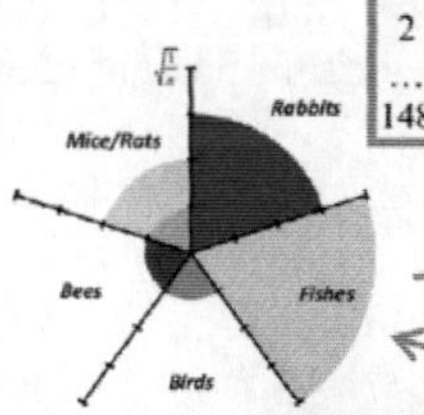

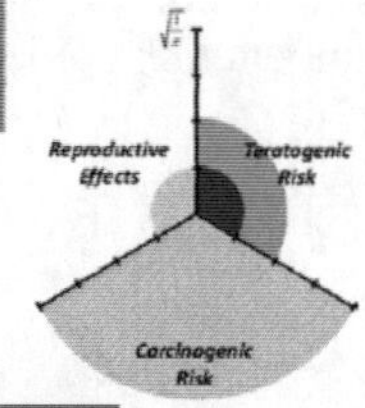

Case 1 Case 1 Case 1 Case 1

Acute Toxicity (AT)

#	*Mice/Rats*	*Rabbits*	*Fishes*	*Birds*	*Bees*
1	Slightly/ /Moderate	High	⊥	Slightly	Slightly
2	Slightly/ /Moderate	⊥	Moderate	Slightly	Slightly
...	...	...	...	...	...
148	Slightly	Slightly	Moderate	Slightly	Slightly

Chronic Toxicity (CT)

#	*Reproductive Effects*	*Teratogenic Risk*	*Carcinogenic Risk*
1	Absence/ /Slightly	Slightly/ /Moderate	⊥
2	Absence/ /Slightly	Slightly/ /Moderate	⊥
...	...	...	...
148	Absence/ /Slightly	Absence/ /Slightly	High

Figure 7. A fragment of the knowledge base for *Environmental and Toxicological Risk Assessment.*

Exemplifying the application of the algorithm presented in [17] in relation to a fictitious term (once a real pesticide assessment is per se not easy) that presents feature vector $ADI = 0.01$, $MCL = \perp$, $EF = [0.39, 0.68]$, $AT = [0.7, 0.85]$, $CT = 0.17$, $DE = 2$, $OE = 2$, one may have:

$$\{ \neg\, tera\left(\left((A_{ADI}, B_{ADI})(QoI_{ADI}, DoC_{ADI})\right), \cdots, \left((A_{OE}, B_{OE})(QoI_{OE}, DoC_{OE})\right)\right)$$

$$\leftarrow not\ tera\left(\left((A_{ADI}, B_{ADI})(QoI_{ADI}, DoC_{ADI})\right), \cdots, \left((A_{OE}, B_{OE})(QoI_{OE}, DoC_{OE})\right)\right)$$

$$tera\underbrace{\left(\left((0.02,\ 0.02)(1,\ 1)\right),\ \cdots,\ \left((1,\ 1)(1,\ 1)\right)\right)}_{\substack{attribute`s\ values\ ranges\ once\ normalized\ and \\ respective\ QoI\ and\ DoC\ values}} :: 1 :: 0.85$$

$$\underbrace{[0, 1] \qquad \cdots \qquad [0, 1]}_{attribute`s\ domains\ once\ normalized} \qquad \} :: 1$$

5. A Case Based approach to Computing

In this section is set the model of the universe of discourse, where the computational part is based on a *CBR* approach to computing. Contrasting with other problem solving tools (e.g., those that use *Decision Trees* or *Artificial Neural Networks*), relatively little work is done offline [24]. Undeniably, in almost all the situations the work is performed at query time. The main difference between this approach and the typical *CB* one relies on the fact that not only all the cases have their arguments set in the interval [0, 1], a situation that is complemented with the prospect of handling incomplete, unknown, or even self-contradictory data or knowledge (Figure 6).

When confronted with a new case, the system is able to retrieve all cases that meet such a structure and optimize, when necessary, such a population, i.e., it considers the attributes *DoC*'s value of each case or of their optimized counterparts when analyzing similarities among them. Thus, under the occurrence of a new case, the goal is to find similar cases in the *Case Base*. Having this in mind, the algorithm described above is applied to a new case, that presents the feature vector ($ADI = \perp$, $MCL = 0.003$, $EF = 0.71$, $AT = [0.65, 0.8]$, $CT = [0.42, 0.58]$, $DE = 1$, $OE = 2$), having in consideration that the cases retrieved from the *Case Base* satisfy the invariant:

$$\cap_{i=1}^{n}(\boldsymbol{B}_i, \boldsymbol{E}_i) \neq \emptyset \tag{6}$$

denoting that the intersection of the attributes range in the cases that make the *Case Base* repository or their optimized counterparts ($\boldsymbol{B}_i$), and the equals in the new case ($\boldsymbol{E}_i$), cannot be empty. Then, the computational process may be continued, with the outcome (once applying the algorithm presented in [17]):

$$tera_{new}\underbrace{\left(\left((0,\ 1)(1,\ 0)\right),\ \cdots,\ \left((1,\ 1)(1,\ 1)\right)\right)}_{new\ case} :: 1 :: 0.85$$

Now, the *new case* may be portrayed on the *Cartesian* plane in terms of its *QoI* and *DoC*, and by using clustering methods [25] it is feasible to identify the cluster(s) that intermingle with the *new one* (epitomized as a square in Figure 8). The *new case* is compared with every retrieved case from the clusters using a similarity function *sim*, given in terms of the average of the modulus of the arithmetic difference between the arguments of each case of the selected cluster and those of the *new case,* which is crucial when different clustering methods are examined. Thus, one may have:

$$\underbrace{\begin{array}{l} tera_1\left(\left((0.04,\ 0.04)(1,\ 1)\right),\ \cdots,\ \left((1,\ 1)(1,\ 1)\right)\right) :: 1 :: 0.99 \\ tera_2\left(\left((0.06,\ 0.06)(1,\ 1)\right),\ \cdots,\ \left((0.5,\ 0.5)(1,\ 1)\right)\right) :: 1 :: 0.84 \\ \vdots \\ tera_j\left(\left((0.03,\ 0.03)(1,\ 1)\right),\ \cdots,\ \left((1,\ 1)(1,\ 1)\right)\right) :: 1 :: 0.85 \end{array}}_{normalized\ cases\ from\ retrieved\ cluster}$$

Assuming that every attribute has equal weight, for the sake of presentation, the dissimilarity between $tera_{new}$ and the $tera_1$, i.e., $tera_{new\rightarrow 1}$, may be computed as follows:

$$tera_{new\rightarrow 1}^{DoC} = \frac{\|0-1\| + \cdots + \|0.99-0.95\| + \|0.99-1\| + \|1-1\| + \|1-1\|}{7} = 0.15$$

Thus, the similarity for $tera_{new\rightarrow 1}^{DoC}$ is set as *1 – 0.15 = 0.85*. Regarding *QoI* the procedure is similar, returning $tera_{new\rightarrow 1}^{QoI} = 1$. Thus, one may have:

$$tera_{new\rightarrow 1}^{QoI,\,DoC} = 1 \times 0.85 = 0.85$$

These procedures should be applied to the remaining cases of the retrieved clusters in order to obtain the most similar ones, which may stand for the possible solutions to the problem.

In order to evaluate the performance of the proposed model the dataset was divided in exclusive subsets through the ten-folds cross validation [21]. In the implementation of the respective dividing procedures, ten accomplishments were performed for each one of them. Table 1 presents the coincidence matrix of the *CB* model, where the values presented denote the average of 30 (thirty) experiments. A perusal to Table 1 shows that the model accuracy was 87.8% (i.e., 130 instances correctly classified in 148 cases). Thus, the predictions made by the *CB* model are satisfactory, attaining accuracy close to 90%. The sensitivity and specificity of the model were 91.7% and 80.8%, while *Positive* and *Negative Predictive Values* were 89.8% and 84.0%, respectively. Thus, the proposed model correctly identified 89.8% of the positive cases, i.e., pesticides with high *potential environmental* or *toxicological risk*. Moreover, it also classified appropriately 84.0% of the negative cases, i.e., pesticides with *low environmental and toxicological risk.*

The present model, beyond to consider the pesticide chemical properties, enables the integration of *Acute* and *Chronic Toxicity* data with other factors such as *Environmental Fate* and *Human Exposure*, being therefore assertive in the prediction of *Pesticide Toxicological Risk*. Thus, it is our claim that the proposed model is able to evaluate the *Environmental and Toxicological Risk* of each pesticide properly, and may be a major contribution to achieve high standards concerning environmental sustainability and public health protection.

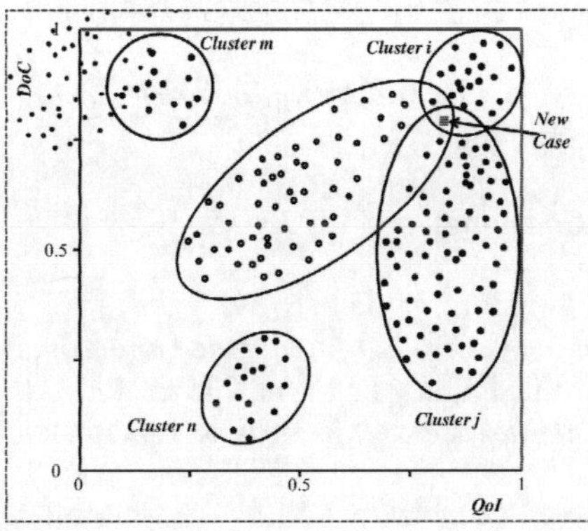

Figure 8. A case's set clusters.

Table 1. The coincidence matrix for *CBR* model.

	Predictive	
Target	True (1)	False (0)
True (1)	88	8
False (0)	10	42

6. Conclusions

The proposed methodology for problem solving is able to give an adequate response to predict the environmental and toxicological risk of pesticide exposure. Nevertheless, it can be considered a hard task since it is necessary to consider different variables and/or conditions with complex relations entwined among them, where the data may be incomplete, self-contradictory, and even unknown. In order to overcome these difficulties the methodology followed in this work was centred on a formal framework based on *LP* for knowledge representation and reasoning, complemented with a *CB* approach to computing. It may set the basis for an overall approach to such systems, susceptive of application in different arenas. Furthermore, under this line of thinking the cases' retrieval and optimization phases were heightened when compared with existing tactics or methods. Additionally, under this approach the users may define the cases weights attributes on-the-fly, letting them to choose the appropriate strategies to address the problem (i.e., it gives the user the possibility to narrow the search space for similar cases at runtime). A possible limitation on its use is not on the model in itself, but on the unavailability of data, information or knowledge but, even in these situations, once it has the capacity to handle incomplete data, information or knowledge, either in its qualitative or quantitative form, its usefulness is assured. Future developments of the model should include the biotransformation pathways and routes of exposure, and consider the contact time and the individual's susceptibility.

Acknowledgments

This work has been supported by COMPETE: POCI-01-0145-FEDER-007043 and FCT – Fundação para a Ciência e Tecnologia within the Project Scope: UID/CEC/00319/2013.

References

[1] Akatar W R, Sengupta D and Chowdhury A 2009 *Interdiscip. Toxicol.* **2** 1–12
[2] Popp J, Pető K and Nagy J 2013 *Agron. Sustain. Dev.* **33** 243–255
[3] Verma J P, Jaiswal D K and Sagar R 2014 *Rev. Environ. Sci Biotechnol.* **13** 429–466
[4] Hodgson E 2012 Introduction to Pesticide Biotransformation and Disposition *Pesticide Biotransformation and Disposition* ed E Hodgson (Amsterdam: Elsevier) chapter 1 pp 1–3
[5] Needham L L, Patterson D G, Barr D B, Grainger J. and Calafat A M 2005 *Anal. Bioanal. Chem.* **381** 397–404
[6] Environmental Protection Agency 2001 *General Principles for Performing Aggregate Exposure and Risk Assessments* (Washington: Office of Pesticide Programs)
[7] Renwick A G 2002 *Pest Manag. Sci.* **58** 1073–82
[8] Antón A, Castells F, Montero J I and Huijbregts M 2004 *Chemosphere* **54** 1225–35
[9] Alister C and Kogan M 2006 *Crop Prot.* **25** 202–211
[10] Juraske R, Antón A, Castells F and Huijbregts M A 2007 *Environ. Int.* **33** 886–893
[11] Kakas A, Kowalski R and Toni F 1998 *Handbook of Logic in Artificial Intelligence and Logic Programming* vol 5, ed D Gabbay *et al* (Oxford: Oxford University Press) pp 235–324
[12] Pereira L and Anh H 2009 *Studies in Computational Intelligence* vol 199, ed K Nakamatsu (Berlin: Springer) pp 51–64
[13] Neves J 1984 *Proc. 1984 Annual Conf. of the ACM on the 5th Generation Challeng* ed R Muller and J Pottmyer (New York: Association for Computing Machinery) pp 50–54

[14] Neves J, Machado J, Analide C, Abelha A and Brito L 2007 *Lecture Notes in Artificial Intelligence* vol 4874, ed J Neves *et al* (Berlin: Springer) pp 160–169
[15] Lucas P 2003 *Proc. of AI-2003 (Research and Developments in Intelligent Systems XX)* ed F Coenen *et al* (London: Springer) pp 309–321
[16] Machado J, Abelha A, Novais P, Neves J and Neves J 2008 *Proc. of the ESM 2008* ed C Bertelle and A Ayesh (Ghent: Eurosis) pp 291–298
[17] Fernandes F, Vicente H, Abelha A, Machado J, Novais P and Neves J 2015 *Proc. of the 2015 Science and Information Conf.* pp 362–370
[18] Aamodt A and Plaza E 1994 *AI Commun.* **7** 39–59
[19] Richter M M and Weber R O 2013 *Case-Based Reasoning: A Textbook* (Berlin: Springer)
[20] Vicente H, Dias S, Fernandes A, Abelha A, Machado J and Neves J 2012 *J. Water Supply Res. T.* **61** 446–459
[21] Haykin S 2009 *Neural Networks and Learning Machines* (New Jersey: Pearson Education)
[22] Mendes R, Kennedy J and Neves J 2003 *Proc. of the 2003 IEEE Swarm Intelligence Symp.* pp 88–94
[23] National Pesticide Information Center http://npic.orst.edu/index.html
[24] Carneiro D, Novais P, Andrade F, Zeleznikow J and Neves J 2013 *Knowl. Inf. Syst.* **36** 789–826
[25] Figueiredo M, Esteves L, Neves J and Vicente, H. 2016 *Chem. Educ. Res. Pract.* **17** 156–171

Development of gas fire detection system using tunable diode laser absorption spectroscopy

Y L Jiang[1, 3], G Li[1], T Yang[1] and J J Wang[2]

[1] College of Civil and Environmental Engineering, Anhui Xinhua University, China
[2] State Key Laboratory of Fire Science, University of Science and Technology of China, China

E-mail: jiangyl@mail.utsc.edu.cn

Abstract. The conventional fire detection methods mainly produce an alarm through detecting the changes in smoke concentration, flame radiation, heat and other physical parameters in the environment, but are unable to provide an early warning of a fire emergency. We have designed a gas fire detection system with a high detection sensitivity and high selectivity using the tunable semiconductor diode laser as a light source and combining wavelength modulation and harmonic detection technology. This system can invert the second harmonic signal obtained to obtain the concentration of carbon monoxide gas (a fire characteristic gas) so as to provide an early warning of fire. We reduce the system offset noise and the background noise generated due to the laser interference by deducting the system background spectrum lines from the second harmonic signal. This can also eliminate the interference of other gas spectral lines to a large extent. We detected the concentration of the carbon monoxide gas generated in smoldering sandalwood fire and open beech wood fire with the homemade fire simulator, and tested the lowest detectable limit of system. The test results show that the lowest detectable limit can reach 5×10^{-6}; the system can maintain stable operation for a long period of time and can automatically trigger a water mist fire extinguishing system, which can fully meet the needs of early fire warning.

1. Introduction

Of all disasters, fire is the most frequent and the most common threat to public safety and social development. It not only causes environmental pollution, destroys ecological balance, but also seriously harms the development of the society. Take America for example, about 1,240,000 fires were reported in 2013. These fires resulted in 3,240 civilian fire fatalities, 15,925 civilian fire injuries and an estimated $11.5 billion in direct property loss [1].Although the cause of fire and combustion processes vary widely, each case of fire has certain physical characteristics, i.e. smoke, flames, and a certain amount of heat generated in the combustion process. By monitoring these physical characteristics we can determine whether a fire is really happening. But traditional physical detection is not always effective in detecting fires. At present, fire detection based on the gas concentrations produced in the combustion process has become an important direction of research. Among the methods for detecting gas concentration, tunable diode laser absorption spectroscopy technology receives increasing attention of the researchers for its characteristics of high selectivity, high sensitivity and high resolution, and has been recognized as a high-sensitivity measurement method with detection limits up to 10-6 or higher.

TDLAS technology uses the wavelength tuning characteristic of tunable diode laser to obtain the characteristic absorption spectra of the gas to be measured for qualitative or quantitative analysis. Compared with other optical methods, TDLAS exhibits significant advantages in both sensitivity and selectivity. The TDLAS measurement of trace gas concentrations is based on the detection of the absorption lines of the gas molecules, and the frequency and linearity of the absorption lines are the inherent, unique characteristics of gas molecules (just like human fingerprints). This feature can be recorded under high-precision and high-resolution conditions.

The spectral parameters of line positions, integrated line strengths, temperature dependence, and line broadening coefficient needed for quantitation of these molecules can be found in the high resolution transmission (HITRAN) molecular absorption database of spectral lines[10].

2. Determination of characteristic gases associated with fire

TDLAS technology for fire detection detects the concentration of the gas generated in the combustion process. Because the composition of the gas generated in the combustion process is very complicated, it is important to determine which gas should be selected as a characteristic gas for fire detection.

The vast majority of combustible materials that naturally exist contain two elements, C and H, which will generate a considerable amount of CO, CO_2 and H_2O in smoldering or flaming conditions. Given the influence of the humidity in the environment, H_2O is usually not deemed as one of the fire detection parameters. As one of the component gases of the atmosphere, CO_2 exists even in non-fire environments at a certain concentration. Because its content in the air is extremely low under normal circumstances and only appears in a considerable amount when fires occur, CO is usually selected as an indicator gas for fire detection. Figure 1 shows a test done in our laboratory which is used to determine whether carbon monoxide (CO) or carbon dioxide (CO_2) gas is more suitable for detecting a fire. When a tester breathes, the concentration of CO_2 gas near the detection point changes and gradually begins to rise from the normal value of 403 ppm (403 ppm for the current concentration of CO_2 gas in the test room) up to the maximum value of 985 ppm, and then begins to decline, while the concentration of CO gas remains unchanged. Only when there is a fire (simulated through the cotton rope smoldering fire test) does the concentration of CO gas gradually begin to rise...

In order to evaluate the adaptability of carbon monoxide detector to each kind of fire, researchers conducted comparative experiments on carbon monoxide detector, smoke detector and other detectors using six types of experimental fire specified by the European Standard EN54. Only carbon monoxide detector responds to all of the six types of fire, showing the adequacy of using CO as an indicator gas for fire detection.

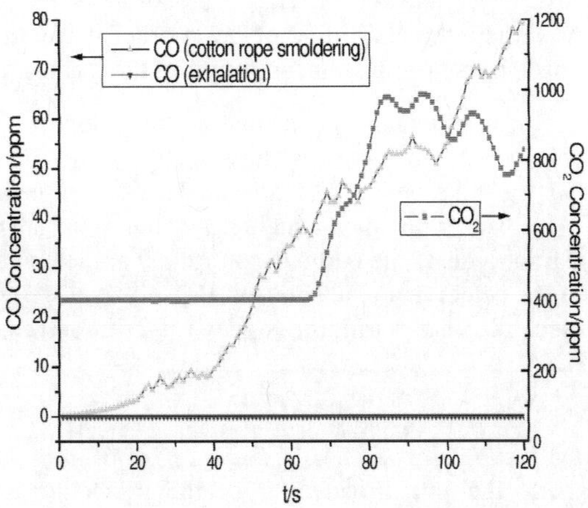

Figure 1. The test experiment for determination of fire characteristic gas.

3. Theory

According to the Lambert-Beer's Law, a beam of light will attenuate as it goes through gas. In case of no saturation, the intensity of transmitted light can be described by the following formula:

$$I(\lambda) = I_0(\lambda)e^{[-\sigma(\lambda)cl]} \tag{1}$$

Where $I_0(\lambda)$ and $I(\lambda)$ represent the input and output light intensities respectively, $\sigma(\lambda)$ means the absorption cross section of the gas molecules to be measured where the wavelength is λ (which also can be expressed as the product of shape and intensity of the molecular absorption lines), c is the concentration of the gas to be measured, and l is the length of absorption path.

In order to reduce the interference of noise signals [11]and improve the signal to noise ratio[12], the wavelength modulation technology is typically used (modulation frequency <1MHz[13]) to modulate the output wavelength of the laser. Assuming the center frequency of the laser light is ν_{cen}, and the modulation amplitude and modulation frequency are δm and ω_m, respectively, we get the instantaneous frequency of the laser emission wavelength as follows:

$$\nu = \nu_{cen} + \delta m\cos(\omega_m t) \tag{2}$$

The transmitted light intensity of the laser can be described by the following Fourier series:

$$I(\nu_{cen}, t) = \sum_{n=0}^{\infty} \Phi_n(\nu_{cen})\cos(n\omega_m t) \tag{3}$$

In Formula (3), $\Phi_n(\nu_{cen})$ means n-order harmonic component. When the gas concentration is very low, $\sigma cl << 1$ .Furthermore,I0 can regarded as a constant in an ideal situation, $\Phi_n(\nu_{cen})$ can be approximated by the following formula:

$$\Phi_n(\nu_{cen}) = \frac{2I_0 cl}{\pi}\int_0^{\pi} -\sigma(\nu_{cen} + \delta_m\cos\theta)\cos(n\theta)d\theta \tag{4}$$

Where $\theta = \omega_m t$. When the modulation amplitude is much smaller than the absorption linewidth, the expansion of $\Phi_n(\nu_{cen})$ according to the Taylor series can be expressed as follows:

$$\Phi_n(\nu_{cen}) = \frac{2^{1-n} I_0 cl}{n!}\delta_m{}^n \frac{d^n\sigma}{d\nu^n}\Big|_{\nu=\nu_{cen}} \tag{5}$$

As can be seen from Formula (5), the amplitude of the n-order harmonic signal is proportional to the gas concentration. Therefore, in theory, the concentration of the gas can be calculated with any one of the harmonic components. But in practice, as n increases, the value of $\Phi_n(\nu_{cen})$ and the harmonic linewidth will increase as well, which would affect the sensitivity and the resolution of the adjacent interference spectral line. On the other hand, the amplitude of even-order harmonics reaches the maximum at the center of the absorption line, which helps the laser stabilize at the center of the absorption spectra, while the amplitude of the odd harmonics is 0 at the center frequency, which works against the laser stabilizing at the center of the absorption lines. Therefore, when the above-mentioned factors are considered, the second order harmonic signal (n=2) is selected in the experiment as a detection signal, i.e.

$$\Phi_2(\nu_{cen}) = \frac{I_0 cl}{4}\delta_m{}^2 \frac{d^2\sigma}{d\nu^2}\Big|_{\nu=\nu_{cen}} \tag{6}$$

Under atmospheric pressure, the line broadening of the molecules mainly takes the form of pressure broadening, which can be expressed with the Lorenz linetype:

$$g_N(\nu,\nu_0) = \frac{\Delta\nu_N / 2\pi}{(\nu-\nu_0)^2 + (\Delta\nu_N/2)^2}) \qquad (7)$$

Where $\Delta\nu_N$ means the half-width of the collision broadening.

4. Experimental setup

The experimental devices are shown in Figure 2. The whole detection system comprises signal generator (TFG3050), laser(NLK1C5EAAA), laser controller (LDC3724B,ILX Lightwave), White absorption cell (homemade), lock-in amplifier (7265, Signal Recovery), fire simulation device (homemade) and data acquisition system (INV303B).

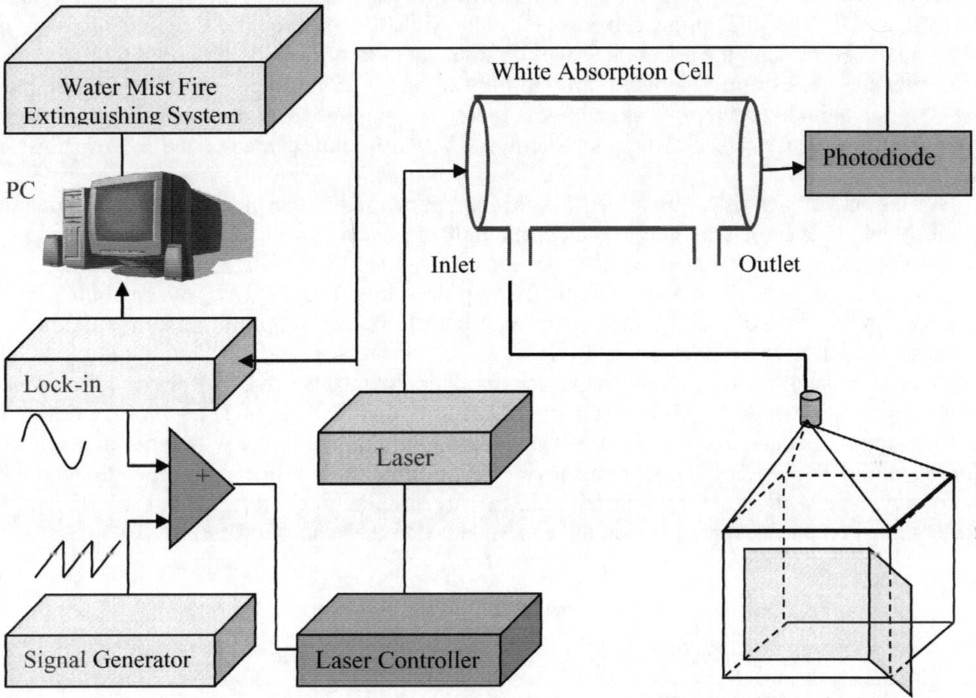

Figure 2. Schematic diagram of the detection system.

For CO detection, the best light source is lasers with wavelength of 4.65 microns. But in fact, the far-infrared laser is expensive to buy and inconvenient to maintain. CO also has absorption line of wavelength 1579.737 nm in its absorption spectrum, which, although has lower absorption intensity than 4.65 microns wavelength, can still be used for fire detection depending on the strong radiation of the laser. Therefore, we choose the continuously tunable DFB diode laser which is relatively affordable. The laser can work at room temperature, with output power of 10 mW, emission wavelength of 1579 nm (the CO characteristic absorption peak) and output frequency coinciding with the absorption lines of carbon monoxide.

The temperature and current of the laser are controlled by a semiconductor laser controller (LDC3724B, ILX LightWave). Controller has a variety of built-in temperature sensors. The change rate of wavelengths with the temperature is 0.1 nm / K.

In order to reduce the interference of the noise spectrum, we superimpose a 5 kHz high-frequency sine wave signal on the 50 Hz low frequency sawtooth signal to modulate the emission wavelength.

The sawtooth wave is generated by the signal generator and the high-frequency sine wave is generated by the lock-in amplifier. These two waveforms, after superimposed by the adder, are injected into the laser controller for laser modulation and scanning.

The beam emitted from the laser enters the absorption cell through the lens after being led out by the optical fiber. The optical fiber is 9/125 μm(1km) single-mode corresponding to the wavelength of about 1579 nm whose transmission loss is below 1dB/km. The White absorption cell is an very important part of the measurement system which based on multiple reflection. The optical path is increased in the cell through the increase of the number of reflections. The total optical path is approximately 100 m, and the temperature uniformity of the cell body is better than ± 0.5℃, and a high vacuum of more than 1.33×10^{-5} Pa can be maintained within the cell. The White multiple reflection cell consists of a spherical mirror, a corner reflector and a field lens. The number of reflections can be adjusted with the spherical mirror in order to change the optical path in the multiple reflection cell. The absorption cell is open, and when fire breaks out, the fire products enter the absorption cell through the opening. Some of the incident laser beam is absorbed by CO that is part of the products, while some is emitted through the exit end after multiple reflections before received by the PIN photodiode on the optical path, which converts the optical signals into electrical signals. The wavelength response range of the photodiode is 1000-1700 nm, characterized by low noise and high sensitivity.

As the useful signals are relatively weak and submerged in the noise signals, the output signals from the photodetector are input to a lock-in amplifier, where a second harmonic detection is used to return second harmonic signals of the gas absorption lines. These signals are collected by the data acquisition card and fed into the computer for processing. The TDLAS, by combining wavelength modulation spectroscopy and second harmonic detection, can suppress noise to a large extent and improve the sensitivity of the system[14-16].

In order to simulate fire, a homemade fire simulation device is used. The device is an iron container whose bottom is a hollow cuboid 50 cm by 50 cm and 80 cm in height. The top of the device is a cone with 45 degrees angle, and the tip is connected with a cylindrical pipe with inner diameter of 2.5 cm for installing PVC pipes. Fire products can hardly enter the White cell depending on its natural buoyancy, so a suction device is installed in the fire simulation device, which can efficiently transfer the combustion products produced in the experiment into the absorption cell.

5. Experiment

5.1. Second harmonic detection

A homemade gas distribution system is used to prepare CO gases of various concentrations required for the experiment. The mass flow ratio method is adopted in the preparation and 99.999% industrial carbon monoxide and 99.999% nitrogen of high purity are used. The gas distribution system is based on accurate mass flowmeter SFC4000 (Sensirion), and produces the mixed gas with a pressure of 1 atm. To avoid interference from other gases, the White cell is vacuumed (2.6×10^{-4} Pa or less) many times prior to the experiment, and then rinsed with 99.999% pure N2 before vacuumed again. This is to rule out the possibility of spectral absorption caused by the presence of other interference gases.

The method where the system background spectra are deducted from the second harmonic signal obtained from the experiment is used to reduce the offset noise caused by modulation and the background noise formed by the optical streaks as a result of laser interference. This not only effectively suppresses the background interference, but also largely eliminates the effects of the spectra of other interference gases.

Figure 3 shows the second harmonic signals of CO at different concentrations. It can be found that the signal is substantially symmetrical on both sides, indicating that the impact of the residual amplitude modulation of the laser is relatively small.

Figure 4 shows the linear relationship between the configured concentration and the measured concentration of the CO gas. As can be seen from Figure 4, the concentration measured by the system has a good linear relationship with the configured concentration, where the linear correlation is 0.99. Under the same conditions, its good reproducibility is proven by a number of tests, with the maximum deviation between the signals controlled within a small range.

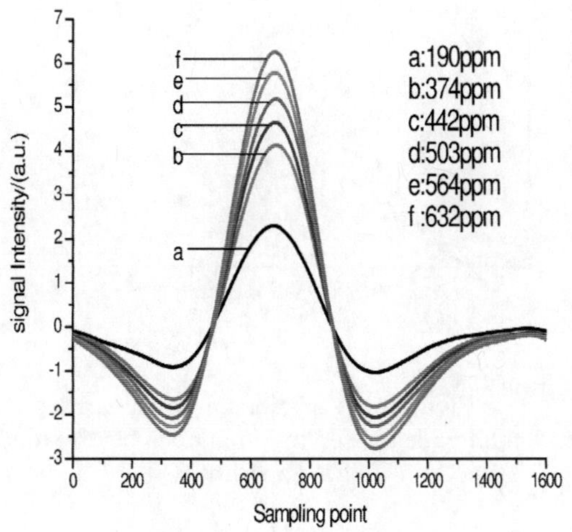

Figure 3. The second harmonic signals of CO at different concentrations.

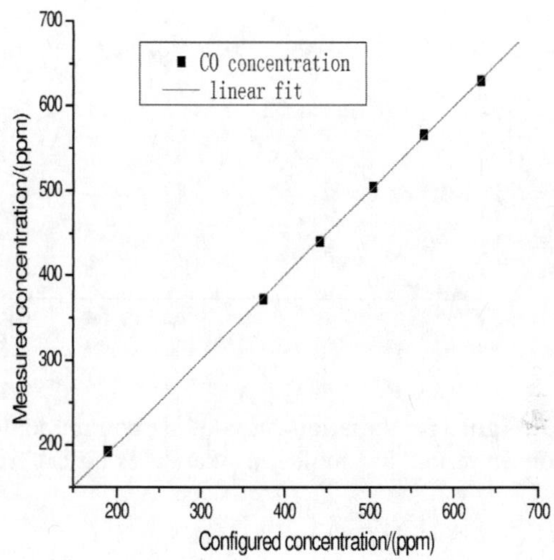

Figure 4. The linear relationship between the measured and the true values of CO

5.2. Fire simulation

After the system is commissioned, sandalwood and beech wood are lit in the combustion chamber to simulate smoldering fires and open fires. At the same time the fan is started and experimental data are collected. The combustion products emitted from the smoke vent in the combustion chamber are transferred to the entrance to the White cell through the PVC pipe. While the data is being collected, the fixed window smoothing function and frequency-domain averaging function are usable to eliminate the impact of random noise.

During the smoldering fire simulation test, the ignited and then put out sandalwood with a diameter of 2.5 mm is placed on the center bottom of the combustion chamber, so as to generate continuous smoke. The change in CO gas concentration is as shown in Figure 5. As can be seen from the figure, the concentration of CO gas continues to increase from the beginning of the test; it reaches 5 ppm at t=21s and then continues to increase, and reaches the maximum value of 79 ppm at t=119s. Later, it starts to decline (The decline curve is not shown in Fig.5).

In the simulation of open flames, pieces of beech wood with dimension of 1cm × 2cm × 2.5 cm (moisture content lower than 3%) are stacked in three tiers and placed in the combustion chamber. On the center bottom of the combustion chamber is a 5cm-diameter dish filled with 5 ml of industrial alcohol which is ignited. The change in the CO concentration generated from the beech burn is shown in Figure 6. As can be seen from the Figure, the CO concentration reaches 5 ppm at t=29s. Although it dips briefly at the first peak, the CO concentration maintains a growing trend in general, and reaches a maximum value of 34 ppm at t=120s before starting to decrease. As the data in Figure 6 only represents a capture of 120 seconds, the downward curve is not shown in the Figure.

Figures 5 and 6 show that both smoldering fires and open flames can lead to continuous rise of CO concentration in air. Whereas the CO content in normal air is typically very low, it is feasible to determine the occurrence of fire depending on the concentration of CO gas. Generally, smoldering generates higher concentrations of CO than open flames. The maximum CO concentration shown in Figure 5 is approximately 2.32 times that in the Figure 6.

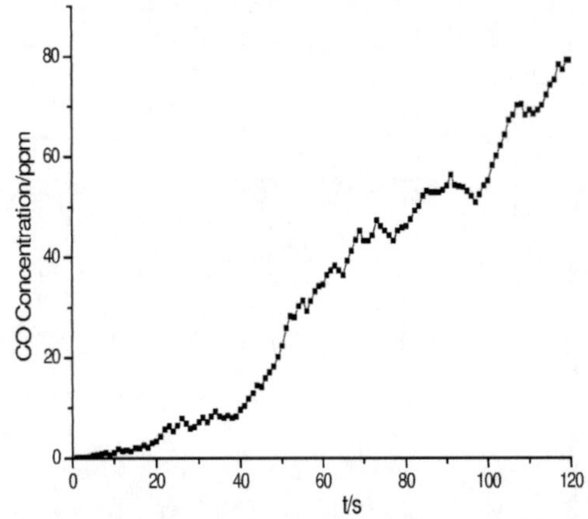

Figure 5. Variation curve of carbon monoxide concentration in smoldering sandalwood experiment.

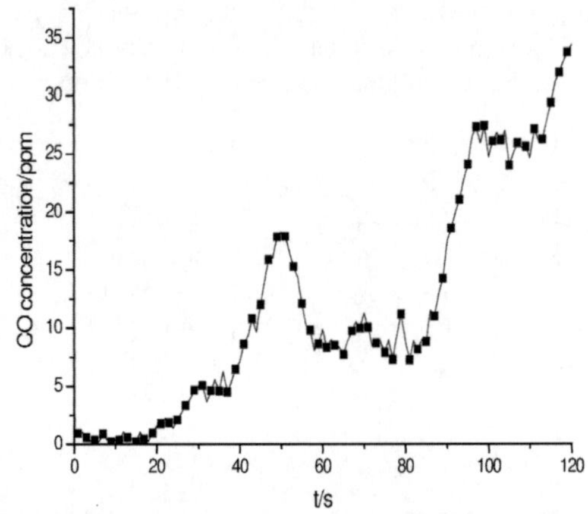

Figure 6. Variation curve of carbon monoxide concentration in open beech wood fire experiment.

5.3. Coordinated fire suppression

A gas fire detection system that works well should be able to sound the alarm and act in conjunction with fire extinguishing systems, in addition to the ability to detect fire. The signals of the CO concentration are judged using the self-made program. When the concentration exceeds the set threshold value, the microcontroller(AT89S51) on the control circuit board issues a command that enables the relay control circuit to connect, which triggers the sound and light alarm (P2475) and starts the water mist fire extinguishing system to put out the fire. The water mist is driven by high-pressure nitrogen, and the 30 L water tank is connected with the pipeline made of stainless steel equipped with a filter. The high-pressure air reaches the required pressure through the relief valve and pressurizes the water in the tank, which flows through the shut-off valve before sprayed through the spray nozzle in the form of fine mist. The alarm threshold usually varies with the importance of places and can be set as required. We set 5 ppm as the threshold value of the alarm signal, meaning any signal showing a concentration greater than or equal to 5 ppm is considered a fire alarm. The experiments show that the alarm works once the experiment starts, and the water mist fire extinguishing system begins to function as well.

6. Conclusion

Tunable diode laser absorption spectroscopy (TDLAS) uses the wavelength tuning characteristic of diode laser for qualitative or quantitative analysis of the target gas. This approach not only has the feature of high precision and good selectivity, but also fast response and has been applied in the monitoring of atmospheric trace gas, gas leaks and other fields.

In the paper, the concentrations of CO generated by the smoldering sandalwood and the open beech wood fire are measured by the least-squares data processing method using narrow linewidth tunable semiconductor laser as a light source in combination with a long optical path White cell, wavelength modulation and the harmonic detection technology. The research results show that the system works well in online measuring the CO concentrations in the environment with real-time displays. The lowest detectable limit is 5×10^{-6}, and the system runs with stability on long-term basis and can meet the requirements of early fire warning. In the experiment, background subtraction is used to effectively reduce the impact of the residual amplitude modulation and the background noised formed by the optical streaks resulted from laser interference. This improves the signal-to-noise ratio and guarantees that the CO concentration in the environment is measured online and accurately obtained through inversion. With a different wavelength diode laser as a light source, the TDLAS technology can also be used for the detection of other gases, and therefore holds great promise for further applications in areas such as environmental protection, industrial control, chemical reaction dynamics and other fields.

Acknowledgment
This study was sponsored by the Key Project of the Outstanding Young Talent Support Program of Anhui Province University in 2016 (gxyqZD2016392).

References

[1] Karter M J 2014 Fire loss in the United States during 2013 *National Fire Protection Association Fire Analysis and Research Division*

[2] Pogány A, Ott O, Werhahn O, et al. 2013 Towards traceability in CO_2 line strength measurements by TDLAS at 2.7 mm *J Quant Spectrosc Ra* **130** 147-57.

[3] Cai T D, Wang G S, Cao Z S, et al. 2014 Sensor for headspace pressure and H_2O concentration measurements in closed vials by tunable diode laser absorption spectroscopy *Opt Laser Eng* **58** 48-53.

[4] Gao Q, Zhang Y G, Yu J, et al. 2013 Tunable multi-mode diode laser absorption spectroscopy for methane detection *Sensor Actuat A-Phys* **199** 106-110.

[5] Melroy H R, Adkins E M, Pause M J, et al. 2015 Species measurements in a nitrogen-diluted, ethylene air diffusion flame using direct sampling mass spectrometry and tunable diode laser absorption spectroscopy *P Combust Inst* **35** 3749-55.

[6] Shemshad J, Aminossadati S M, Kizil M S, et al. 2012 A Review of developments in near infrared methane detection based on tunable diode laser *Sensor Actuat B-Chem* **171-172** 77-92.

[7] Catoire V, Bernard F, Mebarki Y, et al. 2014 A tunable diode laser absorption spectrometer for formaldehyde atmospheric measurements validated by simulation chamber instrumentation *J Environ Sci* **24(1)** 22-33.

[8] Boehm G, Bachmann A, Rosskopf J, et al. 2011 Comparison of InP and GaSb- based VCSELs emitti-ng at 2.3μm suitable for carbon monoxide detection *J Cryst Growth* **323** 442-45.

[9] Krzempek K, Lewicki R, Nähle L, et al. 2012 Continuous wave, distributed feedback diode laser bas-ed sensor for trace gas detection of ethane *Appl Phys B* **106** 251-55.

[10] Wagner S, Klein M, Kathrotia T, et al. 2012 In situ TDLAS measurement of absolute acetylene concentration profiles in a non-premixed laminar counter-flow flame *Appl Phys B* **107** 585-589.

[11] Silver J A 1992 Frequency modulation spectroscopy for trace species detection theory and comparis- on among experimental method *Appl Optics* **31** 707-17.

[12] Harward C N, Thweatt W D, Baren R E, et al. 2006 Determination of molecular line parameters for acrolein (C_3H_4O) using infrared tunable diode laser absorption spectroscopy *Spectrochim Acta A* **63** 970–80.

[13] Hangauer A, Chen J, Strzoda R, et al. 2013 Multi-harmonic detection in wavelength modulation spectroscopy systems *Appl Phys B* **110(2)** 177-85.

[14] Li H J, Wehe S D, McManus K R 2011 Real-time equivalence ratio measurements in gas turbine combustors with a near-infrared diode laser sensor *P Combust Inst* **33** 717–24.

[15] Ray A, Bandyopadhyay A, De S, et al. 2007 A simple scanning semiconductor diode laser source and its application in wavelength modulation spectroscopy around 825nm *Opt Laser Technol* **39** 359–67.

[16] Kluczynski P, Axner O 1999 Theoretical description based on Fourier analysis of wavelength modulation spectrometry in terms of analytical and background signals *Appl Optics* **38** 5803–15.

On the interdisciplinary nature of water-related programs in American public universities

F Li[1] and C L Du

School of Public Administration, Hohai University, Nanjing, 210098, China

E-mail: lfhohai@163.com

Abstract. This paper presents a research on the interdisciplinary nature of 167 water-related programs in 46 selected American public universities. Using keyword coding and Classification method, and the results show that water-related disciplines largely focus on environment, natural resources, engineering, ocean and agricultural studies, which are all applied disciplines. Moreover, water-related disciplines lay great emphasis on the interdisciplinary research and research platforms construction with the research fields of environment, natural resources and agricultural.

1. Introduction

As the material basis for human survival, water has linked thousands of things in nature, and water-related issues are major livelihood issues. In recent years, water-related problems have become more and more complex, and also more difficult to solve. Problems such as water disasters, water shortages, water ecological damage and water pollution occurs everywhere around the world. It's essential for a country to develop education and research programs focusing on water issues. Since the establishment of the People's Republic of China, the development of water-related research and education in China has made great achievements over the past 60 years. Meanwhile, there are still a few problems—the universities establishing water-related disciplines are mainly technological universities, and the interactions and collaborations between water-related disciplines and other disciplines are insufficient. Facing the complexity and flexibility of water-related problems, it is important for China to cultivate high quality water-related interdisciplinary programs, as well as build capacity in both water education and water research, in order to give better solutions to contemporary water problems.

University-based water-related interdisciplinary programs play a main and important role in water research. American universities (especially public universities) are the leaders in water-related research [1]. Among the world's top 11 institutions which published the most research papers in the field of water resource management from 1990 to 2012, 8 institutions were from the United States, and four of them were American public universities, namely, Texas A & M University, University of Arizona, New Mexico State University, and Colorado State University [2]. Therefore, American public universities are a perfect source for the study of water-related programs. We selected a group of 46 American public universities as the sample. By analyzing the water-related degree programs as well as the academic departments and colleges which host those programs in American public universities, we aimed to explore the interdisciplinary nature in water-related programs, and find the patterns of the interaction between both water-related and unrelated disciplines.

2. Definition of Water-related Interdisciplinary

Nowadays, the development of water-related disciplines is no longer limited to "pure" water disciplines, but widely interacts and integrates with other disciplines, resulting in a number of emerging water-related subjects or water-related research fields, such as water economy, water culture, water education and so on [3][4]. According to the classification standards for academic programs and water-related research, we define the water-related interdisciplinary as academic programs or degree programs focusing on the education and research of both water-related problems and direct water problems. Non-degree programs are not included in our study.

3. Sample and Method

3.1 Selecting Research Samples

The research samples were selected from member universities of National Institutes for Water Resources (NIWR). NIWR consists of nationwide water resources institutes based at 54 public universities in the United States [3]. Among these universities, 8 universities without water-related interdisciplinary degree programs were removed. Finally, we selected 46 American public universities as the sample (see Table 1). These public universities belong to 44 different states and major territories. A collection of outstanding public universities are included in our sample, which covers University of California at Davis, Georgia Institute of Technology, University of Minnesota at Twin Cities, Purdue University, etc. Through a thorough search in the universities' official websites, we eventually identified 167 water-related degree programs offered by 46 universities. Each targeted university offers 3.6 water-related degree programs on average, indicating that the sample is ideal for our research since the universities we selected have strong roots in water-related research.

Table 1. 46 American public universities list (in alphabetical order).

Colorado State University	University of Alabama at Huntsville	University of Missouri-Columbia
Cornell University	University of Alaska Fairbanks	University of Nebraska-Lincoln
Georgia Institute of Technology	University of Arizona	University of Nevada, Reno
Iowa State University	University of Arkansas	University of New Hampshire
Mississippi State University	University of California-Davis	University of Rhode Island
Montana State University Bozeman	University of California-Riverside	University of South Carolina
New Mexico State University	University of Connecticut	University of Tennessee, Knoxville
North Dakota State University	University of Delaware	University of the District of Columbia
Ohio State University	University of Florida	University of Vermont
Oklahoma State University	University of Idaho	University of Wisconsin-Madison
Oregon State University	University of Kentucky	University of Wyoming
Pennsylvania State University	University of Maryland ,College Park	Utah State University
Purdue University	University of Massachusetts, Amherst	Virginia Polytechnic Institute and State University
Rutgers, New Jersey State University	University of Minnesota Crookston Campus	Washington State University
South Dakota State University	University of Minnesota Twin Cities Campus	West Virginia University
Texas A&M University		

3.2 Keywords Coding and Classification

The organizational structure for an academic discipline usually consists of three levels in American universities. These three levels are listed as follows in a low-to-high order: the degree program level, the departmental level (sometimes refers to a unit or a school), and the college level (sometimes refers to a division). The same structure applies to water-related disciplines. We name the three organizational levels of water-related disciplines as water-related degree programs, water-related departments and water-related colleges just for convenience. Among them, water-related degree programs are the smallest units where water-related disciplines are developed. The program name reflects the specific research area and specialty of a water-related discipline, while a department or a college usually includes a wider range of research areas. To be accurate, a department usually contains a number of research fields of the same discipline [5]. The department name embodies the range of research fields where water-related research is involved. The establishing of a college is usually based on a broader category of disciplines [6], and a college usually comprises of a group of different disciplines [7]. As a result, the name of a water-related college not only reflects a broad field category where the water-related discipline belongs to, but also reflects the interaction and integration between "pure" water discipline and other related disciplines. In this paper, we used the text analysis methods to classify and code the keywords of the names of water-related degree programs, departments and colleges. By doing so, we are able to identify the range of water-related research areas and examine the interdisciplinary nature of water-related disciplines.

We proposed a guideline to code and group key words appearing in the names of water-related degree programs, departments and colleges (see Table 2). The identification and classification of keywords has been advised and approved by several experts in the fields of water-related research. Taking the coding and grouping of programs as example, we judged whether the program name had the exact words from the keyword list, if yes, then the program was assigned to the category where the keyword belonged. Each program can be assigned to multiple categories. The departments and colleges are assigned to certain categories using the same method as we do to programs. To ensure the accuracy of coding, each item has been coded by two persons separately and cross-checked afterwards.

Table 2. Guideline of keyword coding and grouping.

Category	Levels	Keywords
Arts and sciences (AS)	C[a]	arts and sciences; liberal arts and sciences; letters and science; graduate school/division
Science and engineering (SE)	D[b] /C	science and technology; science and engineering
Science (SCI)	P[c] /D/C	math; physical; science; chemistry
Engineering and application (ENG)	P/D/C	civil; engineering; applied; computer; mine; architecture; house; fire; built; technology
Resource and environment (NRE)	P/D/C	environment; nature; desert; eco-; resource; conservation; natural
Biomedical science (BIO)	P/D/C	life; human; bio-; plant; evolution; organism; wildlife; entomological; health; animal
Agriculture, forestry, animal husbandry and fishery (AGR)	P/D/C	agriculture; food; fishery; crop; soil; forest; rangeland; horticulture; irrigation
Earth science (GEO)	P/D/C	atmosphere; earth; geo-; meteorology; climate; land; atmosphere; air; plane
Economic management and humanities social science (SOC)	P/D/C	sustain; law; design; rural; social; urban; global; history; economic; management; policy; business; journalism
Pure water (WATER)	P/D/C	hydrology; water; limnology; watershed; water resource

[a]. College level.
[b]. Department level.
[c]. Program level.

4. Interdisciplinary Status of Water-related Programs

The scope of water-related research fields and the interdisciplinary status of water-related disciplines can be judged by: a) whether the names of water-related programs, departments and colleges are assigned to multiple categories; b) what categories are mostly likely to appear in the names at the same time. Based on the differences in the meanings of names, the explanations and implications should be different. For example, the more categories a program name has been assigned to, the smaller and more focused research area the program has. In other words, the program will host a specific interdisciplinary research field. However, if a department name is assigned to multiple categories, this department tends to have a broader coverage of water-related research areas. Similarly, if a college name is assigned to multiple categories, this college may host a collection of different disciplines which may lead to the formation of interdisciplinary research fields through interaction and integration between disciplines.

The rate of water-related programs, departments and colleges with assignments to multiple categories are 39.7%, 52.9% and 35.6% respectively. Through the figures, we can see that American public universities tend to educate students with a wide and broad major in water-related programs. More than 60% of the water-related degree programs are assigned to a single discipline category. While for the water- related departments, they have hosted a wider scope of disciplines or research fields. For more than half of the departments are assigned to at least two disciplinary categories. The figure for the water-related colleges matches expectation. Most of the colleges are built based on a single discipline category. Later on, we will focus on the water-related programs, departments and colleges with multiple assignments, and examine the interdisciplinary nature behind these multiple assignments.

4.1 Interdisciplinary research fields

According to multiple assignments of water- related degree programs, the interdisciplinary research fields mainly include:

Firstly, the interdisciplinary fields of natural resources, environment and other academic disciplines are commonly developed in the water-related programs in American public universities. Among all the degree programs with multiple assignments, there are 36 degree programs assigned to "resources and environment" category and other categories at the same time (see Figure 1). For examples, the Civil and Environmental Engineering Program based at Utah State University and the Ecohydrological Engineering Program based at University of Idaho are assigned to both "resources and environment" category and "engineering and application" category; the Environment Science and Management Program based at University of Rhode Island and Environmental and Water Resource Economics Program based at University of Arizona are assigned to both "resources and environment" category and "economic management and humanities social science" category; the Environmental Soil Program based at Utah State University and Forest Resources and Environmental Conservation Program based at Virginia Polytechnic Institute and State University are assigned to both "resources and environment" category and "agriculture, forestry, animal husbandry and fishery" category.

Secondly, the interdisciplinary fields of agriculture, forestry, animal husbandry & fishery and other academic disciplines are also common in American public universities' water-related programs. The number of this kind of degree program is 10, and the best example is Biological & Agricultural Engineering Program based at University of Arkansas, University of Idaho, Pennsylvania State University and Washington State University at the same time; the Environmental Management in Agriculture and Natural Resources Program based at University of Florida and Rangeland Ecology and Watershed Management Program based at University of Wyoming are assigned to "agriculture, forestry, animal husbandry and fishery" category, "resources and environment" category and "economic management and humanities social science" category at the same time.

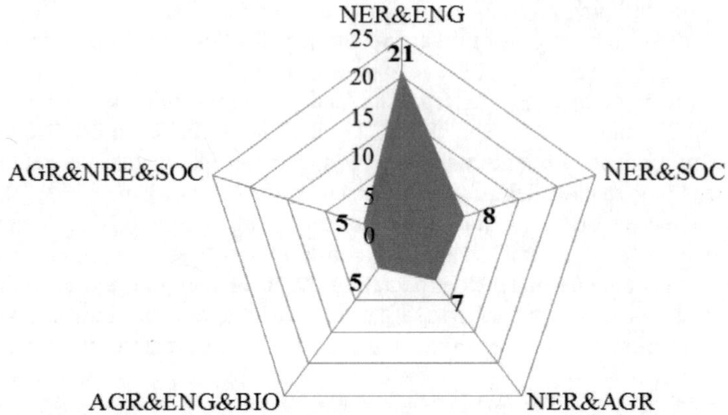

Figure 1. Distribution of water-related degree programs in multiple categories.

4.2 Interdisciplinary research platforms

In the light of multiple assignments of water-related department, the data show that there are 56 two-category department platforms (see Figure 2). And "resources and environment" category discipline does not necessarily based at the "resources and environment" category department platform, for instance, in Oklahoma State University, the Environment and Natural Resources Program is based at the Department of Biosystems and Agricultural Engineering. In addition, there are 11 three-category department platforms which build-up by "agriculture, forestry, animal husbandry and fishery" category, "biomedical science" category and "engineering and application" category, accounting for 12% of the total. For example, the Biosystems and Agricultural Engineering Department provides Environmental Sciences Program in Oklahoma State University. To put it in a nutshell, water-related research platforms are developed with a various construction form.

Based on multiple assignments of water-related colleges, the "agriculture, forestry, animal husbandry and fishery" category is the prime participant, for example, Natural Resources Program based at College of Life Sciences and Agriculture in New Hampshire University, is assigned to both "agriculture, forestry, animal husbandry and fishery" category and "biomedical science" category; Natural Resources Management Program based at College of Agriculture, Food Systems in North Dakota State University is assigned to both "agriculture, forestry, animal husbandry and fishery" category and "resources and environment" category.

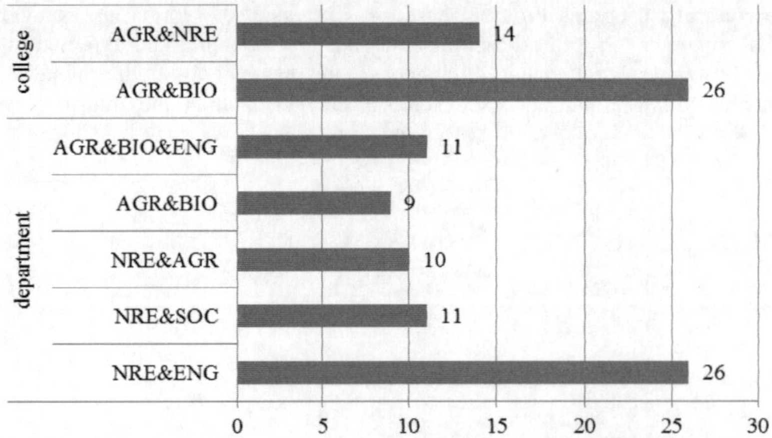

Figure 2. Distribution of water-related departments and colleges with multiple assignments.

5. Conclusions and implications

This study mainly focused on the interdisciplinary nature of water-related programs in American public universities. Through a text analysis of the names of water-related degree programs, departments and colleges, we found that American public universities had paid much attention to the applicability and diversity of water-related programs when developing interdisciplinary research fields. Here are some implications for China's future strengthening interdisciplinary water-related research.

Firstly, it's important to broaden the scope of water-related research through stronger interaction and integration of different disciplines. On the one hand, most of the water-related departments are acting as the platforms that consist of many other discipline categories; on the other hand, water-related colleges are mainly established as comprehensive research platforms. As a result, the water-related departments and colleges have both developed the water-related interdisciplinary in width and depth. Their departments and colleges lay emphasis on the construction of interdisciplinary platforms. For Chinese universities, it's important to broaden the water-related research fields by cultivating new water-related interdisciplinary [7] [8]. Furthermore, the internal and external academic resources must be channeled to establish high-quality inter-disciplinary research platforms [9].

Secondly, strengthen the applicability of water-related interdisciplinary through creating applicative research platforms and cultivating talents with applicable abilities. Most of water-related interdisciplinary in American public universities focused on applied research and applied talents training. Water problems have both universal and national characteristics. As for the development of water-related interdisciplinary in China, we should encourage the cooperation between basic and applied water-related disciplines, such as environmental engineering, geological engineering, and agricultural engineering and so on. Humanities and social sciences should also be included in the cooperation to develop new water-related interdisciplinary such as water law. Most importantly, the development of water-related interdisciplinary in China should attend to solve this country's own water problems.

Finally, break the linear relationship among the disciplines, departments and colleges to establish a diversified water-related interdisciplinary system [10]. First of all, the water-related departments and colleges should be allowed to develop degree programs from a different disciplinary category. For instance, in American public universities, some of the degree programs in "resources and environment" category are not host in the departments or colleges belonged to the same category, but hosted in the "earth science" category colleges. Besides, American public universities have developed several degree programs across the departments or colleges, or even co-hosted the programs with other universities. It is also good for the diversity of water-related inter-disciplinary system. Chinese universities should change and reform the traditional disciplinary system. It is a big step to break the linear relationship between departments and degree programs, and develop a flexible disciplinary system allowing for cultivating water-related interdisciplinary across the departments, colleges and even institutions.

Acknowledgments

This research is supported by Central University Fundamental Research Founds for the Project (No. 2015B40184), Higher Education Scientific Research Program of Hohai University (No. 20141203), and the National Natural Science Foundation of China (No. 71403079).

References

[1] Wang M H, Li J F and Ho Y S 2011 Research articles published in water resources journals: A bibliometric analysis *Desalin Water Treat China* **28** pp 353-365.
[2] Ren J L, Lyu P H, Wu X M, Ma F C, Wang Z Z and Yang G 2013 An informetric profile of Water Resources Management literatures *Water Resour Manage China* **27(13)** pp 4679-96.
[3] The National Institutes For Water Resources. https://niwr.net/public/Migration/about- national-institutes-water-resources, Oct 2015.
[4] Wang H, Wang J H, Qin D Y, Chen C Y, Jiang D and Yao Z J 2002 The study of modern water

resources assessment and water subject system *Earth Sci. Prog. China* **17** pp 12-17.

[5] Yan Y 2005 The connotation of college system and setup of the college *Edu Res.China* **10** pp 76-79.

[6] Song H Y 2006 A comparative study of college set in China and American universities *Edu Dvpt Res.China* **26(7)** pp 65-68.

[7] Yan Y P 2013 Implications of the prevalence and declining of academic fields in U.S *Global Edu China.* **7(7)** pp 95-100.

[8] Noorden R Van 2015 Interdisciplinary research by the numbers *Nature* **525** pp 306-307.

[9] Pei X, Zhang S L and Yu Q 2007 Relying on large-span and multidisciplinary technology platform training high-level innovative talents--Practice concept definite in graduate innovation center constructing *Academic Degrees and Graduate Education China* **6** pp 12-14.

[10] Song H M, Chang S and Dong W C 2014 The exploration and enlightenment of disciplinary cross and fusion in American university *Academic Degrees and Graduate Education China* **9** pp 73-77.

Numerical study on pollutant transport in Dalian bay based on hydrodynamic model

Huiting Qiao[1], Jin Li[2], Hongxing Zhang[1], Kaibin Zhao[1] and Mingliang Zhang[1, 3]

[1] School of Ocean Science and Environment, Dalian Ocean University, Dalian, Liaoning, 116023, China
[2] Yuanyanggou National Ocean Park, Panjin, Liaoning, 124010, China

E-mail: zhmliang_mail@126.com

Abstract: Based on the depth-averaged two-dimensional shallow water and pollutant transport equation, the coupling model of water flow and water quality with explicit scheme is developed in this study. The unstructured triangular grid is adopted to locally refine the mesh around sewage outlet or in high-gradient regions of terrain change for the coupling model. The finite volume method is applied to ensure the conservation of mass for each element. This hydrodynamic model applies the Roe solver approximate Riemann solution with second-order accuracy to compute the water momentum flux on the grid interface. Taking Dalian Bay as the research object, the numerical model established is used to simulate the hydrodynamic characteristics and pollutant transport process. The computed results of the tide level, flow current and flow direction agree well with the measured data in Dalian Bay. The spatial and temporal distribution of pollutant in water are analyzed and discussed in this study. Simulated results show that the two dimensional hydrodynamic and pollutant transport model can accurately simulate the mass transport in coastal waters, and it can provide a scientific basis on coastal water environment protection for the research water.

1. Introduction

Industrial and agricultural waste water and urban sewage discharged into the nearby water body will lead to the pollution of the water body. For the purpose of managing and protecting the ecological environment of water, it is essential to control transport and distribution of pollutants in the water body. Obviously, the process of pollutant transport is determined by the characteristics of the coastal circulation and the properties of pollutant over the entire ocean domain. Under the action of hydrodynamic factors, the numerical simulation method is used to calculate and analyze the movement of pollutants, which has the advantages of economy, flexible and high efficiency. Therefore, more and more scholars dedicate to the study of working on pollutant diffusion - transportation mathematical models. Kong et al (2013) presented a high-resolution numerical method for solving mass transport problems involving advection and anisotropic diffusion in shallow water based on unstructured mesh [1]. Zhang et al (2015) introduced a computationally efficient model based on a simplified adaptive grid system and the numerical scheme was able to provide well-balanced solutions of mass and solute concentration for applications involving wetting and drying over complex domain topographies [2]. Liang et al (2010) developed a numerical model for predicting solute transport in shallow waters, the alternating operator-splitting technique is used to separate the spatial discretization of the convection-diffusion terms in each step of the time marching procedure [3].

About Dalian bay, there is little research on the application on the shallow water model to deal with mass transport problems. The present work focuses on developing a depth-averaged 2D numerical model to simulate the hydrodynamic characteristics and pollutant transport in Dalian bay. The purpose of this application is to verify the capacity of the model in simulating hydrodynamic and mass transport processes in natural shallow water bodies.

2. Governing Equations

The Shallow Water Equation (SWEs) written in conservation and vector form are

$$\frac{\partial \mathbf{U}}{\partial t}+\frac{\partial \mathbf{F}}{\partial x}+\frac{\partial \mathbf{G}}{\partial y}=\frac{\partial \mathbf{F}_d}{\partial x}+\frac{\partial \mathbf{G}_d}{\partial y}+\mathbf{S} \tag{1}$$

In which, t is the time, x and y are the horizontal coordinates, U is the vector of conserved variables, F, G, F_d, and G_d are convection fluxes and diffusion fluxes in the x and y directions, respectively, S is source terms which can be defined respectively as follows:

$$\mathbf{U}=\begin{bmatrix} h \\ uh \\ vh \end{bmatrix}, \quad \mathbf{F}=\begin{bmatrix} hu \\ hu^2 \\ huv \end{bmatrix}, \quad \mathbf{G}=\begin{bmatrix} hv \\ hvu \\ hv^2 \end{bmatrix}$$

$$\mathbf{F}_d=\begin{bmatrix} 0 \\ \nu_t \dfrac{\partial uh}{\partial x} \\ \nu_t \dfrac{\partial vh}{\partial x} \end{bmatrix}, \quad \mathbf{G}_d=\begin{bmatrix} 0 \\ \nu_t \dfrac{\partial uh}{\partial y} \\ \nu_t \dfrac{\partial vh}{\partial y} \end{bmatrix}, \quad \mathbf{S}=\begin{bmatrix} 0 \\ -gh\dfrac{\partial \eta}{\partial x}-\tau_{bx} \\ -gh\dfrac{\partial \eta}{\partial y}-\tau_{by} \end{bmatrix} \tag{2}$$

Where h is the flow depth, u and v are the depth-averaged velocity in the x and y directions, respectively, η is the water level, τ_{bx} and τ_{by} are the friction slopes in the x and y directions, respectively.

Mass transport equation:

$$\frac{\partial(hC)}{\partial t}+\frac{\partial(uhC)}{\partial x}+\frac{\partial(vhC)}{\partial y}=\frac{\partial}{\partial x}(\nu_t h\frac{\partial C}{\partial x})+\frac{\partial}{\partial y}(\nu_t h\frac{\partial C}{\partial y})+S_C \tag{3}$$

3. Finite volume method

Usually, the numerical methods may be classified into three groups: Finite-difference methods (FDM), finite-element methods (FEM) and finite volume methods (FVM). In literature, some 2D hydrodynamic models based on the shallow water equations have been implemented to calculate the flow pattern of ocean using the FEM, FDM or FVM (e.g. Sankaranarayanan et al 2003 [4]; Blain, 2005 [5]; Brière, 2007 [6]; Akbar et al, 2013 [7]; Kuang et al, 2011 [8]; Falcão et al, 2013 [9]).

In this paper, a cell-centered finite volume method is adopted to sole the SWEs. The computational area is divided into a set of triangular meshes. Equation (1) can be integrated over a control volume V as

$$\int_{V_i}\frac{\partial \mathbf{U}}{\partial t}d_V+\int_{V_i}\nabla\cdot\mathbf{E}d_V=\int_{V_i}\frac{\partial \mathbf{F}_d}{\partial x}d_V+\int_{V_i}\frac{\partial \mathbf{G}_d}{\partial y}d_V+\int_{V_i}\mathbf{S}d_V \tag{4}$$

Integrating equation (4) over the area of each control volume, applying Green's theorem to equation (4), a line integral equation can be obtained:

IOP Conf. Series: Earth and Environmental Science 52(2017) 012094 doi:10.1088/1755-1315/52/1/0

$$\frac{\Delta \mathbf{U}_i}{\Delta t} A_i = -\oint_{L_i} \mathbf{E}^* \cdot \mathbf{n} dl + \oint_{L_i} (\mathbf{F}_d \cdot n_x + \mathbf{G}_d \cdot n_y) dl + \int_{V_i} \mathbf{S} d_V \tag{5}$$

where L_i is the boundary of cell V, is the unit outward normal vectors on the cell edges, is angle between outside the normal vector and the positive direction of the x-axis, is the area of the cell i. After discretizing and arranging the line integral equation (5):

$$\Delta \mathbf{U}_i = -\frac{\Delta t}{A_i} \sum_{j=1}^{m} (\mathbf{E}_{ij}^* \cdot \mathbf{n}_{ij}) l_{ij} + \frac{\Delta t}{A_i} \sum_{j=1}^{m} (\mathbf{F}_d \cdot n_x + \mathbf{G}_d \cdot n_y) l_{ij} + \Delta t \cdot \mathbf{S} \tag{6}$$

where l_{ij} is the length of unit edge; $\mathbf{E}_{ij}^* \cdot \mathbf{n}_{ij}$ is the normal vector numerical flux through the j_{th} edge of the i_{th} cell.

Mass transport equation can be also discretized over a triangular control volume, integrating equation (3) over the area of the i_{th} control volume, one can obtain

$$\int_{V_i} \left[\frac{\partial (hC)}{\partial t} + \frac{\partial (uhC)}{\partial x} + \frac{\partial (vhC)}{\partial y} \right] d_V = \int_{V_i} \left[\frac{\partial}{\partial x}(v_t h \frac{\partial C}{\partial x}) + \frac{\partial}{\partial y}(v_t h \frac{\partial C}{\partial y}) + \mathrm{S}_C \right] d_V \tag{7}$$

Applying the Green's theorem to equation (7), discretizing the equation (7) over a time interval, one obtains

$$\Delta C_i = -\frac{\Delta t}{A_i} \sum_{j=1}^{m} (\mathbf{Q}_{ij} \cdot C_{ij} \cdot \mathbf{n}_{ij}) + \frac{\Delta t}{A_i} \sum_{j=1}^{m} \left(\left(v_t h \frac{\partial C}{\partial \mathbf{n}} \right)_{ij} \cdot l_{ij} \right) + \Delta t \cdot \mathbf{S}_c \tag{8}$$

In which, $\mathbf{Q}_{ij}$ is the flux across the element side; $\left(\frac{\partial C}{\partial \mathbf{n}} \right)_{ij}$ is the transport variable gradient of the outward normal direction.

4. Numerical validation and analysis

4.1. Introduction of Dalian Bay

The present model is applied to simulate the current situation of Dalian Bay in the northern Yellow Sea. The computational area is from 121°34′ to 122°10′ in east longitude and from 38°39′ to 39°09′ in north latitude, include Laohutan and Dalian Port, and there are two islands (Dasanshan island and Xiaosanshan island) in computed domain (see Figure 1). The water depth is smaller in the bay, about more than ten meters. While it is larger outside the bay, about forty meters. The water exchange capability in inner bay is poor since the flow velocity of it is lower than that of the bay mouth. Dalian bay is a typical semi-enclosed bay. Along the coast, there are a large number of municipal industrial sewage outfalls and kinds of ship terminals, which carries most of Dalian's domestic sewage and industrial waste water. Considering the poor exchange of water discharge, the marine environment of the Dalian Bay get worse.

4.2. Model application

In this case, the computation domain of Dalian Bay is discretized into 15260 grid cells, because the domain is complex, a fine grid is used near the coastal line, island area and a coarse grid is used in open sea. The water elevations are given by the Tidal Model Driver (TMD). The tidal process covers 216 hours from 13th July, 2010 to 21th July, 2010. The Manning's *n* is set as 0.03. The computational time step is 0.2 s. Figure 2 shows the comparison of simulated and the measured water surface elevation in Dalian port and Laohutan tidal stations. The simulated surface elevations are in good agreement with the observed data. Figure 3 shows the comparison of flow velocities in V1 and V2 monitoring stations. The simulated tide velocities have a slight deviation compared with the measured data, the agreement between them is reasonably good. The flow fields in flood and ebb tides are reasonably well predicted, as shown in figure 4. Around the islands, the tide is in a spiral type, and the rest is basically reciprocating flow. The flood current is in the northeast into the gulf, the southwest direction is ebb current. Whether it is flood or ebb current, the velocity at the mouth of the bay is larger. In total, one can see that this model can applied to simulate the flow phenomena of coastal water, and it has practical application value.

Based on hydrodynamic simulation, the presently developed transport models applied to simulate the solute transport in the Dalian bay mouth. The parameters such as the grid, the open boundary and the time step of the simulation are similar to that of hydrodynamic simulation. The pollutant emission scheme is an imaginary scheme, the release point is shown in figure 1 and the emission rate is 30g/s, lasted for six hours. Figure5 illustrates the movement of the pollutant cloud and the simulated flow fields after 6 hours of the release. From the figure we can see that the pollutant concentration field varies with the trend oscillation. As shown in figure 5, the solute oscillates with the tidal flow. As time progresses, the pollutant has gradually increased its distribution range and becomes less concentrated, which can be seen by comparing the concentration fields of figure 5 (a, b, c, and d). Although there is no measured data to verify this simulation, based on the characteristics of the hydrodynamic and pollutant transport and combined with the simulation results, one can see the concentration distribution is rational and the present solute transport model has the capacity to simulate mass transport processes in natural shallow water bodies.

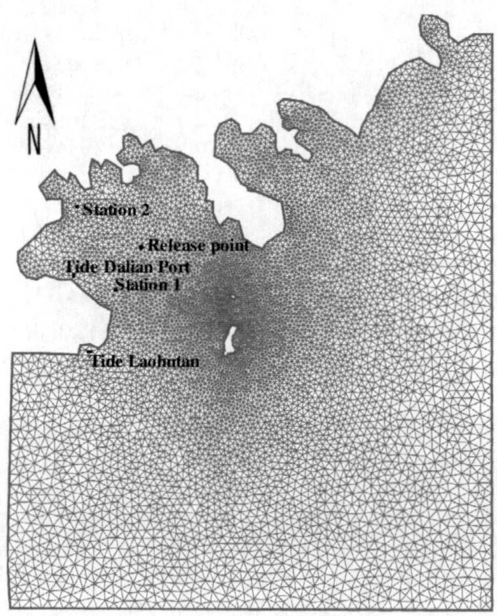

Figure 1. Mesh and measurement stations and pollutant release point at computational area

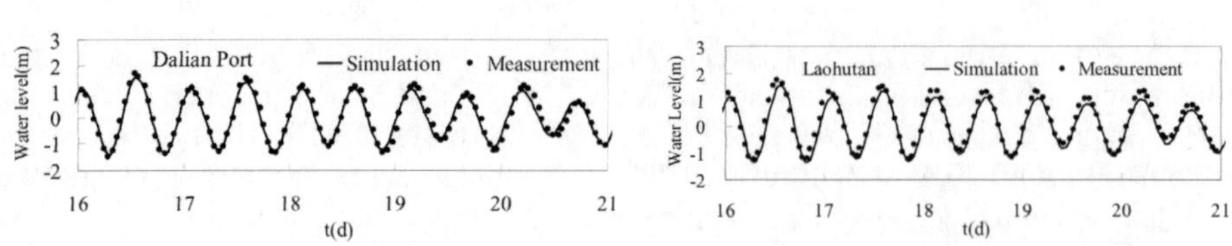

Figure 2. Comparison of simulation and measurement water surface elevation at selected tidal stations

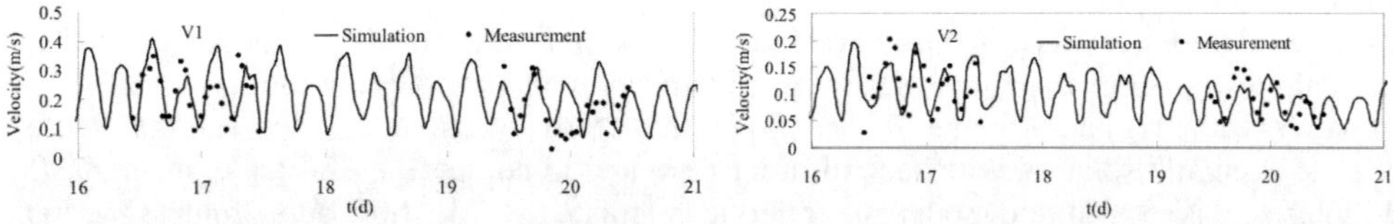

Figure 3. Comparison of flow velocities in V1 and V2 monitoring stations

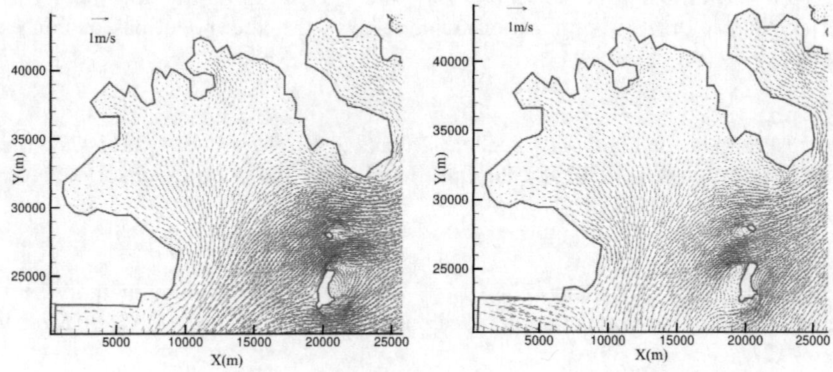

Figure 4. Flood and ebb tide in Dalian bay

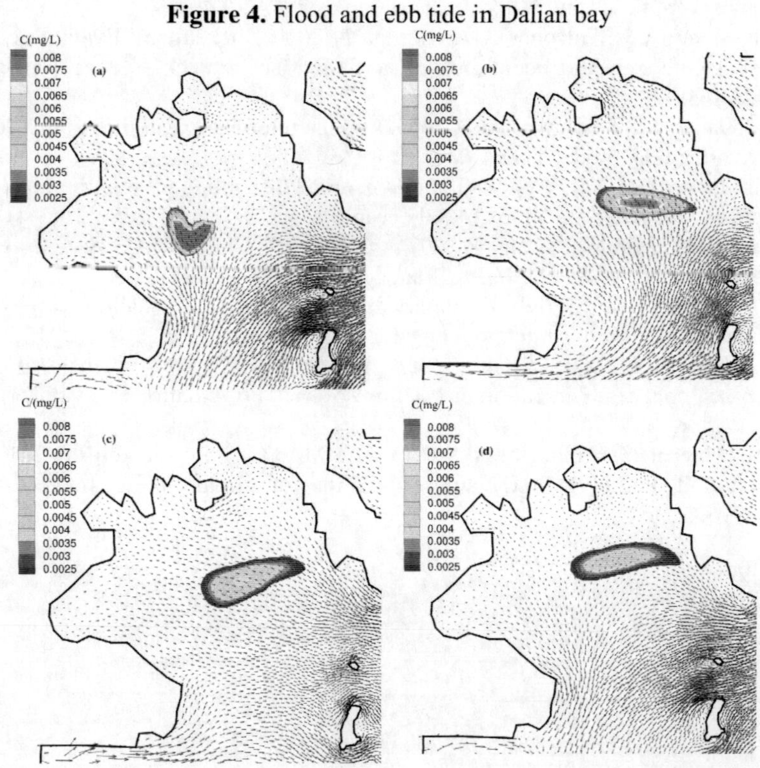

Figure 5. Development of the pollutant concentration field and flow field for the Dalian Bay mouth coastal water

5. Conclusions

Based on the two-dimensional shallow water equations and water quality equation, and the finite volume method with an unstructured triangular grid are adopted, the coupling model of water flow and water quality with explicit scheme is set up in this study. This model applies the Roe solver approximate Riemann solution with second-order accuracy to compute the water momentum flux on the grid interface. And then, the model is applied to calculate the tidal flow and simulate the transport of a fictitious conservative solute in Dalian bay to verify its practicability. Through the example of the pollutant migration and diffusion in the Dalian Bay, the calculation ability and the accuracy of the model are verified in the practical application, which shows that the model has certain engineering application value.

Acknowledgments

This work by the Program for Liaoning Excellent Talents in University (LJQ2013077), the Liaoning Natural Science Foundation (2014020148), the Open Fund of the State Key Laboratory of Hydraulics and Mountain River Engineering (SKHL1517).

References

[1] Kong J, Xin P, Shen C J, Song Z Y, Li L 2013 A high-resolution method for the depth-integrated solute transport equation based on an unstructured mesh *Environ. Modell.& Softw.* **40** 109-127.

[2] Zhang L L, Liang Q H, Wang Y L, Yin J X 2015 A robust coupled model for solute transport driven by severe flow conditions *J. Hydro-environ. Res.* **9** 49-60.

[3] Liang D F, Wang X L, Falconer R A, Bettina N 2010 Bockelmann-Evans Solving thedepth-integratedsolute transport equation with a TVD-MacCormack scheme *Environ. Modell.& Softw.* **25** 1619-29.

[4] Sankaranarayanan S, French McCayD 2003 Application of a two-dimensional depth averaged hydrodynamic tidal model *Ocean Eng.* **30** 1807-32.

[5] Blain C N, Massey T C 2005 Application of a coupled discontinuous–continuous Galerkin finite element shallow water model to coastal ocean dynamics *Ocean Model.* **10(3-4)** 283-315.

[6] Brière C, Abadie S, Bretel P, Lang P 2007 Assessment of TELEMAC system performances, a hydrodynamic case study of Anglet, France *Coast. Eng.* **54** 345–356.

[7] Akbar M, Aliabadi S 2013 Hybrid numerical methods to solve shallow water equations for hurricane induced storm surge modeling *Environ. Modell.& Softw.* **46** 118-128.

[8] Kuang C P, Chen S Y, Zhang Y, Gu J, Pan Y, Huang J 2011 A two-dimensional morphological model based on a next generation circulation solver I: Formulation and validation *Coast. Eng.* **59** 1-13.

[9] Falcão A P, Mazzolari A, Gonçalves A B, Araújo MAVC, Trigo-Teixeira A 2013 Influence of elevation modelling on hydrodynamic simulations of a tidally-dominated estuary *J. Hydrol.* **497** 152-164.

Pollution evaluation of heavy metals in soil near smelting area by index of geoaccumulation (Igeo)

S H Huang[1], Y Yang [2], C Y Yuan[1], Q Li[1], K Ouyang [1,4], B Wang[1] and Z X Wang[3]

[1] Hunan Research Institute of Nonferrous Metals, Changsha 410083, China

[2] College of Resources and Environment, Huazhong Agricultural University, Wuhan 430070, China

[3] South China Institute of Environmental Sciences, Ministry of Environmental Protection, Guangzhou, 510655, China.

E-mail: 153897262@qq.com

Abstract: In order to investigate the heavy metal pollution conditions of soil of smelting area in Zijiang of Chenzhou, Hunan province, 42 samples were studied. The concentrations of heavy metals As, Pb, Cd, Zn and Cu in the soil were determined by using atomic absorption spectrometry(AAS) and atomic fluorescence spectrometry(AFS). Then the potential pollution risks of heavy metal in the soil were evaluated by method of geological acumination index (I_{geo}). The results indicated that the average concentrations of As, Pb, Cd, Zn and Cu were 187.79, 2074.52, 15.72, 2178.89, 39.69 mg/kg respectively. The geological evaluation of the cumulative index results showed that the contamination degree of 5 heavy metals follow the sequence of Cd> Zn >Pb > As >Cu. The results show that Cd reached extremely pollution degree, Zn reached strong pollution-extremely pollution levels, the pollution of Pb in the soil is classified as strong pollution degree, Cu and As of no pollution according to the results of I_{geo} based on the background value of heavy metals in the soil of Hunan Province.

1. Introduction

Recently, soil pollution has received increasing attention. Heavy metals are regarded as serious pollutants of soil because of their environmental persistence, toxicity and ability to be incorporated into food chains [1].

Heavy metals in soil may come from agriculture activities, industrialization, mining and smelting activity. Among these, mining and smelting activity is considered as one of the most dangerous anthropogenic activities [2]. Mining operation, grinding, concentrating ores and disposal of tailings, provide obvious sources of contamination in the environment, along with mine and mill wastewater [3]. This will lead to the release and migration of heavy metals thus cause heavy metals pollution of soil near the mining area.

Heavy metals can be migrated from the soil to other ecosystem components, such as groundwater, plants, thus affecting human health through drinking water and food chain, so the evaluation of heavy metal pollution in soil is very important. The index of geoccumulation (I_{geo}) has been used as a measure of bottom sediment contaminations since 1970s and numerous researchers have employed it to assess the contamination of soils and sediments [4-6]. It not only considers the impact of the background of natural geological process, but also considers the impact of human activities on the heavy metal pollution [7].

This paper studied the heavy metal pollution in soil of Qingjiang lead-zinc mining area in Zixing of Chenzhou, Hunan province. The geological evaluation of the cumulative index is adopted to evaluate the pollution levels of heavy metals As, Pb, Cd, Zn and Cu in the soil. The aim of this study is to assess the level of metal concentrations in soil of smelting mine area and provide a scientific basis for heavy metal pollution control.

2. Materials and methods

2.1. Sample collection and preparation

The geography coordinate of the survey site is longitude 113°17′17.76″E~113°17′36.07″E, latitude 25°45′39.29″N~25°46′1.13″N,located at Qingjiang county, Chenzhou City, Hunan Province.42 soil samples were collected according to the correlated regulations and standards (GB/T15618-1995). Soil samples were taken every 20 cm from the top soil to a depth of 100 cm.

Each sample picked out from a mixture of 3~5 subsamples for 1 kg or so by quarter method. The contents of heavy metals including Cu, As, Zn, Pb and Cd are analyzed by ICP-MS. And Cold Atomic Fluorophotometry is employed to analyze the contents of As.

2.2. The index of geoaccumulation

The geoaccumulation index (I_{geo}) is put forward by Muller [8]. The value of the geoaccumulation index is calculated by the following equation:

$$I_{geo} = \log_2\left[\frac{C_n}{1.5B_n}\right]$$

C_n in the formulas is the measured concentration of the examined heavy metal in the soil, B_n is the geochemical background concentration of metal '*n*', and factor 1.5 is the background matrix correction factor due to lithogenic effects.

Table 1. Index of geoaccumulation(I_{geo}) and contamination level.

I_{geo}	Class	Contamination level
≤0	0	Unpolluted
0-1	1	Unpolluted to moderately polluted
1-2	2	Moderately polluted
2-3	3	Moderately to strongly polluted
3-4	4	Strongly polluted
4-5	5	Strongly to extremely polluted
≥5	6	Extremely polluted

Muller has distinguished seven classes of geo accumulation index [8]. Table 1 shows Index of geoaccumulation (I_{geo}) and contamination level.

3. Results and discussion

The heavy metals considered of concern in the investigation are As, Pb, Cd, Zn and Cu. To assess the environmental pollution of the soil, the concentrations of above heavy metals were determined. Statistical characteristics of heavy metals content in the samples are given in Table 2.

Table 2. The concentrations of heavy metals in samples.

Samples	Concentration of heavy metals (mg/kg)				
	As	Pb	Cd	Zn	Cu
1	64.24	399.58	5.73	738.08	19.59
2	71.87	1019.23	5.38	816.18	27.77
3	750.10	1532.42	38.37	3105.00	50.88
4	73.21	542.83	4.24	610.06	22.03
5	81.38	416.84	8.25	843.78	19.00
6	66.84	421.92	8.69	945.41	26.67
7	306.08	851.93	53.50	2761.18	36.97
8	40.59	331.22	6.05	528.30	23.66
9	32.38	295.84	7.69	621.90	23.43
10	1375.41	1769.45	113.40	29282.50	203.39
11	63.33	115.13	1.89	133.83	22.80
12	50.71	40.66	0.64	95.09	22.18
13	49.50	58.54	1.35	144.64	25.59
14	70.33	302.00	4.31	451.35	30.29
15	745.72	8794.18	84.60	12743.07	105.18
16	431.87	10501.53	52.63	7670.18	78.65
17	459.22	8609.23	34.22	4003.80	59.41
18	352.06	7701.51	31.14	3446.44	48.93
19	52.36	337.02	6.21	564.55	37.64
20	40.41	229.60	4.18	547.13	18.10
21	753.06	3837.55	85.44	6303.06	111.76
22	478.90	16651.57	6.36	3131.40	102.87
23	39.58	693.76	3.13	510.44	21.96
24	38.71	923.39	3.42	591.02	24.24
25	45.75	285.86	3.84	380.49	26.16
26	54.74	138.66	3.84	184.89	23.66
27	286.30	6224.85	7.60	2199.32	65.90
28	370.39	12221.02	24.47	4345.99	104.46
29	39.12	225.30	23.98	1328.05	26.26
30	30.31	66.13	14.42	999.27	22.67
31	34.46	63.56	0.85	90.89	16.93
32	45.97	57.09	0.46	92.25	16.50
33	42.98	61.88	0.13	72.01	17.56
34	47.59	45.76	1.21	93.19	21.59
35	36.96	79.13	0.67	124.13	22.35
36	46.16	89.85	1.45	82.72	20.02
37	35.84	73.06	0.45	92.44	19.26
38	58.62	71.91	0.45	92.81	23.78
39	57.03	314.75	0.41	210.93	12.34
40	46.82	236.17	1.37	185.46	21.11
41	60.86	248.78	2.43	172.08	20.99
42	59.41	249.33	1.57	178.17	22.54
Average	187.79	2074.52	15.72	2178.89	39.69
Background	27.95	89.20	0.60	103.16	17.89

Table 2 shows that the concentration of heavy metals in the soil ranges as follows: Cd (0.13-113.40 mg/kg), Cu (12.34-203.39 mg/kg), As (30.31-1375.41 mg/kg), Pb (40.66-16651.57 mg/kg), Zn (72.01-29282.50 mg/kg). The mean concentrations of Cd, Cu,,As, Pb, Zn in soil were 15.72, 39.69 , 187.79, 2074.52, 2178.89 mg/kg, respectively. It can be found that the average concentrations of heavy metals in soil are much higher than the background concentration of heavy metals in Hunan province.

The calculated I_{geo} of heavy metals in soil samples are shown in Figure.1.

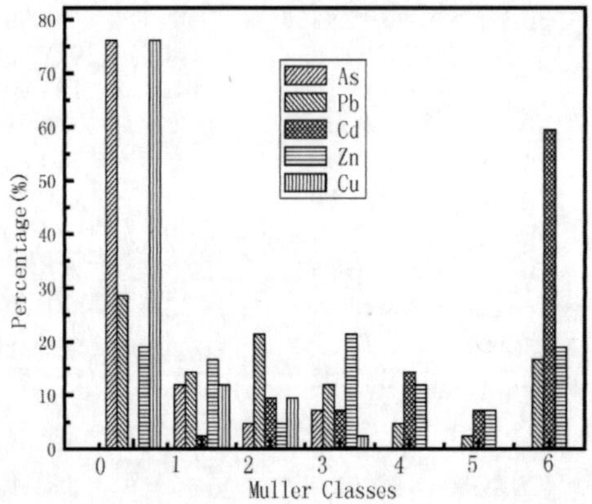

Figure 1. Percentage of samples in Muller class.

Figure.1 displays sample percentages in Müller classes for As, Pb, Cd, Zn and Cu. As and Cu concentrations fall mainly in classes 0 and 1. For Cd, 59.52% of the samples fall in class 6, 7.15% in class 5, 14.28% in class 4, 7.15% in class 3, 9.52% in class 2 and 2.38% in class 1. For Pb, 28.57% and 21.43% of the samples are included in classes 0 and 2, 14.29% in class 1, 11.90% in class 3, 4.76% in class 4, 2.38% in class 5 and the remaining 16.67% fall in class 6. For Zn, 19.05% of the samples fall in class 0, 16.67% in class 1, 4.76% in class 2, 21.43% in class 3, 11.90% in class 4, 7.14% in class 5 and 19.05% in class 6.

Table 3. The average values of I_{geo} values for each metal.

Heavy metals	I_{geo} value	Pollution level
As	-0.37557	unpolluted
Pb	3.964368	strongly polluted
Cd	7.033816	extremely polluted
Zn	4.48427	strongly to extremely polluted
Cu	-0.04501	unpolluted

Table 3 shows the average values of I_{geo} values for each metal and their pollution level. The results indicate that the soil of the study area can be categorized as follows: unpolluted with Cu and As, strongly polluted with Pb, strongly to extremely polluted with Zn and extremely polluted with Cd. The contamination degree from strong to weak in soil is: Cd > Zn >Pb > As >Cu.

4. Conclusions

The average concentrations of Cd, Cu, As,Pb, Zn in soil were 15.72, 39.69, 187.79 ,2074.52, 2178.89 mg/kg respectively. It can be concluded that the average concentrations of heavy metals in soil are much higher than the background concentration of heavy metals in Hunan province.

The geoaccumulation index of heavy metals in the soil decreases as follows: Cd>Zn>Pb >As>Cu. I_{geo} results reveal that the study area is not contaminated with respected to As and Cu, strongly contaminated with Pb, strongly to extremely contaminated with Zn， and extremely contaminated with Cd.

References

[1] Liao G L, Liao D X and Li Q M 2008 Heavy metals contamination characteristics in soil of different mining activity zone *Trans. Nonferrous Met. Soc. China* **18** 207–211

[2] Jung M C, Thornton I 1996 Heavy metal contamination of soils and plants in the vicinity of a lead zinc mine, Korea *Appl Geochem* **11** 53-59

[3] Chopin E I B, Alloway B J 2007 Distribution and mobility of trace elements in soils and vegetation around the mining and smelting areas of Tharsis, Rıotinto and Huelva, Iberian Pyrite Belt, SW Spain *Water air soll poll* **182** 245-261

[4] Khaled A, ElNemr A and ElSikaily A 2006 An assessment of heavy metal contamination in surface sediments of the Suez Gulf using geoaccumulation indexes and statistical analysis *Chem ecol* **22** 239-252

[5] Mico C, Recatala L and Peris M 2006 Assessing heavy metal sources in agricultural soils of an European Mediterranean area by multivariate analysis *Chemosphere* **65** 863-872

[6] Sin S N, Chua H, Lo W, Ng L M 2001 Assessment of heavy metal cations in sediments of Shing Mun River, Hong Kong *Environint* **26** 297-301

[7] Gowd S S, Reddy R M and Govil P K 2010 Assessment of heavily metal contamination in soils at Jajmau (Kanpur) and Unnao industrial areas of the Ganga Plain, Uttar Pradesh, India *J Hazard mater* **174** 113-121

[8] Muller G 1969 Index of geoaccumulation in sediments of the Rhine river *Geojournal* **2** 108-118

Prospects for energy efficiency improvement and reduction of emissions and life cycle costs for natural gas vehicles

A V Kozlov[1], A S Terenchenko, V A Luksho and K E Karpukhin

Federal State Unitary Enterprise Central Scientific Research Automobile and Automotive Institute "NAMI" (FSUE «NAMI»), 125438, Avtomotornaya Street, 2, Moscow, Russia

E-mail: a.kozlov@nami.ru

Abstract. This work is devoted to the experimental investigation of the possibilities to reduce greenhouse gas emissions and to increase energy efficiency of engines that use natural gas as the main fuel and the analysis of economic efficiency of use of dual fuel engines in vehicles compared to conventional diesel. The results of experimental investigation of a 190 kW dual-fuel engine are presented; it is shown that quantitative and qualitative working process control may ensure thermal efficiency at the same level as that of the diesel engine and in certain conditions 5...8% higher. The prospects for reduction of greenhouse gas emissions have been assessed. The technical and economic evaluation of use of dual fuel engines in heavy-duty vehicles has been performed, taking into account the total life cycle. It is shown that it is possible to reduce life cycle costs by two times.

1. Introduction

The Natural gas (NG) is the most prospective alternative fuel used under the conditions of the Russian Federation. Use of natural gas allows obtaining high operating and economic parameter values, as well as fighting against noise pollution [1-3]. This fuel with the methane content of over 90% may be designated as environmentally safe, as it contains almost no aromatic hydrocarbons or sulfur compounds, which is of high importance in the context of reduction of smoke level and emissions of particulate matters with the exhaust gas [4-7].

The issues of using natural gas in motor vehicles are strategic and being successfully solved all over the world [8], particularly, in the Russian Federation as well [9].

Focusing on wider use of natural gas in motor vehicles will allow increasing the efficiency of fuel and energy conversion and reducing negative environment impact [10].

Currently, gas engines for motor vehicles with the gross weight of more than 3.5 t are being built on a diesel basis, because there are no gasoline engines with spark ignition of such displacement, which could be used as a base for conversion [11].

The main focus area of engineering solutions is adaptation of the diesel design for operation powered by alternative types of fuel (natural gas, etc.) and their combinations with conventional fuel in order to reduce fuel consumption, improve environmental, technical and economic characteristics of engines and to increase their competitiveness [12].

There are several ways to convert diesel engines for operation powered by gas fuel. One of the solutions for the task of extension of gas fuel use in motor vehicles is dual-fuel engines, which are capable of operation powered by conventional diesel fuel (DF) and alternative (gas) fuel [13]. It is the most efficient way to use such engines in heavy-duty trucks and vehicles servicing inter-city and long-distance transport, especially, for vehicles operating under severe conditions, with the possibility to recover the expenses for their conversion and to widely use them both in close vicinity to natural gas vehicle refill stations and far away from the latter [14].

The peculiar properties of the dual-fuel engine working process enable obtaining the best economic effect at medium and high loads and provide the possibility to operate by diesel fuel when using up all the gas in the tank. The diesel engine operating according to the dual-fuel cycle is also a prospective solution in terms of reduction of nitrogen oxide and particulate matter emissions, while requiring marginal design updates [15, 16].

Therefore, dual-fuel engines may be considered as one of the efficient solutions for the use of compressed natural gas in motor vehicles with diesel engines [17].

Acceptable economic and environmental parameters of the dual-fuel engine may be reached by extension of the maximum optimizing control range (air/fuel ratio, pilot portion fuel amount, injection timing), application of phased multipoint gas injection and mixed quantitative and qualitative power control. In this regard, investigations with the purpose of optimizing these parameters have been performed.

2. Investigation subject and objectives

The investigation subject was diesel engine YaMZ-6566 with common rail fuel system, which operates according to the dual-fuel cycle with diesel fuel pilot portion and is equipped with the system of multipoint injection of gas into intake manifold developed by FSUE "NAMI". The characteristics of these engines are given in Table 1.

Table 1. Investigation subjects

Basic characteristics	Values	
Engine code designation given by the manufacturer	YaMZ-6566	
ICE version	Diesel engine	Dual-fuel engine
Engine displacement, l	11.15	
Operating cycle	Four-stroke	
Engine cylinder number and arrangement	V6	
Compression ratio	17.5	
Cylinder bore, mm	130	
Piston stroke, mm	140	
Fuel	Diesel	Diesel and Natural gas
Maximum power, kW	200 (at 1900 min^{-1})	197 (at 1900 min^{-1})
Maximum permissible engine speed, min^{-1}	2100	
Maximum torque, N·m	1170 (at 1100-1500 min^{-1})	1150 (at 1100-1500 min^{-1})

A gas supply system and a control system, ensuring operation in accordance with the dual-fuel cycle with ignition of the air-fuel mixture by the diesel fuel pilot portion, were installed onto the engine. The gas supply is phased multiport (6 gas injectors), the quantitative control of the air supply was performed by an electronic throttle.

The peculiar feature of the dual fuel engine is that two fuels, which differ in their physical and chemical properties and require different conditions for the optimized operation within the cylinder, are used in the working process. Given the fact that in terms of maximum economic efficiency natural gas (methane) shall be the main fuel in the dual-fuel process, the working process conditions shall be formed considering the optimization of combustion of this gas fuel.

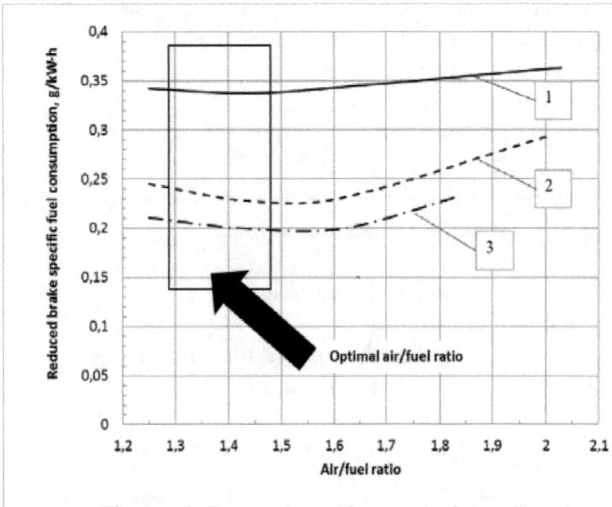

Figure 1. Dependence between the reduced specific fuel consumption and the air-fuel equivalence ratio (where Mk is the torque: 1 - Mk=150 N·m; 2 - Mk=350 N·m; 3 - 550 N·m)

The objective of this investigation was the comparative analysis of technical and economic parameters of the engine when operating using diesel fuel and according to the dual-fuel cycle during the engine bench tests and the total life cycle assessment.

3. Results of engine investigations on test bench

The optimized and efficient natural gas combustion is known to be reached at certain air-fuel ratio. Therefore, the works on adjustment of the optimized value of the air-fuel equivalence ratio (α) ensuring full and efficient methane combustion on the one hand and stable ignition of the minimum diesel pilot portion on the other hand, were performed in the first phase.

The dependence between the reduced specific fuel consumption (specific diesel and natural gas consumption, reduced to the diesel fuel lowest heat of combustion) and the air-fuel equivalence ratio α is shown in Figure 1.

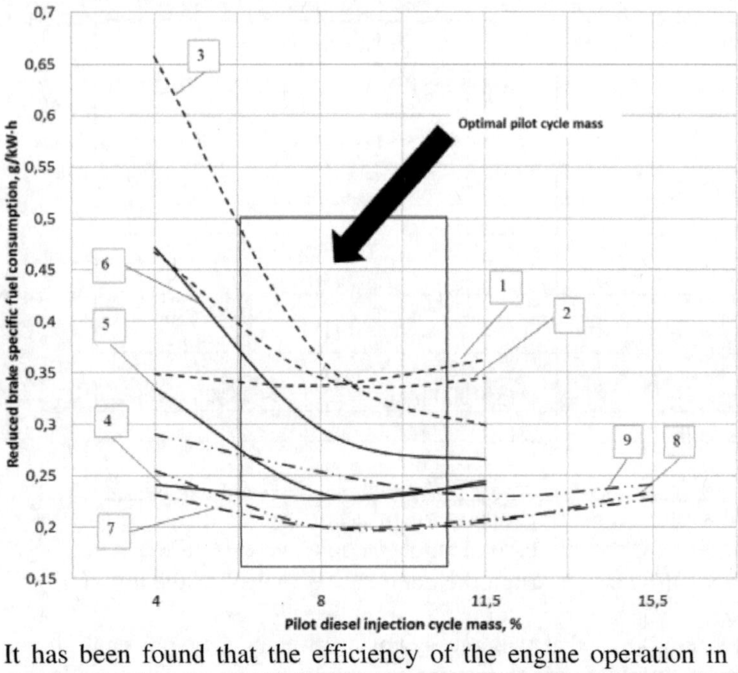

Figure 2. Definition of optimized amounts of the diesel fuel pilot portion in percentage of the maximum supply at the full load (where α is the air-fuel equivalence ratio, Mk is the torque:

1 - α=1.4, Mk=150 N·m;
2 - α=1.65, Mk=150 N·m;
3 - α=2, Mk=150 N·m;
4 - α=1.4, Mk=350 N·m;
5 - α=1.65, Mk=350 N·m;
6 - α=2, Mk=350 N·m;
7 - α=1.4, Mk=550 N·m;
8 - α=1.65, Mk=550 N·m;
9 - α=2, Mk=550 N·m.)

It has been found that the efficiency of the engine operation in dual-fuel cycle with the minimum values of reduced specific fuel consumption is reached when α is maintained within the range from 1.3 to 1.45.

When assessing the parameters that influence the efficiency of the dual fuel process, it is necessary to consider the impact of the amount of the pilot portion of diesel fuel injected in the cylinder. Figure 2 shows the dependencies between the reduced effective fuel consumption and the amount of the diesel fuel pilot portion for different loads and different air-fuel equivalence ratio. The diesel fuel pilot portion was defined as a fraction of the maximum supply of diesel fuel (in diesel mode) at the given speed (n=1200 min^{-1}).

As a result of these investigations a pilot portion amount within the range of 6-11 mg/cycle (5-10%) has been defined for further tests.

The selected pilot portion has been defined based on the conditions of reaching the highest fuel efficiency. However, with such values of the cycle supply a problem of high heat density of the nozzle emerges at high loads. There are two ways to reduce the heat load at the nozzle. The first one is to increase the cycle supply of diesel fuel (in this case, the method is not applicable). The second one is to increase the injection time of the minimum amount of the cycle fuel supply. The second method may be performed by reducing the pressure in the fuel rail.

The change of the fuel rail pressure is known to have a significant influence on the jet structure of the injected fuel and its spray fineness. Consequently, when optimizing the parameters of the fuel supply in order to increase the dual-fuel process efficiency, it is necessary to consider already two factors. The first one is setting up the least pressure in the fuel rail in order to increase the time of injection and create the best conditions for the spray nozzle cooling. Moreover, the second one is maintaining the pressure above the level, at which an evidence of the working process deterioration appears.

Due to the abovementioned, the next test series have been performed in order to study the influence of the fuel rail pressure on the working process of the engine (Figure 3) operating according to the dual-fuel cycle. The tests have been performed at the engine speed n=1200 min^{-1} at three torque levels: 135, 350 and 550 N·m, and the fuel pressure in the rail have increased from 30 to 90 MPa.

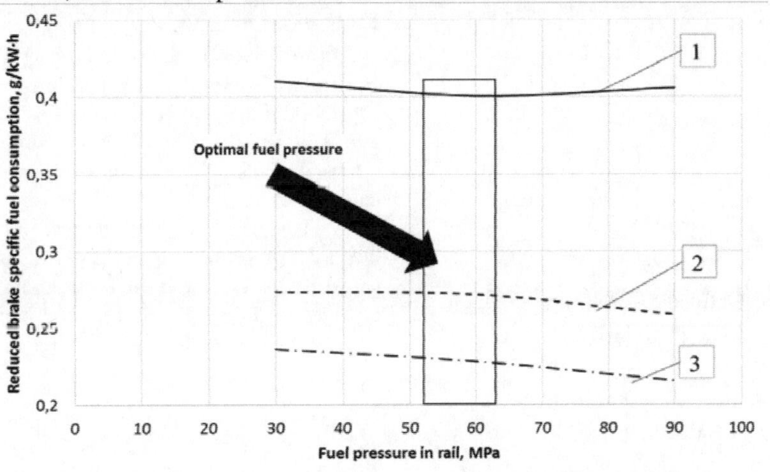

Figure 3. Dependence between the reduced specific fuel consumption and the fuel pressure in the rail (1 - Mk=135 N·m; 2 - Mk=350 N·m; 3 - 550 N·m)

The diagram shows that the increase of the diesel fuel pressure in the rail does not contribute to the improvement of the engine work efficiency parameters in the dual-fuel mode. Therefore, the optimized pressure may be considered as 50-60 MPa.

As a result of the optimization of the air-fuel equivalence ratio (which was calculated taking into account pilot diesel fuel and natural gas consumption), diesel fuel pilot portion and the diesel fuel pressure in the fuel rail, the settings of the control system have been adjusted and the load characteristics of the engine operating according to the dual-fuel cycle with the mixed quantitative and qualitative power control have been obtained. The load characteristic of the dual-fuel engine in certain operation mode (at the engine speed of 1200 min^{-1}) is shown in Figure 4.

The results of the performed engine experimental tests show that use of the optimized algorithms of the control of the engine working process according to the engine load characteristics ensures increase of the engine thermal efficiency by 5-8% in certain modes in comparison to the diesel engine.

To meet the requirements of the modern standards for the emissions of harmful substances in exhaust gases, an engine should be equipped with an oxidation catalyst and selective catalytic reduction converters of nitrogen oxides (SCR). A diagram of the aftertreatment system is shown in Figure 5.

The installation of a complex aftertreatment system of exhaust gases allows refining HC, CO and NO_x up to 65-88%, which ensures meeting the EURO-5 requirements by a dual-fuel engine, and after further improvement of the control and aftertreatment systems – meeting the requirements of EURO-6.

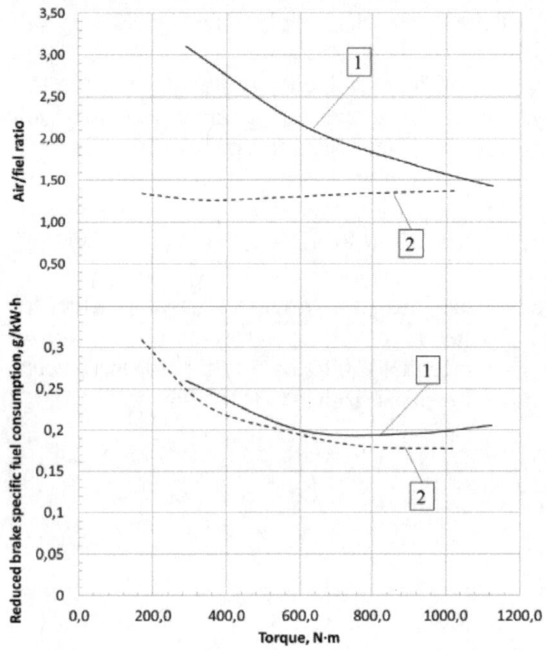

Figure 4. Comparison of the reduced specific fuel consumption with regard to the load characteristic for the dual-fuel and diesel engines (n=1200 min^{-1}, 1 - Diesel engine, 2 – Dual-fuel engine)

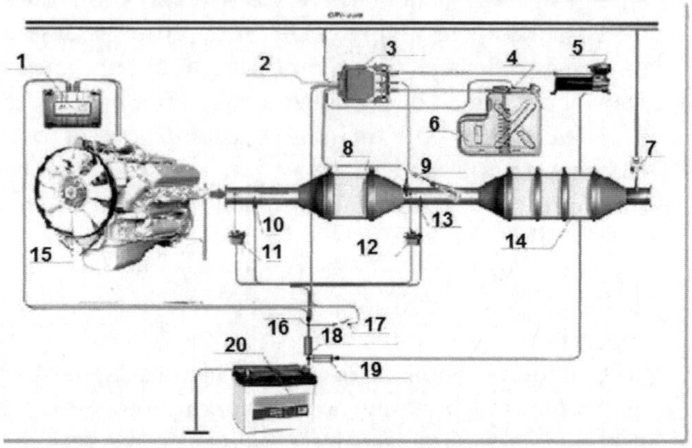

Figure 5. Diagram of the complex aftertreatment system of exhaust gases SNOGMET (1 – engine controller; 2 – ambient temperature sensor; 3 – SCR supply module with controller; 4 – fluid level sensor and temperature sensor AdBlue; 5 – compressor (0.6 MPa); 6 – tank AdBlue ; 7 – pressure sensor; 8 – thermofor catalytic cracking (TCC) unit; 9 – injector AdBlue (0.5 MPa); 10 – TCC unit inlet temperature sensor; 11 – TCC unit inlet pressure sensor; 12 – TCC unit outlet pressure sensor; 13 – TCC unit inlet temperature sensor; 14 –SCR catalyst module; 15 – engine; 16 – contact +24V; 17 – key; 18,19 – fuse 20 A; 20 – battery

4. Assessment of parameters in total life cycle

For the assessment of the efficiency of use of natural gas in engines operating according to the dual-fuel cycle, the costs at all stages of the total life cycle (TLC) are assessed. The assessment has been performed according to the TLC analysis procedure for a vehicle with a powertrain, which operates using diesel fuel and according to the dual-fuel cycle.

The investigation subject was a heavy-duty vehicle with the gross weight of 23,300 kg with engine YaMZ-6566 with two versions (Table 2): diesel and dual-fuel.

For the calculations, a complex procedure for technical, economic and environmental assessment of prospects of use of alternative fuel in vehicles within the total life cycle from the raw material extraction to its use in vehicle operation and disposal, which was developed by FSUE "NAMI", has been used [18-20]. The procedure takes into consideration the consumption of natural resources and energy, emissions of hazardous substances and environmental damage caused by their negative impact, as well as money expenditures for the performance of the life cycle of alternative fuel and powertrains.

IOP Conf. Series: Earth and Environmental Science **52** (2017) 012096 doi:10.1088/1755-1315/52/1/012096

Table 2. Input data for total life cycle cost assessment

Parameter	Values	
Price of DF, €/l	0.50	0.50
Price of NG, €/m^3	-	0.24
Running time per service life, km	1,000,000	
Annual mileage, km	100,000	
Engine price, €	9182	10164
Specific DF consumption, minimum kg/kWh	0.192	0.021
Specific NG consumption, minimum kg/kWh	-	0.168
DF consumption per hour, kg/h	19.95	1.86
NG consumption per hour, kg/h	-	15.4
DF consumption, l/100 km	30.51	2.85
NG consumption, m^3/100 km	-	28.2

For the analysis of the dual-fuel engine economic efficiency in comparison to the diesel engine, the costs during its total life cycle and economic effect per individual components and in general have been calculated, which is shown in Table 3.

In general the positive effect of 70691 € has been obtained in comparison of the diesel engine to the engine operating according to the dual-fuel cycle. This result can be explained by lower natural gas price. Values of costs during the total life cycle on production and operation stages is shown in Figure 6.

Table 3. Total life cycle costs, €

	Diesel engine	Dual-fuel engine
Fuel	DF	DF + NG
Cumulative costs	161741	91050
Economic effect		70691

The figure shows that the main part of expenditures, about 90%, is accounted for the operation stage.

According to the performed assessment it can be noted that use of compressed natural gas as a fuel allows reducing costs for the total life cycle almost by two times.

Carbon dioxide emissions during the vehicle total life cycle have been calculated, which is shown in Figure 7. It has been shown that dual-fuel engine allows decreasing CO_2 emissions by 22.6% in comparison to diesel variant.

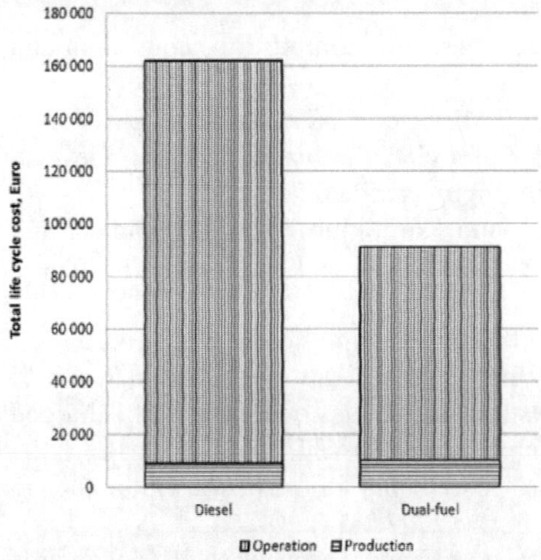

Figure 6. Costs during the total life cycle of the heavy-duty vehicle with engine YaMZ-6566 operating according to the diesel and dual-fuel operation mode

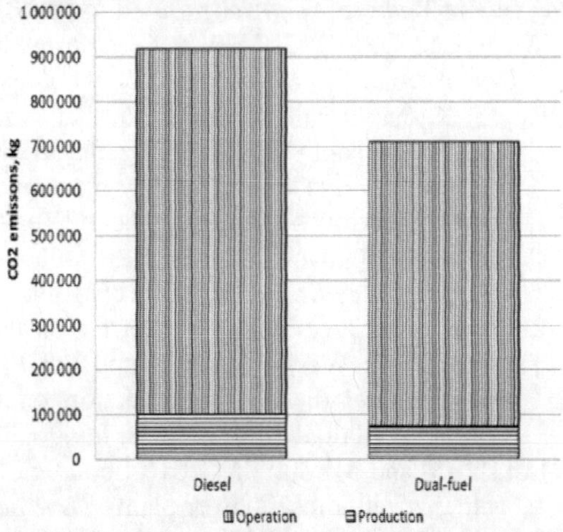

Figure 7. CO_2 emissions during the total life cycle of the heavy-duty vehicle with engine YaMZ-6566 operating according to the diesel and dual-fuel operation mode

5. Conclusions

It has been found that, when converting the diesel engine YaMZ-6566 to the dual-fuel cycle operation, the minimum values of the specific effective fuel consumption are reached as follows: at the air-fuel equivalence ratio equal to 1.3...1.45, which is maintained in the engine operating modes of 10% to 100% of the full load; at the optimized value of the diesel pressure in the fuel rail – 50-60 MPa; at the amount of the diesel pilot portion equal to 6-11 mg/cycle (5-10% of the maximum fuel supply).

Use of the optimized algorithms of the control of the dual-fuel engine working process according to the engine load characteristics ensures improvement of thermal efficiency of the engine by 5-8% in certain modes (at low and medium loads) in comparison to diesel mode of operation.

As a result of the analysis of the economic efficiency of the dual-fuel engine in comparison to the diesel engine it has been found that use of natural gas as the main fuel allows reduction of costs for the total life cycle almost by two times (in Russian Federation conditions). Dual-fuel engine has total life cycle CO_2 emission by 22.6% lower than diesel one. Emission legislation requirements Euro-5 or Euro-6 for both diesel and dual-fuel modes can be fulfilled with using of complex aftertreatment system.

Therefore, the obtained results allow considering natural gas used in dual-fuel engines, as the most prospective alternative fuel in the short run.

Acknowledgments

The paper was prepared under the agreement No. 14.624.21.0005 with the Ministry of Education and Science of the Russian Federation (unique project identifier RFMEFI62414X0005) to create an experimental model of the aftertreatment system of toxic components for gas engines.

References

[1] Abedin M J, Imran A, Masjuki H H, Kalam M A, Shahir S A, Varman M and Ruhul A M 2016 An overview on comparative engine performance and emission characteristics of different techniques involved in diesel engine as dual-fuel engine operation *Renewable and Sustainable Energy Reviews* **60** 306-316.

[2] Zhang Q, Li M and Shao S 2015 Combustion process and emissions of a heavy-duty engine fueled with directly injected natural gas and pilot diesel *Applied Energy* **157** 217-228.

[3] Luksho V 2010 Diesel engine converting into the gas engine with a controlled thermodynamic cycle *Transport on Alternative Fuel* **6(18)** 44-50.

[4] Wei L and Geng P 2016 A review on natural gas/diesel dual fuel combustion, emissions and performance *Fuel Processing Technology* **142** 264-278.

[5] Roy S, Kumar D A, Banerjee R and Kumar B P 2014 A TMI based CNG dual-fuel approach to address the soot–NOx–BSFC trade-off characteristics of a CRDI assisted diesel engine – an EPA perspective *Journal of Natural Gas Science and Engineering* **20** 221-240.

[6] Luksho V, Kozlov A, Terenchenko A, Ter-Mkrtichian J and Karpukhin K 2015 Technical and economic analysis of vehicles pollutant emissions reduction technologies *Biosciences Biotechnology Research Asia* **12(2)** 1867-72.

[7] Lukanin V and Trofimenko Yu 2001 Industrial Transport Ecology: Proc. for high schools 273.

[8] Yang B, Xi C, Wei X, Zeng K and Lai M 2015 Parametric investigation of natural gas port injection and diesel pilot injection on the combustion and emissions of a turbocharged common rail dual-fuel engine at low load *Applied Energy* **143** 130-137.

[9] Luksho V and Mironov M 2010 Economic efficiency of use of natural gas as a motor fuel in vehicles with diesel power plants *Transport on Alternative Fuel* **2(14)** 20-26.

[10] Lounici M S, Loubar K, Tarabe L, Balistrou M, Niculescu D-C and Tazerout M 2014 Towards improvement of natural gas-diesel dual fuel mode: An experimental investigation on performance and exhaust emissions *Energy* **64** 200-211.

[11] Wang Z, Zhao Z, Wang D, Tan M, Han Y, Liu Z and Dou H 2016a Impact of pilot diesel ignition mode on combustion and emissions characteristics of a diesel/natural gas dual fuel heavy-duty engine *Fuel* **167** 248-256.

[12] Panchishny V, Luksho V, Kozlov A, Terenchenko A and Karpukhin K 2015 The problem of toxicity of gas engines and their solutions *Biosciences Biotechnology Research Asia* **12** (Spl. Edn. 2) 217-224.
[13] Lim O, Iida N, Cho G and Narankhuu J 2012 The research about engine optimization and emission characteristic of dual fuel engine fueled with natural gas and diesel *SAE* 2012-32-0008.
[14] Yang B, Wei X, Xi C, Liu Y, Zeng K and Lai M-C 2014 Experimental study of the effects of natural gas injection timing on the combustion performance and emissions of a turbocharged common rail dual-fuel engine *Energy Conversion and Management* **87** 297-304.
[15] Li W, Liu Z and Wang Z 2016 Experimental and theoretical analysis of the combustion process at low loads of a diesel natural gas dual-fuel engine *Energy* **94** 728-741.
[16] Wang B, Li T, Ge L and Ogawa H 2016b Optimization of combustion chamber geometry for natural gas engines with diesel micro-pilot-induced ignition *Energy Conversion and Management* **122** 552-563.
[17] Cameretti M C, Tuccillo R, De Simio L, Iannaccone S and Ciaravola U 2016 A numerical and experimental study of dual fuel diesel engine for different injection timings *Applied Thermal Engineering* **101** 630-638.
[18] Zvonov V, Kozlov A and Kutenev V 2001 Environmental Safety of the Cars in the Full Life Cycle 248.
[19] Luksho V, Kozlov A and Terenchenko A 2011 Total life cycle assessment of natural gas as motor fuel *Transport on Alternative Fuel* **3(21)** 4-9.
[20] Vershkov L, Groshev V, Gavrilov V et al 1999 Temporary methods of definition of the prevented ecological damage 68.

The research on the impact of energy- environment policy on regional development—based on CGE model

Y M Jin

State Grid Energy Research Institute, Beijing 102209, RPC

E-mail: jinyanming@sgeri.sgcc.com.cn

Abstract: In this study, we will focus on study the impact of energy and environment regulatory on the inter-regional coordinated development from the regional level. Making use of multi-regional energy-economy-environmental computable general equilibrium(CGE) model, we will analysis the effect of regional energy regulation in the future. We will research the impact of the carbon emissions trading and other environmental policy on regional economic development and industrial structure. The results show that the regulation of energy and environment can promote regional industry to upgrade and different policies of energy and environmental are needed to implement to mitigate the negative impact on the economy of different regions. In the paper, we expand the traditional regional energy-environmental CGE model tools for providing new quantitative methods to study regional energy and environmental problems in China.

1. Introduction

China's economic and social development faces many prominent problems, such as the restriction of environment and the unbalanced development of regional economics. China's energy resources are mainly concentrated in the north and the west, while economic growth areas are mainly concentrated in the southeast areas. The reverse distribution of economy and resources leads to the pressure of energy transportation. At present, many areas of our country are shortage of water and energy resources. The solution to these issues needs fundamentally change the mode of the regional economy and energy development. The impact of energy and environmental regulation on the economy is not only reflected in the growth rate, but also related to the industrial structure, the people's welfare and other aspects.

Computable general equilibrium model has been widely used in energy and environmental policy simulation. Energy-environmental CGE model is based on a typical economic system, but the energy elements and pollution emissions need to be carefully handled. For the energy-environmental CGE model, it is the core problem that how to embed the energy environment factor into the CGE model. The most of the energy environment CGE model is built up based on the standard CGE model framework by exogenous environment factor,which chang the production function or utility function for the feedback the environmental impact of emissions. Roy et.al (1995) [1] broke down the energy-intensive sectors in the model for estimating environmental losses due to carbon emissions from all sectors. He et al. (2010) [2] directly calculated emissions of pollution using fixed emission factors for energy ,in order to simulation the impact of environmental taxes on our energy reduction and economic development. In recent years, some scholars have tried to change the model of the basic data set to establish structure-derived type of energy and environmental CGE model. For instance, Gao

Ying et al (2009) [3] transformed the traditional social accounting matrix framework, which is introduced the new accounting accounts, such as resource recovery, pollution control and natural endowment factor, etc. Xie et al(2000) [4] increased the description of the environmental pollution controlling activities in the model,which established pollution control department to simulate the pollution reduction behavior and charactered emission reduction costs of the production department and environmental protection investmen, etc..From the regional level, the influence of various environmental regulations and energy regulation on interregional industry, capital and labor transfer is seldom considered.The few studies systematically consider the effect of the multiple energy regulations and environmental regulations on regional economic coordinated development.Niu Yujing et al (2012) [5] simulated and analyzed the impact of multi-regional carbon tariffs on carbon leakage using the CGE model of the eight regions in the world,introducting the international banking and international transport services and other virtual sectors, etc.

Compared with previous research on the national level of energy and environmental policies [6-10], this paper focuses on the impact of environmental regulation on regional energy trade, financial and human capital flows from the cross-regional level.We simulates the effect of environmental regulation and energy mechanism on regional economic coordinated development. In addition, the paper will enrich and expand the traditional regional energy-environmental CGE model tools for providing new quantitative methods to study regional energy and environmental problems in China.

2. The model and Scenario design

2.1. Multi-regional energy-environment-economy CGE model

We construct a multi-regional resource-economy- environment CGE model,which based on the social accounting matrix of Guangdong, Guizhou and other regions of China in 2007.Three regions have their own production and demand structure, and the regions are interrelated through commodity trade and the flow of production elements. The model includes 28 production departments,which are three resource recovery sector, four pollution control department, four resource sectors and 17 traditional productive sectors, and two types of groups are namely urban and rural , and six kinds of production factors are land, physical capital, environmental capacity of capital, agricultural labor, productive workers and professional and technical personnel.

In order to better characterize all kinds of energy and environmental regulations in the model, considering the energy consumption related to pollutant emissions, the energy and environmental capacity assets are combined to energy - asset beam with the Leontief function. It should be pointed out that any one of the production department has two kinds of treatment methods for pollutants, namely direct control (the goods of the pollution control department as an intermediate input) or buy emission rights. Environmental capacity assets can become capital to bring benefits, which the right of using environmental capacity can be transferred with compensation in the case of emissions trading. The environmental capacity assets become the capital with expected return, which also can be viewed as emission rights of an enterprise through the government auction or free distribution.

The model assumes full employment of labor, capital mobility, exogenously given regional environmental capacity capital. Environmental capacity assets can be assumed incomplete or complete flow between the sectors or regions based on the scenario setting, and liquidity depends on the rate of environmental capacity capital return, namely price of emission rights and elasticity of substitution. When the emission trading is not implemented, the model assumes that the supply of environmental capacity is in the form of a constant elasticity function. When the market of emission trading is established, equilibrium equation on emission rights between departments among regions is as follows.

The price of emission rights in various regions,

$$\begin{cases} Pev_r = \left(\sum_i \alpha_{i,r}^{es} PENV_{i,r}^{1+\sigma^{es}} \right)^{\frac{1}{1+\sigma^{es}}} & if \quad \sigma^{es} \neq \infty \\ TEN_r^s = \sum_i EN_{i,r}^d \quad if \quad \sigma^{es} = \infty & \end{cases}$$

Emission rights price of production department in various regions,

$$\begin{cases} EN^{s}_{i,r} = \alpha^{es}_{i,r} TEN^{s}_{r} \left(\dfrac{PENV_{i,r}}{Pev_{r}} \right)^{\sigma^{es}} & if \quad \sigma^{es} \neq \infty \\ PENV_{i,r} = Pev_{r} \quad if \quad \sigma^{es} = \infty \end{cases}$$

Equilibrium conditions of the emission rights market,

$$EN^{s}_{i,r} = EN^{d}_{i,r}$$

Pev_r	Return on environmental assets of region r
TEN^s_r	The total amount of all environmental assets of region r
$PENV_{i,r}$	Prices environmental assets of region r department i
$EN^d_{i,r}$	Demand for environmental assets of region r department i
$EN^s_{i,r}$	All environmental assets of region r department i
$\alpha^{es}_{i,r}$	Share parameters of environmental assets supply of region r department i
σ^{es}	The sectorial elastic parameters of environmental assets

2.2. Scenario Design

According to IPCC fifth report, in order to achieve the sustainable development of human, the average global temperature rise must be controlled within 2 ℃ in 2100. At present, China has become the world's largest carbon emitter. China's carbon emission in 2014 is about 94 billion tons, accounting for 29% of global carbon emissions. To assume the responsibility of emissions reductions, China commits that carbon emissions per unit of GDP in 2030 will be reduced by more than 60% in 2005. This study mainly set four scenarios to analyze the impact of energy and environmental regulations on regional economic development mode. The first scenario of environmental regulation is mainly the establishment of the carbon tax and carbon emissions trading. The second scenario is mainly the energy tax and the limitation energy consumption. Specific design scenarios are shown in Table 1.

Table 1 . The scenarios of energy and environmental regulations

Category	Scenario description
A1: Energy tax	The tax rate of electric power, coal and refined oil is 10%.
A2: Limit energy consumption	Reduce energy consumption by 1% of the electricity, coal and refined oil.
B1: Carbon tax	The carbon tax is 10% in the final consumption goods of sectors..
B2: Carbon tax and Carbon emissions trading	The final goods consumption is levied 10% carbon tax, and establish regional carbon emissions trading market.

3. The Result

3.1. The impact of environment policy on the regional economy

When the emissions trading market is not established, all regions that levied on the carbon tax, are more dependent on the output effect to achieve the emission reduction targets, so that the actual GDP decreases more than that in the case of the implementation of emissions trading scenarios. Guangdong and Guizhou are affected differently in both environmental policy scenarios due to the different industrial structure. Guangdong Province is based on the manufacturing, where the long-term extensive economic growth makes the local environmental capacity of capital increasingly scarce. If the emission reduction targets are achieved only through a carbon tax (B1), real GDP decline 0.14%, mainly due to carbon tax that increases production costs, reduce profits, and result in shrinking of production scale, the decline in rate of wages, reduce household disposable income. Residents reduce consumption under the double impact of commodity price and income. Guangdong foreign imports and transferred area decreased, which are affected by the regional demand.

The real GDP fell 0.11 percent in Guangdong Province when the government levies on carbon tax. Guangdong Province can purchase emission reduction in other provinces to reduce costs due to the establishment of trans-regional emissions trading market, so that Guangdong can alleviate the negative impact of increasing production costs caused by the high local abatement costs. Compared to only impose a carbon tax (B1) scenario, the workers' wages and capital returns in various regions have increased, while the consumption of energy has increased. Under the restraint reduction, trans regional emissions trading enables manufacturers to more use of labor and capital to replace environmental capacity of capital. Under the B1 scenario, since the non-regional emissions trading, manufacturers can only achieve the effect of emissions by reducing energy consumption and production under the restrictions of limited environmental capacity of capital. The results of model are shown in Table 2.

Table 2 . The effect of environmental regulations on the regional economy (%)

	Carbon tax (B1)			Carbon tax + Carbon emissions trading (B2)		
	Guizhou	Guangdong	China	Guizhou	Guangdong	China
Real GDP	-0.03	-0.14	-0.06	-0.03	-0.11	-0.05
Total output	-0.12	-0.24	-0.10	-0.10	-0.19	-0.08
Resident consumption	-0.04	-0.03	-0.05	-0.03	-0.02	-0.04
Investment	0.00	-0.29	-0.01	0.00	-0.23	-0.01
export	-0.64	-0.19	-0.17	-0.52	-0.16	-0.14
Import	-0.10	-0.14	-0.01	-0.08	-0.11	-0.01
regional transfer-out	-0.21	-0.15	-	-0.17	-0.12	-
regional transfer-in	-0.08	-0.25	-	-0.07	-0.21	-
Disposable income of rural residents	-0.40	0.10	-	-0.33	0.08	-
Disposable income of urban residents	-0.20	-0.55	-	-0.16	-0.45	-
Rural residents' welfare benefits	-0.02	0.00	-	-0.02	0.00	-
Urban residents' welfare benefits	-0.01	-0.32	-	-0.01	-0.26	-
The average wage rate of agricultural labor	-0.42	-0.75	-	-0.34	-0.62	-
The average wage rate of industrial workers	-0.50	-0.72	-	-0.41	-0.59	-
Technical personnel average wage rate	-0.29	-0.63	-	-0.24	-0.52	-
Return on capital	0.42	0.42	0.42	0.35	0.35	0.35
Coal consumption	-1.16	-1.45	-0.48	-0.95	-1.18	-0.39
Oil and Gas consumption	-0.62	-1.58	-0.34	-0.50	-1.29	-0.28
Thermal power consumption	-1.34	-1.19	-0.86	-1.09	-0.97	-0.71
Hydroelectric power consumption	0.18	0.09	-0.53	0.15	0.07	-0.44

From the change of industrial structure, the output of the key industries in all regions decreased due to the carbon tax. But the supply output of hydroelectric power increases because the demand for clean energy such as hydroelectric power is increasing due to the decreasing output scale of thermal power supply, and so substitution effect hydropower and thermal power occurs. In the whole country and Guizhou Province, the output of coal mining industry is of the sharpest decline, followed by thermal power supply, petrochemical and other carbon intensive sectors, while hydroelectricity and other electricity supply industry, social services and other less carbon-intensive manufacturing industries, labor-intensive industries have little impact. In Guangdong province, the decrease of each sector output is smaller than that of Guizhou province, the most significant decline in output is still the coal mining industry.

3.2. The impact of energy policy on the regional economy

Because of the energy tax, the price of commodity in Guangdong is lower than one of the Guizhou and other regions due to its industrial structure. Exports and regional transfer-out continue to grow, making the local GDP not fall but a slight rise. Guizhou affected more by heavy taxation than Guangdong, and Guizhou's real GDP falls 0.23% due to the influence by exports and domestic transfer-out. The main reason is that the local economic base is weak, which any shock would have a greater impact on commodity prices and factor returns, and hinders the circulation of commodities and factor mobility between Guizhou and other areas, leading local real GDP subject to greater negative impact.

Under the restriction of total energy consumption, the GDP of Guangdong, Guizhou and the whole country decrease respectively by 1.4%, 5.5% and 3.2%, and the consumption of the residents in these regions decreased respectively by 2.5%, 3.6% and 4.8%. In view of this, mandatory policies of limiting the total amount of energy consumption, such as pulling a switch power brownouts, will lead to the destruction of economic growth and reduction of income. In the production sector, the electricity production, coal mining and supply industry, the oil and gas industry, chemical industry and metal rolling processing industry, belonging to high energy-consuming sector, the outputs decrease respectively by 0.21%, 0.20 %, 0.19%, 0.17% and 0.14%. The results of model are shown in Table 3. In addition, in a general equilibrium framework, limiting the supply of products in the energy sector will directly or indirectly lead to reduce the output amount of upstream and downstream industries, which affect the tertiary industry, such as the retail trade industry, etc.

Table 3 .The impact of energy regulation on the regional economy (%)

	Energy tax(A1)				Limit energy consumption(A2)			
	Guizhou	Guangdong	ROC	China	Guizhou	Guangdong	ROC	China
Real GDP	-0.002	0.010	-0.014	-0.012	-0.032	-0.014	-0.049	-0.055
Resident consumption	-0.019	0.012	-0.035	-0.030	-0.036	-0.025	-0.050	-0.048
Investment	-0.009	-0.012	-0.022	-0.021	0.002	-0.287	0.010	-0.007
Export	0.002	-0.238	0.006	-0.008	-0.642	-0.190	-0.160	-0.170
Import	-0.233	0.061	-0.027	-0.003	-0.096	-0.135	0.042	-0.014
Transfer-out	0.018	0.021	-0.005	0.003	-0.210	-0.148	-0.222	-
Transfer-in	-0.014	0.084	-0.047	—	-0.080	-0.253	-0.148	-
Disposable income of rural residents	-0.014	-0.046	0.072	—	-0.398	0.096	-0.003	-
Disposable income of urban residents	-0.014	0.004	-0.072	-0.076	-0.195	-0.548	-0.003	-
Welfare income of rural residents	-0.119	-0.204	-0.151	-0.161	-0.018	0.004	-0.614	-
Welfare income of urban resident	-0.001	0.000	-0.157	-0.158	-0.007	-0.317	-1.420	-
The average wage rate (industrial workers)	0.023	-0.240	-0.030	—	-0.495	-0.719	-0.006	-
The average wage rate (technical staff)	-0.082	-0.283	-0.183	—	-0.294	-0.633	-0.005	-

4. The conclusions and recommendations

All kinds of energy and environmental regulations have some negative effect on the economic development of various regions. The establishment of emissions trading and other market mechanisms has a certain positive effect on alleviating the negative impact on the economy due to emission constraints. We prove that the emission trading is a lower cost pollution control method, which is helpful to achieve the overall control objectives in the more economical way. In view of the characteristics of high energy consumption and heavy pollution in China, The combination of government measures and market mechanism which establish the perfect emissions trading market, is conducive to economic and social sustainable development.

Our country should balance the local economy with a different environmental tax rate. Due to the unbalanced regional economic situation, Guizhou and other underdeveloped areas in China provide raw materials and primary industrial products with the low added value. While Guangdong and other developed areas in China have developed manufacturers .At the same time, the consumption market is small in Guizhou due to the difference of income, so any shock of tax has a significant negative impact on Guizhou's economy. Therefore, the transfer payment was employed, reduce tax rates and the differential tax system were carried out in western areas, while the central government increase financial aid to the West. Under environmental taxation leverage, the environment is improved due to reduce emissions, while economic disparities between regions are narrowed.

The energy tax and energy consumption will have an important impact on the upstream and downstream industries as well as the income of the residents and the price of consumer goods. Implementation of energy tax reform will increase sector production costs, reduce the total energy consumption, but also indirectly increase the consumer price. In the control of total energy consumption, the government should avoid the phenomenon of capacity constraints such as blackouts, to avoid damage to economic growth.

Reference

[1] Roy Boyd, Kerry Krutilla, W Kip 1995 Viscusi Energy Taxation as a Policy Instrument to Reduce CO_2 Emissions:A Net Benefit Analysis *Journal of Environmental Economics and Management* **29** 1-24.

[2] He Jianwu, Li Shantong 2010 Carbon Reduction and Regional Economy *Management Review* **22** 9-16.

[3] Gao Ying, Li Shan Tong 2009 Study on the Construction of Dynamic Recursive CGE Model within a Sustainable Development Framework *Future and Development* **11** 12-20.

[4] Xie, Sidney Saltzman 2000 Environmental Policy Analysis: An Environmental Computable General-Equilibrium Approach for Developing Countries *Journal of Policy Modeling* **22** 453-489.

[5] Niu Yu Jing, Chen Wen Yin, Wu Zong Xin 2012 Construction of Global Multi-regional CGE Model and Simulation Analysis of Carbon Leakage *The Journal of Quantitative & Technical Economics* **11** 34-50.

[6] Wei Weixian 2009 An Analysis of China Energy and Environmental Policies Based on CGE Model *Statistical Reseach* **7** 3-12.

[7] Budy P, Resosudarmo, Erik Thorbecke 1996 The impact of environmental policies on household incomes for different socio-economic classes: The case of air pollutants in Indonesia *Ecological Economics* **17** 83-94.

[8] Rutherford T F, Montgomery W D, Bernstein P M, CETM 1997 A Dynamic General Equilibrium Model of Global Energy Markets, Carbon Dioxide Emissions and International Trade Working Paper 97-3, Boulder: University of Colorado

[9] Jin Yan Ming, Lei Ming 2012 Research on SO_2 Emission Trade-Based on Resource- Economy-Environment CGE Model *Joumal of Shanxi Finance and Economics University* **34** 1-10.

[10] Dufournaud M C, J Harrington and P Rogers 1988 Leontief's Environmental repercussions and the economic structure revised: A general equilibrium formation *Geographical Analysis* **20** 318-327.

Study on conversion relationships of compressive strength indexes for recycled lightweight aggregate concrete

Xiang-gang Zhang[1, 2, 3], Jian-hui Yang[1, 2, 4] and Xiao-mei Kuang[2]

[1] Henan Polytechnic University, Henan Province Engineering Laboratory of Eco-architecture and the Built Environment, Jiaozuo 454000, China;
[2] Henan Polytechnic University, School of Civil Engineering, Jiaozuo 454000, China;
[3] Guangxi Key Lab of Road structure and materials, Nanning 530000, China.

E-mail: xgzhang@hpu.edu.cn

Abstract: In order to study cube compressive strength and axial compressive strength of recycled lightweight aggregate concrete(RLAC), and conversion relationship between the two, with the replacement rate of recycled lightweight coarse aggregate as change parameters, 15 standard cube test specimens and 15 standard prism test specimens were produced to carry out the test. Then compressive strength of test specimens were measured, and the law of different replacement rate of recycled lightweight coarse aggregate influencing compressive strength of RLAC was analyzed, as the method of statistical regression adopted, the conversion relationships between of cube compressive strength and axial compressive strength of RLAC was obtained. It is shown that compressive strength of RLAC are lower than compressive strength of ordinary concrete; and that compressive strength of RLAC gradually decreases as replacement rate of recycled lightweight coarse aggregate increases; as well as, the conversion relationship between axial compressive strength and cube compressive strength of RLAC is different from ordinary concrete; based on the experimental data, conversion relationship formula between compressive strength indexes of RLAC was established. It is suggested that the replacement rate of recycled lightweight aggregate should be controlled within 25%.

1. Introduction

Construction waste is increasing in China with the rapid development of urbanization in recent years, how to deal with the construction waste has become a difficult problem for scientific research workers [1,2]. According to the statistics [3], at present, construction waste has reached 30%~40% of the city rubbish, among which 50%~60% is the waste concrete in China. It will seriously pollute the environment if these constructions aren't utilized scientifically and effectively. Through those steps of classifying, crushing, washing and drying, the recycling of construction waste will be mixed to be the recycled aggregate formation according to the design of the mixed proportion and gradation. Sand and other natural coarse aggregate were replaced partially or even totally by the recycling of construction waste, and it can be configured to be recycled concrete after adding water and cement [4-6]. Recycling those wastes of concrete can decrease the consumption of natural aggregate in the construction industry and alleviate the pressure of the increasing lack of natural aggregate, meanwhile, the economic and social benefits are obvious. Above all, it has more important practical significance for protecting ecological environment [7, 8].

At present, studying on conversion relationships between compression strength indexes of recycled aggregate concrete is in a preliminary stage. Among them, Xiao Jianzhuang et al [9]studied comprehensively on relationships between strength indexes of recycled aggregate concrete, tested the strength of the specimens, and initially obtained the conversion relationships between cube compressive strength of recycled concrete and other strength indexes through statistical regression of experimental data. Experiments showed that recycled concrete was similar with ordinary concrete in flexural strength, but the conversion relationship between the strength indexes of recycled concrete and ordinary concrete is different. Chai Yuanyuan et al [10]studied on conversion relationships between basic mechanical properties indexes of recycled aggregate concrete and proposed the conversion relationships of some basic mechanical properties indexes of recycled concrete. Lu Shengwu et al[11] studied on the compressive strength of recycled concrete using waste ceramic as fine aggregate, it is shown that the conversion relationship between the compressive strength and the tensile strength of recycled concrete was different from ordinary concrete. A formula on the conversion by compressive strength to tensile strength of recycled concrete was established.

In the past studies, it has been done mostly on recycled concrete derived from ordinary waste concrete, but little research on new recycled concrete from waste lightweight aggregate concrete, and no relevant literature has been found yet. It is very necessary to study the advantages of natural aggregate and waste lightweight aggregate concrete in a certain proportion, which can reduce the weight of the structure, save the natural aggregate and so on.

In this paper, the compressive strength of RLAC of 30 test specimens were carried out, and analyzed law that the effect of replacement rate of recycled lightweight aggregate on the compressive strength of RLAC. The conversion relationship between the axial compressive strength and the cube compressive strength was established, and can provide a reference for the research of RLAC.

2. Experiment

2.1. Raw materials

Natural coarse aggregate adopted continuous gradation of gravel, as shown in Figure 1 (a); The recycled lightweight aggregate (design strength grade LC20) was prepared by crushing, screening, washing and drying the lightweight concrete specimens after being tested at a structure hall of university in Henan Province, as shown in Figure 1 (b). The sand gradation was medium; ordinary tap water was used as mixing water; the cement was PO42.5 produced by a Jiaozuo cement limited. Basic physical properties of the coarse aggregate and recycled coarse aggregate are shown in Table 1. It can be concluded that the water absorption rate of the recycled lightweight aggregate is much larger than that the natural coarse aggregate, which can reach 9.46%. The reason is that the rough surface of recycled coarse aggregate, and more internal voids.

(a) Natural coarse aggregate (b) Recycled coarse lightweight aggregate

Figure 1. Coarse aggregate

Table 1. The basic physical property indexes of coarse aggregate

Aggregate Category	Grain size /mm	Bulk density (kg/m^3)	Apparent density /(kg/m^3)	Water absorption rate/%	Voidage /%
Natural coarse aggregate	5.00-31.50	1481.00	2500.00	0.62	40.76
Recycled lightweight coarse aggregate	5.00-31.50	776.00	1389.00	9.46	44.13

2.2. *Mix proportions*

In this experiment, 5 types (0%, 25%, 50%, 75%, 100%) of replacement rate were designed, the replacement rate was designed according to the mass percentage of the recycled lightweight aggregate to the total coarse aggregate. The replacement rate of 0% of the test specimen as a natural test specimen, and the remaining test specimens as the contrast test specimens. 15 cube test specimens and 15 prism test specimens were produced, in each group of replacement rate was 3 cube specimens and 3 prism specimens. Water absorption rate was tested according to the 《Light aggregate and its test method》 (GB/T17431.2-2010), considering the rough surface and higher water absorption rate of recycled coarse aggregate, therefore, additional water was added as compensation, and the amount of additional water was obtained by the product of the recycled light coarse aggregate and its water absorption rate. Under the different replacement rate, the mixed proportion of quality of LC20 RLAC is shown in Table 2.

Table 2. Quality mix proportion of RLAC

Replacement rate	Cement /kg	Natural coarse aggregate/kg	Recycled lightweight coarse aggregate/kg	Sand /kg	Net duty of water/kg	Additional water/kg	Effective water to binder ratio
0%	440.00	1066.32	0.00	634.08	208.00	0.00	0.47
25%	440.00	799.74	139.68	634.08	208.00	13.21	0.47
50%	440.00	533.16	279.36	634.08	208.00	26.43	0.47
75%	440.00	266.58	419.04	634.08	208.00	39.64	0.47
100%	440.00	0.00	558.72	634.08	208.00	52.85	0.47

Note: Effective water-cement ratio is the ratio of net duty of water to cement quality.

2.3. *Preparation of specimens and loading*

The size of the test specimens was designed to be 150mm×150mm×150mm standard cube test specimen and 150mm×150mm×300mm standard prism specimen. Standard size of steel mould were used for the test, and the recycled concrete was molded according to the 《Standard test method for the performance of ordinary concrete》 (GB/T50080-2002). The moulds were removed from the forming specimens after 24 h in the structure hall, to be the standard curing 28d after loading, measured its compressive strength.

Figure 2. Loading setup

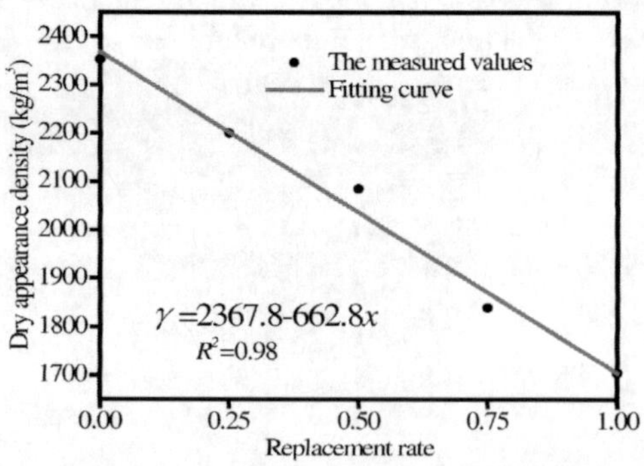

Figure 3. Relation curve between dry appearance density and replacement rate

This test was carried out according to the loading method of the 《Ordinary concrete mechanical property test method》 (GB50081-2002), the loading setup as shown in Figure 2 adopted a SYE-2000 pressure testing machine and the maximum range was 2000kN . The loading rate of the test specimens was kept constant 9kN/s in this experiment.

3. Test results and influence factors

3.1. Dry apparent density

According to the study [12], the dry apparent density has a direct relationship between the compressive strength of test specimen and replacement rate. The three groups of cube test specimens had produced under the different replacement rate, and its size is100 mm×100 mm×100 mm. After the standard curing, test specimens were dried to constant weight by using the model of SY101-2 electric blast oven in the range of 105 ℃, and then the dry apparent density of RLAC with different replacement rates was calculated after measuring its quality as shown in Table 3.

Table 3. Dry appearance density

Replacement rate	0%	25%	50%	75%	100%
Dry appearance density/(kg/m^3)	2353	2200	2085	1839	1705

It can be seen from Table 3 that the dry apparent density of RLAC is lower than that of ordinary concrete, and the dry apparent density of RLAC decreases with the increasing of replacement rate. The magnitude of the reduction (this is the difference between the latter and the former accounting for the former) were 6.50%, 5.23%, 11.80% and 7.29% respectively. With the gradual increase of the replacement rate, the dry apparent density of the specimen have a linear relationship with the replacement rate, as shown in Figure 3.

Accord to the data in Table 3, the least square method is used to fit the relationship between the apparent density and the replacement rate, as shown in Figure 3.

Obviously, the mathematical expression of the relationship between the dry apparent density and the replacement rate curve is

$$\gamma=2367.8-662.8x \tag{1}$$

Where γ is the drying apparent density, x is the replacement rate of recycled lightweight concrete, formula (1) has fitting precision of 0.98.

3.2. Process and pattern of failure

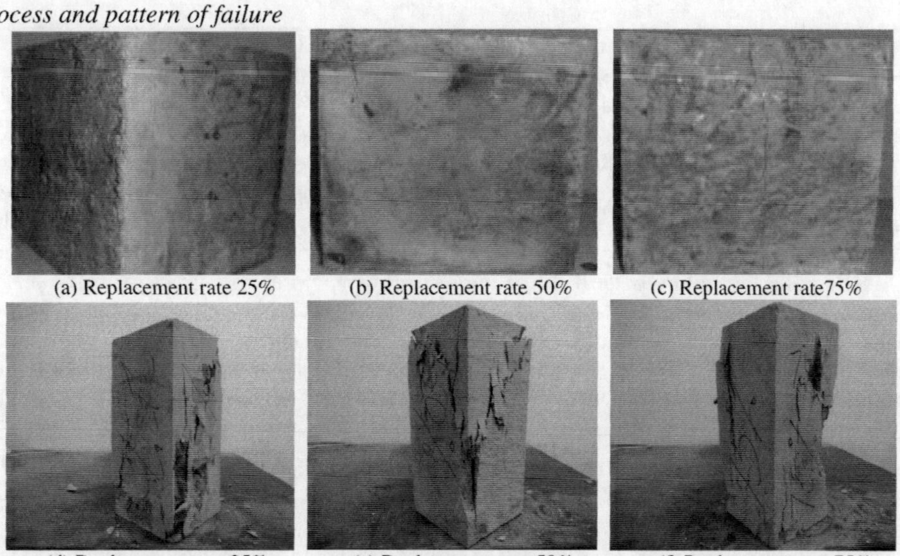

(a) Replacement rate 25% (b) Replacement rate 50% (c) Replacement rate75%

(d) Replacement rate 25% (e) Replacement rate 50% (f) Replacement rate 75%

Figure 4. Failure modes

The failure process of RLAC test specimen and ordinary concrete test specimen is approximately similar. At the initial stage of loading, there was no significantly change in the surface of the test specimens; the internal stress of test specimens increased with the increase of compression loading; as the load continues to increase, specimen's surface began to appear some smaller cracks and micro-cracks with the generating sound of faint crackling, and gradually expand, through; test specimens were failure when the load increased to the ultimate load; the failure cracks of all test specimens were basically along the vertical direction, which was inconsistent with the inclination angle of the crack of the ordinary recycled concrete[13]. But the destructive pattern of RLAC was different from ordinary concrete, the features of porous and brittle texture of recycled lightweight aggregate made the cube specimens damage along the direction parallel to the crack, and was different from the damage of the inverted cone of ordinary concrete. The failure of the prism first appeared from the surface of micro-cracks, and then began to expand and through, the test specimens were damaged due to a large area of the occurrence of massive spalling in the end. The failure patterns of some cube test specimens and prism test specimens are shown in Figure 4(a), (b), (c), (d), (e) and (f), by observing the failure of the test specimens, the damage area of RLAC is found mainly in the aggregate interface and the recycled light aggregate itself.

3.3. Compressive strength

The ultimate load of all the test specimens was measured, and test specimens were known actual bearing area. Methods provided according to the《Mechanical properties of ordinary concrete test method》(GB50081-2002) , the cube compressive strength and axial compressive strength of RLAC were calculated, as shown in Table 4.

Table 4. Test results of RLAC

Replacement rate	f_{cu}/MPa	f_c/MPa	f_c/f_{cu}
0%	37.62	25.55	0.68
25%	31.82	24.76	0.78
50%	27.72	24.53	0.88
75%	26.68	22.54	0.84
100%	20.57	21.30	1.04

Note: fcu is the cube compressive strength; fc is the axial compressive strength.

From Table 4, the cube compressive strength and axial compressive strength of RLAC meet the design strength, and the ratio of axial compressive strength to cube compressive strength of RLAC is higher than the ordinary concrete. When the replacement rate of recycled lightweight aggregate is 100%, the compressive strength of the cube test specimens are lower than the axial compressive strength, this is because the coarse aggregate of RLAC was composed entirely of the regenerated lightweight aggregate at this time; it is the lowest that the dry apparent density of RLAC with sparse and brittle materials, and then the restraining range of the bearing surface of the test specimen is reduced, which make the failure pattern of the cube specimens closes to the prism specimens (Figure 4); it is limited that the increase of cube compressive strength, coupled with the dispersion of the larger concrete itself, and then led to the occurrence of the phenomenon.

Obvious, the law of the compressive strength and the replacement rate is shown in Figure 5 and Figure6.

It can be seen from Figure 5 that the cube compressive strength of RLAC is linear with the replacement rate of recycled lightweight aggregate. The measured compressive values were fitted to the expression by using the principle of least square method:

$$y = ax + b \tag{2}$$

Where the values of a and b are 15.70 and 36.73 respectively, and the fitting precision is 0.95.

It can be seen from Figure 6 that the axial compressive strength of RLAC has a linear relationship with the replacement rate of recycled lightweight aggregate. The least square method is used to fit the expression:

$$y = ax + b \tag{3}$$

Where the values of a and b are respectively -4.29 and 25.88, and the fitting precision is 0.91.

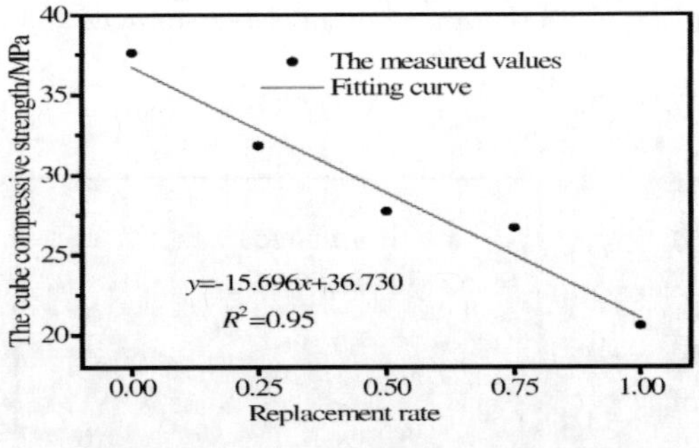

Figure 5. Relation curve between cube compressive strength and replacement rate

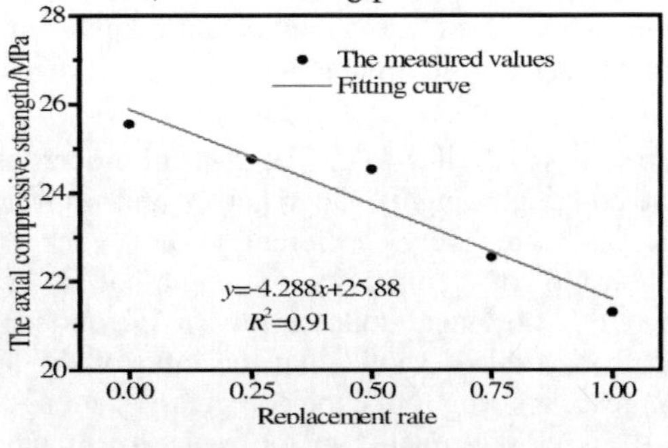

Figure 6. Relation curve between axial compressive strength and replacement rate

3.4. Influence factor analysis

Under different replacement rates, the cube specimens of recycled lightweight aggregate and the axial compressive strength are shown in Figure 7 and Figure 8.

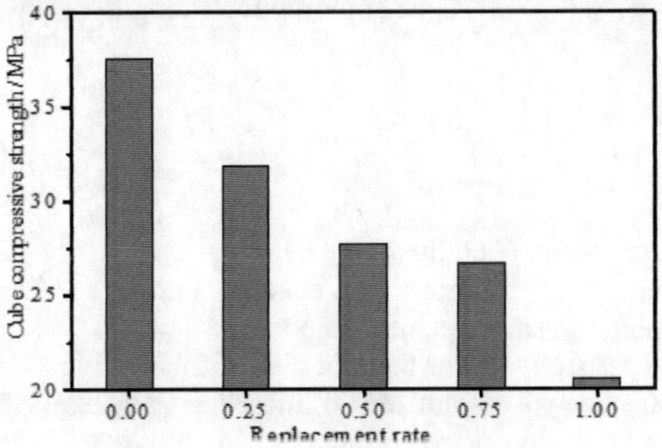

Figure 7. Cube compressive strength under different replacement rate

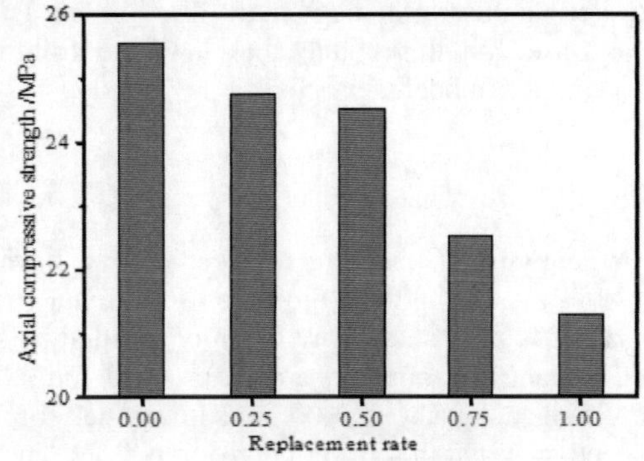

Figure 8. Axial compressive strength under different replacement rate

From Figure 7 and 8 show that the compressive strength of RLAC is lower than the compressive strength of ordinary concrete. With the replacement rate increasing, the cube compressive strength of RLAC and the axial compressive strength is decreased, which is due to the interior of recycled lightweight aggregate accumulates more cracks and micro cracks. The initial damage increase gradually with the increase of replacement rate, and the strength decreases gradually at the macro level. With the increase of replacement rate, the reduction of the compressive strength of the cube specimen is 15.42%, 12.88%, 3.75%, 22.90% respectively, the magnitude of decrease of axial compressive strength is 3.09%, 0.93%, 8.11%, 5.50% respectively.Thus, only when the replacement rate of recycled lightweight aggregate from 50% to 75%, the change of cube compressive strength is not obvious; only when the replacement rate of recycled lightweight aggregate from 25% to 50%, the change of axial compressive strength is not obvious; When the replacement rate of the recycled lightweight aggregate is changed in turn, the decrease of the cube and axial compressive strength is obvious. But when the replacement rate from 0% to 25% to 50%, the magnitude of change of axial compressive strength is far less than 5%, In order to facilitate the popularization and application of RLAC, it is suggested that the replacement rate of recycled light aggregate should be controlled within 25% in the mix proportion of recycled lightweight aggregate.

3.5. The conversion relationship of compressive strength

For ordinary concrete, the conversion relationship[14]between the axial compressive strength and the cube compressive strength is

$$f_c / f_{cu} = 0.76 \tag{4}$$

For the RLAC, due to the mechanical properties of recycled lightweight aggregate and ordinary aggregate are very different. The conversion relationship of compressive strength of ordinary concrete is no longer suitable for RLAC. As can be seen from Table 4 shows that the ratio of the axial compressive strength to cube compressive strength of RLAC is closely related to the replacement rate of recycled lightweight aggregate. In general, with the increase of the replacement rate. the ratio of axial compressive strength to cube compressive strength increases gradually. Figure 9 shows the changeable law of the ratio of axial compressive strength and cube compressive strength to the replacement rate of recycled lightweight aggregate through measured data. Based on this changeable law, the following mathematical model is proposed:

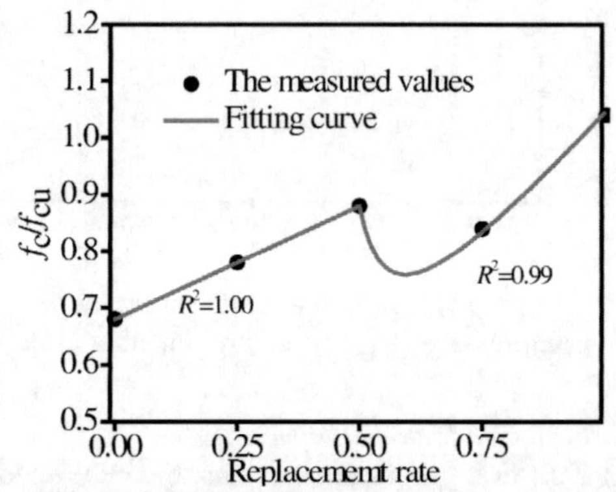

Figure 9. Relation curve between axial and cube compressive strength

$$y = \begin{cases} ax + b & 0 \le x \le 0.5 \\ x + \dfrac{1}{cx + d} & 0.5 < x \le 1 \end{cases} \tag{5}$$

Where y is f_c/f_{cu}, x is the replacement rate of recycled lightweight aggregate.

After measuring data, the ratio of the axial compressive strength to the cube compressive strength of RLAC is shown in Figure 9, which is fitted by using the least square method.

The control parameters a and b is 0.40 and 0.68 respectively in the upper half of the fitting curve, the fitting accuracy is 1.00, the lower half of the control parameters c and d is 36.5 and -15.6 respectively, and the fitting precision is 0.99. The relationship between the axial compressive strength and the cube compressive strength of RLAC is obtained by the method of statistical regression:

$$f_c = \begin{cases} (0.4x + 0.68) f_{cu} & 0 \le x \le 0.5 \\ (x + \dfrac{1}{36.5x - 15.6}) f_{cu} & 0.5 < x \le 1 \end{cases} \tag{6}$$

It can be seen in the Table 5 that the comparison between the calculated values and the measured values by using equation (6).

Table 5. The calculated value compared with the measured value

Replacement rate	Measured value f_{cu}	Measured value f_c	Calculated value f_c	The ratio of calculated value to measured value
0	37.62	25.55	25.58	1.00
25%	31.82	24.76	24.82	1.00
50%	27.72	24.53	24.39	0.99
75%	26.68	22.54	22.28	0.99
100%	20.57	21.30	21.55	1.01

It can be concluded from Table 5 that the ratio of measured value to calculated values is 1.00 about the fc, the variance is 0.00 and the coefficient of variation is 0.01, the calculated values by using equation (6) is in good agreement with the measured values.

4. Conclusions

- The ratio of axial compressive strength to cube compressive strength of RLAC is higher than the ordinary concrete;
- With the gradual increase of the replacement rate of recycled lightweight aggregate, the compressive strength of RLAC cube and axial are gradually decrease, the regularity is more obvious, and the strength is lower than the ordinary concrete;
- The conversion formula of compressive strength of ordinary concrete is not suitable for RLAC any more. Based on the measured data, it can establish the conversion relationships between compressive strength indexes of RLAC.
- In order to facilitate the popularization and application of RLAC, it is suggested that the replacement rate of recycled light aggregate should be controlled within 25% in the mix proportion of recycled lightweight aggregate.

Acknowledgments
The authors would like to thank 2017 Science and Technology Research Projects of Henan Province, 2016 Science and Technology Project of Prevention and Control of the Key Technology on Safety Production Major Accident (henan-0006-2016AQ), 2015 Jiaozuo Science and Technology Plan Project, Key Scientific Research Project of Colleges and Universities in Henan province (15A560008), Open Topic of Guangxi Key Lab of Road structure and materials(2015gxjgclkf-004), for their generous support to the research projects.

References

[1] Zhang X G, Chen Z P, Xue J Y, et al. 2012 Experimental study and mechanical behavior analysis of recycled aggregate concrete filled steel tubular long columns under axial compression *Journal of Building Structures* **33**(9) 12-20.
[2] Xiao Jianzhuang 2008 Recycled concrete Beijing:China Building Industry Press.
[3] Zhao J, Liu Q X, Lin L Q, et al. 2013 Evolution and comparision of construction waste of large cities in china *Joural of Central South University (Science and Technology)* **44**(3) 1297-1304.
[4] Zhang Xianggang, Chen Zongping, Xue Jianayng 2015 Physical mechanical performance of recycled aggregate concrete *Bulletin of the Chinese Ceramic Society* **34**(6) 1684-89.
[5] Chen Z P, Xu J J, Zheng HH 2013 Basic mechanical properties test and stress-strain constitutive relations of recycled coarse aggregate concrete *Journal of Building Materials* **16**(1) 24-32.
[6] Gu Song, Su Yan, Hu Hongjun 2015 Basic mechanical properties and development research of recycled concrete *Material and Adminicle* **4** 90-94.
[7] Monalisa B, Bhattacharyya S K, Minocha A K, et al. 2014 Recycled aggregate from C&D waste & its use in concrete–A breakthrough towards sustainability in construction sector: A review *Construction and Building Materials* **68** 501-516.
[8] Liu S H, Leng F G 2007 Recycled concrete technology Beijing:China Building Industry Press.
[9] Xiao Jianzhuang, Li Jiabin 2005 Study on relationships between strength indexes of recycled aggregate concrete *Journal of Building Materials* **8**(2) 197-201.
[10] Chai Yuanyuan, Liu Feng, Guo Shuiping 2012 Study on conversion relationships between basic mechanical properties indexes of recycled aggregate concrete *Building Technique Development* **39**(2) 41-44.
[11] Lu Shengwu, Zeng Zhixing 2013 Experimental study on the compressive strength of recycled concrete using waste ceramic as fine aggregate *Concrete* **4** 56-60.
[12] Ma Y P, Liu J J, Yang X J, et al. 2014 Effect of apparent density on the plastic shrinkage cracking and strength of the cement-based materials *Materials Reviews* **28**(1) 134-138.
[13] Xiao Jianzhuang, Li Janzhuang, Zhang Ch. 2005 Mechanical properties of recycled aggregate concrete under uniaxial loading *Cement and concrete research* **35** 1187-94.
[14] Guo Zhenhai 2013 Reinforenced concrete principle Beijing: Tsinghua University Press.

Numerical simulation study on drift and diffusion of Dalian Oil Spill

Huan Li [1], Yan Li [1, 3], Cheng Li [1], Guosong Wang [1], Shanshan Xu [1], Jun Song [1] and Song Zhang [2]

[1] National Marine Data and Information Service, Tianjin, China;
[2] National Ocean Technology Center, Tianjin, China

E-mail: ly_nmdis@163.com

Abstract. Marine oil spill has long-term harmful impact on both marine ecosystem and economics. Recently as the increase in China's rapid economic growth, the demand for energy is increasing, leading to the high risk of marine oil spill pollution. So it is essential that we improve emergency response capacity in marine oil spill pollution and develop oil spill prediction and early warning in China. In this study, based on Lagrange tracking approach, we have developed an oil spill model. Combining with high-resolution meteorological and hydrodynamic model, the oil spill model was applied to predict the drift and diffusion processes of Dalian oil spill. The predicted results are well agreed with the analyzed synthetic aperture radar (SAR) image, and provided effective oil spill behaviour prediction to Shandong Maritime Safety Administration.

1. Introduction

Marine oil spill has long-term harmful impact on both marine ecosystem and economics. In recent years, as the increase in China's rapid economic growth, the demand for energy is increasing, leading to the high risk of marine oil spill pollution. On July 16, 2010, an oil spill occurred in Dalian Xingang Port, leading to more than 1500 tons of oil spilled into the water. The spill has seriously affected local fisheries, aquaculture, tourism and shipping industry and caused huge economic losses to Dalian. Therefore, it is essential that we improve emergency response capacity in marine oil spill pollution and develop oil spill prediction and early warning in China.

Most of studies on the oil spill prediction and early warning have been focused on the development of oil spill model. It's the fact that many different oil spill models have been developed during the past 20 years [1-11] and the state of the art in oil spill modeling has been reviewed by various authors [12, 13]. The publicly available oil spill trajectory model GNOME, which was developed by the Hazardous Materials Response Division (HAZMAT) of the National Oceanic and Atmospheric Administration Office of Response and Restoration (NOAA OR&R), can be used to calculate a 'best guess' of an oil spill's trajectory and the associated uncertainty in that trajectory during marine oil spill response [14]. In addition, another sophisticated operational oil spill model systems e.g. OSIS [15], OSCAR [16], OILMAP [13], MOTHY [17], BOOM [18] and TESESO [19] have played a key role in the oil spill response for different countries. Generally, as a focused process, the transport of oil spill is affected by marine dynamic environment changes as well as the physicochemical properties of oil slicks. The marine dynamic environment information e.g. winds, water currents which can be forecasted by numerical models. Therefore, the accuracy of marine dynamic environmental forecasting is extremely essential for the oil spill prediction. In other words, it has become apparent that the prediction accuracy of oil spill model depends on the selection of meteorological or hydrodynamic model.

In this paper, based on Lagrange tracking approach, we have developed an oil spill model. Combining with high-resolution meteorological and hydrodynamic model, the oil spill model was applied to predict the drift and diffusion process of Dalian oil spill on July 16, 2010. By comparing the model results with the Radasat-2 satellite image provided by Shandong Maritime Safety Administration, the model showed good performance in oil spill behavior prediction. Details of the model are described in Section 2, and details of model application are described in Section 4.

2. Models and methods

In this paper, the meteorological and hydrodynamic models have been coupled for high efficient information exchange. The outputs from the meteorological model can provide free surface boundary conditions to drive the hydrodynamic model. Moreover, the outputs from meteorological and hydrodynamic models can provide marine dynamic environment information to the oil spill model for prediction. Figure 1 showed the workflow of the models.

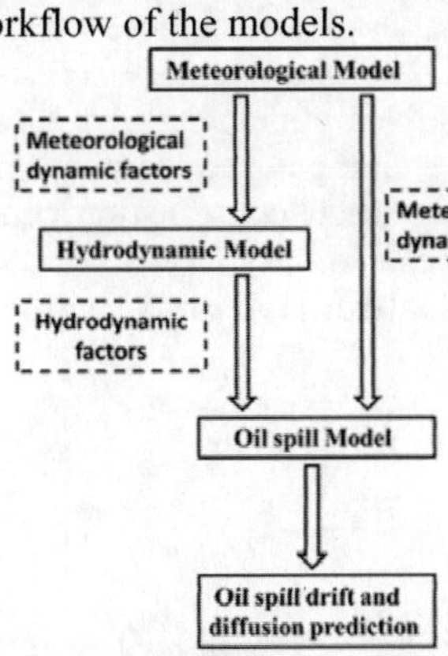

Figure 1. The workflow of the models.

2.1. Meteorological model

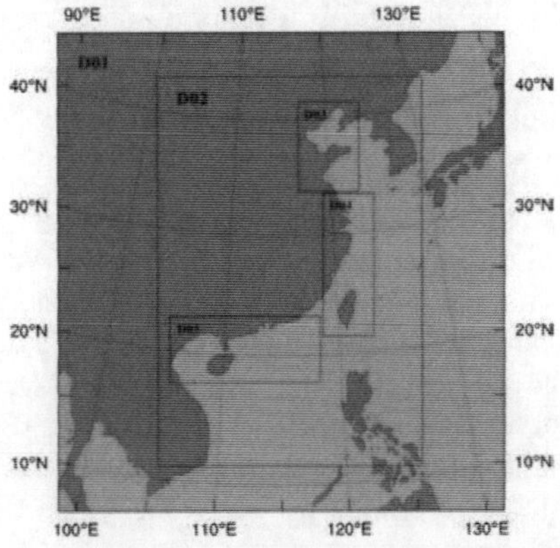

Figure 2. A triple-nested simulation scheme in WRF

In this paper, WRF (Weather Research and Forecasting Model) was selected as the meteorological forecast component. This model can accurately describe the physical processes in weather system cha--nges. In addition, due to its open source and quick update performance, WRF is widely used in domestic and international institutes and universities [20].

In this work, a triple-nested simulation scheme is adopted for high-resolution weather forecast. As shown in Figure 2, the first nested area D01 with coarse grid resolution (36km) was designed to forecast the large-scale weather phenomenon over the whole East Asia and its adjacent seas. The second nested area D02 with higher grid resolution (12km) was designed to forecast the weather phenomenon over the China Seas. Finally, the third nested areas D03, D04 as well as D05 with highest grid resolution (4km) was designed to the forecast the weather phenomenon over the key waters in the China Seas, among which, D03 is for the Bohai Sea and North Yellow Sea. Local high-resolution meteorological forecast has been realized for harbors, waterways and areas with high-risk of oil spill.

2.2. Hydrodynamic model

In this paper, FVCOM (Finite Volume Coastal Ocean Model) was selected as the hydrodynamic forecast component. As we know, the forecasted hydrodynamic information are commonly provided by structured grid ocean models based on the finite difference method. These systems are generally adequate in the open sea or shelf waters, where the geometry is regular or smooth[21]. However, for irregular and complex geometries, e.g. coastal areas, these operational tools are not sufficient to provide information concerning the fundamental variables controlling the oil spill event and to forecast the oil dispersion. FVCOM is solved numerically by a second-order accurate discrete flux calculation in the integral form of the governing equations over an unstructured triangular grid. The ability of the model in accurately solving scalar conservation equations in addition the topological flexibility provided by unstructured meshes and the simplicity of the coding structure have made this model ideally suited for many coastal and interdisciplinary scientific applications[22].

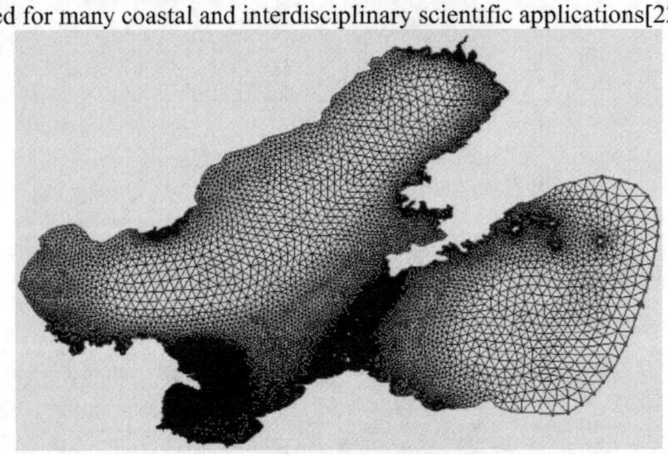

Figure 3. The grid configuration for the Bohai Sea domain

In this work, the grid resolution in most of computing areas was 10km. But in the irregular and prioritized areas, e.g. coastal areas, harbours and the adjacent waters around oil drilling platforms, the grid resolution was refined to 100m and further to 50m (in Figure 3). The refined grid could reflect the topography in the Bohai Sea. The atmospheric forcing at sea surface was provided from WRF, while the tide forcing at open boundaries consisted of tidal elevations and barotropic velocities for 8 major tide constituents, which are obtained from the Oregon State University Tidal Data Inversion Software (TPXO7.2) and global barotropic tide model[23].

2.3. Oil spill model

In this paper, the spreading empirical formula developed by Lehr et al.[24] has been adopted. In this formula, the area of oil slick can be represented as:

$$S = 1 / 4\pi MN \tag{1}$$

where S is the area of oil slick, M and N are the lengths of the minor and major ellipse axis respectively, given by:

$$M = 1.13\left[\frac{\rho_w - \rho_o}{\rho_o}\right]^{1/3} V_0^{1/3} t^{1/4} \tag{2}$$

$$N = M + 0.0034 U_{wind}^{4/3} t^{3/4} \tag{3}$$

where ρ_w is the water density, ρ_0 is the oil density, V_0 is the initial volume of oil spill, t is the time after the oil slick commences spreading (min) and U_{wind} is the wind speed. Let the concentric and similar ellipse on which the particle is located have major and minor axes n and m, with $\frac{m}{M} = \frac{n}{N}$. If the coordinates of the particle relative to the principal axes of the ellipse are (X,Y), whose x-axis is selected in the direction of the wind, here write $X = n\cos\theta$ and $Y = m\cos\theta$. Then the oil particle is displaced outwards with the same elliptical angle θ, as follows[24]:

$$\Delta X = \Delta n \cos\theta = \Delta n\left(\frac{X}{n}\right) = X\left(\frac{\Delta N}{N}\right) \tag{4}$$

$$\Delta Y = \Delta m \sin\theta = \Delta m\left(\frac{Y}{m}\right) = Y\left(\frac{\Delta M}{M}\right) \tag{5}$$

In this paper, based on Lagrange tracking approach, we have developed an oil spill model. The drift and diffusion of oil is solved by tracking a large number of oil particles equivalent to oil slicks. The position of each particle is affected by surface winds, currents and turbulent dispersion. The numerical model solves the following equation:

$$\frac{d\vec{x}}{dt} = \vec{V}_A(\vec{X}_i,t) + \vec{V}_D(\vec{X}_i,t) \tag{6}$$

where $\vec{V}_A$ is the advective velocity; $\vec{V}_D$ is the diffusive velocity; $\vec{X}_i$ is the particle position. $\vec{V}_A$ is calculated as the linear combination of current velocity and wind speed as follows:

$$\vec{V}_A = \vec{V}_{Current} + C_D \vec{V}_{Wind} \tag{7}$$

where $\vec{V}_{Current}$ is the surface current velocity; $\vec{V}_{Wind}$ is the wind speed at 10m height over sea surface;

C_D is the wind drag coefficient. The turbulent diffusive velocity is calculated by Monte Carlo sampling in the range of velocities [$-V_D$, V_D], which are proportional to the diffusion coefficients .The velocity fluctuation at each time step Δt is defined as:

$$|V_D| = \sqrt{\frac{6D}{\Delta t}} \tag{8}$$

where D is the diffusion coefficient. It is worth mentioning that there is no weathering process was considered in this oil spill model, so a more advanced version will be developed in the future.

3. Model Validation

3.1. Validation of meteorological model

In this work, the wind speed observation from the marine observation station near Dalian XingangPort were adopted to validate the meteorological model results. As shown in Figure 4, statistical analysis showed that the root mean square error (RMSE) of wind speed was less than 1.85 m/s (Figure 4a), and the RMSE of the wind direction was less than 31 degrees (Figure 4b). Both the model results and observation showed a similar variation trend and strength especially in the first 24th hours. It was concluded that the meteorological inputs were reliable.

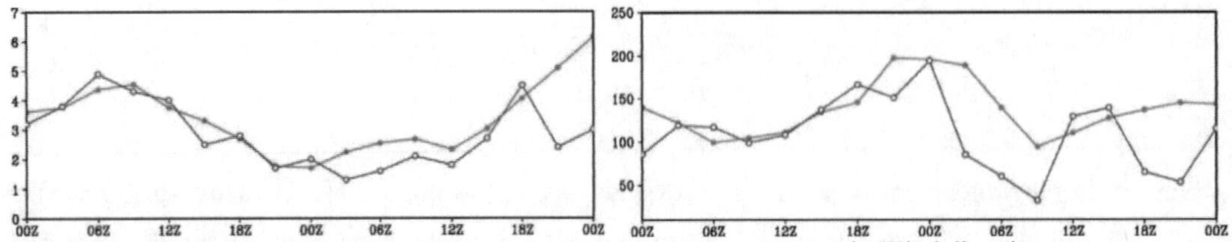

a. Wind speed b. Wind direction

Figure 4. Statistical analysis of the predicted wind and observations from the marine observation station near Dalian Xingang Port (Blue line: observation; Red line: model results)

3.2. Validation of hydrodynamic model

In this work, observations from three stations including the surface current velocities and direction as well as tidal elevations, were adopted to validate the model results. Table 1 showed the related information of each station.

Table 1. The station information

Number	Name	Longitude	Latitude
1	LTS	121° 05′	38° 27′ ′
2	CSJ	122° 44′	37° 25′
3	LQS	123° 48′	34° 02′

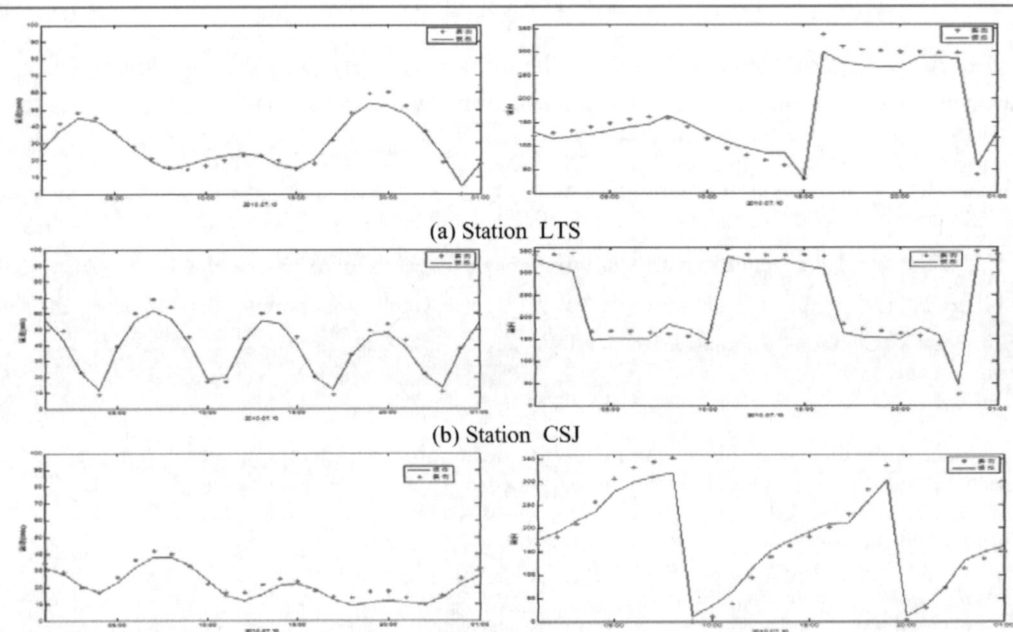

(a) Station LTS

(b) Station CSJ

(c) Station CSJ

Figure 5. The current velocities (left) and direction (right) from the model and observations at different station

(Blue solid line: observation; Red point: model outputs)

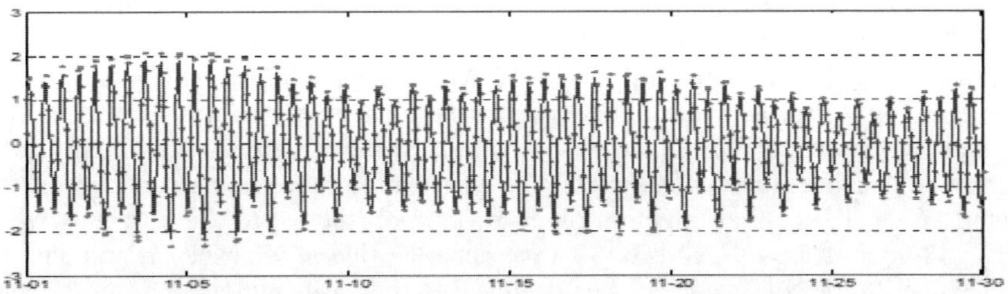

Figure 6. The tidal elevation comparison between the model outputs and observation
(Blue solid line: observation; Red point: model outputs)

As shown in Figure 5, from 0:00 July 10, 2010 to 1:00 July 11, 2010 the surface current velocities (left) and direction (right) from the numerical model (red point) were in good agreement with the observation (blue solid line) at different stations. It was obvious that the model could produce reliable prediction of the current velocity and direction.

Moreover, the simulated tidal elevations on November, 2010 were selected for the further validation. As shown in Figure 6, the elevation from the model outputs (blue solid line) were also in good agreement with observation (red point).

3.3. Surface water currents

It is showed the simulated surface water currents forced by winds and tides near Dalian at different times in Figure 7. Obviously, the essential formation of water current is reversing currents, and the direction changes of water current occur at the time of high or low tide. The maximum current velocity is rapidly generated when the rate of tidal level rising or ebb reached maximum.

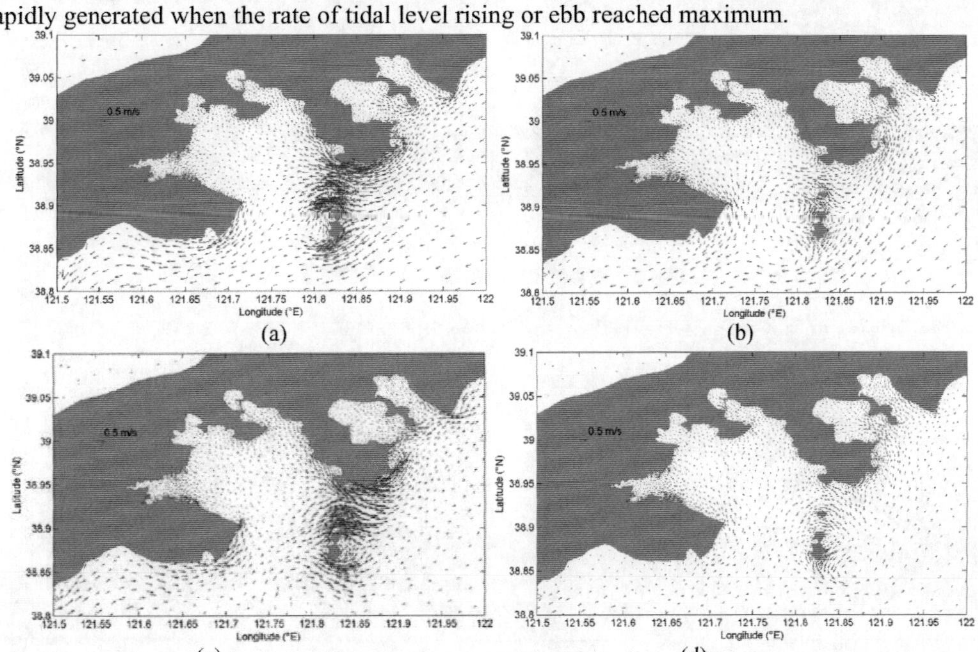

Figure 7. The simulated surface water currents near Dalian at different times

4. Model application

In this paper, the oil spill model was used to predict the oil drift and diffusion after Dalian oil spill occurred. And the prediction were compared with the Radasat-2 synthetic aperture radar (SAR) image.

SAR data have a significant role in remote sensing-based slick detection. the visibility concept of the oil slick in a SAR image is a function of the reduced radar backscatter from the surface[25, 26]. Recently, some new approaches for utilizing RADARSAT data for oil slick detection have been explored and effectively improve the emergency response capacity in marine oil spill[26, 27].

Figure 8 showed the drift and diffusion of spilled oil at different times. After the oil spill occurred, under the impact of surface currents and winds, the oil began to move to the southwest (Figure 8 (a)). 12 hours later, due to the current direction reversion, the oil moved to the northeast with diffusion (Figure 8 (b)). 18 hours later, the oil drifted by the northern side of Xiaoshan Island and entered into Dalian Bay (Figure 8 (c)).28 hours later, the oil moved to the north and entered into Dayao Bay with oil-shoreline interaction (Figure 8 (d)). 48 hours later, as shown in Figure 8 (e), a large amount of oil passed through Sanshan Channel and began to invade Dalian Bay. Meanwhile, the oil reached to the coast of Dashan and Sanshan Island.72 hours later, shown in Figure 8 (f), after long time of drift and diffusion, the oil pollution distributed throughout Dayao Bay, Xiaoyao Bay and the waters near Dalian Bay.

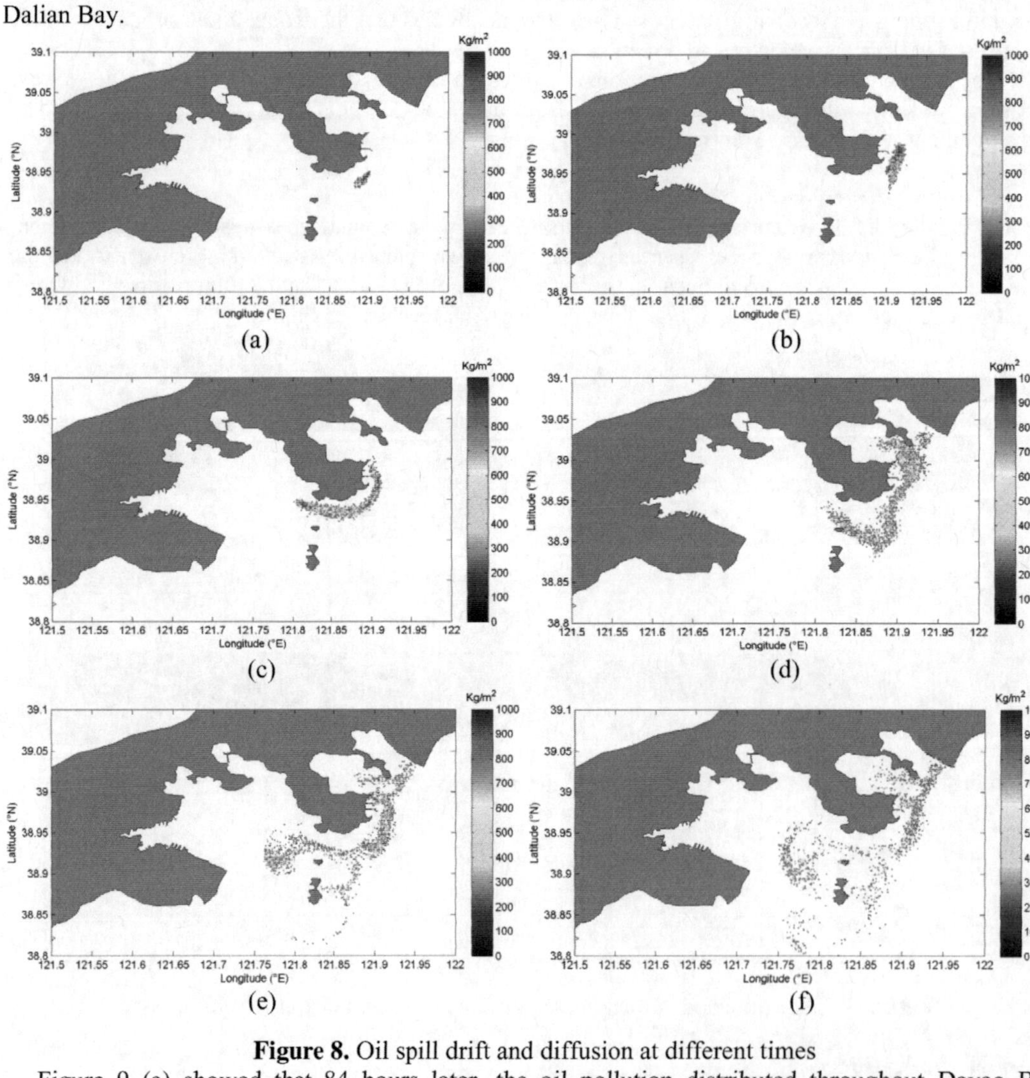

Figure 8. Oil spill drift and diffusion at different times

Figure 9 (a) showed that 84 hours later, the oil pollution distributed throughout Dayao Bay, Xiaoyao Bay and the waters near Dalian Bay, and the calculated pollution area rearched $184 km^2$. In

Figure 9 (b), the Radasat-2 SAR image provided by Shandong Maritime Safety Administration showed the extracted oil spill distribution near Dalian waters and the calculated pollution area was 195km^2. The comparison displayed that the prediction results are well agreed with the satellite observation.

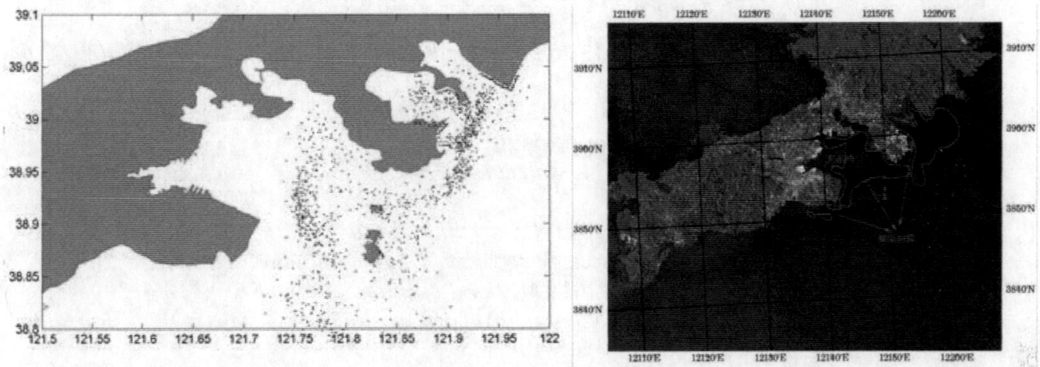

(a) (b)

Figure 9. The predicted oil spill distribution after 84 hours (a) and the Radasat-2 SAR image (b)

5. Conclusions and outlook

In this paper, based on Lagrange tracking approach, we have developed an oil spill model. Combining with high-resolution meteorological and hydrodynamic model, the oil spill model was applied to predict the drift and diffusion processes of Dalian oil spill. The prediction results showed after 72 hours of drift and diffusion, the oil pollution distributed throughout Dayao Bay, Xiaoyao Bay and the waters near Dalian Bay. The prediction is well agreed with the analyzed SAR image, and provided effective oil spill behavior information to Shandong Maritime Safety Administration. However, in this oil spill model no weathering equation has been coupled, so we will develop a more advanced version in the future.

Acknowledgements

This study was jointly supported by the program of National Key Technology Research and Development Program of the Ministry of Science and Technology of China (No.2014BAB12B02) and Tianjin Key Technology Research and Development Program (No.14ZCZDSF00012).We would like to express our gratitude to Shandong Maritime Safety Administration who provided the analyzed Radasat-2 SAR image to this study.

References

[1] Aamo O M, Reed M, Downing K 1995 *Proceedings of the 1995 Oil Spill Conference,* American Petroleum Institute pp 429-438.
[2] Al Rabeh A H, Lardner R W, Gunay N, et al. 1995 *Fourth Saudi Engineering Conference*, King Abdul-Aziz University Press, Jeddah, pp 415-427.
[3] Reed M, French D, Rines H, et al. 1995 *Proceedings of the 1995 Oil Spill Conference*, American Petroleum Institute pp 61-66.
[4] Al Rabeh A H, Lardner R W, Gunay N 2000 *Environmental Modelling and Software* **15** 425-442.
[5] Lehr W J, Jones R, Evans M, et al. 2002 *Environmental Modelling and Software* **17** 191-199.
[6] Chao X, Jothi Shankar N, Wang S Y 2003 *Journal of Hydraulic Engineering* **129** 495-503.
[7] French-McCay D P 2004 *Environmental Toxicology and Chemistry* **23** 2441-56.
[8] Carracedo P, Torres-Lopez S, Barreiro M, et al. 2006 *Marine Pollution Bulletin* **53** 350-360.

[9] Ambjorn C 2007 *Engineering and Management* **3** 60-66.
[10] Guo W J, Wang Y X 2009 *Marine Pollution Bulletin* **58** 726-734.
[11] Wang J H, Shen Y M 2010 *Science China Technological Series* **53** 2423-34.
[12] ASCE 1996 *Journal of Hydraulic Engineering* **122** 594–609.
[13] De Dominicis M, Pinardi N, Odiatis G, et al. 2013 *Geoscientific Model Development* **6(1)** 1851-69.
[14] Yang Y, Chen Z L, Ying L, et al. 2013 *Applied Mechanics and Materials*, 295/296/297/298: 1535-1542.
[15] Leech M, Walker M, Wiltshire M, et al. 1993 *Proceedings of 16th Arctic and Marine Oil Spill Program, Canada:Technical Seminar* pp 549-572.
[16] Reed M, Aamo O M, Daling P S 2011 *Molecular Plant-Microbe Interactions: MPMI* **14(5)** 639-652.
[17] Daniel P, Marty F, Josse P, et al. 2003 *Proceedings of the 2003 International Oil Spill Conference, Washington, DC*: American Petroleum Institute pp 1-6.
[18] Cucco A, Sinerchia M, Ribotti A, et al. 2012 *Marine Pollution Bulletin* **64(6)** 1186-1200.
[19] Abascal A J, Castanedo S, Fernandez V, et al. 2011 *Proceedings of Oceans 2011 IEEE-Spain, Santander:IEEE* pp 1-8.
[20] Michalakes J 1998 World Scientific, River Edge, New Jersey, pp 117-124.
[21] Chen C, Huang H, Robert R C, et al. 2007 *Journal of Geophysical Research Oceans* **112(C3)** 83-87.
[22] Chen C, Robert R C and Cowles G 2006 *Special Issue entitled "Advance in Computational Oceanography", Oceanography* **19(1)** 78-89.
[23] Egbert G D, Erofeeva S Y 2002 *Journal of Atmospheric Oceanic Technology* **19** 183-204.
[24] Lehr W J, Fraga R J, Belen M S, et al. 1984 *Marine Pollution Bulletin* **15** 326-329.
[25] Bern T I, Whal T, Anderssen T, et al. 1993 *Photogramm Eng Remote Sensing* **59** 423-428.
[26] Marghany M 2004 *Environmental Modelling & Software* **19(5)** 473-483.
[27] Marghany M 2014 *Marine pollution bulletin* **89** 20-29.

Conversion of carbon dioxide to carbon monoxide by pulse dielectric barrier discharge plasma

Taobo Wang[1], Hongxia Liu[1, 3], Xiang Xiong[2] and Xinxin Feng[1]

[1] Department of Environmental Science and Engineering, Xi'an Jiaotong University, Xi'an 710049, China.

[2] School of Chemistry and Molecular Engineering, East China Normal University, Shanghai 200241, China

E-mail: hxliu72@xjtu.edu.cn

Abstract: The conversion of carbon dioxide (CO_2) to carbon monoxide (CO) was investigated in a non-thermal plasma dielectric barrier discharge (DBD) reactor, and the effects of different process conditions on the CO_2 conversion were investigated. The results showed that the increase of input power could optimize the conversion of CO_2 to CO. The CO_2 conversion and CO yield were negatively correlated with the gas flow rate, but there was an optimum gas flow rate, that made the CO selectivity best. The carrier gas (N_2, Ar) was conducive to the conversion of CO_2, and the effect of N_2 as carrier gas was better than Ar. The conversion of CO_2 to CO was enhanced by addition of the catalyst (5A molecular sieve).

1. Introduction

The International Energy Agency states that the concentration of CO_2, the main greenhouse gas (GHG), has been increasing significantly in the atmosphere over the past century, while the impact of GHG on climate change is now admitted[1-4]. As an alternative to catalytic dry reforming reaction, non-thermal plasma appears as a promising technology, since highly energetic electrons generated by plasma at room temperature are able to initiate chemical processes[5-7]. Plasma-catalysis has been widely investigated for gaseous pollutant abatement[8,9] and dry reforming of methane[10-13]. The conversion of CO_2 using the plasma method can not only reduce its greenhouse effect, but also produce CO and other chemicals[14]. Up to now, different plasma sources have been used for CO_2 conversion, including DBD[15], corona discharge[16], glow discharge[17], microwave discharge[18], and gliding arc discharge[19]. Advantages of pulsed DBD treatment are: high destruction efficiency, no demands on temperature and pressure, insensitive to contamination, and no damage from high loads, simply installed, compact, little service, small scale. However, the plasma CO_2 dissociation process would be attractive only under specific conditions, which is, in the scenario of cheap[20] and clean (low carbon emissions) electricity production.

In this paper, in order to optimize the operating conditions for the conversion of CO_2 using DBD plasma reactor, we investigated the effects of discharge parameters, gas flow rate and different diluent gases. The CO_2 dissociation with or without molecular sieves type 5A as catalyst filled in the discharge area of the reactor was also investigated. The conversion of CO_2, as well as the yield efficiency and selectivity of CO, will be discussed with respect to these parameters on the plasma conversion of CO_2. Additionally, spectrum diagnosis studies have provided insight into the mechanism of CO_2 activation and the role of the catalyst in the reaction.

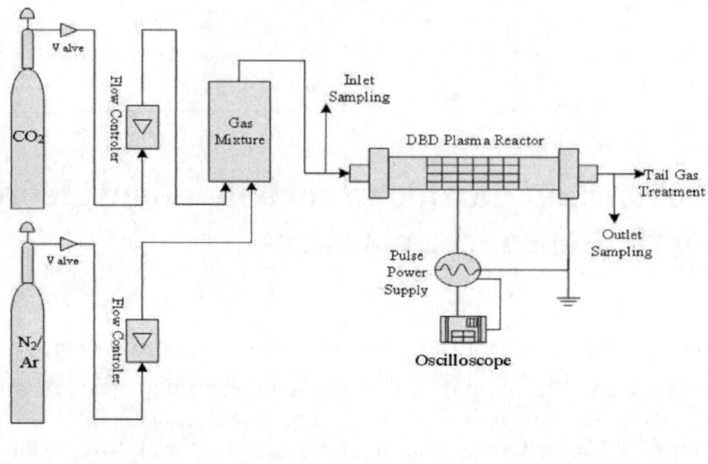

Figure 1. Schematic diagram of the experimental setup

2. Experimental

Figure 1 shows the main components of the experimental equipment including a DBD reactor and a high voltage pulse generator. The homemade DBD reactor is a classical coaxial dielectric barrier discharge plasma type. A stainless steel wire mesh was wrapped around the outside of a corundum tube, with an external diameter of 25 mm and an inner diameter of 20 mm, and acted as a high voltage electrode. A stainless steel rod with an outer diameter of 10 mm was placed in the center of the corundum tube and used as a low-field electrode. The length of the discharge region and after glow region is 120 mm with a discharge gap fixed at 2 mm. The DBD reactor was supplied by a high voltage pulse power supply with a peak-to-peak voltage of 30 kV and a pulse frequency of 0-1000 Hz. The voltage across the external capacitor (47 nF) was measured to determine the charge passing through the DBD. All the electrical signals were sampled by a two-channel digital oscilloscope (DS1102E).

To obtain the desired concentration, the CO_2 and diluent gas flow rates were controlled by independent flow control valves. The flow rate was also controlled by the same specification of flow control valves. For the experiment in the reaction of CO_2 and diluent gases, the CO_2 was mixed with N_2 or Ar. The mixed gases stream then entered the DBD reactor. For each experiment, the conversion of CO_2 (X_{CO2}) was defined as

$$X_{CO_2} = (n_{\text{in. }CO_2} - n_{\text{out. }CO_2}) / n_{\text{in. }CO_2} \tag{1}$$

the yield of CO (Y_{CO}) was defined as

$$Y_{CO} = n_{\text{out. }CO} / n_{\text{in. }CO_2} \tag{2}$$

and the selectivity of CO (S_{CO}) was defined as

$$S_{CO} = n_{\text{out. }CO} / (n_{\text{in. }CO_2} - n_{\text{out. }CO_2}) \tag{3}$$

where n is the total amount of molecules (mol). The active particles and reaction intermediates were analyzed by fiber-optic spectrometer (AvaSpec-2048FT-4-DT)[21]. The catalyst samples were measured by Fourier Transform Infrared Spectrometer (FTIR)[22].

3. Results and discussion

3.1. Effects of input power

Increasing the input power by only changing the applied voltage does not change the average electric field of the plasma, since the gas voltage and breakdown voltage of the CO_2 in the DBD reactor is almost constant with the increase in the input power[23] (calculated from the Lissajous figure[24]). The conversion of CO_2, yield and selectivity of CO at 150 ml/min flow rate and 700 Hz pulse frequency power supply as a function of input power were investigated for the pure CO_2 (black lines in Figure 2) and for CO_2 mixed with either Ar (red lines) or N_2 (blue lines) in a 1:1 molar ratio. Figure 2 (a) (b) shows that when the input power is increased, the conversion of CO_2 and yield of CO increase. The conversion of CO_2 and yield of CO are higher in the presence of diluent gases and N_2 is more conducive to enhancing CO_2 conversion rate compared to Ar.

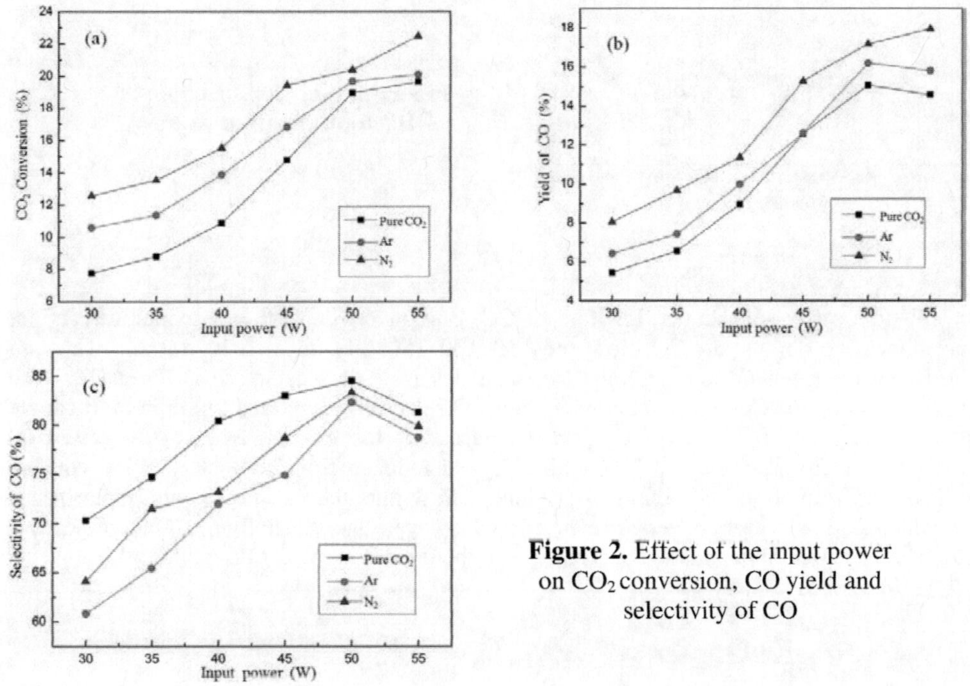

Figure 2. Effect of the input power on CO_2 conversion, CO yield and selectivity of CO

This effect may be attributed to the fact that electrical energy can produce electrons with much higher average kinetic energies than the surrounding gas-phase ions and molecules. These energetic electrons can interact with the CO_2 gas to produce highly reactive species (i.e., radicals, anions, cations, and secondary electrons) that will preferentially destroy CO_2[25]. With the increase in the input power, the density of high-energy electrons increases and the electron energy distribution function varies. Thus, the opportunity for collisions between CO_2 and reactive species is increased. This assumption can explain the increase in CO_2 conversion and CO production with the increase in the input power. However, the selectivity of CO increases to a maximum at 50W and then decreases with the increasing of input power, as we can see from Figure 2 (c). This result could due to CO_2 and generated CO being partially ionized to generate carbon in the reactor with the increase of input power. The results of the emission spectrum showed that the reaction produced CO_2^+ (290 nm), C (312 nm), O (800–820 nm), CO (561 nm), C_2 (512 nm) and other particles in the presence of N_2 and plotted in Figure 3.Because conversion of CO_2 in a DBD reactor generally follow two reaction pathways involving electron impact[26],

$$CO_2 + e^- \rightarrow CO + O + e \quad (4)$$

$$CO_2 + e^- \rightarrow C + O_2 + e \quad (5)$$

The formation of solid carbon on the corundum tube wall and the inner surface of electrode can also result from carbon monoxide or carbon dioxide dissociation[5] (Eqs. (5) and (6)):

$$CO + e^{-} \rightarrow C + O + e \tag{6}$$

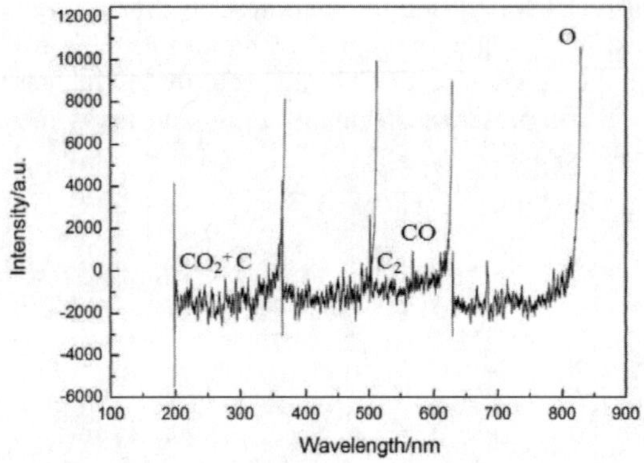

Figure 3. Emission spectrum of DBD for CO_2 mixed with N_2

3.2. Effects of gas flow rate

In order to probe the effects of the gas flow rate of CO_2 on conversion of the dissociation reaction, data were taken for six different pure CO_2 flow rates: 50, 100, 150, 200, 250 and 300 ml/min. The data were obtained with input power 50W and a pulse frequency of 700Hz using the DBD reactor. The effect of gas flow rate on the CO_2 conversion, yield and CO selectivity is plotted in Figure 4. It can be shown that the conversion of CO_2 and CO yield decrease with the increase of gas flow rate. CO_2 conversion reduced from 23.31% to 13.11% and CO yield reduced from 16.48% to 8.2% with an increase in flow rate from 50 to 300 ml/min. With higher flow rate, the mean residential time of gas was smaller, which leads to lower conversion efficiency. However, the selectivity of CO increases to a maximum 79.51% at 150 ml/min and then decreases with the increasing of gas flow rate.

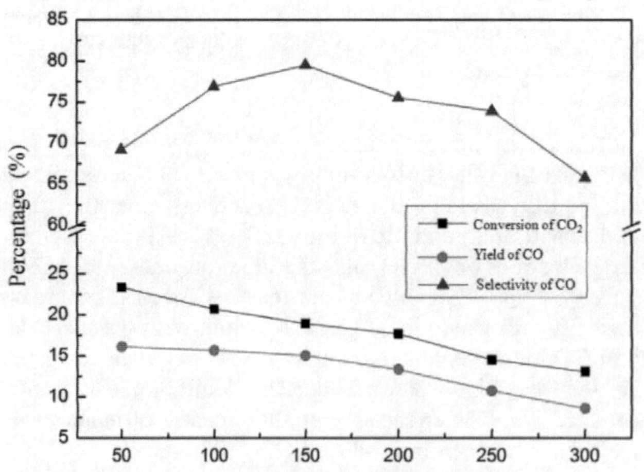

Figure 4. Effect of flow rate on CO_2 conversion, CO yield and selectivity of CO

The results may be explained as: the increase of flow rate having a negative effect on the main reaction and side reaction respectively. With higher flow rate, the mean residential time of gas was smaller, which leads to lower conversion efficiency[27]. When the flow rate is less than 150 ml/min, the experiment is dominated by the main reaction, which generates CO. As the flow rate increases above 150 ml/min, the side reaction, which generates carbon, becomes more dominant.

3.3. Effect of carrier gas

Figure 5 shows the CO_2 conversion, yield and selectivity of CO of the plasma processing of CO_2 with different carrier gases at an input power of 50 W and a total flow rate of 200 ml/min. It is evident that when carrier gas concentration increases, the conversion of CO_2 becomes gradually higher. With the increase in the N_2 and Ar concentration from 0 to 50%, the CO_2 conversion increased to 21.32% and 19.37%, respectively. When the ratio of the carrier gas and CO_2 is less than 50%, the CO yield with Ar is slightly lower than the yield without it. However, when increasing N_2 reaches a certain percentage, the yield of CO will be higher than the yield without N_2.

S. L. Brock[28] et al. showed that the carrier gas is more easily activated than CO_2. The use of diluent gases has a function of optimizing discharge conditions so that CO_2 molecules are more easily converted to active oxygen atoms and CO. Besides, adding a carrier gas leads to the reduction of the absolute flow rate of CO_2, so the opportunity for collisions between CO_2 and energetic electrons is increased at the same input power. It is interesting to note that N_2 as the diatomic molecule has a larger collision area than Ar, which is a monoatomic molecule, so an N_2 activated molecule may have a greater collision probability with CO_2 molecules and thus promote the conversion of CO_2 and yield of CO. The reason for the decrease of CO yield with Ar carrier gas may be due to a change in the discharge status in the reactor.

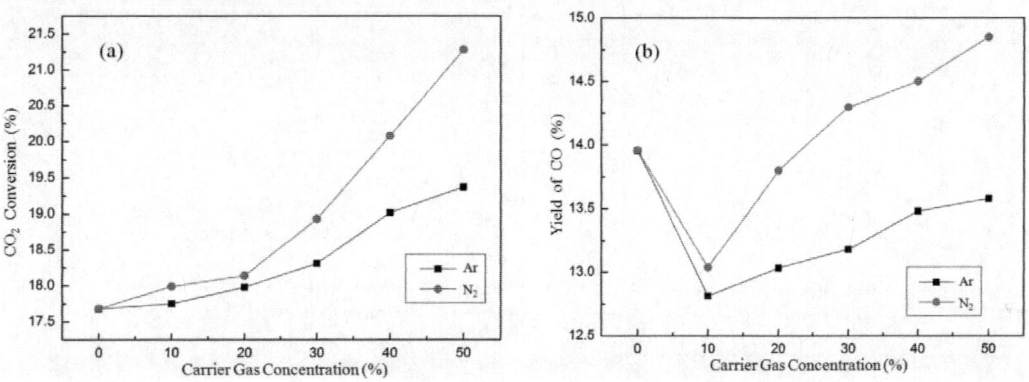

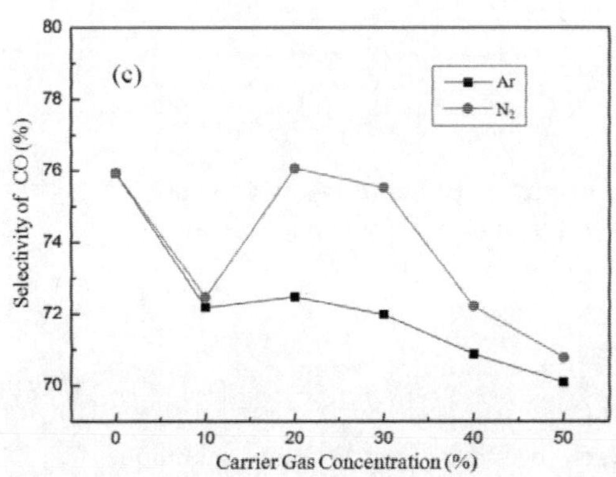

Figure 5. Effect of carrier gas on CO_2 conversion, CO yield and selectivity of CO

As can be seen from Figure 5 (c), it shows a maximum selectivity of CO at 20% carrier gas concentration. After the addition of the carrier gas, the active substance increased to promote the conversion of CO_2 to CO in the reaction system. However, when the carrier gas is added excessively, a large number of active particles will collide with CO, which leads to further conversion of CO molecules.

3.4. *Effect of catalyst*

Figure 6 shows the conversion of CO_2 by high-voltage pulse discharges combined with 3 g of 5A molecular sieve at a pure CO_2 flow rate of 150 ml/min and a pulse frequency of 700 Hz in the DBD reactor. It can be observed that the conversion of CO_2 is apparently enhanced in the presence of catalyst with the increase in the input power. Although the yield of CO also increased, the increment is small. Compared with no packing catalyst, CO selectivity decreased. Also, in both cases selectivity increases first and then decreases with the increasing input power.

These apparent synergistic effects may be attributed to the following reasons: (1) In the DBD reactor packed with molecular sieve, the high frequency of surface collisions involving reactive and or energetic species may suggest that surface mediated processes could also affect the overall conversion efficiency. (2) The high dielectric constant of the catalyst serves to enhance the electric field in the interstitial regions throughout the entire reactor volume. During the course of a voltage cycle, irregular regions around the molecular sieve pass through the Paschen limit and a corona forms. The molecular sieve in effect spreads the discharges that are dispersed from the central electrode to the molecular sieve, ensuring that the entire gas flow is exposed to the electric field. (3) The 5A molecular sieve shows a significant interaction with CO. Molecular sieves have a certain degree of CO adsorption.

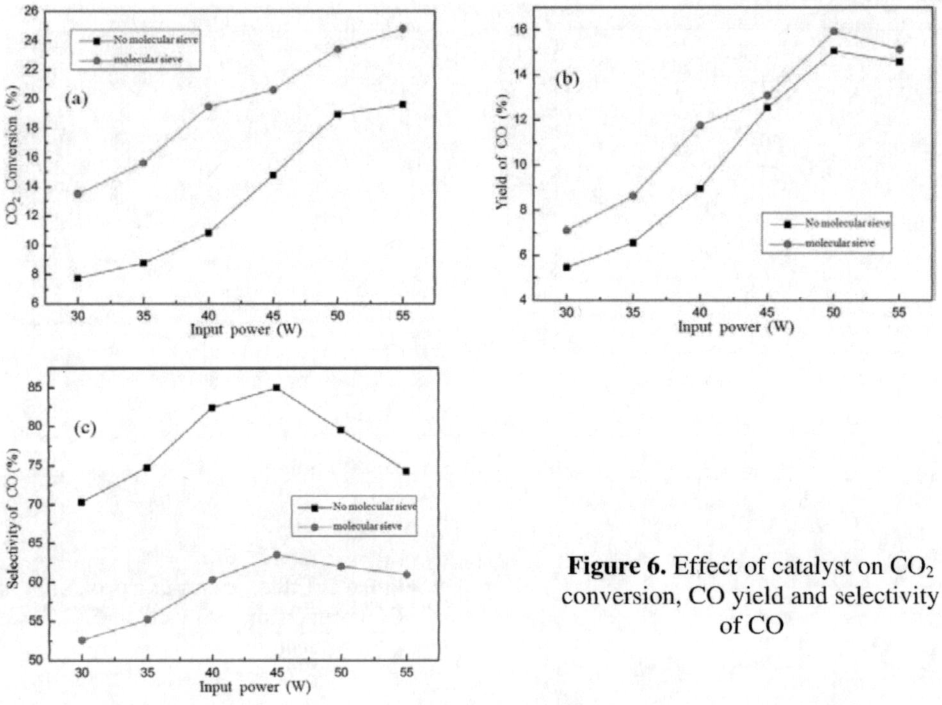

Figure 6. Effect of catalyst on CO_2 conversion, CO yield and selectivity of CO

4. Conclusion

Plasma conversion of CO_2 may prove to be an effective and energy efficient technique for CO_2 remediation if the parameters governing the plasmochemical reactions can be optimized. In this study, the effects of the key plasma process parameters (input power, gas flow rate, carrier gas and catalyst) and their interactions on plasma conversion of CO_2 in a coaxial DBD reactor has been investigated. The results show that the conversion of CO_2 increases with increasing input power and added carrier gas, but decreases with the increase of gas flow rate. The yield of CO was negatively correlated with the gas flow rate, but there was an optimum value, which made the CO selectivity best. The carrier gas (N_2, Ar) was conducive to the conversion of CO_2, and the effect of N_2 as carrier gas was better than Ar. The CO_2 conversion and CO yield were enhanced by addition of molecular sieve. In conclusion, this study demonstrates the feasibility of utilizing pulsed dielectric barrier discharges combined with different process parameters to convert CO_2 to CO. More studies are underway to further improve the efficiency of plasma CO_2 conversion.

Acknowledgements:
This work was supported financially by the Fundamental Research Funds for the Central Universities (NO. xjj2015132) and National Natural Science Foundation of China (NO. 21077084). Additionally, we thank our colleagues involved in the Lab of the Non-Thermal Plasma Technology and Lab of Air Pollution Control & Recycling for support of this research.

References

[1] IEA, CO_2 emissions from fuel combustion, Highlights. 2014 ed.
[2] Chen G X, Georgieva V, Godfroid T, Snyders R, Delplancke-Ogletree M P 2016 *Appl. Catal. B: Environmental.* **190** 115-124.
[3] Hu B, Guild C, Suib S L 2013 *J. CO_2 Util.* **1** 18–27.
[4] Hussaina M, Akhter P, Saracco G, Russo N 2015 *Appl. Catal.* ***B*** 170–171 53–65.
[5] Yap D, Tatibouët J M, Batiot-Dupeyrat C 2015 *J. CO_2 Util.* **12** 54.
[6] Yu Q, Kong M, Liu T, Fei J, Zheng X 2012 *Plasma Chem. Plasma Process* **32** 153–163.
[7] Paulussen S, Verheyde B, Xin T, Bie C D, Martens T, Petrovic D, Bogaerts A, Sels B 2010 *Plasma Sources Sci. Technol.* **19** 034015.
[8] Chen H L , Lee H M, Chen S H, Chao Y, Chang M B 2008 *Appl. Catal. B* **85** 1–9.
[9] Van Durme J, Dewulf J, Leys C, Van Langenhove H 2008 *Appl. Catal. B* **78** 324–333.
[10] Heintze M, Pietruszka B 2004 *Catal. Today.* **89** 21–25.
[11] Long H, Shang S, Tao X, Yin Y, Dai X 2008 *Int. J. Hydrogen Energy* **33** 5510–5515.
[12] Zhang A J, Zhu A M , Guo J, Xu Y, Shi C 2010 *Chem. Eng. J.* **156** 601–606.
[13] Tu X, Whitehead J C 2012 *Appl. Catal. B.* **125** 439–448.
[14] Liu C, Xu G, Wang T 1999 *Fuel Process. Technol.* **58** 119–134.
[15] Ramakers M, Michielsen I, Aerts R, Meynen V, Bogaerts A 2015 *Plasma Process. Polym.* **12** 755.
[16] Horvath G , Skalny J D, Mason N J 2008 *J. Phys. D.* **41** 225207.
[17] Yamamoto A, Mori S, Suzuki M 2007 *Thin Solid Films.* **515** 4296.
[18] Spencer L F, Gallimore A D 2013 *Plasma Sources Sci. Technol.* **22** 015019.
[19] Nunnally T, Gutsol K, Rabinovich A, Fridman A, Gutsol A, Kemoun A 2011 *J. Phys. D Appl. Phys.* **44** 274009.
[20] Samukawa S, Hori M, Rauf S, Tachibana K, Bruggeman P, Kroesen G, Whitehead J C, Murphy A B, Gutsol A F, Starikovskaia 2012 *S. J. Phys. D: Appl. Phys.* **45** 253001.
[21] Imani H, Sarkhani S 2015 Optik - International Journal for Light and Electron Optics, **126** (15-16):1419–1422.
[22] O'Brien P G, Sandhel A Wood T E, et al. 2014 *Advanced Science* **1** **(1)**.
[23] Mei D H, He Y L, Liu S Y, Yan J, Tu X 2016 *Plasma Process and Polymers* **13** 550.
[24] Tu X, Gallon H J, Twigg M V, Gorry P A, Whitehead J C 2011 *J. Phys. D* **44** 274007.
[25] Wen Y Z, Jiang X Z, Liu W P 2002 *Journal of fuel chemistry and technology* **3** 193–197.
[26] Brock S L , Marquez M, Suib S L , et al. 1998 *J Catal* **180** **(2)** 225–233.
[27] Wen Y Z, Jiang X Z 2001 *Plasma Chem. Plasma Process* **21** 665–678.
[28] Brock S L, Shimojo T, Marquez M, Marun C, Suib S L, Matsumoto H, Hayashi Y 1999 *Journal of Catalysis* **184** 123–133.

Study on the preparation of straw activated carbon and its phenol adsorption properties

Liping Chen

College of Chemistry and Environment Science, Inner Mongolia Normal University, Hohhot 010022, China

E-mail: clp@imnu.edu.cn

Abstract. Using sunflower straw as raw materials to manufacture straw activated carbon-modified by phosphoric acidand adsorption isotherm of phenol on straw activated carbon was studied in a batch reactor. The physical properties of the prepared straw activated carbons were characterized by scanning electron microscopy. The effect of various parameters, adsorbent dose, pH and temperature, were studied on optimum conditions. The results have shown that the absorbent was efficient, the removal ratio of phenol up to 99.36% with an adsorbent dosage of 16 g·L-1, a pH of 6.0-8.0, at 25 ℃. The experimental adsorption data fitted reasonably well to the Langmuir isotherm, the maximum adsorption capacity was 109.89 mg/g. The process of adsorption is a exothermic process.

1. Introduction

Phenol is a highly toxic organic pollutant, which is harmful to environment and health. Phenol can be absorbed by skin, the digestion tract and the respiratory tract so that short-term exposure of phenol can cause central nervous system, kidney, cardiac muscle, blood and other organ system more acute damage. And phenol also has chronic toxicity for algae, fish and shrimps, which can inhibit the growth of the aquatic life, and even lead to death. Phenol has been registered as priority pollutants in the water quality standards for drinking water by China and Europe and the United States and other countries.

Several methods have been developed to remove phenol from wastewater, main including microbial degradation [1], chemical oxidation [2], photocatalytic degradation [3], membrane separation [4], solvent extraction [5] and adsorption [6]. Yet, still the adsorption technique using activated carbon in dealing with wastewater is the most favorable method due to high adsorption capacity, huge BET surface area, big pore structure and stable chemical properties. At present, activated carbon adsorption method has good removal effect, high adsorption capacity and low operational cost. Commercial activated carbon is not suitable for promotion because of its high cost. So it is necessary to develop activated carbon raw material of low cost and easy acquisition for mass production.

Sunflower straw is a kind of a large amount of waste agricultural production in China, the sunflower has been planted for 1720 km^2 each year only in Hetao plain, Inner Mongolia, China, and its straw production has reached to one million tons each year. Using sunflower straw as raw material to prepare activated carbon not only wide prospect of straw development but also offers low-cost new activated carbon raw material. The paper is about using the sunflower straw of low cost agricultural production to make the straws activated carbon with the activator of phosphoric acid. Researches have been done for the adsorptive performances to solve phenol pollution problems of wastewater and provided basic data for research and application of taking sunflower straw as raw material to make activated carbon.

2. Materials and methods

2.1. Preparation and characterization of carbon

The sunflower straw is from Hetao plain, Inner Mongolia, China. The surface dust of sunflower straw was cleaned and then cut into about 2 cm short sections, washed by tap water again, rinsed twice or three times by distilled water at last, baked to dry in the temperature 120 ℃ in oven, powdered, screened by 20 mesh sieve. The samples were immersed in phosphoric acid solution about 2 h with the ratio 1.5:1 (weight) of phosphoric acid to sunflower straw, baked to dry in the temperature 110 ℃ in oven, put into tube type furnace, heated under the protection of nitrogen flow to rise temperature with 10 ℃/min until to 300 ℃, activated for 90 min in this constant temperature, again immersed in 5% HCl for 8 h, washed to about 7 of water's pH value with distilled water, dried in 110 ℃, powdered, screened by 60 mesh sieve, weighed to calculate recovery. In this experiment, dry sunflower straw is 100 g, powder of activated carbon manufactured is 54.2 g, recovery is about 54.2%.

2.2. Measuring adsorbent properties

The morphology of the adsorbents was observed by scanning electron microscopy (Hitachi S-3400 series.). The powder of adsorbents were dispersed and coated on the support film of tape, then used to be determined with test voltage of 20 kV and beam of 5×10^{-9} mA.

2.3. Batch Adsorption Studies

Some amounts of adsorbent had accurately been weighted and put into 250 mL conical flask, then 50 mL simulated wastewater containing phenol was added into adsorbent and oscillated for 24 h in the thermostatic water bath oscillators, then taken out, filtered by 0.145 μm microfiltration membrane. The phenol concentration was then measured by UV spectrophotometer (DR-5000, Germany). These three experiments were carried out in duplicates and the average values were reported. The blank experiments' results showed that the total concentration of phenol changed less than 1%, so the effects of these operating processes can be omitted.

The equilibrium adsorption capacity and adsorption efficiency were calculated as follows:

$$q_e = \frac{(C_0 - C_e)V}{m} \tag{1}$$

$$\eta = \frac{C_0 - C_e}{C_o} \times 100\% \tag{2}$$

Where, q_e is the adsorption capacity of phenol on adsorbent ($mg\cdot g^{-1}$), C_0 the initial concentration of phenol solution ($mg\cdot L^{-1}$), C_e the equilibrium phenol concentration in solution ($mg\cdot L^{-1}$), m the mass of adsorbent used (g) and V is the volume of phenol solution (L), η the adsorption efficiency (%).

3. Results and discussion

3.1. Scanning electron microscopy

Figure 1 shows Scanning electron microscopy of activated carbon.

IOP Conf. Series: Earth and Environmental Science **52** (2017) 012101 doi:10.1088/1755-1315/52/1/012101

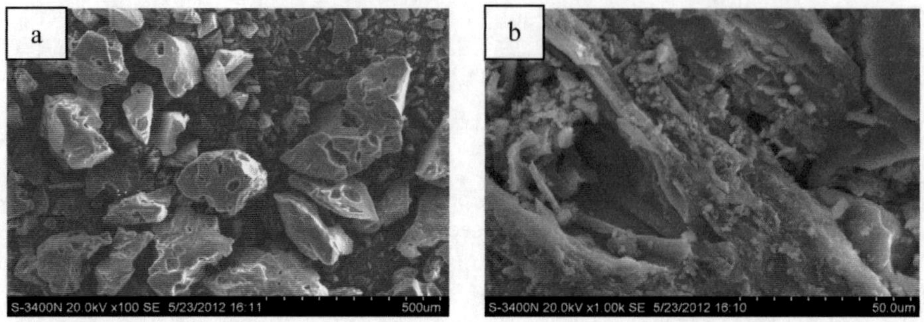

Figure 1. SEM micrographs of activated carbon at: (a) 300 X; (b) 3000X

From the SEM image can be seen that the morphological characteristics of activated carbon used in the experiment has the uneven surface, porous structure, and pore unobstructed, so as to ensure easy accesses to more phenol activated carbon mesoporous and microporous, also increase the adsorption capacity.

3.2. Effect of adsorbent dosage

The adsorption of the phenol on activated carbon was studied by varying the activated carbon concentration (0.2–1.0 g/50 mL) for phenol concentration of 250 mg/L, adsorption temperature 25 ℃, oscillated 120 r/min for 24 h. The percentage of adsorption increased with increase in the activated carbon concentration (Figure 2). This was attributed to increased carbon surface area and the availability of more adsorption sites [7]. When the dosage of activated carbon is 0.4 g, phenol removal rate reached the maximum value of 99.36%.

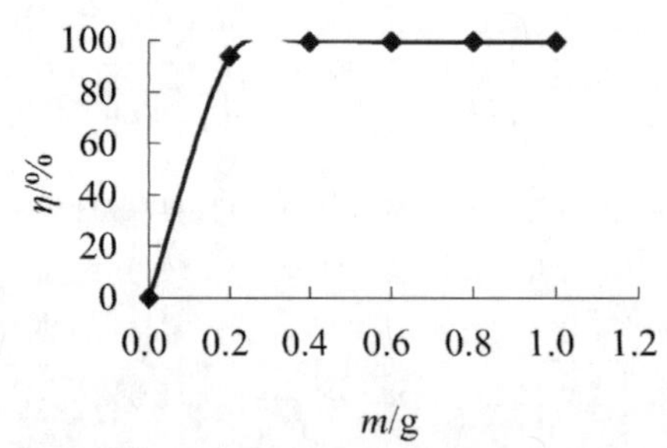

Figure 2. Effect of adsorbent dose on the adsorption of phenol

3.3. Effect of solution pH

The dosage of adsorbent at 16 $g \cdot L^{-1}$, the concentration of phenol at 500 $mg \cdot L^{-1}$, adjusted for pH value by 0.1 $mol \cdot L^{-1}$ HCl and NaOH, the adsorption temperature at 25 ℃, and oscillating time at 24 h. The relationship between pH value and the removal rate of phenol or equilibrium adsorption amount may be seen in Figure 3.

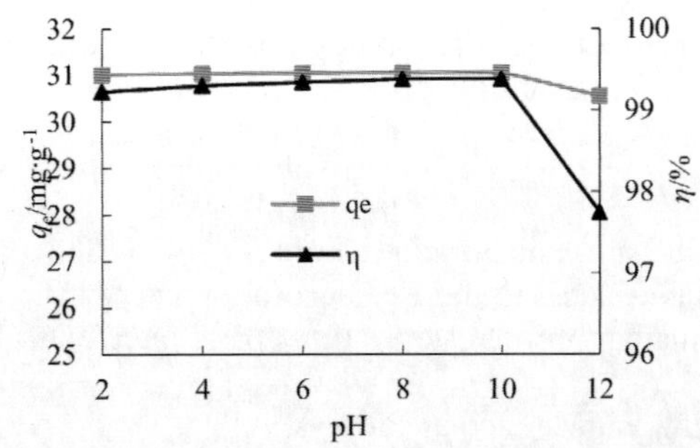

Figure 3. Effect of pH on removal of phenol by activated carbon

Figure 3 shows that the adsorption capacity of phenol increased with increase in the pH when pH< 10, and phenol removal rate and adsorption capacity of phenol decrease obviously when pH> 10. When pH is 6-10, phenol removal rate is above 99% and adsorption capacity of phenol is about 30 mg/g, that is to say the removal efficiency is good. The reason may be that phenol exists in molecular state so that the increase adsorption capacity of phenol on activated carbon due to great affinity of straw activated carbon surface. The pK_a of phenol is 9.96, the decrease in the adsorption capacity of phenol is attributed to both greater solubility of dissociated phenol at pH>pK_a and increased repulsion forces between the dissociated form of the adsorbate and the carbon surface. Because phenol is weak acidic, the most appropriate pH should be 6-8 in subsequent experiments.

3.4. Adsorptive isotherms

The adsorptive isotherm curves may be seen in Figure 4 when the dosage of adsorbent is at 16 g·L^{-1}, with the initial concentration of phenol from 500 to 900 mg·L^{-1}, pH=6.0-8.0, and oscillating time 24 h.

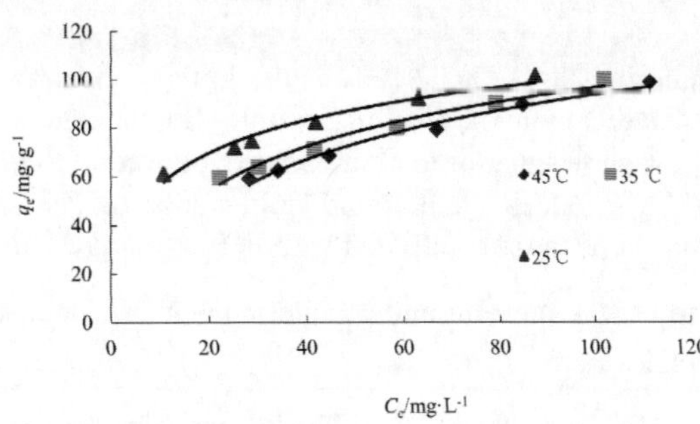

Figure 4. Equilibrium isotherm data for the adsorption of phenol at different Temperatures for phenol wastewater

Figure 4 shows that the greater the concentration of phenol in the same temperature, the greater the amount of activated carbon adsorption. The higher the temperature, the less the equilibrium adsorption amount, implying an exothermic nature of the adsorption process. And under the conditions of high temperature, dynamic energy of theirs is larger so the adsorbed substances get more easily rid of the control of adsorbent.

3.5. Adsorption model

The distribution of phenol between the solid-solution interface equilibrium has been described by the Langmuir equation was linearized as follows:

$$\frac{C_e}{q_e} = \frac{1}{q_m b} + \frac{C_e}{q_m} \tag{3}$$

Where C_e is the equilibrium concentration (mg·L^{-1}), q_m is the monolayer adsorption capacity (mg·g^{-1}) and b is the constant related to the free adsorption energy (Langmuir constant, L·mg^{-1}). The qm and *b* values were calculated from the slopes ($1/q_m$) and intercepts ($1/bq_m$) of linear plots of C_e/q_e versus C_e(Table 1).

The Freundlich equation was linearized as follows:

$$\ln q_e = \ln K_F + \frac{1}{n} \ln C_e \tag{4}$$

Where K_F and n are constants. The n and K_F values were calculated from the slopes ($1/n$) and intercepts ($\ln K_F$) of linear plots of lnqe versus $\ln C_e$(Table 1). It is considered that the $1/n$ value between 0.1 and 1 indicates a beneficial adsorption, 1/n value is more than 2 indicates unfavourable adsorption [8]. Table 1 shows that the Langmuir isotherm model turned out to be extremely satisfactory with the highest R^2 value (>0.99), compared to the other model. Langmuir equation is based on thus theories: the surface of adsorbent is homogeneous and monolayer adsorption, that is to say a adsorption site corresponds to a adsorbent molecule. Langmuir maximal adsorption capacity is 129.87 mg/g at 45 ℃. Statistical analysis suggested that variance analysis significance probability is found to be significant ($p<0.001$) and the temperature has significant influence on adsorption of phenol.

The essential features of a Langmuir isotherm can be expressed in terms of a dimensionless constant separation factor or equilibrium parameter, R_L that is used to predict if an adsorption system is favourable or unfavourable [9]. R_L is defined by:

$$R_L = \frac{1}{1+bC_0} \tag{5}$$

Where b is the Langmuir constant (L·mg^{-1}), C_0 is the initial concentrations of the phenol(mg·L^{-1}). R_L values between 0 and 1 indicate favourable adsorption. The values of R_L at different temperatures are given in Table 1. Table 1 shows that at the temperature between 25-45℃, initial concentrations of phenol are 500-900 mg/L, R_L values were found quite close to zero. This situation can also be interpreted as the adsorption of phenol on activited carbon is a relatively irreversible reaction [10].

Table 1. Langrnuir and Freundlich isotherms constants

t(℃)	Freundlich isotherm			Langmuir isotherm			
	b	$1/n$	R^2	q_m(mg/g)	K_F (L/mg)	R^2	R_L
25	33.6126	0.2432	0.9904	129.8701	0.1959	0.9952	0.0056～0.0101
35	20.3552	0.3398	0.9868	125.0000	0.3398	0.9937	0.0032～0.0059
45	16.7065	0.3755	0.9957	109.8901	0.3755	0.9990	0.0030～0.0053

4. Conclusions

The present study showed that the sunflower straw activated carbon is a promising adsorption for the removal of phenol from aqueous solution. The highest removal rate of phenol on straw activated carbon is 99.37% when the concentration of phenol at 500 mg·L^{-1}, adsorbent dosage of 16 g·L^{-1}, oscillating time 24 h, pH=6-8, and the adsorption temperature 25 ℃. The adsorptive isotherm curves can be fitted with Langmuir equations, and the maximum adsorption capacity of Langmuir is at 109.89 mg/g at 25 ℃. The adsorption of phenol is a monolayer chemical adsorption and endothermic process.

Acknowledgements
This work was financially supported by the Hohhot Natural Science Foundation (2014MS0221).

References

[1] El-Naas M H, Al-Zuhair S, Makhlouf S 2010 Batch degradation of phenol in a spouted bed bioreactor system *J. Ind. Eng. Chem.* **16** 267–272.

[2] Yavuz Y, Koparal S 2006 Electrochemical oxidation of phenol in a parallel plate reaction using ruthenium mixed metal oxide electrode *J. Hazard. Mater* **B136** 296–302.

[3] Svetlana Ivanova, Anna Penkova, María del Carmen Hidalgo 2015 Synthesis and application of layered titanates in the photocatalytic degradation of phenol *Applied Catalysis B: Environmental* **163** 23-29.

[4] Busca G, Berardinelli S, Resini C, Arrighi L 2008 Technologies for the removal of phenol from fluid streams: a short review of recent developments *J. Hazard. Mater* **160** 265–288.

[5] Yang C, Qian Y, Zhang L, Feng J 2006 Solvent extraction process development and on-sitetrial-plant for phenol removal from industrial coal-gasification wastewater *Chem. Eng. J.* **117** 179–185.

[6] Tulaphol S, Bunsan S, Kanchanatip E 2016 Influence of chlorine substitution on adsorption of gaseous chlorinated phenolics on multi-walled carbon nanotubes embedded in SiO_2 *International Journal of Environmental Science and Technology* **13**(6) 1465-74.

[7] Namasivayam C, Yamuna R T 1995 Adsorption of direct red by biogas residual slurry *Environ. Pollut* **89** 1.

[8] Namasivayam C, Kanchana N 1992 Waste banana pith as adsorbent for colour removal from wastewaters *Chemosphere* **25** 1691-1706.

[9] McKay G, Blair H S, Garden J R 1982 Adsorption of dyes on chitin. 1. equilibrium studies *J. Appl. Polym. Sci.* **27** 3043-57.

[10] Ghosh D, Bhattacharyya G K 2002 Adsorption of methylene blue on kaolinite *Appl. Clay Sci.* **20** 295–300.

Environment-friendly drilling operation technology

Huaidong Luo[1], Ning Jing[1], Yanna Zhang[2, 4], Hongjun Huang[3], Jun Wei[1]

[1] CNPC international (Chad) co., LTD, China
[2] China petroleum exploration and development research institute, China;
[3] CNPC Drilling research institute, China

E-mail: zh-yanna@qq.com

Abstract: Under the circumstance that international safety and environmental standards being more and more stringent, drilling engineering is facing unprecedented challenges, the extensive traditional process flow is no longer accepted, the new safe and environment-friendly process is more suitable to the healthy development of the industry. In 2015,CNPCIC adopted environment-friendly drilling technology for the first time in the Chad region, ensured the safety of well control, at the same time increased the environmental protection measure, reduced the risk of environmental pollution what obtain the ratification from local government.This technology carries out recovery and disposal of crude oil, cuttings and mud without falling on the ground. The final products are used in road and well site construction, which realizes the reutilization of drilling waste, reduces the operating cost, and provides a strong technical support for cost-cutting and performance-increase of drilling engineering under low oil price.

1. Introduction

As people's emphasis on environmental protection, National Medium and Long-Term Science and Technology Development Plan Outline, Twelve Five-Year Plan Outline, Five-Year Science and Technology Development Planning all give priority to develop environmental protection technology [1, 2]. A large amount of drilling waste, including drilling sewage, drilling fluid, drilling cuttings and waste oil, causes great pollution to the environment during the oil drilling process. Currently, digging operation is used in most of the drilling sites, and anti-seepage cloth is placed in the mud pool [3-5]. But the cloth damage often occurs, which leads to environmental accidents such as leakage of mud and mud-contaminated soil and water. At the same time, drilling waste is discharged into the mud pool. Although air-dry buried, these wastes are still soluble, and will cause serious environmental risks to the surrounding soil, water, crops and air once soaked in the rain or washed by rivers [6].

In 2013, due to the local workers' labor insurance events of handling drilling mud pool, it led to environmental disputes between Chad's government and CNPCIC and caused a number of unfavorable conditions for construction in Chad. Meanwhile drilling waste discharge in the mud pit for dry burial, but these wastes still have soluble organic pollution, heavy metal pollution, inorganic hydrocarbon pollution, so they would have serious environmental risks on the surrounding soil, water, crops, and the air once immersed in rain and eroded by river . Therefore, China International (Chad) Co., Ltd. carried out the environment-friendly drilling operation technology for chad project region H.

The drilling fluid system used in the Chad region since 2007 is the Bio-Pro system. This system's single-agent and the system itself both passed environmental standards check and met the local environmental protection requirements. Therefore, there is no pollution source in the drilling fluid system itself. But because the highly saturated oil in this region, and formation pore of high pressure, so the overflow is caused frequently during the process of drilling, resulting in crude oil's invasion of drilling fluid, cuttings mixed in, or return of oil-contained cuttings, causing serious environmental pollution. So the main feature of drilling waste in this region is oil-contained, so the difficulty in the treatment of oil wastewater, oil sludge, oil cuttings is without falling on the ground.

2. Methods

In order to thoroughly solve the waste mud governance problem, Chad project department is phasing out past passive centralized way of waste mud "end-treatment" way-landfill method, actively seek pollution thoroughly, advanced way of sludge treatment in line with the "reduction, harmless and recycling" cleaner production principle, so that to fundamentally solve waste mud pollution risk.

2.1. Oil water separation technique

When the crude oil mixed with water-based drilling fluid, it will form an oil-in-water emulsion with water as a continuous phase, use water-soluble inorganic polymer demulsifier - polymerized aluminum sulfate (PAS) [7-9], and added amount is 50mg / L with an obvious layer between oil and water and a short time, and the addition of demulsifier will not cause pollution to the environment.

2.2. Oil sludge separation technology

At home and abroad about the oil (crude oil, lubricating oil) aqueous waste drilling mud solid-liquid separation process are rare, generally thermal desorption is adopted inonly a few research report of the oil waste mud solid-liquid separation, while the waste oil of solid-liquid separation research has not been reported under the normal temperature [10-12].

According to the characteristics of waste in Chad region, By using "oil-displacement" principle, appropriate amount of oil-displacement agent can effectively realize the displacing of oil and mud particles, and by the treatment of coagulant and flocculanting agent [13,14], oil water can be separated from sediment effectively. After treatment the mud mixture bypasses the oil separation device where sedimentation and floatation oil is conducted. The lower layer of separation device waste mud is pumped into the vacuum belt filter for dehydration by a mud pump, and recover the upper layer oil.

2.3. On-site waste mud treatment process

The drilling waste mud treatment system of this technology adopts skid-mounted, modular assembly equipment, the whole process of the treatment of drilling waste mud assures collection without falling or reuse or compliance treatment; Throughout the treatment process, according to drilling design scheme, drilling depth and changes of the mud system, the treatment process is adjusted timely.

3. Results

The key of the environment-friendly drilling operation is to abandon the traditional extensive drilling method and adopt the no-falling receiving and treatment technology to realize the closed-loop recovery and utilization of the drilling waste [15,16]. It is not only with the stringent environmental requirements, but also can be better adapted to the international security situation increasingly stringent environmental standards.

In 2014, Kazakhstan company contractsthe transportation of waste drilling mud and debris to specialized contractors resultfrom the process of drilling and workover, to centralized disposal and realize the environment friendly, but the local government taxes burden is too high to cost.To the three important steps of the cuttings and mud receiving, transporting, and treatment on-site, scientific demonstration and design have been carried out, thus assures the smooth implementation of the environment-friendly operation technology in the field, the operation procedure is shown in Figure 1 and Figure 2.

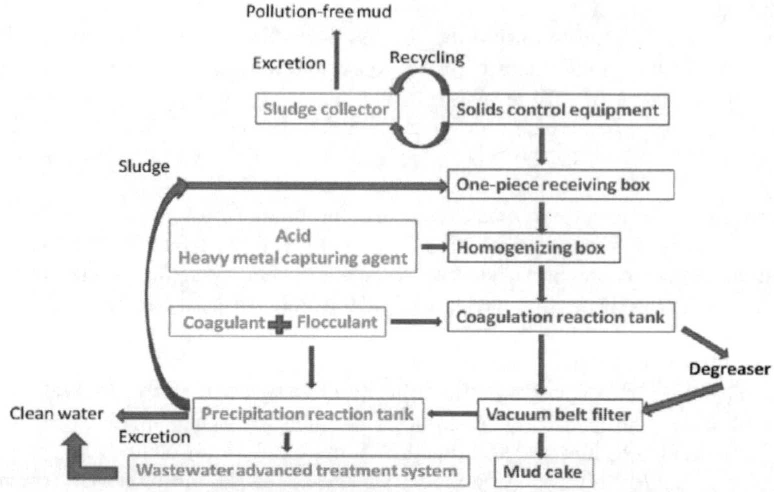

Figure1 Environment friendly operation technology

Figure 2 Processing equipment of drilling waste at the scene

4. Discussion

In the Chad Baobab region, because the well C1-4 has a narrow formation security density window during the drilling process of the three open buried hill formation, overflow and well leakage and other complex situations occurred many times, including 17 times overflows, crude oil spillage of 520m^3, mud leakage about 670 m^3. In view of the coexistence of gushing and leaking and the high risk of environmental pollution, the operation department takes the advantages of environment-friendly drilling operation to completely control the spilled crude oil in the equipment, and to realize the closed-loop recovery of the wellbore output fluid to ensure that the crude oil does not fall and reduce the environmental risks. It saves about USD 240,000 than traditional operation (See Table 1), and at the same time it shortens the construction cycle and accumulates valuable experience in dealing with the complexity of the buried hill formations, and has been endorsed by the Chadian government's on-site inspection team.

This well using environmentally friendly operation, not only effectively prevent environmental pollution risk, but also avoid the mud pit excavation, backfilling, laid impervious cloth, the most important is to reduce the amount of single well waste processing. Return of cuttings and waste pulp waste are collected at the scene of the two steel tank or buffer tank, so the waste need cleaning up and transfer is actuallyonly the drilling returns solid and liquid wastes, therefore directly save disposal charge about $150000. The second open completion mud is collected in the buffer tank, part will be used in two wells under construction. In theory, the well drilling fluid can be reused 3 times, sothe average single well mud cost and handling fees will save $70000. As a result, the relative of pit operation, single well save $240000 on average. The detail can be seen in the table below.

Table 1 Environment friendly operation cost savings

Project	cost (USD)	remark
Avoid mud pit excavation	6000	Average mud pit excavation cost of single well
Avoid backfilling slurry pit	15000	Average mud pit backfilling cost of single well
Save the impervious cloth	14000	Impervious cloth, 50m*50m,14000USD
Reduce pollution produced by the digging soil disposal	150000	Reduce the contaminated soil in the mud pit wall and bottom about 500m^3.Solid processing costs about $300 per square , Single well solid handling to reduce $150000.
The main part hole completion mud recycling	70000	Second mud 150 square, average every time a new mud can be reused, three times the average per well save two open mud 100 square, mud costs an average of $500/square, total average single well save materials and handling about $70000
Add Device	15000	Increase equipment costs about $15000 for average single well
Single well average direct cost savings	**240000**	

5. Conclusion

Environment-friendly drilling operation technology improves on-site operating efficiency, reduces environmental protection risks, and provides a strong technical support for cost-cutting and performance-increase of drilling engineering under low oil price and stringent international safety and environmental standards of the host country of the resources. And it can adapt more suitably to the tendency of more stringent international safety and environmental standards.

References

[1] Mo Yuting, Wu Zhenlin 2010 Analysis of Obstacles and Countermeasures of Contract Project in Kazakhstan under the Financial Crisis *Foreign Investment in China* **223** 106-107.
[2] Peng Yuan, Yang Xu, Sun Changjian 2007 Research on Harmless Treatment of Waste Mud *Environmental Science and Management* **04**
[3] Song Zhanpei 2012 Solidification Treatment Technology of Abandoned Drilling Fluid in Iran Y Oil Field *Environmental Science and Management* **03**
[4] Liu Yucheng, Wu Mian 2010 Chen Mingyan. Research Progress and Prospect of Solidification Treatment Technology for Drilling Waste Mud *Environmental Science and technology* **S1**
[5] Lu Guangliang, Wang Jian, Lu Daosu, Liu Changlong 2009 Research on Photo Fenton Oxidation / Flocculation Combined Treatment of Weak Gel Containing Oil Field Sewage Technology *Journal of Chongqing University of Science and Technology (Natural Sciences Edition)* **05**
[6] Cong Xiaoqiang, Zhang Yi, Wang Haiyan, Jiang Yi 2012 Advances in Treatment Technology of Oilfield Drilling Fluid Waste *Guangzhou Chemical Industry* **09**
[7] Cao Hongyu, Wang Jinliang 2011 Brief Discussion about Harmless Treatment Technology of Drilling Mud *China Petroleum and Chemical Standard and Quality* **07**
[8] Veloso J and Dos Santos G B 2013 The Challenges for the Treatment of Drilling Fluid Wastes Generated by E&P Industry in Brazil *Society of Petroleum Engineers*.
[9] Antle G, Gover P and Pruett J O 2003 Integrated Waste Management: Successful Implementations of Thermal Phase Separation Technology for Oil- and Synthetic-Based Cuttings and Drilling Fluid Waste *Society of Petroleum Engineers*
[10] Morillon A, Vidalie J F, Hamzah U S, Suripno S and Hadinoto E K 2002 Drilling and Waste Management *Society of Petroleum Engineers*
[11] Svensen T and Taugbol K 2011 Drilling Waste Handling in challenging Offshore Operations *Society of Petroleum Engineers*
[12] Denney D 2011 Holistic Drilling-Fluid and Waste Management in the Fayetteville Shale *Society of Petroleum Engineers*
[13] Priatna R, Sjahroezah A and Bledoeg R I 2000 Waste Management: An Approach to the Management of Drilling Waste *Society of Petroleum Engineers*
[14] Kakadjian S, Thompson J E, Torres J R and Quintero H 2014 Stable Fracturing Fluids From Waste Water *International Petroleum Technology Conference*
[15] Svensen T and Taugbol K 2011 Drilling Waste Handling in Challenging Offshore Operations (Russian) *Society of Petroleum Engineers*
[16] Veil J A 2002 Drilling Waste Management: Past, Present, and Future *Society of Petroleum Engineers*

Effect of heavy-metal on synthesis of siderophores by *Pseudomonas aeruginosa* ZGKD3

Peili Shi[1], Zhukang Xing[2], Yuxiu Zhang[1,4], Tuanyao Chai[3]

[1] Department of Environmental & Biological Engineering, School of Chemical & Environmental Engineering, China University of Mining & Technology (Beijing), D11 Xueyuan Road, Beijing 100083;
[2] Huozhou Secondary School of Shanxi, Huozhou 031400, China;
[3] College of Life Science, University of Chinese Academy of Sciences, A19 Yuquan Road, Beijing 100049, PR China.

E-mail: zhangyuxiu@cumtb.edu.cn

Abstract. Most siderophore-producing bacteria could improve the plant growth. Here, the effect of heavy-metal on the growth, total siderophore and pyoverdine production of the Cd tolerance *Pseudomonas aeruginosa* ZGKD3 were investigated. The results showed that ZGKD3 exhibited tolerance to heavy metals, and the metal tolerance decreased in the order $Mn^{2+}>Pb^{2+}>Ni^{2+}>Cu^{2+}>Zn^{2+}>Cd^{2+}$. The total siderophore and pyoverdine production of ZGKD3 induced by metals of Cd^{2+}, Cu^{2+}, Zn^{2+}, Ni^{2+}, Pb^{2+} and Mn^{2+} were different, the total siderophore and pyoverdine production reduced in the order $Cd^{2+}>Pb^{2+}>Mn^{2+}>Ni^{2+}>Zn^{2+}>Cu^{2+}$ and $Zn^{2+}>Cd^{2+}>Mn^{2+}>Pb^{2+}>Ni^{2+}>Cu^{2+}$, respectively. These results suggested that ZGKD3 could grow in heavy-metal contaminated soil and had the potential of improving phytoremediation efficiency in Cd and Zn contaminated soils.

1. Introduction

Soil contamination with heavy metals has become increasingly prominent with the rapid development of human industry and agriculture which causes great hazard to natural environment [1]. Heavy metal ions could be absorbed by plant root and then transport from roots to shoots and cause a considerable threat to human health through food chain due to its high toxicity [2]. Nowadays, in comparison with conventional methods such as physical separation, washing and stabilization, more attention has been paid to plant-microorganism combined bioremediation phytoremediation [3-4]. Various studies have reported that heavy-metal tolerant microbes can directly improve the efficiency of phytoremediation, the possible strategies including metal mobilization, metal chelation and oxidation/reduction reactions [5-7]. In particular, many metal resistant bacteria are capable of synthesize siderophores which provides benefits to plants [8-9]. Siderophores play an important role in phytoremediation due to its strong affinity for Fe^{3+}, Cu^{2+}, Zn^{2+}, $Ni^{2+}$and Cd^{2+}. Siderophore-producing *Pseudomonas aeruginosa* increased the concentrations of Cr and Pb in maize by mobilizing Cd and Pb in soils [10]. Sharma reported that siderophore-producing *P. aeruginosa* GRP_3 alleviated the chorotic symptoms and significantly enhanced chlorophyll content and biomass of *Vigna radiate* L [11]. Ni tolerant endophytic bacteria isolated from *Alyssum bertolonii* were capable of producing siderophores, enhanced the biomass and Ni accumulation of inoculated plants [12]. Similarly, Dimkpa found that siderophores produced by *Streptomyces acidiscabies* E13 alleviated oxidative stress induced by heavy metal of cowpea [13]. A large amount of siderophore-priducing bacterial were screened and applied to improve phytoremediation efficiency, and it has been evidenced that siderophores produced by bacteria can protect microbes against the toxicity of heavy metals [14-15], whereas the mechanism of heavy-metal tolerance of siderophore-producing bacteria is still unknown.

In the present study, *P. aeruginosa* ZGKD3 isolated from soil contaminated by gangue pile of coal area in our laboratory exhibited high tolerance to Cd^{2+}, Cu^{2+}, Zn^{2+}, Ni^{2+}, Pb^{2+} and Mn^{2+}. The objectives of this study were to investigate the effect of different concentrations of multiple heavy metals on synthesis of siderophores of *P. aeruginosa* ZGKD3. Furthermore, determine the potential of ZGKD3 in improving phytoremediation efficiency in heavy-metal contaminated soils and these results may provide a base for revealing the heavy metal tolerance mechanism of bacteria.

2. Materials and Methods

2.1. Bacteria and media.

The heavy-metal resistant *P. aeruginosa* ZGKD3 were isolated from soil contaminated by gangue pile of coal area of Shanxi province in our laboratory. ZGKD3 were cultivated in broth medium which contained 3.0 g beef extract, 10.0 g peptone, and 5.0 g sodium chloride per liter with an initial pH of 7.2 and MSA (sugar-Asp) medium which contained 20.0 g sucrose, 2.0 g aspartic acid, 1.0 g K_2HPO_4 and 0.5 g $MgSO_4$ per liter with an initial pH of 7.2 [16].

2.2. The growth of ZGKD3 under heavy-metal stress.

The growth of *P. aeruginosa* ZGKD3 under different concentrations of heavy-metal stress was assayed in MSA. ZGKD3 were grown on nutrient broth medium for 16 h, 2 ml cells of ZGKD3 were inoculated into 100 ml Erlenmeyer flasks containing 50 mL of sterile MSA medium with $CdCl_2$, $CuCl_2$, $ZnCl_2$, $NiCl_2$, $Pb(NO_3)_2$ and $MnCl_2$ (0, 200, 400 and 1000 μM), respectively, and incubated in an rotary shaker (150 rpm) at 37 °C for 24 h, the effect of heavy-metal on growth and the ability of alkaline production of ZGKD3 were investigated at 16 h. The biomass of bacterial cell was determined by a UV-Visible spectrophotometer, the absorbance was measured at 600 nm (OD_{600}). All of the chemical reagent were analytical reagent.

2.3. Quantitative analysis siderophore synthesis by ZGKD3 under heavy-metal stress.

The effects of different concentrations of heavy-metal on siderophore synthesis of ZGKD3 were test. Bacterial samples were obtained from MSA medium contained $CdCl_2$ (0, 200, 400, 1000 and 3000 μM), $CuCl_2$, $ZnCl_2$, $NiCl_2$, $Pb(NO_3)_2$ and $MnCl_2$ (0, 200, 400 and 1000 μM), respectively. Detection of total siderophores and pyoverdine production by ZGKD3 were carried out, the total siderophore production was assayed by chromo azurol S (CAS) plate assay [17]. For quantification of siderophore and pyoverdine production were investigated at 12, 24 and 48 h by a UV-Visible spectrophotometer at the absorbance of 630 and 400 nm, respectively.

2.4. Statistical analysis.

All data were analyzed by SPSS 16.0 for significant differences ($P<0.05$). Statistical analyses were performed by one-way ANOVA.

3. Results

3.1. The growth of ZGKD3 in MSA medium under heavy metal stress

The effect of six heavy metals on growth of ZGKD3 was assayed. As shown in Figure 1, the concentrations of Cd^{2+}, Cu^{2+}, Zn^{2+}, Ni^{2+}, Pb^{2+} and Mn^{2+} (200, 400, 1000 μM) inhibited the growth of ZGKD3 and as a consequence of the increasing dose of heavy meal, the stronger inhibitive effects on growth of bacterial were observed, suggested that there is a significant negative correlation between the heavy-metal concentration and the growth of ZGKD3. Moreover, ZGKD3 could tolerate Ni^{2+}, Pb^{2+}

and Mn^{2+} at the concentrations of 200, 400 and 1000 μM, and the growth of ZGKD3 was significantly inhibited by Cd^{2+}, Zn^{2+} and Cu^{2+} at the concentration of 200 μM. The strongest inhibitory effect on the growth of ZGKD3 was observed at the high concentrations of Cd^{2+}, the biomass (OD_{600}) of ZGKD3 dreased from 0.792 to 0.098 with the increasing dose of Cd^{2+}. Thus, the effects of various heavy metals on growth of ZGKD3 were different.

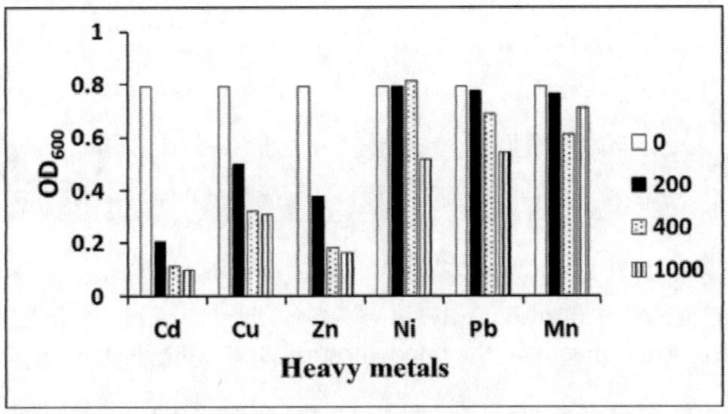

Figure 1. Effect of heavy metals on the growth of strain ZGKD3 in MSA medium contained heavy metals at the concentration of 0, 200, 400 and 1000 μM at 16 h

3.2. The total siderophore and pyoverdine production of ZGKD3 in MSA medium under heavy metal stress

Siderophore production of ZGKD3 was confirmed by adding the culture supernatant into the holes on CAS agar plate and orange haloes were observed. Cd^{2+}, Cu^{2+}, Zn^{2+}, Ni^{2+}, Pb^{2+} and Mn^{2+} exhibited different effects on siderophore production of bacteria (Figure 2). The largest orange haloes was found in the Zn^{2+} group, next is Cd^{2+}. Significant increase in the total siderophore and poverdine production of strain ZGKD3 was found, with the increase of 8-90% for 200-1000 μM of Cd^{2+} and 35-242% for 200-1000 μM Zn^{2+}. However, Cu^{2+} significantly inhibited the total siderophore and poverdine production of strain ZGKD3, and the reduction was range from 26-88%. In comparison with control, Ni^{2+}, Pb^{2+} and Mn^{2+} had no markedly difference in total siderophore and poverdine production (Figure 3 and 4), the total siderophore and pyoverdine production reduced in the order $Cd^{2+}>Pb^{2+}>Mn^{2+}>Ni^{2+}>Zn^{2+}>Cu^{2+}$ and $Zn^{2+}>Cd^{2+}>Mn^{2+}>Pb^{2+}>Ni^{2+}>Cu^{2+}$, respectively.

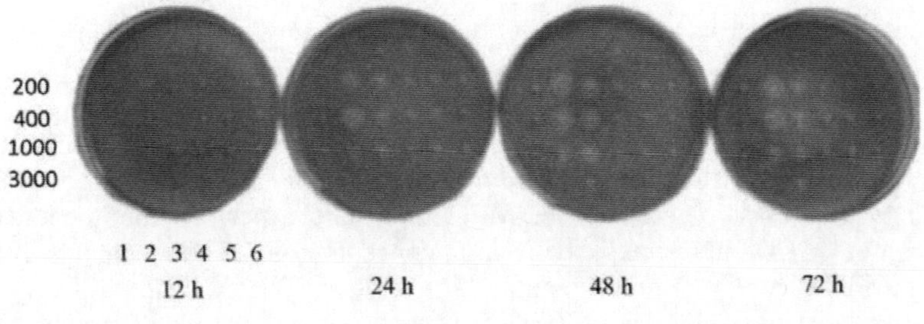

Figure 2. Pyoverdine production by ZGKD3 exposed to heavy metals at the concentrations of 200, 400, 1000 and 3000 μM. The orange haloes on each plate are: A and B in No.1 row indicate negative control and positive control, respectively. The numbers from 1to 6 indicate Cu^{2+}, Zn^{2+}, Cd^{2+}, Ni^{2+}, Pb^{2+} and Mn^{2+}, respectively.

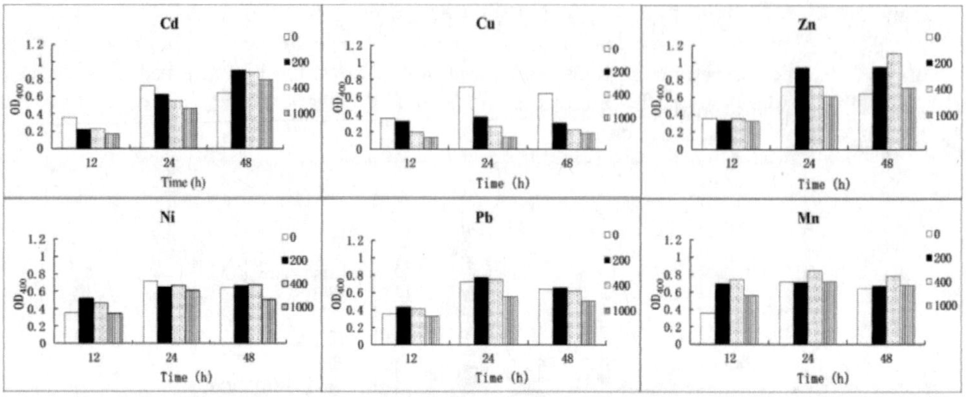

Figure 3. Effects of heavy metals on pyoverdine production by strain ZGKD3 in MSA medium contained heavy metals at the concentration of 0, 200, 400 and 1000 μM.

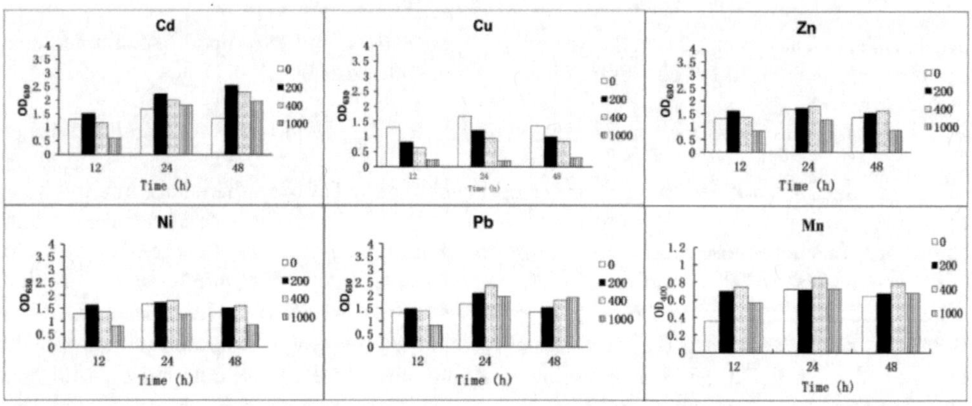

Figure 4. Effects of heavy metals on total siderophore production by strain ZGKD3 in MSA medium contained heavy metals at the concentration of 0, 200, 400 and 1000 μM.

4. Discussion

4.1. The heavy metals inhibited the growth of ZGKD3

Various studies have evidenced that heavy metals could induce inhibitory effects on growth of bacteria. For instance, Zhang found that the biomass of *Bacillus subtillus* decreased by 96.1% at 0.2 mM of Cd [18]. Jiang reported that the growth of *Bacillus subtillus* was inhibited and decreased by 95% at the concentration of 0.25 mM Cd compared to control [19]. In this paper, a experiment was used to determine the growth of ZGKD3 in response to different concentrations of heavy metals. According to the growth curves, ZGKD3 present a great variance in tolerance towards different heavy metals. In this study, an increase in bacteria biomass of ZGKD3 was observed under Ni^{2+} stress of 200 and 400 μM, the result was similar with the reports previously [20]. Thus, low concentration of heavy metal might promote the growth of bacteria, in the contrary, high concentrations of Ni inhibited the growth of ZGKD3. For the Cd^{2+}, Cu^{2+}, Zn^{2+}, Pb^{2+} and Mn^{2+}, all the concentrations of heavy metals exhibited inhibitory effect on bacteria. Therefore, the role of heavy metals on bacteria was dependent on their respective concentration and the kind of heavy metals, suggested that ZGKD3 had different tolerance to multiple heavy metals, ZGKD3 is more sensitive to Cd^{2+} than other matals and the metal tolerance of ZGKD3 decreased in the order $Mn^{2+} > Pb^{2+} > Ni^{2+} > Cu^{2+} > Zn^{2+} > Cd^{2+}$.

4.2. The heavy metals induced the siderophore production of ZGKD3

In the present study, Cd^{2+} and Zn^{2+} were the heavy metal that showed the greater toxicity than that of other heavy metals. In this sense, most of bacteria couldn't grow at high concentrations of heavy metals. Furthermore, bacteria could remove heavy metals from their growth environment with intracellular and surface accumulation of Cu, Zn, Pb and Cd [21]. A large number of bacteria can produce diffusible light-green pigment on a CAS agar plate, and the light-green pigment has been characterized previously to be siderophores [22]. Most of *P. aeruginosa* could produce siderophores, including pyoverdine and pyochelin. It has been suggested that Al, Cu, Mn, Ga and Ni could induce synthesis of siderophores [10]. Similarly, Sinha demonstrated that Cd-resistant strain KUCd1 induced siderophore production maximally at 1.75 mM of Cd concentration [23]. Dao found that the presence of 0.125-1 mM of Cd could stimulate pyoverdine production of *P. aeruginosa* strain PAO1 [24]. Dimlpa observed that Al, Cd, Cu and Ni induced three hydroxamate siderophores by *Streptomyces* sp. Strains [25]. Furthermore, it has been confirmed that siderophores produced by bacteria could chelate many heavy metals, such as Al, Cd, Zn, Cu and Pb, and it was different in chelate ability of bacteria for every heavy metal [26]. Some studies demonstrated that siderophore could increase or decrease the toxicity of heavy metals in bacteria. Pyochelin produced by *P. aeruginosa* increased the toxicity of vanadium to bacteria [27]. However, Braud found synthesis of siderophores decreased the toxicity of multiple heavy metals to *P. aeruginosa* [28]. According to our results, the effects of heavy metals on growth of ZGKD3 were different, the ability of producing siderophore of ZGKD3 varied with different heavy metals might be one of the reasons. The toxicity of heavy metals on bacteria dependent on the amount of heavy metal accumulation in bacteria cells. Hence, the more siderophore produced, the more heavy metals were absorbed by ZGKD3 through the chelation of siderophres for heavy metals, which found to be high toxic to bacteria. In the present study, Cd^{2+} and Zn^{2+} significantly induced the synthesis of pyoverdine, and remarkably inhibited the growth of ZGKD3, indicated that the absorption of ZGKD3 for Cd^{2+} and Zn^{2+} was more than that for other metals. Therefore, *P. aeruginosa* ZGKD3 might have the potential of improving the phytoextraction efficiency in Cd and Zn contaminated soils.

5. Conclusions

P. aeruginosa ZGKD3 exhibited different tolerance to multiple heavy metals. In comparison with Cd^{2+}, Zn^{2+}, Ni^{2+}, Pb^{2+} and Mn^{2+}, and the metal tolerance decreased in the order $Mn^{2+} > Pb^{2+} > Ni^{2+} > Cu^{2+} > Zn^{2+} > Cd^{2+}$. Moreover, Cd^{2+} and Zn^{2+} significantly stimulated the total siderophore and poverdine production of strain ZGKD3. Therefore, *P. aeruginosa* ZGKD3 can act as siderophore-producing bacteria and applied to microbe and plants combined remediation in Cd and Zn contaminated soils.

Acknowledgements

This work was supported by the National Natural Science Foundation of China under Grant No. 31370281 and Fundamental Research Funds for the Central Universities under Grant No. 2010YH05.

References

[1] Kermani J N, Ghasemi F, Khosravan, A. Farahmand, et al. 2010 *Iran. J. Environ. Health. Sci. Eng.* **7** 279-286

[2] Nawrot T S, Staessen J A, Roels H A, Munters E, Cuypers A, Richart T, Ruttens A, Smeets K, Clijsters H, Vangronsveld J 2010 *Biometals* **23** 769-782

[3] Chehregani A, Noori M, Yazdi H L 2009 *Ecotoxicol Environ Saf* **72** 1349–53

[4] Haque N, Peralta-Videa J R, Jones G L, Gill T E, Gardea-Torresdey J L 2008 *Environ Pollut* **153** 362-368

[5] Ma Y, Prasad M N V, Rajkumar M, Freitas H 2011 *Biotechnol Adv* **29** 248–258

[6] Rajkumar M, Ae N, Prasad MNV, Freitas H 2010*Trends Biotechnol* **28** 142–149

[7] Kidd P, Barcelo J, Bernal M P, Navari-Izzo F, Poschenrieder C, Shilev S,Clemente R, Monterroso C 2009 *Environ Exp Bot* **67** 243–59

[8] Kuffner M, Puschenreiter M, Wieshammer G, Gorfer M, Sessitsch A 2008 *Plant Soil* **304** 35-44
[9] Rajkumar M, Ae N, Prasad M N V, Freitas H 2010 *Trends Biotechnol* **28** 142-149
[10] Braud A, Jézéquel K, Bazot S, Lebeau T 2009 *Chemosphere* **74** 280–286
[11] Sharma A, Johri B N, Sharma A K, Glick B R 2003 *Soil Biol. Biochem* **35** 887–894
[12] Barzanti R, Ozino F, Bazzicalupo M, Gabbrielli R, Galardi F, Gonnelli C, Mengoni A 2007 *Microb. Ecol* **53** 306-316
[13] Dimkpa C O, Merten D, Svatoš A, Büchel G, Kothe E 2009 *Soil Biol Biochem* **41** 154–162
[14] Cortese M S, Paszczynski A, Lewis T A, Sebat J L, Borek V, Crawford I 2002 *Biometals* **15** 103-120
[15] Fiedler H P, Krastel P, Müller J, Gebhardt K, Zeeck A 2001 FEMS *Microbiol. Lett* **196** 147-151
[16] Shin S H, Lim Y, Lee S E, Yang N W, Rhee J H. 2001 *Journal of Microbiological Methods* **44** 89-95
[17] Schwyn B, Neiland J B 1987 *Anal Biochem* **160** 47–56
[18] Zhang W, Chen G H, Guo J, Zhao Y J, Gao Y C, Wang J N 2013 *Industrial safety and environmental protection* **39** 42-45
[19] Ma G F, Wang J, Zhang C B 2009 *Microbiology* **35** 882-887
[20] Cabrero A, Fernandez S, Fernando M, Julian G 1998 *Water research* **32** 1355-62
[21] María V C, Antonio C H, Antonio F V, María V M T 2014 *Advances in Microbiology* **4** 644-655
[22] Meyer J M, Abdallah M A 1978 *J Gen Microbiol* **107** 319–328
[23] Sinha S, Mukherjee S K 2008 *Curr Microbiol* **56** 55-60
[24] Dao K H T, Hamer K E, Clark C L, Harshman L G 2001 *Ecol Appl* **9** 441–448
[25] Dimkpa C O, Svatos A, Dabrowska P, Schmidt A, Boland W, Kothe E 2008 *Chemosphere* **74** 19-25
[26] Schalk I J, Hannauer M, Braud A 2011 *Microbial* **13** 2844-54
[27] Baysse C, Vos D D, Naudet Y, Vandermonde A, Ochsner U, Meyer J M 2000 *Microbiol-SGM* **146** 2425-34
[28] Braud A, Geoffroy V, Hoegy F, Mislin G L A, Schalk I J 2010 *Environ Microbiol Report* **2** 419-425

Synchronous rapid start-up of the methanation and anammox processes in two-stage ASBRs

Y Duan[1], W R Li and Y Zhao

Dept. of Environmental Science and Engineering, Taiyuan University of Technology, Taiyuan, PRC

E-mail: duanyun@tyut. edu.cn

Abstract. The "methanation + anaerobic ammonia oxidation autotrophic denitrification" method was adopted by using anaerobic sequencing batch reactors (ASBRs) and realized a satisfactory synchronous removal of chemical oxygen demand (COD) and ammonia-nitrogen (NH_4^+-N) in wastewater after 75 days operation. 90% of COD was removed at a COD load of 1.2 kg/(m^3•d) and 90% of TN was removed at a TN load of 0.14 kg/(m^3•d). The anammox reaction ratio was estimated to be 1: 1.32: 0.26. The results showed that synchronous rapid start-up of the methanation and anaerobic ammonia oxidation processes in two-stage ASBRs was feasible.

1. Introduction

Nowadays, the main pollutants are organic matters and nitrogenous substances in wastewater, which not only intensified the contradiction of water shortage in our country, but also seriously restricted the implementation of the sustainable development strategy. Therefore, efficient removal of organic matters and nitrogenous substances in wastewater would be one of the most important research directions in the field of controlling water pollution and improving water environment.

At present, the treatment processes for the wastewater contained organic matters and nitrogenous substances mainly focus on the removal of COD and NH_4^+-N separately. More and more pollutants like organics, ammonium were discharged together in wastewater from pharmacy [1], food [2], papermaking industry [3] etc. A renewable process that might be capable of simultaneously removing organics and ammonium was received a primary attention. Among all the biological technology in the field of wastewater treatment, the methanation and anaerobic ammonia oxidation methods [4, 5]were more efficient ways. These biochemical processes were produced by different functional biological flora under the anaerobic conditions [6, 7]. As we all know, the doubling time of anammox bacteria was reported to be 11 days with maximum specific growth rate of 0.0027•h^{-1}[1,8]. A disadvantage of this low growth rate was the long start-up period required for anammox process which limited its widespread application. Therefore, reducing the duration of methanation and anammox start-up processes was one of the main challenges for environmental engineering. Yinyin [9] achieved the simultaneous anaerobic ammonia oxidation, methanation and denitrification in UASB reactor. In her study, various activated sludge was inoculated in the same UASB reactor and the reactor which started successfully after about 48 days. The denitrification bacteria and anaerobic ammonium oxidation bacteria could be cultured and enriched in the UASB reactor. However, the results showed that the removal rates of COD and NH_4^+-N were not ideal. Zubo [10] studied on integration of methanogenesis with simultaneous anaerobic ammonium oxidation and denitrification in EGSB reactor, and the granular sludge was formed in EGSB reactor inoculated with the aerobic activated sludge for 120 days' start-up operation. Then, denitrification bacteria and anaerobic ammonium oxidation bacteria were cultured and enriched for 270 days in EGSB reactor with nitrite and ammonium. In Zubo's study, the removal rate was increased, but the start-up time was a little long, which was up to 120 days. There were various wastewater treatment technologies such as anaerobic baffled reactor (ABR [11]), membrane bioreactor (MBR [12]), upflow anaerobic sludge blanket (UASB [13]), biological aerated filter (BAF [14]) and anaerobic sequencing batch reactor (ASBR[1]) et al., which had been used for the cultivation of methanogenic and anammox bacteria. As the third generation of all anaerobic reactors, anaerobic sequencing batch reactor (ASBR) had some advantages: the rapid start-up time, the simple structure, low operation cost, good biological flocculation ability and solid-liquid separation efficiency, little demand for the alkalinity, and efficient anti-shock loading with the formation of A granular sludge, etc [15,16].

The experiment, which aimed at reducing the COD and NH_4^+-N in two-stage ASBRs by cultivating the methanogenic and anammox activated sludge, achieved synchronous rapid start-up of methanation and anaerobic ammonia oxidation.

2. Materials and methods

2.1 ASBR start-up and operation

Two identical anaerobic sequencing batch reactors (R1 and R2) were utilized to investigate the rapid start-up processes in this study. The working volume of each reactor was all approximately 2.5 L with an internal diameter of 15 cm and a height of 28.5 cm. The two reactors were operated at 35°C, and the initial pH of R1 and R2 was 7.4 and 8.1, respectively. When the two reactors were separately operated, the fill and draw were operated with 1000 mL influent and 1000 mL effluent every cycle for hydraulic retention time (HRT) 24 hours. Every ASBR operation cycle was consisted of five steps: filling (4 min), reaction (1200 min), settling (120 min), discharge (4 min) and idle (112 min). Each device was intermittently stirred for 8 minutes per hour [17] by biogas produced by itself, the amount of biogas was collected by a biogas bag and measured by a glass syringe. The whole reaction process was controlled by time switch and miniature vacuum pump. When the two reactors were combined by a latex drainage tubing, the effluent of R1 was as the influent of R2, the the fill and draw flow of linked-up state was the same as the separate state. The HRT of R1 and R2 was respectively reduced to 12 hours in every cycle and the HRT of the whole cycle was 24 hours. In the two-stage anaerobic reactors, the influent concentration could be increased after the effluent concentrations of COD and NH_4^+-N met the requirements of the National Discharge Standard of China (GB8978-1996). An opaque, black plastic film enclosure was served to inhibit the growth of photosynthetic bacteria [18].

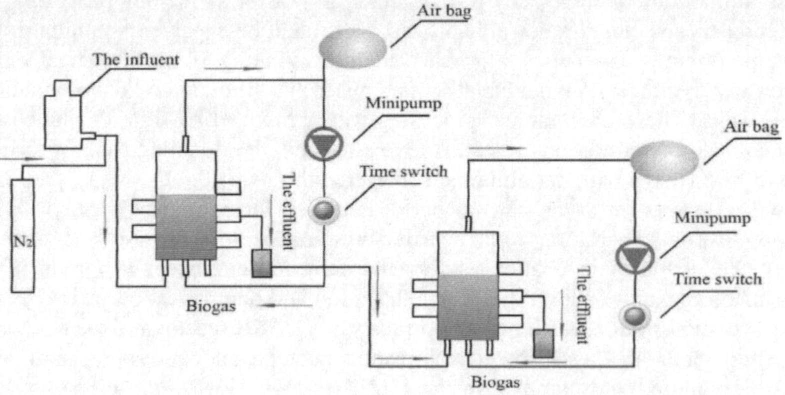

Figure 1. Schematic diagram of anaerobic sequenced batch reactor.

2.2 Wastewater and inoculum

Anaerobic sludge obtained from a wastewater treatment plant in Jinzhong city was used as the inoculum. The anaerobic methanogenic bacteria were sourced from a labscale mature anaerobic reactor and anammox bacteria were sourced from an upflow microaerobic sludge reactor cultured for over 2 years. That $C_6H_{12}O_6$ was added into the wastewater to enhance COD loading rate, NH_4Cl was added to enhance NH_4^+-N loading rate, and $NaHCO_3$ was added to keep the pH of methanation process at 7.4 and the pH of anammox process at 8.1. The initial biomass concentration (expressed as volatile suspended solids, SS) was 13.6g SS/L and VSS/SS was 80%. The methanogenic and anammox bacteria were 15.14 g/L and 16.89 g/L, respectively. R1 was seeded with 1000 mL anaerobic sludge and 250 mL methanogenic bacteria, R2 was also seeded with1000 mL anaerobic sludge and 250 mL anammox bacteria. The biomass ratios of anaerobic sludge and anaerobic bacteria for R1 and R2 were all 4:1.

The reactors were fed with synthetic wastewater containing (per liter) showed in Table 1: $C_6H_{12}O_6$ 300-1500mg, NH_4Cl 50-120mg, $NaNO_2$ 65-120mg, $NaHCO_3$ 1500mg, KH_2PO_4 30mg, $MgSO_4{\cdot}7H_2O$ 300mg, $CaCl_2{\cdot}7H_2O$ 230mg, trace element solution I (including EDTA 5000mg and $FeSO_4$5000mg) 2mland trace element solution II (including $ZnSO_4{\cdot}7H_2O$ 430mg, $CuSO_4{\cdot}5H_2O$ 250mg, $MnCl_2{\cdot}4H_2O$ 990mg, $NiCl_2{\cdot}6H_2O$ 190mg, $CoCl_2{\cdot}6H_2O$ 240mg, H_3BO_4 14mg) 2ml. Filling the N_2 in the synthetic wastewater to remove dissolved oxygen (DO), which kept the DO below 0.3mg per liter.

Table 1.Components of synthetic wastewater.

Components	Concentration(mg/L)
$C_6H_{12}O_6$	300-1500
NH_4Cl	50-120
$NaNO_2$	65-120
$NaHCO_3$	1500
KH_2PO_4	30
$MgSO_4{\cdot}7H_2O$	300
$CaCl_2{\cdot}7H_2O$	300
Trace element I[a]	2ml/L
Trace element II[b]	2ml/L

[a] Trace element I (mg/L) : EDTA 5000 $FeSO_4$ 5000

[b] Trace element □ (mg/L) :$ZnSO_4{\cdot}7H_2O$ 430, $CuSO_4{\cdot}5H_2O$ 250, $MnCl_2{\cdot}4H_2O$ 990, $NiCl_2{\cdot}6H_2O$ 190, $CoCl_2{\cdot}6H_2O$ 240, H_3BO_4 14

2.3 Analytical methods

COD, NH_4^+-N, NO_2^--N, NO_3^--N, MLSS and MLVSS were detected according to the Standard Methods [19]. DO and pH were measured with a dissolved oxygen meter (Taiwan Hengxin, AZ8403) and a pH meter (Switzerland Mettler Toledo, DELTA320), respectively.

3. Results and discussion

3.1 The rapid start-up of the methanation process

In this stage, the synthetic organic wastewater was added into the sludge-feeding reactor. The initial COD concentration of influent was 300mg/L and the COD removal rate was 65%. In the 3rd day, the COD concentration of influent was increased to 350mg/L, and the removal rate was slightly decreased. The reason was that the activity of the initial sludge was low and the ability adapted to new environment was weak, which resulted in the accumulation of volatile fatty acids. Then the COD concentration of influent was remained at 350mg/L for a period of time (approximately one week), the activity of the sludge was gradually recovered and the removal rate of COD was increased. The COD concentration of the influent was increased at the rate of 50mg/L during the 11th day to the 33rd day.

When the COD concentration of the influent was 1000mg/L, the removal rate of COD reached to 80%. In order to shorten the start-up time and improve the anti-shock load, the COD concentration of the influent was increased at the rate of 100mg/L from the 35th day. When the COD concentration of the influent was 1500mg/L, the removal rate reached to 85%. In the 43rd day, the designed COD load and removal rate were reached ahead of time. The COD concentration of the influent and the effluent in the methanation reactor were shown in Figure 2.

In the 45th day, the actual wastewater contained organic matter and ammonia-nitrogen was simulated. An appropriate amount of NH_4Cl was added into the influent to maintain the NH_4^+-N concentration of 50mg/L, when the COD concentration was still 1500mg/L. Then, the COD removal rate was slightly decreased. It might be that the NH_4^+-N reacted with a small amount of dissolved oxygen, and the ability that methanogenesis consumed organic matter was decreased. In order to shorten the start-up time and improve the removal load of COD, the hydraulic retention time was reduced to 12 hours from the 51st day. After 4 days, the COD concentration of influent was remained at 1500 mg/L and the removal rate of COD reached to 85% in 55th day. The removal rate of COD was not inhibited for the addition of NH_4^+-N in the influent, on the contrary, it was slightly increased. It was estimated that the COD was consumed not only by the methanogenic bacteria, but also the partial nitrifying bacteria and denitrifying bacteria. In the 55th day, the COD concentration of influent was increased to 1500mg/L, while the COD concentration of effluent was less than 50mg/L with the COD removal rate at 95%. Finally, the methanation reactor was in stable operation and the anarobic sludge was in good state.

3.2 The rapid start-up of the anammox process

The anammox reactor was started up with a low NH_4^+-N load. A right amount of $NaNO_2$ was added into the wastewater contained NH_4^+-N, and the substrates were provided for the anammox reaction. In order to ensure the anaerobic reaction condition, filling the N_2 into the influent for 20 minutes and controlling the pH of R2 at 8.1 with $NaHCO_3$. The anammox process could be divided into three stages, and the variations of NH_4^+-N, NO_2^--N, and NO_3^--N in the anammox reactor were shown in Figure 3, 4 and 5, respectively.

The first stage (approximately 11days) was the sludge adaptation stage. The NH_4^+-N concentration of influent was 50mg/L and the NO_2^--N concentration was 65mg/L. After an anaerobic reaction cycle (24 hours), the NH_4^+-N concentration of effluent was 45-55 mg/L and the NO_2^--N concentration was 1-3 mg/L. With the organic matter utilized by a number of denitrifying bacteria [20] from the sludge, NH_4^+-N was produced, which caused higher NH_4^+-N concentration of effluent than that influent. In the period of this stage, the anammox reaction was not discovered.

The second stage (approximately 20 days) was that the sludge showed anammox activity stage. In this stage, the NH_4^+-N concentration of the influent was 60 mg/L and the NO_2^--N concentration was 80 mg/L. After an anaerobic reaction cycle (approximately 24 hours), the NH_4^+-N concentration of effluent was 20-30 mg/L and the NO_2^--N concentration was 10-20 mg/L. In this stage, the removal rate of NH_4^+-N concentration was significantly increased and a small amount of NO_2^--N was produced, which indicated that the anammox bacteria appeared in the sludge. However, the anammox bacteria was not dominant in the activated sludge [21], the TN load should be further improved to enrich the anammox bacteria in the following stage.

In the third stage (approximately 20 days), the activity of the anammox sludge was improved and the reactor was more stably operated. The NH_4^+-N concentration of influent was 80 mg/L and the NO_2^--N concentration was 100 mg/L. During the first 10 days of the stage, the NH_4^+-N concentration of effluent was 25 to 30 mg/L, the NO_2^--N concentration was 10 to 20 mg/L, and the NO_3^--N concentration was 15 to 20 mg/L. The removal rate of NH_4^+-N concentration was slightly increased, but the NO_2^--N was significantly improved. Then the hydraulic retention time was reduced to 12 hours, but the NH_4^+-N and NO_2^--N concentration of influent were remained. After an anaerobic reaction cycle (12 hours), the removal rates of NH_4^+-N and NO_2^--N were close to 100%, respectively. At the same time, the NO_3^--N [22] concentration of effluent was reached to 20 mg/L. Finally, the removal ratio of

NH_4^+-N, NO_2^--N, and NO_3^--N concentration was approximately to 1:1. 32:0. 26 [23, 24] in the anammox reactor.

3.3 Treatment for wastewater in two-stages ASBRs

From the 55th day, the methanation reactor (R1) and anammox reactor (R2) were linked up. The simulative wastewater contained COD and NH_4^+-N was filled to the methanation reactor, and then was discharged into the anammox reactor. In this operation, the effluent of R1 was as the influent of R2. The whole anaerobic reaction cycle was 24 hours, including the methanation period (12 hours) and the anammox period (12hours). During the first 10 days of the linked-up state, the COD concentration of influent was constant 1500mg/L, and the NH_4^+-N concentration was 100mg/L. After 12 hours, the COD removal rate of R1 was 95%, but the NH_4^+-N concentration in the effluent of R1 was nearly invariable. Then, the effluent of R1 with nearly 50mg/L COD concentration and 100mg/L NH_4^+-N flowed into the R2. Over a period of anammox process (12 hours), the effluent COD concentration of R2 was about 0mg/L, and the effluent NH_4^+-N concentration was about 1-3 mg/L. In order to improve the removal load of NH_4^+-N, the influent NH_4^+-N concentration of the system was increased to 120mg/L, while the influent COD concentration of R1 was still remained 1500mg/L, to prevent higher COD concentration from influencing the mechanism of removing process of TN in R2. An appropriate amount of $NaNO_2$ (100mg/L-120mg/L) matched with the NH_4^+-N concentration was added into the R2 by sampling opening, which provided the substrates for the anammox reaction. Finally, when the COD, NH_4^+-N and NO_2^--N concentration of the two-stage reactors were remained 1500mg/L, 120mg/L and 120mg/L. The removal rates of COD, NH_4^+-N and NO_2^--N were respectively as high as 95%, 90% and 100%. Thus, the steady-state condition was reached and the desired data were obtained after approximately 20 days. The removal ratio of NH_4^+-N, NO_2^--N and NO_3^--N concentration was approximately close to 1:1.32:0.26[23,24]. According to some relevant references[20,22,24,25], the synchronous start-up of methanation and anammox processes were successfully achieved after 75 days.

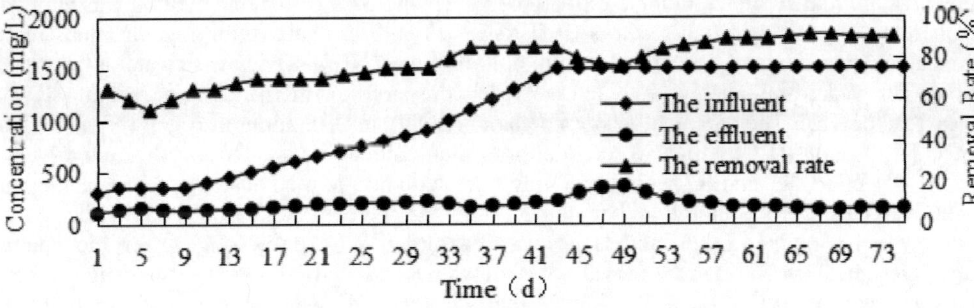

Figure 2. The variation of COD concentration.

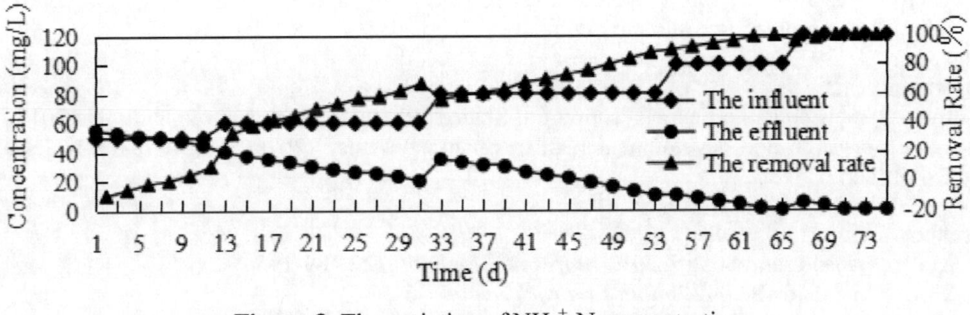

Figure 3. The variation of NH_4^+-N concentration.

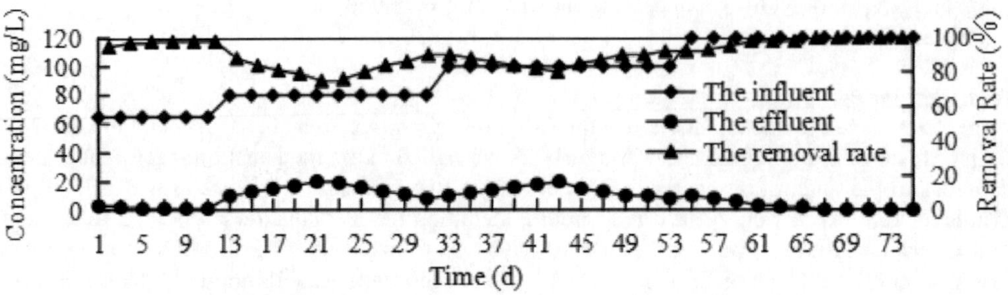

Figure 4. The variation of NO_2^--N concentration.

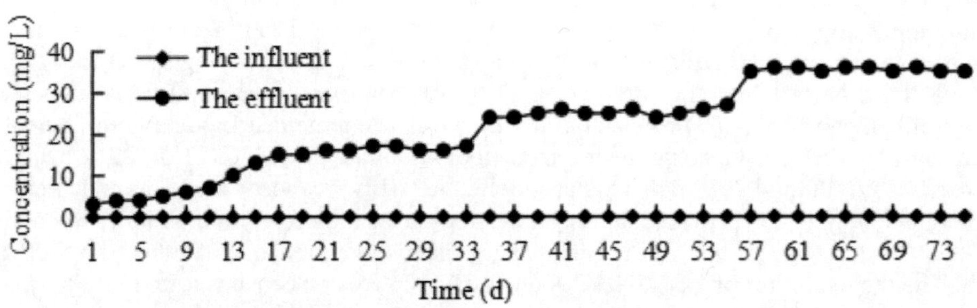

Figure 5. The variation of NO_3^--N concentration.

4. Conclusion

The "methanation + anaerobic ammonia oxidation autotrophic denitrification" method was adopted by using anaerobic sequencing batch reactors (ASBRs) and realized a satisfactory synchronous removal of chemical oxygen demand (COD) and ammonia-nitrogen (NH_4^+-N) in wastewater after 75 days operation. The linked-up anaerobic reactors were consisted of methanogenic reactor (R1) and anammox reactor (R2). The chemical oxygen demand (COD) and ammonia-nitrogen (NH_4^+-N) was in form of $C_6H_{12}O_6$ and NH_4Cl which were added into the synthetic water. During the first 54 days of whole start-up period, R1 and R2 were separately operated and the methanation process achieved the COD removal rate at 95% in R1. When the R1 and R2 were in linked-up station, the synthetic wastewater which contained COD and NH_4^+-N was discharged into the two-stage anaerobic reactors in sequence. Designed synchronous removal rates of COD and NH_4^+-N in the two-stage anaerobic reactors was reached after 75 days. It achieved that 95% of COD was removed with a COD load of 1.2kg/(m^3•d) and over 90% of TN was removed at a TN load of 0.14 kg/(m^3•d). The removal ratio of NH_4^+-N, NO_2^--N and NO_3^--N concentration was approximately close to 1:1.32:0.26. Compared with other anaerobic reactors, the stably operated ASBRs had many advantages in the field of controlling water pollution and protecting water environment.

Acknowledgments

The authors acknowledge the financial support of Shanxi Scholarship Council of China (No.2012-034) and focus on research and development plan in Shanxi Province (201603D221028-1), which have facilitated the carried work.

References

[1] Xin W, Baoan L and Song Z 2012 *Bioresour. Technol.* **122** 189-195
[2] Ming X and Peide S 2002 *Pollut. Control Technol.* **15** 25-27
[3] Shuai F, Can W and Baoting C 2016 *Desalin. Water Treat.* **57** 12755-62
[4] Strous M, Heijnen J.J and Kuenen J.G 1998 *Appl. Microbiol. Biotechnol* **50** 589-596

[5] Fux C, Boehler M and Huber P 2002 *J. Biotechnol* **99** 295-306
[6] Daijun Z 2003 *J.Environ. Sci.* **15** 423-432
[7] Wang J and Jing K 2005 *Process Biochem.* **40** 1973-78
[8] Wouter R.L, Vander S and Wiebe R.A 2007 *Water Res.* **41** 4149-63
[9] Yin Y, Wenyan Z and Jinbao W 2012 *Jiangxi Sci.* **30** 584-589
[10] Zu B, Daijun Z and Yuhua B 2007 *Re.Environ.Sci.* **20** 51-57
[11] William P. B and David C.S 1999 *Water Res.* **33** 1559-78
[12] Vander Star W, Muyzer G and Van Loosdrecht M 2008 *Biotechnol. Bioeng.* **101** 286-294
[13] Schmidt J. E and Ahring B.K 1996 Biotechnol. and Bioeng. **49** 229-246
[14] Gilmore K.R, Husovitz K.J and Holst T 1999 *Water Sci. and Technol.* **39** 227-234
[15] Linyong L and Yaxin L 2011 *Sci. Technol.* **27** 431-435
[16] Yaxin L and Xiuping Y 2008 *China Water. Wastewater* **24** 27-30
[17] Shaoqing Z, Liqiu Z and Chaosheng Z 2013 *gdchem.* **40** 96-97
[18] Qiaoling H, Yibo W and Di W 2008 *Indus. Safety Environ. Prot.* **34** 17-19
[19] Gilcreas F W. 1998 *A.P.H.A.* 113-X.
[20] Shaoqi Z and Junqin Y 2005 *J. Food Sci. Biotechnol.* **24** 1-5
[21] Chamchoi N and Nitisoravut S. 2007 *Chemosphere* **66** 2225-32
[22] Mingshi Z, Shaoqi Z and Wu Z 2008 *Chin. J. Environ. Eng.* **2** 11-15
[23] Dapena-Mora A, Campos J.L and Mosquera-Corral A 2004 *J. Biotechnol.* **110** 159-170
[24] Zhiquan Y, Shaoqi Z and Yanbo S 2009 *J. Hazard Mater.* **169** 113-118
[25] Mike J and Marc S 1998 *FEMS Microbiol. Rev.* **22** 421-437

Determining light requirements of groundcover plants from subtropical natural forest using hemispherical photography

Yi Zhang, Yonglin Zhong, Mingfeng Xu and Zhiyao Su[1]

College of Forestry and Landscape Architecture, South China Agricultural University, Guangzhou 510642, China

E-mail: zysu@scau.edu.cn

Abstract. In order to determine light requirements of indigenous groundcover plants for potential use in urban landscaping, we conducted a plant census in Yinpingshan Nature Reserve, Dongguan, China, and measured canopy structure and understory light regimes using hemispherical photography. We found that canopy openness, transmitted direct solar radiation, and transmitted diffuse solar radiation exhibited highly significant spatial heterogeneity. Species composition and diversity of groundcover plants were highly dependent on canopy structure and understory light condition. Greater diversity and more stems of groundcover plants were associated with greater canopy openness and understory radiation in most cases. Highly significant differences in species composition were detected along canopy openness, transmitted direct solar radiation, and transmitted diffuse solar radiation gradients, respectively. We also detected indicator species for specific understory light regimes, which will provide useful information when applying such species in urban greening under various light environments.

1. Introduction

Groundcover plants play an essential role in plant diversity maintenance, energy flow, and nutrient cycling in forest ecosystems [1]. Understanding the mechanisms of species composition and diversity patterns of groundcover plants is beneficial to the delivery of forest ecosystem services and to selecting plants for urban greening. Groundcover plants respond sensitively to environmental changes [2]. Understory light regime is regarded as one of the most important environmental factors driving groundcover plant distributions [3, 4], as light is the primary element in plant photosynthesis and directly affect the growth and development of plants [5]. In addition, light indirectly affect plant distributions through changing soil properties, including soil temperature, humidity, and nutrients [6]. Thus, heterogeneity in understory light contributes to the forming of species composition and diversity patterns of groundcover plants [2]. However, effects of understory light on the species composition and diversity patterns of groundcover plants remain unclear.

Understory light regimes are affected by many factors, particularly canopy structure and the distribution of overstory plants [7]. Forest canopy intercepts and redistributes light resources in the understory. Greater canopy openness results in greater availability and intensity of understory light. Therefore, canopy structure indirectly affects groundcover plants through altering understory light regimes [8]. Thus, effects of canopy structure and understory light on the species composition and diversity patterns of groundcover plants require further studies.

In this study, we collected data on canopy structure of overstory plants, understory light regime, and analyzed species composition and diversity patterns of groundcover plants in Yinpingshan Nature Reserve in Dongguan city, China. We assessed the effect of canopy structure, understory light regime on species composition and diversity patterns of groundcover plants.

2. Methods

2.1. Study area

The research was conducted in Yinpingshan Nature Reserve (22°52'N~22°56'N, 114°10'E~114°15'E), Dongguan city, China. The study site is located in the central south of Guangdong province and lies within the south subtropical monsoon climatic zone. The reserve receives abundant precipitation and solar radiation. The mean annual temperature and mean annual precipitation range from 21 to 22°C and from 1500 to 2400 mm, respectively. The mean relative humidity is 79%. The vegetation is largely evergreen monsoon forests and evergreen broadleaved forests. The soil type is largely mountain red soil, latosolic red soil, and yellow soil. The parent rock is granites and sand shales [9].

2.2. Data collection

We established two 1-ha permanent forest plots in the reserve. Each plot was a square of 100 m × 100 m and was divided into 25 subplots (20 m × 20 m). Five 2 m × 2 m quadrats were established in each subplot. In each quadrat, we recorded all groundcover plants by species name, number of stems, and coverage.

In each quadrat, a hemispherical image was obtained using a Nikon CoolPix 4500 digital camera connected with a Nikkor FC-E8 fisheye lens. Hemispherical images were analyzed using the free software Gap Light Analyzer 2.0 [2].

2.3. Statistical analysis

We used Kruskal-Wallis test to determine the differences in canopy openness, transmitted direct solar radiation, and transmitted diffuse solar radiation among plots. Canopy openness, transmitted direct solar radiation, and transmitted diffuse solar radiation were classified into three classes, respectively. We used Multi-response Permutation Procedures (MRPP) to test for the differences in species composition along a canopy openness gradient and each understory light gradient. We also detected indicator species in communities along a canopy openness gradient and each understory light gradient, using Indicator Species Analysis (ISA). Kruskal-Wallis test was performed using Statistica 8.0, while MRPP and ISA analyses were carried out using PC-ORD 6.0.

3. Results

3.1. Understory light regimes

Canopy openness varied strongly among plots, indicating highly significant spatial heterogeneity ($P < 0.0001$). Canopy openness of plot 26 to plot 50 was greater than those of plot 1 to plot 25 (Figure 1). Both transmitted direct solar radiation ($P < 0.0001$; Figure 2) and transmitted diffuse solar radiation ($P < 0.0001$; Figure 3) showed highly significant spatial heterogeneity and differed consistently with canopy openness.

3.2. Species composition and diversity

Number of stems, number of species, and Shannon-Wiener index of groundcover plants strongly fluctuated among plots. Plots with greater canopy openness and understory light intensity had more stems and species of groundcover plants in most cases (Figures 4-5). However, Shannon-Wiener index did not vary with canopy openness and understory light (Figure 6).

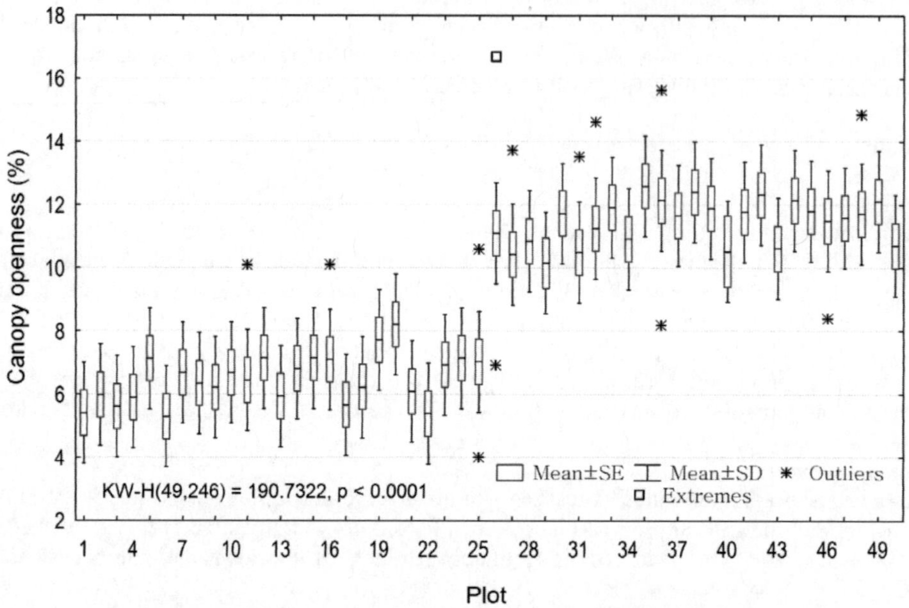

Figure 1. Spatial variations in canopy openness across plots.

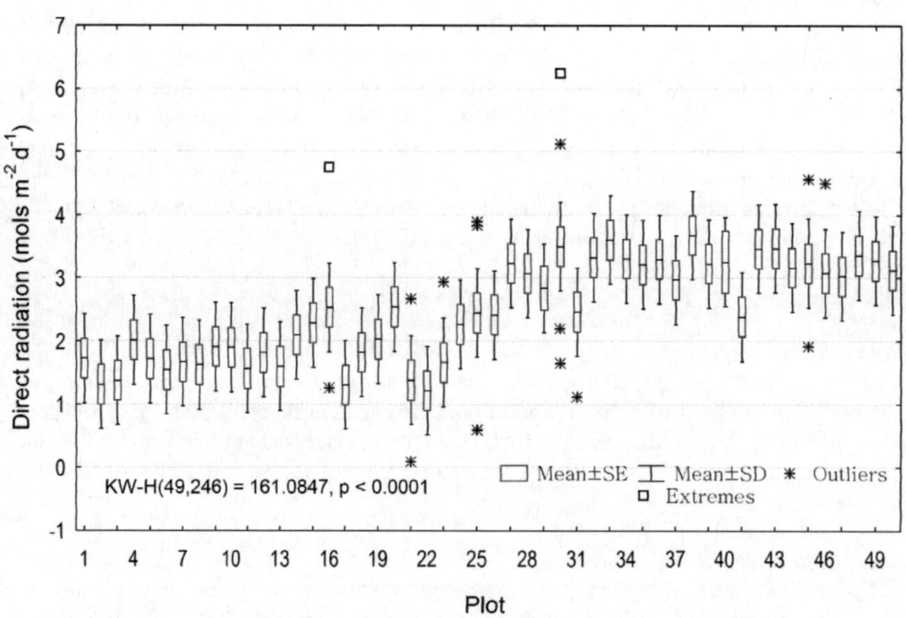

Figure 2. Spatial variations in transmitted direct radiation across plots.

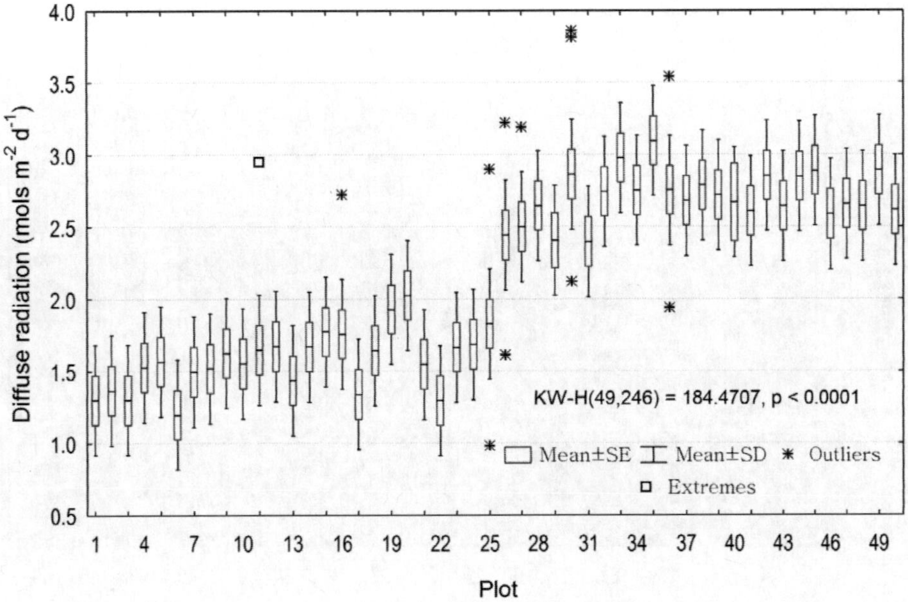

Figure 3. Spatial variations in transmitted diffuse radiation across plots.

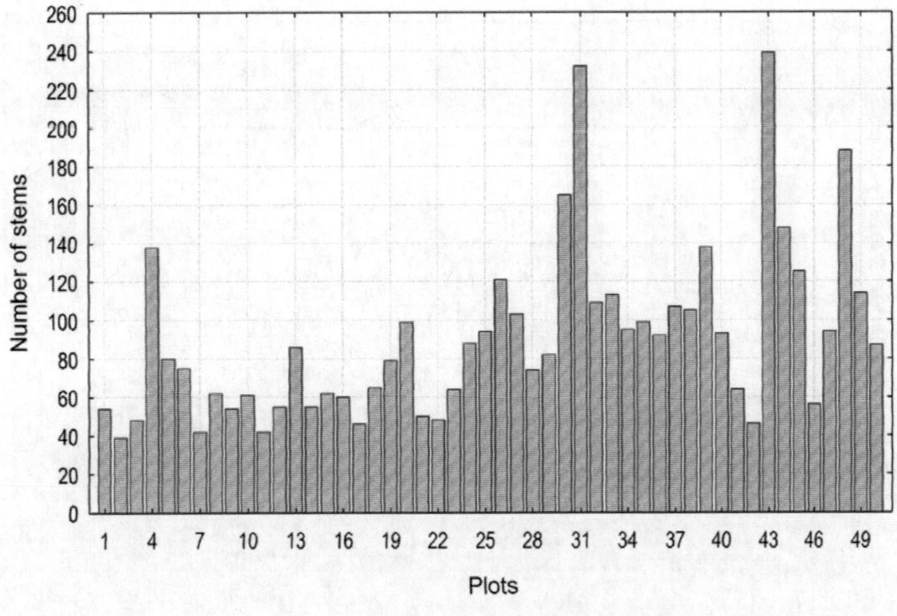

Figure 4. Number of species recorded in plots.

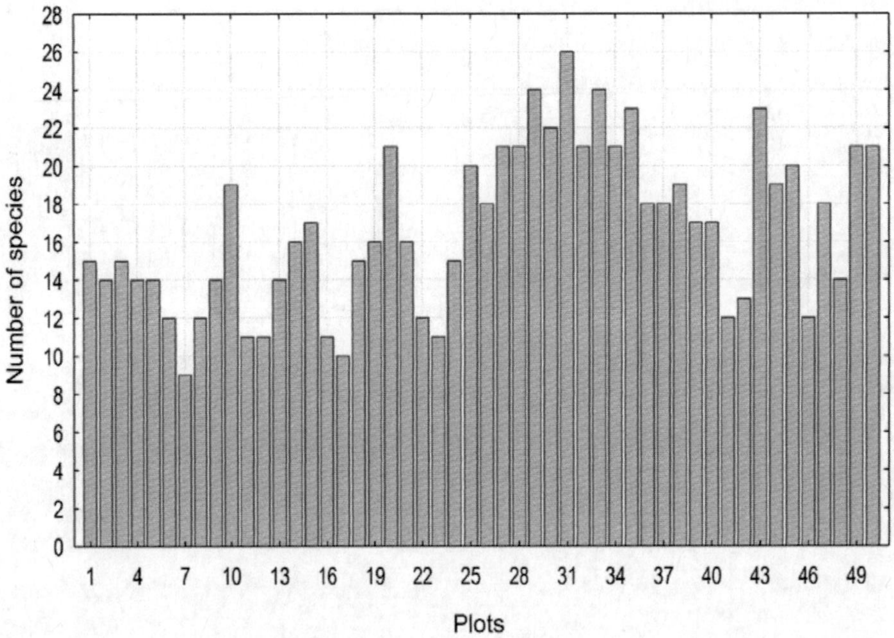

Figure 5. Number of stems recorded in plots.

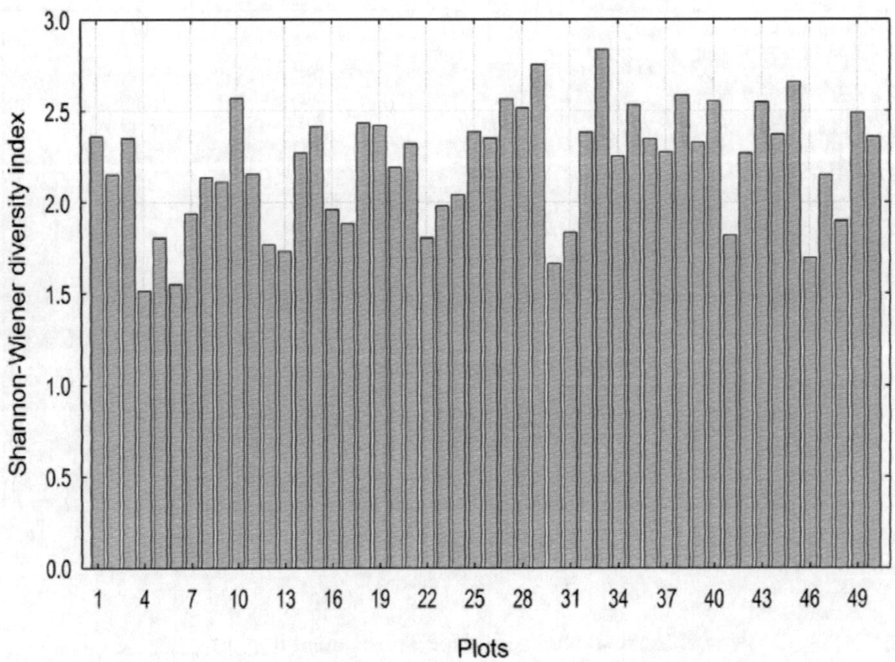

Figure 6. Shannon-Wiener diversity index for the sampling plots.

The species composition of groundcover plants was highly dependent on understory light regime, showing highly significant differences ($P < 10^7$) along the canopy openness, the transmitted direct solar radiation, and the transmitted diffuse solar radiation gradients, as assessed by multi-response permutation procedures (MRPP), respectively (Table 1).

Table 1. Multi-response permutation procedures used to test for the significance of variation in species composition across gap light-related gradients.

Grouping variable	Observed delta	Expected delta	*T*	*A*	*p*
Canopy openness	0.69	0.79	-13.75	0.13	$< 10^{-7}$
Direct radiation	0.67	0.79	-17.03	0.16	$< 10^{-7}$
Diffuse radiation	0.67	0.79	-16.63	0.15	$< 10^{-7}$

3.3. *Indicator species of light gradient*

Indicator species analysis revealed the preference of plant species for habitats [10]. Seven indicators were detected in the habitats with low canopy openness, transmitted direct solar radiation, or transmitted diffuse solar radiation. However, only two species, i.e., *Millettia dielsiana* and *Smilax china*, were the shared indicators (Table 2-4). Five indicators were found in the habitats with medium openness, transmitted direct solar radiation, or transmitted diffuse solar radiation, of which *Dalbergia hancei* and *Gnetum parvifolium* were the shared indicators (Table 2-4). Eleven indicators were found in the habitats with high canopy openness, transmitted direct solar radiation, or transmitted diffuse solar radiation, and only *Rhododendron simsii* was the shared indicator (Table 2-4).

Table 2. Groundcover plants as indicators of canopy openness gradient.

Canopy openness	Species	Indicator value	IV from randomization		*P*
			Mean	S.D.	
Low	*Camellia oleifera*	23.5	8.8	4.22	0.028
Low	*Castanopsis fissa*	29.3	15.4	5.57	0.029
Low	*Cunninghamia lanceolata*	36.6	13.9	5.63	0.004
Low	*Millettia dielsiana*	47.1	13	5.43	0.000
Low	*Smilax china*	48.3	18.5	5.74	0.001
Medium	*Dalbergia hancei*	25	8.8	4.4	0.010
Medium	*Gnetum parvifolium*	25.8	12	5.19	0.020
High	*Alyxia sinensis*	34	12.8	5.58	0.004
High	*Elaeagnus loureirii*	23.5	8.9	4.36	0.032
High	*Rhododendron simsii*	39	15.9	5.92	0.004

4. Conclusions

Highly significant spatial heterogeneity existed in canopy openness, transmitted direct solar radiation, and transmitted diffuse solar radiation. The number of stems and number of species also varied spatially and showed similar trends with canopy structure and understory light regimes.

Plants can be used as indicators for identifying special habitats. *Millettia dielsiana* and *Smilax china* were indicative of low light and low canopy openness environment; *Dalbergia hancei* and *Gnetum parvifolium* were indicative of medium light and medium canopy openness environment; and *Rhododendron simsii* was indicative of high light and high canopy openness environment. Thus, we regarded *Millettia dielsiana* and *Smilax china* as shade-tolerant plants and *Rhododendron simsii* as light-demanding plant. These indicators were dominant species in their specific habitats and had useful information for plant application in urban greening and landscaping in various light environments.

Table 3. Groundcover plants as indicators of direct radiation gradient.

Direct radiation	Species	Indicator value	IV from randomization		*P*
			Mean	S.D.	
Low	*Camellia oleifera*	23.5	8.9	4.52	0.014
Low	*Castanopsis fissa*	27.5	15.5	5.66	0.041
Low	*Cunninghamia lanceolata*	35.7	14	5.54	0.005
Low	*Evodia lepta*	35.2	21.7	6.43	0.038
Low	*Liquidambar formosana*	17.6	7.7	4.2	0.048
Low	*Millettia dielsiana*	33.6	13.1	5.46	0.004
Low	*Smilax china*	41	18.6	5.79	0.005
Medium	*Dalbergia hancei*	30.8	8.8	4.43	0.004
Medium	*Gnetum parvifolium*	42.9	12.1	5.21	0.001
Medium	*Litsea cubeba*	25.8	10.9	4.96	0.016
Medium	*Viburnum fordiae*	27.1	15.3	5.46	0.040
High	*Cayratia japonica*	43.2	18.3	6.78	0.003
High	*Clematis uncinata*	25	10.1	4.91	0.021
High	*Elaeagnus loureirii*	20	8.9	4.61	0.040
High	*Pandanus austrosinensis*	25.2	11.7	5.07	0.023
High	*Rhododendron simsii*	45.4	15.9	5.93	0.001
High	*Rhodoleia championii*	22.2	11.2	5.14	0.050
High	*Tetrastigma hemsleyanum*	46.6	16.4	5.69	0.000
High	*Tricalysia dubia*	25.6	11.9	5.11	0.020

Table 4. Groundcover plants as indicators of diffuse radiation gradient.

Diffuse radiation	Species	Indicator Value	IV from randomization		*P*
			Mean	S.D.	
Low	*Evodia lepta*	38	21.8	6.61	0.026
Low	*Millettia dielsiana*	61.5	13.2	5.57	0.000
Low	*Smilax china*	31.9	18.7	5.9	0.035
Medium	*Dalbergia hancei*	25	8.9	4.46	0.010
Medium	*Gnetum parvifolium*	33.4	12.2	5.42	0.006
Medium	*Pithecellobium lucidum*	34.6	14.5	5.45	0.005
Medium	*Viburnum fordiae Hance*	48.2	15.4	5.53	0.000
High	*Cayratia japonica*	46.2	18.4	6.86	0.002
High	*Clematis uncinata*	23.8	10.1	4.88	0.022
High	*Pandanus austrosinensis*	33.3	11.7	5.05	0.003
High	*Rhododendron simsii*	43.9	16	6.07	0.002
High	*Rhodoleia championii*	21.5	11.2	5.18	0.048
High	*Schefflera octophylla*	32.3	15.6	5.74	0.017
High	*Smilax hypoglauca*	23.8	10	4.81	0.023
High	*Tricalysia dubia*	24.9	11.9	5.24	0.030

Acknowledgments

This research is supported by the Science and Technology Planned Programs of Guangdong Province (Grant No. 2013B020305008) and a grant from Dongguan Forestry Administration of Guangdong (DCD-2013-115).

References:

[1] Bartels S F and Chen H Y H 2010 Is understory plant species diversity driven by resource quantity or resource heterogeneity? *Ecology* **91** 1931-38

[2] Ou Y, Su Z, Ke X and Li Z 2012 Vascular ground flora in relation to topography, canopy structure and gap light regimes in a subtropical broadleaved forest (south china) *Pol. J. Ecol.* **60** 463-478

[3] Ou Z, Cao J, Shen W, Tan Y, He Q and Peng Y 2015 Understory flora in relation to canopy structure, soil nutrients, and gap light regime: A case study in southern china *Pol. J. Environ. Stud.* **24** 2559-68

[4] Marialigeti S, Tinya F, Bidlo A and Odor P 2016 Environmental drivers of the composition and diversity of the herb layer in mixed temperate forests in Hungary *Plant Ecol.* **217** 549-563

[5] Selzer L J and Busso C A 2016 Pigments and photosynthesis of understory grasses: Light irradiance and soil moisture effects *Russ. J. Plant Physl.* **63** 224-234

[6] Abdallah F, Michalet R, Maalouf J, Ouled-Dhaou S, Touzard B, Noumi Z and Chaieb M 2016 Disentangling canopy and soil effects of a savanna tree species on its understorey *J. Veg. Sci.* **27** 771-779

[7] Abella S R and Springer J D 2015 Effects of tree cutting and fire on understory vegetation in mixed conifer forests *Forest Ecol. Manag.* **335** 281-299

[8] Cook J E 2015 Structural effects on understory attributes in second-growth forests of northern Wisconsin, USA *Forest Ecol. Manag.* **347** 188-199

[9] Cui J, Zeng H, Wang Y, Zhang Y, Hu Y and Su Z 2015 Canopy structure and radiation attenuation effects of broad-leaved forest in yinpingshan national nature reserve *Journal of Northwest Forestry University* **30** 45-49

[10] Su Z, Ke X and Zhang S 2012 Vascular plants as indicators of organic carbon gradient in subtropical forested soils *Pol. J. Environ. Stud.* **21** 1393-98

Growth process and model simulation of three different classes of *Schima superba* in a natural subtropical forest in China

Hui Wei[1], Xiangwen Deng [1, 2, 4], Shuai Ouyang [1, 2], Lijun Chen[3] and Yonghe Chu[3]

[1] Faculty of Life Science and Technology; Central South University of Forestry and Technology, Changsha, 410018, China
[2] National Engineering Laboratory for Applied Technology of Forestry & Ecology in South China, Changsha, 410018, China
[3] Forestry Bureau of JingZhou Miao and Dong autonomous county, HuaiHua, 418413, China

E-mail: dxwfree@126.com

Abstract. *Schima superba* is an important fire-resistant, high-quality timber species in southern China. Growth in height, diameter at breast height (DBH), and volume of the three different classes (overtopped, average and dominant) of *S. superba* were examined in a natural subtropical forest. Four growth models (Richards, edited Weibull, Logistic and Gompertz) were selected to fit the growth of the three different classes of trees. The results showed that there was a fluctuation phenomenon in height and DBH current annual growth process of all three classes. Multiple intersections were found between current annual increment (CAI) and mean annual increment (MAI) curves of both height and DBH, but there was no intersection between volume CAI and MAI curves. All selected models could be used to fit the growth of the three classes of *S. superba*, with determinant coefficients above 0.9637. However, the edited Weibull model performed best with the highest R^2 and the lowest root of mean square error (RMSE). *S. superba* is a fast-growing tree with a higher growth rate during youth. The height and DBH CAIs of overtopped, average and dominant trees reached growth peaks at ages 5–10, 10–15 and 15–20 years, respectively. According to model simulation, the volume CAIs of overtopped, average and dominant trees reached growth peaks at ages 17, 55 and 76 years, respectively. The biological rotation ages of the overtopped, average and dominant trees of *S. superba* were 29, 85 and 128 years, respectively.

1. Introduction

Schima superba is a large evergreen species of the Theaceae family, also known as lotus wood or the lotus tree. It has fire-resistant qualities and produces high-grade timber, and is mainly distributed at an elevation of 400~1500 m to the south of the Yangtze River in China. With current global warming and the threat of forest fires [1], it is useful to study the growth process and characteristics of this fire-resistant tree species. Cheng examined the relationship between the height and diameter at breast height (DBH) of *S. superba* [2]; however, the growth process of *S. superba* has seldom been reported. Due to the differentiation within a forest, trees in the same forest stand have different growth characteristics [3]. Thus, being an excellent fire-resistant species, it is important to investigate the growth process and model simulations of *S. superba*, especially for the three different classes of trees (overtopped, average and dominant).

This study aimed to investigate the growth process of the different classes of *S. superba* using stem analysis data and four growth models (Richards, edited Weibull, Logistic and Gompertz) in natural subtropical forest. We suggest that our results will provide important information for the cultivation of forest fire prevention belts and will help to optimize forestry operations and management.

2. Materials and Methods

2.1. Site description

The study site was located in the Paiyashan state-owned forest farm, Jingzhou Miao and Dong autonomous county, Hunan province, China (latitude 26°24'-26°35'N; longitude 109°27'-109°38'E, elevation 400–800 m above sea level; figure 1).

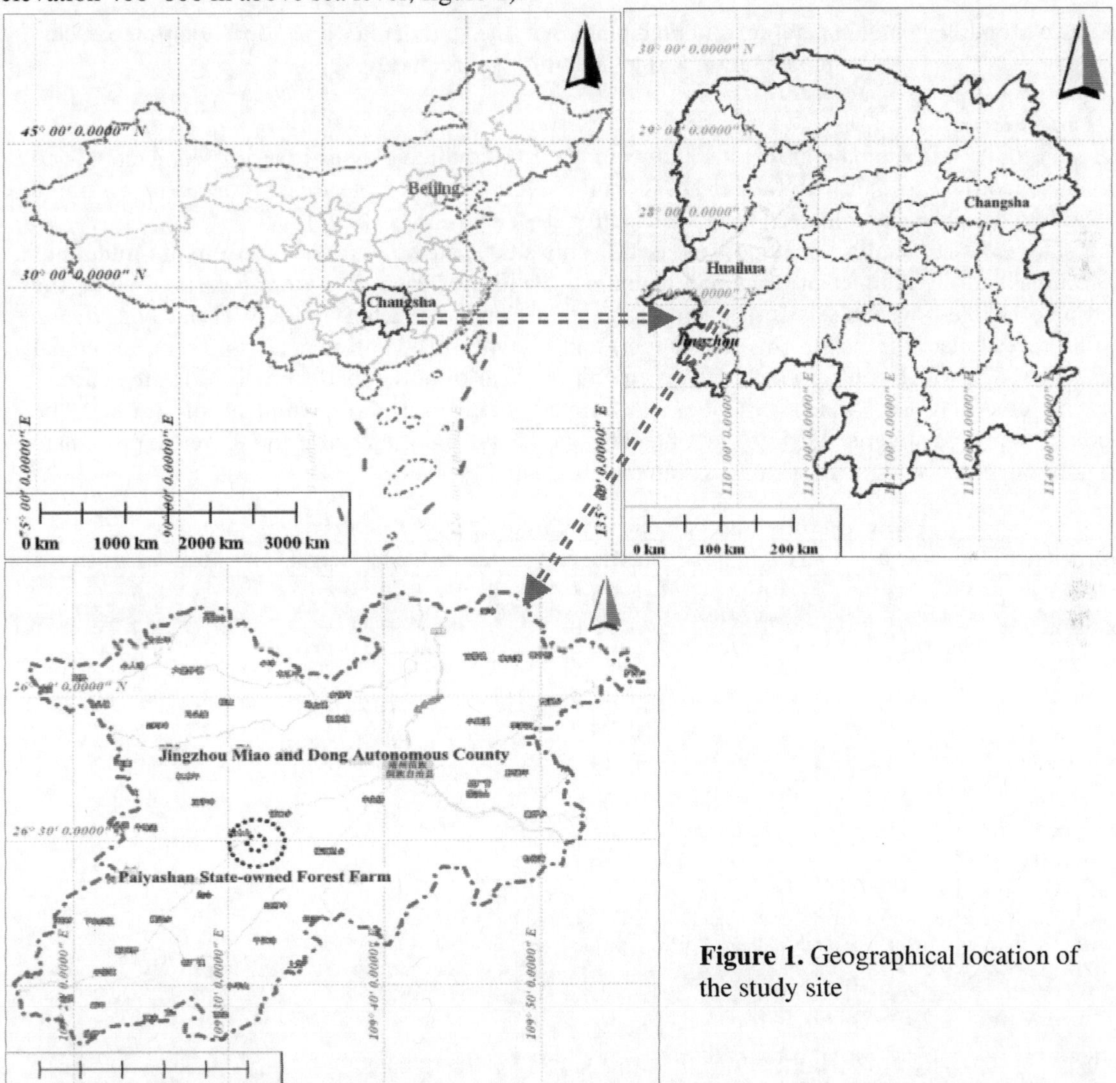

Figure 1. Geographical location of the study site

The region has a humid mid-subtropical monsoon climate with a mean annual temperature of 16.7 °C. Mean annual precipitation amounts to 1250 mm, occurring mostly between April and August (Figure 2). The soil parent material is purple sandshale and with a red soil or a mountain yellow soil. The dominant vegetation is plantations and secondary forests that are composed of many native tree species.

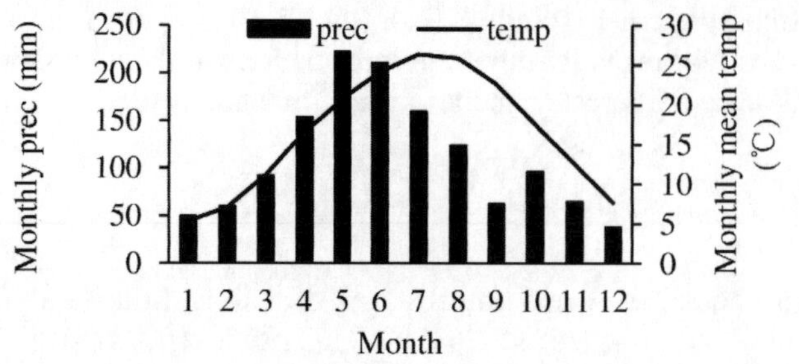

Figure 2. Monthly mean temperature and precipitation in the study area. 'Prec' and 'temp' represent precipitation and temperature, respectively.

2.2. Data collection

All trees in the *S. superba* natural forest stand were placed into three classes (overtopped, average and dominant) according to their height and DBH. Three overtopped sample trees, four average sample trees and two dominant sample trees were selected for stem analysis (Table 1).

The standard stem analysis method was used in this study, a disc was cut out from the middle of each section, and the number of rings and annual ring width of each disc were measured using the WinDENDRO tree-ring analysis instrument (Regent Instruments Canada INC.). Visual FoxPro 9.0 was used to calculate the height (m), DBH (cm) and volume (m^3). Formula (1) is the mean annual increment (MAI), which refers to the average growth per year of a tree; and formula (2) is the current annual increment (CAI), which describes the annual change in the increment of a tree. The interpolation method was used to calculate height increments, and the central basal area approximate quadrature method was used to calculate volume increments [4].

Table 1. Information on classified trees (*Schima superba*)

Classified trees	Age	Height (m)	DBH (cm)	Under-branch height (m)	Mean crown (m)
Dominant tree [a]	56	18.2	28.5	5.4	9.0
Dominant tree	65	18.5	33.8	6.7	8.0
Average tree [b]	30	14.6	13.0	4.6	3.9
Average tree	30	15.6	17.8	2.9	4.5
Average tree	52	16.4	24.3	7.6	6.8
Average tree	56	12.5	19.9	4.4	9.0
Overtopped tree [c]	31	9.0	7.4	5.6	3.5
Overtopped tree	34	10.0	11.1	3.5	4.0
Overtopped tree	55	11.9	14.1	2.0	6.1

[a]: Growth best, no pests, with the largest crown, and occupy the upper canopy;
[b]: Growth better, no pests, with a narrow crown, and located in the middle canopy;
[c]: Growth worst, no pests, with a severely squeezed crown, and in a state of obvious suppression.

$$MAI = Y_t/t \tag{1}$$

$$CAI = Y_t - Y_{t-1} \tag{2}$$

Where Y is the height, DBH or volume increment, and t is the tree age.

2.3. Model description and selection criteria

Tree growth models express the regularity of tree growth increments that change along with age using mathematical equations [5]. In the current study, we chose the Richards, edited Weibull, Logistic and Gompertz growth models (Table 2) according to the principles of model building, design flexibility, good biological explanation and superior applicability [6]. Model parameters were fit using the

Levenberg–Marquardt optimization algorithm [7] of the 1stOpt program (7D-Soft High Technology INC.).

Table 2. Growth models selected for comparison of classified trees

Model	Formula	Variables	Parameters *
Richards	$y = a(1 - e^{-bx})^c$	x, y	a, b, c
Edited Weibull	$y = a(1 - e^{-bx^c})$	x, y	a, b, c
Logistic	$y = a/(1 + be^{-cx})$	x, y	a, b, c
Gompertz	$y = ae^{-be^{-cx}}$	x, y	a, b, c

*'a' is the parameter of biological extreme point; 'b' is a parameter indicating the speed of tree growth; 'c' is a parameter concerned with anabolism.

Model evaluation and comparisons were based on graphical and numerical analyses of the values of the following statistics: formula (3), root of mean square error (RMSE) [8], which is frequently used to measure the differences between the values (sample and population values) predicted by a model or an estimator and the values actually observed (the smaller the value, the smaller the value of the error); formula (4), coefficient of determination (R^2) [9], which reflects the consistency between simulated and measured values (the closer to 1 the value, the better the fit). The expressions for these statistics are as follows:

$$RMSE = \sqrt{\frac{\sum_{i=1}^{n}(y_i - \hat{y}_i)^2}{n - p - 1}} \quad (3)$$

$$R^2 = 1 - \frac{\sum_{i=1}^{n}(y_i - \hat{y}_i)^2}{\sum_{i=1}^{n}(y_i - \bar{y}_i)^2} \quad (4)$$

Where, y_i, $\hat{y}_i$ and $\bar{y}_i$ are the observed, predicted and mean values of height, DBH or volume increment, respectively; n is the total number of data used in fitting the model; and p is the number of independent variables.

3. Results

3.1. Growth in height

The height CAI and MAI curves for the three classes (dominant, average and overtopped) of *S. superba* are presented in figure 3. The height MAI curves of dominant and average trees remained constant with time. However, the height MAI curve of overtopped trees gradually decreased with time and levelled off from the second year. The height CAI curves for the three tree classes showed a multiple-peak phenomenon. Height CAI curve peaks for dominant trees appeared at age 16 years, 23 years and 51 years; for average trees at age 6 years and 23 years; and for overtopped trees in years 1, 3 and 9. There were multiple intersections between height CAI and MAI curves, which indicated there were multiple distorted relationships between height CAI and MAI curves for the three classes of trees. It was therefore necessary to investigate the distortion, and regression methods and model simulations were used to reduce it.

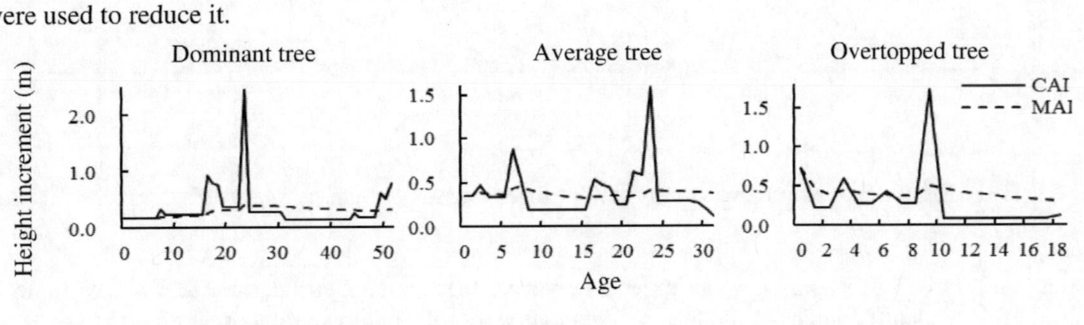

Figure 3. Height growth process of three classes of *Schima superba*

3.2. DBH growth process

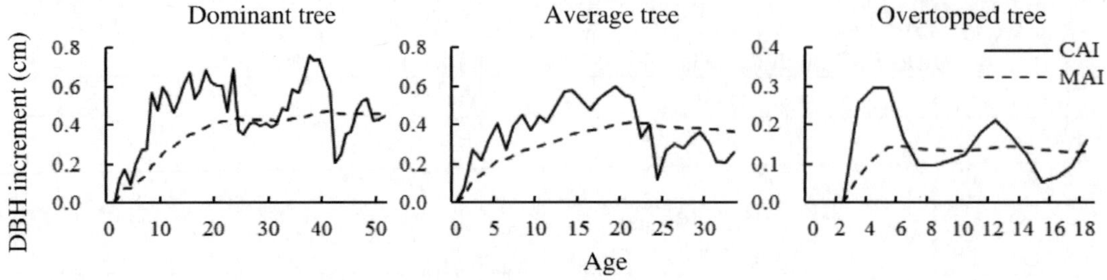

Figure 4. DBH growth process of three classes of *Schima superba*

Figure 4 shows the DBH CAI and MAI curves for the three classes of *S. superba*. The DBH MAIs increased with time during early life. Subsequently, the MAI curves of the dominant, average and overtopped trees started to stabilize in the 25th year, 22nd year and 6th year, respectively. The DBH CAI curves of dominant and average trees took on a fluctuant increasing process during their early life. Later on, the DBH CAI curve for dominant trees had two troughs at 24–31 years and 42 years; but there were 2 troughs for average trees at 24 years and 32 years. The DBH CAI for overtopped trees increased to a maximum during the first two years when the height of overtopped trees reached breast height; the trough values of the DBH CAI curve for overtopped trees appeared at age 7–10 years and 15 years, meanwhile, a peak appeared at 12 years. There were multiple intersections between DBH CAI and MAI curves, but less than that of the tree height curves, especially in the average class. In other words, the DBH was less influenced than tree height by random factors such as environment, tending and thinning.

3.3. Volume growth process

Figure 5 presents the volume CAI and MAI curves for the three classes of *S. superba*. Growths in volume for all types were similar; CAIs and MAIs increased year by year, but the CAI curve of dominant trees showed some fluctuation. Furthermore, volume CAI was always greater than volume MAI in the three classes. The volume of dominant and average trees increased slowly during the early growth stage; later on, dominant trees at 5 years and average trees at 7 years began to enter a period of fast-growth. Moreover, both volume CAI and MAI curves showed no downward trend and did not intersect. In other words, the age of all sample trees did not achieve maximum volume MAI, as confirmed by the models below.

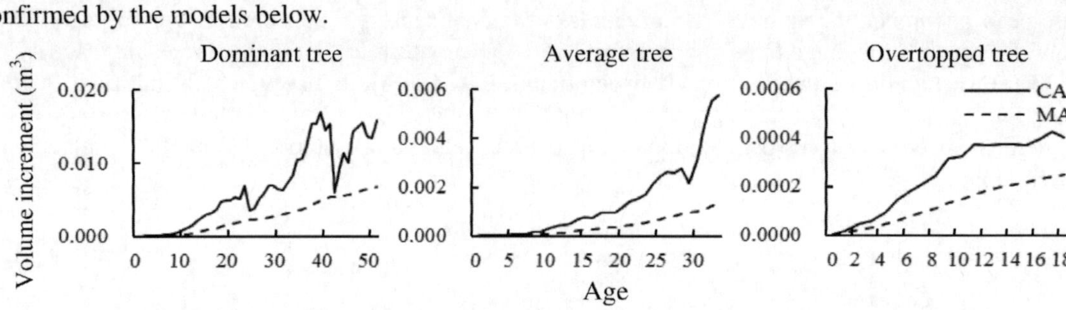

Figure 5. Volume growth process of three classes of *Schima superba*

3.4. Model fit and selection

In current study, tree age was used as an independent variable, and annual height, DBH and volume were used as dependent variables in the four models that were selected to simulate tree growth. The R^2 values of the four models were higher and similar among the four models (Table 3). The minimum value of R^2 was 0.9637 when using the Richards model to simulate growth in height of overtopped *S. superba*. The RMSE values ranged from 0.0001 to 0.8496, and the edited Weibull model showed the best fit to the growth of all three tree classes (with the highest R^2 and the lowest RMSE).

3.5. Model simulation of height

The most suitable model to simulate the height increment was the edited Weibull model. The height CAI and MAI fitted curves of the three classes of *S. superba* are presented in figure 6. As can be seen from the graph, the relationship between the height CAI and MAI is very clear. During the youthful stage, the height CAI and MAI increased with age, but the CAI curves increased faster and reached maximum values earlier than that of the MAI curves. The CAI and MAI curves crossed when the MAI reached its maximum; from then on, the CAI always remained less than the MAI.

The height CAI of dominant trees reached its maximum value of 0.54 $m \cdot a^{-1}$ in the 17th year, and the MAI reached its maximum value of 0.37 $m \cdot a^{-1}$ in the 29th year. The height CAI and MAI of average trees reached their maximum values earlier than the dominant trees, which were 0.43 $m \cdot a^{-1}$ in the 10th year and 0.39 $m \cdot a^{-1}$ in the 17th year, respectively. The height CAI and MAI of overtopped trees reached their maximum values at the earliest age, which were 0.53 $m \cdot a^{-1}$ in the 4th year and 0.45 $m \cdot a^{-1}$ in the 8th year, respectively.

Table 3. Parameters and related statistics for the four growth models

Model	Tree class	Growth factor	Parameters a	b	c	R^2	RMSE
Richards	Dominant	Height	15.746	0.081	3.927	0.9879	0.5852
		DBH	41.293	0.026	1.904	0.9977	0.3677
		Volume	2.390	0.017	3.682	0.9984	0.0043
	Average	Height	17.973	0.041	1.448	0.9849	0.5573
		DBH	15.614	0.074	2.636	0.9988	0.1479
		Volume	0.618	0.037	7.608	0.9975	0.0030
	Overtopped	Height	6.812	0.147	1.756	0.9637	0.3586
		DBH	3.039	0.130	2.410	0.9893	0.0871
		Volume	0.012	0.077	3.553	0.9999	0.0001
Edited Weibull	Dominant	Height	14.734	0.001	2.136	0.9870	0.5926
		DBH	34.292	0.002	1.623	0.9974	0.3952
		Volume	0.704	0.000	3.031	0.9985	0.0041
	Average	Height	16.609	0.011	1.334	0.9856	0.5452
		DBH	13.540	0.003	1.922	0.9992	0.1204
		Volume	0.618	0.037	7.608	0.9975	0.0030
	Overtopped	Height	6.252	0.034	1.551	0.9680	0.3418
		DBH	2.779	0.012	1.746	0.9881	0.0933
		Volume	0.007	0.001	2.606	0.9998	0.0000
Logistic	Dominant	Height	14.341	32.207	0.161	0.9876	0.5774
		DBH	24.872	20.708	0.104	0.9885	0.8496
		Volume	0.457	164.536	0.118	0.9980	0.0051
	Average	Height	14.503	10.871	0.121	0.9867	0.5220
		DBH	12.449	25.208	0.190	0.9965	0.2595
		Volume	0.244	523.444	0.143	0.9972	0.0032
	Overtopped	Height	6.007	11.704	0.347	0.9790	0.2686
		DBH	2.481	20.172	0.322	0.9755	0.1343
		Volume	0.006	80.377	0.313	0.9977	0.0001
Gompertz	Dominant	Height	15.366	5.646	0.095	0.9887	0.5549
		DBH	29.599	4.197	0.055	0.9949	0.5531
		Volume	0.991	7.906	0.039	0.9985	0.0042
	Average	height	15.720	3.085	0.072	0.9874	0.5059
		DBH	14.058	4.707	0.105	0.9991	0.1242
		volume	0.485	12.085	0.049	0.9976	0.0029
	Overtopped	Height	6.412	3.206	0.210	0.9717	0.3123
		DBH	2.712	4.485	0.192	0.9851	0.1030
		Volume	0.008	6.816	0.130	0.9998	0.0001

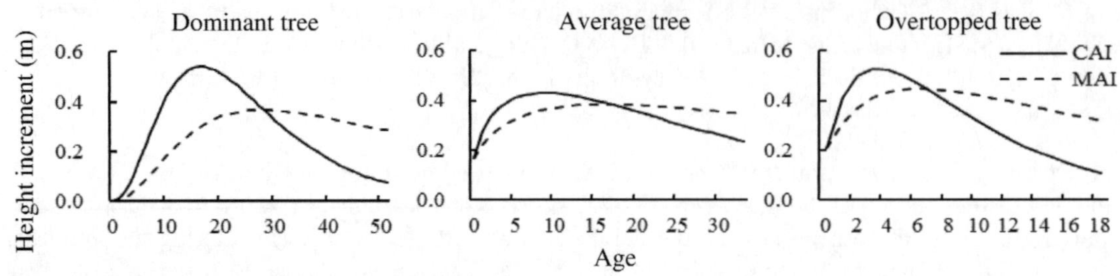

Figure 6. Height CAI and MAI fitted curves of the three classes of *S. superba* using edited Weibull model

3.6. Model simulation of DBH

Figure 7 shows the DBH CAI and MAI fitted curves according to the best fit model (edited Weibull model) for the three classes of *S. superba*. The process of growth in DBH is similar to that of height. A higher CAI rate than MAI rate appeared during the youthful stage. Since the two curves crossed at the time when the MAI reached its maximum, the CAI remained less than the MAI. From dominant trees to average trees to overtopped trees, the age of these intersections appeared earlier and earlier. The DBH CAI of dominant trees reached its maximum value of 0.55 $cm \cdot a^{-1}$ in the 34th year, and the MAI reached its maximum value of 0.46 $cm \cdot a^{-1}$ in the 45th year. The DBH CAI of average trees reached a maximum of 0.53 $cm \cdot a^{-1}$ in the 14th year, with a maximum MAI of 0.40 $cm \cdot a^{-1}$ in the 24th year. The DBH CAI of overtopped trees reached its maximum value of 0.18 $cm \cdot a^{-1}$ in the 9th year, with a maximum MAI of 0.14 $cm \cdot a^{-1}$ in the 13th year.

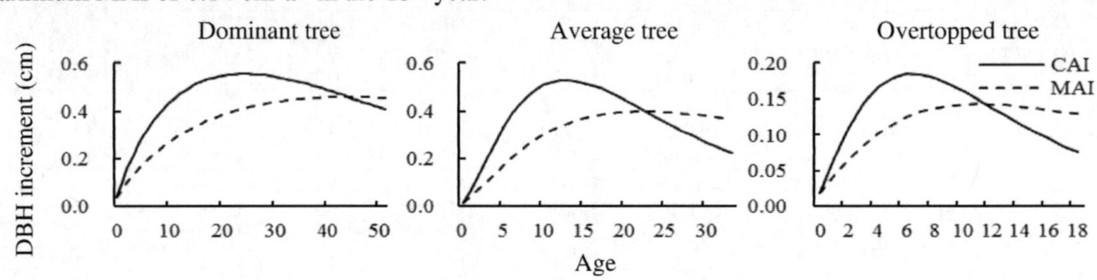

Figure 7. DBH CAI and MAI fitted curves of the three classes of *S. superba* using edited Weibull model

3.7. Model simulation of volume

Figure 8 presents the volume CAI and MAI fitted curves according to the edited Weibull model for the three *S. superba* classes. The relationship between volume CAI and MAI curves was similar to that of the height and DBH. The volume CAI of dominant trees reached a maximum value of 0.0177 $m^3 \cdot a^{-1}$ in the 76th year, and the two curves crossed when the MAI of volume reached a maximum value of 0.0123 $m^3 \cdot a^{-1}$ in the 128th year, which indicated that the biological rotation age of the average trees was 128 years. The biological rotation age is the age at which a tree or stand is harvested if the management objective is to maximize long-term yield [10]. The volume CAI of average trees reached a maximum value of 0.0091 $m^3 \cdot a^{-1}$ in the 55th year, and the two curves crossed when the MAI reached a maximum value of 0.0053 $m^3 \cdot a^{-1}$ in the 85th year, the biological rotation age of the average trees was therefore 85 years. The volume CAI of overtopped trees reached its maximum value of 0.0004 $m^3 \cdot a^{-1}$ in the 17th year, the MAI reached its maximum value of 0.0003 $m^3 \cdot a^{-1}$ in the 29th year, and the biological rotation age of overtopped trees was therefore 29 years.

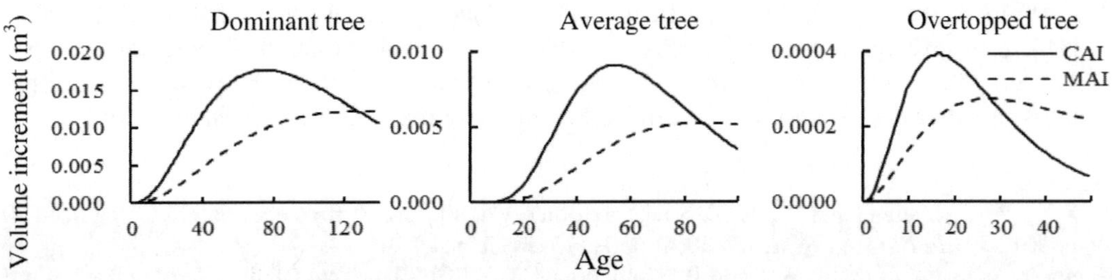

Figure 8. Volume CAI and MAI fitted curves of the three classes of *S. superba* using edited Weibull model

4. Discussion

The current study discussed the growth of three classes (overtopped, average and dominant) of *S. superba* in a natural forest based on stem analysis data. The results revealed that there were many fluctuations during the current annual growth process of the three tree classes. Previous studies show that the height CAI of *Fokienia hodginsii* (Dunn) has four peak values and the DBH CAI has two peak values [11]; multiple intersections have been found between the height or DBH CAI and MAI curves of *Quercus acutissima* [12]. These conclusions are corroborated by the current study on *S. superba,* indicating that the fluctuation phenomenon during the growth process is not species-typical, and may be due to climatic factors or human disturbance, such as tending and cutting. There have been many studies on the relationships between tree growth and climatic factors. For example, Zhang *et al.* found that the productivity of *Pinus massoniana* is significantly affected by relative humidity, moisture, air temperature, sunshine duration and so on [13].

The effectiveness of using *S. superba* as a biological fire prevention belt relate to the height under branch, the size of the crown, leaf water content and the volume of the trunk [14]. To understand the fire prevention features of *S. superba*, this study simulated its growth process using four growth models: Richards, edited Weibull, Logistic and Gompertz. The results showed that all four growth models could fit the height, DBH and volume increments of the three classes of *S. superba*. Wang *et al.* studied the growth process and models of *Quercus veriablilis*; also, found that the Weibull, Logistic, and Gompertz models closely fit the DBH growth in natural forests [15]. The edited Weibull model produced the best fit in the current study, and these results suggest that it is a good predictor of tree growth too.

Tree growth varies due to the impact of genetics and environment (climate factors, site condition, competition, management and so on) [16]. Ma *et al.* found that growth differs among individual *Pinus elliottii* trees, but the overall growth trends are consistent [17]. In the current study, differences between the growth process of the three classes of *S. superba* were found; however, all the CAI and MAI curves of height and DBH intersected early (Figure 6, 7), indicating that *S. superba* is a fast-growing tree with a high growth rate during youth and that this rate decreases with time. Previous studies have shown that, for *S superba*, maximum growth rate in height and DBH take place during the first 20 years [18]; similar results were found in the average and overtopped trees in the current study. Therefore, to produce good timber of large diameter, it is important to apply fertilizer and practice thinning during this fast-growing period.

5. Conclusions

In this study, growth in height, DBH and volume of dominant, average and overtopped trees of *S. superba* were investigated. Four growth models (Richards, edited Weibull, Logistic, and Gompertz) were selected to fit the growth process. The results revealed a fluctuation phenomenon during the current annual height and DBH growth processes for the three tree classes. All models could be used to fit the growth of the three classified trees of *S. superba*, with determinant coefficients above 0.9637. However, the edited Weibull model performed best with the highest R^2 and the lowest RMSE. *S. superba* is an early fast-growing tree with a higher growth rate during youth. CAIs in tree height and

DBH of overtopped, average and dominant trees reached growth peaks at ages of 5–10, 10–15 and 15–20 years [18], respectively. However, the CAIs of the volume of overtopped, average and dominant trees reached growth peaks at 17, 55 and 76 years, respectively. The mature ages of the overtopped, average and dominant trees of *S. superba* were 29, 85 and more than 100 years, respectively.

References

[1] Di X Y, Chu X, Yang G and Wu H 2015 Distribution of summer forest fires occurred in China from 2000 to 2012 *World Forestry Research* **28(4)** pp 72-75 (in Chinese)

[2] Cheng Y N, She J Y, Meng W Deng B P and Wang W X 2014 The *Schima superba* diameter at breast height (DBH) and height relation model research *Journal of Fujian Forestry Science and Technology* **41(2)** pp 109-114 (in Chinese)

[3] Yu B, Wu J S G L, Wang B T and Wang L M 2008 Analysis on classified stem transformation characteristic of *Larix gmelinii* natural forest at Daxingan mountains *Acta Ecologica Sinica* **28(11)** pp 5750-57 (in Chinese)

[4] Meng X Y 2006 Holzmesslehre (Beijing: China Forestry Publishing House) pp 171-196 (in Chinese)

[5] Yang B G, Feng Z K, Zhang B G, Han G S and Quan M Y 2008 Growth model for diameter at breast height of Chinese white poplar at Wenyuhe riverside in Chaoyang district of Beijing *Journal of Beijing Forestry University* **30(1)** pp 202-207 (in Chinese)

[6] Munro D D 1974 Forest growth models - a prognosis. In J. Fries (Ed.) Growth models for tree and stand simulation Sweden: Royal College of Forestry Stockholm pp 7-12

[7] Marquardt D W 1963 An algorithm for least-squares estimation of nonlinear inequalities *SIAM Journal on Applied Mathematics* **11** pp 431-441

[8] Li Y Q, Deng X W, Huang Z H, XIANG W H, YAN W D, LEI P F, ZHOU X L and PENG C H 2015 Development and evaluation of models for the relationship between tree height and diameter at breast height for Chinese-fir plantations in subtropical China *PloS one* **10(4)**

[9] Lee R J and Nicewander W A 1988 Thirteen ways to look at the correlation coefficient *The American Statistician* **42(1)** pp 59-66

[10] Husch B, Miller C L and Beers T W 1982 Forest Mensuration Wiley. New York. p 402

[11] Huang Y P, Fan F R, Zhou D X, Su S J, Wang B F and Shen Q T 2014 Research on the growth process of *Fokienia hodginsii* (Dunn) variant types plantation in Sanming city, Fujian *Chinese Agricultural Science Bulletin* **30(4)** pp 12-16 (in Chinese)

[12] Wei G Y, Qin D W, Sun C Y, Qin W M and Yang C 2014 Simulation and research on the growth regularity of *Quercus acutissima* plantation *Journal of Northwest Forestry University* **29(4)** pp 145-150 (in Chinese)

[13] Zhang L Y, Deng X W, Lei X D, Zhao Z H, and Xiang W H 2013 *Pinus massoniana* productivity at different age stages in relation to climatic factors *Chinese Journal of Ecology* **32(5)** pp 1104-10 (in Chinese)

[14] Zhou Y F, Zhou G M, Yu S Q, Xu X J and Jin W 2008 Spatial distribution of combustible substance of *Schima superba* stands in Zhejiang province, eastern China *Journal of Beijing Forestry University* **30(6)** pp 99-106 (in Chinese)

[15] Wang X M, Lu Y C, Ning J K, Ren Y M, Wang Q F and Wang H 2009 Research on growth process and growth models of *Quercus veriablilis* in Beijing region *Forest Research* **22(6)** pp 860-864 (in Chinese)

[16] Wu X F and Hu Y L 1999 Study of dynamics models of forest growth and nutrition III. Basic assumption and precondition of growth equation *Journal of Central south Forestry University* **19(2)** pp 1-10 (in Chinese)

[17] Ma Z Q, Liu Q J, Wang H M and Guo Z W 2011 The growth pattern of *Pinus elliottii* plantation in central subtropical China *Acta Ecologica Sinica* **31(6)** pp 1525-37 (in Chinese)

[18] Xu G F, Yu S Q, Zhao J, Huang W X and Mao X F 2009 Investigation on *Schima superba* fire-resistant forest belts communities and its growth features in different sites *Journal of Fujian Forestry Science and Technology* **36(3)** pp 88-91 (in Chinese)

Tracing the evolution of degraders in activated sludge during the sludge's acclimation to a xenobiotic organic

N M Chong[1], C H Fan and Y C Yang

Department of Environmental Engineering, DaYeh University, No. 168, University Road, Dacun, Changhua, Taiwan 51591, R.O.C.

E-mail: chong@mail.dyu.edu.tw

Abstract. The molecular biology method of high-throughput pyrosequencing was employed to examine the change of activated sludge community structures during the process in which activated sludge was acclimated to and degraded a target xenobiotic. The sample xenobiotic organic compound used as the activated sludge acclimation target was the herbicide 2,4-dichlorphenoxyacetic acid (2,4-D). Indigenous activated sludge microorganisms were acclimated to 2,4-D as the sole carbon source in both the batch and the continuous-flow reaction modes. Sludge masses at multiple time points during the course of acclimation were subjected to pyrosequencing targeting the microorganisms' 16S rRNA genes. With the bacterial 16S rRNA sequencing results the genera that increased in abundance were checked with degradative pathway databases or literature to confirm that they are commonly seen as potent degraders of 2,4-D. From this systematic examination of degrader changes at time points during activated sludge acclimation and degradation of the target xenobiotic, the trend of degrader evolution in activated sludge over the sludge's acclimation process to a xenobiotic was traced.

1. Introduction

Activated sludge microorganisms generally grow on substrates made of organic pollutants that are biogenic in nature. Under this natural situation, activated sludge microorganisms are not reactive with (indigenous to) any xenobiotic compound and are unable to degrade the xenobiotic. Fortunately, activated sludge can develop xenobiotic degradation capability after an acclimation process in which indigenous sludge is placed in reaction with the target xenobiotic for a sufficient period until degradation starts and proceeds to completion [1, 2]. Through the process of acclimation, some indigenous microorganisms, mainly bacteria, would acquire xenobiotic degradation capability and are converted degraders, while some would remain incapable (non-degraders) [2, 3]. How degraders evolve is an interesting and pressing question; determination of degrader evolution, in amount and its variation over time, serves the correct description of activated sludge's degradative strength which dictates the behaviour and performance of the xenobiotic treatment sludge [4, 5].

It has long been recognized that the requirement of suitable methods for the identification and still more preferably, quantification of xenobiotic degrader be readily available. So far, there is seldom a claim for the most useful method to achieve such a requirement. The review of Widada et al. [6] has highlighted all previous molecular techniques for monitoring xenobiotic degrader bacteria, including the relatively recent methods of DGGE and rt-PCR. However, tracing degrader all through the process in which degradability develops is scantly reported, if any, in the literature. High-throughput pyrosequencing provides readings that can be translated to bacterial taxa to show microbial community structures at any time point [7]. When specific genus can be examined, by off-line methods currently, for its potential xenobiotic degradability, these pyrosequencing readings may be used to identify degraders that are contained within microbial communities such as those in activated sludge. The purposes of this study were to analyze activated sludge community structure using high-throughput pyrosequencing, and with the feasibility of off-line degrader identification, to trace the trend and dynamics of degrader evolution during activated sludge acclimation and degradation of a xenobiotic. Trends of degrader evolution in the common acclimation reactions of batch and continuous-flow modes were determined which revealed the difference in amount of degrader obtainable due to different acclimation pressure exerted by the xenobiotic on the activated sludge biomass [3, 8]. In addition to devising a novel method that traces degrader evolution, this study also disclose the actual differences and changes in amount of degraders that happen in reactors that are commonly used in practical applications.

2. Materials and methods

2.1. Activated sludge and sample xenobiotic

The sample xenobiotic compound used as the activated sludge acclimation/degradation target was the herbicide 2,4-dichlorphenoxyacetic acid (2,4-D).

The initial activated sludge seed was obtained from soil that did not have any record of 2,4-D nor metal contamination. The mixed culture from soil was grown to a suitable amount on Nutrient Broth (NB Difco 234000) for multiple (at least three) subcultures in multiple shake-flasks. The shake-flask microorganisms were considered indigenous and were used as seed to a separate fed-batch reactor for cultivation of sludge biomass.

2.2. Reactor operations

Fed-batch reactor cultivated and maintained indigenous activated sludge that was used at different times or batches that were to acclimate the biomass to 2,4-D. Acclimation reactions were performed in two (2) types of reactors: 1) Batch reactors for acclimation of indigenous activated sludge to 2,4-D; and 2) continuous flow reactors (CSTR) fed solely 2,4-D for acclimating indigenous activated sludge to 2,4-D that was input continuously. Shake-flasks were 1000 mL conical flasks, each contained a medium volume of 400 mL initially. CSTRs had built-in clarifier and a reactor volume of 3 L. Feeds to all reactors also contained minerals of $FeCl_3$, 1.2 mg/L; $CaCl_2$, 12 mg/L; $MgSO_4.7H_2O$, 65 mg/L; NH_4Cl, 125 mg/L; K_2HPO_4, 200.0 mg/L; and KH_2PO_4, 156.6 mg/L. These minerals are typically used in synthetic wastewater with which pH is buffered and essential nutrients are provided in quantities not limiting bacterial growth. Reactors used for acclimation activated sludge to 2,4-D were operated such that acclimation was completed, as shown from the non-detectable level of 2,4-D in the effluent. For both the batch and CSTR acclimation reactors, sludge was harvested from each reactor in regular interval (typically one day). The sludge masses were analyzed for their community structures by the process of high throughput pyrosequencing.

Measurement of activated sludge concentration was by suspended solids (SS) according to 'Total Suspended Solids Dried at 103-105°C of Standard Methods [8]; measurement of 2,4-D was performed by HPLC [9].

2.3. Pyrosequencing

Activated sludge biomasses that were to acclimate to 2,4-D in batch and CSTR reactors were subjected to 16S rRNA gene pyrosequencing analyses. 16S rRNA gene is present in almost all bacteria and is currently used as a standard genetic marker for bacteria identification. Before pyrosequencing procedures the batch reactor sludge from multiple shake-flasks and the CSTR sludge from duplicate reactors were mixed in equal mass proportions. Equal masses of these sludge mixtures were subjected to pyrosequencing. Pyrosequencing was performed by a local biotechnology laboratory, using 454 GS Junior System following protocols recommended by the manufacturer.

2.4. *Degrader identification*

The abundances of all bacteria at the family level were first studied about their 2,4-D degradability. All bacteria present in the sludge, each at the genus level, were separately verified for their true 2,4-D degrader characteristics with references firstly to gene and pathway data aggregator websites, such as MetaCyc [10, 11], and when a particular species' 2,4-D degradability was uncertain from the pathways databases, confirmation of such degradability was extensively searched from published literature.

3. Results and discussion

3.1. *Degrader identification*

Activated sludge biomasses, starting at their indigenous states, were acclimated and after which degraded 2,4-D in the batch and continuous flow reactors. From the high-throughput sequencing results, the relative abundances (% of all readings) of all bacteria under the taxa level of genus were recorded. Each genus was searched for its 2,4-D degradative characteristics from degradation pathway databases. Degraders within each of the sludge masses were identified and their relative abundances were determined. Simultaneously, non-degraders and their abundances were similarly determined.

Using the result of the CSTR acclimated sludge as an example, Figure 1 shows the courses of rise and fall of some prominent degraders, as well as those of some non-degraders. Genus names for the degraders and non-degraders are shown in the legend. The sum of the total number of degrader bacteria is shown as the curve indicated (and coloured red). In contrast, the black curve is the relative amount (number) of non-degraders. The degrader curve shows the evolution dynamics of degraders during activated sludge acclimation in the continuous flow reaction. Degrader and non-degrader followed a harmonious succession with each other.

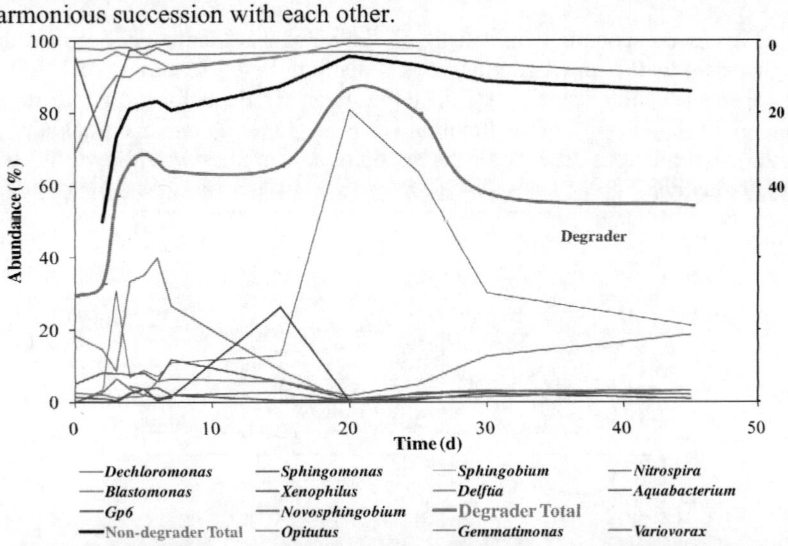

Figure 1. Total degrader summed up of individual genus that was verified 2,4-D degradative, while the rest were non-degraders. Legends show genera most prominent as degrader or non-degraders

3.2. *Mapping degradation and degrader trends*

Figure 2a shows the growth of sludge (represented by SS) and the disappearance of 2,4-D during the batch acclimation process. In the batch reaction, acclimation of sludge to 2,4-D was achieved on about the third day (typical of most batch acclimation of activated sludge to 2,4-D), after which 2,4-D was degraded in a noticeably high rate. Figure 2b shows the evolution of degrader (and also the corresponding non-degrader changes) in the batch-acclimation reaction. Mapping the time of total degrader variation with that of 2,4-D degradation, it is shown that the degraders increased immediately at the start of acclimation reaction, and rose to a highest quantity on about the 4th day. When

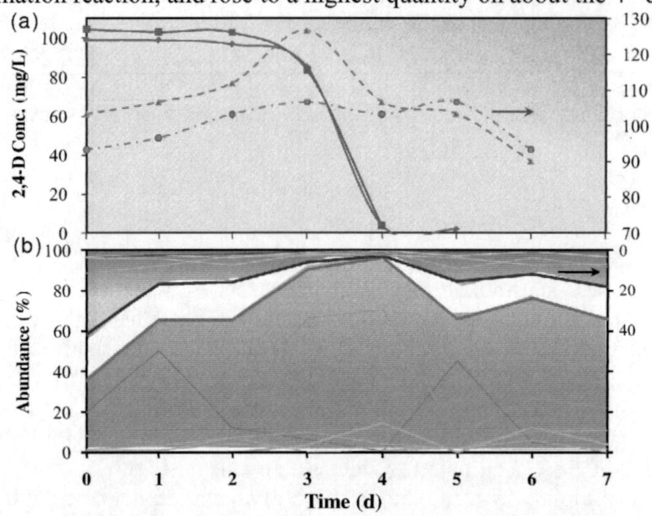

Figure 2. 2,4-D degrader evolution trends corresponding to 2,4-D degradation in batch acclimation and degradation reaction. (a) 2,4-D degradation (left axis) and SS growth (right axis) of duplicate tests; (b) degrader evolution (left axis).

degradation of 2,4-D was complete (after Day-5), the number of degraders declined to a level that was slightly higher than that at the beginning of acclimation. At the last stage of 2,4-D degradation, sludge microorganisms are able to metabolize the downstream products for growth. With the depletion of 2,4-D and therefore diminishing of acclimation force, degrader abundance gradually gave way to non-degraders. At the continuous absence of the xenobiotic, degraders would revert to non-degraders (de-acclimation) gradually [5].

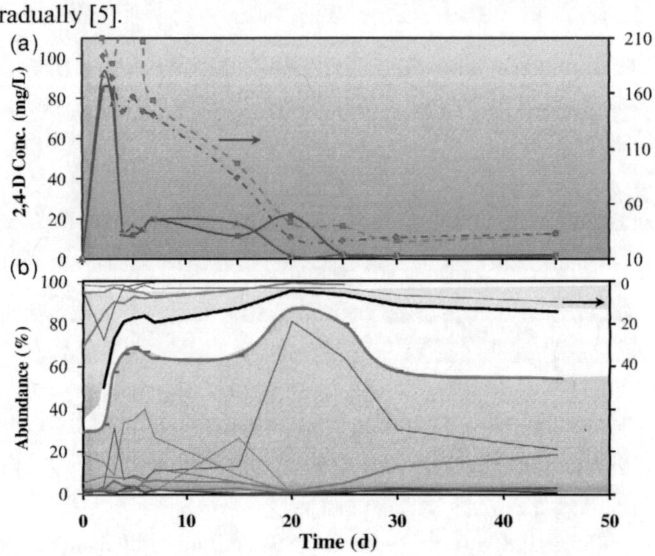

Figure 3. 2,4-D degrader evolution trend corresponding to 2,4-D degradation in continuous-flow acclimation and degradation reaction. (a) 2,4-D degradation (left axis) and SS growth (right axis) of duplicate tests; (b) degrader evolution (left axis).

Figure 3a shows the growth of sludge and the disappearance of 2,4-D during the process of continuous flow acclimation. In a continuous flow reaction, influent 2,4-D, as a non-reactive compound, must reach a concentration theoretically equal to that of the influent after a hydraulic residence time. Only then the high 2,4-D concentration puts the sludge under substantial acclimation pressure. After this delay of 2,4-D concentration build-up, continuous flow acclimation was completed on about the 4th day. The rise of degraders in continuous-flow acclimation (Figure 3b) is similar to that of the batch acclimation with rapid increase when acclimation was the main activity. Also, degrader number reached the highest point when degradation of 2,4-D was complete (2,4-D effluent as low as non-detectable). After degradation was achieved, 2,4-D was still being continuously input so that mass of 2,4-D was continuously present in spite its concentration was significantly diluted in the reaction tank. The need for degrader bacteria kept degrader abundance at a certain high level. When the continuous flow reaction reached steady state, degrader content also reached a steady level that was a little lower than that of the maximum.

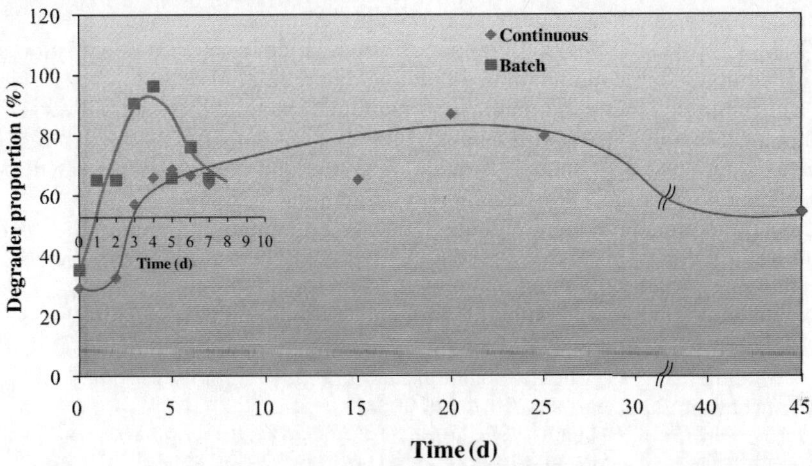

Figure 4. Comparison of batch and CSTR degradation evolution trends.

Figure 4 displays in parallel the evolutionary dynamics of degraders during the acclimation of activated sludge to a xenobiotic organics in batch and continuous flow reactors. In batch reaction, the starting concentration of 2,4-D is high to exert a high acclimation pressure to the sludge. The sum of degrading genera increased quickly during batch acclimation, in close concurrent to the degradation of the xenobiotic as a growth substrate. Approaching the end of acclimation where the target was depleted, degrader number decreased to certain level pending further reduction under de-acclimation. Acclimation in CSTR suffers a lag due to an initial dilution. Rise of degraders after the lag had a rate similar to batch acclimation, but the proportion of degrader was at all times lower than that attainable by batch acclimation. CSTR acclimation maintained a high degrader level (at approximately the maximum) and remained at that level for a length of time until the steady state when degrader was lowered. Compared to CSTR acclimation, batch acclimation yielded higher amount of degrader, but could maintain this amount for shorter period of time.

4. Conclusions

In this study, the evolution dynamics of xenobiotic degraders during the acclimation of activated sludge to the xenobiotic matter was obtained using an up-to-date molecular biological technology. Conclusions can be drawn:

- Xenobiotic degraders, due to the acclimation pressure, rise at a fast rate.

- The overall amount of degrader present in the reactor depends on the concentration of the acclimating xenobiotic. Degrader amounts decline rapidly at low xenobiotic concentration, but this is only more accurate after depletion of the xenobiotic than when xenobiotic is being diluted yet its mass is being continuously input.

Acknowledgment
The work reported in this presentation was supported in parts by the research grant MOST 103-2221-E-212-001-MY3 provided by the Ministry of Science and Technology of Taiwan, the Republic of China.

References
[1] Chong N M 2005 Development of a tool for measuring the degradation capacity of microorganisms for a xenobiotic *Enzyme. Microb. Technol.* **37** 467-471.
[2] Chong N M and Lin T Y 2007 Measurement of the degradation capacity of activated sludge for a xenobiotic organic *Bioresour. Technol.* **98** 1124-27.
[3] Chong N M and Chang H W 2009 Plasmid as a measure of microbial degradation capacity for 2,4-dichlorophenoxyacetic acid *Bioresour. Technol.* **100** 1174-79.
[4] Bali U and Şengül F 2002 Performance of a fed-batch reactor treating a wastewater containing 4-chlorophenol *Proc. Biochem.* **37** 1317-23.
[5] Sahinkaya E and Dilek F B 2002 Effects of 2,4-dichlorophenol on activated sludge *Appl. Microbial. Biotechnol.* **59** 361-367.
[6] Widada J, Nojiri H and Omori T 2002 Recent developments in molecular techniques for identification and monitoring of xenobiotic-degrading bacteria and their catabolic genes in bioremediation *App.l Microbiol. Biotechnol.* **60** 45-59.
[7] Wang X, Hu M, Xia Y, Wen X and Ding K 2012 Pyrosequencing analysis of bacterial diversity in 14 wastewater treatment systems in China *Appl. Environ. Microbiol.* **78** 7042-47.
[8] Chong N M and Huang J C 2012 Production of xenobiotic degrader for potential application in bioaugmentation *Bioresour. Technol.* **125** 305-311.
[9] APHA (American Public Health Association) 1998 *Standard methods for the examination of water and wastewater* (Washington D.C. USA.)
[10] Nguyen L H and Chong N M 2015 Development of an ATP measurement method suitable for xenobiotic treatment activated sludge biomass *J. Chromatogr. B* **1000** 69-76.
[11] MetaCyc metabolic pathways database: http://www.metacyc.org/META/
[12] UniProt taxonomy database: http://www.uniprot.org/taxonomy/

New collection of *Acrogenospora spharerocephala* from freshwater habitats of Northern Thailand

X J Liu, F P Li, W Liu, X L Huang and H Zhang[1]

Faculty of Environmental Science & Engineering, Kunming University of Science & Technology, Kunming 650500, People´s Republic of China

E-mail: zhanghuang2002113@163.com

Abstract. During investigations of freshwater fungi in Thailand, we collected dematiaceous hyphomycetes *Acrogenospora spharerocephala*n in lotic freshwater habitat. The description and morphological photographs of new collection is illustrated. *Acrogenospora spharerocephala* was characterized by unbranched conidiophores and producing solitary non-septate conidia acrogenously from recurrently proliferating conidiogenous cells. The conidia are dark brown to black when mature, paler at young, spherical to subspherical.

1. Introduction

So far, 537 hyphomycetes species were reported from freshwater habitats [1]. This number is almost as same as the freshwater ascomycota (579 species). More hyphomycetes were reported in many fungal diversity studies. [2], [3] Classification of freshwater hyphomycetes is based on morphological features. [4], [5]

During investigations of freshwater fungi in Thailand [6], we collected dematiaceous hyphomycetes *Acrogenospora spharerocephala* in lotic freshwater habitat. The description and morphological photographs of new collection are illustrated. The differences between similar species are provided.

2. Methods

Submerged wood was randomly collected from the lotic freshwater habitats in Chiang Mai and Chiang Rai Province, Thailand, in January and April 2010, following the reported procedures described [2]. Samples were placed separately in Zip lock plastic bags with sterile moist tissue paper. The woody substrates were examined under a dissecting microscope for fruiting bodies after one week of incubation and until 2 months [7].

Observations and photomicrographs were made from material mounted in water or lactic acid (85%) using a Nikon ECLIPSE 80i microscope. Melzer's reagent (MEZ; 0.5 g iodine, 1.5 g IKI, 20 g chloral hydrate, 20 ml distilled water) and aqueous cotton blue were added to determine staining reactions of the ascus apical apparatus [8]. Measurements were made with the Tarosoft (R) Image Frame Work. The herbarium specimens are deposited at Mae Fah Luang University (MFLU), Chiang Rai, Thailand.

3. Results and discussions

Acrogenospora sphaerocephala (Berk. & Broome) M.B. Ellis, Dematiaceous Hyphomycetes (Kew): 114 (1971)

Basionym. – *Monotospora sphaerocephala* Berk. & Broome, *Ann. Mag. nat. Hist.*, Ser. 3 3: 361 (1859)

Synomym. – *Halysium sphaerocephalum* (Berk. & Broome) Vuill.

Monosporella sphaerocephala (Berk. & Broome) S. Hughes, *Can. J. Bot.* 31: 654 (1953)

Monotosporella sphaerocephala (Berk. & Broome) S. Hughes, *Can. J. Bot.* 36: 787 (1958)

Figure 1

Asexual morph – unknown.

Description – *Colonies* sparse, scattered, black, glistening, hairy. *Mycelium* mostly immersed, consisting of septate, thin-walled, smooth, pale brown to brown hyphae. *Conidiophores* macronematous, mononematous, 300–400 μm long, 8–10 μm wide at the base, 4–6 μm wide at the tip, solitary or in some groups, annellations at the apex of a conidiophore resulting from successive percurrent proliferations, septate, erect, unbranched, septate, long cylindrical, slightly tapering towards the apex, pale to mid brown, slightly paler at the apex, smooth, thick-walled at the base, slightly flexuous. *Conidiogenous cells* monoblastic, integrated, terminal, 70–80 × 4–6 μm, light brown, cylindrical, proliferate percurrently. *Conidia* acrogenous 24–32 μm ($\bar{x}$ = 28.2, n =15) in diam., holoblastic, with young conidium developing at the apex of the conidiophore and mature conidia pushed sidewise in a sympodial manner after successive percurrent proliferations of the conidiophores, solitary, dry, dark brown to black when mature, paler at young, spherical to subspherical, with a truncate base, unicellular, smooth, thick-walled.

Habitat – On submerged decaying wood.

Known freshwater distribution – Brunei, Hong Kong, Hungary, Philippines, South Africa, the Seychelles, Thailand, U.S.A.

Material examined. – Thailand, Chiang Mai, Doi suthep Pui National Park, on submerged wood, 25 Apr. 2010, Huang Zhang, MFLU11-1064; ibid MFLU11-1068; ibid MFLU11-1132.

Notes – The genus *Acrogenospora* was established to accommodate the species transferred from *Monotosporella* which have unbranched conidiophores and producing solitary non-septate conidia acrogenously from percurrently proliferating conidiogenous cells [9]. *Monotosporella* species are similar to *Acrogenospora* in many respects, the main differentiating characters being conidia: septate vs. unicellular [10].

Ten species were reported so far, in which six were from freshwater habitats. They are *A. sphaerocephala* (Berk. & Broome) M.B. Ellis, *A. ellipsoidea* D.M. Hu, L. Cai & K.D. Hyde, *A. gigantospora* S. Hughes, *A. ovalis* Goh, K.D. Hyde & K.M. Tsui, *A. subprolata* Goh, K.D. Hyde & K.M. Tsui and *A. verrucispora* Hong Zhu, L. Cai & K.Q. Zhang. They are distinguished by the color of conidiophores and conidial color, shape and dimension. The asexual morph of *A. altissima* and *A. megalospora* are associated with *Farlowiella*. *A. sphaerocephala* is collected in this study.

A. sphărerocephala, the type species of *Acrogenospora*, is characterized by dark brown to black when mature, paler at young, spherical to subspherical conidia [9]. It differs from *A. verrucospora* by the smooth and larger conidia [11]. *A. spharerocephal* is frequently found on submerged wood. We tried to do single spore isolation but the conidia do not germinate on WA or PDA.

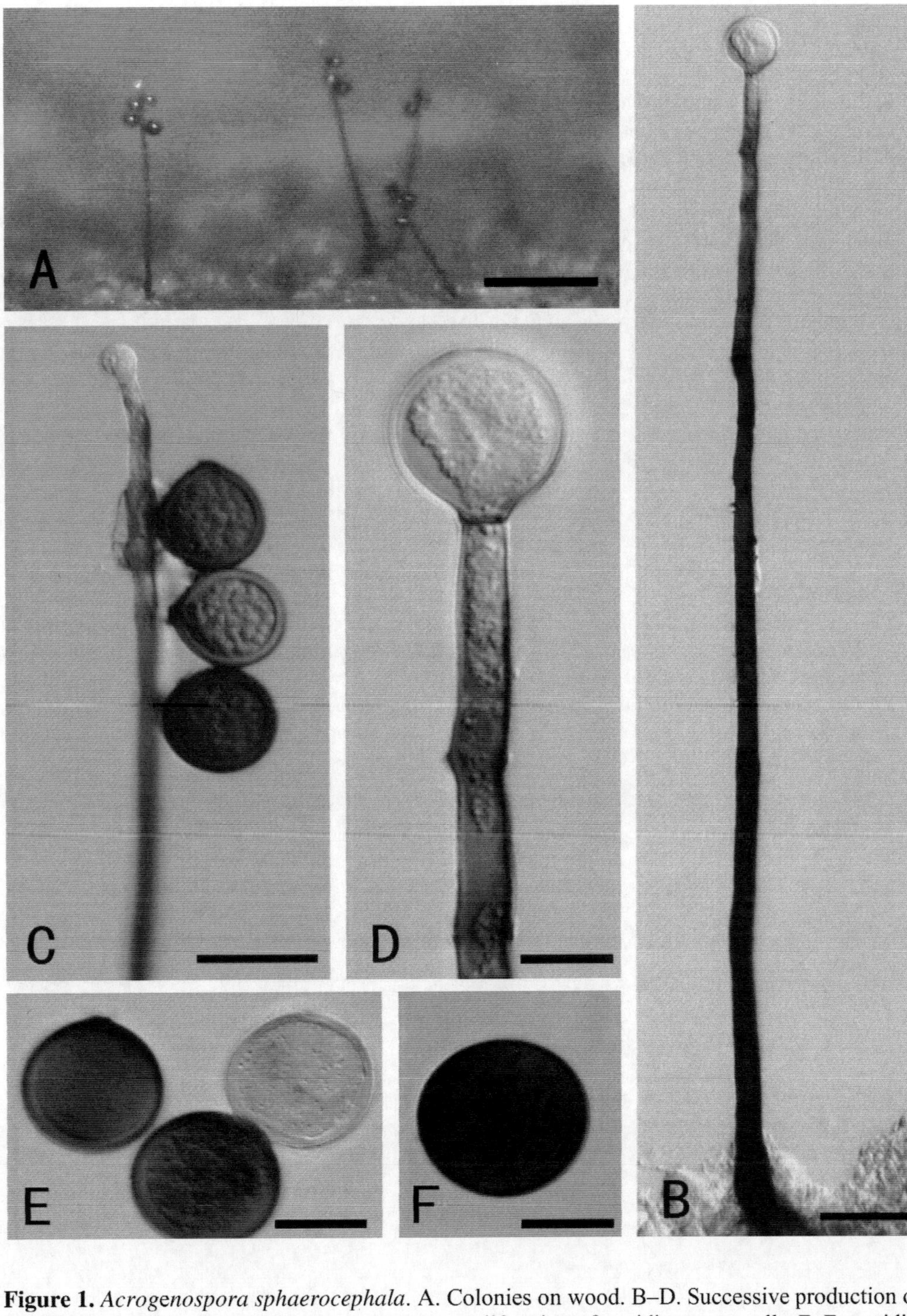

Figure 1. *Acrogenospora sphaerocephala*. A. Colonies on wood. B–D. Successive production of holoblastic conidiophores. Note the percurrent proliferation of conidiogenous cells. E–F. conidia. Scale bars: A = 300 μm, B–C = 30 μm, D–F = 5 μm.

References

[1] Shearer C A and Raja H A. Freshwater Ascomycetes Database: http://fungi.life.illinois.edu/ (Accessed on Aug. 2013)

[2] Kurniawati E, Zhang H, Chukeatirote E, Sulistyowati L, Moslem M A and Hyde K D 2010 Diversity of freshwater ascomycetes in freshwater bodies at Amphoe Mae Chan, Chiang Rai. *Cryptogamie Mycol* **31(3)** pp 323–331

[3] Cai L, Zhang K Q, McKenzie E H C and Hyde K D 2003 Freshwater fungi from bamboo and wood submerged in the Liput River in the Philippines. *Fungal Divers* **13** pp 1–12

[4] Zhao G Z, Yu P and Liu X Z 2013 *Cancellidium* and *Canalisporium* (Hyphomycetes) from China. *Nova Hedwigia* **96(1-2)** pp 221–236

[5] Cai L, McKenzie E and Hyde K 2004 New species of *Cordana* and *Spadicoides* from decaying bamboo culms in China. *Sydowia* **56(2)** pp 6–12

[6] Zhang H, Hyde K D, Abdel-Wahab M A, Abdel-Aziz F A, Ariyawansa H A, KoKo T W, Zhao R L, Alias S A, Bahkali A H and Zhou D Q 2013 A modern concept for *Helicascus* with a *Pleurophomopsis*-like asexual state. *Sydowia* **65(1)** pp 147–166

[7] Raja H A, Fournier J, Shearer C A and Miller AN 2012 Freshwater ascomycetes: *Coniochaeta gigantospora* sp. nov. based on morphological and molecular data. *Mycoscience* **53(5)** pp 373–380

[8] Raja H A and Shearer C A 2008 Freshwater ascomycetes: new and noteworthy species from aquatic habitats in Florida. *Mycologia* **100(3)** pp 467–489

[9] Ellis M B (Ed) 1971 *Dematiaceous Hyphomycetes* Commonwealth Mycological Institute, Surrey, England

[10] Tsui C K, Goh TK, Hyde K D and Hodgkiss IJ 2001 New species or records of *Cacumisporium*, *Helicosporium*, *Monotosporella* and *Bahusutrabeeja* on submerged wood in Hong Kong streams. *Mycologia* **93(2)** pp 389–397

[11] Zhu H, Cai L, Hyde K D and Zhang K Q 2005 A new species of *Acrogenospora* from submerged Bamboo in Yunnan, China. *Mycotaxon* **92** pp 383–386

Effect on the operation properties of DMBR with the addition of GAC

Jizhi Lin[1], Qian Zhang[2], Junming Hong[2, 3]

[1] Department of Science & Technology Research, Huaqiao University, Xiamen, Fujian, China

[2] College of Chemical Engineering,Huaqiao University, Xiamen 361021, Fujian, China

E-mail: jmhong@hqu.edu.cn

Abstract. The membrane bioreactor and dynamic membrane bioreactor were used to examine the effect of granular activated carbon (GAC) on the treatment of synthetic wastewater. After the addition of different volume fractions GAC in the DMBR, the operation parameters, effluent COD, NH_4^+-N, NO_3^--N, TN concentrations and sludge viscosity of the bioreactor was investigated. The results showed that the addition of GAC could relieve the membrane fouling and improve the removal efficiencies of pollutants in the DMBR. The effluent concentrations of pollutants were linear correlation with the addition of volume fractions of GAC in the bioreactor. The value of R2 of each modulation was almost more than 0.9. The sludge viscosity was almost not affected by the volume fractions of GAC in the bioreactor. The best volume fractions of GAC were 20% in the DMBR.

1. Introduction

Energy conservation and environmental industry is one of the fastest growing industries in the world and the major industry of Xiamen. Based on the cogentinvestmentand development in Fujian province, a series of new techniques was developed, such as the dynamic membrane bioreactor process. The widespread application of membrane technology to wastewater treatment has assisted pollutants removal performance [1]. Although membrane bioreactor (MBR) process is reliable in terms of perfect capture of solids in the system, membrane fouling has restricted its practical application and increased operational costs [2]. Various approaches have been adopted to prevent fouling such as increasing fluid velocity [3], increasing fluid pumping or gas sparging rate, relaxation, backwashing [4], modifying the membrane surface [5], subcritical flux operation [6] and so on.

A widely used method of fouling control in MBRs involves the use of additives such as powdered activated carbon (PAC) [7]. The membrane fouling was relieved by the low dose powdered activated carbon addition at high sludge retention times [8]. The application of PAC was effective in decreasing the content of extracellular polymeric substances (EPS) [9].

A new kind of membrane, made up of common silk and fouling cake, has been developed to actively utilize the separation ability of fouling cake. Based on the fact that the silk with large pores of 50-500 μm has no separation ability, while sludge in reactor is actually rejected by the fouling cake which looks like a new separation membrane added onto the silk, this kind of membrane is thus called dynamic membrane[10]. The dynamic membrane process would also be an effective method to settle the problem of membrane fouling. Dynamic membrane could be formed and re-formed in-situ. When the membrane was severely fouled, the dynamic layer can be replaced by a new deposited layer. Dynamic membrane not only has almost all characteristics belonging to common membrane, but also has some distinct features, such as low expenditure, low energy consumption and high flux and so on. The dynamic membrane bioreactor (DMBR) is a kind of submerge membrane bioreactor in which the separation membrane is substituted by dynamic membrane [11].

The studies of selection of membrane materials and the pollutants removal effects was mainly focused in the DMBR [12, 13]. The research on the addition of granular activated carbon (GAC) was not much, which was used to improve the effluent water qualities in the DMBR.

The purpose of this study is, therefore, to investigate the operation characteristics in the addition of GAC in the DMBR. The operation parameters, effluent concentrations of pollutants and sludge viscosity were investigated in the DMBR, when the addition of volume of GAC was changed.

2. Materials and methods

2.1. Materials

The commercial GAC with 10~20 mesh size was used in this study. The volume fractions of GAC were 0, 10%, 20%, and 30% in the DMBR.

The components of synthetic wastewater were 278 mg•L^{-1} glucose, 278 mg•L^{-1} starch, 13.16mg•L^{-1} potassium dihydrogen phosphate, 132.8 mg•L^{-1} ammonium sulfate, 66.0 mg•L^{-1} magnesium sulfate, 111.0 mg•L^{-1} sodium bicarbonate, 6.0 mg•L^{-1} manganese sulfate, 6.0 mg•L^{-1}calcium chloride, 2.0 mg•L^{-1} ferrous sulfate [14]. The concentrations of influent COD, total nitrogen, ammonia and total phosphorus were 237.6 ~ 311.4 mg•L^{-1}, 28.0 ~ 28.5 mg•L^{-1}, 27.5 ~ 28.0 mg•L^{-1}, 2.37 ~ 3.28 mg•L^{-1}, respectively. The pH of influent was 6.5 ~ 7.5.

The activity sludge of the bioreactor was from an oxidation ditch in a municipal wastewater treatment plant. The activity sludge was cultivated by a fill-and-draw process for one month with synthetic wastewater.

2.2. Experiment

The effective volume of DMBR was 10.0 L. The schematic of the reactor setup was shown in Figure 1. The DMBR had a submerged 37 μmterylene filter module with a surface area of 0.10 m^2. The total aeration rates of DMBR and MBR were 0.16 m^3•h^{-1}. The concentrations of dissolved oxygen (DO) were between 3~4 mg•L^{-1}. Hydraulic retention times (HRTs) were 4 h.

The influent flow of DMBR were 2.5 L•h^{-1}. When the transmembrane pressure of DMBR increased to 4000 Pa, 45000 Pa, respectively, the membrane module was taken out and cleaned by rinsing with tap water. During this period, GAC was added three times in DMBR. Each time the volume of GAC was 1000 mL.

2.3. Analytical methods

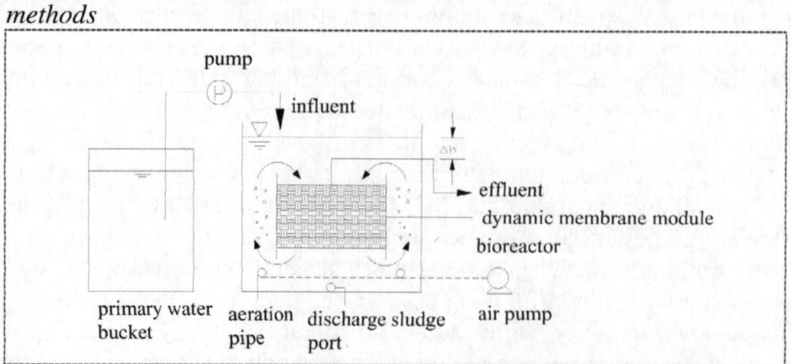

Figure 1. Diagram of an integrated dynamic membrane bioreactor

The influent and effluent were simultaneously sampled every day. The routine monitoring items included pH, the influent and effluent concentrations of COD, NH_4^+-N, NO_3^--N and TN. These items were measured by using Chinese NEPA Standard Methods[Chinese NEPA, 2002].DO was measured using DO meter. Sludge viscosity was measured by the rotational viscometer (NDJ-5S).

3. Results and discussion

3.1. Operation parameters

Figure 2 show the transmembrane pressure of DMBR with different GAC volume fractions. The value of transmembrane pressure of DMBR was about 3.5×10^3Pa. The addition of GAC hadnot effect on the transmembrane pressure. When the value of volume fractions of GAC was more than 10%, the transmembrane pressurehavenot changed.

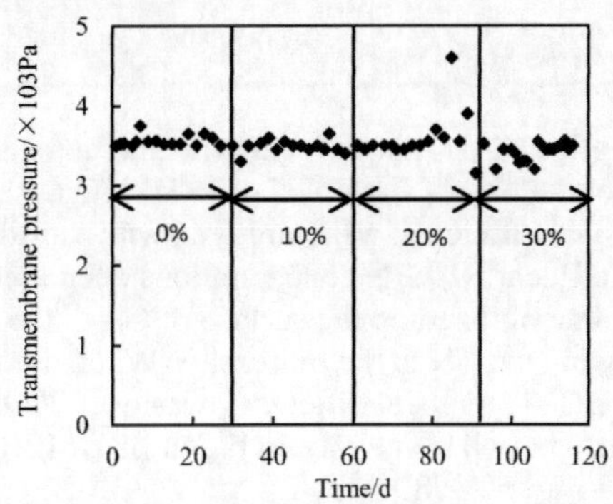

Figure 2. The transmembrane pressure of DMBR with different GAC volume fractions

3.2. Removal of pollutants

3.2.1. COD. Variations of effluent COD concentrations with the volume fractions of GAC are shown in Figure 3. Variations of the COD removal efficiency with the volume fractions of GAC are shown in Table 1. The removal efficiencies of COD were more than 90% in the bioreactor. Effluent COD concentration decreased from 26.1 mg•L^{-1} to 12.0 mg•L^{-1} in the DMBR, when the volume fractions of GAC was raised from 0 to 30%. Because the addition of GAC can immobilize microorganisms and absorb refractory organic compounds in the bioreactor, increasing in the volume fractions of GAC could improve the removal efficiency of COD. When the value of volume fractions of GAC was more than 20%, COD removal efficiencies were over 95% in the bioreactor.

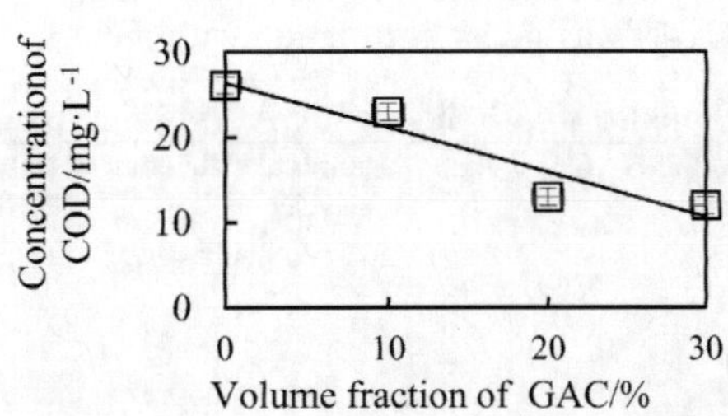

Figure 3. Effluent COD concentrations fordifferent GAC volume fractions in the bioreactor

According to Figure 3, the effluent COD concentration of DMBR was correlated with a linear equation in the following form (r_{DMBR}^2=0.938):

$$E_{COD-DMBR} = 26.36 - 0.522(V) \quad (1)$$

WhereE is the effluent COD concentration in the DMBR and (V) is the volume fraction of GAC.

Table 1.COD removal efficiencies with different GAC volume fractions in the bioreactor

Volume fraction of GAC	**Removal efficiency of COD**
	DMBR
0%	91.30%
10%	92.34%
20%	95.67%
30%	96.00%

3.2.2. NH_4^+-N. Variations of effluent NH_4^+-N concentrations with the volume fractions of GAC are shown in Figure 4. Variations of the NH_4^+-N removal efficiency with the volume fractions of GAC are shown in Table 2. The removal efficiency was improved with the addition of the volume fraction of GAC in the bioreactor. The effluent NH_4^+-N concentrations decreased with the addition of GAC. Because microorganisms were immobilized on the surface of GAC, the addition of GAC could improve the removal efficiency of NH_4^+-N in the bioreactor. When the value of the volume fraction of GAC was more than 20%, NH_4^+-N removal efficiency was over 70% in the DMBR.

According to Figure 4, the effluent NH_4^+-N concentration of DMBR was correlated with a linear equation in the following form (r_{DMBR}^2=0.998) :

$$E_{NH_4^+-N-DMBR} = 13.64 - 0.299(V) \quad (2)$$

WhereE is the effluent NH_4^+-N concentration in the DMBR and (V) is the volume fraction of GAC. The concentration of NH_4^+-N was linear correlation with the volume fraction of GAC.

3.2.3. NO_3^--N. Variations of effluent NO_3^--N concentrations with the volume fractions of GAC are shown in Figure 5. The effluent NO_3^--N concentrations were quickly increased with the addition of GAC in the DMBR. Then, the denitrification ability was not changed in the DMBRandthe concentration of effluent nitrate in the DMBR increased.

According to Figure 4, the effluent NO_3^--N concentration of DMBR was correlated with a linear equation in the following form (r_{DMBR}^2=0.743):

$$E_{NO_3^--N-DMBR} = 1.489 + 0.202(V) \quad (3)$$

Where E is the effluent NO_3^--N concentration in the DMBR and (V) is the volume fraction of GAC. The concentration of NO_3^--N was linear correlation with the volume fraction of GAC.

Table2.COD removal efficiencies with different GAC volume fractions in the bioreactor

Volume fraction of GAC	**Removal efficiency of COD**
	DMBR
0%	51.62%
10%	63.14%
20%	73.61%
30%	83.14%

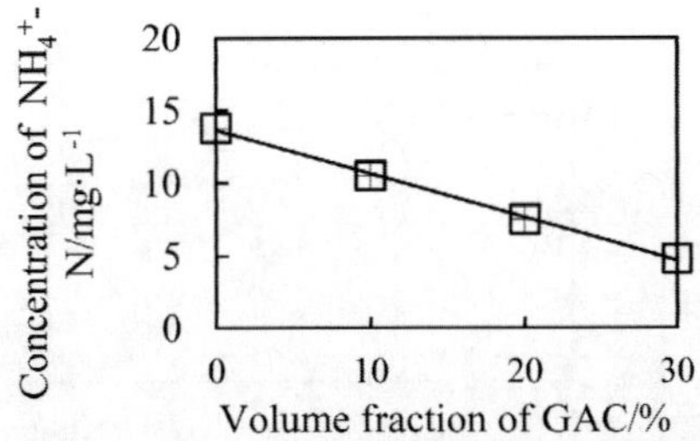

Figure 4. Effluent NH_4^+-N concentrations for different GAC volume fractions in the bioreactor

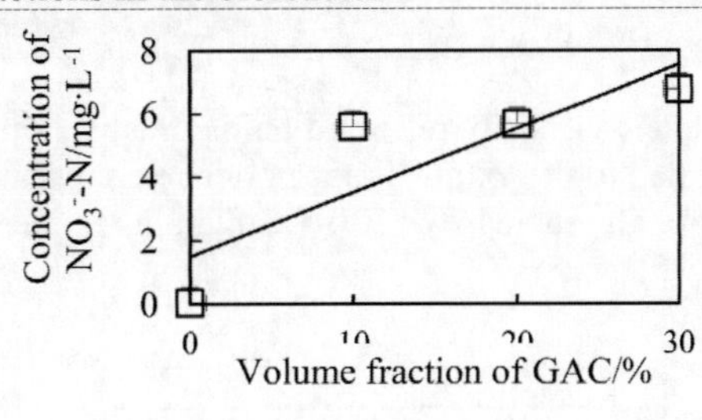

Figure 5. Effluent NO_3^--N concentrations with different GAC volume fractions in the bioreactor

3.2.4. TN. Variations of effluent TN concentrations with the volume fractions of GAC are shown in Figure 6. Variations of the TN removal efficiency with the volume fractions of GAC are shown in Table 3. The removal efficiency of TN was almost not changed in the DMBR. The addition of GAC did not changed the denitrification ability in the DMBR. The effluent TN concentration decreased with the addition of GAC. Because microorganisms were immobilized in the surface of GAC, the addition of GAC could improve the removal efficiency of TN in the bioreactor. When the value of the volume fraction of GAC was more than 20%, TN removal efficiency was over 49% in the DMBR.

According to Figure 6, the effluent TN concentration of DMBR was correlated with a linear equation in the following form (r_{DMBR}^2=0.935): :

$$E_{TN-DMBR} = 13.84 - 0.054(V) \qquad (4)$$

Where E is the effluent TN concentration in the DMBR and V is the volume fraction of GAC. The concentration of TN had alinear correlation with the volume fraction of GAC.

Table3.COD removal efficiencies with different GAC volume fractions in the bioreactor

Volume fraction of GAC	**Removal efficiency of COD**
	DMBR
0%	44.38%
10%	45.91%
20%	49.15%
30%	49.65%

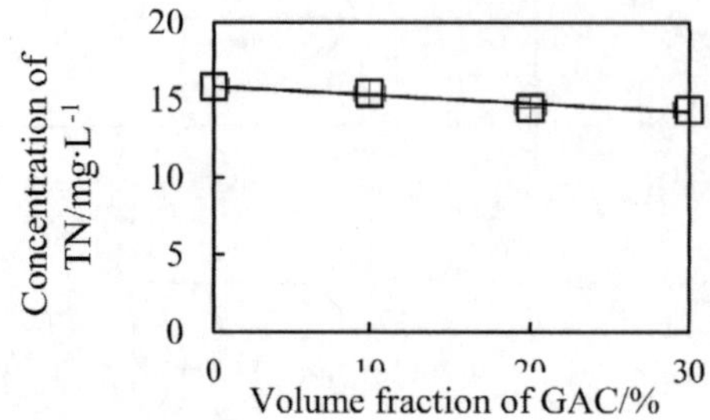

Figure 6. Effluent TN concentrations with different GAC volume fractions in the bioreactor

3.2.5. The sludge properties. Sludge viscosity reflected on the relative amount of micelle in the bioreactors. The micelle was related to the extracellular polymeric substances (EPS), such as polysaccharide, protein and so on. The increase of EPS resulted in the increase of the sludge viscosity in the sludge mixture [15].

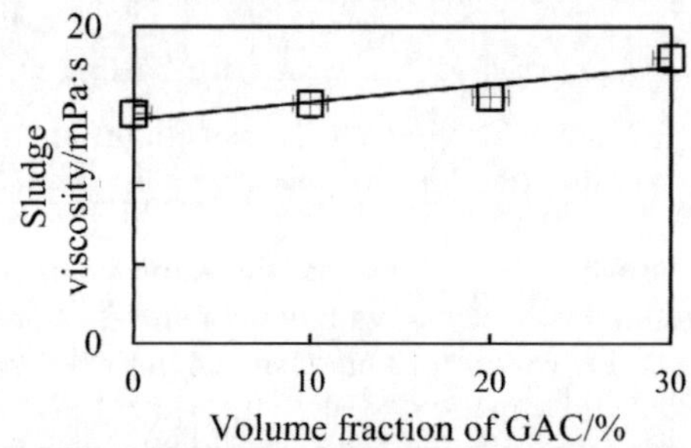

Figure 7. Change of sludge viscosity with different GAC volume fractions in the bioreactor

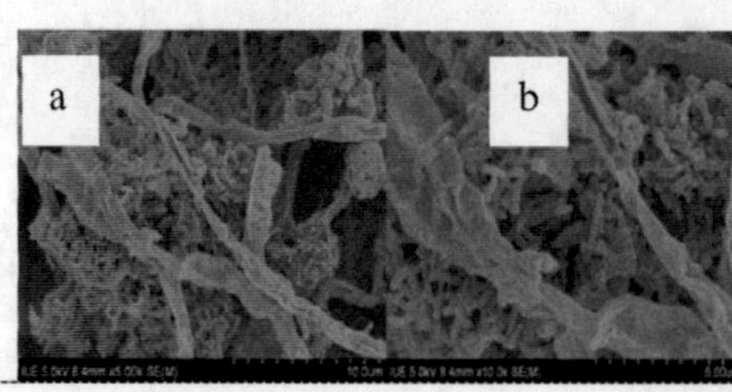

Figure 8. SEM of activated sludge in the bioreactor

Variations of sludge viscosity concentrations with the volume fractions of GAC areshown in Figure 7. The value of sludge viscosity has almost not changed in the bioreactor, whatever GAC was added or not. The value of DMBR was about 16 mPa•s. Therefore, the concentration of EPS did not changed in the bioreactor.

The SEM of activated sludge of two bioreactors is shown in Figure 8. From Figure 8-b, many kinds of microorganisms are found in the bioreactor, such as nematodes, bacteria and filamentous. From

Figure 8-a, as microorganisms were arranged densely in the DMBR, the oxygen could not be easily transferred to the interior of the microorganisms. An anoxic environment was easily formed.

4. Conclusion

This study compares the operation characteristics, the removal efficiencies of pollutants and sludge properties with different volume fractions of GAC in the DMBR treating synthetic wastewater. The addition of GAC can improve the removal pollutants efficiencies of DMBR. The best volume fractions of GACwere 20% in the DMBR. The concentrations of effluent pollutants had alinear correlation with the volume fractions of GAC in the bioreactor. The sludge viscosity was almost not affected by the volume fractions of GAC in the bioreactor.

Acknowledgements

The authors gratefully acknowledge the financial support fromXiamen Technology project Foundation (3502Z20153025, 3502Z20151256).

References

[1] Monti A, Hall ER, Dawson RN, Husain H, Kelly HG 2006 *Bioengineering* **94** 740
[2] Le C P, Chen V, Fane G 2006 *J. Membrance Sci.***284** 17.
[3] Brockmann M, Seyfried CF 1996 *Water Sci. Technol.* **34** 205.
[4] Bouhabila EH, Aim RB, Buisson H 2001 Sep. *Purif. Technol.***22** 123.
[5] Ma H, Bowman CN, Davis RH 2000 *J. Membrane Sci.***173** 191.
[6] Jeison D, Van J B 2006 Biochem. *Eng. J.***29** 204.
[7] Park H, Choo KH, Lee CH 1999 Sep. *Sci. Technol.***34** 2781.
[8] Maxime R, Perry M, Arie Z, Wim R, Hardy T 2009 *Water Res.***43** 345.
[9] Zhao Y, Gu P 2006 Sep. *Purif. Technol.***52** 154.
[10] Lee J, Ahn WY, Lee CH 2001 *Water Res.***35** 2435.
[11] Hongbo L, Chang Y, Wen P 2009 *Chem. Eng. J.* **148** 290.
[12] Kiso Y, Jung YJ, Ichinari T, Park M 2000 *Water Res.***34** 4143.
[13] Zhang XY, Wang ZW, Wu ZC, Wei TY, Lu FH, Tong J, Mai SH 2011 *Process Biochem.* **4** 9219.
[14] Liu R, Hunag X, Yu Q 2001 *Environ. Sci.***22** 20.
[15] Chang IS, Bag SO, Lee CH 2001 *Process Biochem.***36** 855.

Remote sensing monitoring land use change in Donglutian coal mine, Shuozhou City

Baoying Ye[1], Ling Liu

China University of Geosciences, Beijing, 100083, China

E-mail: yebaoying@cugb.edu.cn

Abstract: This paper monitored the coal mine exploitation in Donglutian coal mine, Shuozhou city, Shanxi Province. Landsat satellite images from 2008 to 2016 were selected, and then 15m color composite images were obtained through data processing and image fusion. On this basis, the land use map from 2008 to 2016 was obtained using visual interpretation method. Results showed that the main land use type in this area was cropland, unused land and coalmine. Area of cropland and unused land kept decreasing year by year, while coal mine expanded rapidly. The expansion of coal mine concentrated on two time periods: from 2009 to 2010 and from 2012 to 2013. During these two time periods, topsoil stripping was the main exploitation type,while deep mining was the main type for other times. Results also presented that the exploitation number of small coals kept increasing year by year, from the initial number of 26 at 2008 to 42 at 2016.

1. Introduction

Remote sensing and GIS are the main techniques to study land use/land cover change [1]. Many scholars have conducted dynamic monitoring of environmental change in the mine area by using remote sensing data [2-4], and received ideal effect. Remote sensing also provides the possibility for long-term monitoring of land use [5], provide supervision measures to disordered mining, and also provide accurate and reliable data basis for the study of ecological environmental change.

Located in the east of loess plateau, Donglutian coal mine is a ecological environment system which are sensitive to the environmental change with small plasticity to maintain its stability. This paper chose Donglutian coal mine as the study area, and used Landsat7/8 remote sensing image as the main data sources. Visual interpretation was conducted to determine type, timing, quantity of land deterioration in the area from 2008 to 2016. The obtained results can provide scientific basis for the making of land reclamation planning [6].

2. Study area

Pingshuo coal mine is one of the large coal bases in the north of Shanxi province included in the 13 large national planning coal base. It was Located in the north of Ningwu coal mine, administrative belonging to Pinglu district, Shuozhou city. With the area of 380 km^2, the mine has geological reserves of 11221 Mt. In general, there are 16 fields (3 for open pit and 13 for wells). The total construction scale are 93.50Mt/a, among which the construction scale of three open pit mines are15.00Mt/a for

Antaibao open pit coal mine, 10.00Mt/a for Anjialing open pit mine, and 20.00Mt/a for Donglutian mine. The mine service life is 75a. Donglutian coal mine locates in the northeast of Pingshuo mine, 10km from the northeast of Pinglu district, 28km from Shuozhou city, administrative belonging to

Shuozhou Pinglu area of jurisdiction. The geographical coordinates are longitude 112 ° 23 '58' ~ 112 ° 27 '52 " and latitude 39 ° 29' 58 "~ 39 ° 35 '33" (Figure 1). With 4.42 ~ 5.47 km long in the east-west direction and 6.53 ~ 10.3 km width in north-south direction, covering an area of 48.73 km^2, the mine had geological resource of 1848.92 Mt. A lot mines are in production and small mines prevails. By the end of May, 2004, there had 10 small coal mines with production life 3-5 years (Figure 2). Besides, the border of 5 small mines outside extended to the mine [7]. Small coal mine production capacity are below 0.15 Mt/a. Furthermore, the recovery rate is rather low, generally less than 30%. Disordered mining and resource waste is very serious. In order to put an end to combat phenomena of the disordered mining of small mines, Donglutian coal mine started officially in 2009.

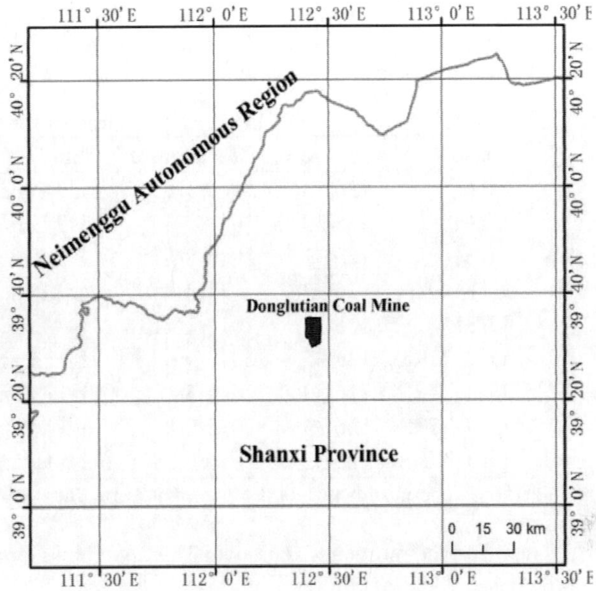

Figure 1. Location of Donglutian coal mine

3. Data source and processing

3.1. Remote sensing data

Remote sensing data used are mainly landsat7（ETM）/8(OLi) from 2008 to 2016 (Table 1). In total, 9 images downloaded from USGS websites were used, including 30m multiband spectral bands and 15m panchromatic band. Data acquired are mainly from May to October during which cloud cover was less than 1%.

3.2. Data processing and procedure

Envi5.2 are used to conduct the preprocessing work of remote sensing data, including atmospheric correction, radiation correction, geometry correction, and image fusion. Standard false color composite data are adopted, 432 band for ETM, 853 bands for OLi.

(1) The Landsat 7 scan-line corrector (SLC), a mechanism designed to correct the under sampling of the primary scan mirror, failed on May 31, 2003. With the SLC now permanently

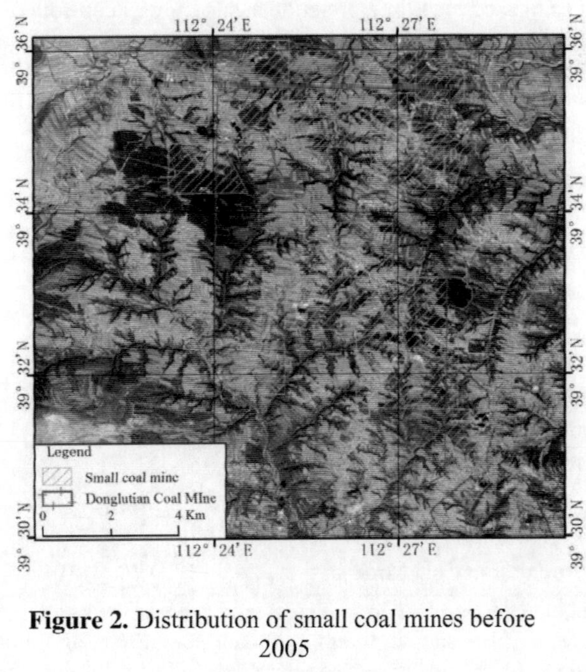

Figure 2. Distribution of small coal mines before 2005

turned off, the ETM+ is losing approximately 22% of the data due to the increased scan gap. So before using landsat 7, it is necessary to fill the gap. The specific method was to use the patches of each band and conducted interpolation processing.(http://landsat.usgs.gov/sci_an.php)

Table 1. Data source of remote sensing

sensor	path/row	date acquired	bands	resolution	source
ETM	126/33	2009/10/5	1-7,Panchromatic	30m/15m	USGS
ETM	126/33	2008/8/31	1-7,Panchromatic	30m/15m	USGS
ETM	126/33	2010/8/5	1-7,Panchromatic	30m/15m	USGS
ETM	126/33	2011/8/8	1-7,Panchromatic	30m/15m	USGS
ETM	126/33	2012/10/13	1-7,Panchromatic	30m/15m	USGS
OLI_TIRS	126/33	2013/9/6	1-11,Panchromatic	30m/15m	USGS
OLI_TIRS	126/33	2014/5/20	1-11,Panchromatic	30m/15m	USGS
OLI_TIRS	126/33	2015/6/8	1-11,Panchromatic	30m/15m	USGS
OLI_TIRS	126/33	2016/7/28	1-11,Panchromatic	30m/15m	USGS

(2) Firstly, to conduct atmospheric correction and radiation correction for 30m ETM multiband data.

(3)Then fuse the multiband data with 15 m pan band, and Gram-Schmidt Pan sharpening algorithm was selected.

(4) Conduct geometric correction to the fused images. 1:50 000 topographic maps in 1985 was selected as the control map, and using quadratic polynomial method to select more than 20 control points with RMS error o < 1. The corrected 2008 ETM image was chosen as the control image when doing the geometric correction to images from other years.

3.3 Remote sensing interpretation and extraction of change information

After image correction and interpretation key establishment, visual interpretation was done in ArcGIS 10.1 platform. The data was divided into Cropland, woodland, grass land, resident and coalmine, as well as unused land. Interpretation accuracy was controlled. The minimum interpretation chart spot for planar feature was 6*6 pixel (a pixel equals 15 meters), 2 pixel for linear feature. After interpretation, conduct topology processing and errors checking, and ultimately the land use map was obtained. Use the high resolution images to check the image, evaluate the result of the interpretation, and correct the wrong patches.

4. Results and analyzes

4.1. Area change of land use

Table 2. Area of every land use type from 2008 to 2016(km^2)

classes	2008	2009	2010	2011	2012	2013	2014	2015	2016
cropland	59.52	57.28	54.31	53.13	51.89	49.67	48.27	47.08	46.36
woodland	0.58	0.18	0.12	0.05	0.00	0.00	0.00	0.00	0.00
road	0.32	0.32	0.32	0.31	0.31	0.31	0.31	0.31	0.31
resident	3.99	3.96	3.57	3.93	3.77	3.31	3.36	3.19	3.23
coalmine	2.04	6.32	11.38	12.33	14.20	17.50	19.28	21.30	22.11
unused land	20.71	19.09	17.46	17.40	16.98	16.36	15.93	15.28	15.13
Donglutian Coal Mine		4.44	9.29	10.12	11.71	15.47	17.15	18.08	18.79

The land use map and area of each land type were shown in Figure 3, Table 2.The land use type in this area was rather simple, mainly cropland, coalmine, and unused land. The change trend was shown in Figure 4. Cropland and unused land reduced year by year, and the coal mine increased gradually. Area of cropland decreased by 13.16km^2 in 8 years' time with average decrease rate 1.64km^2 per year, generally remained stable. Unused land decreased by 5.58 km^2 , with the average decrease rate 0.7km^2, remained rather stable too.

Area of coal mine increased by 20.07$km^2$with the average growth rate 2.51km^2. The increase is more significant from 2008 to 2010, as it is the preliminary stage of mine exploration, and the increase rate began decreasing from 2011 to 2016, which had closed linked with fall of domestic coal prices.

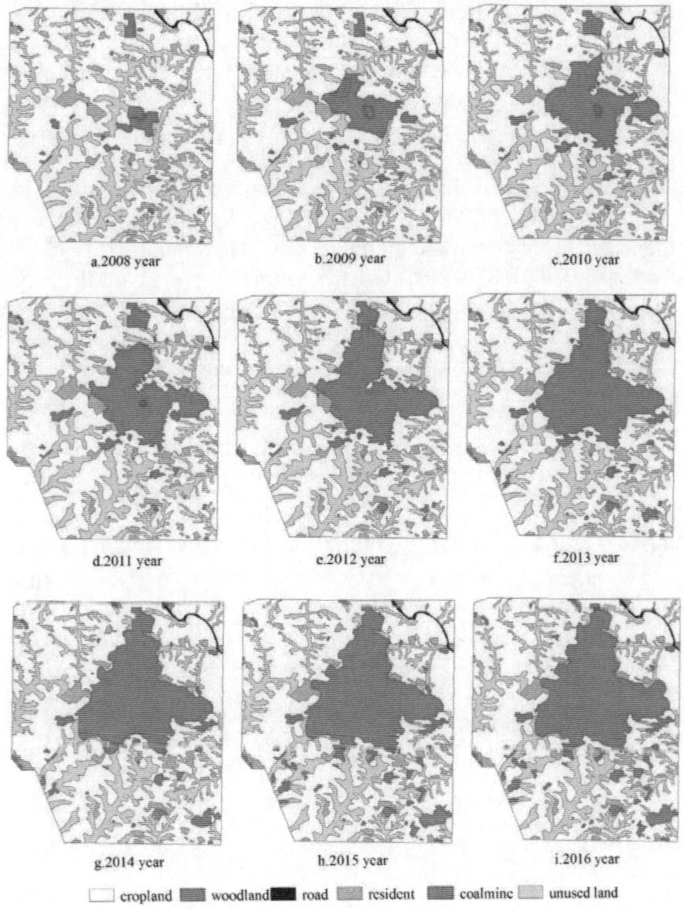

Figure 3. Land use map of the study area from 2008 to 2016

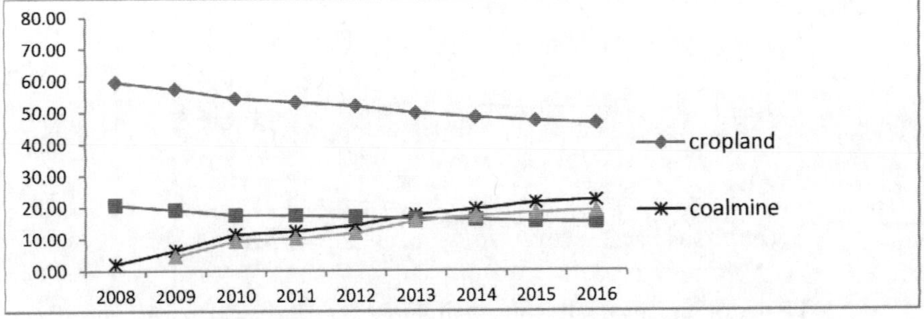

Figure 4. The area change of main land use type from 2008 to 2016

4.2. The monitoring of mining progress

Donglutian coal mine started from January, 2009, and covered an area of 4.44km^2, mainly for industrial construction. In 2010, the area reached 9.29 km^2, topsoil stripping work conducted except for the industrial construction. In 2011, coal mining started but the expansion was small. In 2012, the expansion remained small, and mainly for deep coal mining. In 2013, however, the expansion became obvious, increased by 3.75 km^{2}compared with area in 2012 with the southwest topsoil stripping the main cause. From 2014 to 2016, the mining activities were not as significant as before, and the area remain stable, mainly because the fall of international coal prices and national policies that reduce the coal production.

4.3. Monitoring the small coal mines

Through interpretation results, it was found that the number of small mine increased instead of decreasing. It increased from 26 in 2008 to 42 in 2016. The minimum mines of various periods covers an area of 0.003 km, and the largest is less than 0.773 km^2. The total area was 2.040- 3.326 km2 with the average size between 0.064- 0.079 km^2. On the whole, the area of the small mines gradually increased (table 3 and Figure 5). In spatial distribution (Figure 6), the latest mining of small coal mines mainly concentrated in the southeast and southwest side with no new small mines in the north. Therefore, data showed that the occurrence of Donglutiaon didn't changed the disordered mining of small mines in the area.

Table 3. Statistics of small mines from 2008 to 2016

	2008	2009	2010	2011	2012	2013	2014	2015	2016
Count	26	31	28	32	34	32	32	42	42
Minimum(km^2)	0.003	0.003	0.004	0.003	0.004	0.004	0.004	0.004	0.003
Maximum(km^2)	0.750	0.439	0.773	0.773	0.773	0.256	0.500	0.696	0.682
Sum(km^2)	2.040	2.171	2.085	2.209	2.487	2.034	2.121	3.211	3.326
Mean(km^2)	0.078	0.070	0.074	0.069	0.073	0.064	0.066	0.076	0.079

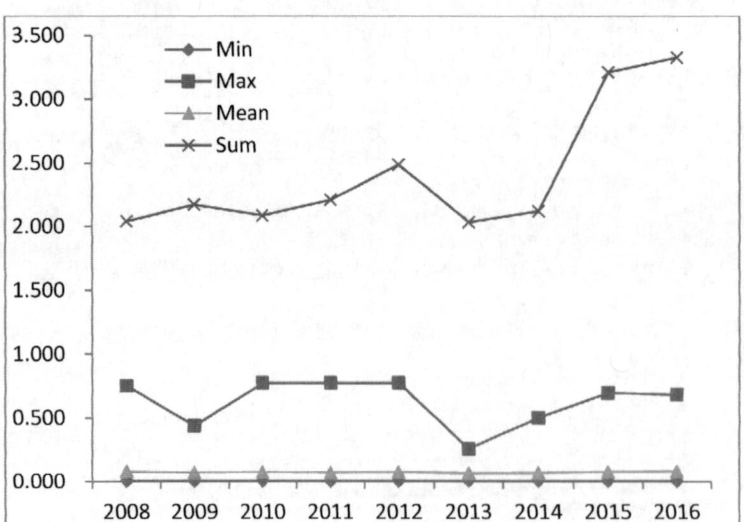

Figure 5. Statistical curve of small mine from 2008-2016

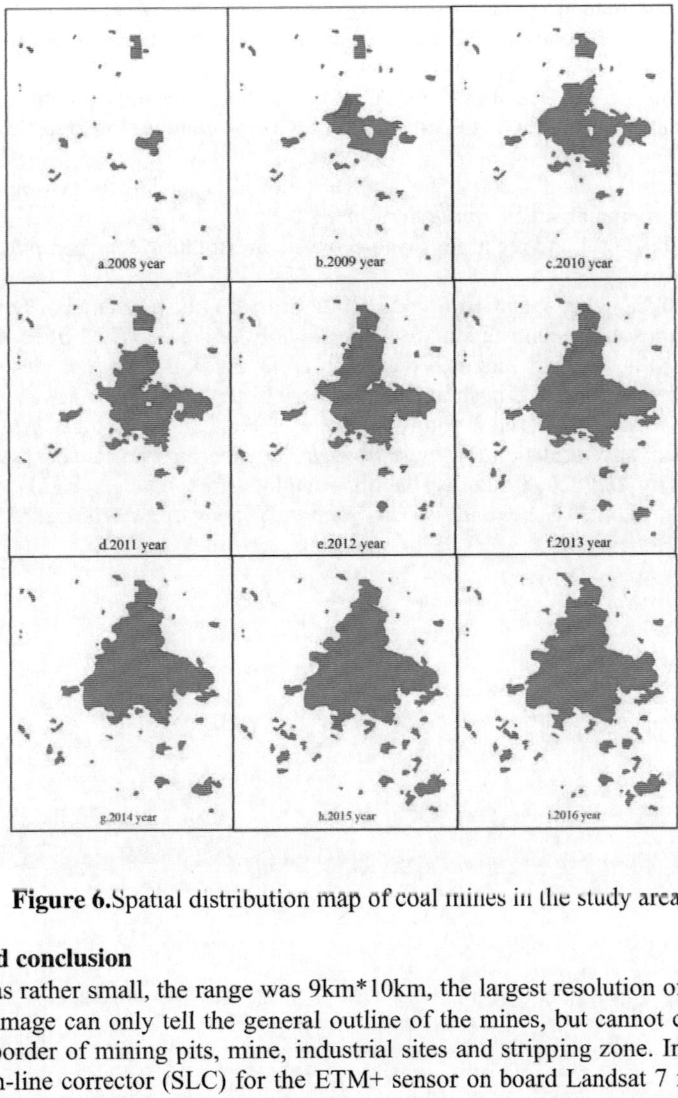

Figure 6. Spatial distribution map of coal mines in the study area

5. Discussion and conclusion

The study area was rather small, the range was 9km*10km, the largest resolution of landsat ETM(OLI) was 15m, so the image can only tell the general outline of the mines, but cannot clearly interpret the location and the border of mining pits, mine, industrial sites and stripping zone. In addition, On May 31, 2003, the scan-line corrector (SLC) for the ETM+ sensor on board Landsat 7 failed permanently. The SLC compensates for the forward motion of the satellite, and without an operating SLC, images have wedge-shaped gaps that range from a single pixel in width near the image-nadir, to about 12 pixels towards the edges of the scene. Missing pixels comprise about 22% pixels of these images, and loss about the width of 7 pixels [8-9].In this paper, The interpretation accuracy of this paper was 6 * 6 pixel. Therefore, the ground feature type less than 7*7 pixel, if in the position of the missing stripe, cannot be interpreted. Through the above statistics, it was found that there are some small mines with the area less than 7*7 pixel, so the result may be influenced to some extent.

Remote sensing technique is a effective way to monitor the long-term coal mining activities in the study area. Thus, it can provide the effective support for the exploitation of small mines, and also provide a favorable method for government to conduct mine reorganization and environmental protection.

References

[1] Chen H L, Xu X D, Liu Y J 2005 Review of researches on remote sensing monitoring and

impact on environment of land use/cover change*MeteorolSciTechnol* **33(4)** 289

[2] Bian Z F, Zhang Y P 2006 Land use changes in Xuzhou coal mining area *ActaGeogr Sin* **61(4)** 349

[3] Chen L Q, Guo D Z, Hu ZL, et al. 2004 A study on remote sensing monitoring land use change and reclamation measures of subsided land in Xuzhou coal mining area *ProgGeogr* **23(2)** 10-12

[4] Chen H L, Chen G, Guo J Z 2004 The application of Landsat TM for ecological environmental monitoring in mining area *Remote SensInf* **1** 31-32.

[5] Li J C, Bai Z K 2000 Land reclaiming and ecosystemrestoration for open pit coal mine *Beijing: Science Press.*

[6] Zhou Wei, BaiZhongke, YuanCHun, et al. 2007 Stimation of influence of Dong open-cast coal mine area on soil erosion and land use *Transactions of the CSAE* **23(3)** 55-61.

[7] China Coal Xi'an Design Engineering Co., LTD 2009 The initial design instruction of Donglutian coal mine of Pingshuo Coal Industry Company.

[8] Arvidson T, Goward S, Gasch J and Williams D 2006 Landsat-7 long-term acquisition plan: Development and validation *Photogrammetric Engineering and Remote Sensing* **72** 1137–46.

[9] Ju J C and Roy D P 2008 The availability of cloud-free Landsat ETM plus data over the conterminous United States and globally *Remote Sensing of Environment* **112** 1196–1211.

Measuring urban sprawl in China by night time light images

Lu Liu[1, 2, 3], **Lin Tang**[2]

[1]School of Urban Planning and Management, Harbin Institute of Technology, Shenzhen Graduate School, Shenzhen, China

[2]Center for Assessment and Development of Real Estate, Shenzhen, China

E-mail: faceocean@qq.com

Abstract. In the process of urbanization, a phenomenon called "urban sprawl" usually occurs. This phenomenon may exaggerated the negative effects of urbanization on environment, public and social health, energy efficiency, and maintenance of farmland. Therefore, the understanding of this phenomenon is urgently required for us to achieve sustainable development. This study proposed a group of night time lights (NTL) indicators of urban sprawl, which intend to use the distribution of lightness to quantify urban sprawl. These measures are proved to be efficient in describing urban sprawl. In addition, they are consistent and easy calculating, making comparison analysis easy to be done. These indicators are used to study urban sprawl in China during the year 2000 to 2010, the results show that in the last ten years, metropolitan areas in the northern part of China have undergone a more sprawl-like urban growth compared with other parts of China.

1. Introduction

In the process of urbanization, a phenomenon called "urban sprawl" usually occurs. Urban sprawl refers to spontaneous urban development pattern, which is scattered, fragmented, and with low density. This phenomenon may be caused by fragmented local government, poor planning, or exclusionary zoning. Many previous studies that urban sprawl has exaggerated the negative effects of urbanization on environment, public and social health, energy efficiency, and maintenance of farmland. A profound understanding of this phenomenon is therefore highly demanded.

Therefore, the definition of sprawl and the indicators selected to measure it are mainly based on the perspective and objectives of a specific study. In this study, we refer to urban sprawl as an unplanned, fragmented, low-density process of urban growth. Moreover, the urban sprawl here should be deemed a phenomenon to describe, rather than to define [1, 2]. In other words, the measures of urban sprawl should focus on depicting this phenomenon from some specific aspects based on the research problem rather than trying to find a comprehensive measuring method. Furthermore, relative measurement, instead of absolute measurement should be adopted to study urban sprawl, which means that the degree of sprawl is assessed based on the comparison analysis among different regions (relatively more sprawl-like, or less sprawl-like), rather than giving an absolute black and white definition to this phenomenon(sprawled or not sprawled).

The previous studies of 'urban sprawl' have focused primarily on the US [3, 4], where new growth has often been discontinuous and extensive in the past few years; very few studies have focused on China. This research gap may be attributed to the usually high population density of urban areas in China, causing people to ignore the occurrence and possible effects of urban sprawl. However, the dramatic expansion of urban land in the last few decades has been accompanied by serious urban sprawl. As proven by several previous studies, the urban land in China has increased exponentially to become more than two times its size in the last twenty years, and some places are very inefficiently developed. Therefore, a comprehensive understanding of urban sprawl in China, which may raise the concern of making better policy and plan on urban land development, is required.

Therefore, this study aims to introduce a set of night time lights measures of urban sprawl, which are consistent and easy to measure. Also, the urban sprawl in China is scrutinized based on the indicators introduced in this study.

2. Data and methodology

2.1. Study Areas

To measure the degree of urban sprawl, two difficulties typically arise. The first is how to select the appropriate scale to measure the phenomenon. The second is how to identify the spatial extent of "urban area".

In previous studies, a scale to study urban sprawl is usually selected based on the objective of the research and the limitation of datasets. In this study, the metropolitan scale is used, because NTL imagery has coarse resolution (30 arc), and is thus inappropriate for the analysis at small scales, such as the local and city scales. Moreover, lit-up areas in the NTL imagery can reveal the extent of various metropolitan areas, which are composed of one or several cities, because the lit-up areas of spatially related cities are always continuous.

Second, previous research often adopts political boundaries to delineate the spatial extents of metropolitan areas. For instance, in the "Report on Development of Metropolitan Areas in China in 2010" [5], 23 metropolitan areas in China have been determined based on the economic and political relationships among cities, and the extents are delineated by the administrative boundaries of these cities. However, this approach appears to be an unreliable means of delineating the metropolitan area, because it may overestimate or underestimate actual urban extent. Meanwhile, the extracted extents are incomparable among nations[6], thus leading difficulty to conduct a comparative analysis. Therefore, we propose a more reasonable and consistent strategy to delineate the extent of the metropolitan areas using NTL images.

In this study, the extents of metropolitan areas are extracted from NTL imagery based on the continuous lit-up areas. First, dim lights (pixels with a DN value lower than 12), which are more probably emitted by rural areas with sparse population, are excluded. Each continuously lit-up area is then deemed as an individual urban patch, and urban patches larger than 1000 km^2 are identified as metropolitan areas. Moreover, urban patches within a distance of 10 km^2 from these metropolitans are merged to them, given that these small urban patches may be largely influenced by the metropolitan areas and are thus a part of them. We do not consider other small urban patches outside the metropolitan areas in this research, because they are not included in any metropolitan area. Moreover, to divide the urban area into city center and suburban area, we set a single threshold of DN value 54 to discriminate urban and suburban areas, this threshold is obtained by overlaying the TM derived urban center with NTL images.

2.2. NTL Imagery

In this study, we use the distribution of DN value in NTL imagery to describe urban sprawl, because brightness is highly related to urban activities on the Earth's surface[7,8]. In many previous studies, the summed lights and the lit-up areas for a region on NTL imagery have been proven to be highly related to various indicators of urbanization, including economic activities, population and area of urban land. Furthermore, the brightness of each pixel has also been proven to represent population density to some extent and has thus been used as an assistive data set for mapping the world population density in the Global Rural Urban Mapping Project (GRUMP)[9]. Therefore, describing the distribution of urban development using NTL imagery is reasonable. The NTL imagery is obtained from the data sharing Infrastructure of Earth System Science Web site (http://www.geodata.cn/, assessed on 15 May, 2014).

2.3. Temporal Measures of Urban Sprawl

We define sprawl like urban growth as a process during which the extents of the urban areas expand significantly outward in a specific period of time. Meanwhile, the expanded regions have relatively low intensity of development, and insignificant urban development has occurred to fill the existing urban or sub-urban area. Table 1 lists the indicators for the three dimensions that are considered to describe this process, and their NTL-based indicators are explained as follows.

2.3.1. Expansion of urban extent The expansion of urban extent measures the degree to which the whole metropolitan area expands its urban land. The NTL-based indicator for this dimension is calculated by:

$$EU = (A_{t2} - A_{t1})/A_{t1} \quad (1)$$

where *EU* refers to the expansion of urban extent, and A_t represents the area of the metropolitan area at the time point t. *t1* represents the previous point of time, and *t2* represents the latter point of time.

A very high *EU* value indicates that the metropolitan area has significantly expanded its urban land. However, a sprawl like urban growth cannot be signified only based on a high value of EU. The intensity of expansion should also be further examined to verify the efficiency of such development.

2.3.2. Intensity of Expansion The intensity of expansion tends to measure the intensity of development at places that were originally rural lands, but were later transformed to parts of urban or suburban areas. As illustrated in Fig.1, the blue regions are the expanded regions of metropolitan areas from year 2000 to year 2010.

The indicator of this dimension is obtained by calculating the mean *DN* value of expanded regions at time point *t2*. A low value may imply that, the expanded regions are not efficiently developed with low intensity, that is, the regions exhibit more sprawl-like urban growth.

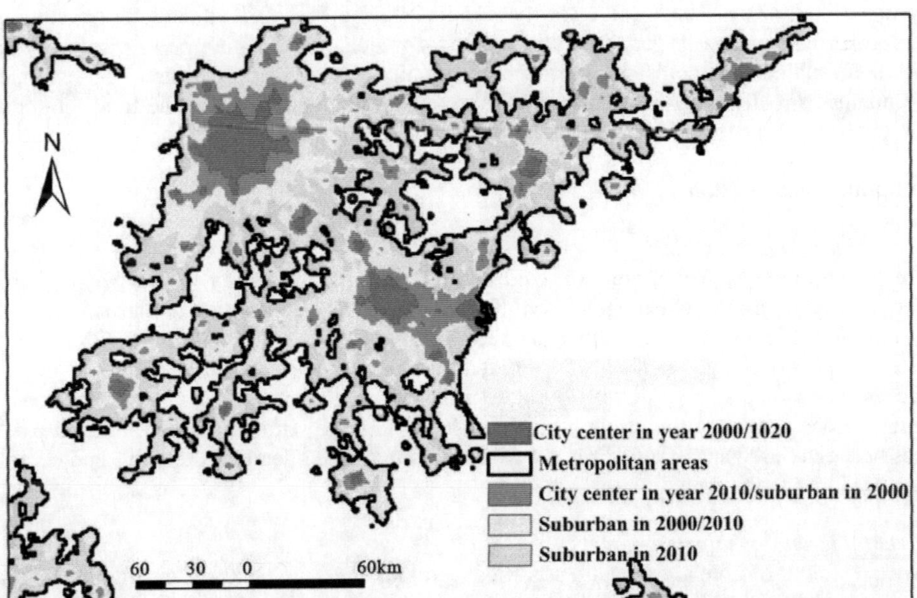

Figure 1. The change of urban extent during 2000 to 2010 in Beijing-Tianjin-Tangshan area.

2.3.3. Infilling of suburban area This dimension measures the degree to which development is concentrated in the already existing suburban area. A compact development should generally focus on efficiently utilizing the already existing urban land, rather than developing new urban land. Therefore, much infilling development in suburban area may indicate a less sprawl like urban growth.

The indicator for this dimension is obtained by calculating the change of intensity (intensity is represented by mean DN value on NTL imagery) in the suburban area. A higher value of intensity change in the region may indicate that significant development has occurred in the already existing suburban region, and a less sprawl-like growth can be signified.

Table 1. The temporal measures of urban sprawl

Dimension	Definition	Operationalization	Measure of sprawl
Expansion of urban extent	The degree to which the urban extent has been expanded	$EU=(A_{t2}-A_{t1})/A_{t1}$	Higher value signifies more sprawl like urban growth
Intensity of expansion	The degree to which the expanded region are intensely developed	Mean DN value in t2, in the expanded urban areas (IE)	Higher value signifies less sprawl like urban growth
Infilling of suburban	The degree to which the original suburban areas are developed	Change of Mean DN value during t1 and t2, in places where are suburban area in both t1 and t2 (INS)	Higher value signifies less sprawl like urban growth

2.4. Analysis Method

The urban sprawl in China is then examined based on the measures introduced above. The analysis is processed in three steps.

The temporal measures of urban sprawl are used to scrutinize the historical trend of urban sprawl during years 2000 to 2010. The Z-score for these indicators are first derived, and the negative number is calculated for EU for consistency with the other two indicators. The summed value of all these Z-scores is then obtained to represents the degree to which the metropolitan area is developed in a sprawled manner. Finally we use a distribution map to examine and compare the trend of sprawl like growth in the last decade in various metropolitan areas of China.

3. Result and interpretation

3.1. Metropolitan Areas in China

The metropolitan areas in China are extracted based on the methodology introduced in Subsection 7.2.1. The metropolitan extents extracted from NTL imagery successfully include urban lands that are spatially closed and related to one another. Moreover, compared with the extent of BTT metropolitan areas used in previous studies, the NTL-derived metropolitan extents can delineate the densely urbanized area more accurately, and can separate urban areas that are not spatially continuous. Based on this method, we obtained 50 metropolitan areas, including two with areas of over 10,000km^2, eight with areas between 5,000 and 10,000 km^2, 14 with areas of approximately 2,000 km^2, and 23 with area between 1000 km^2 to 2000 km^2.

3.2. Correlation relationship of measures

As shown in Table 2**Table** , correlation is weak among these indicators, given that all pairs of indicators show low *r* value. This result may indicate that these three measures are independent in measuring sprawl. Moreover, the area of the metropolitan area shows a weak relationship with all other indicators, implying that these indicators are also independent of size.

Table 2. Correlation analysis among temporal measures of sprawl

Correlation r	EU	IE	INS	Area
EU	1	.559**	-0.046	-0.44
IE		1	.485**	-0.107
INS			1	0.56
Area				1

**. Correlation is significant at the 0.01 level (2-tailed), N=50

3.3. Urban Sprawl in China

Fig.2 depicts the map of these measures for 50 metropolitan areas in China. In the northern part of China, the metropolitan areas have undergone a more sprawl like urban growth in the last 10 years compared with other parts of China.

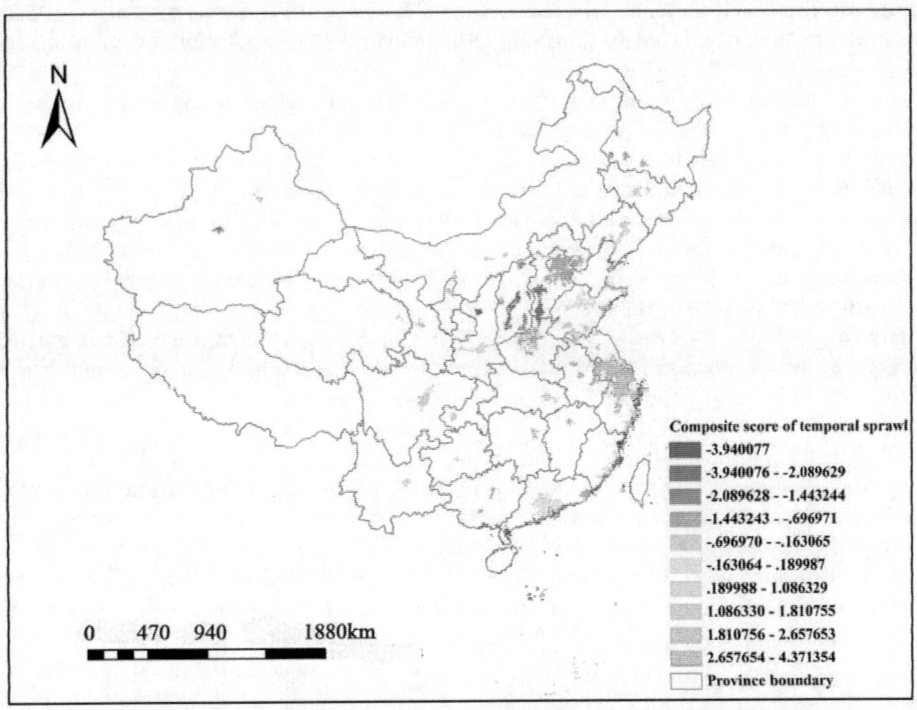

Figure 2. Urban sprawl during year 2000 to 2010 by temporal indicators

4. Discussion and conclusion

This study proposed a group of NTL indicators of urban sprawl, which intend to use the distribution of DN values to quantify urban sprawl. These NTL indicators can provide an overall idea of the level urban sprawl, and are very easy to be calculated, especially for a group of large regions. These measures are proved to be efficient in describing urban sprawl. In addition, they are consistent and easy calculating, making comparison analysis easy to be done. These indicators are used to study urban sprawl in China during the year 2000 to 2010, the results show that in the last ten years, metropolitan areas in the northern part of China have undergone a more sprawl-like urban growth compared with other parts of China.

However, NTL indicators have several limitations. First, the saturation of NTL may influence the analysis results, especially for densely developed urban areas. Second, the response of NTL to urban growth may vary across regions and years, and this variation causes bias in the analysis results. Third, the NTL indicators of urban sprawl cannot comprehensively reflect urban sprawl, such as the mixed use of urban functions. Therefore, these indicators are better combined with other datasets in future studies, to comprehensively understand urban sprawl.

Moreover, given that these indicators are calculated based on the distribution and changes in the DN value on the NTL imagery, more information should be obtained on how these DN values are correlated to urban activities on the earth surface when appropriate datasets are accessible. Moreover, although we have discovered the regional discrepancy in the urban sprawl of various metropolitan areas in China, the underlying causes are still unknown. Therefore, various socio-economic factors and government policies should be studied to identify these reasons. Finally, the manner by which urban sprawl affects the lives of the people should be further explored

References

[1] Emily Hoffhine Wilson, James D Hurd, Daniel L Civco 2003 Development of a Geospatial Model to Quantify, Describe and Map Urban Growth *Remote Sensing of Environment* **86**(3) 275-285.

[2] Bhatta B, Saraswati S, Bandyopadhyay D 2010 Urban Sprawl Measurement from Remote Sensing Data *Applied Geography* **30**(4) 731-740.

[3] Shima Hamidi, Reid Ewing A 2014 longitudinal study of changes in urban sprawl between 2000 and 2010 in the United States *Landscape and Urban Planning* **128** 72-82.

[4] George Galster, Royce Hanson, Michael R Ratcliffe 2001 Wrestling sprawl to the ground: defining and measuring an elusive concept *Housing policy debate* **12**(4) 681-717.

[5] Chuanglin Fang, Shimou Yao, Shenghe Liu 2011 Report on Development of Metropolitan Areas in China in 2010 Beijing: *Science Press.*

[6] Annemarie Schneider, Curtis E. Woodcock 2008 Compact, dispersed, fragmented, extensive? A comparison of urban growth in twenty-five global cities using remotely sensed data, pattern metrics and census information *Urban Studies* **45**(3) 659-692.

[7] Christopher D Elvidge, Edward H Erwin, Kimberly E Baugh 2009 Overview of DMSP Nightime Lights and Future Possibilities 1-5.

[8] Ting Ma, Chenghu Zhou, Tao Pei 2012 Quantitative estimation of urbanization dynamics using time series of DMSP/OLS nighttime light data: A comparative case study from China's cities *Remote Sensing of Environment* **124** 99-107.

[9] Balk D L, Deichmann U, Yetman G. 2006 'Determining Global Population Distribution: Methods, Applications and Data', in *Advances in Parasitology*, ed. by Alastair Graham Simon I. Hay and J. Rogers DavidAcademic Press,) pp 119-156.

IOP Conf. Series: Earth and Environmental Science **52** (2017) 012112 doi:10.1088/1755-1315/52/1/012112

Thinking on Sichuan-Chongqing gas pipeline transportation system reform under market-oriented conditions

Yanzhi Duan

[1] School of Economics and Management, Southwest Jiaotong University, Chengdu, Sichuan 610031, China

[2] Natural Gas Economic Research Institute of PetroChina Southwest Oil & Gas Field Company, Chengdu, Sichuan 610051, China

E-mail: duanyanz@petrochina.com.cn

Abstract. The gas pipeline networks in Sichuan and Chongqing (Sichuan-Chongqing) region have formed a fully-fledged gas pipeline transportation system in China, which supports and promotes the rapid development of gas market in Sichuan-Chongqing region. In the circumstances of further developed market-oriented economy, it is necessary to carry out further the pipeline system reform in the areas of investment/financing system, operation system and pricing system to lay a solid foundation for improving future gas production and marketing capability and adapting itself to the national gas system reform, and to achieve the objectives of multiparty participated pipeline construction, improved pipeline transportation efficiency and fair and rational pipeline transportation prices. In this article, main thinking on reform in the three areas and major deployment are addressed, and corresponding measures on developing shared pipeline economy, providing financial support to pipeline construction, setting up independent regulatory agency to enhance the industrial supervision for gas pipeline transportation, and promoting the construction of regional gas trade market are recommended.

1. Introduction

Sichuan Basin is the cradle of China's gas industry. Thanks to the development in more than 50 years, a fully-fledged gas pipeline network system has been formed in this region, and a matched gas pipeline transportation system adaptable to the integrated development of regional production, transportation, storage and sale has been structured. In recent years, the development of gas industry in this region has become further market-oriented, and so there exists a demand for market-oriented reform in the areas of investment system, operation system and pricing system. In this article, the main thinking on Sichuan-Chongqing pipeline transportation system reform under market-oriented conditions and the realization approach are addressed by referring to the features of Sichuan-Chongqing pipeline networks and production and consumption market.

It is necessity to reform Sichuan-Chongqing gas pipeline transportation system. Firstly, increased pipeline requirements resulted from adjustment to supply and demand market necessitates pipeline investment system reform. With the large scale development of high sulphur gas in Longwangmiao and Northeast Sichuan, the import of external source gas into Sichuan, the export of internal source gas out of Sichuan, and the unbalanced development of internal and external gas markets, it becomes more urgent to improve the adaptability of existing gas pipeline networks and construct new pipelines. In addition, both the return on investment and the recovery of capital are slow, and a great sum of money will be occupied. In this case, continuous reliance on the two central oil enterprises for investment and construction is hard to satisfy the demand for gas supply and consumption. Secondly, rational pricing and improving the transportation efficiency of gas pipelines call for pipeline operation system reform.

Sichuan-Chongqing gas pipeline networks mainly feature interwoven gathering pipelines for production and transportation/distribution pipelines for consumption. It is hard to distinguish source gas cost from pipeline transportation cost, which has brought difficulties to pricing due to the lack of definite basis for adjusting source gas price and pipeline transportation price. This problem is originated from the system that fails to separate gas production from gas transportation. Furthermore, PetroChina and Sinopec have different clients, and so there exist differences in their supply and demand requirements. It is no surprise to have the situation that gas is imported into and exported out of Sichuan simultaneously, which as a whole will give rise to the waste of pipeline transportation cost and the decrease of pipeline transportation efficiency. Thirdly, construction of regional trade market requires pipeline transportation price reform. China is carrying forward actively a market-oriented gas price reform, and both Sichuan and Chongqing are making separate feasibility research for the setup of regional gas trade market. For the construction of gas trade market, it is necessary to put in place a price system that is "tolerant with both ends but strict with the middle", which means while the access to upstream and sales markets is deregulated, the prices for pipelining different source gas to different clients should be defined. According to the existing pricing mechanism, pipeline transportation price is included in portal station price. Such a pricing system is too simple, and its structure is unfledged. With this mechanism, it is unable to identify the actual cost for supplying gas from its source to the client, which is unfair to both the seller and the client. Besides, the construction of new pipelines cannot change the price level of portal stations, which is unfavourable to the investment and construction of gas pipelines. Consequently, it is the inevitable requirements for constructing regional trade market in Sichuan-Chongqing region to further pursue gas pipeline transportation price reform.

2. Main Features of Sichuan-Chongqing Gas Market and Pipeline Transportation

2.1. A fully-fledged gas market in China

In 2015, more than 34 BCM (short for *billion cubic meters*) of gas was produced, while 23.3 BCM of gas was consumed in Sichuan-Chongqing region. Both the gas production and consumption of this region were among the top ones of China. Gas consumption has accounted for more than 12% in the primary energy consumption structure of this region for a long time, which is above the average level of 5.9% in this country. In this region, there are 159 cities using gas, which means 72% of the total cities in this region are gas consumers. By 2020, the gas production will be about 60 BCM and consumption will also be greatly improved. Moreover, with the further development of market-oriented domestic gas industry, the local governments are planning to build a regional gas trade platform and reforming the gas sales model to promote market prosperity.

2.2. With a well-developed gas pipeline network system

2.2.1. Cobwebbed pipeline network. In Sichuan-Chongqing region, an integrated pipeline system with the functions of gas production, transportation, sales, and dispatching has been constructed. Besides, the market is connected to other major markets outside of Sichuan Basin via Zhong-Gui gas pipeline, Zhong-Wu gas pipeline and Sichuan-East gas pipeline. By the end of 2015, about 9,000km gas pipelines with the transportation capacity of 30 BCM had been constructed in this region. By 2020, its gas transportation capacity will exceed 35 BCM [1].

2.2.2. Complicate investment and operation system. For a long time, the main loop network and most short branch pipeline networks have been operated by Southwest Oil and Gas Field Company of PetroChina. Sinopec has also constructed its own small regional gas supply pipeline networks in West and East Sichuan. Since 2014, both the two major oil companies have carried out a diversified ownership reform in the pipeline transportation sector. As a result, Sichuan Gas Investment Company and Chongqing Gas Pipeline Company were founded successively to take charge of the construction of branch Sichuan-Chongqing gas pipelines. As gas production area near to consumption area, PetroChina sets up a 3-tier pipeline transportation management system: The production, operation and pipeline management departments are responsible for the management of gas dispatch, pipeline transportation and pipeline development; the gas transportation and mining area departments are responsible for executing gas pipeline transportation and operation; and the subordinated operation sectors and transportation/distribution stations are responsible for specific site operation.

2.2.3. A leader in transportation price reform Sichuan-Chongqing region has led three reforms in gas pipeline transportation prices. The first reform resulted in a unitary pipeline transportation price. It was first originated in Sichuan Basin in 1963. Later, the operators of Chang-Na gas pipeline, Wei-Wu gas pipeline and other gas pipelines began to charge their clients at this unitary price[2]. The second reform resulted in a pricing method based on transportation distance. In 1976, this pricing method was first implemented in Sichuan with the issuance of a provision on charging gas consumers based on pipeline transportation distance by the Ministry of Petroleum and Chemical Industry[3]. The third reform has resulted in the current stamp pricing system. In 2012, a portal station price reform was carried out in Sichuan and Chongqing. According to this new pricing system, the pipeline transportation cost is included in the portal station price, which means 0.17rmb yuan/m^3 and 0.16 rmb yuan/m^3 major pipeline network transportation fee respectively and 0.11 rmb yuan/m3 administrative fee for gas distribution. In addition, Sichuan province has also stipulated the transportation prices for 22 short inter-provincial gas pipelines.

3. Thinking on Sichuan-Chongqing Gas Pipeline Transportation System Reform

Sichuan-Chongqing gas pipeline transportation system reform should take market development trend into consideration. Initiatives should be taken to meet the requirements for allowing diversified ownership of pipeline transportation industry, separating gas pipeline networks from their operation, improving energy transportation efficiency and developing regional trade market. The system reform should focus on the areas of investment, operation and pricing to attract more investment. Sichuan-Chongqing pipeline networks should be subject to overall planning, operation and dispatching. More fair and rational pipeline transportation pricing system should be worked out to promote the further development and high efficiency of Sichuan-Chongqing gas market.

3.1. Pipeline investment system reform

Main ideas

- Deepen pipeline investment decision-making system reform;
- Build a consolidated investment and financing platform for Sichuan-Chongqing gas pipeline construction;
- Improve the investment and financing mode for pipeline construction;
- Expand continually the financing channels for pipeline construction;
- Encourage private capital's access to Sichuan-Chongqing branch pipeline construction; and
- Minimize pipeline asset volume to share development.

3.1.1. Develop jointly the pipeline construction plan to guide investment and decision-making Local governments of Sichuan and Chongqing may unite PetroChina, Sinopec, joint venture pipeline companies and major sales companies to establish a pipeline construction planning consultation committee, whose major functions are developing jointly five-year and ten-year Sichuan-Chongqing gas pipeline construction plans based on production and market change trends and under the premise of guaranteeing return on investment and supply security, making adjustment to and executing them stage by stage, ensuring the guidance of these plans to investment decision-making, simplifying approval procedures for planned projects, and offering necessary support.

3.1.2. Set up consolidated pipeline investment and financing platform Based on the existing diversified investment entities in Sichuan-Chongqing region, gas pipeline network investment companies invested respectively by PetroChina, Sinopec, government capitals and private capitals should be consolidated to found a subsidiary pipeline company jointly with local clients or local governments to attract more funds for pipeline construction. The consolidated pipeline network investment company can serve as a regional subsidiary of the state pipeline network company in the future.

3.1.3. Define investment interfaces for various pipeline companies Pipelines constituting Sichuan-Chongqing gas pipeline networks can be divided into two categories in terms of application, i.e. transportation/distribution pipelines and client branch pipelines. The former is defined as the inter-area gas transportation/distribution pipelines inside the Basin to serve the five areas of Northwest Sichuan, Middle Sichuan, South Sichuan, Northeast Sichuan and Chongqing, while the latter refers to the independent sales pipelines connected directly to clients inside each area[4]. It is recommended that construction and operation of the former should be executed solely by the central enterprises, while construction and operation of the latter should be executed by existing joint venture pipeline companies themselves or by their joint ventures with local clients. In this way, the investment enthusiasm of joint venture pipeline companies at all tiers can be aroused to share development, and functions of different investors involved can be defined clearly[5]. The central oil enterprises are mainly responsible for securing safety and supply, while other investors can participate in pipeline construction in a more market-oriented way. Based on the principle of interface dividing, client branch pipelines owned by the central enterprises should be subjected to restructuring or stripping to realize the reduction of pipeline assets by means of property right transfer or acquisition.

3.2. Pipeline operation system reform

Main ideas

- Make down-to-earth efforts to improve pipeline operation, so that the existing 3-tier management system (administrative office management + mining area operation) adopted by oil and gas field companies for their own can be gradually transited to operation by entities;
- Identify pipeline operation cost composition;
- Simulate market-oriented operation and consolidate the operation agencies of Sichuan-Chongqing pipeline networks at a proper time in the future, so that the independent operation of Sichuan-Chongqing pipeline networks can be achieved gradually; and
- Improve the entire pipeline transportation efficiency.

3.2.1. Carry out experiment on the independence and professionalization of pipeline operation teams Professional operation companies (as tier-2 units or wholly-owned subsidiaries) responsible for pipeline operation should be founded at a proper time for such experiment. They are supposed to provide professional dispatching service for the operation of gas transportation/distribution pipelines. Operation of the professional companies will cover the main loop network (excluding gathering pipelines) and client branch pipelines in Sichuan-Chongqing region. In the long run, the gas transportation/distribution pipeline operation teams of PetroChina and Sinopec should be consolidated to establish a Sichuan-Chongqing gas pipeline network operation company. Oil and gas field companies should only reserve their ownership of pipelines and assign their operation and management rights to the operation company. The trend is expressed in Figure 1. In doing so, repeated investment in pipeline assets could be avoided, operation cost composition could be identified clearly, professionalization level of operation could be improved gradually, implementation of third party access system could be facilitated, and pipeline operation efficiency could be improved.

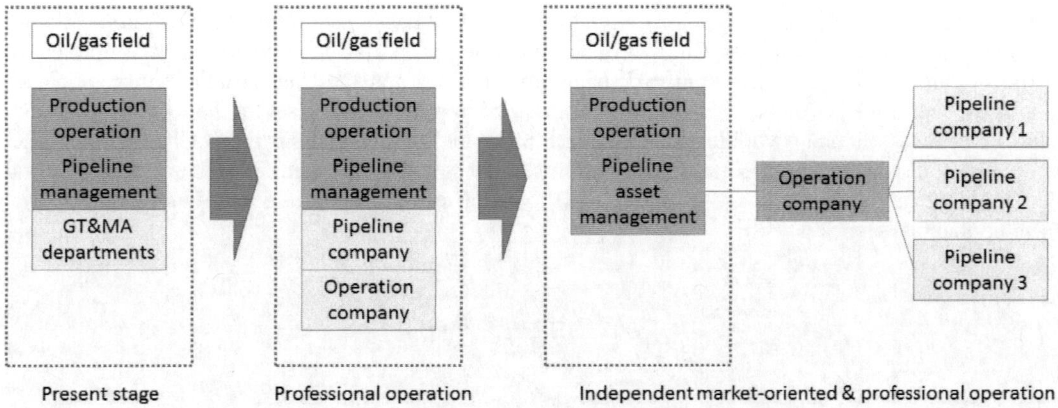

Figure1. The trend of operation system reforming

3.2.2. Set high efficiency and low cost operation objectives Firstly, gas production status should be fully taken into account. Under same price conditions, nearby areas should become the first choice for gas sales. Surplus gas could be sent into the main Sichuan-Chongqing pipeline networks for transportation and dispatching. Secondly, gas sales status should be taken into account as well. Unloading from the loop network and transportation of source gas should be arranged in a way to achieve the lowest pipeline transportation cost. In the future, if operation becomes independent, operation companies will be able to balance the gas production and sales status of oil companies and other shale gas enterprises and realize the overall planning for gas transportation and dispatching inside the Basin to avoid the situation of gas importing into and exporting out of Sichuan simultaneously, reduce the entire transportation cost, and improve the overall efficiency.

3.2.3. Adopt market-oriented operation mechanism Firstly, a market-oriented pipeline transportation sales mechanism should be put in place. Clients can make a reservation with the operation companies for pipeline transportation capacity whose market transaction should be allowed inside Sichuan-Chongqing region. Secondly, economic levers should be used to balance gas supply. If clients' actual gas demand is more or less than the transportation capacity reserved by them, operation companies can give them economic punishment [6]. Thirdly, profit models of operation companies should be closely related to their sales volume (i.e. They can get certain percentage of dispatching and operation fee for every cubic meters of gas sale.), or pipeline companies should procure operation companies' operation and dispatching service by competitive bidding. In the long run, pipeline companies may consider sharing the added values of pipeline transportation income with oil and gas companies. For example, new benefits could be created by optimizing operation management and selling surplus pipeline transportation capacity. A certain percentage should be taken from such benefits as the incomes of operation companies. Fourthly, cost accounting should be independent. Labour cost, repair cost, loss and all other costs incurred during operation and dispatching should be accrued to operation companies, and be used as key indicators for performance evaluation and supervision.

3.3. Pipeline transportation pricing system reform
Main ideas

- Build rationally price structure for Sichuan-Chongqing pipelines of different properties;
- Adopt a service cost-based pricing method and provide strict regulation to reflect the difference in pipeline transportation costs for different clients in different areas; and
- Guide the market-oriented allocation of gas resource.

3.3.1. Stipulate regional price ranges for loop network, and set prices for client branch pipelines based on transportation distance Sichuan-Chongqing gas pipeline transportation price is mainly comprised of the transportation price of transportation/distribution pipelines and the transportation price of client branch pipelines. Five gas transportation/distribution price ranges should be set for the five consumption areas of Northwest Sichuan, Middle Sichuan, South Sichuan, Northeast Sichuan and Chongqing, while the pricing of client branch pipeline transportation should be distance-based. See Figure 2. Besides, a pricing method based on season and gas supply quality (such as discontinuous or continuous gas supply) is also applicable.

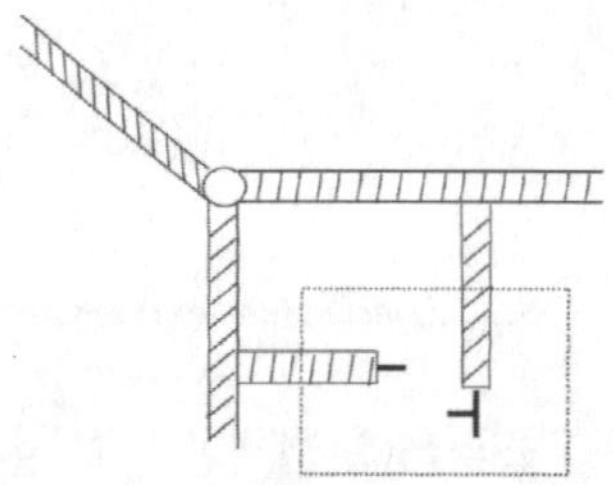

Figure 2. Combing of region-based price and distance-based price.

3.3.2. Regulate and set prices for pipeline companies of independent legal entities based on service costs In Sichuan-Chongqing region, the two major oil companies, joint venture pipeline companies and subsidiary branch pipeline companies organized by joint venture pipeline companies are all independent legal entity companies. Maximum allowable incomes or total service costs should be determined for pipeline companies, whose services should be subject to separate pricing. The maximum allowable incomes of each pipeline company mainly include pipeline operation expenditure, pipeline depreciation, income tax expenditure, other tax expenditure and income from investment calculated from multiplying the net pipeline assets by the regulatory return rate [7]. A pipeline transportation price adjustment mechanism should be put in place to adjust the overall price level every two or three years based on the service cost and profit change of pipeline companies. If the design for operation system reform is executed, pipeline operation expenditures will be the total costs paid to operation companies, and so strict regulation will be necessary [8].

3.3.3. Set entrance/exit pipeline transportation prices for purchased/sold gas at major hub points According to the tie-in conditions of Sichuan-Chongqing pipeline networks with main national pipeline networks, Zhongxian (Zhong-Wu gas pipeline), Guangyuan, Nanbu, Nanchong, Tongliang, jiangjin (Zhong-Gui gas pipeline) and Dazhou (Chuan-East gas pipeline) are all major junction points with external markets. With the development of market-oriented gas trade in Sichuan-Chongqing region, there will be more connections between the internal and external markets. The workload for source gas dispatching at these connection points will be increased significantly to impact the production, transportation and dispatch of source gas inside the Basin and add to operation difficulties and cost. Because the increased cost should be borne by clients of these hub points, it is necessary to set pipeline transportation prices separately for the gas flowing into or out of these hub points. To be specific with price form, entrance and exit pipeline transportation prices for purchased and sold gas can be set separately at these hub points. Such pricing should be mainly based on labour cost, loss and the depreciation cost of supporting facilities associated with the special service.

For example, if a client of Nanbu (Sichuan province) directly purchases gas from external market through Zhongwei-Guiyang gas pipeline, and the purchased gas enters Sichuan-Chongqing pipeline network via Nanbu hub and is finally supplied to the client, the pipeline transportation fee incurred by this client will be the total of Nanbu hub entrance fee, Chuanzhong price range gas transportation/distribution fee and the distance-based client branch pipeline transportation fee.

3.3.4. Learn from European management model for gas pipeline transportation pricing under market-oriented conditions With the deepening of market-oriented development and the further adjustment of Sichuan-Chongqing gas sales system, gas sales business will become more market-oriented, and independent gas retailers or profiteers may emerge. In this case, Sichuan-Chongqing gas market structure can be guided to develop in the direction of British gas market structure[9,10], and the pipeline transportation price should be subject to corresponding adjustment. An entrance/exit pricing system can be switched to, and such important junction points as Yucheng, Suining and Jiangjin can be selected for entrance/exit pipeline transportation pricing. The pipeline transportation pricing system for regional gas distribution should be subjected to further subdivision based on price ranges.

4. Conclusions and suggestions

As to the fast development of gas industry in Sichuan-Chongqing region, it is an important acting point to speed up the reform of pipeline transportation system. Furthermore, from the viewpoint of new situation and new requirements facing the gas industry in this region, it is urgent to implement a reform for the gas pipeline transportation system. In this article, we recall the history and the feature of natural gas transmission industry of Sichuan-Chongqing region, and then we study the ways of pipeline transportation system reform, including investment system reform, operation system reform and pricing system reform. In general, the pipeline transportation system should be more liberalised.

Furthermore, there are some recommendations on guarantee measures for reforming. At first, streamline the administration, deregulate pipeline markets and strive to develop shared pipeline economy. Secondly, provide financial support to pipeline construction. Meanwhile, we should establish independent regulatory agency to strengthen industrial regulation for gas pipeline transportation. Finally, regional trade market is necessary and important, as it is obedient to the market-oriented future of gas industry in China [11].

Acknowledgments
The author thanks Petroleum and Natural gas Development Research Center of Sichuan for financial supporting by the program No.SKB16-02. Besides, the author thanks Professor Jiang ziang for helpful discussions.

References

[1] Jiang Ziang, Wang Fuping, Duan Yanzhi, et al. 2016 The development trend of China natural gas market and coping strategies under the new situation in Sichuan Chongqing Gas District as an example *Natural Gas Industry* **34**(4) 123-130.
[2] Hu Aolin 2014 How to build China's natural gas market *Natural Gas Industry* **34**(9) 18-23.
[3] Lü Miao 2010 Natural gas storage construction and service pricing *Urban Gas* **12**(7) 30-35.
[4] Jiang Ziang, Feng Meng, Zhang Hong 2015 Thoughts on promoting China's natural gas energy revolution *Natural Gas Industry* **34**(3) 120-124.
[5] Zhou Zhibin, Bai Lanjun, Jiang Ziang, et al. 2008 Study on the economic development of natural gas in China *Beijing:Petroleum Industry Press*
[6] Duan Yanzhi, Shi Yufeng, He Runmin, Li Zizi 2015 Characteristics and Enlightenment of the European natural gas market *Natural Gas Industry* **34**(5) 123-130.
[7] Niu Yubin, Ling Zhongguo, Zhu Xiaohai, et al. 2013 Natural gas market and pricing *Beijing:China Market Press*
[8] Sumeet Gupta, Arvind Kumar Jain 2012 Analytical Study of Natural Gas Pipeline Tariff *Asian Journal of Business and Economics* **2** 1-16.
[9] Harm van Goor, Bert Scholtens 2014 Modeling natural gas price volatility: The case of the UK gas market *Energy* **72** 126-134.
[10] European Commission. Gas Network Codes & Guidelines [EB/OL].
[11] Sergey Paltsev, Danwei Zhang 2015 Natural gas pricing reform in China: Getting closer to a market system? *Energy Policy* **85** 43-56.

Institute of Physics
Dirac House, Temple Back
Bristol BS1 6BE UK

ISSN: 1755-1307
ISBN 978-1-5108-3662-4

9 781510 836624

CURRAN ASSOCIATES INC.
proceedings.com